Apoptosis and Beyond

Apoptosis and Beyond

The Many Ways Cells Die

Volume 1

Edited by James Radosevich
University of Chicago, United States

This edition first published 2018
© 2018, John Wiley & Sons Inc

All rights reserved. No part of this publication may be reproduced, stored in a retrieval system, or transmitted, in any form or by any means, electronic, mechanical, photocopying, recording or otherwise, except as permitted by law. Advice on how to obtain permission to reuse material from this title is available at http://www.wiley.com/go/permissions.

The right of James Radosevich to be identified as the author of the editorial material in this work has been asserted in accordance with law.

Registered Office(s)
John Wiley & Sons, Inc., 111 River Street, Hoboken, NJ 07030, USA

Editorial Office
The Atrium, Southern Gate, Chichester, West Sussex, PO19 8SQ, UK

For details of our global editorial offices, customer services, and more information about Wiley products visit us at www.wiley.com.

Wiley also publishes its books in a variety of electronic formats and by print-on-demand. Some content that appears in standard print versions of this book may not be available in other formats.

Limit of Liability/Disclaimer of Warranty
While the publisher and authors have used their best efforts in preparing this work, they make no representations or warranties with respect to the accuracy or completeness of the contents of this work and specifically disclaim all warranties, including without limitation any implied warranties of merchantability or fitness for a particular purpose. No warranty may be created or extended by sales representatives, written sales materials or promotional statements for this work. The fact that an organization, website, or product is referred to in this work as a citation and/or potential source of further information does not mean that the publisher and authors endorse the information or services the organization, website, or product may provide or recommendations it may make. This work is sold with the understanding that the publisher is not engaged in rendering professional services. The advice and strategies contained herein may not be suitable for your situation. You should consult with a specialist where appropriate. Further, readers should be aware that websites listed in this work may have changed or disappeared between when this work was written and when it is read. Neither the publisher nor authors shall be liable for any loss of profit or any other commercial damages, including but not limited to special, incidental, consequential, or other damages.

Library of Congress Cataloging-in-Publication Data

Names: Radosevich, James A. (James Andrew), editor.
Title: Apoptosis and beyond : the many ways cells die / edited by James Radosevich.
Description: Hoboken, NJ : Wiley-Blackwell, 2018. | Includes bibliographical
   references and index. |
Identifiers: LCCN 2018022533 (print) | LCCN 2018023112 (ebook) | ISBN
   9781119432357 (Adobe PDF) | ISBN 9781119432432 (ePub) | ISBN 9781119432425 (cloth)
Subjects: | MESH: Cell Death
Classification: LCC QH671 (ebook) | LCC QH671 (print) | NLM QU 375 | DDC
   571.9/36–dc23
LC record available at https://lccn.loc.gov/2018022533

Cover image: © ImageJournal-Photography/Getty images
Cover design by Wiley

Set in 10/12 pt WarnockPro-Regular by Thomson Digital, Noida, India

Printed in Singapore by C.O.S. Printers Pte Ltd

Contents

List of Contributors *ix*

Volume 1

1   **General View of the Cytoplasmic and Nuclear Features of Apoptosis** *1*
    Humberto De Vitto, Juan P. Valencia, and James A. Radosevich

2   **Mitochondria in Focus: Targeting the Cell-Death Mechanism** *13*
    Humberto De Vitto, Roberta Palorini, Giuseppina Votta, and
    Ferdinando Chiaradonna

3   **Microbial Programmed Cell Death** *49*
    Neal D. Hammer

4   **Autophagy** *71*
    Mollie K. Rojas, Juel Chowdhury, Khatja Batool, Zane Deliu,
    and Abdallah Oweidi

5   **Cell Injury, Adaptation, and Necrosis** *83*
    Sarah G. Fitzpatrick and Sara C. Gordon

6   **Necroptosis** *99*
    Ben A. Croker, James A. Rickard, Inbar Shlomovitz, Arshed Al-Obeidi,
    Akshay A. D'Cruz, and Motti Gerlic

7   **Ferroptosis** *127*
    Ebru Esin Yoruker and Ugur Gezer

8   **Anoikis Regulation: Complexities, Distinctions, and Cell Differentiation** *145*
    Marco Beauséjour, Ariane Boutin, and Pierre H. Vachon

9   **Cornification** *183*
    Leopold Eckhart

10 Excitotoxicity *197*
    Julie Alagha, Sulaiman Alshaar, and Zane Deliu

11 Molecular Mechanisms Regulating Wallerian Degeneration *205*
    Mohammad Tauseef and Madeeha Aqil

12 Pyronecrosis *225*
    Maryam Khalili and James A. Radosevich

13 Phenoptosis: Programmed Death of an Organism *237*
    M.V. Skulachev and V.P. Skulachev

14 Molecular Mechanisms Underlying Oxytosis *289*
    Amalia M. Dolga, Sina Oppermann, Maren Richter, Birgit Honrath, Sandra Neitemeier, Anja Jelinek, Goutham Ganjam, and Carsten Culmsee

15 Pyroptosis *317*
    Kate E. Lawlor, Stephanie Conos, and James E. Vince

16 Paraptosis *343*
    Maryam Khalili and James A. Radosevich

17 Hematopoiesis and Eryptosis *367*
    Mollie K. Rojas, Chintan C. Gandhi, and Lawrence E. Feldman

Volume 2

18 Cyclophilin D-Dependent Necrosis *375*
    Jatin Mehta and Chandi Charan Mandal

19 Role of Phospholipases in Cell Death *395*
    Manikanda Raja, Juel Chowdhury, and James A. Radosevich

20 TRIAD (Transcriptional Repression-Induced Atypical Death) *411*
    Takuya Tamura and Hitoshi Okazawa

21 Alkylating-Agent Cytotoxicity Associated with O^6-Methylguanine *427*
    Latha M. Malaiyandi, Lawrence A. Potempa, Nicholas Marschalk, Paiboon Jungsuwadee, and Kirk E. Dineley

22 Entosis *463*
    Jamuna A. Bai and Ravishankar Rai V.

23 Mitotic Catastrophe *475*
    Raquel De Souza, Lais Costa Ayub, and Kenneth Yip

24 NETosis and ETosis: Incompletely Understood Types of Granulocyte Death and their Proposed Adaptive Benefits and Costs *511*
Marko Radic

25 Parthanatos: Poly ADP Ribose Polymerase (PARP)-Mediated Cell Death *535*
Amos Fatokun

26 Methuosis: Drinking to Death *559*
Madeeha Aqil

27 Oncosis *567*
Priya Weerasinghe, Sarathi Hallock, Robert Brown, and L. Maximilian Buja

28 Autoschizis: A Mode of Cell Death of Cancer Cells Induced by a Prooxidant Treatment *In Vitro* and *In Vivo* *583*
J. Gilloteaux, J.M. Jamison, D. Arnold, and J.L. Summers

29 Programmed Death 1 (PD1)-Mediated T-Cell Apoptosis and Cancer Immunotherapy *695*
Chandi Charan Mandal, Jatin Mehta, and Vijay K. Prajapati

List of Contributors

Julie Alagha
Department of Oral Medicine and
Diagnostic Sciences
University of Illinois at Chicago
Chicago, IL, USA

Arshed Al-Obeidi
Dana-Farber Boston's Children Cancer
and Blood Disorder Center
Harvard Medical School
Boston, MA, USA

Sulaiman Alshaar
Department of Oral Medicine and
Diagnostic Sciences
University of Illinois at Chicago
Chicago, IL, USA

Madeeha Aqil
Department of Oral Medicine and
Diagnostic Sciences
University of Illinois at Chicago
Chicago, IL, USA

D. Arnold
Anesthesiology Unit
Belfast, ME, USA

Lais Costa Ayub
Department of Molecular, Genetic and
Structural Biology
University of Ponta Grossa
Ponta Grossa, PR, Brazil

Jamuna A. Bai
Department of Studies in Microbiology
University of Mysore
Mysore, India

Khatja Batool
Department of Oral Medicine and
Diagnostic Sciences
University of Illinois at Chicago
Chicago, IL, USA

Marco Beauséjour
Department of Anatomy and Cellular
Biology
University of Sherbrooke
Sherbrooke, QC, Canada

Ariane Boutin
Department of Anatomy and Cellular
Biology
University of Sherbrooke
Sherbrooke, QC, Canada

Robert Brown
Department of Pathology and Laboratory
Medicine
University of Texas Health Science
Center at Houston
Houston, TX, USA

L. Maximilian Buja
Department of Pathology and Laboratory Medicine
University of Texas Health Science Center at Houston
Houston, TX, USA

Ferdinando Chiaradonna
Department of Biotechnology and Biosciences
University of Milano-Bicocca
Milan, Italy

Juel Chowdhury
Department of Oral Medicine and Diagnostic Sciences
University of Illinois at Chicago
Chicago, IL, USA

Stephanie Conos
The Walter and Eliza Hall Institute of Medical Research
Bundoora, VIC, Australia

Ben A. Croker
Dana-Farber Boston's Children Cancer and Blood Disorder Center
Harvard Medical School
Boston, MA, USA

Carsten Culmsee
Institute of Physiological Chemistry
Philipps University of Marburg
Marburg, Germany

Akshay A. D'Cruz
Dana-Farber Boston's Children Cancer and Blood Disorder Center
Harvard Medical School
Boston, MA, USA

Zane Deliu
Department of Oral Medicine and Diagnostic Sciences
University of Illinois at Chicago
Chicago, IL, USA

Raquel De Souza
University Health Network, Department of Radiation Physics, Pharmaceutical Sciences
University of Toronto
Toronto, ON, Canada

Humberto De Vitto
Center of Health and Science
Federal University of Rio de Janeiro
Rio de Janeiro, Brazil

Kirk E. Dineley
Chicago College of Osteopathic Medicine
Midwestern University
Downers Grove, IL, USA

Amalia M. Dolga
Department of Molecular Pharmacology, Groningen Research Institute of Pharmacy (GRIP)
University of Groningen
Groningen, The Netherlands

Leopold Eckhart
Department of Dermatology
Medical University of Vienna
Vienna, Austria

Amos Fatokun
School of Medical Sciences, Faculty of Life Sciences
University of Bradford
Bradford, UK

Lawrence E. Feldman
Department of Hematology and Oncology
University of Illinois at Chicago
Chicago, IL, USA

Sarah G. Fitzpatrick
Department of Oral and Maxillofacial Diagnostic Sciences
University of Florida
Gainesville, FL, USA

Chintan C. Gandhi
Department of Hematology and Oncology
University of Illinois at Chicago
Chicago, IL, USA

Goutham Ganjam
Institute of Pharmacology and Clinical Pharmacy, Biochemisch-Pharmakologisches Centrum Marburg
Philipps University of Marburg
Marburg, Germany

Motti Gerlic
Department of Clinical Microbiology and Immunology, Sackler Faculty of Medicine
Tel Aviv University
Tel Aviv, Israel

Ugur Gezer
Department of Basic Oncology, Oncology Institute
Istanbul University
Istanbul, Turkey

J. Gilloteaux
Department of Anatomical Sciences
St. Georges' University International School of Medicine, KB Taylor Global Scholar's Program at Northumbria University
Newcastle upon Tyne, UK

Sara C. Gordon
School of Dentistry
Oral Medicine University of Washington
Seattle, WA, USA

Sarathi Hallock
Memorial University of Newfoundland, AMC Cancer Research Center
University of Texas Health Science Center at Houston
Houston, TX, USA

Neal D. Hammer
Department of Microbiology and Molecular Genetics
Michigan State University
East Lansing, MI, USA

Birgit Honrath
Institute of Pharmacology and Clinical Pharmacy, Biochemisch-Pharmakologisches Centrum Marburg
Philipps University of Marburg
Marburg, Germany

and

Department of Molecular Pharmacology, Groningen Research Institute of Pharmacy (GRIP)
University of Groningen
Groningen, The Netherlands

J.M. Jamison
Department of Urology and Apatone Development Laboratory
Summa Health System
Akron, OH, USA

Anja Jelinek
Institute of Pharmacology and Clinical Pharmacy, Biochemisch-Pharmakologisches Centrum Marburg
Philipps University of Marburg
Marburg, Germany

Paiboon Jungsuwadee
School of Pharmacy
Fairleigh Dickinson University
Florham Park, NJ, USA

Maryam Khalili
Department of Oral Medicine and Diagnostic Sciences
University of Illinois at Chicago
Chicago, IL, USA

Kate E. Lawlor
The Walter and Eliza Hall Institute of Medical Research
Bundoora, VIC, Australia

Latha M. Malaiyandi
Chicago College of Osteopathic Medicine
Midwestern University
Downers Grove, IL, USA

Chandi Charan Mandal
Department of Biochemistry
Central University of Rajasthan
Rajasthan, India

Nicholas Marschalk
Chicago College of Osteopathic Medicine
Midwestern University
Downers Grove, IL, USA

Jatin Mehta
National Institute of Pathology, ICMR
Safdarjang Hospital
New Delhi, India

Sandra Neitemeier
Institute of Pharmacology and Clinical Pharmacy, Biochemisch-Pharmakologisches Centrum Marburg
Philipps University of Marburg
Marburg, Germany

Hitoshi Okazawa
Department of Neuropathology, Medical Research Institute
Tokyo Medical and Dental University
Tokyo, Japan

Sina Oppermann
German Cancer Research Center (DKFZ) and National Center of Tumordiseases (NCT)
Heidelberg, Germany

Abdallah Oweidi
Department of Oral Medicine and Diagnostic Sciences
University of Illinois at Chicago
Chicago, IL, USA

Roberta Palorini
SYSBIO Center for Systems Biology, Department of Biotechnology and Biosciences
University of Milano-Bicocca
Milan, Italy

and

Luxembourg Centre for Systems Biomedicine
Esch-sur-Alzette, Luxembourg

Lawrence A. Potempa
Roosevelt University College of Pharmacy
Schaumburg, IL, USA

Vijay K. Prajapati
Department of Biochemistry
Central University of Rajasthan
Rajasthan, India

Marko Radic
Department of Microbiology, Immunology and Biochemistry
University of Tennessee Health Science Center
Memphis, TN, USA

James A. Radosevich
Department of Oral Medicine and Diagnostic Sciences
University of Illinois at Chicago
Chicago, IL, USA

Manikanda Raja
Department of Oral Medicine and Diagnostic Sciences
University of Illinois at Chicago
Chicago, IL, USA

Maren Richter
Institute of Pharmacology and Clinical Pharmacy, Biochemisch-Pharmakologisches Centrum Marburg
Philipps University of Marburg
Marburg, Germany

James A. Rickard
Department of Biochemistry
La Trobe University
Melbourne, VIC, Australia

Mollie K. Rojas
Department of Oral Medicine and
Diagnostic Sciences
University of Illinois at Chicago
Chicago, IL, USA

Inbar Shlomovitz
Department of Clinical Microbiology
and Immunology, Sackler Faculty
of Medicine
Tel Aviv University
Tel Aviv, Israel

M.V. Skulachev
Belozersky Institute of Physico-Chemical
Biology, Institute of Mitoengineering,
Faculty of Bioengineering and
Bioinformatics
Lomonosov Moscow State University
Moscow, Russia

V.P. Skulachev
Belozersky Institute of Physico-Chemical
Biology, Institute of Mitoengineering,
Faculty of Bioengineering and
Bioinformatics
Lomonosov Moscow State University
Moscow, Russia

J.L. Summers
Department of Urology and Apatone
Development Laboratory
Summa Health System
Akron, OH, USA

Takuya Tamura
Department of Neuropathology, Medical
Research Institute
Tokyo Medical and Dental University
Tokyo, Japan

Mohammad Tauseef
Department of Pharmaceutical Sciences
College of Pharmacy
Chicago State University
Chicago, IL, USA

Ravishankar Rai V.
Department of Studies in Microbiology
University of Mysore
Mysore, India

Pierre H. Vachon
Department of Anatomy and Cellular
Biology
University of Sherbrooke
Sherbrooke, QC, Canada

Juan P. Valencia
University of Rio de Janeiro
Rio de Janeiro, Brazil

James E. Vince
The Walter and Eliza Hall Institute of
Medical Research
Bundoora, VIC, Australia

Giuseppina Votta
SYSBIO Center for Systems Biology,
Department of Biotechnology and
Biosciences
University of Milano-Bicocca
Milan, Italy

Priya Weerasinghe
Department of Pathology and Laboratory
Medicine
University of Texas Health Science
Center at Houston
Houston, TX, USA

Kenneth Yip
Department of Biology
University of Toronto
Toronto, ON, Canada

Ebru Esin Yoruker
Department of Basic Oncology,
Oncology Institute
Istanbul University
Istanbul, Turkey

1

General View of the Cytoplasmic and Nuclear Features of Apoptosis

Humberto De Vitto,[1] Juan P. Valencia,[2] and James A. Radosevich[3]

[1]Center of Health and Science, Federal University of Rio de Janeiro, Rio de Janeiro, Brazil
[2]University of Rio de Janeiro, Rio de Janeiro, Brazil
[3]Department of Oral Medicine and Diagnostic Sciences, University of Illinois at Chicago, Chicago, IL, USA

Abbreviations

AIF	apoptosis-inducing factor
Apaf-1	apoptotic protease activating factor-1
ATP	adenosine triphosphate
BA	bongkrekic acid
Bcl-2	B-cell lymphoma-2
BID	BH3-interacting domain death agonist
bp	base pair
CAD	caspase-activated DNase
c-FLIP	cellular FLICE-inhibitory protein
Chx	cyclohexamide
CsA	cyclosporine A
CTL	cytotoxic T lymphocyte
cyt c	cytochrome c
DISC	death-inducing signal complex
DED	death effector domain
endo D	endonuclease D
endo G	endonuclease G
ER	endoplasmic reticulum
FADD	Fas-associated death-domain protein
FasL	fatty acid synthetase ligand
FasR	fatty acid synthetase receptor
Gzm-A	granzyme-A
Gzm-B	granzyme-B
ICAD	inhibitor of caspase-activated DNase
Kb	kilobase
MEF	mouse embryonic fibroblast
MOMP	mitochondrial outer-membrane permeabilization
NK	natural killer
OMM	outer mitochondrial membrane
PCD	programmed cell death
PFN	Perforin

Apoptosis and Beyond: The Many Ways Cells Die, First Edition. Edited by James Radosevich.
© 2018 John Wiley & Son Inc. Published 2018 by John Wiley & Son Inc.

PT	pore transition
RIP	receptor-interacting protein
ROCK I	Rho effector protein
ROS	reactive oxygen species
SET	stress-response complex
tBID	truncated BID
TNF	tumor necrosis factor
TNFR1	tumor necrosis factor receptor 1
TRADD	TNF receptor-associated death domain
TUNEL	terminal deoxynucleotide tranferase dUTP nick end labeling

1.1 Introduction

The normal development of a cell and the life cycles of the multicellular organism rely on a finely tuned balance between cell survival and death. In a biological context, cells need to grow, divide, and die. In regard to the latter process, cells have developed a very precisely regulated means of programmed cell death (PCD), which contributes to the maintenance of normal cell turnover, leading to reduced impact on tissues, organs, and the organism itself. Some cells have evolved a PCD process called **apoptosis**. Apoptosis can be simply defined as a set of biochemical cytoplasmic and mitochondrial events that may lead to the execution phase of nuclear events.

A wide array of stress stimuli can trigger the apoptotic process, and the biochemical signal can then be amplified in the cytoplasm and mitochondria by both extrinsic and intrinsic pathways. The convergence of the apoptotic signal is considered the activation of a family of **c**ysteine **asp**artyl-specific prote**ases** (**caspases**), composed of 12 proteins strictly involved in the apoptotic cell death process. The dying cells activate the execution pathway that leads to the appearance of blebs and to the "pinching off" of many of them, forming "apoptotic bodies," which may be rounded and retracted from their own tissue. Subsequently, the immune system cells are able to eliminate the apoptotic bodies through an engulfment cell process. The morphological and biochemical features during the apoptotic process are not fully understood.

At the nuclear level, it is well established that endonucleases and exonucleases may hydrolyze the DNA into small fragments (200 pb) [1]. The nuclear events depend on caspase activation. Caspase 3 is considered the most important protease of the executioner pathway, and is activated by different initiator caspases. For instance, caspase 8 is activated from the death receptor, caspase 9 is involved in the mitochondrial apoptotic process, and caspase 10 is involved in the Perforin/granzyme (PFN/Gzm) pathways. The cleaved caspase 3 cleaves the endonuclease caspase-activated DNase (CAD), degrading the DNA at nucleosomal linkers [2,3], which generates small DNA fragments (~50–300 kb). The subsequent processing of the DNA by exonucleases and endonucleases leads to the formation of 200 bp fragments. Many organelles, such as the Golgi apparatus, endoplasmic reticulum (ER), lysosomes, and mitochondria, can be recycled or eliminated, depending on the apoptotic stimuli. It is important to note that mitochondria play a pivotal role in apoptosis, since they can release cytochrome c (cyt c) and endonuclease D (endo D), leading to cell death [4,5].

One of the apoptotic pathways is the extrinsic or death-receptor pathway. It depends for its activation on a death domain and a death ligand, such as tumor necrosis factor alpha (TNFα) and tumor necrosis factor receptor 1 (TNFR1). The ligand represents the external death signal, leading to the intracellular signaling of the effector pathway. The main receptors recruit adaptor proteins like Fas-associated death-domain protein (FADD), TNF receptor-associated death domain (TRADD), and receptor-interacting protein (RIP) [6–8], which in turn recruit other molecules such as pro-caspase 8. The dimerization of the death effector domain (DED) leads to the formation of a death-inducing signal complex (DISC), triggering the subsequent process of autocatalysis of pro-caspase 8 to an activated protein (caspase 8) [9]. Caspase 8 activation is considered the main feature that starts the extrinsic pathway, leading to cell death. In many cases, depending on the apoptotic stimuli, the extrinsic pathway can crosstalk with the intrinsic pathway through proteolysis of the BH3-only protein, BH3-interacting domain death agonist (BID), which is what promotes the release of cyt c from the mitochondria into the cytoplasm. In the cytoplasm, cyt c may be assembled with the adaptor protein apoptotic protease activating factor-1 (Apaf-1) and ATP, generating in the cytosol the multi-molecular holoenzyme complex called the "apoptosome" (Figure 1.1) [10].

The PFN/Gzm pathway is considered part of the extrinsic pathway. It is activated when cells are infected by viruses and/or bacteria. Mechanistically, cytotoxic T lymphocytes (CTLs) and natural killer (NK) cells produce granzymes. The granzymes, together with the PFN, may facilitate the pore formation and action of Gzms that lead to cell death. Two types of Gzm, type A (GmzA) and type B (GmzB), have been described. GmzA is considered the most important serine-protease present in CTL and NK granules. GmzB relies on the mechanism of oligomerization of PFN to enter into the cell, and depends for its action on the activation of caspases, principally caspase 3 [11,12].

It is important to note that different stimuli and players are enlisted in the various apoptotic processes. Moreover, it is well known that immune-system cells, mitochondria, and nuclear events are involved in apoptosis. The entire biochemical process is still elusive. Accordingly, the current chapter will focus on cytoplasmic and nuclear events. We describe and highlight an overview of cytoplasmic events, including the extrinsic pathway and perforin/granzyme pathways (GmzA and GmzB), as well as the main nuclear features of the current process, by correlating these events with the intrinsic apoptotic pathway. We then use such pathway characteristics to explore in more detail the mechanisms of the regulation of important human diseases, including cancer and neurodegenerative diseases.

1.2 Cytoplasmic Events

During apoptosis, the dying cells become rounded and retracted from the tissue. Defined "blebs" appear within the cells. The process culminates in the "pinching off" of many blebs, producing "apoptotic bodies." These bodies recruit phagocytes, which engulf them to recycle some of the molecules. The immune system represents the major mechanism capable of eliminating the apoptotic bodies, apparently in the same way that phagocytes eliminate "non-self" particles. These morphological and biochemical features of apoptosis have been extensively studied, but the whole mechanism is not fully understood.

It is well established that during the apoptotic process, the nuclear DNA is condensed and fragmented. Several proteins, such as endonucleases and exonucleases, may degrade the DNA chain by hydrolyzing it in small fragments, with approximately 200 bp [1]. In

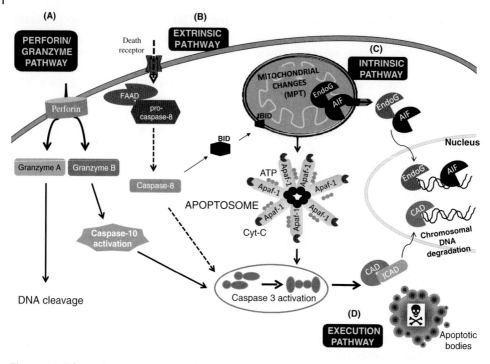

Figure 1.1 Schematic representation of the cytoplasmic and nuclear events of apoptosis. The Perforin/Granzyme pathway, extrinsic pathway, and intrinsic pathway represent the three main pathways of apoptosis. Through a vast array of death signals, all three pathways can be triggered. (A) The Perforin/Granzyme pathway is a unique pathway that partially works in a caspase-independent fashion (granzyme A branch), leading directly to DNA cleavage and cell death. However, the activation of the granzyme B branch can trigger initiator caspase 10, which activates executioner caspase 3. (B) The extrinsic pathway, when activated, can cleave pro-caspase 8 to caspase 8 by FAAD, then activate executioner caspase 3. Caspase 8 plays an important role in the activation of a truncated BID (tBID) protein, leading to the release of mitochondria proteins like cyt c. (C) Upon receiving incoming signals, the intrinsic pathway induces MPTP opening, leading to the release from the mitochondria of proteins such as cyt c, endo G/AIF, and Htra2/Omi. On the cytosol, cyt c forms the apoptosome, which cleaves pro-caspase 3, triggering the execution pathway. (D) The execution pathway is characterized by cell shrinkage, chromatin condensation, and the formation of cytoplasmic blebs and apoptotic bodies. cyt c, cytochrome c; FAAD, Fas-associated death domain; MTPT, mitochondrial permeability transition pore.

the cytoplasm, the Golgi apparatus, ER, lysosomes, and mitochondria can be eliminated or recycled, depending on the apoptotic stimuli. For instance, during oxidative stress involving an increase of reactive oxygen species (ROS), several mitochondrial proteins, such as cyt c and endo D, may be released from the mitochondrial intermembrane space into the cytoplasm and nucleus, leading to cell death [4,5]. Interesting, it has been indentified that the mitochondria, ER, and nucleus are targets of the Gzm pathway, which may trigger apoptosis, facilitating the PCD process.

1.2.1 The Extrinsic Pathway

The extrinsic pathway, also called the death receptor pathway, is involved in transmembrane receptor-mediated interactions, including those of the TNF family [13], which shares the features of the cysteine-rich extracellular domains and a cytoplasmatic domain (death domain) [14]. The death ligand represents the external death signal from the cell surface to the intracellular signaling and effector pathways. The mechanism requires the binding of the extracellular death ligands to the transmembrane cell receptors. The best-characterized ligands and their corresponding death receptors have been identified: (i) TNFα and TNFR1; (ii) fatty acid synthetase ligand and fatty acid synthetase receptor (FasL and FasR); (iii) Apo2 ligand and death receptor 4 (Apo2L and DR4); (iv) Apo2 ligand and death receptor 5 (Apo2L and DR5); and (v) Apo3 ligand and death receptor 3 (Apo3L and DR3) [14–18]. The receptors form clusters and can bind with their cognate trimeric ligands, leading to the recruitment of adaptor proteins, including FADD, TRADD, and RIP [6–8]. In turn, FADD or TRADD can recruit several molecules, such as pro-caspase 8, binding to them via dimerization of the DED, leading to DISC formation and subsequent autocatalysis of pro-caspase 8 and its active form (caspase 8) [9]. Caspase 8 activation is considered the main feature that triggers the extrinsic pathway, leading to cell death. Activated caspase 8 is involved in many proteolytic processes, including the activation of caspases 3, caspase 6, and caspase 7. These enzymes help induce the execution phase of apoptosis (Figure 1.1).

Depending on the apoptotic stimuli, the extrinsic pathway can crosstalk with the intrinsic pathway through proteolysis of the BH3-only protein, BID. The truncated BID (tBID) protein promotes release of mitochondrial cyt c into the cytoplasm, where it can assemble with the apoptosome complex, leading to cell death [10]. However, death receptor-mediated apoptosis can be inhibited by cellular FLICE-inhibitory protein (c-FLIP), which binds to both FADD and caspase 8, inactivating the autocatalytic effect of the caspase 8 complex [19,20]. Different mechanisms of inhibition of the extrinsic apoptosis pathway have also been described, including via the protein Toso, which blocks Fas-induced apoptosis, inhibiting the processing of caspase 8 in immune cells [21].

1.2.1.1 The Perforin/Granzyme Pathway

To eliminate potential dangerous cells like tumor cells or cells infected by viruses or bacteria, the immune system relies on CTLs and NK cells, both of which are produced by the action of Gzms. PFN, a protein capable of binding the membrane of the target cell, facilitate the pore formation that permits the action of Gzms. Gzms are considered specific serine-proteases involved in cell death. They are produced as inactive precursor molecules, designed to avoid the self-destruction of CTLs and NK cells. In human cells, five different Gzms have been reported. GzmA and GzmK are located on chromosome 5 and act as tryptases that cleave proteins following arginine or lysine (basic) residues. GzmB and GzmH are located on chromosome 14. GzmM is located on chromosome 19 and cleaves following methionine or leucine basic residues [22]. There are two Gzm-dependent pathways involved in cell death.

1.2.1.1.1 The Granzyme A Pathway

The GzmA is considered the most important serine-protease mechanism described in CTL and NK granulles. Unlike the GzmB pathway, which relies on the oligomerization of the PFN to enter into the target cell, the GzmA can activate a parallel pathway in a caspase-independent manner, leading to DNA degradation, such as single-stranded DNA damage [23]. Intracellular GzmA substrates have been found in the cytoplasm (Pro-IL-1β) [24], mitochondria (NDUFS3) [25], ER, and nucleus (SET1, APE1, HMGB2) [26–29], and are associated with histone H1, core histones, lamin A, B, and C, Ku70, and PARP1 [30–33]. Various stimuli can trigger the GzmA pathway, such as ROS generation, the loss of membrane potential, and mitochondrial swelling. This can lead to the disruption of the nuclear envelope, inhibition of DNA repair, and activation of cytokines, as a consequence of the accumulation of GzmA in the nucleus [34]. Between mitochondrial changes (within minutes) and phospathidyl serine externalization (30 minutes to 1 hour), dying cells can recruit the macrophage scavenger system [23].

GzmA is less cytotoxic than the GzmB pathway, which is active at micromolar-range concentrations [35]. At 2 hours after the stimulation of apoptosis, several features are present. This cell-death pathway does not activate caspases, because cell-death GzmA activation is known as a non-apoptotic death [36]. Moreover, GzmA does not permebilize the outer mitochondrial membrane (OMM), avoiding the releasing of mitochondrial apoptotic mediators like cyt c. The entry of GzmA into the mitochondria can be partially inhibited by cyclosporine A (CsA) and bongkrekic acid (BA), suggesting a role of permeability for the transition pore (PT) in GzmA mitochondrial damage [23,35].

The oxidative damage drives the ER to make the ER-associated oxidative stress response complex (SET), which contains two endonucleases (Ape1 and NM23-H1) and a $5'$-$3'$ exonuclease (Trex1), chromatin modifying proteins (SET1 and pp32), and DNA-binding proteins that protect against DNA distortion (HMGB2) [23,27,29,37]. GzmA enters into the nucleus and cleaves SET1, which inhibits NM23-H1 endonuclease activity, causing this complex to nick the DNA, and allowing Trex1 to act as an endonuclease [38]. In the same way, GzmA cleaves and inactivates HGMB2 and Ape1 [26], cleaves the linker histone H1, and removes the tail of core histones, allowing the nucleases to attack [30]. GzmA then cleaves and inactivates Ku70 and PARP-1, both of which are involved in DNA repair through the recognition of single- or double-strand breaks [32,33].

1.2.1.1.2 The Granzyme B Pathway

GzmB is produced by CTL and NK cells, which release it via granules. It binds its receptor, the mannose-6-phosphate/insulin-like growth factor II receptor, and is endocytosed but remains arrested in endocytic vesicles until it is released by PFN. The GzmB pathway relies on caspase activation, unlike the GzmA pathway.

The proteolitic activity of GzmB is similar to caspase activity, cleaving substrates after the aspartate (basic) residues. Caspase 3, 6, 7, 8, 9, and 10 have been found to serve as GzmB substrates *in vitro* [39–46], but only caspase 3 is believed to be important *in vivo* [11,12]. As a further mechanism, GzmB can process BID, promoting cyt c release, SMAC/Diablo activation, formation of apoptosis inducing factor (AIF), and release of Omi from the mitochondria. It does this by recruiting the inhibitor of the anti-apoptotic B-cell lymphoma 2 (Bcl-2) family member, especially the Bax protein, to the mitochondrial membrane, leading to apoptosome formation [45–47]. GzmB can also process caspase 3 and 7, initiating the apoptotic process [48]. It has been demonstrated that pro-apoptotic caspase activation happens within minutes of target-cell recognition by CTLs. Unexpectedly, there is a rapid rate of caspase 3/7 biosensor activation following GzmB cversus Fas-mediated signal induction in murine CTLs. This Fas-mediated induction is detected after 90–120 minutes in porcine, murine, and human CTLs, consistent with FasL/Fas-induced activity [49]. Recently, key roles for GzmB have been described, positioning it as an allergic inflammatory response of NK [50]. It has also been shown that the major NK cell-activating receptor NKG2D and the NK cell effector are both mediated by GzmB.

1.2.2 Nuclear Features of Apoptosis

The first description of the apoptotic process as a basic biological phenomenon different from necrosis (based not only on morphological criteria) was given by Kerr et al. [51]. The authors described two characteristics of apoptosis: (i) cytoplasmic and nuclear condensation and the disruption of the cell into a number of membrane-bound, well-fragmented pieces; and (ii) formation of apoptotic bodies that are taken up by other cells for degradation. This study shed new light on the apoptotic mechanism as an important process of PCD that regulates several biological processes, including embryogenic development and aging, cell turnover in different tissues, and the control of the immune system. Inappropriate control of apoptosis appears in many human disorders, leading biologists to seek a better understanding of the entire process. Intriguingly, a wide variety of stimuli – both physiological and pathological conditions – can trigger apoptosis. In this section, we address the main biochemical features of apoptosis that focus on nuclear events, including the activation of the execution caspases (i.e., caspase 3, caspase 8, caspase 9, and caspase 10), chromatin condensation, DNA fragmentation, and the formation of apoptotic bodies.

Early evidence described DNA fragmentation as a key feature of apoptosis. Using low concentrations of an exogenous agent like γ-irradiation to induce cell death, it was shown that the DNA of lymphocytes was completely degraded into oligonucleosomal fragments. Further, cells induced with near-physiological concentrations of glucocorticoid hormones showed chromatin condensation as an early structural change. In fact, this particular nuclear morphological change was associated with excision of the nucleosome chains from nuclear chromatin through activation of an intracellular, but non-lysosomal, endonuclease [52]. At this time, it was already known that some members of the caspase family, comprising 12 proteins, are strictly involved in the apoptotic cell death process [53,54]. These are the signals after mitochondrial outer-membrane permeabilization (MOMP) that activate the caspase pathway. However, the interconnection between the nuclear and cytoplasmic events involved in apoptosis became better appreciated when a nuclease protein (Nuc-1), a homolog of mammalian DNAase II, which plays a role in DNA degradation in the nematode *Caenorhabditis elegans*, was identified as acting downstream of Ced-3 and Ced-4 [56]. In particular, attempts have long been made to understand the link between the executioner caspases and subsequent nuclear apoptotic events, since a variety of death stimuli can activate these proteases, which amplify the signal of cell death.

Caspase 3 is considered the most important protease of the executioner pathway, and can be activated by any of the initiator caspases. Caspase 3 can be activated by caspase 8, which is activated from the death receptor; by caspase 9, which is involved in the mitochondrial apoptotic process; or by caspase 10, which is involved in the PFN/GzmB pathway. Each of these pathways is responsive to a wide range of stimuli capable of amplifying the cellular death signal in an energy-dependent manner. The cleavage of caspase 3 results in the activation of the endonuclease CAD. In apoptotic cells, activated caspase 3 cleaves inhibitor of caspase-activated DNase (ICAD) to dissociate the CAD:ICAD complex, allowing CAD to cleave chromosomal DNA. The CAD:ICAD complex inhibits the CAD activity as DNase. When CAD is cleaved by caspase 3, it can degrade chromosomal DNA like a scissor-like homodimer, cleaving double-strand DNA at nucleosomal linkers [2,3].

CAD is able to condense chromatin and to fragment chromosomal DNA in an irreversible manner that compromises DNA replication and gene transcription, leading to cell death. Accompanied by chromatin condensation, chromosomal DNA is cleaved into high-molecular-weight fragments of 50–300 kb, which are subsequently processed into low-molecular-weight fragments of approximately 180 bp. Several models have been designed to study the role of CAD in PCD. These studies have shown that the inhibition of CAD activity – for instance, by inducing degradation by a chaperon – can abolish internucleosomal DNA fragmentation. However, the inefficient DNA degradation activity detected in CAD-deficient cells suggests the existence of additional nuclease(s) during apoptosis. An interesting example that links the extrinsic and intrinsic pathways is related to the mammalian endonuclease G (endo G). Endo G is a nuclease that was first identified in the mitochondrial intermembrane space; upon apoptotic stimuli, it may be released from the mitochondria and translocated to the nucleus. The endonuclease activity is responsible for cleaving nucleic acids, representing a caspase-independent apoptotic pathway initiated from mitochondria. In mouse embryonic fibroblast (MEF) cells, taken from a DFF45/ICAD-knockout (KO) mouse, there was no detectable caspase 3-dependent activity, and it was shown that there was minimal DNA fragmentation. Moreover, the induction of apoptosis by ultraviolet irradiation or treatment with cyclohexamide (Chx) led to the release of both endo G and cyt c from the mitochondria to the cytosol and nuclei. The identification of DNA fragmentation has been used as a fundamental biological marker of apoptosis. The main method for detecting apoptotic PCD is known as terminal deoxynucleotidyl transferase dUTP nick end labeling (TUNEL) [57].

There have been several further reports providing evidence for caspase-independent death programming *in vitro* and *in vivo* by cathepsins B and D, calpains, and serine proteases. Some of these death routines become evident only when the caspase-dependent pathway is inhibited, particularly in the case of ATP depletion or when using caspase inhibitors.

The evolutionarily conserved execution phase of apoptosis is characterized by cell morphology changes, including cell shrinking, plasma-membrane blebbing, and separation of cell fragments into apoptotic bodies. It is known that the actin–myosin system plays a key role in bleb formation through the activity of the Rho effector protein (ROCK I), which leads to the phosphorylation of myosin light-chain ATPase activity and coupling of actin–myosin filaments to the plasma membranes. Apoptotic bodies consist of cytoplasm-packed organelles that contain nuclear fragment. The integrity of the apoptotic bodies is maintained in order to avoid the release of their cellular constituents into the surrounding interstitial tissue, which would block activation of the inflammatory reaction; this permits the apoptotic bodies to be degraded efficiently within phagolysosomes by macrophages and various surrounding cells.

Although the evolutionarily conserved execution phase of apoptosis has been the theme of many studies, a full understanding of apoptosis at the molecular level is needed if we are to gain deeper insights into its basic and applied biology, particularly regarding new therapeutic strategies.

References

1 Williams JR, Little JB, Shipley WU. Association of mammalian cell death with a specific endonucleolytic degradation of DNA. *Nature* 1974;**252**:754–5.

2. Kothakota S, Azuma T, Reinhard C, Klippel A, Tang J, Chu K, et al. Caspase-3-generated fragment of gelsolin: effector of morphological change in apoptosis. *Science* 1997;**278**:294–8.
3. Enari M, Sakahira H, Yokoyama H, Okawa K, Iwamatsu A, Nagata S. A caspase-activated DNase that degrades DNA during apoptosis, and its inhibitor ICAD. *Nature* 1998;**391**:43–50.
4. Frank S, Gaume B, Bergmann-Leitner ES, Leiner WW, Robert EG, Catez F, et al. The role of dynamin-related protein 1, a mediator of mitochondrial fission, in apoptosis. *Dev Cell* 2001;**1**:515–25.
5. Lane JD, Lucocg J, Pryde J, Barr FA, Woodman PG, Allan VJ, Lowe M. Caspase-mediated cleavage of the stacking protein GRASP65 is required for Golgi fragmentation during apoptosis. *J Cell Biol* 2002;**156**:495–509.
6. Hsu H, Xiong J, Goeddel DV. The TNF receptor 1-associated protein TRADD signals cell death and NF-kappa B activation. *Cell* 1995;**81**:495–504.
7. Grimm S, Stanger BZ, Leder P. RIP and FADD: two "death domain"-containing proteins can induce apoptosis by convergent, but dissociable, pathways. *Proc Natl Acad Sci USA* 1996;**93**(10):923–7.
8. Wajant H. The Fas signaling pathway: more than a paradigm. *Science* 2002;**296**:1635–6.
9. Kischkel FC, Hellbardt S, Behrmann I, Germer M, Pawlita M, Krammer PH, Peter ME. Cytotoxicity-dependent APO-1 (Fas/CD95)-associated proteins form a death-inducing signaling complex (DISC) with the receptor. *Embo J* 1995;**14**:5579–88.
10. Honglin L, Zhu H, Xu C, Juan Y. Cleavage of Bid by caspase 8 mediates the mitocondrial damage in the Fas pathway of apoptosis. *Cell* 1998;**94**:491–501.
11. Darmon AJ, Ley TJ, Nicholson DW, Bleackley RC. Cleavage of CPP32 by granzyme B represents a critical role for granzyme B in the induction of target cell DNA fragmentation. *J Biol Chem* 1996;**271**(21):709–12.
12. Atkinson EA, Barry M, Darmon AJ, Shostak I, Turner PC, Moyer RW, Bleackley RC. Cytotoxic T lymphocyteassisted suicide. Caspase 3 activation is primarily the result of the direct action of granzyme B. *J Biol Chem* 1998;**273**(21):261–6.
13. Locksley RM, Killeen N, Lenardo MJ. The TNF and TNF receptor superfamilies: integrating mammalian biology. *Cell* 2001;**104**:487–501.
14. Ashkenazi A, Dixit VM. Death receptors: signaling and modulation. *Science* 1998;**281**:1305–8.
15. Chicheportiche Y, Bourdon PR, Xu H, Hsu YM, Scott H, Hession C, et al. TWEAK, a new secreted ligand in the tumor necrosis factor family that weakly induces apoptosis. *J Biol Chem* 1997;**272**(32):401–10.
16. Peter ME, Krammer PH. Mechanisms of CD95 (APO-1/Fas)-mediated apoptosis. *Curr Opin Immunol* 1998;**10**:545–51.
17. Suliman A, Lam A, Datta R, Srivastava RK. Intracellular mechanisms of TRAIL: apoptosis through mitochondrial-dependent and -independent pathways. *Oncogene* 2001;**20**:2122–33.
18. Rubio-Moscardo F, Blesa D, Mestre C, Siebert R, Balasas T, Benito A, et al. Characterization of 8p21.3 chromosomal deletions in B-cell lymphoma: TRAIL-R1 and TRAIL-R2 as candidate dosage-dependent tumor suppressor genes. *Blood* 2005;**106**:3214–22.
19. Kataoka T, Schroter M, Hahne M, Schneider P, Irmler M, Thome M, et al. FLIP prevents apoptosis induced by death receptors but not by perforin/granzyme B, chemotherapeutic drugs, and gamma irradiation. *J Immunol* 1998;**161**:3936–42.

20. Scaffidi C, Schmitz I, Krammer PH, Peter ME. The role of c-FLIP in modulation of CD95-induced apoptosis. *J Biol Chem* 1999;**274**:1541–8.
21. Hitoshi Y, Lorens J, Kitada SI, Fisher J, LaBarge M, Ring HZ, et al. Toso, a cell surface, specific regulator of Fas-induced apoptosis in T cells. *Immunity* 1998;**8**:461–71.
22. Garcia-Sanz JA, MacDonald HR, Jenne DE, Tschopp J, Nabholz M. Cell specificity of granzyme gene expression. *J Immunol* 1990;**145**:3111–18.
23. Martinvalet D, Zhu P, Lieberman J. Granzyme A induces caspase-independent mitochondrial damage, a required first step for apoptosis. *Immunity* 2005;**22**:355–70.
24. Irmler M, Hertig S, MacDonald HR, Sadoul R, Becherer JD, Proudfoot A, et al. Granzyme A is an interleukin 1 beta-converting enzyme. *J Exp Med* 1995;**181**:1917–22.
25. Martinvalet D, Dykxhoorn DM, Ferrini R, Lieberman J. Granzyme A cleaves a mitochondrial complex I protein to initiate caspase-independent cell death. *Cell* 2008;**133**:681–92.
26. Fan Z, Beresford PJ, Zhang D, Lieberman J. HMG2 interacts with the nucleosome assembly protein SET and is a target of the cytotoxic T-lymphocyte protease granzyme A. *Mol Cell Biol* 2002;**22**:2810–20.
27. Fan Z, Beresford PJ, Oh DY, Zhang D, Lieberman J. Tumor suppressor NM23-H1 is a Granzyme A-activated DNase during CTL-mediated apoptosis, and the nucleosome assembly protein SET is its inhibitor. *Cell* 2003;**112**:659–72.
28. Fan Z, Beresford PJ, Zhang D, Xu Z, Novina CD, Yoshida A, et al. Cleaving the oxidative repair protein Ape1 enhances cell death mediated by granzyme A. *Nat Immunol* 2003;**4**:145–53.
29. Beresford PJ, Kam CM, Powers JC, Lieberman J. Recombinant human granzyme A binds to two putative HLA-associated proteins and cleaves one of them. *Proc Natl Acad Sci USA* 1997;**94**(17):9285–90.
30. Zhang D, Pasternack MS, Beresford PJ, Wagner L, Greenberg AH, Lieberman J. Induction of rapid histone degradation by the cytotoxic T lymphocyte protease granzyme A. *J Biol Chem* 2001;**276**:3683–90.
31. Zhang D, Beresford PJ, Greenberg AH, Lieberman J. Granzymes A and B directly cleave lamins and disrupt the nuclear lamina during granule-mediated cytolysis. *Proc Natl Acad Sci USA* 2001;**98**:5746–51.
32. Zhu P, Zhang D, Chowdhury D, Martinvalet D, Keefe D, Shi L, Lieberman J. Granzyme A, which causes single-stranded DNA damage, targets the double-strand break repair protein Ku70. *EMBO Rep* 2006;**7**(4):431–7.
33. Zhu P, Martinvalet D, Zhang D, Schlesinger A, Chowdhury D, Lieberman J. The cytotoxic T lymphocyte protease granzyme A cleaves and inactivates poly(adenosine 5'-diphosphate-ribose) polymerase-1. *Blood* 2009;**114**:1205–16.
34. Jans DA, Briggs LJ, Jans P, Froelich CJ, Parasivam G, Kumar S, et al. Nuclear targeting of the serine protease granzyme A (fragmentin-1). *J Cell Sci* 1998;**111**:2645–54.
35. Martinvalet D, Walch M, Jensen DK, Lieberman J. Response: Granzyme A: cell death-inducing protease, pro-inflammatory agent, or both? *Blood* 2009;**114**:3969–70.
36. Beresford PJ, Xia Z, Greenberg AH, Lieberman J. Granzyme A loading induces rapid cytolysis and a novel form of DNA damage independently of caspase activation. *Immunity* 1999;**10**:585–94.
37. Beresford PJ, Zhang D, Oh DY, Fan Z, Greer EL, Russo ML, et al. Granzyme A activates an endoplasmic reticulum-associated caspase-independent nuclease to induce single-stranded DNA nicks. *J Biol Chem* 2001;**276**(43):285–93.

38 Chowdhury D, Beresford PJ, Zhu P, Zhang D, Sung JS, Demple B, et al. The exonuclease TREX1 is in the SET complex and acts in concert with NM23-H1 to degrade DNA during granzyme A-mediated cell death. *Mol Cell* 2006;**23**:133–42.
39 Chinnaiyan AM, Hanna WL, Orth K, Duan H, Poirier GG, Froelich CJ, Dixit VM. Cytotoxic T-cellderived granzyme B activates the apoptotic protease ICE-LAP3. *Curr Biol* 1996;**6**:897–9.
40 Gu Y, Sarnecki C, Fleming MA, Lippke JA, Bleackley RC, Su MS. Processing and activation of CMH-1 by granzyme B. *J Biol Chem* 1996;**271**(10):816–20.
41 Srinivasula SM, Fernandes-Alnemri T, Zangrilli J, Robertson N, Armstrong RC, Wang L, et al. The Ced-3/interleukin 1beta converting enzyme-like homolog Mch6 and the lamin-cleaving enzyme Mch2alpha are substrates for the apoptotic mediator CPP32. *J Biol Chem* 1996;**271**(27):099–106.
42 Orth K, Chinnaiyan AM, Garg M, Froelich CJ, Dixit VM. The CED-3/ICE-like protease Mch2 is activated during apoptosis and cleaves the death substrate lamin A. *J Biol Chem* 1996;**271**(16):443–6.
43 Duan H, Orth K, Chinnaiyan AM, Poirier GG, Froelich CJ, He WW, Dixit VM. ICE-LAP6, a novel member of the ICE/Ced-3 gene family, is activated by the cytotoxic T cell protease granzyme B. *J Biol Chem* 1996;**271**(16):720–4.
44 Fernandes-Alnemri T, Armstrong RC, Krebs J, Srinivasula SM, Wang L, Bullrich F, et al. In vitro activation of CPP32 and Mch3 by Mch4, a novel human apoptotic cysteine protease containing two FADD-like domains. *Proc Natl Acad Sci USA* 1996;**93**:7464–9.
45 Heibein JA, Goping IS, Barry M, Pinkoski MJ, Shore GC, Green DR, Bleackley RC. Granzyme B-mediated cytochrome c release is regulated by the Bcl-2 family members bid and Bax. *J Exp Med* 2000;**192**:1391–402.
46 Susin SA, Lorenzo HK, Zamzami N, Marzo I, Snow BE, Brothers GM, et al. Molecular characterization of mitochondrial apoptosis-inducing factor. *Nature* 1999;**397**:441–6.
47 Suzuki Y, Imai Y, Nakayama H, Takahashi K, Takio K, Takahashi R. A serine protease, HtrA2, is released from the mitochondria and interacts with XIAP, inducing cell death. *Mol Cell* 2001;**8**:613–21.
48 Sutton VR, Davis JE, Cancilla M, Johnstone RW, Ruefli AA, Sedelies K, et al. Initiation of apoptosis by granzyme B requires direct cleavage of bid, but not direct granzyme B-mediated caspase activation *J Exp Med* 2000;**192**:1403–14.
49 Li J, Figueira SK, Vrazo AC, Binkowski BF, Butler BL, Tabata Y, et al. Real-time detection of CTL function reveals distinct patterns of caspase activation mediated by Fas versusgranzyme B. *J Immunol* 2014;**193**(2):519–28.
50 Farhadi N, Lambert L, Triulzi C, Openshaw PJ, Guerra N, Culley FJ. Natural killer cell NKG2D and granzyme B are critical for allergic pulmonary inflammation. *J Allergy Clin Immunol* 2014;**133**(3):827–35.e3.
51 Kerr JF, Wyllie AH, Currie AR. Apoptosis: a basic biological phenomenon with wide-ranging implications in tissue kinetics. *Br J Cancer* 1972;**26**(4):239–57.
52 Wyllie AH, Kerr JFR, Currie AR. Cell death: the significance of apoptosis. *Int Rev Cytol* 1980;**68**:251–306.
53 Daniel N, Korsmeyer S. Cell death: critical control point. *Cell* 2004;**116**:205–19.
54 Yuan J, Horvitz HR. The *Caenorhabditis elegans* cell death gene ced-4 encodes a novel protein and is expressed during the period of extensive programmed cell death. *Development* 1992;**116**:309–20.

55 Li H, Zhu H, Xu CJ, Yuan J. Cleavage of BID by caspase 8 mediates the mitochondrial damage in the Fas pathway of apostosis. *Cell* 1998;**94**:491–501.
56 Ellis HM, Horvitz HR. Genetic control of programmed cell death in the nematode *C. elegans*. *Cell* 1986;**44**(6):817–29.
57 Gavrieli Y, Sherman Y, Ben-Sasson SA. Identification of programmed cell death in situ via specific labeling of nuclear DNA fragmentation. *J Cell Biol* 1992;**119**:493–501.

2

Mitochondria in Focus: Targeting the Cell-Death Mechanism

Humberto De Vitto,[1] Roberta Palorini,[2,3] Giuseppina Votta,[2] and Ferdinando Chiaradonna[4]

[1]*Center of Health and Science, Federal University of Rio de Janeiro, Rio de Janeiro, Brazil*
[2]*SYSBIO Center for Systems Biology, Department of Biotechnology and Biosciences, University of Milano-Bicocca, Milan, Italy*
[3]*Luxembourg Centre for Systems Biomedicine, Esch-sur-Alzette, Luxembourg*
[4]*Department of Biotechnology and Biosciences, University of Milano-Bicocca, Milan, Italy*

Abbreviations

$\Delta\Psi m$	mitochondrial inner transmembrane potential
acetyl-CoA	acetyl coenzyme A
ACL	ATP-citrate lyase
ACS	acetyl-CoA synthetase
AD	Alzheimer's disease
ADP	adenosine diphosphate
AIDS	acquired immunodeficiency syndrome
AIF	apoptosis-inducing factor
ALS	amyotrophic lateral sclerosis
ANT	adenosine nucleotide translocator
Apaf-1	apoptotic protease activating factor-1
APP	amyloid precursor protein
ATP	adenosine triphosphate
BA	bongkrekic acid
BIR	baculoviral IAP repeat
CARD	caspase-associated recruitment domain
CCCP	carbonyl cyanide *m*-chlorophenylhydrazone
CsA	cyclosporine A
CypD	cyclophilin D
cyt c	cytochrome c
DDR	DNA damage response
ECM	extracellular matrix
endo G	endonuclease G
ER	endoplasmic reticulum
ETC	electron transport chain
FAK	focal adhesion kinase
GPX	glutathione peroxidase
GR	glutathione reductase

Apoptosis and Beyond: The Many Ways Cells Die, First Edition. Edited by James Radosevich.
© 2018 John Wiley & Son Inc. Published 2018 by John Wiley & Son Inc.

HD	Huntington's disease
H2B	histone 2B
HIF-1α	hypoxia-inducible factor 1α
HK	hexokinase
IAP	inhibitor of apoptosis protein
IMM	inner mitochondrial membrane
KILLER	death receptor-4 and -5
MAPK	mitogen-activated protein kinase
MLKL	mixed-lineage kinase domain-like
MMP	matrix metalloproteinase
MOMP	mitochondrial outer-membrane permeabilization
MPT	mitochondrial permeability transition
MPTP	mitochondrial permeability transition pore
MST1	mammalian sterile-20
mtDNA	mitochondrial DNA
NAD^+	oxidized nicotinamide adenine dinucleotide
NADH	reduced nicotinamide adenine dinucleotide
$NADP^+$	oxidized nicotinamide adenine dinucleotide phosphate
NADPH	reduced nicotinamide adenine dinucleotide phosphate
NLS	nuclear localization signal
OMM	outer mitochondrial membrane
OXPHOS	oxidative phosphorylation
PBR	peripheral benzodiazepine receptor
PCD	programmed cell death
PD	Parkinson's disease
PGAM5	phosphoglycerate mutase family member 5
PT	permeability transition
PUMA	p53-upregulated modulator of apoptosis
RNS	reactive nitrogen species
RIPK1	receptor-interacting protein kinase 1
ROS	reactive oxygen species
SOD	superoxide dismutase
TCA	tricarboxylic acid
TRAIL	tumor necrosis factor-related apoptosis-inducing ligand
tRNA	transfer RNA
TRXR2	thioredoxin reductase
UPR	unfolded protein response
VDAC	voltage-dependent anion channel

2.1 Introduction

As a natural control, multicellular animals must tune their cell fate between life and death. Apoptosis, one of the most important and regulated processes of destruction, is a complex, energy-dependent biochemical network [1–3]. In organisms, apostosis is used to tightly control cell numbers and tissue size, and to protect against potentially dangerous cells and environmental signals that threaten homeostasis [4]. The term "apoptosis" was definitively inserted in the scientific literature by J.F. Kerr, A.H. Wyllie, and A.R. Currie [5]. In their seminal paper, these authors described a type of programmed cell death (PCD) observed in various tissues and cell types that had distinct morphological features as compared to cells undergoing pathological necrotic cell death [5].

In the 1980s, new insights were gained into apoptosis by the identification of the *Bcl-2* gene as an important determinant for the development and maintenance of B-cell leukemia [6,7]. In particular, some authors described the ability of the Bcl-2 protein to promote cell immortalization and the survival of B-lymphoid cells, otherwise destined to die, by cooperating with the oncogene *c-myc* [8]. Along with these first indications regarding the role of Bcl-2 in cell survival, immunolocalization experiments clearly demonstrated that Bcl-2 was an outer mitochondrial membrane protein, suggesting a relationship between the mitochondria and apoptotic cell death [9]. However, the first indication regarding the connection between apoptosis and the mitochondria must be assigned to D.D. Neymeyer and colleagues, who showed that the apoptotic process could be blocked by the Bcl-2 protein [10]. Intensive efforts were promptly made to understand the key regulatory mechanisms linking the mitochondria and apoptosis, which permitted the identification of other proteins, including Bcl-xL, E1B 19K, CED-9, and Mcl-1, which, like Bcl-2, could act as inhibitors of cell death [11–13]. However, the exact mechanism by which Bcl-2 inhibited dependent cell-death protease activation was not established. Two scientific teams, advised by D.D. Newmeyer and X. Wang [14,15], simultaneously published papers characterizing, for the first time, the role of mitochondrial cytochrome c (cyt c) protein as a key element for apoptosis in *Xenopus* egg extract and in a human cell line. This protein normally resides in the space between the inner and outer mitochondrial membranes (IMM and OMM), where it functions to transfer electrons as part of oxidative phosphorylation (OXPHOS). However, upon certain stimuli triggering cell death, cyt c is released from the mitochondrion into the surrounding cytoplasm. The authors suggested that such a cytoplasmic localization permits cyt c to activate a cascade of events, especially characterized by specific proteases activation, which together lead to apoptotic death. Most interestingly, cyt c translocation, and consequently protease activation, was prevented by Bcl-2 expression and mitochondrial localization, underlining its main role in the inhibition of mitochondria-dependent apoptosis [14,15]. As the role of mitochondria in the apoptotic program was further clarified, other details regarding the cytoplasmic mechanisms were identified. Indeed, since cyt c is released into the cytosolic compartment, a dATP-dependent complex, called the "apoptosome," forms through the action of apoptotic protease activating factor-1 (Apaf-1) and cyt c, which proceeds to activate a latent protease, namely procaspase 9, into caspase 9 [16]. The latter is a member of a family of **c**ysteine **asp**artyl-specific prote**ases** (**caspases**), comprising 12 proteins strictly involved in the apoptotic cell death process.

In 1998, D.R. Green and J.C. Reed published a special issue addressing the relation between mitochondria and apoptosis [17]. The authors identified and discussed the main mechanisms altering mitochondrial physiology and leading to this suicide event. They suggested that some signals coming to the mitochondria, such as calcium (Ca^{++}), oxidants, proapoptotic proteins (e.g., Bax/Bak), and even caspase proteins, can modulate mitochondrial physiology and produce outgoing signaling, leading to PCD. In their seminal work, the authors hypothesized an apoptotic mechanism called the "mitochondria swelling route," based on an alteration of the mitochondrial membrane features. In particular, this speculative mechanism is characterized by a mitochondrial inner transmembrane potential ($\Delta\Psi m$) collapse, leading to mitochondrial channels opening and ultimately to mitochondrial permeability transition pore (MPTP) opening. This results in OXPHOS uncoupling, due to H^+ gradient dissipation, and mitochondrial volume dysregulation. As a consequence, the mitochondrial matrix volume expansion can eventually induce outer-membrane rupture, releasing caspase-activating proteins into the cytosol and causing PCD via apoptosis [17]. On the other hand, pharmacological inhibitors of the MPTP can act as potent inhibitors of cyt c release, and hence prevent apoptosis [18]. Taken together, it was reasonably attractive to postulate that the mitochondria are a critical control point not only in cell death, but also in cell survival.

Given the advance of apoptosis issues, different mechanisms have been hypothesized to explain the role of the mitochondria in PCD. In fact, it has been proposed that a mitochondrion can act either as a passive organelle that releases cytotoxic proteins into the cytoplasm by passive leakage following the action of the proapoptotic BH-3-only proteins (e.g., Bid or Bim) or as an active organelle that induces, for instance, Ca^{++} mobilization from the endoplasmic reticulum (ER) to the mitochondria, leading to mitochondrial outer-membrane permeabilization (MOMP). In addition, morphological changes in the mitochondria, such as fission, induced by proteins like Drp1 and Fis, promote the production of reactive oxygen species (ROS), and inner-membrane changes may induce cyt c release across the outer membrane. This implies that, somehow, cytosolic proteins (such as Bid and other members of the Bcl-2 family that target the outer membranes of mitochondria) transduce signals into the interior of the organelle [19–21]. Nevertheless, as D.D. Newmeyer and S. Ferguson-Miller have said, the "mitochondria and apoptosis scenario was turning diverse and controversial." In this regard, the authors have suggested that the next challenges to significantly improving our understanding will be to determine if these different proposed mechanisms may be utilized simultaneously, sequentially, and/or in a specific cellular context [19].

During the 2000s, some points of critical importance to understanding the complex and uncertain apoptotic process were addressed. Studies explaining the diverse array of positive and negative signals acting on the mitochondria, the diverse functions of pro- and antiapoptotic members of the Bcl-2 family, and the biochemical aspects of caspase activation must be considered the milestones [3,22,23]. All these aspects will be more extensively illustrated in later sections; here, we want just to underline that the identification of critical control points, such as the role of incoming signals, Bcl-2-family proteins, and caspase, has provided important clues to the role of mitochondrial-controlled apoptosis in maintaining physiological homeostasis, either during organism development [24–26] or in many pathological processes, for instance cancer and neurodegenerative diseases [27]. Otherwise, such dual roles are not unexpected, since mitochondria acting as a central hub of different stimuli (e.g., Ca^{++} or oxygen radicals), coming from different intracellular organelles (e.g., nucleus or ER) and with a different intensity (e.g., cytoplasmic Ca^{++} concentration) and duration, may control an array of cell outcomes that span from full survival to excessive apoptosis. All of these stimuli contribute to various human diseases, such as cancer, autoimmune diseases, viral infections, acquired immunodeficiency syndrome (AIDS), ischemia injury, and neurodegenerative diseases [27]. Cancer is a convincing example of human disease associated with increased cell survival; one for which evidence unquestionably exists in support of the idea that mitochondrial bioenergetics and cell metabolism are strongly associated with cell death/survival as critical features for tumor progression [28–32]. Thus, such identification has permitted the development of new therapeutic strategies addressing the relation between mitochondria and apoptosis (see Section 2.3).

In line with this, the current chapter is focused on the role of the mitochondria in apoptosis. We will describe and highlight an overall view of mitochondrial morphology, mitochondrial metabolism, and mitochondrial bioenergetics by correlating these features with the intrinsic apoptotic pathway. Such mitochondrial characteristics will then be used to explore in more detail the fascinating issue of how regulation of mitochondrial function can result either in cell accumulation or cell loss, especially in human diseases, including cancer and neurodegenerative diseases.

2.2 Mitochondria: Overview

2.2.1 Mitochondrial Morphology

In a somatic mammalian cell, it is well established that each mitochondrion harbors thousand of copies of mitochondrial DNA (mtDNA) and houses the OXPHOS system (see Figure 2.1) [33]. Mitochondrial morphology in mammalian cells has received a great

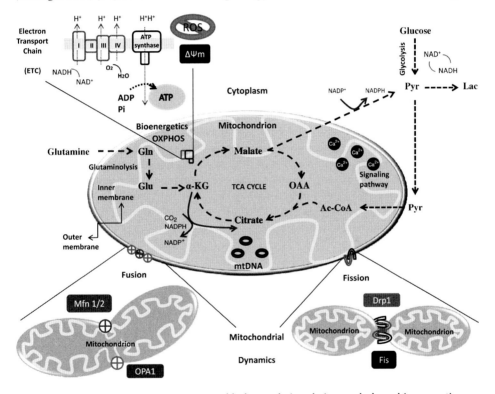

Figure 2.1 Mitochondria operate as a central hub, regulating their morphology, bioenergetics, and biosynthetic activity in order to respond to the metabolic demands of the cell. *Bioenergetics:* OXPHOS machinery uses electron transport chains (ETCs; complexes I–IV) to transport electrons to oxygen (the final electron acceptor), which is reduced to H_2O. The electron transport generates a $\Delta\Psi m$, which is required to activate the ATP synthase used to power the synthesis of the energy carrier molecule ATP from ADP + Pi. *Metabolism:* Glucose (glycolysis) and glutamine (glutaminolysis) are important nutrients required to govern catabolic and anabolic mitochondrial pathways (the tricarboxylic acid (TCA) cycle). This metabolic network shows as rate-limiting steps, with enzymes regulated by the oxidized/reduced nicotinamide adenine dinucleotide (NAD^+/NADH) and nicotinamide adenine dinucleotide phosphate ($NADP^+$/NADPH) ratio, the ATP/ADP ratio, and the acetyl-CoA/ CoA ratio. *Morphology:* Mitochondria harbor multiple copies of mtDNA, divided between two different compartments: the intermembrane space and the mitochondrial matrix (inner and outer membranes). They also harbor a complex enzymatic machinery of fusion (Mnf1/2 and Opa1) and fission (Drp1 and Fis), required to control mitochondrial dynamics in a vast spectrum of stimuli (e.g., ROS and Ca^{2+}).

deal of interest, especially after early findings that the mitochondrial network dynamic may be critical for their function and for cell homeostasis [34]. Although the molecular mechanisms regulating mitochondrial motility, fusion, and fission are not yet totally understood, they have together been related to a protective role in mitochondrial integrity and function. A weakness of the mitochondrial fusion and fission process has been linked to human diseases, including cancer and neurodegenerative disorders [35,36]. Mitochondria change their morphology in response to various stimuli. For example, during the early stages of apoptosis, mitochondrial fragmentation and cristae remodeling are recognized as important morphological alterations leading to cell death. Most interestingly, some proteins that control mitochondrial network dynamics appear to participate in the apoptotic process [37]. The mitochondrial fusion and fission machinery includes a set of outer-membrane proteins (e.g., Drp1 and Fis1) and one of inner-membrane proteins (e.g. Mfn1, Mnf2, and Opa1), which play a key role in apoptosis [38]. In fact, several studies performed in mammalian cells have shown that either decreasing Drp1 or Fis1 activity or increasing Mfn1 or Mfn2 activity may interfere with cyt c release and hence block PCD. On the other hand, it has also been shown that a reduction of Mfn1 or Mfn2 activity increases cell responsiveness to apoptotic stimuli [38]. Moreover, increased expression of the proapoptotic proteins Bax and Bak induces mitochondrial fragmentation and fusion of mitochondrial cristae, resulting in cell death [39]. In contrast, overexpression of the antiapoptotic Bcl-2 protein increases mitochondrial size and structural complexity, preventing cell death [40].

It is tempting to speculate that under metabolic stress, mitochondrial fusion may be an efficient mechanism by which to eliminate and recycle damaged mitochondria, avoiding the loss of an essential compartment and preserving energy production through OXPHOS. Conversely, mitochondrial fission has been described as an mtDNA and/or protein quality-control system that leads, in stressful conditions, to the establishment of healthy mitochondria. However, during high levels of stress (e.g., nutrient deprivation; see Figure 2.2), the fission machinery can trigger apoptosis [41]. As suggested by Youle and van der Bliek [41], in normal cells, the fusion and fission machinery helps to mitigate stress and to create new mitochondria, but disruptions in these processes affect normal development. Accordingly, further understanding of mitochondrial dynamics could help better elucidate the process of mitochondria cell death.

2.2.2 Mitochondria and Cell Metabolism

In the last 20 years, several findings have shown that cell metabolism and cell death are strictly associated [42]. In general, the mitochondrion is believed to act as a sensor capable of determining cell fate under several types of stress: whether to adapt and survive or die. The ability of mitochondria to keep cells alive or to lead to an irreversible catastrophic event is programmed into an array of mitochondrial metabolic pathways that are used to control the major metabolic signals. In mammalian cells, the most important mechanism that supports the mitochondria and cell metabolism is the adenosine triphosphate/adenosine diphosphate (ATP/ADP) ratio, coordinated by OXPHOS machinery. This complex metabolic network relies on nutrient availability (e.g., glucose, glutamine, lipids) and oxygen supply. Furthermore, OXPHOS activity is considered the mayor source of production of ROS, suggesting that mitochondria may sense oxidative stress leading to cell metabolic control and apoptosis [42].

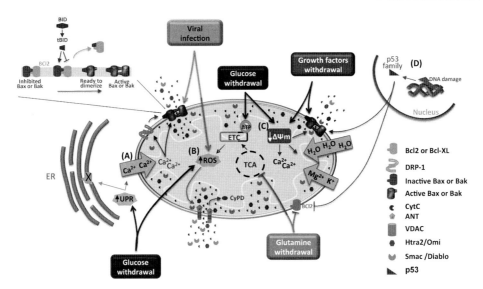

Figure 2.2 Representation of incoming death signals. Increased matrix metalloproteinase (MMP) is the "point of no return" in the cascade of events leading to apoptosis, allowing the release of apoptogenic factors. Different events lead to different MMP changes during apoptosis induction, such as an increase of Ca^{++} uptake (A), ROS formation (B), a drop in $\Delta\Psi m$ (C), or increased expression of Bax due to p53 activated by DNA damage (D). These events are driven by a wide range of stress conditions, including glucose and glutamine withdrawal (which activates almost all of them), growth-factor withdrawal, and viral infection, which are able to act in a direct way on Bax or Bak activation and through $\Delta\Psi m$ and ROS changes, respectively.

Another class of metabolic signal that emerges with a substantial amount of interest in studies of metabolic control and apoptosis is the balance between oxidized and reduced nicotinamide adenine dinucleotide (NAD^+/NADH) and nicotinamide adenine dinucleotide phosphate ($NADP^+$/NADPH). It is remarkable that catabolic, anabolic, and waste-disposal mitochondrial pathways act as rate-limiting steps, with enzymes regulated by the NAD^+/NADH and $NADP^+$/NADPH ratio. For instance, in order to sustain metabolic pathways, those that synthetize macromolecules (anabolism), those that degrade molecules to release energy (catabolism), and those that eliminate toxic product (waste disposal) all depend on redox homeostasis, which may impact on cell-death pathways. Interestingly, some mitochondrial byproducts are found in the cytosolic environment, such as $NADP^+$/NADPH plus ATP. The enzyme glutathione synthetase generates reduced glutathione (GSH). The redox balance between reduced and oxidized glutathione (GSSG) represents one of the most important antioxidant systems, playing a pivotal role in mitigating ROS and reactive nitrogen species (RNS) production. This mechanism emerges as a potentially key pathway that could be exploited for the development of novel strategies in human diseases (see Section 2.3.2) [43].

The third class of metabolic signals, the acetyl coenzyme A (acetyl-CoA)/CoA ratio is regulated by two enzymes: acetyl-CoA synthetases (ACS) and ATP-citrate lyase (ACL). This set of reactions provides acetyl-CoA for the lipogenesis and synthesis of other macromolecules, and for histone acetylation reactions used to regulate gene expression and enzyme function. Interestingly, interference in the acetyl-coenzymeA (acetyl-CoA)/CoA ratio may be sufficient to trigger the apoptotic process [44].

2.2.3 Mitochondria and Bioenergetics Metabolism

In the 1940s, mitochondria were described as housing OXPHOS machinery in eukaryotes cells, creating interest in the study of mitochondria as a biological component that unites many diseases. Structurally, mitochondria are organelles with an outer membrane and an inner membrane. Thus, they can be divided into two different compartments: the intermembrane space and the mitochondrial matrix. These compartments are structurally and functionally distinct. The particular biochemical anatomy of mitochondria provided underlying insights that allowed Peter Mitchell to introduce his chemiosmotic theory. The author hypothesized that biological oxidation reactions generate a proton gradient across the IMM that is transduced into a more useful form of chemical energy: the molecule of ATP [45]. In line with this observation, energy is generated in mitochondria by oxidizing hydrogen derived from our dietary carbohydrates (glycolysis), amino acids such as glutamine (glutaminolysis), and fatty acid (β-oxidation) through the tricarboxylic acid (TCA) cycle. These catabolic pathways can fuel the mitochondrial redox center, represented by universal acceptors (NADH and FADH2) that donate protons through the electron transport chain (ETC), and involve the reduction of O_2 to H_2O, generating ATP as a final product (see Section 2.3.1 for a detailed description) [46]. Mitochondria also encompass various others metabolic pathways, participating in lipogenesis, the urea cycle, and the synthesis of pirimidines, heme, and some amino acids [33]. Therefore, mitochondria must be considered organelles strongly responsible for the bioenergetics mechanisms controlling cell life and death. The entire role of mitochondrial dynamics and mitochondrial bioenergetics and metabolism is summarized in Figure 2.1.

Notably, nutrient shortage, hypoxia, oncogenes expression, ROS production, Ca^{2+} signaling, and other incoming and outgoing signals can influence cell metabolism, interfering in mitochondrial energy production and hence directly impacting on the rewiring of metabolic pathways. One important method of controlling cell fate involves regulation by Bcl-2 family proteins, which undergo MOMP and release cyt c and other mitochondrial proteins into the cytosol, triggering apoptosis. We address some examples in the next section.

2.3 Mitochondrial Apoptotic Pathways

2.3.1 Incoming Signals: Cytoplasmic and Nuclear Events

2.3.1.1 Cytoplasmic Events

Mitochondria have been described as playing a central role in the apoptotic process [17] at several levels: maintenance of ATP production [47], $\Delta\Psi m$ and mitochondrial permeability transition (MPT) for release of certain apoptogenic factors from the intermembrane space into the cytosol [17], and ROS production [48]. A major point of discussion, however, is whether mitochondria are the central decision maker or simply amplify, as a fundamental hub, the signaling pathways that link the detection of intracellular danger to adaptive responses by housing crucial signal transducers. This section is dedicated to the second aspect of mitochondrial involvement in apoptosis.

Mitochondria contain two membranes: the OMM fully surrounds the IMM, with a small intermembrane space between them. The IMM has restricted permeability and is loaded with proteins involved in electron transport and ATP synthesis. It surrounds the mitochondrial matrix, where the TCA cycle produces the electrons that are pumped from one protein complex to the next within the IMM. At the end of this ETC, the final electron acceptor is oxygen, and this ultimately forms H_2O. At the same time, the ETC produces ATP. During electron transport, the participating protein complexes drive protons from the matrix out to the intermembrane space. This creates a concentration gradient of protons that then flows back into the mitochondria through another protein complex (ATP synthase) to power synthesis of the energy carrier molecule ATP [49]. The total force driving protons into the mitochondria (i.e., Δp) is a combination of $\Delta\Psi m$, a charge or electrical gradient, and mitochondrial pH gradient (ΔpH_m, an H^+ chemical or concentration gradient).

The OMM has protein-based pores that allow the passage of ions and proteins through the MPTP. The MPTP consists of the voltage-dependent anion channel (VDAC), the adenine nucleotide translocator (ANT), and cyclophilin D (CypD) [50]. These proteins cooperate to form a large conductance channel, and they are responsible for the permeability transition (PT) of mitochondria. The increase in PT is one of the major causes of cell death in a variety of conditions, and it may also play a role in mitochondrial autophagy [51].

In physiological conditions, mitochondria maintain a high $\Delta\Psi m$, which is used to generate ATP and for the import of proteins. In this setting, ROS are kept to a minimum [52]. When the IMM becomes permeable to positively charged ions (K^+, Mg^{++}, and Ca^{++}), which flow en masse into the mitochondrial matrix, driven by its electronegative nature, there is a dissipation of $\Delta\Psi m$ [17], acidification of the matrix (if the pH drops below 7.0) [53], abolition of mitochondrial ATP synthesis, and massive entry of H_2O into the mitochondrial matrix, causing an osmotic imbalance that results in MOMP and the release into the cytosol of several factors that are normally confined within the intermembrane space, leading to apoptosis (see later). Most interestingly, several studies have shown that CyPD or ANT inhibition by cyclosporine A (CsA) and bongkrekic acid (BA), respectively, may block apoptosis in some systems, suggesting the involvement of the MPTP in apoptosis. However, Bossy-Wetzel et al. [54] have described a noncanonical system whereby cyt c release and caspase activation can occur before any detectable loss of MOMP, with the opening of the MPTP occurring downstream of apoptosome formation. Nonetheless, whether caspases can induce MPTP opening requires future elucidation [54]. However, MOMP represents an event that marks the point of no return of multiple signal-transduction cascades leading to cell death. Indeed, defects that alter the capacity of mitochondria to undergo MOMP are associated with a large array of human pathologies, including infectious diseases, ischemic conditions, neurodegenerative disorders, and cancer [55,56], and the dissipation of the $\Delta\Psi m$ constitutes an early and irreversible step in the cascade of events that leads to apoptotic cell death [57].

It is of note that the MPTP can operate in three distinct states: (i) a closed state, characterized by a high $\Delta\Psi m$; (ii) a low-conductance state, characterized by a partly open pore, with a permeability to molecules <300 Da and a reversible decrease of the $\Delta\Psi m$; (iii) the classic high-conductance state, characterized by an irreversible pore flip from the low-conductance state to the high-conductance state, with a permeability to molecules <1.5 kDa and an irreversible decrease of the $\Delta\Psi m$ collapse [58]. Maintenance of the $\Delta\Psi m$ is essential for cell survival; indeed, the MPTP remains tightly closed during normal mitochondrial function, in order to drive ATP synthesis and maintain OXPHOS activity [58]. Nonetheless, disruption of the $\Delta\Psi m$ by direct methods, such as the mitochondrial uncoupler carbonyl cyanide *m*-chlorophenylhydrazone (CCCP), does not induce PT per se [59], suggesting that PT requires more than just dissipation of the $\Delta\Psi m$ in order to occur.

Mitochondrial membrane integrity is tightly controlled through interactions between pro- and antiapoptotic members of the Bcl-2 protein family [56]. Antiapoptotic Bcl-2 family members interact with and close the VDAC, whereas some but not all, proapoptotic members interact with the VDAC to open a protein-conducting pore through which apoptogenic factors pass (see later). The VDAC interacts not only with Bcl-2 family members but also with proteins such as gelsolin, an actin-regulatory protein, which is able to converge a variety of cell-survival and cell-death signals to MPTP status [60].

The Bcl-2 family resides immediately upstream of the mitochondria, and can be divided into two classes with antagonistic properties: antiapoptotic (Bcl-2, Bcl-xL, Bcl-W, A1, Bag-1, and Mcl-1) and proapoptotic (Bax, Bak, Bad, Bid, Bcl-s, and Bok) [2]. The ratio between the antiapoptotic and proapoptotic Bcl-2 [52] members determines the susceptibility of cells to a death signal. All these proteins share in common at least a BH-3 domain: an amphipathic α-helical that serves as a critical death domain.

Proapoptotic Bcl-2 members can be subgrouped into the "Bax" family, which has several domains homologous to domains of Bcl-2, while the BH-3-only domain subgroup has only the BH-3 domain in common with Bcl-2. Members of the Bax family normally reside in an inactive form in the OMM or in the cytosol, whereas inactive members of the BH-3-only-domain family are normally localized into the cytosol. Both subgroups, upon activation by proapoptotic signals, need to pass from inactive monomeric conformation to active oligomeric complexes. BH-3-only-domain proteins are further subdivided into two groups: molecules typified by Bid, which can bind and activate proapoptotic Bax and Bak, and molecules such as Bim, Bad, and Noxa, which preferentially bind and inhibit the antiapoptotic Bcl-2/Bcl-xL proteins. Following multiple death stimuli, for instance, Bax translocates as a large homo-oligomer at the outer surface of the mitochondria, where it participates, in an almost unclear mode, in causing channel opening and mitochondrial fragmentation. Bak, when it activates the cell-death process, is arranged in large homo-oligomer complexes, apparently representing the active conformation [61].

Several ER-membrane proteins have been reported to interact with Bcl-2 family members and influence the apoptotic process. For example, Bax inhibitor 1 protein, a mammalian apoptosis suppressor, is localized specifically to the ER membrane [62]. Moreover, a member of the BH3-only family, Spike, localized in the ER, inhibits the formation of a complex between BAP31 and Bcl-xL [63]; in this way, BAP31 can be cleaved, and its amino-terminal fragment integrated in the ER, causing an early release of Ca^{++} from the ER, with an associated uptake of Ca^{++} into mitochondria and mitochondrial recruitment of a dynamin-related protein that mediates rupture of the OMM [64]. Mitochondria take up Ca^{++} electrophoretically from the cytosol through a uniport transporter. The energy-dependent Ca^{++} uptake, coupled to the release mediated by the exchange systems, constitutes an energy-dissipating mitochondrial Ca^{++} cycle. The affinity for Ca^{++} of the uniporter is low, and the size of the mitochondrial Ca^{++} pool is small under physiological conditions. However, under pathological conditions, intracellular Ca^{++} concentrations rise, becoming capable of stimulating numerous pathways, including activation of calcium-dependent proteases. Calpain-family cysteine proteases, implicated in the activation of Bax and Bid and in the inhibition of Bcl-2 and Bcl-xL [65], stimulate ROS production [66]. In turn, Bcl-2 resides in the ER, where it can regulate Ca^{++} homeostasis [67], and thus the induction of MPT (see Figure 2.2). In fact, MPT is also a Ca^{++}-linked process, and many chemicals and radicals promote MPT formation. Typically, the effect of such inducers is to decrease the threshold amount of mitochondrial Ca^{++} needed to cause PT opening. In pathological settings, where the MPT contributes to cell killing, Ca^{++} may have several roles. First, increased Ca^{++} alone, and its uptake into mitochondria, may cause MPT onset. Second, other stressors may decrease the threshold for Ca^{2+}-induced MPT, such that Ca^{2+} need not change but is still permissive for MPT onset. Lastly, stressors and increased Ca^{2+} may act synergistically to induce MPT.

Furthermore, Ca^{++} activates the calcium-sensitive mitochondrial fission protein Drp1, which has been implicated in Bax-induced channel opening and the release of cyt c [68].

Another intrinsic source leading to cell death by apoptosis is ROS. The most abundant and common ROS is the superoxide anion (O_2^-), which may react with other radicals, including nitric oxide (NO), to produce RNS, or else may undergo to dismutation to produce another ROS: hydrogen peroxide (H_2O_2). The latter, when not metabolized, can be further transformed to hydroxyl radical (OH) in the presence of metal ions by the Fenton reaction. OH is one of the strongest oxidants in nature, is highly reactive, and generally acts essentially as a damaging molecule [69].

Physiological levels of the respiratory chain produce ROS mainly from complex I (NADH/ubiquinone oxidoreductase); lower levels produce them from complex III (ubiquinol/cyt c oxidoreductase) [69]. At low levels, ROS have been described in cell hypoxia adaptation, where they participate by regulating the stability of hypoxia-inducible factor 1α (HIF-1α); moderate levels of ROS have been involved in regulating the production of proinflammatory cytokines by directly activating the inflammasome and mitogen-activated protein kinase (MAPK); on the other hand, an excess of mitochondrial ROS has been associated with apoptosis and autophagy induction as a consequence of mitochondrial pore opening and autophagy-specific gene 4 (ATG4) oxidation [70]. Activation of the structurally related protein complex termed "the apoptosome" also requires ROS generated by mitochondria [70]. Moreover, increased levels of ROS may cause oxidative stress, with an enhanced activity of the antioxidant defense system and mitochondrial damage, including protein carbonylation, lipid peroxidation, or mtDNA damage [70].

Experiments using antioxidants have shown that ROS act upstream of mitochondrial membrane depolarization [71], Bax relocalization, and cyt c release, executing caspase activation and nuclear fragmentation. It has been established that both ROS accumulation and PCD require the presence of a functional mitochondrial respiratory chain in most ROS-dependent cell-death systems [71].

Mitochondrial ROS are characterized by a lack of biological specificity, or even by an extreme reactivity; these are features contrary to the requirements of a specific signaling role. The major effect of an increased ROS production is an imbalance of intracellular redox status; that is, an enhancement of the cellular oxidative tonus that is the central common effector of PCD.

Likewise, mitochondria also control non-apoptotic types of PCD, including regulated necrosis. The execution of regulated necrosis has been associated with an oxidative and metabolic boost that is mainly mediated by mitochondria [72,73]. It has recently been shown that pronecrotic signals transduced by the receptor-interacting protein kinase 1 (RIPK1)–RIPK3-containing complex, commonly known as the "necrosome," activate a mechanism that depends on mixed-lineage kinase domain-like (MLKL) and phosphoglycerate mutase family member 5 (PGAM5), resulting ultimately in mitochondrial fragmentation and necrotic cell death [74].

These death signals are activated by a wide range of stimuli, such as nutrient and growth-factor deprivation, unfolded protein response (UPR), and viral infection.

Oxygen and glucose deprivation due to insufficient blood circulation can decrease cancer and neuronal cell survival through distinct mechanisms [75–77]. The mechanism of cell death seems to be dependent on ATP depletion, which in turn activates the mitochondrial death cascade [78,79]. Alternatively, oxidative stress due to mitochondrial oxygen free radicals can mediate glucose withdrawal-induced cell death and trigger Bax-associated events, including the JNK/MAPK signaling pathways [80]. A combination of oxidative stress and ATP depletion has also been reported to precede cell death after glucose withdrawal [75,81]. Basically, glucose deprivation activates a metabolic signaling amplification loop, leading to an early increase of intracellular ATP level followed by a progressive decrease associated with a quick $\Delta\Psi m$ drop [82].

Moreover, glucose deprivation, as well as ER Ca^{++} pool overload or depletion, can result in a harmful accumulation of unfolded proteins as a consequence of protein glycosylation reduction, which leads to a UPR-dependent cell-death mechanism [77]. This pathway splits into two streams according to the severity and longevity of the ER stress, where the cell must make a choice for the good of the organism between survival and PCD. The Bcl-2 family of proteins is central to the cell-death arm of the UPR pathway. In addition, Bax/Bak [83], as well a number of BH3-only proteins, including Bad [84], Bid [85], and Bim, have been linked to ER stress-initiated cell death. In this light, the reduction of protein synthesis can help to solve ER stress and increase cell survival, although a prolonged inhibition of mitochondrial protein synthesis can induce ROS production [86].

Glutamine starvation also induces cell death involving the mitochondria. It can be rescued by the overexpression of Bcl-2 or Bcl-xL. Glutamine substitution with a membrane-permeable form of α-ketoglutarate, pyruvate, or oxaloacetate also rescues cell survival [87]. Nevertheless, glutamine depletion of Myc-transformed cells leads to a profound reduction in the levels of TCA-cycle intermediates despite abundant extracellular availability of glucose, supporting the importance of glutamine in the maintenance of mitochondrial anaplerosis [88].

Growth-factor withdrawal is associated with a metabolic arrest that can result in apoptosis. Cell death is preceded by a loss of OMM integrity and by cyt c release. These mitochondrial events appear to follow a relative increase in $\Delta\Psi m$. This change in membrane potential results from the failure of the ANT/VDAC complex to maintain ATP/ADP exchange, while physiological exchange of ADP for ATP is promoted by Bcl-xL expression, permitting OXPHOS to be regulated by cellular ATP/ADP levels and allowing the mitochondria to adapt to changes in metabolic demand [31,89,90].

It is important to note the observation that patients infected with RNA viruses are under chronic oxidative stress. Oxidative stress via RNA virus infections can contribute to several aspects of viral disease pathogenesis, including apoptosis of the host cell with a strong increase in ROS production, which may play an important signaling role in the regulation of viral replication and organelle function [91]. Viruses either induce or inhibit various mitochondrial processes in a highly specific manner, so that they can replicate and produce progeny. Some viruses encode Bcl-2 homolog proteins capable of counteracting the cellular proapoptotic routes activated by cytoplasmic and/or mitochondrial proteins. Others viruses modulate the PTP and either prevent or induce the release of the apoptotic proteins from the mitochondria (see Figure 2.2) [92].

2.3.1.2 Nuclear Events

As previously described, the apoptotic process may be initiated by several signals, including genomic damage. DNA damage results from multiple environmental and intrinsic sources, such as radiation, x-rays, ultraviolet (UV), and alkylating agents, as well as byproducts of endogenous processes such as ROS and RNS from metabolism and errors from DNA replication [93]. Eukaryotes have evolved, and activate upon DNA damage a complex set of reactions that safeguard the genome: the so-called "DNA damage response" (DDR). Depending on the cell type and the severity and extent of DNA damage, different cellular responses may be elicited. Weak DNA damage normally leads to the induction of cell-cycle arrest, whereas severe and irreparable damage causes the induction of senescence or cell-death programs, such as apoptosis, mitotic catastrophe, autophagy, and necrosis. In this section, we will delineate the mechanisms leading to apoptosis, which is activated in response to cellular stress and is executed when repair of DNA damage is slow or incomplete. Several studies have clearly shown the key role of the p53 protein in the crosstalk between DNA damage and apoptosis. Indeed, complete transcriptional activation of p53 following DNA damage induces expression of several apoptotic target genes, such as p53-upregulated modulator of apoptosis (PUMA), p53AIP1, Bax, and Apaf-1, regulating in this way the mitochondrial intrinsic apoptotic pathway [94]. Also, the extrinsic apoptotic pathway is activated by p53 through transcriptional regulation of tumor necrosis factor-related apoptosis-inducing ligand (TRAIL) receptors, death receptor-4 and -5 (KILLER), and the CD95 receptor (Fas/Apo-1) and CD95 (Fas/Apo-1) ligand [95].

Even if p53 acts mostly as a nuclear transcription factor, it has also been shown to trigger apoptosis in the cytoplasm by transcription-independent mechanisms. It has been found to translocate to the cytoplasm in response to different stress signals, where it stimulates MOMP and caspase activation and antagonizes the antiapoptotic proteins Bcl-2 and Bcl-xL through direct binding (see Figure 2.2) [96]. In addition, the two other members of the p53 family, p73 and p63, are also involved in apoptotic responses to DNA damage. Indeed, the combined absence of p63 and p73 has been shown to strongly impair the induction of p53-dependent apoptosis in response to DNA damage [97]. However, p73 can be proapoptotic itself, even in the absence of p53. In fact, p73 may induce apoptosis via the mitochondrial pathway using PUMA and Bax as mediators, followed by the release of cyt c. In addition, p73 regulates many proapoptotic genes also known to be target sof p53, such as Noxa, caspase 6, and CD95 [98].

There are still many gaps that must be filled in order to provide insight into the association between DDR and intrinsic and extrinsic apoptotic pathways, as well as the relationship between apoptosis and the other cell-death mechanisms, such as mitotic catastrophe and autophagy.

2.3.2 Outgoing Signals: Signals after MOMP (Caspase Activation Pathway)

MPTP opening leads to the loss of the $\Delta\Psi m$ and to the release from the mitochondria into the cytosol of some proapoptotic proteins, normally sequestered into the mitochondrial intermembrane space [99]. There are two groups of apoptotic proteins that move from mitochondria to the cytosol after MOMP: an early group, consisting of cyt c,

Smac/DIABLO, and HtrA2/Omi [100–102], which is engaged for the initial steps of apoptosis, and a later group, consisting of apoptosis-inducing factor (AIF) and endonuclease G (endo G), which is released from the mitochondria when the cells have irreversibly committed to die.

The release of cyt c from mitochondria directly initiates apoptotic events through the formation of the cyt c/Apaf-1/caspase 9-containing apoptosome complex. Apaf-1 is a cytosolic protein with an N-terminal caspase-associated recruitment domain (CARD), a nucleotide-binding domain, and a C-terminal domain containing 12–13 WD40 repeats. The binding of cyt c to Apaf-1 induces the hydrolysis of dATP to dADP and its subsequent replacement by exogenous dATP, an event that is required for apoptosome formation (see Figure 2.3) [103]. In fact, it permits the exposure of CARD on Apaf-1, which can then oligomerize and become a platform on which caspase 9 is recruited and activated through a CARD–CARD interaction [104,105]. The apoptosome is physiologically regulated by several factors at different levels; both the interaction between cyt c and Apaf-1 and the activity of caspase 9 can be modulated [56,106,107]. For instance, physiological levels of nucleotides or transfer RNA (tRNA) may inhibit apoptosis by directly binding to cyt c and consequently by preventing Apaf-1–cyt c interaction [108]. Further regulation of apoptosome assembly results from an alteration of the intracellular

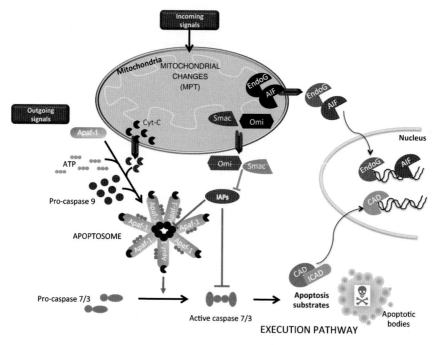

Figure 2.3 Schematic representation of the outgoing signals of the intrinsic apoptotic pathway. Upon the receipt of lethal incoming signals, MPTP opening leads to the release from the mitochondria of some proteins (cyt c, Smac/DIABLO, and Htra2/Omi), which participate in the activation of the caspase cascade. The effector caspases cleave numerous targets that permit the execution of apoptosis, including the endonuclease ICAD/CAD, shown in the figure. Other endonucleases (e.g., AIF and endo G) are released from the mitochondria and translocate to the nucleus to mediate DNA degradation in a caspase-independent manner. All cellular events following the execution pathway will participate in the formation of the apoptotic bodies later degraded by phagocytosis. See text for the details.

levels of K^+ and Ca^{++} [109,110]. Also, the intracellular redox balance may control cyt c function. Indeed, it has been shown that cyt c oxidation or reduction is associated with its pro- and antiapoptotic functions, respectively [111]. Apoptosome function has also been shown to be negatively influenced by different heat-shock proteins [112]. Moreover, caspase 9 activity is inhibited by phosphorylation of its Thr125 site by the action of some protein kinases, including extracellular signal-regulated kinase 1 and 2 (ERK1 and ERK2) and cyclin-dependent kinase 1 (CDK1) [113,114], due to their positive role in cell proliferation [115,116].

Significantly, apoptosome-mediated activation of caspase 9 works as an apoptotic "molecular timer"; this leads to caspase 9 autoprocessing, and will reduce either its affinity for the apoptosome or its activity in relation to its main substrate, caspase 3 [117]. Since such a mechanism has been shown also to be influenced by the caspase 9 level, it has been argued that caspase activity post-MOMP is strongly regulated by caspase expression levels [56].

Released together with cyt c, Smac/DIABLO and HtrA2/Omi have the function of inhibiting inhibitors of apoptosis proteins (IAPs). IAPs make up a family of antiapoptotic proteins that currently has eight members in mammalian cells: NAIP (BIRC1), c-IAP1 (BIRC2), c-IAP2 (BIRC3), XIAP (BIRC4), survivin (BIRC5), Apollon/Bruce (BIRC6), ML-IAP (BIRC7 or livin), and ILP-2 (BIRC8). Among these, only c-IAP1-2, XIAP, and ML-IAP are directly involved in the regulation of apoptosis and interfere with caspase activity; the others regulate cell survival in different ways [118]. In general, IAPs are characterized by the baculoviral IAP repeat (BIR) domain, which can be present either in a single copy or in an array of two or three repeats in the N-terminal portion. Most IAPs contain other functional regions, such as a RING domain conferring E3-ubiquitin ligase activity or a CARD. The only IAP that directly inhibits the activity of caspases is XIAP, which is normally present in the apoptosome complex. It binds caspases 9 at two different sites and inhibits its activity, both avoiding caspase 9 dimerization and hiding the catalytic residue [117,119]. The other IAPs cannot directly inhibit caspase activity, and most work through protein-target ubiquitination [120]. In fact, it has been proved that both XIAP itself and cIAPs can regulate the stability of the different caspases through their ubiquitination and subsequent degradation by proteasome [121,122]. cIAP2 is also able to mono-ubiquitinate caspase 3 and 7 [123], suggesting a post-translational mechanism for caspase regulation, independent of the proteasome-mediated degradation.

It has been demonstrated that Smac/DIABLO can interact with IAPs, particularly XIAP, which avoids the interaction of IAPs with caspase 9 and thus promotes the proteolytic activation and enzymatic activity of caspase 3 [124,125]. On the other hand, it has been suggested that the inhibitory action of Omi/HtrA2 on IAPs is executed through their direct cleavage. This mechanism of action has been observed *in vitro*, particularly with cIAP1, the cleavage of which reduces its ability to inhibit and ubiquitinate caspases. Omi/HtrA2 inhibition of IAP is catalytic and irreversible, and thus more efficient than the stoichiometric anti-IAP activity of Smac/DIABLO [126]. Independent of mechanism, both proteins have an important role in promoting and sustaining apoptosis. This first group of mitochondrial-released proteins (through participation in the activation and recruitment to the apoptosome of caspase 9 and, hence, caspases 3 and 7) is known to have a fundamental role in apoptosis [16,127].

Before discussing the cellular and biochemical effects of caspase 3 and 7 activation, we will briefly present the characteristics, classification, and activation mechanisms of these caspases. Caspases are proteins that play a central role in the execution of apoptosis. They are specific proteases that cleave their substrates after aspartic acid residues in a specific amino acidic context [11]. They can be subdivided into two groups: the initiator caspases, which prefer (I, L, V)ExD motifs, and the effector caspases, which preferentially cleave at DExD motifs. Caspases are generally synthesized as inactive proenzymes or zymogens, which are activated by proteolytic cleavage. For initiator caspases, it has been observed that cleavage is neither required nor sufficient for their activation. In fact, zymogens require dimerization in order to assume an active conformation, and this process is independent of their cleavage [3–5]. The dimerization event occurs at multiprotein activating complexes, as already described for caspase 9 that is recruited by apoptosome. The effector caspases are activated only through a proteolytic cut within their interdomain linker, which is between the large and small subunits, normally caused by an initiator caspase or occasionally by other proteases. The functional significance of the mechanism of caspase activation is not known, but it has been suggested that the cleavage allows the small and large subunits to separate and rearrange to form the catalytic site [128].

Caspases 3 and 7 cleave specific targets that permit the execution of apoptosis. Although a redundant mechanism between these two caspases was long proposed, a recent study has opened the possibility that they have distinct roles during apoptosis. While caspase 3 is required for the efficient execution of apoptosis, making it the dominant effector caspase, caspase 7 seems mainly to be required for apoptotic cell detachment, a typical feature of the apoptotic cells that is probably useful in their removal by phagocytes [129]. Accordingly, caspase 7 can cleave different actin and cytoskeleton factors, and it appears as the principal caspase involved in focal adhesion kinase (FAK) proteolysis [130]. Further, only the elimination of caspase 3 can delay cell death after an apoptotic stimulus; this does not happen in the absence of caspase 7 [129]. Thus, the action of effector caspases is explosive, leading to great change in the cell. Such a cellular caspase-mediated rearrangement has been well reviewed in Taylor et al. [131]. First, caspases act on several key cellular structures, leading to a general contraction of the organelles and contributing to membrane blebbing, a distinctive feature of apoptosis. Therefore, caspases cleave many constituents of the cytoskeleton, such as components of actin microfilaments (including actin itself), actin-associated proteins (myosin, spectrins, actinin, gelsolin, and filamin), microtubular proteins (tubulins and microtubule-associated proteins), and intermediate filament proteins (vimentin, keratins, and nuclear lamins). Caspases also cleave lamins A, B, and C, contributing to the disintegration of the nuclear lamins and the collapse of the nuclear envelope. This event is linked to the DNA fragmentation that occurs during apoptosis as a result of the different endonucleases released by mitochondria (see later), due partly to caspase activity. In particular, caspases permit the activation of CAD [132], which contributes to the fragmentation of the chromatin. DNA fragmentation is associated with the chromatin condensation permitted by histone 2B (H2B) and H2AX phosphorylation. The kinase responsible for this modification, mammalian sterile-20 (MST1) [133,134], is cleaved and activated by caspase 3. As mentioned earlier, another characteristic of the apoptotic cells is their detachment from the extracellular matrix (ECM), in a process involving the caspase-dependent dismantling of cell matrix focal-adhesion sites and cell–cell adhesion complexes. Caspases can also mediate the packaging of the organelles into apoptotic bodies (see later). In fact, caspases are directly involved in the reorganization of the Golgi apparatus and ER, and while they do not contribute directly to mitochondrial morphological rearrangement, they do influence mitochondrial respiratory function. Moreover, during apoptosis, the processes of transcription and translation have to be regulated. Accordingly, caspases have transcription factors, translation initiation factors, and ribosomal proteins as targets.

As mentioned earlier, beyond the first group (constituting cyt c, Smac/DIABLO, and Omi/HtrA2), there is a second group of proteins released from mitochondria during apoptosis. This includes AIF and endo G, which are only released after the cell has been irreversibly committed to die [3]. AIF protein is encoded with a mitochondrial localization signal at the N-terminus, which is removed by a peptidase after its import into the mitochondria [10] and localizes at the OMM. During apoptosis, AIF can be cleaved from its membrane anchor by proteases, generating a soluble fragment that can be released into the cytosol upon permeabilization of the OMM. Subsequently, as a result of two nuclear localization signals (NLSs), AIF translocates into the nucleus, where it contributes to large-scale DNA fragmentation into ~50–300 kb pieces and to condensation of peripheral nuclear chromatin [135]. This early form of nuclear condensation is referred to as "stage I" condensation. Although AIF is involved in an essential passage of apoptosis, it is clear that it is not necessarily involved in all forms of apoptosis and assumes an important role only in certain cell types [136]. After its release from the mitochondria, endo G also moves to the nucleus, where it cleaves nuclear chromatin to produce oligonucleosomal DNA fragments [137]. Endo G works like AIF, in a caspase-independent manner. Together with the aforementioned endonucleases, DEF40/CAD also contributes to the fragmentation of DNA during apoptosis, but it works in a caspase-dependent manner. In healthy cells, it is sequestered in the cytosol by its chaperone and inhibitor DFF45/ICAD, while during apoptosis, DFF45/ICAD is cleaved by caspases, allowing the liberation and localization of DFF40/CAD to the nucleus and the subsequent fragmentation of the chromatin [132]. The action of caspase-activated nuclease CAD and endo G may help ensure that apoptosis is irreversible. This later and more pronounced chromatin condensation performed by these two endonucleases is referred to as "stage II" condensation.

The late and last events of apoptosis include apoptotic body production and the release of attractants for phagocytes by apoptotic cells. Both of these events are important in avoiding the leakage of potentially toxic or immunogenic contents from dying cells and in preventing an inflammatory response [138]. The apoptotic bodies are vesicles formed following nuclear and cytosol fragmentation and the extrusion of cytosol bound to the plasma membrane. They are removed via phagocytosis. To permit this removal, the apoptotic cells release a variety of molecules with attractive properties [139]. First, they establish a chemotactic gradient in order to attract phagocytes. Several signals released by dying cells have been reported: fractalkine (CX3CL1), lysophosphatidylcholine (LPC), sphingosine-1-phosphate (S1P), and the nucleotides ATP and UTP [140–143]. Next, the apoptotic cells present on their surfaces transmit signals that permit their recognition by the phagocytes. Again, numerous signals have been identified, including the exposure of the phospholipid phosphatidylserine, changes in the glycosylation of surface proteins, changes in surface charge, binding of serum proteins, and expression of specific intercellular molecules. Ultimately, after engulfment, apoptotic cells are digested into their basic cellular building blocks.

2.4 Mitochondria between Cellular Life and Death

2.4.1 How Do these Organelles Keep Cells Alive?

2.4.1.1 Tumor Resistance to Apoptosis

One of the hallmarks of human cancers is their ability to evade cell death; this represents a key cause of resistance to current tumor treatments [144].

This cancer cell feature is a result of various mechanisms that interfere at different levels with the apoptotic signaling. First, in many tumors, there is an abnormal expression of anti- and proapoptotic proteins. In particular, it has been observed that there is enhanced expression of antiapoptotic Bcl-2 family members, both Bcl-2 itself and other components such as Bcl-xL and Mcl-1 (see review [145,146]). Also, the expression of IAPs is deregulated in cancer (see also [147]). Genetic evidence indicates that *cIAP1* and *cIAP2* are proto-oncogenes that are affected by chromosomal aberrations in cancers. Although there are diverging reports, the overexpression of XIAP has been associated with tumor aggressiveness and poor prognosis. Moreover, the expression of both survivin and ML-IAP has been identified as tumor-specific. In fact, both are expressed at low or undetectable levels in most normal human tissues but show high expression levels in various cancers. Together with the increased expression of the antiapoptotic proteins, it is evident in tumors that the inactivation of proapoptotic genes leads to the downregulation or mutation of proapoptotic molecules. An example is Bax, whose mutations resulting in its loss of function are common in several tumors and are associated with poor prognosis. Also, the silencing of Apaf-1 has been observed in tumors, particularly in melanoma, where the *APAF1* locus often shows allelic loss. The aberrant expression of the genes coding for proteins involved in the apoptosis has been linked to the impairment of DNA methylation, which is another feature of cancer cells [148].

Mitochondria and the cancer energetic shift seem to confer additional elements of resistance to apoptosis. In fact, the activity of OXPHOS has been demonstrated to be specifically required for the execution of cell death [149,150]. On the other hand, the impairment of OXPHOS contributes to cancer cell resistance to cell death and chemotherapy, in particular avoiding the overproduction of ROS following the proapoptotic stimulation and consequent oxidation and ROS-mediated modification of mitochondrial proteins that favors the opening of PTP [151]. In this context, it has been proven that the downregulation of the H^+-ATP synthase, which is a common characteristic of cancer cells that rely mainly on glycolysis as an energy source, is a molecular strategy for avoiding ROS-mediated apoptosis [152,153].

Among the tumor-specific changes that contribute to metabolic reprogramming, the enzyme hexokinase (HK) has a peculiar role. Cancer cells overexpress HK isoforms that associate with the OMM protein VDAC. This association, tighter than in normal cells, guarantees HK preferential access to mitochondria-generated ATP via the ANT and protection by its product, glucose-6-phosphate, which at high concentration normally inhibits the enzyme [154]. The interaction of HK and VDAC inhibits MOMP; in fact, HK can induce the protein kinase C isoform ε (PKCε)-catalyzed phosphorylation of VDAC, which inhibits PTPC opening. Accordingly, inactivation of HK facilitates the induction of MOMP and subsequent apoptosis [155].

Mitochondrial fusion and fission cover an important role in apoptosis; in fact, while mitochondrial fusion can counteract the apoptotic event, mitochondrial fission promotes mitochondrial membrane depolarization, cyt c release, and apoptosis [156]. In this context, the GTPase family of proteins, which regulates mitochondrial dynamics, also modulates apoptosis, through activities involving members of the Bcl-2 family; on the other hand, pro- and antiapoptotic proteins can regulate mitochondrial fusion and fission [157]. With this a relationship, the aberrant expression or regulation of the proteins involved can leave the mitochondria unable to respond to stimuli, modifying their morphology, and therefore can induce apoptosis deregulation and resistance. This condition has been found in many cancer cell types [158,159].

2.4.1.2 Neurodegeneration: A "Too Much" Cell-Death Disease

Deregulation of apoptosis is associated with several pathologies, including neurodegenerative disorders (Figure 2.4) [160]. Apoptosis, normally, is a controlled mechanism, especially in multicellular organisms. In fact, during their normal development and at adulthood, such organisms show active apoptotic processes that dynamically participate in removing old and damaged cells and in maintaining tissue homeostasis without causing injury to contiguous cells [161]. Therefore, with the exception of post-mitotic cells such as differentiated neurons and muscle cells, which are usually highly resistant to apoptosis, the great part of the body's cells are commonly renewed, particularly within epithelia, endothelia, and the blood. With regard to the different neurodegenerative diseases – Alzheimer's disease (AD), Parkinson's disease (PD), amyotrophic lateral sclerosis (ALS), and Huntington's disease (HD) – we can state that they are all characterized by increased levels of expression or activity of key apoptotic molecules. Nevertheless, despite the heterogeneity of these diseases, an important common feature among them is the involvement of mitochondria with increased apoptosis. For instance, in AD, extensive data show that abnormalities in the metabolism of amyloid precursor protein (APP), an important causative factor, result in mitochondrial dysfunction and eventually cell death. This toxic effect is associated with ROS increase, induction of apoptosis, and, hence, weakened memory [162]. Although most deleterious effects of APP are attributed to accumulation of its intracellular metabolite, Aβ [163,164], other causes (also leading to apoptosis) must be determined. For instance, extracellular Aβ (i.e., released from dying cells) can enter into other cells and cause mitochondrial dysfunction [165,166]. *In vivo* and *in vitro* experiments have shown that soluble Aβ also impairs mitochondrial metabolism by decreasing cytochrome oxidase activity and increasing H_2O_2 generation [167]. Another causative factor associated with AD is ER stress due to the accumulation of unfolded proteins and alteration of ER Ca^{++} regulation [166,168]. Also, in PD, the apoptotic process is considered the main mechanism of cell death associated with this neurodegeneration. An important cause of cell death is the accumulation of unfolded α-syn [169]. Such an accumulation results from mutations in the PARK2 gene that encodes Parkin protein, which works as an ubiquitin E3 ligase for α-syn [170]. Importantly, it has been suggested that the disease may be spread from one nerve cell to another by the malformed protein (e.g., α-syn), rather than arising spontaneously in the cells [171]. In addition, in PD, Parkin and its partner, PINK1, a mitochondrial kinase, cooperate to control mitochondrial morphology and functionality [172,173]. In particular, recent evidence indicates PINK1 protein as a regulator of Ca^{++} efflux in mitochondria, since its deficiencies result in the accumulation of Ca^{++} in the mitochondria and in increased generation of ROS followed by opening of mPTP and elevation in the rate of cell death [174–176]. Another protein involved in the onset of PD is DJ-1, a mitochondrial protein with protective effects partly ascribed to its function as

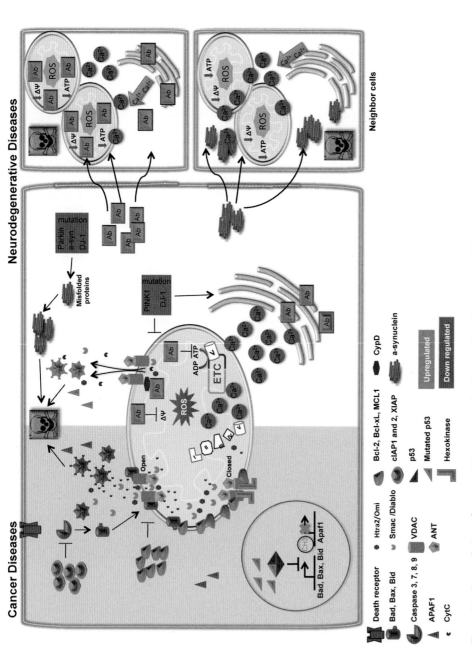

Figure 2.4 Deregulation of apoptosis is associated with several pathologies, including cancer and neurodegenerative disorders. Cancer cells, as shown in the left part of the figure, not only are able to grow uncontrollably, but also can evade cellular death pathways. In fact, as is shown, cancer cells can increase the expression/stability of anti- over proapoptotic proteins. In addition, in almost half of all tumors studied, the apoptotic gatekeeper – the protein p53 – has been found mutated. Other mitochondrial defects, such as the reduced expression and activity of complex I and V, also play a role in the evasion of apoptosis by cancer cells. On the other hand, neurodegenerative diseases are characterized by an increased rate of cell death, especially through the mitochondrial pathway. As shown in the right part of the figure – and fully described throughout the text – oxidative stress, perturbed calcium homeostasis, mitochondrial dysfunction, and accumulation of misfolded proteins all contribute to the aberrant regulation of apoptosis in neuronal cells that leads to the neurodegenerative disorders.

an oxidative stress-induced chaperon [177]. The mitochondrial localization of DJ-1, PINK1, and Parkin strongly supports the link between PD and mitochondrial dysfunction.

Since a full description of the neurodegenerative diseases is beyond the scope of this chapter, we will simply note that much attention in recent years has been given to experimental drugs that are capable of modulating apoptosis. In fact, "cause-directed" trials involving antioxidants and enhancers of mitochondrial function – as well as agents capable of blocking excitotoxicity or apoptosis – are currently under study.

2.4.2 Can We Use Mitochondria as a Therapeutic Target?

Studies of the apoptotic process have focused on the development of novel therapeutic targets [178]. In particular, they have improved our understanding of mitochondrial physiology. In this regard, since mitochondria are required to maintain cellular bioenergetics and regulate the point of no return of the apoptotic process, it is feasible to speculate that mitochondria might be a suitable tool by which to exploit therapeutic approaches to many human disorders (Figure 2.5). However, there are still some mechanisms to be elucidated in the relationship between cell-death regulation, mitochondrial metabolism, and morphology [42]. As previously addressed, the ratio of pro- to antiapoptotic Bcl-2 proteins determines the cell fate between life and death. Regardless, certain of these proteins can induce the oligomerization of Bax and/or Bak, resulting in MOMP, widespread proteolysis, and cell death; in contrast, other proteins can interrupt signaling upstream of Bax/Bak oligomerization, preventing cell death [179,180]. From this perspective, in this section we will discuss emerging data describing how there can be so many ways to exploit mitochondrial bioenergetics, redox balance, and mitochondrial dynamics as promising targets for therapeutic approaches.

It is important to note that dozens of signal-transducer pathways have been targeted in mitochondria in order to induce or prevent cell death. Some of these are inducers of MOMP, Bcl-2-targeted drugs, BH3 mimetics, redox-active compounds, peripheral-type benzodiazepine receptor ligands, ANT ligands, steroid analogs, cationic ampholytes, and mitochondrial proapoptotic factors such as chemotherapeutic agents. Generally, each class of these agents triggers events upstream of MOMP, using Bcl-2 proteins as putative targets. However, the often high concentration of these drugs required in conventional chemotherapies has led to some doubts regarding the precise mechanism of cell death [178].

Recent insights have demonstrated that MPTP represents a potential drug target. Structurally, the MPTP or so-called "PTPC" constitutes VDAC, ANT, and CyPD. Mechanistically, this dynamic supramolecular complex, when in a low-conductance state, can interact with Bcl-2 proteins, contributing to the exchange of small molecules between the mitochondrial matrix and the cytosol. The complex structure of PTPC relies on an interaction with the peripheral benzodiazepine receptor (PBR) and with HK, which uses mitochondrial ATP to catalyze the rate-limiting step of glycolysis. Proapoptotic signals, including Ca^{++} fluxes from the ER to the mitochondria and the collapse of mitochondrial energy demand, may alter the PTPC conductance state, allowing the entry of small solutes into the mitochondrial matrix. Hence, due to the dissipation of the $\Delta\Psi m$ and the osmotic swelling of the mitochondrial matrix, both events can trigger MOMP, leading to apoptosis. In line with this notion, several ANT ligands and VDAC-opening compounds that interact with anti- and proapoptotic members of the Bcl-2 family protein may induce and/or prevent apoptosis. Lonidamine, 4-(N-(S-glutathionylacetyl)

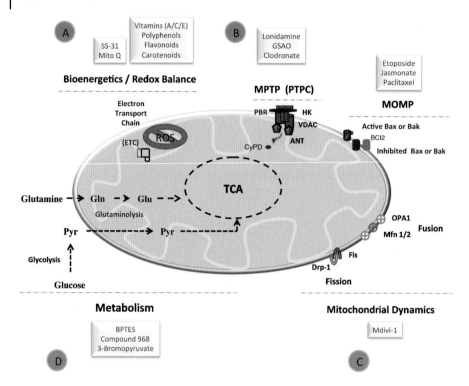

Figure 2.5 Targeting mitochondrial functions to exploit therapeutic approaches for cancer and neurodegeneration. (A) Mitochondrial-target antioxidants: mitochondrial ROS are required for cancer progression and are involved in the etiology and pathogenesis of neuronal disorders. (B) MPTP-target compounds: these can directly or indirectly lead to or prevent MOMP and cell death. (C) Mitochondrial dynamics: Mdivi-1 is a selective inhibitor of mitochondrial division and apoptosis. It has shown efficacy in models of neurodegenerative diseases. (D) Mitochondrial-target metabolism: many cancer cell types use glutamine and glucose for growth and survival. Inhibition of key enzymes of glycolysis and glutaminolysis has shown efficacy in models of cancer. Compound 968 and BPTES are specific inhibitors of mitochondrial glutaminase (Gls), and 3-Bromopyruvate is a potent antiglycolytic agent.

amino) (GSAO), and clodronate are examples of agents designed to interact with MTPT and so cause cell death. Lonidamine has been used in a phase II clinical study containing patients with recurrent glioblastoma multiforme. In parallel, another clinical study showed decreasing tumor growth and low cytotoxicity effects when lonidamine was combined with diazepam (a specific PBR ligand) [181,182]. The second compound, so-called "GSAO," showed direct inhibition of ATP/ADP antiporter activity, ROS overproduction, cytosolic ATP depletion, and PT induction. Accordingly, GSAO inhibited angiogenesis and tumor growth in mice with low toxicity at efficacious doses [183]. Finally, a long-term follow-up in patients with primary breast cancer and metastases to the bone marrow showed improved overall survival after oral administration of clodronate [184]. Other recent insights have demonstrated that MOMP represents a promising target for triggering cell death. Intensive efforts have been carried out to develop specific agents that can directly bind Bax and/or Bak undergoing conformational modifications or antagonizing Bcl-2, Bcl-xL, and Mcl-1 activity, and so contribute to MOMP-dependent apoptosis. A variety of these compounds (e.g., etoposide, jasmonate, and paclitaxel) are in various clinical trial-phase studies of tumors characterized by Bcl-2-mediated apoptosis resistance, including leukemia and lymphoma [185–187]. Thus, targeting mitochondrial MOMP and MPTP to trigger apoptosis could benefit cancer therapy. However, the exact efficacy of these compounds remains to be elucidated in clinical studies [188].

The mitochondria have also been pointed to as the main source of oxygen and nitrogen radicals. In fact, overproduction or incorporation of free radicals may cause oxidative damage of macromolecules (e.g., DNA, proteins, and lipids), which alter the normal cell physiology, leading to oxidative stress and cell death. In healthy cells, two important mechanisms for coping with oxidative stress have been described: non-enzymatic and enzymatic antioxidant systems. Non-enzymatic antioxidants are considered the vitamins A, C, and E and natural compounds such as polyphenols, flavonoids, and carotenoids [189]. Conversely, the most important intracellular enzymatic antioxidant systems are superoxide dismutase (SOD), catalase, glutathione reductase (GR), glutathione peroxidase (GPX), and thioredoxin reductase (TRXR2) [190]. Notably, both systems are required in order to maintain cell physiology homeostasis, through regulation of reactive oxygen and nitrogen molecules levels. Nonetheless, a range of disorders, including AD, PD, ALS, and other neuronal disorders can be associated with inherited genetic lesions of mtDNA and/or nuclear DNA that encode mitochondrial enzymes (e.g., DJ-1, PINK1, and SOD1). Despite this, there are standing disagreements about mitochondrial dysfunction, and there can be implications in the etiology and pathogenesis of neuronal disorders. For instance, either *in vivo* studies, performed by using transgenic mice harboring mutated mitochondrial antioxidant enzyme MnSOD, or *in vitro* studies, performed by using astrocytes treated with the mitochondrial uncoupler CCCP in order to mimic the amyloidogenic APP, have shown a higher incidence of phenotypes associated with AD-related pathology [191,192]. Moreover, chronic infusion of rotenone or other mitochondrial complex I inhibitors, such as 1-methyl 4-phenyl-1,2,3,6-tetrahydropyridine, has resulted in a parkinsonian phenotype in laboratory animals [193,194]. In this regard, it is plausible to speculate that imbalanced antioxidant mechanisms fail to detoxify the overproduction and accumulation of free radicals, leading to oxidative stress, excitotoxicity, and apoptosis. Thus, there has been great interest in targeting basic mitochondrial redox balance processes in various models of neuronal disorders. In fact, some observations have focused on mitochondrial dysfunction as a main casual role in neuronal diseases, using simplistic therapeutic approaches. Antioxidant therapy aimed at decreasing ROS generation (e.g., by expressing uncoupling proteins) and restoring mitochondrial and intracellular antioxidant systems has shown moderate success in clinical trials [190]. For example, the antioxidants SS tetrapeptide (SS-31) and Mito Q have been shown to prevent oxidative stress-related mitochondrial dysfunction. Upon analysis of mitochondrial morphology and activity, either in a mouse model of AD or in a mouse neuroblastoma cell line (N2a) incubated with the Aβ peptide, it was found that Mito Q and SS31 prevented Aβ toxicity in mitochondria and may thus have potential for use against neuronal disorders [195]. Mito Q is currently being used in phase II trials of AD, and it has been considered an important candidate for other oxidative stress-related pathologies [196].

Finally, mitochondrial dynamics perturbation has recently been recognized as a pathological feature of many human disorders, suggesting mitochondrial dynamics targeting may be an appealing idea. It is known that Bax and Bak (proapoptotic) appear to promote mitochondrial fission, leading to MOMP and cell death. In contrast, Bcl-xL (antiapoptotic) appears to regulate mitochondrial fusion by interacting with Mfn2 or by blocking the activity of proapoptotic proteins [197–199]. Interesting, a recent study using yeast screens of chemical libraries identified one compound capable of uncoupling mitochondrial fission and apoptosis: Mdivi-1 blocked Bid-activated Bax/Bak-dependent cyt c release from mitochondria, MOMP, and apoptosis. This novel compound may thus represent a new class of therapeutic approaches for neurodegenerative diseases (Figure 2.5) [200]. In addition, further understanding of the mitochondrial dynamics machinery could shed light on the development of novel therapeutic strategies for fighting cancer.

2.5 Conclusion

In this chapter, we have highlighted some emerging insights into mitochondria and apoptosis. Mitochondria are required for the regulation of an important means of apoptotic cell death. Bioenergetics and metabolism have been extensively investigated over the past decade, although some mechanisms are still to be elucidated. Work in this field has provided insights into the targeting of mitochondrial apoptotic pathways (e.g., MOMP, MPTP, and Bcl-2 family protein) as eligible candidates for the exploitation of therapeutic approaches to many human diseases, including cancer and neurodegeneration. A caveat to the current approaches to the induction or prevention of cell death is that the proteins and metabolites involved in mitochondrial bioenergetics, redox balance, and mitochondrial dynamics are also involved in steady-state cell physiology in a wide array of cell types. These include neuronal cells, proliferative cells, immune cells, and stem cells. Furthermore, targeting mitochondria to induce or prevent cell death via apoptotic processes should be linked with other types of cell-death program, such as mitotic catastrophe, autophagy, necrosis, and other mechanisms addressed in this book. Studies targeting mitochondria will enhance our knowledge, avoid misinterpretations of ambiguous outcomes, and enable the development of novel therapies capable of exploiting the peculiar features of mitochondrial bioenergetics, metabolism, and morphology in a specific cell context.

References

1. Ellis RE, Yuan JY, Horvitz HR. Mechanisms and functions of cell death. *Annu Rev Cell Biol* 1991;7:663–98.
2. Elmore S. Apoptosis: a review of programmed cell death. *Toxicol Pathol* 2007;**35**(4):495–516.
3. Riedl SJ, Shi Y. Molecular mechanisms of caspase regulation during apoptosis. *Nat Rev Mol Cell Biol* 2004;**5**(11):897–907.
4. Hengartner MO. The biochemistry of apoptosis. *Nature* 2000;**407**(6805):770–6.
5. Kerr JF, Wyllie AH, Currie AR. Apoptosis: a basic biological phenomenon with wide-ranging implications in tissue kinetics. *Br J Cancer* 1972;**26**(4):239–57.
6. Tsujimoto Y, Finger LR, Yunis J, Nowell PC, Croce CM. Cloning of the chromosome breakpoint of neoplastic B cells with the t(14;18) chromosome translocation. *Science* 1984;**226**(4678):1097–9.
7. Tsujimoto Y, Ikegaki N, Croce CM. Characterization of the protein product of bcl-2, the gene involved in human follicular lymphoma. *Oncogene* 1987;**2**(1):3–7.

8 Vaux DL, Cory S, Adams JM. Bcl-2 gene promotes haemopoietic cell survival and cooperates with c-myc to immortalize pre-B cells. *Nature* 1988;**335**(6189):440–2.
9 Hockenbery D, Nunez G, Milliman C, Schreiber RD, Korsmeyer SJ. Bcl-2 is an inner mitochondrial membrane protein that blocks programmed cell death. *Nature* 1990;**348**(6299):334–6.
10 Newmeyer DD, Farschon DM, Reed JC. Cell-free apoptosis in Xenopus egg extracts: inhibition by Bcl-2 and requirement for an organelle fraction enriched in mitochondria. *Cell* 1994;**79**(2):353–64.
11 Chinnaiyan AM, Orth K, O'Rourke K, Duan H, Poirier GG, Dixit VM. Molecular ordering of the cell death pathway. Bcl-2 and Bcl-xL function upstream of the CED-3-like apoptotic proteases. *J Biol Chem* 1996;**271**(9):4573–6.
12 Nunez G, Clarke MF. The Bcl-2 family of proteins: regulators of cell death and survival. *Trends Cell Biol* 1994;**4**(11):399–403.
13 Reed JC. Bcl-2 and the regulation of programmed cell death. *J Cell Biol* 1994;**124**(1–2):1–6.
14 Kluck RM, Bossy-Wetzel E, Green DR, Newmeyer DD. The release of cytochrome c from mitochondria: a primary site for Bcl-2 regulation of apoptosis. *Science* 1997;**275**(5303):1132–6.
15 Yang J, Liu X, Bhalla K, Kim CN, Ibrado AM, Cai J, et al. Prevention of apoptosis by Bcl-2: release of cytochrome c from mitochondria blocked. *Science* 1997;**275**(5303):1129–32.
16 Li P, Nijhawan D, Budihardjo I, Srinivasula SM, Ahmad M, Alnemri ES, Wang X. Cytochrome c and dATP-dependent formation of Apaf-1/caspase-9 complex initiates an apoptotic protease cascade. *Cell* 1997;**91**(4):479–89.
17 Green DR, Reed JC. Mitochondria and apoptosis. *Science* 1998;**281**(5381):1309–12.
18 Loeffler M, Kroemer G. The mitochondrion in cell death control: certainties and incognita. *Exp Cell Res* 2000;**256**(1):19–26.
19 Newmeyer DD, Ferguson-Miller S. Mitochondria: releasing power for life and unleashing the machineries of death. *Cell* 2003;**112**(4):481–90.
20 Scorrano L, Ashiya M, Buttle K, Weiler S, Oakes SA, Mannella CA, Korsmeyer SJ. A distinct pathway remodels mitochondrial cristae and mobilizes cytochrome c during apoptosis. *Dev Cell* 2002;**2**(1):55–67.
21 Zong WX, Lindsten T, Ross AJ, MacGregor GR, Thompson CB. BH3-only proteins that bind pro-survival Bcl-2 family members fail to induce apoptosis in the absence of Bax and Bak. *Genes Dev* 2001;**15**(12):1481–6.
22 Cory S, Adams JM. The Bcl2 family: regulators of the cellular life-or-death switch. *Nat Rev Cancer* 2002;**2**(9):647–56.
23 Yip KW, Reed JC. Bcl-2 family proteins and cancer. *Oncogene* 2008;**27**(50):6398–406.
24 Kroemer G, Galluzzi L, Brenner C. Mitochondrial membrane permeabilization in cell death. *Physiol Rev* 2007;**87**(1):99–163.
25 Kushnareva Y, Newmeyer DD. Bioenergetics and cell death. *Ann NY Acad Sci* 2010;**1201**:50–7.
26 Susin SA, Zamzami N, Kroemer G. Mitochondria as regulators of apoptosis: doubt no more. *Biochim Biophys Acta* 1998;**1366**(1–2):151–65.
27 Thompson CB. Apoptosis in the pathogenesis and treatment of disease. *Science* 1995;**267**(5203):1456–62.
28 DeBerardinis RJ, Thompson CB. Cellular metabolism and disease: what do metabolic outliers teach us? *Cell* 2012;**148**(6):1132–44.

29. Gaglio D, Metallo CM, Gameiro PA, Hiller K, Danna LS, Balestrieri C, et al. Oncogenic K-Ras decouples glucose and glutamine metabolism to support cancer cell growth. *Mol Syst Biol* 2011;**7**:523.
30. Metallo CM, Vander Heiden MG. Understanding metabolic regulation and its influence on cell physiology. *Mol Cell* 2013;**49**(3):388–98.
31. Vander Heiden MG, Cantley LC, Thompson CB. Understanding the Warburg effect: the metabolic requirements of cell proliferation. *Science* 2009;**324**(5930):1029–33.
32. Viale A, Pettazzoni P, Lyssiotis CA, Ying H, Sánchez N, Marchesini M, et al. Oncogene ablation-resistant pancreatic cancer cells depend on mitochondrial function. *Nature* 2014;**514**(7524):628–32.
33. Larsson NG. Somatic mitochondrial DNA mutations in mammalian aging. *Annu Rev Biochem* 2010;**79**:683–706.
34. Chan DC. Mitochondrial fusion and fission in mammals. *Annu Rev Cell Dev Biol* 2006;**22**:79–99.
35. Detmer SA, Chan DC. Functions and dysfunctions of mitochondrial dynamics. *Nat Rev Mol Cell Biol* 2007;**8**(11):870–9.
36. Rossignol R, Gilkerson R, Aggeler R, Yamagata K, Remington SJ, Capaldi RA. Energy substrate modulates mitochondrial structure and oxidative capacity in cancer cells. *Cancer Res* 2004;**64**(3):985–93.
37. Karbowski M, Youle RJ. Dynamics of mitochondrial morphology in healthy cells and during apoptosis. *Cell Death Differ* 2003;**10**(8):870–80.
38. Rolland S, Conradt B. The role of mitochondria in apoptosis induction in *Caenorhabditis elegans*: more than just innocent bystanders? *Cell Death Differ* 2006;**13**(8):1281–6.
39. Nechushtan A, Smith CL, Lamensdorf I, Yoon SH, Youle RJ. Bax and Bak coalesce into novel mitochondria-associated clusters during apoptosis. *J Cell Biol* 2001;**153**(6):1265–76.
40. Kowaltowski AJ, Cosso RG, Campos CB, Fiskum G. Effect of Bcl-2 overexpression on mitochondrial structure and function. *J Biol Chem* 2002;**277**(45):42 802–7.
41. Youle RJ, van der Bliek AM. Mitochondrial fission, fusion, and stress. *Science* 2012;**337**(6098):1062–5.
42. Green DR, Galluzzi L, Kroemer G. Cell biology. Metabolic control of cell death. *Science* 2014;**345**(6203):1250256.
43. Trachootham D, Lu W, Ogasawara MA, Nilsa RD, Huang P. Redox regulation of cell survival. *Antioxid Redox Signal* 2008;**10**(8):1343–74.
44. Wellen KE, Hatzivassiliou G, Sachdeva UM, Bui TV, Cross JR, Thompson CB. ATP-citrate lyase links cellular metabolism to histone acetylation. *Science* 2009;**324**(5930):1076–80.
45. Mitchell P, Moyle J. Stoichiometry of proton translocation through the respiratory chain and adenosine triphosphatase systems of rat liver mitochondria. *Nature* 1965;**208**(5006):147–51.
46. Wallace DC. Mitochondria and cancer: Warburg addressed. *Cold Spring Harb Symp Quant Biol* 2005;**70**:363–74.
47. Shchepina LA, Pletjushkina OY, Avetisyan AV, Bakeeva LE, Fetisova EK, Izyumov DS, et al. Oligomycin, inhibitor of the F0 part of H+-ATP-synthase, suppresses the TNF-induced apoptosis. *Oncogene* 2002;**21**(53):8149–57.

48 Wong CH, Iskandar KB, Yadav SK, Hirpara JL, Loh T, Pervaiz S. Simultaneous induction of non-canonical autophagy and apoptosis in cancer cells by ROS-dependent ERK and JNK activation. *PLoS One* 2010;**5**(4):e9996.
49 Wang H, Oster G. Energy transduction in the F1 motor of ATP synthase. *Nature* 1998;**396**(6708):279–82.
50 Halestrap AP, McStay GP, Clarke SJ. The permeability transition pore complex: another view. *Biochimie* 2002;**84**(2–3):153–66.
51 Wrighton KH. Autophagy: from one membrane to another. *Nat Rev Mol Cell Biol* 2010;**11**(7):464.
52 Sena LA, Chandel NS. Physiological roles of mitochondrial reactive oxygen species. *Mol Cell* 2012;**48**(2):158–67.
53 Crompton M. Mitochondrial intermembrane junctional complexes and their role in cell death. *J Physiol* 2000;**529**(Pt. 1):11–21.
54 Bossy-Wetzel E, Newmeyer DD, Green DR. Mitochondrial cytochrome c release in apoptosis occurs upstream of DEVD-specific caspase activation and independently of mitochondrial transmembrane depolarization. *EMBO J* 1998;**17**(1):37–49.
55 Kroemer G, Dallaporta B, Resche-Rigon M. The mitochondrial death/life regulator in apoptosis and necrosis. *Annu Rev Physiol* 1998;**60**:619–42.
56 Tait SW, Green DR. Mitochondria and cell death: outer membrane permeabilization and beyond. *Nat Rev Mol Cell Biol* 2010;**11**(9):621–32.
57 Zamzami N, Larochette N, Kroemer G. Mitochondrial permeability transition in apoptosis and necrosis. *Cell Death Differ* 2005;**12**(Suppl. 2):1478–80.
58 Henry-Mowatt J, Dive C, Martinou JC, James D. Role of mitochondrial membrane permeabilization in apoptosis and cancer. *Oncogene* 2004;**23**(16):2850–60.
59 Ly JD, Grubb DR, Lawen A. The mitochondrial membrane potential (deltapsi(m)) in apoptosis; an update. *Apoptosis* 2003;**8**(2):115–28.
60 Tsujimoto Y. Bcl-2 family of proteins: life-or-death switch in mitochondria. *Biosci Rep* 2002;**22**(1):47–58.
61 Letai AG. Diagnosing and exploiting cancer's addiction to blocks in apoptosis. *Nat Rev Cancer* 2008;**8**(2):121–32.
62 Orrenius S, Zhivotovsky B, Nicotera P. Regulation of cell death: the calcium-apoptosis link. *Nat Rev Mol Cell Biol* 2003;**4**(7):552–65.
63 Mund T, Gewies A, Schoenfeld N, Bauer MK, Grimm S. Spike, a novel BH3-only protein, regulates apoptosis at the endoplasmic reticulum. *FASEB J* 2003;**17**(6):696–8.
64 Breckenridge DG, Stojanovic M, Marcellus RC, Shore GC. Caspase cleavage product of BAP31 induces mitochondrial fission through endoplasmic reticulum calcium signals, enhancing cytochrome c release to the cytosol. *J Cell Biol* 2003;**160**(7):1115–27.
65 Chen M, He H, Zhan S, Krajewski S, Reed JC, Gottlieb RA. Bid is cleaved by calpain to an active fragment in vitro and during myocardial ischemia/reperfusion. *J Biol Chem* 2001;**276**(33):30 724–8.
66 Bernardi P, Krauskopf A, Basso E, Petronilli V, Blachly-Dyson E, Di Lisa F, Forte MA. The mitochondrial permeability transition from in vitro artifact to disease target. *FEBS J* 2006;**273**(10):2077–99.
67 Distelhorst CW, Shore GC. Bcl-2 and calcium: controversy beneath the surface. *Oncogene* 2004;**23**(16):2875–80.

68 Szabadkai G, Simoni AM, Chami M, Wieckowski MR, Youle RJ, Rizzuto R. Drp-1-dependent division of the mitochondrial network blocks intraorganellar Ca2+ waves and protects against Ca2+-mediated apoptosis. *Mol Cell* 2004;**16**(1):59–68.

69 Murphy MP. How mitochondria produce reactive oxygen species. *Biochem J* 2009;**417**(1):1–13.

70 Li X, Fang P, Mai J, Choi ET, Wang H, Yang XF. Targeting mitochondrial reactive oxygen species as novel therapy for inflammatory diseases and cancers. *J Hematol Oncol* 2013;**6**:19.

71 Park WH, Han YW, Kim SH, Kim SZ. An ROS generator, antimycin A, inhibits the growth of HeLa cells via apoptosis. *J Cell Biochem* 2007;**102**(1):98–109.

72 Vandenabeele P, Galluzzi L, Vanden Berghe T, Kroemer G. Molecular mechanisms of necroptosis: an ordered cellular explosion. *Nat Rev Mol Cell Biol* 2010;**11**(10):700–14.

73 Vitale I, Galluzzi L, Castedo M, Kroemer G. Mitotic catastrophe: a mechanism for avoiding genomic instability. *Nat Rev Mol Cell Biol* 2011;**12**(6):385–92.

74 Wang Z, Jiang H, Chen S, Du F, Wang X. The mitochondrial phosphatase PGAM5 functions at the convergence point of multiple necrotic death pathways. *Cell* 2012;**148**(1–2):228–43.

75 Chiaradonna F, Sacco E, Manzoni R, Giorgio M, Vanoni M, Alberghina L. Ras-dependent carbon metabolism and transformation in mouse fibroblasts. *Oncogene* 2006;**25**(39):5391–404.

76 Moley KH, Mueckler MM. Glucose transport and apoptosis. *Apoptosis* 2000;**5**(2):99–105.

77 Palorini R, Cammarata FP, Balestrieri C, Monestiroli A, Vasso M, Gelfi C, et al. Glucose starvation induces cell death in K-ras-transformed cells by interfering with the hexosamine biosynthesis pathway and activating the unfolded protein response. *Cell Death Dis* 2013;**4**:e732.

78 Fujita R, Ueda H. Protein kinase C-mediated cell death mode switch induced by high glucose. *Cell Death Differ* 2003;**10**(12):1336–47.

79 Xu RH, Pelicano H, Zhou Y, Carew JS, Feng L, Bhalla KN, et al. Inhibition of glycolysis in cancer cells: a novel strategy to overcome drug resistance associated with mitochondrial respiratory defect and hypoxia. *Cancer Res* 2005;**65**(2):613–21.

80 Blackburn RV, Spitz DR, Liu X, Galoforo SS, Sim JE, Ridnour LA, et al. Metabolic oxidative stress activates signal transduction and gene expression during glucose deprivation in human tumor cells. *Free Radic Biol Med* 1999;**26**(3–4):419–30.

81 Isaev NK, Stelmashook EV, Dirnagl U, Plotnikov EY, Kuvshinova EA, Zorov DB. Mitochondrial free radical production induced by glucose deprivation in cerebellar granule neurons. *Biochemistry (Mosc)* 2008;**73**(2):149–55.

82 Liu Y, Song XD, Liu W, Zhang TY, Zuo J. Glucose deprivation induces mitochondrial dysfunction and oxidative stress in PC12 cell line. *J Cell Mol Med* 2003;**7**(1):49–56.

83 Scorrano L, Oakes SA, Opferman JT, Cheng EH, Sorcinelli MD, Pozzan T, Korsmeyer SJ. BAX and BAK regulation of endoplasmic reticulum Ca2+: a control point for apoptosis. *Science* 2003;**300**(5616):135–9.

84 Elyaman W, Terro F, Suen KC, Yardin C, Chang RC, Hugon J. BAD and Bcl-2 regulation are early events linking neuronal endoplasmic reticulum stress to mitochondria-mediated apoptosis. *Brain Res Mol Brain Res* 2002;**109**(1–2):233–8.

85 Gao Z, Shao Y, Jiang X. Essential roles of the Bcl-2 family of proteins in caspase-2-induced apoptosis. *J Biol Chem* 2005;**280**(46):38 271–5.

86 Ramachandran A, Moellering DR, Ceaser E, Shiva S, Xu J, Darley-Usmar V. Inhibition of mitochondrial protein synthesis results in increased endothelial cell susceptibility to nitric oxide-induced apoptosis. *Proc Natl Acad Sci USA* 2002;**99**(10):6643–8.

87 Yuneva M, Zamboni N, Oefner P, Sachidanandam R, Lazebnik Y. Deficiency in glutamine but not glucose induces MYC-dependent apoptosis in human cells. *J Cell Biol* 2007;**178**(1):93–105.

88 Wise DR, Thompson CB. Glutamine addiction: a new therapeutic target in cancer. *Trends Biochem Sci* 2010;**35**(8):427–33.

89 Kristiansen M, Ham J. Programmed cell death during neuronal development: the sympathetic neuron model. *Cell Death Differ* 2014;**21**(7):1025–35.

90 Vander Heiden MG, Chandel NS, Schumacker PT, Thompson CB. Bcl-xL prevents cell death following growth factor withdrawal by facilitating mitochondrial ATP/ADP exchange. *Mol Cell* 1999;**3**(2):159–67.

91 Reshi ML, Su YC, Hong JR. RNA viruses: ROS-mediated cell death. *Int J Cell Biol* 2014;**2014**:467452.

92 Anand SK, Tikoo SK. Viruses as modulators of mitochondrial functions. *Adv Virol* 2013;**2013**:738794.

93 Jackson SP, Bartek J. The DNA-damage response in human biology and disease. *Nature* 2009;**461**(7267):1071–8.

94 Brady CA, Jiang D, Mello SS, Johnson TM, Jarvis LA, Kozak MM, et al. Distinct p53 transcriptional programs dictate acute DNA-damage responses and tumor suppression. *Cell* 2011;**145**(4):571–83.

95 Hock AK, Vousden KH. Tumor suppression by p53: fall of the triumvirate? *Cell* 2012;**149**(6):1183–5.

96 Jin Z, El-Deiry WS. Overview of cell death signaling pathways. *Cancer Biol Ther* 2005;**4**(2):139–63.

97 Flores ER, Tsai KY, Crowley D, Sengupta S, Yang A, McKeon F, Jacks T. p63 and p73 are required for p53-dependent apoptosis in response to DNA damage. *Nature* 2002;**416**(6880):560–4.

98 Stiewe T, Putzer BM. p73 in apoptosis. *Apoptosis* 2001;**6**(6):447–52.

99 Saelens X, Festjens N, Vande Walle L, van Gurp M, van Loo G, Vandenabeele P. Toxic proteins released from mitochondria in cell death. *Oncogene* 2004;**23**(16):2861–74.

100 Du C, Fang M, Li Y, Li L, Wang X. Smac, a mitochondrial protein that promotes cytochrome c-dependent caspase activation by eliminating IAP inhibition. *Cell* 2000;**102**(1):33–42.

101 Cai J, Yang J, Jones DP. Mitochondrial control of apoptosis: the role of cytochrome c. *Biochim Biophys Acta* 1998;**1366**(1–2):139–49.

102 van Loo G, van Gurp M, Depuydt B, Srinivasula SM, Rodriguez I, Alnemri ES, et al. The serine protease Omi/HtrA2 is released from mitochondria during apoptosis. Omi interacts with caspase-inhibitor XIAP and induces enhanced caspase activity. *Cell Death Differ* 2002;**9**(1):20–6.

103 Kim HE, Du F, Fang M, Wang X. Formation of apoptosome is initiated by cytochrome c-induced dATP hydrolysis and subsequent nucleotide exchange on Apaf-1. *Proc Natl Acad Sci USA* 2005;**102**(49):17 545–50.

104 Adrain C, Slee EA, Harte MT, Martin SJ. Regulation of apoptotic protease activating factor-1 oligomerization and apoptosis by the WD-40 repeat region. *J Biol Chem* 1999;**274**(30):20 855–60.

105 Cain K, Bratton SB, Langlais C, Walker G, Brown DG, Sun XM, Cohen GM. Apaf-1 oligomerizes into biologically active approximately 700-kDa and inactive approximately 1.4-MDa apoptosome complexes. *J Biol Chem* 2000;**275**(9):6067–70.

106 Cain K, Bratton SB, Cohen GM. The Apaf-1 apoptosome: a large caspase-activating complex. *Biochimie* 2002;**84**(2–3):203–14.

107 Schafer ZT, Kornbluth S. The apoptosome: physiological, developmental, and pathological modes of regulation. *Dev Cell* 2006;**10**(5):549–61.

108 Mei Y, Yong J, Liu H, Shi Y, Meinkoth J, Dreyfuss G, Yang X. tRNA binds to cytochrome c and inhibits caspase activation. *Mol Cell* 2010;**37**(5):668–78.

109 Bao Q, Lu W, Rabinowitz JD, Shi Y. Calcium blocks formation of apoptosome by preventing nucleotide exchange in Apaf-1. *Mol Cell* 2007;**25**(2):181–92.

110 Karki P, Seong C, Kim JE, Hur K, Shin SY, Lee JS, Cho B, Park IS. Intracellular K(+) inhibits apoptosis by suppressing the Apaf-1 apoptosome formation and subsequent downstream pathways but not cytochrome c release. *Cell Death Differ* 2007;**14**(12):2068–75.

111 Brown GC, Borutaite V. Regulation of apoptosis by the redox state of cytochrome c. *Biochim Biophys Acta* 2008;**1777**(7–8):877–81.

112 Takayama S, Reed JC, Homma S. Heat-shock proteins as regulators of apoptosis. *Oncogene* 2003;**22**(56):9041–7.

113 Allan LA, Clarke PR. Phosphorylation of caspase-9 by CDK1/cyclin B1 protects mitotic cells against apoptosis. *Mol Cell* 2007;**26**(2):301–10.

114 Allan LA, Morrice N, Brady S, Magee G, Pathak S, Clarke PR. Inhibition of caspase-9 through phosphorylation at Thr 125 by ERK MAPK. *Nat Cell Biol* 2003;**5**(7):647–54.

115 Santamaria D, Barriere C, Cerqueira A, Hunt S, Tardy C, Newton K, et al. Cdk1 is sufficient to drive the mammalian cell cycle. *Nature* 2007;**448**(7155):811–15.

116 Zhang W, Liu HT. MAPK signal pathways in the regulation of cell proliferation in mammalian cells. *Cell Res* 2002;**12**(1):9–18.

117 Bratton SB, Walker G, Srinivasula SM, Sun XM, Butterworth M, Alnemri ES, Cohen GM. Recruitment, activation and retention of caspases-9 and -3 by Apaf-1 apoptosome and associated XIAP complexes. *EMBO J* 2001;**20**(5):998–1009.

118 Vaux DL, Silke J. Mammalian mitochondrial IAP binding proteins. *Biochem Biophys Res Commun* 2003;**304**(3):499–504.

119 Shiozaki EN, Chai J, Rigotti DJ, Riedl SJ, Li P, Srinivasula SM, et al. Mechanism of XIAP-mediated inhibition of caspase-9. *Mol Cell* 2003;**11**(2):519–27.

120 Berthelet J, Dubrez L. Regulation of apoptosis by inhibitors of apoptosis (IAPs). *Cells* 2013;**2**(1):163–87.

121 Choi YE, Butterworth M, Malladi S, Duckett CS, Cohen GM, Bratton SB. The E3 ubiquitin ligase cIAP1 binds and ubiquitinates caspase-3 and -7 via unique mechanisms at distinct steps in their processing. *J Biol Chem* 2009;**284**(19):12 772–82.

122 Suzuki Y, Nakabayashi Y, Takahashi R. Ubiquitin-protein ligase activity of X-linked inhibitor of apoptosis protein promotes proteasomal degradation of caspase-3 and enhances its anti-apoptotic effect in Fas-induced cell death. *Proc Natl Acad Sci USA* 2001;**98**(15):8662–7.

123 Huang H, Joazeiro CA, Bonfoco E, Kamada S, Leverson JD, Hunter T. The inhibitor of apoptosis, cIAP2, functions as a ubiquitin-protein ligase and promotes in vitro monoubiquitination of caspases 3 and 7. *J Biol Chem* 2000;**275**(35):26 661–4.

124 Adrain C, Creagh EM, Martin SJ. Apoptosis-associated release of Smac/DIABLO from mitochondria requires active caspases and is blocked by Bcl-2. *EMBO J* 2001;**20**(23):6627–36.

125 Chai J, Du C, Wu JW, Kyin S, Wang X, Shi Y. Structural and biochemical basis of apoptotic activation by Smac/DIABLO. *Nature* 2000;**406**(6798):855–62.

126 Yang QH, Church-Hajduk R, Ren J, Newton ML, Du C. Omi/HtrA2 catalytic cleavage of inhibitor of apoptosis (IAP) irreversibly inactivates IAPs and facilitates caspase activity in apoptosis. *Genes Dev* 2003;**17**(12):1487–96.

127 Srinivasula SM, Ahmad M, Fernandes-Alnemri T, Alnemri ES. Autoactivation of procaspase-9 by Apaf-1-mediated oligomerization. *Mol Cell* 1998;**1**(7):949–57.

128 Boatright KM, Salvesen GS. Mechanisms of caspase activation. *Curr Opin Cell Biol* 2003;**15**(6):725–31.

129 Brentnall M, Rodriguez-Menocal L, De Guevara RL, Cepero E, Boise LH. Caspase-9, caspase-3 and caspase-7 have distinct roles during intrinsic apoptosis. *BMC Cell Biol* 2013;**14**:32.

130 Wen LP, Fahrni JA, Troie S, Guan JL, Orth K, Rosen GD. Cleavage of focal adhesion kinase by caspases during apoptosis. *J Biol Chem* 1997;**272**(41):26 056–61.

131 Taylor RC, Cullen SP, Martin SJ. Apoptosis: controlled demolition at the cellular level. *Nat Rev Mol Cell Biol* 2008;**9**(3):231–41.

132 Enari M, Sakahira H, Yokoyama H, Okawa K, Iwamatsu A, Nagata S. A caspase-activated DNase that degrades DNA during apoptosis, and its inhibitor ICAD. *Nature* 1998;**391**(6662):43–50.

133 Cheung WL, Ajiro K, Samejima K, Kloc M, Cheung P, Mizzen CA, et al. Apoptotic phosphorylation of histone H2B is mediated by mammalian sterile twenty kinase. *Cell* 2003;**113**(4):507–17.

134 Wen W, Zhu F, Zhang J, Keum YS, Zykova T, Yao K, et al. MST1 promotes apoptosis through phosphorylation of histone H2AX. *J Biol Chem* 2010;**285**(50):39 108–16.

135 Joza N, Susin SA, Daugas E, Stanford WL, Cho SK, Li CY, et al. Essential role of the mitochondrial apoptosis-inducing factor in programmed cell death. *Nature* 2001;**410**(6828):549–54.

136 Norberg E, Orrenius S, Zhivotovsky B. Mitochondrial regulation of cell death: processing of apoptosis-inducing factor (AIF). *Biochem Biophys Res Commun* 2010;**396**(1):95–100.

137 Li LY, Luo X, Wang X. Endonuclease G is an apoptotic DNase when released from mitochondria. *Nature* 2001;**412**(6842):95–9.

138 Savill J, Dransfield I, Gregory C, Haslett C. A blast from the past: clearance of apoptotic cells regulates immune responses. *Nat Rev Immunol* 2002;**2**(12):965–75.

139 Hochreiter-Hufford A, Ravichandran KS. Clearing the dead: apoptotic cell sensing, recognition, engulfment, and digestion. *Cold Spring Harb Perspect Biol* 2013;**5**(1):a008748.

140 Lauber K, Bohn E, Kröber SM, Xiao YJ, Blumenthal SG, Lindemann RK, et al. Apoptotic cells induce migration of phagocytes via caspase-3-mediated release of a lipid attraction signal. *Cell* 2003;**113**(6):717–30.

141 Truman LA, Ford CA, Pasikowska M, Pound JD, Wilkinson SJ, Dumitriu IE, et al. CX3CL1/fractalkine is released from apoptotic lymphocytes to stimulate macrophage chemotaxis. *Blood* 2008;**112**(13):5026–36.

142 Elliott MR, Chekeni FB, Trampont PC, Lazarowski ER, Kadl A, Walk SF, et al. Nucleotides released by apoptotic cells act as a find-me signal to promote phagocytic clearance. *Nature* 2009;**461**(7261):282–6.

143 Gude DR, Alvarez SE, Paugh SW, Mitra P, Yu J, Griffiths R, et al. Apoptosis induces expression of sphingosine kinase 1 to release sphingosine-1-phosphate as a "come-and-get-me" signal. *FASEB J* 2008;**22**(8):2629–38.

144 Arlt A, Muerkoster SS, Schafer H. Targeting apoptosis pathways in pancreatic cancer. *Cancer Lett* 2013;**332**(2):346–58.

145 Fulda S. Tumor resistance to apoptosis. *Int J Cancer* 2009;**124**(3):511–15.

146 Igney FH, Krammer PH. Death and anti-death: tumour resistance to apoptosis. *Nat Rev Cancer* 2002;**2**(4):277–88.

147 Fulda S, Vucic D. Targeting IAP proteins for therapeutic intervention in cancer. *Nat Rev Drug Discov* 2012;**11**(2):109–24.

148 Hervouet E, Cheray M, Vallette FM, Cartron PF. DNA methylation and apoptosis resistance in cancer cells. *Cells* 2013;**2**(3):545–73.

149 Tomiyama A, Serizawa S, Tachibana K, Sakurada K, Samejima H, Kuchino Y, Kitanaka C. Critical role for mitochondrial oxidative phosphorylation in the activation of tumor suppressors Bax and Bak. *J Natl Cancer Inst* 2006;**98**(20):1462–73.

150 Dey R, Moraes CT. Lack of oxidative phosphorylation and low mitochondrial membrane potential decrease susceptibility to apoptosis and do not modulate the protective effect of Bcl-x(L) in osteosarcoma cells. *J Biol Chem* 2000;**275**(10):7087–94.

151 Wu CC, Bratton SB. Regulation of the intrinsic apoptosis pathway by reactive oxygen species. *Antioxid Redox Signal* 2012;**19**(6):546–58.

152 Martinez-Reyes I, Cuezva JM. The H(+)-ATP synthase: a gate to ROS-mediated cell death or cell survival. *Biochim Biophys Acta* 2014;**1837**(7):1099–112.

153 Santamaria G, Martinez-Diez M, Fabregat I, Cuezva JM. Efficient execution of cell death in non-glycolytic cells requires the generation of ROS controlled by the activity of mitochondrial H+-ATP synthase. *Carcinogenesis* 2006;**27**(5):925–35.

154 Mathupala SP, Ko YH, Pedersen PL. Hexokinase II: cancer's double-edged sword acting as both facilitator and gatekeeper of malignancy when bound to mitochondria. *Oncogene* 2006;**25**(34):4777–86.

155 Galluzzi L, Kepp O, Tajeddine N, Kroemer G. Disruption of the hexokinase-VDAC complex for tumor therapy. *Oncogene* 2008;**27**(34):4633–5.

156 Suen DF, Norris KL, Youle RJ. Mitochondrial dynamics and apoptosis. *Genes Dev* 2008;**22**(12):1577–90.

157 Modica-Napolitano JS, Singh KK. Mitochondrial dysfunction in cancer. *Mitochondrion* 2004;**4**(5–6):755–62.

158 Alirol E, Martinou JC. Mitochondria and cancer: is there a morphological connection? *Oncogene* 2006;**25**(34):4706–16.

159 Chiche J, Rouleau M, Gounon P, Brahimi-Horn MC, Pouyssegur J, Mazure NM. Hypoxic enlarged mitochondria protect cancer cells from apoptotic stimuli. *J Cell Physiol* 2010;**222**(3):648–57.

160 Agostini M, Tucci P, Melino G. Cell death pathology: perspective for human diseases. *Biochem Biophys Res Commun* 2011;**414**(3):451–5.

161 Ghavami S, Shojaei S, Yeganeh B, Ande SR, Jangamreddy JR, Mehrpour M, et al. Autophagy and apoptosis dysfunction in neurodegenerative disorders. *Prog Neurobiol* 2014;**112**:24–49.

162 Lustbader JW, Cirilli M, Lin C, Xu HW, Takuma K, Wang N, et al. ABAD directly links Abeta to mitochondrial toxicity in Alzheimer's disease. *Science* 2004;**304**(5669):448–52.

163 Abramov AY, Canevari L, Duchen MR. Beta-amyloid peptides induce mitochondrial dysfunction and oxidative stress in astrocytes and death of neurons through activation of NADPH oxidase. *J Neurosci* 2004;**24**(2):565–75.

164 Picone P, Carrotta R, Montana G, Nobile MR, San Biagio PL, Di Carlo M. Abeta oligomers and fibrillar aggregates induce different apoptotic pathways in LAN5 neuroblastoma cell cultures. *Biophys J* 2009;**96**(10):4200–11.

165 Anandatheerthavarada HK, Biswas G, Robin MA, Avadhani NG. Mitochondrial targeting and a novel transmembrane arrest of Alzheimer's amyloid precursor protein impairs mitochondrial function in neuronal cells. *J Cell Biol* 2003;**161**(1):41–54.

166 Resende R, Moreira PI, Proença T, Deshpande A, Busciglio J, Pereira C, Oliveira CT. Brain oxidative stress in a triple-transgenic mouse model of Alzheimer disease. *Free Radic Biol Med* 2008;**44**(12):2051–7.

167 Manczak M, Anekonda TS, Henson E, Park BS, Quinn J, Reddy PH. Mitochondria are a direct site of A beta accumulation in Alzheimer's disease neurons: implications for free radical generation and oxidative damage in disease progression. *Hum Mol Genet* 2006;**15**(9):1437–49.

168 Resende R, Ferreiro E, Pereira C, Resende de Oliveira C. Neurotoxic effect of oligomeric and fibrillar species of amyloid-beta peptide 1–42: involvement of endoplasmic reticulum calcium release in oligomer-induced cell death. *Neuroscience* 2008;**155**(3):725–37.

169 Protter D, Lang C, Cooper AA. αSynuclein and mitochondrial dysfunction: a pathogenic partnership in Parkinson's disease? *Parkinsons Dis* 2012;**2012**:829207.

170 Yamamoto A, Friedlein A, Imai Y, Takahashi R, Kahle PJ, Haass C. Parkin phosphorylation and modulation of its E3 ubiquitin ligase activity. *J Biol Chem* 2005;**280**(5):3390–9.

171 Luk KC, Kehm V, Carroll J, Zhang B, O'Brien P, Trojanowski JQ, Lee VM. Pathological alpha-synuclein transmission initiates Parkinson-like neurodegeneration in nontransgenic mice. *Science* 2012;**338**(6109):949–53.

172 Geisler S, Holmstrom KM, Skujat D, Fiesel FC, Rothfuss OC, Kahle PJ, Springer W. PINK1/Parkin-mediated mitophagy is dependent on VDAC1 and p62/SQSTM1. *Nat Cell Biol* 2010;**12**(2):119–31.

173 Lazarou M, Jin SM, Kane LA, Youle RJ. Role of PINK1 binding to the TOM complex and alternate intracellular membranes in recruitment and activation of the E3 ligase Parkin. *Dev Cell* 2012;**22**(2):320–33.

174 Akundi RS, Huang Z, Eason J, Pandya JD, Zhi L, Cass WA, et al. Increased mitochondrial calcium sensitivity and abnormal expression of innate immunity genes precede dopaminergic defects in Pink1-deficient mice. *PLoS One* 2011;**6**(1):e16038.

175 Gandhi S, Wood-Kaczmar A, Yao Z, Plun-Favreau H, Deas E, et al. PINK1-associated Parkinson's disease is caused by neuronal vulnerability to calcium-induced cell death. *Mol Cell* 2009;**33**(5):627–38.

176 Heeman B, Van den Haute C, Aelvoet SA, Valsecchi F, Rodenburg RJ, Reumers V, et al. Depletion of PINK1 affects mitochondrial metabolism, calcium homeostasis and energy maintenance. *J Cell Sci* 2011;**124**(Pt. 7):1115–25.

177 Zhou W, Zhu M, Wilson MA, Petsko GA, Fink AL. The oxidation state of DJ-1 regulates its chaperone activity toward alpha-synuclein. *J Mol Biol* 2006;**356**(4):1036–48.
178 Galluzzi L, Larochette N, Zamzami N, Kroemer G. Mitochondria as therapeutic targets for cancer chemotherapy. *Oncogene* 2006;**25**(34):4812–30.
179 Certo M, Del Gaizo Moore V, Nishino M, Wei G, Korsmeyer S, Armstrong SA, Letai A. Mitochondria primed by death signals determine cellular addiction to antiapoptotic BCL-2 family members. *Cancer Cell* 2006;**9**(5):351–65.
180 Letai A, Bassik MC, Walensky LD, Sorcinelli MD, Weiler S, Korsmeyer SJ. Distinct BH3 domains either sensitize or activate mitochondrial apoptosis, serving as prototype cancer therapeutics. *Cancer Cell* 2002;**2**(3):183–92.
181 Oudard S, Carpentier A, Banu E, Fauchon F, Celerier D, Poupon MF, et al. Phase II study of lonidamine and diazepam in the treatment of recurrent glioblastoma multiforme. *J Neurooncol* 2003;**63**(1):81–6.
182 Ravagnan L, Marzo I, Costantini P, Susin SA, Zamzami N, Petit PX, et al. Lonidamine triggers apoptosis via a direct, Bcl-2-inhibited effect on the mitochondrial permeability transition pore. *Oncogene* 1999;**18**(16):2537–46.
183 Don AS, Kisker O, Dilda P, Donoghue N, Zhao X, Decollogne S, et al. A peptide trivalent arsenical inhibits tumor angiogenesis by perturbing mitochondrial function in angiogenic endothelial cells. *Cancer Cell* 2003;**3**(5):497–509.
184 Lehenkari PP, Kellinsalmi M, Näpänkangas JP, Ylitalo KV, Mönkkönen J, Rogers MJ, et al. Further insight into mechanism of action of clodronate: inhibition of mitochondrial ADP/ATP translocase by a nonhydrolyzable, adenine-containing metabolite. *Mol Pharmacol* 2002;**61**(5):1255–62.
185 Kidd JF, Pilkington MF, Schell MJ, Fogarty KE, Skepper JN, Taylor CW, Thorn P. Paclitaxel affects cytosolic calcium signals by opening the mitochondrial permeability transition pore. *J Biol Chem* 2002;**277**(8):6504–10.
186 Robertson JD, Gogvadze V, Zhivotovsky B, Orrenius S. Distinct pathways for stimulation of cytochrome c release by etoposide. *J Biol Chem* 2000;**275**(42):32 438–43.
187 Rotem R, Heyfets A, Fingrut O, Blickstein D, Shaklai M, Flescher E. Jasmonates: novel anticancer agents acting directly and selectively on human cancer cell mitochondria. *Cancer Res* 2005;**65**(5):1984–93.
188 Fulda S, Galluzzi L, Kroemer G. Targeting mitochondria for cancer therapy. *Nat Rev Drug Discov* 2010;**9**(6):447–64.
189 Uttara B, Singh AV, Zamboni P, Mahajan RT. Oxidative stress and neurodegenerative diseases: a review of upstream and downstream antioxidant therapeutic options. *Curr Neuropharmacol* 2009;**7**(1):65–74.
190 Lin MT, Beal MF. Mitochondrial dysfunction and oxidative stress in neurodegenerative diseases. *Nature* 19 2006;**443**(7113):787–95.
191 Busciglio J, Pelsman A, Wong C, Pigino G, Yuan M, Mori H, Yankner BA. Altered metabolism of the amyloid beta precursor protein is associated with mitochondrial dysfunction in Down's syndrome. *Neuron* 2002;**33**(5):677–88.
192 Ohyagi Y, Yamada T, Nishioka K, Clarke NJ, Tomlinson AJ, Naylor S, et al. Selective increase in cellular A beta 42 is related to apoptosis but not necrosis. *Neuroreport* 2000;**11**(1):167–71.
193 Betarbet R, Porter RH, Greenamyre JT. GluR1 glutamate receptor subunit is regulated differentially in the primate basal ganglia following nigrostriatal dopamine denervation. *J Neurochem* 2000;**74**(3):1166–74.

194 Fornai F, Schlüter OM, Lenzi P, Gesi M, Ruffoli R, Ferrucci M, et al. Parkinson-like syndrome induced by continuous MPTP infusion: convergent roles of the ubiquitin-proteasome system and alpha-synuclein. *Proc Natl Acad Sci USA* 2005;**102**(9):3413–18.
195 Manczak M, Mao P, Calkins MJ, Cornea A, Reddy AP, Murphy MP, et al. Mitochondria-targeted antioxidants protect against amyloid-beta toxicity in Alzheimer's disease neurons. *J Alzheimers Dis* 2010;**20**(Suppl. 2):S609–31.
196 Smith RA, Hartley RC, Murphy MP. Mitochondria-targeted small molecule therapeutics and probes. *Antioxid Redox Signal* 2011;**15**(12):3021–38.
197 Llambi F, Green DR. Apoptosis and oncogenesis: give and take in the BCL-2 family. *Curr Opin Genet Dev* 2011;**21**(1):12–20.
198 Sheridan C, Delivani P, Cullen SP, Martin SJ. Bax- or Bak-induced mitochondrial fission can be uncoupled from cytochrome C release. *Mol Cell* 2008;**31**(4):570–85.
199 Karbowski M, Lee YJ, Gaume B, Jeong SY, Frank S, Nechushtan A, et al. Spatial and temporal association of Bax with mitochondrial fission sites, Drp1, and Mfn2 during apoptosis. *J Cell Biol* 2002;**159**(6):931–8.
200 Cassidy-Stone A, Chipuk JE, Ingerman E, Song C, Yoo C, Kuwana T, et al. Chemical inhibition of the mitochondrial division dynamin reveals its role in Bax/Bak-dependent mitochondrial outer membrane permeabilization. *Dev Cell* 2008;**14**(2):193–204.

3

Microbial Programmed Cell Death

Neal D. Hammer

Department of Microbiology and Molecular Genetics, Michigan State University, East Lansing, MI, USA

Abbreviations

ALD	apoptotic-like death
ECM	extracellular matrix
EDF	extracellular death factor
eDNA	extracellular DNA
ETC	electron transport chain
PCD	programmed cell death
PG	peptidoglycan
PMF	proton motive force
ROS	reactive oxygen species
SDP	sporulation-delaying protein
SKF	sporulation killing factor
TA system	toxin–antitoxin system

3.1 Introduction

Programmed cell death (PCD) is the response of a cell to a stimulus that induces the expression of effectors that kill that cell. The molecular details that differentiate the distinct PCD pathways in cells of eukaryotic origin have been an enormous focus of study; however, the PCD pathways active in bacterial cells are a relatively new discovery – one that is gaining appreciation. Microbes colonize nearly every environment on the planet. Therefore, understanding the role of PCD pathways in the shaping of microbial communities will have tremendous impact. Furthermore, targeting PCD pathways could be a useful therapeutic strategy for combating bacterial pathogens.

There are many differences between bacterial cells and the cells that make up multicellular eukaryotic organisms. Bacterial cells are often smaller, and are protected from the environment by a cell envelope [1]. The molecular components of the envelope differ depending on the species of the bacteria. In some species, the envelope is composed of two membrane lipid bilayers, separated by the periplasmic space. This membrane organization is similar to that observed in mitochondria. The periplasmic

Apoptosis and Beyond: The Many Ways Cells Die, First Edition. Edited by James Radosevich.
© 2018 John Wiley & Son Inc. Published 2018 by John Wiley & Son Inc.

space contains a polymer comprising glycan threads crosslinked by peptides [1]. This polymer is called peptidoglycan (PG). In other species, a thick layer of PG is the only component of the cell envelope, and therefore the only structure protecting the bacterial cell from the external environment. The tinctoral properties of the outer membrane and the thick PG can be exploited to differentiate between these two types of bacteria using a staining technique known as Gram's method [1]. Light microscopy can then be used to observe the cells' Gram stain phenotype. Double membrane-containing bacteria are Gram negative, while bacteria that contain a thick PG layer are Gram positive.

Another difference between bacterial cells and their eukaryotic counterparts is that bacterial cells do not contain membrane-enclosed organelles such as the nucleus [1]. The nucleus of eukaryotes harbors multiple linear chromosomes that contain the DNA of the organism. In most but not all bacteria, the DNA is contained within a single, circular chromosome [1]. Consistent with this, bacteria contain only a single copy of their genes. Often, the bacterial chromosomal DNA is supplemented by additional genetic information, contained on small pieces of circular DNA called plasmids [1]. In some cases, plasmids contain the genetic information that allows the bacterium harboring them to resist antibiotics or to produce toxins and other virulence factors.

Perhaps the biggest difference between bacterial cells and the eukaryotic cells that make up a multicellular organism, as it pertains to PCD, is that the bacterial cells are autonomous, whereas each eukaryotic cell is a component of the larger organism (with the obvious exception of unicellular eukaryotes). PCD in multicellular organisms that develop or change makes inherent sense: it is a function of ontogenesis. As the structure of the multicellular organism changes over time, different cells are required to maintain its viability or structure. The elimination of cells that are infected with pathogens is another function of eukaryotic PCD. However, the fact that PCD is required for the development and immunity of a multicellular organism leads to an interestingly question: Why would an autonomous bacterial cell want to kill itself, putting the entire species at peril? A predominant feature of microbial life likely provides an answer this question.

In nature, bacteria often live within large, multicellular, multispecies communities called biofilms. Biofilms are found in diverse environments, including the thermal vents at the bottom of the ocean and as a major component of dental plaques [2–5]. The networks of cells that compose the biofilm are held together by a complex matrix of polymers. These polymers include cellulose, proteins, and extracellular DNA (eDNA). There are dedicated genetic programs that facilitate the secretion of these components; however, in some cases, eDNA results from cell lysis. Therefore, PCD-mediated cell lysis leads to the production of eDNA and the biosynthesis of an extracellular matrix (ECM), which helps the bacterial population persist within its environment [6,7]. Biofilms also facilitate increased resistance to toxic molecules, such as antibiotics, by limiting their diffusion and reducing the exposure of some of the bacterial cells within the biofilm community. Consequently, biofilms that have developed on medical implants often require surgical removal, because treatment with antibiotics is not sufficient to clear the bacterial cells residing within them [8]. In these instances, perturbing the ability of the bacterial cells to form a biofilm by interfering with the PCD pathways that facilitate the production of eDNA will provide a clinical benefit.

Another example of a bacterial developmental stage that leads to PCD is the formation of a spore. This example will be discussed in Section 3.2. In addition to specific developmental processes such as biofilm formation and spore development, PCD may also provide a mechanism for genetic exchange or for increasing the genetic stability of a population by eradicating cells that experience severe DNA damage [9]. Furthermore, bacterial PCD can protect populations of bacteria from predation by bacteriophages: viruses that replicate within bacterial cells. In some cases, phage infection induces specific types of PCD pathway, as discussed in Section 3.3 [10]. Evidence suggests that bacteria undergo an apoptosis-like cell death in response to excessive oxidative stress and DNA damage [11,12]. The molecular signatures shared by these "terminally stressed" bacterial cells and apoptotic eukaryotic cells are extracellular exposure of phosphatidylserine, decreased membrane potential and cell division, and DNA condensation and fragmentation [13]. This finding implies that some degree of conservation exists between eukaryotic and bacterial PCD. Interestingly, the ectopic expression of the eukaryotic apoptosis effector molecules Bax and Bak in bacterial cells has revealed the molecular mechanism of their killing effect [14]. This provides further evidence that the PCD pathways of bacteria and of eukaryotic organisms are conserved. The study of bacterial PCD is relatively recent, but it is clear that this field is gaining interest.

3.2 Sporulation: A Tale of Two Bacterial Cell-Death Pathways

Some species of bacteria are endowed with the ability to form a dormant spore upon nutrient deprivation. The durability of the spore protects the genetic material residing within it until favorable growth conditions are encountered. Because sporulation is a committed and energetically expensive process, the soil-dwelling Gram-positive bacterium *Bacillus subtilis* has evolved an interesting "last ditch" effort to stall sporulation [15,16]. The observation that the number of viable cells significantly decreases just prior to the onset of sporulation suggests that not all the cells within the population produce spores. In order to resolve this finding, it was proposed that a subpopulation of the bacilli dies before sporulation begins [16]. This decrease was subsequently attributed to toxins produced by sister cells within the population that kill susceptible cells. Killing of the sensitive sister cells results in nutrient release, providing an additional boost of nutrients for the living cells [16,17]. Accordingly, this process was termed "cannibalism" [16]. Both sporulation and canniblism will be discussed at the molecular level in the context of *B. subtilis*, although many species of bacteria are capable of sporulation and likely undergo a similar PCD.

3.2.1 Cannibalism

Only half of the *B. subtilis* cells growing in an environment that induces sporulation activate the transcriptional response necessary for this survival strategy [16]. The transcriptional response regulator Spo0A controls the expression of the genes required to form a spore [18]. Spo0A also controls the expression of two toxins that kill the *B. subtilis* cells that have not activated Spo0A (Spo0A-OFF) [16,19]. Therefore, Spo0A-ON cells kill Spo0A-OFF cells (Figure 3.1A). The killed Spo0A-OFF cells release nutrients and stall the sporulation process in the Spo0A-ON cells by delaying entry into the irreversible stages of sporulation. This example of cannibalism allows *B. subtilis* cells to delay sporulation until every possible reservoir of nutrients has been exhausted.

The two molecules that are secreted by the Spo0A-ON bacilli, sporulation killing factor (SKF) and sporulation-delaying protein (SPD), mediate the killing process (Figure 3.1A) [16,19,20]. Both SKF and SDP are hydrophobic peptides, comprising 26

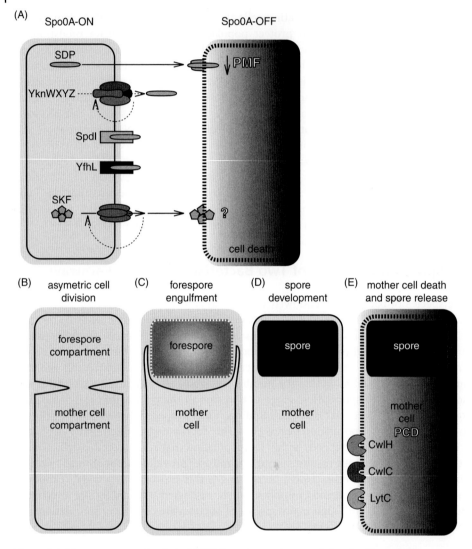

Figure 3.1 The programmed cell-death (PCD) pathways in sporulating *Bacillus subtilis*.
(A) Cannibalism of Spo0A-OFF sister bacilli is induced by toxins produced by Spo0A-ON sister cells. Two toxins are produced by the Spo0A-ON cells: sporulation-delaying protein (SDP) and sporulation killing factor (SKF). In the Spo0A-OFF cells, SDP disrupts the proton motive force (PMF) and induces PCD. The mechanism by which SKF kills cells is unknown. The Spo0A-ON cells that produce SDP and SKF are immune to these toxins. Immunity to SDP is provided in part by SpdI and YfhL, proteins that bind SPD and sequester its toxic effect. SDP can also be pumped out of the Spo0A-ON cells by YknWXYZ. Similarly, SKF is pumped out of Spo0A-ON cells. After the nutrients provided by this cannibalism are exhausted, the Spo0A-ON cells will induce spore formation. (B) Sporulation begins with an asymmetric cell division, which is followed by an engulfment of the forespore by the mother cell. (C,D) The mother cell nurtures the forespore as it matures into a spore. (E) The mother cell then undergoes PCD and releases the spore via the activation of several PG hydrolases: CwlH, CwlC, and LytC.

and 42 amino acids, respectively [20,21]. *B. subtilis* mutant cells that produce only SDP or SKF are able to kill cells grown in close proximity, demonstrating that each toxin has the capacity to mediate cell death. In accordance with this, purified SDP or SKF induce bacterial cell death when added exogenously. Interestingly, purified and exogenously added SDP demonstrates a more robust killing effect than purified and exogenously added SKF [20]. The ability of SDP to kill bacterial cells extends beyond *B. subtilis*, as this toxin has been shown to induce cell death in other species of bacteria [20,22]. This feature of SDP is likely explained by its mechanism of action: SDP induces cell death of neighboring cells by rapidly collapsing their proton motive force (PMF) [23]. The ability to kill other bacterial species is hypothesized to help *B. subtilis* biofilms compete for nutrients within the environment. In support of this idea, when grown in a biofilm, *B. subtilis* produces high levels of SDP, which is toxic to other microorganisms within the environment [22,24]. To date, the mechanism by which SKF kills is unknown. One hypothesis implies that the mechanism by which this toxin induces cell death is analogous to that of another class of bacterial-produced toxins called bacteriocins [25]. Some bacteriocins kill cells by forming a pore in the outer membrane, permeabilizing the bacterial cell [26].

SDP and SKF can selectively kill SpoA-OFF cells because SpoA-ON cells have mechanisms for resisting SDP and SKF toxicity. SKF is secreted in a pumplike fashion outside of the Spo0A-ON cells. This pump presumably provides the resistance to SKF, as mutants capable of producing SKF but not secreting it succumb to its cytotoxic effects [16]. Therefore, Spo0A-ON cells are resistant to the effects of SKF because they can pump out the toxin (Figure 3.1A). Spo0A-OFF cells do not express the pump and are sensitive to SKF cytotoxicity [16]. Immunity to SDP, on the other hand, is mediated by a specific protein, SpdI [19]. *B. subtilis* expressing SpdI are protected from SDP because SpdI sequesters SDP, keeping it from perturbing the PMF [19]. In addition to SpdI, YfhL and the multiprotein complex, YknWXYZ, also provide resistance to SDP [27]. YknWXYZ is predicted to protect bacilli from SDP in a pumplike fashion, in a manner similar to the mechanism of SKF resistance (Figure 3.1A). YfhL is a paralog of SpdI and is hypothesized to provide resistance to SDP in a similar manner [27,28]. Expression of all the immunity proteins, SpdI, YfhL, and YknWXYZ, is repressed in the Spo0A-OFF cells, providing a molecular explanation for the sensitivity of the Spo0A-OFF cells to SDP toxicity [19,27,29,30].

3.2.2 Sporulation

The process of sporulation in *B. subtilis* can be divided into several stages [31]. It begins with an asymmetric division of the cell, resulting in a large mother cell and a smaller forespore [32]. Next, the mother cell engulfs the forespore. In the final stage, the forespore matures into a spore and the mother cell undergoes lysis, resulting in the release of the spore (Figure 3.1B–E). The molecular details underlying the asymmetric cell division, mother-cell engulfment of the forespore, and maturation of the forespore to a spore are beyond the scope of this chapter, but they have been reviewed extensively elsewhere [33,34]. The focus of this section is on the lysis of the mother cell, which facilitates release of the mature spore.

Three proteins degrade the cell wall of the mother cell, allowing for release of the mature spore: LytC, CwlC, and CwlH (Figure 3.1E) [35–39]. Interestingly, genetic inactivation of any of these three proteins individually does not impair spore release, highlighting the redundancy built into this developmental process. However, a *lytC cwlC* double mutant and a *lytC cwlC cwlH* triple mutant are impaired for mother-cell lysis and release of the spore [40]. This defect in mother-cell lysis does not impact spore development, as the *lytC cwlC* double-mutant spore is able to resist high temperatures and chemical insult [38,41]. Mother-cell death can also be observed using time-lapse microscopy [40]. In studies using this technique, rupture of the mother-cell membrane occurred prior to cell-wall degradation. Membrane rupture of the mother cell was also observed upon inactivation of the triple mutant *lytC cwlC cwlH*, providing support for the idea that membrane rupture occurs prior to cell-wall lysis [40]. These results point to additional membrane-rupturing factors involved in the PCD of mother cells. Further analysis is needed to genetically define the proteins and the mechanism involved in rupture of the mother-cell membrane.

The finding that a spore develops normally under laboratory conditions when cell wall-degrading enzymes are inactivated implies that release of the spore is more important in environmental settings. Further analysis of the double and triple mutants under environmentally relevant conditions is needed to determine how spore release impacts the fitness of *B. subtilis* populations over time. Given that sporulating cells reside at the air interface within a *B. subtilis* biofilm, it is tempting to speculate that a defect in mother-cell lysis and spore release may impact the ability of the spore to interact with environmentally relevant factors that come in contact with the *B. subtilis* biofilm [42].

Many species of bacteria have the capacity to produce spores during times of excessive stress. It is likely that the cannibalism and PCD of the mother cell during the sporulation process in *B. subtilis* also occur in other species as well. For example, in the Gram-negative predatory bacterium *Myxococcus xanthus*, PCD is involved in the development of a specialized multicellular structure that contains spores, called a "fruiting body." Just as in the cell death that occurs just prior to *B. subtilis* sporulation, only a small percentage of *M. xanthus* go on to sporulate [43]. This altruistic social behavior of the vegetative cells that undergo PCD feeds the cells within the fruiting body and helps the population of *M. xanthus* cells to persist in the environment through the production spores [9].

3.3 Toxin–Antitoxin Systems

3.3.1 Introduction to TA Systems

Perhaps the most persuasive evidence for the existence of bacterial PCD is the discovery of toxin–antitoxin (TA) systems. TA systems are gene pairs that encode a toxin and its cognate, labile antitoxin. In most cases, a physical interaction between the toxin and the antitoxin proteins renders the toxin inactive. Therefore, any condition that reduces the amount of antitoxin results in active toxin and cell death. TA systems were first discovered on circular autonomous-replicating foreign DNA elements called plasmids. The TA system encoded on the plasmid maintained the foreign DNA within the bacterial cell, because loss of the plasmid would reduce the amount of antitoxin and active the toxin, resulting in cell death. Therefore, the presence of the TA system ensured the plasmid was maintained within the bacterial population. TA systems can be found in many species of bacteria and archaea, and the number of TA systems harbored by a single bacterium can be as few as zero or as many as 88 [44–50]. The finding that some species of bacteria do not contain TA systems supports the notion that TA systems are not an essential feature of bacterial physiology. Additionally, some laboratory strains develop inactivating mutations within TA systems upon their passage, suggesting that the TA systems are not essential for growth under laboratory conditions [11,12,51]. These findings imply that TA systems function primarily under environmental conditions. Within the environment, many species of bacteria compete for limited resources. The fact that the small peptides produced by some species of bacteria induce TA-system activation and PCD of neighboring bacteria supports this idea [52]. Nevertheless, the prevalence of TA systems throughout bacterial species and within microbial genomes indicates that these systems are important modulators of bacterial PCD.

3.3.2 TA-System Classification and Functional Roles

There are five types of TA systems [23,53]. The defining feature that distinguishes the TA system is the molecular nature of the TA interaction. In type I TA systems, the toxin synthesis is inhibited because the antitoxin mRNA anneals to the toxin mRNA. Formation of an mRNA–mRNA duplex blocks the ribosome binding site, limiting expression of the toxin. The TA RNA duplex also induces degradation of the toxin transcript, resulting in a further reduction of toxin expression [54–56]. Type I toxins are often small hydrophobic proteins that have been demonstrated to disrupt the membrane potential of the bacterial cell [56,57].

The type II TA systems interact via protein–protein interactions. In the type II systems, both the toxin and the antitoxin are small proteins, and proteolysis of the antitoxin activates the toxin [25]. Toxin activation can also be stimulated by transcriptional repression of its cognate antitoxin, which reduces the antitoxin–toxin ratio [58–60]. This transcriptional repression, in combination with proteolysis of the labile antitoxin, leads to a reduction of the antitoxin protein and activation of the toxin. It has recently been established that an additional mechanism is required to activate some type II TA systems; this mechanism will be discussed in further detail in Section 3.3.3. Briefly, the MazEF type II TA system requires the extracellular death factor (EDF) for activation of the toxin [61]. The EDF is a secreted linear pentapeptide that is imported by bacterial cells and physically interacts with the MazF toxin. The EDF–MazF interaction increases the endoribonuclease activity of MazF [62], which degrades mRNA and arrests translation, resulting in cell death [63,64]. Interestingly, the EDF also increases the activity of another type II TA-system endoribonuclease toxin, ChpBK [62,65]. It remains to be determined whether there are additional EDFs that impact the other types of TA systems, beyond type II, but it has been established that other species of bacteria outcompete MazEF-expressing bacteria by producing unique EDFs capable of killing them [52]. MazF elicits cytotoxicity by degrading mRNA and inhibiting translation, but the mechanism of action for type II TA toxins is diverse and also includes obstruction of translational machinery, inactivation of DNA gyrase, and inhibition of cell-wall synthesis [53].

The type III TA system functions as an RNA–protein interaction complex. The antitoxin RNA molecule neutralizes the toxin by binding directly to it, keeping it inactive. Changes in the transcription of the antitoxin RNA reduce the antitoxin–toxin ratio, resulting in active toxin. In the ToxIN type III system, *toxI* encodes the antitoxin and *toxN* encodes the toxin – another endoribonuclease that degrades RNA molecules. The ToxIN system protects bacteria from phage infection. Upon infection, the phage alters transcription of the bacterial cell, resulting in a reduced *toxI*–ToxN ratio, which leads to activated ToxN endoribonuclease activity and PCD. This altruistic suicide preserves the other bacterial cells within the population and halts additional phage infections [66].

Type IV and type V TA systems have recently been discovered [25,53,67]. In the type IV TA systems, the antitoxin occludes the target of the toxin. In the only described case of such a TA system, the toxin, CbtA, inhibits polymerization of the bacterial cytoskeleton proteins FtsZ (bacterial tubulin ortholog) and MreB (bacterial actin ortholog), which maintain cell shape and are required for cell division. The antitoxin in this system, CbeA, promotes the polymerization and stability of these cytoskeletal proteins [68]. Under stress-inducing conditions, the antitoxin is degraded, and collapse of the bacterial cytoskeleton induces PCD [67]. In the type V TA system, the antitoxin is an endoribonuclease that cleaves the mRNA encoding the toxin. As with type IV, only one of these type V systems has been observed, but other bacteria are proposed also to contain them [69]. In *Escherichia Coli*, GhoS is a sequence-specific endoribonuclease that degrades the GhoT mRNA. Translation of GhoT in the absence of GhoS leads to membrane damage, leakage of the cytoplasmic contents, and formation of ghost cells [69]. Interestingly, regulation of GhoS and GhoT is controlled by a type II TA system, MqsR/MqsA [70]. Under conditions of stress, GhoS mRNA is degraded by MqsR, resulting in increased GhoT protein expression and cell death [70]. This is the first example of a TA system regulating another TA system. However, TA systems have been found to impact common physiological processes, suggesting that crosstalk between them is likely more common than has been observed.

The diversity that underscores the molecular mechanisms of the TA interaction is consistent with the various functional roles of the TA systems. The first TA systems to be discovered were encoded within extra-chromosomal foreign DNA, and the function attributed to these systems was the maintenance of the plasmid within a bacterium [58]. The discovery that TA systems can be encoded within the genome and that genomically encoded TA systems also induce bacterial PCD greatly increased interest in such systems [59]. The studies that resulted from this increased interest found that TA systems have many functions within a bacterial cell. For example, type I, II, and III TA systems can protect populations of bacteria from phage infection [66,71,72]. In this context, the function of TA systems is analogous to the apoptosis of eukaryotic cells infected with viruses. Eukaryotic PCD pathways also protect the organism from aberrant cell growth that occurs as a result of DNA damage, which can lead to cancer. In bacteria, some TA systems are induced when the cell becomes overly burdened with DNA damage, and it has been proposed that this protects the genomic integrity of the species [73]. One type of DNA damage-inducing stressor is the production of reactive oxygen species (ROS) when bacterial cells are treated with antibiotics [11]. The capacity of antibiotics to induce oxidative stress is controversial [12,74–76]; however, TA systems have been demonstrated to increase the ability of bacterial cells to resist antibiotics via two mechanisms. First, TA systems play a crucial role in biofilm formation, and bacteria growing within the biofilm are recalcitrant to antimicrobials [77]. Second, TA systems can induce the persister phenotype [69–80]: slow-growing, metabolically inert cells that arise stochastically within the population [9,81]. The reduced metabolism of persister cells allows them to resist antibiotics that kill actively dividing cells. The TA system-mediated induction of the persister phenotype may provide a molecular explanation for the failure of some antibiotic therapies [81]. Therefore, understanding the mechanistic links between TA systems, induction of the persister phenotype, and antibiotic resistance could be of therapeutic value. For example, chemical inactivation of the TA systems that induce the persister phenotype has the potential to generate more robust antibiotic-mediated cell killing through the elimination of the persister cells. Additionally, small molecules that perturb the toxin–antitoxin interaction could be used to induce aberrant PCD [46]. Induction of aberrant PCD through the targeting of TA systems could also aid in the clearance of disease-associated biofilm formation or of biofilms that disrupt industrial processes. The fact that many bacterial pathogens encode multiple TA systems further establishes TA systems as potential antimicrobial targets. In sum, TA systems play significant roles in multiple aspects of bacterial physiology, including stabilization of foreign DNA elements, gene regulation, protection from phage infection, resistance to antibiotics, biofilm formation, and control of growth [82]. Further characterization of the interplay between TA systems and bacterial PCD is of clinical and industrial value.

3.3.3 MazEF

The *mazEF* type II TA system was the first such system found to be encoded on the bacterial chromosome [59]. This was a significant finding, because TA systems were previously thought to be required only for the maintenance of extra-chromosomal DNA elements. The fact that the *mazEF* TA system was encoded within the chromosome and that this system induced PCD implied that bacterial cells have an inherent mechanism for the promotion of cell death. To date, *mazEF* is the most-studied TA system, primarily in the microbiome-associated Gram-negative bacterium *E. coli* [46]. Therefore, the focus of this section will predominantly be on studies conducted in *E. coli*, although the role of MazEF in other species of bacteria will be briefly discussed.

In *E. coli*, the *mazEF* genes are located adjacently on the chromosome and, as is common in bacterial cells, are co-transcribed as a single mRNA molecule [59]. MazE is the labile antitoxin and MazF is the stable toxin. Consistent with *mazEF* being characterized as a type II TA system, the MazE and MazF proteins physically interact within the *E. coli* cytoplasm. This physical interaction sequesters MazF and neutralizes its cytotoxicity [59,83]. MazF elicits toxicity via an endoribonuclease activity that specifically targets ACA nucleotide motifs within single-stranded mRNA molecules [63,64]. Global degradation of cellular mRNA by MazF is proposed to lead to inhibition of protein synthesis and cell death, although a revised model of MazF activity and its role in PCD has recently been suggested (see later).

Disruption of the MazE–MazF interaction can occur through several mechanisms, and many stressors can induce MazF cytotoxicity. The stressors that promote activation of MazF include nutritional starvation sensed through increased levels of the bacterial signaling molecule ppGpp [59,84], antibiotics that inhibit transcription or translation [60], elevated temperatures [85], phage infection [72], DNA damage [72], and the oxidative stress that occurs upon exposure to H_2O_2 [72]. In response to these stressors, MazF cytotoxicity becomes active via reduced transcription of *mazE* and degradation of the MazE protein by the serine protease ClpAP (Figure 3.2) [59].

Production of the EDF is also required for the induction of MazF-mediated PCD [61]. The EDF is a linear pentapeptide that is secreted by *E. coli* in a population-dependent manner [61]. Cultures of *E. coli* grown to high density produce EDF, and when this high-density cell culture is stressed with antibiotics or any of the other conditions listed in Figure 3.2, the cells induce MazF PCD [61]. The amino-acid composition of the EDF is as follows: asparagine, asparagine, tryptophan, asparagine, asparagine (single-letter amino acid sequence: NNWNN) [61]. This small peptide overcomes the antitoxin function of MazE by physically binding MazF, leading to increased endoribonuclease activity [62]. It has been proposed that EDF directly competes with MazE to bind to MazF, but further

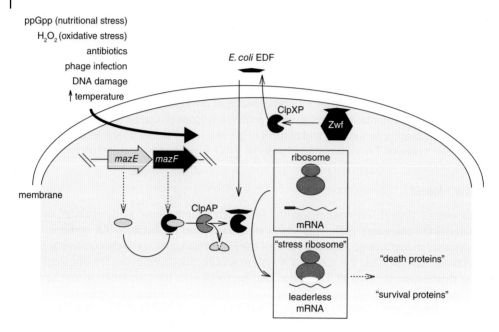

Figure 3.2 The MazEF PCD pathway. In *Escherichia coli*, the MazEF TA system is activated by nutritional stress, oxidative stress, antibiotics, phage infection, DNA damage, and increasing temperature. MazE inhibits MazF endoribonuclease activity. MazF becomes active when *E. coli* encounters any of the conditions just listed and is exposed to EDF. When these two requirements are met, MazE is degraded by ClpAP. EDF interacts with MazF, activating MazF. The endoribonuclease activity of MazF specifically modifies the translational machinery by producing "stress ribosomes" that translate leaderless mRNA. Such leaderless mRNA is also a product of MazF. This results in the translation of a select group of "death proteins," as well as of "survival proteins." It is currently unknown how the cell regulates the differential production of the "death proteins" and "survival proteins." Thick lines indicate positive transcriptional regulation; dashed lines indicate translation of genes into proteins.

molecular and structural characterization of these protein interactions is required [52]. Synthetically manufactured EDF induces PCD in a MazF-dependent manner, highlighting the fact that *E. coli* cells exposed to both stress and EDF undergo PCD [61]. Within the cell, production of EDF occurs through proteolysis of the NNWNN-containing protein Zwf by the ClpXP protease [86]. In agreement with this, the extracellular fraction of *E. coli* mutants inactivated for *zwf* or *clpXP* does not contain EDF [86]. In the current model, ClpXP cleaves Zwf, resulting in secretion of NNWNN (Figure 3.2). The EDF is then imported by other cells within the population, and once internalized, EDF binds and activates MazF. Interestingly, the soil-dwelling bacteria *B. subtilis* and *Pseudomonas aeruginosa* produce distinct EDFs that induce MazEF-dependent PCD [52]. The *B. subtilis* EDF is composed of six amino acids (RGQQNE) and activates PCD in *B. subtilis* via the *mazEF* homolog *ydcDE* [52]. The EDFs produced by *B. subtilis* (RGQQNE) and *E. coli* (NNWNN) induce PCD not only in the species that produced them but also in the other species. For example, the *B. subtilis* EDF activates PCD in *E.coli* in a MazEF-dependent manner, while the *E. coli* EDF induces cell death in *B. subtilis* by activating YdcDE [52]. Despite the fact that the *E. coli* EDF and the *B. subtilis* EDF are composed of different amino acids, the finding that they induce PCD in other species of bacteria through a MazEF or a MazEF homolog implies that some degree of conservation exists between these two TA systems. Intriguingly, *P. aeruginosa* produces three unique EDFs: one is nine amino acids long and the other two are sixteen amino acids long. The two sixteen-amino-acid EDFs have distinct amino-acid sequences [52,87]. All three of the *P. aeruginosa* EDFs stimulate PCD in *E. coli* and *B. subtilis* by activating MazEF and YdcDE, respectively. This result is surprising, given that *P. aeruginosa* is one of the few bacterial species that does not encode a MazEF homolog [52]. The production of three EDFs by *P. aeruginosa* likely gives it a competitive advantage when co-cultured with *E. coli* or *B. subtilis*. In general, the intraspecies killing activity of the EDFs suggests that induction of a neighboring microbe's TA systems is one way of eliminating competing microorganisms during growth under stress-inducing conditions.

The dual requirement of EDF and MazEF for PCD induction underscores the multifaceted but highly controlled nature of this PCD pathway. Induction of MazEF via supplementation with EDF can alter the effect stressors have on bacterial physiology and viability. For example, supplementation of EDF to cultures of *E. coli* exposed to antibiotics that normal only arrest bacterial growth (bacterostatic) results in killing of the bacterial cells (bactericidal) [51]. The shift from a bacteriostatic mechanism of action to a bactericidal mechanism of action occurs through MazF- and EDF-dependent increases in ROS [51]. This result implies that supplementation of a bacteriostatic antibiotic regimen with EDF could provide a more robust antimicrobial effect. The finding that EDF and MazF dictate the response of the cell to antibiotic treatment provides support for the hypothesis that MazF regulates multiple PCD pathways within the cell. This hypothesis predicts that MazF promotes cell death in some circumstances and cell survival in others. In support of this hypothesis, a small population of *E. coli* cells that activate MazF survive, dependent on the expression of a small subset of "survival proteins" [88]. Accordingly, inactivating the genes that encode these proteins causes a further reduction in the number of cells that survive MazEF PCD. Translation of this select group of "survival proteins" occurs because MazF can modify the translational machinery to direct the synthesis of a select group of proteins. Specifically, MazF endoribonuclease activity cleaves the 16S rRNA, resulting in a subpopulation of "stress ribosomes." Additionally, MazF cuts mRNA at specific sites, generating leaderless mRNA that is subsequently transcribed by the "stress ribosomes" [89]. The stress-induced translational machinery leads to the synthesis of a select group of proteins. The "survival proteins" are one class of this group, but the MazF-altered translational machinery also promotes the production of "death proteins" (Figure 3.2) [88,89]. These data suggest that MazF activity can tailor the protein population within a cell to support either survival or death. It remains to be elucidated whether these two groups of proteins, the "survival proteins" and "death proteins," are selectively synthesized within individual cells or if both populations are translated within the same cell [88]. The latter implies a more complex mechanism for the induction of MazF-mediated PCD.

Another layer was recently added to the multidimensional role of MazEF in PCD. Cells inactivated for *mazEF* challenged with stress that induces severe DNA damage undergo an apoptotic-like death (ALD) [74]. In cells that express MazEF, the ALD is masked, because MazEF inhibits the ALD [12,74]. The MazEF-dependent translation of the "survival proteins" promotes the inhibition of the ALD. The molecular signature of the ALD is distinct from that of MazEF-induced cell death, and includes membrane depolarization, degradation of rRNA, upregulation of a specialized group of proteins involved in repairing excessive DNA damage, a decrease in electron transport chain

(ETC) activity, and formation of high levels of hydroxyl radicals via the Fenton reaction [12,74]. Why does *E. coli* have two PCD pathways that seemingly perform the same function? One hypothesis is that the ALD is a selective pressure to maintain the EDF–MazEF PCD [12]. This hypothesis is supported by the fact that only cells inactivated for EDF–MazED undergo ALD. Without the ALD, the *mazEF* mutant cells would no longer respond to EDF and thus no longer undergo PCD. Therefore, the presence of an ALD pathway reduces the survival of *mazEF* mutant "cheaters." Like MazEF PCD, the ALD pathway also leads to a small population of survivors. A comparison between the MazEF PCD survivors and the ALD survivors will likely identify the characteristics that distinguish these two pathways.

The ability of MazEF to control multiple PCD pathways has led to the postulation that MazEF function is analogous to the function of tumor-suppressor protein p53 [90]. In moderately stressed cells, p53 activity allows for repair of the cell, but in cells that have undergone excessive damage, p53 induces PCD [91]. Like that of p53, the role of MazEF is multifaceted, and includes the following: (i) induction of PCD in a majority of *E. coli* within a population through the expression of "death proteins"; (ii) survival of a small population of these cells via the expression of "survival proteins"; and (iii) inhibition of the ALD when the cell encounters severe stress.

MazEF plays a major role in dictating the PCD pathways in *E. coli*, but this TA system is also found in many other species of bacteria. It is likely that the MazEF homologs in other bacteria also play significant roles in the PCD pathways in these species. For instance, *Mycobacterium tuberculosis* contains 88 putative TA systems [45]. This cohort of TA systems includes at least seven MazF homologs [92]. When ectopically expressed in *E. coli*, some of these induce PCD [92]. Several of the *M. tuberculosis* MazFs have been experimentally verified to play a role in antibiotic tolerance and virulence [93]. Like *E. coli* MazF, one of the *M. tuberculosis* MazF homologs, MazF-mt6, has recently been demonstrated to degrade 23S rRNA molecules. However, instead of simply altering the activity of the ribosome, the cleavage of the 23S rRNA by MazF-mt6 inactivates protein synthesis globally and arrests growth [94]. Accordingly, MazF-mt6 aids in *M. tuberculosis* persistence [94–96]. Another fascinating aspect of the *M. tuberculosis mazF* genes is the absence of cognate *mazE* antitoxin genes. This finding implies that the *M. tuberculosis* MazF proteins are controlled differently that the canonical type II TA systems. The mechanisms that control the *M. tuberculosis* MazF toxins are the focus of current study, and could lead to the discovery of novel control pathways involved in bacterial PCD.

One example of an alternative MazF control mechanism can be found in the formation of fruiting bodies in *M. xanthus*. The *M. xanthus* MazF homolog (MazF-mx) promotes PCD during fruiting-body formation [97]. Interestingly, the cognate antitoxin for MazF-mx is a transcriptional regulator, MrpC, which binds MazF-mx. The phosphorylation of MrpC leads to activation of MazF-mx and PCD in some *M. xanthus* isolates [97,98]. These findings demonstrate that other species of bacteria have alternative MazF control mechanisms. Given the conservation of MazF homologs throughout bacteria, the molecular means by which they induce PCD are likely as diverse as the bacterial species themselves.

3.4 PCD in *S. aureus* Biofilm Development

Staphylococcus aureus is a Gram-positive pathogen of significant clinical relevance, because it is the leading cause of endocarditis, oestomylitis, skin and soft-tissue infection, and bacteremia in the United States [99,100]. *S. aureus* asymptomatically colonizes a third of the US population, highlighting its prevalence [101]. Its ability to cause considerable morbidity and mortality is a result of its dynamic colonization capacity and the rapid rate at which it develops resistance to antibiotics. One feature of its physiology that enables *S. aureus* to both persistently colonize its host and resist antimicrobials is its ability to form biofilms.

Biofilms are multicellular structures that arise on both biotic and abiotic surfaces. These large communities of bacterial cells are held together within an ECM, which in many cases is composed of protein, cellulose, and eDNA. eDNA is a major component of biofilms produced by *S. aureus* [7,102]. The production of eDNA is dependent upon the PCD of a subpopulation within the community of *S. aureus* cells. This subpopulation responds to specific cues that lead to PCD, lysis, and eDNA release [7,103]. In accordance with this, the PCD pathway is a highly regulated and tightly controlled process. Biofilm formation is thought to be an important aspect of many *S. aureus* infections [104]; therefore, elucidation of the molecular details that control the PCD of *S. aureus* and the release of eDNA is of considerable clinical value.

The proteins CidAB mediate the PCD of *S. aureus* by regulating a PG-hydrolase activity that induces cell lysis (Figure 3.3) [102,105,106]. CidAB-induced PCD leads to eDNA release and enhanced biofilm development [102]. In the absence of CidAB, biofilm development is impaired, as the cells produce fivefold less eDNA [102]. The mechanism by which CidAB controls PG-hydrolase activity is unknown, but the current model proposes that CidAB acts as a holin: a pore-forming protein through which hydrolases can transit to the cell wall and degrade the PG [107]. Holins are utilized by bacteriophages as a way of controlling lysis of the bacterial host cell [107]. Premature induction of hydrolase activity is detrimental to phage infection and can lead to abortive infections. Therefore, holin activity is controlled in order to maintain the structure of the host cell until phage development has been completed. This control is achieved by a class of proteins called antiholins, which block the formation of the holin pore and, therefore, transit of the phage hydrolase to the bacterial cell wall. The occlusion of the hydrolase from the bacterial cell wall keeps the host cell intact until the phage is ready to induce host-cell lysis. Activation is achieved by a conformational change that allows the holin to oligomerize and form a pore through which the PG-hydrolase can pass [107]. In the CidAB PCD pathway, the LgrAB protein is proposed to function as the antiholin that controls CidAB pore formation (Figure 3.3) [108]. In accordance with this, both LgrAB and CidAB localize to the membrane, where they form higher-ordered oligomers [108]. Additionally, the phenotypes of a *cidAB* mutant are antithetical to an *lgrAB* mutant [105,109]. The *cidAB* mutant is impaired in terms of PG-hydrolase activity, but such activity in the *lgrAB* mutant is enhanced [105,109]. The molecular mechanisms that support LgrAB and CidAB protein–protein interaction are unknown, as is the mechanism by which the inhibitor effect of LgrAB is relieved from CidAB. However, elegant control of PCD by the holin–antiholin system is proposed to occur in higher eukaryotes. In support of this, Bax and Bcl-2 exhibit holin–antiholin properties when expressed in bacteria [14].

The physical nature of the LgrAB and CidAB interaction is unknown, but the transcriptional regulation of *lrgAB* and *cidAB* provides insight into the physiological processes that can lead to CidAB activation. LgrAB protein expression is under the control of the transcriptional regulator LytRS (Figure 3.3). The genes encoding *lytRS* are directly upstream of the *lrgAB* locus. In bacteria, a transcriptional regulator is often located in close proximity to the genes it regulates within the genome. In accordance with this, inactivation of *lytRS* decreases *lrgAB* transcription and increases CidAB-

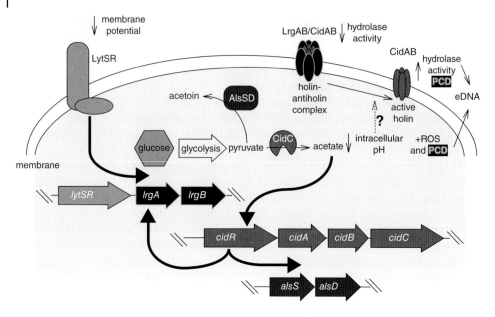

Figure 3.3 PCD during *Staphylococcus aureus* biofilm development. *S. aureus* biofilm development is a tightly regulated PCD pathway that results in the production of eDNA. Cell lysis leading to the production of eDNA is controlled by two transcriptional regulators: LytSR and CidR. LytSR regulates the expression of LrgAB, while CidR regulates the expression of both LrgAB and CidAB. LytSR induces transcription of LrgAB in response to perturbations to the membrane potential. The LrgAB proteins control the activity of the holin-like proteins CidAB. This control is likely mediated through a direct protein–protein interaction that inhibits the holin activity of CidAB. Upon induction of this PCD pathway, activation of CidAB leads to increased PG-hydrolase activity and eDNA release. Another CidR-regulated protein, CidC, also contributes to this PCD pathway. CidC causes the acidification of the cell cytoplasm by converting pyruvate to the weak acid, acetate. Control of this metabolic induction of PCD is afforded by the proteins AlsSD, which produce the neutral product acetoin. Thick lines indicate positive transcriptional regulation.

dependent hydrolase activity [110,111]. Reduced membrane potential is the signal that triggers this LytRS regulatory cascade [111]. This result suggests that altering the transcription of *lrgAB* through LytRS is sufficient to activate CidAB and PCD. The ability of LytRS to sense and respond to alterations in the membrane potential implies that environmental factors that reduce the membrane potential inhibit the CidAB PCD pathway. Upstream of the *cidAB* genes is another transcriptional regulator that impacts PCD, *cidR*. CidR controls the expression of *cidAB* and another gene located downstream of *cidAB*, *cidC* [112]. The signal that activates CidR-dependent transcription is acetate [106,112]. In accordance with this, CidC is a pyruvate oxidase that produces acetate from pyruvate and provides a feedforward regulatory loop that potentiates PCD through acidification of the cytoplasm [113]. This acidification occurs through excessive production of acetate, the product of the CidC-catalyzed reaction [114]. Acetate is a weak acid, and when produced it lowers the pH of the cell cytoplasm and arrests respiration. CidC potentiates CidAB PCD, and the cells that undergo CidC-mediated PCD share the molecular hallmarks of eukaryotic PCD, including ROS generation, DNA damage that can be detected in a TUNEL assay, and respiratory dysfunction [114]. It remains to be determined how the acidification of the cytoplasm impacts the antiholin and holin activities of the LrgAB and CidAB, respectively. However, it has been proposed that acidification of the cytoplasm might activate CidAB, through a similar mechanism to that by which acidification promotes oligomerization and insertion of Bax into the membrane (Figure 3.3) [115]. This model proposes that CidC acidification of the cytoplasm directly activates CidAB, but acidification of the cytoplasm might also alleviate LrgAB inhibition of CidAB [114]. Alternatively, CidR might respond to increased acetate levels and induce *cidAB* transcription [112]. The resulting increase in CidAB proteins could be enough to shift the LrgAB:CidAB ratio to a predominant CidAB-activated holin state.

The CidC acidification of the cytoplasm and the PCD that follows can be controlled by a metabolic pathway that competes for the excess pyruvate (Figure 3.3). The enzyme AlsSD also uses pyruvate as a substrate, but converts it to acetoin. Acetoin has no effect on the pH of the cell, and therefore protects against acidification of the cytoplasm and resultant PCD. In accordance with this, *S. aureus* mutants inactivated for *alsSD* experience exacerbated PCD due to uncontrolled rapid acidification of the cytoplasm in a CidC-dependent manner [114]. Interestingly, the genes encoding AlsSD are under the control of CidR [116]. CidR control of AlsSD provides another safeguard against uncontrolled PCD. In total, two independent but overlapping transcriptional networks, one controlled by LytRS and one controlled by CidR, provide a means of regulation that protects against aberrant PCD and functions to promote biofilm development.

The fact that CidR induces the expression of both pro-PCD genes (*cidAB* and *cidC*) and anti-PCD genes (*lrgAB* and *alsSD*) suggests a complex and multifaceted regulatory network is required to induce Cid-dependent PCD. A clue to the identity of these additional factors is found in the expression profiles of *cidAB* and *lrgAB* within a biofilm. *cidAB* expression predominantly occurs within the centers of the large "tower-like" structures that make up the biofilm [117]. Within the biofilm, oxygen is limited, suggesting that reduced oxygen levels might impact *cidAB* expression and PCD. In agreement with this notion, *cidAB* expression can also be induced by hypoxia [103,117]. These findings demonstrate the temporal nature of *cidAB* expression and will help in the identification of other factors that induce Cid-mediated PCD [103]. The identification of these factors is important, because this PCD pathway is active in models of biofilm formation that mimic clinically relevant biofilm-dependent disease [114].

The CidAB/LrgAB regulatory network is complex, and further experimentation is required in order to obtain a more comprehensive model of staphylococcal PCD. The fact that CidR regulates a pyruvate oxidase and responds to acetate levels within the cell highlights the importance of pyruvate metabolism to the control of *S. aureus* PCD during biofilm development. The association between pyruvate metabolism, the Cid proteins, and PCD extends to other species of bacteria. The pyruvate oxidase, SpxB, in *Streptococcus pneumonia* induces H_2O_2 production and a PCD that provides a selective advantage during infection [118]. In another Streptococcal species, *Streptococcus mutans*, CidAB and LrgAB homologs are regulated by glucose and impact oxidative stress tolerance [119]. Finally, in *Bacillus anthracis*, the causative agent of anthrax, CidR, regulates CidAB, LrgAB, and hydrolase activity [120]. These examples of pyruvate-regulated PCD pathways highlight the importance of central metabolism to bacterial PCD.

3.5 Conclusion

PCD pathways play many different but important roles in bacteria, including development of spores and biofilms, response to stressful growth conditions, reduction of susceptibility to infection by phage, exchange of genetic information, and provision of resistance to antimicrobials. As such pathways are discovered in new bacterial species, this list of functions is certain to grow. The three examples of bacterial PCD described in this chapter share many similarities with eukaryotic PCD. The PCD in both classes of organism is complex and highly regulated. Additionally, there is a striking degree of conservation between some of the molecular signatures of bacterial and eukaryotic PCD pathways. Within different bacterial PCD pathways, generation of ROS, perturbation of the cell membrane or cell wall by effector proteins, depolarization of the membrane, and degradation of DNA have all been established. This high degree of conservation between eukaryotic and bacterial PCD certainly suggests a common evolutionary link between these two systems.

The bacterial PCD pathways described here represent the leading edge of a growing field. Many more such pathways are just beginning to be characterized. For example, a PCD pathway is activated in several pathogenic bacteria, including *S. pneumonia* and *Haemophilus influenza*, in response to the human milk and lipid-protein complex HAMLET [121]. The freshwater Gram-negative bacterium *Caulobacter crescentus* induces PCD via the protein BapE, an endonuclease involved in DNA fragmentation. Membrane depolarization is also induced in *C. crescentus* [122]. Finally, the Gram-negative plant pathogen *Xanthomonas campestris* initiates PCD in response to growth in rich media, leading to increased ETC activity, generation of ROS, and DNA degradation [123,124]. The continued study of PCD in bacteria will inevitably elucidate a variety of new mechanisms by which cell death is induced and controlled in these microorganisms. This insight will lead to a better understanding of what determines whether a bacterial cell lives or dies – an insight that has profound clinical and industrial implications. A greater understanding of the molecular mechanisms that support bacterial PCD may also allow us to categorize the different types of death experienced by bacteria, much like the distinct PCD pathways that are active in eukaryotes.

References

1. Madigan MT, Martinko JM, Brock TD. *Brock Biology of Microorganisms*, 11th edn. Upper Saddle River, NJ: Pearson Prentice Hall. 2006.
2. Lyons NA, Kolter R. On the evolution of bacterial multicellularity. *Curr Opin Microbiol* 2015;24C:21–8.
3. Vlamakis H, Chai Y, Beauregard P, Losick R, Kolter R. Sticking together: building a biofilm the *Bacillus subtilis* way. *Nat Rev Microbiol* 2013;11:157–68.
4. Costerton JW, Cheng KJ, Geesey GG, Ladd TI, Nickel JC, Dasgupta M, Marrie TJ. Bacterial biofilms in nature and disease. *Annu Rev Microbiol* 1987;41:435–64.
5. Costerton JW, Lewandowski Z, Caldwell DE, Korber DR, Lappin-Scott HM. Microbial biofilms. *Annu Rev Microbiol* 1995;49:711–45.
6. Bayles KW. The biological role of death and lysis in biofilm development. *Nat Rev Microbiol* 2007;5:721–6.
7. Mann EE, Rice KC, Boles BR, Endres JL, Ranjit D, Chandramohan L, et al. Modulation of eDNA release and degradation affects *Staphylococcus aureus* biofilm maturation. *PLoS ONE* 2009;4:e5822.
8. Jacques M, Marrie TJ, Costerton JW. Review: Microbial colonization of prosthetic devices. *Microb Ecol* 1987;13:173–91.

9 Lewis K. Programmed death in bacteria. *Microbiol Mol Biol Rev* 2000;**64**:503–14.
10 Engelberg-Kulka H, Sat B, Reches M, Amitai S, Hazan R. Bacterial programmed cell death systems as targets for antibiotics. *Trends Microbiol* 2004;**12**:66–71.
11 Kohanski MA, Dwyer DJ, Hayete B, Lawrence CA, Collins JJ. A common mechanism of cellular death induced by bactericidal antibiotics. *Cell* 2007;**130**:797–810.
12 Erental A, Kalderon Z, Saada A, Smith Y, Engelberg-Kulka H. Apoptosis-like death, an extreme SOS response in *Escherichia coli*. *MBio* 2014;**5**: e01426–14.
13 Dwyer DJ, Camacho DM, Kohanski MA, Callura JM, Collins JJ. Antibiotic-induced bacterial cell death exhibits physiological and biochemical hallmarks of apoptosis. *Mol Cell* 2012;**46**:561–72.
14 Pang X, Moussa SH, Targy NM, Bose JL, George NM, Gries C, et al. Active Bax and Bak are functional holins. *Genes Dev* 2011;**25**:2278–90.
15 Dworkin J, Losick R. Developmental commitment in a bacterium. *Cell* 2005;**121**:401–9.
16 Gonzalez-Pastor JE, Hobbs EC, Losick R. Cannibalism by sporulating bacteria. *Science* 2003;**301**:510–13.
17 Claverys JP, Havarstein LS. Cannibalism and fratricide: mechanisms and raisons d'etre. *Nat Rev Microbiol* 2007;**5**:219–29.
18 Sonenshein AL. Control of sporulation initiation in *Bacillus subtilis*. *Curr Opin Microbiol* 2000;**3**:561–6.
19 Ellermeier CD, Hobbs EC, Gonzalez-Pastor JE, Losick R. A three-protein signaling pathway governing immunity to a bacterial cannibalism toxin. *Cell* 2006;**124**:549–59.
20 Liu WT, Yang YL, Xu Y, Lamsa A, Haste NM, Yang JY, et al. Imaging mass spectrometry of intraspecies metabolic exchange revealed the cannibalistic factors of *Bacillus subtilis*. *Proc Natl Acad Sci USA* 2010;**107**:16 286–90.
21 Watrous JD, Phelan VV, Hsu CC, Moree WJ, Duggan BM, Alexandrov T, Dorrestein PC. Microbial metabolic exchange in 3D. *ISME J* 2013;**7**:770–80.
22 Gonzalez DJ, Haste NM, Hollands A, Fleming TC, Hamby M, Pogliano K, et al. Microbial competition between *Bacillus subtilis* and *Staphylococcus aureus* monitored by imaging mass spectrometry. *Microbiology* 2011;**157**: 2485–92.
23 Lamsa A, Liu WT, Dorrestein PC, Pogliano K. The *Bacillus subtilis* cannibalism toxin SDP collapses the proton motive force and induces autolysis. *Mol Microbiol* 2012;**84**:486–500.
24 Lopez D, Vlamakis H, Losick R, Kolter R. Cannibalism enhances biofilm development in *Bacillus subtilis*. *Mol Microbiol* 2009;**74**:609–18.
25 Allocati N, Masulli M, Di Ilio C, De Laurenzi V. Die for the community: an overview of programmed cell death in bacteria. *Cell Death Dis* 2015;**6**:e1609.
26 Snyder AB, Worobo RW. Chemical and genetic characterization of bacteriocins: antimicrobial peptides for food safety. *J Sci Food Agric* 2014;**94**:28–44.
27 Butcher BG, Helmann JD. Identification of *Bacillus subtilis* sigma-dependent genes that provide intrinsic resistance to antimicrobial compounds produced by Bacilli. *Mol Microbiol* 2006;**60**:765–82.
28 Yamada Y, Tikhonova EB, Zgurskaya HI. YknWXYZ is an unusual four-component transporter with a role in protection against sporulation-delaying-protein-induced killing of *Bacillus subtilis*. *J Bacteriol* 2012;**194**:4386–94.
29 Ellermeier CD, Losick R. Evidence for a novel protease governing regulated intramembrane proteolysis and resistance to antimicrobial peptides in *Bacillus subtilis*. *Genes Dev* 2006;**20**:1911–22.

30. Qian Q, Lee CY, Helmann JD, Strauch MA. AbrB is a regulator of the sigma(W) regulon in *Bacillus subtilis*. *FEMS Microbiol Lett* 2002;**211**:219–23.
31. Errington J. Regulation of endospore formation in *Bacillus subtilis*. *Nat Rev Microbiol* 2003;**1**:117–26.
32. Stragier P, Losick R. Molecular genetics of sporulation in *Bacillus subtilis*. *Annu Rev Genet* 1996;**30**:297–41.
33. Tan IS, Ramamurthi KS. Spore formation in *Bacillus subtilis*. *Environ Microbiol Rep* 2014;**6**:212–25.
34. Higgins D, Dworkin J. Recent progress in *Bacillus subtilis* sporulation. *FEMS Microbiol Rev* 2012;**36**:131–48.
35. Vollmer W, Joris B, Charlier P, Foster S. Bacterial peptidoglycan (murein) hydrolases. *FEMS Microbiol Rev* 2008;**32**:259–86.
36. Foster SJ. Analysis of the autolysins of *Bacillus subtilis* 168 during vegetative growth and differentiation by using renaturing polyacrylamide gel electrophoresis. *J Bacteriol* 1992;**174**:464–70.
37. Kuroda A, Asami Y, Sekiguchi J. Molecular cloning of a sporulation-specific cell wall hydrolase gene of *Bacillus subtilis*. *J Bacteriol* 1993;**175**:6260–8.
38. Smith TJ, Foster SJ. Characterization of the involvement of two compensatory autolysins in mother cell lysis during sporulation of *Bacillus subtilis* 168. *J Bacteriol* 1995;**177**:3855–62.
39. Nugroho FA, Yamamoto H, Kobayashi Y, Sekiguchi J. Characterization of a new sigma-K-dependent peptidoglycan hydrolase gene that plays a role in *Bacillus subtilis* mother cell lysis. *J Bacteriol* 1999;**181**:6230–7.
40. Hosoya S, Lu Z, Ozaki Y, Takeuchi M, Sato T. Cytological analysis of the mother cell death process during sporulation in *Bacillus subtilis*. *J Bacteriol* 2007;**189**:2561–5.
41. Smith TJ, Blackman SA, Foster SJ. Autolysins of *Bacillus subtilis*: multiple enzymes with multiple functions. *Microbiology* 2000;**146**(Pt. 2):249–62.
42. Vlamakis H, Aguilar C, Losick R, Kolter R. Control of cell fate by the formation of an architecturally complex bacterial community. *Genes Dev* 2008;**22**:945–53.
43. Rosenbluh A, Rosenberg E. Role of autocide AMI in development of *Myxococcus xanthus*. *J Bacteriol* 1990;**172**:4307–14.
44. Yamaguchi Y, Park JH, Inouye M. Toxin–antitoxin systems in bacteria and archaea. *Annu Rev Genet* 2011;**45**:61–79.
45. Ramage HR, Connolly LE, Cox JS. Comprehensive functional analysis of *Mycobacterium tuberculosis* toxin–antitoxin systems: implications for pathogenesis, stress responses, and evolution. *PLoS Genet* 2009;**5**:e1000767.
46. Engelberg-Kulka H, Amitai S, Kolodkin-Gal I, Hazan R. Bacterial programmed cell death and multicellular behavior in bacteria. *PLoS Genet* 2006;**2**:e135.
47. Gronlund H, Gerdes K. Toxin–antitoxin systems homologous with *relBE* of *Escherichia coli* plasmid P307 are ubiquitous in prokaryotes. *J Mol Biol* 1999;**285**:1401–15.
48. Mittenhuber G. Occurrence of *mazEF*-like antitoxin/toxin systems in bacteria. *J Mol Microbiol Biotechnol* 1999;**1**:295–302.
49. Pandey DP, Gerdes K. Toxin–antitoxin loci are highly abundant in free-living but lost from host-associated prokaryotes. *Nucleic Acids Res* 2005;**33**:966–76.
50. Rice KC, Bayles KW. Molecular control of bacterial death and lysis. *Microbiol Mol Biol Rev* 2008;**72**:85–109, table of contents.

51 Kolodkin-Gal I, Sat B, Keshet A, Engelberg-Kulka H. The communication factor EDF and the toxin–antitoxin module *mazEF* determine the mode of action of antibiotics. *PLoS Biol* 2008;**6**:e319.
52 Kumar S, Kolodkin-Gal I, Engelberg-Kulka H. Novel quorum-sensing peptides mediating interspecies bacterial cell death. *MBio* 2013; 4: e00314–13.
53 Schuster CF, Bertram R. Toxin–antitoxin systems are ubiquitous and versatile modulators of prokaryotic cell fate. *FEMS Microbiol Lett* 2013; 340: 73–85.
54 Vogel J, Argaman L, Wagner EG, Altuvia S. The small RNA IstR inhibits synthesis of an SOS-induced toxic peptide. *Curr Biol* 2004;**14**:2271–6.
55 Darfeuille F, Unoson C, Vogel J, Wagner EG. An antisense RNA inhibits translation by competing with standby ribosomes. *Mol Cell* 2007;**26**:381–92.
56 Fozo EM, Kawano M, Fontaine F, Kaya Y, Mendieta KS, Jones KL, et al. Repression of small toxic protein synthesis by the Sib and OhsC small RNAs. *Mol Microbiol* 2008;**70**:1076–93.
57 Gerdes K, Bech FW, Jorgensen ST, Lobner-Olesen A, Rasmussen PB, Atlung T, et al. Mechanism of postsegregational killing by the *hok* gene product of the *parB* system of plasmid R1 and its homology with the *relF* gene product of the *E. coli relB* operon. *Embo J* 1986;**5**:2023–9.
58 Engelberg-Kulka H, Glaser G. Addiction modules and programmed cell death and antideath in bacterial cultures. *Annu Rev Microbiol* 1999;**53**:43–70.
59 Aizenman E, Engelberg-Kulka H, Glaser G. An *Escherichia coli* chromosomal "addiction module" regulated by guanosine [corrected] 3′,5′-bispyrophosphate: a model for programmed bacterial cell death. *Proc Natl Acad Sci USA* 1996;**93**:6059–63.
60 Sat B, Hazan R, Fisher T, Khaner H, Glaser G, Engelberg-Kulka H. Programmed cell death in *Escherichia coli*: some antibiotics can trigger *mazEF* lethality. *J Bacteriol* 2001;**183**:2041–5.
61 Kolodkin-Gal I, Hazan R, Gaathon A, Carmeli S, Engelberg-Kulka H. A linear pentapeptide is a quorum-sensing factor required for *mazEF*-mediated cell death in *Escherichia coli*. *Science* 2007;**318**:652–5.
62 Belitsky M, Avshalom H, Erental A, Yelin I, Kumar S, London N, et al. The *Escherichia coli* extracellular death factor EDF induces the endoribonucleolytic activities of the toxins MazF and ChpBK. *Mol Cell* 2011;**41**:625–35.
63 Zhang J, Zhang Y, Inouye M. Characterization of the interactions within the mazEF addiction module of *Escherichia coli*. *J Biol Chem* 2003;**278**:32 300–6.
64 Zhang J, Zhang Y, Zhu L, Suzuki M, Inouye M. Interference of mRNA function by sequence-specific endoribonuclease PemK. *J Biol Chem* 2004;**279**:20 678–84.
65 Masuda Y, Miyakawa K, Nishimura Y, Ohtsubo E. *chpA* and *chpB*, Escherichia coli chromosomal homologs of the pem locus responsible for stable maintenance of plasmid R100. *J Bacteriol* 1993;**175**:6850–6.
66 Fineran PC, Blower TR, Foulds IJ, Humphreys DP, Lilley KS, Salmond GP. The phage abortive infection system, ToxIN, functions as a protein-RNA toxin–antitoxin pair. *Proc Natl Acad Sci USA* 2009;**106**:894–9.
67 Masuda H, Tan Q, Awano N, Wu KP, Inouye M. YeeU enhances the bundling of cytoskeletal polymers of MreB and FtsZ, antagonizing the CbtA (YeeV) toxicity in *Escherichia coli*. *Mol Microbiol* 2012;**84**:979–89.
68 Tan Q, Awano N, Inouye M. YeeV is an *Escherichia coli* toxin that inhibits cell division by targeting the cytoskeleton proteins, FtsZ and MreB. *Mol Microbiol* 2011;**79**:109–18.

69 Wang X, Lord DM, Cheng HY, Osbourne DO, Hong SH, Sanchez-Torres V, et al. A new type V toxin–antitoxin system where mRNA for toxin GhoT is cleaved by antitoxin GhoS. *Nat Chem Biol* 2012;**8**:855–61.

70 Wang X, Lord DM, Hong SH, Peti W, Benedik MJ, Page R, Wood TK. Type II toxin/antitoxin MqsR/MqsA controls type V toxin/antitoxin GhoT/GhoS. *Environ Microbiol* 2013;**15**:1734–44.

71 Pecota DC, Wood TK. Exclusion of T4 phage by the *hok/sok* killer locus from plasmid R1. *J Bacteriol* 1996;**178**:2044–50.

72 Hazan R, Engelberg-Kulka H. *Escherichia coli mazEF*-mediated cell death as a defense mechanism that inhibits the spread of phage P1. *Mol Genet Genomics* 2004;**272**:227–34.

73 Rowe-Magnus DA, Guerout AM, Biskri L, Bouige P, Mazel D. Comparative analysis of superintegrons: engineering extensive genetic diversity in the Vibrionaceae. *Genome Res* 2003;**13**:428–42.

74 Erental A, Sharon I, Engelberg-Kulka H. Two programmed cell death systems in *Escherichia coli*: an apoptotic-like death is inhibited by the *mazEF*-mediated death pathway. *PLoS Biol* 2012;**10**:e1001281.

75 Keren I, Wu Y, Inocencio J, Mulcahy LR, Lewis K. Killing by bactericidal antibiotics does not depend on reactive oxygen species. *Science* 2013;**339**:1213–16.

76 Liu Y, Imlay JA. Cell death from antibiotics without the involvement of reactive oxygen species. *Science* 2013;**339**:1210–13.

77 Wen Y, Behiels E, Devreese B. Toxin–antitoxin systems: their role in persistence, biofilm formation, and pathogenicity. *Pathog Dis* 2014;**70**:240–9.

78 Moyed HS, Broderick SH. Molecular cloning and expression of *hipA*, a gene of *Escherichia coli* K-12 that affects frequency of persistence after inhibition of murein synthesis. *J Bacteriol* 1986;**166**:399–403.

79 Black DS, Kelly AJ, Mardis MJ, Moyed HS. Structure and organization of *hip*, an operon that affects lethality due to inhibition of peptidoglycan or DNA synthesis. *J Bacteriol* 1991;**173**:5732–9.

80 Tripathi A, Dewan PC, Siddique SA, Varadarajan R. MazF-induced growth inhibition and persister generation in *Escherichia coli*. *J Biol Chem* 2014;**289**:4191–205.

81 Lewis K. Persister cells. *Annu Rev Microbiol* 2010;**64**:357–72.

82 Magnuson RD. Hypothetical functions of toxin–antitoxin systems. *J Bacteriol* 2007;**189**:6089–92.

83 Kamada K, Hanaoka F, Burley SK. Crystal structure of the MazE/MazF complex: molecular bases of antidote-toxin recognition. *Mol Cell* 2003;**11**:875–84.

84 Engelberg-Kulka H, Reches M, Narasimhan S, Schoulaker-Schwarz R, Klemes Y, Aizenman E, Glaser G. rexB of bacteriophage lambda is an anti-cell death gene. *Proc Natl Acad Sci USA* 1998;**95**:15 481–6.

85 Hazan R, Sat B, Engelberg-Kulka H. *Escherichia coli mazEF*-mediated cell death is triggered by various stressful conditions. *J Bacteriol* 2004;**186**:3663–9.

86 Kolodkin-Gal I, Engelberg-Kulka H. The extracellular death factor: physiological and genetic factors influencing its production and response in *Escherichia coli*. *J Bacteriol* 2008;**190**:3169–75.

87 Kumar S, Engelberg-Kulka H. Quorum sensing peptides mediating interspecies bacterial cell death as a novel class of antimicrobial agents. *Curr Opin Microbiol* 2014;**21**:22–7.

88 Amitai S, Kolodkin-Gal I, Hananya-Meltabashi M, Sacher A, Engelberg-Kulka H. *Escherichia coli* MazF leads to the simultaneous selective synthesis of both "death proteins" and "survival proteins." *PLoS Genet* 2009;**5**:e1000390.

89 Vesper O, Amitai S, Belitsky M, Byrgazov K, Kaberdina AC, Engelberg-Kulka H, Moll I. Selective translation of leaderless mRNAs by specialized ribosomes generated by MazF in *Escherichia coli*. *Cell* 2011;**147**:147–57.

90 Bayles KW. Bacterial programmed cell death: making sense of a paradox. *Nat Rev Microbiol* 2014;**12**:63–9.

91 Fridman JS, Lowe SW. Control of apoptosis by p53. *Oncogene* 2003;**22**:9030–40.

92 Zhu L, Zhang Y, Teh JS, Zhang J, Connell N, Rubin H, Inouye M. Characterization of mRNA interferases from *Mycobacterium tuberculosis*. *J Biol Chem* 2006;**281**:18 638–43.

93 Tiwari P, Arora G, Singh M, Kidwai S, Narayan OP, Singh R. MazF ribonucleases promote *Mycobacterium tuberculosis* drug tolerance and virulence in guinea pigs. *Nat Commun* 2015;**6**:6059.

94 Schifano JM, Edifor R, Sharp JD, Ouyang M, Konkimalla A, Husson RN, Woychik NA. Mycobacterial toxin MazF-mt6 inhibits translation through cleavage of 23S rRNA at the ribosomal A site. *Proc Natl Acad Sci USA* 2013;**110**:8501–6.

95 Betts JC, Lukey PT, Robb LC, McAdam RA, Duncan K. Evaluation of a nutrient starvation model of *Mycobacterium tuberculosis* persistence by gene and protein expression profiling. *Mol Microbiol* 2002;**43**:717–31.

96 Han JS, Lee JJ, Anandan T, Zeng M, Sripathi S, Jahng WJ, et al. Characterization of a chromosomal toxin–antitoxin, Rv1102c-Rv1103c system in Mycobacterium tuberculosis. *Biochem Biophys Res Commun* 2010;**400**:293–8.

97 Nariya H, Inouye M. MazF, an mRNA interferase, mediates programmed cell death during multicellular *Myxococcus* development. *Cell* 2008;**132**:55–66.

98 Boynton TO, McMurry JL, Shimkets LJ. Characterization of *Myxococcus xanthus* MazF and implications for a new point of regulation. *Mol Microbiol* 2013;**87**:1267–76.

99 Klevens RM, Morrison MA, Nadle J, Petit S, Gershman K, Ray S, et al. Invasive methicillin-resistant *Staphylococcus aureus* infections in the United States. *JAMA* 2007;**298**:1763–71.

100 Lowy FD. *Staphylococcus aureus* infections. *N Engl J Med* 1998;**339**:520–32.

101 Kuehnert MJ, Kruszon-Moran D, Hill HA, McQuillan G, McAllister SK, Fosheim G, et al. Prevalence of *Staphylococcus aureus* nasal colonization in the United States, 2001–2002. *J Infect Dis* 2006;**193**:172–9.

102 Rice KC, Mann EE, Endres JL, Weiss EC, Cassat JE, Smeltzer MS, Bayles KW. The cidA murein hydrolase regulator contributes to DNA release and biofilm development in *Staphylococcus aureus*. *Proc Natl Acad Sci USA* 2007;**104**:8113–18.

103 Moormeier DE, Bose JL, Horswill AR, Bayles KW. Temporal and stochastic control of *Staphylococcus aureus* biofilm development. *MBio* 2014;**5**:e01341–14.

104 Kwon AS, Park GC, Ryu SY, Lim DH, Lim DY, Choi CH, et al. (2008). Higher biofilm formation in multidrug-resistant clinical isolates of *Staphylococcus aureus*. *Int J Antimicrob Agents;* **32**:68–72.

105 Rice KC, Firek BA, Nelson JB, Yang SJ, Patton TG, Bayles KW. The *Staphylococcus aureus* cidAB operon: evaluation of its role in regulation of murein hydrolase activity and penicillin tolerance. *J Bacteriol* 2003;**185**:2635–43.

106 Rice KC, Nelson JB, Patton TG, Yang SJ, Bayles KW. Acetic acid induces expression of the *Staphylococcus aureus* cidABC and lrgAB murein hydrolase regulator operons. *J Bacteriol* 2005;**187**:813–21.

107 Wang IN, Smith DL, Young R. Holins: the protein clocks of bacteriophage infections. *Annu Rev Microbiol* 2000;**54**:799–825.

108 Ranjit DK, Endres JL, Bayles KW. *Staphylococcus aureus* CidA and LrgA proteins exhibit holin-like properties. *J Bacteriol* 2011;**193**:2468–76.

109 Groicher KH, Firek BA, Fujimoto DF, Bayles KW. The *Staphylococcus aureus lrgAB* operon modulates murein hydrolase activity and penicillin tolerance. *J Bacteriol* 2000;**182**:1794–801.

110 Brunskill EW, Bayles KW. Identification of LytSR-regulated genes from *Staphylococcus aureus*. *J Bacteriol* 1996;**178**:5810–12.

111 Patton TG, Yang SJ, Bayles KW. The role of proton motive force in expression of the *Staphylococcus aureus cid* and *lrg* operons. *Mol Microbiol* 2006;**59**:1395–404.

112 Yang SJ, Rice KC, Brown RJ, Patton TG, Liou LE, Park YH, Bayles KW. A LysR-type regulator, CidR, is required for induction of the *Staphylococcus aureus cidABC* operon. *J Bacteriol* 2005;**187**:5893–900.

113 Patton TG, Rice KC, Foster MK, Bayles KW. The *Staphylococcus aureus cidC* gene encodes a pyruvate oxidase that affects acetate metabolism and cell death in stationary phase. *Mol Microbiol* 2005;**56**:1664–74.

114 Thomas VC, Sadykov MR, Chaudhari SS, Jones J, Endres JL, Widhelm TJ, et al. A central role for carbon-overflow pathways in the modulation of bacterial cell death. *PLoS Pathog* 2014;**10**:e1004205.

115 Xie Z, Schendel S, Matsuyama S, Reed JC. Acidic pH promotes dimerization of Bcl-2 family proteins. *Biochemistry* 1998;**37**:6410–18.

116 Yang SJ, Dunman PM, Projan SJ, Bayles KW. Characterization of the *Staphylococcus aureus* CidR regulon: elucidation of a novel role for acetoin metabolism in cell death and lysis. *Mol Microbiol* 2006;**60**:458–68.

117 Moormeier DE, Endres JL, Mann EE, Sadykov MR, Horswill AR, Rice KC, et al. (2013) Use of microfluidic technology to analyze gene expression during *Staphylococcus aureus* biofilm formation reveals distinct physiological niches. *Appl Environ Microbiol*; **79**:3413–24.

118 Regev-Yochay G, Trzcinski K, Thompson CM, Lipsitch M, Malley R. SpxB is a suicide gene of *Streptococcus pneumoniae* and confers a selective advantage in an in vivo competitive colonization model. *J Bacteriol* 2007;**189**:6532–9.

119 Ahn SJ, Rice KC, Oleas J, Bayles KW, Burne RA. The *Streptococcus mutans* Cid and Lrg systems modulate virulence traits in response to multiple environmental signals. *Microbiology* 2010;**156**:3136–47.

120 Ahn JS, Chandramohan L, Liou LE, Bayles KW. Characterization of CidR-mediated regulation in *Bacillus anthracis* reveals a previously undetected role of S-layer proteins as murein hydrolases. *Mol Microbiol* 2006;**62**:1158–69.

121 Hakansson AP, Roche-Hakansson H, Mossberg AK, Svanborg C. Apoptosis-like death in bacteria induced by HAMLET, a human milk lipid-protein complex. *PLoS ONE* 2011;**6**:e17717.

122 Bos J, Yakhnina AA, Gitai Z. BapE DNA endonuclease induces an apoptotic-like response to DNA damage in *Caulobacter*. *Proc Natl Acad Sci USA* 2012;**109**:18 096–101.

123 Gautam S, Sharma A. Involvement of caspase-3-like protein in rapid cell death of *Xanthomonas*. *Mol Microbiol* 2002;**44**:393–401.

124 Gautam S, Sharma A. Rapid cell death in *Xanthomonas campestris* pv. glycines. *J Gen Appl Microbiol* 2002;**48**:67–76.

4

Autophagy

Mollie K. Rojas, Juel Chowdhury, Khatja Batool, Zane Deliu, and Abdallah Oweidi

Department of Oral Medicine and Diagnostic Sciences, University of Illinois at Chicago, Chicago, IL, USA

Abbreviations

ACD	autophagic cell death
AMPK 5′	AMP-activated protein kinase
ATG	autophagy-related gene
ATP	adenosine triphosphate
BECN1	beclin-1
CMA	chaperone-mediated autophagy
HF	heart failure
I/R	ischemia/reperfusion
LAP	LC3-associated phagocytosis
mTOR	mammalian target of rapamycin kinase
NCCD	Nomenclature Committee on Cell Death
ROS	reactive oxygen species
VSMC	vascular smooth-muscle cell
WPB	Weibel-palade body

4.1 Introduction

In order to maintain homeostasis, cellular components must continually be recycled. Unnecessary or dysfunctional macromolecular components must be disassembled and degraded. Degradation of old cellular components is balanced by new macromolecular synthesis. Eukaryotic cells have adapted a variety of strategies to maintain homeostasis, including protein- and vesicle-based degradation and quality control systems. One important homeostatic pathway is macroautophagy (or, commonly, autophagy), which targets aged or damaged organelles, protein aggregates, or long-lived proteins for degradation and recycling. Through autophagy, intracellular substrates are sequestered into autophagosome vesicles. Autophagosomes fuse with lysosomes in order to utilize their digestive enzymes to degrade internalized cargo. By this mechanism, the autophagy pathway allows eukaryotic cells to harness the degradative power of lysosomes to

Apoptosis and Beyond: The Many Ways Cells Die, First Edition. Edited by James Radosevich.
© 2018 John Wiley & Son Inc. Published 2018 by John Wiley & Son Inc.

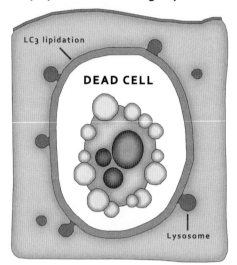
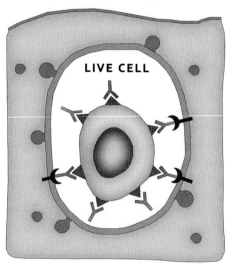

"Role of LC3 lipidation to cell-containing vacuoles:

- Corpse degradation
- Anti-inflammatory response
- Antigen presentation?

- Tumor cell death and degradation
- Cytokine secretion?
- Antigen presentation?

Figure 4.1 Predicted consequences of LC3 lipidation to different cell-containing vacuoles. (A) The lipidation of LC3 to phagosomes (referred to as "LAP") facilitates lysosome fusion. For phagosomes harboring dead or dying cells, LC3 lipidation allows for efficient corpse degradation, and it is required for an anti-inflammatory response mediated by secreted cytokines. For immunogenic forms of cell death, LC3 lipidation may also facilitate antigen presentation. (B) Vacuoles harboring viable cells engulfed by the cell cannibalism mechanism "entosis" also exhibit LC3 lipidation, which facilitates lysosome fusion and the death of engulfed cells. FcγR-mediated phagocytosis of live tumor cells could also induce LC3 lipidation, which could facilitate tumor cell killing, modulate cytokine secretion, or influence antigen presentation. A viable engulfed tumor cell with antibodies bound to its surface is depicted [1]. *Source:* S.N. Odeh and O.N. Fathallah. Reproduced with permission from S.N. Odeh and O.N. Fathallah.

turnover bulk and long-lived intracellular substrates and recycle their building blocks for use in macromolecular synthesis.

In this chapter, we will discuss the three types of known autophagy: macroautophagy, microautophagy, and chaperone-mediated autophagy (CMA) (Figure 4.1). We will also discuss the role of autophagy in nutrient starvation, infection, cancer, and cardiovascular disease.

4.2 Types of Autophagy

Three types of autophagy exist in eukaryotic cells: macroautophagy ("autophagy"), microautophagy, and CMA. Macroautophagy and microautophagy work together to recycle and degrade cellular components. CMA is a special type of autophagy that is very selective in regards to the proteins that it degrades.

4.2.1 Macroautophagy

Macroautophagy, or simply autophagy, is the most common and well-described type of autophagy. It most frequently occurs in cells that are stressed, and involves the synthesis of autophagosomes. Autophagosomes are double-membraned vacuoles that sequester proteins and other cellular components before fusing with lysosomes. Once the autophagosome fuses with a lysosome, the cellular contents are degraded by hydrolases housed within the lysosome. This fusion allows for recycling of cellular components [2].

Autophagy mostly functions in cellular survival, but a role in cell death has been studied. Autophagy functions to turn over old or underutilized proteins and cytoplasmic materials (Figure 4.2). It helps maintain homeostasis, and protects cells by eliminating harmful organelles or protein components [2].

4.2.2 Microautophagy

Microautophagy also functions to recycle cellular components [4]. It is very active and helpful during periods of starvation. Eukaryotic cells can cope with starvation by recycling and reutilizing their own components.

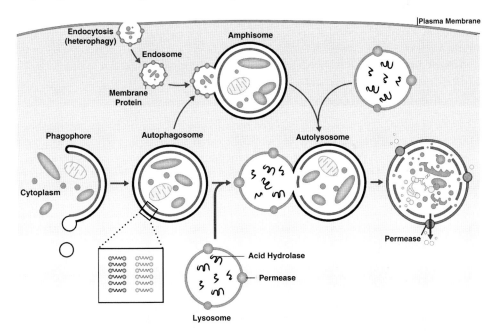

Figure 4.2 The formation of phagolysosomes [3]. During autophagy, sequestration begins with the formation of phagophore, which expands into double-membrane autophagosome while surrounding a portion of the cytoplasm. The autophgosome may fuse with an endosome (the product of endocytosis), in a form of heterophagy. (Heterophagy occurs when the cell internalizes and degrades material that originates outside of it. In contrast, autophagy occurs when the cell consumes part of itself.) The product of the endosome–autophagosome fusion is called an amphisome. The completed autophagosome or amphisome fuses with a lysosome, which supplies acid hydrolases. The enzymes in the resulting compartment, an autolysosome, break down the inner membrane from the autophagosome and degrade the cargo. The resulting macromolecules are released and recycled in the cytosol. *Source:* Adapted from Klinosky 2007 [3] with permission from Macmillan Publishers Ltd. Adapted by S.N. Odeh and O.N. Fathallah. Reproduced with permission from S.N. Odeh and O.N. Fathallah.

Microautophagy differs from macroautophagy in terms of the processing of cellular components. Autophagosomes are critical to the process of macroautophagy, but are not critical in the process of microautophagy. Microautophagy involves direct engulfment of cellular material by the lysosome.

Macroautophagy and microautophagy do not compete with each other; rather, they function together to recycle and degrade cellular components [4].

4.2.3 Chaperone-Mediated Autophagy

CMA does not involve autophagosomes. Rather, it is a selective process whereby proteins are shuttled directly across the lysosomal barrier for degradation [5]. It functions in conjunction with macroautophagy during periods of cellular starvation [5], working mainly to recycle amino acids from degraded proteins [5].

4.3 Functions of Autophagy

"Autophagy is induced in response to many stresses that ultimately lead to apoptosis, such as organelle dysfunction, metabolic stress, atrophy and chemotherapeutic intervention, pathogen infection and starvation" [6]. It helps maintain cellular homeostasis and is initiated in response to cellular stressors. It is critical for cellular survival and destruction in situations involving aging, infection, cancer, cardiovascular disease, nutrient starvation, and repair mechanisms. It is known to exhibit both cytoprotective and cytotoxic effects [7].

4.3.1 Autophagic Cell Death

The role of autophagy in cell death has been recently elucidated. Autophagy has long been known to serve a "pro-survival" role; newer research suggests that it may also play a "causative role in cell death" [2]. It generally functions to restore homeostasis; however, if cellular stressors are not adequately resolved, cellular death may result [6]. It tends to act in a pro-survival manner, but "both cytoprotective and cytotoxic functions of autophagy have been reported" [7]. Extensive research into the functions of autophagy has revealed that "enforced overactivation of autophagy can lead to ACD (autophagic cell death)" [7]. ACD utilizes the autophagosome–lysosomal pathway to render cells incapable of survival [7]. It is theorized that ACD may be threshold-dependent. The pro-survival versus pro-death functionality of autophagy may depend on the duration and extent of signaling [7].

Many cell-death mechanisms overlap with each other and/or contain features of one another [6]. There are several different mechanisms of programmed cell death (PCD), including apoptosis or type I cell death [7]. It is often difficult to completely separate or distinguish apoptotic mechanisms of cell death from one another [6]. "Programmed cell death is an evolutionarily conserved intrinsic mechanism enabling damaged and unwanted cells to commit suicide. This cellular suicide may occur either via apoptosis or via activation of alternate death programs" [7].

In 2012, leading researchers in cell death formally proposed a classification of cell-death mechanisms via the Nomenclature Committee on Cell Death (NCCD): "Five relatively well-characterized different modes of (programmed) cell death (include): (1) extrinsic apoptosis, (2) intrinsic apoptosis, (3) regulated necrosis, (4) mitotic catastrophe (mitosis), and (5) autophagic cell death (ACD)" [7].

ACD is a form of cell death that involves autophagy and that differs from apoptosis, necrosis, and pyroptosis (Figure 4.3). Apoptosis is the "predominant cell death pathway"

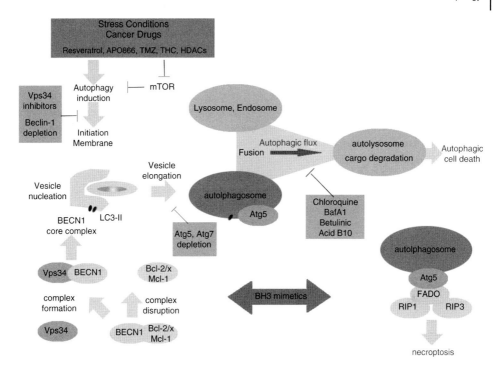

Figure 4.3 Different stages of autophagy and definition of autophagic cell death (ACD). Autophagy can be induced by multiple stimuli, including metabolic stress, organelle dysfunction, protein aggregation, and several cancer drugs, many of which target the central autophagy regulator mTOR (see Table 4.1). The different stages of this process are tightly regulated by the core autophagy proteins encoded by autophagy-related genes (ATGs). Autophagosome biogenesis starts with the formation of an initiation membrane, which can be derived from the endoplasmic reticulum (ER) and several other cellular membrane sources. Vesicle nucleation is promoted by a large macromolecular complex containing the lipid kinase Vps34 (BECN1 core complex). BECN1 (ATG6) serves to activate Vps34, leading to formation of PtdIns3P, which is required for this stage of the autophagic pathway. Vesicle elongation is regulated by two ubiquitin-like conjugation systems involving several ATG proteins: (1) a large protein complex containing ATG5 (and ATG12/ATG16) and (2) ATG7/ATG3-driven attachment of phosphatidylethanolamine to LC3-1, leading to the generation of LC3-II, which is inserted into the autophagosomal membrane. Following vesicle closure, mature autophagosomes fuse with lysosomes or endosomes to autolysosomes in which the autophagosomal content is digested by lysosomal proteases. Excessive activation of the autophagy pathway can lead to an autophagy-dependent cell death in several paradigms. The term "autophagic cell death" should be exclusively limited to cases of cell death that are mediated and not simply accompanied by autophagy. Therefore, only cases where inhibition of the autophagic pathway suppresses cell death can be considered as true ACD. The effects of autophagy inhibition on cell death can be experimentally addressed at different stages of autophagy, either by Vps34 inhibitors or by knockout/knockdown of core autophagic modulators, such as ATG5, ATG7, or BECN1, mediated by an enhanced autophagic flux rather than by alternative forms of cell death, such as necroptosis. In light of this controversy and the cytotoxicity of the drugs inhibiting the autophagic flux from autophagosomes to lysosomes (chloroquine, Bafilomycin A1), interference at this stage of the autophagy pathway is currently not a generally accepted approach to the analysis of ACD. BH3 mimetics have been implicated in several paradigms of ACD. They are capable of inducing the release of BECN1 from its inhibitory interaction with BCL-2/BCL-X_L, and have been shown to recruit necrosome components to the autophagosomal membrane, thereby inducing ACD (for details, please refer to the main text) [7]. *Source:* S.N. Odeh and O.N. Fathallah. Reproduced with permission from S.N. Odeh and O.N. Fathallah.

Table 4.1 Example of autophagic cell death (ACD)-inducing anticancer drugs. *Source:* Data from Fulda and Kogel 2015 [7].

Substances	Cancer types	Autophagy inhibition used
BH3 mimetics		
Obatoclax	ALL	BECN1 and ATG siRNA 3-MA
Obatoclax	RM5	ATG5 shRNA, ATG7 siRNA, BafA1
Obatoclax	ALL	BECN1 and ATG5 siRNA 3-MA
Obatoclax	Colon carcinoma	BECN1 and ATG5 siRNA
Obatoclax	Breast carcinoma	BECN1 siRNA, 3-MA
Obatoclax	Pancreatic carcinoma	BECN1 siRNA
Obatoclax	AML	CQ
Obatoclax	B-cell lymphoma	BECN1 siRNA
Gosspyol	Prostate carcinoma	BECN1 and ATG5 siRNA, 3-MA
Gosspyol	GBM	BECN1 and ATG5 siRNA, BafA1
Gosspyol	Peripheral nerve-sheath tumor	3-MA
Natural products		
Betulinic acid derivative B10	GBM	BECN1 and ATG5, ATG7 siRNA
Betulinic acid	Cervical carcinoma	ATG5 siRNA, ATG5$^{-/-}$ MEFs
Resveratrol	Breast Carcinoma	BECN1 and hVP34, ATG7 siRNA
Resveratrol	CML	ATG5, p62, and LC3 siRNA
HDACs		
SAHA	Chondrosarcoma	3-MA
SAHA	Cervical carcinoma	BECN1 and ATG7 siRNA
SAHA, OSU-HDAC 42	HCC	ATG5 siRNA, 3-MA
Chemotherapeutics		
TMZ	GBM	3-MA
TMZ	GBM	BECN1 and ATG5 siRNA
Cannabinoids		
THC	GBM	ULK1, ATG5 and Ambra-1 siRNA, ATG5$^{-/-}$ MEFs
THC, OWH-015	HCC	ATG5 siRNA, 3-MA
Others		
APO866	Leukemia, lymphoma cells	ATG5 and ATG7 siRNA, 3-MA
Lapatinib	HCC	BECN1 and ATG5, ATG7 shRNA, 3-MA

and is caspase-dependent [2]. It is commonly referred to as "programmed cell death," but it is not the only cell-death mechanism that involves PCD. Apoptosis and ACD share many signaling proteins, and both are forms of PCD. There is significant mechanistic overlap between apoptosis and ACD – so much so, in fact, that it is often difficult to distinguish between the processes [6].

Due to the significant mechanistic overlap between ACD and apoptosis, the lines between the two processes may be blurred. According to Booth, if cellular stresses persist and "autophagy is no longer able to support cell survival, cells can respond by activating the processes of apoptosis in order to ensure their controlled and efficient elimination, without triggering local inflammation" [6]. It is suggested that autophagy and apoptosis may work synergistically to ensure survival of the individual, whether by promoting cellular death or survival [6].

4.3.2 Nutrient Starvation

Autophagy functions to maintain homeostasis. In cells experiencing hypoxia, low cellular energy, or a limitation of nutrients, autophagy is rapidly induced [8,9]. The induction of autophagy at this critical time in a cell's life promotes survival of the cell [9]. In this situation, autophagy functions to degrade and recycle cellular proteins to be used for the synthesis of new proteins or substrates in the production of ATP [9]. "The induction of autophagy promotes a shift from aerobic respiration to glycolysis and allows cellular components of the autophagosome to be hydrolyzed to energy substrates. Increased levels of autophagy are typical in activated immune cells and are a mechanism for the disposal of ROS and phagocytosed debris" [8].

4.3.3 Role in Infection

Data from recent research suggests that autophagy "is a protective mechanism whereby the cell can regulate the levels of cytokine production" [8]. In regulating cytokine production, autophagy can be used to "sequester and degrade" proinflammatory cytokines [8]. If autophagy is inhibited, proinflammatory cytokine levels are increased, leading to increased mortality [8].

4.3.4 Role in Cancer

Autophagy plays a role in defending cells against tumors and enhancing the efficacy of anticancer treatments [9]. "The ability to induce autophagy appears to be beneficial in solid tumors experiencing metabolic stress, as it allows cells at the hypoxic and nutrient-limited core of a tumor to persist for extended periods of time" [9]. Some studies have also suggested that if autophagy is inhibited, the efficacy of anticancer drugs is increased [9]. "In contrast, autophagy has also been invoked as a second mechanism of cell death, and a number of chemotherapeutic agents have been shown to induce cell death accompanied by the upregulation of autophagy" [9].

4.3.5 Role in Cardiovascular Disease

4.3.5.1 Cardiovascular Development

Autophagy is known to play a role in cardiac development via regulation of cardiac progenitor cells. Various research studies have shown that "knockdown of [autophagy-related gene] ATG5, ATG7, and BECN1 result[s] in abnormal heart structure, including defects in cardiac looping, abnormal chamber morphology, and aberrant valve development" [10]. Research has also shown that knockdown of ATG5 in mice results in "defects in heart valve development and chamber septation" [10].

4.3.5.2 Cardiovascular Diseases

The role of autophagy in cardiovascular disease is complex. In order to maintain cellular homeostasis, autophagy is required [10]. However, excessive fluctuations in the level of autophagy can result in alterations of normal cardiac functions. "In ischemia/reperfusion, heart failure, hypertrophy, diabetes mellitus, atherosclerosis, plaque destabilization, lesional thrombosis, and vascular smooth-muscle cell (VSMC) proliferation, autophagic flux is abnormally elevated, contributing to cardiac and vessel dysfunction. In *Atg5* and *Becn1* knockout animals and during aging, autophagic activity is decreased, perturbing cellular homeostasis and contributing to cardiovascular diseases, such as postoperative atrial fibrillation (POAF), ischemia-induced damage, hypertrophy, heart failure, and vascular endothelial cell dysfunction" [10].

Autophagy may play a role in multifactorial cardiovascular disease. "Hypertension is one of the largest contributors to the worldwide burden of cardiovascular diseases, and its prevalence is close to 30% in the world population" [10]. Hypertension is regulated by several mechanisms, including "the sympathetic, parasympathetic, renin–angiotensin–aldosterone and antidiuretic hormone systems" [10]. If any of these hormone systems becomes dysregulated, an individual may develop hypertension [10]. "Because most of the peptides and hormones belonging to these systems are capable of regulating autophagy, it is possible to speculate that dysregulation of autophagy could be associated with hypertension, obesity, diabetes mellitus, and end organ damage" (Figure 4.4) [10].

4.3.5.3 Ischemia/Reperfusion

Autophagy is known to play a role under certain conditions of cardiac stress, including ischemia/reperfusion (I/R) and heart failure (HF) [10]. In both I/R and HF, cardiomyocytes undergo nutrient starvation. "Under those conditions, autophagy is important for the turnover of organelles and protein aggregate degradation at low basal levels under normal conditions" [10].

Ischemia triggers autophagy due to an adaptive mechanism that helps to provide nutrients and to eliminate damaged organelles, including mitochondria. Failure to remove damaged mitochondria can result in formation of ROS and initiation of apoptosis [10].

Several studies have shown that pharmacological intervention of autophagy results in cardiomyocyte death, "indicating that autophagy functions as a prosurvival mechanism" [10]. In cases of chronic ischemia, "autophagy can inhibit apoptosis and diminish tissue damage" [10].

4.3.5.4 Heart Failure

When the heart is stressed, it responds via hypertrophy. Chronic hypertrophy is followed by HF [10]. "In a model of pressure overload, the degree of autophagic activity correlates with the magnitude of hypertrophic growth and the rate of transition to HF.

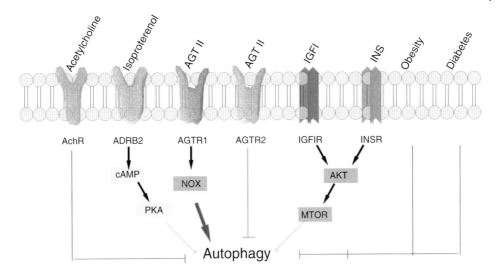

Figure 4.4 Autophagy regulators in cardiomyocytes. A wide variety of stimuli can regulate autophagy in cardiomyocytes. Some of them are autophagy activators and are associated with cardiovascular diseases. However, acetylcholine, catecholamines, aging, insulin-like growth factor 1 (IGF1), and insulin (INS) are capable of inhibiting autophagy. INS and IGF1 are well-known cardioprotective agents. AchR, acetylcholine receptor; AGT-II, angiotensin II; AGTR, angiotensin II receptor type 1; IGF1R, IGF1 receptor; INSR1, INS receptor 1; PKA, protein kinase A. *Source:* S.N. Odeh and O.N. Fathallah. Reproduced with permission from S.N. Odeh and O.N. Fathallah.

Cardiomyocyte-specific overexpression of BECN1 amplifies the pathological remodeling response. Conversely, BECN1 haploinsufficiency partially rescues HF. These data suggest that autophagy can be maladaptive under conditions of severe pressure overload. Analysis of human samples has supported additional evidence that autophagic cell death contributes to HF pathogenesis" [10].

4.3.5.5 Autophagy and Vascular Aging

Recent research suggests that autophagy may contribute to a variety of vascular diseases, including atherosclerosis [11]. It has been suggested that autophagy may modulate rates of vascular aging. It has long been known that age is a major risk factor for cardiovascular diseases, and "autophagy is increasingly being implicated as a modulator of longevity. It is tempting to, therefore, speculate that a decline in vascular autophagy might contribute to what we consider to be the phenotype of the aging vasculature [Figure 4.5]. Such properties include increased stiffening of the large arteries and impaired endothelial-dependent relaxation. The latter property is thought to reflect the effect of higher levels of ROS produced within the aging vasculature. Defects in these processes have been increasingly linked to cardiovascular disease exacerbated by oxidative stress" [11].

Murine models suggest that autophagic flux is reduced in aging vasculature [11]. "Analysis of endothelial cells from old versus young mice, as well as analysis of endothelial cells from old versus young patients, reveals that older endothelial cells have lower levels of BECN1 and higher levels of SQSTM1/p62 (Sequestosome-1).

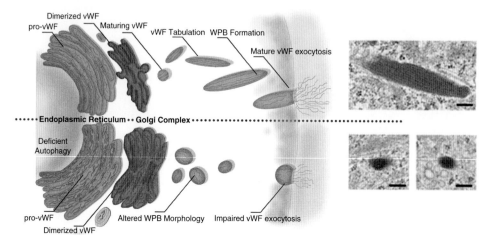

Figure 4.5 Endothelial disruption of autophagic flux impairs Weibel-palade body (WPB) formation. Graphical illustration of the normal maturation and secretion of von Willebrand factor (vWF) within endothelial cells (top). After deletion of Atg7 or Atg5 within the endothelium (bottom), there is no change in the level of Vwf, but more is found in the endoplasmic reticulum (ER) and Golgi, and less within the mature vWF. Moreover, perhaps because of impaired ER and Golgi hemostasis, the pH of the WPB is altered in the absence of autophagy, leading to a shorter and more rounded morphology. Electron micrographs of these different WPB morphologies are shown [11]. *Source:* S.N. Odeh and O.N. Fathallah. Reproduced with permission from S.N. Odeh and O.N. Fathallah.

Because SQSTM1/p62 is largely cleared through autophagy, these results are at least consistent with the notion that the aging vasculature has reduced autophagic flux. The notion that this might be more than a correlation is supported by pharmacological interventions that attempt to increase autophagy. The addition of trehalose or spermidine, two agents that stimulate autophagy, seems to reverse aspects of arterial aging. The parameters studied in these models included both measures of arterial stiffening and endothelial-dependent function" (see Figure 4.6 and Table 4.2) [11].

Table 4.2 Some known alterations to endothelial, vascular smooth-muscle cell (VSMC), and macrophage biology following positive or negative manipulation of autophagic flux. *Source:* Data from Fulda and Kogel 2015 [7].

Cell Type	Increase or Decrease of Autophagy
Endothelial cell	↑ autophagy: ↑ NO production; ↑ protection from APP/β-amyloid, Ang-II, and hyperglycemia; and ↓ angiogenesis, ↓ autophagy: ↓ vWF secretion, and ↓ NO availability
VSMC	↑ autophagy: ↑ protection from oxidized lipids, ↓ BMPR2 expression with ↑ PAH, ↑ VSMC phenotype switching, ↓ vascular calcifications
Macrophage	↓ autophagy: ↑ atherosclerosis

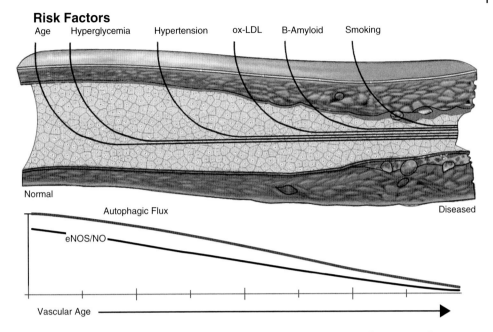

Figure 4.6 **Convergence of autophagic flux and vascular aging**. A variety of intrinsic and extrinsic factors (age, glucose, lipids, etc.) can regulate vascular autophagy. The induction of autophagy may be important in minimizing the damage that these factors can induce. With increasing age, vascular autophagic flux seems to decline, and this decline may contribute to an impairment in endothelial function (with decreased nitric oxide (NO) signaling) and to an increase in vascular stiffness. Efforts to maintain and augment vascular autophagy may prove to be therapeutically beneficial. eNOS, endothelial nitric oxide synthase; ox-LDL, oxidized low-density lipoprotein [11]. *Source:* Nussenzwieg et al. 2015 [11]. Reproduced with permission from Wolters Kluwer Health, Inc.

4.4 Conclusion

The role of autophagy in many cellular processes is constantly being elucidated. As more research is conducted, we will be able to further ascertain the mechanisms involved in the promotion of cellular survival via autophagy versus the promotion of cellular death. Further research is required to help differentiate cellular death promoted by autophagy from cellular death carried out via autophagy.

References

1 Kim S, Overholtzwer M. Autophagy proteins regulate cell engulfment mechanisms that participate in cancer. *Semin Cancer Biol* 2013;**23**(5):329–36.
2 Duprez L, Wirawan E, Vanden Berghe T, Vandenabeele, P. Major cell death pathways at a glance. *Microbes Infect* 2009;**11**:1050–62.
3 Klinosky DJ. Autophagy: from phenomenology to molecular understanding in less than a decade. *Nat Rev Mol Cell Biol* 2007;**8**:931–7.

4 Castro-Obregon S. The discovery of lysosomes and autophagy. *Nat Edu* 2010;3(9):49.
5 Kaushik S, Cuervo AM. Chaperone-mediated autophagy: a unique way to enter the lysosome world. *Trend Cell Biol* 2012;**22**(8):407–17.
6 Booth LA, Tavallai S, Hamed HA, Cruickshanks N, Dent P. The role of cell signalling in the crosstalk between autophagy and apoptosis. *Cell Signal* 2014;**26**:549–55.
7 Fulda S, Kögel D. Cell death by autophagy: emerging molecular mechanisms and implications for cancer therapy. *Oncogene* 2015;**34**(40):1–9.
8 Brunicardi FC, Andersen DK, Billiar TR, Dunn DL, Hunter JG, Matthews JB, et al. Oncology. In: Belval B, Naglieri C (eds.) *Schwartz's Principles of Surgery*, 10th edn. New York: McGraw-Hill Education. 2015.
9 Lichtman MA, Kipps TJ, Seligsohn U, Kaushansky K, Prchal JT. Apoptosis. In: Kaushansky K, Lichtman M, Beutler E, Kipps T, Prchal J, Seligsohn U (eds.) *Williams Hematology*, 8th edn. New York: McGraw-Hill. 2010.
10 Gatica D, Chiong M, Lavandero S, Klionsky DJ. Molecular mechanisms of autophagy in the cardiovascular system. *Circ Res* 2015;**116**(3):456–67.
11 Nussenzweig SC, Verma S, Finkel T. The role of autophagy in vascular biology. *Circ Res* 2015;**116**(3):480–8.

5

Cell Injury, Adaptation, and Necrosis

Sarah G. Fitzpatrick[1] and Sara C. Gordon[2]

[1]Department of Oral and Maxillofacial Diagnostic Sciences, University of Florida, Gainesville, FL, USA
[2]School of Dentistry, Oral Medicine University of Washington, Seattle, WA, USA

Abbreviations

AD	Alzheimer's disease
ATP	adenosine triphosphate
CNS	central nervous system
ER	endoplasmic reticulum
FFA	free fatty acid
NCCD	Nomenclature Committee on Cell Death
NPC1	Niemann–Pick C1
PCD	programmed cell death
RA	rheumatoid arthritis

5.1 Introduction: Reversible vs. Irreversible Cell Injury

When a cell is injured, it may or may not be capable of recovering, leading to the concepts of reversible and irreversible cell injury. The cell's survival may depend on the type of damage sustained, its severity, and its duration. Cells have a remarkable ability to recover from injury by adapting to the changes around them, but if the injury is too severe and a cell cannot adapt or repair itself, then cell death may occur. Necrosis is a kind of irreversible cell injury that leads to the death of part of the cell or of the cell in its entirety. Necrotic cell death is usually a result of an external injury (environmental, infective, ischemic), in contrast to programmed cell death (PCD) or apoptosis, which results from intrinsic factors such as genetic signaling [1]. Injuries such as loss of blood flow (ischemia), excess heat, hazardous toxins, and physical trauma frequently result in necrosis. Cells that have undergone apoptosis may undergo secondary necrosis (apoptotic necrosis) to clear the cellular debris [2].

In recent years, the view of necrosis as a random or uncontrolled process has been modified, as it has been shown that it may be both inhibited and controlled, leading to the potential for the development of treatment therapies that could target necrotic diseases [3].

Apoptosis and Beyond: The Many Ways Cells Die, First Edition. Edited by James Radosevich.
© 2018 John Wiley & Son Inc. Published 2018 by John Wiley & Son Inc.

5.2 Reversible Cell Injury: Adaptation and Intracellular Accumulations

If an injury does not kill a cell – in other words, if the duration and severity of the injury is survivable – then reversible cell injury may occur, and the cell may exhibit a number of adaptive changes. It may change its shape, its form, or its function; it may get larger or smaller; it may become more or less numerous. The adaptation or adaptations exhibited depend on the type of cell, its location in the body, and the type of stress or injury experienced. The net result is a change in the growth and/or differentiation of the affected cell that enables it to survive in an altered environment.

5.2.1 Hydropic Swelling

When a cell is initially damaged, it accumulates fluid due to disruption of the cellular mechanisms for maintaining fluid balance (cell membrane, sodium pump, and adenosine triphosphate (ATP) supply). This may be seen microscopically as **hydropic swelling**. Affected cells look enlarged and pale, with normal-appearing nuclei. Under electron microscopy, the cisternae of the endoplasmic reticulum (ER) are swollen with fluid, and polysomes may detach from the rough ER. The mitochondria may also swell. The cell membrane may develop protrusions (blebbing). This is generally agreed to be the first observable change in response to cell injury. It is usually reversible; if the cause of the injury goes away, the cell generally reverts to its normal appearance. If the injury is overwhelming, necrosis will occur [4].

5.2.2 Decrease in Size: Atrophy

Perhaps the simplest form of cellular adaptation is **atrophy**, or a reduction in cell size (Figure 5.1). Atrophy usually results from a decrease in function or usage of the cell, due either to a disruption of the nutrients or blood supply needed for cell survival, a decrease

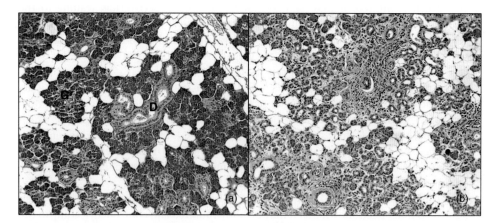

Figure 5.1 Atrophy of salivary glands. (A) Normal parotid salivary gland with ducts (D) and acinar gland cells (G). (B) Parotid salivary gland in the same patient, compressed by a nearby salivary tumor. The ductal structures and acinar gland cells are reduced in size in the area of compression, and many of the acinar gland cells have disappeared entirely. Hematoxylin and eosin stain ×20 magnification.

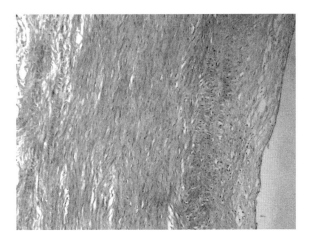

Figure 5.2 Hypertrophy. Smooth-muscle hypertrophy in the wall of an artery in a hypertensive patient. Hematoxylin and eosin stain ×10 magnification.

in the hormonal signals instructing the cell to be active, or a case of chronic inflammation. A common example: When a patient breaks his or her arm, it is immobilized in a cast for many weeks in order to promote healing. The muscles of the affected arm are not used as much as the opposing limb. When the patient's cast is removed, he or she will notice that the healed arm has become noticeably smaller in diameter. This is because the unused myocytes (muscle cells) have shrunk in size, in a process called "disuse atrophy" [5].

5.2.3 Increase in Size: Hypertrophy and Hyperplasia

When there are increased demands on a group of cells, they may adapt by increasing in individual cell size (**hypertrophy**) (Figure 5.2), by becoming more numerous (**hyperplasia**) (Figure 5.3), or through a combination of both. Hypertrophy and hyperplasia are

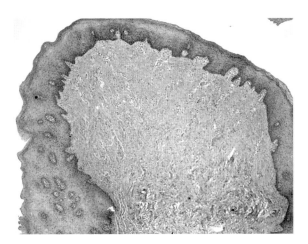

Figure 5.3 Hyperplasia. Hyperplasia of the fibrous connective tissue in response to injury of the oral mucosal tissue. The hyperplastic fibrous tissues have formed a raised bump called a fibroma. Hematoxylin and eosin stain ×10 magnification.

usually triggered by increased functional demands, loss of other functional tissue, or increased hormonal signals. They can also occur due to chronic inflammation.

Some types of cell are not capable of further division; these are called "permanent tissues." Neural and muscular tissues are prime examples. Other tissues are capable of cell proliferation in response to injury; these are called "stable tissues." Examples are glands, liver, and kidney. Other tissues are constantly reproducing and regenerating; these are called "labile tissues." Examples are skin, mucosa, and bone marrow.

Hyperplasia occurs when the cells are capable of further proliferation (i.e., in stable and labile tissues). For example, during the proliferative phase of the menstrual cycle, the uterine endometrium undergoes hyperplasia in response to a hormonal trigger [6]. Older men frequently develop benign prostatic hyperplasia: enlargement of the prostate gland due to hyperplasia triggered by alteration of hormonal signaling. If a patient moves to a high altitude, their bone marrow may respond to lowered oxygen levels by manufacturing additional erythrocytes (red blood cells) to carry oxygen.

When there is increased demand on permanent tissues, hypertrophy will occur. For example, strength training builds an increase in muscle mass by increasing the size of each individual myocyte via hypertrophy [5]. If a patient is morbidly obese, they may develop cardiac enlargement because their cardiac myocytes individually enlarge in response to an increased workload.

Frequently, hypertrophy and hyperplasia work together to meet an increased functional demand on a group of cells or an organ. One example of this is the increase in size of the uterus during pregnancy, in which the uterine muscle undergoes hypertrophy and the endometrium proliferates [6].

Hyperplasia usually occurs in response to normal cell signaling. However, if cells proliferate in the absence of normal cell signaling, this is not called hyperplasia, but instead is termed **neoplasia**. Continued neoplastic cell proliferation can give rise to cancer [5].

5.2.4 Change in Differentiation: Metaplasia

In **metaplasia** (Figure 5.4), the injured cells transform from their original mature cell type to another mature type that is better suited to withstanding the changed

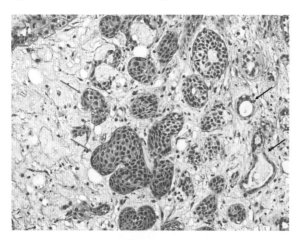

Figure 5.4 Metaplasia. Metaplasia of salivary ducts in response to trauma. The normal ductal-type epithelium (right arrows) has been replaced by squamous-type epithelium (left arrows). Hematoxylin and eosin stain ×20 magnification.

environment. For example, the epithelial lining of the respiratory tract is normally made up of respiratory-type epithelium, a specialized form of tall vertically aligned cells with tiny brush-like cilia on the surface. This specialized epithelium assists in "housecleaning" the respiratory tract by removing irritants and mucus. When faced with a chronic irritant such as cigarette smoke, this epithelium may lose these features and change to a hardier form made up of multiple horizontal layers of cells (stratified squamous epithelium).

Like hyperplasia, metaplasia may also lead to the development of neoplasia in some circumstances, such as in a condition called Barrett's esophagus, where the squamous or horizontally layered epithelium in the esophagus is replaced by glandular epithelium. This condition carries an increased risk of the development of esophageal cancer [5].

5.2.5 Intracellular Accumulations

In some cases, microscopic signs of a prior injury can be seen in the form of intracellular deposits of different forms of cellular material or pigmentation. Some of these deposits include normal cellular or tissue components or blood products [5].

5.2.5.1 Lipids

In injury or chronic disease, **fat (lipid)** may accumulate in tissues where it is not normally found. This generally occurs because of an overabundance of lipid, or due to impaired lipid processing. Vesicles of lipid accumulate in the affected cell's cytoplasm. The most commonly affected organ is the liver. If the liver is acutely exposed to toxins such as alcohol, the hepatocytes can accumulate lipids, including triglycerides: a condition called "fatty liver" (Figure 5.5). This can progress to inflammation and focal necrosis (steatohepatitits) [7]. Intracellular fat accumulation can occur in macrophages, in lesions called xanthomas; this is believed to be a reaction to epithelial damage, as in the verruciform xanthoma shown in Figure 5.6, or else an overabundance of systemic lipids, as in xanthelasma [8,9].

Cholesterol, a specific lipid utilized in cell membrane production, may become trapped inside cells and tissues in response to cell injury or chronic inflammation (Figure 5.7). Abnormal intracellular cholesterol accumulations also occur in some genetic diseases such as lysosomal storage diseases, including Niemann–Pick disease.

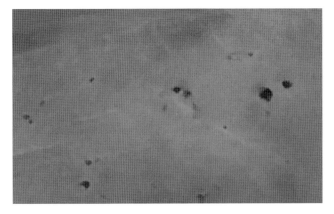

Figure 5.5 Fatty liver. Accumulation of excess lipid in the liever of a patient with obesity. The liver has a pale yellow appearance due to macrovesicles of lipid within the hepatocytes.

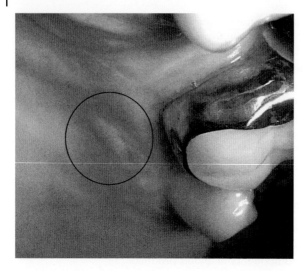

Figure 5.6 Lipid-laden macrophages. Verruciform xanthoma, presenting as a rough yellowish area on the palate. This results from local accumulation of lipid-filled macrophages just under the mucosal surface. It is believed to be triggered by epithelial damage.

Patients with Niemann–Pick disease type C develop cholesterol accumulations in the liver and spleen, as well as in the central nervous system (CNS), leading to early death [10]. Niemann–Pick C1 (NPC1) is an endosomal protein that helps to regulate the processing of cholesterol by cells. Its normal function appears to play a role in protection against atherosclerosis [11]. Atherosclerosis is one of the most common underlying causes of death in humans; cholesterol accumulates in the walls of blood vessels, leading to pathologic lesions called atheromas, which can become inflamed, degenerate, and cause blood-vessel blockages [12].

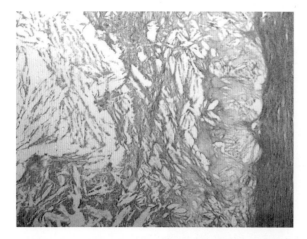

Figure 5.7 Cholesterol accumulation. Cholesterol trapped in chronically inflamed connective tissue in an atherosclerotic plaque in the wall of an artery, presenting as slit-like spaces. Hematoxylin and eosin stain ×10 magnification.

5.2.5.2 Proteins

Protein deposits may be aggregations of normal cellular products, or they may signal a systemic disease. **Amyloidosis** (Figure 5.8) is the deposition of beta-pleated proteins, usually in soft tissues. These accumulations can occur in conjunction with several different systemic diseases, such as multiple myeloma, a malignancy of plasma cells, or a sustained systemic inflammatory disease such as rheumatoid arthritis (RA) [13]. They can also occur locally. For example, amyloid deposits interfere with heart function in the genetic disease cardiac amyloidosis [14] and are accompanied by inflammation in the brain in Alzheimer's disease (AD) [15].

5.2.5.3 Glycogen

Glycogen deposits normally supply a readily accessible form of glucose. Abnormal cellular deposits of glycogen may occur within the framework of systemic disease.

Glycogen deposits within the epithelial cells of the kidney are associated with renal disease, which often secondarily complicates diabetes mellitus [16].

Patients with hereditary glycogen storage diseases accumulate glycogen in tissues because of errors in glycogen synthesis or breakdown. For example, in glucose-6-phosphate deficiency (von Gierke's disease), glycogen and lipid accumulate in the liver and kidneys, causing many systemic problems, including hepatomegaly and renomegaly [17,18]. Patients with Pompe's disease have a deficiency of lysosomal acid α-glucosidase and accumulate glycogen in the lysosomes of their cardiac and skeletal muscles. In the classic form of infantile-onset Pompe's disease, cardiomegaly develops and the child frequently dies of heart failure. In some other forms of Pompe's disease, the skeletal muscle is more prominently involved, and death may occur from respiratory insufficiency [19].

5.2.5.4 Pigments

Exogenous (foreign) or cell-produced pigmentation of cells can sometimes indicate a past history of injury. **Anthracosis** is the deposition of carbon particles from air pollution in the macrophages or scavenger cells within the lungs [20]. Other products may be formed by the breakdown of normal tissue components. Examples include

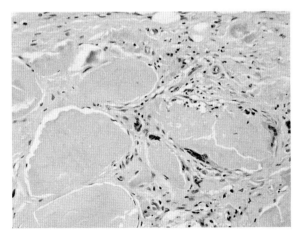

Figure 5.8 Amyloidosis. Amyloid deposits in the fibrous connective tissue appear as large pink amorphous areas. Foreign-body giant cells surround the proteinaceous material. Hematoxylin and eosin stain ×40 magnification.

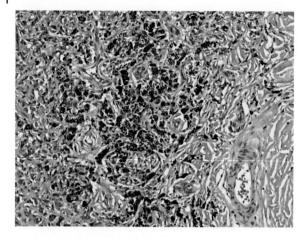

Figure 5.9 Hemosiderin. Hemosiderin is a pigment created by the breakdown of blood products. It is seen here trapped in the connective tissue in an area of inflammation. Hematoxylin and eosin stain ×10 magnification.

lipofucsin, a byproduct of phospholipid membrane breakdown, and **hemosiderin** (Figure 5.9), a blood-product breakdown substance [5]. Both may appear as brown pigmentations, and they are often found in chronically inflamed tissues. **Melanin**, the protective pigmentation usually found in the basal layer of epithelium, may be produced in response to injury or chronic inflammation, and post-inflammatory deposits may be seen where melanin is ingested by macrophages called melanophages. Current research is adding to our understanding of the immune functions of melanocytes [21].

5.3 Irreversible Cell Injury and Cell Death

If injury to the cell is sufficiently severe, or is sustained for a long enough period to pass the cell threshold of recovery, cell death will occur. Changes indicating cell death take place on a biochemical, a cellular, and a tissue level. In 2009, the Nomenclature Committee on Cell Death (NCCD) gave the following criteria for considering a cell dead: "(1) the cell has lost the integrity of its plasma membrane, as defined by the incorporation of vital dyes (e.g., PI) in vitro; (2) the cell, including its nucleus, has undergone complete fragmentation into discrete bodies (which are frequently referred to as 'apoptotic bodies'); and/or (3) its corpse (or its fragments) has been engulfed by an adjacent cell in vivo" [22].

5.3.1 Nuclear Changes in Cell Death

In irreversible cell death, changes to the nuclear material include pyknosis, karyorrhexis, and karyoysis (Figure 5.10) [2]. Cells affected by apoptosis usually undergo pyknosis (nuclear shrinkage) followed by karyorrhexis (nuclear fragmentation), followed by "apoptotic necrosis" and karyolysis, or dissolution of the nuclear material [2]. On the other hand, cells undergoing necrosis usually first show cellular swelling (oncosis), followed by karyolysis [2,23]. However, the differing forms of nuclear change are not specific to either form of cell death, and examples may be found in both types. In cells

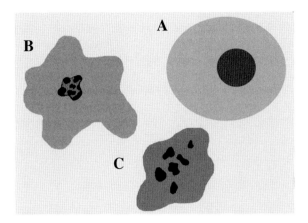

Figure 5.10 Nuclear changes in cell death. (A) Illustration of a normal cell, with central nucleus and surrounding cytoplasm. (B) In the first stage of cell death, the nuclear material has become condensed, darker in color, and smaller in size (pyknosis). The cytoplasm has also become more eosinophilic (pink) because of cross-linking of cellular proteins. (C) In the second stage of cell death, the nuclear material fragments within the cytoplasm (karyorrhexis). The cytoplasm often shows more pronounced eosinophilia. In karyolysis, the final stage (not shown), the nuclear material is digested and disappears, so only the cytoplasm remains.

that have received chemotherapy for cancer treatment, all three forms of nuclear change have been noted [24].

5.3.1.1 Pyknosis

Pyknosis can occur as a result of necrosis or apoptosis (Figure 5.10B). Pyknotic cells will show a condensation of the nucleus. This phenomenon may be seen in multiple types of clinical situation. For example, after a myocardial infarction, pyknotic nuclei of the necrotic cardiac muscle cells are noted about 24–48 hours after the ischemic event [25]. In other forms of cell injury, pyknosis may be observed when cell death has been slow, rather than rapid [26].

5.3.1.2 Karyorrhexis

Karyorrhexis refers to the fragmentation of cell nuclei into "nuclear dust" (Figure 5.10C). Karyorrhexis often follows pyknosis; following a myocardial infarction, for example, karyorrhexis is noted extensively in the cardiac muscle about 3–5 days after the ischemic event [25].

5.3.1.3 Karyolysis

In **karyolysis**, the cell nucleus is dissolved by the automatic responses of the cell to eliminate damaged structures. This is generally the final stage of cell change following individual cell death [27].

5.3.2 Tissue Patterns of Necrosis

When a certain threshold of cell death has occurred, necrosis is evident on a tissue-wide basis, rather than just in individual cells. Multiple patterns of tissue necrosis can be seen, and they are often suggestive of the underlying disease process or pattern.

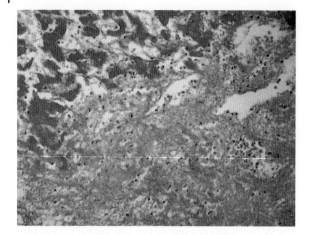

Figure 5.11 Coagulative necrosis. Coagulative necrosis in heart muscle after a myocardial infarction. Hematoxylin and eosin stain ×20 magnification.

Some important differences may be seen in cell death resulting from apoptosis when compared to cell death resulting from a necrotic process. Generally, necrotic cell processes usually activate a robust inflammatory response, whereas apoptosis does not result in extensive inflammation [3]. This is important because some types of injury are most effectively curtailed by a robust immune response (e.g., viral and bacterial infections) [28]. The small-intestine cells have been shown to employ a combination of apoptotic and necrotic processes to renew themselves after a pathogenic infection such as a virus [23]. In the past, cancer therapy often focused on targeting apoptosis of cells, but some newly developing therapies seek to cause necrosis of the tumor cells in order to elicit a greater immune response, which may destroy additional tumor cells [23,29]. Further, necrosis can kill tumor cells even when they have developed mechanisms to avoid apoptosis by the body's own surveillance system [28].

5.3.2.1 Coagulative Necrosis
Coagulative necrosis (Figure 5.11) suggests a sudden loss of oxygen to the tissues and is seen in the necrosis of the cardiac muscle after a myocardial infarction. In coagulative necrosis, the cellular outline persists but the nuclear material is lost from the destroyed cells [4].

5.3.2.2 Liquefactive Necrosis
Liquefactive necrosis (Figure 5.12) suggests an acute bacterial infection. Enzymatic destruction of the dead cells results in dead-cell fragments and neutrophils, clinically seen as purulent exudate (pus) [4].

5.3.2.3 Caseous Necrosis
Caseous necrosis (Figure 5.13) is often seen in chronic or longstanding bacterial infections – characteristically, those caused by mycobacteria such as tuberculosis – and occasionally in deep fungal infections such as histoplasmosis. A zone of dead-cell fragments is surrounded by a granuloma (Figure 5.14). The term "caseous" refers to the pale cheese-like appearance of the central necrotic zone. A granuloma consists of a

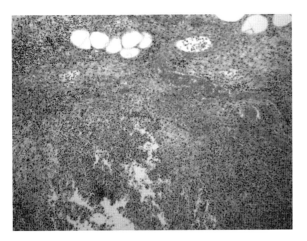

Figure 5.12 Liquefactive necrosis. In a case of acute suppurative appendicitis, large collections of neutrophils and debris line the appendix wall. Here, they have destroyed much of the surface lining. Dead cells are fragmented. Hematoxylin and eosin stain ×10 magnification.

central nidus of scavenger macrophages (sometimes termed "epithelioid" macrophages, because of their casual resemblance to plump epithelial cells) and special multinucleated giant cells formed by the fusion of several macrophages. Chronic inflammatory cells such as lymphocytes permeate the periphery of the granuloma [30]. Granulomas are an attempt to wall off an infection, while the macrophages serve to attack and kill the infection.

5.3.2.4 Gummatous Necrosis

Gummatous necrosis is characteristic of the late (tertiary) stage of the bacterial infection syphilis. It is a destructive type of coagulative necrosis seen in the centers of granulomas, which generally have an inflamed fibrotic outermost zone [4].

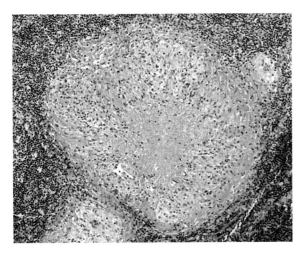

Figure 5.13 Caseous necrosis. Caseating granuloma in a patient with tuberculosis, exhibiting central caseous necrosis surrounded by giant cells and histiocytes. Hematoxylin and eosin stain ×10 magnification.

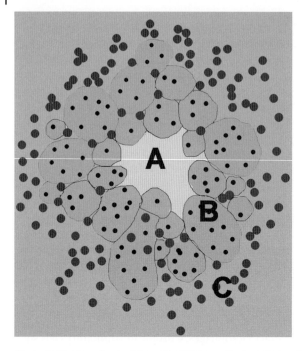

Figure 5.14 Granuloma. Schematic diagram of a granuloma showing a central zone (A) surrounded by a zone of plump macrophages (B), which may take the form of multinucleated giant cells. The outermost zone (C) is dominated by lymphocytes. The central zone is not always present, but it can contain necrotic material, which may be a clue to the etiology. Caseous necrosis in the central zone suggests certain infections, particularly tuberculosis, while gummatous necrosis is a strong but not infallible indicator that syphilis is the causative agent.

5.3.2.5 Fat Necrosis

Fat necrosis is a nonspecific destruction of adipose or fat tissue. It usually occurs when an acute infection or inflammation in nearby glandular tissue causes the release of enzymes, especially lipases, that break down the contents of the adipocytes, releasing free fatty acids (FFAs). It is especially common when there is extensive inflammation in the pancreas. Cations react with these FFAs, resulting in the production of calcium and magnesium soaps, in a process called "saponification." Saponification, if extensive, can lower a patient's serum calcium levels. Under the microscope, fat necrosis can be recognized by the presence of adipose cells without nuclei, surrounded by inflammation. Chalky-looking calcium deposits may be formed by the reaction between the enzymes and fatty acids released by the damaged fat cells [5]. In the breast, fat necrosis may be caused by injury; this must be distinguished from malignancy on imaging and biopsy [31].

5.3.2.6 Fibrinoid Necrosis

Fibrinoid necrosis (Figure 5.15) involves destruction of blood-vessel walls secondary to inflammation in the vessel wall. Plasma proteins are deposited in the smooth-muscle layer surrounding blood vessels, leading to an intensely pink appearance under the microscope [4]. This is seen in temporal arteritis (also known as giant-cell arteritis), a painful granulomatous inflammatory disease of the arteries. It is more common in

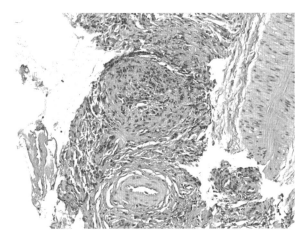

Figure 5.15 Fibrinoid necrosis. Vasculitis, or inflammation of the blood vessel, damages the vessel wall, causing necrosis. Pink fibrinous deposits appear interspersed within the wall. Hematoxylin and eosin stain ×10 magnification.

elderly patients, and is most common in the temporal artery, where the damage sustained may lead to loss of vision in some cases [32].

5.3.3 Intratissue Accumulations: Dystrophic Calcification

Just as cells may accumulate deposits of exogenous or internal products after injury, tissue that has been damaged by necrosis may also show deposits of material that hint at past injury. Dystrophic calcifications (Figure 5.16) are the most common such finding. Unlike metastatic calcifications, which occur due to a systemic calcium derangement, dystrophic calcifications are localized deposits of calcium salts in areas that have been injured. The calcified deposits may be as small as a grain of sand, or more extensive. They may occur in tuberculous involvement of lymph nodes (scrofula).

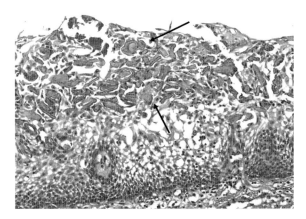

Figure 5.16 Dystrophic calcification. The formation of abnormal calcifications (arrows) in the lining of a longstanding and chronically inflamed odontogenic cyst of the jaws. Hematoxylin and eosin stain ×20 magnification.

Dystrophic calcification is especially common in the cardiovascular system, where it may increase the severity of preexisting disease [5]. When it occurs in aging or damaged heart valves, it can impede their function, because the valve leaflets lose flexibility and the valve orifice becomes restricted. Calcific aortic-valve stenosis is usually a disease of the elderly. Dystrophic calcification is commonly seen in severe atherosclerosis, and cardiac calcium-scoring CT tests are sometimes used to assess the risk of coronary artery disease.

Localized calcification can also occur in settings of tumors and cysts. Such calcifications have such a well-established association with breast cancer that they form the basis of mammography; clustered, linear, or segmental calcifications may be suspicious signs of breast cancer. However, they are also associated with benign processes such as the odontogenic cyst in Figure 5.16.

5.4 Conclusion

Cells may be injured for a variety of reasons. While they possess a number of adaptive capabilities to survive certain forms of injury, if they experience severe or sustained injury they may die. When a sufficient number of cells die, tissue damage and tissue necrosis can be observed. The patterns of cell death and tissue damage may give clues to the source of the destruction and aid in determining the cause of injury or the disease diagnosis.

References

1 Gamrekelashvili J, Geten TF, Korangy F. Immunogenicity of necrotic cell death. *Cell Mol Life Sci* 2015;**72**(2):273–83.
2 Majno G, Joris I. Apoptosis, oncosis, and necrosis, an overview of cell death. *Am J Pathol* 1995;**146**(1):3–15.
3 McCall K. Genetic control of necrosis – another type of programmed cell death. *Curr Opin Cell Biol* 2010;**22**:882–8.
4 Stevens A, Lowe J, Scott I. *Core Pathology*, 3rd edn. Marylands Heights, MO: Mosby Elsevier. 2009.
5 Kumar V, Cotran RS, Robbins SL. *Robbins Basic Pathology*, 9th edn. Philadelphia, PA: Saunders Elsevier. 2013.
6 Burkitt HG, Stevens A, Lowe JS, Young B. *Wheater's Basic Histopathology, a Color Atlas and Text*, 4th edn. Edinburgh: Churchill Livingstone. 2003.
7 Sears D. Fatty liver. Emedicine by Medscape. August 16, 2012. Available from http://emedicine.medscape.com/article/175472-overview (last accessed March 22, 2018).
8 Fair KP. Xanthomas. Emedicine by Medscape. October 23, 2014. Available from http://emedicine.medscape.com/article/1103971-overview (last accessed March 22, 2018).
9 Lambert CW. Verruciform xanthoma. Emedicine by Medscape. October 31, 2014. Available from http://emedicine.medscape.com/article/1061445-overview (last accessed March 22, 2018).
10 Cianciola NL, Carlin CR. Molecular pathways for intracellular cholesterol accumulation: common pathogenic mechanisms in Niemann–Pick disease type C and cystic fibrosis. *Arch Biochem Biophys* 2011: **515**(1–2); 54–63.
11 Zhang JR, Coleman T, Langmade SJ, Scherrer DE, Lane L, Lanier MH, et al. Niemann-Pick C1 protects against atherosclerosis in mice via regulation of macrophage intracellular cholesterol trafficking. *J Clin Invest* 2008;**118**(6):2281–90.

12 Ladich ER. Atherosclerosis pathology. Emedicine by Medscape. October 4, 2012. Available from http://reference.medscape.com/article/1612610-overview (last accessed March 22, 2018).
13 Holmes RO. Amyloidosis. Emedicine by Medscape. November 26, 2014. Available from http://emedicine.medscape.com/article/335414-overview (last accessed March 22, 2018).
14 Quarta CC, Kruger JL, Falk RH. Cardiac amyloidosis. *Circulation* 2012;**126**:e178–82.
15 Latta CH, Brothers HM, Wilcock DM. Neuroinflammation in Alzheimer's disease; a source of heterogeneity and target for personalized therapy. *Neurosci* 2014;**pii**: S0306-4522(14)00820-3.
16 Seidiu I, Torffvit O. Decreased urinary concentration of Tamm-Horsfall protein is associated with development of renal failure and cardiovascular death within 20 years in type 1 but not in type 2 diabetic patients. *Scand J Urol Nephrol* 2008;**42**(2):168–74.
17 Froissart R, Piraud M, Boudjemline AM, Viarey-Saban C, Petit F, Hubert-Buron A, et al. Glucose-6-phosphate deficiency. *Orphanet J Rare Dis* 2011;**6**:27.
18 Bali DS, Chen YT, Goldstein JL. Glycogen storage disease type I. April 19, 2006 [updated September 19, 2013]. GeneReviews. Available from http://www.ncbi.nlm.nih.gov/books/NBK1312/ (last accessed March 22, 2018).
19 Leslie N, Tinkle BT. Glycogen storage disease type II (Pompe disease). August 31, 2007 [updated May 9, 2013]. GeneReviews. Available from http://www.ncbi.nlm.nih.gov/books/NBK1261/ (last accessed March 22, 2018).
20 Khan FJ. Coal worker's pneumoconiosis. Emedicine by Medscape. March 27, 2014. Available from http://emedicine.medscape.com/article/297887-overview#a0104 (last accessed March 22, 2018).
21 Feller L, Masilana A, Khammissa RAG, Altini M, Jadwat Y, Lemmer J. Melanin: the biophysiology of oral melanocytes and physiological oral pigmentation. *Head Face Med* 2014;**10**:8.
22 Kroemer G, Galluzzi L, Vandenabeele P, Abrams J, Alnemri E, Baehrecke E, Melino G. Classification of cell death: recommendations of the Nomenclature Committee on Cell Death 2009. *Cell Death Differ* 2009;**16**(1); 3–11.
23 Proskuryakov SY, Konoplyannikov AG, Gabai VL. Necrosis: a specific form of programmed cell death? *Exp Cell Res* 2003;**283**:1–16.
24 Sethi D, Sen R, Parshad S, Khetarpal S, Garg M, Sen J. Histopathologic changes following neoadjuvant chemotherapy in various malignancies. *Int J App Basic Med Res* 2012;**2**:111–16.
25 Burke AP. Pathology of acute myocardial infarction. Emedicine by Medscape. July 17, 2013. Available from http://emedicine.medscape.com/article/1960472-overview (last accessed March 22, 2018).
26 Burgoyne LA. The mechanisms of pyknosis: hypercondensation and death. *ExpCell Res* 1999;**248**:214–22.
27 Trump BE, Berezesky IK, Chang SH, Phelps PC. The pathways of cell death: oncosis, apoptosis, and necrosis. *Toxicol Pathol* 1997;**25**:82–7.
28 Festjens N, Vanden Berghe T, Vandenabeele P. Necrosis, a well-orchestrated form of cell demise: signaling cascades, important mediators and concomitant immune response. *Biochim Biophys Acta* 2006;**1757**:1371–87.
29 Zhang J. Lou X, Jin L, Zhou R, Liu S, Xu N, Liao DJ. Necrosis, and then stress induced necrosis-like cell death, but not apoptosis, should be the preferred cell death mode for chemotherapy: clearance of a few misconceptions. *Oncoscience* 2014;**1**(6):407–22.

30 Yeldandi AV. Pathology of pulmonary infectious granulomas. Emedicine by Medscape. April 3, 2013. Available from http://emedicine.medscape.com/article/2078678-overview (last accessed March 22, 2018).

31 Amin AL, Purdy AC, Mattingly JD, Kong AL, Termuhlen PM. Benign breast disease. *Surg Clin N Am* 2013;**93**:299–308.

32 Flood TA. Temporal arteritis pathology. Emedicine by Medscape. May 9, 2013. Available from http://emedicine.medscape.com/article/1612591-overview (last accessed March 22, 2018).

6

Necroptosis

Ben A. Croker,[1] James A. Rickard,[2] Inbar Shlomovitz,[3]
Arshed Al-Obeidi,[1] Akshay A. D'Cruz,[1] and Motti Gerlic[3]

[1] Dana-Farber Boston's Children Cancer and Blood Disorder Center, Harvard Medical School, Boston, MA, USA
[2] Department of Biochemistry, La Trobe University, Melbourne, VIC, Australia
[3] Department of Clinical Microbiology and Immunology, Sackler Faculty of Medicine, Tel Aviv University, Tel Aviv, Israel

Abbreviations

AIM2	absent in melanoma 2
BMDC	bone marrow-derived dendritic cells
c-FLIP	cellular FLICE (FADD-like IL-1β-converting enzyme)-inhibitory protein
cIAP	cellular inhibitor of apoptosis protein
CMV	human cytomegalovirus
CYLD	cylindromatosis
DAI	DNA-dependent activator of IFN-regulatory factors
DAMP	damage-associated molecular pattern
EAE	experimental autoimmune/allergic encephalomyelitis
FADD	Fas-associated death-domain protein
HD	Huntington's disease
HMGB1	high-mobility group protein B1
HSC	hematopoietic stem cells
HSV	herpes simplex virus
IDO	indoleamine-2,3-dioxygenase
IL-1	interleukin 1
I/R	ischemia/reperfusion
IFN	interferon
LPS	lipopolysaccharide
LT-HSC	long-term hematopoietic stem cell
LUBAC	linear ubiquitin chain assembly complex
MAPK	mitogen-activated protein kinase
MCMV	murine cytomegalovirus
MLKL	mixed lineage kinase domain-like
MS	multiple sclerosis
mtDNA	mitochondrial DNA
Nec-1	necrostatin-1
NLRs	nucleotide-binding domain and leucine-rich repeat-containing gene family

Apoptosis and Beyond: The Many Ways Cells Die, First Edition. Edited by James Radosevich.
© 2018 John Wiley & Son Inc. Published 2018 by John Wiley & Son Inc.

PARP	poly ADP ribose polymerase
PCD	programmed cell death
PGAM5	phosphoglycerate mutase family member 5
PKR	protein kinase R
RHIM	receptor-interacting protein homotypic interaction motif
RIG-I	retinoic acid-inducible gene 1
RIPK1	receptor-interacting protein kinase-1
RIPK3	receptor-interacting protein kinase-3
ROS	reactive oxygen species
SMAC	small mitochondria-derived activator of caspases
TLR	Toll-like receptor
TNF	tumor necrosis factor
TNFR1	TNF receptor 1
TRPM7	transient receptor potential melastatin-related 7
XIAP	X-linked inhibitor of apoptosis protein

6.1 Introduction

Programmed and regulated forms of cell death serve diverse functions in multicellular organisms. The balance between cellular differentiation, proliferation, and programmed cell death (PCD) is critical for embryonic development, organ function, tumorigenesis, and immune responses during life. When apoptotic cells are not cleared by phagocytic cells, an accidental form of cell death traditionally viewed as necrosis ensues, resulting in the nonspecific release of cellular contents. This can drive inflammation and organ damage. Necrosis has therefore been viewed as a sign of injury, stress, or infection.

6.1.1 Defining Necroptosis

For over 4 decades, it was believed that two forms of cell death exist: apoptosis, a programmed form, and necrosis, a non-programmed and accidental form (Figure 6.1). Recent discoveries have overturned this simplistic model to reveal molecular machineries controlling programmed or regulated necrosis. Many experimental approaches support the existence of non-apoptotic forms of cell death, including genetic, biochemical, and pharmacological studies.

The best-characterized forms of non-apoptotic cell death are pyroptosis (see Chapter 15) and necroptosis. Although originally described in 2000 as a serine-threonine kinase receptor-interacting protein kinase-1 (RIPK1)-dependent and caspase-independent form of cell death [1], necroptosis is now known to proceed independently of RIPK1 in some settings. Necroptosis is therefore currently best defined as a caspase-independent form of cell death that requires two proteins: the serine-threonine kinase receptor-interacting protein kinase-3 (RIPK3) and the pseudokinase mixed lineage kinase domain-like (MLKL) (Figure 6.2); these will be discussed in detail later in the chapter.

6.2 Inducing Necroptosis

Numerous stimuli can engage the necroptosis cell-death cascade, including death receptors, Toll-like receptors (TLRs), and intracellular receptors. These activation

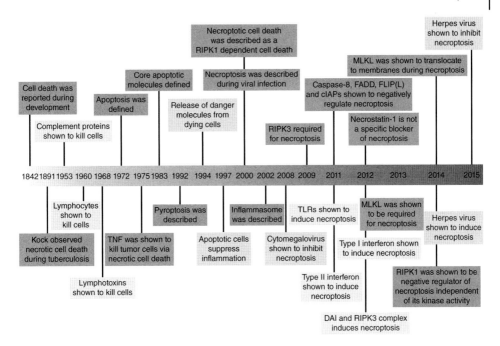

Figure 6.1 The history of necroptosis. Key historical events in the necroptosis field.

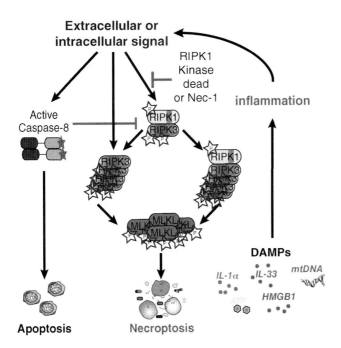

Figure 6.2 The key players in necroptotic cell death. When caspase 8 activity is inhibited, stimulation of extra- or intracellular receptors induces necroptosis via receptor-interacting protein kinase-1 (RIPK1)-dependent or independent pathways. Both scenarios depend on the phosphorylation of receptor-interacting protein kinase-3 (RIPK3), which leads to aggregation and phosphorylation of mixed-lineage kinase domain-like (MLKL) by RIPK3. Phosphorylated MLKL then leads to necroptosis and the release of damage-associated molecular patterns (DAMPs).

pathways have one common feature: the inhibition of caspase 8 by genetic or pharmacological means.

6.2.1 Cell Death Receptors

In 1975, tumor necrosis factor (TNF) was found to induce the necrotic death of some tumor cell lines (Figure 6.1), spurring hope that it would be suitable for the induction of tumor cell death [2]. However, in most circumstances TNF induces inflammation and promotes cell survival. One of the key factors controlling signaling downstream of the TNF receptor 1 (TNFR1) is RIPK1, which coordinates pro-survival and pro-death signaling to ensure appropriate biological outcomes during infection and inflammation [3–8].

RIPK1 was first discovered as a binding partner of Fas, and its overexpression was found to induce cell death [9]. It was the first intracellular player identified in the necroptosis pathway, and was shown to be essential for cell death in the presence of caspase inhibitors and the death receptor ligand FasL [1]. The death domain of RIPK1 can interact with TNFR1 in a complex called complex I (Figure 6.3). Complex I contains TRADD (a death domain-containing protein), TRAF2, cellular inhibitor of apoptosis (cIAP)1, cIAP2, and the linear ubiquitin chain assembly complex (LUBAC), comprising SHARPIN, HOIL-1, and HOIP [10–14]. Here, RIPK1 is ubiquitinated by several proteins, including cIAPs [15–18]. RIPK1 ubiquitylation by cIAP1, cIAP2, and LUBAC promotes TAK1, IKK, and NEMO recruitment, which impairs complex II assembly and induces pro-survival and pro-inflammatory signaling (Figure 6.3). Caspase 8 can cleave cylindromatosis (CYLD), a tumor suppressor that can de-ubiquitinate RIPK1; the cleavage products of CYLD inhibit necroptosis by an undetermined mechanism [18]. Complex I drives cytokine production and induces the expression of the anti-apoptotic protein cellular FLICE-inhibitory protein (c-FLIP) via MAP kinase activation and NF-κB (Figure 6.3). However, RIPK1 is not required for the activation of NF-κB in response to TNF, at least in mouse embryonic fibroblasts [19]. Generation of cytoplasmic complex II, containing RIPK1, Fas-associated death-domain protein (FADD), and caspase 8, initiates apoptosis – an outcome that depends on the autoprocessing of caspase 8 and is inhibited by c-FLIP$_L$ [11,20–23]. Expression of the short c-FLIP isoform c-FLIP$_S$ can inhibit apoptosis, but fails to block necroptosis [11,20].

Inhibition of caspase 8 can change the type of death from apoptosis to necroptosis. If cIAP1, cIAP2, and X-linked inhibitor of apoptosis protein (XIAP) [11,12] functions are inhibited by small mitochondria-derived activator of caspases (SMAC) [24] or SMAC mimetics, or if RIPK1 is de-ubiquitinated by the TRIM27/USP7 complex [25,26], RIPK1 then mediates apoptosis as part of a cytoplasmic complex together with caspase 8 and FADD (Figure 6.3). Caspase 8 and FADD are essential for receptor-induced apoptosis [27], yet they also negatively regulate RIPK3-dependent necroptosis (see later) [21,28,29]. In addition to caspase 8, FADD, and RIPK1, protein phosphatase 1b (PPM1B) also negatively regulates TNF-induced necroptosis [30]. PPM1B was identified in a RIPK3 immunocomplex and was found to dephosphorylate Thr231 and Ser232 on RIPK3, thereby preventing RIPK3 autophosphorylation and MLKL recruitment to RIPK3.

6.2.2 Toll-Like Receptors

Activation of TLR3 and the subsequent recruitment of TRIF to the receptor has several effects: (i) activation of NF-κB signaling, which depends in part on an interaction with

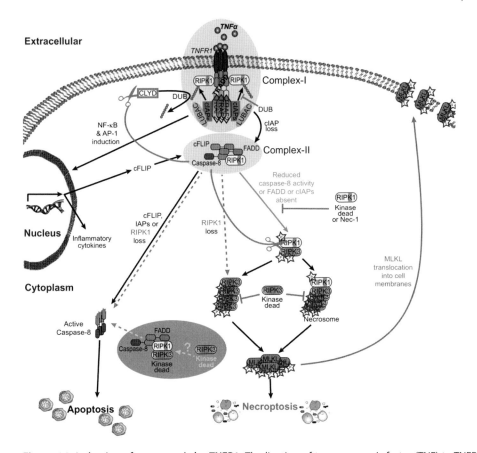

Figure 6.3 Induction of necroptosis by TNFR1. The ligation of tumor necrosis factor (TNF) to TNFR1 recruits the TNF receptor-associated death domain (TRADD) and receptor-interacting protein kinase 1(RIPK1) via their respective death domains. TRADD recruits TRAF2 and thereby cellular inhibitor of apoptosis proteins (cIAPs), which in turn ubiquitinate complex I components that recruit the linear ubiquitin-chain assembly complex (LUBAC) E3 ligase complex that generates linear ubiquitin chains to induce NF-κB nuclear translocation and AP-1 signaling. When complex I activity is impaired (e.g., when cIAPs are lost), the subsequent formation of the cytosolic complex II can induce either apoptotic or necroptotic cell death. c-FLIP is induced in an NF-κB-dependent manner, and binds the cytosolic complex II to inhibit caspase 8 autoprocessing and thus apoptosis, while enabling caspase 8 to restrain necroptosis, likely by cleaving RIPK1. RIPK3/MLKL-dependent necroptosis can be triggered by the depletion of cIAPs and concurrent caspase inhibition, RIPK1 de-ubiquitination by CYLD, or the reduction of of RIPK1, FADD, or caspase 8. Inhibition of RIPK1 or RIPK3 kinase activity by genetic or pharmacological means blocks necroptosis. Inhibition of RIPK3 kinase activity can lead to the formation of a RIPK3–RIPK1–FADD–caspase 8 complex that triggers apoptosis.

RIPK1 via the receptor-interacting protein homotypic interaction motif (RHIM) domain; (ii) activation of caspase 8-dependent apoptosis; and (iii) activation of RIPK1/RIPK3-dependent necroptosis when caspase 8 is absent (Figure 6.4). All TLRs have the capacity to induce necroptotic cell death indirectly via production of cytokines such as TNF. TLR3 and TLR4 can initiate necroptosis directly by coupling the RHIM domains of TRIF and RIPK3, but only when caspase 8 is inhibited (Figure 6.4) [31]. The necroptotic effects of TLR3 are independent of TNF production, whereas TLR4 stimulation provides an intermediate phenotype due to the ability of MyD88 to drive

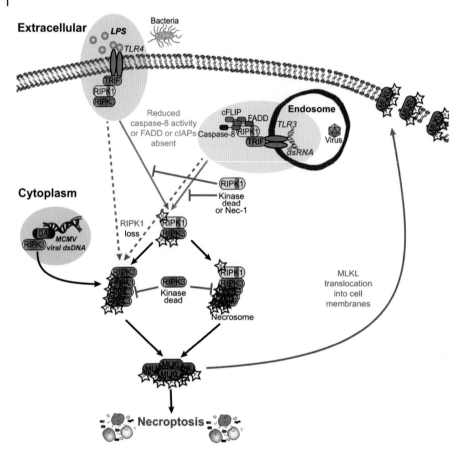

Figure 6.4 Induction of necroptosis by Toll-like receptors 3 and 4 (TLR3 and TLR4) and DNA-dependent activator of interferon (IFN)-regulatory factors (DAI). Necroptosis can be induced directly by TLR3, which recognizes viral dsRNA, via a TRIF–RIPK1–FADD–c-FLIP–caspase 8 complex; by TLR4, which recognizes bacterial lipopolysaccharide (LPS), via a TRIF–RIPK1–RIPK3 complex; or by the intracellular receptor DAI, which recognizes viral dsDNA. Both TLR pathways induce RIPK1 and RIPK3 phosphorylation, which leads to necroptosis, as shown in Figure 6.2. During murine cytomegalovirus (MCMV) infection, the intracellular receptor DAI can directly recruit RIPK3 via a RHIM–RHIM interaction to induce RIPK3 phosphorylation, oligomerization, and necroptosis.

TNF production downstream of TLR4. The kinetics of the response also varies, with TLR3 and TLR4 coupling directly and rapidly (4–6 hours) to the necroptotic machinery and TNF-dependent necroptosis coupling via indirect TLR1/2/5/9 activity over a longer time frame (12–18 hours) [31]. TRIF/RIPK3-dependent necroptosis occurs independently of the role of TRIF in NF-κB- and IRF3-dependent cytokine production downstream of TLR3 and TLR4 [31].

6.2.3 Interferon Receptors

Interferons (IFNs) are associated with an increase in cell death in the presence of caspase inhibitors [32,33]. In 2011, IFNγ was shown to induce RIPK1-dependent necroptosis in a

JAK-STAT-dependent manner when NF-κB signaling was inhibited (Figure 6.1) [32]. A year later, IFNα/β were shown to induce necroptosis during *Salmonella* infection (Figure 6.1) [33]. Both IFNα/β and IFNγ upregulate protein kinase R (PKR) expression via JAK-STAT signaling, and phosphorylated PKR is thought to interact with RIPK1/RIPK3 and induce necroptosis when FADD is absent or phosphorylated, or when caspase 8 is inhibited (Figure 6.5) [31,34]. PKR may also be responsible for phosphorylating – and thereby inhibiting – FADD. Macrophages deficient in the IFNα/β receptor

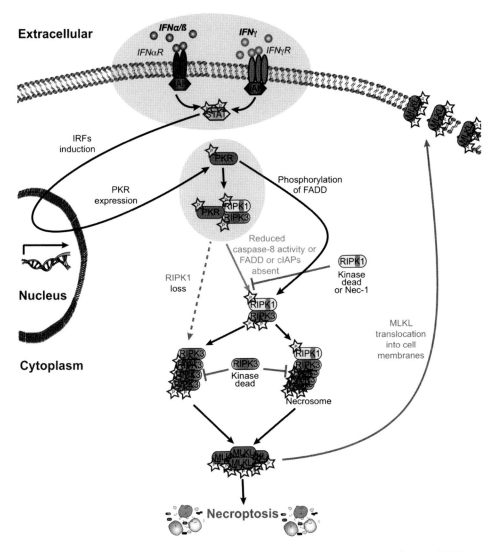

Figure 6.5 Induction of necroptosis by interferons (IFNs). IFNs type 1 (IFNα/β) and type 2 (IFNγ) induce necroptosis in a protein kinase R (PKR)-dependent manner. The binding of interferons to their receptors (IFNαR or IFNγR) induces phosphorylation of JAK and the receptor, leading to STAT recruitment and phosphorylation. Phosphorylated STAT further induces IRF signaling, promoting PKR expression and the formation of a PKR–RIPK1–RIPK3 complex. A reduction in caspase 8 activity (such as TNF- and TLR-induced necroptosis) and phosphorylation of FADD by PKR lead to RIPK1–RIPK3–MLKL-dependent necroptosis.

are resistant to lipopolysaccharide (LPS), polyinosinic-polycytidylic acid, TNFα, and IFNβ-induced necroptosis, suggesting that IFNα/β are the dominant factors controlling macrophage necroptosis [35].

6.2.4 Intracellular Receptors

While some viruses encode factors with direct cytopathic effects, the majority appear to induce cell death via the activation of host PCD. *Sendai virus* or transfected poly(I:C) can induce recruitment of caspase 8 and RIPK1 to a retinoic acid-inducible gene 1 (RIG-I) complex to control the magnitude and duration of RIG-I signaling, but it is not thought to regulate necroptosis in the process. In the absence of caspase 8, virus-induced IRF3 activation is enhanced by the ubiquitination of RIPK1 at Lys377, and antagonized by caspase 8-mediated cleavage of RIPK1 to a 38 kDa isoform [36]. The cytosolic DNA sensor DNA-dependent activator of IFN-regulatory factors (DAI) has the capacity to bind RIPK3 via RHIM interactions, thereby acting as a sensor for viral DNA and a direct activator of RIPK3-dependent necroptosis, independently of RIPK1 (Figure 6.4) [37]. Understanding the nature of these interactions at a structural level may help define modes of RIPK3 oligomerization and activation.

6.3 Mechanisms of Necroptosis

Like apoptosis, several inputs can trigger the induction of regulated necroptotic cell death, although the downstream events are largely conserved (Figures 6.2–6.5). Inhibition of caspase 8 activity is a requisite feature of necroptosis. Caspase 8 cleaves and inactivates RIPK3, but numerous genetic mutations can disrupt this regulatory mechanism (see later) to initiate a downstream cascade, leading to permeabilization of the cell membranes by MLKL.

6.3.1 RIPK1-Dependent Necroptosis

RIPK1 was the first intracellular protein identified as a key component of the necroptosis pathway (Figure 6.1) [1], but RIPK1-independent forms of necroptosis are now well recognized (see later). It was first suggested that RIPK1 kinase activity is necessary for the induction of necroptosis, because blocking it using necrostatin-1 (Nec-1) prevents necroptosis (Figure 6.1) [38]. This earlier *in vitro* work using Nec-1 was confirmed using mice expressing $Ripk1^{D138N}$ and $Ripk1^{K45A}$ kinase dead alleles, which are resistant to many necroptotic stimuli [6,39]. However, the identity of RIPK1 kinase substrates remains to be defined.

6.3.2 RIPK3

It took almost 9 years to identify the key protein downstream of RIPK1 in the necroptosis pathway (Figure 6.1). Two groups demonstrated the importance of RIPK3 in necroptosis, and suggested that phosphorylation of RIPK1 and RIPK3 drives the assembly of a RIPK1–RIPK3 complex (necrosome) to initiate necroptosis [40,41]. RIPK3 was shown to phosphorylate RIPK1 and itself (autophosphorylation). RIPK1–RIPK3 phosphorylation and necrosome formation lead to the recruitment and phosphorylation of the pseudokinase MLKL by RIPK3, a key step in commitment to necroptotic cell death (Figures 6.2–6.5) [42–44]. RIPK3 has the capacity to drive caspase 1 and caspase 8 activation, leading to apoptotic death in some settings; this therefore cannot be solely relied on as a definition of necroptosis.

6.3.3 MLKL: The End or Just the Beginning?

MLKL has recently emerged as a central player in the execution of RIPK3-dependent necroptotic death (Figure 6.1) [43], but its precise role in this cell-death pathway is not yet fully understood. Phosphorylation of the activation loop of MLKL induces a conformational change, disrupting an autoinhibitory interaction between the pseudokinase domain and the four-helix bundle. This promotes an interaction with phosphatidylinositol phosphates, promoting membrane localization (Figures 6.3–6.5) [45,46]. It is not clear if MLKL membrane localization and pore formation represent the end of this pathway [47], or if additional players such as phosphoglycerate mutase family member 5 (PGAM5) and transient receptor potential melastatin-related 7 (TRPM7) function downstream to modulate mitochondrial stability and calcium current across the plasma membrane to kill the cell [47,48]. Dimerized RIPK3 and MLKL function can induce necroptosis independently of the mitochondrial fission factor Drp1 that is reported to be downstream of PGAM5 [49]. Recently, human cytomegalovirus (CMV) was shown to block necroptosis despite phosphorylation of MLKL, suggesting that other players downstream of MLKL may exist [50]. Phosphorylated MLKL was shown to interact with ESCRTIII machinery to promote phosphatidylserine (PS) exposure and release of PS-positive extracellular vesicles – "necroptotic bodies" – during necroptosis, promoting "eat me" and "find me" signals similar to those seen in apoptosis [51–53].

6.3.4 RIPK1-Independent Necroptosis

The use of Nec-1 to inhibit RIPK1 kinase activity and block necroptosis has made RIPK1 and RIPK1 kinase activity synonymous with necroptosis. It is therefore a surprise that unlike RIPK3-deficiency, RIPK1-deficiency fails to fully rescue the embryonic lethality associated with caspase 8 deficiency (Figures 6.6 and 6.7) [21,28,29].

RIPK1-deficiency is neonatal-lethal within minutes of birth (Figure 6.6) [54]. The partial rescue of RIPK1-deficient neonates by combined loss of RIPK3 and MLKL is also counterintuitive [3–7]. $Ripk1^{-/-}Ripk3^{-/-}$ and $Ripk1^{-/-}Mlkl^{-/-}$ mice survive longer than $Ripk1^{-/-}$ mice (Figure 6.6) but succumb to an intestinal defect characterized by extensive apoptosis [3–5]. Intestinal-specific deletion of RIPK1 phenocopies the disease seen in $Ripk1^{-/-}Ripk3^{-/-}$ mice (Figure 6.6) [5,55]. The loss of RIPK1 kinase activity has no effect on mouse development (Figure 6.6) [6,7], indicating that RIPK1 negatively regulates RIPK3 and MLKL independently of its kinase activity. In support of this, RIPK1 can inhibit an artificially dimerizable form of RIPK3 that drives necroptosis [56]. A deficiency of RIPK1 increases sensitivity to necroptosis that is triggered by TNF and TLR [57].

6.4 Outcomes of Necroptosis: Good or Bad?

Necroptosis is a morphologically lytic form of regulated cell death resulting in the release of the contents of the cell (Figure 6.2). These factors may act to modulate inflammation

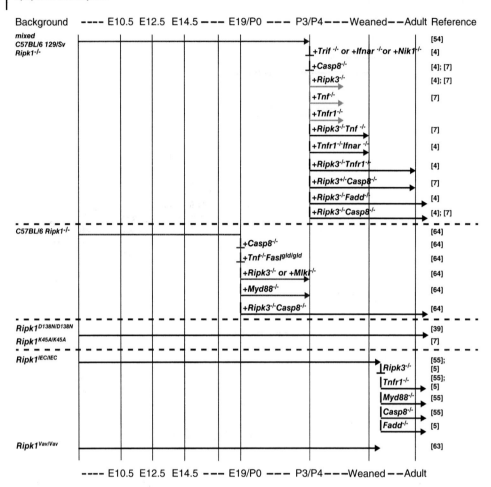

Figure 6.6 The role of receptor-interacting protein kinase 1 (RIPK1) in the necroptotic pathway during embryonic development and neonatal life. The loss of RIPK1 induces morbidity and mortality that is dependent on IFN, TRIF, MyD88, caspase 8, RIPK3, and MLKL signaling. The findings outlined demonstrate the importance of RIPK1 as a regulator of apoptosis and necroptosis.

during infection or chronic inflammatory disease. Therefore, depending on the context, necroptotic cell death-driven inflammation may help to combat infection or may delay resolution, causing tissue damage. Excessive necroptotic cell death has also been implicated in the pathogenesis of autoimmune and autoinflammatory diseases.

6.4.1 Embryonic Developmental Checkpoints are Regulated by Members of the Necroptotic and Apoptotic Cell-Death Family

One of the earliest suggestions for a non-apoptotic role of caspase 8 came from the analysis of $Casp8^{-/-}$ and $Fadd^{-/-}$ embryos. $Casp8^{-/-}$, $Fadd^{-/-}$, $Casp8^{DED/DED}$, and mutants lacking caspase 8 catalytic activity all die at embryonic day (E) 10.5–12.5 (Figure 6.7) [27,58–60]. TNFR1 signaling contributes to the lethality, because $Casp8^{-/-}Tnfr1^{-/-}$

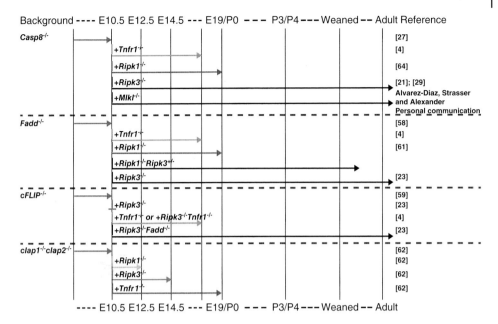

Figure 6.7 Negative regulation of the necroptosis pathway by caspase 8 during embryonic development. Caspase 8 and FADD are central negative regulators of RIPK3/MLKL-dependent necroptosis during embryonic development. Without caspase 8 and FADD, TNFR1 signaling induces necroptotic cell death in embryonic day (E) 10.5–12.5 embryos.

embryos survive longer than $Casp8^{-/-}$ embryos (Figure 6.7) [4]. Finally, $Casp8^{-/-}Ripk3^{-/-}$ and $Fadd^{-/-}Ripk3^{-/-}$ mice are viable (Figure 6.7), indicating that caspase 8 regulates RIPK3 activity at this embryonic developmental checkpoint [4,21,29,61]. Compound cIAP1 and cIAP2 deficiency is also lethal at E10.5, and is also rescued by loss of RIPK3, RIPK1, or TNFR1 (Figure 6.7), suggesting that cIAP1 and cIAP2 are involved in this developmental checkpoint [62]. c-FLIP can negatively regulate caspase 8, but deficiency in c-FLIP phenocopies caspase 8 deficiency (Figure 6.7), suggesting that a caspase 8/c-FLIP heterodimer and not a caspase 8 homodimer inhibits RIPK3 activity [22]. Caspase 8 enzymatic activity requires cleavage of the caspase 8 zymogen, but its role in embryonic development and the regulation of RIPK3 activity does not require cleavage [22].

In the absence of caspase 8, RIPK3 autophosphorylation (which can also be seen in the absence of RIPK1) promotes the formation of RIPK3 oligomers and the recruitment and phosphorylation of MLKL to drive necroptosis.

6.4.2 Necroptosis in the Hematopoietic System

RIPK3-dependent necroptotic cell death limits the self-renewal capacity of $Ripk1^{-/-}$ long-term hematopoietic stem cell (LT-HSC) in lethally irradiated recipient mice in a TNF-dependent manner [3,63]. This finding implicates non-apoptotic cell death as a key biological process restricting HSC "self-renewal," and could have clinical implications for improving the engraftment potential of HSC in transplantation settings or in patients with life-threatening infections.

6.4.3 Inflammatory Diseases

The hallmarks of inflammation include redness, heat, swelling, pain, and loss of function. Both pathogen- and host-derived molecules can be detected by components of the innate immune system, including nucleotide-binding domain and leucine-rich repeat-containing-proteins (NLR), TLRs, MAVS/RIG-I, and AIM2. Sensing of these "danger" or "damage" molecules leads to the production of cytokines and chemokines, which recruit innate and adaptive immune cells. A circumscribed response can be beneficial, but a systemic uncontrolled response will cause morbidity or even mortality.

A clear example of the inflammation that can be induced by necroptosis comes from the study of $Ripk1^{-/-}$ mice. These mice have sterile multi-organ inflammation, which is dependent on the key necroptotic proteins RIPK3 and MLKL (see earlier), and on MyD88, a protein that transduces signals downstream of TLRs and interleukin 1 (IL-1)-family cytokine receptors [3]. Disruption of the apoptotic pathway fails to prevent skin inflammation and lethality, suggesting that inflammation is driven in large part by necroptosis. Consistent with this, epidermis-specific RIPK1-deficient mice develop skin inflammation in a manner dependent on MLKL and RIPK3 (necroptosis), but not on FADD (apoptosis) (Table 6.1) [5]. Several "damage" signal molecules, including HMGB1, IL-33, IL-1α, and mitochondrial DNA, signal via MyD88-dependent receptors. However, only IL-1α and IL-33 were detected in the plasma of $Ripk1^{-/-}$ neonates – and not in $Ripk1^{-/-}Ripk3^{-/-}$ or $Ripk1^{-/-}Mlkl^{-/-}$ neonates [64]. IL-33 is constitutively expressed in healthy cells, but is released during necroptosis to trigger MyD88-dependent inflammation, thereby serving as a "damage" signal molecule. During apoptosis, IL-33 is inactivated by caspases 3 and 7 [65,66]. $Ripk1^{-/-}Tnfr1^{-/-}Trif^{-/-}$ survive longer than $Ripk1^{-/-}Tnfr1^{-/-}$ littermates (Figure 6.6), suggesting important roles for TLR3/TLR4 signaling in the pathology of mice lacking RIPK1. $Ripk1^{-/-}Tnfr1^{-/-}Ifnar^{-/-}$ knockout mice are partially protected compared to $Ripk1^{-/-}Tnfr1^{-/-}$ littermates (Figure 6.6) [4], further indicating important roles for type I IFN signaling.

The role of RIPK1 as a negative regulator of RIPK3 is supported by studies of RIPK1 kinase inhibition and a dimerizable form of RIPK3. The authors propose that the binding of Nec-1 to RIPK1 may act in a dominant negative fashion on RIPK3, preventing its oligomerization and activation [56,57]. Spontaneous RIPK3 oligomerization may also be prevented by the recruitment of RIPK1, caspase 8, and FADD, and by subsequent degradation by cIAP proteins.

6.4.4 Necroptosis in Sterile Tissue Damage

Necroptosis has been implicated in morbidity and mortality associated with heart attack, atherosclerosis, pancreatitis, and liver, retinal, and renal injury (as summarized in Table 6.1). However, the use of Nec-1 to investigate the role of necroptosis introduces significant caveats due to the possibility of indoleamine-2,3-dioxygenase (IDO) inhibition and its short *in vivo* half-life [67]. The use of $Ripk1$ kinase dead ($Ripk1^{D138N/D138N}$ and $Ripk1^{K45A/K45A}$) mice and $Ripk3^{-/-}$ and $Mlkl^{-/-}$ mice, all of which are viable (Figures 6.6 and 6.8), is currently a preferred alternative to help define the role of necroptosis *in vivo* [7,39,42,44,68].

Nec-1 and RIPK3 deficiency prevent cone cell death in models involving $Rd10^{-/-}$ mice [69], dsRNA-induced retinal degeneration [70], and retinitis pigmentosa [71]. Nec-1 affords protection in models of ischemic brain injury [38,72], controlled cortical impact (brain trauma) [73], Huntington's disease (HD) [74], myocardial infarction, and cardiac hypoxia [75], although all of these await confirmation with genetic models. Nec-1 and RIPK3 deficiency increases survival in an ischemia/reperfusion (I/R) model, but the lack of effect of MLKL deficiency suggests that morbidity may be independent of cell death [76–78]. In the skin, RIPK3 or MLKL deficiency prevents epidermal hyperplasia in mice lacking FADD [79], caspase 8 [80], or RIPK1 [3–5]. Interestingly, loss of RIPK1 kinase activity (using $Ripk1^{K45A/K45A}$ mice) prevents the cutaneous inflammatory disease in SHARPIN mutant mice [81], although this phenotype is driven by apoptosis rather than necroptosis [64,82]. A RIPK3 deficiency is protective against atherosclerosis in $Ldlr^{-/-}$ and $Apoe^{-/-}$ mice [83]. Cerulein-induced pancreatitis is alleviated in the absence of RIPK3 and MLKL, but exacerbated by Nec-1 [41,44,84]. Ethanol-induced liver injury can be reduced by RIPK3 deficiency; Nec-1 has no effect [85], whereas RIPK3 deficiency reduces inflammatory hepatocarcinogenesis but exacerbates jaundice and cholestasis [86]. Inhibition of RIPK1 kinase activity and RIPK3 deficiency protects oligodendrocytes from necroptosis, leading to reduced demyelination, motor dysfunction, and microglial activation in mouse models of experimental autoimmune/allergic encephalomyelitis (EAE) [87]. Herpesviruses have been implicated in the etiology of multiple sclerosis (MS), and the expression of a viral homolog of c-FLIPs (v-FLIPs) suggests that v-FLIPs-mediated caspase 8 inhibition and activation of necroptosis may contribute to MS-associated inflammation [88].

6.4.5 Necroptosis During Infection

One of the fundamental responses to infection is now recognized to be the induction of cell death. This response is conserved across kingdoms, playing key roles in organisms as

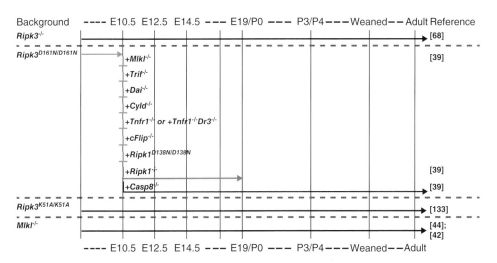

Figure 6.8 Receptor-interacting protein kinase 3 (RIPK3) and mixed-lineage kinase domain-like (MLKL) are activated during embryonic development when caspase 8 is inhibited. Inhibition of caspase 8 leads to arrest of development at E10.5–12.5 due to excessive RIPK3/MLKL-dependent cell death (see Figure 6.7). Inhibition of RIPK3 kinase activity by mutation of D161 on RIPK3 drives caspase 8-dependent apoptosis and embryonic lethality, also at E10.5–12.5, demonstrating the high plasticity of these pathways.

Table 6.1 Contribution of receptor-interacting protein kinase 1 (RIPK1), RIPK3, and mixed-lineage kinase domain-like (MLKL) to inflammation and tissue damage.

Tissue	Investigation	Necrostatin-1 (Ripk1 Inhibition)	Ripk3 Deletion	Mlkl Deletion	Comments
Heart	Myocardial infarction/cardiac hypoxia	Protects [75,89,90]			Nec-1+Z-VAD protects more than Nec-1 alone [90]
	Cardiac allograft		Improved graft survival [91]		HMGB1 release in graft tissue [91]
Skin	Epidermal Fadd deletion: skin inflammation		Prevents [79]		$Tnf^{-/-}$, $Myd88^{-/-}$, and $Tnfr1^{-/-}$ delays phenotype, HMGB1 release in skin [79] Loss of RIPK1 kinase activity ($Ripk1^{D138N/D138N}$) or epidermal Ripk1 deletion delays phenotype [5]
	Tamoxifen-inducible Casp8 deletion (CreER): skin inflammation		Protects [80]		
	$Ripk1^{-/-}$ mice: skin inflammation		Protects [3]	Protects [3]	Protection with $Myd88^{-/-}$ but not $Tnfr1^{-/-}$ Increased IL-33 expression
	Epidermal Ripk1 deletion: skin inflammation		Prevents [5]	Prevents	Tnfr1 and Trif deletion reduces inflammation
	$Sharpin^{cpdm/cpdm}$ dermatitis		Mild protection [64,82]	No change [64]	Tnf deletion prevents[a] [92] $Ripk1^{K45A/K45A}$ kinase dead cross prevents[a] [81] Casp8- and Fadd-dependent [64,82]
	Toxic epidermal necrolysis (TEN)				Increased RIPK3 and p-MLKL expression in skin of TEN patients [93]
Pancreas	Cerulein-induced pancreatitis	Worsens [94]	Protects [41]	Protects [44]	
Liver	Ethanol-induced liver injury	No change [85]	Protects [85]		Increased RIPK3 expression in ethanol-fed mice and human alcoholic liver disease patients [85,95]

Organ	Condition	Nec-1 effect	Ripk3−/− effect	Comments
	Fas-induced hepatitis		No change [29]	Phenotype driven by caspase 8
	LPS/GalN-injected mice: hepatitis		No change [29]	
	Liver parenchymal cell TAK1 deletion: inflammatory hepatocarcinogenesis		Worsens: promotes tumor growth [86]	RIPK3 augments cholestasis and jaundice
	Bile-duct ligation-induced liver injury		Protects against acute liver inflammation [96]	RIPK3 deletion worsens cholestasis. Increased RIPK3 and p-MLKL in primary biliary cholangitis human liver samples
	αGalCer immune-mediated liver injury	No change [97]	No change	RIPK1 knockdown significantly worsened injury, but no exacerbation in $Ripk1^{D138N/D138N}$ mice – points to kinase-independent protective role of RIPK1
	Concanavalin A-induced liver injury	Protects [98,99]	No change [80]	
	Non-alcoholic steatohepatitis (NASH) mouse model and chronic liver disease human samples		Protects [100]	Increased RIPK3 expression in NASH and alcoholic steatohepatitis patient livers. Increased p-MLKL in NASH patient livers
Eye	dsRNA-induced retinal degeneration		Protects [70]	HMGB1 release [70]
	Retinitis pigmentosa (*Rd10* mice): cone cell death	Protects [101]	Protects [101]	*Ripk3*−/− mice not protected from rod degeneration
	Retinitis pigmentosa (*Irbp*−/− mice)	Both Nec-1 and Nec-1s protect [71]		Increased TNF, RIPK1, and RIPK3 expression in *Irbp*−/− retinas
	Retinal detachment injury	No change [102]	Protects [102]	Increased RIPK3 expression following retinal detachment. Protection with Nec-1+Z-VAD
Ear	Spiral ganglion neuron injury	Minimal effect [103]		Increased RIPK3 expression Nec-1+Z-VAD protects more than Nec-1 or Z-VAD alone [103]
Kidney	Renal ischemia reperfusion injury	Protects [84]	Protects [78,104]	Increased HMGB1 in wild-type but not *Ripk3*−/− kidneys [104]. *Cypd*−/− mice also have improved survival, *Cypd*−/− *Ripk3*−/− survive [78]

(*continued*)

Table 6.1 (Continued)

Tissue	Investigation	Necrostatin-1 (Ripk1 Inhibition)	Ripk3 Deletion	Mlkl Deletion	Comments
	Renal transplantation		Improved graft survival [104]		HMGB1 reduced in $Ripk3^{-/-}$ kidneys [104]
	Subtotal nephrectomy rat model of chronic kidney disease	Protects [105]			Increased RIPK3 expression in subtotal nephrectomy kidney tissue Nec-1+Z-VAD protects more than Nec-1 alone
Brain	Ischemic brain injury	Protects [38,72]			
	Trauma (controlled cortical impact)	Protects [73]			
	Huntington's disease model (R6/2 transgenic mouse)	Protects [74]			
	Oligodendrocyte cuprizone-induced inflammation and degeneration, and experimental autoimmune encephalomyelitis (EAE) multiple sclerosis models	Protects [87] (Necrostatin 7N-1)	Protects [87]		$Ripk3$ deletion and 7N-1 treatment protect against demyelination [87]
	$Optn^{-/-}$ (Optineurin) mice: dysmyelination and axonal degeneration amyotrophic lateral sclerosis model	Protects (Nec-1s) [106]	Protects		$Optn^{-/-}Ripk1^{D138N/D138N}$ mice also protected Increased expression of RIPK1, p-RIPK1, RIPK3, MLKL, and p-MLKL in human ALS spinal cord samples
Intestine	Intestinal epithelial cell (IEC) $Fadd$ deletion: intestinal damage		Prevents [28]		Tnf deletion attenuates and $Myd88$ deletion prevents colitis, but not small-intestine Paneth cell loss and enteritis Increased intestinal RIPK3 expression in $Fadd^{-/-}$ IECs [28] Crossing to $Ripk1^{-/-}$ or $Ripk1^{D138N/D138N}$ mice does not prevent phenotype [5]
	Tamoxifen-inducible $Casp8$ deletion (CreER): intestinal damage		Prevents [80]		
	$Ripk1^{-/-}$ mice: intestinal phenotype		No change [3,4]	No change [3]	Protection with $Tnfr1^{-/-}$ but not $Myd88^{-/-}$ Casp8-dependent [3]

	IEC *Ripk1* deletion: intestinal damage	No change [5,55]		Combined loss of *Ripk3* and intestinal epithelial *Fadd* prevents *Myd88* deletion or germ-free environment does not prevent phenotype [5] *Casp8* deletion prevents, deletion of *Tnfr1* or *Myd88* or antibiotic treatment protects [55]
	Inducible IEC *cFlip* deletion: intestinal damage and lethality	No change [107]		Viability of *Fadd*$^{-/-}$ *Flip*$^{-/-}$ *Ripk3*$^{-/-}$ mice indicates apoptosis drives pathology in mice lacking intestinal cFLIP and RIPK3 [23]
	Dextran sodium sulfate (DSS)-induced colitis	Worsens [108]		RIPK3 has beneficial role in damaged intestine repair [108]
	Crohn's disease (CD) and ulcerative colitis (UC) patients			Increased intestinal MLKL/RIPK3 expression [109] Paneth cell necroptosis in CD patients [110] TNF inhibitors effective in CD and UC [111,112]
	Intestinal ischemia reperfusion injury	Protects [113]		Nec-1+Z-VAD protects more than Nec-1 alone HMGB1 release in oxygen-glucose-deprived intestinal cell line *in vitro*
Vascular	*Ldlr*$^{-/-}$ and *Apoe*$^{-/-}$ mice: atherosclerosis	Protects [83]		*Ripk3* deletion protected against advanced atherosclerosis in *Ldlr*$^{-/-}$ mice, but no benefit at earlier time point
Systemic	TNF-injected mice: systemic inflammatory response syndrome (hypothermia and lethality)	Protects[b] [114,115]/ worsens [94] Nec-1s also protects [115]	Protects [6,39,94,114]	Associated with increased plasma mtDNA (possible DAMP) that was prevented by *Ripk3* deletion *Ripk3* deletion or Nec-1 treatment did not prevent intestinal damage [114] *Ripk1*$^{D138N/D138N}$ have enhanced survival and reduced hypothermia [6,39]

(*continued*)

Table 6.1 (Continued)

Tissue	Investigation	Necrostatin-1 (Ripk1 Inhibition)	Ripk3 Deletion	Mlkl Deletion	Comments
	Caecal ligation puncture: sepsis	Worsens [116]	Protects [114,117]/no change [44]	No change [44]	
	LPS-injected mice: inflammation		Similar cytokine expression [44,68,80]	Similar cytokine expression [44]	
	LPS/Z-VAD-injected mice: inflammation		Reduced cytokine expression [118]		
	TNF/Z-VAD-injected mice: acute shock (hypothermia and lethality)	Protects [114]/worsens [94]	Protects [6,94,114]		$Ripk1^{D138N/D138N}$ protected [6]
	$Ripk1^{-/-}$ mice: inflammatory phenotype		Protects [3]	Protects [3]	Protection with $Myd88^{-/-}$ but not $Tnfr1^{-/-}$ Plasma IL-1α and IL-33 [3]
Obesity	Obese mouse model (choline-deficient high-fat diet)		Worsens white adipose tissue inflammation [119]		RIPK3 expression increased in white adipose tissue Ripk3 deletion induced white adipose tissue apoptosis and inflammation with impaired glucose tolerance; $Ripk3^{-/-}$ $Casp8^{-/-}$ mice were protected from this

a) *Tnf* deletion does not prevent the systemic *Sharpin*$^{cpdm/cpdm}$ phenotype, including the development of splenomegaly or loss of Peyer's patches [92], whereas *Sharpin*$^{cpdm/cpdm}$ $Ripk1^{K45A/K45A}$ RIPK1 kinase dead mice do not have the multi-organ pathology seen in *Sharpin*$^{cpdm/cpdm}$ mice [81].

b) Nec-1, Nec-1s (more specific than Nec-1), and Nec-1i ("inactive") are all protective at high doses, but at low doses Nec-1 and Nec-1i (but not Nec-1s) accelerate lethality. This may be due to off-target effects of Nec-1 and Nec-1i, i.e., targeting indoleamine 2,3-dioxygenase (IDO) [115]. This dose-dependent effect may explain why Nec-1 increases lethality in the model, as reported by Linkermann et al. [94]. Nec-1, necrostatin-1 (RIPK1 kinase inhibitor); Z-VADn Z-VAD-fmk (pan-caspase inhibitor); p-MLKL/p-RIPK1, phosphorylated MLKL/RIPK1.

diverse as plants and mammals. Cell death can reduce pathogen burden by eliminating infected cells. It does not require input from innate or adaptive immune cells, but can be influenced by the presence of inflammatory cytokines (see earlier). Pathogens that interfere with host cell-death pathways can inadvertently trigger alternative inflammatory forms of regulated cell death. These regulated pathways, which can be activated systemically during severe infection, appear to play key roles in diverse disease processes.

6.4.5.1 Viral Infection

Host cells activate cell death to limit viral replication, often via RIPK1/caspase 8-dependent pathways. Vaccinia virus encodes proteins that can inhibit apoptosis [120,121] and pyroptosis [122], but these attempts to hijack the cell result in the induction of RIPK1/RIPK3-dependent necroptosis [6,122–124]. Murine cytomegalovirus (MCMV) encodes two suppressors of this machinery: (i) a M36-encoded caspase 8 inhibitor; and (ii) a RHIM inhibitory protein (M45), which blocks DAI/RIPK3 interactions and thereby RIPK3-dependent necroptosis, independently of RIPK1. Human CMV, which lacks a M45 homolog, adopts a different strategy to inhibit necroptosis downstream of RIPK3 and MLKL phosphorylation [50]. Herpes simplex virus 1 (HSV-1) protein ICP6 and HSV-2 protein ICP10 interact with RIPK1 and RIPK3 via an RHIM interaction to induce RIPK3/MLKL-dependent necroptosis in mouse cells, leading to a host antiviral response [125,126]. In human cells, as opposed to in mouse cells, these HSV proteins have been found to inhibit TNF-induced necroptosis, indicating the importance of host factors in response to virus infection [125–128].

6.4.5.2 Bacterial Infection

Nec-1 can block the kinase activity of RIPK1 to interfere with killing induced by caspase inhibitors, IAP antagonists, or pathogens such as *Yersinia pestis*. But what pathways are downstream of RIPK1 kinase activity, and are they restricted to RIPK3/MLKL-dependent necroptosis? Some bacteria have evolved to inhibit RIPK1 signaling, including *Porphyromonas gingivalis*, which cleaves RIPK1 via its lysine-specific (Kgp) protease [129]. The YopJ protein of *Y. pestis* (also known as YopP in *Yersinia enterocolitica*) is an acetyltransferase and de-ubiquitinase that can inhibit NF-κB and mitogen-activated protein kinase (MAPK). These proteins prevent pro-survival signaling and c-FLIP production, leading to caspase 8 activation and cell death [130,131]. RIPK1 inhibition or deficiency also prevents caspase 8 processing and activity, suggesting that RIPK1 is required for caspase 8-triggered cell death [130,131]. In these settings, the kinase activity of RIPK1 may change the conformation of RIPK1 to expose the RHIM and death domain (DD), enabling interaction with FADD and RIPK3. It is unclear whether the negative regulatory roles of RIPK1 in necroptosis and the heightened sensitivity of RIPK1-deficient cells to TNF alter the response of cells to *Y. pestis*. It is also unclear what is the role of caspase 1 in the regulation of cell death, and it seems possible that both pathways can contribute to it (see later). The outcome may be cell type- and pathogen dose-specific. Where RIPK1 sits in this pathway during *Y. pestis* infection remains to be elucidated: is it upstream of caspase 8, or can the inhibition of one pathway of cell death enable activation of an alternative one?

6.5 Crosstalk with Other Forms of Cell Death (Apoptosis and Pyroptosis)

An emerging theme in the field of inflammatory cell death is that the apoptotic and non-apoptotic cell-death pathways communicate extensively. When cIAPs are depleted with SMAC mimetics, RIPK3 activation triggers caspase 8 activation, as well as NLRP3 inflammasome and caspase 1 activation, leading to reactive oxygen species (ROS) production and IL-1β secretion, independently of the kinase activity of RIPK3 [132–135]. Similarly, caspase 8-deficient bone marrow-derived dendritic cells have increased RIPK3-dependent NLRP3 inflammasome activity [136]. $Fadd^{-/-}Ripk3^{-/-}$ and $Casp8^{-/-}Ripk3^{-/-}$ BMDCs display defective NLRP3 inflammasome activation, confirming the key role of RIPK3 in this setting of caspase 8 deficiency. It should also be appreciated that caspase 8 can directly process pro-IL-1β downstream of FAS/CD95 in macrophages and neutrophils, indicating it plays a multitude of roles in apoptosis, necroptosis, and the inflammasomes.

A20, a de-ubiquitinating enzyme, has dual functions in the negative regulation of the NLRP3/caspase 1 machinery [137] and the necroptotic machinery. A20 is an NF-κB-inducible gene that can remove lysine 63-linked ubiquitin chains and add K48-linked ubiquitin chains to RIPK1 [15]. A20 also prevents spontaneous NLRP3 activity via a RIPK3-dependent process involving K63-linked ubiquitination of Lys133 on pro-IL-1β in a complex comprising caspase 1, RIPK1, RIPK3, pro-IL-1β, and caspase 8. Thus, A20 serves as a critical negative regulator of cytokine production by controlling the activation of this key inflammasome complex [137].

RIPK1-dependent, RIPK3-independent forms of death also exist, as exemplified by the YopJ protein of *Y. pestis* [117,118] and by mice with mutations in SHARPIN, a component of complex I. SHARPIN mutant mice ($Sharpin^{cpdm/cpdm}$) develop a disease characterized by excessive caspase 8-dependent cell death [64,82]. $Sharpin^{cpdm/cpdm} Ripk1^{K45A/K45A}$ mice, but not $Sharpin^{cpdm/cpdm} Mlkl^{-/-}$ mice, are protected from disease (Table 6.1), indicating a central role for RIPK1 kinase activity independently of necroptosis.

As previously discussed, RIPK3 activation can drive caspase 8-dependent apoptosis, MLKL-dependent necroptosis, or caspase 1-dependent pyroptosis. Mutations affecting the kinase domain of RIPK3 unexpectedly induce caspase 8-dependent lethality (Figure 6.8), but not MLKL-mediated cell death. Cook and colleagues propose that the availability of substrates – namely caspase 8, MLKL, and FADD – determines the outcome to dimerization of RIPK1 and RIPK3 [138]. A RIPK3 construct that can be induced to dimerize was used to show that the kinase domain of RIPK3 drives MLKL-mediated necroptosis in the absence of RIPK1, caspase 8, and FADD. In contrast, in the absence of MLKL, dimerized RIPK3 induces caspase 8-dependent apoptosis and cleavage of caspase 3 and poly ADP ribose polymerase (PARP). This process is enhanced by RIPK1, and occurs independently of RIPK3 kinase function [138]. It is therefore possible that more than one pathway might be activated in a cell to contribute to the loss of cell viability [138].

6.6 Targeting Necroptosis (RIPK1/RIPK3/MLKL Inhibitors)

Inhibition of necroptosis may have therapeutic value in many contexts. Pharmacological or genetic manipulation of RIPK1 and RIPK3 alters the morbidity and mortality of cerulein-induced pancreatitis, retinal degeneration, atherosclerosis in *Apoe* and *Ldlr* mutant mice, TNF-induced inflammation in mice, I/R injury of the kidneys, steatohepatitis, myocardial infarction, and hepatic injury induced by ethanol (Table 6.1) [28,41,70,77,83–85,104,110]. The clinical relevance of these findings is only now being tested.

The lethality of RIPK1-deficient mice, together with the normal development of RIPK3-deficient mice (Figures 6.6 and 6.8), suggested that RIPK3 might be a more suitable target for drug development. However, $Ripk3^{D161N/D161N}$ mice display increased caspase 8 activity, leading to apoptosis (Figure 6.8). In agreement with this, compounds targeting D161 on RIPK3 induce high levels of apoptosis via RHIM interactions with RIPK1 and activation of caspase 8 (Figure 6.3, gray circle). In contrast to RIPK3-D161 N, the K51A, D143 N, and D161G mutations on RIPK3 do not induce a similar response, and have instead been shown to inhibit necroptosis in response to TLR3, TLR4, DAI, TNFR1, and IFNβ (Figure 6.8) [133]. Therefore, RIPK3 antagonists, which inhibit necroptosis while promoting induction of apoptosis, may find a more useful application in the setting of oncology.

6.7 Conclusion

Necroptosis has emerged from the shadow of apoptosis as a key player in regulated cell death. As discussed in this chapter, the RIPK3- and MLKL-dependent caspase-independent form of cell death known as necroptosis is kept under strict control by several negative regulators, including caspase 8 and RIPK1. Loss of negative regulation by insidious biochemical processes during disease, or by pathogen-derived molecules, triggers the activation of this inflammatory form of cell death, which can alter immunological responses and cause tissue damage. As this is an emerging field, numerous questions remain. What are the substrates for RIPK1? Is MLKL the final player in this pathway to cell death? If so, how does it function? What are the specific danger molecules that are released during necroptosis? Incorporating future findings in this novel area of cell death will result in new strategies to interfere in acute and chronic inflammatory disease, and alleviate organ damage and immune suppression associated with life-threatening systemic infection.

References

1 Holler N, Zaru R, Micheau O, Thome M, Attinger A, Valitutti S, et al. Fas triggers an alternative, caspase-8-independent cell death pathway using the kinase RIP as effector molecule. *Nat Immunol* 2000;**1**(6):489–95.

2 Carswell EA, Old LJ, Kassel RL, Green S, Fiore N, Williamson B. An endotoxin-induced serum factor that causes necrosis of tumors. *Proc Natl Acad Sci USA* 1975;**72**(9):3666–70.

3 Rickard JA, O'Donnell JA, Evans JM, Lalaoui N, Poh AR, Rogers T, et al. RIPK1 regulates RIPK3-MLKL driven systemic inflammation and emergency hematopoiesis. *Cell* 2014;**157**(5):1175–88.

4 Dillon CP, Weinlich R, Rodriguez DA, Cripps JG, Quarato G, Gurung P, et al. RIPK1 blocks early postnatal lethality mediated by caspase-8 and RIPK3. *Cell* 2014;**157**(5):1189–202.

5 Dannappel M, Vlantis K, Kumari S, Polykratis A, Kim C, Wachsmuth L, et al. RIPK1 maintains epithelial homeostasis by inhibiting apoptosis and necroptosis. *Nature* 2014;**513**(7516):90–4.
6 Polykratis A, Hermance N, Zelic M, Roderick J, Kim C, Van TM, et al. Cutting edge: RIPK1 Kinase inactive mice are viable and protected from TNF-induced necroptosis in vivo. *J Immunol* 2014;**193**(4):1539–43.
7 Kaiser WJ, Daley-Bauer LP, Thapa RJ, Mandal P, Berger SB, Huang C, et al. RIP1 suppresses innate immune necrotic as well as apoptotic cell death during mammalian parturition. *Proc Natl Acad Sci USA* 2014;**111**(21):7753–8.
8 O'Donnell MA, Legarda-Addison D, Skountzos P, Yeh WC, Ting AT. Ubiquitination of RIP1 regulates an NF-kappaB-independent cell-death switch in TNF signaling. *Curr Biol* 2007;**17**(5):418–24.
9 Stanger BZ, Leder P, Lee TH, Kim E, Seed B. RIP: a novel protein containing a death domain that interacts with Fas/APO-1 (CD95) in yeast and causes cell death. *Cell* 1995;**81**(4):513–23.
10 Lavrik IN, Mock T, Golks A, Hoffmann JC, Baumann S, Krammer PH. CD95 stimulation results in the formation of a novel death effector domain protein-containing complex. *J Biol Chem* 2008;**283**(39):26 401–8.
11 Feoktistova M, Geserick P, Kellert B, Dimitrova DP, Langlais C, Hupe M, et al. cIAPs block Ripoptosome formation, a RIP1/caspase-8 containing intracellular cell death complex differentially regulated by cFLIP isoforms. *Mol Cell* 2011;**43**(3):449–63.
12 Tenev T, Bianchi K, Darding M, Broemer M, Langlais C, Wallberg F, et al. The ripoptosome, a signaling platform that assembles in response to genotoxic stress and loss of IAPs. *Mol Cell* 2011;**43**(3):432–48.
13 Haas TL, Emmerich CH, Gerlach B, Schmukle AC, Cordier SM, Rieser E, et al. Recruitment of the linear ubiquitin chain assembly complex stabilizes the TNF-R1 signaling complex and is required for TNF-mediated gene induction. *Mol Cell* 2009;**36**(5):831–44.
14 Tokunaga F, Sakata S, Saeki Y, Satomi Y, Kirisako T, Kamei K, et al. Involvement of linear polyubiquitylation of NEMO in NF-kappaB activation. *Nat Cell Biol* 2009;**11**(2):123–32.
15 Wertz IE, O'Rourke KM, Zhou H, Eby M, Aravind L, Seshagiri S, et al. De-ubiquitination and ubiquitin ligase domains of A20 downregulate NF-kappaB signalling. *Nature* 2004;**430**(7000):694–9.
16 Bertrand MJ, Milutinovic S, Dickson KM, Ho WC, Boudreault A, Durkin J, et al. cIAP1 and cIAP2 facilitate cancer cell survival by functioning as E3 ligases that promote RIP1 ubiquitination. *Mol Cell* 2008;**30**(6): 689–700.
17 Dynek JN, Goncharov T, Dueber EC, Fedorova AV, Izrael-Tomasevic A, Phu L, et al. c-IAP1 and UbcH5 promote K11-linked polyubiquitination of RIP1 in TNF signalling. *EMBO J* 2010;**29**(24):4198–209.
18 Moquin DM, McQuade T, Chan FK. CYLD deubiquitinates RIP1 in the TNFalpha-induced necrosome to facilitate kinase activation and programmed necrosis. *PLoS ONE* 2013;**8**(10):e76841.
19 Wong WW, Gentle IE, Nachbur U, Anderton H, Vaux DL, Silke J. RIPK1 is not essential for TNFR1-induced activation of NF-kappaB. *Cell Death Differ* 2010;**17**(3):482–7.

20 Geserick P, Hupe M, Moulin M, Wong WW, Feoktistova M, Kellert B, et al. Cellular IAPs inhibit a cryptic CD95-induced cell death by limiting RIP1 kinase recruitment. *J Cell Biol* 2009;**187**(7):1037–54.
21 Oberst A, Dillon CP, Weinlich R, McCormick LL, Fitzgerald P, Pop C, et al. Catalytic activity of the caspase-8-FLIP(L) complex inhibits RIPK3-dependent necrosis. *Nature* 2011;**471**(7338):363–7.
22 Pop C, Oberst A, Drag M, Van Raam BJ, Riedl SJ, Green DR, Salvesen GS. FLIP(L) induces caspase 8 activity in the absence of interdomain caspase 8 cleavage and alters substrate specificity. *Biochem J* 2011;**433**(3):447–57.
23 Dillon CP, Oberst A, Weinlich R, Janke LJ, Kang TB, Ben-Moshe T, et al. Survival function of the FADD-CASPASE-8-cFLIP(L) complex. *Cell Rep* 2012;**1**(5):401–7.
24 Garrison JB, Correa RG, Gerlic M, Yip KW, Krieg A, Tamble CM, et al. ARTS and Siah collaborate in a pathway for XIAP degradation. *Mol Cell* 2011;**41**(1):107–16.
25 Zaman MM, Nomura T, Takagi T, Okamura T, Jin W, Shinagawa T, et al. Ubiquitination-deubiquitination by the TRIM27-USP7 complex regulates tumor necrosis factor alpha-induced apoptosis. *Mol Cell Biol* 2013;**33**(24):4971–84.
26 Zaman MM, Shinagawa T, Ishii S. Trim27-deficient mice are susceptible to streptozotocin-induced diabetes. *FEBS Open Bio* 2013;**4**:60–4.
27 Varfolomeev EE, Schuchmann M, Luria V, Chiannikulchai N, Beckmann JS, Mett IL, et al. Targeted disruption of the mouse caspase 8 gene ablates cell death induction by the TNF receptors, Fas/Apo1, and DR3 and is lethal prenatally. *Immunity* 1998;**9**(2):267–76.
28 Welz PS, Wullaert A, Vlantis K, Kondylis V, Fernández-Majada V, Ermolaeva M, et al. FADD prevents RIP3-mediated epithelial cell necrosis and chronic intestinal inflammation. *Nature* 2011;**477**(7364):330–4.
29 Kaiser WJ, Upton JW, Long AB, Livingston-Rosanoff D, Dalye-Bauer LP, Hakem R, et al. RIP3 mediates the embryonic lethality of caspase-8-deficient mice. *Nature* 2011;**471**(7338):368–72.
30 Chen W, Wu J, Li L, Zhang Z, Ren J, Liang Y, et al. Ppm1b negatively regulates necroptosis through dephosphorylating Rip3. *Nat Cell Biol* 2015;**17**(4):434–44.
31 Kaiser WJ, Sridharan H, Huang C, Mandal P, Upton JW, Gough PJ, et al. Toll-like receptor 3-mediated necrosis via TRIF, RIP3, and MLKL. *J Biol Chem* 2013;**288**(43):31268–79.
32 Thapa RJ, Basagoudanavar SH, Nogusa S, Irrinki K, Mallilankaraman K, Slifker MJ, et al. NF-kappaB protects cells from gamma interferon-induced RIP1-dependent necroptosis. *Mol Cell Biol* 2011;**31**(14):2934–46.
33 Robinson N, McComb S, Mulligan R, Dudani R, Krishnan L, Sad S. Type I interferon induces necroptosis in macrophages during infection with *Salmonella enterica* serovar *Typhimurium*. *Nat Immunol* 2012;**13**(10):954–62.
34 Thapa RJ, Nogusa S, Chen P, Maki JL, Lerro A, Andrake M, et al. Interferon-induced RIP1/RIP3-mediated necrosis requires PKR and is licensed by FADD and caspases. *Proc Natl Acad Sci USA* 2013;**110**(33):E3109–18.
35 McComb S, Cessford E, Alturki NA, Joseph J, Shutinoski B, Startek JB, et al. Type-I interferon signaling through ISGF3 complex is required for sustained Rip3 activation and necroptosis in macrophages. *Proc Natl Acad Sci USA* 2014;**111**(31):E3206–13.

36 Rajput A, Kovalenko A, Bogdanov K, Yang SH, Kang TB, Kim JC, et al. RIG-I RNA helicase activation of IRF3 transcription factor is negatively regulated by caspase-8-mediated cleavage of the RIP1 protein. *Immunity* 2011;**34**(3):340–51.

37 Upton JW, Kaiser WJ, Mocarski ES. DAI/ZBP1/DLM-1 complexes with RIP3 to mediate virus-induced programmed necrosis that is targeted by murine cytomegalovirus vIRA. *Cell Host Microbe* 2012;**11**(3):290–7.

38 Degterev A, Huang Z, Boyce M, Li Y, Jagtap P, Mizushima N, et al. Chemical inhibitor of nonapoptotic cell death with therapeutic potential for ischemic brain injury. *Nat Chem Biol* 2005;**1**(2):112–19.

39 Newton K, Dugger DL, Wickliffe KE, Kapoor N, de Almagro MC, Vucic D, et al. Activity of protein kinase RIPK3 determines whether cells die by necroptosis or apoptosis. *Science* 2014;**343**(6177):1357–60.

40 Cho YS, Challa S, Moquin D, Genga R, Ray TD, Guildford M, Chan FK. Phosphorylation-driven assembly of the RIP1-RIP3 complex regulates programmed necrosis and virus-induced inflammation. *Cell* 2009;**137**(6):1112–23.

41 He S, Wang L, Miao L, Wang T, Du F, Zhao L, Wang X. Receptor interacting protein kinase-3 determines cellular necrotic response to TNF-alpha. *Cell* 2009;**137**(6):1100–11.

42 Murphy JM, Czabotar PE, Hildebrand JM, Lucet IS, Zhang JG, Alvarez-Diaz S, et al. The pseudokinase MLKL mediates necroptosis via a molecular switch mechanism. *Immunity* 2013;**39**(3):443–53.

43 Sun L, Wang H, Wang Z, He S, Chen S, Liao D, et al. Mixed lineage kinase domain-like protein mediates necrosis signaling downstream of RIP3 kinase. *Cell* 2012;**148**(1–2):213–27.

44 Wu J, Huang Z, Ren J, Zhang Z, He P, Li Y, et al. Mlkl knockout mice demonstrate the indispensable role of Mlkl in necroptosis. *Cell Res* 2013;**23**(8):994–1006.

45 Hildebrand JM, Tanzer MC, Lucet IS, Young SN, Spall SK, Sharma P, et al. Activation of the pseudokinase MLKL unleashes the four-helix bundle domain to induce membrane localization and necroptotic cell death. *Proc Natl Acad Sci USA* 2014;**111**(42):15 072–7.

46 Dondelinger Y, Declercq W, Montessuit S, Roelandt R, Goncalves A, Bruggeman I, et al. MLKL compromises plasma membrane integrity by binding to phosphatidylinositol phosphates. *Cell Rep* 2014;**7**(4):971–81.

47 Cai Z, Jitkaew S, Zhao J, Chiang HC, Choksi S, Liu J, et al. Plasma membrane translocation of trimerized MLKL protein is required for TNF-induced necroptosis. *Nat Cell Biol* 2014;**16**(1):55–65.

48 Wang Z, Jiang H, Chen S, Du F, Wang X. The mitochondrial phosphatase PGAM5 functions at the convergence point of multiple necrotic death pathways. *Cell* 2012;**148**(1–2):228–43.

49 Moujalled DM, Cook WD, Murphy JM, Vaux DL. Necroptosis induced by RIPK3 requires MLKL but not Drp1. *Cell Death Dis* 2014;**5**:e1086.

50 Omoto S, Guo H, Talekar GR, Roback L, Kaiser WJ, Mocarski E. Suppression of RIP3-dependent Necroptosis by Human Cytomegalovirus. *J Biol Chem* 2015;**290**(18):11 635–48.

51 Gong YN, Guy C, Olauson H, Becker JU, Yang M, Fitzgerald P, Linkermann A, Green DR. ESCRT-III acts downstream of MLKL to regulate necroptotic cell death and its consequences. *Cell* 2017;**169**(2):286–300.

52 Yoon S, Kovalenko A, Bogdanov K, Wallach D. MLKL, the protein that mediates necroptosis, also regulates endosomal trafficking and extracellular vesicle generation. *Immunity* 2017;**47**(1):51–65.

53 Zargarian S, Shlomovitz I, Erlich Z, Hourizadeh A, Ofir-Birin Y, Croker BA, Regev-Rudzki N, Edry-Botzer L, Gerlic M. Phosphatidylserine externalization, "necroptotic bodies" release, and phagocytosis during necroptosis. *PLoS Biol* 2017;**15**(6):e2002711.

54 Kelliher MA, Grimm S, Ishida Y, Kuo F, Stanger BZ, Leder P. The death domain kinase RIP mediates the TNF-induced NF-kappaB signal. *Immunity* 1998;**8**(3):297–303.

55 Takahashi N, Vereecke L, Bertrand MJ, Duprez L, Berger SB, Divert T, et al. RIPK1 ensures intestinal homeostasis by protecting the epithelium against apoptosis. *Nature* 2014;**513**(7516):95–9.

56 Orozco S, Yatim N, Werner MR, Tran H, Gunja SY, Tait SW, et al. RIPK1 both positively and negatively regulates RIPK3 oligomerization and necroptosis. *Cell Death Differ* 2014;**21**(10):1511–21.

57 Kearney CJ, Cullen SP, Clancy D, Martin SJ. RIPK1 can function as an inhibitor rather than an initiator of RIPK3-dependent necroptosis. *FEBS J* 2014;**281**(21):4921–34.

58 Yeh WC, de la Pompa JL, McCurrach ME, Shu HB, Elia AJ, Shahinian A, et al. FADD: essential for embryo development and signaling from some, but not all, inducers of apoptosis. *Science* 1998;**279**(5358):1954–8.

59 Yeh WC, Itie A, Elia AJ, Ng M, Shu HB, Wakeham A, et al. Requirement for Casper (c-FLIP) in regulation of death receptor-induced apoptosis and embryonic development. *Immunity* 2000;**12**(6):633–42.

60 Sakamaki K, Tsukumo S, Yonehara S. Molecular cloning and characterization of mouse caspase-8. *Eur J Biochem/FEBS* 1998;**253**(2):399–405.

61 Zhang H, Zhou X, McQuade T, Li J, Chan FK, Zhang J. Functional complementation between FADD and RIP1 in embryos and lymphocytes. *Nature* 2011;**471**(7338):373–6.

62 Moulin M, Anderton H, Voss AK, Thomas T, Wong WW, Bankovacki A, et al. IAPs limit activation of RIP kinases by TNF receptor 1 during development. *EMBO J* 2012;**31**(7):1679–91.

63 Roderick JE, Hermance N, Zelic M, Simmons MJ, Polykratis A, Pasparakis M, Kelliher MA. Hematopoietic RIPK1 deficiency results in bone marrow failure caused by apoptosis and RIPK3-mediated necroptosis. *Proc Natl Acad Sci USA* 2014;**111**(40):14 436–41.

64 Rickard JA, Anderton H, Etemadi N, Nachbur U, Darding M, Peltzer N, et al. TNFR1-dependent cell death drives inflammation in Sharpin-deficient mice. *eLife* 2014;**3**: doi: 10.7554/eLife.03464

65 Luthi AU, Cullen SP, McNeela EA, Duriez PJ, Afonina IS, Sheridan C, et al. Suppression of interleukin-33 bioactivity through proteolysis by apoptotic caspases. *Immunity* 2009;**31**(1):84–98.

66 Cayrol C, Girard JP. The IL-1-like cytokine IL-33 is inactivated after maturation by caspase-1 *Proc Natl Acad Sci USA* 2009;**106**(22):9021–6.

67 Vandenabeele P, Grootjans S, Callewaert N, Takahashi N. Necrostatin-1 blocks both RIPK1 and IDO: consequences for the study of cell death in experimental disease models. *Cell Death Differ* 2013;**20**(2):185–7.

68 Newton K, Sun X, Dixit VM. Kinase RIP3 is dispensable for normal NF-kappa Bs, signaling by the B-cell and T-cell receptors, tumor necrosis factor receptor 1, and Toll-like receptors 2 and 4. *Mol Cell Biol* 2004;**24**(4):1464–9.

69. Murakami T, Ockinger J, Yu J, Byles V, McColl A, Hofer AM, Horng T. Critical role for calcium mobilization in activation of the NLRP3 inflammasome. *Proc Natl Acad Sci USA* 2012;**109**(28):11 282–7.
70. Murakami Y, Matsumoto H, Roh M, Giani A, Kataoka K, Morizane Y, et al. Programmed necrosis, not apoptosis, is a key mediator of cell loss and DAMP-mediated inflammation in dsRNA-induced retinal degeneration. *Cell Death Differ* 2014;**21**(2):270–7.
71. Sato K, Li S, Gordon WC, He J, Liou GI, Hill JM, et al. Receptor interacting protein kinase-mediated necrosis contributes to cone and rod photoreceptor degeneration in the retina lacking interphotoreceptor retinoid-binding protein. *J Neurosci* 2013;**33**(44):17 458–68.
72. Northington FJ, Chavez-Valdez R, Graham EM, Razdan S, Gauda EB, Martin LJ. Necrostatin decreases oxidative damage, inflammation, and injury after neonatal HI. *J Cereb Blood Flow Metab* 2011;**31**(1):178–89.
73. You Z, Savitz SI, Yang J, Degterev A, Yuan J, Cung GD, et al. Necrostatin-1 reduces histopathology and improves functional outcome after controlled cortical impact in mice. *J Cereb Blood Flow Metab* 2008;**28**(9):1564–73.
74. Zhu S, Zhang Y, Bai G, Li H. Necrostatin-1 ameliorates symptoms in R6/2 transgenic mouse model of Huntington's disease. *Cell Death Dis* 2011;**2**:e115.
75. Oerlemans MI, Liu J, Arslan F, den Ouden K, van Middelaar BJ, Doevendans PA, Sluijter JP. Inhibition of RIP1-dependent necrosis prevents adverse cardiac remodeling after myocardial ischemia-reperfusion in vivo. *Basic Res Cardiol* 2012;**107**(4):270.
76. Linkermann A, De Zen F, Weinberg J, Kunzendorf U, Krautwald S. Programmed necrosis in acute kidney injury. *Nephrol Dial Transplant* 2012;**27**(9):3412–19.
77. Linkermann A, Heller JO, Prókai A, Weinberg JM, De Zen F, Himmerkus N, et al. The RIP1-kinase inhibitor necrostatin-1 prevents osmotic nephrosis and contrast-induced AKI in mice. *J Am Soc Nephrol* 2013;**24**(10):1545–57.
78. Linkermann A, Bräsen JH, Darding M, Jin MK, Sanz AB, Heller JO, et al. Two independent pathways of regulated necrosis mediate ischemia-reperfusion injury. *Proc Natl Acad Sci USA* 2013;**110**(29):12 024–9.
79. Bonnet MC, Preukschat D, Welz PS, van Loo G, Ermolaeva MA, Bloch W, et al. The adaptor protein FADD protects epidermal keratinocytes from necroptosis in vivo and prevents skin inflammation. *Immunity* 2011;**35**(4):572–82.
80. Weinlich R, Oberst A, Dillon CP, Janke LJ, Milasta S, Lukens JR, et al. Protective roles for caspase-8 and cFLIP in adult homeostasis. *Cell Rep* 2013;**5**(2):340–8.
81. Berger SB, Kasparcova V, Hoffman S, Swift B, Dare L, Schaeffer M, et al. Cutting edge: RIP1 kinase activity is dispensable for normal development but is a key regulator of inflammation in SHARPIN-deficient mice. *J Immunol* 2014;**192**(12):5476–80.
82. Kumari S, Redouane Y, Lopez-Mosqueda J, Shriaishi R, Romanowska M, Lutzmayer S, et al. Sharpin prevents skin inflammation by inhibiting TNFR1-induced keratinocyte apoptosis. *eLife* 2014;**3**: doi: 10.7554/eLife.03422
83. Lin J, Li H, Yang M, Ren J, Huang Z, Han F, et al. A role of RIP3-mediated macrophage necrosis in atherosclerosis development. *Cell Rep* 2013;**3**(1):200–10.
84. Linkermann A, Bräsen JH, Himmerkus N, Liu S, Huber TB, Kunzendorf U, Krautwald S. Rip1 (receptor-interacting protein kinase 1) mediates necroptosis and contributes to renal ischemia/reperfusion injury. *Kidney Int* 2012;**81**(8):751–61.

85 Roychowdhury S, McMullen MR, Pisano SG, Liu X, Nagy LE. Absence of receptor interacting protein kinase 3 prevents ethanol-induced liver injury. *Hepatology* 2013;**57**(5):1773–83.

86 Vucur M, Reisinger F, Gautheron J, Janssen J, Roderburg C, Cardenas DV, et al. RIP3 inhibits inflammatory hepatocarcinogenesis but promotes cholestasis by controlling caspase-8- and JNK-dependent compensatory cell proliferation. *Cell Rep* 2013;**4**(4):776–90.

87 Ofengeim D, Ito Y, Najafov A, Zhang Y, Shan B, DeWitt JP, et al. Activation of necroptosis in multiple sclerosis. *Cell Rep* 2015;**10**(11):1836–49.

88 Virtanen JO, Jacobson S. Viruses and multiple sclerosis. *CNS Neurol Disord Drug Targets* 2012;**11**(5):528–44.

89 Smith CC, Davidson SM, Lim SY, Simpkin JC, Hothersall JS, Yellon DM. Necrostatin: a potentially novel cardioprotective agent? *Cardiovasc Drugs Ther* 2007;**21**(4):227–33.

90 Koshinuma S, Miyamae M, Kaneda K, Kotani J, Figueredo VM. Combination of necroptosis and apoptosis inhibition enhances cardioprotection against myocardial ischemia-reperfusion injury. *J Anesth* 2014;**28**(2):235–41.

91 Pavlosky A, Lau A, Su Y, Lian D, Huang X, Yin Z, et al. RIPK3-mediated necroptosis regulates cardiac allograft rejection. *Am J Transplant* 2014;**14**(8):1778–90.

92 Gerlach B, Cordier SM, Schmukle AC, Emmerich CH, Rieser E, Haas TL, et al. Linear ubiquitination prevents inflammation and regulates immune signalling. *Nature* 2011;**471**(7340):591–6.

93 Kim SK, Kim WJ, Yoon JH, Ji JH, Morgan MJ, Cho H, Kim YC, Kim YS. Upregulated RIP3 expression potentiates MLKL phosphorylation-mediated programmed necrosis in toxic epidermal necrolysis. *J Invest Dermatol* 2015;**135**(8):2021–30.

94 Linkermann A, Bräsen JH, De Zen F, Weinlich R, Schwendener RA, Green DR, et al. Dichotomy between RIP1- and RIP3-mediated necroptosis in tumor necrosis factor-alpha-induced shock. *Mol Med* 2012;**18**:577–86.

95 Roychowdhury S, Chiang DJ, Mandal P, McMullen MR, Liu X, Cohen JI, et al. Inhibition of apoptosis protects mice from ethanol-mediated acceleration of early markers of CCl4-induced fibrosis but not steatosis or inflammation. *Alcohol Clin Exp Res* 2012;**36**(7):1139–47.

96 Afonso MB, Rodrigues PM, Simão AL, Ofengeim D, Carvalho T, Amaral JD, et al. Activation of necroptosis in human and experimental cholestasis. *Cell Death Dis* 2016;**7**(9):e2390.

97 Suda J, Dara L, Yang L, Aghajan M, Song Y, Kaplowitz N, Liu ZX. Knockdown of RIPK1 markedly exacerbates murine immune-mediated liver injury through massive apoptosis of hepatocytes, independent of necroptosis and inhibition of NF-kappaB. *J Immunol* 2016;**197**(8):3120–9.

98 Zhou Y, Dai W, Lin C, Wang F, He L, Shen M, et al. Protective effects of necrostatin-1 against concanavalin A-induced acute hepatic injury in mice. *Mediators Inflamm* 2013;**2013**:706156

99 Jouan-Lanhouet S, Arshad MI, Piquet-Pellorce C, Martin-Chouly C, Le Moigne-Muller G, Van Herreweghe F, et al. TRAIL induces necroptosis involving RIPK1/RIPK3-dependent PARP-1 activation. *Cell Death Differ* 2012;**19**(12):2003–14.

100 Afonso MB, Rodrigues PM, Carvalho T, Caridade M, Borralho P, Cortez-Pinto H, et al. Necroptosis is a key pathogenic event in human and experimental murine models of non-alcoholic steatohepatitis. *Clin Sci (Lond)* 2015;**129**(8):721–39.

101 Murakami Y, Matsumoto H, Roh M, Suzuki J, Hisatomi T, Ikeda Y, et al. Receptor interacting protein kinase mediates necrotic cone but not rod cell death in a mouse model of inherited degeneration. *Proc Natl Acad Sci USA* 2012;**109**(36):14 598–603.

102 Trichonas G, Murakami Y, Thanos A, Morizane Y, Kayama M, Debouck CM, et al. Receptor interacting protein kinases mediate retinal detachment-induced photoreceptor necrosis and compensate for inhibition of apoptosis. *Proc Natl Acad Sci USA* 2010;**107**(50):21 695–700.

103 Wang X, Wang Y, Ding ZJ, Yue B, Zhang PZ, Chen XD, et al. The role of RIP3 mediated necroptosis in ouabain-induced spiral ganglion neurons injuries. *Neurosci Lett* 2014;**578**:111–16.

104 Lau A, Wang S, Jiang J, Haig A, Pavlosky A, Linkermann A, et al. RIPK3-mediated necroptosis promotes donor kidney inflammatory injury and reduces allograft survival. *Am J Transplant* 2013;**13**(11):2805–18.

105 Zhu Y, Cui H, Xia Y, Gan H. RIPK3-mediated necroptosis and apoptosis contributes to renal tubular cell progressive loss and chronic kidney disease progression in rats. *PLoS One* 2016;**11**(6):e0156729.

106 Ito Y, Ofengeim D, Najafov A, Das S, Saberi S, Li Y, et al. RIPK1 mediates axonal degeneration by promoting inflammation and necroptosis in ALS. *Science* 2016;**353**(6299):603–8.

107 Wittkopf N, Günther C, Martini E, He G, Amann K, He YW, et al. Cellular FLICE-like inhibitory protein secures intestinal epithelial cell survival and immune homeostasis by regulating caspase-8. *Gastroenterology* 2013;**145**(6):1369–79.

108 Moriwaki K, Balaji S, McQuade T, Malhotra N, Kang J, Chan FK. The necroptosis adaptor RIPK3 promotes injury-induced cytokine expression and tissue repair. *Immunity* 2014;**41**(4):567–78.

109 Pierdomenico M, Negroni A, Stronati L, Vitali R, Prete E, Bertin J, et al. Necroptosis is active in children with inflammatory bowel disease and contributes to heighten intestinal inflammation. *Am J Gastroenterol* 2014;**109**(2):279–87.

110 Günther C, Martini E, Wittkopf N, Amann K, Weigmann B, Neumann H, et al. Caspase-8 regulates TNF-alpha-induced epithelial necroptosis and terminal ileitis. *Nature* 2011;**477**(7364):335–9.

111 Targan SR, Hanauer SB, van Deventer SJ, Mayer L, Present DH, Braakman T, et al. A short-term study of chimeric monoclonal antibody cA2 to tumor necrosis factor alpha for Crohn's disease. Crohn's Disease cA2 Study Group. *N Engl J Med* 1997;**337**(15):1029–35.

112 Järnerot G, Hertervig E, Friis-Liby I, Blomquist L, Karlén P, Grännö C, et al. Infliximab as rescue therapy in severe to moderately severe ulcerative colitis: a randomized, placebo-controlled study. *Gastroenterol* 2005;**128**(7):1805–11.

113 Wen S, Ling Y, Yang W, Shen J, Li C, Deng W, et al. Necroptosis is a key mediator of enterocytes loss in intestinal ischaemia/reperfusion injury. *J Cell Mol Med* 2017;**21**(3):432–43.

114 Duprez L, Takahashi N, Van Hauwermeiren F, Vandendriessche B, Goossens V, Vanden Berghe T, et al. RIP kinase-dependent necrosis drives lethal systemic inflammatory response syndrome. *Immunity* 2011;**35**(6):908–18.

115 Takahashi N, Duprez L, Grootjans S, Cauwels A, Nerinckx W, DuHadaway JB, et al. Necrostatin-1 analogues: critical issues on the specificity, activity and in vivo use in experimental disease models. *Cell Death Dis* 2012;**3**:e437.

116 McNeal SI, LeGolvan MP, Chung CS, Ayala A. The dual functions of receptor interacting protein 1 in fas-induced hepatocyte death during sepsis. *Shock* 2011;**35**(5):499–505.

117 Sharma A, Matsuo S, Yang WL, Wang Z, Wang P. Receptor-interacting protein kinase 3 deficiency inhibits immune cell infiltration and attenuates organ injury in sepsis. *Crit Care* 2014;**18**(4):R142.

118 He S, Liang Y, Shao F, Wang X. Toll-like receptors activate programmed necrosis in macrophages through a receptor-interacting kinase-3-mediated pathway. *Proc Natl Acad Sci USA* 2011;**108**(50):20 054–9.

119 Gautheron J, Vucur M, Schneider AT, Severi I, Roderburg C, Roy S, et al. The necroptosis-inducing kinase RIPK3 dampens adipose tissue inflammation and glucose intolerance. *Nat Commun* 2016;**7**:11 869.

120 Dobbelstein M, Shenk T. Protection against apoptosis by the vaccinia virus SPI-2 (B13R) gene product. *J Virol* 1996;**70**(9):6479–85.

121 Wasilenko ST, Stewart TL, Meyers AF, Barry M. Vaccinia virus encodes a previously uncharacterized mitochondrial-associated inhibitor of apoptosis. *Proc Natl Acad Sci USA* 2003;**100**(24):14 345–50.

122 Gerlic M, Faustin B, Postigo A, Yu EC, Proell M, Gombosuren N, et al. Vaccinia virus F1L protein promotes virulence by inhibiting inflammasome activation. *Proc Natl Acad Sci USA* 2013;**110**(19):7808–13.

123 Aoyagi M, Zhai D, Jin C, Aleshin AE, Stec B, Reed JC, Liddington RC. Vaccinia virus N1L protein resembles a B cell lymphoma-2 (Bcl-2) family protein. *Protein Sci* 2007;**16**(1):118–24.

124 Cooray S, Bahar MW, Abrescia NG, McVey CE, Bartlett NW, Chen RA, et al. Functional and structural studies of the vaccinia virus virulence factor N1 reveal a Bcl-2-like anti-apoptotic protein. *J Gen Virol* 2007;**88**(Pt. 6): 1656–66.

125 Wang X, Li Y, Liu S, Yu X, Li L, Shi C, et al. Direct activation of RIP3/MLKL-dependent necrosis by herpes simplex virus 1 (HSV-1) protein ICP6 triggers host antiviral defense. *Proc Natl Acad Sci USA* 2014;**111**(43): 15 438–43.

126 Huang Z, Wu SQ, Liang Y, Zhuo X, Chen W, Li L, et al. RIP1/RIP3 binding to HSV-1 ICP6 initiates necroptosis to restrict virus propagation in mice. *Cell Host Microbe* 2015;**17**(2):229–42.

127 Guo H, Kaiser WJ, Mocarski ES. Manipulation of apoptosis and necroptosis signaling by herpesviruses. *Med Microbiol Immunol* 2015;**204**(3):439–48.

128 Guo H, Omoto S, Harris PA, Finger JN, Bertin J, Gough PJ, et al. Herpes simplex virus suppresses necroptosis in human cells. *Cell Host Microbe* 2015;**17**(2):243–51.

129 Madrigal AG, Barth K, Papadopoulos G, Genco CA. Pathogen-mediated proteolysis of the cell death regulator RIPK1 and the host defense modulator RIPK2 in human aortic endothelial cells. *PLoS pathogens* 2012;**8**(6):e1002723.

130 Philip NH, Dillon CP, Snyder AG, Fitzgerald P, Wynosky-Dolfi MA, Zwack EE, et al. Caspase-8 mediates caspase-1 processing and innate immune defense in response to bacterial blockade of NF-kappaB and MAPK signaling. *Proc Natl Acad Sci USA* 2014;**111**(20):7385–90.

131 Weng D, Marty-Roix R, Ganesan S, Proulx MK, Vladimer GI, Kaiser WJ, et al. Caspase-8 and RIP kinases regulate bacteria-induced innate immune responses and cell death. *Proc Natl Acad Sci USA* 2014;**111**(20):7391–6.

132 Vince JE, Wong WW, Gentle I, Lawlor KE, Allam R, O'Reilly L, et al. Inhibitor of apoptosis proteins limit RIP3 kinase-dependent interleukin-1 activation. *Immunity* 2012;**36**(2):215–27.

133 Mandal P, Berger SB, Pillay S, Moriwaki K, Huang C, Guo H, et al. RIP3 induces apoptosis independent of pronecrotic kinase activity. *Mol Cell* 2014;**56**(4):481–95.

134 Yabal M, Müller N, Adler H, Knies N, Gross CJ, Damgaard RB, et al. XIAP restricts TNF- and RIP3-dependent cell death and inflammasome activation. *Cell Rep* 2014;**7**(6):1796–808.

135 Hockendorf U, Yabal M, Jost PJ. RIPK3-dependent cell death and inflammasome activation in FLT3-ITD expressing LICs. *Oncotarget* 2016;**7**(36):57 483–4.

136 Kang TB, Yang SH, Toth B, Kovalenko A, Wallach D. Caspase-8 blocks kinase RIPK3-mediated activation of the NLRP3 inflammasome. *Immunity* 2013;**38**(1):27–40.

137 Duong BH, Onizawa M, Oses-Prieto JA, Advincula R, Burlingame A, Malynn BA, Ma A. A20 restricts ubiquitination of pro-interleukin-1beta protein complexes and suppresses NLRP3 inflammasome activity. *Immunity* 2015;**42**(1):55–67.

138 Cook WD, Moujalled DM, Ralph TJ, Lock P, Young SN, Murphy JM, Vaux DL. RIPK1- and RIPK3-induced cell death mode is determined by target availability. *Cell Death Differ* 2014;**21**(10):1600–12.

7

Ferroptosis

Ebru Esin Yoruker and Ugur Gezer

Department of Basic Oncology, Oncology Institute, Istanbul University, Istanbul, Turkey

Abbreviations

AD	Alzheimer's disease
AIF	apoptosis-inducing factor
ATP	adenosine triphosphate
CPX	ciclopirox olamine, iron chelator
DFX	deferoxamine, iron chelator
ECM	extracellular matrix
endo G	endonuclease G
ER	endoplasmic reticulum
FDA	US Food and Drug Administration
Fe^{++}	ferrous iron
Fe^{+++}	ferric iron
GPX	glutathione-dependent peroxidase
H_2O_2	hydrogen peroxide
HCC	hepatocellular carcinoma
HD	Huntington's disease
IRE	iron-responsive element
IRP	iron-regulatory protein
NADPH	reduced nicotinamide adenine dinucleotide phosphate
NOX	nicotinamide adenine dinucleotide phosphate oxidase
$O_2^{\bullet-}$	superoxide
$OH^{\bullet}$	hydroxil radical
PCD	programmed cell death
PD	Parkinson's disease
PDGFR-β	platelet-derived growth-factor receptor β
PUFA	polyunsaturated fatty acid
RA	rheumatoid arthritis
Rb	retinoblastoma
RBC	red blood cell
RNS	reactive nitrogen species
$ROO^{\bullet}$	peroxyl radical
ROS	reactive oxygen species
SOD	superoxide dismutase

Apoptosis and Beyond: The Many Ways Cells Die, First Edition. Edited by James Radosevich.
© 2018 John Wiley & Son Inc. Published 2018 by John Wiley & Son Inc.

TNF	tumor necrosis factor
VDAC	voltage-dependent anion channel
VEGFR	vascular endothelial growth-factor receptor

7.1 Introduction: The Balance of Cell Death

Homeostatic mechanisms depend for their effectiveness on the balance between cell survival and cell death. The extracellular matrix (ECM) is very important to cell survival and cell life quality. Cells are constantly protecting their own structure. They tend to maintain their intracellular environment within certain physiological limits; this is known as "maintaining normal homeostasis." As cells encounter physiologic stresses or pathologic conditions, they can undergo adaptation for the preservation of viability and function. Severe or persistent stress results in irreversible injury, and as a consequence, the affected cells die. Cell death is also a normal and crucial process in embryogenesis, the development of organs, and the maintenance of homeostasis mechanism [1]. This mechanism is crucial for development of the fertilized egg into an adult organism. The benefits of cell death include the deletion of unnecessary structures, the adjustment of cell numbers, and the elimination of abnormal, misplaced, nonfunctional, or harmful cells. The balance between cell survival and cell death is broken in the pathogenesis of many diseases, including cancer autoimmunity, ischemic injury, and neurodegenerative disorders [2,3].

Apoptosis (type I cell death), autophagy (type II), and necrosis (type III or accidental) are recognized as major forms of cell death. For 2 decades, apoptosis was the only known cell-death mechanism during infection, homeostasis, and development. In addition to necrosis, it was considered an accidental cell-death pathway that occurred in response to physicochemical conditions [4–6]. In recent years, studies have shown that there are in fact several pathways regulated by necrosis. These forms of non-apoptotic cell death are classified according to their underlying pathways, each characterized by a particular aspect. Ferroptosis is one of the forms that is included in regulated necrosis. It is defined as a non-apoptotic cell-death mechanism that is morphologically, biochemically, and genetically distinct from apoptosis, necrosis, and autophagy [7]. Understanding of alternative regulated cell-death mechanisms may be beneficial for the treatment of diseases, development of new therapeutics, and overcoming of drug resistance to chemotherapeutic agents for a wide range of diseases.

7.2 Overview of Free Radicals

7.2.1 Free Oxygen Radicals

Free radicals are important for their toxic effects. Obtaining a better understanding of ferroptosis is thus the first step to understanding reactive oxygen species (ROS) in cellular metabolism. ROS can be defined as molecules or molecular fragments containing one or more unpaired electrons [8]. Most have a few microseconds' half-life within an organism. Radicals derived from oxygen represent the most important class of radical species generated in living systems. Examples include superoxide ($O_2^{\bullet-}$), hydrogen peroxide (H_2O_2), hydroxyl radical ($OH^{\bullet}$), peroxyl radical ($ROO^{\bullet}$), and a related class of

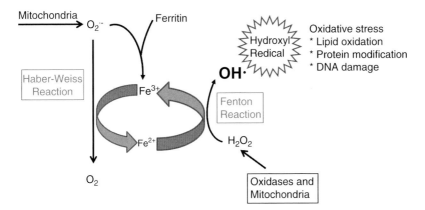

Figure 7.1 Fenton and Haber–Weiss reactions in a cell. *Source:* Adapted from Kell 2010 [11] under the terms of the Creative Commons Attribution Licence, CC-BY, via Springer.

reactive nitrogen species (RNS) derived from the reaction of nitric oxide (NO) with oxygen or superoxide [9,10]. H_2O_2 is not a free radical, but due to its chemical effects and consequent activity, it is classified as a highly reactive oxygen compound. Hydroxyl radical is the most damaging free-oxygen radical and is derived from $O_2^{\bullet-}$ and H_2O_2. The two sources of $OH^{\bullet}$ are the Fenton and Haber–Weiss reactions in cells (Figure 7.1) [11]. The Fenton reagent is a mixture of Fe(II) and hydrogen peroxide. The Fe(II) reacts with the peroxide to form Fe(III), a hydroxyl radical, and a hydroxide anion. In the Haber–Weiss reaction, $OH^{\bullet}$ is generated from an interaction between $O2^{\bullet-}$ and H_2O_2. This reaction proceeds very slowly, but if it is catalyzed by iron it occurs much more rapidly. It regenerates the Fe(II) used via oxidation with a peroxyl radical to give oxygen gas as a product [12–14].

ROS are known to play a significant role in biological systems, since they can be either harmful or beneficial to vitality. Cellular signaling systems and immune-system defenses against infectious agents experience beneficial effects from ROS. Low concentrations of ROS induce mitogenic responses in cells, while excessive levels may lead to oxidative stress, unbalancing of vital functions, and apoptosis or necrosis. ROS are toxic, mutagenic, and carcinogenic, due to their high reactivity. DNA and cell structures such as lipids and proteins are major targets of ROS in living organisms. ROS also play important roles in enzyme activity, cell signaling, regulation of transcription factor activities, adaptation to diverse growth conditions, and host defense in multicellular organisms [15,16]. ROS are formed by several different mechanisms, including irradiation by ultraviolet (UV) light, x-rays, gamma rays, and products of metal-catalyzed reactions. They are also produced by air pollution in the atmosphere, responses by neutrophils and macrophages during inflammatory cell activation, and byproducts of mitochondria-catalyzed electron-transport reactions [17].

7.2.2 Sources of ROS within the Cell

Free radicals and other ROS are accidentally produced during metabolism in the organism. Inflammation and aging can be included among ROS resources. ROS can be produced from both endogenous and exogenous substances in living organisms. Mitochondria, peroxisomes, cytochrome P450 metabolism, and inflammatory cell responses are endogenous sources of ROS [18]. The most important *in vivo* source is the mitochondrion. Adenosine triphosphate (ATP) is formed by oxidative phosphorylation, and $O_2^{\bullet-}$, H_2O_2, and $OH^{\bullet}$ are released from this reaction. Hydrogen peroxide is released from the mitochondria to the cytoplasm, but $O_2^{\bullet-}$ is not. Intramitochondrial superoxide dismutase (SOD) converts $O_2^{\bullet-}$ to H_2O_2 and thereby preserves the cell from oxidative damage. It is known that up to 2% of total mitochondrial oxygen consumption "normally" goes toward the production of ROS. The hydrogen peroxide molecule is not a radical species, because it does not contain an unpaired electron. Free-radical production in the nuclear membrane and endoplasmic reticulum (ER) results from the oxidation of membrane-bound cyctochrome [18–21].

Another source of superoxide radical is the auto-oxidation reaction of compounds such as thiols, flavins, catecholamines, and tetrahydrofolate. Hydrogen peroxide and superoxide radicals occur during the catalytic cycles of many enzymes. Xanthine oxidase, amino acid oxidase, aldehyde oxidase, reduced nicotinamide adenine dinucleotide phosphate (NADPH) oxidase (NOX), tryptophane dioxygenase, and dehydrogenase flavoprotein are all examples of these enzymes. NOX are multi-subunit transmembrane proteins that transfer electrons across biological membranes and catalyze the production of ROS. The NOX family uses oxygen as an electron acceptor to produce superoxide anion radicals. These enzymes have been found in all tissues and have been shown to have several mechanisms, including cell growth and apoptosis, angiogenesis, and the regulation of the ECM [22–25].

Transition metals such as iron and copper are specially involved in physiological oxidation-reduction reactions. Because of this, they act as catalysts for various reactions. In particular, copper and iron catalyze the synthesis of hydroxyl radicals from hydrogen peroxide and superoxide radicals [13].

Arachidonic acid metabolism is another important endogenous source of reactive oxygen metabolites. Arachidonic acid is released from the plasma membrane as a result of stimulation of phagocytic cells, and a variety of free radicals are produced by its enzymatic free oxidation [26].

There are various exogenous agents that can cause the production of ROS. Radiation and environmental agents (air pollution, pesticides, solvents, cigarette smoke, anesthetics, and aromatic hydrocarbons) cause free-radical formation. Some of the antineoplastic agents, such as bleomycin and doxorubicin, generate lipid peroxidation in cells. Oxidation of catecholamines also produces ROS. An increase in catecholamin, induced by neural activation, may play a role in the stress-related pathogenesis of various diseases [27].

7.2.3 The Impact of ROS

ROS may damage important biological materials, such as DNA, protein, carbohydrates, and lipids. Therefore, the elimination of ROS is important for the defense of the cell.

Lipids are the most vulnerable biomolecules to be affected by free radicals. Membrane lipids and fatty acids interact readily with free radicals and generate peroxidation products. Oxidative degradation of polyunsaturated fatty acids (PUFAs) with free radicals is known as "non-enzymatic lipid peroxidation"; this event continues as a chain reaction. Lipid peroxidation causes direct irreversible damage to membrane structures; it is also indirectly destructive to other cell components, as aldehydes are formed. Molecular oxygen is its basic molecule, facilitated by Fe^{++} ions. Lipid peroxidation occurs not only in the plasma membrane but also in peroxisomes, microsomes, and mitochondria. Changes in the permeability and fluidity of the membrane lipid bilayer can dramatically alter cell integrity as a result of lipid peroxidation. Thus, the toxicity of lipid peroxidation products in mammals generally causes neurotoxicity, nephrotoxicity, and hepatotoxicity [28,29].

Proteins with sulfur-containing amino acids, such as methionine and cystein, and amino acids with unsaturated bonds, such as tryptophan, tyrosine, phenylalanine, and histidine, are easily affected by ROS. This is especially true for sulfur-centered radicals and carbon-centered organic radicals. Very active ROS species impair the function of peptides and amino acids through hydroxylation.

In summary, ROS cause denaturation of proteins, breaks in the peptide chain, crosslinking, inhibition of enzymes, and permeability changes in tissues and cells [30,31].

It has been estimated that one human cell is exposed to approximately 1.5×10^5 oxidative adducts a day from hydroxyl radicals and other such reactive species. Free oxygen radicals are accepted as mediators of mutagenesis and carcinogenesis and are responsible for DNA strand breaks [32]. We know that free radical-mediated DNA damage occurs in various cancers. The hydroxyl radical is known to react with all components of the DNA molecule (whether single- or double-strand DNA, and whether with purine or pyrimidine bases). The guanine base is particularly sensitive to oxidation, so it is used for oxidative injury detection. Analytical methodology can currently detect as little as 25 fmol of 8-hydroxy-deoxyguanosine [33]. Permanent modification of genetic material resulting from this "oxidative damage" is thought to be the first step in mutagenesis, carcinogenesis, and aging. DNA damage can cause either repression or induction of transcription, induction of signaling pathways, and genomic instability – all of which are associated with carcinogenesis.

Free radicals impact carbohydrates through monosaccharide auto-oxidation or depolymerization of the carbohydrates directly. Superoxides and oxaloaldehydes are produced during auto-oxidation of monosaccharides. These molecules have the ability to bind to DNA, RNA, and proteins and play a significant role in cancer and aging. In inflammatory diseases such as rheumatoid arthritis (RA), it has been shown that the hyaluronic acid that is the major type of mucopolysaccharide is depolymerized by oxygen-derived free radicals [34].

In order to prevent damage by ROS, detoxifying cells have an anti-oxidant defense system that involves both enzymatic (catalase, SOD, andglutathione peroxidase) and non-enzymatic (glutathione, ascorbic acid, beta carotene, melatonin, and cysteine) detoxification of various ROS. At higher ROS concentrations, hydrogen peroxide can inhibit caspases in the cell; this leads to a switch from apoptosis to necrosis [35].

7.3 Molecular Mechanism of Ferroptosis

Dolma et al. [36] identified in *in vitro* studies that small molecules like the chemical compound erastin selectively kill cells carrying an active Ras oncogene. They suggested that the genetically targeted small molecules may serve as anticancer drugs for personalized treatment [36]. Scott J. Dixon and his colleagues [37] were the first to use the term "ferroptosis" and describe the principle behind this cell-death mechanism, in 2012. Occurrence of oxidative stress, membrane lipid peroxidation, and loss of cell viability are the most basic characteristic of ferroptosis [38].

7.3.1 Discovery of the Small Molecule Erastin

Erastin is a cell-permeable piperazinyl–quinazolinone compound that exhibits oncogene-selective lethality in cells with H-Ras mutations. Erastin has been shown to bind mitochondrial voltage-dependent anion channels (VDACs) and alter their gating; it rapidly induces an oxidative, non-apoptotic cell death in several human tumors harboring activating mutations in RAS–RAF–MEK signaling. The Ras/Raf/**m**itogen-activated protein kinase/ERK **k**inase (MEK)/**e**xtracellular signal-**r**egulated kinase (ERK) cascade communicates a signal from cell-surface receptors to DNA in the nucleus of the cell (Figure 7.2) [39,40].

Depending upon the stimulus and cell type, cells respond via signal prevention, induction of apoptosis, or cell-cycle progression. Thus, the ferroptotic pathway is an appropriate target for therapeutic intervention [41–43].

In a paper by Brent Stockwell's laboratory at Columbia University [37], the term "ferroptosis" is proposed to describe iron-dependent accumulation of ROS. Stockwell's team studied the role of ROS in ferroptosis in order to understand the molecular events leading to this cell-death pathway. How ROS are generated and what roles they may play in the cell-death process are important questions for this topic. The researchers had

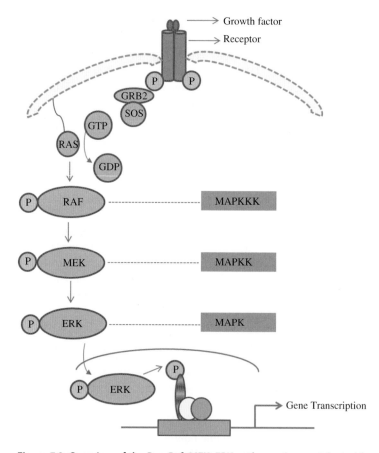

Figure 7.2 Overview of the Ras–Raf–MEK–ERK pathway. *Source:* Adapted from Chang et al. 2003 [40] with permission from Macmillan Publishers Ltd.

previously published on two anticancer compounds: erastin and RSL3 (Ras selective lethal) [36,44]. They showed that VDAC was a potential target for erastin and that the cells were directed to a cell-death pathway, but observed that the mechanism of cell death was different from that of apoptosis. The increase of ROS is noteworthy, and the death is prevented by iron chelation or inhibition of iron uptake [39,44]. The Stockwell laboratory showed that inhibitors of apoptosis, autophagy, and necrosis did not prevent erastin-induced death, but that antioxidants and iron chelators did [37].

7.3.2 Erastin–VDAC–Cystine/Glutamate Antiporter Interaction

Erastin-induced death by ROS in cells is suppressed by iron chelators [44]. Ferroptosis can be initiated by structurally diverse small molecules such as sulfasalazine, RSL3, and RSL5. Erastin has been shown to trigger ferroptosis by inhibiting cystine uptake in cells via a cystine/glutamate antiporter known as system x_c^-. The reduced form of cystine is known as cysteine. Glutathione (created by cysteine) goes from glycine to glutamic acid. Inhibition of cystine uptake results in the depletion of glutathione, and glutathione depletion leads to iron-dependent accumulation of ROS [45]. Glutathione is a tripeptide with a thiol group. It prevents or reduces the destructive effects of free radicals, and it serves as a substrate for many enzymes, including transferases and peroxidases. Because of this, it protects biological membranes against lipid peroxidation in an enzyme-dependent reaction. Gluathione is oxidized to the oxidized glutathion form (GSSG) by the glutathion peroxidase (GSHPx) enzyme [46,47].

$$2GSH + H_2O_2 \rightarrow GSSG + 2H_2O$$

Glutathione also reacts with extremely harmful ROS such as $O_2^{\bullet-}$ and $OH^{\bullet}$ without enzyme catalysis, protecting the cell [48].

The mitochondrial VDAC2 and VDAC3 also have affinity for the lethal effects of erastin. Mitochondrial function and cell viability require flux of metabolites, which is achieved by VDACs. The VDACs are found in the mitochondrial outer membrane and lead to permeabilization of small ions and metabolites [49,50]. Interaction of erastin with VDACs triggers the release of iron and iron-binding compound (Figure 7.3). This process leads to accumulation of ROS by the promotion of Fenton chemistry and consequent lipid peroxidation, resulting ultimately in plasma-membrane permeabilization and cell death. RSL3 is not dependent on VDACs or system X_c^-. Its molecular target is still unknown [37]. It can be initiated by selecting the target glutathione-dependent peroxidase 4 (GPX4). The biochemical function of GPXs is to reduce lipid hydroperoxides and hydrogen peroxide to their corresponding alcohols or water. Unlike other GPXs, GPX4 can catalyze the reduction of lipid peroxides. There are eight isoforms of GPX in humans [51]. A study on this subject by Yang et al. [38] reported that GPX4 deletion caused lethality or cell loss; the authors thought this was via ferroptosis. Their experimental design showed that RSL3 binds and inhibits GPX4, so that ferroptosis is induced by RSL3 and GPX4 is a central regulator of this cell-death pathway.

7.3.3 The Importance of Iron in Ferroptosis

Homeostasis of iron is essential for human organisms. Iron is also necessary for growth, development, normal cellular functioning, and the synthesis of some hormones and connective tissue. The total quantity of iron in the human body is 2–5 g (35–45 mg/kg).

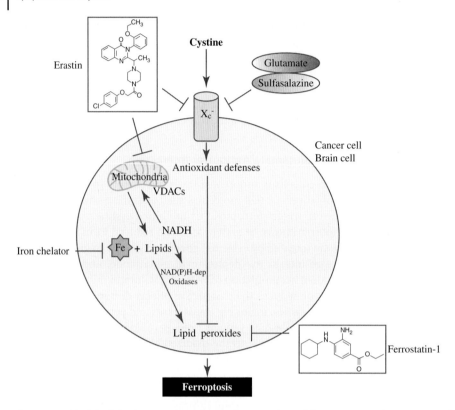

Figure 7.3 Model of the ferroptosis pathway. *Source:* Adapted from Dixon et al. 2012 [37] with permission from Elsevier.

Most of the body contains iron, and control of this element is critical because it can potentially be toxic [52].

Metal ion-binding proteins (e.g., hemoglobin, ferritin, transferrin, lactoferrin, myoglobin) are considered basic antioxidant defense systems. Metal ions are important to many enzymes, including cytchromes, cyctchrome oxidase, peroxidase, and catalase (Table 7.1). Oxidant damage caused by iron is indirectly accelerated by ferrous iron (Fe^{++}). Iron and copper ions may transform some less reactive compounds into more

Table 7.1 Iron distribution in the human body.

Iron-Binding Proteins	Percentage of Total
Hemoglobin	66.0
Myoglobin	3.0
Fe-containing enzymes	0.1
Ferritin and others storage forms	30.0
Transferrin	0.1
Intracellular labile iron	1.0

reactive ones within a short span of time. Therefore, they are kept bound to carrier proteins and storage proteins in the organism [52,53].

Iron is a major component of hemoglobin, a metalloprotein found in red blood cells (RBCs) that carries oxygen to all parts of the body. About 70% of the human body's iron is found in the hemoglobin. Another 30% is stored in ferritin, a protein found in nearly all tissues, and a few per cent in myoglobin, a protein specifically utilized by muscle cells. Iron contained in blood is normally bound to the protein transferrin. Each molecule of transferrin can transport two molecules of iron to those areas of the body that require it [54].

Human homeostasis of iron is regulated by iron absorption, iron recycling, and mobilization of stored iron. Cellular iron homeostasis is controlled post-transcriptionally by the iron-regulatory proteins (IRP)/iron-responsive element (IRE) regulatory system. Recent findings suggest that the IRP/IRE regulatory network is involved in cancer progression and metastasis [55].

Ferroptosis is described as an iron-dependent cell-death pathway, but the precise role of iron in ferroptosis is unclear. Inhibition of heme-dependent enzymes such as NADPH oxidase can protect cells from ferroptosis, while the use of cobalt chloride inhibits ferroptosis due to the displacement capability of iron. Consequently, Dixon et al. [7,37] have suggested that ferroptosis is dependent on intracellular iron but not on other metals. In addition, iron-containing enzymes are essential for ferroptosis, which is inactivated by iron chelation [7,37].

7.3.4 Differences from Other Cell-Death Mechanisms

Ferroptosis is morphologically, biochemically, and genetically distinct from apoptosis, autophagy, and necrosis [37,39,44].

In multicellular organisms, cells that are unwanted are eliminated by energy-dependent apoptosis. Apoptosis occurs normally during development and aging as a homeostatic mechanism. However, it also occurs in cells damaged by disease, chemotherapeutic agents, and irradiation. This programmed cell death (PCD) is mediated by proteolytic enzymes known as caspases. The induction of apoptosis occurs by three different mechanisms:

1) Intrinsic or mitochondrial pathways are induced using caspases (trigger by internal signals).
2) FasL and tumor necrosis factor (TNF) death activators bind to their receptor and initiate caspases, leading to cell death (trigger by external signals).
3) Apoptosis inducing factor (AIF) and endonuclease G (endo G) are released from mitochondrial membranes, leading to apoptosis without caspase activity [56].

Autophagy is a catabolic process and is adopted as a means of recycling intracellular components. Autophagic cell death is a caspase-independent process that exhibits an extensive autophagic degradation of dysfunctional organelles such as the Golgi apparatus, polyribosomes, ER, and cytoplasmic macromolecules [57,58].

Necrotic cell death is considered an unregulated, energy-independent form of cell death. An example is ischemia: drastic depletion of oxygen or glucose by the cell, leading to necrosis, which may occur in the centers of large malignant tumors [59].

Dixon et al. [37] examined how cells die via the ferroptotic cell-death pathway by comparing HRAS-mutant BjeLR-engineered tumor cells treated with erastin, staurosporine, and H2O2/rapamycin for induced ferroptosis, apoptosis, necrosis, and autophagy, respectively. Ferroptosis was defined by smaller-than-normal mitochondrial morphology and increased membrane density. They also showed that erastin-treated cells did not have bioenergetic failure. They concluded that there is another cell-death mechanism separate from apoptosis, necrosis, and autophagy [37].

A different study on this subject showed that the classic features of apoptosis are not observed in ferroptosis. Ferroptosis is independent of the proapoptotic Bcl-2 family members BAX and BAK and from the release of cytochrome C from mitochondria [39,44]. The researchers reported that caspases are not activated during ferroptosis.

7.3.5 Inducers and Inhibitors of Ferroptosis

Hundreds of lethal compounds have been screened in several studies in order to identify the molecular mechanism of cancer cell death. Understanding the pathway of cell death is important not only to treating cancer, but also to the prevention of emerging diseases related to pathological cell death. Although it is still a controversial topic whether or not ferroptosis occurs *in vivo*, this cell-death type can be initiated with a number of small molecules (e.g., erastin, sulfasalazine, RSL3, RSL5) in *in vitro* studies (Figure 7.4). Erastin and RSL3 have been found to function selectively only in RAS-mutant cell lines [36,44]. Experimental studies show that ferroptosis inhibitors like ferrostatin-1 (Fer-1) and liproxstatin-1 hold promise against human diseases such as renal failure, stroke, cardiovascular diseases, and neurodegenerative diseases. Skouta et al. [61] investigated this topic and suggested that inhibition of ferroptosis by Fer-1 in cellular models of

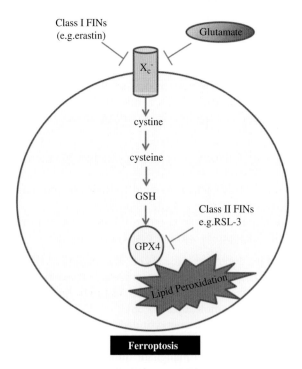

Figure 7.4 Class I and II inducers of ferroptosis. *Source:* Adapted from Yang et al. 2014 [38] with permission from Elsevier.

kidney dysfunction, periventricular leukomalacia, and Huntington's disease (HD) is crucial for inhibition of these diseases. Dixon et al. [37] showed that erastin-induced death in cancer cells and glutamate-induced death in rat brain slices share a molecular mechanism, and suggested that glutamate and erastin may bind to the same target.

7.4 Clinical Significance of Ferroptosis

Ferroptosis is a recognized form of regulated necrosis that also occurs in pathologic nontransformed conditions. Activation of ferroptosis has considered as a therapeutic approach to the destruction of cancer cells.

7.4.1 Ferroptosis and Cancer

Activation of alternative cell-death pathways may be beneficial for previously untreatable diseases such as cancer. Evasion of apoptosis is one of the hallmarks of cancer [62]. The mechanism of resistance to apoptosis has encouraged the development of new drugs that act against non-apoptotic cell-death mechanisms. Today, various agents are in clinical use as chemotherapeutic anticancer drugs aimed at tumor cell death. Most lethal compounds damage tumors via induction of apoptosis or necrosis [63,64]. Some lead to cancer cell death by unknown mechanisms.

Sorafenib is a US Food and Drug Administration (FDA)-approved small molecule that inhibits tumor-cell proliferation and tumor angiogenesis and increases the rate of cell death in advanced hepatocellular carcinoma (HCC) and thyroid and kidney cancer. It acts by inhibiting the serine–threonine kinases Raf-1 and B-Raf and the receptor tyrosine kinase activity of vascular endothelial growth-factor receptors (VEGFRs) 1, 2, and 3 and platelet-derived growth-factor receptor β (PDGFR-β) [65]. Dixon et al. [66] have shown that sorafenib inhibits system x_c^- and that inhibition of cystine import can lead to induction of ferroptotic cell death. Louandre et al. [67] reported that using the iron chelator deferoxamine (DFX) in the treatment of HCC with sorafenib protected the cells from the cytotoxic effects of sorafenib [67]. Another study showed that the retinoblastoma (Rb)-negative status of HCC cells promotes ferroptosis. Rb-negative cancer cells are more cytotoxic to treatment with sorafenib because of the increased levels of oxidative stress [68].

In conclusion, understanding the underlying molecular mechanism of ferroptosis and the function of different cell-death mechanisms could provide the rationale for the cell-death discovery of novel drugs. Potential targets could be developed for new personalized drugs of use in the treatment of cancer.

7.4.2 Ferroptosis and Neurodegeneration

Alzheimer's disease (AD) and Parkinson's disease (PD) are progressive neurodegenerative diseases that are associated with the regional accumulation and abnormal deposition of specific proteins in the brain [69]. The precise pathways that lead to neuronal death are not well established. It is known that Fe^{++} is highly reactive, and an excess of Fe^{2+} may stimulate the overproduction of reactive chemical species, such as the hydroxyl radical ($OH^•$). Such free radicals are responsible for oxidative stress and are considered to be a primary contributing factor to neurodegeneration.

Why is iron accumulated during the aging process? That is unknown, but some studies show that the elevated levels of iron observed in neuronal and cancer cells may lead to susceptibility to ferroptotic cell death. Inhibitors of ferroptosis could be used to prevent oxidative iron-dependent cell death in the nervous system. Dixon et al. [37] found ferrostatin-1 and iron chelators such as ciclopiroxolamine (CPX) to be potent inhibitors of glutamate-induced death in brain-cell populations. Skouta et al. [70] showed that a decrease in cell death is associated with ferrostatin in an HD rat brain-slice model. These results suggest that oxidative glutamate toxicity and HD may involve ferroptosis and that inhibition of this cell death type can be targeted for the prevention of neurodegenerative diseases in the future.

In the past, researchers working on cell-death mechanisms focused on apoptosis when developing new drugs. We now realize that the ferroptotic cell-death mechanism may be a better target. More work is required to provide an understanding of the mechanism of ferroptosis, to help in identifying new pharmacological drug targets.

References

1. Miura M. Active participation of cell death in development and organismal homeostasis. *Develop, Growth Differ* 2011;**53**(2):125–36.
2. Yuan J, Kroemer G. Alternative cell death mechanisms in development and beyond. *Genes Dev* 2010;**24**:2592–602.
3. Tait SWG, Ichim G, Green DR. Die another way – non-apoptotic mechanisms of cell death. *J Cell Sci* 2014;**127**:2135–44.
4. Berghe TV, Linkermann A, Jouan-Lanhouet S, Walczak H, Vandenabeele P. Regulated necrosis: the expanding network of non-apoptotic cell death pathways. *Nat Rev Mol Cell Biol* 2014;**15**:135–47.
5. Galluzzi L, Maiuri MC, Vitale I, Zischka H, Castedo M, Zitvogel L, Kroemer G. Cell death modalities: classification and pathophysiological implications. *Cell Death Differ* 2007;**14**:1237–43.
6. Suzanne M, Steller H. Shaping organisms with apoptosis. *Cell Death Differ* 2013;**20**:669–75.
7. Dixon SJ, Stockwell BR. The role of iron and reactive oxygen species in cell death. *Nat Chem Biol* 2014;**10**:9–17.
8. Halliwell B, Gutteridge JMC. *Free Radicals in Biology and Medicine*, 3rd edn. Oxford: Oxford University Press. 1999.
9. Winterbourn CC. Reconciling the chemistry and biology of reactive oxygen species. *Nat Chem Biol* 2008;**4**(5):278–86.
10. Kirkinezos IG, Moraes CT. Reactive oxygen species and mitochondrial diseases. *Semin Cell Dev Biol* 2001;**12**:449–57.
11. Kell D. Towards a unifying, systems biology understanding of large-scale cellular death and destruction caused by poorly liganded iron: Parkinson's, Huntington's, Alzheimer's, prions, bactericides, chemical toxicology and others as examples. *Arch Toxicol* 2010;**84**:825–89.
12. Haber F, Weiss J. The catalytic decomposition of hydrogen peroxide by iron salts. *Proc R Soc London Ser A* 1934;**147**:332–51.
13. Kehrer JP. The Haber–Weiss reaction and mechanisms of toxicity. *Toxicology* 2000;**149**:43–50.

14 Winterbourn CC. Toxicity of iron and hydrogen peroxide: the Fenton reaction. *Toxicol Lett* 1995;**82–83**:969–74.
15 Hirota K, Murata M, Sachi Y, Nakamura H, Takeuchi J, Mori K, Yodoi J. Distinct roles of thioredoxin in the cytoplasm and in the nucleus a two step mechanism of redox regulation of transcription factor NF-κB. *J Biol Chem* 1999;**274**(24):27 891–7.
16 Jonas NJ, Arnér ESJ. Reactive oxygen species, antioxidants, and the mammalian thioredoxin system. *Free Radic Biol Med* 2001;**31**(11):1287–312.
17 Poljšak B, Fink R. The protective role of antioxidants in the defence against ROS/RNS-mediated environmental pollution. *Oxid Med Cell Longev* 2014;**2014**:671539.
18 Valko M, Izakovic M, Mazur M, Rhodes CJ, Telser J. Role of oxygen radicals in DNA damage and cancer incidence. *Mol Cell Biochem* 2004;**266**:37–56.
19 Loschen G, Azzi A. On the formation of hydrogen peroxide and oxygen radicals in heart mitochondria. *Recent Adv Stud Cardiac Struct Metab* 1975;**7**:3–12.
20 Boveris A, Cadenas E. Mitochondrial production of superoxide anions and its relationship to the antimycin insensitive respiration. *FEBS Lett* 1975;**54**(3):311–14.
21 Toyokuni S, Akatsuka S. Pathological investigation of oxidative stress in the post-genomic era. *Pathol Int* 2007;**57**:461–73.
22 Kuroda J, Ago T, Nishimura A, Nakamura K, Matsuo R, Wakisaka Y, et al. Nox4 is a major source of superoxide production in human brain pericytes. *J Vasc Res* 2014;**51**(6):429–38.
23 Sciarretta S, Yee D, Ammann P, Nagarajan N, Volpe M, Frati G, Sadoshima J. Role of NADPH oxidase in the regulation of autophagy in cardiomyocytes. *Clin Sci (Lond)* 2015;**128**(7):387–403.
24 Fruehauf JP, Meyskens FL Jr. Reactive oxygen species: a breath of life or death? *Clin Cancer Res* 2007;**13**(3):789–94.
25 Bedard K, Krause KH. The NOX family of ROS-generating NADPH oxidases: physiology and pathophysiology. *Physiol Rev* 2007;**87**(1):245–313.
26 Kim C, Kim YJ, Kim JH. Cytosolic phospholipase A2, lipoxygenase metabolites, and reactive oxygen species. *BMB Rep* 2008;**41**(8):555–9.
27 Lodovici M, Bigagli E. Oxidative stress and air pollution exposure. *J Toxicol.* 2011;**2011**:487074.
28 Repetto M, Semprine J, Boveris A. Lipid peroxidation: chemical mechanism, biological implications and analytical determination. Available from http://www.intechopen.com/books/lipid-peroxidation/lipid-peroxidation-chemical-mechanism-biological-implications-and-analytical-determination (last accessed March 22, 2018).
29 Barrera G. Oxidative stress and lipid peroxidation products in cancer progression and therapy. *ISRN Oncol* 2012;**2012**:137289.
30 Davies KJ. Intracellular proteolytic systems may function as secondary antioxidant defenses: an hypothesis. *J Free Radic Biol Med* 1986;**2**(3):155–73.
31 Davies KJ, Goldberg AL. Oxygen radicals stimulate intracellular proteolysis and lipid peroxidation by independent mechanisms in erythrocytes. *J Biol Chem* 1987;**262**(17):8220–6.
32 Beckman KB, Ames BN. Oxidative decay of DNA. *J Biol Chem* 1997;**272**(32):19 633–6.
33 Helbock HJ, Beckman KB, Shigenaga MK, Walter PB, Woodall AA, Yeo HC, Ames BN. DNA oxidation matters: the HPLC-electrochemical detection assay of 8-oxo-deoxyguanosine and 8-oxo-guanine. *Proc Natl Acad Sci USA* 1998;**95**(1):288–93.

34 McNeil JD, Wiebkin OW, Betts WH, Cleland LG. Depolymerisation products of hyaluronic acid after exposure to oxygen-derived free radicals. *Ann Rheum Dis* 1985;**44**(11):780–9.

35 Hampton MB, Fadeel B, Orrenius S. Redox regulation of the caspases during apoptosis. *Ann N Y Acad Sci* 1998;**854**:328–35.

36 Dolma S, Lessnick SL, Hahn WC, Stockwell BR. Identification of genotype-selective antitumor agents using synhetic lethal chemical screening in engineered human tumor cells. *Cancer Cell* 2003;**3**:285–96.

37 Dixon SJ, Lemberg KM, Lamprecht MR, Skouta R, Zaitsev EM, Gleason CE, et al. Ferrotosis: an iron-dependent form of nonapoptotic cell death. *Cell* 2012;**149**(5):1060–72.

38 Yang WS, SriRamaratnam R, Welsch ME, Shimada K, Skouta R, Viswanathan VS, et al. Regulation of ferroptotic cancer cell death by GPX4. *Cell* 2014;**156**:317–31.

39 Yagoda N, von Rechenberg M, Zaganjor E, Bauer AJ, Yang WS, Fridman DJ, et al. RAS-RAF-MEK-dependent oxidative cell death involving voltage-dependent anion channels. *Nature* 2007;**447**(7146):864–8.

40 Chang F, Steelman LS, Lee JT, Shelton JG, Navolanic PM, Blalock WL, et al. Signal transduction mediated by the Ras/Raf/MEK/ERK pathway from cytokine receptors to transcription factors: potential targeting for therapeutic intervention. *Leukemia* 2003;**17**:1263–93.

41 Maldonado EN, Sheldon KL, Dehart DN, Patnaik J, Manevich Y, Townsend DM, et al. Voltage dependent anion channels modulate mitochondrial metabolism in cancer cells: regulation by free tubulin and erastin. *J Biol Chem* 2013;**288**(17):11 920–9.

42 White MK, McCubrey JA. Suppression of apoptosis: role in cell growth and neoplasia. *Leukemia* 2001;**15**(7):1011–21.

43 McCubrey JA, Lee JT, Steelman LS, Blalock WL, Moye PW, Chang F, et al. Interactions between the PI3K and Raf signaling pathways can result in the transformation of hematopoietic cells. *Cancer Detect Prev* 2001;**25**(4):375–93.

44 Yang WS, Stockwell BR. Synthetic lethal screening identifies compounds activating iron-dependent, nonapoptotic cell death in oncogenic-RAS-harboring cancer cells. *Chem Biol* 2008;**15**(3):234–45.

45 Murphy TH, Miyamoto M, Sastre A, Schnar RL, Coyle JT. Glutamate toxicity in a neuronal cell line involves inhibition of cystine transport leading to oxidative stress. *Neuron* 1989;**2**:1547–58.

46 Di Mascio P, Murphy ME, Sies H. Antioxidant defense systems: the role of carotenoids, tocopherols, and thiols. *Am J Clin Nutr* 1991;**53**(1 Suppl.):194S–200S.

47 Townsend DM, Tew KD, Tapiero H. The importance of glutathione in human disease. *Biomed Pharmacother* 2003;**57**(3–4):145–55.

48 Larson RA. The antioxidants of higher plants. *Phytochemistry* 1988;**2**:969–78.

49 Colombini M. VDAC: the channel at the interface between mitochondria and the cytosol. *Mol Cell Biochem* 2004;**256–257**(1–2):107–15.

50 Rostovtseva T, Colombini M. ATP flux is controlled by a voltage-gated channel from the mitochondrial outer membrane. *J Biol Chem* 1996;**271**(45):28 006–8.

51 Brigelius-Flohe R, Maiorino M. Glutathione peroxidases. *Biochim Biophys Acta* 2013;**1830**:3289–303.

52 Valko M, Morris H, Cronin MTD. Metals, toxicity and oxidative stress. *Curr Med Chem* 2005;**12**:1161–20.

53 Brock JH. Iron-binding proteins. *Acta Paediatr Scand Suppl* 1989;**361**:31–43.

54 Aisen P, Listowsky I. Iron transport and storage proteins. *Annu Rev Biochem* 1980;**49**:357–93.
55 Zhang AS, Enns CA. Molecular mechanisms of normal iron homeostasis. *Hematology Am Soc Hematol Educ Program* 2009: 207–14.
56 Reed J. Mechanisms of apoptosis. *Am J Pathol* 2000;**157**(5):1415–30.
57 Kanzawa T, Germano IM, Komata T, Ito H, Kondo Y, Kondo S. Role of autophagy in temozolomide-induced cytotoxicity for malignant glioma cells. *Cell Death Differ* 2004;**11**:448–57.
58 Kondo Y, Kanzawa T, Sawaya R, Kondo S. The role of autophagy in cancer development and response to therapy. *Nat Rev Cancer* 2005;**5**:726–34.
59 You-Zhi L, Chiang JL, Pinto AV, Pardee AB. Release of mitochondrial cytochrome C in both apoptosis and necrosis induced by 8-lapachone in human carcinoma cells. *Mol Med* 1999;**5**(4):232–9.
60 Lefranc F, Kiss R. Autophagy, the Trojan horse to combat glioblastomas. *Neurosurg Focus* 2006;**20**(4):E7.
61 Anheli JPF, Schneider M, Proneth B, Tyurina YY, Tyurin VA, Hammond VJ, et al. *Nat Cell Biol* 2014;**16**:1180–91.
62 Hanahan D, Weinberg RA. The hallmarks of cancer. *Cell* 2000;**100**(1):57–70.
63 Bezabeh T, Mowat MRA, Jarolim L, Greenberg AH, Smith ICP. Detection of drug-induced apoptosis and necrosis in human cervical carcinoma cells using HNMR spectroscopy. *Cell Death Diff.* 2001;**8**(3):219–24.
64 Ocker M, Höpfner M. Apoptosis-modulating drugs for improved cancer therapy. *Eur Surg Res* 2012;**48**(3):111–20.
65 Llovet JM, Ricci S, Mazzaferro V, Hilgard P, Gane E, Blanc JF, et al. Sorafenib in advanced hepatocellular carcinoma. *N Engl J Med* 2008;**359**(4):378–90.
66 Dixon SJ, Patel DN, Welsch M, Skouta R, Lee ED, Hayano M, et al. Pharmacological inhibition of cystine-glutamate exchange induces endoplasmic reticulum stress and ferroptosis. *Elife* 2014;**3**:e02523.
67 Louandre C, Ezzoukhry Z, Godin C, Barbare JC, Mazière JC, Chauffert B, Galmiche A. Iron-dependent cell death of hepatocellular carcinoma cells exposed to sorafenib. *Int J Cancer* 2013;**133**(7):1732–42.
68 Louandre C, Marcq I, Bouhlal H, Lachaier E, Godin C, Saidak Z, et al. The retinoblastoma (Rb) protein regulates ferroptosis induced by sorafenib in human hepatocellular carcinoma cells. *Cancer Lett* 2015;**356**(2 Pt. B):971–7.
69 Jomova K, Vondrakova D, Lawson M, Valko M. Metals, oxidative stress and neurodegenerative disorders. *Mol Cell Biochem* 2010;**345**(1–2):91–104.
70 Skouta R, Dixon SJ, Wang J, Dunn DE, Orman M, Shimada K, et al. Ferrostatins inhibit oxidative lipid damage and cell death in diverse disease models. *J Am Chem Soc* 2014;**136**(12):4551–6.

8

Anoikis Regulation: Complexities, Distinctions, and Cell Differentiation

Marco Beauséjour, Ariane Boutin, and Pierre H. Vachon

Department of Anatomy and Cellular Biology, University of Sherbrooke, Sherbrooke, QC, Canada

Abbreviations

AJ	adherens junction
Akt (PKB)	Ak mouse strain thymoma (protein kinase B)
Apaf-1	apoptosis protease activating factor-1
ASK-1	apoptosis signal-regulating kinase-1
Bad	Bcl-2-associated death promoter
Bak	Bcl-2 homologous antagonist killer
Bax	Bcl-2-associated X protein
Bcl-2	B-cell lymphoma-2
Bcl-X_L	B-cell lymphoma-extra large
BID	BH3-interacting-domain death agonist
Bim	Bcl-2-interacting mediator of cell death
Bit1	Bcl-2 inhibitor of transcription protein 1
Bmf	Bcl-2-modifying factor
Bok	Bcl-2-related ovarian killer
CAD	caspase-activated DNase
CAM	cell-adhesion molecule
Cdk5	cyclin-dependent kinase 5
c-Flip	cellular FLICE-inhibitory protein
cyt c	cytochrome c
DAPK1/2	death-associated protein kinase 1/2
DISC	death-inducing signaling complex
EB-PA	epidermolysis bullosa with pyloric atresia
EBS	epidermolysis bullosa simplex
E-cadherin	epithelial cadherin
ECM	extracellular matrix
EGF	epidermal growth factor
EGFR	EGF receptor
EMT	epithelial–mesenchymal transition
ER	endoplasmic reticulum
Erk	extracellular regulated kinase
FADD	Fas-associated death-domain protein
FAK	focal adhesion kinase

Apoptosis and Beyond: The Many Ways Cells Die, First Edition. Edited by James Radosevich.
© 2018 John Wiley & Son Inc. Published 2018 by John Wiley & Son Inc.

FasL	Fas ligand
Grb2	growth factor receptor-bound protein 2
GSK-3β	glycogen synthase kinase-3β
hIEC	human intestinal epithelial cell
IAP	inhibitor of apoptosis protein
ILK	integrin-linked kinase
jBID	JNK-dependent-phosphorylated BID
ITG7A CMD	congenital muscular dystrophy with integrin alpha 7 deficiency
JEB-H	junctional epidermolysis bullosa–Herlitz
JNK	C-Jun N-terminal kinase
Mcl-1	myeloid leukemia cell differentiation protein 1
MDC1A	merosin-deficient congenital muscular dystrophy type 1A
Mdm2	mouse double minute 2 homolog
MEK1/2	mitogen/extracellular signal-regulated kinase 1/2
MEKK1	mitogen-activated protein kinase kinase kinase 1
MK	mitogen-activated protein kinase-activated protein kinase
MLC	myosin light chain
MSK1/2	mitogen and stress kinase 1/2
mTORC2	mammalian target of rapamycin complex 2
PDK1	phosphoinositide-dependent kinase 1
PI3-K	phosphatidylinositol 3-kinase
PIDD	p53-induced death domain protein 1
PINCH1/2	particularly interesting new cysteine–histidine-rich protein 1/2
PTK6	protein tyrosine kinase 6
Raf	rapidly accelerated fibrosarcoma kinase
RAIDD	RIPK-associated protein with a death domain
Ras	rat sarcoma small GTPase
RCD	regulated cell death
RhoA	Ras homolog gene member A
RICTOR	rapamycin-insensitive companion of mTOR
RIPK1	receptor-interacting protein kinase 1
ROCK1	Rho-associated, coiled coil-containing protein kinase 1
RSK-1	p90 ribosomal S6 kinase-1
RTK	receptor tyrosine kinase
SAM	substrate-adhesion molecule
SAPK	stress-activated MAPK
Shc	Src homology 2 domain-containing transforming protein 1
tBID	truncated BID
TGF-α	transforming growth factor-α
TNF-α	tumor necrosis factor-α
TRADD	TNF receptor-associated death domain
TRAIL	TNF-related apoptosis-inducing ligand

8.1 Anoikis: Basic Forensics

Where multicellular organisms like mammals are concerned, a multiplex array of regulatory systems is required for the embryogenesis and ontogeny, as well as for the maintenance, renewal, and repair, of the multifaceted assortment of organs and tissues that allows them to function, survive, and reproduce. One such critical system is **regulated cell death** (RCD) [1–12], broadly defined as "a sequence of events based on cellular metabolism that lead to cell destruction" [1]. Notwithstanding, such a concise definition, RCD includes a number of mechanistically distinct subroutines of cell death (apoptosis, autophagic cell death, programmed necrosis, etc.), which can be caspase-dependent, caspase-independent, or both, depending on the deathly stimulus, the initiating transduction events that activate the executioner mechanisms, and the molecular makeup of said executioner mechanisms [1,2,7–12]. This chapter will focus on aspects of the "caspase-dependent apoptosis" side of RCD, particularly with regard to integrin-mediated signaling and a distinct subroutine of apoptosis called **anoikis**.

Signifying in the Greek language "homelessness" or "the state of being without a home," the term "anoikis" was first coined by S.M. Frisch in 1994 as an identifier of a then-novel anchorage-dependent form of RCD [13]. Also referred to as "detachment-induced apoptosis" or "integrin-mediated death," anoikis is primarily defined as caspase-dependent apoptosis that is induced by inadequate, inappropriate, or non-existent cell–extracellular matrix (ECM) interactions [10,13–25]. Consequently, an overview of apoptosis and of cell–ECM and cell–cell adhesion signaling is mandatory before we can tackle the basics of anoikis, its regulation, its initiation, and its execution.

8.1.1 Cell Survival: The Balance Between Life and Death

Regardless of their state of being, normal cells are intrinsically wired to enter apoptosis by default [1–12,25]. In other words, cells require survival signals to remain alive and, consequently, to suppress apoptosis – unless apoptosis is required [1–9,12,14–25]. Depending on the tissue and cell type, survival signals will often include those provided by distinct growth factors, hormones, and their specific receptors [3–5,12,25]. In addition to the loss of such survival signals, apoptosis may likewise be induced by proinflammatory cytokines, as well as through various insults that cause cell stress or cell damage, such as high-error DNA replication, plasma-membrane peroxidation, DNA damage, protein misfolding, and mitochondrial dysfunction [1–9,12,25].

B-cell lymphoma-2 (Bcl-2) homologs perform critical roles as decisional arbiters of cell survival and apoptosis [6–9,19,24–33]. In humans, some 20 members of this family have been identified so far [26–33]. The majority of Bcl-2 homologs are ubiquitously expressed [26–33]; however, some can be selectively expressed in a more restricted subset of tissues [26–33]. Furthermore, some homologs can be expressed exclusively following specific apoptotic stimuli [26–33].

Bcl-2 homologs act principally as apoptotic suppressors (e.g., Bcl-2, Bcl-X_L, Mcl-1), effectors (Bax, Bak and Bok), sensitizers (e.g., Bad, Bmf, Noxa), or activators (e.g., BID, Bim, Puma) [6–9,19,24–33]. When a cell is subjected to an apoptotic stimulus, the balance of Bcl-2 homolog expression and activities will be affected, so that sensitizers and activators gain prominence, allowing them to inhibit the suppressors and thus prevent them latter from blocking the effectors, which, in turn, are then free to form pores at mitochondria [6–9,19,24–33]. Alternately, the activators can synergize with the effectors to facilitate their translocation from the cytosol to mitochondria [6–9,19,24–33]. In any case, the pores formed by effectors cause a loss of mitochondrial membrane integrity that leads to the release of cytochrome c (cyt c) [6,24–36]. In the cytosol, cyt c acts as cofactor to apoptosis protease activating factor-1 (Apaf-1) in forming the **apoptosome**, which recruits the precursor form of the initiator caspase 9 ("pro-caspase 9") for the latter's subsequent large-scale activation [6,24–36]. In turn, caspase 9 initiates an irreversible amplifying activation cascade of executioner caspases (caspases 3, 6, and 7), which activates in turn additional apoptotic implementers (e.g., caspase-activated DNase (CAD), which enacts internucleosomal DNA degradation that is characteristic of caspase-dependent apoptosis) and cleave their various substrates (e.g., kinases, actin, etc.) [34–40]. Incidentally, numerous additional molecules implicated in the decision and/or in execution of apoptosis have been identified to date, including inhibitors of apoptosis proteins (IAPs), which inhibit caspases, and IAP inhibitors (e.g., Smac/Diablo, Omni), which are released via the loss of mitochondrial membrane integrity caused by effector Bcl-2 homologs [36,41–44].

Hence, the "life versus death" fate of a cell depends primarily on a closely controlled balancing act between the anti- and proapoptotic activities of Bcl-2 homologs [6–9,19,24–33]. In addition to the direct regulation of gene expression via survival signals or specific apoptotic stimuli (e.g., p53 inducing the expression of Puma and Noxa upon DNA damage), alternate mRNA splicing (e.g., the Bcl-X_L mRNA being alternately spliced as a proapoptotic, Bcl-X_S mRNA), proteolytic processing (e.g., inactive precursor BID being cleaved into activator-functional truncated BID (tBID)), phosphorylation (e.g., phosphorylation of Bad on either of the S112, S136, or S155 residues, inactivating its sensitizer functions), and sequestration (e.g., Bim and Bmf, typically sequestered in microtubules and actin microfilaments, respectively, and thus being prevented from performing their proapoptotic functions unless released) all likewise contribute significantly to establishing, as well as affecting, such a critical balance [6–9,25–33,45–47].

8.1.2 Of Apoptosomes and DISCs

It must be pointed out here that the previously described sequence of events that led to the establishment of the apoptosome (and consequent caspase 9 activation) constitutes what is generally recognized as being the **intrinsic**, or "common," pathway of apoptosis [1–5,7–12,31,32,34–36,44,47]. Conversely, the **extrinsic** (or "death receptor" or "receptor-mediated") pathway [2–4,10–12,25–28,48–53] is induced by the binding of "death ligands" (tumor necrosis factor-α (TNF-α), Fas ligand (FasL), TNF-related apoptosis-inducing ligand (TRAIL)) to their specific "death receptors" [50–53]. Upon activation, these receptors recruit the adaptor Fas-associated death-domain protein (FADD) (TNF receptor-associated death domain (TRADD), in the case of TNF-α and its receptor), which in turn recruits the precursor form of the initiator caspase 8 and, thus, forms a **death-inducing signaling complex** (DISC) [48–53], which drives the activation of caspase 8 (caspase 10, another initiator caspase, may also be activated via a DISC, depending on the death receptor implicated) [2–4,10–12,25–28,48–53]. Incidentally, it is of note that the primary inhibitor of death receptor-mediated apoptotic signaling is the cytosolic c-Flip protein, which blocks the formation of a DISC by binding FADD (or TRADD) [49–53,61]. The antiapoptotic functions of c-Flip to this effect are primarily regulated via its expression levels, like the aforementioned IAPs.

There are two subtypes of extrinsic apoptosis: type I and type II. In the case of type I cell death, cells display a high density of death receptors, and/or a high expression of pro-caspase 8, and/or a low expression of IAPs (and c-Flip); consequently, the activation of caspase 8 will be of sufficient intensity to result in the amplifying activation cascade of executioner caspases, making the process forthwith irreversible [3,4,10–12,48–53]. In type II cell death, however, cells display a low density of death receptors and/or a low expression of pro-caspase 8; hence, the activation of caspase 8 will instead be of low intensity, leading to a low-intensity activation of executioner caspases. Consequently, for the process to be rendered irreversible, caspase 8 and executioner caspases must initially perform tasks that will affect a translocation of the receptor-initiated death signal toward that of the formation of the apoptosome [3,4,10–12,25–28,48–53]. Such additional tasks include the cleavage of BID into tBID and the proteolytic-driven destabilization of actin microfilaments and microtubules (causing the release of Bmf and Bim), which together cause a shift in the balance of anti- and proapoptotic Bcl-2 homolog activities in favor of effectors being allowed to form pores in mitochondria and incite apoptosome formation [3,4,10–12,25–28,48–53].

Although the use of the terms "intrinsic" and "extrinsic" remains popular to this day, in order to discriminate functionally what are commonly referred to as "the two major apoptotic pathways" [2–8,10–12,24–28], it has become evident over the last decade or so that things are not that clear cut (pun intended). For instance, and as just mentioned, it is frequently the case that a death receptor-mediated (i.e., extrinsic) induction of apoptosis necessitates the translocation of the DISC signal to the apoptosome (i.e., the intrinsic pathway) if the process is to be irreversible and final [3,4,10–12,25–28,48–53]. Furthermore, the extrinsic pathway may not lead to apoptosis per se, but instead induce autophagic cell death or regulated necrosis – again, depending on the cell type and the physiological circumstances at play [3,4,10–12,25–33,47–54]. This should not be surprising, considering the intersecting "gray areas" between apoptosis, autophagic cell death, and regulated necrosis, including whether these are executed in a caspase-dependent or -independent manner [2–5,7–12,26–28,30–33,47–54].

It is, however, the activation of caspase 2 (also an initiator caspase) that best illustrates the blurring of lines between the so-called "intrinsic" and "extrinsic" pathways of apoptosis induction. On the one hand, specific apoptotic stimuli such as DNA damage or stress/damage sustained by the endoplasmic reticulum (ER) will induce the scaffold p53-induced death domain protein 1 (PIDD) to recruit the adaptor receptor-interacting protein kinase (RIPK)-associated protein with a death domain (RAIDD), which will in turn recruit pro-caspase 2 in order to form the **PIDDosome**, resulting in the activation of caspase 2 [3,4,7–9,48,49,53,55–58]. Once activated, caspase 2 will (at the very least) cleave BID into tBID, contributing to a translocation of the signal to the apoptosome [2–5,7,8,55–58]. On the other hand, the activation of a death receptor can lead to its direct recruitment of RAIDD, which will then recruit pro-caspase 2, leading again to caspase 2 activation [50,53–55,57–59]. Incidentally, caspase 2 can even be activated in a DISC-activated-caspase 8-dependent manner [57,58,60], presumably in order to accelerate translocation of the deathly signal to the apoptosome.

Hence, RCD pathways can not only crosstalk, but are not restricted to obtusely defined terms such as "extrinsic" and "intrinsic." The relevance of laying out such considerations will become evident later in the chapter.

8.1.3 Integrins and Cell–ECM Interactions: The 4-1-1

Cell–ECM interactions play crucial roles in regulating cell processes, including cell survival [14–25,62,63]. The transmembrane receptors of the integrin family are the primary mediators of the biological and physiological functions of cell–ECM interactions [14,15,18–25,62–69]. Integrins are heterodimeric ($\alpha\beta$); eighteen α and eight β subunits have been identified so far in humans, forming twenty-four distinct receptors with differing – albeit, in some cases overlapping – ligand specificities [64–71]. Integrins that share the $\beta1$ subunit in common constitute most receptors for ECM components and are thus known as the "cell–ECM adhesion integrins" (or "cellular integrins") functional subgroup of the larger integrin family [64–71]. This subgroup also includes the α6β4 integrin, which is expressed exclusively in epithelial cells [70,75,76].

For any given cell type, the repertoire of integrins displayed depends on the specific contexts of ECM composition, tissue type, and/or species concerned [23,25,69,70,75]. Furthermore, post-transcriptional alternative splicing and/or post-translational proteolytic processing can occur for some α and β integrin subunits; this results largely in the production of variants with alterations in their cytoplasmic tails, adding further versatility to their signaling properties and, consequently, to their roles and functions [72–74]. Hence, cellular responses to the ECM mediated by integrins are quite varied and multifaceted [14–25,62–77]. In this respect, the first reports showing that cell anchorage to the ECM constitutes a critical survival factor [13,78] were swiftly followed by observations that distinct ECM components and integrins are distinctively responsible for cell survival according to the cell type, tissue type, and/or species [15,18,21,23,25,63,75–77,79–86].

8.1.4 Integrin Signaling: Anchorage Life Lines

Integrins binding to their ECM ligands not only create direct links between a cell's cytoskeleton and the ECM, but also engage an assortment of transduction signals that direct gene expression, cell shape, and cell behavior (Figure 8.1) [14–25,64–68,70,71,74,75,77,79,87,88]. Although ligand-binding (i.e., activated) integrins lack intrinsic enzymatic activity, they nonetheless mediate the assembly of cytoplasmic scaffolding and signaling networks into **integrin adhesomes** ("outside–in signaling"), thereby functioning as conduits of bidirectional signal transduction across the cell membrane [64–67,70,74,77,87–91]. Conversely, some cytoplasmic interactions can regulate an integrin's affinity for its ECM ligand and dictate its activation state regardless of ECM ligand availability for binding ("inside–out signaling") [14–25,65–70,74–77,87–90]. In addition, some integrins can enact lateral interactions with other cell-surface proteins, such as growth-factor receptors, tetraspanins, nectins, and caveolins, further expanding integrin signaling capabilities and/or modulatory means of integrin activity [14–25,65–70,74–77,87–90]. Hence, a given repertoire of expressed integrins can not only produce distinct signals for a specific cell type, but also perform distinct modulatory roles in the control of cell processes within the same given tissue [14–25,64–68,70,71,74,75,77,79,87–90]. To this effect, the still-growing list of signal-transduction pathways that can be engaged by integrins is rather extensive; for the purposes of the present chapter, it is sufficient to note here that, for a given cell, pathways such as PI3-K/Akt, Ras/Raf/MEK/Erk, or the pathways of C-Jun N-terminal kinase (JNK) and p38 stress-activated mitogen-activated protein kinases (SAPKs) can be activated alone – or in combination – according to the repertoire of integrins expressed [14–25,64–70,74–77,87–91].

Because integrins lack enzymatic activity, they require kinase proxies to enact signal transduction following their activation (Figure 8.1). Hence, signaling by β1 integrins results largely from the recruitment and activation of focal adhesion kinase (FAK) [24,25,66,67,87–89,91–95], which typically then recruits and activates Src [24,25,66,67,87–89,91–102]. Conversely, Src may be the first to be recruited and activated by integrins [89,98,103], then recruit/activate FAK in turn [89,98–100,102]. In any event, integrin-mediated FAK/Src signaling represents a major driver of the assembly of integrin adhesomes, which constitute diverse signaling cassettes consisting of an increasing array of signaling molecules, including adaptors, scaffolders, and other kinases (Figure 8.1) [66,67,77,87–102,104,105]. Likewise, integrin-mediated FAK/Src

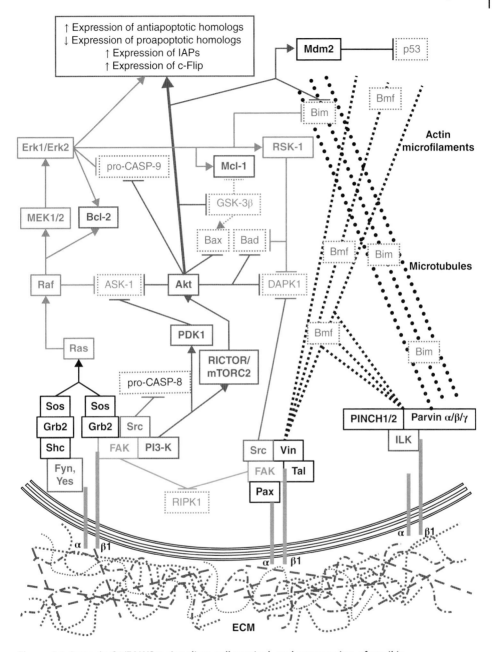

Figure 8.1 Integrin β1/FAK/Src signaling, cell survival, and suppression of anoikis.

signaling contributes to the assembly and formation of focal adhesions via their interactions with a rising number of partners (e.g., paxillin, kindlins, talin, vinculin), leading not only to the creation of a bridge between the ECM and the actin cytoskeleton, but also to the regulation of the stability and organization of actin microfilaments and microtubules (Figure 8.1) [19,22–25,64–69,77,79,87–89,91–102,104–106]. In addition to establishing overall cytoskeletal stability and tensegrity, this enables – or maintains – the sequestration of Bmf and/or Bim (Figure 8.1). The modulation of integrin-mediated signaling and focal adhesion assembly can further be generated by specific α integrin subunits, via the lateral association with membrane proteins (e.g., tetraspanins, caveolins, nectins) or via other adaptors and kinases (Figure 8.1) [66,67,70,76,77,88–91,94,99,102,104–109]. Similarly to β1 integrins, the α6β4 integrin participates in the stimulation of pathways such as PI3-K/Akt and Ras/Raf/MEK/Erk; however, it does so chiefly via its engagement of Src, and not of FAK. Additionally, α6β4 is a key contributor to the formation and maintenance of hemidesmosomes: cell-anchoring complexes that physically link the ECM to keratin intermediate filaments, which differ greatly in composition from focal adhesions and yet further organize/stabilize the cytoskeleton of epithelial cells [70,76,109–114].

It is given that overall signal transduction by integrins is similar to that of receptor tyrosine kinases (RTKs). To this effect, there is extensive crosstalk between RTK- and integrin-mediated signaling, the two often cooperating in the regulation of various cell processes, including survival [17–25,66,67,69,74–77,88,89,91,94–96,98–100,108,109,111,115–118]. Incidentally, integrins can perform the inside–out activation of RTK signaling, and vice versa, largely via FAK, Src, and/or integrin-linked kinase (ILK) [17–25,66,67,69,74–77,88,89, 91–101,108–111,115–118]. Hence, integrin-mediated signaling can be expanded even further through such crosstalk with various growth factor receptors, whether they be RTKs or otherwise [17–25,66,67,69,74–77,88,89,91,94–96,98–100,108,109,111,115–118].

Much remains to be understood of the integrin-mediated promotion of cell survival. In any case, PI3-K/Akt and Ras/Raf/MEK/Erk represent the best-known survival-promoting pathways to be engaged by integrin-mediated FAK/Src signaling (Figure 8.1) [17–25,67,69,87–89,91,94,95,99,100,111,119–142]. Concerning PI3-K/Akt, Akt is well known for its numerous cell-survival functions, such as the inhibitory phosphorylation of Bax, Bad, Bim, and GSK-3β (which is known to positively phosphorylate Bax and negatively phosphorylate Mcl-1), the phosphorylation of pro-caspase 9 (to suppress its activation), and the stimulating phosphorylation of Mdm2 (and consequent inhibitory phosphorylation of p53 by the latter) (Figure 8.1) [24–28,120–122, 125–129,143–145]. Interestingly, Akt can also bind, phosphorylate, and thus lead to the activation of FAK [148]. It is now well established that the PI3-K-dependent phosphorylation of the S473 residue of Akt by rapamycin-insensitive companion of mammalian target of rapamycin (RICTOR)/mammalian target of rapamycin complex (mTORC2) is indicative not only of its activation, but also of its principal commitment to cellular survival functions [120,121,125,127–129,146,147]. However, Akt activation can be independent of PI3-K activity, a situation that appears to be species- and cell type-dependent, as well as isoform-dependent [119–130,149]. To this effect, Src, as well as fellow family members Fyn, Yes, and protein tyrosine kinase 6 (PTK6), can bind and phosphorylate Akt to activate the latter, or at least to potentiate its activity [149–152]. ILK is often still considered part of the PI3-K/Akt pathway [107,108,154–156]. Indeed, and although ILK contributes to the formation of integrin-mediated focal adhesions by binding directly to the cytoplasmic tail of the β subunits of activated receptors (Figure 8.1) [107,108], it has been shown to be PI3-K-dependent in its activation and to act as an upstream activator of Akt [120,153]. However, considering the weight of evidence that has presently established ILK as a pseudokinase [154–156], its previously tacit kinase-mediated signaling roles in promoting integrin-driven cell survival, including Akt activation, should be set aside without further consideration. In any event, the scaffolding functions of ILK in focal adhesion assembly, actin microfilament linkage (with scaffold/adaptor partners particularly interesting new cysteine–histidine-rich protein 1/2 (PINCH1/2) and parvin α, β, γ) [107,108,153,155,156], and microtubule assembly/stability [156,157] are definite contributors to the sequestration of Bmf and Bim (Figure 8.1).

With regard to Ras/Raf/MEK/Erk, it is known that Raf can phosphorylate Bcl-2 in order to potentiate its suppressor functions, whereas p90 ribosomal S6 kinase-1 (RSK-1) (downstream effector of Erk1/Erk2) enacts an inhibitory phosphorylation of Bad (Figure 8.1) [133–137,143,144,158,159]. Erk1/Erk2 can likewise phosphorylate Bcl-2 (to potentiate its suppressor functions), Mcl-1 (to protect it from degradation), Bim (to inhibit its activator functions), and pro-caspase 9 (to prevent its activation) (Figure 8.1) [131–133,137,143,144]. Incidentally, and aside from some obvious overlapping roles, PI3-K/Akt and Ras/Raf/MEK/Erk can crosstalk with each other to cooperate – if not synergize – in the promotion of cell survival. To this effect, Ras can activate PI3-K, leading to the activation of Akt [120–122,126,127,130,139–142,160], whereas phosphoinositide-dependent kinase 1 (PDK1) (another effector of PI3-K) can directly activate RSK-1 [158–161]. In any case, the engagement of PI3-K/Akt and/or Ras/Raf/MEK/Erk by integrin-mediated FAK/Src signaling results overall in the upregulation of the expression of the antiapoptotic Bcl-2 homologs, IAPs and c-Flip, and in the simultaneous downregulation of the expression of proapoptotic Bcl-2 homologs (Figure 8.1) [17–25,120,125–129,133,137].

Hence, depending on the composition of the integrin adhesomes they help assemble and the specific integrin receptors implicated, the importance of FAK and/or Src in the integrin-driven promotion of cell survival is intimately linked with their pivotal role in the engagement of pathways such as PI3-K/Akt and/or Ras/Raf/MEK/Erk (Figure 8.1), according to the cell type and species studied. Incidentally, FAK and Src can furthermore contribute directly to the promotion of cell survival (Figure 8.1). For instance, Src can phosphorylate pro-caspase 8 (to suppress its activation), allowing the latter to recruit and engage PI3-K as a constituent of integrin adhesomes [162,163]. As for FAK, it has been shown that it can bind and sequester RIPK1 in order to prevent its recruitment of FADD and the consequent formation of a DISC [164,165]. Furthermore, Cdk5-phosphorylated FAK plays a role in microtubule organization/stability [166] and, by inference, in contributing to the sequestration of Bim.

In conclusion, a cell's integrin-mediated anchorage to its ECM constitutes a potent, multilayered device whose most significant purpose is to promote and maintain cell survival. Interestingly, the complex nature of the regulation of cell processes, including survival, is strikingly enhanced by the specific implication of isoforms, splicing variants, and/or family members of adaptors, scaffolds, and kinases, often resulting in distinct – if not differing – roles among the pathways in which they engaged. A few examples identified to date in humans include PI3-K isoform complexes (four catalytic subunit isoforms: p110α, β, γ, and δ; three regulatory subunit isoforms: p85α, p85β, and p55γ), Akt isoforms (−1 to −3), and Shc ($p46^{Shc}$, $p52^{Shc}$, $p66^{Shc}$) family members [97,101,119,123,125,126,129,130,134–136,158,159,167,168]. That these can be selectively expressed according to cell type, tissue type, and species, in addition to their specific functions within the same given cell type, further underscores the intricacies involved in the regulation of the various known cell processes – including, of course, integrin-mediated cell survival.

8.1.5 Adherens Junctions: Sticking Together for a Living

For multicellular organisms, a variety of cell-adhesion mechanisms govern the way cells are organized in their tissues and organs. The term "cell adhesion" is, in truth, an umbrella-type one that covers two distinct (yet similar) categories of interaction between cells and their environment: cell–ECM adhesion/interactions and cell–cell adhesion/interactions [169,170]. Both categories comprise the same three basic, functional components: (i) an extracellular ligand; (ii) a transmembrane receptor; and (iii) a linkage with the cytoskeleton – as well as the generation of signal transduction via adhesomes assembled at the cytoplasmic tail of the receptor [169–171]. Receptors for cell–ECM interactions and cell–cell adhesion are named, respectively, "substrate-adhesion molecules" (SAMs) and "cell-adhesion molecules" (CAMs) [169]. Like their cell–ECM counterparts, stable cell–cell interactions are required to maintain the structural integrity of tissues, while dynamic changes in cell–cell interactions are crucial for the morphogenesis, maturation, renewal, and healing/reparation, of tissues [169–171].

In epithelial cells, a prominent subtype of cell–cell interaction takes the form of **adherens junctions** (AJs). AJs are formed by extracellular homotypic interactions between transmembrane epithelial cadherins (E-cadherins) on (typically) the lateral surfaces of adjacent cells. The cytoplasmic domains of adhesion-active E-cadherins form complexes with plaque proteins known as catenins (namely, α-, β-, and p120-catenin) [106,169–174]. In turn, the catenins collaborate to establish a physical link (via vinculin, among others) with actin microfilaments of the cytoskeleton [106,169–174]. Additionally, binding E-cadherins will recruit Src in order to engage the PI3-K/Akt pathway [17,19,24,111,173–176]. Although much remains to be understood of E-cadherin-driven signaling, it is nonetheless established that AJs not only contribute to the regulation of the organization and stability of actin microfilaments [106,169–176] (and presumably to the sequestration of Bmf), but furthermore participate in the promotion of cell survival [17,19,24,111,175,176]. Interestingly, yet perhaps unsurprisingly, E-cadherin-mediated actin organization and cell signaling can crosstalk with the actin organization and cell signaling of RTKs (principally via Src) [17,19,106,111,115,117,175,176] and integrins (via Src and/or FAK) [17,19, 106,111,115,116,175–178].

8.2 Investigation of Anoikis: Who, What, When, Why?

Cells of multicellular organisms must continually monitor their environment by sensing soluble mediators, cell–cell contacts, and mechanical signals, all in order to perform their assigned roles and stay alive while doing so. In other words, it is a critical requirement for cells to integrate biochemical and mechanical information if they are to function within their tissue of residence [179–182]. One important example (for the present chapter, at least) of such environmental sensing is **mechanotransduction** (or "mechanosignaling"). Simply and concisely, the term "mechanotransduction" refers to the molecular mechanisms by which cells convert mechanical stimuli into biochemical activity and/or changes in gene expression, cell shape, or cell behavior. Each individual cell exists within a three-dimensional microenvironment, in which it is exposed to mechanical and physical cues – to this effect, it is principally via cell adhesion (cell–ECM and cell–cell) that a cell is able to sense and recognize not only the biochemical composition of its extracellular neighborhood, but also its physical and topographical characteristics (e.g., immediate neighboring cells, pliability, dimensionality, ligand spacing) [65,179–182].

It is now well recognized that mechanical forces contribute to the regulation of all known cell processes, including proliferation, differentiation, migration, tissue organization, and cell survival [65,179–182]. Hence, in this respect, integrins and E-cadherins function as mechanosensors through the focal adhesions (or hemidesmosomes) and AJs (respectively) that they form [65,68,176,179–182]. Given that anoikis is also known as "detachment-induced apoptosis" or "integrin-mediated death," the obvious question here is: What is the relevance of mechanotransduction for anoikis, and vice versa?

8.2.1 Anoikis: The Monitor of Cell Positioning

Like apoptosis, and RCD in general, anoikis performs important roles during organogenesis, as well as in tissue maintenance and renewal [17,20,22,23,63,75,79,110,111,183]. For example, during vertebrate embryogenesis, inner endodermal cells lose adhesion to their ECM and consequently undergo anoikis, allowing cavitation to occur [184]. As other examples, the induction of anoikis in obsolete cells is the cell process that underlies the involution of mammary glands and the renewal of the epidermis, as well as that of the intestinal epithelium [185–188]. It is now understood that cells possess an innate/default mechanotransduction-driven monitoring system that is meant to ensure that the integrins they express interact with the proper ECM ligands, ensuring RCD of cells when it does not occur naturally – and that monitoring system is nothing other than anoikis. Hence, any cell that strays accidentally (or otherwise) from its assigned position within its given tissue, either by interacting with a "wrong" ECM or by losing anchorage to its own ECM, is targeted for death by anoikis [20–25,183,189–193]. In other words, for multicellular organisms, anoikis constitutes *the* critical defence mechanism that serves to prevent detached cells from readhesion to ECMs that are different than their assigned ones, or to avert straying cells from illicitly wandering towards incorrect locations, and thus ensures a healthy continuity of tissue function and integrity.

8.2.2 Integrins and Anoikis: Letting Go Means the End

Considering the importance of integrin-mediated cell–ECM interactions in establishing a cell's tensegrity, in organizing/stabilizing its cytoskeleton, and in driving signaling for its continued survival, it is not surprising that the disruption or loss of integrin binding induces anoikis [14–25,65,67,68,70,75,179,180,183,189–193]. But how does anoikis act as a terminator of detached or illicitly wandering cells? Much remains to be understood of the regulation, initiation, and execution of anoikis. Nevertheless, it is recognized that the process constitutes a "four-punch hit" [25] against cell survival that is somewhat similar to type II cell death, insofar as it involves elements from both the common and the extrinsic pathways of caspase-dependent apoptosis.

The first "punch" originates from FAK and/or Src deactivation following the loss of integrin binding, leading to a disengagement of survival-promoting pathways such as PI3-K/Akt and Ras/Raf/Mek/Erk (Figure 8.2). Consequently, the pro-survival roles performed by these pathways (Figure 8.1) undergo critical failure. The second "punch" arises from the concomitant disassembly of anchoring focal adhesions (and/or hemidesmosomes) – in large part due to the loss of the integrin-mediated engagement of

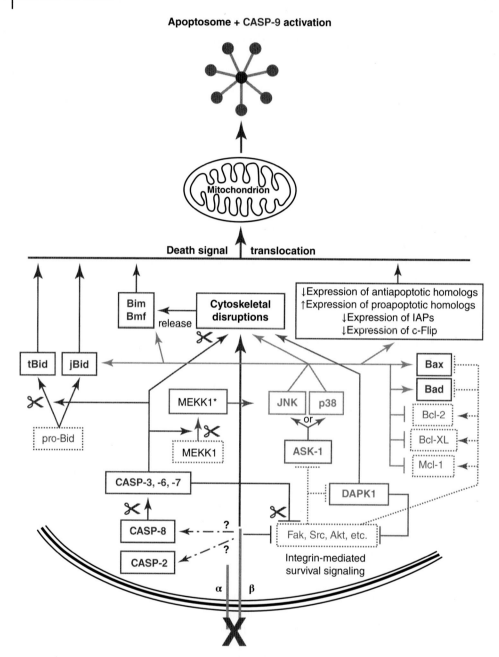

Figure 8.2 Induction of anoikis: basic mechanics. CASP, caspase.

FAK, Src, and/or ILK (e.g., Figure 8.1) – causing the destabilization of the cytoskeleton and consequent release of Bmf and Bim from sequestration (Figure 8.2). In this respect, anoikis can be viewed as the outcome of a constant mechanosensing tension test that has been met with failure, because integrins find themselves unable to support the stable cytoskeletal tensegrity that they largely helped establish and maintain, due to the loss of binding to their proper ECM ligands. Furthermore, p66^{Shc} [168,194] can be found at destabilizing focal adhesions due to integrin-binding failure, and is thought to act as an additional mechanosensor for sound/correct integrin–ECM interactions [20,189, 193,197]. Upon the disruption or loss of integrin–ECM interactions, p66^{Shc} translocates to focal adhesions and induces an elevated/sustained activation of the small GTPase RhoA, which in turn initiates (or contributes further to) the destabilization of the cytoskeleton [20,189,195–197] through signaling events that remain to be fully identified (but are likely to implicate the phosphorylation of myosin light chain (MLC) by the RhoA effector, Rho-associated, coiled coil-containing protein kinase 1 (ROCK1)) [196,197]. Although it remains to be clearly understood how p66^{Shc} is induced to translocate to focal adhesions upon integrin-binding failure, such an "anoikis-inducing mechanosensing" function of p66^{Shc} appears to be independent of its better-known ability to translocate to mitochondria and thereby induce apoptosis [194,195,197].

The third "punch" comes from the activation of "deathly" (i.e., "proapoptotic") kinases [16,17,19,20,22–25,183,198]. The roles of the JNK and p38 SAPKs in cell survival, apoptosis, and anoikis remain ambiguous [25,132,133,137,199–201]. Three isoforms for JNK (−1 to −3) and four for p38 (α, β, γ, and δ) are known – all of which can be selectively expressed, depending on the contexts of cell type and species [132,133,137,199–205]. To this effect, JNK and p38 will contribute to cell survival, apoptosis/anoikis, or neither, not only according to cell type, but also in an isoform-distinct manner [25,132,133,137,199–208]. Furthermore, either a JNK isoform, a p38 isoform, or an isoform of each will contribute to the promotion of cell survival or to driving apoptosis/anoikis [25,82,132,133,137,198–212]. Hence, although the implications of JNK and/or p38 in survival, as well as in apoptosis/anoikis, remain to be fully understood, it is nevertheless recognized that a prolonged/sustained activation of specific JNK and/or p38 isoforms allows them to enact their deadlier functions [16,25,132,133,137,198–210]. In this respect, an elevated and/or sustained activation of an apoptotic JNK isoform, or of a p38 one, arises mostly from the upstream activation of apoptosis signal-regulating kinase-1 (ASK-1) (Figure 8.2) [25,132, 133,137,198–208,213–215]. Interestingly, Akt and PDK1 enact an inhibitory phosphorylation of ASK-1 [216,217], whereas active Raf directly binds ASK-1 [218], keeping the latter's activation in check (Figure 8.1). Turnabout being fair play, activated ASK-1 has been shown to negatively phosphorylate PDK1 [217]. While the precise apoptotic roles (if any) that are enacted by each specific JNK and p38 isoform remain to be fully elucidated, it is nonetheless recognized that apoptosis/anoikis-induced JNK or p38 can generally perform numerous deathly functions, such as: (i) activatation of p53 to facilitate the isoform's own proapoptotic functions (e.g., inducing the expression of Puma, Noxa); (ii) induction/upregulation of the expression of FasL (death ligand of Fas) for autocrine "death stimulation" via DISC; (iii) contribution to the destabilization of cytoskeletal elements (such as microtubules and microfilaments, either directly or via downstream effectors such as mitogen-activated protein kinase (MAPK)-activated protein kinase (MK) 2, 3, and 5) to induce/enhance the freeing of Bim and Bmf from sequestration; (iv) downregulation of the expression of IAPs, c-Flip, and antiapoptotic Bcl-2 homologs and simultaneous upregulation of the expression of proapoptotic ones; and (v) phosphorylation of Bcl-2 homologs, either in the cytosol or at mitochondria (Figure 8.2) [25,116,158,159,203–208]. Proapoptotic phosphorylating functions of Bcl-2 homologs have mainly been characterized in the case of JNK, albeit not for any particular isoform. They include the inhibitory phosphorylation of Bcl-2, Bcl-X$_L$, and Mcl-1, the phosphorylation of BID to produce an active JNK-dependent-phosphorylated BID (jBID) form (instead of the usual tBID one), the phosphorylation of Bax for its activation and mitochondrial translocation, the positive/potentiating phosphorylation of Bad, and the phosphorylation of Bim and Bmf to either prevent their sequestration or cause their release (Figure 8.2) [25,143,144,203,204,206,208]. Note that while similar proapoptotic phosphorylations of Bcl-2 homologs have been reported for p38, the specific residues targeted remain to be firmly identified (Figure 8.2) [25,200,204,205,207,208]. Conversely, mitogen and stress kinase (MSK) 1 and 2, two downstream effectors of p38, have been shown to phosphorylate Bad in order to potentiate its sensitizing functions [25,158,159,200,204,205,207,208]. Additionally, MSK1 and MSK2 can inhibit Akt via phosphorylation [158,159].

Another deathly/proapoptotic kinase family is the death-associated protein kinases (DAPK), principally DAPK1) [219–222]. The exact mode of activation for DAPK1 remains poorly understood, but it has been reported to have been brought about by signaling of death receptors (namely those for TNF-α and FasL) [219–222]. Conversely, Src, Akt, and RSK-1 enact inhibitory phosphorylations on DAPK1 (Figure 8.1) [219–223] and, accordingly, integrin binding and FAK activation likewise suppress its activation [225]. The proapoptotic roles of DAPK1 remain to be fully elucidated. Nonehtless, it is known that it contributes to the destabilization of the actin cytoskeleton (Figure 8.2) [219–224]. Additionally, activated DAPK1 can further deactivate integrins via an inside–out mechanism that involves the displacement and replacement of talin, or by binding paxillin at the cytoplasmic tail of β subunits; as a consequence, increased/accelerated disassembly of focal adhesions occurs (Figure 8.2) [225–227]. Incidentally, DAPK1 has been shown to phosphorylate Beclin-1, an autophagic cell death-driving factor, in order to free (and/or protect) it from inhibitory binding by Bcl-2 and Bcl-X_L [219,228]. In this respect, past experimental attempts at inhibiting intrinsic-pathway apoptotic effectors following the induction of anoikis have often failed to protect cells, causing instead a Beclin-1-mediated autophagic death (23-25,183). Hence, the anchorage loss-induced activation of DAPK1 may constitute a functional molecular bridge between anoikis and autophagic cell death. Aside from DAPK1, DAPK2 has also been shown to be activated upon induction of anoikis [229] – however, the anoikis-driving functions performed by this DAPK family member remain unknown.

The fourth "punch" to hit cell survival due to the loss of integrin anchorage consists in the induction of a type II cell death-like pathway of apoptosis with the activation of caspase 8 (Figure 8.2) [17,19–25,163,183]. While such caspase 8 activation appears to be an early/immediate event following the loss of integrin-mediated cell–ECM anchorage, the precise mechanisms responsible remain elusive. Previous reports have shown that both pro-caspase 8 and activated caspase 8 are found in association with the cytoplasmic tails of β1 and/or β3 subunits of integrins that have lost their binding [163,230,231]. As outlined earlier, pro-caspase 8 is found in adhesomes (and thus associated with the cytoplasmic tails of integrin β subunits) under healthy/adhering conditions [163,231], most likely because pro-caspase 8 complexes with Src following its inhibitory phosphorylation by the latter [163]. Nonetheless, the formation of a DISC at the tails of unbound integrins remains contentious as an explanation for the activation of caspase 8 in anoikis.

On the one hand, some studies from different cell types have shown that caspase 8 activation can be FADD-dependent in anoikis [164,165,232–237] and that c-Flip can inhibit the process (235,238–239). These findings appear to be supported by subsequent observations that, upon FAK and Src being downactivated following detachment from the ECM, RIPK1 is freed from FAK [164,165] and pro-caspase 8 is no longer inhibited by Src [163,240]. Furthermore, it is known that the concomitant downactivation of PI3-K/Akt and/or Ras/Raf/MEK/Erk leads to a downregulation of the expression of c-Flip [17–25,183,190–193]. Therefore, it can be surmised that RIPK1, FADD, and pro-caspase 8 will be free to form a DISC at unligated integrins [164]. However, it has been reported that FAK-freed RIPK1 forms a FADD-dependent DISC at the cytoplasmic domain of the death-receptor Fas (death receptor 6) [164], not integrins. In hindsight, this was to be expected, since the expression of FasL and/or TRAIL can be upregulated to enhance anoikis following the loss of integrin binding in some cell types, and anoikis can be greatly attenuated by the inhibition of the revelant death receptor [17,19,20,22–25,183,189–193,232,233,238,239]. In any event, RIPK1 does not form a DISC at unligated integrins [165].

On the other hand, it has been shown that FADD is not associated with the cytoplasmic tails of β3 subunit-containing unbound integrins in human umbilical-vein endothelial cells undergoing anoikis, even though pro-caspase 8 and activated caspase 8 are [230]. Additionally, it has been reported in human keratinocytes that bound β1 integrins contain the β1A subunit variant, whereas the β1B variant is expressed in the cytosol and apparently is not part of any functional integrin heterodimer [231]; interestingly, and as is to be expected, pro-caspase 8 is associated with the cytoplasmic tails of β1A subunits when those cells are adhering, whereas upon the loss of anchorage, β1A integrins are internalized so as to co-localize with the cytosolic β1B subunits, whereby pro-caspase 8 somehow shuttles from the tails of the former to those of the latter, in order to undergo activation in a FADD-independent manner [231].

To complicate matters further, an immediate/early activation of caspase 2 has also been observed upon the induction of anoikis (Figure 8.2) [241–244]. This should not be surprising, considering that caspase 2 can be activated by caspase 8 [57,58,60]. However, a small number of reports have shown not only that caspase 2 activation can precede that of caspase 8 [241,242], but also that the knockdown of caspase 2, in a similar manner to that of caspase 8, can confer protection against anoikis [244]. The questions of whether caspase 2 is found at the tails of β subunits of unbound integrins, of how it is activated during the induction of anoikis, and of whether caspase 8 activation is actually caspase 2-dependent remain unanswered. Likewise, how exactly pro-caspase 8 and/or activated caspase 8 associate with β subunits of unligated integrins and what are the precise mechanisms responsible for caspase 8 activation at these sites remain open questions. We may find out that the mechanisms of caspase 8 activation upon the induction of anoikis, as well as those of caspase 2, simply happen to differ according to the combined contexts of the cell type, of the integrin repertoire expressed (including variants), and of the species concerned – not unlike everything else that has so far been outlined herein.

In any event, caspase 8 activation following the loss of integrin-mediated anchorage is typically of low intensity. This leads to a likewise low-intensity activation of executioner caspases. Hence, anoikis implicates the following necessary events, many of which are reminiscent of type II cell death: the caspase-mediated cleavage activation of BID into tBID, the caspase-mediated cleavage of pro-survival kinases (e.g., FAK, Akt), and the caspase-mediated destabilization of the cytoskeleton, with subsequent liberation of Bmf and Bim from sequestration (Figure 8.2) [17,20,22,24,25,183]. Additionally, mitogen-activated protein kinase kinase kinase 1 (MEKK1), an upstream activator of the JNK pathway, may be cleaved by caspases (namely caspase 7) to generate a constitutive active form ("MEKK1*") that drives a prolonged/sustained activation of JNK1 [198,209,210], which undertakes its proapoptotic phosphorylating functions (Figure 8.2). Taken all together, these events, along with the three concomitant "punches," effect a translocation of the anchorage loss-induced death signal to mitochondria to bring about the formation of the apoptosome and activation of caspase 9 – consequently rendering the process irreversible (Figure 8.2). Incidentally, it is of note that the proanoikis molecule Bit1 is released from mitochondrial membranes following the loss of integrin-mediated signaling – however, its functions as "anoikis effector" remain to be understood [245].

Given that normal cells need proper anchorage to their ECM to remain alive, and considering the "four-punch hit" following the loss of cell anchorage, integrins are viewed as *suppressors* of anoikis [14–25,70,79,183,187,190–193]. The observation that the forced expression of dominant negative mutants of FAK and/or Src readily induces anoikis, whereas the forced expression of constitutive active mutants of either or both protects against it, greatly encourages such a perception of integrins and integrin-mediated signaling [92–101,246]. To this effect, similar results are obtained when mutants of kinases that are engaged by integrin–FAK/Src-mediated signaling are used, such as Akt [14–25,70,79,119–142,183,187,190–193]. That notwithstanding, there is evidence that some integrins may in fact contribute to *sensitizing* cells to anoikis. For instance, the knockdown of the expression of the α8β1 integrin in human intestinal epithelial crypt cells confers a measure of resistance to anoikis [93,247–249]. Another example is αvβ3 in colon cancer cells [250,251]. The *raison d'être* of anoikis-sensitizing integrins is unclear. However, it is understood that anoikis-sensitizing functions will be enacted by specific integrins depending on the context of cell type (and/or species). Case in point: the αvβ3 integrin performs anoikis-suppressing roles in endothelial cells, as well as in many other cell types [116,252], whereas the α8β1 receptor has been shown to suppress anoikis in myofibroblasts [253].

8.2.3 Adherens Junctions and Anoikis: Going Rogue Warrants Execution

Upon losing anchorage, active AJs in epithelial cells undergo disassembly, and E-cadherin-mediated cell–cell contacts are therefore disengaged. This results in enhancement/precipitation of anoikis, in accordance with the fact that E-cadherin-driven signaling not only organizes/stabilizes the cytoskeleton [106,169–174], but also promotes cell survival via its engagement of Src and the PI3-K/Akt pathway [17,19,24,111,173–176]. Interestingly, JNK has been shown to phosphorylate β-catenin and, consequently, to induce/hasten AJ disassembly [254]. It is thought that the activation of JNK during anoikis may constitute an inside–out mechanism for E-cadherin inactivation, following the loss of cell anchorage [254]. Conversely, it is now established that the disruption, or loss, of E-cadherin-mediated AJs in epithelial cells is sufficient to induce anoikis. In this respect, E-cadherins are recognized as mechanosensors of equal importance to that of integrins [111,174,176,178, 182,183,189–193]. Thus, "E-cadherin-mediated death," like "integrin-mediated death," can also be viewed as the outcome of a constant mechanosensing tension test that has been met with failure, because E-cadherins are unable to support the stable cytoskeletal tensegrity that they contributed to establishing and maintaining, due to the loss of binding to E-cadherins of adjoining cells. In other words, AJs ensure that any epithelial cell that seeks to wander away from its tissue, by losing its adhesive contacts with its neighboring cells, will be targeted for death by anoikis [111,174,176,178,182, 183,189–193]. Therefore, on the one hand, the disruption/loss of AJs leads to the enactment of the first (loss of survival signals) and second (destabilization of the cytoskeleton) "punches" against cell survival, much as described for integrins in Section 8.2.2. On the other, it is unclear how such cell–cell detachment can lead to the activation of apoptotic kinases (the third "punch"). Although the loss of E-cadherin-mediated Akt activation may allow the activation of ASK-1 and/or DAPK1, such an occurrence upon the disruption/loss of AJs remains to be established. Likewise, how caspase 8 activation (the fourth "punch") may occur following the disruption/loss of E-cadherin interactions is unknown.

It must be noted here that the roles of AJs in cell survival and anoikis constitute somewhat of a paradox. For instance, while some studies have shown that the forced maintenance of AJs under absence of integrin–ECM anchorage conditions delays the onset of anoikis [255], others have reported that the presence of active AJs between cells renders them altogether more sensitive to the process [13,256]. Such apparent discrepancy is thought to be related to the actual levels of cytoskeletal organization of cells, meaning that a higher organization of a cell's cytoskeleton translates into a greater sensitivity to anoikis [111,174,176,178,182,183,189–193]. This correlation is likely due, at least in part, to greater amounts of Bmf and/or Bim molecules being sequestered in highly polarized cells, as compared to less polarized ones. Hence, cytoskeletal disruptions – whether due to loss of cell–ECM interactions or to loss of cell–cell contacts – invariably cause greater harm to highly polarized cells [65,68,111,174,176, 178–183,189–193]. In any event, further studies of the mechanosensing functions of AJs will undoubtedly provide a better understanding of the roles of E-cadherin-mediated signaling in the suppression – as well as the induction – of anoikis. In the meantime, it is à propos to revise here the earlier definition of anoikis as being "caspase-dependent apoptosis that is induced by inadequate, inappropriate, or non-existent *cell adhesion*."

8.2.4 Anoikis: A Coroner's Conclusions

It is of interest that anoikis in normal cells is characterized by a delay in the irreversible commitment to the process, following its trigger [13–25,198]. This explains the earlier observations that cells can be rescued by allowing their reattachment, provided this occurs within a limited time span following their loss of anchorage [13–25,198]. Such a "window of anoikis reversibility" varies in length, depending on the cell type and species, between some 15 minutes and 4 hours [13–25,198]. It is thought that the length of a given window of anoikis reversibility represents the time span required for the anchorage loss-induced death signal to reach mitochondria (and for the apoptosome to form), which in turn is primarily determined by the specific integrins (and variants) implicated, the precise determinants of cell survival/death in play (e.g., Bcl-2 homolog expression profiles encountered, survival pathways/isoforms engaged, apoptotic kinases/isoforms involved, etc.), and the degree of cytoskeletal organization/cell polarity imparted not only by integrins (i.e., focal adhesions, hemidesmosomes), but also by cell–cell interactions (e.g., AJs) [25]. Additionally, it has been observed that the activation of pro-survival kinases such as Src, Akt, and/or Erk1/Erk2 experiences a short (5–15-minute), transient upactivation following the loss of integrin binding [13–25,198]. The mechanisms by which such "deathly gasps" occur remain poorly understood, but are considered to protect cells from anoikis during normal cell processes that require transient detachments from the ECM, such as migration and cytokinesis [25,192]. In this respect, they may also contribute to defining the length of a window of anoikis reversibility [13–25,198]. However, considering that an illicit upactivation of Erk1/Erk2 in normal cells can trigger the common pathway of apoptosis [257,258] or activate DAPK1 for consequent autophagic cell death [221,225,257,258], it is a germane question whether the aforementioned "deathly gasps" may not in fact represent "poisoned fruits." Regardless, it remains indisputable that once caspase 9 activation occurs, any window of anoikis reversibility is thereafter closed, and the process becomes irreversible – to the point that experimental attempts to inhibit it will likely result in autophagic cell death mediated by Beclin-1 [23–25,183]. Incidentally, this would explain why the overexpression of Bcl-2 or Bcl-X_L can greatly delay, if not fully prevent, anoikis [14,17–22,25,83,198,209,228].

In conclusion, the preceding considerations together shine a bright spotlight upon the multifaceted intricacies that underlie the regulation and execution of cell survival, apoptosis, and – of particular relevance to this chapter – anoikis. In the latter case, such regulatory complexity includes mechanistic distinctions according to the contexts of ECM composition, the specific integrins implicated, the characteristics of cell–cell adhesion, and the degree of cell polarity (cytoskeletal organization). Furthermore, recent studies have enabled the unveiling of yet another level of complexity in the regulation of cell survival, apoptosis, and anoikis: the existence of distinct control mechanisms according to the state of cell differentiation.

8.3 Differentiation State-Specific Regulation of Anoikis: The Case of Human Intestinal Epithelial Cells

The intestinal epithelium is a classic physiological model system for investigating the functional connections between cell adhesion and cell state [25,259–263]. The continuous and rapid renewal of this simple epithelium occurs along a well-defined morphological unit, the crypt–villus axis. This axis consists generally in two cell populations: the proliferative, immature crypt cells, and the differentiated (nonproliferative) villus ones [261–264]. As part of the process of dynamic renewal of the intestinal epithelium, differentiated enterocytes that are obsolete undergo anoikis as a means of exfoliation upon reaching the apex of villi [25,187,261,262,264,265]. With regard to crypts, occasional apoptosis may occur to remove damaged or defective proliferative/undifferentiated cells [25,261,262,264,265].

It was this very contrast of cell fate between crypt and villus enterocytes, in addition to their respective and specific profiles of Bcl-2 homolog expression (which are established during the differentiation process) [261,262,266–270], that brought about the general concept of distinct regulatory mechanisms for cell survival and apoptosis according to the state of cell differentiation [25,261,262,266]. Accordingly, said concept has been studied and demonstrated predominantly in human intestinal epithelial cells (hIECs) [25,261,262]. For instance, the PI3-K/Akt and MEK/Erk pathways are selectively implicated in the promotion of enterocytic survival according to the state of differentiation, including with regard to their modulation of the expression/activity of individual anti- and proapoptotic Bcl-2 homologs [22,25,268,270–272]. As another example, further concerning the PI3-K/Akt pathway, the Akt-1 isoform performs critical hIEC survival functions regardless of the state of differentiation, whereas the Akt-2 isoform contributes to the maintenance of cell survival in differentiated hIECs only (Akt-3 is not expressed in the human intestinal epithelium [277,278,282,283]; P.H. Vachon, unpublished results) (Figure 8.3). Incidentally, ILK is not required for cell survival in either undifferentiated or differentiated hIECs [297], which raises the question of cell type- and/or species context-dependent pro-survival roles for this pseudokinase.

8.3.1 hIEC Anoikis Regulation: Also a Matter of Cell Differentiation

As would be expected from the normal fate of villus IECs (i.e., exfoliation by anoikis), differentiated enterocytes are more susceptible to anoikis than undifferentiated ones [268,277–279]. This accounts for studies that have shown that mechanical/shearing forces in the intestine occasionally cause incidental anoikis of villus cells only [187,280]. Interestingly, crypt and villus hIECs express distinct profiles of integrins (including splicing and/or proteolytic variants) as they interact with specific ECM components, which in turn are found deposited differentially along the crypt–villus axis [25, 259–262,265,266,273–276]. Although crypt IECs happen to express α8β1 as an anoikis-sensitizing integrin (Figure 8.3A) [247], the differentiated/villus ones are nonetheless highly polarized – in addition to bearing hemidesmosomes and AJs – in stark contrast to their poorly polarized undifferentiated/crypt counterparts (as stylized in Figure 8.3A versus 8.3B) [25,256,273,281]. As an aside, AJs in differentiated IECs have been shown to indeed engage Src and PI3-K/Akt for contributions to cell-survival signaling [255,298,299]. Be that as it may, these observations together constitute the bases that underlie the fact that differentiated/villus enterocytes are substantially more susceptible to anoikis than undifferentiated/crypt ones.

Interestingly, such distinctions in anoikis susceptibility between crypt and villus enterocytes have been shown to translate into differentiation state-specific mechanisms of integrin-mediated regulation of anoikis [25,268,270–272,277–279,282]. For instance, β1 integrins, FAK, and Src distinctively modulate the expression/activity of Bcl-2 homologs, depending on the status of enterocytic differentiation [268,270–272]. Furthermore, α2β1A, α5β1A, and α6Aβ4A^{ctd-} are required for the suppression of anoikis in undifferentiated IECs, while α3Aβ1 and α6Bβ4A perform this role in differentiated ones [268,279] (Figure 8.3). Similarly, the engagement of FAK and Src themselves, of PI3-K (p110 catalytic/p85 regulatory) isoform complexes, of distinct pools of Akt isoforms (namely Akt-1, Akt-2), and of MEK/Erk by integrin β1- and/or β4-mediated signaling is clearly specific according to the state of differentiation, which in turn translates into differentiation state-specific roles in the suppression of anoikis for these signaling molecules/pathways [22,268,270,272,277–279,282,283] (M. Beauséjour et al., unpublished data) (Figure 8.3). Hence, the integrin β1/FAK/Src/PI3-K/Akt-1 and β4/Src pathways antagonize the proapoptotic activation of p38β in undifferentiated enterocytes, whereas the integrin β1/FAK/PI3-K/Akt-1, β1/FAK/Src/MEK/Erk, and β4/Src/Akt-2 pathways contribute together to antagonizing the activation of p38δ in differentiated ones [25,270,272,277,283] (P.H. Vachon, unpublished results) (Figure 8.3). Therefore, the integrin-mediated regulation of cell survival and anoikis is indeed beholden to differentiation state-specific mechanisms.

8.3.2 Victimology: hIECs Only? (Definitely Not)

Of course, the general concept of differentiation state-specific controls of cell survival, apoptosis, and anoikis also applies to tissues other than the intestinal epithelium –

(A)

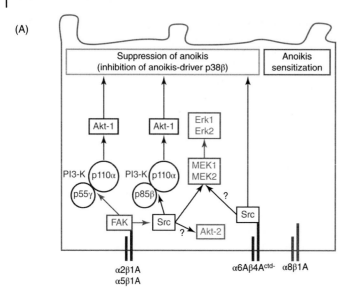

(B)

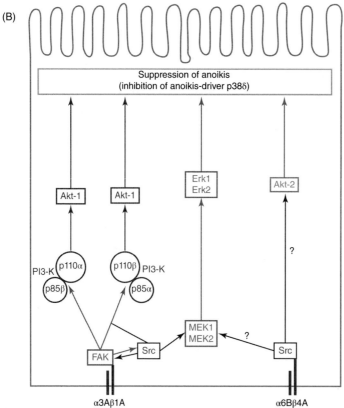

Figure 8.3 Integrin-mediated regulation of anoikis in (A) undifferentiated and (B) differentiated human intestinal epithelial cells (hIECs).

although not without expected cell type- and/or species-dependent distinctions [25]. For example, skeletal muscle myoblasts require fibronectin and α5β1A for survival, while their differentiated counterparts (myocytes/myotubes/myofibers) require laminin-211 and α7Bβ1D instead [80–83]. Furthermore, myoblasts depend on an integrin-driven FAK/Src/MEK/Erk pathway for cell survival/suppression of anoikis, whereas their differentiated counterparts are dependent on an integrin-driven Fyn/PI3-K/Akt-2 one [82,83,284,285]. Three key questions present themselves with regard to our further understanding of such differentiation state-specific mechanisms of cell survival, apoptosis, and anoikis: (i) Why such distinct mechanisms exist according to the state of cell differentiation; (ii) How these mechanisms work (i.e., the further functional identification of the specific extracellular ligands, integrins, and molecules/pathways involved, as well as their respective differentiation-specific roles in the suppression or induction of apoptosis/anoikis); and (iii) In what capacity they contribute to the emergence of tissue/organ-specific diseases when they become dysregulated.

8.4 Anoikis and Disease: Too Much or Too Little, It's Still Unhealthy

As with apoptosis, numerous diseases are characterized by a dysregulation of integrin-mediated survival and anoikis signaling [4,20,23,25,67,69,70,75,79,110,111,113,118, 173,183,186,189–193,245,252,286].(For instance, muscular dystrophies and epidermolysis bullosae constitute paragons of physiopathologies that are caused primarily by a deregulated induction of anoikis, due to genetic-borne defects in any of the three basic components of cell adhesion (the extracellular-side ligand, transmembrane receptor, and cytoskeletal-side adhesomes) [67,70,75,110,111,113,287–289].

Muscular dystrophies are diseases that principally affect skeletal muscles and are characterized by progressive muscle wasting and weakness. The loss of expression of the laminin-211 ("merosin") α2 chain (which causes merosin-deficient congenital muscular dystrophy type 1A, MDC1A) and the loss of expression of the α7 integrin subunit (which causes congenital muscular dystrophy with integrin alpha 7 deficiency, ITG7A CMD) are two examples of the skeletal muscle anchorage-driven tensegrity defects that have been identified for this group of degenerative physiopathologies [287,288].

Conversely, epidermolysis bullosae are characterized by a pronounced skin fragility, which manifests clinically as blisters and erosions. The loss of expression of keratin 14 (which causes epidermolysis bullosa simplex, EBS), the loss of expression of laminin-322 α3 chain expression (which causes junctional epidermolysis bullosa–Herlitz, JEB-H), or the loss of expression of the β4 integrin subunit (which causes epidermolysis bullosa with pyloric atresia, EB-PA) constitute but a few examples of the molecular defects that affect the stable and healthy α6β4/hemidesmosome-driven tensegrity of the epidermis [289].

In the same vein, the necessity of understanding the regulation of cell adhesion-mediated survival has gained considerable importance in cancer research, especially given that the acquisition of a resistance to anoikis represents a critical, as well as a *limiting*, step in tumor progression and, especially, the emergence of metastases [69,189–193,252,290]. Notwithstanding that normal cells undergo anoikis in response to loss of anchorage, cancer cells speedily develop various mechanisms to resist anoikis, and exploit these to progress further in malignancy and, eventually, to metastasize to other tissues. To this effect, cancer cells are well known for exhibiting significant alterations in their repertoire of expressed integrins [69,70,76,111,118, 183,190,192,252,275,291], as well as in their surrounding ECM [23,79,111,183, 275,292], which allows them to prevail over the induction of anoikis during the initial/early phases of oncogenic transformation and, later in the course of tumorigenic progression, during their metastatic colonization of other organs (and thereby facilitating adaptation to the metastatic site). Additionally, exfoliated cancer cells can aggregate to form multicellular spheroids, a cellular arrangement that imparts some measure of resistance to anoikis – albeit by means which remain to be fully understood [301,302].

Accordingly, metastatic cells are notorious for displaying a striking resistance to anoikis [69,111,183,189–192,290–292]. Hence, there is a clear need not only to seek an eventual understanding of the acquisition of resistance to anoikis by cancer cells, but also to identify potential molecular targets that are susceptible to shutting down such problematic resistance to the process. In this respect, much attention has been given to the roles of FAK and Src in tumor progression [67,94–96,98–100], as well as to those pathways that are typically engaged by them, such as PI3-K/Akt and Ras/Raf/MEK/Erk [67,121,122,124–131,133,136–138,140–142,149,160,161,167,183,189–192,290]. Along with these, other known players in the suppression (e.g., antiapoptotic Bcl-2 homologs, IAPs, etc.) and in the induction and/or execution (e.g., DAPK1, proapoptotic homologs, ASK-1, Bit1, etc.) of anoikis are likewise being scrutinized [26–29,36, 41–43,52–54,61,90,200,204,213,214,220,245]. Anoikis-sensitizing integrins may be included in this growing category, given that while anoikis-resistant colon carcinoma cells do not express α8β1, forcing its expression reinstates a good measure of anoikis susceptibility in them [247]. Correspondingly, anoikis-sensitive colon carcinoma cells acquire a notable resistance to the process following their loss of expression of αvβ3 [250].

Still in relation to colon cancer (and carcinomas at large), much interest is being given to E-cadherins – indeed, active AJs similarly sensitize cells to anoikis [111,174,176,178,182,183,189–193,290]. This interest stems largely from a crucial process in tumor progression, termed "epithelial–mesenchymal transition" (EMT) [173,183,189–192,290,293–295]. The hallmarks of EMT are: (i) loss of an epithelioid morphology, principally due to a downregulation of epithelial CAMs such as E-cadherin; (ii) loss of epithelium-specific genes; (iii) gain of mesenchymal-specific genes; and (iv) loss of sensitivity to anoikis [173,183,189–192,290,293–295]. Hence, during EMT, epithelial cells undergo a series of morphological changes, including their detachment from one another. They then concomitantly modify their basal–apical polarity and shape by altering their cytoskeleton, enabling them to migrate and overcome anoikis [173,183,189–192,286,290,293–295]. In other words, EMT allows carcinoma cells to bypass their (formerly) epithelial innate/default mechanotransduction-driven monitoring system, which is constituted by integrins and E-cadherins. Such a major phenotypic change is considered a prerequisite for the acquisition of anchorage-independent growth and resistance to anoikis, leading to the emergence of invasive and metastatic cancer cells [173,183,189–192,290,293–295]. The mechanistic link between autophagic cell death and anoikis may grow in significance, given the increasing evidence for a modulatory role for autophagy in EMT and anoikis resistance [303]. Therefore, gaining a better understanding of the molecular underpinnings of EMT is of importance for cancer research, in order eventually to develop novel therapeutic approaches that will reverse/suppress EMT and thus restore anoikis sensitivity.

However, the crosstalk between integrins, E-cadherins, and growth factor receptors (RTKs or otherwise) complicates matters greatly [115–117,173,175–179,183, 189–192,290,293–295]. For instance, the deregulated activity of an RTK can confer resistance to anoikis by maintaining/enhancing integrin-mediated cell survival after the loss of attachment via inside–out signaling. In this respect, the deregulation of RTK signaling is well known to be able to induce EMT in normal epithelial cells [173,183,189–192,290,293–295]. For instance, the constitutive activation of the MET RTK in IECs results in the latter undergoing EMT (along with a downregulation of E-cadherin, of course) and acquiring a high resistance to anoikis [300]. As another example, colon carcinoma cells that secrete epidermal growth factor (EGF)/transforming growth factor-α (TGF-α) in high numbers display a high resistance to anoikis by, at least in part, sustaining the activation of Src via autocrine stimulation of epidermal growth factor receptor (EGFR) and, thus, allowing an illicit sustainment of functional Src–FAK interactions and consequent FAK/Src-mediated suppression of anoikis [296].

Overall, the task of elucidating the precise mechanisms that are responsible for anoikis resistance in cancer cells has demonstrated itself to be a baffling one – especially when considering the inherent molecular differences between the various types of cancers, as well as the cellular and molecular heterogeneity among patients afflicted by the same type of cancer. There is nonetheless little doubt that further increasing our knowledge of the regulatory determinants of cell survival and anoikis within a given tissue will ultimately improve our mechanistic comprehension of anoikis-linked diseases of said tissue, in addition to clearing the road for the design of molecular therapeutic approaches that are altogether tissue- and disease-specific. In this respect, the acquisition of a greater understanding of differentiation state-specific distinctions in the integrin-mediated control of cell survival and anoikis should not only allow for a more complete knowledge of a given tissue's normal physiology, but also deliver a full accounting of its anoikis-linked physiopathological molecular foundations, including with regard to cancer.

8.5 Conclusion

Cell adhesion, via integrins and E-cadherins, constitutes a crucial contributor of the survival signals required for cells to live. Furthermore, integrin- and E-cadherin-mediated cell survival and suppression of anoikis constitute a complex and comprehensive mechanosensory surveillance mechanism that acts as a watchdog in upholding the correct positions of cells within their respective tissues, thereby sentencing to death any cell that strays from its allocated position – either by losing anchorage to its own ECM, by interacting with an ECM that differs in composition to its own, or by losing its adhesive contacts with its immediate neighbors. Consequently, anoikis embodies a critical administrator of the establishment and maintenance of the functional organization of tissues, by ensuring the disposal of cells that are "going rogue" with regard to their precise cell–ECM and cell–cell adhesion requirements for tensegrity and survival, whether this occurs accidentally during normal physiological processes (e.g., tissue renewal) or otherwise.

The control of anoikis implicates molecular mechanisms that are distinct according to the precise composition of the ECM, the repertoire of integrins expressed, the degree of cytoskeletal organization (i.e., cell polarity) conferred (in large part) by specific integrin–

ECM and cell–cell interactions, the signaling molecules and pathways engaged and/or suppressed by individual integrins, the particular isoforms and/or family members participating in such pathways, the repertoire of expressed growth factor receptors, and the various regulators/effectors of apoptosis that are likewise expressed. While some measure of mechanistic commonality between certain cell types may exist, we should remain aware of the reality that the integrin- and E-cadherin-driven regulation of anoikis is a multilayered and multifaceted affair that depends not only on the contexts of tissue type and species, but also on the state of cell differentiation. Consideration of these caveats should allow for an eventual full elucidation of the selective integrin- and E-cadherin-mediated underpinnings that are implicated in tissue-specific, anoikis-linked pathological disorders – as well as in tumor progression and metastasis.

References

1. Edinger AL, Thompson CB. Death by design: apoptosis, necrosis and autophagy. *Curr Opin Cell Biol* 2004;**16**(6):663–9.
2. Assunção Guimarães C, Linden R. Programmed cell deaths. Apoptosis and alternative deathstyles. *Eur J Biochem* 2004;**271**(9):1638–50.
3. Danial NN, Korsmeyer SJ. Cell death: critical control points. *Cell* 2004;**116**(2):205–19.
4. Zakeri Z, Lokshin RA. Cell death in health and disease. *J Cell Mol Med* 2007;**11**(6):1214–24.
5. Penaloza C, Orlanski S, Ye Y, Entezari-Zaher T, Javdan M, Zakeri Z. Cell death in mammalian development. *Curr Pharm Des* 2008;**14**(2):184–96.
6. Colin J, Gaumer S, Guenal I, Mignotte B. Mitochondria, Bcl-2 family proteins and apoptosomes: of worms, flies and men. *Front Biosci (Landmark Ed)* 2009;**14**:4127–37.
7. Eisenberg-Lerner A, Bialik S, Simon HU, Kimchi A. Life and death partners: apoptosis, autophagy and the cross-talk between them. *Cell Death Differ* 2009;**16**(7):966–75.
8. Heath-Engel HM, Chang NC, Shore GC. The endoplasmic reticulum in apoptosis and autophagy: role of the BCL-2 protein family. *Oncogene* 2008;**27**(50):6419–33.
9. Levine B, Sinha S, Kroemer G. Bcl-2 family members: dual regulators of apoptosis and autophagy, *Autophagy* 2008;**4**(5):600–6.
10. Kroemer G, Galluzzi L, Vandenabeele P, Abrams J, Alnemri ES, Baehrecke EH, et al. Classification of cell death: recommendations of the Nomenclature Committee on Cell Death 2009. *Cell Death Differ* 2009;**16**(1):3–11.
11. Galluzzi L, Vitale I, Abrams JM. Molecular definitions of cell death subroutines: recommendations of the Nomenclature Committee on Cell Death 2012. *Cell Death Differ* 2012;**19**(1):107–20.
12. Galluzzi L, Bravo-San Pedro JM, Vitale I, Aaronson SA, Abrams JM, Adam D, et al. Essential versus accessory aspects of cell death: recommendations of the NCCD 2015. *Cell Death Differ* 2015;**22**(1):58–73.
13. Frisch SM, Francis H. Disruption of epithelial cell-matrix interactions induces apoptosis. *J Cell Biol* 1994;**124**(4):619–26.
14. Frisch SM, Ruoslahti E. Integrins and anoikis. *Curr Opin Cell Biol* 1997;**9**(5):701–6.
15. Meredith JE, Schwartz MA. Integrins, adhesion and apoptosis. *Trends Cell Biol* 1997;**7**(4):146–50.
16. Frisch SM, Screaton RA. Anoikis mechanisms. *Curr Opin Cell Biol* 2001;**13**(5):555–62.

17 Grossmann J. Molecular mechanisms of "detachment-induced apoptosis – anoikis." *Apoptosis* 2002;7(3):247–60.
18 Stupack DG, Cheresh DA. Get a ligand, get a life: integrins, signalling and cell survival. *J Cell Sci* 2002;**115**(Pt. 19):3729–38.
19 Martin SS, Vuori K. Regulation of Bcl-2 proteins during anoikis and amorphosis. *Biochim Biophys Acta* 2004;**1692**(2–3):145–57.
20 Gilmore AP. Anoikis. *Cell Death Differ* 2005;**12**(Suppl. 2):1473–7.
21 Reddig PJ, Juliano RL. Clinging to life: cell to matrix adhesion and cell survival. *Cancer Metastasis Rev* 2005;**24**(3):425–39.
22 Chiarugi P, Giannoni E. Anoikis: a necessary death program for anchorage-dependent cells. *Biochem Pharmacol* 2008;**76**(11):1352–64.
23 Marastoni S, Ligresti G, Lorenzon E, Colombatti A, Mongiat M. Extracellular matrix: a matter of life and death. *Connect Tissue Res* 2008;**49**(3):203–6.
24 Gilmore AP, Owens TW, Foster FM, Lindsay J. How adhesion signals reach a mitochondrial conclusion – ECM regulation of apoptosis. *Curr Opin Cell Biol* 2009;**21**(5):654–61.
25 Vachon PH. Integrin signalling, cell survival, and anoikis: distinctions, differences, and differentiation. *J Signal Transduct* 2011;**2011**:738137.
26 Cory S, Huang DC, Adams JM. The Bcl-2 family: roles in cell survival and oncogenesis. *Oncogene* 2003;**22**(53):8590–607.
27 Adams JM, Cory S. Bcl-2-regulated apoptosis: mechanism and therapeutic potential. *Curr Opin Immunol* 2007;**19**(5):488–96.
28 Adams JM, Cory S. The Bcl-2 apoptotic switch in cancer development and therapy. *Oncogene* 2007;**26**(9):1324–37.
29 Yip KW, Reed JC. Bcl-2 family proteins and cancer. *Oncogene* 2008;**27**(50):6398–406.
30 Youle RJ, Strasser A. The BCL-2 protein family: opposing activities that mediate cell death. *Nat Rev Mol Cell Biol* 2008;**9**(1):47–59.
31 Chi X, Kale J, Leber B, Andrews DW. Regulating cell death at, on, and in membranes. *Biochim Biophys Acta* 2014;**1843**(9):2100–13.
32 Czabotar PE, Lessene G, Strasser A, Adams JM. Control of apoptosis by the BCL-2 protein family: implications for physiology and therapy. *Nat Rev Mol Cell Biol* 2014;**15**(1):49–63.
33 Moldoveanu T, Follis AV, Kriwacki RW, Green D. Many players in BCL-2 family affairs. *Trends Biochem Sci* 2014;**39**(3):101–11.
34 Riedl SJ, Salvesen GS. The apoptosome: signalling platform of cell death. *Nat Rev Mol Cell Biol* 2007;**8**(5):405–13.
35 Yuan S, Akey CW. Apoptosome structure, assembly, and procaspase activation. *Structure* 2013;**21**(4):501–15.
36 Yadav N, Chandra D. Mitochondrial and postmitochondrial survival signalling in cancer. *Mitochondrion* 2014;**16**:18–25.
37 Fuentes-Prior P, Salvesen GS. The protein structures that shape caspase activity, specificity, activation and inhibition. *Biochem J* 2004;**384**(Pt. 2):201–32.
38 Timmer JC, Salvesen GS. Caspase substrates. *Cell Death Differ* 2007;**14**(1):66–72.
39 Poreba M, Strózyk A, Salvesen GS, Drag M. Caspase substrates and inhibitors. *Cold Spring Harb Perspect Biol* 2013;**5**(8):a008680.
40 Connolly PF, Jäger R, Fearnhead HO. New roles for old enzymes: killer caspases as the engine of cell behavior changes. *Front Physiol* 2014;**5**:149.

41 Hunter AM, LaCasse EC, Korneluk RG. The inhibitors of apoptosis (IAPs) as cancer targets. *Apoptosis* 2007;**12**(9):1543–68.
42 Martinez-Ruiz G, Maldonado V, Ceballos-Cancino G, Grajeda JP, Melendez-Zajgla J. Role of Smac/DIABLO in cancer progression. *J Exp Clin Cancer Res* 2008;**27**:48.
43 Dubrez L, Berthelet J, Glorian V. IAP proteins as targets for drug development in oncology. *Onco Targets Ther* 2013;**9**:1285–304.
44 Kilbride SM, Prehn JH. Central roles of apoptotic proteins in mitochondrial function. *Oncogene* 2013;**32**(22):2703–11.
45 Basu A, DuBois G, Haldar S. Posttranslational modifications of Bcl2 family members – a potential therapeutic target for human malignancy. *Front Biosci* 2006;**11**:1508–21.
46 Kutuk O, Letai A. Regulation of Bcl-2 family proteins by posttranslational modifications. *Curr Mol Med* 2008;**8**(2):102–18.
47 Brunelle JK, Letai A. Control of mitochondrial apoptosis by the Bcl-2 family. *J Cell Sci* 2009;**122**(Pt. 4):437–41.
48 Chen M, Wang J. Initiator caspases in apoptosis signalling pathways. *Apoptosis* 2002;**7**(4):313–19.
49 Li J, Yuan J. Caspases in apoptosis and beyond. *Oncogene* 2008;**27**(48):6194–206.
50 Wilson NS, Dixit VM, Ashkenazi A. Death receptor signal transducers: nodes of coordination in immune signalling networks. *Nat Immunol* 2009;**10**(4):348–55.
51 Sessler T, Healy S, Samali A, Szegezdi E. Structural determinants of DISC function: new insights into death receptor-mediated apoptosis signalling. *Pharmacol Ther* 2013;**140**(2):186–99.
52 Fouqué A, Debure L, Legembre P. The CD95/CD95L signalling pathway: a role in carcinogenesis. *Biochim Biophys Acta* 2014;**1846**(1):130–41.
53 Fulda S. Tumor-necrosis-factor-related apoptosis-inducing ligand (TRAIL). *Adv Exp Med Biol* 2014;**818**:167–80.
54 Zhivotovsky B, Orrenius S. Cell death mechanisms: cross-talk and role in disease. *Exp Cell Res* 2010;**316**(8):1374–83.
55 Degterev A, Boyce M, Yuan J. A decade of caspases. *Oncogene* 2003;**22**(53):8543–67.
56 Krumschnabel G, Sohm B, Bock F, Manzl C, Villunger A. The enigma of caspase-2: the laymen's view. *Cell Death Differ* 2009;**16**(2):195–207.
57 Bouchier-Hayes L, Green DR. Caspase-2: the orphan caspase. *Cell Death Differ* 2012;**19**(1):51–7.
58 Janssens S, Tinel A. The PIDDosome, DNA-damage-induced apoptosis and beyond. *Cell Death Differ* 2012;**19**(1):13–20.
59 Kitevska T, Spencer DM, Hawkins CJ. Caspase-2: controversial killer or checkpoint controller? *Apoptosis* 2009;**14**(7):829–48.
60 Olsson M, Vakifahmetoglu H, Abruzzo PM, Högstrand K, Grandien A, Zhivotovsky B. DISC-mediated activation of caspase-2 in DNA damage-induced apoptosis. *Oncogene* 2009;**28**(18):1949–59.
61 Micheau O. Cellular FLICE-inhibitory protein: an attractive therapeutic target? *Expert Opin Ther Targets* 2003;**7**(4):559–73.
62 Hynes RO. The extracellular matrix: not just pretty fibrils. *Science* 2009;**326**(5957):1216–19.
63 Rozario T, DeSimone DW. The extracellular matrix in development and morphogenesis: a dynamic view. *Dev Biol* 2010;**341**(1):126–40.
64 Hynes RO. Integrins: bidirectional, allosteric signalling machines. *Cell* 2002;**110**(6):673–87.

65 Ross TD, Coon BG, Yun S, Baeyens N, Tanaka K, Ouyang M, Schwartz MA. Integrins in mechanotransduction. *Curr Opin Cell Biol* 2013;**25**(5):613–18.
66 Morse EM, Brahme NN, Calderwood DA. Integrin cytoplasmic tail interactions. *Biochemistry* 2014;**53**(5):810–20.
67 Winograd-Katz SE, Fässler R, Geiger B, Legate KR. The integrin adhesome: from genes and proteins to human disease. *Nat Rev Mol Cell Biol* 2014;**15**(4):273–88.
68 Manninen A. Epithelial polarity – generating and integrating signals from the ECM with integrins. *Exp Cell Res* 2015;**334**(2):337–49.
69 Seguin L, Desgrosellier JS, Weis SM, Cheresh DA. Integrins and cancer: regulators of cancer stemness, metastasis, and drug resistance. *Trends Cell Biol* 2014;**25**(4):234–40.
70 van der Flier A, Sonnenberg A. Function and interactions of integrins. *Cell Tissue Res* 2001;**305**(3):285–98.
71 Askari JA, Buckley PA, Mould AP, Humphries MJ. Linking integrin conformation to function. *J Cell Sci* 2009;**122**(Pt. 2):165–70.
72 de Melker AA, Sonnenberg A. Integrins: alternative splicing as a mechanism to regulate ligand binding and integrin signalling events. *Bioessays* 1999;**21**(6):499–509.
73 Armulik A. Splice variants of human β1 integrins: origin, biosynthesis and functions. *Front Biosci* 2002;**7**:d219–27.
74 Gahmberg CG, Fagerholm SC, Nurmi SM, Chavakis T, Marchesan S, Grönholm M. Regulation of integrin activity and signalling. *Biochim Biophys Acta* 2009;**1790**(6):431–44.
75 Danen EH, Sonnenberg A. Integrins in regulation of tissue development and function. *J Pathol* 2003;**201**(4):632–41.
76 Stipp CS. Laminin-binding integrins and their tetraspanin partners as potential antimetastatic targets. *Expert Rev Mol Med* 2010;**12**:e3.
77 Giancotti FG. Complexity and specificity of integrin signalling. *Nat Cell Biol* 2000;**2**(1):E13–14.
78 Meredith J, Fazeli B, Schwartz M. The extracellular matrix as a survival factor. *Mol Biol Cell* 1993;**4**(9):953–61.
79 Berrier AL, Yamada KM. Cell-matrix adhesion. *J Cell Physiol* 2007;**213**(3):565–73.
80 Vachon PH, Loechel F, Xu H, Wewer UM, Engvall E. Merosin and laminin in myogenesis; specific requirement for merosin in myotube stability and survival. *J Cell Biol* 1996;**134**(6):1483–97.
81 Vachon PH, Xu H, Liu L, Loechel F, Hayashi Y, Arahata K, et al. Integrins (α7β1) in muscle function and survival; disrupted expression in merosin-deficient congenital muscular dystrophy. *J Clin Invest* 1997;**100**(7):1870–81.
82 Laprise P, Poirier E-M, Vézina A, Rivard N, Vachon PH. Merosin-integrin promotion of skeletal myofiber cell survival: differentiation state-distinct involvement of p60^{Fyn} tyrosine kinase and p38α stress-activated MAP kinase. *J Cell Physiol* 2002;**191**(1):69–81.
83 Laprise P, Vallée K, Demers M-J, Bouchard V, Poirier EM, Vézina A, et al. Merosin (laminin-2/4)-driven survival signalling: complex modulations of Bcl-2 homologs. *J Cell Biochem* 2003;**89**(6):1115–25.
84 Gawlik KI, Mayer U, Blomberg K, Sonnenberg A, Ekblom P, Durbeej M. Laminin α1 chain mediated reduction of laminin α2 chain deficient muscular dystrophy involves integrin α7β1 and dystroglycan. *FEBS Lett* 2006;**580**(7):1759–65.

85 Rooney JE, Gurpur PB, Yablonka-Reuveni Z, Burkin DJ. Laminin-111 restores regenerative capacity in a mouse model for α7 integrin congenital myopathy. *Am J Pathol* 2009;**174**(1):256–64.

86 Gawlik KI, Akerlund M, Carmignac V, Elamaa H, Durbeej M. Distinct roles for laminin globular domains in laminin α1 chain mediated rescue of murine laminin α2 chain deficiency. *PLoS One* 2010;**5**(7):e11549.

87 Giancotti FG, Ruoslahti E. Integrin signalling. *Science* 1999;**285**(5430):1028–32.

88 Harburger DS, Calderwood DA. Integrin signalling at a glance. *J Cell Sci* 2009;**122**(Pt. 2):159–63.

89 Cabodi S, Di Stefano P, Leal Mdel P, Tinnirello A, Bisaro B, Morello V, et al. Integrins and signal transduction. *Adv Exp Med Biol* 2010;**674**:43–54.

90 Bouvard D, Pouwels J, De Franceschi N, Ivaska J. Integrin inactivators: balancing cellular functions in vivo and in vitro. *Nat Rev Mol Cell Biol* 2013;**14**(7):430–42.

91 Martin KH, Slack JK, Boerner SA, Martin CC, Parsons JT. Integrin connections map: to infinity and beyond. *Science* 2002;**296**(5573):1652–53.

92 Parsons JT. Focal adhesion kinase: the first ten years. *J Cell Sci* 2003;**116**(Pt. 8):1409–16.

93 Cohen LA, Guan JL. Mechanisms of focal adhesion kinase regulation. *Curr Cancer Drug Targets* 2005;**5**(8):629–43.

94 Zhao J, Guan JL. Signal transduction by focal adhesion kinase in cancer. *Cancer Metastasis Rev* 2009;**28**(1–2):35–49.

95 Sulzmaier FJ, Jean C, Schlaepfer DD. FAK in cancer: mechanistic findings and clinical applications. *Nat Rev Cancer* 2014;**14**(9):598–610.

96 Frame MC. Newest findings on the oldest oncogene; how activated src does it. *J Cell Sci* 2004;**117**(Pt. 7):989–98.

97 Parsons SJ, Parsons JT. Src family kinases, key regulators of signal transduction. *Oncogene* 2004;**23**(48):7906–9.

98 Playford MP, Schaller MD. The interplay between Src and integrins in normal and tumor biology. *Oncogene* 2004;**23**(48):7928–46.

99 Mitra SK, Schlaepfer DD. Integrin-regulated FAK-Src signalling in normal and cancer cells. *Curr Opin Cell Biol* 2006;**18**(5):516–23.

100 Brunton VG, Frame MC. Src and focal adhesion kinase as therapeutic targets in cancer. *Curr Opin Pharmacol* 2008;**8**(4):427–32.

101 Ingley E. Src family kinases: regulation of their activities, levels and identification of new pathways. *Biochim Biophys Acta* 2008;**1784**(1):56–65.

102 Reynolds AB, Kanner SB, Bouton AH, Schaller MD, Weed SA, Flynn DC, Parsons JT. SRChing for the substrates of Src. *Oncogene* 2014;**33**(37):4537–47.

103 Arias-Salgado EG, Lizano S, Sarkar S, Brugge JS, Ginsberg MH, Shattil SJ. Src kinase activation by direct interaction with the integrin β cytoplasmic domain. *Proc Natl Acad Sci USA* 2003;**100**(23):13 298–302.

104 Legate KR, Fässler R. Mechanisms that regulate adaptor binding to beta-integrin cytoplasmic tails. *J Cell Sci* 2009;**122**(Pt. 2):187–98.

105 Moser M, Legate KR, Zent R, Fässler R. The tail of integrins, talin, and kindlins. *Science* 2009;**324**(5929):895–9.

106 Han SP, Yap AS. The cytoskeleton and classical cadherin adhesions. *Subcell Biochem* 2012;**60**:111–35.

107 Legate KR, Montañez E, Kudlacek O, Fässler R. ILK, PINCH and parvin: the tIPP of integrin signalling. *Nat Rev Mol Cell Biol* 2006;**7**(1):20–31.

108 Böttcher RT, Lange A, Fässler R. How ILK and kindlins cooperate to orchestrate integrin signalling. *Curr Opin Cell Biol* 2009;**21**(5):670–5.
109 Cabodi S, del Pilar Camacho-Leal M, Di Stefano P, Defilippi P. Integrin signalling adaptors: not only figurants in the cancer story. *Nat Rev Cancer* 2010;**10**(12):858–70.
110 Wilhelmsen K, Litjens SH, Sonnenberg A. Multiple functions of the integrin alpha6beta4 in epidermal homeostasis and tumorigenesis. *Mol Cell Biol* 2006;**26**(8):2877–86.
111 Gilcrease MZ. Integrin signalling in epithelial cells. *Cancer Lett* 2007;**247**(1):1–25.
112 de Pereda JM, Ortega E, Alonso-García N, Gómez-Hernández M, Sonnenberg A. Advances and perspectives of the architecture of hemidesmosomes: lessons from structural biology. *Cell Adh Migr* 2009;**3**(4):361–4.
113 Toivola DM, Boor P, Alam C, Strnad P. Keratins in health and disease. *Curr Opin Cell Biol* 2015;**32C**:73–81.
114 Loschke F, Seltmann K, Bouameur JE, Magin TM. Regulation of keratin network organization. *Curr Opin Cell Biol* 2015;**32C**:56–64.
115 Pugacheva EN, Roegiers F, Golemis EA. Interdependence of cell attachment and cell cycle signalling. *Curr Opin Cell Biol* 2006;**18**(5):507–15.
116 Streuli CH, Akhtar N. Signal co-operation between integrins and other receptor systems. *Biochem J* 2009;**418**(3):491–506.
117 Lemmon MA, Schlessinger J. Cell signalling by receptor tyrosine kinases. *Cell* 2010;**141**(7):1117–34.
118 Deb M, Sengupta D, Patra SK. Integrin-epigenetics: a system with imperative impact on cancer. *Cancer Metastasis Rev* 2012;**31**(1–2):221–34.
119 Vanhaesebroeck B, Ali K, Bilancio A, Geering B, Foukas LC. Signalling by PI3-K isoforms: insights from gene-targeted mice. *Trends Biochem Sci* 2005;**30**(4):194–204.
120 Duronio V. The life of a cell: apoptosis regulation by the PI3-K/PKB pathway. *Biochem J* 2008;**415**(3):333–44.
121 Franke TF. PI3-K/Akt: getting it right matters. *Oncogene* 2008;**27**(50):6473–88.
122 Engelman JA. Targeting PI3K signalling in cancer: opportunities, challenges and limitations. *Nat Rev Cancer* 2009;**9**(8):550–62.
123 Vanhaesebroeck B, Guillermet-Guibert J, Graupera M, Bilanges B. The emerging mechanisms of isoform-specific PI3K signalling. *Nat Rev Mol Cell Biol* 2010;**11**(5):329–41.
124 Vogt PK, Hart JR, Gymnopoulos M, Jiang H, Kang S, Bader AG, et al. Phosphatidylinositol 3-kinase: the oncoprotein. *Curr Top Microbiol Immunol* 2010;**347**:79–104.
125 Hers I, Vincent EE, Tavaré JM. Akt signalling in health and disease. *Cell Signal* 2011;**23**(10):1515–27.
126 Stephens L, Hawkins P. Signalling via class 1A PI3Ks. *Adv Enzyme Regul* 2011;**51**(1):27–36.
127 Martini M, De Santis MC, Braccini L, Gulluni F, Hirsch E. PI3K/AKT signalling pathway and cancer: an updated review. *Ann Med* 2014;**46**(6):372–83.
128 Porta C, Paglino C, Mosca A. Targeting PI3K/Akt/mTOR Signalling in Cancer. *Front Oncol* 2014;**4**:64.
129 Toker A, Marmiroli S. Signalling specificity in the Akt pathway in biology and disease. *Adv Biol Regul* 2014;**55**:28–38.
130 Bauer TM, Patel MR, Infante JR. Targeting PI3 kinase in cancer. *Pharmacol Ther* 2015;**146C**:53–60.

131 Roberts PJ, Der CJ. Targeting the Raf-MEK-ERK mitogen-activated protein kinase cascade for the treatment of cancer. *Oncogene* 2007;**26**(22):3291–310.
132 Raman M, Chen W, Cobb MH. Differential regulation and properties of MAPKs. *Oncogene* 2007;**26**(22):3100–12.
133 Kim EK, Choi EJ. Pathological roles of MAPK signalling pathways in human diseases. *Biochim Biophys Acta* 2010;**1802**(4):396–405.
134 Roskoski R Jr. RAF protein-serine/threonine kinases: structure and regulation. *Biochem Biophys Res Commun* 2010;**399**(3):313–17.
135 Wimmer R, Baccarini M. Partner exchange: protein-protein interactions in the Raf pathway. *Trends Biochem Sci* 2010;**35**(12):660–8.
136 Maurer G, Tarkowski B, Baccarini M. Raf kinases in cancer-roles and therapeutic opportunities. *Oncogene* 2011;**30**(32):3477–88.
137 Kyriakis JM, Avruch J. Mammalian mitogen-activated protein kinase signal transduction pathways activated by stress and inflammation: a 10-year update. *Physiol Rev* 2012;**92**(2):689–737.
138 Campbell PM. Oncogenic Ras pushes (and pulls) cell cycle progression through ERK activation. *Methods Mol Biol* 2014;**1170**:155–63.
139 Goitre L, Trapani E, Trabalzini L, Retta SF. The Ras superfamily of small GTPases: the unlocked secrets. *Methods Mol Biol* 2014;**1120**:1–18.
140 Neuzillet C, Tijeras-Raballand A, de Mestier L, Cros J, Faivre S, Raymond E. MEK in cancer and cancer therapy. *Pharmacol Ther* 2014;**141**(2):160–71.
141 Samatar AA, Poulikakos PI. Targeting RAS-ERK signalling in cancer: promises and challenges. *Nat Rev Drug Discov* 2014;**13**(12):928–42.
142 Stephen AG, Esposito D, Bagni RK, McCormick F. Dragging ras back in the ring. *Cancer Cell* 2014;**25**(3):272–81.
143 Basu A, DuBois G, Haldar S. Posttranslational modifications of Bcl2 family members – a potential therapeutic target for human malignancy. *Front Biosci* 2006;**11**:1508–21.
144 Kutuk O, Letai A. Regulation of Bcl-2 family proteins by posttranslational modifications. *Curr Mol Med* 2008;**8**(2):102–18.
145 Strozyk E, Kulms D. The role of AKT/mTOR pathway in stress response to UV-irradiation: implication in skin carcinogenesis by regulation of apoptosis, autophagy and senescence. *Int J Mol Sci* 2013;**14**(8):15 260–85.
146 Huang J, Manning BD. A complex interplay between Akt, TSC2 and the two mTOR complexes. *Biochem Soc Trans* 2009;**37**(Pt. 1):217–22.
147 Vadlakonda L, Dash A, Pasupuleti M, Anil Kumar K, *Reddanna P. The Paradox of Akt-mTOR Interactions. Front Oncol* 2013;**3**:165.
148 Wang S, Basson MD. Protein kinase B/Akt and focal adhesion kinase: two close signalling partners in cancer. *Anticancer Agents Med Chem* 2011;**11**(10):993–1002.
149 Mahajan K, Mahajan NP. PI3K-independent AKT activation in cancers: a treasure trove for novel therapeutics. *J Cell Physiol* 2012;**227**(9):3178–84.
150 Chen R, Kim O, Yang J, Sato K, Eisenmann KM, McCarthy J, et al. Regulation of Akt/PKB activation by tyrosine phosphorylation. *J Biol Chem* 2001;**276**(34):31 858–62.
151 Zheng Y, Peng M, Wang Z, Asara JM, Tyner AL. Protein tyrosine kinase 6 directly phosphorylates AKT and promotes AKT activation in response to epidermal growth factor. *Mol Cell Biol* 2010;**30**(17):4280–92.
152 Zheng Y, Gierut J, Wang Z, Miao J, Asara JM, Tyner AL. Protein tyrosine kinase 6 protects cells from anoikis by directly phosphorylating focal adhesion kinase and activating AKT. *Oncogene* 2013;**32**(36):4304–12.

153 McDonald PC, Fielding AB, Dedhar S. Integrin-linked kinase – essential roles in physiology and cancer biology. *J Cell Sci* 2008;**121**(Pt. 19):3121–32.
154 Boudeau J, Miranda-Saavedra D, Barton GJ, Alessi DR. Emerging roles of pseudokinases. *Trends Cell Biol* 2006;**16**(9):443–52.
155 Wickström SA, Lange A, Montanez E, Fässler R. The ILK/PINCH/parvin complex: the kinase is dead, long live the pseudokinase! *EMBO J* 2010;**29**(2):281–91.
156 Ghatak S, Morgner J, Wickström SA. ILK: a pseudokinase with a unique function in the integrin-actin linkage. *Biochem Soc Trans* 2013;**41**(4):995–1001.
157 Fielding AB, Dedhar S. The mitotic functions of integrin-linked kinase. *Cancer Metastasis Rev* 2009;**28**(1–2):99–111.
158 Carriere A, Ray H, Blenis J, Roux P.P. The RSK factors of activating the Ras/MAPK signalling cascade. *Front Biosci* 2008;**13**:4258–75.
159 Anjum R, Blenis J. The RSK family of kinases: emerging roles in cellular signalling. *Nat Rev Mol Cell Biol* 2008;**9**(10):747–58.
160 Klinger B, Blüthgen N. Consequences of feedback in signal transduction for targeted therapies. *Biochem Soc Trans* 2014;**42**(4):770–5.
161 Raimondi C, Falasca M. Targeting PDK1 in cancer. *Curr Med Chem* 2011;**18**(18):2763–9.
162 Senft J, Helfer B, Frisch SM. Caspase-8 interacts with the p85 subunit of phosphatidylinositol 3-kinase to regulate cell adhesion and motility. *Cancer Res* 2007;**67**(24):11 505–9.
163 Frisch SM. Caspase-8: fly or die. *Cancer Res* 2008;**68**(12):4491–3.
164 Kurenova E, Xu LH, Yang X, Baldwin A.S. Jr., Craven RJ, Hanks SK, et al. Focal adhesion kinase suppresses apoptosis by binding to the death domain of receptor-interacting protein. *Mol Cell Biol* 2004;**24**(10):4361–71.
165 Kamarajan P, Bunek J, Lin Y, Nunez G, Kapila YL. Receptor-interacting protein shuttles between cell death and survival signalling pathways. *Mol Biol Cell* 2010;**21**(3):481–8.
166 Xie Z, Sanada K, Samuels BA, Shih H, Tsai LH. Serine 732 phosphorylation of FAK by Cdk5 is important for microtubule organization, nuclear movement, and neuronal migration. *Cell* 2003;**114**(4):469–82.
167 Gonzalez E, McGraw TE. The Akt kinases: isoform specificity in metabolism and cancer. *Cell Cycle* 2009;**8**(16):2502–8.
168 Wills MK, Jones N. Teaching an old dogma new tricks: twenty years of Shc adaptor signalling. *Biochem J* 2012;**447**(1):1–16.
169 Gumbiner BM. Cell adhesion: the molecular basis of tissue architecture and morphogenesis. *Cell* 1996;**84**(3):345–57.
170 Gumbiner BM. Regulation of cadherin-mediated adhesion in morphogenesis. *Nat Rev Mol Cell Biol* 2005;**6**(8):622–34.
171 Takeichi M. Dynamic contacts: rearranging adherens junctions to drive epithelial remodelling. *Nat Rev Mol Cell Biol* 2014;**15**(6):397–410.
172 Yap AS, Brieher WM, Gumbiner BM. Molecular and functional analysis of cadherin-based adherens junctions. *Annu Rev Cell Dev Biol* 1997;**13**:119–46.
173 Schackmann RC, Tenhagen M, van de Ven RA, Derksen PW. p120-catenin in cancer – mechanisms, models and opportunities for intervention. *J Cell Sci* 2013;**126**(Pt. 16):3515–25.
174 Capaldo CT, Farkas AE, Nusrat A. Epithelial adhesive junctions. *F1000Prime Rep* 2014;**6**:1.

175 Müller EJ, Williamson L, Kolly C, Suter MM. Outside-in signalling through integrins and cadherins: a central mechanism to control epidermal growth and differentiation? *J Invest Dermatol* 2008;**128**(3):501–16.

176 Bhatt T, Rizvi A, Batta SP, Kataria S, Jamora C. Signalling and mechanical roles of E-cadherin. *Cell Commun Adhes* 2013;**20**(6):189–99.

177 Serrels A, Canel M, Brunton VG, Frame MC. Src/Fak-mediated regulation of E-cadherin as a mechanism for controlling collective cell movement. *Cell Adh Migr* 2011;**5**(4):360–5.

178 Weber GF, Bjerke MA, DeSimone DW. Integrins and cadherins join forces to form adhesive networks. *J Cell Sci* 2011;**124**(Pt. 8):1183–93.

179 DuFort CC, Paszek MJ, Weaver VM. Balancing forces: architectural control of mechanotransduction. *Nat Rev Mol Cell Biol* 2011;**12**(5):308–19.

180 Geiger B, Spatz JP, Bershadsky AD. Environmental sensing through focal adhesions. *Nat Rev Mol Cell Biol* 2009;**10**(1):21–33.

181 Wozniak MA, Chen CS. Mechanotransduction in development: a growing role for contractility. *Nat Rev Mol Cell Biol* 2009;**10**(1):34–43.

182 Huveneers S, de Rooij J. Mechanosensitive systems at the cadherin-F-actin interface. *J Cell Sci* 2013;**126**(Pt. 2):403–13.

183 Horbinski C, Mojesky C, Kyprianou N. Live free or die: tales of homeless (cells) in cancer. *Am J Pathol* 2010;**177**(3):1044–52.

184 Murray P, Edgar D. Regulation of programmed cell death by basement membranes in embryonic development. *J Cell Biol* 2000;**150**(5):1215–21.

185 Houben E, De Paepe K, Rogiers V. A keratinocyte's course of life. *Skin Pharmacol Physiol* 2007;**20**(3):122–32.

186 Muschler J, Streuli CH. Cell-matrix interactions in mammary gland development and breast cancer. *Cold Spring Harb Perspect Biol* 2010;**2**(10):a003202.

187 Bertrand K. Survival of exfoliated epithelial cells: a delicate balance between anoikis and apoptosis. *J Biomed Biotechnol* 2011;**2011**:534139.

188 Glukhova MA, Streuli CH. How integrins control breast biology. *Curr Opin Cell Biol* 2013;**25**(5):633–41.

189 Ma Z, Liu Z, Myers DP, Terada LS. Mechanotransduction and anoikis: death and the homeless cell. *Cell Cycle* 2008;**7**(16):2462–5.

190 Guadamillas MC, Cerezo C, del Pozo MA. Overcoming anoikis – pathways to anchorage-independent growth in cancer. *J Cell Sci* 2011;**124**(Pt. 19):3189–97.

191 Nagaprashantha LD, Vatsyayan R, Lelsani PC, Awasthi S, Singhal SS. The sensors and regulators of cell-matrix surveillance in anoikis resistance of tumors. *Int J Cancer* 2011;**128**(4):743–52.

192 Paoli P, Giannoni E, Chiarugi P. Anoikis molecular pathways and its role in cancer progression. *Biochim Biophys Acta* 2013;**1833**(12):3481–98.

193 Buchheit CL, Weigel KJ, Schafer ZT. Cancer cell survival during detachment from the ECM: multiple barriers to tumour progression. *Nat Rev Cancer* 2014;**14**(9):632–41.

194 Pellegrini M, Pacini S, Baldari CT. p66SHC: the apoptotic side of Shc proteins. *Apoptosis* 2005;**10**(1):13–18.

195 Ma Z, Myers DP, Wu RF, Nwariaku FE, Terada LS. p66^{Shc} mediates anoikis through RhoA. *J Cell Biol* 2007;**179**(1):23–31.

196 Miñambres R, Guasch RM, Perez-Aragó A, Guerri C. The RhoA/ROCK-I/MLC pathway is involved in the ethanol-induced apoptosis by anoikis in astrocytes. *J Cell Sci* 2006;**119**(Pt. 2):271–82.

197 Debnath J. p66^{Shc} and Ras: controlling anoikis from the inside-out. *Oncogene* 2010;**29**(41):5556–8.
198 Frisch SM. Anoikis. *Methods Enzymol* 2000;**322**:472–9.
199 Bode AM, Dong Z. The functional contrariety of JNK. *Mol Carcinog* 2007;**46**(8):591–8.
200 Loesch M, Chen G. The p38 MAPK stress pathway as a tumor suppressor or more? *Front Biosci* 2008;**13**:3581–93.
201 Bogoyevitch MA, Ngoei KR, Zhao TT, Yeap YY, Ng DC. c-Jun N-terminal kinase (JNK) signalling: recent advances and challenges. *Biochim Biophys Acta* 2010;**1804**(3):463–75.
202 Bogoyevitch MA. The isoform-specific functions of the c-Jun N-terminal kinases (JNKs): differences revealed by gene targeting. *Bioessays* 2006;**28**(9):923–34.
203 Bogoyevitch MA, Kobe B. Uses for JNK: the many and varied substrates of the c-Jun N-terminal kinases. *Microbiol Mol Biol Rev* 2006;**70**(4):1061–95.
204 Wagner EF, Nebreda AR. Signal integration by JNK and p38 MAPK pathways in cancer development. *Nat Rev Cancer* 2009;**9**(8):537–49.
205 Cuadrado A, Nebreda AR. Mechanisms and functions of p38 MAPK signalling. *Biochem J* 2010;**429**(3):403–17.
206 Liu J, Lin A. Role of JNK activation in apoptosis: a double-edged sword. *Cell Res* 2005;**15**(1):36–42.
207 Zarubin T, Han J. Activation and signalling of the p38 MAP kinase pathway. *Cell Res* 2015;**15**(1):11–18.
208 Sui X, Kong N, Ye L, Han W, Zhou J, Zhang Q, et al. p38 and JNK MAPK pathways control the balance of apoptosis and autophagy in response to chemotherapeutic agents. *Cancer Lett* 2014;**344**(2):174–9.
209 Frisch SM, Vuori K, Kelaita D, Sicks S. A role for Jun-N-terminal kinase in anoikis; suppression by bcl-2 and crmA. *J Cell Biol* 1996;**135**(5):1377–82.
210 Cardone MH, Salvesen GS, Widmann C, Johnson G, Frisch SM. The regulation of anoikis: MEKK-1 activation requires cleavage by caspases. *Cell* 1997;**90**(2):315–23.
211 Khwaja A, Downward J. Lack of correlation between activation of Jun-NH2-terminal kinase and induction of apoptosis after detachment of epithelial cells. *J Cell Biol* 1997;**139**(4):1017–23.
212 Wang Y, Huang S, Sah VP, Ross J. Jr., Brown JH, Han J, Chien KR. Cardiac muscle cell hypertrophy and apoptosis induced by distinct members of the p38 mitogen-activated protein kinase family. *J Biol Chem* 1998;**273**(4):2161–8.
213 Fujisawa T, Takeda K, Ichijo H. ASK family proteins in stress response and disease. *Mol Biotechnol* 2007;**37**(1):13–18.
214 Hattori K, Naguro I, Runchel C, Ichijo H. The roles of ASK family proteins in stress responses and diseases. *Cell Commun Signal* 2009;**7**:9.
215 Shiizaki S, Naguro I, Ichijo H. Activation mechanisms of ASK1 in response to various stresses and its significance in intracellular signalling. *Adv Biol Regul* 2013;**53**(1):135–44.
216 Kim AH, Khursigara G, Sun X, Franke TF, Chao MV. Akt phosphorylates and negatively regulates apoptosis signal-regulating kinase 1. *Mol Cell Biol* 2001;**21**(3):893–901.
217 Seong HA, Jung H, Ichijo H, Ha H. Reciprocal negative regulation of PDK1 and ASK1 signalling by direct interaction and phosphorylation. *J Biol Chem* 2010;**285**(4):2397–414.

218 Matallanas D, Birtwistle M, Romano D, Zebisch A, Rauch J, von Kriegsheim A, Kolch W. Raf family kinases: old dogs have learned new tricks. *Genes Cancer* 2011;**2**(3):232–60.
219 Bialik S, Kimchi A. Lethal weapons: DAP-kinase, autophagy and cell death: DAP-kinase regulates autophagy. *Curr Opin Cell Biol* 2010;**22**(2):199–205.
220 Michie AM, McCaig AM, Nakagawa R, Vukovic M. Death-associated protein kinase (DAPK) and signal transduction: regulation in cancer. *FEBS J* 2010;**277**(1):74–80.
221 Lin Y, Hupp TR, Stevens C. Death-associated protein kinase (DAPK) and signal transduction: additional roles beyond cell death. *FEBS J* 2010;**277**(1):48–57.
222 Shiloh R, Bialik S, Kimchi A. The DAPK family: a structure-function analysis. *Apoptosis* 2014;**19**(2):286–97.
223 Bialik S, Kimchi A. The DAP-kinase interactome. *Apoptosis* 2014;**19**(2):316–28.
224 Ivanovska J, Mahadevan V, Schneider-Stock R. DAPK and cytoskeleton-associated functions. *Apoptosis* 2014;**19**(2):329–38.
225 Wang WJ, Kuo JC, Yao CC, Chen RH. DAP-kinase induces apoptosis by suppressing integrin activity and disrupting matrix survival signals. *J Cell Biol* 2002;**159**(1):169–79.
226 Kuo JC, Lin LR, Staddon JM, Hosoya H, Chen RH. Uncoordinated regulation of stress fibers and focal adhesions by DAP kinase. *J Cell Sci* 2003;**116**(Pt. 23):4777–90.
227 Kuo JC, Wang WJ, Yao CC, Wu PR, Chen RH. The tumor suppressor DAPK inhibits cell motility by blocking the integrin-mediated polarity pathway. *J Cell Biol* 2006;**172**(4):619–31.
228 Zalckvar E, Berissi H, Eisenstein M, Kimchi A. Phosphorylation of Beclin 1 by DAP-kinase promotes autophagy by weakening its interactions with Bcl-2 and Bcl-XL. *Autophagy* 2009;**5**(5):720–2.
229 Li H, Ray G, Yoo BH, Erdogan M, Rosen KV. Down-regulation of death-associated protein kinase-2 is required for beta-catenin-induced anoikis resistance of malignant epithelial cells. *J Biol Chem* 2009;**284**(4):2012–22.
230 Stupack DG, Puente XS, Boutsaboualoy S, Storgard CM, Cheresh DA. Apoptosis of adherent cells by recruitment of caspase-8 to unligated integrins. *J Cell Biol* 2001;**155**(3):459–70.
231 Lotti R, Marconi A, Truzzi F, Dallaglio K, Gemelli C, Borroni RG, et al. A previously unreported function of β(1)B integrin isoform in caspase-8-dependent integrin-mediated keratinocyte death. *J Invest Dermatol* 2010;**130**(11):2569–77.
232 Frisch SM. Evidence for a function of death-receptor-related, death-domain-containing proteins in anoikis. *Curr Biol* 1999;**9**(18):1047–9.
233 Rytömaa M, Martins LM, Downward J. Involvement of FADD and caspase-8 signalling in detachment-induced apoptosis. *Curr Biol* 1999;**9**(18):1043–6.
234 Rytömaa M, Lehmann K, Downward J. Matrix detachment induces caspase-dependent cytochrome c release from mitochondria: inhibition by PKB/Akt but not Raf signalling. *Oncogene* 2000;**19**(39):4461–8.
235 Marconi A, Atzei P, Panza C, Fila C, Tiberio R, Truzzi F, et al. FLICE/caspase-8 activation triggers anoikis induced by beta1-integrin blockade in human keratinocytes. *J Cell Sci* 2004;**117**(Pt. 24):5815–23.
236 Miyazaki T, Shen M, Fujikura D, Tosa N, Kim HR, Kon S, et al. Functional role of death-associated protein 3 (DAP3) in anoikis. *J Biol Chem* 2004;**279**(43):44 667–72.
237 Bouchentouf M, Benabdallah BF, Rousseau J, Schwartz M, Tremblay JP. Induction of anoikis following myoblast transplantation into SCID mouse muscles requires the Bit1 and FADD pathways. *Am J Transplant* 2007;**7**(6):1491–505.

238 Aoudjit F, Vuori K. Matrix attachment regulates Fas-induced apoptosis in endothelial cells: a role for c-Flip and implication for anoikis. *J Cell Biol* 2001;**152**(3):633–43.

239 Mawji IA, Simpson CD, Hurren R, Gronda M, Williams MA, Filmus J, et al. Critical role for Fas-associated death domain-like interleukin-1-converting enzyme-like inhibitory protein in anoikis resistance and distant tumor formation. *J Natl Cancer Inst* 2007;**99**(10):811–22.

240 Zhao Y, Sui X, Ren H. From procaspase-8 to caspase-8: revisiting structural functions of caspase-8. *J Cell Physiol* 2010;**225**(2):316–20.

241 Grossmann J, Walther K, Artinger M, Kiessling S, Schölmerich J. Apoptotic signalling during initiation of detachment-induced apoptosis ("anoikis") of primary human intestinal epithelial cells. *Cell Growth Differ* 2001;**12**(3):147–55.

242 Sourdeval M, Lemaire C, Deniaud A, Taysse L, Daulon S, Breton P, et al. Inhibition of caspase-dependent mitochondrial permeability transition protects airway epithelial cells against mustard-induced apoptosis. *Apoptosis* 2006;**11**(9):1545–59.

243 Sourdeval M, Boisvieux-Ulrich E, Gendron MC, Marano F. Mitochondrial inside-out signalling during alkylating agent-induced anoikis. *Front Biosci (Landmark Ed)* 2009;**14**:1917–31.

244 Yoo BH, Wang Y, Erdogan M, Sasazuki T, Shirasawa S, Corcos L, et al. Oncogenic ras-induced down-regulation of pro-apoptotic protease caspase-2 is required for malignant transformation of intestinal epithelial cells. *J Biol Chem* 2011;**286**(45):38 894–903.

245 Jenning S, Pham T, Ireland SK, Ruoslahti E, Biliran H. Bit1 in anoikis resistance and tumor metastasis. *Cancer Lett* 2013;**333**(2):147–51.

246 Frisch SM, Vuori K, Ruoslahti E, Chan-Hui PY. Control of adhesion-dependent cell survival by focal adhesion kinase. *J Cell Biol* 1996;**134**(3):793–9.

247 Benoit YD, Larrivée JF, Groulx JF, Stankova J, Vachon PH, Beaulieu JF. Integrin α8β1 confers anoikis susceptibility to human intestinal epithelial crypt cells. *Biochem Biophys Res Commun* 2010;**399**(3):434–9.

248 Subauste MC, Pertz O, Adamson ED, Turner CE, Junger S, Hahn KM. Vinculin modulation of paxillin-FAK interactions regulates ERK to control survival and motility. *J Cell Biol* 2004;**165**(3):371–81.

249 Ziegler WH, Liddington RC, Critchley DR. The structure and regulation of vinculin. *Trends Cell Biol* 2006;**16**(9):453–60.

250 Kozlova NI, Morozevich GE, Chubukina AN, Berman AE. Integrin αvβ3 promotes anchorage-dependent apoptosis in human intestinal carcinoma cells. *Oncogene* 2001;**20**(34):4710–17.

251 Morozevich GE, Kozlova NI, Chubukina AN, Berman AE. Role of integrin αvβ3 in substrate-dependent apoptosis of human intestinal carcinoma cells. *Biochemistry (Mosc)* 2003;**68**(4):416–23.

252 Jin H, Varner J. Integrins: roles in cancer development and as treatment targets. *Br J Cancer* 2004;**90**(3):561–5.

253 Farias E, Lu M, Li X, Schnapp LM. Integrin α8β1-fibronectin interactions promote cell survival via PI3 kinase pathway. *Biochem Biophys Res Commun* 2005;**329**(1):305–11.

254 You H, Lei P, Andreadis ST. JNK is a novel regulator of intercellular adhesion. *Tissue Barriers* 2013;**1**(5):e26845.

255 Hofmann C, Obermeier F, Artinger M, Hausmann M, Falk W, Schoelmerich J, et al. Cell-cell contacts prevent anoikis in primary human colonic cells. *Gastroenterology* 2007;**132**(2):587–600.

256 Escaffit F, Perreault N, Jean D, Francoeur C, Herring E, Rancourt C, et al. Repressed E-cadherin expression in the lower crypt of human small intestine: a cell marker of functional relevance. *Exp Cell Res* 2005;**302**(2):206–20.

257 Mebratu Y, Tesfaigzi Y. How ERK1/2 activation controls cell proliferation and cell death: is subcellular localization the answer? *Cell Cycle* 2009;**8**(8):1168–75.

258 Cagnol S, Chambard JC. ERK and cell death: mechanisms of ERK-induced cell death – apoptosis, autophagy and senescence. *FEBS J* 2010;**277**(1):2–21.

259 Beaulieu JF. Integrins and human intestinal cell functions. *Front Biosci* 1999;**4**: D310–21.

260 Lussier C, Basora N, Bouatrouss Y, Beaulieu JF. Integrins as mediators of epithelial cell-matrix interactions in the human small intestinal mucosa. *Microsc Res Tech* 2000;**51**(2):169–78.

261 Ménard D, Beaulieu JF, Boudreau F, et al. Gastrointestinal tract (GI tract) In: Unsicker K, Krieglstein K (eds.) *Cell Signalling and Growth Factors in Development – Part II*. Weinheim: Wiley-VCH. 2005.

262 Vachon PH. Cell survival: differences and differentiation. *Med Sci (Paris)* 2006;**22**(4):423–9.

263 Lévy E, Delvin E, Ménard D, Beaulieu JF. Functional development of human fetal gastrointestinal tract. *Methods Mol Biol* 2009;**550**:205–24.

264 Edelblum KL, Yan F, Yamaoka T, Polk PB. Regulation of apoptosis during homeostasis and disease in the intestinal epithelium. *Inflamm Bowel Dis* 2006;**12**(5):413–24.

265 Tarnawski AS, Szabo I. Apoptosis-programmed cell death and its relevance to gastrointestinal epithelium: survival signal from the matrix. *Gastroenterology* 2001;**120**(1):294–9.

266 Potten CS. Epithelial cell growth and differentiation. II. Intestinal apoptosis. *Am J Physiol* 1997;**273**(2 Pt. 1):G253–7.

267 Vachon PH, Cardin E, Harnois C, Reed JC, Vézina A. Early establishment of epithelial apoptosis in the developing human small intestine. *Int J Dev Biol* 2000;**44**(8):891–8.

268 Gauthier R, Harnois C, Drolet JF, Reed JC, Vézina A, Vachon PH. Human intestinal epithelial cell survival; differentiation state-specific control mechanisms. *Am J Physiol Cell Physiol* 2001;**280**(6):C1540–54.

269 Vachon PH, Cardin E, Harnois C, Reed JC, Plourde A, Vézina A. Early acquisition of bowel segment-specific Bcl-2 expression profiles during the development of the human ileum and colon. *Histol Histopathol* 2001;**16**(2):497–510.

270 Harnois C, Demers MJ, Bouchard V, Vallée K, Gagné D, Fujita N, et al. Human intestinal epithelial crypt cell survival and death: complex modulations of Bcl-2 homologs by Fak, PI3-K/Akt-1, MEK/Erk, and p38 signalling pathways. *J Cell Physiol* 2004;**198**(2):209–22.

271 Gauthier R, Laprise P, Cardin E, Harnois C, Plourde A, Reed JC, et al. Differential sensitivity to apoptosis between the human small and large intestinal mucosae: linkage with segment-specific regulation of Bcl-2 homologs and involvement of signalling pathways. *J Cell Biochem* 2001;**82**(2):339–55.

272 Bouchard V, Harnois C, Demers MJ, Thibodeau S, Laquerre V, Gautheir R, et al. β1 integrin/Fak/Src signalling in intestinal epithelial crypt cell survival: integration of complex regulatory mechanisms. *Apoptosis* 2008;**13**(4):531–42.

273 Stutzmann J, Bellissent-Waydelich A, Fontao L, Launay JF, Simon-Assmann P. Adhesion complexes implicated in intestinal epithelial cell-matrix interactions. *Microsc Res Tech* 2000;**51**(2):179–90.

274 Teller IC, Auclair J, Herring E, Gauthier R, Ménard D, Beaulieu JF. Laminins in the developing and adult human small intestine: relation with the functional absorptive unit. *Dev Dyn* 2007;**236**(7):1980–90.

275 Beaulieu JF. Integrin α6β4 in colorectal cancer. *World J Gastrointest Pathophysiol* 2010;**1**(1):3–11.

276 Benoit YD, Groulx JF, Gagné D, Beaulieu JF. RGD-dependent epithelial cell-matrix interactions in the human intestinal crypt. *J Signal Transduct* 2012;**2012**:248759.

277 Vachon PH, Harnois C, Grenier A, Dufour G, Bouchard V, Han J, et al. Differentiation state-selective roles of p38 isoforms in human intestinal epithelial cell anoikis. *Gastroenterology* 2002;**123**(6):1980–91.

278 Bouchard V, Demers MJ, Thibodeau S, Laquerre V, Fujita N, Tsuruo T, et al. Fak/Src signalling in human intestinal epithelial cell survival and anoikis: differentiation state-specific uncoupling with the PI3-K/Akt-1 and MEK/Erk pathways. *J Cell Physiol* 2007;**212**(3):717–28.

279 Beauséjour M, Thibodeau S, Demers MJ, Bouchard V, Gauthier R, Beaulieu JF, Vachon PH. Suppression of anoikis in human intestinal epithelial cells: differentiation state-selective roles of α2β1, α3β1, α5β1, and α6β4 integrins. *BMC Cell Biol* 2013;**14**:53.

280 Bullen TF, Forrest S, Campbell F, Dodson AR, Hershman MJ, Pritchard DM, et al. Characterization of epithelial cell shedding from human small intestine. *Lab Invest* 2006;**86**(10):1052–63.

281 Basora N, Herring-Gillam FE, Boudreau F, Perreault N, Pageot L-P, Simoneau M, et al. Expression of functionally distinct variants of the β4A integrin subunit in relation to the differentiation state in human intestinal cells. *J Biol Chem* 1999;**274**(42):29 819–25.

282 Dufour G, Demers MJ, Gagné D, Dydensborg AB, Teller IC, Bouchard V, et al. Human intestinal epithelial cell survival and anoikis: differentiation state-distinct regulation and roles of protein kinase B/Akt isoforms. *J Biol Chem* 2004;**279**(42):44 113–22.

283 Beauséjour M, Noël D, Thibodeau S, Bouchard V, Harnois C, Beaulieu JF, et al. Integrin/Fak/Src-mediated regulation of cell survival and anoikis in human intestinal epithelial crypt cells: selective engagement and roles of PI3-K isoform complexes. *Apoptosis* 2012;**17**(6):566–78.

284 Smythe GM, Rando TA. Altered caveolin-3 expression disrupts PI(3) kinase signalling leading to death of cultured muscle cells. *Exp Cell Res* 2006;**312**(15):2816–25.

285 Rotwein P, Wilson EM. Distinct actions of Akt1 and Akt2 in skeletal muscle differentiation. *J Cell Physiol* 2009;**219**(2):503–11.

286 Jaalouk DE, Lammerding J. Mechanotransduction gone awry. *Nat Rev Mol Cell Biol* 2009;**10**(1):63–73.

287 McNally EM, Pytel P. Muscle diseases: the muscular dystrophies. *Annu Rev Pathol* 2007;**2**:87–109.

288 Wallace GQ, McNally EM. Mechanisms of muscle degeneration, regeneration, and repair in the muscular dystrophies. *Annu Rev Physiol* 2009;**71**:37–57.

289 Uitto J, McGrath JA, Rodeck U, Bruckner-Tuderman L, Robinson EC. Progress in epidermolysis bullosa research: toward treatment and cure. *J Invest Dermatol* 2010;**130**(7):1778–84.

290 Frisch SM, Schaller M, Cieply B. Mechanisms that link the oncogenic epithelial-mesenchymal transition to suppression of anoikis. *J Cell Sci* 2013;**126**(Pt. 1):21–9.

291 Shattil SJ, Kim C, Ginsberg MH. The final steps of integrin activation: the end game. *Nat Rev Mol Cell Biol* 2010;**11**(4):288–300.
292 Barkan D, Green JE, Chambers AF. Extracellular matrix: a gatekeeper in the transition from dormancy to metastatic growth. *Eur J Cancer* 2010;**46**(7):1181–8.
293 Singh A, Settleman J. EMT, cancer stem cells and drug resistance: an emerging axis of evil in the war on cancer. *Oncogene* 2010;**29**(34):4741–51.
294 Lamouille S, Xu J, Derynck R. Molecular mechanisms of epithelial-mesenchymal transition. *Nat Rev Mol Cell Biol* 2014;**15**(3):178–96.
295 Lindsey S, Langhans SA. Crosstalk of oncogenic signalling pathways during epithelial-mesenchymal transition. *Front Oncol* 2014;**4**:358.
296 Demers MJ, Thibodeau S, Noël D, Fujita N, Tsuruo T, Gautheir R, et al. Intestinal epithelial cancer cell anoikis resistance: EGFR-mediated sustained activation of Src overrides Fak-dependent signalling to MEK/Erk and/or PI3-K/Akt-1. *J Cell Biochem* 2009;**107**(4):639–54.
297 Gagné D, Groulx JF, Benoit YD, Basora N, Herring E, Vachon PH, Beaulieu JF. Integrin-linked kinase regulates migration and proliferation of human intestinal cells under a fibronectin-dependent mechanism. *J Cell Physiol* 2010;**222**(2):387–400.
298 Fouquet S, Lugo-Martínez VH, Faussat AM, Renaud F, Cardot P, Chambaz J, et al. Early loss of E-cadherin from cell-cell contacts is involved in the onset of Anoikis in enterocytes. *J Biol Chem* 2004;**279**(41):43 061–9.
299 Lugo-Martínez VH, Petit CS, Fouquet S, Le Beyec J, Chambaz J, Pinçon-Raymond M, et al. Epidermal growth factor receptor is involved in enterocyte anoikis through the dismantling of E-cadherin-mediated junctions. *Am J Physiol Gastrointest Liver Physiol* 2009;**296**(2):G235–44.
300 Pomerleau V, Landry M, Bernier J, Vachon PH, Saucier C. Met receptor-induced Grb2 or Shc signals both promote transformation of intestinal epithelial cells, albeit they are required for distinct oncogenic functions. *BMC Cancer* 2014;**14**:240.
301 Sodek KL, Murphy KJ, Brown TJ, Ringuette MJ. Cell-cell and cell-matrix dynamics in intraperitoneal cancer metastasis. *Cancer Metastasis Rev* 2012;**31**:393–414.
302 Weiswald L-B, Bellet D, Dangles-Marie V. Spherical cancer models in tumor biology. *Neoplasia* 2015;**17**(1):1–15.
303 Mowers EE, Sharifi MN, Macleod KF. Autophagy in cancer metastasis. *Oncogene* 2017;**36**(12):1619–30.

9

Cornification
Leopold Eckhart

Department of Dermatology, Medical University of Vienna, Vienna, Austria

Abbreviations

CAD	caspase-activated DNase
DNase	deoxyribonuclease
EDC	epidermal differentiation complex
ER	endoplasmic reticulum
filaggrin	filament aggregating protein
ICAD	inhibitor of caspase-activated DNase
KLK	kallikrein
KRTAP	keratin-associated protein
LEKTI	lympho-epithelial Kazal-type related inhibitor I
mTORC	mammalian target of rapamycin complex
NMF	natural moisturizing factor
PCD	programmed cell death
SH	sulfhydryl
SOX	sulfhydryl oxidase
SPINK5	serine peptidase inhibitor, Kazal type 5 gene
SPRR	small proline-rich protein
TGase	transglutaminase
TIG3	tazarotene-induced gene 3

9.1 Cornification is a Key Step in the Terminal Differentiation of Epidermal Keratinocytes

9.1.1 Epidermal Keratinocytes Establish the Skin Barrier and Build Up Skin Appendages Such as Hair

The epidermis of humans and other mammals is protected from external noxes by a layer of dead epidermal cells and by hair, which also consists of dead epidermal cells. The coordinated mode of cell death that generates the components of these skin structures is called cornification. The distinctive features of cornification are extensive crosslinking of cellular proteins and the maintenance of dead cell corpses as functional elements of the

Apoptosis and Beyond: The Many Ways Cells Die, First Edition. Edited by James Radosevich.
© 2018 John Wiley & Son Inc. Published 2018 by John Wiley & Son Inc.

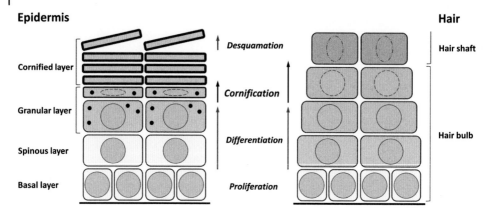

Figure 9.1 Terminal differentiation of keratinocytes in the epidermis and in hair. The nucleus and keratohyalin granules are indicated. Cornification involves the breakdown of the nucleus.

tissue. Cornification differs mechanistically from all other types of cell death and strongly depends on the preceding steps of epidermal cell differentiation, which prepare the cell for its execution [1].

The main cell type of the epidermis is the keratinocyte. Others include melanocytes, Langerhans cells, and Merkel cells, all of which have specific functions in the interaction of the body with the environment. However, only keratinocytes contribute significantly to the skin barrier against mechanical damages and the uncontroled passage of substances through the body surface. The latter functions depend on the unique ability of keratinocytes to cornify and establish resilient intercellular connections [2].

The epidermis is constantly renewed by the proliferation of cells in the basal layer, by differentiation in suprabasal layers, by cornification of differentiated keratinocytes, and by desquamation of superficial dead cells. The sheet-like structure of the epidermis is interrupted by hair follicles, in which the epidermal epithelium is modified to allow for the continuous growth of hair. The formation of the hair shaft involves a special mode of differentiation, in which keratinocytes establish a highly interconnected cytoskeleton and retain intercellular connections. Both interfollicular and hair keratinocytes undergo cornification, but with distinct differences in its mechanism [3]. The basic organization of the epidermis and the hair and the main keratinocyte differentiation events are schematically depicted in Figure 9.1.

9.1.2 The Cornified Layer of the Epidermis is Formed by a Multistep Differentiation Program

The basal layer of the epidermis contains stem cells and so-called "transient amplifying cells," which proliferate via either horizontal or vertical cell division. In the first case, both daughter cells keep contact with the basement membrane and retain their proliferative potential. If, however, the cell division is oriented vertically relative to the basement membrane, one daughter cell detaches from the membrane and consequently ceases proliferation. Instead, this cell starts a differentiation process that involves further movement away from the basal layer due to the pressure of newly formed cells. The differentiation of keratinocytes is controled by transcription factors such as p63, AP1, Ezh2, and KLF4, which repress or activate the expression of distinct sets of genes during the outward passage of keratinocytes through the layers of the epidermis [4,5]. The targets genes of the transcriptional regulators of keratinocyte differentiation code for proteins that ensure the structural integrity of the epidermis and for enzymes that facilitate the remodelling of the cell structure, as well as alterations in cellular metabolism when cells approach the body surface [6].

The cytoskeleton of keratinocytes is mainly made up of keratins. In the basal layer, the cells express keratin 5 (K5), a type II intermediate filament protein, and K14, a type I intermediate filament protein. These keratins heterodimerize and form filaments, which are connected to neighboring cells via desmosomes and to the basement membrane via hemidesmosomes. Intermediate filaments are also attached to the nuclear membrane via plectin, nesprin-3, and perhaps other yet-uncharacterized linkers [7,8]. When keratinocytes move to the suprabasal layers, they cease to express K5 and K14 and start the expression of K1, K2, and K10. In the human epidermis, the expression of K1 precedes that of K2, and both appear to interact with K10. By contrast, K1 and K2 are expressed in a mutually exclusive manner in the mouse, where K1 and K10 form the suprabasal cytoskeleton in the interfollicular epidermis of body regions with a dense hair coat, whereas K2 and K10 are the main components of the cytoskeleton in the epidermis of ears and soles [9]. High levels of keratin gene expression increase the keratin content of differentiating keratinocytes to over 80% [10].

Keratins are controled by post-translational modifications such as phosphorylation and by interactions with so-called "keratin-associated proteins" (KRTAPs). In the epidermis, filaggrin (filament aggregating protein) is the best-characterized keratin regulator. Filaggrin belongs to the S100 fused-type protein family and, accordingly, consists of an amino-terminal S100 domain and a long repetitive carboxy-terminal domain. Expression of the filaggrin gene during late differentiation of keratinocytes results in the accumulation of the proform of filaggrin in the form of keratohyalin granules. These structures are the distinctive feature of the granular layer of the epidermis. Proteolytic processing of pro-filaggrin, which depends on prior desphosphorylation, releases mature filaggrin, which contributes to the aggregation and bundling of keratin filaments [11,12]. Besides interacting with keratins, filaggrin also serves as an important source of urocanic acid, an endogenous sunscreen [13].

The mechanical resilience of the epidermis depends both on keratin filaments within individual cells and on the linkage of neighboring cells via desmosomes. The desmosomal cell–cell contacts are flexibly dissolved and reformed during the outward movement of differentiating cells but become fixed in the outer layers of the epidermis. While cornification is mainly an intracellular remodelling process, the desmosomal cell–cell contacts are also modified to form the co-called "corneodesmosomes." They maintain the compact organization of the cornified layer of the epidermis. Ultimately, corneodesmosomes are, however, cut by targeted proteolysis. Cornified cells are thus detached from the epidermis and shed to the environment in a process named desquamation [14].

9.1.3 Hair Consists of Cornified, Interconnected Keratinocytes

Besides terminal differentiation of keratinocytes in the epidermis, alternative pathways of differentiation are active in the hair follicles. Hair keratinocyte differentiation follows the same basic principles as interfollicular epidermal keratinocyte differentiation. Proliferation occurs in the basal layer and differentiation with transcriptionally driven alterations of the cytoskeleton in the suprabasal cells (Figure 9.1, right panel). However, there are notable differences between epidermal and hair differentiation. Most importantly, a peculiar cell fate-determination program establishes concentric circles of keratinocytes, which follow distinct differentiation pathways, with specific sets of keratins and associated proteins being expressed in each concentric layer [15]. As keratinocytes move away from the basement membrane, the circular arrangement of cell identities is maintained and a tubular organization of the hair follicles is achieved. Thus, the hair follicle comprises a central hair shaft surrounded by the inner and the outer root sheaths. The hair shaft is further divided into a medulla, a cortex, and a cuticle, and likewise, the root sheaths have several sublayers. In the present context, it is important to note that the functionality of the hair follicle depends on differential cornification of keratinocytes in the various layers. Keratinocytes of the hair shaft cornify and maintain intercellular connections so that a mechanically resilient hair fiber is formed. The outgrowth of the fiber from the surrounding epithelium depends on the scaffolding function of the inner root sheath. There, keratinocytes cornify and later dissociate from the hair shaft so that an essential tissue gap is formed around the growing hair fiber. The keratinocytes of the outer root sheath do not cornify, but maintain the integrity of the epithelial connection between the hair bulb and the epidermis. The latter is crucial because the hair follicle undergoes extensive remodelling during the growth and regression phases of the hair cycle [16]. Moreover, the attachment of sebaceous glands to the hair follicle adds to the complexity of this mini-organ. For a comprehensive description of cell differentiation in the hair follicle, the reader is referred to an excellent review by Langbein and Schweizer [15]. As the molecular aspects of keratinocyte differentiation remain to be determined for several compartments of the hair follicle, the differentiation and cornification program will be described here only for the cells of the hair shaft.

9.2 The Mechanism of Cornification

9.2.1 Cellular Remodelling in the Granular Layer Precedes Cornification and Includes the Removal of Mitochondria and Other Organelles

While keratinocytes of all living layers of the epidermis increase the number of cytoskeletal proteins during differentiation, the granular layer shows additional cellular changes. Among these, the breakdown of organelles such as mitochondria, the Golgi apparatus, the endoplasmic reticulum (ER), and lysosomes may be considered part of the actual process of cell death. Consequently, the aforementioned intracellular compartments disappear during the transition of cells to the cornified layer of the epidermis.

The granular layer of the epidermis is defined histologically, and actually consists of several layers of cells that display progressive accumulation of keratohyalin granules and lamellar bodies. Keratohyalin granules are protein aggregates, consisting mainly of profilaggrin. They are not surrounded by a membrane and, therefore, are to be considered part of the cytosolic fraction of keratinocytes. By contrast, lamellar bodies are vesicles with a single membrane border, containing lipid lamellae and proteins. They are formed in an incompletely understood process from precursors that are related to distinct compartments of the ER and/or lysosomes. In fact, many proteins in lamellar bodies are identical to proteins of lysosomes [17]. Eventually, the lamellar bodies fuse with the apical side of granular-layer cells and thereby secrete their content to the

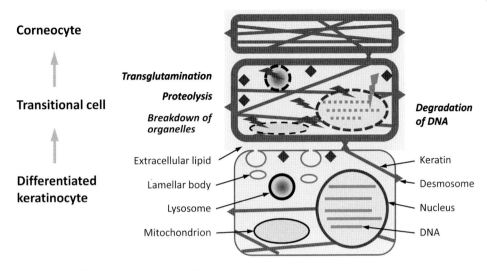

Figure 9.2 Cellular processes of cornification in the epidermis.

extracellular space (Figure 9.2). The secretion process accounts for a significant loss of cellular material, and may therefore contribute to the dying of keratinocytes.

Little is known about the steps involved in organelle removal during cornification. Lysosomes have been reported to become more abundant when mitochondria, the Golgi apparatus, and the ER disappear in the granular layer [18]. In one study, lysosomes of this epidermal layer were found to contain and perhaps degrade mitochondria [19]. These data support a role for macroautophagy: the enclosure and transport of cell organelles to lysosomes for degradation in the stratum granulosum. Accordingly, the expression of the recombinant autophagosome reporter green fluorescence protein light-chain protein 3 in mice has been used to visualize autophagosomes in the outer layers of the epidermis [20]. However, the deletion of essential autophagy proteins does not result in severe disturbances of cornification, but appears to be compatible with the formation of a functional skin barrier [20,21]. Taking this all together, autophagy may contribute to the removal of organelles during terminal differentiation.

Lysosomes may also contribute to cornification via their obligatory role as scaffolds for the mammalian target of rapamycin complex 1 (mTORC1). In line with this hypothesis, the lysosomal adaptor protein p18, which anchors mTORC1 to lysosomes, has been shown to be essential for cornification. Deletion of this protein in epidermal keratinocytes of the mouse results in the accumulation of autophagosomes, glycogen granules, and condensed nuclei in immature but dead cells on the skin surface. Because they have a defective skin barrier function, newborn mutant mice dehydrate fast and die perinatally [22].

Yet another hypothesis suggests that lysosomes do not primarily take up intracellular content during cornification but rather disintegrate to release their content (hydrolytic enzymes such as proteases of the cathesin family) into the cytosol. The lysosomal enzymes require an acidic pH for optimal catalytic activity, which appears to be incompatible with a role in the neutral milieu of the cytosol. However, the pH decreases at the skin surface, and it is speculated that the intracellular pH may already drop during cornification. Experimental evidence for or against this hypothesis is not conclusive at present. The activity of lysosomal cysteine proteases increases during terminal keratinocyte differentiation [23], and lysosomal proteases contribute to distinct subprocesses of cornification [3]. This notion is supported by the epidermal phenotypes of mutations in genes for lysosomal proteases. Ablation of cathepsin D results in decreased activity of the cornification enzyme transglutaminase 1 (TGase 1; see Section 9.2.3), defective cornification, and increased thickness of the cornified layer [24]. Similarly, cathepsin L activates TGase3 [25], and the deletion of the cathepsin L genes causes an aberrant increase in the thickness of the cornified layer [26]. Likewise, deleterious mutations of cathepsin C lead to hyperkeratosis in human patients [27]. Cystatin M/E inhibits cathepsin L and V and is essential for normal cornification, suggesting that the lysosomal protease activities must be tightly controled in order to prevent aberrant proteolysis [28]. Importantly, cystatin M/E localizes to the cytosol, perhaps indicating that it inhibits cathepsins that are prematurely released from lysosomes.

Besides releasing proteases, lysosomes are also a source of enzymes that degrade nucleic acids. The lysosomal endonuclease DNase 2 is present in the cornified layer, where it degrades DNA from external sources, and possibly also residual DNA from keratinocytes [29]. As the breakdown of lysosomes occurs in parallel to the breakdown of the nucleus, an interdependence of these processes is conceivable. Interestingly, autophagolysosomal activities and nuclear breakdown are linked in sebaceous glands, where keratinocytes undergo non-cornifying cell death [21]. Clearly, further experimental studies are necessary to test the physiological relevance of lysosomal enzymes in cornification.

Finally, not only organelles but also the cytosol is processed during cornification. Cornification is associated with a decrease in the water content to approximately 40% in the lower cornified layer [30]. This is accompanied by shrinkage of the cell to approximately 50% of its pre-cornification volume [31]. Interestingly, the cysteine protease caspase-14 requires a milieu that can be mimicked *in vitro* only by high concentrations of kosmotropic salts, which effectively withdraw water from proteins [32,33]. Caspase-14 contributes to the degradation of filaggrin to amino acids, which again affects the physicochemical properties of the keratinocyte cytosol, including its water-retention capability.

9.2.2 The Nucleus is Broken Down during Cornification

The viability of cells generally depends on the integrity of the genetic information in the nucleus. Therefore, the breakdown of the nucleus is a hallmark of most types of cell death. The nucleus is also dismantled during cornification of epidermal keratinocytes. Histologically, terminal differentiation of keratinocytes often involves indentions in the nuclear envelope, apparent compression of the nucleus in the flattened cells of the granular layer, and chromatin condensation [34]. Ultimately, the nucleus disappears from the cornifying cell. While the entire differentiation of an epidermal keratinocyte, from the cell division in the basal layer to desquamation, takes approximately 2 weeks, the process of nuclear breakdown may be executed within a time frame of a few hours [35].

The central step in the degradation of the nucleus is the hydrolysis of DNA. Being a chemically stable molecule, DNA requires enzymatic catalysis for its efficient breakdown. Like apoptotic DNA degradation by caspase-activated DNase (CAD), cornification-associated DNA degradation involves a specific enzyme, DNase1L2. DNase1L2 is a member of the DNase1 family of endonucleases, and therefore it resembles DNase1, the main DNase of the serum, and DNase1L3, a DNase implicated in certain forms of apoptosis [36]. In contrast to all other members of the DNase1 family, DNase1L2 is expressed exclusively in the epidermis, and its expression is transcriptionally upregulated during differentiation of keratinocytes [37]. Accordingly, DNase1L2 is present at high amounts in the granular layer of the interfollicular epidermis, but also in differentiated hair keratinocytes [37,38]. The suppression of DNase1L2 expression in an *in vitro* model of differentiating human keratinocytes has demonstrated its requirement for DNA removal under these conditions.

Deletion of the *DNase1l2* gene in the mouse results in the retention of nuclear DNA in keratinocytes, which undergo cornfication to form skin appendages such as hair, nails, and the scales of the tail [38]. Furthermore, the epithelia on the surface of the tongue and the esophagus of this mouse model are parakeratotic; that is, nuclear remnants are readily detectable on histological sections. In hair fibers, DNA is aberrantly present in the medulla, cortex, and cuticle, suggesting that all types of hair corneocyte require DNase1L2 for DNA removal. In the interfollicular epidermis of DNase1L2 knockout (KO) mice, DNA is degraded during cornification, indicating compensatory activity of other DNase(s) [38].

DNase1L2 KO mice aberrantly retain elevated levels of nuclear DNA in cornifying hair keratinocytes, which alters their ultrastructural organization and decreases the resistance of hair to mechanical stress [38]. The observation that DNase1L2 KO inhibits the normal proteolysis of histones during cornification suggests that DNase1L2 activity is not an endpoint of cornification but a step that is followed by downstream processes such as breakdown of DNA-binding proteins [39].

9.2.3 The Crosslinking of Proteins Stabilizes Corneocytes

A most distinctive feature of cornification is the massive crosslinking of proteins. In epidermal keratinocyte cornification, proteins are mainly crosslinked by transglutamination [1,40], whereas disulfide bond-mediated crosslinking is highly active in the cornification of hair and nail keratinocytes. Transglutamination is the formation of an N-epsilon-(gamma-glutamyl)lysine isopeptide link between a glutamine and a lysine residue, whereby ammonia is released (Figure 9.3, upper panel). Disulfide bonds form between cysteine residues via the oxidation of sulfhydryl (SH) groups (Figure 9.3, lower panel). Both transglutamination and disulfide bridge formation transform target proteins into supramolecular structures that are largely resistant to proteolysis and therefore allow corneocytes to become stable structures.

Transglutamination is catalyzed by TGases. In epidermal keratinocytes, TGases 1, 3, and 5 are expressed [40]. TGase1 is proteolytically activated by cathepsin D [24], linking transglutamination to a lysosomal protease. TGase1 is also regulated by the protein product of the tazarotene-induced gene 3 (TIG3). TIG3 is expressed in differentiating epidermal keratinocytes. The carboxy-terminal membrane-anchoring domain of TIG3 facilitates its attachment to the plasma membrane, where TIG3 interacts with and activates TGase1 [41]. This mechanisms contributes to the preferential localization of transglutamination activity in the cell periphery.

The anchorage of TGase1 to the cell membrane results in the formation of a protein envelope, referred to as the "cornified envelope" (some publications use this term as a synonym for "corneocyte"). The cornified envelope is 5–10 nm thick and largely consists of proteins that are expressed during late differentiation of keratinocytes. Many of these target proteins are encoded by genes within a gene cluster named the "epidermal differentiation complex" (EDC) [42,43]. The evolution of the EDC has been linked to that of epidermal cornification and the transition from an aquatic or amphibian to a fully terrestrial vertebrate lifestyle [44]. Quantitatively important components of the cornified

Transglutamination

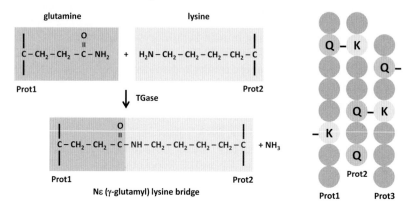

Disulfide bond formation

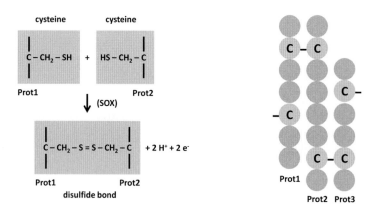

Figure 9.3 Mediation of protein crosslinking during cornification by transglutamination and disulfide bond formation. Transglutamination is the main mode of crosslinking in epidermal keratinocytes, whereas disulfide bond formation predominates in hair keratinocytes. Prot, protein; SOX, sulfhydryl oxidase; TGase, transglutaminase.

envelope are the EDC proteins involucrin, loricrin, and small proline-rich proteins (SPRRs). In addition, filaggrin and keratins are partly transglutaminated [45,46]. The critical role of TGase1 in keratinocyte cornification is demonstrated by the effects of its deletion: lack of TGase1 suppresses the formation of a cornified envelope, the condensation of keratin filaments, and the removal of the nucleus [47,48].

In contrast to the corneocytes of the interfollicular epidermis, those of the hair fibers use predominantly disulfide bonds to crosslink proteins [49]. However, transglutamination by TGase3 is important in the cuticle, and noncovalent protein interactions also contribute to the mechanical properties of hair [50]. The cortex of the hair fiber is essentially filled with keratins and KRTAPs. The keratins of the hair differ from those of the "soft" epidermis in having a higher cysteine-residue content [15,51]. Likewise, most KRTAPs have high cysteine contents, with maximum values beyond 30% of total amino acids [52]. Mechanistically, disulfide bonds are formed either enzymatically by sulfhydryl oxidases (SOX) or non-enzymatically by a change from a reducing to an oxidizing milieu [53].

9.2.4 Cornification is Accompanied by Maintenance and Modification of Cell–Cell Contacts

Keratinocyte cornification is a mode of cell death that is mediated by intracellular events. However, the intracellular processes are intimately linked to changes at the cell surface and in the milieu around the cell, so that the dead cell remnant is integrated into a supracellular functional structure (i.e., either the cornified layer of the epidermis or the hair fiber). In the epidermis, corneocytes are connected by corneodesmosomes and embedded in a lipid matrix.

Corneodesmosomes are modified desmosomes [54]. They are formed by the attachment of the protein corneodesmosin to desmosomes that are preformed in earlier differentiation stages. Corneodesmosin is secreted from granular-layer keratinocytes and enhances the connective force of cell–cell contacts. When the cell is dead, corneodesmosomes link the cell corpses and establish a mechanically resistant sheet (i.e., the cornified layer). The cornified layer itself consists of multiple layers of thin corneocytes. Its thickness depends on the body site, being largest on the palms and soles, and is controlled by the onset of desquamation during the passive outward movement of corneocytes. Desquamation involves the proteolytic cleavage of corneodesmosomes and the shedding of dead corneocytes into the environment [55].

The key proteases of desquamation are serine proteases of the kallikrein (KLK) family [56]. KLK5 and KLK7 are secreted from keratinocytes of the granular layer via lamellar bodies [57]. They require proteolytic maturation, upon which they cleave corneodesmosin, desmocollin, and desmoglein, thereby breaking corneodesmosomes. Lympho-epithelial Kazal-type related inhibitor I (LEKTI) is an essential inhibitor of KLKs. It is encoded by the *SPINK5* (serine peptidase inhibitor, Kazal type 5) gene, the site of mutations in Netherton syndrome [58]. Reduction of LEKTI allows for elevated activities of KLKs and enhanced removal of corneodesmosomal bonds between corneocytes, resulting in premature desquamation, barrier defects, and severe skin inflammation [58]. Interestingly, KLK7 is also implicated in the activation of caspase 14, which is considered an intracellular protein [59]. Recent studies have improved our understanding of the etiology and the diagnosis of ichthyoses [60–62].

9.3 Differences and Commonalities between Cornification and Other Modes of Cell Death

Cornification is a mode of programmed cell death (PCD) that is physiologically limited to keratinocytes. Keratinocytes of epithelia other than the epidermis and hair follicles, such as the corneal epithelium of the eye, cornify under special stress conditions [63]. It is important to note that the program of cornification in the epidermis is modified under conditions of inflammation and wound healing. Typically, an accelerated mode of cornification is associated with incomplete breakdown of the nucleus and retention of nuclear remnants in corneocytes. This condition is known as parakeratosis and is a histological hallmark of several skin diseases, including psoriasis. Although the barrier properties generated by aberrant cornification are inferior to those of the homeostatic epidermis, the modified forms of cornification must still be regarded as types of cell death. Thus, skin-associated defects of cornification are not failures of cells to die but deviations from the ordered sequence of cellular changes. Consequently, cell death in diseased skin loses some features of pure cornification and acquires features of other cell-death modalities. In line with this notion, normal terminal differentiation of keratinocytes involves the suppression of alternative modes of cell death, such as apoptosis and necrosis [64].

Cornification shares some principles with apoptosis, a ubiquitous and evolutionarily ancient type of controled cell death. In mammalian cells, apoptosis is characteristically associated with the activity of caspases. Most prominently, caspase 3 is an executioner protease of apoptosis, and detection of its mature and active form in tissues is a sign of apoptosis. Cornification does not involve the activity of caspase 3 [65]. However, another member of the caspase family, namely caspase 14, *is* activated during cornification [66,67]. Caspase 14 is expressed specifically in differentiating epidermal keratinocytes, undergoes proteolytic maturation in the upper granular layer, and contributes to the process of cornifcation [68]. Deletion of caspase 14 results in aberrant processin of filaggrin, increases the sensitivity of the epidermis to ultraviolet B (UVB) damage, and predisposes the skin to the development of parakeratosis. Moreover, caspase 14-deficient mice develop a slightly abnormal skin microflora. There is experimental evidence that caspase 14 contributes to the proteolytic breakdown of mature filaggrin to amino acids, including histidine, which serves as precursor for the UVB-absorbant urocanic acid [68]. An alternative model proposes a role of caspase 14 in the initiation of nuclear DNA degradation. In the latter scenario, caspase 14 cleaves the inhibitor of caspase-activated DNase (ICAD) and thereby releases CAD [69]. This would be perfectly analogous to the initiation of nuclear DNA degradation in apoptosis. However, caspase 14-deficient mice are able to degrade DNA during epidermal cornification, and keratinocytes of the nails do not express caspase 14 even under normal conditions, but do efficiently remove DNA [70].

Cornification also has some similarity to autophagic cell death [71]. The removal of organelles during cornification is reminiscent of autophagy, and perhaps involves molecular components of autophagy (see earlier). However, besides this role within the death program of cells, autophagy has mostly pro-survival functions, and its role in cornification remains to be further dissected.

The breakdown of intracellular membranes distinguishes cornification from apoptosis but makes it resemble necrosis. Indeed, the putative activity of lysosomal proteases during cornification suggests mechanistic similarities to cell death by lysosomal membrane permeabilization. While necrotic cells disintegrate and release their content, cornifying cells show only a controlled release of cellular content via lamellar bodies and incompletely defined mechanisms of volume loss. Therefore, necrosis is often associated with tissue inflammation, whereas cornification does not trigger inflammatory processes in neighboring living cells. However, the epidermis shows a peculiar tendency to activate programmed necrosis (necroptosis), which is suppressed by caspase 8 [72–74].

As described earlier, cornification depends on massive crosslinking of cellular proteins, and transglutaminases play central roles in the formation of the cornified layer of the epidermis. However, transglutaminases are not exclusive to cornification, because TGase2 is activated in apoptosis [40]. Nevertheless, protein crosslinking in cornification has the unique effect of preserving the integrity of the cell during its death. This allows the integration of the cell corpse into large assemblies of dead cells, such as the cornified layer of the epidermis and the hair. Finally, the location of cornifying cells at the surface of the body is associated with their unique fate of being shed to the environment. Together, these features make cornification an autonomous mode of cell death that ultimately serves to protect the cells of the internal parts of the body.

References

1. Candi E, Schmidt R, Melino G. The cornified envelope: a model of cell death in the skin. *Nat Rev Mol Cell Biol* 2005;**6**:328–40.
2. Proksch E, Brandner JM, Jensen JM. The skin: an indispensable barrier. *Exp Dermatol* 2008;**17**(12):1063–72.
3. Eckhart L, Lippens S, Tschachler E, Declercq W. Cell death by cornification. *Biochim Biophys Acta* 2013;**1833**:3471–80.
4. Rorke EA, Adhikary G, Young CA, Roop DR, Eckert RL. Suppressing AP1 factor signaling in the suprabasal epidermis produces a keratoderma phenotype. *J Invest Dermatol* 2015;**135**:170–80.
5. Wurm S, Zhang J, Guinea-Viniegra J, García F, Muñoz J, Bakiri L, et al. Terminal epidermal differentiation is regulated by the interaction of Fra-2/AP-1 with Ezh2 and ERK1/2. *Genes Dev* 2015;**29**(2):144–56.
6. Koster MI, Roop DR. Mechanisms regulating epithelial stratification. *Annu Rev Cell Dev Biol* 2007;**23**:93–113.
7. Wilhelmsen K, Litjens SH, Kuikman I, Tshimbalanga N, Janssen H, van den Bout I, et al. Nesprin-3, a novel outer nuclear membrane protein, associates with the cytoskeletal linker protein plectin. *J Cell Biol* 2008;**171**(5):799–810.
8. Ketema M, Kreft M, Secades P, Janssen H, Sonnenberg A. Nesprin-3 connects plectin and vimentin to the nuclear envelope of Sertoli cells but is not required for Sertoli cell function in spermatogenesis. *Mol Biol Cell* 2013;**24**(15):2454–66.
9. Fischer H, Langbein L, Reichelt J, Praetzel-Wunder S, Buchberger M, Ghannadan M, et al. Loss of keratin K2 expression causes aberrant aggregation of K10, hyperkeratosis, and inflammation. *J Invest Dermatol* 2014;**134**:2579–88.
10. Feng X, Zhang H, Margolick JB, Coulombe PA. Keratin intracellular concentration revisited: implications for keratin function in surface epithelia. *J Invest Dermatol* 2013;**133**(3):850–3.
11. Sandilands A, Sutherland C, Irvine AD, McLean WH. Filaggrin in the frontline: role in skin barrier function and disease. *J Cell Sci* 2009;**122**(Pt. 9):1285–94.
12. de Veer SJ, Furio L, Harris JM, Hovnanian A. (2014) Proteases: common culprits in human skin disorders. *Trends Mol Med* 2014;**20**(3):166–78.
13. Barresi C, Stremnitzer C, Mlitz V, Kezic S, Kammeyer A, Ghannadan M, et al. Increased sensitivity of histidinemic mice to UVB radiation suggests a crucial role of endogenous urocanic acid in photoprotection. *J Invest Dermatol* 2011;**131**:88–194.
14. Simpson CL, Patel DM, Green KJ. Deconstructing the skin: cytoarchitectural determinants of epidermal morphogenesis. *Nat Rev Mol Cell Biol* 2011;**12**:565–80.
15. Langbein L, Schweizer J. Keratins of the human hair follicle. *Int Rev Cytol* 2005;**243**:1–78.
16. Alonso L, Fuchs E. The hair cycle. *J Cell Sci* 2006;**119**:391–3.
17. Raymond AA, Gonzalez de Peredo A, Stella A, Ishida-Yamamoto A, Bouyssie D, et al. Lamellar bodies of human epidermis: proteomics characterization by high throughput mass spectrometry and possible involvement of CLIP-170 in their trafficking/secretion. *Mol Cell Proteomics* 2008;**7**:2151–75.
18. Lavker RM, Matoltsy AG. Formation of horny cells: the fate of cell organelles and differentiation products in ruminal epithelium. *J Cell Biol* 1970;**44**:501–12.
19. Morioka K, Takano-Ohmuro H, Sameshima M, Ueno T, Kominami E, Sakuraba H, Ihara S. Extinction of organelles in differentiating epidermis. *Acta Histochem Cytochem* 1999;**32**:465–76.

20. Rossiter H, König U, Barresi C, Buchberger M, Ghannadan M, Zhang CF, et al. Epidermal keratinocytes form a functional skin barrier in the absence of Atg7 dependent autophagy. *J Dermatol Sci* 2013;**71**:67–75.
21. Sukseree S, Rossiter H, Mildner M, Pammer J, Buchberger M, Gruber F, et al. Targeted deletion of Atg5 reveals differential roles of autophagy in keratin K5-expressing epithelia. *Biochem Biophys Res Commun* 2013;**430**(2):689–94.
22. Soma-Nagae T, Nada S, Kitagawa M, Takahashi Y, Mori S, Oneyama C, Okada M. The lysosomal signaling anchor p18/LAMTOR1 controls epidermal development by regulating lysosome-mediated catabolic processes. *J Cell Sci* 2013;**126**(Pt. 16):3575–84.
23. Tanabe H, Kumagai N, Tsukahara T, Ishiura S, Kominami E, Nishina H, Sugita H. Changes of lysosomal proteinase activities and their expression in rat cultured keratinocytes during differentiation. *Biochim Biophys Acta* 1991;**1094**:281–7.
24. Egberts F, Heinrich M, Jensen JM, Winoto-Morbach S, Pfeiffer S, Wickel M, et al. Cathepsin D is involved in the regulation of transglutaminase 1 and epidermal differentiation. *J Cell Sci* 2004;**117**:2295–307.
25. Cheng T, Hitomi K, van Vlijmen-Willems IM, de Jongh GH, Yamamoto K, Nishi K, et al. Cystatin M/E is a high affinity inhibitor of cathepsin V and cathepsin L by a reactive site that is distinct from the legumain-binding site. A novel clue for the role of cystatin M/E in epidermal cornification. *J Biol Chem* 2006;**281**:15 893–9.
26. Tobin DJ, Foitzik K, Reinheckel T, Mecklenburg L, Botchkarev VA, Peters C, Paus R. The lysosomal protease cathepsin L is an important regulator of keratinocyte and melanocyte differentiation during hair follicle morphogenesis and cycling. *Am J Pathol* 2002;**160**:1807–21.
27. Toomes C, James J, Wood AJ, Wu CL, McCormick D, Lench N, et al. Loss-of-function mutations in the cathepsin C gene result in periodontal disease and palmoplantar keratosis. *Nat Genet* 1999;**23**:421–4.
28. Zeeuwen PL. Epidermal differentiation: the role of proteases and their inhibitors. *Eur J Cell Biol* 2004;**83**:761–73.
29. Fischer H, Scherz J, Szabo S, Mildner M, Benarafa C, Torriglia A, et al. DNase 2 is the main DNA-degrading enzyme of the stratum corneum. *PloS One* 2011;**6**:e17581.
30. Caspers PJ, Lucassen GW, Carter EA, Bruining HA, Puppels GJ. In vivo confocal Raman microspectroscopy of the skin: noninvasive determination of molecular concentration profiles. *J Invest Dermatol* 2001;**116**:434–42.
31. Al-Amoudi A, Norlen LP, Dubochet J. Cryo-electron microscopy of vitreous sections of native biological cells and tissues. *J Struct Biol* 2004;**148**:131–5.
32. Mikolajczyk J, Scott FL, Krajewski S, Sutherlin DP, Salvesen GS. Activation and substrate specificity of caspase-14. *Biochemistry* 2004;**43**(32):10 560–9.
33. Fischer H, Stichenwirth M, Dockal M, Ghannadan M, Buchberger M, Bach J, et al. Stratum corneum-derived caspase-14 is catalytically active. *FEBS Lett* 2004;**577**(3):446–50.
34. Wier KA, Fukuyama K, Epstein WL. Nuclear changes during keratinization of normal human epidermis. *J Ultrastruct Res* 1971;**37**:138–45.
35. Matoltsy A. Structure and function of the mammalian epidermis. In: Bereiter-Hahn J, Matoltsy AG, Richards KS (eds.). *Biology of the Integument 2*. Berlin: Springer. 1984.
36. Errami Y, Naura AS, Kim H, Ju J, Suzuki Y, El-Bahrawy AH, et al. Apoptotic DNA fragmentation may be a cooperative activity between caspase-activated deoxyribonuclease and the poly(ADP-ribose) polymerase-regulated DNAS1L3, an endoplasmic reticulum-localized endonuclease that translocates to the nucleus during apoptosis. *J Biol Chem* 2013;**288**:3460–8.

37 Fischer H, Eckhart L, Mildner M, Jaeger K, Buchberger M, Ghannadan M, Tschachler E. DNase1L2 degrades nuclear DNA during corneocyte formation. *J Invest Dermatol* 2007;**127**:24–30.
38 Fischer H, Szabo S, Scherz J, Jaeger K, Rossiter H, Buchberger M, et al. Essential role of the keratinocyte-specific endonuclease DNase1L2 in the removal of nuclear DNA from hair and nails. *J Invest Dermatol* 2011;**131**:1208–15.
39 Eckhart L, Fischer H, Tschachler E. Mechanisms and emerging functions of DNA degradation in the epidermis. *Front Biosci (Landmark Ed)* 2012;**17**:2461–75.
40 Eckert RL, Kaartinen MT, Nurminskaya M, Belkin AM, Colak G, Johnson GV, Mehta K. Transglutaminase regulation of cell function. *Physiol Rev* 2014;**94**:383–417.
41 Eckert RL, Sturniolo MT, Jans R, Kraft CA, Jiang H, Rorke EA. TIG3: a regulator of type I transglutaminase activity in epidermis. *Amino Acids* 2009;**36**:739–46.
42 Henry J, Toulza E, Hsu CY, Pellerin L, Balica S, Mazereeuw-Hautier J, et al. Update on the epidermal differentiation complex. *Front Biosci* 2012;**17**:1517–32.
43 Kypriotou M, Huber M, Hohl D. The human epidermal differentiation complex: cornified envelope precursors, S100 proteins and the "fused genes" family. *Exp Dermatol* 2012;**21**:643–9.
44 Strasser B, Mlitz V, Hermann M, Rice RH, Eigenheer RA, Alibardi L, et al. Evolutionary origin and diversification of epidermal barrier proteins in amniotes. *Mol Biol Evol* 2014;**31**:3194–205.
45 Kalinin AE, Kajava AV, Steinert PM. Epithelial barrier function: assembly and structural features of the cornified cell envelope. *Bioessays* 2002;**24**:789–800.
46 Segre J. Complex redundancy to build a simple epidermal permeability barrier. *Curr Opin Cell Biol* 2003;**15**:776–82.
47 Matsuki M, Yamashita F, Ishida-Yamamoto A, Yamada K, Kinoshita C, Fushiki S, et al. Defective stratum corneum and early neonatal death in mice lacking the gene for transglutaminase 1 (keratinocyte transglutaminase). *Proc Natl Acad Sci USA* 1998;**95**:1044–9.
48 Kuramoto N, Takizawa T, Takizawa T, Matsuki M, Morioka H, Robinson JM, Yamanishi K. Development of ichthyosiform skin compensates for defective permeability barrier function in mice lacking transglutaminase 1. *J Clin Invest* 2002;**109**:243–50.
49 Popescu C, Höcker H. Cytomechanics of hair basics of the mechanical stability. *Int Rev Cell Mol Biol* 2009;**277**:137–56.
50 John S, Thiebach L, Frie C, Mokkapati S, Bechtel M, Nischt R, et al. Epidermal transglutaminase (TGase 3) is required for proper hair development, but not the formation of the epidermal barrier. *PLoS One* 2012;**7**:e34252.
51 Eckhart L, Dalla Valle L, Jaeger K, Ballaun C, Szabo S, Nardi A, et al. Identification of reptilian genes encoding hair keratin-like proteins suggests a new scenario for the evolutionary origin of hair. *Proc Natl Acad Sci USA* 2008;**105**:18 419–23.
52 Rogers MA, Langbein L, Praetzel-Wunder S, Winter H, Schweizer J. Human hair keratin-associated proteins (KAPs). *Int Rev Cytol* 2006;**251**:209–63.
53 Saaranen MJ, Ruddock LW. Disulfide bond formation in the cytoplasm. *Antioxid Redox Signal* 2013;**19**:46–53.
54 Jonca N, Leclerc EA, Caubet C, Simon M, Guerrin M, Serre G. Corneodesmosomes and corneodesmosin: from the stratum corneum cohesion to the pathophysiology of genodermatoses. *Eur J Dermatol* 2011;**21**(Suppl. 2):35–42.
55 Singh B, Haftek M, Harding CR. Retention of corneodesmosomes and increased expression of protease inhibitors in dandruff. *Br J Dermatol* 2014;**171**:760–70.

56 Caubet C, Jonca N, Brattsand M, Guerrin M, Bernard D, Schmidt R, et al. Degradation of corneodesmosome proteins by two serine proteases of the kallikrein family, SCTE/KLK5/hK5 and SCCE/KLK7/hK7. *J Invest Dermatol* 2004;**122**:1235–44.

57 Ishida-Yamamoto A, Simon M, Kishibe M, Miyauchi Y, Takahashi H, Yoshida S, et al. Epidermal lamellar granules transport different cargoes as distinct aggregates. *J Invest Dermatol* 2004;**122**:1137–44.

58 Hovnanian A. Netherton syndrome: skin inflammation and allergy by loss of protease inhibition. *Cell Tissue Res* 2013;**351**:289–300.

59 Yamamoto M, Miyai M, Matsumoto Y, Tsuboi R, Hibino T. Kallikrein-related peptidase-7 regulates caspase-14 maturation during keratinocyte terminal differentiation by generating an intermediate form. *J Biol Chem* 2012;**287**:32 825–34.

60 Oji V, Tadini G, Akiyama M, Blanchet Bardon C, Bodemer C, Bourrat E, et al. Revised nomenclature and classification of inherited ichthyoses: results of the First Ichthyosis Consensus Conference in Sorèze 2009. *J Am Acad Dermatol* 2010;**63**(4):607–41.

61 Schmuth M, Martinz V, Janecke AR, Fauth C, Schossig A, Zschocke J, Gruber R. Inherited ichthyoses/generalized Mendelian disorders of cornification. *Eur J Hum Genet* 2013;**21**(2):123–33.

62 Samuelov L, Sprecher E. Peeling off the genetics of atopic dermatitis-like congenital disorders. *J Allergy Clin Immunol* 2014;**134**(4):808–15.

63 Pelegrino FS, Pflugfelder SC, De Paiva CS. Low humidity environmental challenge causes barrier disruption and cornification of the mouse corneal epithelium via a c-jun N-terminal kinase 2 (JNK2) pathway. *Exp Eye Res* 2012;**94**:150–6.

64 Lippens S, Denecker G, Ovaere P, Vandenabeele P, Declercq W. Death penalty for keratinocytes: apoptosis versus cornification. *Cell Death Differ* 2005;**12**(Suppl. 2):1497–508.

65 Fischer H, Rossiter H, Ghannadan M, Jaeger K, Barresi C, Declercq W, et al. Caspase-14 but not caspase-3 is processed during the development of fetal mouse epidermis. *Differentiation* 2005;**73**(8):406–13.

66 Lippens S, Kockx M, Knaapen M, Mortier L, Polakowska R, Verheyen A, et al. Epidermal differentiation does not involve the pro-apoptotic executioner caspases, but is associated with caspase-14 induction and processing. *Cell Death Differ* 2000;**7**:1218–24.

67 Eckhart L, Declercq W, Ban J, Rendl M, Lengauer B, Mayer C, et al. Terminal differentiation of human keratinocytes and stratum corneum formation is associated with caspase-14 activation. *J Invest Dermatol* 2000;**115**:1148–51.

68 Denecker G, Hoste E, Gilbert B, Hochepied T, Ovaere P, Lippens S, et al. Caspase-14 protects against epidermal UVB photodamage and water loss. *Nat Cell Biol* 2007;**9**:666–74.

69 Yamamoto-Tanaka M, Makino T, Motoyama A, Miyai M, Tsuboi R, Hibino T. Multiple pathways are involved in DNA degradation during keratinocyte terminal differentiation. *Cell Death Dis* 2014;**5**:e1181.

70 Jäger K, Fischer H, Tschachler E, Eckhart L. Terminal differentiation of nail matrix keratinocytes involves up-regulation of DNase1L2 but is independent of caspase-14 expression. *Differentiation* 2007;**75**:939–46.

71 Liu Y, Levine B. Autosis and autophagic cell death: the dark side of autophagy. *Cell Death Differ* 2015;**22**:367–76.

72 Lee P, Lee DJ, Chan C, Chen SW, Ch'en I, Jamora C. Dynamic expression of epidermal caspase 8 simulates a wound healing response. *Nature* 2009;**458**:519–23.

73 Vandenabeele P, Galluzzi L, Vanden Berghe T, Kroemer G. Molecular mechanisms of necroptosis: an ordered cellular explosion. *Nat Rev Mol Cell Biol* 2010;**11**:700–14.

74 Dannappel M, Vlantis K, Kumari S, Polykratis A, Kim C, Wachsmuth L, et al. RIPK1 maintains epithelial homeostasis by inhibiting apoptosis and necroptosis. *Nature* 2014;**513**(7516):90–4.

10

Excitotoxicity

Julie Alagha, Sulaiman Alshaar, and Zane Deliu

Department of Oral Medicine and Diagnostic Sciences, University of Illinois at Chicago, Chicago, IL, USA

Abbreviations

AD	Alzheimer's disease
ALS	amyotrophic lateral sclerosis
AMPA	α-amino-3-hydroxy-5-methyl-isoxazole-4-propioate
APP	amyloid precursor protein
Ca^{++}	calcium
CNS	central nervous system
cyt c	cytochrome c
ER	endoplasmic reticulum
glx	combined glutamate and glutamine
HD	Huntington's disease
HORSE	Huntington's Outreach Project for Education at Stanford
Mg^{++}	magnesium
Na^{+}	sodium
NBQX	a glutamate receptor antagonist
nNOS	neuronal nitric oxide synthase
NMDA	N-methyl-D-aspartate
PD	Parkinson's disease
PKC	protein kinase C
VDCC	voltage-dependent calcium channel

10.1 Introduction

In recent years, there has been a marked increase in the prevalence of neurodegenerative diseases. Human neurodegenerative diseases are defined as involving a progressive loss of nerve function caused by certain neurological deficits. This chapter provides information about (i) what excitotoxicity means and the toxic effect it has on neurons, (ii) how excessive activation of glutamate receptors leads to the death of neurons, and (iii) the major role of excitotoxicity in neurodegenerative diseases. It places emphasis on the events that elicit neuronal excitation leading to many neuropathological diseases [1,2].

Apoptosis and Beyond: The Many Ways Cells Die, First Edition. Edited by James Radosevich.
© 2018 John Wiley & Son Inc. Published 2018 by John Wiley & Son Inc.

10.2 Defining Excitotoxicity

"Excitotoxicity" refers to toxic effects that result from excessive activation of excitatory amino acid receptors. Because the amino acid, glutamate, is the major neurotransmitter in the central nervous system (CNS), excitotoxicity is neuronal death caused by excessive stimulation of the amino acid receptors [3].

10.3 The Pathophysiology of Excitotoxicity

10.3.1 The Role of Glutamate

Glutamate is an amino acid that acts as a principal excitatory neurotransmitter in the brain. Amino acids consist of a central carbon atom (α carbon) bonded to a carboxyl group (COOH) and an amino group (NH_3). Normally, the role of glutamate is to carry messages across synapses through the brain and spinal cord. Neuronal toxicity occurs when there is too much glutamate due to overstimulation of neurons [4].

Glutamate plays a major role in synaptic plasticity, memory, and cognitive functions. Too much glutamate in the brain can lead to adverse effects, including severe damage. The signaling effect of glutamate depends on the ability of cells to respond. Cells with glutamate receptors are sensitive to glutamate and are capable of binding to it, activating its signaling function [5].

10.3.2 Glutamate Receptors

The main types of glutamate receptors are ionotropic and metabotropic. Ionotropic receptors are coupled directly to membrane ion channels. They are subdivided into N-methyl-D-aspartate (NMDA), α-amino-3-hydroxy-5-methyl-isoxazole-4-propionate (AMPA), and Kainate. Under normal conditions, glutamate released from nerve terminals acts on postsynaptic receptors to cause depolarization, which causes in turn the release of Mg^{++} from NMDA receptor channels, allowing Ca^{++} to pass through the pore. NMDA receptors activate channels that cause the influx of extracellular calcium and sodium. AMPA and NMDA receptors jointly contribute to the process of synaptic plasticity, which is involved in memory, excitotoxicity, learning, and neuroprotection. When all receptors work together, it causes the release of glutamate. Metabotropic receptors are proteins linked to intracellular G proteins that regulate intracellular second messengers such as cyclic nucleotides, inositol-1,4,5-triphosphate, and calcium [6].

Any changes in the amino-acid sequences of glutamate receptors could cause changes in the permeability of calcium or in other properties that would make excitotoxicity worse. Recent evidence suggests that mutations in postsynaptic proteins of AMPA receptors can cause a channelopathy effect in epilepsy. Knowledge of glutamate receptors and their distribution may lead to the development of receptor antagonists for the treatment of neurological diseases [6].

10.3.3 Calcium Overload

Calcium ions are necessary for neuronal cell function and survival. However, when they reach critical levels, cell injury or death may occur. The first steps in the excitotoxic process are the opening of the ionotropic glutamate receptors and the influx of Na^+ and Ca^{++}. An increase in the concentration of cytoplasmic Ca^{++} is caused by the activation of the glutamate receptors. This occurs through AMPA and NMDA receptors channels. The activation of metabotropic glutamate receptors also causes release of Ca^{++} from the endoplasmic reticulum (ER). The main observed process leading to neuronal death or injury is excessive accumulation of intracellular calcium. Overstimulation of the NMDA receptors causes this neuronal calcium overload [7].

The initial glutamate-receptor opening of the sodium and calcium channels causes influx of calcium and membrane depolarization. This depolarization activates the voltage-dependent calcium channels (VDCCs) that increase the intracellular calcium levels [7].

10.3.4 Excessive Glutamate Accumulation

Excessive glutamate accumulation in the synaptic space is the key process that starts the entire excitotoxic cascade. This can occur through a change in the normal cycling of intracranial glutamate that increases the release of glutamate into the extracellular space or decreases the absorption of glutamate from the synaptic space. This accumulation can also occur through a leakage of glutamate from injured neurons [5].

Normal extracellular glutamate concentration is about 0.6 µmol/L. When there is neuronal injury, glutamate concentrations reach 2–5 µmol/L. Traumatic injury to neurons can cause glutamate concentrations of about 10 µmol/L to the extracellular space. Any mechanical injury to a neuron risks damaging to neighboring neurons. Recent therapy suggests treating anyone who has had injuries to the head or spinal column with glutamate receptor blockers, in order to minimize the spread of neuron death [5].

10.3.5 Intracellular Toxic Effects

The buildup of high intracellular calcium levels causes a surge of membrane, cytoplasmic, and nuclear events, leading to neurotoxicity. The elevated calcium levels overactivate enzymes, including phospholipases, protein kinase C (PKC), proteases, nitric oxide synthase (NOS), phosphatases, and endonucleases. The activation of phospholipase A promotes platelet-activating factor and arachidonic acid and its metabolites. Platelet-activating factor increases glutamate release, leading to excitotoxicity. Arachidonic acid inhibits reabsorption of glutamate, leading to further activation of glutamate receptors, which in turn causes more arachidonic acid formation. An increase in arachidonic acid levels forms oxygen free radicals, which activate phospholipase A and, as mentioned earlier, leads to further formation of arachidonic acid. These enzymes lead to neuronal self-digestion [8].

Nitric oxide synthase, another important activated enzyme, forms nitric oxide. When NMDA receptors are excessively stimulated, they produce abnormally increased levels of nitric oxide and superoxide ions. These substances react to form peroxynitrite, an extremely toxic substance, leading to neuronal death. Nitric oxide can also damage DNA, which creates more free radicals and causes membrane depolarization. In Huntington's disease (HD), this nitric oxide-initiated neurotoxic cascade is an important component of the mechanism of cell death [5].

10.4 Excitotoxicity and Oxidative Stress

Oxidative stress may contribute to the excitotoxic process and make neurons more susceptible to excitotoxicity. Many studies have shown that neurons are killed more often by glutamate when they are under conditions of increased oxidative stress. Oxidative stress disrupts the functions of the key proteins responsible for the maintenance of cellular calcium homeostasis. Along with disrupting membrane transporter functions in neurons, oxidative stress can disrupt glutamate transporter function in astrocytes, which causes excitotoxicity due to the increased concentration of extracellular glutamate. Most neurodegenerative disorders are associated with oxidative stress, and overactivation of glutamate receptors causes the deaths of many neurons in these disorders [9].

10.5 Excitotoxicity and Brain Injuries

Acute brain injuries are characterized by an excessive depolarization of the neurons and reversal of astrocytic glutamate uptake, both of which lead to an excessive amount of extracellular glutamate, which is linked to excitotoxicity as follows: the release of cytochrome c (cyt c) from the mitochondria is considered one of the primary pathways by which to activate caspase-independent cell apoptosis. In acute brain injuries, this process is closely linked to excitotoxicity; the excessive release of glutamate from presynaptic nerve terminals activates NMDA in postsynaptic neurons, and NMDA is one of the glutamate receptor subtypes. This activation opens the cation-operated Ca^{++} channels, leading to a Ca^{++} influx into the cells, which in turn activates two enzymes: calpain I, which causes the release of cyt c from the mitochondria, and neuronal nitric oxide synthase (nNOS) [10].

Activated calpain I blocks the cation exchange between Na+ and Ca^{++} cations across the plasma membrane, leading to further accumulation of Ca^{++}, which in turn feeds into the same mechanism, providing an additional Ca^{++} source other than the activation of (NMDA). Activated (nNOS), on the other hand, produces nitric oxide, which creates free radicals such as (O2−, ONOO−), which cause more damage to the cellular components (membranes, proteins, DNA) [10].

All these mechanisms have cell necrosis as a final result, and that is the basis of all acute brain injuries. This process has been linked to many such injuries, such as worse Glasgow scale comas, damage due to long elevation of intracranial pressure, trauma, and epilepsy [10].

10.6 Excitotoxicity and Neurodegenerative Diseases

Excitotoxicity is a process by which neurological cells go through apoptosis. It can be expected that different disorders will result depending on the location of the affected neurons, including stroke, epilepsy, damage due to elevated intracranial pressure, and cerebral ischemia (Figure 10.1). Excitotoxicity also causes neurodegenerative diseases such as Parkinson's disease (PD), Alzheimer's disease (AD), and HD. This section will provide information on the different processes that take place in neurodegenerative disorders.

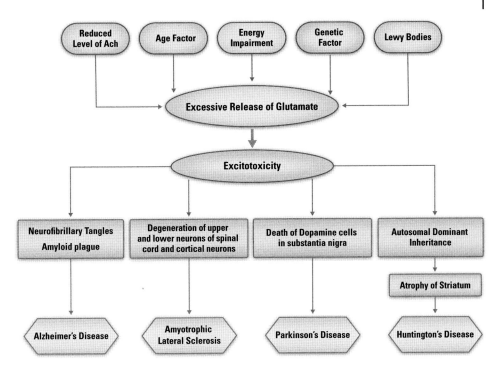

Figure 10.1 Excitotoxic mechanisms in neurodegenerative diseases [11]. *Source:* Mehta et al. 2013 [11]. Reproduced with permission from Elsevier.

10.6.1 Huntington's Disease

HD is an autosomal-dominant inherited disease caused by a mutation in a gene on chromosome 4. The coding region of this gene consists of repeated sections of (CAG) (10–35 times, varying from person to person). The (CAG) codes for glutamine or glutamic acid, and the end result is a protein called huntingtin. In HD, the repeated number increases significantly, to 40 or more, increasing the number of glutamine repeats, which is why HD is one of nine polyglutamine expansion disorders according to the Huntington's Outreach Project for Education at Stanford (HORSE) [12]. The neurodegenerative effect of HD is more severe on the striatal medium spiny neurons, which are more sensitive to energy impairment than other neurons. Studies in rodents show that these neurons die through exposure to the mitochondrial toxin 3-nitropropionic acid, via an excitotoxic mechanism [13].

Symptoms of HD involve uncontrolled movements or twitching and changes in mood and behavior, such as depression. The motor dysfunction keeps on progressing and often results in death. HD affects about 30 000 Americans.

10.6.2 Alzheimer's Disease

AD is one of the most common neurodegenerative diseases, currently affecting over 5 million Americans. It is a progressive neurodegenerative disease that destroys neurons, leading to dementia, memory loss, behavioral changes, and a decrease in language and thinking skills. AD is characterized by an accumulation of beta amyloid protein in neurons, leading to their death. This beta-amyloid accumulation can be seen on autopsy as plaques in the brain tissue.

Some cases have an early age of onset; these are linked to mutations in three different genes: amyloid precursor protein (APP), presenilin-1, and presenilin-2. The altered processing of APP is considered a critical step in AD, leading to the accumulation of beta amyloid in the neurons. This accumulation leaves the cell susceptible to excitotoxic mechanisms by inducing membrane lipid peroxidation and impairing the function of ion-motive ATPases, glucose and glutamate transporters, and ion [14–20].

The therapeutic agents tested currently for AD include (i) glutamate receptor antagonists blocking the excitotoxic process through glutamate and (ii) agents that block calcium, preventing the activation of the critical enzymes necessary for apoptosis: calpain I and nNOS.

10.6.3 Parkinson's Disease

PD is another neurodegenerative disease characterized by a progressive destruction of the dopamine-generating cells in the substantia nigra in the midbrain, affecting the motor system of the brain. The end result is often death: according to parkinsons.org, about 1 million individuals suffer from the disease in the United States, with 50 000–60 000 new cases each year.

The exact mechanism of cells destruction is not fully understood, but studies show increased levels of glutamate and upregulation of glutamate receptors in the dying cells, making them more susceptible to excitotoxicity [21].

The main drug used to manage PD is Levodopa, a known precursor for dopamine; it is used alone or with carbidopa (Sinemet). This kind of treatment is known to be symptomatic, providing relief of symptoms by increasing levels of dopamine, but without addressing the main etiology of the disease. For that reason, other classes of medication are being tested based on the excitotoxic theory behind PD, including include NBQX (a glutamate receptor antagonist) and other types of antioxidants and neurotrophic factors [22]. These are all still in clinical trials. Other ideas for treating PD include the use of stem cells to replace damaged cells in the substantia nigra.

10.6.4 Other Diseases Caused by Excitotoxicity

Amyotrophic lateral sclerosis (ALS) has also been linked to excitotoxicity through a very similar mechanism of oxidative stress and impaired glutamate transport, which results in degradation of upper and lower motor neurons, leading to complete paralysis and death. There is a need to develop more glutamate receptor-antagonist drugs in order to treat neurodegenerative diseases. While several drugs are currently under study, a better approach would be to examine the processes that lead to neurons being vulnerable to excitotoxicity, such as oxidative stress and metabolic compromise, which play key roles in most neurodegenerative disorders [20].

10.7 Neuroradiologic Observations of Excitotoxicity

Routine neuroimaging studies reflect the stages of glutamate excitotoxicity damage. The imaging of the spreading involvement of areas adjacent to the trauma and infarction of one area and of the delayed spread of these conditions provides examples of images tested.

The development of spectroscopy at 0.5 T provides an opportunity to observe the combined glutamate and glutamine (glx) peak *in vivo*. Delineation of this peak is enabled by the coalescence of the multiplets of glx at this field strength. By examining the glx peak, unique clinical observations can be made. Elevated glx peaks in the hippocampi have been seen in patients with glutamate excitotoxicity. Neurodegenerative diseases, such as PD, have been evaluated using this technique, and elevated glx peaks have been seen in the basal ganglia [5].

10.8 Conclusion

Glutamate is the major excitatory neurotransmitter. When found in excessively high levels, it causes excitotoxicity. The influx of calcium through inotropic glutamate-receptor channels causes excitotoxic death. Knowledge concerning excitotoxicity and its mechanism in the human body is limited. Because it is involved in many diseases, further investigations are needed to find new and better therapies for the treatment of excitotoxicity.

References

1. Anderson CM, Swanson RA. Astrocyte glutamate transport: review of properties, regulation, and physiological functions. *Glia* 2000;**32**:1–14.
2. Kim K, Lee SG, Kegelman TP, Su ZZ, Das SK, Dash R, et al. Role of excitatory amino acid transporter-2 (EAAT2) and glutamate in neurodegeneration: opportunities for developing novel therapeutics. 2011;226(10):2484–93.
3. Takei H. Pathology of motor neuron disorders. Emedicine by Medscape. April 28, 2014. Available from https://emedicinemedscapecom/article/2111360-overview (last accessed March 22, 2018).
4. Dong XX, Wang Y, Qin ZH. Molecular mechanisms of excitotoxicity and their relevance to pathogenesis of neurodegenerative diseases. *Acta Pharmacol Sin* 2009;**30**:379–87.
5. Mark LP, Prost RW, Ulmer JL, Smith MM, Daniels DL, Strottmann JM, et al. Pictorial review of glutamate excitotoxicity: fundamental concepts for neuroimaging. *AJNR Am J Neuroradiol* 2001;**22**(10):1813–24.
6. Lau A, Tymianski M. Glutamate receptors, neurotoxicity and neurodegeneration. *Pflugers Arch* 2010;**460**(2):525–42.
7. Matute C, Alberdi E, Ibarretxe G, Sánchez-Gómez MV. Excitotoxicity in glial cells. *Eur J Pharmacol* 2002;**447**:239–46.
8. Gill SS, Pulido OM. Glutamate receptors in peripheral tissues: current knowledge, future research, and implications for toxicology. *Toxicol Pathol* 2001;**29**(2):208–23.
9. Mattson MP, Magnus T. Ageing and neuronal vulnerability. *Nat Rev Neurosci* 2006;**7**(4):278–94.
10. Fujikawa DG. The role of excitotoxic programmed necrosis in acute brain injury. *Comput Struct Biotechnol J* 2015;**13**:212–21.
11. Mehta A, Prabhakar M, Kumar P, Deshmukh R, Sharma P. Excitotoxicity: bridge to various triggers in neurodegenerative disorders. *Eur J Pharmacol* 2013;**698**(1–3):6–18.

12 Liou S. Neurobiology, neurotrophic factors. Available from http://web.stanford.edu/group/hopes/cgi-bin/hopes_test/neurotrophic-factors-and-huntingtons-disease/ (last accessed March 22, 2018).

13 Brouillet E, Conde F, Beal MF, Hantraye P. Replicating Huntington's disease phenotype in experimental animals. *Prog Neurobiol* 1999;**59**:427–68.

14 Mattson MP, Cheng B, Davis D, Bryant K, Lieberburg I, Rydel RE. Beta-amyloid peptides destabilize calcium homeostasis and render human cortical neurons vulnerable to excitotoxicity. *J Neurosci* 1992;**12**:376–89.

15 Mark RJ, Hensley K, Butterfield DA, Mattson MP. Amyloid beta-peptide impairs ion-motive ATPase activities: evidence for a role in loss of neuronal Ca2+ homeostasis and cell death. *J Neurosci* 1995;**15**:6239–49.

16 Mark RJ, Lovell MA, Markesbery WR, Uchida K, Mattson MP. A role for 4-hydroxynonenal in disruption of ion homeostasis and neuronal death induced by amyloid beta-peptide. *J Neurochem* 1997;**68**:255–64.

17 Mark RJ, Pang Z, Geddes JW, Uchida K, Mattson MP. Amyloid beta-peptide impairs glucose uptake in hippocampal and cortical neurons: involvement of membrane lipid peroxidation. *J Neurosci* 1997;**17**:1046–54.

18 Keller JN, Mattson MP. 17beta-estradiol attenuates oxidative impairment of synaptic Na+/K+-ATPase activity, glucose transport and glutamate transport induced by amyloid beta-peptide and iron. *J Neurosci Res* 1997;**50**:522–30.

19 Kruman I, Bruce-Keller AJ, Bredesen DE, Waeg G, Mattson MP. Evidence that 4-hydroxynonenal mediates oxidative stress-induced neuronal apoptosis. *J Neurosci* 1997;**17**:5089–100.

20 Mattson MP. Excitotoxic and excitoprotective mechanisms: abundant targets for the prevention and treatment of neurodegenerative disorders. *Neuromolecular Med* 2003;**3**(2):65–94.

21 Miranda AF, Boegman RJ, Beninger RJ, Jhamandas K. Protection against quinolinic acid-mediated excitotoxicity in nigrostriatal dopaminergic neurons by endogenous kynurenic acid. *Neuroscience* 1997;**78**:967–75.

22 Turski L, Bressler K, Rettig KJ, Loschmann PA, Wachtel H. Protection of substantia nigra from MPP+ neurotoxicity by N-methyl- D-aspartate antagonists. *Nature* 1991;**349**: 414–18.

11

Molecular Mechanisms Regulating Wallerian Degeneration

Mohammad Tauseef[1] and Madeeha Aqil[2]

[1]Department of Pharmaceutical Sciences College of Pharmacy, Chicago State University, Chicago, IL, USA
[2]Department of Oral Medicine and Diagnostic Sciences, University of Illinois at Chicago, Chicago, IL, USA

Abbreviations

ALS	amyotrophic lateral sclerosis
ApoE	apolipoprotein E
BDNF	brain-derived neurotrophic factor
CNS	central nervous system
CNTF	ciliary neurotrophic factor
DRG	dorsal root ganglion
Gal-3	galectin-3
GM-CSF	granulocyte–macrophage colony-stimulating factor
HD	Huntington's disease
IL	interleukin
LIF	leukemia inducible factor
MAG	myelin-associated glycoprotein
MCP-1	chemoattractant protein 1
MIP-1α	macrophage inflammatory protein 1α
MMP9	matrix metalloproteinase 9
NAD^+	oxidized nicotinamide adenine dinucleotide
NGF	nerve growth factor
NLS	nuclear location sequence
NT-3	neurotrophin 3
PD	Parkinson's disease
PNS	peripheral nervous system
TNF-α	tumor necrosis factor-α
TLR	Toll-like receptor
VCP	valosin-containing protein
Wld^s	slow Wallerian degeneration

11.1 Introduction

Axonal injury is one of the most common clinical outcomes following the nerve crush or stretch, often leading to axonal degeneration [1–6]. Neurodegenerative diseases, such as

Apoptosis and Beyond: The Many Ways Cells Die, First Edition. Edited by James Radosevich.
© 2018 John Wiley & Son Inc. Published 2018 by John Wiley & Son Inc.

Parkinson's disease (PD), Huntington's disease (HD), and amyotrophic lateral sclerosis (ALS), are the best-studied causes of axonal injury and axonal loss [1–4,7]. As axons constitute the bulk of the neurons in the nervous system, they maintain the functions of the neuronal circuits [5,6], and axonal cell death thus leads to an alteration of neuronal circuits and pathology. It is therefore a priority in biomedical research to investigate the molecular signaling pathways involved in axonal degeneration and to develop measures to prevent such aberrant processes from occurring [1,3,5,7].

Wallerian degeneration involves the cutting or interruption of a peripheral axon or nerve, causing distal disintegration of the axon [4,5,7]. It was originally defined by Augustus Waller, who demonstrated its effects by severing glossopharyngeal and hypoglossal nerves in frogs [6,8]. He made a microscopic observation of the axon at the distal site of the original injury, finding it was separated out from its cell body in the brain stem [6]. For more than a century, it was believed that the phenomenon of Wallerian degeneration resulted from the cutting of the nutrient supply of the distal axons as a result of the severing of the peripheral axons [6,9,10]. The true details of the process emerged with the discovery of the slow Wallerian degeneration (Wld^s) mouse model [11].

In a Wld^s mouse model, the distal severed axons survive even weeks after the original peripheral axonal injury, rather than the 1.5 days observed *in vivo* [1,11]. This was an unexpected and serendipitous discovery, showing that under certain conditions, the large disconnected fragments of the axons can survive for long periods on their own, without the cell body. It raised many questions and complicated the picture, suggesting that Wallerian degeneration is regulated by a complex array of signaling pathways [9]. In order to develop a better understanding of the pathology and to identify better diagnostic markers of Wallerian degeneration, we need to learn as much as possible about the molecular mechanisms of the disease [9]. This chapter will summarize the current understanding of the signaling mechanisms regulating this process. It will also briefly discuss the role of peripheral neuropathies in the activation of axonal degeneration mechanisms and what ultimately leads to peripheral nerve dysfunction.

11.2 Slow Wallerian Degeneration and the Mechanism of Neuroprotection

The Wld^s mouse strain expresses a tandem triplication that results in the fusion of two genes, nmnat1 and ube4b [12], and generates a 70-amino-acids WLD^s protein [1]. Neuronal expression of Wld^s protects both motor and sensory neurons, as well as the axons of multiple types of central nervous system (CNS) neuron, from granular disintegration [12,13]. Wld^s proteins also offer neuronal protection in other species, such as Drosophila and zebrafish, suggesting that the mechanism is evolutionarily conserved [1,14]. However, delineating the molecular and signaling pathways by which the overexpression of Wld^s leads to suppression of neuronal degeneration is an active area in neuroscience research.

Studies show that the lentivirus mediates overexpression of Nmnat1, a component of the oxidized nicotinamide adenine dinucleotide (NAD$^+$) scavenging pathway, suppressing axonal degeneration of the cultured mammalian dorsal root ganglion neurons [15]. This indicates that Nmnat1, which is a nuclear NAD$^+$ biosynthetic enzyme, is an important regulator in Wld^s-mediated protection of axonal injury mechanism [15]. Moreover, it has been observed that levels of NAD$^+$ in distal severed axons remain high even after 2–4 hours of neuronal injury, before falling again. This drop off in NAD$^+$ levels following axonal injury also leads to a reduction in ATP and prevents axonal fragmentation; it can be maintained for longer periods in animals overexpressing the Wld^s protein in their brain [15–17]. It is implied that neuronal injury could be averted, at least in a preclinical setting, by the addition of NAD$^+$ or its precursor, nicotinamide, even after 3–5 hours of neuronal injury [16,18]. However, when Nmnat1 and NAD$^+$ were expressed in mice versus Wld^s mice, they demonstrated no protection against neuronal injury [16,19]. It is thus clear that the mechanism of Wld^s-mediated neuronal protection is more complex than is currently understood, and may depend on more than Nmnat1, at least under *in vivo* conditions [16–19]. To further explore how Wld^s offers neuroprotection in mice, investigators focused on the key portion of the N-terminal Ube4b molecule: the 16 most N-terminal amino acids (N16) [16,17,19]. These had previously been shown to be associated with valosin-containing protein (VCP) [1,16,20]. Upon deletion of N16, Wld^s failed to offer neuroprotective benefits in a mouse model [20]. Taken together, studies have shown that the N-terminals of both Ube4b and Nmnat1 are required in the protection induced by Wld^s. But how N16 protects neurons still requires thorough investigation. In early studies, it appeared that N16 induced the portion of the cellular pool of Wld^s out of the nuclear site, to the area where it was required [12]. Nmnat1 has a nuclear location sequence (NLS), which is included in the Wld^s fusion protein [12]. This is why Wld^s primarily presents in the nuclear region. Moreover, deletion of the NLS makes the Nmnat a cytoplasmic protein with suppression of axonal injury [21]. Thus, it is the cytoplasmic localization of Nmnat1 that is essential in providing neuroprotection during axonal injury under both *in vivo* and *in vitro* conditions [21–23]. Under basal conditions, Wld^s is present in low amounts in the cytoplasm and in synaptosomal preparations [21,22,24,25]. Finally, a convincing study shows that lentiviral-based delivery of cytoplasmically targeted Nmnat1 protein directly into axons following axonal injury protects them from axonal degeneration [23]. This observation brings strong evidence in favor of a role for Wld^s in preventing axonal injury by acting directly on the axon.

11.3 Induction of Neuronal Protection by Signaling Mechanisms Activated by Wld^s

Research studies suggest that the length of the axon determines how fast the neurons will degenerate [1,6,26–28]. For example, longer detached axons undergo a faster degeneration process [27,28]. Thus, it has been suggested that the stroma might provide some growth factors that continuously transport down the axon [27,28]. Once injured, the survival of the neuron will depend on how soon it will deplete its reserve trophic factors and reach the degenerative pathway. The question that arises is: What are these trophic factors? Their identity will be helpful in the development of any potential therapeutics of neuronal injuries and disorders. A breakthrough came when Gilley and Coleman [29] proposed Nmnat2, a cytoplasmic protein found in the axoplasm, which is actively transported down the axons [28]. Stabilization of Nmnat2 through mutation of its palmitoylation site makes it more protective [30]. On the other hand, knocking down

Nmnat2 from cultured dorsal root ganglion (DRG) neurons causes degeneration of axons [28]. This phenomenon is reversed by Wld^s [28]. *In vivo* studies suggest that Nmnat2 null mice are embryonically lethal and grow axons in the peripheral nervous system (PNS) [31,32]. These axons are shorter than their corresponding wild-type cohorts, both in *in vivo* and *in vitro*. However, overexpression of Wld^s both rescues embryonic lethality and maintains the size of the axons in Nmnat2 null mice [31]. These data indicate that Nmnat2 protects axons from injury and that Wld^s rescues the phenotype in Nmnat2-deleted cultured neuronal cells [32].

How Wld^s induces neuroprotection is still not very clearly understood. The studies discussed here and in Section 11.2 show the convincing roles for Nmnat1, N16, and Nmnat2 in the prevention of axonal injuries, and how Wld^s overexpression rescues the protective functions upon deletion of these proteins in neuronal cell-culture models [1]. This argues that Wld^s mediates axonal protection in rather a complex model in which multiple signaling pathways are operating [1].

11.4 Wallerian Degeneration and the Immune Response to Traumatic Nerve Injury

We will now focus our attention on the pathological processes in which Wallerian degeneration exerts its affects on nerve growth and regeneration [33,34]. For example, traumatic injury to nerves in the PNS results in the loss of neuronal functions [33,34]. To regain those functions, repair is required, which is a multistep process that includes the regeneration of damaged axons and the reinnervation of the target tissue [10]. Here, again, successful functional recovery depends on numerous cellular and molecular processes that reroute their functional signaling away from the injury site toward the denervated target tissue. These are the responses of the PNS following nerve injury, and this pathway is called Wallerian degeneration, as originally discovered by Waller [8,33]. So, an in-depth understanding of the Wallerian degeneration signaling pathways should provide us with the tools to recover the functions of the nerves following axonal injury [10]. Traumatic nerve injury is one of the most common debilitating factors encountered in patients and is among the most unmet medical needs. Thus, a better understanding of the cellular and molecular signaling events of Wallerian degeneration will be helpful in identifying the rational therapeutic treatments of traumatic nerve injury.

Wallerian degeneration is an innate immune response of the PNS following traumatic nerve injury [33,35,36]. As soon as traumatic nerve injury occurs, there is: (i) recruitment of macrophages; (ii) phagocytosis of degenerated myelin; and (iii) generation of various cytokines and chemokines to provide a milieu in which constructive events help regain the loss of functional responses [33,36,37]. Identification of the cells involved in the regulation of these signaling events, along with the timing and magnitude of their activation, is of utmost importance in gaining a better understanding of the role of Wallerian degeneration in the healing of severed nerves. The activation mechanisms of cellular and molecular signaling events are different during the crushing versus the cutting of the nerve. So, the nature of the injury dictates the type of signaling involved in the regaining of physiological functions. Finally, from a research point of view, these events and processes may have different outcomes depending upon which type of animal model is being used [1]. Since Wallerian degeneration encompasses the events that follow traumatic injury of CNS axons, the signaling pathways activated depend on the type of neuronal cell severed [33,38]. For instance, the nature of Wallerian degeneration will differ according to whether Schwann cells or macrophages in the PNS or microglia or oligodendrocytes in the CNS are severed [37,38].

In contrast to neuronal injury, Wallerian degeneration is sometimes used to define events that developed during neuropathies, such as inherited demyelinated disorders [38,39]. These pathways are independent of signaling events activated via injuries induced by Wallerian degeneration. In order to avoid confusion and to make the subject easy to comprehend, we will focus our review on traumatic injury induced by Wallerian degeneration signaling pathways.

11.5 Role of Wallerian Degeneration in the Mechanism of Functional Recovery Following the Traumatic Injury of Peripheral Nerves

Now the major question arises how Wallerian degeneration following traumatic injury leads to the activation of recovery pathways. The sooner the activation of Wallerian degeneration occurs, the sooner it will lead to the regaining of tissue functions. So, if Wallerian degeneration is slowed, as in the Wld^s mouse, it will slow the regeneration of damaged nerves and hamper functional recovery of the target tissue [1,33,40]. Bundles of nerve fibers in the PNS are composed of axons, which are wrapped by Schwann cells. Schwann cells form a myelin sheath around the axons, and microvasculature provides nourishment. During the traumatic injury of PNS nerves, there is sudden tissue damage at the site of physical impact [34,40]. Following the lesion, the nerve located distal to the site of injury (at a location where no direct injury occurred) starts to degenerate by undergoing characteristic cellular and molecular alterations [33,34,40]. This leads to the activation of multistep processes wherein axons break down, demyelination occur, and bone marrow-derived macrophages are recruited [33,40]. These events involve the shedding of myelin and ultimate removal of the degenerated neurons, in order to prime the neurons to start constructive work such as regeneration of nerve fibers [33,34,40].

In the PNS, nerve bundles are composed of axons, around which Schwann cells wrap, forming myelin sheaths. Fibroblasts remain scattered between nerve fibers, and vasculature supplies the required nutrients to the PNS tissue. When a forceful physical impact affects the PNS nerves, as in the case of traumatic injury, there is a sudden loss of tissue function. This is due to the damage that occurs at the immediate area of physical impact. However, nerves lying farther away from the original impact area undergo the cellular changes that characterize Wallerian degeneration. This is the area where injury did not happen, and it is away from the original site of injury. The cellular changes observed during the injury are axon breakdown, rejection of the myelin portion of their membranes by the Schwann cells, and recruitment and activation of bone marrow-derived macrophages. Finally, resident Schwann cells remove degenerated axons and myelin (Figure 11.1). Thus, in order to achieve complete recovery of tissue functions, it is necessary to successfully complete the regeneration of the injured axons at the distal nerve segments that undergo Wallerian degeneration. The characteristic for good functional recovery, at least in human beings, is rapid regeneration of the severed axons [38,39,41]. Research studies show that complete repair is less successful in human

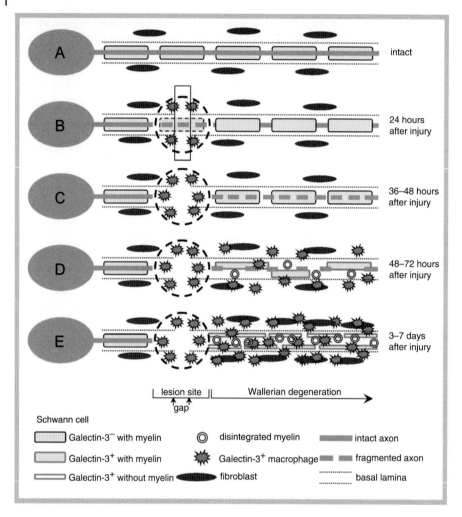

Figure 11.1 Cellular changes in intact and damaged nerves during Wallerian degeneration.
During Wallerian degeneration in peripheral nerve injury, various cellular processes are activated to regenerate the damaged axons. (A) Uninjured myelinating Schwann cells surround an axon, with fibroblasts lying between nerve fibers. (B) Sudden nerve injury damages the tissue at the lesion site (marked by a circle). Galectin-3 (Gal-3)/MAC-2$^+$ macrophages accumulate within 24 hours after the injury. (C) Wallerian degeneration starts 36 hours after the injury. (D) Wallerian degeneration leads to recruitment of Gal-3/MAC-2$^+$ macrophages. Myelin disintegration and Gal-3/MAC-2 expression by Schwann cells begin 48–72 hours after injury. (E) Gal-3/MAC-2$^+$ macrophages and Schwann cells start to remove degenerated myelin. *Source:* Rotshenker 2011 [33]. Reproduced under the terms of the Creative Commons Attribution Licence, CC-BY 2.0, via BioMed Central.

beings than it is in rodent models such as mice and rats [33,34]. These observations suggest that there might be differences in the signaling mechanisms regulating Wallerian degeneration in both humans and rodents. Moreover, there could be differences in the mechanisms involved in the delayed onset of axon destruction, the longer nerve segments that need to be cleared of degenerated myelin, and the longer distances over which regenerating axons need to grow in order to reach their target tissues in humans. Thus, whatever the discrepancy, the most important take-home message is that speeding the Wallerian degeneration process may improve functional recovery.

11.6 Demyelination of Axons and Functional Recovery during Wallerian Degeneration

Why is demyelination of axons the critical step toward the recovery of tissue functions and healing of injured axons? Research studies show that PNS myelin contains molecules, such as myelin-associated glycoprotein (MAG), that inhibits regeneration of severed axons [42–45]. The clearance of myelin is a first step toward axon regeneration, and ultimately functional recovery [33]. For example, functional recovery is delayed considerably in Wld^s mice compared to wild-type [46–48]. To further prove the role of Wallerian degeneration in the functional recovery, researchers knocked out MAG in Wld^s mice; they found that regeneration of severed axons was improved following the knocking out of MAG, even though myelin removal was still slow [45]. Further along this line, PNS myelin and MAG inhibit regeneration *in vitro* [43,44,49]. The prompt removal of degenerated myelin by Wallerian degeneration includes protection of intact axons and myelin after partial injury to PNS nerves, where some but not all axons are axotomized by the impact [33]. For example, injured myelin may activate the complement system to produce membrane attack complexes, which, in turn, inflict damage to the remaining nearby intact axons and myelin [50,51]. Thus, the rapid clearance of degenerated myelin may stop activation of the inflammatory cascade and thereby stop the production of membrane attack complexes and the damage they cause.

11.6.1 Role of Schwann Cells and Macrophages in the Clearance of Myelin during Axonal Injury

During normal Wallerian degeneration, resident Schwann cells and recruited macrophages clear degenerated myelin in wild-type mice [35,52,53]. Thus, they have significant roles in functional recovery. Both *in vivo* and *in vitro* studies suggest that macrophage and Schwann cells have strong potential in the clearance of damaged myelin. Furthermore, following the depletion of macrophages, Schwann cells are capable of effectively completing this task, and vice versa. This suggests that each cell type can remove myelin *in vivo* without the other [52,54–56]. Myelin destruction and removal are delayed considerably during Wallerian degeneration in Wld^s mice, because of axon destruction and macrophage recruitment [52,57–59].

11.6.2 The Role of the Cytokine Network in Wallerian Degeneration-Mediated Functional Recovery

One of the prominent outcomes of traumatic PNS injury is the activation of the immune system [33,34,39,60]. Not just immune cells are involved in the activation of proinflammatory signaling; non-immune cells also produce cytokines at the site of injury, as well as at the distal areas away from the injury [33,34,61]. Thus, a complex network of cytokines actively participates in the normal Wallerian degeneration process [33]. The production of various pro- and anti-inflammatory cytokines and their detailed kinetics, along with their mRNA and protein profiles, have been investigated in depth during

Wallerian degeneration in wild-type as well Wld^s mice [33,62,63]. More importantly, the role of cytokines in degenerated myelin clearance and in macrophage recruitment, as well as in the recovery of functions, has been studied in order to exploit their roles to find better treatments for nerve injury [59,62–64].

The first non-neuronal cell types that respond to injury are the Schwann cells, because they make a close contact with nerve fibers in the PNS [33,65]. They basically express the mRNAs of the proinflammatory cytokines, such as tumor necrosis factor-α (TNFα) and interleukin 1α (IL-1α). Thus, immediately after sensing the injury, Schwann cells respond promptly and produce TNFα and IL-1α [33]. There are thus increased levels of TNFα and IL-1α within 5–6 hours following the nerve injury. Besides proinflammatory cytokine secretions, Schwann cells also produce IL-1β, which is detected between 5 and 10 hours after injury [33]. Next, Schwann cell-derived TNFα and IL-1α induce nearby resident fibroblasts to express and further produce the mRNAs and proteins of the cytokines IL-6 and granulocyte–macrophage colony-stimulating factor (GM-CSF), the secretion of which is detected within 2–5 hours after the injury [33]. Thus, a cascade of cytokine secretion, followed by activation and recruitment of different cell types, including macrophages, takes place. This process will be ongoing for the next 7 days following the traumatic injury of the PNS (Figure 11.2) [33].

What will be the outcomes of the activation of a variety of pro- and anti-inflammatory cytokines? Why is there a need to activate the inflammatory cytokines in the PNS? The production of inflammatory cytokines and chemokines leads to the recruitment of bloodborne macrophages [5,33]. This recruitment begins 2–3 days after the injury and peaks at 7 days. Thus, the levels of secretion of the cytokines TNFα and IL-1β are reduced as macrophage numbers increase [33,34]. The reason for this is that once macrophages are fully recruited to the site of the injury – as well as away from the site of injury – they start to produce TNFα and IL-1β. Recruited macrophages secrete IL-6 and IL-10 proteins, as well. The production of the anti-inflammatory cytokine IL-10 is induced in resident fibroblasts within 5 hours after injury, but levels are low and ineffective, since nerve-resident fibroblasts are poor producers of IL-10 and Schwann cells do not produce it [33,34]. However, recruited macrophages produce IL-10 in abundance quite effectively [33,34]. Therefore, levels of IL-10 initially increase, peaking at day 7 – concomitant with the timing and magnitude of macrophage recruitment. IL-10 then gradually downregulates the production of inflammatory cytokines to the level of normal Wallerian degeneration, which concludes within 2–3 weeks after injury, once the degenerated myelin has already been cleared [66]. Thus, both pro- and anti-inflammatory cytokine networks are essential to activating the recruitment of macrophages, which clear debris and degenerated myelin. The anti-inflammatory cytokine IL-10 will start healing the injury and is helpful in the recovery of various functions (Figure 11.2).

We will now discuss the cytokine profiles in the beginning and later phases of Wallerian degeneration. This will shed further light on how Wallerian degeneration, by using the cytokine network, heals injury and prevents the further loss of axons. In brief, the cytokine profile will provide us with a detailed understanding of what is going on at the signaling level before and after macrophage recruitment. Here, during the first phase, there is a production of the inflammatory cytokines, such as TNFα, IL-1α, IL-1β, GM-CSF, and IL-6 [1,33,34,66]. The second phase is marked by the production of IL-10, IL-6, and a GM-CSF inhibitor molecule, and by the reduced production of TNFα and IL-1β [33,34,39,56]. Therefore, the first phase is inflammatory in nature and the second is more anti-inflammatory.

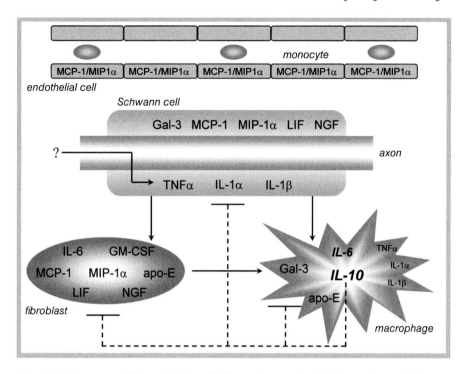

Figure 11.2 Summary of the activation of inflammatory and anti-inflammatory cytokines in Wallerian degeneration. In uninjured and intact nerves, Schwann cells cover the axon to form a myelin sheath. They secrete tumor necrosis factor alpha (TNFα) and interleukin 1α (IL-1α) mRNAs and the TNFα protein. However, once the injury has occurred, normal Wallerian degeneration processes start to kick in. Here, during traumatic injury, there is many-fold upregulation of inflammatory cytokines by Schwann cells as compared to intact axons. Downstream toward the Schwann cells, generation of TNFα and IL-1α induces resident fibroblasts to enhance the production of IL-6 and granulocyte–macrophage colony-stimulating factor (GM-CSF) proteins within 2–5 hours after the injury. Chemokines, chemoattractant protein 1 (MCP-1/CCL2) and macrophage inflammatory protein 1α (MIP-1α/CCL3) are also expressed by TNFα, IL-1β, and IL-6 as of day 1 following the injury to the Schwann cells. Fibroblasts and endothelial cells also participate in the secretion of chemokines and cytokines. The increased production of inflammatory cytokines and chemokines leads to the transmigration and accumulation of monocytes to nerve tissue. This takes place 2–3 days following the injury. Before beginning the infiltration of monocytes, fibroblasts secrete apolipoprotein E (ApoE,) while Schwann cells release galectin-3 (Gal-3)/MAC-2. Later, both ApoE and Gal-3/MAC-2 are involved in the conversion of monocytes into anti-inflammatory M2 macrophages, which further generates ApoE and Gal-3/MAC-2. These M2 macrophages generate anti-inflammatory cytokines, IL-10, and IL-6, and thus inhibit the generation of inflammatory cytokines. *Source:* Rotshenker 2011 [33]. Reproduced under the terms of the Creative Commons Attribution Licence, CC-BY 2.0, via BioMed Central.

Macrophages are immune sentinel cells and are very plastic in nature [67–69]. Recent studies classify macrophages as classical or M1 type and alternative M2 type [68]. When isolated from bone marrow, most macrophages are in M0 phase [68,69]. However, depending on the milieu, once activated into M1 type, they become proinflammatory in nature and thus secrete inflammatory cytokines [68,69]. On the other hand, M2 macrophages are anti-inflammatory, releasing anti-inflammatory cytokines [68,69]. Therefore, it may be possible that recruited macrophages are of the M2 phenotype, as they produce high levels of IL-10 and IL-6, less TNFα and IL-1β, and little or no

GM-CSF [68,69]. They are involved in tissue repair. The question is, what is polarizing these macrophages into become M2 types? Studies show that apolipoprotein E (ApoE) [70] and galectin-3 (Gal-3)/MAC-2 [71] can direct the polarization of recruited macrophages toward the M2 phenotype. ApoE is produced and secreted by resident fibroblasts during normal Wallerian degeneration as of day 2 and later, again via macrophages [72,73], as is Gal-3/MAC-2. Interestingly, both ApoE and Gal-3/MAC-2 are produced in Wld^s mice at injury sites but not during Wallerian degeneration [33,73].

One of the reasons Wld^s mice are resistant to Wallerian degeneration may be a deficient cytokine network activation pathway in those mice. The production of cytokines is very low during slow versus normal Wallerian degeneration, even though the expression of cytokine mRNAs is upregulated [59,62,63,74]. In contrast, cytokine mRNAs are expressed and proteins are produced in injured Wld^s PNS nerves at lesion sites concomitant with macrophage accumulation and activation to phagocytosize myelin. Thus, it is the development of an effective cytokine network that highlights the inflammatory nature of normal Wallerian degeneration. This hastens the clearance of degenerated myelin and quickens functional recovery. The question is why cytokine proteins are not produced during Wld^s even though the expression of their mRNAs was observed [62]. It might be that cytokine mRNAs and proteins are differentially regulated during Wallerian degeneration, and that mRNA expression does not necessarily indicate that the respective protein is produced [33,62]. Thus, to better understand the Wallerian degeneration signaling with respect to cytokine network activation pathways, it is imperative that both the cytokine protein and its mRNA expression during the onset of injury and at later phases be determined.

11.6.3 Role of Chemokines in the Recruitment of Macrophages and Wallerian Degeneration

Like cytokines, chemokines such as chemoattractant protein 1(MCP-1, also known as CCL2, C-C motif ligand 2) and macrophage inflammatory protein 1α (MIP-1α, also known as CCL3) are actively involved in Wallerian degeneration [33,34]. For example, chemokines induce the transendothelial migration of monocytes and help to clear the severed tissue. During normal Wallerian degeneration, Schwann cells produce MCP-1 at the site of injury within hours after injury, and post-day 1 at a distal area [37,75–79]. MCP-1 production is also partly promoted by TNFα and IL-1β upon activation of Toll-like receptors (TLRs) on the surface of Schwann cells. Unlike in normal Wallerian degeneration in wild-type mice, MCP-1 is only generated at the site of injury, not at a distal area, in Wld^s mice. This may be one reason why Wld^s develops in these mice, leading to slow recovery of functions. These events correlate with the occurrence and timing of macrophage recruitment that MCP-1 promotes.

In non-neuronal cells, research findings demonstrate that fibroblasts generate MCP-1, which is induced by IL-6 [80]. Moreover, MIP-1α is involved in the recruitment of macrophages at the site of injury [78]. However, in neuronal tissues, such as Schwann cells, the data show that fibroblasts, endothelial cells, and macrophages may produce MIP-1α/CCL3 upon activation by TNFα, IL-1α, and IL-1β [80,81]. Moreover, macrophage recruitment is further achieved by TNFα-dependent induction of matrix metalloproteinase 9 (MMP9), which is produced by Schwann cells [82–85] and by complement [86–88].

11.6.4 Role of Neurotrophic Factors in the Regulation of Wallerian Degeneration

The immediate outcome of the release of neurotrophic factors is peripheral nerve injury [10,33]. Neurotrophic factors are produced by Schwann cells and fibroblasts during normal Wallerian degeneration, which suggests their potential role in the healing of nerve injury and functional recovery [10,33]. What are neurotrophic factors, and how do they regulate neurological functions during heath and disease? The brief answer is that they are peptides in nature (e.g., nerve growth factor; leukemia inducible factor, LIF). They are involved in: (i) axon growth; (ii) neuronal survival; (iii) synapse formation during normal development; and (iv) recovery following traumatic PNS nerve injury [33,89]. Mechanistically, they provide their neuroprotective benefits by binding to and activating their cognate receptors at the nerve endings and/or after being transported retrograde to neuronal cell bodies [33,89].

We will briefly discuss the neurotrophic roles of the various families of neurotrophic factors. Neurotrophin is one such family, which includes NGF, brain-derived neurotrophic factor (BDNF), neurotrophin 3 (NT-3), and NT-4/5 [90–94]. Studies show that levels of NGF, BDNF, and NT-4 are increased during normal Wallerian degeneration [95–99]. Furthermore, NGF has been found to support neuronal survival and axon growth. At the level of expression, NGF mRNA is transcribed in two phases, both at the site of injury and at distal areas. The first expression of NGF mRNA is detected within a few hours following neuronal injury, and continues to be detectable for 2–3 days. Further, it has been demonstrated that IL-1α, IL-1β, and TNFα increase NGF mRNA upregulation in fibroblasts but not in Schwann cells. However, as with mRNA, increase in the protein concentration of NGF are found only during the second phase of injury [95]. Interestingly, the expression of the NGF expression profile differs depending on the type of injury that has occurred. For example, NGF mRNA expression is marked and prolonged following cut injuries, but only transient levels are observed after crush injuries. This finding suggests that axons that regenerate after crush downregulate NGF expression [97]. In contrast to normal physiological Wallerian degeneration, the upregulation of NGF mRNA expression is hampered in Wld^s mice, which show slow degeneration [48]. These mice produce lower amounts of IL-1β and TNFα cytokines [62].

Cytokines such as LIF, ciliary neurotrophic factor (CNTF), and IL-6 are members of the IL-6 family [91,100,101]. Strong evidence shows increased generation of IL-6 and LIF during normal Wallerian degeneration but not in Wld^s mice. IL-6 is produced by resident fibroblasts that are involved in the recruitment of macrophages at the site of injury [62,74]. LIF is secreted by resident Schwann cells and fibroblasts [102,103]. Apart from being modulators of innate immune functions, IL-6 [104–106] and LIF [107,108] also possess neurotrophic properties, and thus promote neuronal survival and growth [109].

11.7 Role of Wallerian Degeneration in the Development of Neuropathic Pain

Neuropathic pain is the inflammatory painful sensation that occurs spontaneously in response to innocuous stimuli. Studies suggest a role for the innate immune response of injury-induced Wallerian degeneration in the development of neuropathic pain. In this section, we will briefly discuss how Wallerian degeneration regulates the sensation of neuropathic pain in response to pathological stimuli. The exact pathological process involved in the generation of such pain is unclear, due to the involvement of numerous pathological pathways with distinct and as yet undefined molecular mechanisms in its etiology [110–112]. However, studies have identified the role of the innate immune response in Wallerian degeneration and the pathology underlying the development of neuropathic pain. The most important finding using Wld^s mice as a model demonstrates that injury-induced neuropathic pain is delayed and even reduced in such mice [113]. Moreover, a similar observation was recorded in IL-6 deficient mice [114]. It is not only injury that precipitates neuropathic pain, however; it can also be evoked by mere inflammation in absence of injury [115–120]. Thus, inflammatory cytokines such as TNFα, IL-1β, and NGF, which are produced in normal Wallerian degeneration, have a significant pathological role in the generation and sensation of pain. Mechanistically, TNFα and IL-1β sensitize intact axons to produce pain sensation. These cytokines even aggravate the situation in response to mechanical and thermal injury [121]. If the injury remains untreated, IL-1β and TNFα further induce the expression of NGF, which, in turn, sensitizes sensory nerve endings. Thus, a vicious cycle is created, which ultimately leads to the severe form of neuropathic pain sensation [33,121,122]. This phenomenon is generally observed during PNS nerve injury, particularly in a situation of partial axonal injury. Taken together, the mechanism of delayed and reduced neuropathic pain in Wld^s mice may result from the decreased generation of the inflammatory cytokines IL-6, IL-1β, TNFα, and NGF [33,122]. This further suggests a potential therapeutic target for the treatment of neuropathic pain: inflammatory cytokine production [33,122].

11.8 Conclusion

A variety of inflammatory processes and injuries activate Wallerian degeneration to restore nerve function in PNS. Innate immunity has a pivotal role in the injury-mediated activation of Wallerian degeneration. Thus, the sooner the innate immune responses activate, the sooner the repair process begins. In order to turn on the repair pathway, innate immunity helps remove the degenerated myelin so that myelin-mediated inhibitory responses are turned off. This leads to the regeneration of damaged axons. The regeneration processes are carried out efficiently by neurotrophic factors. If the innate immune responses fail to engage, the recovery process will fail. Wallerian degeneration is the earliest step toward axonal recovery during PNS injury induced by inflammation or trauma. The Wallerian degeneration pathways are regulated by innate immunity, and therefore, targeting of the innate immune system will open up new avenues for the treatment of trauma-induced nerve damage and the recovery of neural functions.

References

1 Freeman MR. Signaling mechanisms regulating Wallerian degeneration. *Curr Opin Neurobiol* 2014;**27**:224–31.
2 Liang C, Tao Y, Shen C, Tan Z, Xiong WC, Mei L. Erbin is required for myelination in regenerated axons after injury. *J Neurosci* 2012;**32**:15 169–80.
3 Geden MJ, Deshmukh M. Axon degeneration: context defines distinct pathways. *Curr Opin Neurobiol* 2016;**39**:108–15.

4 Raff MC, Whitmore AV, Finn JT. Axonal self-destruction and neurodegeneration. *Science* 2002;**296**:868–71.
5 Chang B, Quan Q, Lu S, Wang Y, Peng J. Molecular mechanisms in the initiation phase of Wallerian degeneration. *Eur J Neurosci* 2016;**44**(4):2040–8.
6 Koeppen AH. Wallerian degeneration: history and clinical significance. *J Neurol Sci* 2004;**220**:115–17.
7 Gerdts J, Summers DW, Milbrandt J, DiAntonio A. Axon self-destruction: new links among SARM1, MAPKs, and NAD+ metabolism. *Neuron* 2016;**89**:449–60.
8 Waller A. Experiments on the section of glossopharyngeal and hypoglossal nerves of th frog, observations of the alterations produced thereby in the structure of their primitive fibers. *Phil Transact Royal Soc London* 1850;**140**:423–9.
9 Medana IM, Esiri MM. Axonal damage: a key predictor of outcome in human CNS diseases. *Brain* 2003;**126**:515–30.
10 Cashman CR, Hoke A. Mechanisms of distal axonal degeneration in peripheral neuropathies. *Neurosci Lett* 2015;**596**:33–50.
11 Luo L, O'Leary DD. Axon retraction and degeneration in development and disease. *Annu Rev Neurosci* 2005;**28**:127–56.
12 Johnson VE, Stewart W, Smith DH. Axonal pathology in traumatic brain injury. *Exp Neurol* 2013;**246**:35–43.
13 Saxena S, Caroni P. Mechanisms of axon degeneration: from development to disease. *Prog Neurobiol* 2007;**83**:174–91.
14 De Vos KJ, Grierson AJ, Ackerley S, Miller CC. Role of axonal transport in neurodegenerative diseases. *Annu Rev Neurosci* 2008;**31**:151–73.
15 Coleman MP. The challenges of axon survival: introduction to the special issue on axonal degeneration. *Exp Neurol* 2013;**246**:1–5.
16 Hyman BT, Yuan J. Apoptotic and non-apoptotic roles of caspases in neuronal physiology and pathophysiology. *Nat Rev Neurosci* 2012;**13**:395–406.
17 Kole AJ, Annis RP, Deshmukh M. Mature neurons: equipped for survival. *Cell Death Dis* 2013;**4**:e689.
18 Fuchs Y, Steller H. Programmed cell death in animal development and disease. *Cell* 2011;**147**:742–58.
19 Tao J, Rolls MM. Dendrites have a rapid program of injury-induced degeneration that is molecularly distinct from developmental pruning. *J Neurosci* 2011;**31**:5398–405.
20 Sajadi A, Schneider BL, Aebischer P. Wlds-mediated protection of dopaminergic fibers in an animal model of Parkinson disease. *Curr Biol* 2004;**14**:326–30.
21 Riley DA. Ultrastructural evidence for axon retraction during the spontaneous elimination of polyneuronal innervation of the rat soleus muscle. *J Neurocytol* 1981;**10**:425–40.
22 Vanderhaeghen P, Cheng HJ. Guidance molecules in axon pruning and cell death. *Cold Spring Harb Perspect Biol* 2010;**2**:a001859.
23 Bagri A, Cheng HJ, Yaron A, Pleasure SJ, Tessier-Lavigne M. Stereotyped pruning of long hippocampal axon branches triggered by retraction inducers of the semaphorin family. *Cell* 2003;**113**:285–99.
24 Egea J, Klein R. Bidirectional Eph-ephrin signaling during axon guidance. *Trends Cell Biol* 2007;**17**:230–8.
25 Pasterkamp RJ. Getting neural circuits into shape with semaphorins. *Nat Rev Neurosci* 2012;**13**:605–18.

26 Benarroch EE. Acquired axonal degeneration and regeneration: recent insights and clinical correlations. *Neurology* 2015;**84**:2076–85.

27 Parker M, Roberts R, Enriquez M, Zhao X, Takahashi T, Pat Cerretti D, et al. Reverse endocytosis of transmembrane ephrin-B ligands via a clathrin-mediated pathway. *Biochem Biophys Res Commun* 2004;**323**:17–23.

28 Bishop DL, Misgeld T, Walsh MK, Gan WB, Lichtman JW. Axon branch removal at developing synapses by axosome shedding. *Neuron* 2004;**44**:651–61.

29 Gilley J, Coleman MP. Endogenous Nmnat2 is an essential survival factor for maintenance of healthy axons. *PLoS Biol* 2010;**8**:e1000300.

30 Song JW, Misgeld T, Kang H, Knecht S, Lu J, Cao Y, et al. Lysosomal activity associated with developmental axon pruning. *J Neurosci* 2008;**28**:8993–9001.

31 Fuentes-Medel Y, Logan MA, Ashley J, Ataman B, Budnik V, Freeman MR. Glia and muscle sculpt neuromuscular arbors by engulfing destabilized synaptic boutons and shed presynaptic debris. *PLoS Biol* 2009;**7**:e1000184.

32 Boulanger A, Farge M, Ramanoudjame C, Wharton K, Dura JM. Drosophila motor neuron retraction during metamorphosis is mediated by inputs from TGF-beta/BMP signaling and orphan nuclear receptors. *PLoS One* 2012;**7**:e40255.

33 Rotshenker S. Wallerian degeneration: the innate-immune response to traumatic nerve injury. *J Neuroinflammation* 2011;**8**:109

34 Gaudet AD, Popovich PG, Ramer MS. Wallerian degeneration: gaining perspective on inflammatory events after peripheral nerve injury. *J Neuroinflammation* 2011;**8**: 110

35 Stoll G, Jander S, Myers RR. Degeneration and regeneration of the peripheral nervous system: from Augustus Waller's observations to neuroinflammation. *J Peripher Nerv Syst* 2002;**7**:13–27.

36 Coleman MP, Freeman MR. Wallerian degeneration, wld(s), and nmnat. *Annu Rev Neurosci* 2010;**33**:245–67.

37 Martini R, Fischer S, Lopez-Vales R, David S. Interactions between Schwann cells and macrophages in injury and inherited demyelinating disease. *Glia* 2008;**56**:1566–77.

38 Wood MD, Kemp SW, Weber C, Borschel GH, Gordon T. Outcome measures of peripheral nerve regeneration. *Ann Anat* 2011;**193**:321–33.

39 Hoke A. Mechanisms of disease: what factors limit the success of peripheral nerve regeneration in humans? *Nat Clin Pract Neurol* 2006;**2**:448–54.

40 Vargas ME, Barres BA. Why is Wallerian degeneration in the CNS so slow? *Annu Rev Neurosci* 2007;**30**:153–79.

41 Krarup C, Archibald SJ, Madison RD. Factors that influence peripheral nerve regeneration: an electrophysiological study of the monkey median nerve. *Ann Neurol* 2002;**51**:69–81.

42 McKerracher L, David S, Jackson DL, Kottis V, Dunn RJ, Braun PE. Identification of myelin-associated glycoprotein as a major myelin-derived inhibitor of neurite growth. *Neuron* 1994;**13**:805–11.

43 Mukhopadhyay G, Doherty P, Walsh FS, Crocker PR, Filbin MT. A novel role for myelin-associated glycoprotein as an inhibitor of axonal regeneration. *Neuron* 1994;**13**:757–67.

44 Shen YJ, DeBellard ME, Salzer JL, Roder J, Filbin MT. Myelin-associated glycoprotein in myelin and expressed by Schwann cells inhibits axonal regeneration and branching. *Mol Cell Neurosci* 1998;**12**:79–91.

45 Schafer M, Fruttiger M, Montag D, Schachner M, Martini R. Disruption of the gene for the myelin-associated glycoprotein improves axonal regrowth along myelin in C57BL/Wlds mice. *Neuron* 1996;**16**:1107–13.

46 Bisby MA, Chen S. Delayed wallerian degeneration in sciatic nerves of C57BL/Ola mice is associated with impaired regeneration of sensory axons. *Brain Res* 1990;**530**:117–20.

47 Brown MC, Perry VH, Lunn ER, Gordon S, Heumann R. Macrophage dependence of peripheral sensory nerve regeneration: possible involvement of nerve growth factor. *Neuron* 1991;**6**:359–70.

48 Brown MC, Perry VH, Hunt SP, Lapper SR. Further studies on motor and sensory nerve regeneration in mice with delayed Wallerian degeneration. *Eur J Neurosci* 1994;**6**:420–8.

49 Bahr M, Przyrembel C. Myelin from peripheral and central nervous system is a nonpermissive substrate for retinal ganglion cell axons. *Exp Neurol* 1995;**134**:87–93.

50 Ramaglia V, King RH, Nourallah M, Wolterman R, de Jonge R, Ramkema M, et al. The membrane attack complex of the complement system is essential for rapid Wallerian degeneration. *J Neurosci* 2007;**27**:7663–72.

51 Mead RJ, Singhrao SK, Neal JW, Lassmann H, Morgan BP. The membrane attack complex of complement causes severe demyelination associated with acute axonal injury. *J Immunol* 2002;**168**:458–65.

52 Reichert F, Saada A, Rotshenker S. Peripheral nerve injury induces Schwann cells to express two macrophage phenotypes: phagocytosis and the galactose-specific lectin MAC-2. *J Neurosci* 1994;**14**:3231–45.

53 George R, Griffin JW. Delayed macrophage responses and myelin clearance during Wallerian degeneration in the central nervous system: the dorsal radiculotomy model. *Exp Neurol* 1994;**129**:225–36.

54 Perry VH, Tsao JW, Fearn S, Brown MC. Radiation-induced reductions in macrophage recruitment have only slight effects on myelin degeneration in sectioned peripheral nerves of mice. *Eur J Neurosci* 1995;**7**:271–80.

55 Fernandez-Valle C, Bunge RP, Bunge MB. Schwann cells degrade myelin and proliferate in the absence of macrophages: evidence from in vitro studies of Wallerian degeneration. *J Neurocytol* 1995;**24**:667–79.

56 Rotshenker S. Microglia and macrophage activation and the regulation of complement-receptor-3 (CR3/MAC-1)-mediated myelin phagocytosis in injury and disease. *J Mol Neurosci* 2003;**21**:65–72.

57 Lunn ER, Perry VH, Brown MC, Rosen H, Gordon S. Absence of Wallerian degeneration does not hinder regeneration in peripheral nerve. *Eur J Neurosci* 1989;**1**:27–33.

58 David S, Aguayo AJ. Axonal elongation into peripheral nervous system "bridges" after central nervous system injury in adult rats. *Science* 1981;**214**:931–3.

59 Be'eri H, Reichert F, Saada A, Rotshenker S. The cytokine network of Wallerian degeneration: IL-10 and GM-CSF. *Eur J Neurosci* 1998;**10**:2707–13.

60 Chen P, Piao X, Bonaldo P. Role of macrophages in Wallerian degeneration and axonal regeneration after peripheral nerve injury. *Acta Neuropathol* 2015;**130**:605–18.

61 Dubovy P, Jancalek R, Kubek T. Role of inflammation and cytokines in peripheral nerve regeneration. *Int Rev Neurobiol* 2013;**108**:173–206.

62 Shamash S, Reichert F, Rotshenker S. The cytokine network of Wallerian degeneration: tumor necrosis factor-alpha, interleukin-1alpha, and interleukin-1beta. *J Neurosci* 2002;**22**:3052–60.

63. Saada A, Reichert F, Rotshenker S. Granulocyte macrophage colony stimulating factor produced in lesioned peripheral nerves induces the up-regulation of cell surface expression of MAC-2 by macrophages and Schwann cells. *J Cell Biol* 1996;**133**:159–67.
64. Rotshenker S, Aamar S, Barak V. Interleukin-1 activity in lesioned peripheral nerve. *J Neuroimmunol* 1992;**39**:75–80.
65. Mietto BS, Mostacada K, Martinez AM. Neurotrauma and inflammation: CNS and PNS responses. *Mediators Inflamm* 2015;**2015**:251204.
66. Mirski R, Reichert F, Klar A, Rotshenker S. Granulocyte macrophage colony stimulating factor (GM-CSF) activity is regulated by a GM-CSF binding molecule in Wallerian degeneration following injury to peripheral nerve axons. *J Neuroimmunol* 2003;**140**:88–96.
67. Liu G, Yang H. Modulation of macrophage activation and programming in immunity. *J Cell Physiol* 2013;**228**:502–12.
68. Das A, Sinha M, Datta S, Abas M, Chaffee S, Sen CK, Roy S. Monocyte and macrophage plasticity in tissue repair and regeneration. *Am J Pathol* 2015;**185**:2596–606.
69. Giorgio S. Macrophages: plastic solutions to environmental heterogeneity. *Inflamm Res* 2013;**62**:835–43.
70. Khallou-Laschet J, Varthaman A, Fornasa G, Compain C, Gaston A-T, Clement M, et al. Macrophage plasticity in experimental atherosclerosis. *PLoS One* 2010;**5**:e8852.
71. MacKinnon AC, Farnworth SL, Hodkinson PS, Henderson NC, Atkinson KM, Leffler H, et al. Regulation of alternative macrophage activation by galectin-3. *J Immunol* 2008;**180**:2650–8.
72. Aamar S, Saada A, Rotshenker S. Lesion-induced changes in the production of newly synthesized and secreted apo-E and other molecules are independent of the concomitant recruitment of blood-borne macrophages into injured peripheral nerves. *J Neurochem* 1992;**59**:1287–92.
73. Saada A, Dunaevsky-Hutt A, Aamar A, Reichert F, Rotshenker S. Fibroblasts that reside in mouse and frog injured peripheral nerves produce apolipoproteins. *J Neurochem* 1995;**64**:1996–2003.
74. Reichert F, Levitzky R, Rotshenker S. Interleukin 6 in intact and injured mouse peripheral nerves. *Eur J Neurosci* 1996;**8**:530–5.
75. Carroll SL, Frohnert PW. Expression of JE (monocyte chemoattractant protein-1) is induced by sciatic axotomy in wild type rodents but not in C57BL/Wld(s) mice. *J Neuropathol Exp Neurol* 1998;**57**:915–30.
76. Ransohoff RM. Chemokines in neurological disease models: correlation between chemokine expression patterns and inflammatory pathology. *J Leukoc Biol* 1997;**62**:645–52.
77. Siebert H, Sachse A, Kuziel WA, Maeda N, Bruck W. The chemokine receptor CCR2 is involved in macrophage recruitment to the injured peripheral nervous system. *J Neuroimmunol* 2000;**110**:177–85.
78. Perrin FE, Lacroix S, Aviles-Trigueros M, David S. Involvement of monocyte chemoattractant protein-1, macrophage inflammatory protein-1alpha and interleukin-1beta in Wallerian degeneration. *Brain* 2005;**128**:854–66.
79. Boivin A, Pineau I, Barrette B, Filali M, Vallières N, Rivest S, Lacroix S. Toll-like receptor signaling is critical for Wallerian degeneration and functional recovery after peripheral nerve injury. *J Neurosci* 2007;**27**:12 565–76.

80. Chui R, Dorovini-Zis K. Regulation of CCL2 and CCL3 expression in human brain endothelial cells by cytokines and lipopolysaccharide. *J Neuroinflammation* 2010;7:1.
81. Mori K, Chano T, Yamamoto K, Matsusue Y, Okabe H. Expression of macrophage inflammatory protein-1alpha in Schwann cell tumors. *Neuropathology* 2004;**24**:131–5.
82. Siebert H, Dippel N, Mader M, Weber F, Bruck W. Matrix metalloproteinase expression and inhibition after sciatic nerve axotomy. *J Neuropathol Exp Neurol* 2001;**60**:85–93.
83. Shubayev VI, Angert M, Dolkas J, Campana WM, Palenscar K, Myers RR. TNFalpha-induced MMP-9 promotes macrophage recruitment into injured peripheral nerve. *Mol Cell Neurosci* 2006;**31**:407–15.
84. Chattopadhyay S, Myers RR, Janes J, Shubayev V. Cytokine regulation of MMP-9 in peripheral glia: implications for pathological processes and pain in injured nerve. *Brain Behav Immun* 2007;**21**:561–8.
85. La Fleur M, Underwood JL, Rappolee DA, Werb Z. Basement membrane and repair of injury to peripheral nerve: defining a potential role for macrophages, matrix metalloproteinases, and tissue inhibitor of metalloproteinases-1. *J Exp Med* 1996;**184**:2311–26.
86. Bruck W, Friede RL. The role of complement in myelin phagocytosis during PNS wallerian degeneration. *J Neurol Sci* 1991;**103**:182–7.
87. Dailey AT, Avellino AM, Benthem L, Silver J, Kliot M. Complement depletion reduces macrophage infiltration and activation during Wallerian degeneration and axonal regeneration. *J Neurosci* 1998;**18**:6713–22.
88. Liu L, Lioudyno M, Tao R, Eriksson P, Svensson M, Aldskogius H. Hereditary absence of complement C5 in adult mice influences Wallerian degeneration, but not retrograde responses, following injury to peripheral nerve. *J Peripher Nerv Syst* 1999;**4**:123–33.
89. Dekkers MP, Nikoletopoulou V, Barde YA. Cell biology in neuroscience: death of developing neurons: new insights and implications for connectivity. *J Cell Biol* 2013;**203**:385–93.
90. Levi-Montalcini R, Angeletti PU. Nerve growth factor. *Physiol Rev* 1968;**48**:534–69.
91. Silver JS, Hunter CA. gp130 at the nexus of inflammation, autoimmunity, and cancer. *J Leukoc Biol* 2010;**88**:1145–56.
92. Zweifel LS, Kuruvilla R, Ginty DD. Functions and mechanisms of retrograde neurotrophin signalling. *Nat Rev Neurosci* 2005;**6**:615–25.
93. Huang EJ, Reichardt LF. Neurotrophins: roles in neuronal development and function. *Annu Rev Neurosci* 2001;**24**:677–736.
94. Mok SA, Lund K, Campenot RB. A retrograde apoptotic signal originating in NGF-deprived distal axons of rat sympathetic neurons in compartmented cultures. *Cell Res* 2009;**19**:546–60.
95. Heumann R, Korsching S, Bandtlow C, Thoenen H. Changes of nerve growth factor synthesis in nonneuronal cells in response to sciatic nerve transection. *J Cell Biol* 1987;**104**:1623–31.
96. Lindholm D, Heumann R, Meyer M, Thoenen H. Interleukin-1 regulates synthesis of nerve growth factor in non-neuronal cells of rat sciatic nerve. *Nature* 1987;**330**:658–9.
97. Heumann R, Lindholm D, Bandtlow C, Meyer M, Radeke MJ, Misko TP, et al. Differential regulation of mRNA encoding nerve growth factor and its receptor in rat sciatic nerve during development, degeneration, and regeneration: role of macrophages. *Proc Natl Acad Sci USA* 1987;**84**:8735–9.

98 Matsuoka I, Meyer M, Thoenen H. Cell-type-specific regulation of nerve growth factor (NGF) synthesis in non-neuronal cells: comparison of Schwann cells with other cell types. *J Neurosci* 1991;**11**:3165–77.

99 Hattori A, Iwasaki S, Murase K, Tsujimoto M, Sato M, Hayashi K, Kohno M. Tumor necrosis factor is markedly synergistic with interleukin 1 and interferon-gamma in stimulating the production of nerve growth factor in fibroblasts. *FEBS Lett* 1994;**340**:177–80.

100 Bauer S, Kerr BJ, Patterson PH. The neuropoietic cytokine family in development, plasticity, disease and injury. *Nat Rev Neurosci* 2007;**8**:221–32.

101 Murphy M, Dutton R, Koblar S, Cheema S, Bartlett P. Cytokines which signal through the LIF receptor and their actions in the nervous system. *Prog Neurobiol* 1997;**52**:355–78.

102 Banner LR, Patterson PH. Major changes in the expression of the mRNAs for cholinergic differentiation factor/leukemia inhibitory factor and its receptor after injury to adult peripheral nerves and ganglia. *Proc Natl Acad Sci USA* 1994;**91**:7109–13.

103 Curtis R, Scherer SS, Somogyi R, Adryan KM, Ip NY, Zhu Y, et al. Retrograde axonal transport of LIF is increased by peripheral nerve injury: correlation with increased LIF expression in distal nerve. *Neuron* 1994;**12**:191–204.

104 Hirota H, Kiyama H, Kishimoto T, Taga T. Accelerated nerve regeneration in mice by upregulated expression of interleukin (IL) 6 and IL-6 receptor after trauma. *J Exp Med* 1996;**183**:2627–34.

105 Zhong J, Dietzel ID, Wahle P, Kopf M, Heumann R. Sensory impairments and delayed regeneration of sensory axons in interleukin-6-deficient mice. *J Neurosci* 1999;**19**:4305–13.

106 Murphy PG, Borthwick LA, Altares M, Gauldie J, Kaplan D, Richardson PM. Reciprocal actions of interleukin-6 and brain-derived neurotrophic factor on rat and mouse primary sensory neurons. *Eur J Neurosci* 2000;**12**:1891–9.

107 Cheema SS, Richards L, Murphy M, Bartlett PF. Leukemia inhibitory factor prevents the death of axotomised sensory neurons in the dorsal root ganglia of the neonatal rat. *J Neurosci Res* 1994;**37**:213–18.

108 Cafferty WB, Gardiner NJ, Gavazzi I, Powell J, McMahon SB, Heath JK, et al. Leukemia inhibitory factor determines the growth status of injured adult sensory neurons. *J Neurosci* 2001;**21**:7161–70.

109 Dowsing BJ, Morrison WA, Nicola NA, Starkey GP, Bucci T, Kilpatrick TJ. Leukemia inhibitory factor is an autocrine survival factor for Schwann cells. *J Neurochem* 1999;**73**:96–104.

110 Costigan M, Scholz J, Woolf CJ. Neuropathic pain: a maladaptive response of the nervous system to damage. *Annu Rev Neurosci* 2009;**32**:1–32.

111 Woolf CJ. What is this thing called pain? *J Clin Invest* 2010;**120**:3742–4.

112 Zimmermann M. Pathobiology of neuropathic pain. *Eur J Pharmacol* 2001;**429**:23–37.

113 Myers RR, Heckman HM, Rodriguez M. Reduced hyperalgesia in nerve-injured WLD mice: relationship to nerve fiber phagocytosis, axonal degeneration, and regeneration in normal mice. *Exp Neurol* 1996;**141**:94–101.

114 Murphy PG, Ramer MS, Borthwick L, Gauldie J, Richardson PM, Bisby MA. Endogenous interleukin-6 contributes to hypersensitivity to cutaneous stimuli and changes in neuropeptides associated with chronic nerve constriction in mice. *Eur J Neurosci* 1999;**11**:2243–53.

115 Wu G, Ringkamp M, Murinson BB, Pogatzki EM, Hartke TV, Weerahandi HM, et al. Degeneration of myelinated efferent fibers induces spontaneous activity in uninjured C-fiber afferents. *J Neurosci* 2002;**22**:7746–53.

116 Safieh-Garabedian B, Poole S, Allchorne A, Winter J, Woolf CJ. Contribution of interleukin-1 beta to the inflammation-induced increase in nerve growth factor levels and inflammatory hyperalgesia. *Br J Pharmacol* 1995;**115**:1265–75.

117 Woolf CJ, Allchorne A, Safieh-Garabedian B, Poole S. Cytokines, nerve growth factor and inflammatory hyperalgesia: the contribution of tumour necrosis factor alpha. *Br J Pharmacol* 1997;**121**:417–24.

118 Wagner R, Myers RR. Endoneurial injection of TNF-alpha produces neuropathic pain behaviors. *Neuroreport* 1996;**7**:2897–901.

119 Sorkin LS, Doom CM. Epineurial application of TNF elicits an acute mechanical hyperalgesia in the awake rat. *J Peripher Nerv Syst* 2000;**5**:96–100.

120 Zelenka M, Schafers M, Sommer C. Intraneural injection of interleukin-1beta and tumor necrosis factor-alpha into rat sciatic nerve at physiological doses induces signs of neuropathic pain. *Pain* 2005;**116**:257–63.

121 Moalem G, Tracey DJ. Immune and inflammatory mechanisms in neuropathic pain. *Brain Res Rev* 2006;**51**:240–64.

122 Dubovy P. Wallerian degeneration and peripheral nerve conditions for both axonal regeneration and neuropathic pain induction. *Ann Anat* 2011;**193**:267–75.

12

Pyronecrosis

Maryam Khalili and James A. Radosevich

Department of Oral Medicine and Diagnostic Sciences, University of Illinois at Chicago, Chicago, IL, USA

Abbreviations

ASC	apoptosis-associated speck-like protein containing a CARD
CAPS	CIAS1-associated periodic syndromes
CARD	caspase-associated recruitment domain
CIAS1	cold-induced autoinflammatory syndrome 1
CLR	C-type lectin receptor
HIV	human immunodeficiency virus
HMGB1	high-mobility group protein B1
HSP	heat-shock protein
IL	interleukin
IRG	immunity-related guanosine triphosphatase
LT	lethal toxin
MOI	multiplicity of infection
NALP3	NACHT, LRR, and PYD domains-containing protein 3
NCCD	Nomenclature Committee on Cell Death
NLR	NOD-like receptor
NLRP3	NOD-like receptor family, pyrin domain containing 3
NOD	nucleotide-binding domain
PARP	poly ADP ribose polymerase
PCD	programmed cell death
PVL	Panton–Valentine leukocidin
PYPAF1	PYRIN-containing Apaf1-like protein 1
RAGE	receptor for advanced glycation end products
RIP	receptor-interacting protein
TLR	Toll-like receptor
TNF	tumor necrosis factor
TNFR1	tumor necrosis factor receptor 1
TUCAN	tumor upregulated CARD-containing antagonist of caspasenine

Apoptosis and Beyond: The Many Ways Cells Die, First Edition. Edited by James Radosevich.
© 2018 John Wiley & Son Inc. Published 2018 by John Wiley & Son Inc.

12.1 Introduction

The development and homeostasis of organisms are dependent on the balance between cell survival and cell death. It is obvious that there is no life without death, and vice versa. There are many types of cell death, which are variously categorized. Initially, three types of cell-death morphology were described: type I, which is associated with heterophagy ("eating of another"); type II, which is associated with autophagy ("eating of itself"); and type III, which did not involve digestion. Cell deaths by these processes are now referred to as "apoptosis," "cell death associated with autophagy," and "necrosis," respectively. Today, it is considered that cell death is a fundamental process regulated by multiple interconnected (and sometimes overlapping) signaling pathways, and the concept of programmed cell death (PCD) is defined accordingly. In addition to development, cell death is related to a number of other vital processes, including chemotaxis, phagocytosis, regeneration, and immunogenicity [1].

Until recently, apoptosis was considered the hallmark of PCD and the standard form of cell death during development, homeostasis, infection, and pathogenesis – as compared to necrosis, which was mostly considered an "accidental" cell death that occurred in response to physicochemical insults. Today, the concept of "regulated necrosis" is accepted, and recent genetic evidence – as well as the discovery of chemical inhibitors of necrosis – demonstrates multiple pathways for this type of cell death. A genetically controlled form of cell death known as "regulated necrosis" is defined as a process that eventually results in cellular leakage and necrosis. Morphologically, it is characterized by cytoplasmic granulation, as well as organelle and/or cellular swelling ("oncosis"). Multiple modes of cell death share these morphological features, and they need to be examined for common or distinct underlying signaling pathways [2,3].

Regulated necrosis can be induced by several triggers, including alkylating DNA damage, excitotoxins, and the ligation of death receptors, at least under selected circumstances.

Regulated necrosis can be further classified based on its dependence on specific signaling modules, and it should be named accordingly. For instance, cases of regulated necrosis that exhibit receptor-interacting protein 1 (RIP1) activation and can be suppressed by RIP1 inhibitors, including necrostatin-1, should be labeled as "RIP1-dependent regulated necrosis" [2]. Also, it should be noted that the term "necroptosis," which is commonly used as a synonym of regulated necrosis, was originally introduced to indicate a specific case of regulated necrosis that is triggered by tumor necrosis factor receptor 1 (TNFR1) ligation and can be inhibited by the RIP1-targeting chemical necrostatin-1. Different types of cell death, including necroptosis, parthanatos, oxytosis, ferroptosis, ETosis, NETosis, pyronecrosis, and pyroptosis, were defined. These are all considered different forms of regulated necrosis. They share a common mechanistic profile, which consists of four levels: a trigger (level 1) activates an initiator mechanism (level 2), which activates several mediators that propagate the signal (level 3) and ultimately relay it to overlapping biochemical mechanisms (executioners) that cause necrotic cell death (level 4) [3].

Before proceeding to the next section, it is worth explaining that not all the types of cell death mentioned in this section are accepted as separate entities. Recently, the Nomenclature Committee on Cell Death (NCCD) has proposed a functional classification of cell-death subroutines that applies to both *in vitro* and *in vivo* settings. This classification includes extrinsic apoptosis, caspase-dependent or -independent intrinsic apoptosis, regulated necrosis, autophagic cell death, and mitotic catastrophe. Anoikis, parthanatos, pyroptosis, netosis, and cornification are classified as other cell-death modalities and are defined tentatively. Over the past decade, several neologisms have been introduced to indicate very specific signaling pathways that lead to cell death, some of which have a number of different definitions, which has resulted in confusion. Therefore, and with the aim of providing truly functional definitions of cell-death subroutines, the NCCD encourages scientists and authors of scientific publications to use a general versus specific nomenclature, which means (i) the use of general terms that bear functional connotations to specific names, and (ii) to avoid the introduction of neologisms [4]. With this in mind, this chapter will try to focus on pyronecrosis as a subtype of regulated necrosis.

12.2 Definition and Morphology

Pyronecrosis is a recent addition to the terminology of cell death. In 2007, two studies identified a proinflammatory cell-death pathway that was nucleotide-binding domain (NOD)-like receptor (NLR) protein-dependent and had necrotic features [5,6]. This was later termed "pyronecrosis" [6]. Observed in macrophages infected with *S. flexneri*, pyronecrosis is a caspase-independent but cathepsin-dependent cell-death pathway with the morphological characteristics of necrosis [6]. Pyronecrosis is associated with autoinflammatory diseases that have mutations in the NOD-like receptor family, pyrin domain-containing 3 (NLRP3; also known as CIAS1, NALP3, PYPAF1, or cryopyrin) gene. It is also associated with various microbial diseases, such as infections with *Shigella flexneri* [5,6]. It is caspase-independent, whereas pyroptosis is caspase-dependent. This could be because the effector caspase 3 is not cleaved and its substrate poly ADP ribose polymerase (PARP) is not present in pyronecrosis, as cell death proceeds in cells treated with caspase 1-specific inhibitors and/or pancaspase inhibitors [5]. Cell death stops, however, if an inhibitor of the lysosomal protease cathepsin B is used, suggesting that lysosome activity is necessary in this pathway. Pyronecrosis has morphological features characteristic of necrosis and is similar to the latter. It is inherently proinflammatory. Morphological hallmarks of apoptosis are not observed in this process. Pyronecrotic cells demonstrate neither DNA fragmentation nor the loss of mitochondrial membrane potential [6]. As observed by electron microscopy, the morphological changes seen in pyronecrosis are consistent with necrosis. They include membrane degradation, along with uncondensed chromatin. As with classical necrosis, pyronecrosis involves the release of the proinflammatory cytokine high-mobility group protein B1 (HMGB1) [6]. Pyroptosis and pyronecrosis can be stimulated by similar factors to produce similar outcomes. However, pyroptosis, but not pyronecrosis, relies on caspase-1 activity; this represents a significant difference between them [7]. There is an interesting overlap between these cell-death pathways and the NLR proteins. In response to cellular insult, the NLR proteins act as early mediators of inflammation [6]. It remains to be determined what importance NLR proteins have to other cell-death pathways. Although the list of terms used to describe such pathways has expanded over the past decade, future work should continue to refine cell-death terminology and improve classification [7].

12.3 Inflammation and Cell Death

Cell death has an important role in the immune system. Dying cells expose at their surface, molecules that signal to the immune system and release substances that trigger an immune response and inflammation. The inflammatory process can produce necrosis and other types of cell death. Recent progress in cell-death research has demonstrated a tight link between molecularly defined cell death and inflammation. PCD can act in a protective manner in host defense; the death of infected cells may reduce microbial infections, separate uninfected neighboring cells, and alert the host through danger signals and inflammatory mediators [8]. Interestingly, it has been suggested that following an initial regulated necrosis event, cell death and inflammation can induce each other and create a local autoamplification loop that leads to exaggerated cell death and inflammation [9]. Although most studies have focused on apoptosis, recent evidence suggests that additional cell-death pathways are crucial for the triggering of inflammation and immunity. Specifically, two forms of cell death have recently been identified, pyroptosis and pyronecrosis, which appear to exploit the proinflammatory features of necrosis within the context of immunity. However, they differ in the extent to which they resemble apoptosis and necrosis [7].

12.3.1 The Inflammasome

The inflammasome is a multimolecular complex consisting of caspase 1, PYCARD, NALP, and sometimes caspase 5 (also known as caspase 11 or ICH-3). It is a component of the innate immune system and is expressed in myeloid cells. The exact composition of an inflammasome depends on the activator that initiates its assembly. The inflammasome promotes the maturation of the inflammatory cytokines, interleukin 1β (IL-1β) and 18 (IL-18) [10]. It is responsible for activation of inflammatory processes [11] and has been shown to induce pyroptosis, a process of PCD distinct from apoptosis [12]. It is made up of apoptosis-associated speck-like protein containing a CARD (ASC) and tumor upregulated CARD-containing antagonist of caspasenine (TUCAN), as well as cryopyrin (CIAS1, NLRP3). It is the inflammasome that promotes activation of caspase 1/ICE. Pyronecrosis is independent of caspase 1 and caspase 11, but is dependent on the inflammasome component ASC and the lysosomal protein cathepsin B. It leads to the secretion from cells of the proinflammatory mediator HMGB1 [6].

12.3.2 Cryopyrin and NLR

The CATERPILLER or NLR family (NODs or NACHT-LRRs and C-type lectin receptors (CLRs)) is a group of proteins involved in the regulation of innate immunity. Like Toll-like receptors (TLRs), CLRs are most likely intracellular molecules that sense pathogen-derived products. There is increasing evidence for a crucial role of the NLR family in the regulation of immunity. Studies on NLR proteins are largely focused on its ability to mediate the start of an immune response to a pathogenic insult, particularly with regard to inflammation. Recently, it has been shown that NLR proteins also link innate immunity and cell-death signaling. This signaling is not limited to apoptosis, and is linked to both pyroptosis and pyronecrosis [7].

Cryopyrin is a member of the CLR family that is coded by the gene cold-induced autoinflammatory syndrome 1 (CIAS1). CIAS1 is mutated in a condition known as

CIAS1-associated periodic syndrome (CAPS), which has three different patterns of dominantly inherited periodic fevers. Recent reports have pointed out that there is an essential role for IL-1β in the manifestation of mutant CIAS1-associated periodic fevers. It has been shown that mutant CIAS1/NLRP3 causes elevated levels of spontaneous and induced IL-1β both *in vitro* and *in vivo*. Cryopyrin helps to control IL-1β levels through involvement in the multimolecular complex of the inflammasome. As mentioned before, this complex, which also includes ASC and TUCAN, promotes activation of caspase 1/ICE. Caspase 1 cleaves pro-IL-1β to produce mature IL-1β. This is then released from the cell. Mutations in cryopyrin can cause hyperactivation of this pathway, resulting in excessive IL-1β production and severe episodes of inflammation.

Cryopyrin's functions are not limited to autoinflammatory disorders within the immune system. Multiple recent studies have shown that it is an important cellular component that can organize the inflammasome to induce IL-1β release in response to viral, bacterial, and other proinflammatory stimuli. Host response in mammals to microbial pathogens may occur through macrophage/monocyte necrotic-like death, which results in pathogen elimination, but also leads to exacerbated inflammation and sepsis. Cryopyrin can also participate in initiation of necrosis. It has been shown that macrophages that are cryopyrin-deficient are more resistant to cell death in response to the Gram-positive Staphylococcus bacteria [11].

Willingham et al. [6] reported that ASC and cryopyrin are required for pyronecrotic cell death. They also demonstrated that cryopyrin mediates the IL-1β and cell-death response to a Gram-negative bacterium, *S. flexneri*. This causes cellular necrosis and the exacerbation of inflammation. They concluded that Shigella-induced cell death is independent of caspase 1 and IL-1β, suggesting that this process occurs without inflammasome formation. Normal cryopyrin induces cell death only with bacteria or other pathogens, and disease-related cryopyrin represents a hyperactive form of the protein [6].

12.4 Relation of Pyronecrosis to Other Types of Cell Death

Regulated necrosis can be induced by several molecular pathways. The important question is whether these individual pathways are part of an interacting network with common execution mechanisms that result in a similar morphology, or whether multiple programs of regulated necrosis are developed separately in relation to specific stimuli resulting from special processes such as infection or cellular stress [7]. Since apoptosis was long considered the main PCD pathway, it is included in this section as well.

12.4.1 Pyronecrosis and Necrosis

It is well established that the induction of necrotic cell death is a crucial part of the start of an immune response. Necrosis (if not otherwise specified) is an unprogrammed type of cell death that is considered to be passive in nature. It has been known for a long time that necrosis can be seen in monocytes exposed to toxins or infected with intracellular bacteria. In some incidences, pathogen-induced cell death is most likely passive, but the programmed process of pyronecrosis, which is an active one, might be a critical feature of macrophage function. A hallmark of necrosis is the rapid loss of integrity of the plasma membrane and the consequent release of cellular contents into the extracellular medium. These features are central to the process of necrosis in the immune and inflammatory responses.

It is the release of these cellular components that has a significant effect on the local microenvironment. The intracellular components include uric acid, adenine phosphate, purine metabolites, and heat-shock proteins (HSPs), all of which become proinflammatory effectors. Macrophages undergoing necrosis can release proinflammatory cytokines such as tumor necrosis factor (TNF) and IL-1 [13]. Another protein released from necrotic cells that has drawn significant attention is the nuclear DNA-binding HMGB1. When HMGB1 is released, it becomes an agonist for receptor for advanced glycation end products (RAGE), TLR2, and TLR4. These receptors are expressed on the cell surface of monocytes and of other cell types. When they are activated, the inflammation process is heightened through induction of proinflammatory cytokines in the microenvironment [14]. Pyronecrosis (along with pyroptosis; see Section 12.4.2) appears to use the proinflammatory components of necrosis within the context of immunity.

In conclusion, pyronecrosis is a programmed or regulated cell-death type with features of necrosis that is triggered by pathogenic stimuli and dependant on NLRP3, ASC, and cathepsin B.

12.4.2 Pyronecrosis and Pyroptosis

Microbial pathogens, such as *Salmonella* spp. and *Listeria* spp., induce the pyroptosis cell-death pathway. Pyroptosis is like apoptosis in that it is caspase-dependent and involves DNA damage. However, pyroptosis relies on caspase 1, and not on the classical proapoptotic initiators and effector caspases (caspases 3, 8, and 9).

In addition to its apoptotic qualities, pyroptosis exhibits some of the same features as necrosis. Both are characterized by plasma-membrane breakdown, although mitochondrial membrane integrity remains in the case of pyroptosis.

Current studies are directed at defining other molecular mediators involved in pyroptosis [7]. A recent report defines what is called the "pyroptosome": a large complex made up of ASC, which also contains a caspase-associated recruitment domain (CARD). The CARD, as part of the pyroptosome, has been referred to as "PYCARD." Dimers of PYCARD assemble as part of this process. It has also been reported that, *in vivo*, NLR proteins are required for pyroptosome formation [15].

With regard to inflammation, two important aspects of pyroptosis are: (i) the activation of caspase 1 and (ii) the breakdown of the plasma membrane. Pyronecrosis, on the other hand, is an NLR protein-dependent pathway of proinflammatory cell death with primarily necrotic features [5]. It is genetically based, as it involves mutations in NLRP3 (also known as CIAS1, NALP3, PYPAF1, or cryopyrin). It is also associated with microbial pathogens such as *S. flexneri* [5,6]. Unlike pyroptosis, pyronecrosis is caspase-independent. The effector caspase 3 is not cleaved and its substrate PARP is not found in pyronecrosis. Cell death proceeds in the presence of caspase 1-specific inhibitors and pancaspase inhibitors. However, in the presence of an inhibitor of the lysosomal protease cathepsin B, cell death does not occur, suggesting that lysosome activity is necessary in this form of cell death [5,6].

Pathogens that induce cell death, including the Gram-negative bacteria *S. flexneri*, have been investigated in many studies. It has been observed that at low multiplicity of infection (MOI), macrophage cell death induced by *S. flexneri* is caspase 1-dependent [16]. It has also been shown that gene deletion of the NLR family member NLRC4, but not the inflammasome component ASC, reduces the activity of this cell-death pathway. Together, these findings are representative of the pyroptosis pathway. At high MOI, *S. flexneri* induces caspase 1-independent cell death in human monocyte-derived macrophages. This phenomenon is characterized by morphologically necrotic characteristics closely mirroring those of mutant NLRP3-induced cell death. *S. flexneri*-induced necrotic cell death has been shown to be dependent on NLRP3 in both mouse macrophages and the human monocytic cell line THP-1. Like mutant NLRP3-induced cell death, it requires both ASC and the protease cathepsin B [6]. From these findings, one can conclude that NLRC4 mediates *S. flexneri*-induced pyroptosis and that NLRP3 mediates pyronecrosis in response to the same bacteria. This suggests the intriguing possibility that different NLR proteins mediate different forms of cell death, each having a distinct biologic outcome [7].

12.4.3 Pyronecrosis and Apoptosis

Apoptosis is a form of PCD that uses specific molecular mediators, most importantly the apoptotic caspases. For the most part, apoptosis relies on the protease activity of caspases and is a slow and energy-consuming process. It has significant membrane blebbing as a cellular feature, and involves the packaging of cellular material for recycling. Pyronecrosis, on the other hand, is a proinflammatory form of cell death that is triggered by pathogenic stimuli with primarily necrotic features. It is caspase-independent, proceeding when caspase 1-specific inhibitors and/or pancaspase inhibitors are present. However, cell death does not take place in the presence of an inhibitor of the lysosomal protease cathepsin B, which implicates lysosome activity in this pathway [5,6]. Characteristic features of apoptosis are not present. Pyronecrotic cells do not have DNA fragmentation, nor is there a loss of mitochondrial membrane potential [6]. As can be seen using electron microscopy, the morphological changes characteristic of pyronecrosis are consistent with necrosis. They include membrane degradation and the presence of chromatin, which is uncondensed [5,6].

12.5 Molecular Pathway and Mechanism

Despite recent progress in molecular studies on cell death, the mechanism underlying pyronecrosis remains unclear and requires further investigation [8]. Briefly, pyronecrosis is provoked by downstream NLRP3-induced ASC oligomerization but does not require caspase 1 or IL-1β cleavage. Cathepsin B inhibitor inhibits pyronecrosis before ASC oligomerization, while Z-VAD-fmk inhibits the process after ASC oligomerization [17].

Cell-death processes that include a component in which there is a loss of plasma-membrane integrity can lead to the exacerbation of inflammation through the release of intracellular inflammatory cytokines and factors (such as IL-1, TNF, and HMGB1). Willingham et al. [6] showed that NLRP3 and ASC were required for the release of IL-1β and HMGB1 from a human macrophage cell line, THP-1, that was infected with *S. flexneri*. They described pyronecrosis – which is not dependent on IL-1 β or caspase 1, but is dependent on ACS – as a necrotic-like cell death caused by disease-associated mutants of CIAS1. CIAS1 is the gene that encodes cryopyrin and, along with ASC, is required for necrotic-like cell death. Willingham et al. [6] also demonstrated that cryopyrin is capable of mediating both the IL-1β and the cell-death response to

S. flexneri, resulting in cellular necrosis with enhanced inflammation. When Shigella cause pyronecrosis, it is independent of caspase 1 and IL-1β, suggesting that this process occurs without inflammasome formation. As with other necrotic processes, cellular components spill out from the pyronecrotic cell into the microenvironment. These components include the nuclear protein, HMGB1, which is a powerful proinflammatory cytokine when released by the cell. Once released, HMGB1 acts as a potent danger indicator, stimulating the RAGE, TLR2, and TLR4 receptors on neighboring monocytes and macrophages. This triggers the induction of several inflammatory cytokines, including TNF-α and IL-1β [18]. When these receptors are activated, there is further exacerbation of inflammation in the microenvironment through the induction of additional proinflammatory cytokines [6,7]. HMGB1 in the serum of mice is increased during endotoxin exposure, while septic patients who also succumbed to infection have increased levels as well. Neutralization of HMGB1 can significantly reduce inflammation, and improves survival in animal models of established sepsis. The release of HMGB1 elicited by bacterial-induced cell death further supports the use of HMGB1 antagonists to reduce inflammation during sepsis [6]. A therapeutic approach currently under investigation involves neutralizing HMGB1 as a target for the treatment of sepsis, bacteremia, and induced acute respiratory distress syndrome. Broad inhibition of NLRP3-dependent responses may be detrimental to host survival, but neutralization of HMGB1 may help to reduce NLRP3-mediated inflammation without increasing host mortality [18].

12.6 Pyronecrosis-Inducing Factors

Several pathogens are reported to induce cell death that might be interpreted as pyronecrosis [3]. Induction of necrosis by *S. flexneri*, and possibly by other microbial pathogens, may be carried out via secreted toxins.

12.6.1 *Shigella flexneri*

Willingham et al. [6] first described pyronecrosis as a necrotic process in macrophages infected with *S. flexneri*. This process is cryopyrin-dependent, having features similar to the death caused by mutant CIAS1. Such necrotic death is independent of caspase 1 and IL-1β, and therefore independent of the inflammasome. The necrosis seen in primary macrophages requires the presence of virulence gene products from *Shigella*. When cells are at low density (MOI < 10), macrophage cell death induced by *S. flexneri* is seen as being caspase 1-dependent. It has been shown that deletion of NLRC4, but not ASC (a component of the inflammasome), also reduces this process, suggesting pyroptosis as the mechanism of cell death. When the MOI is comparatively high, *S. flexneri* induces caspase 1-independent cell death in human monocyte-derived macrophages. This phenomenon has morphological features similar to necrosis, closely resembling those of mutant NLRP3-induced cell death. Necrotic cell death induced by *S. flexneri* is dependent on NLRP3 in both mouse macrophages and the human monocytic cell line THP-1. Like mutant NLRP3-induced cell death, this process also requires ASC and the protease cathepsin B [6]. It therefore appears that NLRC4 mediates *S. flexneri*-induced pyroptosis and NLRP3 mediates the pyronecrosis cell-death pathway in response to the same bacteria [7].

12.6.2 Neisseria gonorrhoeae

N. gonorrhoeae is a common sexually transmitted pathogen. It has a major impact on transmission of human immunodeficiency virus (HIV), female fertility, and neonatal health worldwide. It induces localized inflammation of the urethra and cervix by causing the production of IL-1β and other inflammatory cytokines. Duncan et al. [19] demonstrated that NLRP3 (cryopyrin, NALP3) is the primary NLR required for IL-1β/IL-18 secretion in response to *N. gonorrhoeae* in monocytes. They also showed that *N. gonorrhoeae* infection induces NLRP3-dependent monocytic cell death via pyronecrosis. Additionally, *N. gonorrhoeae* activates the cysteine protease cathepsin B. By inhibiting cathepsin B, it can be shown that this protease is an apical controlling process in the downstream activities of NLRP3, including IL-1β production, pyronecrosis, and HMGB1 release. Factors capable of initiating the NLRP3-mediated signaling events are present in conditioned media from *N. gonorrhoeae*. Lipooligosaccharide isolated from *N. gonorrhoeae* is a known virulence factor. It is isolated in the form of outer-membrane blebs and activates both NLRP3-induced IL-1β secretion and pyronecrosis [19].

12.6.3 Toxoplasma gondii parasitophorous

T. gondii is a natural intracellular protozoal pathogen of mice and other small mammals. After infection, the parasite freely replicates in a number of cell types and then encysts in muscle and brain tissue. Early immune resistance is mediated by the family of interferon-inducible immunity-related guanosine triphosphatase (IRG) proteins in mice, but little is known of how this resistance takes place. Zhao et al. [20] reported that IRG proteins accumulate on intracellular vacuoles containing the pathogen, and that the vacuolar membrane subsequently ruptures. The rupture of the vacuole is followed by death of the intracellular parasite, which is followed in turn by the death of the infected cell. Cells that show features of death by pyronecrosis include (i) membrane permeabilization and (ii) release of the inflammatory protein HMGB1, but *not* (iii) caspase 1 cleavage. This event only occurs following infection with an avirulent strain of *T. gondii*. Cells infected by virulent strains rarely undergo necrosis. The authors conclude that IRG proteins resist infection by avirulent *T. gondii* using a novel mechanism in which there is disruption of the vacuolar membrane. This results in molecular events that ultimately lead to the necrotic death of the infected cell [20].

12.6.4 Staphylococcus aureus

The *S. aureus* pore-forming toxin Panton–Valentine leukocidin (PVL) is most likely responsible for life-threatening necrotizing infections, characterized by massive tissue inflammation and necrosis. Holzinger et al. [21] showed that PVL binding to monocytes and macrophages leads to release of the caspase 1-dependent proinflammatory cytokines IL-1β and IL-18. PVL also activates the NLRP3 inflammasome. Specific inhibition of this pathway at several steps significantly reduces inflammasome activation and subsequent pyronecrosis. PVL is a strong inducer of IL-1β secretion via a cathepsin B-mediated activation of the NLRP3 inflammasome, which likely contributes to the severe inflammation associated with necrotizing infections. Pore formation of PVL leads to K^+-efflux and the consecutive activation of cathepsin B, which mediates programmed necrosis and activation of NLRP3. The study also showed that PVL is a strong inducer of NLRP3 activation [21].

12.6.5 Bacillus anthracis Lethal Toxin

Recently, *B. anthracis* has been included on the list of pathogens capable of inducing cell death that might be interpreted as pyronecrosis [3]. However, the original article by Averette et al. [22] stated that *B. anthracis* lethal toxin (LT) is recognized by a subset of alleles of the NLR protein, Nlrp1b, resulting in pyroptotic cell death of macrophages and dendritic cells. Macrophages that express an LT-sensitive allele of Nlrp1b undergo pyroptosis in the presence of this toxin, releasing inflammatory cytokines that activate innate immunity. It is not understood how Nlrp1b controls recognition of LT or what downstream events lead to cell death. In this study, the authors used LT to investigate the mechanism of cell death that occurs during pyroptosis [22]. Therefore, the role of LT in pyronecrosis is questionable.

12.6.6 Other Possible Toxins

The induction of necrosis by microbial pathogens may be mediated through toxins. Some of these have been examined directly with respect to caspase 1 activation and cell death.

Streptomyces hygroscopicus produces the toxin nigericin. This is a potassium ionophore that is a potent inducer of both necrosis and IL-1β release in monocytes. As in pyronecrosis, both functions require active cathepsin B [23]. Maitotoxin is another toxin, produced by the dinoflagellate *Gambierdiscus toxicus*. It has been shown that maitotoxin induces necrosis in a manner that is dependent on the calcium-activated cysteine protease calpain. It also induces IL-1β release by mouse macrophages [24]. Like *S. flexneri*, both nigericin and maitotoxin activate the NLRP3 inflammasome. These toxins also change the levels of intracellular potassium. Although molecular mediators like that of nigericin and maitotoxin continue to be identified, the participation of NLRP3 and/or other NLR family members in the macrophage response to these toxins needs to be further studied [7].

12.7 Conclusion

The consequences of pathogen-induced cell death and their relation to immunity have not been studied thoroughly, although recent studies have shed some light on this topic. Obviously, the process denies the invading bacteria an environment in which to replicate. In addition, cell-death programs that have a component in which there is a loss of plasma-membrane integrity also result in an increased inflammatory response through the release of intracellular inflammatory cytokines and factors such as IL-1, TNF, and HMGB1. Two such modes of cell death have recently been identified: pyroptosis and pyronecrosis. There are significant differences between the two pathways, but they both respond to pathogens and promote the proinflammatory response through the release of cellular contents. Pyronecrosis is caspase-independent and cathepsin-dependent. NLR proteins are important regulators of both pathways, and future studies should aim to find additional mediators for these processes. Recently, an inhibitory interaction of mitochondrial antiapoptotic proteins with NLRP1 has been described. Therefore, propyroptotic- and propyronecrotic-associated proteins may be within the collection of proteins recognized to be proapoptotic cell-death mediators [7].

References

1. Kaczmarek A. Vandenabeele P. Krysko D. Necroptosis: the release of damage-associated molecular patterns and its physiological relevance. *Immunity* 2013;**38**:209–23.
2. Cho YS. Challa S. Moquin D. Genga R. Ray TD. Guildford M. et al. Phosphorylation-driven assembly of the RIP1–RIP3 complex regulates programmed necrosis and virus-induced inflammation. *Cell* 2009;**137**:1112–23.
3. Vanden Berghe T. Linkermann A. Jouan-Lanhouet S. Walczak H. Vandenabeele P. Regulated necrosis: the expanding network of non-apoptotic cell death pathways. *Nat Rev Mol Cell Biol* 2014;**15**:135–47.
4. Galluzzi L. Vitale I. Abrams JM. Alnemri ES. Baehrecke EH. Blagosklonny MV. et al. Molecular definitions of cell death subroutines: recommendations of the Nomenclature Committee on Cell Death 2012. *Cell Death Differ* 2012;**19**(1):107–20.
5. Fujisawa A. Kambe N. Saito M. Nishikomori R. Tanizaki H. Kanazawa N. Disease-associated mutations in CIAS1 induce cathepsin B-dependent rapid cell death of human THP-1 monocytic cells. *Blood* 2007;**109**:2903–11.
6. Willingham SB. Bergstralh DT. O'Connor W. Morrison AC. Taxman DJ. Duncan JA. et al. Microbial pathogen-induced necrotic cell death mediated by the inflammasome components CIAS1/cryopyrin/NLRP3 and ASC. *Cell Host Microbe* 2007;**2**:147–59.
7. Ting JP. Willingham SB. Bergstralh DT. NLRs at the intersection of cell death and immunity. *Nature* 2008;**8**:372–9.
8. Yang Y. Jiang G. Zhang P. Fan J. Programmed cell death and its role in inflammation. *Mil Med Res* 2015;**2**:12.
9. Linkerman A. Stockwell BR. Krautwald S. Anders HJ. Regulated cell death and inflammation: an auto-amplification loop causes organ failure. *Nat Rev Immunol* 2014;**14**:759–67.
10. Martinon F. Burns K. Tschopp J. The inflammasome: a molecular platform triggering activation of inflammatory caspases and processing of proIL-beta. *Mol Cell* 2002;**10**(2):417–26.
11. Mariathasan S. Weiss DS. Newton K. McBride J. O'Rourke K. Roose-Girma M. et al. Cryopyrin activates the inflammasome in response to toxins and ATP. *Nature* 2006;**440**:228–32.
12. Fink SL. Cookson BT. Apoptosis, pyroptosis, and necrosis: mechanistic description of dead and dying eukaryotic cells. *Infect Immun* 2005;**73**:1907–16.
13. Chen CJ. Kono H. Golenbock D. Reed G. Akira S. Rock KL. Identification of a key pathway required for the sterile inflammatory response triggered by dying cells. *Nat Med* 2007;**13**(7):851–6.
14. Sunden-Cullberg J. Norrby-Teglund A. Treutiger CJ. The role of high mobility group box-1 protein in severe sepsis. *Curr Opin Infect Dis* 2006;**19**:231–6.
15. Fernandes-Alnemri T. Wu J. Yu JW. Datta P. Miller B. Jankowski W. et al. The pyroptosome: a supramolecular assembly of ASC dimers mediating inflammatory cell death via caspase-1 activation. *Cell Death Differ* 2007;**14**:1590–604.
16. Suzuki T. Franchi L. Toma C. Ashida H. Ogawa M. Yoshikawa Y. et al. Differential regulation of caspase-1 activation, pyroptosis, and autophagy via Ipaf and ASC in Shigella-infected macrophages. *PLoS Pathog* 2007;**3**(8):e111.

17 Satoh T. Kambe N. Matsue H. NLRP3 activation induces ASC-dependent programmed necrotic cell death, which leads to neutrophilic inflammation. *Cell Death Dis* 2013;**4**(5): e644.
18 Willingham SB. Allen IC. Bergstralh DT. Brickey WJ. Huang MT. Taxman DJ. et al. NLRP3 (NALP3, Cryopyrin) facilitates *in vivo* caspase-1 activation, necrosis and HMG≡1 release via inflammasome-dependent and -independent pathways. *J Immunol* 2009;**183**:2008–15.
19 Duncan JA. Gao X. Huang M. O'Connor BP. Thomas CE. Willingham SB. et al. *Neisseria gonorrhoeae* activates the proteinase cathepsin B to mediate the signaling activities of the NLRP3 and ASC-containing inflammasome. *J Immunol* 2009;**182**:6460–9.
20 Zhao Y. Khaminets A. Hunn J. Howard J. Disruption of the *Toxoplasma gondii* parasitophorous vacuole by IFNγ-inducible immunity-related GTPases (IRG proteins) triggers necrotic cell death. *PLoS Pathog* 2009;**5**(2):e1000288.
21 Holzinger D. Gieldon L. Mysore V. Nippe N. Taxman DJ. Duncan JA. et al. *Staphylococcus aureus* Panton-Valentine leukocidin induces an inflammatory response in human phagocytes via the NLRP3 inflammasome. *J Leukocyte Biol* 2012;**92**:1069–81.
22 Averette KM. Pratt MR. Yang Y. Bassilian S. Whitelegge JP. Loo JA. et al. Anthrax lethal toxin induced lysosomal membrane permeabilization and cytosolic cathepsin release is Nlrp1b/Nalp1b–dependent. *PloS ONE* 2009;**4**:e7913.
23 Hentze H. Lin XY. Choi MS. Porter AG. Critical role for cathepsin B in mediating caspase-1-dependent interleukin-18 maturation and caspase-1-independent necrosis triggered by the microbial toxin nigericin. *Cell Death Differ* 2003;**10**:956–68.
24 Zhao X. Pike BR. Newcomb JK. Wang KK. Posmantur RM. Hayes RL. Maitotoxin induces calpain but not caspase-3 activation and necrotic cell death in primary septo-hippocampal cultures. *Neurochem Res* 1999;**24**:371–82.

13

Phenoptosis: Programmed Death of an Organism

M.V. Skulachev and V.P. Skulachev

Belozersky Institute of Physico-Chemical Biology, Institute of Mitoengineering, Faculty of Bioengineering and Bioinformatics, Lomonosov Moscow State University, Moscow, Russia

Abbreviations

AD	Alzheimer's disease
ADP	adenosine diphosphate
ALS	amyotrophic lateral sclerosis
Apaf-1	apoptotic protease activating factor-1
ATP	adenosine triphosphate
ATPase	adenosine triphosphatase
$C_{12}TPP$	dodecyltriphenylphosphonium
CNS	central nervous system
CGRP	calcitonin gene-related peptide
cyt c	cytochrome c
DRG	dorsal root ganglion
FFA	free fatty acid
IMM	inner mitochondrial membrane
LPS	lipopolysaccharide
mROS	mitochondrial ROS
NAC	N-acetylcysteine
NMDA	N-methyl-D-aspartate
NMR	nuclear magnetic resonance
PQ	plastoquinone
ROS	reactive oxygen species
SkQ1	plastoquinonyl decyltriphenylphosphonium
SkQR1	plastoquinonyl decylrhodamine 19
TRPV1	transient receptor potential cation channel subfamily V member 1

13.1 Definition of Phenoptosis and a Short History of the Problem

When we asked the famous philologist and metrician Prof. M.L. Gasparov what to call the programmed death of an *organism* if the programmed death of a cell is already

Apoptosis and Beyond: The Many Ways Cells Die, First Edition. Edited by James Radosevich.
© 2018 John Wiley & Son Inc. Published 2018 by John Wiley & Son Inc.

called "apoptosis," he suggested the term "phenoptosis" [1]. We have used this word in our publications since 1997. In 2012, a special issue of *Biochemistry (Moscow) Phenoptosis* was published (for some review papers devoted to phenoptosis, see references [1–15]).

Phenoptosis is defined as the genetically programmed death of an organism. The death program is usually encoded in the genome of the dying organism, being a chain of biochemical events that ultimately causes its suicide. Less often, death occurs as a result of behavioral responses coded in the genome of the dying individual, its sexual partner, a close relative, or a partner in the ecosystem [2–4,13,16]. At first glance, the existence of phenoptosis is hardly compatible with classical Darwinism. Charles Darwin proclaimed natural selection of individuals to be the basis of biological evolution. Within this paradigm, phenoptosis, as a phenomenon obviously counterproductive for an individual, could not be selected for by the struggle for existence. However in 1818 – that is, in the pre-Darwinian period of biology – the philosopher Arthur Schopenhauer defined the role of the individual in evolution thus: "An individual is not only exposed to destruction in a thousand ways from the most insignificant accidents, but is even destined for death and is led toward it by nature herself, from the moment that individual has served for the maintenance of the species" [17].

Alfred Russell Wallace (who postulated at the same time as Darwin the idea of natural selection) wrote in one of his letters at the turn of the 1870s: "When parents have provided a sufficient number of successors, they themselves, as consumers of nourishment in a constantly increasing degree, are an obstacle to those successors. Natural selection therefore weeds them out, and in many cases favors such races that die almost immediately after they have left successors" (quoted in [18]). Later, in 1881, this principle was independently formulated and developed in detail by another great biologist, August Weismann: "Worn-out individuals are not only valueless to the species, but they are even harmful, for they take the place of those who are sound . . . I consider that death is not a primary necessity, but that it has been secondarily acquired as an adaptation" [18].

Weismann was immediately accused by his contemporaries of being anti-Darwinist, even though Darwin himself was perfectly aware of the limitations of his hypothesis that evolution follows only those directions that are favorable for an individual. That is what he wrote in his second famous book, *The Descent of Man*: "There can be no doubt that a tribe including many members who were always ready to aid one another, and to sacrifice themselves for the common good, would be victorious over most other tribes; and this would be natural selection" [19].

The programmed-death phenomenon attracted the interest of biologists in the second half of the 20th century, when it was discovered that the vast majority of cells die because they actuate their own deadly program. This phenomenon was given the name "apoptosis" in 1972 by J.F. Kerr et al. [20]. They borrowed this term from the Roman scholar and physician Claudius Galen, who noticed that the broken branch of a tree enters winter with leaves dried up, but not fallen from it. According to Galen, defoliation appears to be an *active* process rather than the death of leaves from cold, as was previously believed. The studies of many authors of the late 20th century showed that apoptosis is an extremely widespread phenomenon in all types of multicellular organism. These studies reached their apotheosis when researchers discovered in the worm *Caenorhabditis elegans* genes encoding proteins required for apoptosis. Thus, the first genes of death were identified (the authors of these studies, H.R. Horwitz, J.E. Sulston, and S. Brenner, were awarded the Nobel Prize in Physiology or Medicine in 2002).

It should be noted that the apoptotic program requires energy. If a cell is deprived of energy resources, it will not die as fast as it was normally going to when the apoptotic program initiated the cell suicide. Up to a certain stage of the apoptotic cascade, the deadly program can be stopped and the cell can be saved. By means of modern genetic engineering techniques, it is possible to create a cell incapable of apoptosis. To achieve this, some genes need to be switched off.

13.2 Phenoptosis of Unicellular Organisms

Some organisms consist of a single cell. These are bacteria, protozoa such as amoebae, and unicellular fungi and plants. If these creatures have apoptosis, it means that they possess a suicide program of the *organism*, as in such cases the cell is the organism.

The program of self-destruction has been discovered in yeast; this happened very recently, in the first years of the present century, through the cooperation of Dr. A.A. Hyman's laboratory and Dr. F.F. Severin from our group [21–23]. It was found that a pheromone (a substance secreted by individuals of the opposite sex to attract a partner) could be a signal that triggers the death program. We will not go into the details of sexual reproduction of yeast here; we will only state the main idea. A natural compound (pheromone, a short peptide) that triggers sexual reproduction in yeast kills the yeast cell when the pheromone concentration reaches a certain critical value. It was revealed that the yeast cell dies because pheromone binds to a protein receptor on the surface of the cell. The binding thereby triggers a long cascade of reactions leading to the death of this unicellular organism. The death cascade of yeast has several steps in common with the apoptosis of cells in multicellular organisms.

Apart from yeast, there is a huge world of unicellular microorganisms that are organized in a much simpler way. We mean prokaryotes: eubacteria and archaea. The self-destruction phenomenon has been found to be inherent in them as well, although the mechanisms can differ from those found in eukaryotic cells.

For example, systems of the "long-lived toxin–short-lived antitoxin" type have been found in eubacteria; in this case, a cell *slowly* synthesizes a protein that can potentially kill it. Such suicide does not happen under conditions where there is a sufficient supply of amino acids for protein synthesis: the cell then succeeds in *rapid* synthesis of the protein antidote (antitoxin), which binds to the toxin and neutralizes it. Toxins are not only slowly synthesized, they also disintegrate slowly. At the same time, antitoxins break up very quickly. As a result, "in lean years," when the supply of amino acids is not sufficient for the synthesis of new proteins at a high rate, antitoxin disintegrates and disappears, but the amount of toxin decreases only slightly. The outcome is sad: the toxin is released from its complex with antitoxin and kills the bacterium. Bacteria die, their number diminishes, and therefore the consumption of amino acids decreases. Eventually, the supply of amino acids in the surviving bacteria rises to the level sufficient for the fast rate of protein synthesis, and the lucky survivors once again start synthesizing antitoxin, which binds toxin. In this way, the population of bacteria solves the problem of overpopulation [4,24–26].

It is significant that the lack of not only amino acids, but also respiratory substrates or oxygen, as well as the appearance of pollutants – inhibitors of transcription, translation, or energy metabolism – and other adverse factors inhibiting the biosynthesis of proteins can actuate the "toxin–antitoxin" system as the last line of defense of the bacterial population against total extinction [4]. The mass death of aquatic microorganisms caused by the appearance of viruses in the water may have the same significance: a "scorched earth" tactic used to block the attack of the enemy (here, deadly infection).

Many Gram-negative bacteria have a special enzyme, lysine oxidase, that oxidizes lysine using molecular oxygen, which is reduced not to water (as is the case in the vast majority of other oxidases), but to hydrogen peroxide (H_2O_2) [27]. The latter is poisonous, as it is a precursor of reactive oxygen species (ROS). It is H_2O_2 that kills bacteria in so-called "biofilms" formed by a myriad of agglomerated bacterial cells. Hollow tube-like cavities are formed inside the biofilms, in order to provide nutrients for those bacteria located deep inside them. The same paths are used for the removal of the final products of bacterial metabolism. In the case of Gram-positive bacteria, the same function is performed by pyruvate (rather than lysine) oxidase, which also forms H_2O_2 (see [27]). (On the role of ROS in the programmed death of eukaryotes, see Section 13.12.)

In bacteria, a special signal system has been described that is actuated in response to DNA damage: (i) stimulation of DNA repair; (ii) if repair is not sufficient, the blockade of reproduction; and (iii) for an even greater degree of damage, active lysis of the bacterial wall, leading to cell death (phenoptosis) [4,25]. Such a ruthless principle prevents the loss of genetic heritage due to errors that have crept into the biological text. As K. Lewis writes, "it is quite possible that the main danger waiting for unicellular organisms is not competition, pathogens, or depletion of nutrients, but their own clone turned into a group of 'hopeless monsters' capable of causing death of the entire population" [25]. To avoid this danger, the bacterium with damaged DNA commits suicide long before the genome is deteriorated to such an extent that protein synthesis completely ceases.

There are some other examples of the suicide of unicellular prokaryotes that are worth mentioning:

- lysis of the maternal cell of *Bacillus subtilis* or *Streptomyces* on sporulation, which is required for the release of the spores;
- release of bacteroids from *Rhizobium* cells;
- lysis of some cells of *S. pneumonia* to release DNA into the medium, so that this DNA can later be absorbed by other cells of the same species;
- lysis of colicin-producing *Escherichia coli* cells, so that these poisonous compounds that kill other strains of the same bacterial species can be released to the medium; and
- three mechanisms of *E. coli* suicide caused by cell infection by different bacteriophages. The first includes the opening of channels in the cell membrane; these channels are permeable to protons, and as a result, the cell loses its ability to capture energy released during respiration. The second mechanism is based on the activation of a protease cleaving one of the proteins required for ribosome functioning. The third mechanism includes activation of an RNase, which hydrolyzes one of the transfer RNAs. In the latter two cases, cell death results from the termination of protein synthesis [4,25].

13.3 Phenoptosis of Plants: Discovery of the Genes of Death of a Multicellular Organism

The rapid aging of semelparous (reproducing only once) plants is perhaps the most frequently described example of phenoptosis. According to the prominent plant biochemist and gerontologist L.D. Nooden and his coauthors, "even before the appearance of scientific data on the biochemistry of aging [of such plants], it was seen as an internally programmed process, specific and organized in terms of when, where, and how it proceeds. Although this aging develops with age, it cannot be attributed to the processes of passive aging, since it is controlled by internal and external signals and can be slowed down or speeded up by them. On the contrary, aging as a time-dependent passive accumulation of damages is best illustrated by a gradual decrease in seed viability during storage" [28] (see also [29]).

It is well known that the rapid aging and death of soybean plants can be prevented by depodding [30] or deseeding [31]. A demonstrative experiment was carried out by Nooden and Murray [32], who removed all but a single pod cluster. The plant did not die at the usual time (i.e., during month 3 of life), but remained green, with the exception of one area: the leaves closest to the pod cluster turned yellow. The yellowing occurred even if the petiole contained a zone treated with a jet of steam, killing the phloem, which should prevent the transport of compounds from leaf to pods via the phloem (a living tissue) but not from pods to leaf via xylem (in adult plants, an already dead tissue). This proves that pods induce aging by producing a death signal (or a poison), which kills leaves [6,33]. This observation is in obvious contrast with a statement by Kirkwood and Melov on the impossibility of the existence of an aging program: "There is no evidence that semelparous (capable of only single reproduction) organisms are *actively* destroyed once reproduction is complete" [34].

There are some indications that the phytohormone abscisic acid (from Latin *ab*, meaning "away," and *scindere*, meaning "to cut") plays an important role in triggering the process of leaf aging [35]. Nanomolar concentrations of this organic acid regulate the expression of enzymes that form or destroy ROS in plant leaves [36]. Plant aging is known to be accompanied by a sharp increase in ROS levels [37,38].

Aging of soybeans plants is a rather fast process, taking about 10 days (the maximal life span of this plant is 90 days). Timely removal of ripening pods greatly increases their life span. A similar phenomenon has been described for other semelparous plants, including *Arabidopsis thaliana* (rockcress), the classical model species for plant physiologists and geneticists (for review, see [39]). It is with this plant that S. Melzer and coworkers [39,40], Belgian biologists from Ghent, directly refuted another thesis from the already mentioned article by Kirkwood and Melov: "Yet among the many gene mutations that affect lifespan, often increasing it significantly, none has yet been found that abolishes aging altogether" [34]. Melzer et al. [39,40] have reported that an *A. thaliana* plant with mutations in two (out of 22 000) genes, *soc1* and *full*, switches from sexual to vegetative reproduction. The mutant blooms very late, with a significantly reduced number of flowers and seeds, and at the same time it completely loses the rapid aging caused by these seeds. The life span of mutants increases from 90 days to at least 18 months. The plant acquires cambium, secondary growth, woody stems, and multiple rhizomes. It becomes much larger than wild-type, changing from a small grass to a highly branched shrub with large, thick leaves. Inflorescence meristem is mainly transformed to vegetative.

Melzer et al. [39,40] suggested that the species originally emerged as a perennial shrub that vegetatively reproduced with rhizomes and competed with other shrubs and trees, before later emerging into a grass (as happened with horsetails and ferns). This transition was accompanied by the appearance of sexual reproduction, culminating in the formation of numerous very small seeds, easily carried by wind over long distances. Seeds germinate quickly and, once in open ground, form small shoots of grass that can grow without competing with other plants, which have not yet had time to occupy this part of the ground.

Contemporary rockcress is a short-lived organism that is killed by its own seeds. Early death accelerates the change of generations and, consequently, the plant's evolution. Because it is a semelparous organism, every new portion of seeds is formed by *another* individual, which increases the diversity of offspring. This means that the probability of the appearance of new traits is increased. The change from vegetative to sexual reproduction also accelerates evolution, since in the genome of the progeny is always the hybrid of the genomes of two parents. It is not a coincidence that the transition to sexual reproduction in yeasts is a response to the deterioration of environmental conditions, an attempt to find new properties that might help survive in a changed environment. Apparently, the transformation of *Arabidopsis* from a vegetatively propagating plant into a flowering one occurred relatively recently, since the program for its preceding (vegetative) form is still kept in its genome as a backup.[1]

A. thaliana can be considered a precedent, where inactivation of several genes completely prevents rapid aging and death [6]. It is hardly a rare exception in this respect. Melzer et al. [40] state that "in angiosperms, the perennial woody habit is believed to be the ancestral condition, from which annual herbaceous lineages have evolved several times independently [41]. Conversely, evolution from annual herbaceous ancestors to perennial woody taxa has also repeatedly occurred. For example, in various annual herbaceous lineages, such as *Sonchus* and *Echium*, woody perennial species evolved on isolated islands from their continental annual ancestors" [42–44].

Among perennial plants, there are examples of organisms that propagate vegetatively for many years, then switch over to sexual reproduction and die when their seeds mature. Several species of bamboo have fixed life spans, determined by the time of inflorescence. This time is species-specific, varying from 6 years (blooms in the 6th year) to 120 years (blooms in the 120th year). A similar situation has been described for some other perennial plants, such as agave and the Madagascar palm *Ravenala madagascariensis*, which flourish in their 10th and 100th year of life, respectively, and die immediately after their seeds mature. The plant *Puya reimodii*, found in the Andes, lives even longer, blooming in the 150th years. An experiment conducted in agave showed that regular removal of its seeds caused a large increase in its life span [15,45].

In many cases, the death of a parental plant right after seed maturation has an obvious biological sense: it is necessary to free the space for descendants. In the wild, bamboo thickets are so dense that young shoots cannot compete with the parental plants if they do not free the space, making sunlight, moisture, and soil nutrients available [46]. The death of the tropical tree *Tachigalia versicolor* after fruiting plays the same role, providing a gap in the forest canopy for its descendents [47].

13.4 Phenoptosis in Invertebrates: "Horrible Cruelty of Life"

An article by R. Hanus and his Czech colleagues recently appeared in *Science*, titled "Explosive Backpacks in Old Termite Workers" [48]. Termites – social insects – are

---

1 The short life span of *A. thaliana* provides it with another advantage: it appears in open ground and dies so quickly that other plants simply do not have enough time to develop and compete with it. This is why it spends its entire life producing seeds, in relatively comfortable conditions. It is striking that the case of *A. thaliana* literally confirms the idea of the first explorer of aging as a program, August Weismann, that higher organisms "contain the seeds of death" [18].

known to be divided into castes: the queen and her husbands provide reproduction, big soldiers protect the family from enemies, and small workers gnaw wood, which serves as food for the family. Termites eat this wood by cultivating special cellulose-decomposing bacteria in their intestines. The mandibles used by the workers to gnaw the wood eventually become blunt, and the productivity of the small workaholics steadily declines. The Czech entomologists noticed that in parallel with the blunting of the mandibles, two dark-blue spots appear on the border of the thoracic and abdominal areas of the workers, the size of which increases with age. When the spots reach a certain size, the workers change their profession and become actively involved in battles with enemies: other colonies and invertebrates that feed on termites. When the enemy tries to grab the puny worker "across the belly," the worker explodes, spraying out a poisonous fluid. Biochemical analysis of the blue spots showed that they contain a hemocyanin-type phenol oxidase, which catalyzes the formation of some toxic explosive compound from a harmless precursor produced in the salivary glands of the worker. The glands are reliably separated from the sinuses containing the blue enzyme, so that the explosive mixture is formed only when an enemy attacks the worker and damages its tissues. As a result, both worker and enemy die.

People who live alone or in extreme poverty often save money across their life time to pay for their own funeral. Termite workers save the blue enzyme across their life span so as to die as heroes, sacrificing themselves for their termite mound.

There are many other examples of phenoptosis in invertebrates. Cephalopod mollusks – some species of octopus and squid – reproduce only once (like annual plants). The males of some squid species die right after mating, and females after laying eggs. The example of *Octopus hummelincki* is quite illustrative: females cease to eat and die of starvation after the hatching of their youngsters. As with *Arabidopsis*, biologists can prove that this death is not caused by some inevitable aging of the females, but by a real biological suicide program: a young mother does not lose the ability to eat, live, and repeatedly reproduce if her so-called optical glands are removed [49].

The female of one species of praying mantis decapitates the male at the end of sexual intercourse. For many years, this has been cited as an example of the most sophisticated cruelty in the world of insects, but it turns out that the ejaculation by the male becomes possible only after he is decapitated. The male spiders *Argiope aurantia* and *Latrodectus hasselti* are also killed by the female during mating [50].

The case of mayflies is also striking. Nature has not provided them with a functional mouth. These insects spend most of their lives in the larval state, which has no problems with eating. Then, the larva turns into an adult mayfly: an imago. The imago inherits a small amount of nutrients from the larva, which allows it to fly for a couple of days. If an imago is lucky, it meets a partner of the opposite sex, and the female of the pair lays hundreds of thousands of tiny eggs. After that, the biological function of both mayflies is completed and they die of starvation. This is a typical example of phenoptosis.

The life cycle of cicadas also seems interesting. This is how it is described in *Suicide Genes* by J. Mitteldorf and D. Sagan [15]: Cicada larvae live in soil for years (some species, for up to *17 years*) then turn into an imago, mate with a sexual partner, and die (the male shortly after swarming, the female some time after laying her eggs). Death is caused by acute phenoptosis due to sudden dehydration of the insect's body. Such *modus vivendi* has a deep biological sense, developed over thousands of years of evolution. If cicadas survived after mating and reproduced repeatedly, they would destroy the plants that serve them as food, and at the same time facilitate reproduction of birds, which love to feast on these large invertebrates. Both events are prevented when cicada swarming happens only once in 17 years.

For the latest review on phenoptosis in invertebrates, see [50].

13.5 Phenoptosis in Vertebrates

13.5.1 Phenoptosis of Semelparous Vertebrates

Vertebrates also provide examples of acute phenoptosis following reproduction. Males of the Australian mouse *Antechinus stuartii* and 15 other species of three genera of small Australian marsupials die a couple of weeks after rut, being affected by the same pheromone they originally use to attract females. This pheromone is perceived by the male vomeronasal organ. Prolonged exposure to somehow blocks the controlling functions of the hippocampus with respect to the hypothalamus, leading to severe stress due to increased production of the corticosteroids, adrenaline, and noradrenaline and thus violation of salt metabolism, which results in acute renal failure. Castration of males or keeping them separate from females increases their life span – it then reaches that of females [51]. It is interesting that both Australian mice and yeast (see Section 13.2) use pheromones as a phenoptosis-inducing agent, although the biochemical mechanisms of this tool are completely different in the two cases.

In the case of a small South American marsupial, *Gracilinanus microtarsus*, males die right after mating and females after the end of lactation. The same phenomenon has been described for some other species of American and Australian marsupial, including opossums (A. Vercesi, pers. comm.). Semelparous species can be found among amphibians, reptiles, and fish [15]. All these examples can be regarded as an evolution-accelerating mechanism that increases the diversity of offspring. The diversity will be greater if an organism can become a parent only once (we have already mentioned this in relation to semelparous plants; see Section 13.3).

The fast phenoptoses of annual plants and mammals can be compared with progeria (accelerated aging) of the Pacific salmon. Before spawning, this beautiful fish turns into a ridiculous humpbacked creature with a mouth unsuitable for the eating of food. The humpback salmon dies soon after spawning. This event is reminiscent of an accelerated movie of the entire aging program, occurring across a couple of weeks. It starts with a sharp decrease in immunity and finishes with severe osteoporosis of bones, sarcopenia of skeletal muscles, thinning of skin, and cancerogenesis [52]. Zoologists have long believed that the accelerated aging of salmon results from the hard work they perform when swimming against the current of a river to its headwaters, in some cases traveling thousands of kilometers. This hypothesis collapsed when it was discovered that the same changes take place even when the ocean and spawning site are connected by a strait only several hundred meters long [52]. The change of fish habitat from seawater to freshwater apparently serves as a signal triggering the progeria program. S. Austad believes that salmon progeria and slow aging of higher vertebrates are completely different in nature: accelerated aging is a program, while slow aging results from occasional accumulation of errors [53]. However, this explanation contradicts the previously noted similarity between many signs of the two types of aging.[2]

13.5.2 Acute Phenoptosis in Repeatedly Breeding Vertebrates (Septic Shock, Cancer, Diabetes)

Try asking your friends, "What do you know about lemmings?" The most common answer will be, "Nothing!" Or you might hear an interesting story about mass suicide in the animal kingdom. Lemmings are small rodents, relatives of hamsters, that live in the tundra and forest tundra of Eurasia and North America. Like all other rodents, they are good at two things: eating and reproducing. When their numbers reach a critical level, they eat everything edible around them and begin to starve, and some groups migrate away from their former habitat. In the early 20th century, a bold hypothesis was introduced in a children's encyclopedia: lemming numbers dramatically reduce because the migrating groups do not really search for new habitats, but simply jump off cliffs into the sea. This rather crazy idea would have probably quietly vanished at the next reprinting of the encyclopedia, but the dramatic image of mass suicide of cute hamsters was taken by the Walt Disney Studio, captured on film, and shown in movie theatres to millions of children in the late 1950s. The fact that the shooting was staged, that the lemmings were brought to the shooting site by plane, and that they were dropped from a small cliff by humans remained a secret. This resulted in one the most widespread myths about animals: lemmings became famous for being prone to mass suicide.

However, there are examples of real suicide of repeatedly breeding animals. One of the best known cases concerns much larger and cleverer mammals: whales. Most people have heard news reports of the stranding of groups of whales on beaches. The reasons for such mass suicide remain unclear, but it is hard to doubt that it is indeed suicide – whales can perfectly well distinguish between the shore and the sea depths. However, this sad phenomenon is very rare; it rather belongs to the subject of "Oddities of Nature," together with rains of frogs. Therefore, it is difficult to use it to draw general conclusions about "death genes" pushing living beings to commit suicide.

It is well known that many animals risk their lives when protecting their offspring. This behavioral model – saving children even at the cost of the parent's life – does not seem strange or unnatural to us. Many more examples of such altruism can easily be found, and not only among animals. For example, a favorite object of scientists, the bacterium *Escherichia coli*, sometimes destroys its cell in response to bacteriophage infection well before viral multiplication would cause its death. In this way, phage-infected bacteria prevent the spread of viral infection in the population.

13.5.2.1 Septic Shock

Something similar is also observed in higher animals, including humans. There is a phenomenon called sepsis, where blood infection with a large number of bacteria causes a severe pathology of the entire organism, fraught with death (septic shock). This lethal syndrome is caused by the so-called "endotoxin" of Gram-negative bacteria. Endotoxin is

---

[2] T. Maldonado et al. [54–56] discovered in the humpback salmon brain peptides of amyloid plaques that develop in aging people with Alzheimer's disease (on the similarities between basic principles of progeria and "normal" aging in humans, see [57]).

the lipopolysaccharide (LPS) of the outer layer of the cell wall of such bacteria. Animals do not have compounds of this type, so its appearance in the bloodstream means bacterial infection of the blood. LPS is recognized by the blood protein that binds it. The complex of this protein with LPS is in turn recognized by a special receptor protein of endothelial cells of the blood vessels. The ternary LPS–blood protein–receptor complex activates a chain of events that leads to inflammation and apoptosis [58–60]. At first, this looks like a fight among macro- and microorganisms. Strikingly, however, the role of the microbe in this fight is absolutely passive: the bacterium does not excrete LPS into the medium. Instead, LPS is released only as a result of collapse of the cell wall of the dying bacterium due to the effect of certain factors of the macroorganism. (This is the fundamental difference between LPS and exotoxins, the bacterial chemical weapons ejected by live bacterial cells fighting with the macroorganism.[3]) In fact, the only fault of the bacterium is that it is recognized. Even more strikingly, increasing LPS concentration in the blood causes many-fold stimulation of the processes of inflammation and apoptosis, which, when not being restrained by the macroorganism, cause quick death of the patient.

It is hard to resist the analogy with *E. coli*, which commits suicide when infected by bacteriophage. If the number of bacteria in the bloodstream is relatively low, then the LPS-triggered cascade of events leads to the elimination of infection (this is achieved primarily through the activation of the inflammation system) and recovery. But when the organism cannot cope with the infection, the same LPS becomes its killer: the factor causing phenoptosis [60]. The standard explanation of septic shock (an assumption that the organism can no longer control the reactions that it initiated) is unconvincing. For example, the organism has at its disposal the system of decomposition of many its constituents by mitochondrial oxidative enzymes, which, if uncontrolled, could kill the organism in several minutes – but no physician has seen such a death. It seems more logical to assume that in this case, higher beings follow the principle already invented by prokaryotes: when infected, they try to prevent multiplication of the pathogen, and if they fail, they eliminate themselves. As a result, a program that is deadly harmful for the individual becomes useful for the family, community, or population.

13.5.2.2 Cancer

It is possible that death from cancer is also programmed, since it can lead to the elimination of individuals who have accumulated too many errors in their DNA. It will be beneficial for the population, because otherwise these errors might lead to the appearance in the progeny of the monsters Lewis wrote about in relation to bacteria (see Section 13.2) [25]. The idea of cancer being a mechanism by which the population attempts to eliminate individuals with increased mutational load was formulated by S.S. Sommer in 1994 [63] and developed by V.N. Manskikh [64,65] and A.V. Lichtenshtein [66].

Another possible phenoptotic function of cancer is to shorten the life span, which is, after Weismann, the main evolvability-increasing effect of aging [18]. Certainly, most types of cancer are age-dependent. However, aging as slow phenoptosis seems like a more sophisticated mechanism by which to increase evolvability (see Section 13.6.1). This does not exclude the situation where acute phenoptosis also contributes, just in Weismann's sense, to stimulation of evolution. Moreover, it seems probable that there

---

3 In this fight, the microbe is interested only in restraining the animal's protective systems, and not in the animal's death [61,62].

are some other lethal diseases (first of all, diabetes) that also shorten life span and, hence, accelerate evolution, which operate in parallel to the aging program and cancer.

13.5.2.3 Diabetes

Recently, progress has been made in understanding the molecular mechanism of at least one type of diabetes. In 2014, Riera et al. from the University of California, Berkley, published an important observation in *The Cell* [67]. They showed that knockout of the gene of one of the pain receptors, namely transient receptor potential cation channel subfamily V member 1 (TRPV1), increases the life span of mice. TRPV1 is expressed in afferent sensory neurons of the dorsal root ganglion (DRG) that detect very high temperature and pain stimuli in the dermal and epidermal skin layers, mucous membranes of the mouth and nose, joints, the brain (where it affects hippocampal synaptic plasticity), and a number of other tissues. These neurons transmit signals to the central nervous system (CNS) via the spinal cord. The pancreas also has DRG neurons; their demyelinated C-fibers (expressing TRPV1) densely entwine β-cells of this gland. They excrete two types of neuropeptide: peptide P (stimulates inflammation) and calcitonin gene-related peptide (CGRP) (inhibits insulin production). The level of CGRP peptide has been shown to increase dramatically with age [67,68]. The mechanism of this effect remains unclear, but it is surely directly connected to the development of type 2 diabetes by the blocking of insulin secretion from β-cells [67]. According to Westerterp-Plantenga et al. [69], a diet rich in a substance obtained from hot pepper, capsaicin (an agonist of the TRPV1 pain receptor and an inhibitor of respiratory chain complex I – it is similar in the latter respect to rotenone), causes death of DRG neurons by overstimulation and reduces the number of diabetes cases.

In vivo administration of CGRP8-37 peptide for 6 weeks arrested the entire regulatory chain (CGRP8-37 is a shortened version of CGRP, which acts as an antagonist of CGRP receptor on β-cells). This compound increased the life span of mice and, like TRPV1 gene knockout, slowed the development of a number of senile signs in animals (in particular, it reduced O_2 consumption *in vivo*, caused the disappearance of circadian rhythm in changes of respiratory coefficient, inhibited the expression of the insulin gene, weakened of spatial memory, and impaired coordination of movements) [67].

It may be that age-related deterioration of the state of joints causes pain, which triggers the TRPV1 receptors present therein. Another possibility is that the "master biological clock" that counts our age gives a signal to stop the production of some juvenile hormone that acts as an antagonist of TRPV1 receptor or, conversely, stimulates the production of an aging hormone, an agonist of this receptor. TRPV1 control of the hippocampus might participate in these events. The hippocampus and its calcium homeostasis are known to be very sensitive to aging [70].

Most likely, somewhere on the path leading from the hypothetical "clock" controlling ontogenesis and aging to the receptor, ROS serve as a necessary intermediate. This is true for the geroprotective effect of cold-receptor TRPA-1 acting in the direction opposite to the progeric effect of TRPV1. As we shall note in Section 13.13.3, TRPA-1 activation stimulates a FOXO-type transcription factor, including the transcription of intra- and extracellular antioxidant enzymes and antiapoptotic factors.

As for some events triggered by lack of insulin, they are reversed by the mitochondrial antioxidant plastoquinonyl decyltriphenylphosphonium (SkQ1) (see Section 13.12) [71]. This effect is quite expected, given the amount of data on persistent oxidative stress in diabetes [72].

The more we learn about diabetes, the more it resembles the programs of aging, cancer, and septic shock (i.e., death organized by the organism itself). In particular, the overeating characteristic of advanced age should stimulate both aging and diabetes. The program of septic shock, with its strong increase in the level of fatty acids in the blood [73,74], could use the receptor of these acids to kill the organism via the same cascade as aging, but in its strongly accelerated mode (see Section 13.13.1).

To conclude this section, we would like to emphasize that canceling of aging in the naked mole rat is accompanied by the disappearance of the CGRP peptide (and, together with it, diabetes as one of the causes of death) and desensitization of the organism to capsaicin (see Section 13.7.2) [75]. It is also remarkable that TRPV1 is involved in inflammation. Aging results in changes in TRPV1's activity from anti- to proinflammatory, as shown by Romanovsky et al. [76].

13.6 Aging as Slow (Chronic) Phenoptosis

13.6.1 The Fable of the Fox and the Hares

In the case of annual plants, octopods, Pacific salmon, marsupial mice, and opossums, their rapid death shortly after reproduction is obviously a result of a genetic program. In the case of aging of humans or rats, this is not as obvious. However, there are facts indicating that death can be programmed in these slowly aging organisms, and if such a program does exist, it can be broken and aging will be slowed.

Aging of higher organisms can be defined as *slow and concerted attenuation of all the vital functions with age, eventually resulting in the death of the organism.* We assume that programmed aging was invented by biological evolution to stimulate this evolution.

Imagine two animal species: the first is the usual, aging type, while the second belongs to the category of lucky eternally young ones. Let us look first at aging animals. Young individuals make up the majority of the population. Aging leads to the situation where, starting from a certain age, older individuals lose rivalry with younger ones, and their chances for reproduction decrease. In addition, the chances of successfully escaping dangers, finding food, and resisting diseases and parasites also decrease. In this situation, the older an animal is, the smaller its chances of meeting the next dawn. The old ones do not survive in the wild. As a result, aging leads to the acceleration of change in generations, and hence to higher rates of evolution and population plasticity under rapidly changing conditions.

But let us look more closely at those "iron elders," who still – despite their old age and decrease of general vitality – manage to continue living in the ruthless world of fanged enemies and no-less-fanged young congeners. How do they do it? They surely have some very valuable properties that give them an advantage over the majority of other animals of the same species. Translating this into the language of biology, they have some useful genes that provide a marked evolutionary advantage in the given habitat, which allows them to compete successfully for food and reproduction with young individuals. Thus, due to aging, the population is more rapidly enriched with genes valuable under given conditions. This means that in addition to general acceleration of evolution due to removal of old individuals from the cohort of breeding animals, aging can also support the selection of useful new properties.

What happens in the population of non-aging animals? In this case, age is rather an advantage, since the level of vitality remains the same in older individuals, but the animals have greater experience. Moreover, non-aging animals often continue growing for their entire life time, and size also matters. As a result, in non-aging animals, mortality can decrease with age, and breeding success can increase. Here, we have a population in which a small number of old individuals suppress the majority of young congeners, superseding them in mating competition. Obviously, the genetic diversity of such a population will be considerably lower than that of the population of aging animals. Moreover, since an accumulation of errors in the genetic material takes place – slowly, but inevitably – old non-aging "patriarchs" and "matriarchs" will eventually produce offspring of lesser and lesser quality. Taken together, this means that a non-aging population has much smaller chances of surviving an abrupt change in habitat for the worse.

The fact that the majority of non-aging animals live in highly stable and favorable environmental conditions (see Section 13.7) is consistent with the hypothesis on aging as the mechanism of acceleration of evolution and population plasticity.

A hare will always run away from a fox, as it is a matter of life and death for the hare, while for the fox it is but a question of dinner. This should mean that foxes are not involved in the natural selection of hares. This may be true in relation to young, strong hares, but what about older hares that run slower? Let us consider the following thought experiment. Two young hares, a clever one and a not-so-clever one, upon meeting a fox, have almost equal chances of escaping from the enemy, since they run much faster than the fox. However, with age, the clever hare will gain an advantage over the stupid one (and this advantage might prove decisive), since they will both run slower because of aging. The clever hare, as soon as it notices the fox, will immediately start running and will have a much greater chance of surviving than the stupid one, which will be delayed at the start. This means that only the clever hare will proceed to produce leverets. As a result, the hare population will grow wiser [4].

At this point, we are often asked a question: If aging is such a useful thing, then perhaps gerontologists, trying to cancel aging, can harm humanity as a species by slowing our evolution and thus depriving us of the perspective of development? Should research in this direction be prohibited? Of course not.

The fact is that we are no longer an object of natural selection, which provides the survival of individuals most fit to the environmental conditions. If something in our environment does not suit us, we change the environment instead of changing ourselves. If we feel cold, we do not wait for our descendants to grow thick fur. Instead, we learned long ago to put on animal skins and make fire. If the environment tries to destroy us with microbes, we do not select a cohort of people more resistant to infections, but invent antibiotics. Whether for good or bad, evolution of the human species has practically stopped, and its place has been taken by technical progress. Apparently, in the case of humans, the aging program is doomed to disappear and will sooner or later wither away, just as happened with non-aging animals that have no enemies and therefore do not feel the pressure of natural selection (see Section 13.7). This means that our aging is a harmful atavism, and combating it is reasonable and justified [4].[4]

---

4 Note that if aging is essential for acceleration of evolution, it should start rather early so that the animal has time to age significantly while still being of reproductive age. Aging of the immune system in humans starts somewhere between 10 and 15 years of age. Healing of wounds and accommodation of the visual apparatus start to slow around the same time. Sarcopenia (an age-related decrease in muscle fibers in skeletal muscles) develops after the age of 20, and attenuation of visual acuity after 30 years; decrease of lung volume starts at the age of 35, and that of skin elasticity at 45 years [77].

13.6.2 Turquoise Killifish: "Although My Life is Very Short, I Should Find the Time for Aging"

A remarkable result was obtained in research on aging conducted with various species of killifish, a small African fish of the *Nothobranchius* genus [78]. It was observed that the life span of different species of this genus varies up to fivefold, depending on their habitat in the wild. For instance, *N. furzeri*, which lives in Zimbabwe in puddles formed during the rainy season, which dry up after this period is over, was shown to have a life span of 3 months, which corresponds to the length of the rainy season in this country. *N. rachovii* and *N. kuhntae*, from Mozambique, where the rainfall is fourfold greater, live for about 9.5 months, while *N. guentheri* from Zanzibar, a humid climate with two rainy seasons, lives for over 16 months. In the case of the shortest-lived species, *N. furzeri*, growth and sexual maturation occur very rapidly: within the first month of life. During the next 2 months, females repeatedly spawn until the puddle dries up. The following year, their progeny will hatch from the eggs that have survived the drought. It seems quite remarkable that over the last 2 months of life, the fish manages to age, manifesting by the end of their life a set of typical signs of senescence (reduced mobility, loss of a searching behavior in open space, gibbosity and other manifestations of osteoporosis, accumulation of lipofuscin granules in the liver, a sharp increase of β-galactosidase activity in skin fibroblasts, etc.) [78]. In the case of the longer-living species of the same genus, completion of growth and sexual maturation occur much later. Accordingly, biomarkers of aging also appear later. Notably, the described differences remain even in cases of aquarium breeding, which means that they are determined genetically and are no longer dependent on the current conditions of the habitat [78]. It looks like the short-lived species seek to age within the short life span given to them. It is also important that the aging program has similar manifestations even in such distant vertebrate classes as fish and mammals. This conservatism can be regarded as clear evidence that this program was invented by evolution very early.

13.7 Non-Aging Living Beings

13.7.1 Organisms with Negligible Senescence

The most important conclusion of the newest studies on gerontology is that *aging is not a mandatory attribute of life*. Every living being, from insects to elephants, develops from a single cell, builds its amazing complex body, grows and matures, and finally, having reached sexual maturity, starts reproduction. As Williams noted [79], compared to all the stages of ontogenesis, maintaining the already formed body and preventing it from deterioration should be a much simpler task. Most dangers have been already left behind; the risky periods of infancy, childhood, and adolescence are in the past; the organism is at the peak of its physical state – it seems nothing should prevent it from living and reproducing! And for some animals and plants, nothing does.

Meet the bristlecone pine, *Pinus longaeva*, the longevity record holder on our planet. Five thousand years is rather an old age, is not it? And still there are no signs of senescence, senile decay, or other attributes of old age! Moreover, millennial trees are as active and successful in reproduction as their young centenarian congeners.

You might say that trees are not so interesting, since they differ too much from humans. So, what about the centenarians in the animal kingdom? There are many examples of those as well. The bivalve marine mollusk *Arctica islandica*, inhabiting the cold waters of the North Atlantic and the Arctic Ocean, lives for over 500 years. And here is the gigantic Californian turtle, which reaches a size of half a meter: the oldest individuals live to 100 years, and again, it does not seem that such an old age has any negative effect on their physical state or sexual abilities. Some representatives of rockfish (*Sebastes aleutianus*) live for over 200 years, yet one cannot say that they are in any way inferior to their centenary youngsters, which could be their great-great-grandchildren.

And what about mammals: are there non-aging lucky ones among them? Surely so! Vitality of the bowhead whale apparently does not decrease with age. That is why it is not surprising that this animal is also a record holder in longevity among mammals: it can live for over 200 years [53]. Here, it should be noted that we, people, hold the second place among mammals, with the record of 122.5 years. However, it is possible that this results from a lack of reliable information about many other species of long-lived mammals.

The bowhead whale is not the only non-aging mammal. Another example is found on land: the small hairless rodent of unsightly appearance known as the naked mole rat (*Heterocephalus glaber*). Naked mole rats live in savannahs and semi-savannahs of Ethiopia, Kenya, and Somalia, where they dig long systems of underground passages in soil somehow as hard as concrete. These labyrinths are inhabited by families of up to 300 animals, ruthlessly and brutally ruled by a single queen. The queen bullies all the other females and does not allow them to breed, while herself enjoying the love and attention of one to three husbands and continuously breeding. The rest of the males and females in the family play the role of workers, soldiers, foragers, and nannies. The soldiers are brave: if a dangerous enemy like a snake gets into the burrow, the soldier gives the signal to others, and the tunnel behind them is immured, leaving them alone in deadly battle.

A very interesting fact concerning the naked mole rat is that such age-linked diseases as cancer, diabetes, and cardiovascular, neurological, and inflectional pathologies do not affect this animal. But most remarkably, its mortality is age-independent [80]. Unlike us, naked mole rats do not become weaker with age, do not get covered with wrinkles (not counting those that have appeared in their youth), do not lose teeth, keep their enthusiasm and zeal in matters of procreation, and generally do not suffer a decrease in vitality.

Then why do these mysterious animals die? When in captivity, with no danger of hunger, snakes, or aggressive neighbors, both the queen and her subordinates can live more than 33 years – and it is already clear that this record can be outdone, as laboratory observations of these rodents are not yet finished.

You may say that 33 years is not the best example of longevity. However, it is all a matter of comparison. Mice are in many ways similar to the naked mole rat: they are almost identical in size. But mice live for about 3 years, and by the end of this time their fur turns gray, they start losing hair, and they demonstrate all the other signs of aging. The naked mole rat lives at least 10 times longer, and even after 30 years is still cheerful and happy! Importantly, *the probability of its death does not depend on its age* [80]. We humans may live longer, but the probability of our death increases greatly with age.

Under laboratory conditions, naked mole rats (both the royal family and its sub-ordinates) die very rarely, and death happens for unknown reasons (it is believed that it may be caused by clashes with congeners, which sometimes accidentally have fatal endings [81]). Unlike among such social insects as bees and ants, the naked mole rat queen and her husbands initially have no morphological differences from their subordinate animals. After the death of the queen, she can be replaced by any formerly subordinate female. A striking and obvious difference between naked mole rats and mice is that in the former, reproduction is monopolized by the queen and her husbands, who are well protected against external enemies and hunger by a large team of subordinates. Essentially, the queen and the husbands are exempted from the pressure of natural selection. The absence of enemies is a common feature of other species of non-aging animals: sea urchins are protected with poisonous spines; toads with skin glands that produce strong poisons; large crabs and turtles, as well as pearl mussels and oysters, with their firm shells; huge predators like sharks and large birds with sharp teeth or beaks and powerful claws; and giant whales with their strong herds. Finally, there is the naked mole rat – a small, eusocial animal that has created numerous social groups with a few privileged individuals entrusted with reproduction and, therefore, involvement in evolutionary processes.

The naked mole rat is so interesting because it it is far more convenient for various studies than, for example, the bowhead whale. While bats are apparently also non-aging mammals, the naked mole rat has the further advantage that it is related to aging mice and rats, the well-studied classical research objects of experimental biologists. Bats have no such aging relatives for comparison.

When stating that non-aging animals have no enemies in the wild, we obviously do not mean that it would be enough to protect an animal against enemies for it immediately to become non-aging. Laboratory animals have no problems with predators, but they still age. The probability of death among laboratory mice steadily increases with age – in contrast to naked mole rats under the same laboratory conditions. Apparently, conversion of an animal species from aging to non-aging requires a lot of time, just as with any event mediated by natural selection.[5] Our friend, the Moscow gerontologist Dr. A. Khalyavkin, once mentioned in discussion that naked mole rats, whales, and bats do age, but that it happens *differently* than in other mammals. It is difficult to argue with that. We believe that "differently" means "much slower and without the involvement of a special biological genetic program of aging."

Programmed aging can be easily distinguished from other phenomena of body wear-out on the basis of two required attributes. First, this program causes a *gradual* weakening of *many* key body functions, while a sudden loss of a single vital function is sufficient for dying. This means that biological aging is not a mere suicide, as are cases of acute phenoptosis. Second, it is a *coordinated* weakening of various functions; otherwise, an organism would quickly slide into acute phenoptosis. Biological aging is the realization of the program of slow phenoptosis. Senile phenoptosis should not be confused with age-dependent chemical changes that are not programmed in the genome. For example, L-to-D isomerization of amino acids in the proteins of a whale's crystalline lens is a very slow spontaneous process leading to the situation where about 40% of the L-aspartate in the lens of a 200-year-old whale turns into the D-isomer, which

---

[5] A very interesting study was conducted in the late 20th century by S. Austad [53], who compared the aging rates in two populations of opossums (*Didelphis virginiana*), one living on the continent and one on an island in the barrier reef in the state of Georgia. The main predators that exterminate opossums on the continent disappeared from the islands 4–5 millennia ago, leading to a decrease in the pressure of natural selection on the opossum islanders, which in turn led to the reliable slowing of female aging. In particular, the effect of age on the properties of collagen from the animal tail tendons was shown to develop twice as slowly in the islanders when compared to continental animals.

should have an adverse effect on the properties of this protein, causing lenticular opacity [82]. Perhaps the very fact that crystallins in the whale's lens are not replaced during the life time of this animal represents – due to their isomerization – a mechanism of self-elimination of the oldest individuals in the whale herd. A similar situation is apparently characteristic of elephants, another long-lived mammal. Here, just as in whales, the very large size of the organism minimizes the possibility of natural selection, and there is a special mechanism of self-elimination of the oldest individuals. This mechanism may be a limited number of changes of the teeth: Elephants consume large number of leaves each day, so their teeth wear out rather fast. The teeth are replaced six times during an elephant's life, and the animal may die due to starvation when the last tooth set is worn out.

13.7.2 The Aging Program is Not Initiated in Naked Mole Rats Because of Neoteny (Prolongation of the Neonatal Stage of Ontogenesis)

"Mole rat" is rather misleading. Obviously, it would be better to call this mouse-sized rodent a "mole mouse." *H. glaber* was discovered by the great German naturalist Eduard Rüppell in 1845, so its first description was given in Latin and German [83]. When *H. glaber* was described in English, an opinion was discussed among zoologists that it might be a mammalian example of a **neotenic** animal. Neoteny represents prolongation of the larval state, as in salamanders (*Ambystoma*), where it is called the "axolotl." Axolotls are competent in sexual reproduction, as well as in regeneration of almost all their organs. There are also certain termite castes that are held in larval stage as future replacements of the queen [84]. The naked mole rat (body weight, 39 g [85]) differs from other members of the mole rat (*Bathyergidae*) family in that it is much smaller and has no fur, like any newborn rodent.

In 2009, Nathaniel et al. [86] and Larson and Park [87] described high resistance of hippocampal cells from adult naked mole rats against anoxia and reoxygenation. The group of J. Larson and T.J. Park studied this effect in detail [87–90], describing the molecular mechanism of this resistance, which seems to be related to a high level of one of the subunits of the N-methyl-D-aspartate (NMDA) (glutamate) receptor, namely GluN2D, in the adult state of the naked mole rat. In the mouse, the GluN2D level is high only in newborn animals, and strongly lowers with age. The GluN2D-mediated increase in membrane permeability for cations is more cation-specific, being restricted to only two monovalent cations (Na^+ and K^+); that of GluN2A, B, and C also includes Ca^{++}. Anoxia results in exhaustion of adenosine triphosphate (ATP) and long-term depolarization of a neuron after its excitation, since Na^+/K^+ gradients are not regenerated due to ATP deficiency. Without GluN2D, the situation is even worse, since gradients of not only Na^+ and K^+, but also of Ca^{++}, are not maintained, and intracellular [Ca^{++}] strongly increases. Eventually, the neuron is killed by apoptosis initiated by Ca++-induced stimulation of the production of mitochondrial reactive oxygen species (mROS), causing opening of pores in the inner mitochondrial membrane (IMM) [91,92].

The phenomenon of high brain resistance to hypoxia and oxidative stress is inherent in neonatal mammals. In animals others than the naked mole rat, the resistance strongly decreases with age [93,94]. Similar relationships have been revealed for two other neurophysiological characteristics of adult naked mole rat hippocampus that resemble those of the neonatal rat: (i) absence of pair-pulse synaptic facilitation and (ii) low sensibility of synaptic transmission to extracellular adenosine [87]. Larson and

Park [87,88] explained these relationships within the framework of the neoteny hypothesis concerning the origin of the naked mole rat.

Among other traits described in neonatal mammals and adult naked mole rats are:

1) The similarity of distribution of the Ca^{++}-binding protein calbindin in the CA3 region of the hippocampus in naked mole rats and neonatal primates [95].
2) The capsaicin insensitivity of pain receptors and the absence of related substance P in the mole rat [76]. In rats, such sensitivity increases with age [96].
3) The limited ability of naked mole rats to maintain stable body temperature [75,80].
4) The high retention of brown fat level (which is usually reduced during postnatal development of mammals) in adult naked mole rats (Dr. S. Holtze, pers. comm.).
5) The presence in the intercellular spaces of the adult naked mole rat of large quantities of high-molecular-weight hyaluronan, as discovered by V. Gorbunova, A. Seluanov, and coworkers [97]. In other mammals, this strongly decreases with age [98].
6) The complete absence of auricles from adult naked mole rats [83], just as from neonatal rats. In rats, auricles appear when newborns convert to adults. (In adults of other species of the *Bathyergidae* family, the auricles are usually reduced but are still recognizable.)
7) The absence of captivity infections, cardiovascular and neurological age-related diseases, and cancer and diabetes as reasons for death among adult naked mole rats [80] and very young mammals of other species [81]. This is why mortality does not depend upon age in either adult naked mole rats or the very young of other mammals [80].

A recent study by Dr. S. Holtze at the Leibnitz Institute in Berlin and Dr. M.Y. Vyssokikh in our laboratory at the Belozersky Institute (Moscow) on mitochondria isolated from heart tissue from adult naked mole rats resulted in the description of two more properties common between this animal in the adult state and the neonatal mouse. It was found that (i) the rate of respiration of naked mole rat mitochondria in state 4 after exhaustion of added adenosine diphosphate (ADP) is much higher than the rate before ADP addition (in mouse, these two rates are low and similar) and (ii) the concentration of endogenous nucleotides is much lower in mitochondria of the naked mole rat than in mouse. As shown by Aprille and Asimakis [99], a high rate of state 4 respiration after the addition of ADP and a low level of adenine nucleotides are typical for mitochondria from neonatal animals. In mammals other than the naked mole rat, these unusual features quickly disappear during postnatal development.

In Table 13.1, we list various traits of the neonatal period of mammals that do not disappear in adult naked mole rats. Some of them are usually accounted for by their adaptation to the low oxygen concentration in the subterranean labyrinths where they live. However, this suggestion fails to explain why other members of the *Bathyergidae* family living under the same conditions do not show these traits. For example, they have much lower resistance to anoxia and subsequent oxidative stress. Such resistance changes in the series: naked mole rat >> common mole rat > opossum > Damaraland mole rat, blind mole rat, mouse, rat, gerbil, hamster, prairie vole, rabbit, ferret [88].

Our explanation of the naked mole rat paradox is as follows. This rodent is unique as a eusocial mammal. The queen and one to three husbands are isolated from the pressure of natural selection by an army of subordinates that construct an extensive subterranean labyrinth and fight any potential enemies (snakes and other naked mole rat societies). The subordinates do not reproduce progeny. This social organization makes aging

Table 13.1 Neonatal traits retained in adult naked mole rats. ADP, adenosine diphosphate; NMDA, N-methyl-D-aspartate.

N	Trait
1.	Low body weight (much smaller than in 20 other members of the *Bathyergidae* family) [86]
2.	Absence of fur (unique for the family) [84]
3.	Absence of auricles [84]
4.	Limited ability to maintain constant body temperature [81,95]
5.	Static level of brown fat (Dr. S. Holtze, pers. comm.)
6.	No perception of pain caused by capsaicin due to absence of substance P [76]
7.	Very high resistance of the brain to anoxia and oxidative stress during reoxygenation [88,89]
8.	Continual presence of the GluN2D subunit of NMDA (glutamate) receptor [90]
9.	Absence of Ca++ overload under long-term excitation of neurons [89,91]
10.	Neonatal type of distribution of calbindin in the CA3 region of the hippocampus [96]
11.	Permanent presence of high-molecular-weight hyaluronan in intercellular spaces [98]
12.	Low level of adenine nucleotides in mitochondria isolated from heart muscle (M.Y. Vyssokikh and S. Holtze, in preparation)
13.	Higher rate of state 4 respiration of isolated heart mitochondria after exhaustion of added ADP versus before ADP addition (M. Yu. Vyssokikh and S. Holtze, in preparation)
14.	Absence of such age-linked diseases as cancer, cardiovascular pathologies, brain pathologies, diabetes, infection pathologies among reasons for death [82]
15.	Independence of mortality from age [81]

senseless as a mechanism for increasing evolvability by elevating the pressure of natural selection. As a result, the aging program has most probably been lost through a mutation in a mechanism related to the "master clock" responsible for control of ontogenesis. The "clock" is stopped at a stage where sexual reproduction has already become possible but many traits of the adult – including the aging program – have not yet been elaborated. These include final body weight, the formation of fur and auricles, and other properties listed in Table 13.1. In other words, the naked mole rat, like the axolotl, is a product of neoteny [5,8]. However, unlike the axolotl, the naked mole rat has lost the possibility of completing ontogenesis and forming imago when ambient conditions change.

Alex Comfort wrote that the life span of humans should be 700 years, if only they could maintain, throughout their life, a resistance to diseases as high as that of teenagers. It seems to us that naked mole rats selected just this *modus vivendi* [77].

It is suggested that the master clock initiates the aging program at a certain stage of postnatal development by increasing the level of mROS. Originally, such ROS are involved in the removal of malfunctioning cells or of cells that should be eliminated during ontogenesis (stages 1 and 1a and stage 1b, respectively) [100]. However, the ROS level rises with age, eliminating an increasing proportion of the normal cells required for optimal activity of organs and tissues. This is done by means of multiplication of proapoptotic ROS signals at the level of extracellular ROS. As a result, H_2O_2 produced by an apoptotic cell sends to apoptosis other cells surrounding it (stages 2 and 3) [101]. This decreases cellularity, and hence the functional ability of organs and tissues (stage 4).

Progressive weakening of vital functions (stage 5) leads to slow phenoptosis (i.e., aging) (stage 6).

Aging can be controlled by heavy hyaluronan, an extracellular polysaccharide that quenches extracellular ROS [98,102], inhibits apoptosis via receptors on the cell surface [97], and strongly lowers the rate of diffusion of H_2O_2 in the extracellular space by dramatically increasing the viscosity of the extracellular liquid [97]. At first, the master clock induces synthesis of heavy hyaluronan (initially absent in the naked mole rat embryo), then maintains it at very high level throughout the adult period (stage 2a) [97]. In turn, hyaluronan prevents the ROS-linked aging cascade at stages 3 and 4. In adults of animals other than naked mole rats, the master clock stops stimulating heavy hyaluronan biosynthesis (or activates its hydrolysis), allowing the aging cascade to operate.

13.8 Aging-Stimulated Evolvability is a Function Characteristic of Most Living Beings

In 1989, R. Dawkins [103] introduced the term "evolvability" (ability to evolve). Later, G.P. Wagner and L. Altenberg [104], M. Kirschner and J. Gerhart [105], and T.C. Goldsmith [106,107] suggested that such an ability varies in different species in the following way.

In species that conquer new habitats, natural selection is largely directed toward an increase in evolvability, sometimes at the expense of the interests of the individual. We have already described a thought experiment that illustrates how the gradual decline of vital functions with age can accelerate evolution (see "The Fable of the Fox and the Hares," Section 13.6.1). It is important that the young (i.e., more numerous and more intensely breeding) part of the population does not participate in such an experiment, serving as a guarantee of stability of everything already achieved by evolution (stabilizing selection, postulated by I.I. Schmalhausen) [108,109]. At the same time, the aging part of the population can afford some changes in the genotype, initiating the selection of a new property. If this property proves useful, it will be transmitted to – and fixed in – the progeny. However, if the new feature has adverse side effects that are potentially damaging to the species, the property will not pass the sieve of selection. This experiment will hardly have serious negative consequences for the species, since there are not so many old individuals, they do not breed as actively as the young ones, and they will soon die (D.P. Skulachev, unpublished). Thus, it appears that young individuals are mainly responsible for the conservatism of heredity, and aging ones for its variability. The latter point is facilitated by the fact that mutations – the basis of variability in living organisms – accumulate with age. Such an evolutionary strategy reminds one of the features of sexual dimorphism in many bird species, where females are gray, voiceless, and generally not prone to "sticking out" ("conservative"), while males are motley, vociferous, active, and ready for dangerous situations ("revolutionary") [110,111].

Increasing the diversity of individuals in a population may be the second function of aging as a mechanism facilitating the evolutionary process. According to a witty statement by J. Mitteldorf and D. Sagan, you cannot select larger individuals if all of them are of the same height [15]. Aging alters the properties of the organism. Such processes develop with somewhat different rates in the case of different individuals, and they inevitably result in the divergence of age-dependent traits.

Mitteldorf and Sagan suggest one more possible function of aging: the creation of a cohort of weaker organisms, playing the role of a "demographic buffer" under adverse conditions. Old individuals, being weaker and inferior to the young ones in their reproductive capacity, will be the first to suffer from deteriorating conditions (e.g., the appearance of an enemy), providing some protection for the healthy, strong, and rapidly reproducing population nucleus [15].

It is clear that all these functions of aging can be useful only for future generations, as with everything in evolution. For the individuals of a given current generation, aging is, as a rule, detrimental. Following the words of Osip Mandelstam, they suffer "for the thundering valor of ages to come."

The successful evolution of any species cannot proceed without the "co-evolution" of other species in the same ecosystem [15]. Mitteldorf and Sagan discuss the case of the Rocky Mountain locust (*Melanoplus spretus*) as an example where the interests of ecosystem partners are violated [15]. In 1874, approximately half a million square kilometers (comparable to the entire territory of California) were devastated by these insects. Following this locust plague, nothing green was left throughout the giant territory, and the soil was covered with a layer of eggs ready to produce a new generation of voracious insects the following year. However, this next generation had nothing left to eat, and it died out. The authors stress that this death occurred "not because individual locusts were not adapted enough. It happened because these individuals were too aggressive and too prolific" [15].

Examples exist of the moderation and consideration of a species for the sake of other members of the same ecosystem. There are long-lived birds (penguins, alcidae, condors, vultures, eagles, albatrosses) that lay only one egg. If this egg is broken, the bird lays another one, but it never lays two eggs at once [15]. According to the anthropologist A. M. Carr-Saunders [112], primitive human populations, retaining their number for many thousands of years, used several methods to limit their growth, from abortion to infanticide. M.E. Gilpin [113] describes cases where the evolution of predators in an ecosystem occurs in such a way as to protect victims from complete extermination. On this point, Mitteldorf and Sagan wrote, "you cannot create a stable ecosystem of species if individuals seek to grab as much as they can carry off and multiply as fast as they can" [15].

Lions and buffalo provide a remarkable example of predator–victim co-evolution. Four lionesses search for food in the rearguard of the herd. The first predator might be just a few meters away from the last buffalos. The buffalos do not try to gore and trample their enemy, who is obviously planning to follow the herd. This tactic is understandable, given that a lioness will never attack a buffalo calf. She tracks adult animals weakened by old age, infections, or wounds, and purifies the herd from the rearguard individuals, which (i) systematically constrain the moving tempo of the herd and (ii) can be the source of infections.

In concluding Sections 13.7 and 13.8, there is a fairly large group of non-aging animal species, including some mammals, that have no functional aging program. In the species where such a program is operative (including *Homo sapiens*), it serves to accelerate evolution, increasing the pressure of natural selection on individuals gradually weakening with age. In the case of modern humans, aging is a harmful atavism, because people no longer rely on a slow pace of evolution to adapt to their habitat, and prefer instead to adapt their habitat to their needs.

13.9 Why Natural Selection Could Not Eliminate Phenoptoses

There is no doubt that there are several counterproductive phenoptotic programs that, despite the obvious harm they do to organisms, have become a part of their genomes, being selected and then conserved by biological evolution. These programs include not only death right after breeding, but also rapid aging of many semelparous plants. In the latter case, genes have been found that are required for the killing of the plant after seed ripening; one of the poisons involved in such a killing, abscisic acid, has already been identified.

A possible explanation of the selection of counterproductive programs is that they are carried out by **bifunctional proteins**, which possess two functions: one that is harmful for the organism and one that is extremely useful, and whose disappearance would be lethal. For example, cytochrome c (cyt c) is involved in apoptosis and phenoptosis induced by ROS, which we believe to play the key role in animal aging (see Section 13.12). Cyt c fulfills this counterproductive function (involvement in the aging process) when it is released from mitochondria into the cytosol to interact with apoptotic protease activating factor-1 (Apaf-1) protein. It seems that any mutation inactivating the cyt c gene could prevent aging, thereby providing an advantage for the mutant animal in its struggle for survival. However, cyt c has another, "positive" function: electron transfer along the mitochondrial respiratory chain. The disappearance of cyt c would cause collapse of mitochondrial respiration, much like poisoning with cyanide. Interestingly, binding of cyt c with Apaf-1 and its respiratory-chain partners (cyt c_1 or cytochrome oxidase) is carried out via ionic interaction of the "corolla" of lysine cationic groups of cyt c with anionic groups of dicarboxylic acids of Apaf-1 [114], cyt c_1, and cytochrome oxidase [46]. Perhaps it was cyt c that was selected by evolution to participate in aging, because this small (only 104 amino acids), single-domain protein had already been involved in the vital function of respiration, and a mutation of the corresponding gene affecting cyt c's interaction with Apaf-1 would most likely also lead to the disruption of respiratory function. Nevertheless, in our group, we have shown that replacement of one of the lysines from the cationic corolla of cyt c (K72) by tryptophan leads to the production of a protein fully active with respect to its respiratory function, but which forms an inactive complex with Apaf-1. K72W mutant mice (as well as fibroblasts obtained from them) have proved to be quite viable *in vivo* [115,116]. We are now studying the aging of these mutant mice.

An alternative function has also been discovered in Apaf-1 protein. It is involved in a cascade of processes that leads to the arrest of the cell cycle in response to DNA damage. To perform this function, Apaf-1 is transferred from the cytosol to the cell nucleus, where it interacts with checkpoint kinase 1 [117].

Caspases 9 and 3 are the next enzymes in the same apoptotic (phenoptotic) cascade in which cyt c and Apaf-1 are involved. Besides death programs, they have been found to be essential for the differentiation of stem cells into muscle cells, monocytes, and erythroid cells [118]. Abscisic acid, which causes the death of semelparous plants, is a phytohormone that also regulates a number of vital processes in plants, including embryonic development, reproduction, division and elongation of cells, protection against stress, and so on – so let us not be too quick in quantifying this compound solely as a "lethal poison." Incidentally, abscisic acid has been found in brain and some other mammalian tissues, where its function remains completely obscure [119,120].

The obvious harmfulness of the aging program for the individual thus does not mean that it could not have been selected by evolution. An undeniable fact is that counter-productive programs do exist, and aging might well be one of these programs. (Further discussion of this problem can be found in Section 13.13).

13.10 Why Do We Need an Aging Program if an Organism Will Die in Any Case Due to Accumulation of Occasional Damages?

When discussing this problem, one should take into account the following. It is not at all obvious that the systems selected by billions of years of biological evolution must undergo self-destruction within the short time span of an individual's life. Once synthesized, one and the same molecule of crystallin in the whale's eye lens can exist for over 200 years, and the only change detected during this time is spontaneous L-D-isomerization of its amino acids (for aspartate, this proceeds at the maximal rate: 2% every 10 years) [82].

It should be borne in mind that living systems usually control all the phenomena happening within them, and will attempt to completely avoid spontaneous processes, or at least to minimize them. The question of life and death is absolutely crucial for an organism, and therefore evolution could have hardly left it unattended.

The hypothesis of spontaneous stochastic aging as the cause of death contradicts the fact that even among animals belonging to the most intricate organisms, we can find non-aging species: invertebrates (sea urchin, large crab, oyster, pearl mussel), fish (flounder, sturgeon, pike, shark, northern rockfish species), amphibians (toad), reptiles (giant tortoise, crocodile), birds (albatross,[6] guillemot, raven), and mammals, from the largest (whale) to the smallest (bat, naked mole rat). These examples show that non-aging organisms differ from aging ones not in having greater simplicity, but in some other, quite different quality. For example, the non-aging northern rockfish is hardly organized in a simpler way than its aging southern relative, the life span of which is only about 12 years. The non-aging toad is not simpler than the aging frog. The same can be said with regard to crocodiles versus lizards, bats versus shrews, and naked mole rats versus mice. The latter pair is of particular interest, since the two animals – aging and non-aging – are almost the same in size, are genetically rather close (rodents), and both tolerate captivity quite well.

---

6 Lecomte et al. [121] reported that aged albatrosses (life span over 60 years), nesting on the islands in the Indian Ocean, can fly in search of prey into Antarctic waters, but young and middle-aged individuals never reach the Antarctic Circle (based on observations conducted by a satellite and radio sensors fixed on the bodies of these birds). The authors were searching for signs of aging, which have never previously been seen in these animals. Albatrosses grow throughout their life span, before dying suddenly for unknown reasons. Lecomte and coworkers came to the conclusion that the age-related increase in flight distances is the aging sign for albatrosses, which is of course contrary to the very definition of aging as the *weakening* of vital functions with age, rather than their *enhancing*.

13.11 Accumulation of Stochastic Injuries During Aging

Everything we have learned about living cells and their functioning undoubtedly convinces us of the extreme "bureaucratization" of their management. Long hierarchical chains of controllers have been found in cells. When an enzyme (e.g., muscle adenosine triphosphatase (ATPase) actomyosin) performs some useful work, several other enzymes will control its activity. This chain consists of the controller N1, which directly interacts with actomyosin; controller 2, which controls the work of controller N1; controller 3, which controls controller 2; and controller 4, which controls controller 3. As a rule, all of these controllers are protein kinases or phosphatases. Controller 4 is controlled by a hormone: the commander on a supracellular level. The amount of hormone in the blood is, in turn, controlled by the chain of other controllers. This cumbersome system, when acting cohesively, increases the reliability of cell functioning, in particular by reducing the probability of accumulation of errors in DNA and protein structures. If such errors do occur in DNA, they are recognized by a special controller and corrected by reparation systems. As for damaged proteins, they are recognized and labeled by special enzymes controlling the quality of these polymers. Special polypeptides, ubiquitins, are attached to such damaged proteins. Ubiquitinated proteins are recognized by the "molecular mincer," the proteasome: a minute, tube-like intracellular organelle that binds the proteins to be eliminated and splits them into amino acids [122].

High reliability of the system of DNA and protein quality control is provided by the well-known redundancy of its work: ubiquitination removes not only proteins that have lost their function because of some damage, but also proteins with changes in structure that have not yet affected their functioning. Moreover, on the simultaneous appearance of many proteins with modified structures, the cell commits suicide, undergoing apoptosis or necrosis, followed by decomposition of all its proteins, even if most of them have not changed (i.e. remained native).

Age-related *weakening of the quality control* could save many cells that otherwise would have been destroyed. *Accumulation of cells with random errors in DNA and proteins in the tissues of aging organisms* would be a side effect of such a strategy. Gradual accumulation of errors is indeed observed in the course of aging, a fact usually employed as the main argument by those gerontologists who assume nonprogrammed version of aging in their eternal confrontation with the "programmists."[7]

However, we should not forget that *a decrease in quality control* is likely to have been originally programmed in the genome as a final stage of ontogenesis. Thus, we come to the situation where *aging, having started as the result of the relevant program, is gradually turning into the process of accumulation of random (stochastic) damages to*

---

7 T. Nistrom et al. [123] recently showed that drosophila aging is accompanied by (i) a reduction of proteasome activity, the key mechanism of protein quality control, and (ii) an increase in the number of damaged (carbonylated or hydroxynonenal-bound) proteins. A similar effect has also been described in higher animals (mammals) [125,126], in humans *in vivo* [126,127], and in the culture of human cells [128]. The concentrations of the polypeptide ubiquitin [129] and of the enzymes involved in binding of ubiquitin to the "victim" protein [130,131] have been shown to decrease with age in certain animal tissues. In addition, inactive mutant forms of ubiquitin, which prevent "normal" ones from performing their function as protein quality controllers, have been shown to appear in old animals [132]. Interestingly, food restriction (see Section 13.13.1) significantly reduces the effect of age-related weakening of the quality-control system in rats [133].

biopolymers, which remain unnoticed by the weakened systems of quality control of these polymers.[8]

13.12 An Attempt to Identify a Key Component of the Aging Program and to Abolish It

In 1956, D. Harman suggested that aging is a result of a slow poisoning of an organism by its own ROS [143]. In 1972, he characterized this idea, assuming that mitochondria are organelles that generate the aging-inducing ROS [144]. Next, he hypothesized that an age-linked increase in the effect of mROS is programmed [1,4]. Within the framework of this concept, an attempt was undertaken by our group to interrupt the aging program using mitochondria-targeted antioxidants. To this end, the chloroplast electron carrier plastoquinone (PQ) was conjugated with the decyltriphenylphosphonium cation. The product, SkQ1, proved to be a penetrating cation for black phospholipid membranes. It is specifically accumulated in mitochondria (the only intracellular organelle that is *negatively* charged inside); in its reduced (quinol) form, prevents ROS production by mitochondria, being a very active antioxidant; shows properties of a *rechargeable* antioxidant, since its quinol form is regenerated from quinone by the mitochondrial respiratory chain complex III (this occurs in center *i* of this complex); has a very large window between anti- and prooxidant concentrations; and, at nanomolar concentrations, arrests H_2O_2- or light-induced apoptosis or necrosis in cell culture [145–148]. Further experiments showed that a very dilute (nanomolar) solution of SkQ1 used as an addition to the cultivation medium or to drinking water prolongs the life span of a plants, fungi, crustaceans, insects, a fish, and several mammals [145,149–152]. Under certain conditions, the median life span of mice was doubled [1,149]. Preclinical trials of SkQ1 and its rhodamine analog plastoquinonyl decylrhodamine 19 (SkQR1) showed that, in mammals, these compounds are effective in preventing and/or treating a large group of age-related diseases, including heart infarction [145,153,154], kidney infarction [145,151,153], sarcopenia [155], osteoporosis and kyphosis, heart-mass increase [150], dry-eye syndrome [16], cataract [144,151,156], glaucoma [157], retinopathy [145,151,156,157], amyotrophic lateral sclerosis (ALS) [158], age-linked disappearance of the "research reflex" [159,160], disappearance of estrus in females and of sexual attraction in

---

8 With regard to mitochondria, quality control involves the suicide of damaged mitochondria (mitoptosis) [4,134–138]. If practically all of the cellular mitochondria become damaged, mass mitoptosis takes place: mitochondria first gather near the cell nucleus, then are surrounded by a membrane to form a "mitoptotic body," which is extruded from the cell into the intercellular space [139]. The process of mitophagy – a special type of autophagy specific for defected mitochondria that have lost membrane potential – is responsible for the destruction of individual mitochondria [140]. Such mitochondria are captured by special autophagosomes (called mitophagosomes), which then fuse with lysosomes (or transform into these organelles) and completely digest the trapped mitochondria. If the defect appears in filamentous mitochondria, they are first split into small ellipsoidal mitochondria ("thread–grain" transition) [141]. Recently, C. Franceschi et al. [219] published the results of studies comparing the fibroblasts of 75- and 100-year-old humans. They confirmed that mitophagy weakens with age, finding this weakening was much more pronounced in centenarians, primarily due to the inhibition of the "thread–grain" transition. In another study, published in 2014, H. Maes and P. Agostinis [142] reported a sharp suppression of autophagy in the cells of the most aggressive cancer, melanoma. They assumed that cancer cells use this mechanism to increase the number of their mitochondria and, hence, to mobilize energy resources.

males [149,150], depression before death [149], decrease in resistance to infections [149], inhibition of hematopoiesis and myeloid shift of the blood formula [161], involution of thymus and follicles of spleen [162], deceleration of wound healing [16,71,163], diabetes, alopecia, disappearance of vibrissae in rodents [16,149], and canities of black mice [151]. Moreover, it was found that SkQ1 prevents age-induced stimulation of apoptosis [145,149], inhibits *in vivo* oxidative stress developing with age (e.g., it lowers lipid peroxidation and protein carbonylation in skeletal muscles [145]), increases cardiolipin level in old animals [164], prevents elevation of β-galactosidase and of phosphorylation of histone H2AX with age [145], reduces age-related decrease in growth hormone and IGF-1 and increase in oxidized guanosine in blood [151], lowers the blood level of H_2O_2, and retards chromosome aberrations in epitheliocytes [151].

On the molecular level, prevention of cardiolipin peroxidation proved to be one of the key effects of SkQs. It was found that the chain reaction of lipid decomposition in membranes initiated by specific peroxidation of cardiolipin was interrupted by very low concentrations of SkQ1 added to isolated mitochondria [145,146] (for discussion, see [146,151]). This effect cannot be reproduced by dodecyltriphenylphosphonium (C_{12}TPP), an SkQ analog lacking the quinone residue. Another antioxidant mechanism characteristic of both SkQs and C_{12}TPP has also been described. It consists of cycling of free fatty acids (FFAs) in the mitochondrial membrane, since these penetrating cations are carriers of fatty acid anions [147]. Fatty acid cycling, in turn, mediates H^+ conductance through the IMM, resulting in a decrease in membrane potential and strong inhibition of mROS generation by the respiratory chain [165]. This effect requires a higher concentration of SkQ1 than does the prevention of destruction of cardiolipin. Among other effects of SkQ are prevention *in vivo, in situ* β-amyloid-induced inhibition of long-term potentiation in rat hippocampus [159,166,167], and lowering of β-amyloid accumulation in the OXYS rat strain that is used as a model of Alzheimer's disease (AD) [168].

All these data clearly indicate that aging can be retarded in a predictable way; that is, by a decrease in the level of a key component of the aging program, mROS, caused by a mitochondria-targeted antioxidant. Based on these findings, we tried to retard premature aging in "mutator mice," which have a point mutation in the proofreading domain of mitochondrial DNA polymerase [164]. We found that SkQ1 decelerates aging in these progeric ("mutator") animals and retards the development of many age-linked traits. Independently, similar results were obtained by P. Rabinovich and coworkers [169], who addressed catalase to the mitochondrial matrix (normally, this H_2O_2-decomposing enzyme is absent from the matrix). Like SkQ1, mitochondria-targeted catalase prolonged life span not only in "mutator" but also in control mice [170] and retarded the appearance with age of a large group of pathologies [170–173]. All these effects were much smaller or even completely absent if catalase was addressed to the nucleus or peroxisomes [170].

As to SkQ1, it proved to be efficient in the treatment of not only the slow version of phenoptosis (i.e., aging), but also some cases of death representing fast (cancer) or very fast (sepsis) phenoptotic events.

In one such case, we tried SkQ1 as a tool of chemical therapy of lymphoma in $p53^{-/-}$ mice. As shown by P.M. Chumakov et al. [174], the antioxidant N-acetylcysteine (NAC) prolongs by one-third the life span of $p53^{-/-}$ mutants dying due to lymphomas. Dr. B.P. Kopnin confirmed this result, and compared NAC with SkQ1 in this model. It was found that SkQ1 *can effectively replace NAC at a 1.2×10^6 times lower concentration.* Such high efficacy is not surprising if we look at the scavenging of ROS in the IMM. If the concentration of SkQ1 in the extracellular medium is assumed to be 1 and the membrane potential on the outer cell membrane to be 60 mV, then the concentration of [SkQ1] in cytosol will be equal to 10. If the potential on the IMM is 180 mV, then the concentration of [SkQ1] in the matrix will be 10^4. The distribution coefficient of SkQ1 in the octanol/water system is 10^4. Assuming that this is close to that in the membrane/water system, [SkQ1] in the inner leaflet of the inner membrane should be $\mathbf{10^4 \times 10^4 = 10^8}$ [145].

Independently, in our group, it was shown that cell cultures of several sarcomas are specifically killed by low concentrations of SkQ1 that are without any damaging effect on normal (nonmalignant) cells [175]. Thus, at least some types of this disease malignization are mediated by mROS.

It was mentioned in Section 13.5.2 that septic shock can be considered a very fast form of phenoptosis that kills badly infected individuals to prevent the expansion of an epidemic. Apparently, the death signal is generated by receptors in the blood vessels that monitor the level of a bacteria-specific substance (e.g., LPS) and initiate phenoptosis when the concentration of that substance reaches a certain critical value.

It seems to us that a similar phenoptotic mechanism is actuated when a sudden death occurs soon after a severe crisis state. In such a state, organisms cannot guarantee intactness of the heredity information, so their progeny might be monsters. The following experiment was carried out on rats in our laboratory by Dr. D.B. Zorov's group [145]. One kidney was removed, and the other was subjected to 90-minute ischemia. It was found that all the animals survived the day after the treatment, but that 70% of them died during days 2–5. A single intraperitoneal injection of 1 µmol SkQ1 or SkQR1 per kg of body weight a day before ischemia resulted in the survival of 100 and 90% of the rats, respectively.

We showed the effects of SkQR1 and SkQ1 on the development of septic shock induced by injection of *E. coli* into the bladder of rats. This treatment resulted in severe pyelonephritis and then death of 70% of the animals on days 1–4. Five injections of 100 nmol SkQR1/kg following the injection of *E. coli* decreased the occurrence of death to 10% and normalized the level of the antiapoptotic protein Bcl-2 in the kidney, which was strongly lowered in response to the *E. coli* injection. In kidney cell culture, bacterial lysate or LPS induced strong oxidative stress and apoptotic cell death. Both effects were inhibited by the SkQs [176].

13.13 Regulation of Aging by an Organism

13.13.1 Comparison of Effects of Food Restriction and SkQs

The hypothesis that the aging of an organism represents a case of slow phenoptosis predicts that it should be regulated by the organism itself.

Apparently, the first successful attempt to regulate (to retard) the aging program in animals took place in 1934–43, when C.M. MacCay and, independently, T.B. Robertson and coworkers reported an increase in the life span of rats and mice resulting from some food restriction [177–181]. This restriction, introduced at early stages of life, first led to a delay of the animals' growth. Then, when the animals were given the opportunity to eat freely, their size quickly increased to the norm, but they lived about 70% (males) and 50% (females) longer than animals fed *ad libitum* for their entire life. Lung infections and certain tumors dramatically decreased. Long-lived animals looked young and full of energy throughout their life time, regardless of age. Later, the beneficial effect of food restriction on life expectancy was demonstrated on a variety of eukaryotic species, from yeast to rhesus macaques and humans. On the face of it, Harman's hypothesis on the role of ROS in aging [143] could explain these effects: they might be caused by the decrease in the amount of food oxidized by oxygen – an effect that would lower ROS formation and, hence, prevent the accumulation of occasional oxidative damage to biopolymers. However, experiments have shown that the situation is far more complex than this.

The early works of MacCay's group on food restriction, for example, showed that heat production per kilogram of body weight among food-restricted rats was slightly higher than in the control [180]. Recent studies, meanwhile, directly refute the initial speculation, which had already become generally accepted. First, it was found that only 7 – or even 2 – days of food restriction was enough for young *Drosophila* to become long-lived, the effect being undistinguishable from that achieved by lifelong dietary restriction of this fly [182]. Second, it was discovered that an increase not only in the amount of food consumed, but also in just its *smell*, reduced the geroprotective effect of dietary restriction in *Drosophila* and the nematode *C. elegans* [183,184]. These circumstances are not specific for invertebrates. In 1934, T.B. Robertson et al. found that regular 2-day-long fasts of albino mice were sufficient to increase the life span of these animals by 50–60% [181]. C.J. Carr et al. observed that the reproductive capacity of mice disappeared by the end of the first year of life in animals fed without any restrictions, but was maintained at least until the 21st month if the mice were limited in food consumption during the first 11–15 months, and then received food *ad libitum* [185]. According to the Czech author E. Stuchlikova et al. [186], rats, mice, and golden hamsters restricted in food consumption by 50% for 2 years lived 20% longer than control animals, but those restricted during only the first or only the second year of life lived 40–60 and 30–40% longer, respectively.

Further research has shown that all three major components of food (carbohydrates, proteins, and lipids) contribute to the effect of dietary restriction. The effect of proteins has been studied in detail. It was found that a single amino acid, methionine, is responsible for this effect [187–191]. Methionine is an essential amino acid, since it cannot be synthesized by mammals and must be taken in through food. The prolongation of life – as well as a reduction in ROS generation by mitochondria and a decrease in oxidative damage of the *mitochondrial* DNA – was found to be simulated by a diet in which proteins were replaced with an isocaloric mixture of amino acids without methionine [189–192]. It is important that food restriction has no effect on oxidation of the *nuclear* DNA [193].

Recent observations by G. Barja and coworkers [189] are quite demonstrative. Rats were kept on a methionine-free diet for 7 weeks. This caused a decrease in ROS generation by liver mitochondria of old animals and reduced or even completely reversed age-induced oxidative damage to the mitochondrial DNA, proteins, and lipids.

We think that an animal perceives food restriction as an alarming signal of coming famine. Even partial starvation is known to cause decreased fertility, which threatens the survival of the entire population. To prevent this situation, it seems reasonable to slow the aging program, thereby extending the duration of the reproductive period (i.e., increasing the total number of offspring) [1,7].[9] In other words, the effect of food

---

9 It seems interesting to mention that food restriction has no geroprotective effect on the bee queen – apparently, the aging program is canceled in this organism even if food is not limited [194].

restriction on life expectancy is only indirectly related to ROS,[10] and represents a purely regulatory effect. This is primarily biology, and not chemistry. It's why such clearly signaling effects as short-term fasting (or, conversely, the smell of food) have such a powerful effect on the parameters of the life cycle. It is not a coincidence that a *temporary* food restriction (fasting) is better than a *permanent* one. A relatively short time is sufficient to generate an alarm signal, while prolonged starvation is harmful for an organism. Overeating is more typical for some humans than for animals, which tend not to eat "for the future" when an excess of food is available. It is quite possible that religious fasting is a way of prolonging life through short periods of dietary restriction. On average, people observing religious rites are known to live longer [199].

The hypothesis that the dietary restriction effect is a kind of signal provides a good explanation for the experiments with methionine. Apparently, the organism determines the amount of available proteins in food (and, first, the essential amino acids required for protein biosynthesis) by monitoring, via a special receptor, the amount of one amino acid: methionine.

Such receptors do exist for glucose; the organism can monitor the amount of carbohydrates in food based on the glucose level. It would be logical to assume that regulation of aging by the fat component of food is organized in a similar way. In this case, the organism apparently measures either the level of some essential fatty acids (there are some among the unsaturated fatty acids) or the total concentration of FFAs in blood [73,74].

It seems quite important that dietary restriction not only extends the average life span, but also prolongs youth, as already noted by C.M. MacCay, the discoverer of this phenomenon [177–180]. In this respect, a recently published study by R. Weindruch and coworkers is very illustrative [200,201]. Experiments on 76 macaques, conducted over 20 years (observations started when the animals were from 7 to 14 years old), showed that a long-term 30% food restriction had the following effects: (i) a sharp decrease in age-related death rate (for animals over 30 years old, the death rate was 20% in the food-restricted group versus 50% in the control group fed *ad libitum*); (ii) the absence of diabetes from the causes of death; (iii) a halving of the death rate from cancer (in macaques, this is primarily intestinal adenocarcinoma); (iv) a decrease in death from cardiovascular diseases; (v) a decrease in osteoporosis; (vi) an arrest of the development of such age-related traits as sarcopenia, decline in brain gray matter, alopecia, canities, and so on. By the age of 30 years, 80% of the surviving control macaques showed some traits of aging, whereas only 20% of the experimental animals showed such traits.

This experiment is still far from completion, and, therefore, we can say nothing about the effect of dietary restriction on the maximum life span of primates. However, some data are available for rodents. They show that the median life span in mice, rats, and hamsters increases much more markedly than the maximal life span. Apparently, the aging program controls the median rather than the maximal life span. With regard to the maximal life span, its increase requires first the canceling of all types of cancer – otherwise the organism with a disabled aging program will ultimately "live up to its cancer." Both food restriction and SkQ1 have a protective effect for only some types of cancer; other cancers are not influenced.

The concept outlined in this section can be summed up by the scheme. It is hypothesized that the organism monitors the dietary intake of proteins, carbohydrates,

---

10 A number of observations indicate that food restriction really reduces oxidative stress [195–198].

and fats via special receptors that measure the blood concentrations of methionine, glucose, and fatty acids, respectively. These receptors, when combined with corresponding metabolites, trigger a chain of events that increases the level of mROS. In turn, mROS activate apoptosis, reducing the cellularity of organs and tissues, which weakens the vital functions of the organism (i.e., causes aging). Food restriction reduces the levels of methionine, glucose, and fatty acids, which results in slowing (or even canceling) of the entire deadly cascade of senile phenoptosis.

Rectangularization of survival curves and decreases in early mortality rates have been observed to be characteristic of both food restriction and SkQ1 treatment, so the median life span increases much more than the maximal life span. Both of these factors affect not so much the duration of life as such, but rather the duration of healthy young life (the health span). Both are effective in living beings occupying quite different systematic positions. Food restriction has a geroprotective effect in yeast, worms, insects, and mammals. As for SkQ, it is active in a mycelial fungus, a flowering plant, a crustacean, an insect, fish, and mammals. The effect of both factors is clearly pleiotropic; that is, they cause a response of quite different physiological systems of the organism. Cardiovascular diseases, osteoporosis, vision disorders, and certain types of cancer recede; graying, loss of hair, and age-related depression do not occur.

Contradictory data concerning the effects of food restriction on sarcopenia and immune responses have been reported. Some authors state that such exposure adversely affects both the muscle system and immunity [195,202]. On the other hand, Weindruch et al. [201] reported a lack of sarcopenia in the food-restricted monkeys, and MacCay et al. [177–179] and Robertson et al. [181] observed resistance to pulmonary diseases in rats and mice subjected to food restriction (compare this to the sharp decrease in mortality from infectious diseases and the inhibition of age-dependent involution of thymus and spleen follicles in SkQ1-treated mice and rats [145,151]). However, the effects of dietary restriction and SkQ1 on wound healing have been shown to be opposite: starvation inhibited and SkQ1 stimulated healing [71]. Food restriction could not retard certain aspects of aging of the rat visual apparatus [203], although SkQ1 was effective in these cases [145]. Dietary restriction decreased body temperature and inhibited animal growth, an effect not observed with SkQ1 [7]. These data rule out such a trivial interpretation as the assumption that SkQ1 decreases an animal's food intake due, say, to the loss of appetite: direct measurements on mice treated with SkQ1 revealed no decrease in food intake [150,164].

The fact that food restriction adversely affects some vitally important parameters is not surprising. It has already been mentioned that animals do not tend to overeat even if they are not restricted in food. Usually, they eat as much as is needed to maintain vital functions. Therefore, long-term food restriction entails some disorders. It is also clear that the longer the starvation period, the more probable such disorders become. This may explain the controversy in data on the effects of food restriction on the life span and general state of an organism: in cases where food restriction was not too severe and did not last too long, positive effects were observed, but when gerontologists overdid the restriction, unfavorable side effects occurred. For example, it is commonly accepted that *long-term* food restriction decreases the frequency of estrous cycles (sometimes leading to their complete disappearance) [204], but, as early as 1949, Carr et al. [185] showed that *temporary* food restriction and its subsequent cancellation preserves the estrous cycle until very old age. Recall that the latter effect can also be observed with SkQ1 [145,150].

Another significant circumstance should be taken into account when considering food restriction as a geroprotector for humans. Actually, if food restriction is a warning signal against starvation, then the organism should not respond to it simply by prolonging its life span to compensate for the decline of birth rates in lean years. Other responses also seem quite probable, and some may be not as attractive as the extension of healthy life. For example, it has been noted that in a state of food restriction, people become irritable and short-tempered. Hungry mice placed in squirrel wheels do not want to leave them, and run from 6 to 8 km overnight (with normal feeding, this distance is always shorter than 1 km) [204]. Obviously, this effect cannot be explained by starvation-induced exhaustion and muscle weakness. More likely, we are dealing here with another response to the starvation signal: extreme anxiety and the attempt to scan as large a territory as possible in search for food. Were this effect characteristic of SkQ1, we would observe an enhanced food intake by SkQ1-receiving animals, but this is not the case [150]. An impression arises that the use of SkQ1 is a "purer" way to retard the aging program – one not overburdened with undesirable side effects.

In any case, a clinical trial on humans of the geroprotective effect of food restriction is badly needed. According to L.M. Redman and E. Ravussin [205], the only scientific result on this effect in people was obtained in 1957 by the Argentine researcher, E.A. Vallejo [206]. The experimental group consisted of 60 people fed every other day with a reduced amount of food, so that the average food intake was lowered by 35%. The control group was 60 people who ate without any restrictions. A.J. Stunkard and M. Rockstein published an analysis of the same data 18 years later [207], describing a downward trend in mortality and a halving of the average number of days spent in hospital among people restricted in food intake [14].

Studies of the geroprotective effect of food restriction clearly show that the aging process is under the control of regulatory systems of the organism. We assume that food restriction is just a signal of coming starvation; if this assumption is true, we can expect that other changes in environmental conditions that threaten an individual's existence will also cancel aging as a counterproductive program. Let us consider the available data from such a point of view.

13.13.2 The Geroprotective Effect of Physical Activity

"The Incredible Flying Nonagenarian." That was the title of an article that appeared in the *New York Times* in 2010 [208] on the subject of Olga Kotelko, who set 23 world records for 90-year-olds in different disciplines of athletics and won 17 gold medals in this age group during the Olympic Games for the elderly in Lahti, Finland. It was not merely the fact that Mrs. Kotelko achieved a victory in so many different sport disciplines that was impressive, but also the way in which she defeated her rivals: she threw a javelin more than 8 m farther than her nearest age-group rival; her time in the 100 meters – 23.95 seconds – was faster than that of some finalists in the 80–84-year category, two brackets down; her long-jump record was 1.7 m.

Olga was the seventh of eleven kids in a family of Ukrainian peasants who emigrated to Canada. As a young girl, she helped her family on the farm every day – fed the chickens, slopped the pigs, milked the cows. After that, she would trudge 2 miles to school. Soon after graduating from high school, she got married and had two daughters. Later, she divorced, and moved with her children from Saskatchewan to British Columbia, where she graduated from university with a bachelor's degree. Then Olga became a schoolteacher. The only sport that she was engaged in before the age of 77 years (first as a schoolgirl, then again after retirement) was amateur softball. When she was 77, a local coach suggested she might enjoy track and field. She started training three days a week, and soon excelled not only in sports suitable for her height (1 m 50 cm), such as long jump and run, but also in heavy lifting in the prone position, shot put, and other disciplines that required significant muscle strength. Having gained the fame of the world champion, Olga Kotelko became an object of study by Dr. Tanja Taivassalo, a muscle physiologist from McGill University (Montreal, Canada).

A muscle biopsy carried out in October 2010, when Kotelko was 91, showed no signs of sarcopenia or mitochondrial damage, which usually accompany such old age. According to Taivassalo, in a muscle sample of a person over the age of 65, one would expect to see at least a couple of fibers with some mitochondrial defects. However, in Kotelko's case, not a single fiber with damaged mitochondria was found. Perhaps, persistent, regular, and lengthy training, started by the future Olympic champion at the age of 77, reversed (i.e., *cured*) the aging of her muscle tissue. Her whole life is the story of a very healthy person. She found herself in hospital only once. Professor M.A. Tarnopolsky of McMaster University (Hamilton, Ontario) sums up the Olga Kotelko phenomenon as an example of the extension of youth due to canceling of the aging program: "So you're healthy, healthy, healthy, and then at some point you kick the bucket" [208]. In fact, Tarnopolsky predicted how Olga would die. This happened in July 2014, at the age of 96. She went by herself to the hospital a day before her death.

Olga Kotelko's case may be a precedent for the switching off or strong deceleration of the aging program at the genetic level. The opposite situation is inherent in progeric diseases well known in both animals and humans. In 2011, the groups of M.A. Tarnopolsky from Canada and T.A. Prolla from the USA published a joint work on the effect of physical exercises (three times a week for 45 minutes on a treadmill moving at a speed of 15 m/min) on progeria of "mutator" mice with a defect in the proofreading domain of mitochondrial DNA-polymerase [209,210]. They began these exercises when the mice were about 3 months old, and observed a doubling of their life span when the exercises were concluded after 5 months. Numerous defects accompanying the accelerated aging were reversed, including early graying of hair, alopecia, fatigue, sarcopenia, decrease in total body weight (in particular, brain weight) and skin thickness, increase in the mass of cardiac muscle and spleen, disappearance of fat depots, downsizing of ovaries and testes, a drop in the levels of hemoglobin, erythrocytes, and leukocytes, changes in the number, shape, and size of mitochondria, strong stimulation of apoptosis in various tissues, reduction of the amount of mitochondrial DNA, increase in the number of mutations in this DNA, reduction of the amounts of complexes I–IV in the respiratory chain, and reduction of factor PGC-1α, which regulates mitochondrial biogenesis. All of these changes are also characteristic of normal aging of mice, but they appear much earlier in "mutator" mice. As already noted, experiments carried out on "mutator" mice in the group of B. Cannon in Stockholm showed that SkQ1 significantly increases the life span and normalizes most of the parameters studied by Tarnopolsky and Prolla. Like muscle work, SkQ1 did not completely remove the mutation effect on the life span of "mutator" mice: they lived longer than mice without SkQ1, but still died earlier than wild-type mice. However, strikingly, they died without such signs of premature aging as humpback, baldness, or depression developing about a week prior to death [164].

The "mutator" mouse is not the only model in which serious muscle work partially or completely normalizes the parameters of an aging animal. According to T.O. Stolen et al.

[211], maximal physical activity lasting 3 months in aged mice reverses many pathological changes in their cardiomyocytes and cures such age-related diseases as diabetes.

There arises a question concerning the validity of the data obtained in mice for humans. M.A. Tarnopolsky and coworkers [210] cite numerous observations of the positive effect of physical activity on the elderly.[11] Perhaps Olga Kotelko, whose story opened this section, remains the most striking example. However, we need to remember that a positive result is attained by *prolonged* and rather *heavy* muscle activity. This can hardly be substituted by jogging. The experiment with the treadmill is a rather difficult one; it requires a lot of energy. According to B.P. Yu, a mouse needs to run at least 3 miles a day on a moving walkway in order for the geroprotective effect of muscle activity to be manifested [213]. In 2014, N.T. Broskey et al. [214] published a report stating that regular physical exercises increased the total volume of mitochondria in the skeletal muscles of 80-year old humans, as well as the value of their maximal oxygen consumption (when compared to people of the same age leading a sedentary lifestyle). Physical activity lasting 3 months in "sedentary" elderly increased mitochondrial volume, maximal ATP level in muscles, and the levels of certain transcription factors regulating mitochondrial biogenesis.

Our explanation of the geroprotective effect of physical activity is as follows. We assume that an organism facing the necessity of performing regular intense muscular work slows the aging program, the way it does when responding to the signal of food deficiency. Like dietary restriction, the purpose of this response is to find a reserve of resources to provide for additional energy-consuming function.

In this and preceding sections, we have considered the geroprotective effect of SkQ1, dietary restriction, and intensive muscle work. We have also mentioned data on the targeted delivery of catalase to mitochondria. An unbiased reader will notice that all four of these cases share a number of features:

1) The listed effects are of pleiotropic character. That is, they affect a large group of parameters related to only one thing: all of them are the traits of aging.
2) They have both preventive and therapeutic effects (except catalase, for which relevant analysis has not been carried out). That is, they not only prevent (or at least slow) aging, but also to some extent reverse (cure) certain senile pathologies, if they have not gone too far.
3) The increase of the average or median life span caused by any of the four cases is always greater than that of its maximal value. In the case of SkQ1, where this effect has been specifically studied, these relationships are explained primarily by the fact that some malignant tumors are resistant to geroprotectors.
4) SkQ1, muscular work, and catalase delivery to mitochondria are effective in slowing progeria in "mutator" mice. With regard to food restriction, unfortunately, this has not yet been studied.

In general, all these observations are perfectly consistent with the hypothesis of programmed aging; they could be predicted within the concept suggested by D. Harman back in 1972 [144], according to which it is mROS that play the key role in the aging of an organism.

---

11 Claudius Galen, the Roman physician who introduced the word "apoptosis" into scientific language, often prescribed to his patients digging the ground and mowing the grass. For the latest review on the slowing of brain aging by physical activity, see [212].

13.13.3 Deteriorations in Living Conditions as Nonspecific Geroprotectors

For our logic, it is important that two such different factors as food restriction and physical exercise cause qualitatively similar geroprotective effects. We think that this similarity is not accidental. J. Mitteldorf and D. Sagan [15] believe that any change in conditions that seriously complicates the existence of an organism has a chance to somewhat prolong its life by slowing the aging program. In this way, an organism tries to compensate for the increased energy costs under deteriorated conditions. An impression arises that any deterioration of external or internal parameters significantly affecting the state of an organism can initiate some inhibition of the aging program. These relationships can explain why low doses of radiation increase life span [15], as well as the phenomenon of "hormesis," where small amounts of some poisons (e.g., chloroform [215,216]) have a positive effect on life expectancy. This may be related to the effects of moderate cooling or heating and weak infections causing some stress [15]. It looks like aging is a *facultative* counterproductive program, which can be switched off or decelerated by an organism when deterioration of conditions brings into question an individual's continued existence.

Recent experiments by R. Xiao et al. [217] on the nematode *C. elegans* are particularly demonstrative. An ambient temperature decrease from 25 to 15 °C was shown to increase the median life span of this invertebrate from 9 to 30 days. Mutation of the gene encoding the protein of the calcium channel *activated by cold* (TRPA-1) significantly reduced the cold effect (the nematode lived for only 18 instead of 30 days at 15 °C, versus an unchanged 9 days at 25 °C). It is known that an increase in Ca^{++} levels in the cells of the worm leads to activation of a cascade of enzymes, namely (i) protein kinase C, (ii) kinase of DAF-16 protein, which belongs to FOXO transcription factors, and (iii) DAF-16 protein. The latter regulates the activity of a large group of genes, including intracellular antioxidant proteins (activated by DAF-16) and apoptosis (inhibited), as well as the gene of extracellular superoxide dismutase-3 (activated). Successive knockout of every gene of the proteins in this cascade led to the disappearance of part of the cold effect, which was caused by the impact of TRPA-1 protein. If, instead of knocking out the TRPA-1 protein gene, its activity was artificially increased, the effect of cold also increased: the worms then lived for 36 days at 15 °C.

The following fact is remarkable. When the knocked-out nematode TRPA-1 protein gene was replaced by the gene of a homologous human protein, the normal cold effect on the nematode life span was observed. This finding indicates the universality (from invertebrates to humans) of the cold-linked mechanism of inhibition of the aging program [14].

13.13.4 Psychological Aspects Affecting Human Life Expectancy

When we consider aging as a program regulated by the organism in response to changing external conditions, a question arises concerning the possibility of a human individual actively fighting his or her aging.

In 2013, the prestigious journal *PNAS* published the article "A Functional Genomic Perspective on Human Well-Being" by B.L. Fredrickson et al. from the University of California, Los Angeles [218]. These authors measured the expression of genes in the blood leukocytes of 84 Americans, dividing them into two groups based on the type of pleasure they were getting out of life. One type was *hedonic*. This was based on the simple satisfaction of one's personal needs. The other type represented a deeper *eudaimonic* pleasure, suggesting the awareness of the importance of one's existence not only for oneself, but also for others, as well as self-improvement and pursuit of some higher goals. Surprisingly, the authors discovered statistically significant differences between the two groups in the levels of gene expression of proteins involved in inflammatory reactions of the immune system. In the hedonic type, pleasure significantly increased the expression of inflammation factors, while in the eudaimonic type, this parameter was decreased.

C. Franceschi et al. [219] considered chronic inflammation as one of the main causes of aging ("inflamm-aging"). The role of this process in the development of cancer is well known. Therefore, it would be logical to suggest that those enjoying hedonic pleasures will have shorter life spans than those enjoying eudaimonic pleasures.

Thus, the aging program is controlled by receptor systems that monitor the key parameters of the external and internal environment. At a time when the survival of the population is at stake, the organism can slow the activity of this counterproductive program, which reduces evolvability but increases the chances of individual survival. It would be reasonable if the aging program were slowed in individuals working for the benefit of the community. In fact, B.W. Penninx et al. reports [220] that among many psychological factors, the two that most clearly correlate with human longevity are (i) awareness of one's usefulness to others and (ii) maintenance of a high level of mastery in old age. As to individuals harmful for the community, they might be eliminated by means of acute phenoptosis. In the 1920s, cases were described among Australian aborigines where the tribe, following the shaman's orders, started singing funeral chants for a person who was alive. This provoked disorders in that person's salt metabolism, and within several days the victim died [221].

Homo sapiens, compared to less advanced types of primates, have two fundamental improvements: a huge volume of cerebral cortex and a very sophisticated vocal apparatus. As a result, evolution has produced a being that can transmit information to its descendants not only genetically, in the form of DNA molecules in the sperm and egg, but also verbally, through speech. In other words, if cave bears live behind a particular mountain, there is no need for humans to wait for thousands of years for the innate fear of a mountain of a given shape located in a given place to get fixed in the genes. Instead, an elder can explain to the young: Do not go over that mountain alone and unarmed, because of cave bears.

This invention fundamentally changed the role of old individuals in the state of menopause. They gained the opportunity to work for the improvement of the quality of the young generation by providing them with information acquired throughout their life and preserved not in their genes but in the memory of their brain. The problem is that not all grandparents can provide information in this way. Some – the "useful elders" – really teach youngsters, but others lack the gift of teaching, are tongue-tied or blunt, or have weak memory – or they may be just nasty misanthropes with whom nobody really wants to communicate. In primeval times, such grandparents would have been harmful, as they occupied someone else's place and ate someone else's food while bringing no benefits to the population. In general, they decreased the viability of the *Homo sapiens* population.

If aging is genetically programmed, then most likely it can be regulated (like other genetic programs). We can suggest a rather bold hypothesis: Over many thousands of years of the existence of primitive humans, there appeared mechanisms capable of accelerating or slowing the operation of the aging program according to different factors of human life. In light of the stated concept of "useful" and "harmful" elders, it seems reasonable to assume that aging of the "useful" ones could be slowed so as to give them an opportunity to teach more youngsters. At the same time, aging of "harmful" elders could be accelerated in the interest of other individuals.

13.13.5 Integration of Phenoptotic Signals: Citrate in the Blood as a Predictor of Death

In 2014, Fischer et al. [222] published the results of an analysis of 106 different biomarkers measured with nuclear magnetic resonance (NMR) spectroscopy in the blood plasma of 17 345 persons from Estonia and Finland. The goal of the study was to find such biomarkers as could report on the risk of death during 5 years' observation. Four such biomarkers were identified: three were proteins, and one was a metabolite of low molecular mass, namely citrate.

In fact, the previously mentioned observation that risk of death is proportional to an increase in the plasma level of citrate is in line with a publication by Langley et al. [223] indicating that a citrate level increase in blood is predictive of death from sepsis, as well as with works by Lemon et al. [224] and Garland et al. [225] concerning hypercitricemia in cancer and diabetes, respectively. Most interesting in the study of Fischer et al. [222] is that citrate elevation is inherent in all three types of death classified by the authors: death due to (i) cardiovascular diseases, (ii) cancer, and (iii) all others reasons. This effect is more demonstrative for Estonians (ages 18–103 years) than for Finns (24–74 years); that is, for the elder cohort, where the percentage of age-dependent death should be larger.

The first scheme was published in 1997 by one of the present authors [226]. It is known that $O_2^{\bullet-}$ interacts with the 4Fe-4S cluster of aconitase, the first enzyme of the Krebs cycle. This event leads to a release of Fe^{++} into the medium, causing inhibition of the enzyme, the reactivation of which requires Fe^{++} to be recombined with inactive aconitase [227,228]. As shown by A.D. Vinogradov et al. [229], the 4Fe-4S cluster is not necessary for aconitase to catalyze oxaloacetate *keto-enol* tautomerization, another activity of this protein. The authors concluded that Fe^{++} is hardly involved in the binding and transforming of anionic substrates of aconitase, as was previously assumed [230]. Taking this observation into account, we suggested that Fe^{++} plays a regulatory (rather than catalytic) role in allowing aconitase to be controlled by superoxide ($O_2^{\bullet-}$). The $O_2^{\bullet-}$-induced inhibition at the very beginning of the Krebs cycle results in the exhaustion of substrates of the main dehydrogenase reactions that supply the respiratory chain with reducing equivalents. This must entail oxidation of all the respiratory-chain electron carriers, including those that are capable of one-electron reduction of O_2 to $O_2^{\bullet-}$. As Gardener and Fridovich [231] and Kim Lewis (pers. comm.) have suggested, this effect might operate as an antioxidant mechanism.

Citrate accumulation seems to be another useful consequence of the inhibition of aconitase. Autooxidable "citrate^{3-}–Fe^{++}" complex immediately reacts with O_2 [232]. As a result, Fe^{++} is oxidized to Fe^{+++}, preventing the production of $OH^{\bullet}$, the most aggressive ROS, which requires Fe^{++} to be formed from H_2O_2 (the Fenton reaction). The Fe^{+++} obtained remains bound to citrate^{3-}, since its binding to citrate^{3-} is much stronger than that of Fe^{++} [226].

Within the framework of such reasoning, the positive correlation of the risk of death and the citrate level in the blood might be a result of the involvement of mROS in death-related processes. Moreover, such a correlation could be predicted, assuming that both fast (septic shock, death after crisis, cancer, diabetes) and slow (aging) phenoptoses are mediated by mROS. If this is the case, death should be accompanied by inhibition of aconitase and accumulation of citrate, both of which are attempts by the organism to survive under conditions of mitochondrial oxidative stress.[12]

13.14 General Scheme of Regulation of mROS-mediated Phenoptoses

The scheme suggests the existence of a central mechanism controlling our life span. The final decision to live further or to die depends on the interplay of numerous signals reaching the center. Some are anti- and some prophenoptotic. In the case where prophenoptotic signals are dominating, a phenoptotic mechanism is switched on, resulting in a burst of mROS (acute phenoptoses like septic shock, cancer, and diabetes) or a small but permanent increase (slow age-linked phenoptosis) [11,12].

The roles played by acute and slow phenoptoses in the increasing evolvability of a species seem to be different. Such an acute phenoptosis as septic shock stops an epidemic, defending those evolutionary inventions that have already been made. As to cancer and diabetes, they shorten life span, thereby increasing the frequency of changes of generations, and so the probability of new evolutionary inventions occurring: a function originally imposed by Weismann as a duty to slow aging processes [18]. In our opinion, aging as slow phenoptosis elevates evolvability by increasing the pressure of natural selection on individuals weakening with age (see "The Fable of the Fox and the Hares," Section 13.6.1) [14].

The ability of an organism to temporarily cancel the aging program under conditions critical for its survival may explain why such a counterproductive property as aging has not been eliminated by natural selection. This counterproductiveness is minimized by the very possibility of the partial or even complete cancellation of this process for a certain period of time in response to a signal of upcoming disaster. The organism allows itself to age under conditions not very far from optimal, when the altruistic costs of the increase in evolvability fit into the well-known maxim, "taking a little bit from a lot is not stealing, but simply sharing." However, when the external conditions strongly deteriorate, cancellation of the aging program becomes one of the easily mobilized reserves that can help an individual to win its Battle of Waterloo.

Strikingly, this kind of logic suggests that in a dangerous situation, the organism with a switchable counterproductive program will gain advantages over the organism without one (K.V. Skulachev, unpublished).

---

12 This scheme deals with low citrate concentrations comparable with the concentrations of iron ions. As to high [citrate], it may have some prophenoptotic effects. It has been shown that high [citrate] inhibits phosphofructokinase [233]. This entails inhibition of glycolysis, the only ATP source when the Krebs cycle is not operative due to inactivation of aconitase by superoxide. Moreover, it was shown by B. Zhivotovsky and coworkers [234] that, in cancer cell cultures, 20 mM citrate activates caspases 8 and 2 and induces apoptosis. As found by V. Infantino et al. [235], the citrate/malate antiporter of the inner mitochondrial membrane is induced by LPS (apparently via NFκB), an effect that somehow stimulates ROS production and inflammation under conditions of septic shock.

13.15 Perspectives for the Pharmacological Switching Off of Counterproductive Phenoptotic Programs in Humans: From *Homo sapiens* to *Homo sapiens liberatus*

The genome is the only self-reproducing biological structure whose preservation, evolution, and expansion have taken priority over the well-being of an organism or a group of organisms. In fact, the organism is nothing but a machine serving the genome's interests [46,236,237]. Modern humans do not rely upon the very slow rate of our evolution, so many mechanisms increasing *Homo sapiens'* evolvability, including the aging program, are clearly counterproductive for us. Our attempt to cancel such programs would be a "rise of the machines" against genome tyranny. Our success would symbolize the conversion of humans into *Homo sapiens liberatus*, which would be the highest achievement of medicine in this century [46,237]. SkQs seem to be a promising tool in such an attempt.

13.15.1 Clinical Trials of SkQ1

In Section 13.12, we summarized our studies of various effects of SkQs on animals, which can be regarded as preclinical trials of these new compounds. As to clinical trials of SkQs, they are already in progress. We decided to start using drops of SkQ1 solution as a possible medicine to treat age-related eye diseases (dry-eye syndrome, cataract, and glaucoma). The choice in favor of eye diseases was made, first, due to the probability that any adverse side effects would be much lower for eye drops than for *per os* treatment. The effect of Visomitin eye drops containing 250 nM SkQ1 is compared with that of "Natural Tears" (Alcon), one of the most popular medicines for treating dry-eye syndrome, which (like other such drugs) shows only a symptomatic influence on this disease, considered an incurable pathology by physicians. Visomitin treatment for 3 weeks resulted in the disappearance of all pathological symptoms in 60% of patients, while Natural Tears was effective in only 20%. Among the symptoms in question was tear production. This study was carried out at the Helmholtz Research Institute of Eye Diseases in Moscow by E.V. Yani and coworkers [46,238].

Clinical trials were also performed in eight eye clinics in Russia and two in Ukraine. Drops of Visomitin were used to treat patients suffering from dry-eye syndrome. The duration of treatment was 6 months. The positive effect of the drops was clearly seen, while placebo was ineffective.

In the USA, the ORA Company has already successfully carried out phase 1 and 2 clinical trials of Visomitin on 90 dry-eye-syndrome patients (2014). Phase 3 for Visomitin, as well as phases 1 and 2 for a *per os* introduced SkQ1-containing drug, Plastomitin, are now in preparation.

13.15.2 SkQ-Sensitive and SkQ-Resistant Phenoptoses

It is clear that some kinds of phenoptosis do not require mROS and, hence, are SkQ-resistant. For example, mammary carcinoma and some other malignant transformations are not affected by SkQ1. However, SkQ1 retards the development of lymphomas in $p53^{-/-}$ mice and kills sarcoma cells. Certainly, phenoptoses of the behavioral types can hardly be linked to mROS. However, in each case, special investigation should be carried out.

13.15.3 Can Aging Be Accelerated by Small Doses of Antioxidants?

The whole logic of our review forces us to agree that, in principle, such a danger exists. Indeed, ROS are toxic. However, small doses of poison might slow aging (hormesis, see Section 13.13.3). M. Ristow even introduced the special term "mitohormesis" in relation to poisons formed in mitochondria (e.g., mROS) [239,240]. Recent studies by the Canadian researcher S. Hekimi et al. [241] have shown that inhibition of respiratory-chain complexes I or III causes an increase in the amount of ROS in tissues of the nematode *C. elegans*. Strikingly, in this case, an ROS increase led to some increase in the animal's life span. The addition of rather small (0.1 mM) doses of the powerful prooxidant paraquat gave the same results. Even more surprisingly, it was found that Apaf-1 and caspase 9 were required for such a prolongation of life. Somewhat earlier, H. Khalil et al. [242], in Switzerland, observed a similar effect on mice. They used doxorubicin as a prooxidant; this compound caused the death of 50% of rodents about 50 days after the beginning of intoxication. By this time, *all* the caspase 3-lacking, doxorubicin-treated mutants had died. The researchers discovered the mechanism of the saving effect of caspase 3. They showed that caspase 3 cleaves peptide N from the protein p120 RasGAP. Peptide N, in turn, activates the antiapoptotic protein kinase Akt. Remarkably, the affinity of caspase 3 for p120 RasGAP is so high that even small amounts of the active caspase are sufficient for the appearance of peptide N and activation of Akt. Increased level of active caspase 3 lead to a situation where this caspase cleaves not only p120, but also peptide N. As a result, two shorter peptides, N1 and N2, are formed. These new peptides cannot activate Akt. The affinity of caspase 3 for peptide N is much lower than for p120. Thus, qualitatively the same effect (a ROS increase) first protects cells from death but then – upon a further increase in ROS level – provokes it. It is noteworthy that the mechanisms of cell death are different in wild-type versus caspase 3-deficient mutant mice. Wild-type cells die due to apoptosis, while mutants die due to necrosis.

It is possible that the disastrous consequences of the removal of superoxide from inhaled air can be explained by the cancellation of the hormesis effect in the vomeronasal organ or in lungs following this type of mechanism. With regard to SkQ1, we did not observe any acceleration of aging by low concentrations of this mitochondria-targeted antioxidant. This may be because the mROS level in an aging organism is already above the thresholds at which the effect of mROS-mediated mitohormesis can still be realized [14].

Acknowledgments

This work was supported by Russian Science Foundation 14-50-00029.

References

1. Skulachev VP. What is "phenoptosis" and how to fight it? *Biochemistry (Mosc)* 2012;**77**(7):689–706.
2. Skulachev VP. Aging is a specific biological function rather than the result of a disorder in complex living systems: biochemical evidence in support of Weismann's hypothesis. *Biochemistry (Mosc)* 1997;**62**(11):1191–5.

3. Skulachev VP. Phenoptosis: programmed death of an organism. *Biochemistry (Mosc)* 1999;**64**(12):1418–26.
4. Skulachev VP. Aging and the programmed death phenomena. In: Nystrom T, Osiewacz HD (eds.). *Model Systems in Aging*. Berlin: Springer-Verlag. 2003.
5. Longo VD, Mitteldorf J, Skulachev VP. Programmed and altruistic ageing. *Nat Rev Genet* 2005;**6**(11):866–72.
6. Skulachev VP. Aging as a particular case of phenoptosis, the programmed death of an organism (a response to Kirkwood and Melov "On the programmed/non-programmed nature of ageing within the life history"). *Aging (Albany NY)* 2011;**3**(11):1120–3.
7. Skulachev VP. SkQ1 treatment and food restriction – two ways to retard an aging program of organisms. *Aging (Albany NY)* 2011;**3**(11):1045–50.
8. Libertini G. Classification of phenoptotic phenomena. *Biochemistry (Mosc)* 2012;**77**(7):707–15.
9. Libertini G. Phenoptosis, another specialized neologism, or the mark of a widespread revolution? *Biochemistry (Mosc)* 2012;**77**(7):795–8.
10. Gavrilova NS, Gavrilov LA, Severin FF, Skulachev VP. Testing predictions of the programmed and stochastic theories of aging: comparison of variation in age at death, menopause, and sexual maturation. *Biochemistry (Mosc)* 2012;**77**(7):754–60.
11. Libertini G. The programmed aging paradigm: how we get old. *Biochemistry (Mosc)* 2014;**79**(10):1004–16.
12. Skulachev MV, Severin FF, Skulachev VP. Receptor regulation of senile phenoptosis. *Biochemistry (Mosc)* 2014;**79**(10):994–1003.
13. Skulachev MV, Skulachev VP. New data on programmed aging – slow phenoptosis. *Biochemistry (Mosc)* 2014;**79**(10):977–93.
14. Skulachev MV, Severin FF, Skulachev VP. Aging as an evolvability-increasing program which can be switched off by organism to mobilize additional resources for survival. *Curr Aging Sci* 2015;**8**(1):95–109.
15. Mitteldorf J, Sagan D. *Cracking the Aging Code: The New Science of Growing Old, And What It Means for Staying Young*. Basingstoke: Macmillan. 2016.
16. Skulachev VP, Skulachev MV, Feniouk BA. *Life Without Aging*. Moscow: EKSMO. 2013.
17. Schopenhauer A. *The World as Will and Presentation*. Leipzig: Brauhaus. 1818.
18. Weismann A. *Essays upon Heredity and Kindred Biological Problems*, 2nd edn. Oxford: Clarendon. 1889.
19. Darwin C. *The Descent of Man*. London: John Murray. 1871.
20. Kerr JF, Wyllie AH, Currie AR. Apoptosis: a basic biological phenomenon with wide-ranging implications in tissue kinetics. *Br J Cancer* 1972;**26**(4):239–57.
21. Pozniakovsky AI, Knorre DA, Markova OV, Hyman AA, Skulachev VP, Severin FF. Role of mitochondria in the pheromone- and amiodarone-induced programmed death of yeast. *J Cell Biol* 2005;**168**(2):257–69.
22. Severin FF, Hyman AA. Pheromone induces programmed cell death in S. cerevisiae. *Curr Biol* 2002;**12**(7):R233–5.
23. Skulachev VP. Programmed death in yeast as adaptation? *FEBS Lett* 2002;**528**(1–3):23–6.
24. Engelberg-Kulka H, Amitai S, Kolodkin-Gal I, Hazan R. Bacterial programmed cell death and multicellular behavior in bacteria. *PLoS Genet* 2006;**2**(10):e135.
25. Lewis K. Programmed death in bacteria. *Microbiol Mol Biol Rev* 2000;**64**(3):503–14.

26 Yamaguchi Y, Park JH, Inouye M. Toxin-antitoxin systems in bacteria and archaea. *Annu Rev Genet* 2011;**45**:61–79.
27 Mai-Prochnow A, Lucas-Elio P, Egan S, Thomas T, Webb JS, Sanchez-Amat A, et al. Hydrogen peroxide linked to lysine oxidase activity facilitates biofilm differentiation and dispersal in several gram-negative bacteria. *J Bacteriol* 2008;**190**(15):5493–501.
28 Guiamet JJ, John I, Pichersky E, Nooden LD. Expression of a soybean thiol protease during leaf senescence and nitrogen starvation. *Plant Physiol* 1997;**114**(3):1220.
29 Shahri W. Senescence: concepts and synonyms. *Asian J Plant Sci* 2011;**10**(1):24–8.
30 Leopold AC, Niedergangkamien E, Janick J. Experimental modification of plant senescence. *Plant Physiol* 1959;**34**(5):570–3.
31 Lindoo SJ, Nooden LD. Studies on behavior of senescence signal in anoka soybeans. *Plant Physiol* 1977;**59**(6):1136–40.
32 Nooden LD, Murray BJ. Transmission of the monocarpic senescence signal via the xylem in soybean. *Plant Physiol* 1982;**69**(4):754–6.
33 Neumann PM, Tucker AT, Nooden LD. Characterization of leaf senescence and pod development in soybean explants. *Plant Physiol* 1983;**72**(1):182–5.
34 Kirkwood TB, Melov S. On the programmed/non-programmed nature of ageing within the life history. *Curr Biol* 2011;**21**(18):R701–7.
35 Cutler SR, Rodriguez PL, Finkelstein RR, Abrams SR. Abscisic acid: emergence of a core signaling network. *Annu Rev Plant Biol* 2010;**61**:651–79.
36 Cho D, Shin DJ, Jeon BW, Kwak JM. ROS-mediated ABA signaling. *J Plant Biol* 2009;**52**(2):102–13.
37 Munne-Bosch S, Lalueza P. Age-related changes in oxidative stress markers and abscisic acid levels in a drought-tolerant shrub, Cistus clusii grown under Mediterranean field conditions. *Planta* 2007;**225**(4):1039–49.
38 Munne-Bosch S, Alegre L. Plant aging increases oxidative stress in chloroplasts. *Planta* 2002;**214**(4):608–15.
39 Lens F, Smets E, Melzer S. Stem anatomy supports *Arabidopsis thaliana* as a model for insular woodiness. *New Phytol* 2012;**193**(1):12–17.
40 Melzer S, Lens F, Gennen J, Vanneste S, Rohde A, Beeckman T. Flowering-time genes modulate meristem determinacy and growth form in *Arabidopsis thaliana*. *Nat Genet* 2008;**40**(12):1489–92.
41 Carlquist SJ. *Island Biology*. New York: Columbia University Press. 1974.
42 Groover AT. What genes make a tree a tree? *Trends in Plant Sci* 2005;**10**(5):210–14.
43 Kim SC, Crawford DJ, Francisco Ortega J, Santos Guerra A. A common origin for woody Sonchus and five related genera in the Macaronesian islands: molecular evidence for extensive radiation. *Proc Natl Acad Sci USA* 1996;**93**(15):7743–8.
44 Böhle UR, Hilger HH, Martin WF. Island colonization and evolution of the insular woody habit in *Echium L.* (Boraginaceae). *Proc Natl Acad Sci USA* 1996;**93**(21):11740–5.
45 Libbert A. *Plant Physiology*. Moscow: Mir. 1976.
46 Skulachev VP, Bogachev AV, Kasparinsky FO. *Principles of Bioenergetics*. Berlin: Springer. 2013.
47 Pepper JW, Shelton DE, Rashidi A, Durand PM. Are internal, death-promoting mechanisms ever adaptive? *J Phylogen Evolution Biol* 2013;**1**(3):1000113.
48 Šobotník J, Bourguignon T, Hanus R, Demianová Z, Pytelková J, Mareš M, et al. Explosive backpacks in old termite workers. *Science* 2012;**337**(6093):436.

49 Wodinsky J. Hormonal inhibition of feeding and death in octopus – control by optic gland secretion. *Science* 1977;**198**(4320):948–51.
50 Kartsev VM. Phenoptosis in arthropods and immortality of social insects. *Biochemistry (Mosc)* 2014;**79**(10):1032–48.
51 Bradley AJ, Mcdonald IR, Lee AK. Stress and mortality in a small marsupial (*Antechinus-Stuartii*, Macleay). *Gen Comp Endocr* 1980;**40**(2):188–200.
52 Skulachev VP. Ageing as atavistic program which can be cancelled. *Vestnik RAN* 2005;**75**:831–43.
53 Austad SN. *Why We Age: What Science is Discovering about the Body's Journey through Life.* New York: John Willey and Sons. 1997.
54 Maldonado TA, Jones RE, Norris DO. Distribution of beta-amyloid and amyloid precursor protein in the brain of spawning (senescent) salmon: a natural, brain-aging model. *Brain Res* 2000;**858**(2):237–51.
55 Maldonado TA, Jones RE, Norris DO. Intraneuronal amyloid precursor protein (APP) and appearance of extracellular beta-amyloid peptide (A beta) in the brain of aging kokanee salmon. *J Neurobiol* 2002;**53**(1):11–20.
56 Maldonado TA, Jones RE, Norris DO. Timing of neurodegeneration and beta-amyloid (A beta) peptide deposition in the brain of aging kokanee salmon. *J Neurobiol* 2002;**53**(1):21–35.
57 Kipling D, Davis T, Ostler EL, Faragher RGA. What can progeroid syndrome tell us about human aging? *Science* 2004;**305**(5689):1426–31.
58 Klosterhalfen B, Bhardwaj RS. Septic shock. *Gen Pharmacol* 1998;**31**(1):25–32.
59 Fenton MJ, Golenbock DT. LPS-binding proteins and receptors. *J Leukoc Biol* 1998;**64**(1):25–32.
60 Skulachev VP. Programmed death phenomena: from organelle to organism. *Ann NY Acad Sci* 2002;**959**:214–37.
61 Kirchner JW, Roy BA. The evolutionary advantages of dying young: epidemiological implications of longevity in metapopulations. *Am Nat* 1999;**154**(2):140–59.
62 Kirchner JW, Roy BA. Evolutionary implications of host-pathogen specificity: fitness consequences of pathogen virulence traits. *Evol Ecol Res* 2002;**4**(1):27–48.
63 Sommer SS. Does cancer kill the individual and save the species? *Hum Mutat* 1994;**3**(2):166–9.
64 Manskikh VN. *A Study on Evolution Oncology.* Tomsk: SibSMU. 2004.
65 Manskikh VN. Hypotesis on defence of long-lived species from tumors by phagocytosis of aberrant cells. *Uspekhi Gerontol* 2008;**21**(1):27–33.
66 Lichtenstein AV. Cancer as a programmed death of an organism. *Biochemistry (Mosc)* 2005;**70**(9):1055–64.
67 Riera CE, Huising MO, Follett P, Leblanc M, Halloran J, Van Andel R, et al. TRPV1 pain receptors regulate longevity and metabolism by neuropeptide signaling. *Cell* 2014;**157**(5):1023–36.
68 Melnyk A, Himms-Hagen J. Resistance to aging-associated obesity in capsaicin-desensitized rats one year after treatment. *Obes Res* 1995;**3**(4):337–44.
69 Westerterp-Plantenga MS, Smeets A, Lejeune MP. Sensory and gastrointestinal satiety effects of capsaicin on food intake. *Int J Obes (Lond)* 2005;**29**(6):682–8.
70 Moreno H, Burghardt NS, Vela-Duarte D, Masciotti J, Hua F, Fenton AA, et al. The absence of the calcium-buffering protein calbindin is associated with faster age-related decline in hippocampal metabolism. *Hippocampus* 2012;**22**(5):1107–20.

71 Demyanenko IN, Popova EN, Zakharova VV, Ilyinskaya OP, Vasilieva TV, Romashchenko VP, et al. Mitochondria-targeted antioxidant SkQ1 improves impaired wound healing in old and diabetic mice. *J Investigative Dermatol* 2015;**7**(7):475–85.
72 Rains JL, Jain SK. Oxidative stress, insulin signaling, and diabetes. *Free Radic Biol Med* 2011;**50**(5):567–75.
73 Maitra U, Chang S, Singh N, Li L. Molecular mechanism underlying the suppression of lipid oxidation during endotoxemia. *Mol Immunol* 2009;**47**(2–3):420–5.
74 Schaffler A, Scholmerich J. Innate immunity and adipose tissue biology. *Trends Immunol* 2010;**31**(6):228–35.
75 Park TJ, Comer C, Carol A, Lu Y, Hong HS, Rice FL. Somatosensory organization and behavior in naked mole-rats: II. Peripheral structures, innervation, and selective lack of neuropeptides associated with thermoregulation and pain. *J Comp Neurol* 2003;**465**(1):104–20.
76 Wanner SP, Garami A, Pakai E, Oliveira DL, Gavva NR, Coimbra CC, et al. Aging reverses the role of the transient receptor potential vanilloid-1 channel in systemic inflammation from anti-inflammatory to proinflammatory. *Cell Cycle* 2012;**11**(2):343–9.
77 Comfort A. *The Biology of Senescence*, 3rd edn. New York: Elsevier. 1979.
78 Terzibasi E, Valenzano DR, Cellerino A. The short-lived fish *Nothobranchius furzeri* as a new model system for aging studies. *Exp Gerontol* 2007;**42**(1–2):81–9.
79 Williams GC. Pleiotropy, natural selection, and the evolution of senescence. *Evolution* 1957;**11**(4):398–411.
80 Buffenstein R. The naked mole-rat? A new long-living model for human aging research. *J Gerontol A Biol Sci Med Sci* 2005;**60**(11):1369–77.
81 Delaney MA, Nagy L, Kinsel MJ, Treuting PM. Spontaneous histologic lesions of the adult naked mole rat (*Heterocephalus glaber*): a retrospective survey of lesions in a zoo population. *Vet Pathol* 2013;**50**(4):607–21.
82 George JC, Bada J, Zeh J, Scott L, Brown SE, O'Hara T, et al. Age and growth estimates of bowhead whales (*Balaena mysticetus*) via aspartic acid racemization. *Canad J Zool* 1999;**77**(4):571–80.
83 Ruppell E. Heterocephalus graber (Ruppell). *Abliantllungren aus dem Gebiete der beschreibenden Naturgeschichte* 1845;**3**:99–101.
84 Stedman TL. *Stedman's Medical Dictionary*, 26th edn. Baltimore, MD: Williams & Wilkins. 1995.
85 Mcnab BK. Influence of body size on the energetics and distribution of fossorial and burrowing mammals. *Ecology* 1979;**60**(5):1010–21.
86 Nathaniel T, Umesiric F, Saras A, Olajuyigbe F. Tolerance to oxygen nutrient deprivation in the hippocampal slides of the naked mole rat. *J Cerebr Blood F Met* 2009;**29**:S451.
87 Larson J, Park TJ. Extreme hypoxia tolerance of naked mole-rat brain. *Neuroreport* 2009;**20**(18):1634–7.
88 Larson J, Drew KL, Folkow LP, Milton SL, Park TJ. No oxygen? No problem! Intrinsic brain tolerance to hypoxia in vertebrates. *J Exp Biol* 2014;**217**(7):1024–39.
89 Peterson BL, Larson J, Buffenstein R, Park TJ, Fall CP. Blunted neuronal calcium response to hypoxia in naked mole-rat hippocampus. *PLoS One* 2012;**7**(2):e31568.
90 Peterson BL, Park TJ, Larson J. Adult naked mole-rat brain retains the NMDA receptor subunit GluN2D associated with hypoxia tolerance in neonatal mammals. *Neurosci Lett* 2012;**506**(2):342–5.

91 Skulachev VP. Why are mitochondria involved in apoptosis? Permeability transition pores and apoptosis as selective mechanisms to eliminate superoxide-producing mitochondria and cell. *FEBS Lett* 1996;**397**(1):7–10.

92 Skulachev VP. Role of uncoupled and non-coupled oxidations in maintenance of safely low levels of oxygen and its one-electron reductants. *Q Rev Biophys* 1996;**29**(2):169–202.

93 Bickler PE, Fahlram CS, Taylor DM. Oxygen sensitivity of NMIDA receptors: relationship to NR2 subunit composition and hypoxia tolerance of neonatal neurons. *Neuroscience* 2003;**118**(1):25–35.

94 Cherubini E, Benari Y, Krnjevic K. Anoxia produces smaller changes in synaptic transmission, membrane-potential, and input resistance in immature rat hippocampus. *J Neurophysiol* 1989;**62**(4):882–95.

95 Amrein I, Becker AS, Engler S, Huang SH, Muller J, Slomianka L, et al. Adult neurogenesis and its anatomical context in the hippocampus of three mole-rat species. *Front Neuroanat* 2014;**8**:39.

96 Porseva VV, Shilkin VV, Korzina MB, Korobkin AA, Masliukov PM. [Changes the TRPV1-immunoreactive neurons of the rat spinal nerve sensory ganglia, induced by capsaicin]. *Morfologia* 2011;**139**(3):41–5.

97 Tian X, Azpurua J, Hine C, Vaidya A, Myakishev-Rempel M, Ablaeva J, et al. High-molecular-mass hyaluronan mediates the cancer resistance of the naked mole rat. *Nature* 2013;**499**(7458):346–9.

98 Holmes MWA, Bayliss MT, Muir H. Hyaluronic-acid in human articular-cartilage – age-related-changes in content and size. *Biochem J* 1988;**250**(2):435–41.

99 Aprille JR, Asimakis GK. Postnatal-development of rat-liver mitochondria – state-3 respiration, adenine-nucleotide translocase activity, and the net accumulation of adenine-nucleotides. *Arch Biochem Biophys* 1980;**201**(2):564–75.

100 Vriz S, Reiter S, Galliot B. Cell death: a program to regenerate. *Curr Top Dev Biol* 2014;**108**:121–51.

101 Pletjushkina OY, Fetisova EK, Lyamzaev KG, Ivanova OY, Domnina LV, Vyssokikh MY, et al. Long-distance apoptotic killing of cells is mediated by hydrogen peroxide in a mitochondrial ROS-dependent fashion. *Cell Death Differ* 2005;**12**(11):1442–4.

102 Greenwald RA, Moy WW. Effect of oxygen-derived free radicals on hyaluronic acid. *Arthritis Rheum* 1980;**23**(4):455–63.

103 Dawkins R. The evolution of evolvability. In: Langton C. *Artificial Life Proceedings 6*. Boston, MA: Addison-Wesley. 1989.

104 Wagner GP, Altenberg L. Perspective: complex adaptations and the evolution of evolvability. *Evolution* 1996;**50**(3):967–76.

105 Kirschner M, Gerhart J. Evolvability. *Proc Natl Acad Sci USA* 1998;**95**(15):8420–7.

106 Goldsmith TC. *The Evolution of Aging*, 2nd edn. Annapolis, MD: Azinet Press. 2014.

107 Goldsmith TC. Solving the programmed/non-programmed aging conundrum. *Curr Aging Sci* 2015;**8**(1):34–40.

108 Schmalhausen, II. *Factors of evolution: the theory of stabilizing selection*. Philadelphia, PA: Blakiston. 1949.

109 Levit GS, Hossfeld U, Olsson L. From the "Modern Synthesis" to cybernetics: Ivan Ivanovich Schmalhausen (1884–1963) and his research program for a synthesis of evolutionary and developmental biology. *J Exp Zool B Mol Dev Evol* 2006;**306**(2):89–106.

110 Zahavi A. Mate selection – a selection for a handicap. *J Theor Biol* 1975;**53**(1):205–14.

111 Geodekian VA. Evolutionary theory of sex. *Priroda* 1991;**8**:60–9.
112 Carr-Saunders AM. *The Population Problem: A Study in Human Evolution*. Oxford: Clarendon. 1922.
113 Gilpin ME. *Group Selection in Predator–Prey Communities*. Princeton, NJ: Princeton University Press. 1975.
114 Yu T, Wang X, Purring-Koch C, Wei Y, McLendon GL. A mutational epitope for cytochrome *c* binding to the apoptosis protease activation factor-1. *J Biol Chem* 2001;**276**(16):13 034–8.
115 Sharonov GV, Feofanov AV, Bocharova OV, Astapova MV, Dedukhova VI, Chernyak BV, et al. Comparative analysis of proapoptotic activity of cytochrome c mutants in living cells. *Apoptosis* 2005;**10**(4):797–808.
116 Mufazalov IA, Pen'kov DN, Cherniak BV, Pletiushkina O, Vysokikh M, Kirpichnikov MP, et al. [Preparation and characterization of mouse embryonic fibroblasts with K72W mutation in somatic cytochrome C gene]. *Mol Biol (Mosk)* 2009;**43**(4):648–56.
117 Zermati Y, Mouhamad S, Stergiou L, Besse B, Galluzzi L, Boehrer S, et al. Nonapoptotic role for Apaf-1 in the DNA damage checkpoint. *Mol Cell* 2007;**28**(4):624–37.
118 Murray TV, McMahon JM, Howley BA, Stanley A, Ritter T, Mohr A, et al. A non-apoptotic role for caspase-9 in muscle differentiation. *J Cell Sci* 2008;**121**(Pt. 22):3786–93.
119 Le Page-Degivry MT, Bidard JN, Rouvier E, Bulard C, Lazdunski M. Presence of abscisic acid, a phytohormone, in the mammalian brain. *Proc Natl Acad Sci USA* 1986;**83**(4):1155–8.
120 Bruzzone S, Basile G, Mannino E, Sturla L, Magnone M, Grozio A, et al. Autocrine abscisic acid mediates the UV-B-induced inflammatory response in human granulocytes and keratinocytes. *J Cell Physiol* 2012;**227**(6):2502–10.
121 Lecomte VJ, Sorci G, Cornet S, Jaeger A, Faivre B, Arnoux E, et al. Patterns of aging in the long-lived wandering albatross. *Proc Natl Acad Sci USA* 2010;**107**(14):6370–5.
122 Ciechanover A. Intracellular protein degradation: from a vague idea through the lysosome and the ubiquitin-proteasome system and onto human diseases and drug targeting. *Neurodegener Dis* 2012;**10**(1–4):7–22.
123 Fredriksson A, Johansson Krogh E, Hernebring M, Pettersson E, Javadi A, Almstedt A, et al. Effects of aging and reproduction on protein quality control in soma and gametes of *Drosophila melanogaster*. *Aging Cell* 2012;**11**(4):634–43.
124 Koga H, Kaushik S, Cuervo AM. Protein homeostasis and aging: the importance of exquisite quality control. *Ageing Res Rev* 2011;**10**(2):205–15.
125 Nystrom T. Role of oxidative carbonylation in protein quality control and senescence. *EMBO J* 2005;**24**(7):1311–17.
126 Friguet B, Bulteau AL, Chondrogianni N, Conconi M, Petropoulos I. Protein degradation by the proteasome and its implications in aging. *Ann NY Acad Sci* 2000;**908**:143–54.
127 Shringarpure R, Davies KJ. Protein turnover by the proteasome in aging and disease. *Free Radic Biol Med* 2002;**32**(11):1084–9.
128 Sitte N, Merker K, Von Zglinicki T, Grune T, Davies KJ. Protein oxidation and degradation during cellular senescence of human BJ fibroblasts: part I – effects of proliferative senescence. *FASEB J* 2000;**14**(15):2495–502.
129 Jahngen JH, Lipman RD, Eisenhauer DA, Jahngen E.G. Jr. Taylor A. Aging and cellular maturation cause changes in ubiquitin-eye lens protein conjugates. *Arch Biochem Biophys* 1990;**276**(1):32–7.

130. Ruotolo R, Grassi F, Percudani R, Rivetti C, Martorana D, Maraini G, et al. Gene expression profiling in human age-related nuclear cataract. *Mol Vis* 2003;**9**:538–48.
131. Hawse JR, Hejtmancik JF, Horwitz J, Kantorow M. Identification and functional clustering of global gene expression differences between age-related cataract and clear human lenses and aged human lenses. *Exp Eye Res* 2004;**79**(6):935–40.
132. Tsirigotis M, Zhang M, Chiu RK, Wouters BG, Gray DA. Sensitivity of mammalian cells expressing mutant ubiquitin to protein-damaging agents. *J Biol Chem* 2001;**276**(49):46 073–8.
133. Cui J, Shi S, Sun X, Cai G, Cui S, Hong Q, et al. Mitochondrial autophagy involving renal injury and aging is modulated by caloric intake in aged rat kidneys. *PLoS One* 2013;**8**(7):e69720.
134. Zorov DB, Kinnally KW, Tedeschi H. Voltage activation of heart inner mitochondrial membrane channels. *J Bioenerg Biomembr* 1992;**24**(1):119–24.
135. Skulachev VP. Fatty acid circuit as a physiological mechanism of uncoupling of oxidative phosphorylation. *FEBS Lett* 1991;**294**(3):158–62.
136. Skulachev VP. Chemiosmotic concept of the membrane bioenergetics: what is already clear and what is still waiting for elucidation? *J Bioenerg Biomembr* 1994;**26**(6):589–98.
137. Skulachev MV, Antonenko YN, Anisimov VN, Chernyak BV, Cherepanov DA, Chistyakov VA, et al. Mitochondrial-targeted plastoquinone derivatives. Effect on senescence and acute age-related pathologies. *Curr Drug Targets* 2011;**12**(6):800–26.
138. Skulachev VP. Uncoupling: new approaches to an old problem of bioenergetics. *Biochim Biophys Acta* 1998;**1363**(2):100–24.
139. Lyamzaev KG, Nepryakhina OK, Saprunova VB, Bakeeva LE, Pletjushkina OY, Chernyak BV, et al. Novel mechanism of elimination of malfunctioning mitochondria (mitoptosis): formation of mitoptotic bodies and extrusion of mitochondrial material from the cell. *Biochim Biophys Acta* 2008;**1777**(7–8):817–25.
140. Kotiadis VN, Duchen MR, Osellame LD. Mitochondrial quality control and communications with the nucleus are important in maintaining mitochondrial function and cell health. *Biochim Biophys Acta* 2014;**1840**(4):1254–65.
141. Skulachev VP, Bakeeva LE, Chernyak BV, Domnina LV, Minin AA, Pletjushkina OY, et al. Thread-grain transition of mitochondrial reticulum as a step of mitoptosis and apoptosis. *Mol Cell Biochem* 2004;**256**(1–2):341–58.
142. Maes H, Agostinis P. Autophagy and mitophagy interplay in melanoma progression. *Mitochondrion* 2014;**19**(Pt. A):58–68.
143. Harman D. Aging: a theory based on free radical and radiation chemistry. *J Gerontol* 1956;**11**(3):298–300.
144. Harman D. The biologic clock: the mitochondria? *J Am Ger Soc* 1972;**20**(4):145–7.
145. Skulachev VP, Anisimov VN, Antonenko YN, Bakeeva LE, Chernyak BV, Erichev VP, et al. An attempt to prevent senescence: a mitochondrial approach. *Biochim Biophys Acta* 2009;**1787**(5):437–61.
146. Skulachev VP, Antonenko YN, Cherepanov DA, Chernyak BV, Izyumov DS, Khailova LS, et al. Prevention of cardiolipin oxidation and fatty acid cycling as two antioxidant mechanisms of cationic derivatives of plastoquinone (SkQs). *Biochim Biophys Acta* 2010;**1797**(6–7):878–89.
147. Severin FF, Severina II, Antonenko YN, Rokitskaya TI, Cherepanov DA, Mokhova EN, et al. Penetrating cation/fatty acid anion pair as a mitochondria-targeted protonophore. *Proc Natl Acad Sci USA* 2010;**107**(2):663–8.

148 Agapova LS, Chernyak BV, Domnina LV, Dugina VB, Efimenko AY, Fetisova EK, et al. Mitochondria-targeted plastoquinone derivatives as tools to interrupt execution of the aging program. 3. Inhibitory effect of SkQ1 on tumor development from p53-deficient cells. *Biochemistry (Mosc)* 2008;**73**(12):1300–16.

149 Anisimov VN, Bakeeva LE, Egormin PA, Filenko OF, Isakova EF, Manskikh VN, et al. Mitochondria-targeted plastoquinone derivatives as tools to interrupt execution of the aging program. 5. SkQ1 prolongs lifespan and prevents development of traits of senescence. *Biochemistry (Mosc)* 2008;**73**(12):1329–42.

150 Anisimov VN, Egorov MV, Krasilshchikova MS, Lyamzaev KG, Manskikh VN, Moshkin MP, et al. Effects of the mitochondria-targeted antioxidant SkQ1 on lifespan of rodents. *Aging (Albany NY)* 2011;**3**(11):1110–19.

151 Skulachev MV, Antonenko YN, Anisimov VN, Chernyak BV, Cherepanov DA, Chistyakov VA, et al. Mitochondrial-targeted plastoquinone derivatives. Effect on senescence and acute age-related pathologies. *Curr Drug Targets* 2011;**12**(6):800–26.

152 Dzyubinskaya EV, Ionenko IF, Kiselevsky DB, Samuilov VD, Samuilov FD. Mitochondria-addressed cations decelerate the leaf senescence and death in Arabidopsis thaliana and increase the vegetative period and improve crop structure of the wheat Triticum aestivum. *Biochemistry (Mosc)* 2013;**78**(1):68–74.

153 Bakeeva LE, Barskov IV, Egorov MV, Isaev NK, Kapelko VI, Kazachenko AV, et al. Mitochondria-targeted plastoquinone derivatives as tools to interrupt execution of the aging program. 2. Treatment of some ROS- and age-related diseases (heart arrhythmia, heart infarctions, kidney ischemia, and stroke). *Biochemistry (Mosc)* 2008;**73**(12):1288–99.

154 Khailova LS, Silachev DN, Rokitskaya TI, Avetisyan AV, Lyamsaev KG, Severina II, et al. A short-chain alkyl derivative of Rhodamine 19 acts as a mild uncoupler of mitochondria and a neuroprotector. *Biochim Biophys Acta* 2014;**1837**(10):1739–47.

155 Vays VB, Eldarov CM, Vangely IM, Kolosova NG, Bakeeva LE, Skulachev VP. Antioxidant SkQ1 delays sarcopenia-associated damage of mitochondrial ultrastructure. *Aging (Albany NY)* 2014;**6**(2):140–8.

156 Neroev VV, Archipova MM, Bakeeva LE, Fursova A, Grigorian EN, Grishanova AY, et al. Mitochondria-targeted plastoquinone derivatives as tools to interrupt execution of the aging program. 4. Age-related eye disease. SkQ1 returns vision to blind animals. *Biochemistry (Mosc)* 2008;**73**(12):1317–28.

157 Iomdina EN, Khoroshilova-Maslova IP, Robustova OV, Averina OA, Kovaleva NA, Aliev G, et al. Mitochondria-targeted antioxidant SkQ1 reverses glaucoma lesions in rabbits. *Front Biosci* 2015;**20**:892–901.

158 Lukashev AN, Skulachev MV, Ostapenko V, Savchenko AY, Pavshintsev VV, Skulachev VP. Advances in development of rechargeable mitochondrial antioxidants. *Prog Mol Biol Transl Sci* 2014;**127**:251–65.

159 Skulachev VP. Mitochondria-targeted antioxidants as promising drugs for treatment of age-related brain diseases. *J Alzheimers Dis* 2012;**28**(2):283–9.

160 Stefanova NA, Fursova A, Kolosova NG. Behavioral effects induced by mitochondria-targeted antioxidant SkQ1 in Wistar and senescence-accelerated OXYS rats. *J Alzheimer's Dis* 2010;**21**(2):479–91.

161 Shipounova IN, Svinareva DA, Petrova TV, Lyamzaev KG, Chernyak BV, Drize NI, et al. Reactive oxygen species produced in mitochondria are involved in age-dependent changes of hematopoietic and mesenchymal progenitor cells in mice. A

study with the novel mitochondria-targeted antioxidant SkQ1. *Mech Ageing Dev* 2010;**131**(6):415–21.

162 Obukhova LA, Skulachev VP, Kolosova NG. Mitochondria-targeted antioxidant SkQ1 inhibits age-dependent involution of the thymus in normal and senescence-prone rats. *Aging (Albany NY)* 2009;**1**(4):389–401.

163 Demianenko IA, Vasilieva TV, Domnina LV, Dugina VB, Egorov MV, Ivanova OY, et al. Novel mitochondria-targeted antioxidants, "Skulachev-ion" derivatives, accelerate dermal wound healing in animals. *Biochemistry (Mosc)* 2010;**75**(3):274–80.

164 Shabalina IG, Vyssokikh MY, Gibanova N, Csikasz RI, Edgar D, Hallden-Waldermarson A, et al. Improved health-span and lifespan in mtDNA mutator mice treated with the mitochondrially targeted antioxidant SkQ1. *Aging (Albany NY)* 2017;**9**(2):315–36.

165 Korshunov SS, Skulachev VP, Starkov AA. High protonic potential actuates a mechanism of production of reactive oxygen species in mitochondria. *FEBS Lett* 1997;**416**(1):15–18.

166 Kapay NA, Isaev NK, Stelmashook EV, Popova OV, Zorov DB, Skrebitsky VG, et al. In vivo injected mitochondria-targeted plastoquinone antioxidant SkQR1 prevents beta-amyloid-induced decay of long-term potentiation in rat hippocampal slices. *Biochemistry (Mosc)* 2011;**76**(12):1367–70.

167 Kapay NA, Popova OV, Isaev NK, Stelmashook EV, Kondratenko RV, Zorov DB, et al. Mitochondria-targeted plastoquinone antioxidant SkQ1 prevents amyloid-beta-induced impairment of long-term potentiation in rat hippocampal slices. *J Alzheimers Dis* 2013;**36**(2):377–83.

168 Stefanova NA, Muraleva NA, Skulachev VP, Kolosova NG. Alzheimer's disease-like pathology in senescence-accelerated OXYS rats can be partially retarded with mitochondria-targeted antioxidant SkQ1. *J Alzheimers Dis* 2014;**38**(3):681–94.

169 Dai DF, Chen T, Wanagat J, Laflamme M, Marcinek DJ, Emond MJ, et al. Age-dependent cardiomyopathy in mitochondrial mutator mice is attenuated by overexpression of catalase targeted to mitochondria. *Aging Cell* 2010;**9**(4):536–44.

170 Lee HY, Choi CS, Birkenfeld AL, Alves TC, Jornayvaz FR, Jurczak MJ, et al. Targeted expression of catalase to mitochondria prevents age-associated reductions in mitochondrial function and insulin resistance. *Cell Metab* 2010;**12**(6):668–74.

171 Schriner SE, Linford NJ, Martin GM, Treuting P, Ogburn CE, Emond M, et al. Extension of murine life span by overexpression of catalase targeted to mitochondria. *Science* 2005;**308**(5730):1909–11.

172 Treuting PM, Linford NJ, Knoblaugh SE, Emond MJ, Morton JF, Martin GM, et al. Reduction of age-associated pathology in old mice by overexpression of catalase in mitochondria. *J Gerontol A Biol Sci Med Sci* 2008;**63**(8):813–24.

173 Dai DF, Rabinovitch PS. Cardiac aging in mice and humans: the role of mitochondrial oxidative stress. *Trends Cardiovasc Med* 2009;**19**(7):213–20.

174 Sablina AA, Budanov AV, Ilyinskaya GV, Agapova LS, Kravchenko JE, Chumakov PM. The antioxidant function of the p53 tumor suppressor. *Nat Med* 2005;**11**(12):1306–13.

175 Severina II, Severin FF, Korshunova GA, Sumbatyan NV, Ilyasova TM, Simonyan RA, et al. In search of novel highly active mitochondria-targeted antioxidants: thymoquinone and its cationic derivatives. *FEBS Lett* 2013;**587**(13):2018–24.

176 Plotnikov EY, Morosanova MA, Pevzner IB, Zorova LD, Manskikh VN, Pulkova NV, et al. Protective effect of mitochondria-targeted antioxidants in an acute bacterial infection. *Proc Natl Acad Sci USA* 2013;**110**(33):E3100–8.

177 MacCay CM, Crowell MF. Prolonging the life span. *Sci Mon* 1934;**39**:405–14.
178 MacCay CM, Crowell MF, Maynard LA. The effect of retarded growth upon the length of life span and upon the ultimate body size. *J Nutr* 1935;**10**:63–79.
179 MacCay CM, Maynard LA, Barnes LL. Growth, aging, chronic disease and lifespan in rats. *Arch Biochem* 1943;**2**:469.
180 Will LC, MacCay CM. Ageing, basal metabolism and retarded growth. *Arch Biochem* 1943;**2**:481.
181 Robertson TB, Marston R, Walters JW. Influence of intermittent starvation and of intermittent starvation plus nucleic acid on growth and longevity in white mice. *Aust J Exp Biol Med Sci* 1934;**12**:33.
182 Mair W, Goymer P, Pletcher SD, Partridge L. Demography of dietary restriction and death in *Drosophila*. *Science* 2003;**301**(5640):1731–3.
183 Libert S, Pletcher SD. Modulation of longevity by environmental sensing. *Cell* 2007;**131**(7):1231–4.
184 Libert S, Zwiener J, Chu X, Vanvoorhies W, Roman G, Pletcher SD. Regulation of *Drosophila* life span by olfaction and food-derived odors. *Science* 2007;**315**(5815):1133–7.
185 Carr CJ, King JT, Visscher B. Delay of senescence infertility by dietary restriction. *Proc Fedn Am Soc Exp Biol* 1949;**8**:22.
186 Stuchlikova E, Juricova-Horakova M, Deyl Z. New aspects of the dietary effect of life prolongation in rodents. What is the role of obesity in aging? *Exp Gerontol* 1975;**10**(2):141–4.
187 Richie J.P. Jr. Leutzinger Y, Parthasarathy S, Malloy V, Orentreich N, Zimmerman JA. Methionine restriction increases blood glutathione and longevity in F344 rats. *FASEB J* 1994;**8**(15):1302–7.
188 Miller RA, Buehner G, Chang Y, Harper JM, Sigler R, Smith-Wheelock M. Methionine-deficient diet extends mouse lifespan, slows immune and lens aging, alters glucose, T4, IGF-I and insulin levels, and increases hepatocyte MIF levels and stress resistance. *Aging Cell* 2005;**4**(3):119–25.
189 Caro P, Gomez J, Sanchez I, Garcia R, Lopez-Torres M, Naudi A, et al. Effect of 40% restriction of dietary amino acids (except methionine) on mitochondrial oxidative stress and biogenesis, AIF and SIRT1 in rat liver. *Biogerontology* 2009;**10**(5):579–92.
190 Sanz A, Caro P, Ayala V, Portero-Otin M, Pamplona R, Barja G. Methionine restriction decreases mitochondrial oxygen radical generation and leak as well as oxidative damage to mitochondrial DNA and proteins. *FASEB J* 2006;**20**(8):1064–73.
191 Sanchez-Roman I, Gomez A, Perez I, Sanchez C, Suarez H, Naudi A, et al. Effects of aging and methionine restriction applied at old age on ROS generation and oxidative damage in rat liver mitochondria. *Biogerontology* 2012;**13**(4):399–411.
192 Sanchez-Roman I, Gomez A, Gomez J, Suarez H, Sanchez C, Naudi A, et al. Forty percent methionine restriction lowers DNA methylation, complex I ROS generation, and oxidative damage to mtDNA and mitochondrial proteins in rat heart. *J Bioenerg Biomembr* 2011;**43**(6):699–708.
193 Edman U, Garcia AM, Busuttil RA, Sorensen D, Lundell M, Kapahi P, et al. Lifespan extension by dietary restriction is not linked to protection against somatic DNA damage in *Drosophila melanogaster*. *Aging Cell* 2009;**8**(3):331–8.
194 Remolina SC, Hughes KA. Evolution and mechanisms of long life and high fertility in queen honey bees. *Age (Dordr)* 2008;**30**(2–3):177–85.

195 Sun DX, Muthukumar AR, Lawrence RA, Fernandes G. Effects of calorie restriction on polymicrobial peritonitis induced by cecum ligation and puncture in young C57BL/6 mice. *Clin Diagn Lab Immunol* 2001;**8**(5):1003–11.

196 Garcia AM, Busuttil RA, Calder RB, Dolle ME, Diaz V, McMahan CA, et al. Effect of Ames dwarfism and caloric restriction on spontaneous DNA mutation frequency in different mouse tissues. *Mech Ageing Dev* 2008;**129**(9):528–33.

197 Hamden K, Carreau S, Ayadi F, Masmoudi H, El Feki A. Inhibitory effect of estrogens, phytoestrogens, and caloric restriction on oxidative stress and hepato-toxicity in aged rats. *Biomed Environ Sci* 2009;**22**(5):381–7.

198 Park SK, Kim K, Page GP, Allison DB, Weindruch R, Prolla TA. Gene expression profiling of aging in multiple mouse strains: identification of aging biomarkers and impact of dietary antioxidants. *Aging Cell* 2009;**8**(4):484–95.

199 Eng PM, Rimm EB, Fitzmaurice G, Kawachi I. Social ties and change in social ties in relation to subsequent total and cause-specific mortality and coronary heart disease incidence in men. *Am J Epidemiol* 2002;**155**(8):700–9.

200 Colman RJ, Anderson RM. Nonhuman primate calorie restriction. *Antioxid Redox Signal* 2011;**14**(2):229–39.

201 Colman RJ, Anderson RM, Johnson SC, Kastman EK, Kosmatka KJ, Beasley TM, et al. Caloric restriction delays disease onset and mortality in rhesus monkeys. *Science* 2009;**325**(5937):201–4.

202 Gardner EM. Caloric restriction decreases survival of aged mice in response to primary influenza infection. *J Gerontol A Biol Sci Med Sci* 2005;**60**(6):688–94.

203 Obin M, Halbleib M, Lipman R, Carroll K, Taylor A, Bronson R. Calorie restriction increases light-dependent photoreceptor cell loss in the neural retina of Fischer 344 rats. *Neurobiol Aging* 2000;**21**(5):639–45.

204 Hopkin K. Dietary drawbacks. *Sci Aging Knowledge Environ* 2003;**2003**(8):NS4.

205 Redman LM, Ravussin E. Caloric restriction in humans: impact on physiological, psychological, and behavioral outcomes. *Antioxid Redox Signal* 2011;**14**(2):275–87.

206 Vallejo EA. [Hunger diet on alternate days in the nutrition of the aged]. *Prensa Med Argent* 1957;**44**(2):119–20.

207 Stunkard AJ, Rockstein M. *Nutrition, Longevity and Obesity*. New York: Academic Press. 1976.

208 Grierson B. The incredible flying nonagenarian. *New York Times Magazine*. November 28, 2010.

209 Wanagat J, Dai DF, Rabinovitch P. Mitochondrial oxidative stress and mammalian healthspan. *Mech Ageing Dev* 2010;**131**(7–8):527–35.

210 Safdar A, Bourgeois JM, Ogborn DI, Little JP, Hettinga BP, Akhtar M, et al. Endurance exercise rescues progeroid aging and induces systemic mitochondrial rejuvenation in mtDNA mutator mice. *Proc Natl Acad Sci USA* 2011;**108**(10):4135–40.

211 Stolen TO, Hoydal MA, Kemi OJ, Catalucci D, Ceci M, Aasum E, et al. Interval training normalizes cardiomyocyte function, diastolic Ca++ control, and SR Ca++ release synchronicity in a mouse model of diabetic cardiomyopathy. *Circ Res* 2009;**105**(6):527–36.

212 Marques-Aleixo I, Oliveira PJ, Moreira PI, Magalhaes J, Ascensao A. Physical exercise as a possible strategy for brain protection: evidence from mitochondrial-mediated mechanisms. *Progr Neurobiol* 2012;**99**(2):149–62.

213 Yu BP. *Modulation of Aging Processes by Caloric Restriction*. Boca Raton, FL: CRC Press. 1994.

214 Broskey NT, Greggio C, Boss A, Boutant M, Dwyer A, Schlueter L, et al. Skeletal muscle mitochondria in the elderly: effects of physical fitness and exercise training. *J Clin Endocrinol Metab* 2014;**99**(5):1852–61.

215 Heywood R, Sortwell RJ, Noel PRB, Street AE, Prentice DE, Roe FJC, et al. Safety evaluation of toothpaste containing chloroform. 3. Long-term study in beagle dogs. *J Environm Pathol Toxicol* 1979;**2**(3):835–51.

216 Palmer AK, Street AE, Roe FJC, Worden AN, Vanabbe NJ. Safety evaluation of toothpaste containing chloroform. 2. Long-term studies in rats. *J Envir Pathol Toxicol* 1979;**2**(3):821–33.

217 Xiao R, Zhang B, Dong YM, Gong JK, Xu T, Liu JF, et al. A genetic program promotes *C. elegans* longevity at cold temperatures via a thermosensitive TRP channel. *Cell* 2013;**152**(4):806–17.

218 Fredrickson BL, Grewen KM, Coffey KA, Algoe SB, Firestine AM, Arevalo JM, et al. A functional genomic perspective on human well-being. *Proc Natl Acad Sci USA* 2013;**110**(33):13 684–9.

219 Franceschi C, Bonafe M, Valensin S, Olivieri F, De Luca M, Ottaviani E, et al. Inflamm-aging. An evolutionary perspective on immunosenescence. *Ann NY Acad Sci* 2000;**908**:244–54.

220 Penninx BW, van Tilburg T, Kriegsman DM, Deeg DJ, Boeke AJ, van Eijk JT. Effects of social support and personal coping resources on mortality in older age: the longitudinal aging study. *Am J Epidemiol* 1997;**146**(6):510–19.

221 Eastwell HD. Voodoo death in Australian aborigines. *Psychiatr Med* 1987;**5**:71–3.

222 Fischer K, Kettunen J, Wurtz P, Haller T, Havulinna AS, Kangas AJ, et al. Biomarker profiling by nuclear magnetic resonance spectroscopy for the prediction of all-cause mortality: an observational study of 17 345 persons. *PLoS Med* 2014;**11**(2):e1001606.

223 Langley RJ, Tsalik EL, van Velkinburgh JC, Glickman SW, Rice BJ, Wang C, et al. An integrated clinico-metabolomic model improves prediction of death in sepsis. *Sci Transl Med* 2013;**5**(195):195ra95.

224 Lemon HM, Mueller JH, Looney JM, Chasen WH, Kelman M. Hypercitricemia in human cancer. Factors concerned in pathogenesis and treatment. *Br J Cancer* 1960;**14**:376–96.

225 Garland PB, Randle PJ, Newsholme EA. Citrate as an intermediary in inhibition of phosphofructokinase in rat heart muscle by fatty acids, ketone bodies, pyruvate, diabetes and starvation. *Nature* 1963;**200**(490):169–70.

226 Skulachev VP. Membrane-linked systems preventing superoxide formation. *Biosci Rep* 1997;**17**(3):347–66.

227 Gardner PR. Superoxide-driven aconitase FE-S center cycling. *Biosci Rep* 1997;**17**(1):33–42.

228 Gardner PR, Raineri I, Epstein LB, White CW. Superoxide radical and iron modulate aconitase activity in mammalian cells. *J Biol Chem* 1995;**270**(22):13 399–405.

229 Belikova YO, Kotlyar AB, Vinogradov AD. Identification of the high-molecular-mass mitochondrial oxaloacetate keto-enol tautomerase as inactive aconitase. *FEBS Lett* 1989;**246**(1–2):17–20.

230 Beinert H, Kennedy MC. Aconitase, a two-faced protein: enzyme and iron regulatory factor. *FASEB J* 1993;**7**(15):1442–9.

231 Gardner PR, Fridovich I. Superoxide sensitivity of the *Escherichia coli* aconitase. *J Biol Chem* 1991;**266**(29):19 328–33.

232 Spitsin VI, Martynenko LI. *Inorganic Chemistry*. Moscow: MSU. 1994.

233 Bucay AH. The biological significance of cancer: mitochondria as a cause of cancer and the inhibition of glycolysis with citrate as a cancer treatment. *Med Hypotheses* 2007;**69**(4):826–8.

234 Kruspig B, Nilchian A, Orrenius S, Zhivotovsky B, Gogvadze V. Citrate kills tumor cells through activation of apical caspases. *Cell Mol Life Sci* 2012;**69**(24):4229–37.

235 Infantino V, Iacobazzi V, Menga A, Avantaggiati ML, Palmieri F. A key role of the mitochondrial citrate carrier (SLC25A1) in TNFalpha- and IFNgamma-triggered inflammation. *Biochim Biophys Acta* 2014;**1839**(11):1217–25.

236 Skulachev VP. How to cancel the aging program of organism? *Rus Khim Zh* 2009;**LIII**(3):125–40.

237 Skulachev VP. New data on biochemical mechanism of programmed senescence of organisms and antioxidant defense of mitochondria. *Biochemistry (Mosc)* 2009;**74**(12):1400–3.

238 Yani EV, Katargina LA, Chesnokova NB, Beznos OV, Savchenko AY, Vygodin VA, et al. The first experience of using the drug Vizomitin in the treatment of "dry eyes." *Prakt Med* 2012;**59**:134–7.

239 Schulz TJ, Zarse K, Voigt A, Urban N, Birringer M, Ristow M. Glucose restriction extends *Caenorhabditis elegans* life span by inducing mitochondrial respiration and increasing oxidative stress. *Cell Metab* 2007;**6**(4):280–93.

240 Ristow M, Schmeisser K. Mitohormesis: promoting health and lifespan by increased levels of reactive oxygen species (ROS). *Dose Response* 2014;**12**(2):288–341.

241 Yee C, Yang W, Hekimi S. The intrinsic apoptosis pathway mediates the pro-longevity response to mitochondrial ROS in *C. elegans*. *Cell* 2014;**157**(4):897–909.

242 Khalil H, Peltzer N, Walicki J, Yang JY, Dubuis G, Gardiol N, et al. Caspase-3 protects stressed organs against cell death. *Mol Cell Biol* 2012;**32**(22):4523–33.

14

Molecular Mechanisms Underlying Oxytosis

Amalia M. Dolga,[1] Sina Oppermann,[2] Maren Richter,[3] Birgit Honrath,[1,3] Sandra Neitemeier,[3] Anja Jelinek,[3] Goutham Ganjam,[3] and Carsten Culmsee[4]

[1] Department of Molecular Pharmacology, Groningen Research Institute of Pharmacy (GRIP), University of Groningen, Groningen, The Netherlands
[2] German Cancer Research Center (DKFZ) and National Center of Tumordiseases (NCT), Heidelberg, Germany
[3] Institute of Pharmacology and Clinical Pharmacy, Biochemisch-Pharmakologisches Centrum Marburg, Philipps University of Marburg, Marburg, Germany
[4] Institute of Physiological Chemistry, Philipps University of Marburg, Marburg, Germany

Abbreviations

AD	Alzheimer's disease
AIF	apoptosis-inducing factor
AMPA	α-amino-3-hydroxy-5-methyl-isoxazole-4-propionate
AP-1	activator protein 1
APP	amyloid precursor protein
ATF-4	activating transcription factor-4
ATF-6	activating transcription factor-6
Bak	Bcl-2 homologous antagonist killer
Bax	Bcl-2-associated X protein
Bcl-2	B-cell lymphoma-2
Bcl-X_L	B-cell lymphoma-extra large
BH	B-cell lymphoma protein-2 homology
BID	BH3-interacting domain death agonist
BiP	binding immunoglobulin protein
cBid	caspase 8-cleaved BID
cGMP	cyclic guanosine monophosphate
CHOP	CCAT/enhancer binding protein (C/EBP) homologous protein
CNS	central nervous system
COX	cyclooxygenase
CypA	cyclophilin A
cyt c	cytochrome c
DOPA	dihydroxyphenylalanine
DRP1	dynamin-related protein-1
eIF2α	eukaryotic initiation factor 2
ER	endoplasmic reticulum
ERK	growth factor-regulated extracellular signal-related kinase

Apoptosis and Beyond: The Many Ways Cells Die, First Edition. Edited by James Radosevich.
© 2018 John Wiley & Son Inc. Published 2018 by John Wiley & Son Inc.

FIS1	mitochondrial fission 1 protein
GPX4	glutathione peroxidase-4
GRP78	ER-resident chaperone 78 kDa glucose-regulating protein
GSH	glutathione
GSSG	glutathione disulfide
HETE	hydroxyeicosatetraenoic acid
Hq	Harlequin
HT-22R	glutamate-resistant HT-22 cells
IMM	inner mitochondrial membrane
IMS	intermembrane mitochondrial space
IP_3	inositol triphosphate
IRE1	inositol-requiring enzyme 1
JNK	C-Jun N-terminal kinase
LOX	lipoxygenenase
MAPK	mitogen-activated protein kinase
MAP2K	mitogen-activated protein kinase kinase
MAP3K	mitogen-activated protein kinase kinase kinase
MMP	mitochondrial membrane potential
MOMP	mitochondrial outer-membrane permeabilization
mtDNA	mitochondrial DNA
NAD^+	oxidized nicotinamide adenine dinucleotide
NADH	reduced nicotinamide adenine dinucleotide
NADPH	reduced nicotinamide adenine dinucleotide phosphate
NMDA	N-methyl-D-aspartate
NMDAR	N-methyl-D-aspartate receptor
NOX	nicotinamide adenine dinucleotide phosphate oxidase
Nrf2	nuclear factor (erythroid-derived 2)-like 2
OGD	oxygen-glucose deprivation
OMM	outer mitochondrial membrane
OPA1	optic atrophy-1
ORAI1	calcium release-activated calcium channel protein 1
PD	Parkinson's disease
PERK	PKR-like ER kinase
PUFA	polyunsaturated fatty acid
ROS	reactive oxygen species
SOCE	store-operated calcium entry
tBID	truncated BID
TNFα	tumor necrosis factor alpha
UPR	unfolded protein response
XBP1	X-box-binding protein 1
xCT/X_c^- Cys/Glu	glutamate–cystine antiporter

14.1 Introduction

Oxidative stress is implicated in the pathogenesis and pathophysiology of cardiovascular diseases and age-related neurodegenerative disorders such as Alzheimer's disease (AD), Parkinson's disease (PD), and cerebral ischemia [1,2]. A specific form of oxidative cell death associated with increased extracellular glutamate concentration, inhibition of the glutamate/cysteine (xCT/X_c^- Cys/Glu) antiporter, depletion of glutathione (GSH), and formation of mitochondrial reactive oxygen species (ROS) has been classified as "oxytosis" [3]. The expression of X_c^- Cys/Glu antiporter in the central nervous system (CNS) is essential in regulating extracellular glutamate concentrations, and thereby indirectly controlling neuronal excitotoxicity and cell death. The X_c^- Cys/Glu antiporter acts as a cysteine transporter, using the transmembrane gradient of high- and low-extracellular glutamate concentration as its driving force; high-extracellular glutamate concentration specifically blocks the antiporter, thereby inhibiting the cystine uptake [4]. Blocking of the X_c^- Cys/Glu antiporter by high extracellular glutamate concentrations or glutamate analogs such as homocysteate or quisqualate [5] results in attenuated levels of intracellular cystine, which is otherwise rapidly reduced to cysteine, the rate-limiting amino acid required for GSH synthesis. Since nerve cells are prone to oxidative stress, the activity of the X_c^- Cys/Glu antiporter, by favoring GSH synthesis, is essential to preventing oxidative stress and oxytosis in the CNS [6]. Therefore, understanding how X_c^- Cys/Glu antiporter regulates the cellular redox balance is of utmost importance.

xCT mRNA is expressed in the murine neuronal HT-22 cell line, mouse and rat cortex, primary murine neurons and astrocytes, fibroblasts, and human spinal cord, brain, and pancreas [7]. In cell culture, glutamate elicits toxic effects via the inhibition of X_c^- Cys/Glu antiporter in murine hippocampal HT-22 cells, the rat adrenal medulla pheochromocytoma PC-12 cell line, human neuroblastoma SH-SY5Y, the human teratocarcinoma Ntera/D cell line, and immature primary neurons [8]. Analysis of the murine and human xCT cDNA reveals multiple putative binding sites for the activator protein 1 (AP-1) [9] and four putative antioxidant-response elements regulated by the nuclear factor (erythroid-derived 2)-like 2 (Nrf2) transcription factor [10]. Indeed, upregulation of Nrf2 expression by tert-buthylhydroquinone in neuronal HT-22 cells [7] or by ceftriaxone in astrocytes and stem cell-derived motor neurons [11] strongly increases xCT protein expression and X_c^- Cys/Glu antiporter activity.

The most frequently investigated model systems for oxytosis are the murine hippocampal-derived HT-22 cell line [3,12] and immature primary neurons that do not express ionotropic excitatory receptors at early stages of differentiation within the first days in culture (DIV1–2) [13]. The HT-22 cell line is a subclone of the HT-4 cell line, an immortalized mouse hippocampal cell line [14]. The HT-22 clone was chosen out of 25 clones tested, based on its highest sensitivity to glutamate. The HT-22 subclone has been thoroughly characterized with respect to ionotropic glutamate receptors and monoamine synthesis. Polymerase chain reaction (PCR) and Northern blot analysis demonstrated that N-methyl-D-aspartate (NMDA) receptors, α-amino-3-hydroxy-5-methyl-isoxazole-4-propionate (AMPA) receptor, kainate receptors, quisqualate receptors, and metabotropic glutamate receptors are not expressed in these cells. Furthermore, NMDA, aspartate, AMPA, kainate, and selective metabotropic glutamate receptor agonist (1S,3R)-1-aminocyclopentane-1,3-dicarboxylic acid treatment do not induce HT-22 cell death at concentrations up to 10 mM, and the glutamate receptor antagonists APV and MK-801 do not prevent glutamate toxicity, suggesting neither expression nor functional activity of ionotropic glutamatergic receptors in HT-22 cells. Gas chromatography (GC) analysis shows that HT-22 cells do not produce norepinephrine, dihydroxyphenylalanine (DOPA), dopamine, or epinephrine. However, high concentrations of glutamate (5 mM) and of its analog, quisqualate (0.5 mM), are toxic via competitive inhibition of X_c^- Cys/Glu antiporter, which induces oxytosis. High concentrations of cystine (1 mM) inhibit glutamate toxicity and induction of oxytosis [5].

Oxytosis is characterized by a biphasic increase of ROS levels in neuronal HT-22 cells. Within the first 6–7 hours of glutamate exposure, ROS levels increase in a linear fashion by approximately 10%, then again to up to 200 times the initial ROS levels [15]. The first boost of ROS is associated with GSH depletion, which in turn activates lipoxygenase (LOX, EC 1.13.11) or cyclooxygenase (COX, EC 1.14.99.1) proteins, leading to lipid peroxidation. Increased lipid peroxidation, together with the translocation of the truncated form (truncated BH3-interacting domain death agonist, tBID) of the proapoptotic BID protein, promotes cell death [16,17]. tBID formation leads to the breakdown of the mitochondrial membrane potential (MMP), accumulation of ROS, and calcium overload [17,18]. Calcium influx, even though it is not mediated by NMDA receptors in HT-22 cells, is essential to oxytosis, since cell death can be attenuated by calcium chelators [19]. In consequence, more proapoptotic factors are released, including apoptosis-inducing factor (AIF), which rapidly translocates to the nucleus to induce DNA damage, marking the final steps of neuronal cell death [20,21]. Mitochondrial outer-membrane permeabilization (MOMP) and the release of AIF are accompanied by the second, exponential burst in ROS levels and represent a state of irreversible damage [22–24].

In this chapter, we address the molecular mechanisms underlying oxytosis, with an emphasis on mitochondrial- and endoplasmic reticulum (ER)-dependent pathways.

14.2 The Role of GSH in Oxytosis

The tripeptide γ-L-glutamyl-L-cysteinyl-glycine, known as GSH, is one of the most important antioxidants in eukaryotic cells and the most important one in the CNS [25]. Its reduced form is capable of donating an electron from its thiol group of cysteine to unstable molecules such as free radicals, peroxides, and lipid peroxides. By losing an electron, GSH reacts with another reactive GSH and is converted to its oxidized form, glutathione disulfide (GSSG). Cells can recycle GSSG to GSH by glutathione reductase, but they also depend on a continuous *de novo* synthesis of GSH [26]. Despite its reducing activity, GSH also plays a role in the metabolism of xenobiotics and cellular signaling.

The GSH/GSSG ratio is important for the cellular redox balance, and therefore correlates with the extent of oxidative stress [27]. For instance, in rat brains, a decrease in the GSH/GSSG ratio occurs in an age-related manner and correlates with oxidative damage of the mitochondrial DNA (mtDNA) [28]. A disturbed metabolism of GSH is also described for many neurological disorders associated with oxidative stress, including PD [29] and AD [30].

The rate-limiting step during production of GSH is the reduction of cystine to cysteine. Cysteine should be maintained at low levels in the brain, because of its neurotoxicity [31]. The amount of cystine in the cell is highly regulated by the X_c^- Cys/Glu antiporter, which exchanges intracellular glutamate against extracellular cystine in a ratio of 1:1. Reduced GSH levels result in reduced glutathione peroxidase-4 (GPX4) activity. Under physiological conditions, GPX4 catalyzes the reduction of hydrogen peroxides or lipid peroxides as part of the antioxidant defense [32]. Notably, GPX4 knockout (KO) mice have an embryonic lethal phenotype, and cell lines derived from $GPX4^{+/-}$ mice are more susceptible to oxidative stress than wild-type littermates, emphasizing the pivotal role of GPX4 during development and in redox metabolism. Moreover, neuron-specific GPX4 depletion in mice causes neurodegeneration *in vivo* and *in vitro* [32].

During oxytosis, the initial oxidative damage is attributed to the accumulation of peroxides in consequence of the GSH depletion and subsequent inhibition of GPX4, and is reversible approximately 7–8 hours following glutamate exposure, whereas glutamate exposure for more than 8 hours induces irreversible mitochondrial damage and cell death [24]. Detrimental effects of GSH depletion and subsequent ROS formation have also been shown in different neuronal cell lines [3,33], but a reduction of GSH alone does not lead to maximal ROS production. In fact, protease activities, protein synthesis, and especially the mitochondrial electron transport chain are also highly affected by low levels of cellular GSH [15]. Despite GSH depletion, neuronal cells can be rescued from oxidative damage [15,34]. Although glutamate-resistant HT-22 cells do not display GSH depletion and are protected against oxytosis [35], neuronal cells can be rescued from oxidative damage despite GSH depletion [15,34]. Thus, GSH depletion can trigger and facilitate oxidative stress pathways, but it is not a prerequisite for cell death. This underlines the complex mechanisms of cell death in response to oxidative stress, which also involve, among other things, further activation of LOX enzymes, stimulation of mitogen-activated protein kinases (MAPKs) [36,37], production of mitochondrial ROS [15], and mitochondrial release of AIF [21].

14.3 Upstream Mechanisms of Mitochondrial Dysfunction in Oxytosis

14.3.1 Lipid Peroxidation at the Onset of Oxytosis

In general, COX and LOX enzymes metabolize arachidonic acid to eicosanoids and incorporate molecular oxygen at specific positions in polyunsaturated fatty acids (PUFAs). In healthy organisms, these eicosanoids, comprising prostaglandins, thromboxanes, and leukotrienes, regulate synaptic function, cerebral blood flow, apoptosis, and angiogenesis [38]. There are three known isoforms of COX (COX-1, -2, and -3), which are expressed in the CNS or induced by pathological stimuli [39]. Among the three different mammalian isoforms of LOX (the arachidonate 5-lipoxygenase (5-LOX, EC 1.13.11.34), the arachidonate 12-lipoxygenase (12-LOX, EC 1.13.11.31), and the erythroid cell-specific 15-lipoxygenase (15-LOX, EC 1.13.11.33)), expressed in the brain, the most common isoforms are 12- and 15-LOX. These iron-containing enzymes catalyze the dioxygenation of PUFAs in lipids and generate 12- and 15-hydroxyeicosatetraenoic acids (12- and 15-HETE). Since 12-LOX enzymes generate 12-HETE and to some degree 15-HETE, they are often referred to as "12/15-LOX" [40]. Under pathophysiological conditions, an augmented activity of LOX enzymes leads to the generation of high amounts of ROS, which further contribute to inflammatory processes and mitochondrial damage, thereby promoting aging and apoptosis [38].

In oxytosis, GSH depletion activates 12/15-LOX, which contributes to the first burst of ROS and subsequently leads to mitochondrial damage [24]. Studies have shown that COX-2 KO mice develop less neuronal damage following transient forebrain ischemia [41]. COX-1 KO mice are more susceptible to ischemic brain injury, suggesting a diverse role of COX activity in neuronal cell death *in vivo*. Notably, indomethacin, a broad COX inhibitor, fails to protect HT-22 cells against oxytosis, whereas 12/15-owever,HLOX inhibition with PD146176 prevents glutamate-induced cell death in HT-22 cells [24], indicating a crucial role of 12/15-LOX – but not of COX activity – in oxytosis. This pivotal role of 12/15-LOX activation in other forms of cell death is further supported *in vivo* by its enhanced expression in ischemic mouse brains and *in vitro* by experiments with baicalein, a LOX inhibitor that elicits beneficial effects against glutamate toxicity in primary rat neurons [42]. Moreover, LOX inhibitors are capable of attenuating nitrosative stress in neuronal cells [43], and LOX activity is enhanced in PC12 cells overexpressing mutant amyloid precursor protein (APP) [44]. These findings suggest a contribution of LOX enzymes to neuronal damage in settings of cell death beyond oxytosis. LOX activity initiates further stimulation of growth factor-regulated extracellular signal-related kinases (ERKs), stress-activated C-Jun N-terminal kinases (JNKs), and p38 MAPKs [36,37].

14.3.2 The Function of MAPKs in Oxytosis

The MAPKs are a family of serine/threonine protein kinases consisting of ERKs, JNKs, and p38. They are involved in many cellular processes, including cell growth and differentiation, and also in detrimental mechanisms, such as inflammation and cell death. MAPKs are important transducers of signals from the cell surface to the nucleus, and are activated through a phosphorylation cascade [45]. Briefly, various stimuli (e.g., growth factors, cytokines, oxidative stress) activate MAPK kinase kinases (MAP3K), which in turn stimulate MAPK kinases (MAP2K), which subsequently may phosphorylate the MAPKs ERK, JNK, and p38. Mimicking oxidative stress through the exogenous administration of H_2O_2 has been shown to activate MAPK signaling pathways [46], strongly suggesting the contribution of MAPKs to oxidative stress-induced cell death.

14.3.2.1 Extracellular Signal-Related Kinases (ERKs)

Although the ERK signaling pathway is mostly implicated in cell survival and neuronal differentiation [47], recent studies have shown that it can also play a detrimental role by promoting neuronal injury following cerebral ischemia and brain trauma or in neurodegenerative diseases such as AD and PD [48,49]. The pro-death function of ERK activation in paradigms of oxidative stress has been confirmed in the model system of glutamate-induced cytotoxicity in HT-22 cells [36], as well as in primary cortical neurons [50,51]. Activation of ERK1/2 requires the involvement of the LOX metabolite 20-hydroxyeicosatetraenoic acid [52]. Therefore, ERK1/2 activation can be considered a downstream effect of increased lipid peroxidation [50] and an upstream mechanism of mitochondrial damage, as activated ERK1/2 mediates cytochrome c (cyt c) release [53], decreases mitochondrial respiration, and mediates a loss of the MMP [54]. Furthermore, inhibition of ERK2 phosphorylation using the MEK-inhibitor UO126 protects against ischemic brain damage [55]. Notably, ERK2 inhibition by U0126 also prevents oxidative stress-induced cell death initiated by glutamate exposure or hypoxia, but fails to protect HT-22 neurons against the toxic effects of staurosporine or tumor necrosis factor alpha (TNFα) [51]. Several neuroprotective substances, including α-iso-cubebene and plant extracts from the dried roots of *Polygonum multiflorum* or the leaves of *Chamaecyparis obtusa*, inhibit ERK phosphorylation and prevent oxytosis in HT-22 cells [56–58]. Although ERK activation can mediate pro-survival signals, it appears to contribute to cell death in conditions of oxytosis.

14.3.2.2 C-Jun N-Terminal Kinases (JNKs)

There are three known isoforms of JNK (JNK-1, -2, and -3) which are activated by various extracellular stress stimuli. Oxidative stress can stimulate the JNK pathway via the redox-sensitive MAP3K ASK1 [59], thereby promoting cell-death mechanisms.

ROS activate ASK1 by disrupting its binding to thioredoxin, which leads to the activation of the MAP2Ks MKK4 and MKK7 and thus to stimulation of the JNK pathway [60]. Once activated, JNK phosphorylates transcription factors such as c-Jun and c-Fos [61]. These dimerize to form the transcription factor AP-1, which is involved in several cellular processes, including cell-death pathways. During oxytosis, glutamate activates the c-Jun and c-Fos pathways, which can be inhibited by several neuroprotective compounds, including obovatol, honokiol, and magnolol isolated from the cortex Magnoliae [62].

Another substrate of JNK is the antiapoptotic protein B-cell lymphoma-2 (Bcl-2), which loses its pro-survival properties upon phosphorylation by JNK [63,64], further indicating a regulatory role of JNK in cell-death pathways. The specific JNK inhibitor SP600125 prevents apoptosis in HT-22 cells and increases neuronal viability [37]. This neuroprotective effect has been confirmed by silencing c-Jun with specific siRNA sequences in a model of glutamate-induced oxytosis [65]. In accordance with these findings, c-Jun silencing protects neuronal primary cultures against glutamate exposure in a model of oxygen-glucose deprivation (OGD) [66].

The proapoptotic Bcl-2-associated X protein (Bax) seems to be one of the most important targets of JNK activation during oxidative stress acting upstream of mitochondrial damage. Several findings show that JNK phosphorylation activates Bax *in vitro* [67], and neuronal cell death in the hippocampus *in vivo* has been abolished by the application of the peptide JNK inhibitor Tat-JBD following transient brain ischemia/reperfusion. Tat-JBD is further able to attenuate other apoptotic hallmarks, including the release of Bax from Bcl-2/Bax dimers, the subsequent translocation of Bax to the mitochondria, mitochondrial pore formation, and the release of cyt c from mitochondria [68]. Studies on *Jnk1$^{-/-}$Jnk2$^{-/-}$* primary fibroblasts also reveal a major role for the JNK-induced release of cyt c by Bax activation during ultraviolet (UV)-induced cell death [69]. In conclusion, activation of the JNK cascade contributes to oxidative cell death and mitochondrial demise.

14.3.2.3 p38 Kinase Family
The p38 MAPK family consists of four isoforms: p38α, β, γ, and δ. p38α and β are the only isoforms expressed in the CNS. One of the major MAP3Ks in the p38 pathway is ASK1, which also activates the JNK pathway [70]. Phosphorylation of ASK1 leads to the induction of MKK 3 and 6 and the subsequent activation of p38. MKK 3 and 6 are highly specific for p38, but can also act on JNK [71]. Inhibition of p38 protects HT-22 cells against both glutamate- [72] and ethanol-induced oxidative stress [73]. Enhanced cell viability in primary cortical neurons following p38 inhibition in a model of NMDA-induced excitotoxicity [74] further confirms the detrimental role of p38 in both conditions of neurotoxic N-methyl-D-aspartate receptor (NMDAR) stimulation and ROS-mediated cell death. Moreover, p38 also mediates its proapoptotic effect by activating several pro-death transcription factors, such as CCAT/enhancer binding protein (C/EBP) homologous protein (CHOP) [75]. MAPK activation leads to mitochondrial dysfunction and DNA damage.

14.4 Mitochondrial Dysfunction in Oxytosis

In mammalian cells, signaling pathways of regulated cell death converge at mitochondria, where the permeabilization of the outer mitochondrial membrane (OMM) is considered the point of no return. Since neurons rely on the ability of mitochondria to provide ATP via oxidative phosphorylation, intact mitochondrial morphology and function are a prerequisite for neuronal survival. The energy produced is required for ATP-dependent ion channels or transport proteins and is essential for the mitochondrial calcium buffering capacity and the regulation of cell survival. The inner mitochondrial membrane (IMM) and OMM play fundamental roles in organelle integrity. Transmembrane proteins of the OMM are involved in morphological regulation, while the IMM harbors proteins involved in organelle function. This is of particular importance, as changes in the MMP can affect mitochondrial function. Under conditions of cellular stress, an alteration of the OMM proteins affects membrane integrity and advances cell death. Under pathophysiological conditions, the charge flow is disrupted between the intermembrane mitochondrial space (IMS) and the mitochondrial matrix via the IMM; this is normally maintained by proton pumps in order to balance ion homeostasis (K^+, Ca^{++}, Mg^{++}).

An increase in ROS formation after glutamate exposure has been connected to an impaired regulation of mitochondrial function and dynamics in HT-22 cells. Live cell-imaging studies of HT-22 cells demonstrate a highly dynamic morphology of the mitochondria, which appear as a large interconnected and dynamic network [76]. The mitochondria undergo highly dynamic morphological alterations, reflecting constant fusion and fission: processes required for the efficient function of these organelles and of their role in regulating cellular homeostasis and maintenance [77]. Mitochondrial fusion supports the regeneration of mitochondrial proteins and mtDNA, while mitochondrial fission is crucial for the transport of the organelles throughout the lengths of the axons and dendrites of the neurons [78–80]. Defects in mitochondrial dynamics lead to an increase in oxidative stress, a decrease in energy production, and a complete loss of mtDNA, and therefore to neuronal dysfunction and cell death [76]. During glutamate-induced oxytosis, the mitochondrial morphology shifts from elongated tubulin-like structures to highly fragmented organelles accumulating in the vicinity of the pycnotic nucleus [17].

14.4.1 The Bcl-2 Family Proteins and Their Interaction Partners

An important regulatory mechanism of MOMP and the intrinsic apoptotic pathway is represented by the Bcl-2 family proteins [81–84]. More than 25 Bcl-2 family proteins have been identified, which are subdivided into three groups based on their pro- and antiapoptotic function and the presence of up to four conserved Bcl-2 homology (BH) regions [77]. Antiapoptotic Bcl-2 family proteins (Bcl-2, Bcl-X_L, Bclw, MCL-1, and BFL1) share the BH1–4 domain and prevent MOMP by inhibiting proapoptotic Bcl-2 family members [77]. The proapoptotic multiregion proteins Bax and Bcl-2 homologous antagonist killer (Bak) bear three BH3 domains (BH1–3) that oligomerize to form membrane-permeabilizing pores in the OMM [77,85]. Meanwhile, the proapoptotic BH3 proteins, such as BID, BIM, BAD, NOXA, and PUMA, contain only the BH3 domain. They are activated by various stress stimuli, and directly (activator BH3) or indirectly (sensitizer BH3) activate Bax and Bak [86,87]. While they are common in non-neuronal cells, the expression of the Bcl-2 family members Bcl-X_L, Bcl-2, Bax, Bak, MCL-1, and BID is highly dynamic during neuronal development. In the adult brain, high levels of BID, Bax, and Bcl-X_L are continuously expressed, but significant protein levels of full-length Bak or of the antiapoptotic proteins Bcl-2 and MCL-1 have not been detected [88–92]. Considering these observations, the role of specific Bcl-2 family members in regulating cell-death pathways in neurons might differ from those described in cancer cells or non-neuronal cells. The proapoptotic proteins Bax and Bak are widely accepted as the key players of mitochondrial cell-death pathways in non-neuronal cells, but they seem to be less involved in the mechanisms of neuronal oxytosis. For instance, full-length Bak could not be detected in neurons, yet an unusual splice variant of Bak was detected, which mediated antiapoptotic functions when overexpressed [92].

Increasing evidence highlights a pivotal role for BID in neuronal cell death, by triggering mitochondrial demise and the subsequent acceleration of oxidative stress. Reduced BID expression prevented cell death in a model of OGD in primary cultured neurons, and genetic depletion of BID decreased cell brain damage in models of cerebral ischemia and brain trauma *in vivo* [20,93]. Moreover, BID-deficient neurons are highly resistant to glutamate-induced excitotoxicity and oxidative stress *in vitro* [21,93]. BID cleavage and tBID translocation to mitochondria have been further shown in seizure-induced neuronal death and in human brain tissue after temporal epilepsy [94,95].

The involvement of BID in mitochondrial pathways of neuronal death triggered by oxidative stress has been established in HT-22 neurons, confirming the detrimental role of BID in neurological diseases associated with impairments in mitochondrial integrity and function [17,21,24,96]. In HT-22 cells, glutamate-induced intracellular GSH depletion and the initial formation of ROS were accelerated by BID-dependent mitochondrial damage [17,24]. Pharmacological BID inhibition could not affect early increases in lipid peroxidation after glutamate treatment, but it significantly prevented the secondary pronounced accumulation of lipid peroxides that accompanied the fatal mitochondrial demise [24]. Glutamate-induced transactivation of BID to mitochondria triggers mitochondrial damage and release of AIF, which in turn localizes to the nucleus and induces nuclear condensation and DNA fragmentation [21,23,97–99]. The involvement of BID in oxytosis was confirmed by recent studies using the approach of siRNA-mediated BID gene silencing and pharmacological BID inhibition via small molecules [21,24,96,100]. BID inhibition by BI-6c9 and several novel small-molecule BID inhibitors [96,100] preserved the cell morphology and viability of glutamate-treated HT-22 cells and attenuated the glutamate-induced mitochondrial translocation of BID and the downstream hallmarks of BID-mediated mitochondrial dysfunction [21]. In the model system of oxytosis in HT-22 cells, pharmacological BID inhibition prevented mitochondrial fission, the associated ROS release, and the breakdown of the MMP [96,100]. BID inhibition was sufficient to prevent mitochondrial bioenergetics and respiration by restoring mitochondrial ATP production and oxygen consumption after glutamate treatment [96], thereby confirming BID as a key player in oxytosis. Overall, the aforementioned findings in model systems of neuronal death have led to the discovery of novel BID inhibitors that are able to protect mitochondrial function and have a high relevance for novel therapeutic strategies in neurological diseases where mitochondrial damage plays a prominent role.

The proteins Bax and Bak mediate their proapoptotic activity in their full-length form through conformational changes and oligomerization for pore-formation at the mitochondrial membrane. In contrast, the proapoptotic function of BID is accelerated by the proteolytic cleavage of the cytosolic full-length BID to cBID (caspase 8-cleaved BID). cBID is predominantly localized at the mitochondrial membrane, where the membrane-associated dissociation of the p7 fragment releases the active p15 fragment, termed "tBID" [101]. Cleavage of BID is mainly induced by caspase 8, connecting the extrinsic and intrinsic pathways of apoptosis, but can be also exerted by caspases 3 [102] and 10 [103] and by several non-caspase proteases, such as granzyme B [104], cathepsins [105], and calpains [106,107]. Stimulation of these of non-caspase proteases may represent the mechanism underlying neuronal activation of BID, as oxytosis is mediated independently of caspase 8 or 3 activation [21]. *In vitro* models of hypoxia and *in vivo* studies of cerebral ischemia report a caspase 1-mediated cleavage of BID to its active form, tBID, leading to the subsequent mitochondrial damage and cell death [108]. Finally, calpains, which are activated following ROS-dependent increases in intracellular Ca^{++} levels, are suggested to cleave BID, thereby inducing the tBID-triggered mitochondrial damage [109–111]. However, studies have revealed minor impact of these BID activation pathways in HT-22 cells, as inhibition of caspase 2, 3, or 8, as well as inhibition of calpains, failed to protect HT-22 cells from glutamate-induced cell death [21]. In addition, recent studies show that both full-length BID and tBID are sufficient to induce mitochondrial dysfunction and cell death in neurons [112]. In particular, full-length BID has been shown to initiate apoptosis in cultured rat hippocampal neurons [101,113,114] and to possibly be involved in caspase-independent cell death with slower kinetics, but tBID is suggested to initiate a more rapid mitochondrial damage in a caspase-dependent manner [112]. In the model of oxytosis in HT-22 cells, activated full-length BID and tBID may both contribute to neuronal death, since pharmacological BID inhibition prevents the glutamate-induced translocation of full-length BID to mitochondria and the subsequent mitochondrial damage and cell death [21,96,100]. However, BID cleavage could not be detected in cytosolic and mitochondrial fractions of glutamate-exposed HT-22 cells.

In contrast to the well-established role of BID as a key player in the model of oxytosis in HT-22 cells, the role of Bax in neuronal oxytosis is not fully apparent. Although Bax knockdown failed to protect primary neurons against glutamate toxicity [98], Bax-inhibiting peptides could preserve Bax translocation and glutamate-induced cerebellar granule cell death [115], as well as neuronal cell death following global cerebral ischemia [116]. In addition, knockout of Bax was efficient in providing neuroprotection in mouse models of cerebral ischemia [117,118] and brain trauma [119], and pharmacological inhibition of mitochondrial Bax's pore formation was neuroprotective in a model of global brain ischemia [120].

Besides their direct role in MOMP and the release of death-promoting proteins during oxytosis, the Bcl-2 family members also interact with proteins at the IMM (optic atrophy-1, OPA1) and with dynamin-related proteins (DRP1, Mitofusin-1 and -2) at the OMM, and are thereby also involved in the regulation of mitochondrial dynamics. Defects in the mitochondrial balance between fusion and fission are early events in neurological diseases, as enhanced mitochondrial fragmentation is associated with the induction of neuronal cell death triggered by oxidative stress. Several studies show that mitochondrial fission is mainly mediated by DRP1 and mitochondrial fission 1 protein (FIS1) [121,122], while mitochondrial fusion is mainly regulated by Mitofusin-1 and -2. Under conditions of cellular stress, DRP1 is recruited to the OMM [123,124], where it co-localizes with Bax and Mitofusin-21 [125]. The required function of DRP1 in mitochondrial fission has been reported in several studies, which claim that siRNA-mediated downregulation of DRP1 or the expression of a dominant negative mutant (DRP1K28A) preserves mitochondrial fission, caspase activation, and cell death [123,126,127]. A particular role for DRP1 in oxidative stress-related cell death was concluded from recent studies reporting that a cell line with an endogenous loss-of-activity mutation in DRP1 was resistant to hydrogen peroxide-induced cell death [128]. In the model of oxytosis in HT-22 cells, pharmacological inhibition of DRP1 by small molecules (MdiviA, MdiviB) blocked translocation of DRP1 to mitochondria, which prevented mitochondrial fragmentation and the formation of mitochondrial ROS. Furthermore, DRP1 inhibition or DRP1 gene silencing prevented tBID-induced mitochondrial fragmentation and cell death, suggesting a detrimental interplay between the Bcl-2 family proteins and DRP1. However, the nature of this proposed interaction remains to be elucidated, since a direct interaction between BID and DRP1 has not yet been detected. Global DRP1 KO mice have an embryonic lethal phenotype, and neuronal cell-specific DRP1 knockout led to brain hypoplasia and subsequent death [129]. Interestingly, DRP1 inhibitors demonstrated neuroprotective properties against glutamate-induced excitotoxicity and OGD-related cell death *in vitro* in primary cultured neurons. Moreover, DRP1 inhibitors reduced the infarct size *in vivo* in a model of cerebral ischemia [76], suggesting a critical role of DRP1 in maintaining mitochondrial function and cell-survival pathways.

14.4.2 The Role of Mitochondrial ROS

Under physiological conditions, ROS are produced as byproducts of the oxygen metabolism. These oxygen derivatives appear as peroxides (H_2O_2, $R-O_2$) or free radicals (superoxide anion $O_2^{\bullet-}$, hydroperoxyl $HO_2^{\bullet}$, and hydroxyl $^{\bullet}OH$) and have distinct functions in tissue homeostasis and a variety of redox-signaling cascades [130]. Typically, ROS are produced by different oxidases, such as nicotinamide adenine dinucleotide phosphate (NADPH) oxidase (NOX) and xanthine oxidase [131]. Under physiological conditions, the amount of cellular ROS is tightly controlled and is maintained at low levels. This is achieved by various antioxidant defense mechanisms, including high GSH levels and the activity of superoxide dismutases, catalases, and glutathione peroxidases [131]. If the antioxidative machinery loses its ability to detoxify the cellular ROS, the amount of ROS rises, causing severe cellular damage that ultimately results in cell death. Hallmarks of this damage are the oxidation of unsaturated fatty acids and SH groups of proteins, as well as the formation of protein carbonyls.

Mitochondria are highly dynamic organelles that, under physiological conditions, produce around 2% of their consumed oxygen as ROS. Therefore, they are the main source of detrimental ROS within a cell [131]. Mitochondrial ROS are predominantly produced by the mitochondrial electron transport chain, although other enzymes (e.g., 2-oxoglutarate dehydrogenase, pyruvate dehydrogenase) also contribute to this process [132]. Generation of superoxide anions mainly occurs at mitochondrial complex I and III of the respiratory chain [133]. Superoxide anion production depends on the proton-motive force, reduced/oxidized nicotinamide adenine dinucleotide (NADH/NAD^+) and $CoQH_2/CoQ$ ratios, and local O_2 concentrations [132]. Under physiological conditions, molecular oxygen is reduced to water during ATP generation via the respiratory chain, and only small amounts of superoxides are produced. Most of our knowledge on mitochondrial ROS production arises from investigations in isolated mitochondria, so little is known about the *in vivo* situation or the exact role of mitochondrial ROS during oxidative stress in intact cells [132,133]. There is evidence that both too high and too low amounts of mitochondrial ROS can cause mitochondrial dysfunction and cell death, underlining the pivotal role of ROS balance during cell signaling [134]. The MMP seems to be one of the major regulators of mitochondrial ROS production. An increase in the proton-motive force leads to an enhanced production of ROS [135]. Mitochondrial ROS mainly target mtDNA [136]. Damage of mtDNA leads to impaired synthesis of mitochondrial-complex proteins of the electron transport chain and, thereby, to a disruption of electron transport and to loss of the MMP and a decrease in ATP production (OXPHOS). Furthermore, mitochondria-generated ROS are shown to induce the formation of the mitochondrial permeability transition pore that allows translocation of AIF to the nucleus [99] and, ultimately, cell death.

14.5 Downstream Mechanisms of Mitochondrial Dysfunction

Inhibition of cystine uptake results in GSH depletion, mitochondrial ROS production, and the opening of cyclic guanosine monophosphate (cGMP)-gated channels, allowing the influx of extracellular calcium and the formation of MOMP [7]. Permeabilization of the OMM promotes the release of proapoptotic proteins from the intermembrane space, which brings about the final steps of regulated cell death. AIF is an example of just such a mitochondrial protein that is released from the mitochondria to the nucleus. During oxytosis, AIF undergoes proteolytic cleavage, mitochondrial release, and immediate translocation to the nucleus, where it induces caspase-independent cell death associated with chromatin condensation and large-scale DNA fragmentation (Figure 14.1) [21,24,97,98].

AIF is a flavin adenine dinucleotide-containing, NADH-dependent oxidoreductase anchored to the IMM of the mitochondrial intermembrane space. It is implicated in diverse yet interconnected physiological processes that are important for mitochondrial function and integrity – including, for example, the maintenance of respiratory-chain complex I. Analysis of *in vivo* phenotypes associated with AIF deficiency or defects and the identification of its mitochondrial, cytoplasmic, and nuclear interaction partners have revealed the complex regulation of AIF-mediated signal transduction and suggest an important role of AIF in the maintenance of mitochondrial morphology and energy metabolism. Additionally, the protein is proposed to regulate the respiratory chain indirectly through assembly and/or stabilization of mitochondrial complexes I and III. In mitochondria, AIF plays a pivotal role in the regulation of respiratory-chain super-complexes and their activities, generating and maintaining the MMP, energy supply, ROS formation, and, thus, mitochondrial morphology and integrity [137–141].

Recent studies suggest that AIF itself lacks nuclease activity, suggesting that AIF-mediated DNA damage requires binding to other proteins. For example, AIF has been reported to form complexes with cyclophilin A (CypA) for translocation to the nucleus and DNA degradation [97,142]. More recent studies in cultured mouse hippocampal neurons show that the nuclear AIF–CypA complex also requires histone H_2AX binding for DNA damage and cell death. The AIF–CypA complex requires an interaction between an AIF region and one of the AIF helices, together with a part of the β-barrel of CypA [97]. Recent studies using designed AIF peptides that specifically target AIF–CypA complex formation investigated the potential of these inhibitory peptides in a model of glutamate-induced oxytosis [22]. Inhibition of AIF–CypA complex formation blocked the nuclear translocation of AIF and CypA, reduced the sustained increase in intracellular Ca^{++}, and drastically reduced the sensitivity of glutamate-mediated oxidative stress in neuronal cells [22].

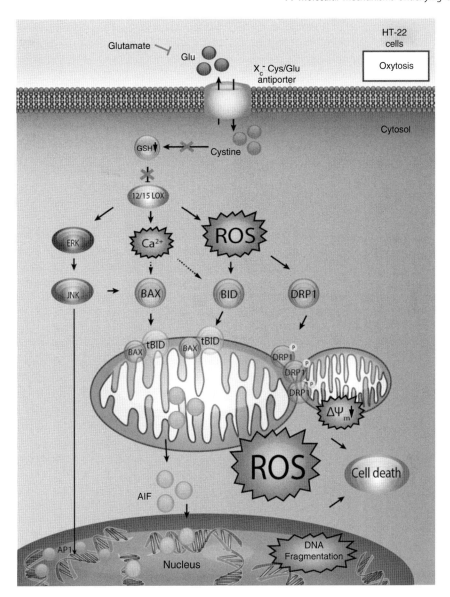

Figure 14.1 Glutamate-induced oxytosis in HT-22 cells. Oxytosis is initiated by high levels of extracellular glutamate that inhibit Xc⁻Cys/Glu antiporter and trigger GSH depletion. A reduction in GSH levels leads to increased 12/15-LOX activity and the formation of reactive oxygen species (ROS). Downstream of this first ROS production, proapoptotic proteins such as BID, Bax, and DRP1 are activated and translocate to the OMM, where Bax can form homo-oligomers, leading to mitochondrial pore formation and mitochondrial outer-membrane permeabilization (MOMP). BID and its active truncated form, tBID, trigger mitochondrial dysfunction through interactions with Bax and DRP1. Upon phosphorylation, DRP1 proteins translocate to the mitochondria and initiate mitochondrial fragmentation/fission. The resulting loss of MMP entails a second burst in ROS and further unbalances the cellular redox levels. Afterwards, apoptosis-inducing factor (AIF) is released from the mitochondria in the cytosol and translocates to the nucleus, promoting large-scale chromatin DNA fragmentation and subsequent cell death. Inhibition of different proteins by small-molecule inhibitors or siRNA can provide neuroprotection and expose novel targets for therapeutics in the treatment of neurological disorders associated with oxytosis.

Since the translocation of mitochondrial AIF to the nucleus is regarded as an initiator event for DNA damage in cell-death pathways, many studies have aimed to reveal the function of AIF under conditions of reduced AIF expression [20,23,141,143]. Indeed, a reduction of AIF expression *in vitro* or *in vivo* prevented neuronal cell death in several models of neurological diseases [23,144,145]. AIF and its interacting partners mediate physiological functions at the level of the mitochondria and determine cellular demise. Beyond its pathological function in neuronal cell death, AIF also plays an important physiological role in mitochondrial function and integrity by stabilizing respiratory chain complex I [146], the rate-limiting step for oxygen consumption in mammalian cells. This is demonstrated by the fact that genetic deletion of AIF is embryonic-lethal [147]. Reduced AIF levels, however, are well tolerated during development and at a young age, as shown when investigating mice carrying the Harlequin (*Hq*) mutation, which leads to a reduction of AIF protein levels by 80–90% [148]. Although *Hq* mice develop cerebellar and retinal neuron degeneration and show increased vulnerability to oxidative stress [148,149] when they are subjected to acute brain-damage paradigms such as cerebral ischemia and brain trauma, they also show significant neuroprotection [20,144]. Furthermore, emerging *in vivo* data identify nuclear AIF translocation and the associated DNA damage as a key feature of neuronal death following neonatal hypoxia–ischemia and traumatic brain injury [144,145].

Alterations in mitochondrial metabolism and function have recently been associated with increased resistance of neurons against stress [150,151], suggesting that moderate AIF depletion may induce mitochondrial preconditioning, a process previously termed "mitochondrial hormesis" or "mitohormesis" [152]. Disturbed AIF expression, processing, and dislocation may consequently impose increasing levels of stress on the cell, which may result in either mitohormesis (mitochondrial adaptation/preconditioning) or the demise of the mitochondria and the cell, depending on the time, strength, and context of the insult. In a model of glutamate-induced oxytosis in HT-22 cells, previous studies showed that AIF depletion prevented mitochondrial fission and perinuclear accumulation, and preserved the cells' tubular shape in the presence of toxic doses of glutamate [22,24]. In HT-22 cells, oxidative stress involves detrimental mitochondrial fission, loss of MMP, and ATP depletion, along with the release of AIF to the nucleus and cell death. Recent studies have demonstrated that AIF gene silencing significantly enhances neuronal survival following a lethal exposure to glutamate. In addition, real-time detection of cell death over a period of 15 hours further supports the neuroprotective effect mediated by AIF silencing. These findings suggest that AIF depletion prevents neuronal death *in vitro* in paradigms of glutamate oxytosis via mitochondrial preconditioning pathways [23].

14.6 Oxytosis and ER Stress: Common Features and Differences

The ER is the organelle involved in the correct assembly and folding of secretory and membrane proteins, as well as some organelle-targeted proteins [153]. Due to the high number of proteins inside the ER lumen, various insults can lead to the accumulation of unfolded proteins and ER stress, including nutrient deprivation, alterations in the oxidation-reduction balance, changes in the intracellular calcium concentration, and failure of post-translational modifications. The accumulation of unfolded proteins activates an adaptive stress response termed the "unfolded protein response" (UPR). The UPR maintains the survival of the cell by employing three different strategies: (i) attenuation of the rate of general translation to reduce the total number of proteins; (ii) expression of ER-resident chaperones to enhance protein folding inside the ER lumen; and (iii) activation of ER-associated degradation to eliminate misfolded proteins [154]. Three ER transmembrane receptors are important for mediation of the UPR in mammalian cells: PKR-like ER kinase (PERK), activating transcription factor-6 (ATF-6), and inositol-requiring enzyme 1 (IRE1). Under physiological conditions, these receptors are associated with the ER-resident chaperone 78 kDa glucose-regulating protein (GRP78; also known as binding immunoglobulin protein, BiP). The accumulation of unfolded proteins inside the ER lumen leads to GRP78 dissociation and the activation of the three ER transmembrane receptors [154].

Active PERK phosphorylates the eukaryotic initiation factor 2 (eIF2α), leading to inhibition of mRNA translation [155]. In addition, certain mRNAs (e.g., ATF-4) avoid the translational block and show intensified translation [156]. Active ATF-6 is a transcription factor for genes containing an ER stress-response element [154]. Active IRE1 exhibits an endoribonuclease activity that enables the removal of an intron in X-box-binding protein 1 (XBP1) mRNA, which is induced by ATF-6, and results in a frame-shift splice variant (sXBP1) [157]. Under conditions of prolonged ER stress, this initially adaptive stress response switches toward proapoptotic signaling [158], which involves in particular prolonged activation of caspase 12 and CHOP [159].

Activation of the UPR has also been detected in glutamate-induced oxytosis. Expression levels of GRP78, CHOP, and caspase 12 were increased in HT-22 cells challenged with toxic concentrations of glutamate [160], while the well-accepted ER stress-inducer tunicamycin, which inhibits N-linked glycosylation in ER, produced enhanced GRP78 and CHOP expression in HT-22 cells [161]. Other frequently used ER stressors, namely brefeldin A, an inhibitor of endomembrane-system transport, and thapsigargin, an irreversible SERCA inhibitor, induced activation of caspase 12 and the expression of a variety of UPR proteins in HT-22 cells [162]. In such cells exposed to ER stress, knockdown of CHOP and caspase 12 only slightly increased cell viability [162]. Furthermore, in a glutamate-induced toxicity model, reduced levels of CHOP and caspase 12 improved cell survival [160]. However, since pan-caspase inhibitors had previously failed to protect against oxytosis in HT-22 cells [21], the impact of caspase 12 activation in glutamate-induced oxidative cell death is under debate.

In HT-22 cells, ERK, p38, and JNK were activated via ER stress induced by brefeldin A and thapsigargin [162]. In paradigms of ER stress, phosphorylation of JNK is mediated by a complex of IRE1, Traf2, and ASK1 [163], while in conditions of oxidative glutamate toxicity, ERK, p38, and JNK phosphorylation are enhanced and inhibitors of MAPKs protect against both ER stress-associated cell death and oxytosis [34,162]. Therefore, the activation of MAPKs seems to be a shared downstream mechanism of oxytosis and ER stress.

The eIF2α/ATF-4 pathway plays a pivotal role in glutamate-induced oxidative cell death. Using genetic screening, Tan and colleagues were the first to identify the involvement of eIF2α in the regulation of intracellular GSH levels and in oxytosis [3]. Both downregulation and constitutive phosphorylation of eIF2α preserved GSH levels, reduced ROS formation and Ca^{++} influx, and attenuated oxytosis in HT-22 cells [3]. In line with these findings, a constitutive repressor of eIF2α dephosphorylation attenuated ROS production and protected against glutamate toxicity. Further, this approach also prevented cell death induced by tunicamycin [164]. On the other hand, pretreatment with thapsigargin and tunicamycin protected against oxytosis, which was attributed to the enhanced eIF2α phosphorylation [165]. The same group generated an artificial eIF2α kinase, which induced constitutive eIF2α phosphorylation and protected against oxytosis and ER stress-associated cell death induced by tunicamycin. Furthermore, salubrinal, a selective inhibitor of eIF2α dephosphorylation, mediated protection against glutamate toxicity in HT-22 cells while promoting an induction of ATF-4 protein [166]. Additionally, ATF-4 overexpression led to enhanced X_c^- Cys/Glu antiporter activity, higher GSH levels, and attenuated oxytosis [166]. In line with these findings, mouse embryonic fibroblasts lacking ATF-4 showed reduced GSH and increased ROS levels [167]. Lewerenz and colleagues provided evidence that a mutated form of ATF-4 is responsible for HT-22 cell resistance against oxidative glutamate toxicity [35]. This ATF-4 mutant is constitutively expressed, uncoupled from eIF2α phosphorylation, and acts as a protective protein. However, the role of ATF-4 in cell death and survival remains controversial, with protective [168] and pro-death [169] effects described in different model systems. Besides its ability to regulate the genes involved in cell-death pathways [169], ATF-4 may also enhance X_c^- Cys/Glu antiporter expression, thereby preventing cell death in settings where it is attributed to GSH depletion.

In different models of focal and global cell death, transient ischemia levels of the ER-related chaperone GRP78 are increased [170–172]. ER-resident chaperones, such as GRP78, regulate protein folding in the ER lumen and prevent aggregation of unfolded proteins [173], suggesting a protective role of chaperones in conditions of disturbed ER function. In fact, ER chaperones play an important role in suppressing oxidative stress and stabilizing calcium in models of neuronal glutamate toxicity *in vitro* [174,175]. Specific inducers of GRP78 protect against ER stress-induced cell death in neuroblastoma cells [176] and show neuroprotective effects following middle cerebral artery occlusion *in vivo* [177]. In HT-22 cells, proteomic analysis reveals downregulation of GRP78 due to glutamate exposure [178], suggesting alterations of the ER protein quality-control process during oxytosis. However, in HT-22 cells, induction of the GRP78 by the ATF-6 pathway protects against tunicamycin-induced cell death but not against glutamate challenge [161]. Thus, although oxytosis seems to be associated with impaired protein folding, ER-resident chaperones are not sufficient to protect against this form of oxidative cell death.

In oxidative glutamate toxicity, 12-LOX produces eicosanoids, which activate soluble guanylate cyclases and thereby increase the concentration of cGMP [7]. Influx of extracellular Ca^{++} via a cGMP-dependent Ca^{++} channel is a necessary step preceding cell death in glutamate-induced oxidative toxicity [179]. In addition to extracellular calcium influx, glutamate can also stimulate inositol triphosphate (IP_3) accumulation by activating group I metabotropic glutamate receptors and triggering Ca^{++} release from the ER via IP_3 receptors in HT-22 cells [180]. Further, thapsigargin mediates ER Ca^{++} store depletion, which results in apoptosis [162]. ER stress induced by tunicamycin in HeLa cells and by glutamate treatment in HT-22 cells leads to calcium overload [22,23,181], suggesting intracellular calcium disturbances are a potential link between ER stress and oxytosis. Therefore, targeting intracellular Ca^{++} regulations may elicit therapeutic strategies in both situations. One prospective approach might be the activation of intracellularly located potassium channels, such as small-conductance calcium-activated potassium channels, which prevent intracellular calcium deregulation and cellular demise in glutamate-induced oxytosis and brefeldin A-mediated ER stress in HT-22 cells (unpublished data) [23]. Recent evidence suggests deregulated store-operated calcium entry (SOCE) as the main mechanism underlying the detrimental Ca^{++} influx in oxytosis [182]. SOCE activation occurs as a result of ER Ca^{++} store depletion via the IP_3 receptor, in order to refill the ER. Stromal interaction molecule 1 activates calcium release-activated calcium channel protein 1 (ORAI1), a calcium channel located at the plasma membrane, which leads to Ca^{++} entry from the extracellular space. Henke and colleagues found reduced ER Ca^{++} content after glutamate stimulation in HT-22 cells. ER Ca^{++} store depletion in these cells (e.g., by the SERCA inhibitor thapsigargin) is sufficient to activate the UPR and results in ER stress-associated apoptosis [162,182]. Potentially, reduced ER Ca^{++} levels may be the reason for UPR activation by glutamate. Glutamate-resistant HT-22 cells (HT-22R) are characterized by reduced levels of ORAI1 and disturbed SOCE signaling [182]. HT-22R are protected not only against oxytosis, but also against ER stress-associated cell death evoked by tunicamycin [183]. However, pharmacological SOCE inhibition is not protective against tunicamycin toxicity in non-resistant HT-22 cells [182].

In conclusion, glutamate-induced oxytosis activates the UPR and shares executors of cell death with the ER stress response, although several effective strategies of protection against ER stress fail in oxytosis, and vice versa. However, Ca^{++} deregulation at the level of the cytosol, the ER, and the mitochondria is a common feature in both situations. Thus, maintaining intracellular calcium homeostasis represents a promising combined strategy in the prevention of oxytosis and ER stress-induced cell death.

14.7 Conclusion

Oxytosis is initiated by very high glutamate concentrations in millimolar range *in vitro* in HT-22 cells and in immature primary neuronal cells. These glutamate concentrations are required for the competitive inhibition of the X_c^- Cys/Glu antiporter, in order to counteract the cystine cell-culture medium concentration of 200 µM necessary for *in vitro* cell survival and proliferation. *In vitro* studies demonstrating that blocking the X_c^- Cys/Glu antiporter with glutamate leads to GSH depletion and cell death are consistent with *in vivo* studies showing substantial increases in extracellular glutamate followed by a significant drop in GSH levels during cerebral ischemia. Moreover, the key hallmarks of oxytosis (including LOX activation and inhibition of GPX4) that led to the mitochondrial transactivation of BID, the loss of mitochondrial integrity and function, the release of AIF to the nucleus, and DNA fragmentation are all detected in various *in vivo* model systems of neuronal death of relevance to human diseases. Furthermore, there is significant evidence from genetic and pharmacological studies that interference with these key mechanisms of oxytosis attenuates neuronal damage *in vivo* (e.g., in models of cerebral ischemia) [184,185]. Overall, oxytosis in HT-22 cells and immature neurons is a well-defined and widely established model system of neuronal oxidative damage that is highly valuable to the identification of novel therapeutic targets for the prevention of mitochondrial dysfunction and neuronal death in oxidative stress conditions related to human diseases.

References

1. Coyle JT, Puttfarcken P. Oxidative stress, glutamate, and neurodegenerative disorders. *Science* 1993;**262**(5134):689–95.
2. Lin MT, Beal MF. Mitochondrial dysfunction and oxidative stress in neurodegenerative diseases. *Nature* 2006;**443**(7113):787–95.

3. Tan S, Schubert D, Maher P. Oxytosis: a novel form of programmed cell death. *Curr Top Med Chem* 2001;**1**(6):497–506.
4. Bannai S, Kitamura E. Transport interaction of L-cystine and L-glutamate in human diploid fibroblasts in culture. *J Biol Chem* 1980;**255**(6):2372–6.
5. Maher P, Davis JB. The role of monoamine metabolism in oxidative glutamate toxicity. *J Neurosci* 1996;**16**(20):6394–401.
6. Albrecht P, Lewerenz J, Dittmer S, Noack R, Maher P, Methner A. Mechanisms of oxidative glutamate toxicity: the glutamate/cystine antiporter system xc- as a neuroprotective drug target. *CNS Neurol Disord Drug Targets* 2010;**9**(3):373–82.
7. Albrecht P, Henke N, Tien ML, Issberner A, Bouchachia I, Maher P, et al. Extracellular cyclic GMP and its derivatives GMP and guanosine protect from oxidative glutamate toxicity. *Neurochem Int* 2013;**62**(5):610–19.
8. Kritis AA, Stamoula EG, Paniskaki KA, Vavilis TD. Researching glutamate-induced cytotoxicity in different cell lines: a comparative/collective analysis/study. *Front Cell Neurosci* 2015;**9**:91.
9. Satoh T, Nakatsuka D, Watanabe Y, Nagata I, Kikuchi H, Namura S. Neuroprotection by MAPK/ERK kinase inhibition with U0126 against oxidative stress in a mouse neuronal cell line and rat primary cultured cortical neurons. *Neurosci Lett* 2000;**288**(2):163–6.
10. Sasaki H, Sato H, Kuriyama-Matsumura K, Sato K, Maebara K, Wang H, et al. Electrophile response element-mediated induction of the cystine/glutamate exchange transporter gene expression. *J Biol Chem* 2002;**277**(47):44 765–71.
11. Lewerenz J, Albrecht P, Tien ML, Henke N, Karumbayaram S, Kornblum HI, et al. Induction of Nrf2 and xCT are involved in the action of the neuroprotective antibiotic ceftriaxone *in vitro*. *J Neurochem* 2009;**111**(2):332–43.
12. Murphy TH, Miyamoto M, Sastre A, Schnaar RL, Coyle JT. Glutamate toxicity in a neuronal cell line involves inhibition of cystine transport leading to oxidative stress. *Neuron* 1989;**2**(6):1547–58.
13. Murphy TH, Schnaar RL, Coyle JT. Immature cortical neurons are uniquely sensitive to glutamate toxicity by inhibition of cystine uptake. *FASEB J* 1990;**4**(6):1624–33.
14. Morimoto BH, Koshland DE Jr. Excitatory amino acid uptake and N-methyl-D-aspartate-mediated secretion in a neural cell line. *Proc Natl Acad Sci USA* 1990;**87**(9):3518–21.
15. Tan S, Sagara Y, Liu Y, Maher P, Schubert D. The regulation of reactive oxygen species production during programmed cell death. *J Cell Biol* 1998;**141**(6):1423–32.
16. Culmsee C, Plesnila N. Targeting Bid to prevent programmed cell death in neurons. *Biochem Soc Trans* 2006;**34**(Pt. 6):1334–40.
17. Grohm J, Plesnila N, Culmsee C. Bid mediates fission, membrane permeabilization and peri-nuclear accumulation of mitochondria as a prerequisite for oxidative neuronal cell death. *Brain Behav Immun* 2010;**24**(5):831–8.
18. Dickey AS, Strack S. PKA/AKAP1 and PP2A/Bβ2 regulate neuronal morphogenesis via Drp1 phosphorylation and mitochondrial bioenergetics. *J Neurosci* 2011;**31**(44):15 716–26.
19. Dolga AM, Netter MF, Perocchi F, Doti N, Meissner L, Tobaben S, et al. Mitochondrial small conductance SK2 channels prevent glutamate-induced oxytosis and mitochondrial dysfunction. *J Biol Chem* 2013;**288**(15):10 792–804.
20. Culmsee C, Zhu C, Landshamer S, Becattini B, Wagner E, Pellecchia M, et al. Apoptosis-inducing factor triggered by poly(ADP-ribose) polymerase and Bid

mediates neuronal cell death after oxygen-glucose deprivation and focal cerebral ischemia. *J Neurosci* 2005;**25**(44):10 262–72.
21. Landshamer S, Hoehn M, Barth N, Duvezin-Caubet S, Schwake G, Tobaben S, et al. Bid-induced release of AIF from mitochondria causes immediate neuronal cell death. *Cell Death Differ* 2008;**15**(10):1553–63.
22. Doti N, Reuther C, Scognamiglio PL, Dolga AM, Plesnila N, Ruvo M, et al. Inhibition of the AIF/CypA complex protects against intrinsic death pathways induced by oxidative stress. *Cell Death Dis* 2014;**5**:e993.
23. Öxler EM, Dolga A, Culmsee C. AIF depletion provides neuroprotection through a preconditioning effect. *Apoptosis* 2012;**17**(10):1027–38.
24. Tobaben S, Grohm J, Seiler A, Conrad M, Plesnila N, Culmsee C. Bid-mediated mitochondrial damage is a key mechanism in glutamate-induced oxidative stress and AIF-dependent cell death in immortalized HT-22 hippocampal neurons. *Cell Death Differ* 2011;**18**(2):282–92.
25. Dringen R, Hirrlinger J. Glutathione pathways in the brain. *Biol Chem* 2003;**384**(4):505–16.
26. Aquilano K, Baldelli S, Ciriolo MR. Glutathione: new roles in redox signaling for an old antioxidant. *Front Pharmacol* 2014;**5**:196.
27. Rodríguez-Rodríguez A, Egea-Guerrero JJ, Murillo-Cabezas F, Carrillo-Vico A. Oxidative stress in traumatic brain injury. *Curr Med Chem* 2014;**21**(10):1201–11.
28. de la Asuncion JG, Millan A, Pla R, Bruseghini L, Esteras A, Pallardo FV, et al. Mitochondrial glutathione oxidation correlates with age-associated oxidative damage to mitochondrial DNA. *FASEB J* 1996;**10**(2):333–8.
29. Bharath S, Hsu M, Kaur D, Rajagopalan S, Andersen JK. Glutathione, iron and Parkinson's disease. *Biochem Pharmacol* 2002;**64**(5–6):1037–48.
30. Schulz JB, Lindenau J, Seyfried J, Dichgans J. Glutathione, oxidative stress and neurodegeneration. *Eur J Biochem* 2000;**267**(16):4904–11.
31. Janáky R, Varga V, Hermann A, Saransaari P, Oja SS. Mechanisms of L-cysteine neurotoxicity. *Neurochem Res* 2000;**25**(9–10):1397–405.
32. Seiler A, Schneider M, Förster H, Roth S, Wirth EK, Culmsee C, et al. Glutathione peroxidase 4 senses and translates oxidative stress into 12/15-lipoxygenase dependent- and AIF-mediated cell death. *Cell Metab* 2008;**8**(3):237–48.
33. Froissard P, Monrocq H, Duval D. Role of glutathione metabolism in the glutamate-induced programmed cell death of neuronal-like PC12 cells. *Eur J Pharmacol* 1997;**326**(1):93–9.
34. Suh HW, Kang S, Kwon KS. Curcumin attenuates glutamate-induced HT22 cell death by suppressing MAP kinase signaling. *Mol Cell Biochem* 2007;**298**(1–2):187–94.
35. Lewerenz J, Sato H, Albrecht P, Henke N, Noack R, Methner A, Maher P. Mutation of ATF4 mediates resistance of neuronal cell lines against oxidative stress by inducing xCT expression. *Cell Death Differ* 2012;**19**(5):847–58.
36. Choi BH, Hur EM, Lee JH, Jun DJ, Kim KT. Protein kinase Cdelta-mediated proteasomal degradation of MAP kinase phosphatase-1 contributes to glutamate-induced neuronal cell death. *J Cell Sci* 2006;**119**(Pt. 7):1329–40.
37. Fukui M, Song JH, Choi J, Choi HJ, Zhu BT. Mechanism of glutamate-induced neurotoxicity in HT22 mouse hippocampal cells. *Eur J Pharmacol* 2009;**617**(1–3):1–11.

38 Phillis JW, Horrocks LA, Farooqui AA. Cyclooxygenases, lipoxygenases, and epoxygenases in CNS: their role and involvement in neurological disorders. *Brain Res Rev* 2006;**52**(2):201–43.

39 Minghetti L. Role of COX-2 in inflammatory and degenerative brain diseases. *Subcell Biochem* 2007;**42**:127–41.

40 Watanabe T, Haeggström JZ. Rat 12-lipoxygenase: mutations of amino acids implicated in the positional specificity of 15- and 12-lipoxygenases. *Biochem Biophys Res Commun* 1993;**192**(3):1023–109.

41 Sasaki T, Kitagawa K, Yamagata K, Takemiya T, Tanaka S, Omura-Matsuoka E, et al. Amelioration of hippocampal neuronal damage after transient forebrain ischemia in cyclooxygenase-2-deficient mice. *J Cereb Blood Flow Metab* 2004;**24**(1):107–13.

42 van Leyen K, Kim HY, Lee SR, Jin G, Arai K, Lo EH. Baicalein and 12/15-lipoxygenase in the ischemic brain. *Stroke* 2006;**37**(12):3014–18.

43 Czubowicz K, Czapski GA, Cieślik M, Strosznajder RP. Lipoxygenase inhibitors protect brain cortex macromolecules against oxidation evoked by nitrosative stress. *Folia Neuropathol* 2010;**48**(4):283–92.

44 Strosznajder JB, Cieslik M, Cakala M, Jesko H, Eckert A, Strosznajder RP. Lipoxygenases and poly(ADP-ribose) polymerase in amyloid beta cytotoxicity. *Neurochem Res* 2011;**36**(5):839–48.

45 Son Y, Cheong YK, Kim NH, Chung HT, Kang DG, Pae HO. Mitogen-activated protein kinases and reactive oxygen species: how can ROS activate MAPK pathways? *J Signal Transduct* 2011;**2011**:792639.

46 Ruffels J, Griffin M, Dickenson JM. Activation of ERK1/2, JNK and PKB by hydrogen peroxide in human SH-SY5Y neuroblastoma cells: role of ERK1/2 in H2O2-induced cell death. *Eur J Pharmacol* 2004;**483**(2–3):163–73.

47 Hetman M, Gozdz A. Role of extracellular signal regulated kinases 1 and 2 in neuronal survival. *Eur J Biochem* 2004;**271**(11):2050–5.

48 Cheung EC, Slack RS. Emerging role for ERK as a key regulator of neuronal apoptosis. *Sci STKE* 2004;**2004**(251):PE45.

49 Zhuang S, Schnellmann RG. A death-promoting role for extracellular signal-regulated kinase. *J Pharmacol Exp Ther* 2006;**319**(3):991–7.

50 Stanciu M, Wang Y, Kentor R, Burke N, Watkins S, Kress G, et al. Persistent activation of ERK contributes to glutamate-induced oxidative toxicity in a neuronal cell line and primary cortical neuron cultures. *J Biol Chem* 2000;**275**(16):12 200–6.

51 Satoh T, Nakatsuka D, Watanabe Y, Nagata I, Kikuchi H, Namura S. Neuroprotection by MAPK/ERK kinase inhibition with U0126 against oxidative stress in a mouse neuronal cell line and rat primary cultured cortical neurons. *Neurosci Lett* 2000;**288**(2):163–6.

52 Muthalif MM, Benter IF, Karzoun N, Fatima S, Harper J, Uddin MR, Malik KU. 20-hydroxyeicosatetraenoic acid mediates calcium/calmodulin-dependent protein kinase II-induced mitogen-activated protein kinase activation in vascular smooth muscle cells. *Proc Natl Acad Sci USA* 1998;**95**(21):12 701–6.

53 Zhang X, Shan P, Sasidhar M, Chupp GL, Flavell RA, Choi AM, Lee PJ. Reactive oxygen species and extracellular signal-regulated kinase 1/2 mitogen-activated protein kinase mediate hyperoxia-induced cell death in lung epithelium. *Am J Respir Cell Mol Biol* 2003;**28**(3):305–15.

54 Cagnol S, Chambard JC. ERK and cell death: mechanisms of ERK-induced cell death - apoptosis, autophagy and senescence. *FEBS J* 2010;**277**(1):2–21.

55 Namura S, Iihara K, Takami S, Nagata I, Kikchi H, Matsushita K, et al. Intravenous administration of MEK inhibitor U0126 affords brain protection against forebrain ischemia and focal cerebral ischemia. *Proc Natl Acad Sci USA* 2001;**98**(20):11 569–74.

56 Park SY, Jung WJ, Kang JS, Kim CM, Park G, Choi YW. Neuroprotective effects of α-iso-cubebene against glutamate-induced damage in the HT22 hippocampal neuronal cell line. *Int J Mol Med* 2015;**35**(2):525–32.

57 Kim HN, Kim YR, Jang JY, Choi YW, Baek JU, Hong JW, et al. Neuroprotective effects of Polygonum multiflorum extract against glutamate-induced oxidative toxicity in HT22 hippocampal cells. *J Ethnopharmacol* 2013;**150**(1):108–15.

58 Jeong EJ, Hwang L, Lee M, Lee KY, Ahn MJ, Sung SH. Neuroprotective biflavonoids of Chamaecyparis obtusa leaves against glutamate-induced oxidative stress in HT22 hippocampal cells. *Food Chem Toxicol* 2014;**64**:397–402.

59 Matsuzawa A, Ichijo H. Redox control of cell fate by MAP kinase: physiological roles of ASK1-MAP kinase pathway in stress signaling. *Biochim Biophys Acta* 2008;**1780**(11):1325–36.

60 Mehan S, Meena H, Sharma D, Sankhla R. JNK: a stress-activated protein kinase therapeutic strategies and involvement in Alzheimer's and various neurodegenerative abnormalities. *J Mol Neurosci* 2011;**43**(3):376–90.

61 Widmann C, Gibson S, Jarpe MB, Johnson GL. Mitogen-activated protein kinase: conservation of a three-kinase module from yeast to human. *Physiol Rev* 1999;**79**(1):143–80.

62 Yang EJ, Lee JY, Park SH, Lee T, Song KS. Neuroprotective effects of neolignans isolated from Magnoliae Cortex against glutamate-induced apoptotic stimuli in HT22 cells. *Food Chem Toxicol* 2013;**56**:304–12.

63 Mielke K, Herdegen T. JNK and p38 stresskinases – degenerative effectors of signal-transduction-cascades in the nervous system. *Prog Neurobiol* 2000;**61**(1):45–60.

64 Maundrell K, Antonsson B, Magnenat E, Camps M, Muda M, Chabert C, et al. Bcl-2 undergoes phosphorylation by c-Jun N-terminal kinase/stress-activated protein kinases in the presence of the constitutively active GTP-binding protein Rac1. *J Biol Chem* 1997;**272**(40):25 238–42.

65 Cardoso AL, Simões S, de Almeida LP, Pelisek J, Culmsee C, Wagner E, Pedroso de Lima MC. siRNA delivery by a transferrin-associated lipid-based vector: a non-viral strategy to mediate gene silencing. *J Gene Med* 2007;**9**(3):170–83.

66 Cardoso AL, Costa P, de Almeida LP, Simões S, Plesnila N, Culmsee C, et al. Tf-lipoplex-mediated c-Jun silencing improves neuronal survival following excitotoxic damage in vivo. *J Control Release* 2010;**142**(3):392–403.

67 Kim BJ, Ryu SW, Song BJ. JNK- and p38 kinase-mediated phosphorylation of Bax leads to its activation and mitochondrial translocation and to apoptosis of human hepatoma HepG2 cells. *J Biol Chem* 2006;**281**(30):21 256–65.

68 Guan QH, Pei DS, Zong YY, Xu TL, Zhang GY. Neuroprotection against ischemic brain injury by a small peptide inhibitor of c-Jun N-terminal kinase (JNK) via nuclear and non-nuclear pathways. *Neuroscience* 2006;**139**(2):609–27.

69 Tournier C, Hess P, Yang DD, Xu J, Turner TK, Nimnual A, et al. Requirement of JNK for stress-induced activation of the cytochrome c-mediated death pathway. *Science* 2000;**288**(5467):870–4.

70 Nagai H, Noguchi T, Takeda K, Ichijo H. Pathophysiological roles of ASK1-MAP kinase signaling pathways. *J Biochem Mol Biol* 2007;**40**(1):1–6.

71. Nakamura K, Johnson GL. Activity assays for extracellular signal-regulated kinase 5. *Methods Mol Biol* 2010;**661**:91–106.
72. Maher P. How protein kinase C activation protects nerve cells from oxidative stress-induced cell death. *J Neurosci* 2001;**21**(9):2929–38.
73. Ku BM, Lee YK, Jeong JY, Mun J, Han JY, Roh GS, et al. Ethanol-induced oxidative stress is mediated by p38 MAPK pathway in mouse hippocampal cells. *Neurosci Lett* 2007;**419**(1):64–7.
74. Park JY, Kim EJ, Kwon KJ, Jung YS, Moon CH, Lee SH, Baik EJ. Neuroprotection by fructose-1,6-bisphosphate involves ROS alterations via p38 MAPK/ERK. *Brain Res* 2004;**1026**(2):295–301.
75. Wang XZ, Ron D. Stress-induced phosphorylation and activation of the transcription factor CHOP (GADD153) by p38 MAP kinase. *Science* 1996;**272**(5266):1347–9.
76. Grohm J, Kim SW, Mamrak U, Tobaben S, Cassidy-Stone A, Nunnari J, et al. Inhibition of Drp1 provides neuroprotection *in vitro* and *in vivo*. *Cell Death Differ* 2012;**19**(9):1446–58.
77. Martinou JC, Youle RJ. Mitochondria in apoptosis: Bcl-2 family members and mitochondrial dynamics *Dev Cell* 2011;**21**(1):92–101.
78. Bereiter-Hahn J, Vöth M. Dynamics of mitochondria in living cells: shape changes, dislocations, fusion, and fission of mitochondria. *Microsc Res Tech* 1994;**27**(3):198–219.
79. Chan DC. Mitochondria: dynamic organelles in disease, aging, and development. *Cell* 2006;**125**(7):1241–52.
80. Jendrach M, Pohl S, Vöth M, Kowald A, Hammerstein P, Bereiter-Hahn J. Morpho-dynamic changes of mitochondria during ageing of human endothelial cells. *Mech Ageing Dev* 2005;**126**(6–7):813–21.
81. Bolaños JP, Moro MA, Lizasoain I, Almeida A. Mitochondria and reactive oxygen and nitrogen species in neurological disorders and stroke: therapeutic implications. *Adv Drug Deliv Rev* 2009;**61**(14):1299–315.
82. Chipuk JE, Bouchier-Hayes L, Green DR. Mitochondrial outer membrane permeabilization during apoptosis: the innocent bystander scenario. *Cell Death Differ* 2006;**13**(8):1396–402.
83. Desagher S, Martinou JC. Mitochondria as the central control point of apoptosis. *Trends Cell Biol* 2000;**10**(9):369–77.
84. Green DR, Kroemer G. The pathophysiology of mitochondrial cell death. *Science* 2004;**305**(5684):626–9.
85. Hardwick JM, Youle RJ. SnapShot: BCL-2 proteins. *Cell* 2009;**138**(2):404, 404.e1.
86. Kuwana T, Bouchier-Hayes L, Chipuk JE, Bonzon C, Sullivan BA, Green DR, Newmeyer DD. BH3 domains of BH3-only proteins differentially regulate Bax-mediated mitochondrial membrane permeabilization both directly and indirectly. *Mol Cell* 2005;**17**(4):525–35.
87. Youle RJ, Strasser A. The BCL-2 protein family: opposing activities that mediate cell death. *Nat Rev Mol Cell Biol* 2008;**9**(1):47–59.
88. Krajewska M, Mai JK, Zapata JM, Ashwell KW, Schendel SL, Reed JC, Krajewski S. Dynamics of expression of apoptosis-regulatory proteins Bid, Bcl-2, Bcl-X, Bax and Bak during development of murine nervous system. *Cell Death Differ* 2002;**9**(2):145–57.

89 Krajewska M, Zapata JM, Meinhold-Heerlein I, Hedayat H, Monks A, Bettendorf H, et al. Expression of Bcl-2 family member Bid in normal and malignant tissues. *Neoplasia* 2002;**4**(2):129–40.

90 Krajewski S, Bodrug S, Krajewska M, Shabaik A, Gascoyne R, Berean K, Reed JC. Immunohistochemical analysis of Mcl-1 protein in human tissues. Differential regulation of Mcl-1 and Bcl-2 protein production suggests a unique role for Mcl-1 in control of programmed cell death in vivo. *Am J Pathol* 1995;**146**(6):1309–19.

91 Krajewski S, Mai JK, Krajewska M, Sikorska M, Mossakowski MJ, Reed JC. Upregulation of bax protein levels in neurons following cerebral ischemia. *J Neurosci* 1995;**15**(10):6364–76.

92 Sun YF, Yu LY, Saarma M, Timmusk T, Arumae U. Neuron-specific Bcl-2 homology 3 domain-only splice variant of Bak is anti-apoptotic in neurons, but pro-apoptotic in non-neuronal cells. *J Biol Chem* 2001;**276**(19):16 240–7.

93 Plesnila N, Zinkel S, Le DA, Amin-Hanjani S, Wu Y, Qiu J, et al. BID mediates neuronal cell death after oxygen/glucose deprivation and focal cerebral ischemia. *Proc Natl Acad Sci USA* 2001;**98**(26):15 318–23.

94 Henshall DC, Bonislawski DP, Skradski SL, Lan JQ, Meller R, Simon RP. Cleavage of bid may amplify caspase-8-induced neuronal death following focally evoked limbic seizures. *Neurobiol Dis* 2001;**8**(4):568–80.

95 Yamamoto A, Murphy N, Schindler CK, So NK, Stohr S, Taki W, et al. Endoplasmic reticulum stress and apoptosis signaling in human temporal lobe epilepsy. *J Neuropathol Exp Neurol* 2006;**65**(3):217–25.

96 Oppermann S, Schrader FC, Elsässer K, Dolga AM, Kraus AL, Doti N, et al. Novel N-phenyl-substituted thiazolidinediones protect neural cells against glutamate- and tBid-induced toxicity. *J Pharmacol Exp Ther* 2014;**350**(2):273–89.

97 Candé C, Vahsen N, Kouranti I, Schmitt E, Daugas E, Spahr C, et al. AIF and cyclophilin A cooperate in apoptosis-associated chromatinolysis. *Oncogene* 2004;**23**(8):1514–21.

98 Cheung EC, Melanson-Drapeau L, Cregan SP, Vanderluit JL, Ferguson KL, McIntosh WC, et al. Apoptosis-inducing factor is a key factor in neuronal cell death propagated by BAX-dependent and BAX-independent mechanisms. *J Neurosci* 2005;**25**(6):1324–34.

99 Ott M, Gogvadze V, Orrenius S, Zhivotovsky B. Mitochondria, oxidative stress and cell death. *Apoptosis* 2007;**12**(5):913–22.

100 Barho MT, Oppermann S, Schrader FC, Degenhardt I, Elsässer K, Wegscheid-Gerlach C, et al. N-acyl derivatives of 4-phenoxyaniline as neuroprotective agents. *Chem Med Chem* 2014;**9**(10):2260–73.

101 Shamas-Din A, Bindner S, Zhu W, et al. tBid undergoes multiple conformational changes at the membrane required for Bax activation. *J Biol Chem* 2013;**288**(30):22 111–27.

102 Billen LP, Shamas-Din A, Andrews DW. Bid: a Bax-like BH3 protein. *Oncogene* 2008;**27**(Suppl. 1):S93–104.

103 Yin XM. Bid, a BH3-only multi-functional molecule, is at the cross road of life and death. *Gene* 2006;**369**:7–19.

104 Sutton VR, Wowk ME, Cancilla M, Trapani JA. Caspase activation by granzyme B is indirect, and caspase autoprocessing requires the release of proapoptotic mitochondrial factors. *Immunity* 2003;**18**(3):319–29.

105 Cirman T, Oresić K, Mazovec GD, Turk V, Reed JC, Myers RM, et al. Selective disruption of lysosomes in HeLa cells triggers apoptosis mediated by cleavage of Bid by multiple papain-like lysosomal cathepsins. *J Biol Chem* 2004;**279**(5):3578–87.

106 Chen M, He H, Zhan S, Krajewski S, Reed JC, Gottlieb RA. Bid is cleaved by calpain to an active fragment *in vitro* and during myocardial ischemia/reperfusion. *J Biol Chem* 2001;**276**(33):30 724–8.

107 Polster BM, Basañez G, Etxebarria A, Hardwick JM, Nicholls DG. Calpain I induces cleavage and release of apoptosis-inducing factor from isolated mitochondria. *J Biol Chem* 2005;**280**(8):6447–54.

108 Zhang WH, Wang X, Narayanan M, Zhang Y, Huo C, Reed JC, Friedlander RM. Fundamental role of the Rip2/caspase-1 pathway in hypoxia and ischemia-induced neuronal cell death. *Proc Natl Acad Sci USA* 2003;**100**(26):16 012–17.

109 Chan SL, Mattson MP. Caspase and calpain substrates: roles in synaptic plasticity and cell death. *J Neurosci Res* 1999;**58**(1):167–90.

110 Goll DE, Thompson VF, Li H, Wei W, Cong J. The calpain system. *Physiol Rev* 2003;**83**(3):731–801.

111 Mattson MP, Duan W, Pedersen WA, Culmsee C. Neurodegenerative disorders and ischemic brain diseases. *Apoptosis* 2001;**6**(1–2):69–81.

112 Ward MW, Rehm M, Duessmann H, Kacmar S, Concannon CG, Prehn JH. Real time single cell analysis of Bid cleavage and Bid translocation during caspase-dependent and neuronal caspase-independent apoptosis. *J Biol Chem* 2006;**281**(9):5837–44.

113 König HG, Rehm M, Gudorf D, Krajewski S, Gross A, Ward MW, Prehn JH. Full length Bid is sufficient to induce apoptosis of cultured rat hippocampal neurons. *BMC Cell Biol* 2007;**8**:7.

114 Li H, Zhu H, Xu CJ, Yuan J. Cleavage of BID by caspase 8 mediates the mitochondrial damage in the Fas pathway of apoptosis. *Cell* 1998;**94**(4):491–501.

115 Iriyama T, Kamei Y, Kozuma S, Taketani Y. Bax-inhibiting peptide protects glutamate-induced cerebellar granule cell death by blocking Bax translocation. *Neurosci Lett* 2009;**451**(1):11–15.

116 Han B, Wang Q, Cui G, Shen X, Zhu Z. Post-treatment of Bax-inhibiting peptide reduces neuronal death and behavioral deficits following global cerebral ischemia. *Neurochem Int* 2011;**58**(2):224–33.

117 Gibson ME, Han BH, Choi J, Knudson CM, Korsmeyer SJ, Parsadanian M, Holtzman DM. BAX contributes to apoptotic-like death following neonatal hypoxia-ischemia: evidence for distinct apoptosis pathways. *Mol Med* 2001;**7**(9):644–55.

118 D'Orsi B, Kilbride SM, Chen G, Perez-Alvarez S, Bonner HP, Pfeiffer S, et al. Bax regulates neuronal Ca++ homeostasis. *J Neurosci* 2015;**35**(4):1706–22.

119 Tehranian R, Rose ME, Vagni V, Pickrell AM, Griffith RP, Liu H, et al. Disruption of Bax protein prevents neuronal cell death but produces cognitive impairment in mice following traumatic brain injury. *J Neurotrauma* 2008;**25**(7):755–67.

120 Hetz C, Vitte PA, Bombrun A, Rostovtseva TK, Montessuit S, Hiver A, et al. Bax channel inhibitors prevent mitochondrion-mediated apoptosis and protect neurons in a model of global brain ischemia. *J Biol Chem* 2005;**280**(52):42 960–70.

121 Montessuit S, Somasekharan SP, Terrones O, Lucken-Ardjomande S, Herzig S, Schwarzenbacher R, et al. Membrane remodeling induced by the dynamin-related protein Drp1 stimulates Bax oligomerization. *Cell* 2010;**142**(6):889–901.

122 Smirnova E, Griparic L, Shurland DL, van der Bliek AM. Dynamin-related protein Drp1 is required for mitochondrial division in mammalian cells. *Mol Biol Cell* 2001;**12**(8):2245–56.

123 Frank S, Gaume B, Bergmann-Leitner ES, Leitner WW, Robert EG, Catez F, et al. The role of dynamin-related protein 1, a mediator of mitochondrial fission, in apoptosis. *Dev Cell* 2001;**1**(4):515–25.

124 Wasiak S, Zunino R, McBride HM. Bax/Bak promote sumoylation of DRP1 and its stable association with mitochondria during apoptotic cell death. *J Cell Biol* 2007;**177**(3):439–50.

125 Karbowski M, Lee YJ, Gaume B, Jeong SY, Frank S, Nechushtan A, et al. Spatial and temporal association of Bax with mitochondrial fission sites, Drp1, and Mfn2 during apoptosis. *J Cell Biol* 2002;**159**(6):931–8.

126 Barsoum MJ, Yuan H, Gerencser AA, Liot G, Kushnareva Y, Gräber S, et al. Nitric oxide-induced mitochondrial fission is regulated by dynamin-related GTPases in neurons. *EMBO J* 2006;**25**(16):3900–11.

127 Lee YJ, Jeong SY, Karbowski M, Smith CL, Youle RJ. Roles of the mammalian mitochondrial fission and fusion mediators Fis1, Drp1, and Opa1 in apoptosis. *Mol Biol Cell* 2004;**15**(11):5001–11.

128 Tanaka A, Kobayashi S, Fujiki Y. Peroxisome division is impaired in a CHO cell mutant with an inactivating point-mutation in dynamin-like protein 1 gene. *Exp Cell Res* 2006;**312**(9):1671–84.

129 Ishihara N, Nomura M, Jofuku A, Kato H, Suzuki SO, Masuda K, et al. Mitochondrial fission factor Drp1 is essential for embryonic development and synapse formation in mice. *Nat Cell Biol* 2009;**11**(8):958–66.

130 Calabrese V, Cornelius C, Mancuso C, Lentile R, Stella AM, Butterfield DA. Redox homeostasis and cellular stress response in aging and neurodegeneration. *Methods Mol Biol* 2010;**610**:285–308.

131 Holmström KM, Finkel T. Cellular mechanisms and physiological consequences of redox-dependent signalling. *Nat Rev Mol Cell Biol* 2014;**15**(6):411–21.

132 Murphy MP. How mitochondria produce reactive oxygen species. *Biochem J* 2009;**417**(1):1–13.

133 Brand MD. The sites and topology of mitochondrial superoxide production. *Exp Gerontol* 2010;**45**(7–8):466–72.

134 Fleury C, Mignotte B, Vayssière JL. Mitochondrial reactive oxygen species in cell death signaling. *Biochimie* 2002;**84**(2–3):131–41.

135 Mailloux RJ, Harper ME. Mitochondrial proticity and ROS signaling: lessons from the uncoupling proteins. *Trends Endocrinol Metab* 2012;**23**(9):451–8.

136 Orrenius S, Nicotera P, Zhivotovsky B. Cell death mechanisms and their implications in toxicology. *Toxicol Sci* 2011;**119**(1):3–19.

137 Daugas E, Nochy D, Ravagnan L, Loeffler M, Susin SA, Zamzami N, Kroemer G. Apoptosis-inducing factor (AIF): a ubiquitous mitochondrial oxidoreductase involved in apoptosis. *FEBS Lett* 2000;**476**(3):118–23.

138 Susin SA, Lorenzo HK, Zamzami N, Marzo I, Snow BE, Brothers GM, et al. Molecular characterization of mitochondrial apoptosis-inducing factor. *Nature* 1999;**397**(6718):441–6.

139 Lorenzo HK, Susin SA. Therapeutic potential of AIF-mediated caspase-independent programmed cell death. *Drug Resist Updat* 2007;**10**(6):235–55.

140 Ferreira P, Villanueva R, Cabon L, Susín SA, Medina M. The oxido-reductase activity of the apoptosis inducing factor: a promising pharmacological tool? *Curr Pharm Des* 2013;**19**(14):2628–36.

141 Chinta SJ, Rane A, Yadava N, Andersen JK, Nicholls DG, Polster BM. Reactive oxygen species regulation by AIF- and complex I-depleted brain mitochondria. *Free Radic Biol Med* 2009;**46**(7):939–47.

142 Zhu C, Wang X, Deinum J, Huang Z, Gao J, Modjtahedi N, et al. Cyclophilin A participates in the nuclear translocation of apoptosis-inducing factor in neurons after cerebral hypoxia-ischemia. *J Exp Med* 2007;**204**(8):1741–8.

143 Joza N, Susin SA, Daugas E, Stanford WL, Cho SK, Li CY, et al. Essential role of the mitochondrial apoptosis-inducing factor in programmed cell death. *Nature* 2001;**410**(6828):549–54.

144 Zhu C, Wang X, Huang Z, Qiu L, Xu F, Vahsen N, et al. Apoptosis-inducing factor is a major contributor to neuronal loss induced by neonatal cerebral hypoxia-ischemia. *Cell Death Differ* 2007;**14**(4):775–84.

145 Slemmer JE, Zhu C, Landshamer S, Trabold R, Grohm J, Ardeshiri A, et al. Causal role of apoptosis-inducing factor for neuronal cell death following traumatic brain injury. *Am J Pathol* 2008;**173**(6):1795–805.

146 Chen Q, Szczepanek K, Hu Y, Thompson J, Lesnefsky EJ. A deficiency of apoptosis inducing factor (AIF) in Harlequin mouse heart mitochondria paradoxically reduces ROS generation during ischemia-reperfusion. *Front Physiol* 2014;**5**:271.

147 Brown D, Yu BD, Joza N, Bénit P, Meneses J, Firpo M, et al. Loss of Aif function causes cell death in the mouse embryo, but the temporal progression of patterning is normal. *Proc Natl Acad Sci USA* 2006;**103**(26):9918–23.

148 Klein JA, Longo-Guess CM, Rossmann MP, Seburn KL, Hurd RE, Frankel WN, et al. The harlequin mouse mutation downregulates apoptosis-inducing factor. *Nature* 2002;**419**(6905):367–74.

149 Laliberté AM, MacPherson TC, Micks T, Yan A, Hill KA. Vision deficits precede structural losses in a mouse model of mitochondrial dysfunction and progressive retinal degeneration. *Exp Eye Res* 2011;**93**(6):833–41.

150 Richter M, Nickel C, Apel L, Kaas A, Dodel R, Culmsee C, Dolga AM. SK channel activation modulates mitochondrial respiration and attenuates neuronal HT-22 cell damage induced by H2O2. *Neurochem Int* 2015;**81**:63–75.

151 Gohil VM, Sheth SA, Nilsson R, Wojtovich AP, Lee JH, Perocchi F, et al. Nutrient-sensitized screening for drugs that shift energy metabolism from mitochondrial respiration to glycolysis. *Nat Biotechnol* 2010;**28**(3):249–55.

152 Correia SC, Carvalho C, Cardoso S, Santos RX, Santos MS, Oliveira CR, et al. Mitochondrial preconditioning: a potential neuroprotective strategy. *Front Aging Neurosci* 2010;**2.pii**:138.

153 Hebert DN, Molinari M. In and out of the ER: protein folding, quality control, degradation, and related human diseases *Physiol Rev* 2007;**87**(4):1377–408.

154 Schröder M, Kaufman RJ. The mammalian unfolded protein response. *Annu Rev Biochem* 2005;**74**:739–89.

155 Harding HP, Zhang Y, Ron D. Protein translation and folding are coupled by an endoplasmic-reticulum-resident kinase. *Nature* 1999;**397**(6716):271–4.

156 Harding HP, Novoa I, Zhang Y, Zeng H, Wek R, Schapira M, Ron D. Regulated translation initiation controls stress-induced gene expression in mammalian cells. *Mol Cell* 2000;**6**(5):1099–108.

157. Yoshida H, Matsui T, Yamamoto A, Okada T, Mori K. XBP1 mRNA is induced by ATF6 and spliced by IRE1 in response to ER stress to produce a highly active transcription factor. *Cell* 2001;**107**(7):881–91.

158. Gorman AM, Healy SJ, Jäger R, Samali A. Stress management at the ER: regulators of ER stress-induced apoptosis. *Pharmacol Ther* 2012;**134**(3):306–16.

159. Rao RV, Ellerby HM, Bredesen DE. Coupling endoplasmic reticulum stress to the cell death program. *Cell Death Differ* 2004;**11**(4):372–80.

160. Jin ML, Park SY, Kim YH, Oh JI, Lee SJ, Park G. The neuroprotective effects of cordycepin inhibit glutamate-induced oxidative and ER stress-associated apoptosis in hippocampal HT22 cells. *Neurotoxicology* 2014;**41**:102–11.

161. Ono Y, Shimazawa M, Ishisaka M, Oyagi A, Tsuruma K, Hara H. Imipramine protects mouse hippocampus against tunicamycin-induced cell death. *Eur J Pharmacol* 2012;**696**(1–3):83–8.

162. Choi JH, Choi AY, Yoon H, Choe W, Yoon KS, Ha J, et al. Baicalein protects HT22 murine hippocampal neuronal cells against endoplasmic reticulum stress-induced apoptosis through inhibition of reactive oxygen species production and CHOP induction. *Exp Mol Med* 2010;**42**(12):811–22.

163. Nishitoh H, Matsuzawa A, Tobiume K, Saegusa K, Takeda K, Inoue K, et al. ASK1 is essential for endoplasmic reticulum stress-induced neuronal cell death triggered by expanded polyglutamine repeats. *Genes Dev* 2002;**16**(11):1345–55.

164. Jousse C, Oyadomari S, Novoa I, Lu P, Zhang Y, Harding HP, Ron D. Inhibition of a constitutive translation initiation factor 2alpha phosphatase, CReP, promotes survival of stressed cells. *J Cell Biol* 2003;**163**(4):767–75.

165. Lu PD, Jousse C, Marciniak SJ, Zhang Y, Novoa I, Scheuner D, et al. Cytoprotection by pre-emptive conditional phosphorylation of translation initiation factor 2. *EMBO J* 2004;**23**(1):169–79.

166. Lewerenz J, Maher P. Basal levels of eIF2alpha phosphorylation determine cellular antioxidant status by regulating ATF4 and xCT expression. *J Biol Chem* 2009;**284**(2):1106–15.

167. Dickhout JG, Carlisle RE, Jerome DE, Mohammed-Ali Z, Jiang H, Yang G, et al. Integrated stress response modulates cellular redox state via induction of cystathionine γ-lyase: cross-talk between integrated stress response and thiol metabolism. *J Biol Chem* 2012;**287**(10):7603–14.

168. Harding HP, Zhang Y, Zeng H, Novoa I, Lu PD, Calfon M, et al. An integrated stress response regulates amino acid metabolism and resistance to oxidative stress. *Mol Cell* 2003;**11**(3):619–33.

169. Lange PS, Chavez JC, Pinto JT, Coppola G, Sun CW, Townes TM, et al. ATF4 is an oxidative stress-inducible, prodeath transcription factor in neurons *in vitro* and *in vivo*. *J Exp Med* 2008;**205**(5):1227–42.

170. Wang S, Longo FM, Chen J, Butman M, Graham SH, Haglid KG, Sharp FR. Induction of glucose regulated protein (grp78) and inducible heat shock protein (hsp70)mRNAs in rat brain after kainic acid seizures and focal ischemia. *Neurochem Int* 1993;**23**(6):575–82.

171. Tajiri S, Oyadomari S, Yano S, Morioka M, Gotoh T, Hamada JI, et al. Ischemia-induced neuronal cell death is mediated by the endoplasmic reticulum stress pathway involving CHOP. *Cell Death Differ* 2004;**11**(4):403–15.

172 Ito D, Tanaka K, Suzuki S, Dembo T, Kosakai A, Fukuuchi Y. Up-regulation of the Ire1-mediated signaling molecule, Bip, in ischemic rat brain. *Neuroreport* 2001;**12**(18):4023–8.
173 Ni M, Lee AS. ER chaperones in mammalian development and human diseases. *FEBS Lett* 2007;**581**(19):3641–51.
174 Yu Z, Luo H, Fu W, Mattson MP. The endoplasmic reticulum stress-responsive protein GRP78 protects neurons against excitotoxicity and apoptosis: suppression of oxidative stress and stabilization of calcium homeostasis. *Exp Neurol* 1999;**155**(2):302–14.
175 Kitao Y, Ozawa K, Miyazaki M, Tamatani M, Kobayashi T, Yanagi H, et al. Expression of the endoplasmic reticulum molecular chaperone (ORP150) rescues hippocampal neurons from glutamate toxicity. *J Clin Invest* 2001;**108**(10):1439–50.
176 Kudo T, Kanemoto S, Hara H, Morimoto N, Morihara T, Kimura R, et al. A molecular chaperone inducer protects neurons from ER stress. *Cell Death Differ* 2008;**15**(2):364–75.
177 Oida Y, Hamanaka J, Hyakkoku K, Shimazawa M, Kudo T, Imaizumi K, et al. Post-treatment of a BiP inducer prevents cell death after middle cerebral artery occlusion in mice. *Neurosci Lett* 2010;**484**(1):43–6.
178 Lee Y, Park HW, Park SG, Cho S, Myung PK, Park BC, Lee DH. Proteomic analysis of glutamate-induced toxicity in HT22 cells. *Proteomics* 2007;**7**(2):185–93.
179 Li Y, Maher P, Schubert D. Requirement for cGMP in nerve cell death caused by glutathione depletion. *J Cell Biol* 1997;**139**(5):1317–24.
180 Sagara Y, Schubert D. The activation of metabotropic glutamate receptors protects nerve cells from oxidative stress. *J Neurosci* 1998;**18**(17):6662–71.
181 Bravo R, Vicencio JM, Parra V, Troncoso R, Munoz JP, Bui M, et al. Increased ER-mitochondrial coupling promotes mitochondrial respiration and bioenergetics during early phases of ER stress. *J Cell Sci* 2011;**124**(Pt. 13):2143–52.
182 Henke N, Albrecht P, Bouchachia I, Ryazantseva M, Knoll K, Lewerenz J, et al. The plasma membrane channel ORAI1 mediates detrimental calcium influx caused by endogenous oxidative stress. *Cell Death Dis* 2013;**4**:e470.
183 Dittmer S, Sahin M, Pantlen A, Saxena A, Toutzaris D, Pina AL, et al. The constitutively active orphan G-protein-coupled receptor GPR39 protects from cell death by increasing secretion of pigment epithelium-derived growth factor. *J Biol Chem* 2008;**283**(11):7074–81.
184 Shivakumar BR, Kolluri SV, Ravindranath V. Glutathione and protein thiol homeostasis in brain during reperfusion after cerebral ischemia. *J Pharmacol Exp Ther* 1995;**274**(3):1167–73.
185 Shih AY, Li P, Murphy TH. A small-molecule-inducible Nrf2-mediated antioxidant response provides effective prophylaxis against cerebral ischemia in vivo. *J Neurosci* 2005;**25**(44):10 321–35.

15

Pyroptosis

Kate E. Lawlor, Stephanie Conos, and James E. Vince

The Walter and Eliza Hall Institute of Medical Research, Bundoora, VIC, Australia

Abbreviations

AD	Alzheimer's disease
AIDS	acquired immunodeficiency syndrome
AIM2	absent in myeloma 2
ALR	AIM2-like receptor
ASC	apoptosis-associated speck-like protein containing a CARD
ATP	adenosine triphosphate
Bak	Bcl-2 homologous antagonist killer
Bax	Bcl-2-associated X protein
BIR	baculoviral IAP repeat
CAPS	CIAS1-associated periodic syndromes
CARD	caspase-associated recruitment domain
cIAP	cellular inhibitor of apoptosis protein
cyt c	cytochrome c
DAMP	damage-associated molecular pattern
FADD	Fas-associated death-domain protein
FCAS	familial cold autoinflammatory syndrome
GBP	guanylate-binding partner
HIV	human immunodeficiency virus
HMGB1	high-mobility group protein B1
IAP	inhibitor of apoptosis protein
IFI16	γ-interferon inducible protein 16
IFNγ	interferon gamma
IFNR	interferon receptor
IL-1	interleukin 1
IL-1R	interleukin 1 receptor
JNK	C-Jun N-terminal kinase
LCMV	lymphocytic choriomeningitis virus
LPS	lipopolysaccharide
LRR	leucine-rich repeat
MD2	myeloid differentiation factor 2
MLKL	mixed-lineage kinase domain-like
NACHT	nucleotide binding and oligomerization

Apoptosis and Beyond: The Many Ways Cells Die, First Edition. Edited by James Radosevich.
© 2018 John Wiley & Son Inc. Published 2018 by John Wiley & Son Inc.

NADPH	reduced nicotinamide adenine dinucleotide phosphate
NAIP	neuronal apoptosis inhibitory protein
NLR	NOD-like receptor
NOD	nucleotide-binding domain
PAMP	pathogen-associated molecular pattern
PARP	poly ADP ribose polymerase
PRR	pattern-recognition receptor
PS	phosphatidylserine
RIG-I	retinoic acid-inducible gene 1
RIPK	receptor-interacting protein kinase
RLR	RIG-1-like receptor
ROS	reactive oxygen species
T3SS	type III secretion system
TNF	tumor necrosis factor
TNFR1	tumor necrosis factor receptor 1
TLR	Toll-like receptor
TRAIL	tumor necrosis factor-related apoptosis-inducing ligand
TRPC1	transient receptor potential channel 1
XIAP	X-linked inhibitor of apoptosis protein

15.1 Introduction

Inflammatory cell death alerts the immune system to a clear and present danger. While "accidental" non-programmed cell death can result from overwhelming physical or toxic stimuli, the last 2 decades of research have uncovered several genetically programmed cell-death pathways that can induce an inflammatory response [1]. These lytic modes of cell death can be triggered by microbial infections, host-derived molecules, or cytokines and environmental toxins. Inflammatory cell death, in particular pyroptosis, can act in a beneficial manner to release intracellular bacteria from their immunotolerant niches within cells, exposing them to neutrophil-mediated clearance [2]. On the other hand, infection-induced pyroptosis may also be detrimental to the host, as it can reduce the hematopoietic progenitor pool during viral infection to limit host recovery [3], and it has recently been documented to be the major cell-death pathway that causes CD4 T-cell depletion in abortive human immunodeficiency virus (HIV) infection [4]. Excessive pyroptosis, or pyroptosis caused by autoactivating mutations in components of the caspase 1 oligomerization machinery, the inflammasomes, may also contribute to autoinflammatory diseases [3,5–8]. Hence, there is a strong interest in understanding under what circumstances inhibiting or promoting inflammatory cell-death pathways may be therapeutically beneficial.

The rapid breakdown of the plasma membrane is thought to dictate whether a cell's demise triggers an inflammatory response via the release of intracellular molecules, collectively termed "damage-associated molecular patterns" (DAMPs), into the external environment. These include cytokines (e.g., IL-1α, IL-33) and cellular constituents or metabolites such as uric acid, adenosine triphosphate (ATP), mitochondrial DNA, high-mobility group protein B1 (HMGB1), S100 proteins, and heat-shock proteins [9]. When released, DAMPs are detected by immune cells, and may thereby initiate or propagate inflammation to stimulate tissue healing or pathogen clearance, or to contribute to diseases such as inflammatory arthritis (gout) and sepsis [10,11]. Indeed, it is now well established that exacerbated or chronic inflammatory responses are a common pathological trait in a diverse number of autoimmune and autoinflammatory diseases, including rheumatoid arthritis, atherosclerosis, diabetes, Alzheimer's disease (AD), inflammatory bowel disease, and cancer [12,13]. In this regard, identifying and neutralizing disease-specific inflammatory signaling components such as the cytokines tumor necrosis factor (TNF) and interleukin 1beta (IL-1β), DAMPs, and their receptors, holds great promise for the development of new therapeutics. Notably, anti-TNF and anti-IL-1 therapies have already revolutionized the treatment of diseases such as rheumatoid arthritis and CIAS1-associated periodic syndromes (CAPS), respectively [13], while HMGB1 neutralization and uric acid depletion have been used successfully to reduce lytic cell death-induced inflammation *in vivo* [14–16].

The innate inflammatory process is mediated by evolutionary conserved innate-immune pattern-recognition receptors (PRRs), which detect a range of DAMPs and pathogen-associated molecular patterns (PAMPs). These include the Toll-like receptor (TLR) family, the retinoid-inducible gene 1 (RIG-1)-like receptors (RLRs), the C-type lectin receptors, and the nucleotide-binding domain (NOD)-like receptor (NLR) family. Historically, PRRs have principally been studied for their ability to induce the rapid transcription and activation of inflammatory cytokines. However, it is now apparent that an additional key function of PRR stimulation can be cell death itself. Specific PAMPs (e.g., lipopolysaccharide (LPS), bacterial toxins, DNA) and DAMPs (e.g., HMGB1, ATP) can induce genetically programmed lytic cell death, including caspase 1-dependent pyroptosis [17,18] and receptor-interacting protein kinase 3 (RIPK3)–mixed-lineage kinase domain-like (MLKL) protein-dependent necroptosis [19].

Inflammasome-mediated caspase 1 activation and pyroptotic cell death are initiated by a subset of NLRs, absent in myeloma 2 (AIM2)-like receptors, or the tripartite-motif family member pyrin [20,21]. A defining feature of pyroptosis is its dependence on caspase 1, which mediates the processes involved in plasma-membrane rupture, cell death, and the release of DAMPs. Additionally, activated caspase 1 is responsible for processing the precursor forms of the inflammatory cytokines IL-1β and IL-18 to trigger their activation, and, alongside IL-1α, directs their secretion out of the cell. IL-β and IL-1α signal via the interleukin 1 receptor (IL-1R) to induce the transcription of a plethora of inflammatory mediators. As such, IL-1β is required for the induction of systemic and local responses to infection and injury by causing flu-like symptoms (fever, nausea, headache, fatigue), activating lymphocytes and epithelial cells, and promoting immune-cell infiltration. IL-18 activates IL-18R, which can lead to interferon gamma (IFNγ) production and Th1 cell polarization. Use of inflammasome-specific and IL-1R-, IL-1β-, IL-1α-, and IL-18-deficient mice has demonstrated the importance of inflammasome-induced cytokine activation and secretion in host protection against a wide range of viral, bacterial, fungal, and protozoan pathogens [13]. Although these pathogens also invariably induce pyroptosis following inflammasome activation and cytokine secretion (at least *in vitro*), the physiological function of pyroptosis in most cases has yet to be determined. In contrast to caspase 1, caspase 11 does not efficiently process IL-1β and IL-18. However, its ability to induce pyroptotic cell death upon binding to cytosolic LPS aids bacterial clearance, likely via DAMP-induced inflammation and/or destruction of the pathogens immune-protected replicative niche, and therefore cell-death induction (not cytokine processing) may be its principle function. In this chapter, we highlight the cardinal features of pyroptotic cell death, how it is signaled through inflammasome receptors, its physiological significance, and its crosstalk with other programmed cell death (PCD) pathways.

15.2 Caspases

Caspases (**c**ysteine-dependent **asp**artate-specific prote**ases**) are evolutionary conserved cytosolic enzymes that cleave protein substrates at defined sites after an aspartate residue to modulate specific cell-signaling pathways. While they have been most intensively studied for their role in PCD, they also regulate such diverse processes as inflammatory cytokine expression and activation [13,22], and cellular proliferation, differentiation, and migration [23]. Typically, the cytosolic inactive caspase enzyme precursor (zymogen) undergoes activation following an appropriate stimulus, which results in precursor processing to a large and small catalytic subunit, which can act to stabilize the caspase homodimer and expose the proteolytic active site. While, in general, caspase processing is associated with increased activity, recent experiments have documented important catalytic functions for non-processed caspase dimers. These include essential roles for pro-caspase 8 in the suppression of necroptotic cell death [24] and a role for full-length caspase 1 in executing pyroptosis following bacterial infection [25].

15.2.1 Apoptotic Caspases

Historically, research has focused on caspase-dependent cell death that is associated with apoptotic caspase activity. Apoptosis can occur following tumor necrosis factor-related apoptosis-inducing ligand (TRAIL), TNF, or Fas death-receptor ligation, where death domain-containing adaptor proteins (e.g., Fas-associated death-domain protein, FADD) bind directly to activated death receptors and then recruit and activate initiator caspases 8 and 10. In contrast, mitochondrial apoptosis occurs in response to DNA damage or to cellular stresses such as cytokine withdrawal, and is tightly regulated by the levels of Bcl-2 family members. When pro-survival Bcl-2 family members are depleted, or when they are inhibited by proapoptotic BH3-only proteins, Bcl-2 homologous antagonist killer (Bak) and Bcl-2-associated X protein (Bax) are activated to cause irreversible mitochondrial membrane damage. This allows the release of cytochrome c (cyt c) from the mitochondria into the cytosol, where it nucleates the formation of the APAF1 and caspase 9 apoptosome cell-death complex. When activated, the death receptor (caspases 8 and 10) and mitochondrial (caspase 9) initiator caspases cleave and activate effector caspases (caspases 3, 6 and, 7), which process defined cellular substrates to bring about the apoptotic phenotype. Initiator caspases exist as monomers, and following their recruitment to death-receptor or mitochondrial apoptotic signaling complexes, they undergo proximity-induced autocatalytic processing and activation via pro-domain dimerization. In contrast, effector caspases, the executioners of cell death, exist as dimers and are predominantly activated by initiator-caspase cleavage.

The hallmark morphological and biochemical features resulting from effector-caspase activity are distinct, and include cell shrinkage (pyknosis), nuclear condensation, DNA fragmentation, and membrane blebbing of apoptotic bodies. Effector-caspase substrates are numerous, but they include the cleavage-induced inactivation of poly ADP ribose polymerase (PARP), which can otherwise induce a necrotic phenotype if overactivated, and flippases, which are required for phosphatidylserine (PS) exposure on the cell surface [26]. The exposure of "find me" and "eat me" signals on apoptotic cells and bodies, such as PS, generally results in their clearance prior to the breakdown of membrane structures and the release of proinflammatory molecules. As such, apoptosis has traditionally been classified as non-inflammatory. However, recent studies highlight how the genetic deletion of key apoptotic repressors can result in severe apoptotic-induced inflammation *in vivo* [27–29], and apoptotic caspase 8 has been demonstrated to directly process and activate inflammatory cytokines such as IL-1β, much like caspase 1 [30–32]. Therefore, it appears that the cell type, tissue type, extracellular environment, expression levels of cell-death components, apoptotic stimulus strength, and timing are all likely to dictate whether apoptosis remains immunologically silent or triggers an inflammatory response.

15.2.2 Inflammatory Caspases

In contrast to the aforementioned apoptotic caspases, caspase 1, caspase 11 (caspases 4 and 5 in humans), and the rather enigmatic caspase 12 have been classified as inflammatory caspases, which are typically expressed in innate immune cells such as monocytes, neutrophils, and macrophages [33]. Inflammatory caspases all contain an N-terminal caspase-associated recruitment domain (CARD) and, in mammals, are located on a single locus (Figure 15.1). Like apoptotic initiator caspase activation following recruitment to the apoptosome, caspase 1 is activated upon dimerization within a large cytosolic protein complex, termed the inflammasome (see later). Sequence

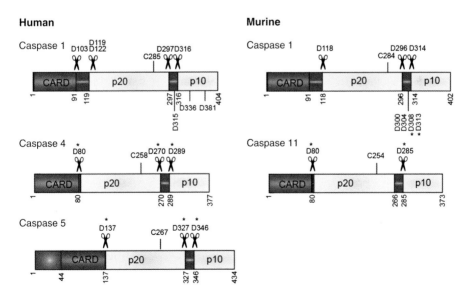

Figure 15.1 Domain structures of the human and murine inflammatory caspases, caspase 1, caspase 4 (caspase 11 mouse), and caspase 5 (caspase 12, not depicted). Common domains within caspases are the caspase-associated recruitment domain (CARD), which facilitates their interaction with inflammasomes or the adaptor protein apoptosis-associated speck-like protein containing a CARD (ASC) (see Figure 15.2), and the catalytically active p20 and p10 subunits. The aspartic acid residues (scissors) indicated represent predicted (*) or identified cleavage sites for autoproteolysis and activation where the preferred substrate sequence has Asp at position P1. The cysteine residues detailed are the catalytic sites for the individual caspases.

analysis suggests that human caspases 4 and 5 were derived from gene duplication of mouse caspase 11 [34]; this is supported by recent functional data showing conserved mechanisms of action – including their ability to bind and be activated by cytosolic LPS and induce pyroptotic cell death [35–37].

Caspase 1 was initially characterized for its ability to cleave inactive precursor IL-1β and IL-18 into their biologically active fragments [38–40]. Caspase 1 activity also induces the secretion of active IL-1β and IL-18 into the extracellular environment through a nonconventional secretory pathway that has yet to be clearly defined [41]. Subsequently, caspase 1 was demonstrated to induce cell death upon infection with pathogenic bacteria, including *Salmonella* and *Shigella*, independent of apoptotic cell-death signaling [42–47]. Just as apoptotic signaling is dispensable for pyroptosis, apoptotic cell death also precedes normally in caspase 1- and caspase 11-deficient mice [48,49]. Caspase 1 killing is associated with rapid cell swelling (oncosis), plasma-membrane pore formation (~1.0–2.4 nm), and eventual osmotic lysis [50], which are characteristic features also observed during both programmed (e.g., necroptosis) and non-programmed necrotic cell death [51]. Whether pyroptosis and other lytic forms of cell death such as necroptosis differ in their inflammatory potential and/or release different subsets of DAMPs remains to be determined.

Unlike in apoptosis, mitochondrial cyt c is not released prior to caspase 1-mediated cell death [42]. Although caspase 1-dependent DNA fragmentation and PARP activity are observed during pyroptosis, these processes are not required for caspase 1 killing or for the release of DAMPs such as HMGB1 [50,52]. Apoptosis-induced PS from the inner plasma-membrane leaflet to the outer allows apoptotic cells to be stained by the PS-binding molecule annexin V. While annexin V is therefore frequently used as a marker of apoptotic cell death, pyroptotic cells also stain positive for it. This may be because caspase 1-mediated pore formation allows annexin V access to the inner membrane, or it may be due to specific pyroptotic-induced PS flipping to the cell-surface outer membrane [53]. As such, it has been suggested that pyroptosis-induced ATP release and PS membrane flipping act as "find me" and "eat me" signals: the same mechanism by which apoptotic cells are detected and removed from the external environment [54]. Therefore, although several morphological features have been proposed to distinguish apoptotic from pyroptotic cell death, it is increasingly apparent that, depending on the stimulus, timing, and signal strength, the features of these and other PCD pathways may overlap significantly; hence, the importance of performing both biochemical and genetic experiments to determine which cell-death pathway is activated following a particular stimulus [1].

15.3 Canonical Inflammasomes: Caspase 1-Activating Platforms

Humans harbor 22 NLR genes [55], among which NLRP1 [56,57], NLRP3 [58,59], NLRP6 [60,61], NLRP7 [62], NLRP12 [63], and NAIP/NLRC4 [64,65] have been reported to sense various danger molecules and form a functional caspase 1-containing inflammasome. There is some evidence for pyroptosis playing a physiological role following NLRP1, NLRP3, and NAIP/NLRC4 activation (Figure 15.2). AIM2-like receptors (ALRs) capable of activating caspase 1 include AIM2 [66,67] and γ-interferon inducible protein 16 (IFI16) [68], which respond to DNA molecules found in the cytosol (AIM2/IFI16) and nucleus (IFI16). The evidence for caspase 1-dependent pyroptosis

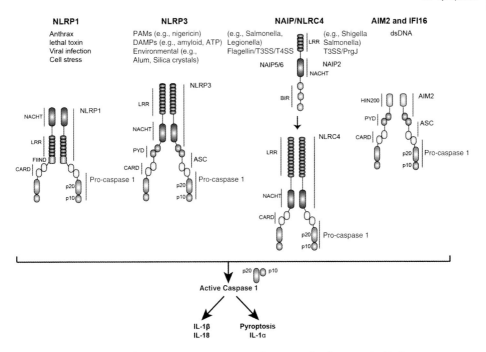

Figure 15.2 Canonical inflammasome activation and pyroptosis. The canonical inflammasomes, including the nucleotide-binding domain (NOD)-like receptors (NLPR1, NLRP3, NLRC4) and absent in myeloma 2 (AIM2), all contain a pyrin domain (PYD) or caspase-associated recruitment domain (CARD), which allows specific interactions with adaptor apoptosis-associated speck-like protein containing a CARD (ASC) or with caspase 1 directly. With the exception of AIM2, all NLRs and NAIP contain leucine-rich repeats (LRRs), which may function as intracellular sensors, and a nucleotide binding and oligomerization (NACHT) domain, which enables adenosine triphosphate (ATP)-dependent oligomerization and activation of the complex. NLRP3 senses a diverse range of triggers, including extracellular ATP, the pore-forming toxin nigericin, and monosodium urate crystals. NLRP1 and NLRC4 are more restricted sensors, where NLRP1 is activated by anthrax lethal toxin, viral or *Toxoplasma* infection, or cellular stress, and NLRC4 is activated by cytosolic bacterial flagellin or T3SS components that are detected by NAIP receptors. AIM2 features a unique DNA-binding HIN200 domain, which it uses to sense cytosolic DNA. In the case of apoptosis-associated speck-like protein containing a CARD (ASC)-containing inflamamsomes, following binding of the NLR to ASC via the PYD domain, ASC foci ("specks") form to recruit caspase 1 via its CARD. Caspase 1 subsequently undergoes dimerization and autoproteolysis to trigger pyroptosis, as well as IL-1β and IL-18 activation.

playing a physiological role via these NLR and ALR inflammasome-sensor proteins is discussed in this section.

Mutation of the *MEFV* gene, which encodes for pyrin, is responsible for the auto-inflammatory disease familial Mediterranean fever [69]. Pyrin is activated following pathogen-mediated inactivation of Rho GTPases and nucleates formation of a pyrin-ASC-caspase 1 inflammasome complex [20]. A physiological role for pyrin-induced pyroptosis has not yet been demonstrated.

Despite the unequivocal data showing that probably all inflammasomes can activate caspase 1 to induce pyroptosis *in vitro*, the critical caspase 1 substrates required for pyroptotic death have yet to be identified. However, given that ion flux and osmotic lysis are characteristic of rapid pyroptotic cell death, it is postulated that caspase 1 may process ion channels or pumps – or their essential regulators – required to maintain cellular ionic homeostasis. Consistent with this notion, the cationic channel subunit transient receptor potential channel 1 (TRPC1) has been reported to be a caspase 11 substrate and to modulate IL-1β secretion [70]. In the following subsection, we summarize how inflammasome-sensor proteins recruit ASC and caspase 1 through pyrin and CARD interactions, ask whether caspase 1 killing and cytokine activation are distinct, separable events, and highlight known physiological functions for inflammasome-mediated pyroptosis.

15.3.1 Inflammasome Structural Features

Several NLR family members, as well as ALRs and pyrin, can form inflammasome complexes to induce caspase 1 activation (Figure 15.2). Other receptors, such as retinoic acid-inducible gene 1 (RIG-I) [71], have been suggested to form inflammasome complexes with the capacity to bind and activate caspase 1 – but the validity of this has been questioned [72]. NLR-forming inflammasomes contain C-terminal leucine-rich repeats (LRRs) that may act to autoinhibit NLR function until stimulated [73]. A central NACHT domain oligomerizes upon inflammasome activation, binds ATP, and is required for inflammasome activity. The N-terminal domain can consist of either a CARD, a pyrin domain, or a baculoviral inhibitor of apoptosis protein (IAP) repeat (BIR) domain. The pyrin domain and CARD mediate homotypic protein interactions to facilitate inflammasome formation. In the case of pyrin domain-containing inflammasomes (e.g., NLRP3, AIM2), these interact with the pyrin domain of ASC, which then recruits caspase 1 through CARD–CARD homotypic interactions (Figure 15.2). Studies suggest that ASC prion-like filaments or clusters, nucleated by ASC phosphorylation [74] and subsequent pyrin–pyrin-domain oligomerization, form a critical scaffold required for caspase 1 recruitment and inflammasome assembly [75,76]. These can be observed visually by flow cytometry or immunomicroscopy as ASC puncta or "specks" [77,78]. Recent studies have suggested that ASC specks released during pyroptosis can be taken up by neighboring phagocytes to promote caspase 1 dimerization and activation, and thereby act like DAMPs to propagate cell death and proinflammatory responses [79,80].

Inflammasomes lacking a pyrin domain, but containing a CARD (e.g., NLRP1, NLRC4), can directly bind and activate caspase 1, although they may also utilize ASC for optimal function [56]. Regardless, the recruitment of pro-caspase 1 into all inflammasomes facilitates proximity-induced oligomerization and activation, resulting in the autoprocessing of caspase 1 into the catalytic p20 and p10 subunits that process precursor IL-1β and IL-18. Although caspase 1 is considered the predominant enzyme that activates IL-1β and IL-18, a number of proteases (e.g., granzyme A, elastase, proteinase 3, caspase 8) may also process precursor IL-1β under specific conditions, including during *Mycobacterium tuberculosis* infection, following tissue damage, or upon TLR or death-receptor ligation [30–32,81].

Because caspase 1-dependent cytokine activation and pyroptosis often go hand-in-hand, and it has often been assumed that pyroptosis may simply represent an efficient mechanism for releasing activated cytokines, we first examine if any evidence exists to suggest that these events – pyroptosis and cytokine activation – can be functionally separated.

15.3.2 Are Caspase 1 Pyroptosis and Cytokine Activation Separable?

Caspase 1 has several proinflammatory substrates other than IL-1β. As such, there is significant experimental difficulty in validating the physiological contribution of caspase 1-dependent pyroptosis. Studies have utilized IL-1R and IL-18 double-knockout (KO) mice to remove the confounding effects of caspase 1-dependent IL-1β, IL-1α, and IL-18 activation. However, not unexpectedly, caspase 1 mediates the secretion of a number of other leaderless proteins involved in inflammation and tissue repair [82], and it has more recently been documented to cause the release of proinflammatory eicosanoids: signaling lipids such as prostaglandins and leukotrienes [83].

Despite these complications, there are emerging data to suggest that the caspase 1-dependent events of cytokine activation and pyroptosis are functionally distinct. Several studies have reported that the caspase 1-dependent activation and secretion of IL-1β occurs at an earlier time point than the death of these same cells [31,84–86]. The act of pyroptosis can also limit intracellular replication of *Salmonella* and expose it to immune attack independent of IL-1 or IL-18 signaling [2]. Further to this, recent studies have documented that *Salmonella*- and *Streptococcus*-infected neutrophils activate NLRC4 and NLRP3 inflammasome-associated caspase 1 to induce IL-1β secretion in the absence of cell death, *in vitro* and *in vivo* [87,88]. It has also been proposed that bacterial-induced caspase 1-dependent pyroptosis is mediated by full-length caspase 1 catalytic activity, while processed casapase 1 is much more efficient at activating IL-1β and IL-18 [25]. In particular, the NLRC4 inflammasome directly binds and activates pro-caspase 1 via CARD–CARD interactions to induce pyroptosis, and under these conditions caspase 1 is localized diffusely throughout the cytosol. In contrast, when NLRC4 complexes with the inflammasome adaptor ASC, caspase 1 recruitment into ASC "specks" induces caspase 1 proteolysis to the p20 and p10 fragments, which efficiently cleaves IL-1β and IL-18, in addition to causing cell death. Whether pyroptosis and cytokine activation can be similarly separated following the activation of other inflammasome-sensor proteins remains undetermined. Regardless, these data imply that caspase 1-dependent pyroptosis does not exist simply to allow the release of cytokines such as IL-1β; in fact, they suggest that in some cell types, such as neutrophils, in which sustained cytokine production or neutrophil engulfment of microbes is required to fight pathogenic infections, pyroptotic cell death may be actively repressed by a yet-to-be-defined mechanism, or else that neutrophils may lack the critical caspase 1 effector substrates required for pyroptosis. Alternatively, in the studies in question [87,88], neutrophil caspase 1 activity may be limited by inflammasome expression levels, and therefore fail to pass the activation threshold necessary to commit the cell to death.

15.3.3 The NLRP3 Inflammasome

Among the inflammasome-sensor proteins, NLRP3 is arguably the most important, for the following reasons: (i) it is mutated and drives pathological caspase 1 activation in several rare heritable diseases collectively known as the CAPS [12]; (ii) pathological activation of NLRP3 has been implicated in a diverse number of disease states, affecting millions of people, including atherosclerosis, type 2 diabetes, heart failure, arthritis, and gout [89]; and (iii) numerous studies have documented how NLRP3-deficient mice are more susceptible to infection by a variety of microbial pathogens such as *Streptococcus pneumoniae* (pneumonia), *Neissseria gonorrhoeae* (gonorrhea), *Plasmodium* sp. (malaria), *Candida albicans* (thrush), and *Influenza* (flu) [13,90].

NLRP3 senses a number of chemically and structurally diverse molecules and cellular stresses, including bacterial pore-forming toxins (e.g., nigericin) [59], different types of bacterial and viral RNA [91], asbestos and silica particles [92], ER stress [93], ultraviolet B (UVB) light [94], and host-derived danger molecules such as islet amyloid polypeptide [95], amyloid β [96], ATP [59], uric acid crystals [97], and hyaluronan [98]. Therefore, it is unlikely that NLRP3 directly interacts with any of these activators. While much research has focused on determining the mechanism of NLRP3 activation, and several efficient NLRP3 inhibitors have been discovered which block NLRP3-dependent cytokine activation and pyroptosis [99–101], the biochemical mechanism by which NLRP3 is activated, and how these inhibitors work, remains enigmatic.

It is well established that NLRP3, like precursor IL-1β, is transcriptionally induced in an NF-κB-dependent manner following TLR stimulation, known as "inflammasome priming." However, NLRP3 activation only takes place once the cell detects a second event triggered by an NLRP3 stimulus. Several models have been proposed for the signaling pathway that leads to NLRP3 oligmerization and inflammasome formation, including ion-pore formation, lysosomal rupture, and reactive oxygen species (ROS) generation. For example, ATP binding to the P2X7 receptor-ion channel and nigericin-induced pore formation are associated with K^+ efflux, while particulates such as alum and silica are postulated to cause lysosomal rupture and the release of proteolytic enzymes, such as cathepsin B. Alternatively, it has been suggested that mitochondrial ROS, or the release of mitochondrial DNA or cardiolipin and their binding to NLRP3, may be the unifying mechanism leading to NLRP3 inflammasome formation. However, all these mechanisms have been disputed to varying degrees [102–104], and, to date, the only reproducible cellular event that appears necessary and sufficient for NLRP3 activity, and which is conserved amongst all known NLRP3 stimuli, is K^+ efflux [104]. The membrane channels or pumps that allow K^+ efflux following stimulation with various NLRP3 activators, and how K^+ efflux triggers NLRP3 oligomerization, remain unknown.

Although NLRP3 activation in macrophages and dendritic cells triggers pyroptosis, it has yet to be determined *in vivo*, using either bacterially or host-derived NLPR3 activators, whether this has a physiological function. However, using mice harboring CAPS-associated mutations in NLRP3, it was shown that deletion of both IL-18 and IL-1R did not recapitulate the deletion of caspase 1 [6]. Specifically, around 50% of mice expressing mutant NLRP3 associated with familial cold autoinflammatory syndrome (FCAS) on an IL-1R and IL-18 null background failed to thrive and died before reaching maturity, while those crossed on to a caspase 1-deficient background had normal survival and no evidence of disease. Importantly, bone marrow-derived cells isolated from FCAS mice displayed enhanced pyroptosis that was only rescued by loss of caspase 1. Therefore, NLRP3-induced pyroptosis may promote disease pathogenesis in the context of murine FCAS, and NLRP3-induced caspase 8 activation and apoptosis is unlikely to play a role in this context (see later). Whether the same holds true for human CAPS disease remains unclear, as CAPS-associated mutations cause more severe disease symptoms in mice, and human CAPS patients can be successfully treated by directly blocking IL-1.

15.3.4 The NLRP1 Inflammasome

NLRP1 can be activated by *Bacillus anthracis* lethal factor-mediated cleavage [105,106], following *Toxoplasma* infection [107–109], viral infection, or chemotherapeutic treatment [3]. As with NLRP3, the host signaling pathways activated by *Toxoplasma*, viral infection, or cellular stresses that are required for the promotion of NLRP1 activity remain unclear. In mice, NLRP1 has undergone two gene-duplication events and encodes for three NLRP1 paralogs, *Nlrp1a*, *Nlrp1b*, and *Nlrp1c*, while in humans, only a single NLRP1 paralog is found. In rodents, the *Nlrp1b* locus controls susceptibility to lethal toxin-induced pyroptosis [105], whereas human NLRP1 is not activated by lethal toxin.

Forward genetic screening has identified a constitutively active *Nlrp1a* mutant mouse (encoding for a Q953P substitution) with a neutrophilic phenotype, which died prematurely at around 3–6 months [3]. While the premature death and systemic inflammation of these mice was rescued by crossing them on to an IL-1R background, hematopoietic stem and progenitor cells underwent pyroptosis that was only rescued following deletion of caspase 1. Consequently, emergency hematopoiesis induced by chemotherapy and cytopenia resulting from lymphocytic choriomeningitis virus (LCMV) infection are compromised and enhanced, respectively, in NLRP1a-activating mice. Importantly, it was also demonstrated that the converse is true: genetic deletion of the NLRP1 locus (*Nlrp1a*, *Nlrp1b*, *Nlrp1c*) results in reduced cytopenia following LCMV infection and improves hematopoietic cell recovery following chemotherapeutic treatment. Therefore, NLPR1-mediated pyroptosis plays an important role in modulating hematopoietic cell death, and can induce immune suppression following infection or periods of cellular stress. The mechanisms leading to NLRP1 activation in these circumstances will be important to determine.

15.3.5 The NAIP/NLRC4 Inflammasome

The cytosolic NAIP/NLRC4 inflammasome responds to bacterial infection and is activated following detection of flagellin or the bacterial type III secretion system (T3SS) rod-and-needle proteins. The ability of NLRC4 to be activated by such diverse proteins was uncovered when these ligands were shown to directly bind distinct NLR NAIP family members in order to mediate direct NAIP–NLRC4 interactions, and thus nucleate NLRC4 inflammasome formation [65,110]. While C57BL/6 mice encode for four functional and two non-coding *Naip* genes that directly bind bacterial flagellin (NAIP5/6), the T3SS rod (NAIP2), or the T3SS needle (NAIP1), humans only harbor one *Naip* gene, which responds only to the T3SS needle protein [110]. NLRC4-activating mutations in humans causes neonatal-onset enterocolitis, periodic fever, and macrophage-activation syndrome [7,111], with isolated monocytes and macrophages showing spontaneous ASC speck formation, enhanced IL-1β and IL-18 secretion, and increased rates of pyroptosis. IL-1 receptor antagonist treatment normalizes markers of systemic inflammation, with no clinical fevers or flares recorded after 7 months of treatment. However, although serum IL-1β levels are dramatically reduced, IL-18 levels remain elevated and patient macrophages remain hyper-responsive. Whether patients respond better to IL-18 and/or caspase 1 inhibition remains to be determined.

The *in vivo* role of pyroptosis in clearing *S. typhimurium* infection was elegantly demonstrated by the creation of a strain that constitutively expresses flagellin [2]. While mice succumbed to wild-type *S. typhimurium* infection within 6–8 days, *S. typhimurium*-expressing flagellin survived infection with dramatically reduced bacterial loads, and this was dependent on NLRC4 but not ASC, suggesting that *in vivo*, ASC is dispensable for murine NLRC4 activity. Importantly, bacterial clearance was dependent on caspase 1 but independent of IL-1R and IL-18. These results suggest that pyroptosis *in vivo* can act to expose intracellular bacteria to neutrophil killing. Similar results were obtained using *L. monocytogenes* expressing *Legionella* flagellin, whereby virulence was attenuated in a NLRC4-dependent, but IL-1β- and IL-18-independent, manner [112]. The central role of inflammasome activation in host defense against bacterial pathogenesis is highlighted by the fact that both *Salmonella* and *Listeria* suppress flagellin expression during systemic infection [18]. Given the importance of IL-1 and IL-18 activation in many infection models, and the role played by caspase 1 in mediating eicosanoid release, deciphering the exact contribution of caspase 1-dependent pyroptosis in endogenous infections remains a significant challenge.

15.3.6 The AIM2 and IFI16 Inflammasomes

AIM2 directly binds double-stranded DNA through a HIN200 domain [66,67,113,114]. It thus recognizes exposed cytosolic DNA from a range of viruses [115], as well DNA released following the cytosolic lysis of microbes such as *Francisella tularensis* [116,117], *Listeria monocytogenes* [118–121], and *Aspergillus fumigates* [122] in order to mediate IL-1 activation, pyroptosis, and pathogen resistance. Although AIM2 invariably induces pyroptosis *in vitro*, the functional consequences of AIM2-dependent pyroptosis in the absence of IL-1R and IL-18 signaling have yet to be documented. The human ALR IFI16 has been shown to detect viral double-stranded DNA in the nucleus or cytosol and activate caspase 1 [68,123]. Recent work has established that an IFI16 caspase 1 inflammasome induces pyroptosis of $CD4^+$ T cells following abortive HIV infection [124,125]. IFI16 binds reverse-transcribed ssHIV DNA that accumulates in the cytosol to trigger ASC recruitment and caspase 1 activation. These groundbreaking results show pyroptosis as a key host pathway that depletes $CD4^+$ T cells and is likely to contribute to the pathogenesis of acquired immunodeficiency syndrome (AIDS).

15.4 The Non-canonical Caspase 11 Inflammasome

LPS binds to and oligomerizes caspase 11 (or the human orthologs, caspases 4 and 5) to induce pyroptotic cell death [35]. It can also activate the NLRP3–caspase 1 inflammasome to modulate IL-1β and IL-18 secretion [126] (Figure 15.3). Because LPS is only found on the surface of Gram-negative bacteria, Gram-positive bacterial infection does not result in caspase 11 activation [127]. It is also important to note that although pyroptosis induced by caspase 11 (or caspases 4 and 5) is assumed to resemble that mediated by caspase 1, a proper, detailed comparison has yet to be made.

LPS is a large molecule that contains a hydrophobic region, lipid A (endotoxin), and a polysaccharide comprising O-antigen – but it is the hexa-acyl lipid A moiety that specifically activates caspase 11 [128,129]. LPS was originally demonstrated to signal through a cell-surface TLR4 and myeloid differentiation factor 2 (MD2) complex to induce the production of inflammatory cytokines, chemokines, and type I IFNs, and, in some circumstances, to cause apoptotic or necroptotic cell death. LPS-TLR4-TRIF induction of type I IFN, or stimulation of IFNγ, activates STAT1 to prime cells for non-canonical inflammasome activation by inducing the expression of caspase 11 [127,130,131]. ROS have also been suggested to enhance caspase 11 expression by increasing C-Jun N-terminal kinase (JNK) activation [132]. However, the non-canonical inflammasome is only activated

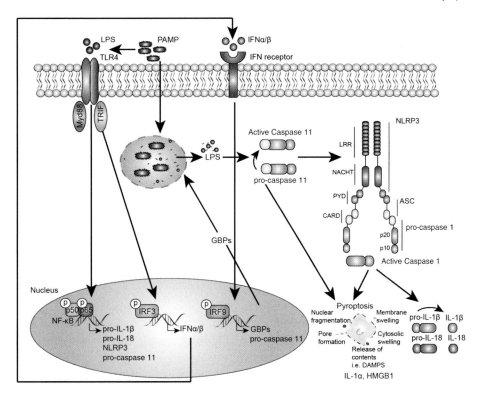

Figure 15.3 Non-canonical inflammasome activation and pyroptosis in response to cytosolic lipopolysaccharide (LPS). Gram-negative bacterial infections, or exposure to LPS, lead to increased NF-κB activity and the expression of pro-IL-1β, pro-IL-18, NLRP3, and pro-caspase 11, and via TLR4-TRIF, can activate a type 1 interferon pathway to activate interferon receptor (IFNR)-induced expression of pro-caspase 11 and guanylate-binding partners (GBPs). Upon caspase 11's sensing the escape of cytosolic LPS from pathogen-harboring vacuoles, in a GBP-dependent manner, it oligomerizes and activates. Active caspase 11 can trigger pyroptosis directly, or else trigger the NLRP3 inflammasome via an unknown mechanism.

if LPS accesses the host cytosol where caspase 11 is located, and as long as caspase 11 is expressed, this occurs independently of TLR4-MD2. Caspase 11 activation has been demonstrated to occur following the escape of a vacuole containing bacteria into the host cytosol, with the subsequent activation of caspase 11 being key to the host restriction of bacterial pathogenesis [133]. However, Gram-negative bacteria that reside in vacuolar membranes, such as *Legionella* and *Salmonella*, only escape into the cytosol in significant numbers if they are genetically engineered to do so [133], and therefore caspase 11 loss has no effect on their virulence *in vivo* [127,130,134,135]. In contrast, bacteria that naturally invade the cytosol, such as *Burkholderia*, will potently activate the caspase 11 inflammasome, and in this case caspase 11 is critical to mediating host resistance to lethal infection [133].

Pyroptosis mediated by caspase 11 exposes intracellular bacteria for neutrophil clearance. For example, host resistance to *Salmonella* that are genetically engineered to escape into the cytosol is critically dependent on caspase 11, but is independent of canonical inflammasome activity, as well as of IL-1β and IL-18. Neutrophil ROS generation is required for bacterial clearance, and consistent with this, reduced nicotinamide adenine dinucleotide phosphate (NADPH) oxidase-deficient mice are defective in mutant *Salmonella* clearance, much as with the loss of caspase 11 [133]. Therefore, it is likely that caspase 11-mediated pyroptosis is required, at least in part, for the restriction of Gram-negative bacteria that access the host cytosol.

The remarkable finding that the original caspase 1 KO mice were also deficient in caspase 11 (due to the knockouts being made on the 129 mouse strain deficient for caspase 11 expression) has revolutionized our understanding of endotoxic shock [135], and it has potentially significant implications for the pathogenesis of sepsis [10]. While the LPS model of sepsis was originally reported to be dependent on caspase 1, the use of bona fide caspase 1 and 11 null mice demonstrates that LPS-induced lethality is caspase 11-dependent but caspase 1-independent [135]. Further to this, if caspase 11 expression is first induced through priming (e.g., TLR3 activation), then subsequent LPS challenge and endotoxic shock can be triggered in the absence of TLR4 [129]. Moreover, because caspase 11 activates IL-1β and IL-18 via NLRP3-caspase 1, this implies pyroptotic cell death – and not NLRP3, caspase 1, or IL-1 – is responsible for endotoxic shock pathogenesis. In this regard, it has been documented that HMGB1 neutralization protects against LPS challenge, suggesting that caspase 11-mediated cell death and DAMP release contribute to pathology [136].

15.5 Crosstalk between Pyroptosis, Necroptosis, and Apoptosis

Caspase 8 is the initiator caspase required for death-receptor apoptosis. However, several studies using immortalized cell lines indicate that ASC can also trigger apoptotic cell death and/or NF-κB activation via this caspase [137,138]. Recently, two landmark papers built upon this work and detailed the mechanism by which the AIM2 and NLRP3 inflammasomes can trigger caspase 8-dependent apoptosis in addition to pyroptosis in primary cells (Figure 15.4) [139,140].

In particular, ASC recruited to AIM2 can trigger caspase 8-mediated apoptosis upon *Francisella tularensis* infection in the absence of caspase 1. Biochemical studies indicate that caspase 8 is recruited to the inflammasome by the interaction between the pyrin domain of ASC and the death-effector domain of caspase 8. Notably, this interaction is specific for caspase 8 and not for other caspases (e.g., caspase 2, 9, 11, and 12). Similar studies using the NLRP3 activator nigericin also show a clear interaction of caspase 8 with ASC-associated NLRP3, and the triggering of apoptotic cell death independent of caspase 1. While rapid caspase 1 pyroptosis appears to be the primary mediator of cell death using high doses of NLRP3 (nigericin) or AIM2 (double-stranded DNA) activators, at lower concentrations there is a preference for caspase 8-mediated apoptosis. The physiological significance of these observations requires further study, but they suggest that the infectious dose may dictate whether a cell dies a pyroptotic or an apoptotic death. Notably, however, an examination of ASC KO (required for both caspase 8 and caspase 1 activation) versus caspase 1/11 double-KO mice (where ASC can still promote caspase 8 activation) revealed no differences in infectious burden or lethality following *F. tularensis* infection. However, IL-18 and IFNγ levels were blunted upon ASC loss when compared to caspase 1/11 deletion, suggesting ASC may interact with caspase 8 to activate IL-18. It has been speculated that AIM2- or NLRP3-induced apoptosis may be beneficial to the host under circumstances where pathogens, such as *Yersinia*, employ strategies to directly inhibit caspase 1 [141].

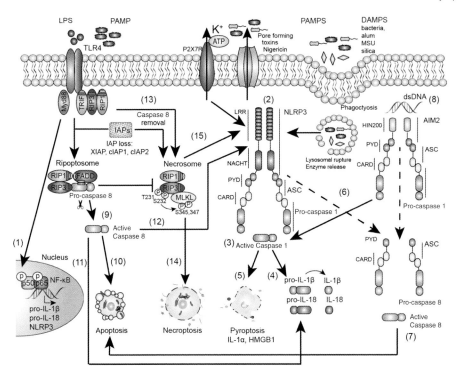

Figure 15.4 Crosstalk between necroptosis and apoptosis signaling with inflammasome activation. Following Toll-like receptor (TLR) (or tumor necrosis factor, TNF) priming (signal 1) in macrophages (1) (namely, pro-IL-1β and NLRP3 expression), exposure to a range of diverse stimuli (signal 2) can activate NLRP3 inflammasome assembly (2) via a number of potential mechanisms, including lysosomal rupture, K+ efflux, and sensing of ROS. NLRP3 inflammasome assembly results in the activation of caspase 1 (3), which can cleave pro-IL-1β and IL-18 to their active mature forms (4) and cause pyroptotic cell death (5). NLRP3 can also interact with pro-caspase 8 via ASC (6), leading to caspase 8 activation and apoptotic cell death (7). Likewise, the AIM2 inflammasome (8) can interact with caspase 1 or 8, depending on the level of DNA, to activate it. When macrophages are exposed to TLR ligands in the absence of all three inhibitor of apoptosis proteins (XIAP, cIAP1, cIAP2), a RIP1-RIP3-FADD-caspase 8 ripoptosome complex can form (9), leading to the activation of caspase 8 (10) and further to a number of other events: caspase 8-mediated apoptosis, caspase 8 cleavage and activation of IL-1β and IL-18 (11), and caspase 8 activation of NLRP3 by an unknown mechanism (12). When IAPs are absent and/or caspase 8 is inactive, a second complex can form (13), namely the necrosome, which can trigger necroptotic cell death (14) and activate the NLRP3 inflammasome (15). It is possible, but not yet proven, that all these cell-death pathways promote NLRP3 inflammasome formation through membrane damage and the release of potassium ions (not depicted).

Recent research has also uncovered novel roles for caspase 8 in activating the NLRP3 inflammasome or in directly cleaving and thereby activating IL-1β and IL-18 (Figure 15.4). TLRs and death receptors can signal caspase 8 activation in innate immune cells such as macrophages; this occurs rapidly upon the loss of caspase 8 repressor proteins such as IAPs or the caspase 8 inhibitor cFLIP. Notably, when either of these proteins is genetically deleted or its synthesis is inhibited, TLR stimulation alone is sufficient to induce direct caspase 8-mediated cleavage of IL-1β and IL-18 in the absence of caspase 1, as with caspase 8 cleavage of IL-1β following Fas ligand stimulation [30–32,142]. Moreover, it has also been reported that an ASC-caspase 8 inflammasome complex forms in response to fungal or mycobacterial stimulation of dectin-1 to modulate caspase 8 processing of IL-1β [143,144]. Although the physiological significance of these observations *in vivo* is not yet clear, as with ASC-caspase 8-induced apoptosis, caspase 8 may play a backup role in activating inflammatory cytokines upon pathogen-mediated inhibition of caspase 1 activity.

TLR stimulation in the absence of IAPs has also been observed to trigger caspase 8-dependent NLRP3-caspase 1 activation (Figure 15.4) [145]. This appears specific for certain stimuli, because caspase 8 is dispensable for canonical NLRP3 activation by nigericin. In a similar manner, it has been reported that caspase 8 may directly promote caspase 1 activity following *Yersinia* infection [145] or NLRP3-caspase 1 activation as a result of chemotherapeutic treatment [31,146]. In the former case, it is proposed that *Yersinia* mediates the inhibition of host-cell NF-κB activity via the bacterial YopJ protein. As a result, YopJ activates caspase 1 in a caspase 8-dependent manner, and mice lacking both RIPK3 and caspase 8 (caspase 8 deletion alone is lethal) are highly susceptible to *Yersinia* infection due to a defect in caspase 1-dependent cytokine activation. However, loss of RIPK3 and caspase 8 also impairs TLR-induced transcription of inflammatory cytokines, including precursor IL-1β [103,147,148], indicating that the defect in monocyte and neutrophil cytokine production following *Yersinia* infection may also result from a general lack of TLR-induced transcriptional responses [147].

Apart from the crosstalk between apoptotic caspase 8 and caspase 1, several studies have documented how necroptotic cell-death signaling can also trigger NLRP3-caspase 1 inflammasome activation in innate immune cells, such as macrophages and dendritic cells. Caspase 8 is required to suppress RIPK3- and MLKL-mediated necroptosis. Hence, when caspase 8's function is inhibited (e.g., by viral CrmA or vFLIP), tumor necrosis factor receptor 1 (TNFR1) and TLRs induce necroptotic cell death, which requires RIPK1, RIPK3 kinase activity, and the RIPK3 substrate MLKL. Studies have reported that when activated, MLKL binds to and disrupts membrane lipids, including the plasma membrane, to cause a lytic cell death and the release of DAMPs [149]. Remarkably, it has been observed that deletion of caspase 8 in dendritic cells produces endotoxin lethality *in vivo* that is dependent on RIPK3-induced IL-1 activation [150]. Biochemical and functional studies *in vitro* have since shown that RIPK3 and MLKL are required for NLRP3-caspase 1 activity following caspase 8 and IAP removal [151]. Hence, multiple cell-death pathways, including apoptosis and necroptosis, can converge on NLRP3 inflammasome activation, and may have evolved to counteract the many mechanisms microbes employ to manipulate and inhibit cell-death signaling and innate-immune responses.

References

1 Galluzzi L, Bravo-San Pedro JM, Vitale I, Aaronson SA, Abrams JM, Adam D, et al. Essential versus accessory aspects of cell death: recommendations of the NCCD 2015. *Cell Death Differ* 2015;**22**(1):58–73.
2 Miao EA, Leaf IA, Treuting PM, Mao DP, Dors M, Sarkar A, et al. Caspase-1-induced pyroptosis is an innate immune effector mechanism against intracellular bacteria. *Nat Immunol* 2010;**11**(12):1136–42.
3 Masters SL, Gerlic M, Metcalf D, Preston S, Pellegrini M, O'Donnel JA, et al. NLRP1 inflammasome activation induces pyroptosis of hematopoietic progenitor cells. *Immunity* 2012;**37**(6):1009–23.

4 Doitsh G, Galloway NL, Geng X, Yang Z, Monroe KM, Zepeda O, et al. Cell death by pyroptosis drives CD4 T-cell depletion in HIV-1 infection. *Nature* 2014;**505**(7484):509–14.
5 Chang W, Lin J, Dong J, Li D. Pyroptosis: an inflammatory cell death implicates in atherosclerosis. *Med Hypotheses* 2013;**81**(3):484–6.
6 Brydges SD, Broderick L, McGeough MD, Pena CA, Mueller JL, Hoffman HM. Divergence of IL-1, IL-18, and cell death in NLRP3 inflammasomopathies. *J Clin Invest* 2013;**123**(11):4695–705.
7 Canna SW, de Jesus AA, Gouni S, Brooks SR, Marrero B, Liu Y, et al. An activating NLRC4 inflammasome mutation causes autoinflammation with recurrent macrophage activation syndrome. *Nat Genet* 2014;**46**(10):1140–6.
8 Kitamura A, Sasaki Y, Abe T, Kano H, Yasutomo K. An inherited mutation in NLRC4 causes autoinflammation in human and mice. *J Exp Med* 2014;**211**(12):2385–96.
9 Bianchi ME. DAMPs, PAMPs and alarmins: all we need to know about danger. *J Leukoc Biol* 2007;**81**(1):1–5.
10 Croker BA, O'Donnell JA, Gerlic M. Pyroptotic death storms and cytopenia. *Curr Opin Immunol* 2014;**26**:128–37.
11 Foell D, Wittkowski H, Roth J. Mechanisms of disease: a "DAMP" view of inflammatory arthritis. *Nat Clin Pract Rheumatol* 2007;**3**(7):382–90.
12 Masters SL, Simon A, Aksentijevich I, Kastner DL. Horror autoinflammaticus: the molecular pathophysiology of autoinflammatory disease (*). *Annu Rev Immunol* 2009;**27**:621–68.
13 Menu P, Vince JE. The NLRP3 inflammasome in health and disease: the good, the bad and the ugly. *Clin Exp Immunol* 2011;**166**(1):1–15.
14 Huebener P, Pradere JP, Hernandez C, Gwak GY, Caviglia JM, Mu X, et al. The HMGB1/RAGE axis triggers neutrophil-mediated injury amplification following necrosis. *J Clin Invest* 2015;**125**(2):539–50.
15 Harris HE, Andersson U, Pisetsky DS. HMGB1: a multifunctional alarmin driving autoimmune and inflammatory disease. *Nat Rev Rheumatol* 2012;**8**(4):195–202.
16 Kono H, Chen CJ, Ontiveros F, Rock KL. Uric acid promotes an acute inflammatory response to sterile cell death in mice. *J Clin Invest* 2010;**120**(6):1939–49.
17 Croker BA, O'Donnell JA, Gerlic M. Pyroptotic death storms and cytopenia. *Curr Opin Immunol* 2014;**26**:128–37.
18 Jorgensen I, Miao EA. Pyroptotic cell death defends against intracellular pathogens. *Immunol Rev* 2015;**265**(1):130–42.
19 Khan N, Lawlor KE, Murphy JM, Vince JE. More to life than death: molecular determinants of necroptotic and non-necroptotic RIP3 kinase signaling. *Curr Opin Immunol* 2014;**26**:76–89.
20 Xu H, Yang J, Gao W, Li L, Li P, Zhang L, et al. Innate immune sensing of bacterial modifications of Rho GTPases by the Pyrin inflammasome. *Nature* 2014;**513**(7517):237–41.
21 Schroder K, Tschopp J. The inflammasomes. *Cell* 2010;**140**(6):821–32.
22 Monie TP, Bryant CE. Caspase-8 functions as a key mediator of inflammation and pro-IL-1beta processing via both canonical and non-canonical pathways. *Immunol Rev* 2015;**265**(1):181–93.
23 Pop C, Salvesen GS. Human caspases: activation, specificity, and regulation. *J Biol Chem* 2009;**284**(33):21 777–81.

24 Oberst A, Dillon CP, Weinlich R, McCormick LL, Fitzgerald P, Pop C, et al. Catalytic activity of the caspase-8-FLIP(L) complex inhibits RIPK3-dependent necrosis. *Nature* 2011;**471**(7338):363–7.

25 Broz P, von Moltke J, Jones JW, Vance RE, Monack DM. Differential requirement for Caspase-1 autoproteolysis in pathogen-induced cell death and cytokine processing. *Cell Host Microbe* 2010;**8**(6):471–83.

26 Segawa K, Kurata S, Yanagihashi Y, Brummelkamp TR, Matsuda F, Nagata S. Caspase-mediated cleavage of phospholipid flippase for apoptotic phosphatidylserine exposure. *Science* 2014;**344**(6188):1164–8.

27 Weinlich R, Oberst A, Dillon CP, Janke LJ, Milasta S, Lukens JR, et al. Protective roles for caspase-8 and cFLIP in adult homeostasis. *Cell Rep* 2013;**5**(2):340–8.

28 Rickard JA, Anderton H, Etemadi N, Nachbur U, Darding M, Peltzer N, et al. TNFR1-dependent cell death drives inflammation in Sharpin-deficient mice. *Elife* 2014;**3**: doi: 10.7554/eLife.03464.

29 Panayotova-Dimitrova D, Feoktistova M, Ploesser M, Kellert B, Hupe M, Horn S, et al. cFLIP regulates skin homeostasis and protects against TNF-induced keratinocyte apoptosis. *Cell Rep* 2013;**5**(2):397–408.

30 Maelfait J, Vercammen E, Janssens S, Schotte P, Haegman M, Magez S, Beyaert R. Stimulation of Toll-like receptor 3 and 4 induces interleukin-1beta maturation by caspase-8. *J Exp Med* 2008;**205**(9):1967–73.

31 Vince JE, Wong WW, Gentle I, Lawlor KE, Allam R, O'Reilly L, et al. Inhibitor of apoptosis proteins limit RIP3 kinase-dependent interleukin-1 activation. *Immunity* 2012;**36**(2):215–27.

32 Bossaller L, Chiang PI, Schmidt-Lauber C, Ganesan S, Kaiser WJ, Rathinam VA, et al. Cutting edge: FAS (CD95) mediates noncanonical IL-1beta and IL-18 maturation via caspase-8 in an RIP3-independent manner. *J Immunol* 2012;**189**(12):5508–12.

33 Fernandez DJ, Lamkanfi M. Inflammatory caspases: key regulators of inflammation and cell death. *Biol Chem* 2015;**396**(3):193–203.

34 Martinon F, Tschopp J. Inflammatory caspases: linking an intracellular innate immune system to autoinflammatory diseases. *Cell* 2004;**117**(5):561–74.

35 Shi J, Zhao Y, Wang Y, Gao W, Ding J, Li P, et al. Inflammatory caspases are innate immune receptors for intracellular LPS. *Nature* 2014;**514**(7521):187–92.

36 Kajiwara Y, Schiff T, Voloudakis G, Gama Sosa MA, Elder G, Bozdagi O, Buxbaum JD. A critical role for human caspase-4 in endotoxin sensitivity. *J Immunol* 2014;**193**(1):335–43.

37 Knodler LA, Crowley SM, Sham HP, Yang H, Wrande M, Ma C, et al. Noncanonical inflammasome activation of caspase-4/caspase-11 mediates epithelial defenses against enteric bacterial pathogens. *Cell Host Microbe* 2014;**16**(2):249–56.

38 Cerretti DP, Kozlosky CJ, Mosley B, Nelson N, Van Ness K, Greenstreet TA, et al. Molecular cloning of the interleukin-1 beta converting enzyme. *Science* 1992;**256**(5053):97–100.

39 Fantuzzi G, Dinarello CA. Interleukin-18 and interleukin-1 beta: two cytokine substrates for ICE (caspase-1). *J Clin Immunol* 1999;**19**(1):1–11.

40 Thornberry NA, Bull HG, Calaycay JR, Chapman KT, Howard AD, Kostura MJ, et al. A novel heterodimeric cysteine protease is required for interleukin-1 beta processing in monocytes. *Nature* 1992;**356**(6372):768–74.

41 Lopez-Castejon G, Brough D. Understanding the mechanism of IL-1beta secretion. *Cytokine Growth Factor Rev* 2011;**22**(4):189–95.

42. Jesenberger V, Procyk KJ, Yuan J, Reipert S, Baccarini M. Salmonella-induced caspase-2 activation in macrophages: a novel mechanism in pathogen-mediated apoptosis. *J Exp Med* 2000;**192**(7):1035–46.
43. Hersh D, Monack DM, Smith MR, Ghori N, Falkow S, Zychlinsky A. The Salmonella invasin SipB induces macrophage apoptosis by binding to caspase-1. *Proc Natl Acad Sci USA* 1999;**96**(5):2396–401.
44. Brennan MA, Cookson BT. Salmonella induces macrophage death by caspase-1-dependent necrosis. *Mol Microbiol* 2000;**38**(1):31–40.
45. Chen Y, Smith MR, Thirumalai K, Zychlinsky A. A bacterial invasin induces macrophage apoptosis by binding directly to ICE. *EMBO J* 1996;**15**(15):3853–60.
46. Hilbi H, Chen Y, Thirumalai K, Zychlinsky A. The interleukin 1beta-converting enzyme, caspase 1, is activated during Shigella flexneri-induced apoptosis in human monocyte-derived macrophages. *Infect Immun* 1997;**65**(12):5165–70.
47. Hilbi H, Moss JE, Hersh D, Chen Y, Arondel J, Banerjee S, et al. Shigella-induced apoptosis is dependent on caspase-1 which binds to IpaB. *J Biol Chem* 1998;**273**(49):32 895–900.
48. Kuida K, Lippke JA, Ku G, Harding MW, Livingston DJ, Su MS, Flavell RA. Altered cytokine export and apoptosis in mice deficient in interleukin-1 beta converting enzyme. *Science* 1995;**267**(5206):2000–3.
49. Wang S, Miura M, Jung YK, Zhu H, Li E, Yuan J. Murine caspase-11, an ICE-interacting protease, is essential for the activation of ICE. *Cell* 1998;**92**(4):501–9.
50. Fink SL, Cookson BT. Caspase-1-dependent pore formation during pyroptosis leads to osmotic lysis of infected host macrophages. *Cell Microbiol* 2006;**8**(11):1812–25.
51. Vanden Berghe T, Linkermann A, Jouan-Lanhouet S, Walczak H, Vandenabeele P. Regulated necrosis: the expanding network of non-apoptotic cell death pathways. *Nat Rev Mol Cell Biol* 2014;**15**(2):135–47.
52. Nyström S, Antoine DJ, Lundbäck P, Lock JG, Nita AF, Högstrand K, et al. TLR activation regulates damage-associated molecular pattern isoforms released during pyroptosis. *EMBO J* 2013;**32**(1):86–99.
53. MacKenzie A, Wilson HL, Kiss-Toth E, Dower SK, North RA, Surprenant A. Rapid secretion of interleukin-1beta by microvesicle shedding. *Immunity* 2001;**15**(5):825–35.
54. Wang Q, Imamura R, Motani K, Kushiyama H, Nagata S, Suda T. Pyroptotic cells externalize eat-me and release find-me signals and are efficiently engulfed by macrophages. *Int Immunol* 2013;**25**(6):363–72.
55. Ting JP, Lovering RC, Alnemri ES, Bertin J, Boss JM, Davis BK, et al. The NLR gene family: a standard nomenclature. *Immunity* 2008;**28**(3):285–7.
56. Faustin B, Lartigue L, Bruey J-M, Luciano F, Sergienko E, Bailly-Maitre B, et al. Reconstituted NALP1 inflammasome reveals two-step mechanism of caspase-1 activation. *Molecular Cell* 2007;**25**(5):713–24.
57. Martinon F, Burns K, Tschopp J. The inflammasome: a molecular platform triggering activation of inflammatory caspases and processing of proIL-beta. *Mol Cell* 2002;**10**(2):417–26.
58. Agostini L, Martinon F, Burns K, McDermott MF, Hawkins PN, Tschopp J. NALP3 forms an IL-1beta-processing inflammasome with increased activity in Muckle-Wells autoinflammatory disorder. *Immunity* 2004;**20**(3):319–25.
59. Mariathasan S, Weiss DS, Newton K, McBride J, O'Rourke K, Roose-Girma M, et al. Cryopyrin activates the inflammasome in response to toxins and ATP. *Nature* 2006;**440**(7081):228–32.

60 Elinav E, Strowig T, Kau AL, Henao-Mejia J, Thaiss CA, Booth CJ, et al. NLRP6 inflammasome regulates colonic microbial ecology and risk for colitis. *Cell* 2011;**145**(5):745–57.

61 Anand PK, Kanneganti TD. NLRP6 in infection and inflammation. *Microbes Infect* 2013;**15**(10–11):661–8.

62 Khare S, Dorfleutner A, Bryan NB, Yun C, Radian AD, de Almeida L, et al. An NLRP7-containing inflammasome mediates recognition of microbial lipopeptides in human macrophages. *Immunity* 2012;**36**(3):464–76.

63 Vladimer GI, Weng D, Paquette SW, Vanaja SK, Rathinam VA, Aune MH, et al. The NLRP12 inflammasome recognizes Yersinia pestis. *Immunity* 2012;**37**(1):96–107.

64 Mariathasan S, Newton K, Monack DM, Vucic D, French DM, Lee WP, et al. Differential activation of the inflammasome by caspase-1 adaptors ASC and Ipaf. *Nature* 2004;**430**(6996):213–18.

65 Kofoed EM, Vance RE. Innate immune recognition of bacterial ligands by NAIPs determines inflammasome specificity. *Nature* 2011;**477**(7366):592–5.

66 Roberts TL, Idris A, Dunn JA, Kelly GM, Burnton CM, Hodgson S, et al. HIN-200 proteins regulate caspase activation in response to foreign cytoplasmic DNA. *Science* 2009;**323**(5917):1057–60.

67 Hornung V, Ablasser A, Charrel-Dennis M, Bauernfeind F, Horvath G, Caffrey DR, et al. AIM2 recognizes cytosolic dsDNA and forms a caspase-1-activating inflammasome with ASC. *Nature* 2009;**458**(7237):514–18.

68 Kerur N, Veettil MV, Sharma-Walia N, Bottero V, Sadagopan S, Otageri P, Chandran B. IFI16 acts as a nuclear pathogen sensor to induce the inflammasome in response to Kaposi Sarcoma-associated herpesvirus infection. *Cell Host Microbe* 2011;**9**(5):363–75.

69 French FMF Consortium. A candidate gene for familial Mediterranean fever. *Nat Genet* 1997;**17**(1):25–31.

70 Py BF, Jin M, Desai BN, Penumaka A, Zhu H, Kober M, et al. Caspase-11 controls interleukin-1beta release through degradation of TRPC1. *Cell Rep* 2014;**6**(6):1122–8.

71 Poeck H, Bscheider M, Gross O, Finger K, Roth S, Rebsamen M, et al. Recognition of RNA virus by RIG-I results in activation of CARD9 and inflammasome signaling for interleukin 1 beta production. *Nat Immunol* 2009;**11**(1):63–9.

72 Rajan JV, Rodriguez D, Miao EA, Aderem A. The NLRP3 inflammasome detects encephalomyocarditis virus and vesicular stomatitis virus infection. *J Virol* 2011;**85**(9):4167–72.

73 Hu Z, Yan C, Liu P, Huang Z, Ma R, Zhang C, et al. Crystal structure of NLRC4 reveals its autoinhibition mechanism. *Science* 2013;**341**(6142):172–5.

74 Hara H, Tsuchiya K, Kawamura I, Fang R, Hernandez-Cuellar E, Shen Y, et al. Phosphorylation of the adaptor ASC acts as a molecular switch that controls the formation of speck-like aggregates and inflammasome activity. *Nat Immunol* 2013;**14**(12):1247–55.

75 Cai X, Chen J, Xu H, Liu S, Jiang QX, Halfmann R, Chen ZJ. Prion-like polymerization underlies signal transduction in antiviral immune defense and inflammasome activation. *Cell* 2014;**156**(6):1207–22.

76 Lu A, Magupalli VG, Ruan J, Yin Q, Atianand MK, Vos MR, et al. Unified polymerization mechanism for the assembly of ASC-dependent inflammasomes. *Cell* 2014;**156**(6):1193–206.

77. Sester DP, Thygesen SJ, Sagulenko V, Vajjhala PR, Cridland JA, Vitak N, et al. A novel flow cytometric method to assess inflammasome formation. *J Immunol* 2015;**194**(1):455–62.
78. Stutz A, Horvath GL, Monks BG, Latz E. ASC speck formation as a readout for inflammasome activation. *Methods Mol Biol* 2013;**1040**:91–101.
79. Franklin BS, Bossaller L, De Nardo D, Ratter JM, Stutz A, Engels G, et al. The adaptor ASC has extracellular and "prionoid" activities that propagate inflammation. *Nat Immunol* 2014;**15**(8):727–37.
80. Baroja-Mazo A, Martin-Sanchez F, Gomez AI, Martínez CM, Amores-Iniesta J, Compan V, et al. The NLRP3 inflammasome is released as a particulate danger signal that amplifies the inflammatory response. *Nat Immunol* 2014;**15**(8):738–48.
81. Netea MG, van de Veerdonk FL, van der Meer JW, Dinarello CA, Joosten LA. Inflammasome-independent regulation of IL-1-family cytokines. *Annu Rev Immunol* 2015;**33**:49–77.
82. Keller M, Ruegg A, Werner S, Beer HD. Active caspase-1 is a regulator of unconventional protein secretion. *Cell* 2008;**132**(5):818–31.
83. von Moltke J, Trinidad NJ, Moayeri M, Kintzer AF, Wang SB, van Rooijen N, et al. Rapid induction of inflammatory lipid mediators by the inflammasome in vivo. *Nature* 2012;**490**(7418):107–11.
84. Brough D, Rothwell NJ. Caspase-1-dependent processing of pro-interleukin-1beta is cytosolic and precedes cell death. *J Cell Sci* 2007;**120**(Pt. 5):772–81.
85. Silveira TN, Zamboni DS. Pore formation triggered by Legionella spp. is an Nlrc4 inflammasome-dependent host cell response that precedes pyroptosis. *Infect Immun* 2010;**78**(3):1403–13.
86. Gross O, Yazdi AS, Thomas CJ, Masin M, Heinz LX, Guarda G, et al. Inflammasome activators induce interleukin-1alpha secretion via distinct pathways with differential requirement for the protease function of caspase-1. *Immunity* 2012;**36**(3):388–400.
87. Chen KW, Gross CJ, Sotomayor FV, Stacey KJ, Tschopp J, Sweet MJ, Schroder K. The neutrophil NLRC4 inflammasome selectively promotes IL-1beta maturation without pyroptosis during acute Salmonella challenge. *Cell Rep* 2014;**8**(2):570–82.
88. Karmakar M, Katsnelson M, Malak HA, Greene NG, Howell SJ, Hise AG, et al. Neutrophil IL-1beta processing induced by pneumolysin is mediated by the NLRP3/ASC inflammasome and caspase-1 activation and is dependent on K+ efflux. *J Immunol* 2015;**194**(4):1763–75.
89. Dinarello CA, Simon A, van der Meer JW. Treating inflammation by blocking interleukin-1 in a broad spectrum of diseases. *Nat Rev Drug Discov* 2012;**11**(8):633–52.
90. Hoffman HM, Brydges SD. Genetic and molecular basis of inflammasome-mediated disease. *J Biol Chem* 2011;**286**(13):10 889–96.
91. Xiao TS. The nucleic acid-sensing inflammasomes. *Immunol Rev* 2015;**265**(1):103–11.
92. Dostert C, Pétrilli V, Van Bruggen R, Steele C, Mossman BT, Tschopp J. Innate immune activation through Nalp3 inflammasome sensing of asbestos and silica. *Science* 2008;**320**(5876):674–7.
93. Menu P, Mayor A, Zhou R, Tardivel A, Ichijo H, Mori K, Tschopp J. ER stress activates the NLRP3 inflammasome via an UPR-independent pathway. *Cell Death Dis* 2012;**3**:e261.

94. Feldmeyer L, Keller M, Niklaus G, Hohl D, Werner S, Beer H-D. The inflammasome mediates UVB-induced activation and secretion of interleukin-1beta by keratinocytes. *Curr Biol* 2007;**17**(13):1140–5.
95. Masters SL, Dunne A, Subramanian SL, Hull RL, Tannahill GM, Sharp FA, et al. Activation of the NLRP3 inflammasome by islet amyloid polypeptide provides a mechanism for enhanced IL-1beta in type 2 diabetes. *Nat Immunol* 2010;**11**(10):897–904.
96. Harrison C. Neurological disease: inflammasome activation in Alzheimer's disease. *Nat Rev Drug Discov* 2013;**12**(2):102.
97. Martinon F, Pétrilli V, Mayor A, Tardivel A, Tschopp J. Gout-associated uric acid crystals activate the NALP3 inflammasome. *Nature* 2006;**440**(7081):237–41.
98. Yamasaki K, Muto J, Taylor KR, Cogen AL, Audish D, Bertin J, et al. NLRP3/cryopyrin is necessary for interleukin-1beta (IL-1beta) release in response to hyaluronan, an endogenous trigger of inflammation in response to injury. *J Biol Chem* 2009;**284**(19):12 762–71.
99. Coll RC, Robertson AA, Chae JJ, Higgins SC, Muñoz-Planillo R, Inserra MC, et al. A small-molecule inhibitor of the NLRP3 inflammasome for the treatment of inflammatory diseases. *Nat Med* 2015;**21**(3):248–55.
100. Lamkanfi M, Mueller JL, Vitari AC, Misaghi S, Fedorova A, Deshayes K, et al. Glyburide inhibits the Cryopyrin/Nalp3 inflammasome. *J Cell Biol* 2009;**187**(1):61–70.
101. He Y, Varadarajan S, Muñoz-Planillo R, Burberry A, Nakamura Y, Nuñez G. 3,4-methylenedioxy-beta-nitrostyrene inhibits NLRP3 activation by blocking assembly of the inflammasome. *J Biol Chem* 2014;**289**(2):1142–50.
102. Lawlor KE, Vince JE. Ambiguities in NLRP3 inflammasome regulation: is there a role for mitochondria? *Biochim Biophys Acta* 2014;**1840**(4):1433–40.
103. Allam R, Lawlor KE, Yu EC, Mildenhall AL, Moujalled DM, Lewis RS, et al. Mitochondrial apoptosis is dispensable for NLRP3 inflammasome activation but non-apoptotic caspase-8 is required for inflammasome priming. *EMBO Rep* 2014;**15**(9):982–90.
104. Muñoz-Planillo R, Kuffa P, Martínez-Colón G, Smith BL, Rajendiran TM, Núñez G. K+ efflux is the common trigger of NLRP3 inflammasome activation by bacterial toxins and particulate matter. *Immunity* 2013;**38**(6):1142–53.
105. Boyden ED, Dietrich WF. Nalp1b controls mouse macrophage susceptibility to anthrax lethal toxin. *Nat Genet* 2006;**38**(2):240–4.
106. Chavarria-Smith J, Vance RE. The NLRP1 inflammasomes. *Immunol Rev* 2015;**265**(1):22–34.
107. Cirelli KM, Gorfu G, Hassan MA, Printz M, Crown D, Leppla SH, et al. Inflammasome sensor NLRP1 controls rat macrophage susceptibility to *Toxoplasma gondii*. *PLoS Pathog* 2014;**10**(3):e1003927.
108. Ewald SE, Chavarria-Smith J, Boothroyd JC. NLRP1 is an inflammasome sensor for *Toxoplasma gondii*. *Infect Immun* 2014;**82**(1):460–8.
109. Gorfu G, Cirelli KM, Melo MB, Mayer-Barber K, Crown D, Koller BH, et al. Dual role for inflammasome sensors NLRP1 and NLRP3 in murine resistance to *Toxoplasma gondii*. *MBio* 2014;**5**(1): pii:e01117-13.
110. Zhao Y, Yang J, Shi J, Gong YN, Lu Q, Xu H, et al. The NLRC4 inflammasome receptors for bacterial flagellin and type III secretion apparatus. *Nature* 2011;**477**(7366):596–600.

111 Romberg N, Al Moussawi K, Nelson-Williams C, Stiegler AL, Loring E, Choi M, et al. Mutation of NLRC4 causes a syndrome of enterocolitis and autoinflammation. *Nat Genet* 2014;**46**(10):1135–9.

112 Sauer JD, Pereyre S, Archer KA, Burke TP, Hanson B, Lauer P, Portnoy DA. Listeria monocytogenes engineered to activate the Nlrc4 inflammasome are severely attenuated and are poor inducers of protective immunity. *Proc Natl Acad Sci USA* 2011;**108**(30):12 419–24.

113 Fernandes-Alnemri T, Yu J-W, Datta P, Wu J, Alnemri ES. AIM2 activates the inflammasome and cell death in response to cytoplasmic DNA. *Nature* 2009;**458**(7237):509–13.

114 Bürckstümmer T, Baumann C, Blüml S, Dixit E, Dürnberger G, Jahn H, et al. An orthogonal proteomic-genomic screen identifies AIM2 as a cytoplasmic DNA sensor for the inflammasome. *Nat Immunol* 2009;**10**(3):266–72.

115 Lupfer C, Malik A, Kanneganti TD. Inflammasome control of viral infection. *Curr Opin Virol* 2015;**12**:38–46.

116 Rathinam VAK, Jiang Z, Waggoner SN, Sharma S, Cole LE, Waggoner L, et al. The AIM2 inflammasome is essential for host defense against cytosolic bacteria and DNA viruses. *Nat Immunol* 2010;**11**(5):395–402.

117 Fernandes-Alnemri T, Yu J-W, Juliana C, et al. The AIM2 inflammasome is critical for innate immunity to Francisella tularensis. *Nat Immunol* 2010;**11**(5):385–93.

118 Wu J, Fernandes-Alnemri T, Alnemri ES. Involvement of the AIM2, NLRC4, and NLRP3 inflammasomes in caspase-1 activation by listeria monocytogenes. *J Clin Immunol* 2010;**30**(5):693–702.

119 Warren SE, Armstrong A, Hamilton MK, Mao DP, Leaf IA, Miao EA, Aderem A. Cutting edge: cytosolic bacterial DNA activates the inflammasome via Aim2. *J Immunol* 2010;**185**(2):818–21.

120 Sauer J-D, Witte CE, Zemansky J, Hanson B, Lauer P, Portnoy DA. Listeria monocytogenes triggers AIM2-mediated pyroptosis upon infrequent bacteriolysis in the macrophage cytosol. *Cell Host Microbe* 2010;**7**(5):412–19.

121 Kim S, Bauernfeind F, Ablasser A, Hartmann G, Fitzgerald KA, Latz E, Hornung V. Listeria monocytogenes is sensed by the NLRP3 and AIM2 inflammasome. *Eur J Immunol* 2010;**40**(6):1545–51.

122 Karki R, Man SM, Malireddi RK, Gurung P, Vogel P, Lamkanfi M, Kanneganti TD. Concerted activation of the AIM2 and NLRP3 inflammasomes orchestrates host protection against aspergillus infection. *Cell Host Microbe* 2015;**17**(3):357–68.

123 Unterholzner L, Keating SE, Baran M, Horan KA, Jensen SB, Sharma S, et al. IFI16 is an innate immune sensor for intracellular DNA. *Nat Immunol* 2010;**11**(11):997–1004.

124 Monroe KM, Yang Z, Johnson JR, Geng X, Doitsh G, Krogan NJ, Greene WC. IFI16 DNA sensor is required for death of lymphoid CD4 T cells abortively infected with HIV. *Science* 2014;**343**(6169):428–32.

125 Jakobsen MR, Bak RO, Andersen A, Berg RK, Jensen SB, Tengchuan J, et al. IFI16 senses DNA forms of the lentiviral replication cycle and controls HIV-1 replication. *Proc Natl Acad Sci USA* 2013;**110**(48):E4571–80.

126 Stowe I, Lee B, Kayagaki N. Caspase-11: arming the guards against bacterial infection. *Immunol Rev* 2015;**265**(1):75–84.

127 Rathinam VA, Vanaja SK, Waggoner L, Sokolovska A, Becker C, Stuart LM, et al. TRIF licenses caspase-11-dependent NLRP3 inflammasome activation by gram-negative bacteria. *Cell* 2012;**150**(3):606–19.

128 Kayagaki N, Wong MT, Stowe IB, Ramani SR, Gonzalez LC, Akashi-Takamura S, et al. Noncanonical inflammasome activation by intracellular LPS independent of TLR4. *Science* 2013;**341**(6151):1246–9.

129 Hagar JA, Powell DA, Aachoui Y, Ernst RK, Miao EA. Cytoplasmic LPS activates caspase-11: implications in TLR4-independent endotoxic shock. *Science* 2013;**341**(6151):1250–3.

130 Gurung P, Malireddi RK, Anand PK, Demon D, Vande Walle L, Liu Z, et al. Toll or interleukin-1 receptor (TIR) domain-containing adaptor inducing interferon-beta (TRIF)-mediated caspase-11 protease production integrates Toll-like receptor 4 (TLR4) protein- and Nlrp3 inflammasome-mediated host defense against enteropathogens. *J Biol Chem* 2012;**287**(41):34 474–83.

131 Schauvliege R, Vanrobaeys J, Schotte P, Beyaert R. Caspase-11 gene expression in response to lipopolysaccharide and interferon-gamma requires nuclear factor-kappa B and signal transducer and activator of transcription (STAT) 1. *J Biol Chem* 2002;**277**(44):41 624–30.

132 Lupfer CR, Anand PK, Liu Z, Stokes KL, Vogel P, Lamkanfi M, Kanneganti TD. Reactive oxygen species regulate caspase-11 expression and activation of the non-canonical NLRP3 inflammasome during enteric pathogen infection. *PLoS Pathog* 2014;**10**(9):e1004410.

133 Aachoui Y, Leaf IA, Hagar JA, Fontana MF, Campos CG, Zak DE, et al. Caspase-11 protects against bacteria that escape the vacuole. *Science* 2013;**339**(6122):975–8.

134 Broz P, Ruby T, Belhocine K, Bouley DM, Kayagaki N, Dixit VM, Monack DM. Caspase-11 increases susceptibility to Salmonella infection in the absence of caspase-1. *Nature* 2012;**490**(7419):288–91.

135 Kayagaki N, Warming S, Lamkanfi M, Vande Walle L, Louie S, Dong J, et al. Non-canonical inflammasome activation targets caspase-11. *Nature* 2011;**479**(7371):117–21.

136 Lamkanfi M, Sarkar A, Vande Walle L, Vitari AC, Amer AO, Wewers MD, et al. Inflammasome-dependent release of the alarmin HMGB1 in endotoxemia. *J Immunol* 2010;**185**(7):4385–92.

137 Hasegawa M, Imamura R, Kinoshita T, Matsumoto N, Masumoto J, Inohara N, Suda T. ASC-mediated NF-kappaB activation leading to interleukin-8 production requires caspase-8 and is inhibited by CLARP. *J Biol Chem* 2005;**280**(15):15 122–30.

138 Hasegawa M, Kawase K, Inohara N, Imamura R, Yeh WC, Kinoshita T, Suda T. Mechanism of ASC-mediated apoptosis: bid-dependent apoptosis in type II cells. *Oncogene* 2007;**26**(12):1748–56.

139 Sagulenko V, Thygesen SJ, Sester DP, Idris A, Cridland JA, Vajjhala PR, et al. AIM2 and NLRP3 inflammasomes activate both apoptotic and pyroptotic death pathways via ASC. *Cell Death Differ* 2013;**20**(9):1149–60.

140 Pierini R, Juruj C, Perret M, Jones CL, Mageot P, Weiss DS, Henry T. AIM2/ASC triggers caspase-8-dependent apoptosis in Francisella-infected caspase-1-deficient macrophages. *Cell Death Differ* 2012;**19**(10):1709–21.

141 Aachoui Y, Sagulenko V, Miao EA, Stacey KJ. Inflammasome-mediated pyroptotic and apoptotic cell death, and defense against infection. *Curr Opin Microbiol* 2013;**16**(3):319–26.

142 Wu YH, Kuo WC, Wu YJ, Yang KT, Chen ST, Jiang ST, et al. Participation of c-FLIP in NLRP3 and AIM2 inflammasome activation. *Cell Death Differ* 2014;**21**(3):451–61.

143 Gringhuis SI, Kaptein TM, Wevers BA, Theelen B, van der Vlist M, Boekhout T, Geijtenbeek TB. Dectin-1 is an extracellular pathogen sensor for the induction and processing of IL-1beta via a noncanonical caspase-8 inflammasome. *Nat Immunol* 2012;**13**(3):246–54.

144 Ganesan S, Rathinam VA, Bossaller L, Army K, Kaiser WJ, Mocarski ES, et al. Caspase-8 modulates dectin-1 and complement receptor 3-driven IL-1beta production in response to beta-glucans and the fungal pathogen, *Candida albicans*. *J Immunol* 2014;**193**(5):2519–30.

145 Philip NH, Dillon CP, Snyder AG, Fitzgerald P, Wynosky-Dolfi MA, Zwack EE, et al. Caspase-8 mediates caspase-1 processing and innate immune defense in response to bacterial blockade of NF-kappaB and MAPK signaling. *Proc Natl Acad Sci USA* 2014;**111**(20):7385–90.

146 Antonopoulos C, El Sanadi C, Kaiser WJ, Mocarski ES, Dubyak GR. Proapoptotic chemotherapeutic drugs induce noncanonical processing and release of IL-1beta via caspase-8 in dendritic cells. *J Immunol* 2013;**191**(9):4789–803.

147 Weng D, Marty-Roix R, Ganesan S, Proulx MK, Vladimer GI, Kaiswer WJ, et al. Caspase-8 and RIP kinases regulate bacteria-induced innate immune responses and cell death. *Proc Natl Acad Sci USA* 2014;**111**(20):7391–6.

148 Gurung P, Anand PK, Malireddi RK, Vande Walle L, Van Opdenbosch N, Dillon CP, et al. FADD and caspase-8 mediate priming and activation of the canonical and noncanonical Nlrp3 inflammasomes. *J Immunol* 2014;**192**(4):1835–46.

149 Murphy JM, Silke J. Ars Moriendi; the art of dying well – new insights into the molecular pathways of necroptotic cell death. *EMBO Rep* 2014;**15**(2):155–64.

150 Kang TB, Yang SH, Toth B, Kovalenko A, Wallach D. Caspase-8 blocks kinase RIPK3-mediated activation of the NLRP3 inflammasome. *Immunity* 2013;**38**(1):27–40.

151 Lawlor KE, Khan N, Mildenhall A, Gerlic M, Croker BA, D'Cruz AA, et al. RIPK3 promotes cell death and NLRP3 inflammasome activation in the absence of MLKL. *Nat Commun* 2015;**6**:6282.

16

Paraptosis

Maryam Khalili and James A. Radosevich

Department of Oral Medicine and Diagnostic Sciences, University of Illinois at Chicago, Chicago, IL, USA

Abbreviations

1-NP	1-nitropyrene
A0	thioxotriazole copper (II) complex
ACTH	adrenocorticotropic hormone
AIF	apoptosis-inducing factor
Alix-CT	C-terminal half of AIP1/Alix
APC	antigen-presenting cell
ATP	adenosine triphosphate
Bc	bromocriptine
BK	big potassium
BKCa	big potassium calcium
cAMP	cyclic adenosine monophosphate
CisPt	cisplatin
CNS	central nervous system
CO	carbon monoxide
CP	phosphine copper (I) complex
EGF	epidermal growth factor
ENOA	alpha-enolase
ER	endoplasmic reticulum
ERK	growth factor-regulated extracellular signal-related kinase
GBM	glioblastoma multiforme
GC	Golgi complex
HCC	hepatocellular carcinoma
HCNP	hippocampal cholinergic neurostimulating peptide
HMGB1	high-mobility group protein B1
HNK	honokiol
IGF1R	insulin-like growth factor 1 receptor
IMM	inner mitochondrial membrane
JNK	C-Jun N-terminal kinase
K+	potassium ion
LMP	lysosomal membrane permeabilization
mAb	monoclonal antibody
MAPK	mitogen-activated protein kinase

Apoptosis and Beyond: The Many Ways Cells Die, First Edition. Edited by James Radosevich.
© 2018 John Wiley & Son Inc. Published 2018 by John Wiley & Son Inc.

mM-CSF	membrane macrophage colony-stimulating factor
mNCX	mitochondrial Na^+/Ca^{++} exchanger
NADPH	reduced nicotinamide adenine dinucleotide phosphate
NCCD	Nomenclature Committee on Cell Death
NK1R	neurokinin-1 receptor
OP-A	ophiobolin A
OxPt	oxaliplatin
PAH	polycyclic aromatic hydrocarbon
PCD	programmed cell death
PEBP	phosphatidylethanolamine-binding protein
PNS	peripheral nervous system
PPD	protopanaxadiol
PRL	prolactin
RER	rough endoplasmic reticulum
RKIP	Raf kinase-inhibitor protein
ROS	reactive oxygen species
RPE	retinal pigment epithelium
TNFR	tumor necrosis factor receptor
UPR	unfolded protein response
UTP	uridine triphosphate
VR1	vanilloid receptor subtype 1
YTX	yessotoxin
zVAD-fmk	benzyloxycarbonyl-valyl-alanyl-aspartyl fluoromethyl ketone

16.1 Introduction

In the process of growth and development, each cell has three choices: to divide, to specialize, and to die. In the language of biology, these are translated to mitosis, differentiation, and cell death [1]. The balance between these choices is critical for the development and maintenance of multicellular organisms. There are several types of death, which can be classified based on different characteristics of the cells. The morphological appearance, enzymatic profile, and functional and immunological aspects of the cells are considered in these classifications [2].

With regard to the morphological appearance, three main types of cell death were initially defined in developing tissues, based fundamentally on the role of the lysosomes [3]. Type 1 was apoptosis and type 2 autophagic degeneration. Type 3 was non-lysosomal vesiculate degradation, which was further subdivided into two subtypes: type 3A, non-lysosomal disintegration, and type 3B, "cytoplasmic" degeneration.

The problem with this classification was that some dying cells showed combined features of more than one of the types, and some seemed not to match any of them [3].

Cell death can also be classified according to enzymological criteria, based on the involvement of nucleases or of distinct classes of proteases, mainly caspases [2]. Hence, "caspase-dependent" and "caspase-independent" cell-death pathways have been proposed [4]. Usually, cell death is considered to be caspase-dependent when it is inhibited by broad-spectrum caspase inhibitors. Caspase-independent cell death can occur despite the effective inhibition of caspases and with the morphological features of apoptosis, autophagy, or necrosis. Apoptosis does not equal caspase activation, but caspase activation could be a major predictor of death and a defining feature of apoptosis [5–7].

With regard to functional aspects, cell death can be classified as programmed or accidental and as physiological or pathological [2]. Some researchers advocate the distinction between developmental/physiological and pathological kinds of cell death. Since many types occur in both situations, any rigid distinction between the two processes should be discouraged. This would also encourage interaction between pathologists and developmental biologists [3]. Programmed cell death (PCD) is a process whereby cell death is genetically programmed; this is generally contrasted with "accidental cell death" or necrosis, which is induced by pathological stimuli [6]. In recent decades, PCD was accepted as synonymous with apoptosis, which is the inherent, controlled cellular death program, but there are in fact various types of PCD [6,8].

Recently, the Nomenclature Committee on Cell Death (NCCD) set up a forum in which names describing distinct modalities of cell death could be critically evaluated and recommendations on their definition and use could be formulated. This was intended to provide uniform and non-rigid nomenclatures that would ensure better understanding and communication among scientists [2]. The first round of recommendations was formulated in 2005 [6] and updated in 2009 [2].

Based on the NCCD's guidelines, there are four dinstinct modalities of cell death: apoptosis, autophagy, cornification and necrosis. Several other modalities are classified as atypical cell death, including mitotic catastrophe, anoikis, exocitotoxicity, Wallerian degeneration, pyroptosis, pyronecrosis, entosis, and paraptosis [2]. It should be kept in mind that there are numerous examples of cell death that display mixed features (e.g., with signs of both apoptosis and necrosis, or synchronization of autophagic vacuolization with signs of apoptosis). Therefore, a clear-cut and absolute distinction between different forms of cell death based on morphological criteria may not be possible [2].

16.2 Definition

Paraptosis (*para* "beside" or "related to" + *ptōsis* "a falling" or apoptosis) [9] is classified as a caspase-independent PCD that is distinct from apoptosis morphologically, biochemically, and in its response to apoptotic inhibitors [10]. It seems that the first use of the term "paraptosis" was by Sperandio et al. [10], who used insulin-like growth factor 1 receptor (IGF1R) to induce cell death in 293T cells and mouse embryonic fibroblasts. Ultrastructurally, the cells lacked the features of apoptosis, including nuclear fragmentation, apoptotic body formation, and chromatin condensation. Instead, cytoplasmic vacuolation was present, with vacuoles derived predominantly from the endoplasmic reticulum (ER). Mitochondrial swelling was also present, but autophagic vacuoles were not. This form of cell death shared some morphological features with necrosis, including cytoplasmic vacuolation and mitochondrial swelling, in addition to the lack of effect of caspase inhibitors. Sperandio et al. concluded that IGF1R may induce a form of PCD with features distinctive from apoptosis, which they called "paraptosis" [10].

While this was the first time that the term appeared in published literature, patterns of cell death with features of paraptosis had been reported in previously published articles [3,4,11]. A similar pattern originally observed in dying neurons and described as "cytoplasmic" or "type 3B" cell death was characterized by initial dilatation of rough endoplasmic reticulum (RER) and, sometimes, the swelling of mitochondria. This type was different from type 1 (apoptosis) in that there were no signs of early condensation of the nucleus, no detectable fragmentation of the cell and, no loss of ribosomes from cisternae. Interestingly, several features of necrosis, such as mitochondrial swelling, dilation of ER, and late karyolysis, were observed [3]. Caspase-independent PCD with necrotic-like morphology characterized by cytoplasmic vacuolation and minimal nuclear changes may represent another example of paraptosis [11]. Interestingly, paraptosis-like PCD has been observed in both plants and protists. Apoptosis does not occur in plants, because their cell walls prevent phagocytosis. Paraptosis has been shown in the unicellular alga, *Dunaliella viridis*, and it has been suggested that this type of PCD may be an ancestral form that is conserved across different forms of life [12].

16.3 Morphology

The old morphological classification of cell death defined type 3, or more specifically type 3B or cytoplasmic, as involving dilation of organelles and cytoplasmic vacuolation [3]. This process begins with dilation of the cisternae of RER, and sometimes with the swelling of mitochondria. The ribosomes remain attached to the dilated cisternae and there is no dispersion of polysomes into free ribosomes [3]. This type has some characteristics of both apoptosis and necrosis. The hallmark of paraptosis is extensive cytoplasmic vacuolation [13] without significant cell membrane blebbing, nuclear shrinkage, or pyknosis [14]. The cells are involved in a process of swelling and vacuolization that begins with physical enlargement of the ER and swelling and clumping of the mitochondria [10,15]. These changes can be observed through light microscopy in the form of rounded, vacuolated, and swollen cells. This swollen appearance could indicate ionic dysregulation followed by water retention. Eventually, disturbance in intracellular ion homeostasis results in osmotic lysis, which releases so-called "danger signals" [15]; these are endogenous (released by tissues undergoing stress, damage, or abnormal death) or exogenous (caused by pathogens) substances that promote massive inflammatory response and stimulate cell-mediated immunity [16]. As mentioned before, the main features of apoptosis (type 1 cell death), including chromatin condensation, nuclear fragmentation, and the formation of apoptotic bodies – as well as membrane blebbing – are not observed in paraptosis [10].

Cytoplasmic vacuolation, which is the main characteristic of paraptosis, is also observed in other types of cell death, such as necrosis and autophagy (type 2 cell death). According to Cagle and Allen, "the most studied cytoplasmic vacuolation-induced cell death is autophagy" [13]. It should be noted that autophagic vacuoles are not observed in paraptosis [10]. This is considered the main difference between paraptosis and autophagic cell death. Also, paraptosis cannot be inhibited by inhibitors of autophagic cell death (e.g., 3-methyladenine) [14]. Vacuolation can occur spontaneously in several cell lines, or it may be induced by various stimuli. The extent of vacuolization depends on the cell type, and the number and size of vacuoles increase gradually. Although cells undergoing vacuolization can escape death and recover, beyond a certain threshold a point of no return is reached and the recovery becomes impossible, leading to cell death [13].

Morphological changes similar to paraptosis have been observed in neural development and in some cases of neurodegeneration. These changes can also be induced by several substances, including cancer drugs. Figure 16.1 and Table 16.1 compare the morphological features of paraptosis with those of some other types of cell death.

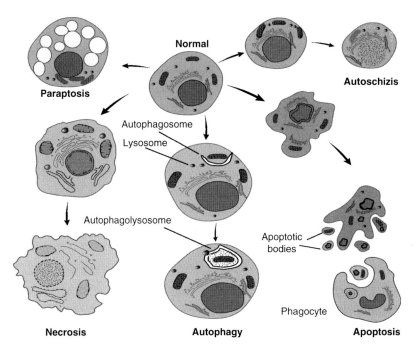

Figure 16.1 Morphological features of different types of cell death. Note the extensive cytoplasmic vacuolization and swelling and clumping of mitochondria in paraptosis. Nuclear fragmentation, membrane blebbing, and apoptotic body formation are absent. *Source:* Cagle and Allen 2009 [13]. Reproduced with permission from Springer.

Table 16.1 Morphological features of different types of cell death. *Source:* Data from Sperandio et al. 2000 [10].

Morphology	Apoptosis	Necrosis	Paraptosis
Cell body	Shrinking	Swelling	Enlargement[a]
Cell membrane	Blebbing	Broken	Intact
Cytoplasm	Compact and condensed	Vacuolized	Vacuolized
Cytoskeleton	Collapse before death	Collapse after death	Collapse before death
Nucleus	Shrinking and fragmented	Enlarged and broken	No fragmentation
Chromatin	Condensation	Clumped	With/without condensation
Lysosomes	Intact	Leaking	Intact
Endoplasmic reticulum	Dilated	Dilated	Dilated[a]
Mitochondria	Intact, swelling[a]	Broken, swelling	Late swollen
Apoptotic bodies	Common	Absent	Absent

a) Sometimes

16.4 Mechanisms

It was long believed that caspases play a general and essential role in PCD. These are a family of cysteine proteases that are activated in apoptosis and inhibited by a broad spectrum of caspase inhibitors, most commonly benzyloxycarbonyl-valyl-alanyl-aspartyl fluoromethyl ketone (zVAD-fmk) and p53. Although caspases are critical determinants of apoptosis, it is now established that caspase activation is not the sole indicator of life and death in PCD. Many studies have shown that while inhibition of caspase activities results in inhibition of apoptosis, it does not prevent cell death [11]. Therefore, the term "caspase-independent PCD" was introduced. Paraptosis is considered a caspase-independent PCD, but several other types have been described, with many overlapping features. Also, many characteristics of different death types can be present at the same time [8]. Although the molecular mechanisms of apoptosis are well established, several aspects of paraptosis remain to be investigated. As already stated, paraptosis is defined by swelling and vacuolization that involves the ER and the mitochondria. The swollen morphology is the result of ionic imbalance, water retention, and eventually osmotic lysis. Intracellular contents such as adenosine triphosphate (ATP), uridine triphosphate (UTP), high-mobility group protein B1 (HMGB1), heat-shock proteins, and various proteases are released and act as "danger signals," promoting extensive inflammation and cell-mediated immunity. Hoa et al. [15] studied the mechanism of paraptosis induction by monocytes in human glioma cells through disruption of ionic homeostasis and big potassium (BK) channel activation. They found that osmotic dysregulation in tumor cells induced by BK-channel activation provides a possible mechanism through which monocyte-mediated cytotoxicity can occur [15].

BK channels, also known as Maxi-K, hSlo, mslo, calcium-dependant (BKCa), large conductance-, and voltage-activated channels, are potassium ion (K+)-specific channels characterized by their large conductance of K+ through cell membranes. They regulate cell volume by controlling intracellular levels of K+, and are themselves indirectly regulated by oxygen. Potassium channels are controlled by two enzymatic proteins, reduced nicotinamide adenine dinucleotide phosphate 450 (NADPH450) reductase and hemoxygenase-2, which produce a secondary messenger, carbon monoxide (CO), in response to oxygen. CO, in turn, allows for the opening of the BK channels. BK channels are found at several cellular locations, including the cell membrane and intracellular sites. They have been observed in the ER and mitochondria, which could explain why these two organelles are specifically targeted in paraptosis. BK may be activated by BK-channel activators such as phloretin and pimaric acid. Forced BK-channel opening results in disruption of normal homeostasis as potassium is expelled from the cell. In order to keep the ionic balance, sodium cations enter the cell, followed by water, resulting in organelle and overall cellular swelling.

Hoa et al.'s study [15] showed that rodent and human mononuclear phagocytes inhibit the growth of U251 glioma cells expressing membrane macrophage colony-stimulating factor (mM-CSF) and that the cells could be identified as paraptotic. Within 30 minutes of treatment with two different BK-channel activators (phloretin and pimaric acid), the U251 glioma cells began swelling and forming vacuoles. Interestingly, the swelling and vacuolization effects induced by BK-channel activators were inhibited by the use of a

BK-channel inhibitor. Iberiotoxin also prevented human monocytes from killing the mM-CSF-expressing U251 cells [15].

The molecular mechanism of cell death leading to paraptosis involves several steps. The initiating event is the production of reactive oxygen species (ROS) by monocytes after contacting mMCSF-expressing U251 glioma cells. Before that, normal homeostasis is present in the tumor cells and ATP, potassium, and sodium are at baseline physiological levels, which means low sodium and high potassium intracellular concentration. Following contact between the monocytes and mM-CSF, ROS bring about CO production by hemoxygenase and P450 reductase. CO causes the opening of BK channels on the cell membrane, ER, and mitochondria, resulting in potassium efflux. In order to maintain cellular electroneutrality, Na enters the cell, followed by water, which produces cellular swelling and, eventually, vacuolization due to swelling of the ER and mitochondria. The elevated intracellular sodium is removed through the ATP-dependant Na/H antiporter to maintain cellular homeostasis [15].

Interestingly, Schneider et al. [17] showed that inhibition of the Na/H exchanger induces caspase-independent death in cerebellar neurons, much like paraptosis.

To remove excess Na ions, the cell needs more ATP, and since the mitochondria are the target organelle in paraptosis, the ability of the cell to produce the level of ATP required to maintain ionic homeostasis is reduced. Loss of ATP production prevents the cell from maintaining a proper volume, and eventually osmotic rupture occurs. It should be noted that cell swelling and vacuolization does not induce immediate death, but that further sequential events are required, which is consistent with the slow depletion of ATP previously described.

Several monocyte/macrophage-derived mediators can induce cell swelling, including H_2O_2, peroxynitrite, arachidonic acid and its metabolites, leukotrienes, and prostaglandins. The exact macrophage-derived mediator responsible for paraptosis induction in Hoa et al's study [15] was not clearly defined, and it was concluded that the mechanisms of cellular death in the mM-CSF expressing glioma cells are multidimensional and have the potential for synergistic interactions.

The ER and mitochondria also store Ca^{++} ions. In fact, other groups of researchers have reported that Ca^{++} ion dysregulation induces paraptosis. Jambrina et al. [18] described a distinct model of rapid cell death triggered by Ca^{++} overload and involving mitochondrial damage leading to paraptotic cell death in Jurkat cells. BK channels are known as Ca^{++}-dependent–voltage-dependent K channels, so the study of Hoa et al. [15] is consistent with that of Jambrina et al [18]. Thus, it is possible that there are common effector mechanisms involving disrupted ionic homeostasis that can lead to paraptosis.

The model presented by Hoa et al. [15] for the mechanism of paraptosis induction can be summarized into six steps (Figure 16.2):

- Step1: Macrophage contact with mM-CSF-expressing cells, production of ROS.
- Step2: Production of CO by hemoxygenase and P450 reductase.
- Step3: Opening of BK channels mediated by CO.
- Step4: Release of K+, Na+ cations enter the cell to maintain the electroneutrality of the cell.
- Step5: Na+ enters the cell, followed by water, to produce cellular swelling and vacuolization.
- Step6: Activation of cellular homeostatic mechanisms to expel excess Na+ [19].

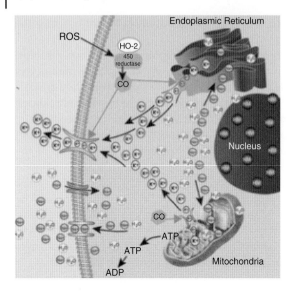

Figure 16.2 Mechanism of paraptosis induction by ionic dysregulation via prolonged BK-channel activation. *Source:* Hoa et al. 2009 [19]. Reproduced under the terms of the Creative Commons Public Domain declaration, CC-PD, via *PLoS One*.

16.5 Pathways

16.5.1 IGF1R-Induced Paraptosis

Paraptosis requires both transcription and translation. Sperandio et al. [10] showed that actinomycin D restores cell viability after transfection of 293T cells with IGF1R. Also, IGF1R-induced cell death was blocked by cycloheximide. None of the apoptosis inhibitors (e.g., the caspase inhibitors zVAD.fmk, BAF, p53, and X-chromosome-linked inhibitor of apoptosis (xiap) or Bcl-XL, from the Bcl2 family) inhibited 293T cell death induced by IGF1R. Despite the lack of effect of caspase inhibitors, a catalytic mutant of caspase-9 zymogen blocked IGF1R-induced cell death. Therefore, it could be suggested that caspase 9 (or caspase 9 zymogen) is involved in the apoptotic as well as the non-apoptotic form of PCD [10]. It should be noted that the term "caspase-independent," used in older publications, was meant to indicate the lack of effect of caspase inhibitors on cell death and was described mainly in contrast to apoptosis.

In another study, the same authors showed that IGF1R-induced paraptosis is mediated through two signaling pathways: mitogen-activated protein kinase (MAPK) and C-Jun N-terminal kinase (JNK)-1 [20]. They found that inhibition of MAPK and downregulation of MEK-2 – as well as of JNK1 – inhibited paraptosis. Therefore, it could be suggested that MAP kinase family members, especially MEK-2 and to a lesser extent JNK1, should be considered paraptosis mediators. The authors also described the first inhibitor of paraptosis, AIP-1/Alix. Activation of both JNK and MAPK by IGF1R was reduced by the overexpression of AIP1/Alix [20].

As already mentioned, paraptosis was inhibited by a catalytic mutant of caspase 9. Allen et al. [21] showed that caspase 9 is a direct target of MAPK, and that the growth factor-regulated extracellular signal-related kinase 2 (ERK2) MAPK pathway inhibits the proapoptotic activity of caspase 9 by direct phosphorylation. It seems that phosphorylation of caspase 9 could function as a switch from a proapoptotic to proparaptotic activity. JNK activation has also been implicated in apoptosis and, based on the findings of this study, JNK1 can mediate paraptosis. Therefore, the early suggestion that there is crosstalk between apoptosis and paraptosis is supported. In fact, paraptosis and apoptosis may be complementary cell-death programs. An example is the withdrawal of a trophic factor versus hyperactivation of its receptor. Trophic factors, or trophins, are agents that stimulate differentiation and survival of cells, and their absence usually causes cell death. Withdrawal of IGF-1 may lead to apoptosis, whereas hyperactivation of IGF1R inactivates cellular apoptotic pathways but may induce the alternative cell-death program of paraptosis. Signal transduction by trophic-factor receptors such as IGF1R could lead to cell survival, proliferation, and differentiation or to cell-death induction, depending on quantitative (hyperstimulation that exceeds a threshold may induce cell death) and qualitative (cell state or costimulatory signals) effects [20].

The function of AIP1/Alix is not completely defined, but it may cooperate in promoting apoptosis and have a potential role in regulating paraptosis. The C-terminal half of AIP1/Alix (also known as Alix-CT) inhibits apoptosis, and it has been shown that overexpression of this region can induce cytoplasmic vacuolization, which is greatly enhanced upon coexpression with endophilins [22]. Based on these findings, the full-length AIP1/Alix facilitates apoptosis but inhibits both paraptosis and paraptosis-dependent cytoplasmic vacuolation. On the other hand, the C-terminally deleted form of AIP1/Alix prevents apoptosis and induces vacuolation. AIP1/Alix is a multifunctional protein that is important in signal transduction and, ultimately, cell integrity [20]. IGF1R- induced paraptosis requires transcription and translation and is mediated by MEK2 and JNK. In addition to the initial IGF1R-induced paraptosis, this cell death type can be induced by many other factors.

16.5.2 VR1-Mediated Paraptosis

As mentioned earlier, cytoplasmic calcium overload can be a death trigger in Jurkat cells through the expression of vanilloid receptor subtype 1 (VR1), which is a receptor-operated specific Ca^{++} channel. Calcium uptake through the channel increases intracellular calcium enough to cause mitochondrial dysfunction and damage, ultimately leading to cell death. The death process displays mixed features of apoptosis and necrosis, without caspase activation, DNA cleavage and degradation, or release of cytochrome c (cyt c) or apoptosis-inducing factor (AIF) from the mitochondria. In conclusion, calcium influx triggers a distinct program of mitochondrial dysfunction, leading to paraptotic cell death (calcium-induced paraptotic PCD) [18].

16.5.3 TAJ/TROY-Mediated Paraptosis

Another paraptosis-inducing pathway is via the overexpression of TAJ/TROY, which is a member of the tumor necrosis factor receptor (TNFR) superfamily and is highly expressed during embryonic development. TAJ/TROY induces paraptosis-like cell morphology in HEK293, HeLa, and 293T cells. TEM studies show excessive cytoplasmic vacuolization and mitochondrial swelling, with no condensation or fragmentation of the nuclei. TAJ/TROY overexpression results in significant upregulation of cellular endogenous PDCD5, an apoptosis-promoting protein that can also promote TAJ/TROY-induced paraptotic cell death. Therefore, PDCD5 is an important regulator of both apoptosis and paraptosis [23].

16.5.4 NK1R-Mediated Paraptosis

Castro-Obregon et al. [24] described a type of cell death in hippocampal, striatal, and cortical neurons mediated by neurokinin-1 receptor (NK1R), the receptor for neuropeptide substance P. NK1R and substance P are considered the first ligand–receptor pair to mediate a non-apoptotic form of PCD. They are widely distributed in the central and peripheral nervous systems (CNS and PNS), and have been involved in pain mediation and depression, as well as in other effects. Their binding activates a PCD that requires gene expression. It does not resemble apoptosis morphologically, and shows vacuolation and phagolysosomes ultrastructurally, with intact plasma and nuclear membranes. This system is not prevented by caspase inhibitors or by overexpression of Bcl-xL. Therefore, NK1R induces a non-apoptotic form of cell death that is similar to paraptosis in its morphology, caspase-independence, and need for gene transcription and translation [24]. This type of cell death is mediated by the MAPK activation pathway, which involves Raf-1, MEK2, and ERK2. Activation of ERK2 leads to phosphorylation of Nur77, which is essential to NK1R-mediated cell death [25].

16.5.5 EGF-Induced Paraptosis

Fombonne et al. [26] reported a form of cell death triggered by epidermal growth factor (EGF) in the pituitary cell line (GH4C1), which has some features of apoptosis (DNA fragmentation) and some of paraptosis (caspase independence and vacuolization). They further studied the endonuclease involved in EGF-induced DNA fragmentation, and showed that EGF triggers the activation and nuclear translocation of L-DNase II. This endonuclease is known to be involved in caspase-independent DNA fragmentation. The paraptosis inhibitor AIP-1/Alix, but not the antiapoptotic Alix-CT, can block EGF-induced cell death. The authors concluded that EGF-induced cell death could be defined as a novel, L-DNase II-mediated form of paraptosis [27]. However, it should be noted that in contrast to classical paraptosis, EGF-induced PCD involves internucleosomal DNA fragmentation but not MAPK [26].

16.5.6 Corticosteroid-Induced Paraptosis

A similar paraptosis cell-death mechanism was reported by *in vitro* incubation of rat retinal cells with corticosteroids. This corticosteroid-induced cell death was dose-dependent, and showed extensive cytoplasmic vacuolization in the retinal pigment epithelial (RPE) cell. No caspase activity was observed, and overexpression of AIP-1/Alix inhibited RPE cell death [28].

16.5.7 1-NP-Induced Paraptosis

Several studies have investigated the toxic effects of nitro-polycyclic aromatic hydrocarbons (nitro-PAHs). These environmental pollutants are generated during incomplete combustion of organic material. 1-nitropyrene (1-NP) is the predominant nitrated PAH emitted in diesel exhaust and is capable of inducing paraptosis in Hepa1c1c7 cells.

Following exposure to 1-NP, the cells show excessive vacuolization, which has been revealed to be partly due to mitochondrial swelling via electron microscopy. The ERK1/2 inhibitor PD 98057 completely blocks the induced vacuolization, while the other MAPK inhibitors have minor effects on this process [29]. 1-NP can also induce apoptosis [30]. Therefore, the toxicity of 1-NP involves different death mechanisms.

16.5.8 Taxol-Induced Paraptosis

Recently, a paraptotic cell-death process induced by Taxol in human lung carcinoma cell lines (ASTC-a-1) has been reported. Taxol (paclitaxel) is a potent anticancer drug that can induce different types of cell death. It has been shown that high concentrations of Taxol result in paraptotic cell death characterized by caspase independence and cytoplasmic vacuolization due to ER and mitochondrial swelling [31]. Further studies have shown that Taxol-induced paraptosis is different from IGF1R-induced paraptosis. Both pathways have similar characteristics: mainly cytoplasmic vacuolization due to ER and mitochondrial swelling, caspase independence, absence of apoptotic bodies, and DNA fragmentation. However, IGF1R-induced paraptosis requires transcription and translation [10], whereas Taxol-induced paraptosis does not require protein synthesis, suggesting that it could be mediated by previously existing proteins in cells [32]. IGF1R-induced paraptosis is mediated by MEK-2 and JNK [20], while Taxol-induced paraptosis is not mediated by MEK but is partially inhibited by the JNK inhibitor, SP600125, although the percentage of vacuolated cells does not decrease, indicating that Taxol-induced cytoplasmic vacuolization may not require JNK activation. This vacuolization is derived predominantly from the ER and may be a result of Taxol-induced ER stress. Stress induction in the ER due to high concentrations of Taxol have been reported. Molecules that cause ER stress may be involved in the vacuolization process. Also, accumulation of misfolded proteins within the ER could possibly lead to swelling, which could be enhanced by ER stress to form significant vacuolization. However, these proteins are not detected in ER vacuolization. Therefore, the role of ER stress in Taxol-induced paraptosis remains unclear. In conclusion, these findings show that paraptotic PCD may be regulated through a number of different pathways [32].

16.5.9 Yessotoxin-Induced Paraptosis

Yessotoxin (YTX) is a marine algal toxin that has been shown to induce apoptosis in several cell types. It has also been shown that the BC3H1 myoblast cell line, when exposed to YTX, undergoes a type of PCD with features similar to paraptosis, rather than to apoptosis. These cells show swelling of the ER and mitochondria and extensive cytoplasmic vacuolation without DNA fragmentation or alterations in chromatin and the cytoskeleton. YTX-induced cell death also has MAPK pathway involvement [33].

16.5.10 Honokiol-Induced Paraptosis

Honokiol (HNK), a pure compound derived from *Magnolia officinalis*, can induce paraptosis-like cell death in human leukemia NB4 and K562 cells. Low concentrations of HNK induce cell death mainly characterized by cytoplasmic vacuolization that is time- and concentration-dependent, and the amount and size of vacuoles are correlated with the cell type. Cytoplasmic vacuolization may be partially associated with the increased generation of ROS [34,35].

Cytoplasmic vacuolization in rat T9 glioma cells involving BK channels and induced by ROS has been described [19]. Cells treated with HNK also generated ROS. Therefore, it could be suggested that HNK-mediated ROS are involved in the process of vacuolation, and ROS may be promoters of HNK-induced cytoplasmic vacuolization. Ultrastructural examinations revealed the presence of ribosomes along the edges of the vacuoles, indicating that the cytoplasmic vacuoles are derived from the swollen ER. TEM studies failed to show double-membraned autophagosomes, indicating that the cell death is not related to autophagy, which is characterized by the formation of phagocytic vacuoles called autophagosomes containing cytoplasm and organelles. Mitochondria in HNK-induced paraptosis were normal ultrastructurally. Dysfunction of the mitochondrial respiratory chain resulted in overproduction of ROS and cell death. Considering that HNK-treated cells generated ROS, it could be suggested that although mitochondrial swelling did not occur in HNK-induced cells, mitochondrial function may have been impaired. Also, the absence of mitochondrial swelling could possibly be related to the relatively low concentration of HNK used and the effects of short-term exposure [35].

16.6 Proteome Profile

As already discussed, the first proteomic analysis of paraptosis was described by Sperandio et al. in 2010 [10]. Their study focused on the relative abundance of each individual protein, and the mechanism of protein regulation (transcriptional or post-transcriptional) was not determined. They showed that during paraptosis, alterations occur in different proteins as follows:

- cytoskeletal or structural proteins;
- mitochondrial proteins;
- signal-transduction protein; and
- some metabolic proteins [14].

16.6.1 Structural Proteins

In paraptotic cells, three structural proteins were altered, namely α-tubulin, β-tubulin, and tropomyosin. A- and β-tubulin are the major constituents of microtubules. Microtubules are one of the main structures of the cytoskeleton, and provide the structural support of the cell, as well as participating in different cellular movements, such as organellar transport, cell locomotion, and chromosome separation during mitosis. Tropomyosin is a major component of the thin filaments in striated and smooth muscles, and participates in the regulation of the contractile process. In non-muscle cells, tropomyosin is associated with the cytoskeleton and the actin-based motor system. Alterations in structural proteins in paraptotic cells are quite expected. During paraptosis, a major structural rearrangement could be observed almost immediately, even by light microscopy; it consisted in the rounded-cell reorganization of the cytoplasm by the appearance of vacuolation. In paraptotic cells, α-tubulin and tropomyosin were both more abundant in the light membrane/organelle fractions (endosomes and Golgi) and less abundant in the cytosol and the dark membrane (composed of mitochondria and lysosomes). β-tubulin levels were decreased in paraptotic cells. The decreased abundance of a protein could suggest its targeting for degradation, such as via the proteasome pathway, or simply via decreased expression. The altered relative abundance of structural proteins observed in subcellular fractions of paraptotic cells is suggestive of an intracellular redistribution of these proteins during paraptosis. Confocal microscopy following immunofluorescence staining of cytoskeletal proteins revealed major alterations in the microtubular network, along with peripheralization of both α- and β-tubulins and tropomyosin staining. In addition, a subpopulation of paraptotic cells, possibly representing a more advanced stage of the cell-death process, displayed a dramatic decrease in both α and β-tubulin levels. The relationship between the disruption of the microtubular network in paraptotic cells and the formation of the cytoplasmic vacuoles remains to be established. The early disruption of the microtubular network observed in paraptosis could provide a distinguishing feature for this type of cell death as compared to type 2 or autophagic cell death. Currently, a functional – or, at least, a facilitating – role has been proposed for microtubules in the process of autophagy. Staining of tubulins along with already known autophagy markers could be a potential tool for differentiating between these two types of cell death.

16.6.2 Mitochondrial Proteins

The greatest changes in protein concentration were found in the two mitochondrial proteins, prohibitin and ATP-synthase-β subunit, indicating a previously unsuspected mitochondrial involvement in this cell-death process.

ATP synthase is an important enzyme with multiple subunits that catalyzes the synthesis of ATP and is located in the inner mitochondrial membrane (IMM). The ATP-synthase β-subunit was more abundant in the P20 mitochondrial fractions of paraptotic cells. Increased levels of an enzyme fundamentally involved in oxidative phosphorylation indicate a need for ATP production for the execution of paraptosis.

Prohibitin is the other mitochondrial protein that showed alteration in paraptotic cells. It is involved in cell-cycle regulation, replicative senescence, tumor suppression, and cell immortalization. Paraptotic cells showed increased levels of prohibitin by almost 3.4-fold compared to controls. These cells also showed an intense and altered mitochondrial staining, with condensed appearance in certain areas. This suggests a reorganization or alteration of the mitochondrial network. It has been shown that increased levels of prohibitin, when associated with a paraptotic stimulus, result in a cell death that is not inhibited by caspase inhibitors (and is not apoptotic). Therefore, the increased levels of prohibitin could have a functional role in the execution of paraptosis [14].

16.6.3 Signal-Transduction Proteins

Phosphatidylethanolamine-binding protein (PEBP) is another protein shown to be decreased in paraptotic cells. Also known as Raf kinase-inhibitor protein (RKIP), PEBP is a multifunctional protein involved in the biogenesis of the hippocampal cholinergic neurostimulating peptide (HCNP), inhibition of the Raf kinase, inhibition of apoptosis, and inhibition of serine proteases. A decreased level of PEBP was observed in all subcellular fractions of paraptotic cells [14]. As mentioned earlier, IGF1R-induced paraptosis requires signaling through the MAPK and JNK pathways [20]. Downregulation of PEBP and other kinase inhibitors would allow MAPK and JNK to accumulate in sufficient levels to induce cell death. In addition to its role in paraptosis, PEBP can also function as an apoptotic inhibitor. Therefore, PEBP can be considered a critical survival factor with the ability to prevent cell death induced by multiple pathways [14].

16.6.4 Metabolic Proteins

Another altered protein in paraptotic cells is alpha enolase. Basically, it is a glycolytic enzyme, but it has many other functions as well. Antibodies against alpha-enolase (ENOA) have been detected in several diseases, and it has been shown that they are capable of inducing cell death through an apoptotic process. Decreased levels of alpha enolase have been reported in paraptotic cells [14].

16.7 Potential Medical Implications

Several substances and agents are reported to have anticancer effects through induction of different death pathways in human cancer cells. Many compounds from naturally derived sources, as well as metal complexes, have been shown to induce paraptotic cell death in cancer cells, and these have potential as cancer therapeutic agents. Traditional chemotherapy could have many harmful side effects; therefore, the use of naturally derived compounds would provide a safer approach to cancer management. Drug resistance is another problem in cancer therapy, and development of new drugs that can induce multiple death pathways is of great interest. It should be noted that multiple cell-death pathways can be activated in the same cell, and the dying cells may show characteristic features of different death processes. The relative speed of the available cell-death programs determines the dominant death phenotype. As mentioned before, cancer cells often develop some resistance to apoptosis, and it is possible that alternative forms of PCD play a critical role in cancer therapy. In fact, chemotherapeutic drugs like paclitaxel, arsenic trioxide, and Doxorubicin, can all induce caspase-independent PCD [8].

A number of recently reported substances with paraptotic effect that have potential as anticancer drugs are presented in this section.

16.7.1 Taxol

Taxol (paclitaxel) is a potent anticancer drug, currently used in the treatment of ovarian, breast, and non-small-cell lung carcinoma. It is an antitubulin drug and can induce several types of cell death, including apoptosis, paraptosis, necrosis, mitotic catastrophe, and autophagy [32]. Although the exact mechanism of Taxol toxicity is still not clear, it has been shown that different concentrations can induce different effects on both the cellular microtubule network and biochemical pathways. Therefore, Taxol's effects on cancer cells are concentration-dependent. While a low concentration of Taxol induces apoptosis, a high concentration induces a caspase-independent cell death with cytoplasmic vacuolization [31,36]. This paraptosis-like PCD does not require protein synthesis. Taxol-induced paraptosis may have implications in cancer therapy, especially for apoptosis-resistant tumors. When a high concentration of Taxol (70 µM) was injected into hypodermic tissue, no obvious side effect was observed, and it has potential as a treatment for skin cancer [36].

16.7.2 Bromocriptine

Bromocriptine (Bc) is an ergot-derivative dopaminergic agonist that binds to the dopamine D2 receptors of anterior pituitary cells, especially on lactotrophs, and reduces prolactin (PRL) secretion by the inhibition of cyclic adenosine monophosphate (cAMP) production and/or cytosolic Ca^{++} elevation. Bc inhibits PRL gene transcription, synthesis, and release. Cytotoxic and antiproliferative effects on pituitary cells have also been reported. Because of its hormone-lowering effect, Bc has been used as a therapeutic agent in patients with hyperprolactinemia, prolactinoma, acromegaly, and adrenocorticotropic hormone (ACTH)-secreting adenoma. It has been shown that prolonged administration of Bc results in remarkable tumor shrinkage, with apoptosis as the classical cell death type responsible for tumor regression. However, it can trigger non-apoptotic cell-death mechanisms as well, which represents the main pattern of cell death. Recently, Palmeri et al. [37] investigated the ultrastructural effects of Bc on estrogen-induced pituitary tumors. They described the morphological changes of Bc-treated cells finally resulting in so-called "dark cell" degeneration. "Dark cells" were considered to be an irreversible stage of degeneration and cell death. Marked alterations of cytoplasmic organelles were observed, with grossly swollen mitochondria and late disruption of the IMM. The RER and Golgi complex (GC) presented progressive and intense swelling, characteristic of non-lysosomal cytoplasmic vacuolization. Nuclear chromatin became more electron-dense, but no nuclear fragmentation was observed in the dark cells, and the plasma membrane did not show signs of disruption. Interestingly, the ultrastructural observations of dark cells were consistent with those of the non-apoptotic cell-death type 3 (cytoplasmic) death of Clarke's classification [3] and of the non-apoptotic cell death described by Sperandio et al. [10] in 293T cells. The authors concluded that paraptosis is the predominant cell death type involved in the regression of pituitary tumors in response to Bc treatment [37].

16.7.3 Copper Complexes

Over the past 40 years, several metal-based compounds have been tested for their potential antitumor properties. Cisplatin (CisPt), a platinum-based agent, is the most widely used anticancer drug. CisPt induces apoptosis, but its use is significantly limited by increasingly high rates of resistance and by cases of nonsensitive cancers. The development of therapies directed through non-apoptotic cell deaths may be a promising approach to overcoming apoptosis resistance in multidrug-resistant cancers [38].

Recently, copper (I,II) complexes have been investigated for their antitumor effects. A phosphine copper (I) complex, [Cu (thp) 4][PF6] (CP), was reported to have a strong antiproliferative effect and remarkable killing ability in cancer cells. CP is a monocationic copper (I) complex with high solubility and stability in water solutions. CP's cytotoxicity is about 40 times greater than that of CisPt, and it is able to overcome multidrug resistance. The great cytotoxicity of CP may be a result of its ability to induce paraptotic cell death characterized by extensive cytoplasmic vacuolization. As explained earlier, vacuolization is a morphological change resulting from disruption of the ER–proteasome functional link. Copper complexes inhibit proteasome activity and may act as antiproliferative agents by accumulating misfolded proteins and disrupting ER homeostasis.

Colorectal cancer is one of the leading causes of cancer-related death in the world, mainly due to its poor response to conventional chemotherapy. Also, resistance to therapy is commonly observed in colon cancer cells, which could be related to the failure to execute apoptosis. The effect of CP on human colon carcinoma cell lines has recently been studied [39]. It showed a strong effect on cell viability but no harmful effect on normal colon fibroblasts. It was more active than both CisPt and oxaliplatin (OxPt), the main drug currently used for the treatment of colorectal cancers. It also efficiently overcame acquired resistance to OxPt. The CP-induced paraptotic cell death in colon cancer cells is characterized by a massive formation of cytoplasmic vacuoles. Evidence shows that ER stress induced by CP is associated with the inhibition of the proteasome 26S and the resulting intracellular accumulation of misfolded proteins, which activates a signaling cascade called the unfolded protein response (UPR). Accordingly, a potential mechanism of CP-induced cell death can be explained. When a specific amount of ubiquitinated unfolded proteins has been accumulated, ER stress and swelling occur, and cells undergo UPR. This cytoprotective pathway temporarily suspends protein production to restore ER homeostasis. If the offending substance is eliminated, the UPR pathway shuts down and normal protein translation and folding resume. Conversely, if the UPR pathway remains functionally active, this brings about the inhibition of bulk protein synthesis, cell-cycle arrest, and eventual cell death. CP is capable of inhibiting all three catalytic activities of 26S proteasome, which could explain its high selectivity against tumor cells compared with non-tumor cells. CP might be considered a potential new therapeutic tool for selectively killing colon cancer cells through paraptosis [39].

Another copper complex with known cytotoxic effect through paraptosis is thioxotriazole copper (II) complex (A0). The cytotoxicity of A0 has been compared with that of CisPt. It was found that A0 upregulates the genes involved in the UPR and the response to heavy metals. Its cytotoxic effects are associated with inhibition of the ubiquitin–proteasome system and with accumulation of ubiquitinylated proteins, in a manner dependent on protein synthesis. It inhibits caspase 3 activity and induces a death program lacking typical apoptotic features. A0-induced cell death is characterized by extensive vacuolization derived from ER dilatation, a classic hallmark of paraptosis. In other words, by inhibiting caspase 3, A0 cancels apoptosis and shifts cells to alternative death programs. Recent evidence shows that paraptosis becomes predominant when apoptosis is inhibited. The mechanism of A0-induced cell death can be explained by a model in which the copper complex induces a massive metal overload in cancer cells. This is considered the primary cause of cell injury followed by accumulation of polyubiquitinated proteins. The accumulation of misfolded proteins triggers ER stress and the UPR. Therefore, A0-treated cells show massive vacuolization, altering the ER morphology. Inhibition of the proteasome prevents the degradation of misfolded proteins from committing the cell to death trough the pro-death arm of the UPR. The ability of A0 to induce a dramatic ER stress and at the same time inhibit caspase 3 forces the cell toward a paraptotic-like death [38].

16.7.4 American Ginseng

American ginseng (*Panax quinquefolius*) is a commonly used herb in the United States. It has several biological effects, including immunomodulatory, anti-inflammatory, and anti-tumor activities. The primary effective components of American ginseng are the ginsenosides. It has been reported that steamed American ginseng extract could potently kill colorectal cancer; the main ginsenosides in the extract are protopanaxadiol (PPD), Rg3, and Rh2. Rh2 is a more potent killer of colorectal cancer cells than Rg3. It can induce two different cell-death pathways in these cells: caspase-dependent apoptosis and caspase-independent paraptosis-like cell death. Both types are p53-dependent. Rh2 induces p53 transcription activity, and p53 inactivation decreases vacuole formation and cell death. Interestingly, Rh2 activates NF-κB signaling through ROS generation in colorectal cancer cells, which partially counteracts the killing activities of Rh2. Accordingly, inhibition of ROS or of the NF-κB pathway increases the toxicity of Rh2. In conclusion, Rh2 treatment protects cells from death by NF-κB pathway activation and at the same time activates the p53 pathway, leading to the induction of both apoptosis and paraptosis-like cell death [40].

16.7.5 Tocotrienols

The tocotrienols are one of the most important components of vitamin E. They are the main bioactive component in palm oil and are also abundant in rye, barley, and oats. Several therapeutic activities have been reported for tocotrienols, including antioxidant, anticancer, antiangiogenic, and immunomodulatory effects. Tocotrienols are isomers of vitamin E and are necessary for human health, rather than a pharmaceutical drug. They contain α-, β-, γ-, and δ-homologs with different numbers of methyl groups and different biological activities. Many studies have reported tocotrienols to induce apoptosis in many types of cancer cell. They can also induce a non-apoptotic caspase-independent PCD in several cancer cells. It has been shown that γ-tocotrienol induces a paraptosis-like cell death in human colon carcinoma SW620 cells [41]. Recently, the effect of γ-tocotrienol on two colon cancer cell lines, SW620 and HCT-8, was investigated [43]. It was shown that γ-tocotrienol induces a paraptosis-like cell death with extensive cytoplasmic vacuole formation. This type of cytotoxicity may be associated with the suppression of the Wnt signaling pathway. Interestingly, activation of the Wnt/β-catenin signal pathway was observed in most colon cancer patients, and one of the mechanisms for transition of colitis to cancer in the human colon could be explained by high activities of this pathway [42]. It was concluded that γ-tocotrienol may be used as a natural substance for the prevention and treatment of apoptosis-resistant colon cancer [43].

16.7.6 Curcumin

Curcumin (diferuloylmethane) is a major component of turmeric (*Curcuma longa*) and is used as a popular dietary spice and herbal medicine. It has shown anti-tumor activities in animal models and in high-risk patients and individuals with precancerous lesions. It is a pharmacologically safe agent that can be used in cancer prevention and treatment because its selects for cancer cells while sparing normal cells. There is epidemiological evidence showing a low incidence of colorectal cancer in India, which could be related to diets high in curcumin. Curcumin has the ability to induce different cell-death pathways, including apoptosis, mitotic catastrophe, and autophagy, in several types of cancer cell. Recent studies have shown that through induction of paraptosis, curcumin is preferentially cytotoxic to malignant breast cancer cells versus normal breast cells. Apoptotic characteristics such as caspase dependency, cell shrinkage, apoptotic bodies, and cell-death inhibition by antiapoptotic proteins were rarely observed in breast cancer cells treated with curcumin. Curcumin induced vacuolation in breast cancer cells, which resulted from the swelling and fusion of mitochondria and/or ER, not from autophagy. Therefore, curcumin-treated cells showed a small number of megamitochondria and an expanded ER. Ultimately, irreversible functional loss of these organelles led to cell death. Other paraptotic features, such as protein synthesis, inhibition by AIP-1/Alix, and involvement of the ERK and JNK pathways, were also observed in curcumin-treated malignant breast cancer cells. Significant proteasomal dysfunction and ER stress were noted in curcumin-treated cancer cells, but not in normal cells. As mentioned earlier, inhibition of proteasome activity increases the accumulation of ubiquitinated proteins in the ER, leading to ER stress. It has been shown that proteasome impairment by curcumin contributes to paraptotic changes in the ER. Proteasomal dysfunction may be necessary but not sufficient for curcumin-induced paraptosis, probably because of other signals responsible for mitochondrial paraptotic changes. It has been suggested that mitochondrial superoxide, rather than H_2O_2, has a critical early role in the curcumin-induced paraptotic changes in both mitochondria and the ER [44].

Recently, the role of Ca^{++} in curcumin-induced paraptosis has been investigated. It has been well established that loss of Ca^{++} homeostasis, cellular Ca^{++} overload, and changes in Ca^{++} distribution within intracellular compartments can cause cytotoxicity through various cell-death modes. Both mitochondria and the ER are major reservoirs for intracellular Ca^{++}. It has been reported that Ca2+ influxes into the mitochondria act as an initial and critical signal for curcumin-induced paraptosis, which triggers dilation of both mitochondria and the ER. In addition, simultaneous inhibition of the mitochondrial Na^+/Ca^{++} exchanger (mNCX) and proteasomes could result in a sustained mitochondrial Ca^{++} overload, inducing paraptosis in malignant breast cancer cells [45].

Inhibition of ERK blocks both mitochondria and ER dilation, and also inhibits the accumulation of polyubiquitinated proteins and the ER stress response. Therefore, ERKs (specifically ERK2) may act as downstream mediators of mitochondrial superoxide in curcumin-induced paraptotic signaling. JNK inhibition mainly attenuates ER dilation without affecting mitochondrial dilation and partially blocks curcumin-induced accumulation of polyubiquitinated proteins and ER stress responses. It can be concluded that the ERK signaling cascade is important for the induction of mitochondrial vacuolation and that JNK may primarily contribute to curcumin-induced ER dilation. Finally, curcumin induces mixed modes of cell death, including apoptosis, necrosis, and paraptosis, in many other cancer cells, such as colon and hepatoma cells. Curcumin is currently in human clinical trials for colon and pancreatic cancer and multiple myeloma [44]. However, it should be noted that it showed a tumor-promoting activity in the lung in a recent study [46]. Therefore, given the potential organ-specific effects of curcumin, extensive preclinical studies are required to verify its clinical use as an anticancer therapeutic agent [44].

16.7.7 Ophiobolin A

Ophiobolin A (OP-A) is a fungal phytotoxin found in *Bipolaris* species. It has been reported to cause ion leakage, block hexose transport in higher plants, and inhibit calmodulin activity. It induces apoptosis in mouse leukemia cells and can inhibit human cancer cell growth. The effect of OP-A on human glioblastoma multiforme (GBM) cells has recently been investigated. GBM is the most common adult primary brain cancer, and the deadliest of all forms of brain tumors. Because of its invasive nature, complete resection is almost impossible. Also, GBM cells have an intrinsic resistance to radiation- and chemotherapy-induced apoptosis. It has been shown that OP-A induces a paraptosis-like cell death in GBM cells. This type of PCD is mediated by inhibition of the

BKCa channels by OP-A, which leads to the disruption of internal K+ homeostasis and, eventually, cell death. Biopsy specimens of malignant gliomas show significant expression of BKCa-channel proteins, and studies of human glioma cell lines have shown that functional BKCa channels – the predominant K channel type – are highly expressed in these cells. As already mentioned, the blockade of BKCa channels with OP-A results in decreased cell proliferation and migration, and an increased level of non-apoptotic cell death. Therefore, the potential use of OP-A as a novel therapeutic agent that can overcome the intrinsic resistance of glioblastoma cells to proapoptotic stimuli may be suggested [47].

16.7.8 Implications for a Non-Genetically Modified Tumor Vaccine

It has been shown that glioma cells bearing mM-CSF can be killed by rodent and human monocytes/macrophages through BK-channel activation. It has also been shown that tumor cells killed by ionic dysregulation caused by the BK channel may be used as a non-genetically modified glioma cell vaccine. The presence of mM-CSF allows prolonged conjugation between tumor cells and myeloid cells, resulting in cytotoxicity, which naturally facilitates anti-tumor immunity. The resulting systemic immunity is elicited through multiple danger-signal production. Since injecting living tumor cells into humans as a vaccine is not possible, the mechanism of mM-CSF tumor cell death might be considered in designing clinical trials in which tumor cells are killed through activation of paraptosis-inducing pathways (i.e., BK-channel activation) or release of ROS (Figure 16.3). The advantage of this process is that tumor cells could easily be killed

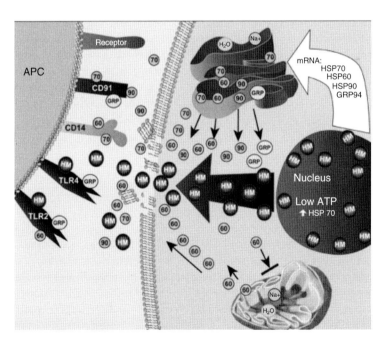

Figure 16.3 Mechanism of induction of paraptosis and immunity by prolonged BK-channel activation. The results of the ionic dysregulation that induces immune responses by stimulating antigen-presenting cells (APCs) are shown. *Source:* Hoa et al. 2009 [19]. Reproduced under the terms of the Creative Commons Public Domain declaration, CC-PD, via *PLoS One*.

by prolonged BK activation *in vitro*, then used as a functionallly killed vaccine, with no need for further genetic manipulation. These findings could have immediate and significant clinical potential for the treatment of cancer and for the development of anticancer vaccines [19].

16.7.9 Anticancer Immunotherapy

Recently, a targeted immunotherapy through monoclonal antibodies (mAbs) specific to a tumor-associated antigen has been introduced for the treatment of hepatocellular carcinoma (HCC). Hep88 is a novel mAB that has been shown to have tumoricidal activity against HCC (HepG2 cell line) while being harmless to the normal liver cell line (Chang liver) [48]. Hep88 mAb proteins are found not only on the cell membrane, but also in the cytoplasm. It has been shown that after 3 days' incubation of HepG2 cells with Hep88 mAb, intracellular vacuolization and ER and mitochondrial dilatation are observed, suggesting that Hep88 mAb induces paraptosis-like PCD in HCC cells. This killing ability is triggered by binding with specific proteins, mortalin (HSPA9), and ENOA. Mortalin is known to act as an antiapoptotic factor by preventing p53 activation in cytoplasm. The paraptosis-like PCD that follows the formation of a Hep88 mAb–mortalin complex in the cytoplasm can be explained by the release of p53 from the p53–mortalin complex, so that it can act through a transcription-independent pathway via the induction of lysosomal membrane permeabilization (LMP). This event results in the release of various lysosomal proteases (granzyme B, cathepsins B and D, etc.), which finally activate the cells into a state of paraptosis. α-enolase is believed to be involved in cell migration and cancer metastasis, through its action as a plasminogen-binding receptor. In addition, a correlation between tumor progression and the upregulation of this enzyme has been reported. It has been shown that a Hep88 mAb–ENOA complex might initiate energy exhaustion in HepG2 cells through glycolysis-pathway obstruction. This synergistic effect of the ATP-depletion phenomenon, along with the effect of the Hep88 mAb–mortalin complex, could finally reduce the capacity of the HCC to remain alive. Therefore, Hep88 mAb may be a promising agent for the development of an effective treatment against HCC in the next decade [48].

16.8 Conclusion

It should be noted that many of the substances discussed in this chapter are capable of inducing more than one type of cell-death program. IGF1R [10], Bc [37], ginsenoside RH2 [40], YTX [33], and 1-NP [29] can all induce both paraptosis and apoptosis under different circumstances. Various factors, including cell type, the hierarchy of intercellular organelles, and the type and severity of insult, are involved in the activation of a particular cell-death pathway. Multiple pathways can interact to form a complex system in order to execute cell death. The ultimate morphological appearance will be determined by: (i) the metabolic situation; (ii) intracellular protein localization; (iii) transport inside the cells; (iv) activation or suppression of individual subroutines; and (v) the relative speed of their execution in each cell type. Crosstalk between different cell-death subroutines can occur under stress conditions and may result in the activation of multiple lethal mechanisms, which can exhibit different degrees of overlap. Receptors involved in cell death may also execute the apoptotic or the paraptotic pathway, or both.

For example, activation of the apoptotic and paraptotic pathways in BC3H1 myoblast cell lines under YTX insult may be related to individual receptors in such cell lines and the interaction of YTX with different cellular components involved in the expression and function of these receptors. It has been suggested that YTX concentration and cell specificity may be critical for the induction of different cell-death mechanisms. Induction of multiple pathways may be necessary to ensure the removal of injured cells or to prevent the survival of unhealthy cells. Simultaneous activation of different cell-death programs may occur under the same insult and in the same cell population. Insults may produce synergistic effects, which are mediated by regulators of several pathways. The induction apparently depends on cell type, type of insult, exposure time, type of cellular receptor, and potentially unknown factors. Multiple cell-death programs can be activated in the same cell population. The dominant cell-death phenotype is determined by the relative speed of the available programs. Although the characteristics of several cell-death pathways can be present, only the fastest and most effective pathway is usually evident. However, it is possible that a cell might switch back and forth between different death-signaling pathways. This could produce an apparent display of characteristics from several different pathways. The most important advantage of the activation of multiple cell-death pathways could be a more efficient and selective elimination of damaged cells, in order to protect the organism against the development of malignant diseases. This fact may have general medical interest for the development of new therapeutic applications. Screening for novel molecules capable of simultaneously triggering different PCDs is a current pharmacological challenge. Therapeutic selectivity, or the preferential killing of malignant cancer cells without significant toxicity to normal cells, should be considered one of the most desirable properties of any potential cancer chemotherapeutic agent [49].

References

1   Melino G. The sirens' song. *Nature* 2001;**412**:23.
2   Kroemer G, Galluzzi L, Vandenabeele P, Abrams J, Alnemri ES, Baehrecke EH, et al. Classification of cell death: recommendations of the Nomenclature Committee on Cell Death 2009. *Cell Death Differ* 2009;**16**(1):3–11.
3   Clarke PG. Developmental cell death: morphological diversity and multiple mechanisms. *Anat Embryol (Berl)* 1990;**181**:195–213.
4   Borner C, Monney L. Apoptosis without caspases: an inefficient molecular guillotine? *Cell Death Differ* 1999;**6**:497–507.
5   Leist M, Jäättelä M. Four deaths and a funeral: from caspases to alternative mechanisms. *Nat Rev Mol Cell Biol* 2001;**2**:1–10.
6   Kroemer G, El-Deiry WS, Golstein P, Peter ME, Vaux D, Vandenabeele P, et al. Classification of cell death: recommendations of the Nomenclature Committee on Cell Death. *Cell Death Differ* 2005;**12**:1463–7.
7   Chipuk JE, Green DR. Do inducers of apoptosis trigger caspase-independent cell death? *Nat Rev Mol Cell Biol* 2005;**6**(3):268–75.
8   Bröker LE, Kruyt FAE, Giaccone G. Cell death independent of caspases: a review. *Clin Cancer Res* 2005;**11**:3155–62.
9   Oxford Living Dictionaries. Paraptosis. Available from http://www.oxforddictionaries.com/definition/english/paraptosis (last accessed March 22, 2018).

10. Sperandio S, de Belle I, Bredesen DE. An alternative, nonapoptotic form of programmed cell death. *Proc Natl Acad Sci USA* 2000;**97**:14 376–81.
11. Kitanaka C, Kuchino Y. Caspase-independent programmed cell death with necrotic morphology. *Cell Death Differ* 1999;**6**(6):508–15.
12. Jimenez C, Capasso JM, Edelstein CL, Rivard CJ, Lucia S, Breusegem S, et al. Different ways to die: cell death modes of the unicellular chlorophyte *Dunaliella viridis* exposed to various environmental stresses are mediated by the caspase-like activity DEVDase. *J Exp Botany* 2009;**60**(3):815–28.
13. Saikumar P, Venkatachalam MA. Apoptosis and cell death. In: Cagle PT, Allen TC (eds.). *Basic Concepts of Molecular Pathology*. Berlin: Springer Science+Business Media. 2009.
14. Sperandio S, Poksay KS, Schilling B, Crippen D, Gibson BW, Bredesen DE. Identification of new modulators and protein alterations in non-apoptotic programmed cell death. *J Cell Biochem* 2010;**111**:1401–12.
15. Hoa NT, Zhang JG, Delgado CL, Myers MP, Callahan LL, Vandeusen G, et al. Human monocytes kill M-CSF-expressing glioma cells by BK channel activation. *Lab Invest* 2007;**87**:115–29.
16. Gallucci S, Matzinger P. Danger signals: SOS to the immune system. *Curr Opin Immunol* 2001;**13**:114–19.
17. Schneider D, Gerhardt E, Bock J, Muller MM, Wolburg H, Lang F, Schulz JB. Intracellular acidification by inhibition of the Na/H-exchanger leads to caspase-independent death of cerebellar granule neurons resembling paraptosis. *Cell Death Differ* 2004;**11**:760–70.
18. Jambrina E, Alonso R, Alcalde M, del Carmen Rodríguez M, Serrano A, Martínez AC, et al. Calcium influx through receptor-operated channel induces mitochondria-triggered paraptotic cell death. *J Biol Chem* 2003;**278**(16):14 134–45.
19. Hoa N, Myers MP, Douglass TG, Zhang JG, Delgado C, Driggers L, et al. Molecular mechanisms of paraptosis induction: implications for a non-genetically modified tumor vaccine. *PLoS One* 2009;**4**(2):e4631.
20. Sperandio S, Poksay K, De Belle I, Lafuente MJ, Liu B, Nasir J, Bredesen DE. Paraptosis: mediation by MAP kinases and inhibition by AIP-1/Alix. *Cell Death Differ* 2004;**11**:1066–75.
21. Allan LA, Morrice N, Brady S, Magee G, Pathak S, Clarke PR. Inhibition of caspase-9 through phosphorylation at Thr 125 by ERK MAPK. *Nat Cell Biol* 2003:**5**(7):647–54.
22. Chatellard-Causse C, Blot B, Cristina N, Torch S, Missotten M, Sadoul R. Alix (ALG-2-interacting protein X), a protein involved in apoptosis, binds to endophilins and induces cytoplasmic vacuolization. *J Biol Chem* 2002;**277**:29 108–15.
23. Wang Y, Li X, Wang L, Ding P, Zhang Y, Han W, Ma D. An alternative form of paraptosis-like cell death triggered by TAJ/TROY and enhanced by PDCD5 overexpression. *J Cell Sci* 2004;**117**:1525.
24. Castro-Obregon S, del Rio G, Chen SF, Swanson RA, Frankowsk H, Rao RV, et al. A ligand-receptor pair that triggers a non-apoptotic form of programmed cell death. *Cell Death Differ* 2002;**9**:807–17.
25. Castro-Obregón S, Rao RV, del Rio G, Chen SF, Poksay KS, Rabizadeh S, et al. Alternative, nonapoptotic programmed cell death: mediation by Arrestin2, ERK2 and Nur77. *J Biol Chem* 2004;**279**:17 543–53.

26 Fombonne J, Reix S, Rasolonjanahary S, Danty E, Thirion S, Laforge-Anglade G, et al. EGF triggers an original caspase-independent pituitary cell death with heterogenous phenotype. *Mol Biol Cell* 2004;**15**:4938–48.
27 Fombonne J, Padron L, Enjalbert A, Krantic S, Torriglia A. A novel paraptosis pathway involving LEI/L-DNaseII for EGF-induced cell death in somato-lactotrope pituitary cells. *Apoptosis* 2006;**11**(3):367–75.
28 Valamanesh F, Torriglia A, Savoldelli M, Gandolphe C, Jeanny JC, BenEzra D, Behar-Cohen F. Glucocorticoids induce retinal toxicity through mechanisms mainly associated with paraptosis. *Mol Vis* 2007;**13**:1746–57.
29 Asare N, Landvik NE, Lagadic-Gossmann D, Rissel M, Tekpli X, Ask K, et al. 1-nitropyrene (1-NP) induces apoptosis and apparently a non-apoptotic programmed cell death (paraptosis) in Hepa1c1c7 cells. *Toxicol Appl Pharmacol* 2008;**230**:175–86.
30 Landvik NE, Gorria M, Arlt VM, Asare N, Solhaug A, Lagadic-Gossman D, Holme JA. Effects of nitrated-polycyclic aromatic hydrocarbons and diesel exhaust particle extracts on cell signaling related to apoptosis: possible implications for their mutagenic and carcinogenic effects. *Toxicology* 2006;**231**:159–74.
31 Chen TS, Wang XP, Sun L, Wang LX, Xing D, Mok M. Taxol induces caspase independent cytoplasmic vacuolization and cell death through endoplasmic reticulum (ER) swelling in ASTC-a-1 cells. *Cancer Lett* 2008;**270**:164–72.
32 Sun Q, Chen T, Wang X, Wei X. Taxol induces paraptosis independent of both protein synthesis and MAPK pathway. *J Cell Physiol* 2010;**222**:421–32.
33 Korsnes MS, Espenes A, Hetland DL, Hermansen LC. Paraptosis-like cell death induced by yessotoxin. *Toxicol Vitro* 2011;**25**:1764–70.
34 Wang Y, Yang Z, Zhao X. Honokiol induces paraptosis and apoptosis and exhibits schedule-dependent synergy in combination with imatinib in human leukemia cells. *Toxicol Mech Methods* 2010;**20**:234–41.
35 Wang Y, Zhu X, Yang Z, Zhao X. Honokiol induces caspase-independent paraptosis via reactive oxygen species production that is accompanied by apoptosis in leukemia cells. *Biochem Biophys Res Comm* 2013;**430**:876–82.
36 Guo WJ, Chen TS, Wang XP, Chen R. Taxol induces concentration dependent apoptotic and paraptosis like cell death in human lung adenocarcinoma (ASTCa1) cells. *J Xray Sci Technol* 2010;**18**:293–308.
37 Palmeri C, Petiti J, Sosa LDV, Gutíerrez S, Paul AD, Mukdsi J, Torres A. Bromocriptine induces paraptosis as the main type of cell death responsible for experimental pituitary tumor shrinkage. *Toxicol Appl Pharmacol* 2009;**240**:55–65.
38 Tardito S, Isella C, Medico E, Marchio L, Bevilacqua E, Hatzoglou M, et al. The thioxotriazole copper (II) complex A0 induces endoplasmic reticulum stress and paraptotic death in human cancer cells. *J Biol Chem* 2009;**284**(36):24 306–19.
39 Gandin V, Pellei M, Tisato F, Porchia M, Santini C, Marzano C. A novel copper complex induces paraptosis in colon cancer cells via the activation of ER stress signaling. *J Cell Mol Med* 2012;**16**(1):142–51.
40 Li B, Zhao J, Wang CZ, Searle J, He TC, Yuan CS, Du W. Ginsenoside Rh2 induces apoptosis and paraptosis-like cell death in colorectal cancer cells through activation of p53. *Cancer Lett* 2011;**301**:185–92.
41 Zhang JS, Li DM, He N, Liu YH, Wang CH, Jiang SQ, et al. A paraptosis-like cell death induced by delta-tocotrienol in human colon carcinoma SW620 cells is associated with the suppression of the Wnt signaling pathway. *Toxicology* 2011;**285**:8–17.

42 Shenoy AK, Fisher RC, Butterworth EA, Pi L, Chang LJ, Appelman HD, et al. Transition from colitis to cancer: high Wnt activity sustains the tumor-initiating potential of colon cancer stem cell precursors. *Cancer Res* 2012;**72**(19):5091–100.

43 Zhang JS, Li DM, Ma Y, He N, Gu Q, Wang FS, et al. γ-tocotrienol induces paraptosis-like cell death in human colon carcinoma SW620 cells. *PLoS One* 2013;**8**(2):e57779.

44 Yoon MJ, Kim EH, Lim JH, Kwon TK, Choi KS. Superoxide anion and proteasomal dysfunction contribute to curcumin-induced paraptosis of malignant breast cancer cells. *Free Radic Biol Med* 2010;**48**:713–26.

45 Yoon MJ, Kim EH, Kwon TK, Park SA, Choi KS. Simultaneous mitochondrial Ca2+ overload and proteasomal inhibition are responsible for the induction of paraptosis in malignant breast cancer cells. *Cancer Lett* 2012;**324**:197–209.

46 Dance-Barnes ST, Kock ND, Moore JE, Lin EY, Mosley LJ, D'Agostino RB, et al. Lung tumor promotion by curcumin. *Carcinogenesis* 2009;**30**:1016–23.

47 Bury M, Girault A, Mégalizzi V, Spiegl-Kreinecker S, Mathieu V, Berger W, et al. Ophiobolin A induces paraptosis-like cell death in human glioblastoma cells by decreasing BKCa channel activity. *Cell Death Dis* 2013;**4**:e561.

48 Rojpibulstit P, Kittisenachai S, Puthong S, Manochantr S, Gamnarai P, Jitrapakdee S, Roytrakul S. Hep88 mAb-initiated paraptosis-like PCD pathway in hepatocellular carcinoma cell line through the binding of mortalin (HSPA9) and alpha-enolase. *Cancer Cell Int* 2014: **14**:69.

49 Korsnes MS. Yessotoxin as a tool to study induction of multiple cell death pathways. *Toxins* 2012;**4**:568–79.

17

Hematopoiesis and Eryptosis

Mollie K. Rojas,[1] Chintan C. Gandhi,[2] and Lawrence E. Feldman[2]

[1]*Department of Oral Medicine and Diagnostic Sciences, University of Illinois at Chicago, Chicago, IL, USA*
[2]*Department of Hematology and Oncology, University of Illinois at Chicago, Chicago, IL, USA*

Abbreviations

APC	antigen-presenting cell
B cell	B lymphocyte
CFU-E	colony-forming unit–erythrocyte
CFU-GM	colony-forming unit–granulocyte–macrophage
CSF	colony-stimulating factor
HSC	hematopoietic stem cell
MHC	major histocompatibility complex
NK	natural killer
PCD	programmed cell death
RBC	red blood cell
T cell	T lymphocyte
WBC	white blood cell

17.1 Introduction

The life span of an erythrocyte is approximately 120 days, and is governed by the inability to synthesize new organelles. The erythrocyte plasma membrane deteriorates as a result of squeezing into and out of capillaries and the narrow spaces of the spleen. Negative feedback tightly regulates erythrocyte production, which occurs at the same rate as erythrocyte destruction. The kidneys stimulate erythrocyte production by releasing erythropoietin. When erythrocytes become damaged and near the end of their life span, they may undergo a specific type of suicidal death known as eryptosis. Eryptosis differs from apoptosis in that it occurs in enucleated erythrocytes. It effectively disposes of defective erythrocytes without rupturing the cell membrane or releasing intracellular material.

Apoptosis and Beyond: The Many Ways Cells Die, First Edition. Edited by James Radosevich.
© 2018 John Wiley & Son Inc. Published 2018 by John Wiley & Son Inc.

17.2 Hematopoiesis

Hematopoiesis (hemopoiesis) is the process by which red blood cells (RBCs), white blood cells (WBCs), and platelets are formed. It begins in the embryonic yolk sac. After 12 weeks' gestation, it relocates to the liver, spleen, thymus, and lymph nodes of the fetus, and after around 28 weeks' gestation, to the red bone marrow. All bone marrow is red at birth, and therefore contains active hematopoietic stem cells (HSCs), which produce the formed components of blood. However, as one ages, some red bone marrow is replaced with yellow bone marrow, which is primarily made up of fat cells. Hematopoiesis in adults is limited to the proximal femur, vertebrae, ribs, skull, and pelvis, because these areas are rich in red bone marrow [1,2].

HSCs are multipotent or pluripotent. Multipotent stem cells are also known as hemocytoblasts and are capable of differentiating into many types of blood cell. HSCs initially differentiate into lymphoid stem cells and myeloid stem cells. Myeloid stem cells then differentiate into myeloid progenitor cells or directly into precursor/blast cells. Lymphoid stem cells differentiate directly into precursor/blast cells. Both progenitor cells and precursor/blast cells are limited in terms of the types of cell that they can produce; they are committed to specific cell lineages. Lymphoid precursor/blast cells further differentiate into different types of WBC or leukocyte. The myeloid progenitor cells differentiate into various types of leukocyte (monocytes and neutrophils), platelet, RBC, and erythrocyte. Myeloid precursor/blast cells differentiate into two types of leukocyte (eosinophils and basophils) [1,2].

The formation of erythrocytes is called erythropoiesis. During erythropoiesis, HSCs differentiate into myeloid stem cells, which further differentiate into progenitor cells and then precursor/blast cells. The specific progenitor cell associated with erythropoiesis is called a colony-forming unit–erythrocyte (CFU-E).

Assorted cytokines, hormones, and growth factors stimulate differentiation into various cell types. The hormones responsible for the "differentiation and proliferation of particular progenitor cells" are called hemopoietic growth factors [3]. Erythropoietin, a hemopoietic growth factor, is essential in the differentiation of erythrocytes. An increased production of erythropoietin increases the quantity of erythrocyte precursor cells. Erythropoietin synthesis is stimulated by the hormone testosterone. Since testosterone levels are higher in males, the level of erythropoietin produced by males is higher than that of females. Erythropoietin is produced primarily in the kidneys [1,2].

17.3 Erythrocytes

Erythrocytes contain hemoglobin, which carries oxygen and provides pigmentation. The life span of an erythrocyte is approximately 120 days. To keep up with the destruction of erythrocytes, an equal number – approximately 2 million mature cells per second – must be produced [2].

17.3.1 Structure and Function of Erythrocytes

Erythrocytes transport oxygen and carbon dioxide in blood. Mature erythrocytes lack a nucleus; this optimizes the space available in the cell for oxygen transport [3]. Since RBCs lack a nucleus as well as organelles, they are limited to carrying oxygen and carbon dioxide. Erythrocytes are not capable of reproduction or other complex metabolic processes.

They are very small in size, measuring 7–8 μm in diameter. They are shaped like biconcave discs and have a strong and flexible plasma membrane, which allows them to squeeze through narrow capillaries without rupturing [2].

Erythrocytes generate oxygen anaerobically, which preserves the oxygen that they are carrying for the metabolic functions of other cells. According to Tortora and Derrickson, "Each RBC contains about 280 million hemoglobin molecules" [2]. Every hemoglobin molecule can transport four oxygen molecules or four carbon dioxide molecules. Oxygen is transported from the lungs to the tissues, while carbon dioxide is transported from the tissues to the lungs and then exhaled.

17.3.2 Life Cycle of Erythrocytes

Erythrocytes live approximately 120 days. Their life span is limited by their inability to synthesize new organelles. They are incapable of reconstructing their plasma membrane, which deteriorates due to squeezing into and out of capillaries and the narrow spaces of the spleen. Old erythrocytes are more likely to burst than are newly synthesized ones, due to their damaged plasma membranes. Ruptured erythrocytes are removed from circulation and destroyed in the spleen and liver. Their components are then recycled [1,2].

17.3.3 Erythrocyte Production

The precursor/blast cell known as the proerythroblast initiates erythrocyte production. Proerythroblasts reside in red bone marrow. According to Tortora and Derrickson, "The proerythroblast divides several times, producing cells that begin to synthesize hemoglobin" [2]. It eventually expels its nucleus, becoming a reticulocyte, then – following release from red bone marrow (about 1–2 days) – matures into an erythrocyte.

17.3.4 Erythropoiesis Regulation

Erythrocyte production is tightly regulated by a negative-feedback system. Typically, the production and destruction of erythrocytes occurs in a steady state. The negative-feedback system is responsible for controlling the quantity of erythrocytes produced by stimulation of the kidneys to release erythropoietin. Erythropoietin speeds the differentiation of proerythroblasts into reticulocytes. If the synthesis of erythrocytes decreases so much that the oxygen-carrying capacity of the blood is affected, synthesis of new erythrocytes will be triggered by the negative-feedback system. Erythropoiesis is mainly triggered by hypoxia: decreased cellular oxygen [2].

17.4 Structure of Leukocytes

17.4.1 Granulocytes

Granulocytes can be distinguished from one another by counting the number of nuclear lobes each possesses and differentiating between the stained colors of the cytoplasmic granules:

- Neutrophils contain small, evenly distributed granules that stain purple. The nucleus of a neutrophil can contain two to five lobes.

- Basophils stain blueish-purple using basic dyes. They contain two-lobed nuclei that are often obscured by round granules, which vary in size.
- Eosinophils contain granules that stain red when acidic dyes are used. Their two-lobed nuclei are usually apparent. The granules of eosinophils are large and uniform in size [2].

17.4.2 Agranulocytes

Agranulocytes contain cytoplasmic granules, but they are very small in size and do not stain well using dyes.

Lymphocytes (T lymphocytes, B lymphocytes, and natural killer (NK) cells) vary in size and diameter. The larger the lymphocyte, the more cytoplasm it contains. Lymphocyte nuclei are round in shape and are surrounded by a cytoplasmic rim that appears sky blue in color.

Monocytes contain kidney-shaped nuclei and stain blueish-gray due to the presence of lysosomes. Monocytes are transported into tissues from the blood. They differentiate into macrophages, which can either remain fixed in tissues or migrate to areas fighting infection via the bloodstream [2].

17.5 Function of WBCs

Leukocytes predominantly participate in the innate and adaptive immune systems. Innate immunity is present from birth and is rapid in its response to foreign invaders. This type of immunity is considered nonspecific because the innate immune system responds in the same way to all foreign pathogens. Innate immunity functions without a memory; the cells of the innate immune system react without recalling any previous exposure to the pathogen in question. It involves first- and second-line defenses. The first line of defense against foreign pathogens entails "physical and chemical barriers of the skin and mucous membranes" [2]. NK cells, neutrophils, and macrophages play a large role in the second line of defense [2]. Adaptive or specific immunity involves immune responses based on memory. Defenses of the adaptive immune system respond to specific pathogens. B and T lymphocytes are major players in the adaptive immune system.

Antigen presentation by phagocytes is necessary for the adaptive immune system to function. All nucleated cells in the body, including leukocytes (granulocytes and agranulocytes), display surface proteins called major histocompatibility complex (MHC) antigens. MHC antigens are required for antigen processing, the first step of the adaptive immune response. There are two types of MHC antigen: class I and class II. Class I MHC antigens are present on all cells of the body except erythrocytes. Class II MHC antigens are present on the cell surfaces of antigen-presenting cells (APCs).

The main function of leukocytes is to respond to foreign pathogens via phagocytosis or other immune responses. In order to perform various immune functions, some leukocytes are transported out of the bloodstream into the tissues.

Neutrophils respond quickly to foreign invaders and phagocytize their pathogens. They first engulf the pathogen and then destroy the invader by releasing chemicals such as lysozyme, superoxide ion, hydrogen peroxide, and hypochlorite ion. They also contain defensins, which destroy pathogens by poking holes in their cell membranes.

Monocytes differentiate into macrophages once they reach the site of infection. Macrophages are considered APCs, which are effective at phagocytizing foreign invaders and then "presenting" a piece of the foreign antigen on their MHC class II complex for recognition and processing by T cells.

Basophils and mast cells release chemical mediators such as histamine at inflammatory sites. Other chemical mediators, such as heparin and serotonin, are also released by basophils and mast cells. These mediators are important to increasing the inflammatory response of the host.

Eosinophils release enzymes such as histaminase, which opposes the effect of histamine. Histaminase is primarily released in response to inflammation produced during allergic reactions. Eosinophils also combat parasitic worms and phagocytize antigen–antibody complexes [2].

B lymphocytes are an important component of the adaptive immune system. They differentiate into several types, including B memory cells and plasma cells, based on cytokine stimulation. B lymphotcytes are APCs that are effective in identifying and phagocytizing foreign invaders and inactivating their toxic products. B memory cells remain in circulation for extended periods of time and are important in the adaptive immune response. Plasma cells produce antibodies in response to foreign invaders.

T lymphocytes are required for the activation of B lymphocytes. After foreign invaders are engulfed by B lymphocytes, a portion of their antigen is presented on the MHC class II complex located on their cell membrane. T lymphocytes recognize the antigen presented on the MHC class II complex as foreign and mount an immune response. They also function in eliminating viruses and cancer cells. If a cell is infected by a virus or affected by cancer, it will digest the foreign proteins and present the antigen on MHC class I complexes for recognition and elimination by T lymphocytes. T lymphocytes are also implicated in host rejection of transplanted tissue.

NK cells protect against a variety of viral infections and tumor cells [2].

17.5.1 WBC Production and Maturation

Granulocytes and monocytes develop from the myeloid stem-cell lineage. Pluripotent stem cells that differentiate into myeloid stem cells further differentiate into different progenitor cells and precursor/blast cells. Eosinophils develop directly from the precursor/blast cell known as an eosinophilic myeloblast. Basophils develop directly from the basophilic myeloblast. Neutrophils and monocytes share a common progenitor cell, the colony-forming-unit–granulocyte–macrophage (CFU-GM), which differentiates into myeloblasts and then into neutrophils and monoblasts. Monocytes and ultimately macrophages develop from monoblasts.

B and T lymphocytes develop from pluripotent stem cells. Pluripotent stem cells differentiate into lymphoid stem cells. Lymphoid stem cells differentiate into precursor/blast cells. The precursor/blast cells called T lymphoblasts further differentiate into T lymphocytes, while those called B lymphoblasts differentiate into B lymphocytes.

NK cells develop in the lymphoid stem cell lineage from the precursor/blast cell called NK lymphoblast.

Production of leukocytes is regulated by the release of cytokines. Cytokines are glycoproteins that are produced by many different types of cell. Interleukins and colony-stimulating factors (CSFs) are specific types of cytokine that are important in the regulation of leukocyte formation.

17.6 Eryptosis

The average life span of an erythrocyte is about 120 days. It is ultimately limited by senescence. Erythrocytes may experience survival-threatening injury, which shortens cellular life span. Eryptosis, a specialized form of programmed cell death (PCD), is responsible for eliminating worn-out enucleated erythrocytes. It removes cells without rupturing their membranes or releasing intracellular material. Eryptosis differs from apoptosis in that it only occurs in enucleated erythrocytes. Apoptosis, in contrast, occurs in cells that contain nuclei and mitochondria. Eryptosis lacks mitochondrial depolarization and nuclear condensation, which are hallmark signs of apoptosis. It shares several features of apoptosis, including cell-membrane blebbing and cell shrinkage. Erythrocytes that undergo eryptosis are eventually engulfed and degraded by macrophages. Eryptosis allows for the removal of immature, infected, or potentially harmful cells.

17.7 Triggers and Signaling of Eryptosis

Several mechanisms have been credited with triggering eryptosis. According to Lang and Qadri, "Eryptosis is stimulated by an increase in cytosolic Ca2+ activity, ceramides, hyperosmotic shock, oxidative stress, energy depletion, hyperthermia, and a wide variety of xenobiotics and endogenous substances" (see Figure 17.1) [4].

As erythrocytes age, they become more susceptible to oxidative stress and, ultimately, eryptosis. Eryptosis is further stimulated by a myriad of xenobiotics and a wide variety of endogenous substances and challenges.

17.8 Physiological Significance of Eryptosis

Eryptosis removes defective erythrocytes prior to hemolysis. The presence of phosphatidylserine on the surface of eryptotic cells is important for their adherence to endothelial cells and macrophages. The eryptotic cells are subsequently engulfed and degraded intracellularly.

Cellular swelling is a response to energy depletion or exposure to toxins. Toxin exposure leads to a cellular gain of sodium and chloride and – through osmosis – of water, which produces the swelling. Cellular swelling causes the cell membrane to rupture, leading to the release of intracellular hemoglobin, which can be troublesome. Hemoglobin can be filtered by the glomerula, but it occludes renal tubules. Eryptosis is stimulated by increasing cytosolic calcium activity, which leads to the hyperpolarization of cells (see Figure 17.1).

Eryptosis also leads to the removal of infected erythrocytes. Microorganisms that are capable of entering erythrocytes include malarial parasites and Babesia microti. These intracellular microorganisms cause oxidative stress, which activates cation channels and triggers eryptosis. Therefore, erythrocytes that are infected with microorganisms will undergo eryptosis, decreasing the risk of spread via the blood. Several genetic disorders, such as sickle-cell trait, beta-thalassemia trait, and glucose 6-phosphate dehydrogenase (G6PD) deficiency, trigger early eryptosis via abnormal erythrocytes. The early eryptosis that is triggered confers protection against severe courses of infections.

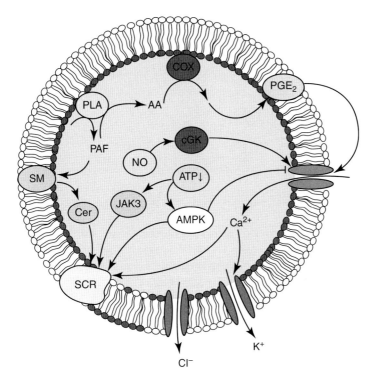

Figure 17.1 Signaling involved in the stimulation and inhibition of eryptosis. AA: arachidonic acid; AMPK: AMP activated protein kinase; ATP: adenosine triphosphate; cGK: cyclic GMP dependent preotein kinase type I; COX: cylcoxygenase; NO: nitric oxide; NSC: nonselective cation channel; PAF: platelet activating factor; PGE2: prostaglandin E_2; PLA: phospholipase A; PS: phosphatidylserine; ROS: reactive oxygen species; SCR: scrambling enzyme; SM: spingomyelinase. *Source:* Lang et al. 2012 [3]. Reproduced with permission from Elsevier.

17.9 Pathophysiological Significance of Eryptosis

If accelerated eryptosis is not adequately compensated by erythropoiesis, it may lead to anemia. The presence of phosphatidylserine on the surface of eryptotic cells may cause problems with microcirculation due to adherence to endothelial cells. It may also stimulate clot formation.

Iron-deficiency anemia causes decreased erythrocyte volume and defective erythrocytes, which may sensitize erythrocytes to trigger eryptosis. Erytposis is further enhanced in patients with diabetes, renal insufficiency, hyperthermia, sepsis, and infections with mycoplasma or malaria. Eryptosis is also seen in several genetic disorders, such as sickle-cell trait, beta-thalassemia trait, glucose 6-phosphate dehydrogenase (G6PD) deficiency, hereditary spherocytosis, Wilson's disease, paroxysmal nocturnal hemoglobinuria, and myelodysplastic syndrome.

17.10 Conclusion

Recently, much has been uncovered regarding the cellular mechanisms triggering and inhibiting erytposis. Eryptosis is stimulated by a myriad of xenobiotics and a wide variety of endogenous substances and challenges. Nonetheless, our understanding of the exact physiological mechanism of eryptosis is still incomplete and requires further research. Eryptosis most likely plays a very important role in a variety of clinical conditions. Understanding its complexities will provide greater understanding of the pathophysiology of a variety of clinical conditions.

References

1 Kindt T, Goldsby, RA, Osborne, BA, Kuby J. *Kuby Immunology*, 6th edn. Basingstoke: Macmillan. 2007.
2 Tortora GJ, Derrickson B. *Principles of Anatomy and Physiology*, 12th edn. Chichester: John Wiley & Sons. 2009.
3 Lang E, Qadri SM, Lang F. Killing me softly – suicidal erythrocyte death. *Int J Biochem Cell Biol* 2012;**44**(8):1236–43.
4 Lang F, Qadri SM. Mechanisms and significance of eryptosis, the suicidal death of erythrocytes. *Blood Purif* 2012;**33**(1–3):125–30.

Apoptosis and Beyond

Apoptosis and Beyond

The Many Ways Cells Die

Volume 2

Edited by James Radosevich
University of Chicago, United States

This edition first published 2018
© 2018, John Wiley & Sons Inc

All rights reserved. No part of this publication may be reproduced, stored in a retrieval system, or transmitted, in any form or by any means, electronic, mechanical, photocopying, recording or otherwise, except as permitted by law. Advice on how to obtain permission to reuse material from this title is available at http://www.wiley.com/go/permissions.

The right of James Radosevich to be identified as the author of the editorial material in this work has been asserted in accordance with law.

Registered Office(s)
John Wiley & Sons, Inc., 111 River Street, Hoboken, NJ 07030, USA

Editorial Office
The Atrium, Southern Gate, Chichester, West Sussex, PO19 8SQ, UK

For details of our global editorial offices, customer services, and more information about Wiley products visit us at www.wiley.com.

Wiley also publishes its books in a variety of electronic formats and by print-on-demand. Some content that appears in standard print versions of this book may not be available in other formats.

Limit of Liability/Disclaimer of Warranty
While the publisher and authors have used their best efforts in preparing this work, they make no representations or warranties with respect to the accuracy or completeness of the contents of this work and specifically disclaim all warranties, including without limitation any implied warranties of merchantability or fitness for a particular purpose. No warranty may be created or extended by sales representatives, written sales materials or promotional statements for this work. The fact that an organization, website, or product is referred to in this work as a citation and/or potential source of further information does not mean that the publisher and authors endorse the information or services the organization, website, or product may provide or recommendations it may make. This work is sold with the understanding that the publisher is not engaged in rendering professional services. The advice and strategies contained herein may not be suitable for your situation. You should consult with a specialist where appropriate. Further, readers should be aware that websites listed in this work may have changed or disappeared between when this work was written and when it is read. Neither the publisher nor authors shall be liable for any loss of profit or any other commercial damages, including but not limited to special, incidental, consequential, or other damages.

Library of Congress Cataloging-in-Publication Data

Names: Radosevich, James A. (James Andrew), editor.
Title: Apoptosis and beyond : the many ways cells die / edited by James Radosevich.
Description: Hoboken, NJ : Wiley-Blackwell, 2018. | Includes bibliographical
    references and index. |
Identifiers: LCCN 2018022533 (print) | LCCN 2018023112 (ebook) | ISBN
    9781119432357 (Adobe PDF) | ISBN 9781119432432 (ePub) | ISBN 9781119432425 (cloth)
Subjects: | MESH: Cell Death
Classification: LCC QH671 (ebook) | LCC QH671 (print) | NLM QU 375 | DDC
    571.9/36–dc23
LC record available at https://lccn.loc.gov/2018022533

Cover image: © ImageJournal-Photography/Getty images
Cover design by Wiley

Set in 10/12 pt WarnockPro-Regular by Thomson Digital, Noida, India

Printed in Singapore by C.O.S. Printers Pte Ltd

Contents

List of Contributors *ix*

Volume 1

1. **General View of the Cytoplasmic and Nuclear Features of Apoptosis** *1*
   Humberto De Vitto, Juan P. Valencia, and James A. Radosevich

2. **Mitochondria in Focus: Targeting the Cell-Death Mechanism** *13*
   Humberto De Vitto, Roberta Palorini, Giuseppina Votta, and Ferdinando Chiaradonna

3. **Microbial Programmed Cell Death** *49*
   Neal D. Hammer

4. **Autophagy** *71*
   Mollie K. Rojas, Juel Chowdhury, Khatja Batool, Zane Deliu, and Abdallah Oweidi

5. **Cell Injury, Adaptation, and Necrosis** *83*
   Sarah G. Fitzpatrick and Sara C. Gordon

6. **Necroptosis** *99*
   Ben A. Croker, James A. Rickard, Inbar Shlomovitz, Arshed Al-Obeidi, Akshay A. D'Cruz, and Motti Gerlic

7. **Ferroptosis** *127*
   Ebru Esin Yoruker and Ugur Gezer

8. **Anoikis Regulation: Complexities, Distinctions, and Cell Differentiation** *145*
   Marco Beauséjour, Ariane Boutin, and Pierre H. Vachon

9. **Cornification** *183*
   Leopold Eckhart

10 Excitotoxicity   *197*
    Julie Alagha, Sulaiman Alshaar, and Zane Deliu

11 Molecular Mechanisms Regulating Wallerian Degeneration   *205*
    Mohammad Tauseef and Madeeha Aqil

12 Pyronecrosis   *225*
    Maryam Khalili and James A. Radosevich

13 Phenoptosis: Programmed Death of an Organism   *237*
    M.V. Skulachev and V.P. Skulachev

14 Molecular Mechanisms Underlying Oxytosis   *289*
    Amalia M. Dolga, Sina Oppermann, Maren Richter, Birgit Honrath, Sandra Neitemeier, Anja Jelinek, Goutham Ganjam, and Carsten Culmsee

15 Pyroptosis   *317*
    Kate E. Lawlor, Stephanie Conos, and James E. Vince

16 Paraptosis   *343*
    Maryam Khalili and James A. Radosevich

17 Hematopoiesis and Eryptosis   *367*
    Mollie K. Rojas, Chintan C. Gandhi, and Lawrence E. Feldman

Volume 2

18 Cyclophilin D-Dependent Necrosis   *375*
    Jatin Mehta and Chandi Charan Mandal

19 Role of Phospholipases in Cell Death   *395*
    Manikanda Raja, Juel Chowdhury, and James A. Radosevich

20 TRIAD (Transcriptional Repression-Induced Atypical Death)   *411*
    Takuya Tamura and Hitoshi Okazawa

21 Alkylating-Agent Cytotoxicity Associated with O^6-Methylguanine   *427*
    Latha M. Malaiyandi, Lawrence A. Potempa, Nicholas Marschalk, Paiboon Jungsuwadee, and Kirk E. Dineley

22 Entosis   *463*
    Jamuna A. Bai and Ravishankar Rai V.

23 Mitotic Catastrophe   *475*
    Raquel De Souza, Lais Costa Ayub, and Kenneth Yip

24 **NETosis and ETosis: Incompletely Understood Types of Granulocyte Death and their Proposed Adaptive Benefits and Costs** *511*
    Marko Radic

25 **Parthanatos: Poly ADP Ribose Polymerase (PARP)-Mediated Cell Death** *535*
    Amos Fatokun

26 **Methuosis: Drinking to Death** *559*
    Madeeha Aqil

27 **Oncosis** *567*
    Priya Weerasinghe, Sarathi Hallock, Robert Brown, and L. Maximilian Buja

28 **Autoschizis: A Mode of Cell Death of Cancer Cells Induced by a Prooxidant Treatment *In Vitro* and *In Vivo*** *583*
    J. Gilloteaux, J.M. Jamison, D. Arnold, and J.L. Summers

29 **Programmed Death 1 (PD1)-Mediated T-Cell Apoptosis and Cancer Immunotherapy** *695*
    Chandi Charan Mandal, Jatin Mehta, and Vijay K. Prajapati

Index *723*

List of Contributors

Julie Alagha
Department of Oral Medicine and
Diagnostic Sciences
University of Illinois at Chicago
Chicago, IL, USA

Arshed Al-Obeidi
Dana-Farber Boston's Children Cancer
and Blood Disorder Center
Harvard Medical School
Boston, MA, USA

Sulaiman Alshaar
Department of Oral Medicine and
Diagnostic Sciences
University of Illinois at Chicago
Chicago, IL, USA

Madeeha Aqil
Department of Oral Medicine and
Diagnostic Sciences
University of Illinois at Chicago
Chicago, IL, USA

D. Arnold
Anesthesiology Unit
Belfast, ME, USA

Lais Costa Ayub
Department of Molecular, Genetic and
Structural Biology
University of Ponta Grossa
Ponta Grossa, PR, Brazil

Jamuna A. Bai
Department of Studies in Microbiology
University of Mysore
Mysore, India

Khatja Batool
Department of Oral Medicine and
Diagnostic Sciences
University of Illinois at Chicago
Chicago, IL, USA

Marco Beauséjour
Department of Anatomy and Cellular
Biology
University of Sherbrooke
Sherbrooke, QC, Canada

Ariane Boutin
Department of Anatomy and Cellular
Biology
University of Sherbrooke
Sherbrooke, QC, Canada

Robert Brown
Department of Pathology and Laboratory
Medicine
University of Texas Health Science
Center at Houston
Houston, TX, USA

List of Contributors

L. Maximilian Buja
Department of Pathology and Laboratory Medicine
University of Texas Health Science Center at Houston
Houston, TX, USA

Ferdinando Chiaradonna
Department of Biotechnology and Biosciences
University of Milano-Bicocca
Milan, Italy

Juel Chowdhury
Department of Oral Medicine and Diagnostic Sciences
University of Illinois at Chicago
Chicago, IL, USA

Stephanie Conos
The Walter and Eliza Hall Institute of Medical Research
Bundoora, VIC, Australia

Ben A. Croker
Dana-Farber Boston's Children Cancer and Blood Disorder Center
Harvard Medical School
Boston, MA, USA

Carsten Culmsee
Institute of Physiological Chemistry
Philipps University of Marburg
Marburg, Germany

Akshay A. D'Cruz
Dana-Farber Boston's Children Cancer and Blood Disorder Center
Harvard Medical School
Boston, MA, USA

Zane Deliu
Department of Oral Medicine and Diagnostic Sciences
University of Illinois at Chicago
Chicago, IL, USA

Raquel De Souza
University Health Network, Department of Radiation Physics, Pharmaceutical Sciences
University of Toronto
Toronto, ON, Canada

Humberto De Vitto
Center of Health and Science
Federal University of Rio de Janeiro
Rio de Janeiro, Brazil

Kirk E. Dineley
Chicago College of Osteopathic Medicine
Midwestern University
Downers Grove, IL, USA

Amalia M. Dolga
Department of Molecular Pharmacology, Groningen Research Institute of Pharmacy (GRIP)
University of Groningen
Groningen, The Netherlands

Leopold Eckhart
Department of Dermatology
Medical University of Vienna
Vienna, Austria

Amos Fatokun
School of Medical Sciences, Faculty of Life Sciences
University of Bradford
Bradford, UK

Lawrence E. Feldman
Department of Hematology and Oncology
University of Illinois at Chicago
Chicago, IL, USA

Sarah G. Fitzpatrick
Department of Oral and Maxillofacial Diagnostic Sciences
University of Florida
Gainesville, FL, USA

Chintan C. Gandhi
Department of Hematology and Oncology
University of Illinois at Chicago
Chicago, IL, USA

Goutham Ganjam
Institute of Pharmacology and Clinical Pharmacy, Biochemisch-Pharmakologisches Centrum Marburg
Philipps University of Marburg
Marburg, Germany

Motti Gerlic
Department of Clinical Microbiology and Immunology, Sackler Faculty of Medicine
Tel Aviv University
Tel Aviv, Israel

Ugur Gezer
Department of Basic Oncology, Oncology Institute
Istanbul University
Istanbul, Turkey

J. Gilloteaux
Department of Anatomical Sciences
St. Georges' University International School of Medicine, KB Taylor Global Scholar's Program at Northumbria University
Newcastle upon Tyne, UK

Sara C. Gordon
School of Dentistry
Oral Medicine University of Washington
Seattle, WA, USA

Sarathi Hallock
Memorial University of Newfoundland, AMC Cancer Research Center
University of Texas Health Science Center at Houston
Houston, TX, USA

Neal D. Hammer
Department of Microbiology and Molecular Genetics
Michigan State University
East Lansing, MI, USA

Birgit Honrath
Institute of Pharmacology and Clinical Pharmacy, Biochemisch-Pharmakologisches Centrum Marburg
Philipps University of Marburg
Marburg, Germany

and

Department of Molecular Pharmacology, Groningen Research Institute of Pharmacy (GRIP)
University of Groningen
Groningen, The Netherlands

J.M. Jamison
Department of Urology and Apatone Development Laboratory
Summa Health System
Akron, OH, USA

Anja Jelinek
Institute of Pharmacology and Clinical Pharmacy, Biochemisch-Pharmakologisches Centrum Marburg
Philipps University of Marburg
Marburg, Germany

Paiboon Jungsuwadee
School of Pharmacy
Fairleigh Dickinson University
Florham Park, NJ, USA

Maryam Khalili
Department of Oral Medicine and Diagnostic Sciences
University of Illinois at Chicago
Chicago, IL, USA

Kate E. Lawlor
The Walter and Eliza Hall Institute of Medical Research
Bundoora, VIC, Australia

Latha M. Malaiyandi
Chicago College of Osteopathic Medicine
Midwestern University
Downers Grove, IL, USA

Chandi Charan Mandal
Department of Biochemistry
Central University of Rajasthan
Rajasthan, India

Nicholas Marschalk
Chicago College of Osteopathic Medicine
Midwestern University
Downers Grove, IL, USA

Jatin Mehta
National Institute of Pathology, ICMR
Safdarjang Hospital
New Delhi, India

Sandra Neitemeier
Institute of Pharmacology and Clinical Pharmacy, Biochemisch-Pharmakologisches Centrum Marburg
Philipps University of Marburg
Marburg, Germany

Hitoshi Okazawa
Department of Neuropathology, Medical Research Institute
Tokyo Medical and Dental University
Tokyo, Japan

Sina Oppermann
German Cancer Research Center (DKFZ) and National Center of Tumordiseases (NCT)
Heidelberg, Germany

Abdallah Oweidi
Department of Oral Medicine and Diagnostic Sciences
University of Illinois at Chicago
Chicago, IL, USA

Roberta Palorini
SYSBIO Center for Systems Biology, Department of Biotechnology and Biosciences
University of Milano-Bicocca
Milan, Italy

and

Luxembourg Centre for Systems Biomedicine
Esch-sur-Alzette, Luxembourg

Lawrence A. Potempa
Roosevelt University College of Pharmacy
Schaumburg, IL, USA

Vijay K. Prajapati
Department of Biochemistry
Central University of Rajasthan
Rajasthan, India

Marko Radic
Department of Microbiology, Immunology and Biochemistry
University of Tennessee Health Science Center
Memphis, TN, USA

James A. Radosevich
Department of Oral Medicine and Diagnostic Sciences
University of Illinois at Chicago
Chicago, IL, USA

Manikanda Raja
Department of Oral Medicine and Diagnostic Sciences
University of Illinois at Chicago
Chicago, IL, USA

Maren Richter
Institute of Pharmacology and Clinical Pharmacy, Biochemisch-Pharmakologisches Centrum Marburg
Philipps University of Marburg
Marburg, Germany

List of Contributors

James A. Rickard
Department of Biochemistry
La Trobe University
Melbourne, VIC, Australia

Mollie K. Rojas
Department of Oral Medicine and
Diagnostic Sciences
University of Illinois at Chicago
Chicago, IL, USA

Inbar Shlomovitz
Department of Clinical Microbiology
and Immunology, Sackler Faculty
of Medicine
Tel Aviv University
Tel Aviv, Israel

M.V. Skulachev
Belozersky Institute of Physico-Chemical
Biology, Institute of Mitoengineering,
Faculty of Bioengineering and
Bioinformatics
Lomonosov Moscow State University
Moscow, Russia

V.P. Skulachev
Belozersky Institute of Physico-Chemical
Biology, Institute of Mitoengineering,
Faculty of Bioengineering and
Bioinformatics
Lomonosov Moscow State University
Moscow, Russia

J.L. Summers
Department of Urology and Apatone
Development Laboratory
Summa Health System
Akron, OH, USA

Takuya Tamura
Department of Neuropathology, Medical
Research Institute
Tokyo Medical and Dental University
Tokyo, Japan

Mohammad Tauseef
Department of Pharmaceutical Sciences
College of Pharmacy
Chicago State University
Chicago, IL, USA

Ravishankar Rai V.
Department of Studies in Microbiology
University of Mysore
Mysore, India

Pierre H. Vachon
Department of Anatomy and Cellular
Biology
University of Sherbrooke
Sherbrooke, QC, Canada

Juan P. Valencia
University of Rio de Janeiro
Rio de Janeiro, Brazil

James E. Vince
The Walter and Eliza Hall Institute of
Medical Research
Bundoora, VIC, Australia

Giuseppina Votta
SYSBIO Center for Systems Biology,
Department of Biotechnology and
Biosciences
University of Milano-Bicocca
Milan, Italy

Priya Weerasinghe
Department of Pathology and Laboratory
Medicine
University of Texas Health Science
Center at Houston
Houston, TX, USA

Kenneth Yip
Department of Biology
University of Toronto
Toronto, ON, Canada

Ebru Esin Yoruker
Department of Basic Oncology,
Oncology Institute
Istanbul University
Istanbul, Turkey

18

Cyclophilin D-Dependent Necrosis

Jatin Mehta[1] and Chandi Charan Mandal[2]

[1] National Institute of Pathology, ICMR, Safdarjang Hospital, New Delhi, India
[2] Department of Biochemistry, Central University of Rajasthan, Rajasthan, India

Abbreviations

Aβ	amyloid beta
AD	Alzheimer's disease
ALS	amyotrophic lateral sclerosis
ANT	adenosine nucleotide translocator
Bak	Bcl-2 homologous antagonist killer
Bax	Bcl-2-associated X protein
BBB	blood–brain barrier
cGMP	cyclic guanosine monophosphate
CsA	cyclosporine A
CypA	cyclophilin A
CypD	cyclophilin D
cyt c	cytochrome c
FKBP	FK506-binding protein
GTP	guanosine triphosphate
IMM	inner mitochondrial membrane
MEF	mouse embryonic fibroblast
MPTP	mitochondrial permeability transition pore
MOMP	mitochondrial outer-membrane permeability
NAD^+	oxidized nicotinamide adenine dinucleotide
NADH	reduced nicotinamide adenine dinucleotide
$NADP^+$	oxidized nicotinamide adenine dinucleotide phosphate
NADPH	reduced nicotinamide adenine dinucleotide phosphate
NO	nitric oxide
OMM	outer mitochondrial membrane
OSCP	oligomycin sensitivity-conferring protein
PCD	programmed cell death
PD	Parkinson's disease
PiC	phosphate carrier
PKA	protein kinase A
PKGI	cGMP-dependent protein kinase I
PPIase	protein prolyl isomerase

Apoptosis and Beyond: The Many Ways Cells Die, First Edition. Edited by James Radosevich.
© 2018 John Wiley & Son Inc. Published 2018 by John Wiley & Son Inc.

ROS   reactive oxygen species
VDAC   voltage-dependent anion channel

18.1 Introduction

Cell death regulates a large number of biological processes and plays a crucial role in animal and plant development. Cell death or loss of cell viability occurs in response to many stimuli, including hypoxia, ischemia, exposure to toxic chemicals, and withdrawal of growth factors. Following ischemia, cells are known to die through a passive form of cell death called necrosis. Necrosis is defined as a passively regulated, uncontrolled cell death lacking the features of both apoptosis and autophagy. It is often involved in many pathological conditions, and it leads to inflammation due to the release of many factors from the dead cells. It affects the neighboring cells, as the release of the intracellular contents of the dying cell causes damage to bystanders. The chronological and molecular order of events are poorly defined, but early loss of plasma-membrane integrity, release of intracellular components, and dilation of cytoplasmic organelles such as mitochondria are the most characteristic features. Cyclophilin D (CypD) is a family member of the cyclosporine-binding proteins, which have been implicated in some but not all types of necrotic cell death. CypD is a mitochondrial matrix protein that may play an important role in the opening of the mitochondrial permeability transition pore (MPTP), leading to necrosis-mediated cell death. Recent genetic studies have revealed that mitochondria isolated from CypD knockout (KO) mice are more resistant to Ca^{++}-induced mitochondrial permeability transition as compared to wild-type. Moreover, studies have shown that targeted disruption of CypD in mice leads to normal cell death in many different cell types upon apoptotic stimuli, whereas resistance leads to necrotic cell death [1,2]. CypD has pleiotropic effects apart from MPTP regulation, as CypD KO hearts show an alteration in branched-chain amino-acid metabolism [3]. Mice that are CypD-deficient are resistant to ischemia/reperfusion-induced cell death *in vivo*. CypD-dependent necrosis is known to play an important role in a variety of diseases, including ischemia-reperfusion injury of the heart and brain, muscular dystrophies, Parkinson's disease (PD), Alzheimer's disease (AD), and cancers [4]. It has also been observed that mitochondrial permeability transition due to opening of the MPTP is inhibited by cyclosporine A (CsA), a CypD blocker. Therefore, inhibition of CypD-dependent mitochondrial pore transition may provide a novel therapeutic target for myocardial infarction and many other diseases. However, targeting of CypD is associated with undesirable immunosuppressive side effects, and there is a need for the development of novel CypD inhibitors. These should ideally be smaller molecules with greater bioavailability, higher absorption, and fewer unwanted side effects than current inhibitors. Therefore, a better understanding of the pathophysiology and role of CypD in CypD-dependent diseases is an important future research goal.

18.2 Cyclophilins

Cyclophilins are ubiquitous proteins that are highly conserved across evolution. They are present throughout animal and bacterial genomes and are identified by the presence of their characteristic peptidyl-prolyl cis-trans-isomerase (PPIase) activity. There are eight

Table 18.1 Roles of major cyclophilins in humans. Cyp, cyclophilin; ER, endoplasmic reticulum; MPTP, mitochondrial permeability transition pore.

Gene/protein name	MW (kDa) of full-length protein	Cellular location	Cellular functions	References
PPIA/CypA	18	Cytoplasm, nucleus, and extracellular space	Protein folding, intracellular protein transport, assembly immune modulation, and cell signaling	[5–7]
PPIB/CypB	23.74	ER, nucleus, and extracellular space	ER redox homeostasis, protein folding, ribosome biogenesis, calcium homeostasis, chemotaxis, and prolactin signaling	[6–11]
PPIC/CypC	22	ER, Golgi, and extracellular space	ER redox homeostasis, DNA degradation, and immune suppression	[6,11]
PPIF/CypD	22	Mitochondria	Opening of MPTP and Ca^{2+} homeostasis	[1,12]
PP1E/Cyp33	33	Nucleus	RNA splicing	[13]
PPIL3/CypF	18	Nucleus and cytoplasm	Cancer progression and resistance to chemotherapy	[14]
PPID/Cyp40	40.76	Nucleus and cytoplasm	Protein folding, ligand binding, and glucocorticoid-, estrogen/progesterone-, and aryl-receptor signalling	[15–17]
CypNK	150	Cell surface and cytoplasm	Tumor recognition	[18]

major cyclophilins in humans, with sixteen unique cyclophilin families, each member of which has a unique subcellular compartment, diverse cellular roles, and functional specialization (Table 18.1).

All cyclophilins share a unique 109-amino-acid domain called the cyclophilin-like domain, surrounded by domains unique to each, which help in subcellular compartmentalization and unique functions. Functionally, cyclophilins belong to the PPIase family of enzymes, which catalyze the isomerization of peptide bonds from trans to cis form at proline residues and which aid in the process of protein folding. Cyclophilins are distributed in almost all cellular compartments and occur in both free and membrane-bound forms. They are regulated by diverse posttranslational modifications, including glycosylation, N-terminal modification, and phosphorylation. Cyclophilin A (CypA) was the first family member to be discovered and is the most abundantly expressed, accounting for 0.1–0.6% of the total cytoplasmic protein. Together with FK506-binding proteins (FKBPs) and parvulins, cyclophilins are collectively known as the immunophilins [19]. Human cyclophilins are involved in a variety of cellular functions and play an important role in various human diseases, including heart failure, arrhythmias, vascular stenosis, endothelial dysfunction, atherosclerosis, and hypertension. A better understanding of the roles cyclophilin plays in the molecular mechanisms underlying these diseases will help us develop novel pharmacological therapies.

Figure 18.1 Schematic diagram of the various domains and motifs of human cyclophilin D (CypD). MTS, mitochondrial targeting sequence; CLD, cyclophilin-like domain.

18.2.1 CypD and its Physiological Roles

CypD, also known as Cyp3, is a unique mitochondrial isoform of the mammalian cyclophilin, encoded by the *PPIF* gene. It was first identified during the screening of the human cDNA library using the human CyP cDNA (encoding hCyP1) as a probe. Western-blot analysis using hCypD-specific antiserum has shown that the expression of CypD/Cyp3 is most abundant in U937 (monocyte-macrophage cells) and PANC-1 (pancreatic carcinoma cells) extracts, followed by Jurkat (T-cell lymphoma), 293 (renal epithelial cells), and DU 145 (prostate cancer cells); it is barely detectable in HT29 (colon carcinoma cells). The full-length CypD protein has 207 amino acids (22 kDa) and includes an N-terminal 29-amino-acid mitochondrial targeting sequence, which provides it with target specificity (Figure 18.1). The *PPIF* gene is located in the 10q21-q23 region and contains six exons.

The structure of human CypD is made up of eight β-strands, two α-helices, and one 3_{10} helix (Figure 18.2) [20]. Moreover, cyclophilin structures are quite conserved, with CypD having 75% structural homology to CypA. CypD has been shown to be the mitochondrial target of the drug CsA, which is known to affect calcium flux and to induce MPTP in mitochondria. The crystal structure of CypD in complex with CsA shows that the drug inhibits PPIase enzymatic activity by binding just above the active site of CypD (Figure 18.2). CypD uses its isomerase activity for most of its mitochondrial functions, as isomerase-deficient CypD mutants are unable to rescue mitochondrial permeability transition and reactive oxygen species (ROS)-induced cell death [1,4,5]. It is postulated that CypD contributes to cell necrosis by opening the MPTP, as mitochondria isolated from CypD KO mice are resistant to mitochondrial permeability transition *in vitro*. Recent investigations have shown that CypD plays an indispensable role in

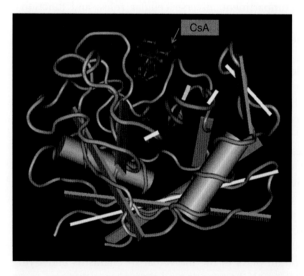

Figure 18.2 Ribbon-and-stick model showing the overall structure of cyclophilin D (CypD) in complex with cyclosporine A (CsA).

ischemia/reperfusion injury of the heart, as the size of the infarcts is reduced in mice with CypD deletion [1,21]. Thus, CypD is one of the major structural components of MPTP, and it is known to trigger the opening of MPTP in response to various stimuli, such as calcium overload and oxidative stress.

18.3 Mechanism of CypD-Dependent Necrosis

18.3.1 Mitochondria and Cell Death

Mitochondria are membranous organelles that are involved in the processes of energy production, pyrimidine synthesis, amino-acid biosynthesis, lipid biosynthesis, and calcium homeostasis, as well as other metabolic activities. Normally, they are referred to as the "powerhouse of the cell," as they are involved in the process of ATP generation through oxidative phosphorylation. However, upon ischemia – calcium overload accompanied by oxidative stress – mitochondria play a central role in the process of cell death. A typical mitochondrion contains an outer mitochondrial membrane (OMM), which is permeable to all solutes (e.g., ions and sugars), and an inner mitochondrial membrane (IMM), which is devoid of porins and is impermeable to all molecules (Figure 18.3).

Various transporters present in the IMM maintain solute homeostasis by preventing a free exchange of solute molecules between the IMM and the cytosol. Major metabolic signals that regulate cell death through mitochondrial pathways are the ATP/ADP ratio, the acetyl-CoA/CoA ratio, the oxidized/reduced nicotinamide adenine dinucleotide (NAD^+/NADH) and nicotinamide adenine dinucleotide phosphate ($NADP^+$/NADPH) ratios, glycosylated proteins, and ROS and Ca^{++} overload [22,23]. Mitochondria also harbor holocytochrome c, which is sequestered in the intermembrane space and plays an important role in many types of cell death (Figure 18.3).

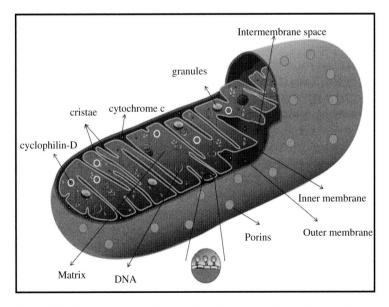

Figure 18.3 Structure of a mitochondrion. Inset shows the F_1–F_0 complexes lining the inner structure of the cristae.

Mitochondria play a key role in determining the cell fate in cases of necrosis and apoptotic cell death. Necrosis is an unorderly, fast-occurring type of cell death triggered by mechanical injury, ischemia/reperfusion injury, calcium overload, and oxidative stress. Apoptosis, on the other hand, is a highly coordinated programmed cell death (PCD) induced by intrinsic pathways or by death ligands from outside the cell (extrinsic pathway). During apoptosis, a lethal stimulus may trigger the oligomerization of Bcl-2-associated X protein (Bax) and Bcl-2 homologous antagonist killer (Bak), the main players in the process of mitochondrial outer-membrane permeability (MOMP) (Figure 18.4). On receiving this

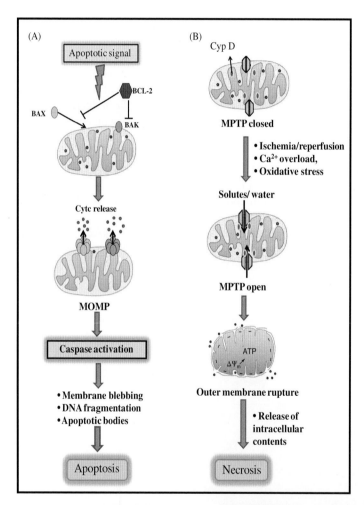

Figure 18.4 Comparative changes in mitochondria during necrosis and apoptotic cell death.
(A) During apoptotic cell death, an apoptotic stimulus induces Bax and Bak oligomerization (whereas Bcl-2 antagonizes this process). Bcl-2-associated X protein (Bax)- and Bcl-2 homologous antagonist killer (Bak)-mediated oligomerization leads to mitochondrial outer-membrane permeability (MOMP)-mediated release of cytochrome c (cyt c) from the intermembrane space into the cytosol. Activation of caspases by this cyt c release triggers apoptotic cell death. (B) Necrotic cell death is characterized by mitochondrial permeability transition pore (MPTP) opening by cyclophilin D (CypD), leading to an influx of water and solutes into the mitochondria. This disruption of ion and solute balance leads to necrotic cell death.

lethal stimulus, mitochondria undergo a MOMP-mediated release of cytochrome c (cyt c), a process regulated by the Bcl-2 family of proteins (Figure 18.4), leading to an amplification of the caspase cascade. Importantly, the functions of the IMM are not affected during MOMP. Activation of caspases culminates in an apoptotic cell death involving membrane blebbing, DNA fragmentation, and the formation of apoptotic bodies. Most importantly, in apoptotic cell death, there is no plasma membrane rupture and the apoptotic bodies are cleared off by phagocytes.

During mitochondrial permeability transition involving necrotic cell death, there is an osmotic swelling of the mitochondrial matrix, which causes mechanical breakdown of the mitochondrial outer membrane (Figure 18.4) [4]. During necrosis, opening of the MPTP leads to flocculent mitochondria and disruption of ion and solute homeostasis, which further leads to disintegration of the cell and the release of intracellular components into its surroundings (Figure 18.4). This release can damage bystander cells. Although the molecular composition of the MPTP is not clearly defined, CypD is one of the most prominent players in mitochondrial permeability transition during necrosis.

18.3.2 Mitochondrial Permeability Transition during Necrosis

Mitochondrial permeability transition is defined as an abrupt increase in the permeability of the IMM, resulting in the dissipation of the chemiosmotic gradient across it. This is mediated by the opening of a channel-like large conductance pore spanning both the IMM and the OMM, known as the MPTP (Figures 18.4 and 18.5). The MPTP is a channel in the IMM that allows molecules less than 1.5 kDa to pass through it in a nonspecific manner. Calcium overload and oxidative stress, along with inducers like inorganic phosphate and thiols, trigger the opening of the MPTP, resulting in the collapse of the transmembrane potential (Figure 18.5) [24–26]. Such conditions are known to prevail during ischemic/reperfusion injury, leading to the opening of the MPTP [27]. Moreover, opening of the MPTP leads to dissipation of the pH gradient and the membrane potential, culminating in osmotic expansion of the mitochondrial matrix. In the absence of transmembrane potential, mitochondria are unable to synthesize ATP

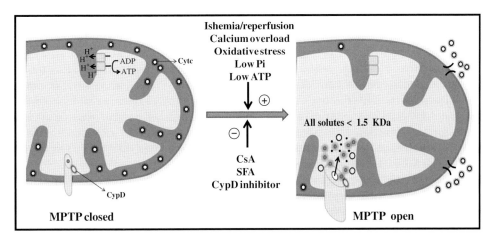

Figure 18.5 The mitochondrial permeability transition pore (MPTP) and the consequences of its opening. CsA, cyclosporine A; CypD, cyclophilin D; cyt c, cytochrome c; Pi, inorganic phosphate; SFA, sanglifehrin A.

and the F_0–F_1 ATPase runs in reverse, hydrolyzing ATP in a futile effort to restore the potential (Figure 18.5). This rapid breakdown of ATP generated by glycolysis and the remaining competent mitochondria results in ATP depletion, affecting the activity of the Na+/K+ ATPase, and disrupts the integrity of the plasma membrane. The opening of the MPTP leads to disruption of ion and solute homeostasis across the plasma membrane, culminating in necrotic cell death distinct from apoptosis (Figure 18.5).

18.3.3 Role of CypD in the Opening of the MPTP

CypD protein is encoded by the *PPIF* gene, with the full-length protein having 22 kDa molecular weight and 207 amino acid residues [28]. Upon its import into the mitochondria, the targeting sequence is cleaved, resulting in an 18 kDa mature isoform residing in the mitochondrial matrix, which accounts for all the CypD functions [12]. CypD is the most studied component of the MPTP, and plays a defining role as a regulator of both MPTP and cell death. Along with Ca^{++}, it plays an important role in the process of mitochondrial permeability transition, opening of MPTP and necrotic cell death. However, in cancer cells, CypD suppresses apoptosis by stabilizing the binding of hexokinase-II to the OMM. This stabilized binding of hexokinase-II to the outer membrane is postulated to interfere with Bax-mediated apoptotic cascades [29]. Thus, CypD-induced mitochondrial permeability transition plays an important role in necrotic cell death but not in apoptotic cell death [1]. The cyclophilin inhibitor drug CsA has been found to inhibit Ca^{++}-induced formation of MPTP [30]. Further studies have shown that CsA prevents MPTP formation by binding to CypD, inhibiting its peptidyl proline cis-trans-isomerase activity [31,32]. Various pathophysiological conditions involving ischemic and hyperfusion injury cause the opening up of the MPTP, and it has been shown that CypD KO mice have resistance to ischemia/reperfusion-induced cell death *in vivo*. CypD-induced MPTP is required for Ca^{++} overload and ROS-induced cell death, but not for the apoptotic cell death regulated by the Bcl-2 family members. It is postulated that CypD regulates necrotic cell death by regulating the threshold for the opening of the putative MPTP [1,2]. Studies involving CypD-interacting proteins show that it interacts with adenosine nucleotide translocator (ANT) in both detergent extracts and purified form. It was hypothesized in the initial working models that ANT from the IMM and voltage-dependent anion channel (VDAC) from the OMM, together with other mitochondrial matrix proteins, constitute the MPTP, where CypD plays a regulatory role. However, genetic knockout studies have shown that MPTP is found in mitochondria deficient in VDAC and ANT [33–38]. Therefore, VDAC is not a constituent of the MPTP – but there is still doubt about ANT, because all its isoforms were not knocked out and it may be possible that some other isoform can execute the pore-forming function [39]. Studies show that phosphate carrier (PiC) in the IMM interacts with CypD and constitutes an important component of the MPTP. The role of PiC in the MPTP is further supported by the finding that phosphate is required for the inhibition of MPTP opening by CsA or CypD knockdown [40]. Moreover, co-immunoprecipitation and GST-CypD pulldown experiments confirm the CsA-dependent binding of PiC to CypD. However, experiments in cell lines and mice involving PiC deletion throw into doubt the role of PiC in MPTP opening [41,42].

The F_0–F_1 ATPase is a 600 kDa complex tethered to the mitochondrial inner membrane. Most recent evidence suggests that the MPTP pore is formed by dimers of F_0–F_1 ATP synthase [43]. CypD binds the oligomycin sensitivity-conferring protein

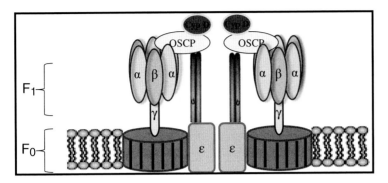

Figure 18.6 Interaction between oligomycin sensitivity-conferring protein (OSCP) and cyclophilin D (CypD).

(OSCP) subunit of the F_0-F_1 ATPase, which aids in the opening of the MPTP (Figure 18.6). Importantly, this interaction of CypD with the ATPase complex is favored by inorganic phosphate and competed by CsA [4,44]. However, even in the presence of CsA or CypD ablation, MPTP pore opening can be induced by higher Ca^{++} loading. The composition of the MPTP is not comprehensively known, and it is suggested that there may be numerous membrane proteins involved in its formation [45]. Thus, CypD facilitates the opening of the MPTP by means of a conformational change in a membrane protein in the presence of calcium and oxidative stress. It also plays a comprehensive role in the regulation of MPTP opening and the resultant necrosis by means of its enzymatic activity. It has pleiotropic effects in addition to its role in MPTP regulation. A recent metabolomic study of hearts from CypD KO mice found changes in branched-chain amino-acid metabolism, pyruvate metabolism, and the citric-acid cycle. These changes can be explained by the finding that alterations in mitochondrial acetylome are observed in CypD KO mice. Importantly, changes in the acetylation patterns in the mitochondria are known to affect mitochondrial metabolism [3,43,46]. Recently, CypD has been shown to interact with mitochondrial transcription factors and regulate the expression of mitochondrial genes [43]. It is speculated that some of the changes in mitochondrial metabolism are directly associated with the CypD alteration of the metabolic genes. In summary, CypD has pleiotropic effects in the regulation of mitochondrial metabolism apart from its central role in mitochondrial permeability transition aided by the opening of the MPTP. All the processes regulated by CypD play a significant part in mitochondrial permeability transition-aided necrotic cell death. Apart from myocardial infarction, CypD is also important in a plethora of diseases, such as neurodegeneration, muscle dystrophy, AD, PD, multiple sclerosis, and aging [47–53].

18.4 Role of CypD in Diseases

18.4.1 Myocardial Infarction

Mitochondria play a pivotal role in determining the survivability of cardiomyocytes exposed to ischemic/reperfusion injury. In many pathological conditions, including myocardial infarction and cardiac surgery, there is a deprivation of blood to the heart (or part of it), leading to an ischemic injury. Ischemia leads to oxygen deprivation of the cardiac tissue, which affects the production of ATP by the oxidative phosphorylation pathway in the mitochondria. Moreover, cardiac mycocytes then switch to anaerobic glycolysis to make up for the reduction in ATP, but rapid accumulation of H^+ and reduced nicotinamide adenine dinucleotide NADH inhibits the activity of glyceraldehyde phosphate dehydrogenase. Reduced ATP production affects the Na^+ and Ca^{++} homeostasis and increased osmolarity in the cardiac myocytes. Mostly, prolonged ischemic injury to the heart leads to the development of necrosis due to mitochondrial dysfunction caused by the opening of the MPTP. Therefore, restoration of the blood supply to the ischemic tissue is required in order to salvage the heart. However, the restoration of the blood supply further exacerbates damage due to the phenomena of reperfusion injury. Return of oxygen after ischemia tends to resume respiration, but at the same time it induces bursts of ROS production by complex I, complex III, and monoamine oxidases [54]. Moreover, damaged mitochondria during ischemia/reperfusion injury produce more ROS [25,55,56], which can mediate Ca^{++} overloading by causing extracellular Ca^{++} influx due to lipid peroxidation and to opening of voltage-sensitive Ca^{++} channels and release of Ca^{++} sequestered from the intracellular stores. Therefore, multiple factors in ischemia/reperfusion injury facilitate MPTP opening, including Ca^{++} overloading, increased inorganic phosphate levels, ATP depletion, and increased ROS generation. Apart from ROS and Ca^{++}, nitric oxide (NO) is another very significant player in reperfusion injury. Its primary target is guanylyl cyclase, which converts guanosine triphosphate (GTP) into secondary messenger cyclic guanosine monosphate (cGMP). Major actions of NO are mediated by two protein kinases: cGMP-dependent protein kinase I (PKGI) and protein kinase A (PKA). Physiologically, NO at lower concentrations exhibits a protective effect, but at higher concentration during reperfusion injury it mediates a damaging role [57,58]. NO reacts with superoxide to form peroxynitrite and contribute to oxidative stress. Peroxynitrite formation is known to inhibit oxidative phosphorylation and activate mitochondrial permeability transition [59–61]. The alteration of the metabolic and ion homeostasis during ischemia/reperfusion injury is known to favor MPTP opening. However, reperfusion rather than ischemia is known to favor the opening of MPTP [27]. Therefore, protection of cardiac myocytes against ischemic/reperfusion injury by inhibition of MPTP formation via inhibition of CypD is a potential therapy. GSK-3β is a potential candidate for such inhibition of MPTP, as GSK-3β inhibition increases the MPTP threshold in isolated cardiomyocytes [62]. CypD is one of the most important factors involved in the MPTP opening during ischemia/reperfusion injury. The role of CypD in myocardial necrosis following ischemia/reperfusion injury is validated by the finding from animal studies that CypD inhibitors such as CsA, NIM811, and genetic ablation of CypD reduce infarct size [1,63]. It has been shown that CypD KO mice have greater cardiac hypertrophy and fibrosis and lesser myocardial function in response to pressure-overload stimulation as compared to control mice. Additionally, the symptoms of progressive cardiac disease were rescued upon the transgene expression of CypD in CypD KO mice [12,64]. It was observed that the cardiomyocytes of CypD KO mice had considerable metabolic reprogramming, where the shift from oxidative phosphorylation to glycolysis was one of the reasons for the progressive heart disease. Moreover, studies involving primary fibroblast and hepatocytes isolated from CypD-null mice have shown increased resistance of the cells to Ca^{++} overload and hydrogen peroxide-induced oxidative stress. However, this resistance is circumvented by even higher Ca^{++} overload. Inhibition of CypD is considered one of the therapeutic approaches in palliating the effects of myocardial infarction and cardiac heart diseases, while CsA and its derivatives are possible medications for the treatment and alleviation of the symptoms associated with myocardial infarction (Figure 18.7).

18.4.2 Neurological Disorders

CypD plays an instrumental role in the mitochondrial permeability transition-mediated changes in the mitochondrial structure of the neurons in the brain. Over time, these alterations can result in changes in mitochondrial function and lead to the development of aging and neurodegenerative diseases like AD, PD, and amyotrophic lateral sclerosis (ALS). CypD KO mice develop normally and show no defects in the brain or cerebrovasculature [65], validating the idea that CypD does not play a role in apoptotic cell death, since apoptosis plays an indispensable role in the process of embryonic development. However, CypD KO C57BL/6J mice showed higher levels of anxiety and emotionality as compared to wild-type. Moreover, it was observed that CypD ablation in the mice resulted in better learning and avoidance of memory tasks, such as active and

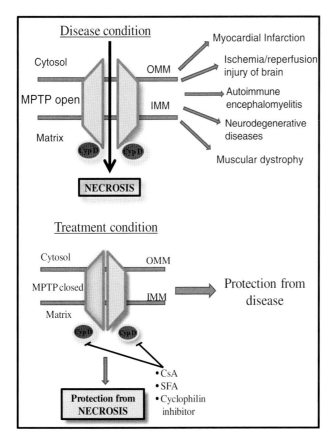

Figure 18.7 Proposed model showing the role of CypD in MPTP pore opening and the resultant necrosis in the progression of many diseases. Inhibition of CypD and various CypD inhibitors can be used to prevent necrosis-mediated damage and to protect against neurodegenerative disease and cell death in myocardial infarction.

passive avoidance, while increased incidence of diabetes was observed in mice having CypD deletion [53]. Thus, CypD plays a fundamental role in the basal neurological functions, the control of calorie balance, and adipogenesis. It mediates mitochondrial dysfunction by opening the MPTP, and it contributes to synaptic and neuronal damage. Therefore, it is a potential target for the treatment of many neurodegenerative disorders.

AD is an adult-onset neurodegenerative disorder exhibiting progressive deterioration in cognitive function and delusions. An age-related progressive deposition of amyloid beta (Aβ) protein in the mitochondria, linked to mitochondrial dysfunction and neuronal stress, has been observed in AD mouse models. Studies show that Aβ protein mediates neurotoxicity effects by means of oxidative stress, mitochondrial swelling, and intracellular calcium perturbations [66,67]. Moreover, it tends to accumulate within mitochondria and to contribute to mitochondrial stress and mitochondrial damage by dysregulation of the MPTP. MPTP opening is aided by oxidative stress and Ca^{++} overload, and once formed, MPTP can further aggravate oxidative stress and calcium perturbations, leading to mitochondrial and neuronal damage. CypD is a prominent player in MPTP, opening and its levels are found to be elevated in AD-affected regions. An interaction of CypD with Aβ has been observed by surface-plasmon resonance using *in vitro* protein–protein studies. Importantly, such an interaction has been confirmed in the brain tissues of patients affected with AD by co-immunoprecipitation [49]. Moreover, CypD null mice were protected from Aβ- and oxidative stress-induced cell death and had better behavioral and synaptic function as compared to controls.

Blockade of CypD in even aged mouse models confers protection against Aβ-mediated toxicity [50]. Similar results have been observed in the mitochondria of an AD mouse model treated with CsA (a CypD inhibitor).

PD is another prominent neurodegenerative disease in which CypD is thought to play an important role. It is a chronic neurodegenerative disorder characterized by the progressive death and degeneration of the dopaminergic neurons in the substantia nigra, with many mutations observed in the mitochondrial DNA of these neurons. The neuronal death and degeneration in PD are associated with Ca^{++} overload in the mitochondria and further opening of the MPTP. Experiments with isolated brain mitochondria show that CypD ablation results in reduced injury to the mitochondria by Ca^{++} overload. Moreover, CypD KO mice show better handling of Ca^{++} overload, lower ROS production, and a higher threshold for mitochondrial permeability transition [51]. Therefore, targeting of CypD is a potential therapeutic approach for the treatment of neurodegenerative diseases (Figure 18.7).

18.5 CypD Inhibitors

Genetic knockout studies using mice have proved a critical role of CypD in necrotic cell death. Therefore, CypD blocking for therapeutic purposes can be beneficial for many diseases where oxidative stress and necrosis play a significant role in mitochondrial dysfunction and concomitant cell death. Drugs like CsA, isolated from the fungus *Tolypocladium inflatum*, are the most potent inhibitors of CypD. They work by inhibiting its PPIase activity [64,68]. However, there are limitations to the use of CsA, as it is known to inhibit calcineurin and is unable to cross the blood–brain barrier (BBB) (Table 18.2) [37]. Moreover, it is active only within a narrow concentration range, with a decline in its protective effects observed at higher concentrations [27,69,70]. The

Table 18.2 **Cyclophilin D (CypD) inhibitors.** CsA, cyclosporine A; BBB, blood–brain barrier; HCV, hepatitis C virus.

Drug	Company	Nature	Clinical trials	Side effects/ limitations of usage	References
CsA	Sandoz/ hexal, Abbott	Cyclic peptide	Phase II/III in traumatic brain injury, cardiac hypertrophy	Immunosuppressant, shows nephrotoxicity and hepatotoxicity, unable to cross BBB, shows activity within a narrow cocncentration range only	[37,78,79]
Alisporivir/ Debio025/ DEB025	Debiopharm group	CsA derivative	Phase I/II trials for HCV treatment, Duchhene muscular dystrophy	Abdominal pain, hotness, vomiting, fatigue, pyrexia	[80,81]
NIM811	Novartis	CsA derivative	Trials for HCV treatment	Nephrotoxicity, hepatotoxicity	[82]
Scy-635	Scynexis	CsA derivative	Phase II trials for HCV treatment	Nephrotoxicity, hepatotoxicity	[83]
Sanglifehrin A	Sigma, Novartis	Nonribosomal polyketide	Animal models for ischemic heart diseases	Low bioavailability, low solubility, immunosuppresant	[73,75]
F680 and F684	INSERM (Institut national de la santé et de la recherche médicale), France	Peptide-based inhibitors	HCV treatment	-	[71,72]

immunosuppressive effect of calcineurin inhibition by CsA prevents any meaningful usage in the clinical setting for the treatment of diseases. Therefore, more selective cyclophilin inhibitors with non-immunosuppressive effects and a slight modification to the CsA backbone have been produced, such as alisporivir (DEB025), NIM811, and SCY-635 [71,72]. Alisporivir, for example, binds to CsA and forms a complex that doesn't bind to calcineurin due to the presence of the N-ethyl-valine at position 4 in its cyclic peptide (Table 18.2). None of these inhibitors has any immunosuppressive effects associated with calcineurin binding, and they can induce MPTP opening and cell death via CypD binding. Currently, however, CsA and its derivatives have limited use in clinical settings because they produce severe side effects, such as nephrotoxicity, neurotoxicity, and hepatotoxicity. Therefore, there is a need to develop new molecules with higher bioavailability and only limited side effects.

Recently, new cyclophilin inhibitors called sanglifehrins have been discovered that are not derived from CsA. Sanglifehrins are naturally occurring, cyclophilin-binding mixed nonribosomal peptides/polyketides produced by the actinomycetes strain *Streptomyces* A92-308110 [73]. High-resolution crystallization data have revealed that sanglifehrins embed deeply into the same site as CsA and enjoy direct interactions with cyclophilins. They exert their immunosuppressive effects in a manner independent of the calcineurin pathway. Importantly, their mode of action is different from that of all other known immunophilins [74,75]. Therefore, new non-immunosuppressive analogs of sanglifehrin A have been synthesized to make use of sanglifehrins in clinical settings [76]. Moreover, preclinical characterization of sanglifehrins has demonstrated their low oral bio-availability and low solubility, stressing the need for novel analogs with better solubility and pharmacokinetic characteristics [73]. In order to address these shortcomings, novel peptide-based inhibitors have been identified using a fragment-based drug-design approach. Inhibitors like F680 and F684 have shown promise, as they inhibit a spectrum of cyclophilins and show limited cytotoxicity [71]. Novel CypD inhibitors derived from quinoxaline have been shown to inhibit mitochondrial swelling and can be used as potential MPTP inhibitors [77]. CypD is a promising target for the purpose of therapeutic intervention in a plethora of diseases. However, there are many challenges in the development of novel analogues and the discovery of new drug molecules that are non-immunosuppressive, exhibit lesser side effects, have better pharmacokinetics, and have higher bioavailability.

18.6 Future Prospects

CypD is a known regulatory component of the MPTP that regulates mitochondrial membrane potential, as catastrophic opening of the MPTP results in necrotic cell death. Although all the constituents of the MPTP are not yet defined, it is widely accepted that inhibition or genetic deletion of CypD prevents cell death in the setting of ischemia/reperfusion injury and other pathological disorders. As and when other components of the MPTP are better defined in the experimental setting, the scope of therapeutic intervention can be expanded, but for now, CypD is the most promising potential candidate for the purpose of therapeutic intervention strategies. Studies have revealed that genetic knockout of CypD results in pleiotropic changes in mitochondrial metabolism. Cardiomyocytes from CypD KO mice show changes in branched-chain amino-acid metabolism, pyruvate metabolism, and the Krebs cycle. Metabolic reprogramming and the shift from oxidative phosphorylation to glycolysis upon CypD ablation may be other facets of cyclophilin function under various clinicopathological conditions. It has been reported that a reduction in the levels of CypD in mouse embryonic fibroblasts (MEFs) results in increased cell proliferation and enhanced cell migration and invasion [84]. It is postulated that CypD may have a potential role in mitochondria-to-nuclei interorganelle signaling, as it influences STAT3-dependent increased cell proliferation and chemokine-directed cell motility. Therefore, CypD may play other important roles in addition to its regulation of MPTP opening in the mitochondria. Importantly, recent studies suggest that in addition to its role in necrotic cell death, CypD might potentially be involved in the processes of tumorigenesis and autophagy.

There is an urgent need for novel drugs targeting CypD function in many pathological conditions. Moreover, there is a need for a better understanding of the other components of the MPTP, so that effective therapies can be designed. Currently, the CypD inhibitors cannot be used for the purpose of therapeutic intervention because of their associated side effects. Most importantly, there are concerns over bioavailability and immunosuppression that render them ineffective for any meaningful use in clinics. Therefore, novel analogs with better pharmacokinetics, greater bioavailability, and reduced side effects need to be designed. Moreover, most CypD inhibitors have a large molecular weight and are unable to cross the BBB. There is thus a need for small, low-molecular-weight compounds that can be used to target CypD and improve mitochondrial function in many associated disorders.

References

1 Baines CP, Kaiser RA, Purcell NH, Blair NS, Osinska H, Hambleton MA, et al. Loss of cyclophilin D reveals a critical role for mitochondrial permeability transition in cell death. *Nature* 2005;**434**(7033):658–62.
2 Nakagawa T, Shimizu S, Watanabe T, Yamaguchi O, Otsu K, Yamagata H, et al. Cyclophilin D-dependent mitochondrial permeability transition regulates some necrotic but not apoptotic cell death. *Nature* 2005;**434**(7033):652–8.
3 Menazza S, Wong R, Nguyen T, Wang G, Gucek M, Murphy E. CypD−/− hearts have altered levels of proteins involved in Krebs cycle, branch chain amino acid degradation and pyruvate metabolism. *J Mol Cell Cardiol* 2013;**56**:81–90.
4 Giorgio V, Soriano ME, Basso E, Bisetto E, Lippe G, Forte MA, Bernardi P. Cyclophilin D in mitochondrial pathophysiology. *Biochim Biophys Acta* 2010;**1797**(6–7):1113–18.
5 Perrucci GL, Gowran A, Zanobini M, Capogrossi MC, Pompilio G, Nigro P. Peptidyl-prolyl isomerases: a full cast of critical actors in cardiovascular diseases. *Cardiovasc Res* 2015: **106**(3):353–64.
6 Stocki P, Chapman DC, Beach LA, Williams DB. Depletion of cyclophilins B and C leads to dysregulation of endoplasmic reticulum redox homeostasis. *J Biol Chem* 2014;**289**(33):23 086–96.
7 Bram RJ, Crabtreet GR. Calcium signalling in T cells stimulated by a cyclophilin B-binding protein. *Nature* 1994;**371**(6495):355–8.
8 Barnes AM, Carter EM, Cabral WA, Weis M, Chang W, Makareeva E, et al. Lack of cyclophilin B in osteogenesis imperfecta with normal collagen folding. *N Engl J Med* 2010;**362**(6):521–8.
9 Rycyzyn MA, Reilly SC, O'Malley K, Clevenger CV. Role of cyclophilin B in prolactin signal transduction and nuclear retrotranslocation. *Mol Endocrinol* 2000;**14**(8):1175–86.
10 Sherry B, Yarlett N, Strupp A, Cerami A. Identification of cyclophilin as a proinflammatory secretory product of lipopolysaccharide-activated macrophages. *Proc Natl Acad Sci USA* 1992;**89**(8):3511–15.
11 Montague JW, Hughes FM, Cidlowski JA. Native recombinant cyclophilins A, B, and C degrade DNA independently of peptidylprolyl cis-trans-isomerase activity potential roles of cyclophilins in apoptosis. *J Biol Chem* 1997;**272**(10):6677–84.
12 Elrod JW, Molkentin JD. Physiologic functions of cyclophilin D and the mitochondrial permeability transition pore. *Circ J* 2013;**77**(5):1111–22.

13 Anderson M, Fair K, Amero S, Nelson S, Harte PJ, Diaz MO. A new family of cyclophilins with an RNA recognition motif that interact with members of the trx/MLL protein family in Drosophila and human cells. *Dev Genes Evol* 2002;**212**(3):107–13.

14 Chen S, Zhang M, Ma H, Saiyin H, Shen S, Xi J, et al. Oligo-microarray analysis reveals the role of cyclophilin A in drug resistance. *Cancer Chemother Pharmacol* 2008;**61**(3):459–69.

15 Pratt WB, Galigniana MD, Harrell JM, DeFranco DB. Role of hsp90 and the hsp90-binding immunophilins in signalling protein movement. *Cell Signal* 2004;**16**(8):857–72.

16 Ratajczak T, Ward BK, Minchin RF. Immunophilin chaperones in steroid receptor signalling. *Curr Top Med Chem* 2003;**3**(12):1348–57.

17 Luu TC, Bhattacharya P, Chan WK. Cyclophilin-40 has a cellular role in the aryl hydrocarbon receptor signaling. *FEBS Lett* 2008;**582**(21):3167–73.

18 Anderson SK, Gallinger S, Roder J, Frey J, Young HA, Ortaldo JR. A cyclophilin-related protein involved in the function of natural killer cells. *Proc Natl Acad Sci USA* 1993;**90**(2):542–6.

19 Wang P, Heitman J. The cyclophilins. *Genome Biol* 2005;**6**(7):226.

20 Kajitani K, Fujihashi M, Kobayashi Y, Shimizu S, Tsujimoto Y, Miki K. Crystal structure of human cyclophilin D in complex with its inhibitor, cyclosporin A at 0.96-Å resolution. *Proteins* 2008;**70**(4):1635–9.

21 Basso E, Fante L, Fowlkes J, Petronilli V, Forte MA, Bernardi P. Properties of the permeability transition pore in mitochondria devoid of Cyclophilin D. *J Biol Chem* 2005;**280**(19):18 558–61.

22 Halestrap A. Calcium, mitochondria and reperfusion injury: a pore way to die. *Biochem Soc Trans* 2006;**34**(2):232–7.

23 Miura T, Tanno M. Mitochondria and GSK-3β in cardioprotection against ischemia/reperfusion injury. *Cardiovasc Drugs Ther* 2010;**24**(3):255–63.

24 Juhaszova M, Wang S, Zorov DB, Nuss HB, Gelichmann M, Mattson MP, Sollott SJ. The identity and regulation of the mitochondrial permeability transition pore. *Ann NY Acad Sci* 2008;**1123**(1):197–212.

25 Halestrap AP, Pasdois P. The role of the mitochondrial permeability transition pore in heart disease. *Biochim Biophys Acta* 2009; 1787 (11):1402–15.

26 Crompton M. The mitochondrial permeability transition pore and its role in cell death. *Biochem J* 1999;**341**:233–49.

27 Halestrap AP, Clarke SJ, Javadov SA. Mitochondrial permeability transition pore opening during myocardial reperfusion – a target for cardioprotection. *Cardiovasc Res* 2004;**61**(3):372–85.

28 Gutiérrez-Aguilar M, Baines CP. Structural mechanisms of cyclophilin D-dependent control of the mitochondrial permeability transition pore. *Biochim Biophys Acta* 2015;**1850**(10):2041–7.

29 Machida K, Ohta Y, Osada H. Suppression of apoptosis by cyclophilin D via stabilization of hexokinase II mitochondrial binding in cancer cells. *J Biol Chem* 2006;**281**(20):14 314–20.

30 Crompton M, Ellinger H, Costi A. Inhibition by cyclosporin A of a Ca2+-dependent pore in heart mitochondria activated by inorganic phosphate and oxidative stress. *Biochem J* 1988;**255**:357–60.

31 Tanveer A, Virji S, Andreeva L, Totty NF, Hsuan JJ, Ward JM, Crompton M. Involvement of cyclophilin D in the activation of a mitochondrial pore by Ca2+ and oxidant stress. *Eur J Biochem* 1996;**238**(1):166–72.

32. Connern CP, Halestrap AP. Purification and N-terminal sequencing of peptidyl-prolyl cis-trans-isomerase from rat liver mitochondrial matrix reveals the existence of a distinct mitochondrial cyclophilin. *Biochem J* 1992;**284**:381–5.
33. Zheng Y, Shi Y, Tian C, Jiang C, Jin H, Chen J, et al. Essential role of the voltage-dependent anion channel (VDAC) in mitochondrial permeability transition pore opening and cytochrome c release induced by arsenic trioxide. *Oncogene* 2004;**23**(6):1239–47.
34. Pestana CR, Silva CH, Pardo-Andreu GL, Rodrigues FP, Santos AC, Uyemura SA, Curti C. Ca 2+ binding to c-state of adenine nucleotide translocase (ANT)-surrounding cardiolipins enhances (ANT)-Cys 56 relative mobility: a computational-based mitochondrial permeability transition study. *Biochim Biophys Acta* 2009;**1787**(3):176–82.
35. Halestrap A. Biochemistry: a pore way to die. *Nature* 2005;**434**(7033):578–9.
36. Kokoszka JE, Waymire KG, Levy SE, Sligh JE, Cai J, Jones DP, et al. The ADP/ATP translocator is not essential for the mitochondrial permeability transition pore. *Nature* 2004;**427**(6973):461–5.
37. Rao VK, Carlson EA, Yan SS. Mitochondrial permeability transition pore is a potential drug target for neurodegeneration. *Biochim Biophys Acta* 2014;**1842**(8):1267–72.
38. Woodfield K, Ruck A, Brdiczka D, Halestrap A. Direct demonstration of a specific interaction between cyclophilin-D and the adenine nucleotide translocase confirms their role in the mitochondrial permeability transition. *Biochem J* 1998;**336**:287–90.
39. Borutaite V. Mitochondria as decision-makers in cell death. *Environ Mol Mutagen* 2010;**51**(5):406–16.
40. Leung AW, Varanyuwatana P, Halestrap AP. The mitochondrial phosphate carrier interacts with cyclophilin D and may play a key role in the permeability transition. *J Biol Chem* 2008; 283 (39):26 312–23.
41. Varanyuwatana P, Halestrap AP. The roles of phosphate and the phosphate carrier in the mitochondrial permeability transition pore. *Mitochondrion* 2012;**12**(1):120–5.
42. Kwong JQ, Davis J, Baines CP, Sargent MA, Karch J, Wang X, et al. Genetic deletion of the mitochondrial phosphate carrier desensitizes the mitochondrial permeability transition pore and causes cardiomyopathy. *Cell Death Differ* 2014;**21**(8):1209–17.
43. Radhakrishnan J, Bazarek S, Chandran B, Gazmuri RJ. Cyclophilin-D: a resident regulator of mitochondrial gene expression. *FASEB J* 2015: **29**(7):2734–48.
44. Giorgio V, von Stockum S, Antoniel M, Fabbro A, Fogolari F, Forte M, et al. Dimers of mitochondrial ATP synthase form the permeability transition pore. *Proc Natl Acad Sci USA* 2013;**110**(15):5887–92.
45. He L, Lemasters JJ. Regulated and unregulated mitochondrial permeability transition pores: a new paradigm of pore structure and function? *FEBS Lett* 2002;**512**(1):1–7.
46. Nguyen TTM, Wong R, Menazza S, Sun J, Chen Y, Wang G, et al. Cyclophilin D modulates mitochondrial acetylome. *Circ Res* 2013;**113**(12):1308–19.
47. Millay DP, Sargent MA, Osinska H, Baines CP, Barton ER, Vuagniaux G, et al. Genetic and pharmacologic inhibition of mitochondrial-dependent necrosis attenuates muscular dystrophy. *Nat Med* 2008;**14**(4):442–7.
48. Palma E, Tiepolo T, Angelin A, Sabatelli P, Maraldi NM, Basso E, et al. Genetic ablation of cyclophilin D rescues mitochondrial defects and prevents muscle apoptosis in collagen VI myopathic mice. *Hum Mol Genet* 2009;**18**(11):2024–31.

49 Du H, Guo L, Fang F, Chen D, Sosunov AA, McKhann GM, et al. Cyclophilin D deficiency attenuates mitochondrial and neuronal perturbation and ameliorates learning and memory in Alzheimer's disease. *Nat Med* 2008;**14**(10):1097–105.

50 Du H, Guo L, Zhang W, Rydzewska M, Yan S. Cyclophilin D deficiency improves mitochondrial function and learning/memory in aging Alzheimer disease mouse model. *Neurobiol Aging* 2011;**32**(3):398–406.

51 Thomas B, Banerjee R, Starkova NN, Zheng SF, Calingasan NY, Yang L, et al. Mitochondrial permeability transition pore component cyclophilin D distinguishes nigrostriatal dopaminergic death paradigms in the MPTP mouse model of Parkinson's disease. *Antioxid Redox Signal* 2012;**16**(9):855–68.

52 Forte M, Gold BG, Marracci G, Chaudhary P, Basso E, Johnsen D, et al. Cyclophilin D inactivation protects axons in experimental autoimmune encephalomyelitis, an animal model of multiple sclerosis. *Proc Natl Acad Sci USA* 2007;**104**(18):7558–63.

53 Luvisetto S, Basso E, Petronilli V, Bernardi P, Forte M. Enhancement of anxiety, facilitation of avoidance behavior, and occurrence of adult-onset obesity in mice lacking mitochondrial cyclophilin D. *Neuroscience* 2008;**155**(3):585–96.

54 Sullivan LB, Chandel NS. Mitochondrial reactive oxygen species and cancer. *Cancer Metab* 2014;**2**(1):17.

55 von Knethen A, Callsen D, Brüne B. Superoxide attenuates macrophage apoptosis by NF-κB and AP-1 activation that promotes cyclooxygenase-2 expression. *J Immunol* 1999;**163**(5):2858–66.

56 Waypa GB, Marks JD, Mack MM, Boriboun C, Mungai PT, Schumacker PT. Mitochondrial reactive oxygen species trigger calcium increases during hypoxia in pulmonary arterial myocytes. *Circ Res* 2002;**91**(8):719–26.

57 Schulz R, Kelm M, Heusch G. Nitric oxide in myocardial ischemia/reperfusion injury. *Cardiovasc Res* 2004;**61**(3):402–13.

58 Davidson SM, Duchen MR. Effects of NO on mitochondrial function in cardiomyocytes: pathophysiological relevance. *Cardiovasc Res* 2006;**71**(1):10–21.

59 Brookes PS, Salinas EP, Darley-Usmar K, Eiserich JP, Freeman BA, Darley-Usmar VM, Anderson PG. Concentration-dependent effects of nitric oxide on mitochondrial permeability transition and cytochrome c release. *J Biol Chem* 2000;**275**(27): 20 474–9.

60 Radi R, Cassina A, Hodara R, Quijano C, Castro L. Peroxynitrite reactions and formation in mitochondria. *Free Radic Biol Med* 2002;**33**(11):1451–64.

61 Brookes PS, Darley-Usmar VM. Role of calcium and superoxide dismutase in sensitizing mitochondria to peroxynitrite-induced permeability transition. *Am J Physiol Heart Circ Physiol* 2004;**286**(1):H39–46.

62 Juhaszova M, Zorov DB, Kim SH, Pepe S, Fu Q, Fishbein KW, et al. Glycogen synthase kinase-3β mediates convergence of protection signaling to inhibit the mitochondrial permeability transition pore. *J Clin Invest* 2004;**113**(11):1535.

63 Argaud L, Gateau-Roesch O, Muntean D, Chalabreysse L, Loufouat J, Robert D, Ovize M. Specific inhibition of the mitochondrial permeability transition prevents lethal reperfusion injury. *J Mol Cell Cardiol* 2005;**38**(2):367–74.

64 Elrod JW, Wong R, Mishra S, Vagnozzi RJ, Sakthievel B, Goonasekera SA, et al. Cyclophilin D controls mitochondrial pore-dependent Ca(2+) exchange, metabolic flexibility, and propensity for heart failure in mice. *J Clin Invest* 2010; **120**(10):3680.

65 Schinzel AC, Takeuchi O, Huang Z, Fisher JK, Zhou Z, Rubens J, et al. Cyclophilin D is a component of mitochondrial permeability transition and mediates neuronal cell death after focal cerebral ischemia. *Proc Natl Acad Sci USA* 2005;**102**(34):12 005–10.

66 Tanaka S, Takehashi M, Matoh N, Iida S, Suzuki T, Futaki S, et al. Generation of reactive oxygen species and activation of NF-κB by non-Aβ component of Alzheimer's disease amyloid. *J Neurochem* 2002;**82**(2):305–15.

67 Brzyska M, Elbaum D. Dysregulation of calcium in Alzheimer's disease. *Acta Neurobiol Exp* 2003;**63**(3):171–84.

68 Takahashi N, Hayano T, Suzuki M. Peptidyl-prolyl cis-trans isomerase is the cyclosporin A-binding protein cyclophilin. *Nature* 1989;**337**(6206):473–5.

69 Di Lisa F, Menabò R, Canton M, Barile M, Bernardi P. Opening of the mitochondrial permeability transition pore causes depletion of mitochondrial and cytosolic NAD+ and is a causative event in the death of myocytes in postischemic reperfusion of the heart. *J Biol Chem* 2001;**276**(4):2571–5.

70 Griffiths EJ, Halestrap AP. Mitochondrial non-specific pores remain closed during cardiac ischaemia, but open upon reperfusion. *Biochem J* 1995;**307**:93–8.

71 Gallay PA. Cyclophilin inhibitors: a novel class of promising host-targeting anti-HCV agents. *Immunol Res* 2012;**52**(3):200–10.

72 Gallay PA. Cyclophilin inhibitors. *Clin Liver Dis* 2009;**13**(3):403–17.

73 Gregory MA, Bobardt M, Obeid S, Chatterji U, Coates NJ, Foster T, et al. Preclinical characterization of naturally occurring polyketide cyclophilin inhibitors from the sanglifehrin family. *Antimicrob Agents Chemother* 2011;**55**(5):1975–81.

74 Kallen J, Sedrani R, Zenke G, Wagner J. Structure of human cyclophilin A in complex with the novel immunosuppressant sanglifehrin A at 1.6 Å resolution. *J Biol Chem* 2005;**280**(23):21 965–71.

75 Zhang L-H, Youn H-D, Liu JO. Inhibition of cell cycle progression by the novel cyclophilin ligand sanglifehrin A is mediated through the NFκB-dependent activation of p53. *J Biol Chem* 2001;**276**(47):43 534–40.

76 Sedrani R, Kallen J, Martin Cabrejas LM, Papageorgiou CD, Senia F, Rohrbach S, et al. Sanglifehrin-cyclophilin interaction: degradation work, synthetic macrocyclic analogues, X-ray crystal structure, and binding data. *J Am Chem Soc* 2003;**125**(13):3849–59.

77 Wang F, Chen J, Shen X, Jiang HL. Novel cyclophilin D inhibitors derived from quinoxaline exhibit highly inhibitory activity against rat mitochondrial swelling and Ca2+ uptake/release. *Acta Pharmacol Sin* 2005;**26**(10):1201–11.

78 Schreiber SL, Crabtree GR. The mechanism of action of cyclosporin A and FK506. *Immunol Today* 1992;**13**(4):136–42.

79 Sullivan PG, Thompson M, Scheff SW. Continuous infusion of cyclosporin A postinjury significantly ameliorates cortical damage following traumatic brain injury. *Exp Neurol* 2000;**161**(2):631–7.

80 Reutenauer J, Dorchies O, Patthey-Vuadens O, et al. Investigation of Debio 025, a cyclophilin inhibitor, in the dystrophic mdx mouse, a model for Duchenne muscular dystrophy. *Br J Pharmacol* 2008;**155**(4):574–84.

81 Gallay PA, Lin K. Profile of alisporivir and its potential in the treatment of hepatitis C. *Drug Des Devel Ther* 2013;**7**:105–15.

82 Goto K, Watashi K, Murata T, Hishiki T, Hijikata M, Shimotohno K. Evaluation of the anti-hepatitis C virus effects of cyclophilin inhibitors, cyclosporin A, and NIM811. *Biochem Biophys Res Commun* 2006;**343**(3):879–84.

83 Hopkins S, Scorneaux B, Huang Z, Murray MG, Wring S, Smitley C, et al. SCY-635, a novel nonimmunosuppressive analog of cyclosporine that exhibits potent inhibition of hepatitis C virus RNA replication *in vitro*. *Antimicrob Agents Chemother* 2010;**54**(2):660–72.

84 Tavecchio M, Lisanti S, Lam A, Ghosh JC, Martin NM, O'Connell M, et al. Cyclophilin D extramitochondrial signaling controls cell cycle progression and chemokine-directed cell motility. *J Biol Chem* 2013;**288**(8):5553–61.

19

Role of Phospholipases in Cell Death

Manikanda Raja, Juel Chowdhury, and James A. Radosevich

Department of Oral Medicine and Diagnostic Sciences, University of Illinois at Chicago, Chicago, IL, USA

Abbreviations

AA	arachidonic acid
AD	Alzheimer's disease
CNS	central nervous system
cPLA$_2$	cytosolic phospholipase A$_2$
cyt c	cytochrome c
ECM	extracellular matrix
EL	endothelial lipase
ER	endoplasmic reticulum
ERK	extracellular signal-regulated kinase
FAK	focal adhesion kinase
FasL	Fas ligand
FasR	Fas receptor
FFA	free fatty acid
GI	gastrointestinal
GPCR	G-protein-coupled receptor
HDL	high-density lipoprotein
HL	hepatic lipase
IL	interleukin
IP$_3$	inositol 1,4,5-trisphosphate
iPLA$_2$	cytosolic Ca++ independent phospholipase A$_2$
JNK	C-Jun N-terminal kinase
LC3	light-chain 3
LDL	low-density lipoprotein
LP	lysosomal permeabilization
LPL	lipoprotein lipase
MAPK	mitogen-activated protein kinase
mitoPLD	mitochondrial phospholipase D
mPA-PLA$_1$	membrane-bound phosphatidic acid-preferring phospholipase A$_1$
mTOR	mammalian target of rapamycin
NMDAR	N-methyl D-aspartate receptor
OMM	outer mitochondrial membrane
PA	phosphatidic acid

Apoptosis and Beyond: The Many Ways Cells Die, First Edition. Edited by James Radosevich.
© 2018 John Wiley & Son Inc. Published 2018 by John Wiley & Son Inc.

PAF-AH	platelet-activating factor – acetyl hydrolases
PA-PLA$_1$	phosphatidic acid-preferring phospholipase A$_1$
PCD	programmed cell death
PD	Parkinson's disease
PI3-K	phosphoinositide 3-kinase
PKC	protein kinase C
PL	human pancreatic lipase
PLA	phospholipase A
PLB	phospholipase B
PLC	phospholipase C
PLD	phospholipase D
PLRP2	pancreatic lipase-related protein 2
PS-PLA$_1$	phosphatidyl serine-specific phospholipase A$_1$
RBC	red blood cell
ROS	reactive oxygen species
RTK	receptor tyrosine kinase
sn	site-specific number
sPLA$_2$	secreted phospholipase A$_2$
TAG	troacylglycerol
TNFα	tumor necrosis factor alpha
TNFR	tumor necrosis factor receptor

19.1 Introduction

The phospholipases are enzymes that cleave at different stereospecific sites of phospholipids. The major superfamilies are PLA, PLB, PLC, and PLD, each of which has several subfamily members. PLA can cleave a phospholipid's acyl ester bond at site-specific number 1 or 2 (sn-1 or sn-2). PLB cleaves at both sn-1 and sn-2. PLC cleaves a glycerophosphate bond, while PLD cleaves the polar head group. Each subfamily member has its own enzymatic activity, tissue distribution, physiological use, and so on.

19.1.1 PLA

PLAs are enzymes that can hydrolyze acyl ester bonds at sn-1 or sn-2. The PLAs that hydrolyze at sn-1 are designated PLA$_1$ and those that hydrolyze at sn-2, PLA$_2$.

19.1.1.1 PLA$_1$

PLA$_1$ hydrolyzes phospholipids to produce 2-acyl-lysophospholipids and fatty acids. The members of this subfamily are phosphatidyl serine-specific phospholipase A$_1$ (PS-PLA$_1$), membrane-bound phosphatidic acid-preferring phospholipase A$_1$ (mPA-PLA$_1$)a,b, hepatic lipase (HL), endothelial lipase (EL), pancreatic lipase-related protein 2 (PLRP2), phosphatidic acid-preferring phospholipase A$_1$ (PA-PLA$_1$), human pancreatic lipase (PL), and lipoprotein lipase (LPL). Generally, PA-PLA$_1$, p125, and KIA0725p are found intracellularly, while the other members are found extracellularly. From crystallographic data on extracellular PLA$_1$s, they have a N and a C terminal domain. The N terminal has the Ser-Asp-His catalytic triad, which is essential for catalytic activity. This set of lipases has three loops in its active sites – lid, β5, and β9 loop – and plays an important role in an enzyme's specificity. These lipases belong to the pancreatic lipase gene family. The N terminal is very important for lipase activity, and similar proteins with only the N terminal have been found in *Drosophila melanogaster* and other species that possess lipase activity [1].

PA-PLA$_1$ is a phosphatidic acid (PA)-sensitive phospholipase. PA is a cone-shaped phospholipid that plays a vital role in the fertility of the sperm, with data showing that it participates in mitochondrial dynamics. It has been shown that mitoPLD produces PA from cardiolipin [2]. p125 and XIAA0725p belong to the subfamily PA-PLA$_1$ and show PLA$_1$ activity in a PA-specific manner. P125 is a sec23p-interacting protein that exhibits remarkable sequence similarity with KIAA0725p, but KIAA0725p lacks the sec23p-interacting region at the N terminal of the sequence. p125 possesses the ability to produce vesicles from the endoplasmic reticulum (ER), while KIAA0725p can alter the morphology of organelles [3].

PS-PLA$_1$ and mPA-PLA$_1$α,β are different to the other enzymes in terms of their substrate specificity. They exhibit only PLA$_1$ activity, while the rest also show triacylglycerol (TAG)-hydrolyzing activity. PS-PLA$_1$ helps in blood coagulation and during phagocytosis by macrophages. mPA-PLA$_1$α participates in muscle contraction and stimulation for cell proliferation. mPA-PLA$_1$β participates in sperm fertility; it is found only in mammals. mPA-PLA$_1$β seems to participate in the progression of tumors. EL predominantly exhibits PLA$_1$ activity, whereas HL exhibits PLA$_1$ and TAG-hydrolyzing activities. As shown in Table 19.1, HL and EL participates in uptake of high-density lipoprotein (HDL), while LPL plays an important role in metabolism and the supply of lipoproteins to the endothelial lines of adipose and heart tissue. PL and PLRP2 are important in the absorption of fatty acids, by catalyzing TAGs and galactolipids from plants [1].

Table 19.1 Different types of extracellular phospholipase A$_1$ (PLA$_1$). LPL, lipoprotein lipase; HL, hepatic lipase; HDL, high-density lipoprotein; EL, endothelial lipase; PL, human pancreatic lipase; PLRP2, pancreatic lipase-related protein 2; TAG, troacylglycerol; PS-PLA$_1$, phosphatidyl serine-specific phospholipase A$_1$; mPA-PLA$_1$α, membrane-bound phosphatidic acid-preferring phospholipase A$_1$ alpha; mPA-PLA$_1$β, membrane-bound phosphatidic acid-preferring phospholipase A$_1$ beta.

Phospholipase A$_1$ (extracellular)	Tissue distribution	Function
LPL	Capillary endothelial, adipose tissue and heart	Plays role in metabolism of lipoproteins
HL	Hepatocytes	Helps in uptake of HDL
EL	Endothelial cells	Helps in uptake of HDL
PL and PLRP2	In pancreatic juices, produced by pancreatic acinar cells	Plays role in intestinal absorption of fatty lipids by hydrolyzing TAGs, helps in hydrolyzing galactolipids from plant membranes
PS-PLA$_1$	Heart, lung, and poorly expressed in platelets	Helps in blood coagulation, phagocytosis by macrophages
mPA-PLA$_1$α	Prostate, testis, ovary, colon, pancreas, kidney, and lung	Stimulation of cell proliferation, muscle contraction
mPA-PLA$_1$β	Found only in testis	Helps in sperm fertility

Source: Aoki et al. 2007 [1]. Reproduced with permission from Elsevier.

19.1.1.2 PLA$_2$

The PLA$_2$ subfamily consists of 15 groups of lipases. The most studied types are secreted phospholipase A$_2$ (sPLA$_2$), cytosolic phospholipase A$_2$ (cPLA$_2$), platelet-activating factor – acetyl hydrolase (PAF-AH), lysosomal PLA$_2$, and cytosolic Ca++ independent phospholipase A$_2$ (iPLA$_2$). As already discussed, this group of enzymes hydrolyzes the sn-2 position of phospholipids. It participates in membrane homeostasis by recycling fatty acid moieties through the liberation of arachidonic acid (AA). AA is used in the production of eicosanoids and PAF, which is a biologically active molecule.

Phospholipases have 10 groups of sPLA$_2$, of which groups I, II, III, V, X, and XII belong to mammals and the rest to reptiles and other species. These lipases are Ca^{++}-dependent and have a His/Asp dyad at the catalytically active site. This site gives the enzyme the ability to perform lipase action intracellularly and in distant cells (via transport to other cells through channels). This particular type of subfamily plays an important role in inflammatory diseases, coagulation, and exocytosis, among other things.

iPLA$_2$ belongs to group VI and has two phenotypes: group VIA and group VIB. VIA participates in membrane homeostasis via deacylation and reacylation reactions. It is found in the cytoplasm, whereas VIB is generally membrane-bound. When a cell is subjected to oxidative stress, VIA causes the release of free fatty acids (FFAs), including AA. VIA and VIB show lysophospholipase and phospholipids transacylase activity in addition to the PLA$_2$ activity, which is relatively higher. VIA have three splice variants with different anykrin repeats. iPLA$_2$ is found in tetramer form, and these anykrin repeats allow oligomerization. VIA has no substrate specificity for the fatty acid in the sn-2 or sn-3 of phospholipids, and it behaves more like PAF-AH. It has been proven that VIA can bind to calmodulin in a Ca^{++}-dependent manner and become inactivated. Caspases can cleave the sequence and enhance the biological function. For VIA activation, additional kinase is required. As a whole, iPLA$_2$ participates in cell growth, apoptosis, and chemotaxis, among other things [4].

cPLA$_2$ belongs to group IV. It has transacylase, lysophospholipase, and PLA$_2$ activity. It is regulated by Ca^{++} for translocation to the intracellular membrane, and participates in inflammation, brain injury, and polyposis, among other things. It has six variants, one of which – group IVC – lacks a C2 domain and behaves in a Ca^{++}-independent manner.

PAF-AHs are intercellular enzymes with three variants: group VIIA, group VIIB, and group VIII A and B. These lipases have the capability to hydrolyze the acetyl group from the sn-2 position of PAF. Group VIIA has the Ser-His-Asp catalytic triad, which gives a wide range of substrate specificities and provides PLA$_1$ activity. Group VIIB shares 43% identity with Group VIIA and has wide substrate specificity, but it also protects cells from undergoing apoptosis under exposure to oxidative stress [5]. Group VIII has two catalytic and one regulatory domain. Catalytic domains α1 and α2 were found to be catalytically active homo- or heterodimers. The specificity of the enzyme depends upon the type of dimer formation: if the dimer is made of two α2, then the enzyme exhibits action on PAF and 1-O-alkyl-2-acetyl-sn-glycero-3-phosphoethanolamine; if it is made of two α1 or has homodimer form it shows less activity [6]. Plasma PAF-AH has the ability to catabolize oxidized phospholipids associated with circulating low-density lipoprotein (LDL) particles [5].

Lysosmal PLA$_2$ belongs to group XV and exhibits activity similar to that of lecithin-cholesterol acyltransferase due to the catalytic triad Ser-His-Asp. It is optimally active at pH 4.5 and is co-localized with lysosomes [6].

19.1.2 PLB

PLB is a type of phospholipase that exhibits hydrolyzing capability on both sn-1 and sn-2. The gene responsible is present in the 2nd chromosome and is denoted *PLB1*. PLB has both phospholipase A_2 and lysophospholipase activity. Usually, it is synthesized as 170 kDa proenzyme, but later it turns into active 97 kDa enzyme via papain treatment. It is found in the intestine, epidermis, and sperm [7]. The *PLB* gene is one of the risk genes for rheumatoid arthritis, and much research shows that PLB1 is important for acrosome exocytosis in sperm, in order to maintain fertility [8,9].

19.1.3 PLC

The PLC family consists of six subfamilies: PLCβ, γ, δ, ε, ζ, and η [10]. The G-protein-coupled receptor (GPCR), receptor tyrosine kinase (RTK), seems to participate in the activation of PLC family members in the presence of hormones and neurotransmitters [11]. Each subfamily has more than one variant. The PLC gene, *PLCB1* in neurons can cause malignant migrating partial seizures of infancy, schizophrenia, and so on in a mutated state [12,13]. From crystallographic data, PLCs usually have X and Y domain for enzyme activity. Apart from the catalytic domains, these subfamilies usually have a Ca^{++}-dependent phoshpolipid-binding domain, as well as Src domains, which have been identified in protein–protein interaction. PLCβ has been shown to participate in differentiation by regulating cyclin D3 and other cellular processes, such as proliferation, via the protein kinase C (PKC)α-mediated pathway [14,15].

19.1.4 PLD

The PLD family consists of three types: PLD1, 2, and mito. MitoPLD was only recently discovered, and it plays an important role in mitochondrial fission and fusion reactions. PLD 1 and 2 have been shown to take part in cellular processes such as apoptosis, proliferation, cell division, cell motility, and cell survival. PLD has a domain that can make interactions with actin, which is how it helps in cell motility in its various phases (cell adhesion, contractility, actin polymerization, spread to substratum, etc.) [16]. It can hydrolyze phospholipids such as phosphatidylcholine to generate choline and PA; this reaction is highly regulated, as all cells are made of phospholipids. PLD1 localizes in the Golgi complex, secretory granules, and endosomes, PLD2 is concentrated in the plasma membrane, and mitoPLD is found on the mitochondrial surface [17]. MitoPLD plays an important role in generating PA and in suppressing transposon mobilization during meiosis in spermatogenesis [18].

19.2 Apoptosis

19.2.1 Role of PLA in Apoptosis

The major types of PLA that play a role in apoptosis are lysosomal PLA_2, $sPLA_2$, $cPLA_2$, $iPLA_2$ β, Palatin-like PLA_2, and $PS-PLA_1$, all of which have been extensively studied. Other subfamilies of PLA mechanisms are still unclear, and some do not participate in cell death.

Apoptosis induced by bacteria is mostly seen in the gastrointestinal (GI) and respiratory tracts. *Escherichia coli*-like bacteria have the ability to induce apoptosis in epithelial cells. They do this by recognizing the phosphatidyl ethanolamine on the plasma membrane and intensifying outer-leaflet expression of phosphatidyl serine. These initial steps lead to an increase in the accumulation of ceramide in the mitochondria. Ceramides are lipids, and are usually found in the intermembrane space. They cleave into sphingnosine and fatty acid, and are produced in excess when the cell undergoes apoptosis. The pathways that lead to an accumulation of ceramide involve the activation of *de novo* synthesis, inhibition of ceramide breakdown, or activation of acidic sphingomyelinase, which is an enzyme that hydrolyzes sphingomyelin to ceramide [19]. In bacterial-induced apoptosis, ceramide levels are increased due to inhibition of ceramide acylation by PLA_2-like transacylase from phosphatidyl ethanolamine; simultaneously, phosphatidyl ethanolamine acts as a substrate to PLD to catalyze the formation of PA. PA seems to equilibrate the balance between mitogenic and apoptotic responses [20]. With a greater concentration of PA in the cytosol, the balance shifts toward an apoptotic response, and vice versa. In later stages of bacterial-induced apoptosis, margination, nuclear condensation, and membrane/apoptotic blebbing are seen. In this type of bacterial-induced cellular death, the cells undergo more apoptosis, rather than necrosis; while doing *in vitro* studies, even apoptosis can look like necrosis. Thus, proper analysis is required before jumping to a conclusion regarding the kind of death pathway is being evoked.

During ischemic shock or seizures, transient release of AA from the brain is seen in the nervous system; this is termed the "Bazan effect." $cPLA_2$ is the common intercellular form of PLA_2, found extensively in the central nervous system (CNS) [21]. Prions (or excess glutamate production) are the trigger that induce $cPLA_2$ activation and neural death. For example, prions induce neurodegenerative diseases like Jakob disease and spongiform encephalitis. When they come in contact with neurons, $cPLA_2$ gets activated, and this mediates the induction of apoptosis in neural cells. The activated $cPLA_2$ is relocated to cellular neurites through β-III-tubulin and catalyzes the conversion of AA to prostaglandins [22]. Prostaglandins help in caspase 3 activation and apoptosis. $cPLA_2$ plays an important role in excitotoxicity induced by glutamate and in oxidative stress-mediated apoptosis of neurons. Oxidative agents activate $cPLA_2$, $sPLA_2$, and $iPLA_2$ in the CNS, which mediate oxidative stress-mediated cell death. $iPLA_2$ is mainly associated with the ER and outer mitochondrial membrane (OMM), which helps in release of cytochrome c (cyt c) [23]. N-methyl D-aspartate receptor (NMDAR) subtype activation triggers apoptosis in neurons through activation of NO_2 synthase and mitochondrial superoxide production. Group IIA $sPLA_2$ plays important roles in mitochondrial superoxide production, which will help in amplification of the death signal triggered by NMDAR [24].

Tumor necrosis factor alpha (TNFα) and Fas ligand (FasL) can trigger apoptosis in almost all types of cells that possess tumor necrosis factor receptor (TNFR)1/2 or Fas receptor (FasR) on their membrane; this mechanism is mediated through activation of $iPLA_2$. The apoptosis trigger produced by TNFα or FasL should have a fully functional death domain to initiate the signaling mechanism. Caspase 3 gets activated and cleaves $iPLA_2$ at Asp^{183}. Truncated $iPLA_2$ is responsible for the release of AA and the production of lysophosphatidycholine, a chemoattractant that invites phagocytic cells to remove apoptotic cells. Caspase 3 can cleave at additional sites as well, leading to different variants of $iPLA_2$ [25]. Biologically active $iPLA_2$ exists in homotetramer form, and oligomerization is an important mechanism for the regulation of iPLA$_2$. iPLA$_2$ can be regulated by oxidants and lipid metabolism, which inactivate iPLA$_2$ by altering the oligomerization of monomers. iPLA$_2$ also facilitates repair of membrane phospholipids like cardiolipin in neurons, but the repair mechanism is susceptible to reactive oxygen species (ROS)-mediated peroxidation. iPLA$_2$ produced both survival and apoptotic effects in rat adrenal medulla cells. Similarly, when the ER of β-cells was subjected to stress, it led to depletion of ER Ca++ stores and to activation of caspase 12. Activated caspase 12 activates caspase 3 and releases cyt c and AA from mitochondria. Cells with highly developed ER show greater vulnerability to ER stress, and iPLA$_2$ plays a vital role in β cell-like cells, but not in neurons. In β cell-like cells, ceramide expression keeps iPLA$_2$ in the "ON" state, instead of inactivating it in the presence of ROS [23]. iPLA$_2$ mediates ceramide production through acidic sphingomyelinase, but when iPLA$_2$VIA loses its C-terminal region, it turns violent toward the cellular membrane.

The deciding event that orchestrates a dying cell to undergo apoptosis or necrosis is totally dependent on LPLA$_2$ activation and lysosomal permeabilization (LP). LP is involved in apoptosis induction, hastening the process. It is triggered by sphingnosine or through inhibition of sphingnosine kinase 1, which leads to lysosomal damage. During lysosomal damage, sPLA$_2$ plays an important role in making the cell osmosis-sensitive through the production of AA. LP releases cathespin D and activates caspases. It can also trigger caspase-independent cell death, where the cell loses its adhesion molecules, resulting in anoikis. LP and apoptotic traits are proportionally related [26].

The sPLA$_2$ family paves the path to membrane permeablization and leads apoptotic cells to undergo phagocytosis. Healthy cells are resistant to hydrolysis by sPLA$_2$. This can be the early step that invites macrophages toward apoptotic cells. While PS-PLA$_1$ is activated during the late stages of apoptosis, in which phosphatidylserine expression is upregulated, this sends the "eat me" signal to phagocytes [25].

In normal cells, cell–extracellular matrix (ECM) interactions can lead to various different cell fates. These interactions are formed by integrin in the normal cell. Integrin's focal adhesion kinase (FAK) signaling complex has been implicated in the regulation of anchorage-dependent cell survival. When FAK is inhibited or downregulated, or when the correct ECM is absent, cells enter apoptosis through anoikis via the p53-dependent pathway, activated by PKCλ and cPLA$_2$ [27].

Cells undergo senescence over time; sPLA$_2$ plays a major role in senescence through activation of the DNA-p53-dependent pathway, which is induced by ROS accumulation [28].

19.2.2 Role of PLC in Apoptosis

The major types of PLC that play a role in apoptosis are PC-PLC, PLCγ1, PLCγ2, PLCδ1, PLCδ3, and PLCβ3. Ca^{++} transfer from the ER to the mitochondria triggers apoptotic pathways to induce the release of mitochondrial proapoptotic factors. PLC activation is required for ER Ca++ depletion and the increase of cytosolic Ca^{++} [29].

During tissue homeostasis, remodeling, and cytotoxic T-cell killing, TNFR1 and Fas/Apo-1 mediate apoptosis. When TNFR1/Apo-1 crosslinking occurs, PC-PLC gets activated and hydrolyzes phosphtidylcholine into DAG and choline. Acidic sphingnomyelinase needs DAG in order to become activated and hydrolyze membrane sphingomyelin. This leads to generation of ceramide via *de novo* synthesis. Accumulated ceramide activates mitogen-activated protein kinase (MAPK) and C-Jun N-terminal kinase (JNK), and leads to the generation of AA by activated PLA_2. Downstream molecular signals like PLA_2 (from the death domain) activate protein phosphatase 2A and cause the downregulation of c-myc. When c-myc becomes downregulated, the cell dies of apoptosis. TNFα-induced apoptosis is mediated by PC-PLC (this was quantified from acute myelogenous leukemia [30,31]). Cytotoxic T cells use this mechanism to trigger apoptosis as a humoral response.

After apoptosis, macrophages phagocytose the apoptotic bodies. In the initial stages of phagocytosis, phospahtidyl serine and RTK-like MerTK recognize the apoptotic body and tyrosine phosphorylates PI-PLC in the macrophage. Activated PI-PLC hydrolyzes phosphatidyl inositol 4,5 phosphate into DAG and inositol 1,4,5-trisphosphate (IP_3). DAG and Ca^{+2} activate PLCγ2 and become associated with MerTK. Association of PLCγ2 with MerTk is the main step in phagocytosis [32]. When this association is not achieved, apoptotic bodies can cause injury to the surrounding cells and induce autoimmunity. PLCγ1 is found in the FAK signaling complex, and its loss results in anoikis (i.e., the cell loses its ECM or adhesion molecules from the surrounding cells or the support) [33].

Loss of PLCδ1 and 3 causes severe apoptosis and dysfunctional vascularisation of the placenta in the developmental stages of the embryo [34]. Embryos with loss of PLCδ1and 3 trigger apoptosis in the cardiomyocytes of the developing fetus, which shows increased heart weight/tibial length ratios [35].

19.2.3 Role of PLD in Apoptosis

PLD comes in three types: PLD1, PLD2, and mitoPLD. It usually cleaves phosphatidylcholine to PA. Phosphatidy ethanolamine is also a substrate of PLD, as discussed earlier, and it plays a major role in giving the final tip of the balance between mitogenic and apoptotic response through PA level. PLD hydrolyzes PC into the secondary messengers, choline and PA.

When shear stress is applied to cells, they react in the following order: (i) PLD1 activation; (ii) stimulated mammalian target of rapamycin (mTOR) signaling; (iii) hypertrophy; and, later, (iv) apoptosis. Increased shear stress on plasma membrane causes hypertrophy. Cells in hypertrophy show increased c-src phosphorylation and PLD1 activation, resulting in apoptosis. This kind of apoptosis is seen in renal and endothelial cells that are in direct and regular contact with shear force exerted by fluid motion [36].

MitoPLD was only recently discovered and is found near the mitochondrial surface. It generates PA during mitochondrial fusion and plays a major role in the meiosis of spermatocytes. Mice lacking mitoPLD show atypical meiotic arrest of apoptosis. MitoPLD is also important in apoptosis and neurodegenerative diseases. High expression results in the juxtaposition of mitochondria during mitochondrial fusion and when RNAi (specific RNA of spermatocyte) causes fragmentation of mitochondria. The mitochondria of the cell that undergoes apoptosis show increased amounts of fission, rather than fusion. Mitochondrial fission plays an important role in apoptosis upstream of cyt c. Dynanin-related protein1 is a mediator of mitochondrial fission and is required for apoptosis. This kind of apoptosis usually occurs during developmental stages. MitoPLD is also the terminator of mitochondrial fusion, via recruitment of Lipin1b-like molecules that promote mitochondrial fission by converting PA into DAG.

However, in the same scenario, high expression of PLD1 is seen in mitochondrial fragments. The mechanism by which PLD1 helps in mitochondrial fragmentation is not really known. MitoPLD and PLD2 help in cell survival, while PLD1 helps in apoptosis [37,38].

19.2.4 Opposite Role: Savior of the Cell

Cells with high replication capability do not have similar effects to phospholipase in cell death. When umbilical endothelial cells are irradiated, signaling occurs from the membrane to the DNA, which induces DNA strand breaks, followed by initiation of apoptosis. On the other hand, membrane-derived signals traveling through the extracellular signal-regulated kinase (ERK) and phosphoinositide 3-kinase (PI3-K)/Akt signaling pathways enhance cell viability. cPLA$_2$ generates secondary messengers to activate the Akt and ERk1/2 pathways of cell survival. When cPLA$_2$ is inhibited, mitotic catastrophe results [39]. The iPLA$_2$δ variant is highly important during development, and its absence could trigger massive apoptosis in the developing embryo [34].

Phospholipases behave differently under the same death stimuli in cancer cells. Secretory PLA$_2$ seems to resist apoptosis in breast cancer cell lines by leading to the formation of lipid droplets. Activation of sPLA$_2$ leads to the production of lipid droplets in response to apoptosis in cancer cells [40]. PLD have been indentified to help in cancer survival by activating mTOR, stabilizing mutant p53, and contributing to the "Warburg effect." PLC- and PI3-K-dependent mechanisms are important for cancer migration and metastasis [41]. Many therapies capable of targeting PLD in order to control cancer are currently being designed.

19.3 Role of Phospholipases in Autophagy

The phospholipases that play a major role in autophagy are PLD1, iPLA2γ, cPLA$_2$, PLCγ1, PLCε, and PC-PLC. Autophagy occurs during starvation, development, and differentiation of cells. Its main purpose is to reduce the production of ROS and the release of toxic intramitochondrial proteins. Aging is the major cause, followed by physical exercise, which can trigger autophagy in muscles, β cells, and liver as a response to glucose tolerance. Autophagy is an evolutionarily conserved trait of eukaryotes that prevents apoptosis by mitochondria during starvation. Autophagic degradation of cytoplasmic components occurs in development, differentiation, and oocyte fertilization. In red blood cells (RBCs), autophagy remodels the cells and makes them ready for differentiation [42]. The best studied form of autophagy is macroautophagy, in which phospaholipase plays a central role. Phospholipase also mediates microautopahgy. Macroautophagy is induced by 30 known autophagy genes, which contribute to the initiation and completion of the autophagic vacuole [43].

During macroautophagy, Arf-6(G-protein) generates PIP$_2$ and induces PLD activity through phosphatidyl 4-phosphate 5-kinase at the plasma membrane. PLD1 hydrolyzes PC and produces PA. PLD1 is targeted to autophagosomes alongside endosomal membranes, which get excluded from preautophagosomal structures and are found in later stages of the autophagic process. PLD1 localizes to the outer membrane of amphisomes: intermediate organelles formed during autophagy through the fusion of autophagosomes and endosomes. When PLD1 is inhibited, a decrease in light-chain 3 II

(LC3-II) levels is seen [44]. Genetic destruction of iPLA2γ seems to result in enlarged and degenerated mitochondria, leading to autophagy [45]. In innate and adaptive immunity, and during the early stages of ischemia and atherosclerosis, triggered autophagy is independent of changes in mTOR and autophagic flux. PC-PLC- and cPLA$_2$-induced autophagy are ATG5-dependent. cPLA$_2$ hydrolyzes AA from the membrane phospholipids, resulting in an increase in LC3-II, which is membrane-associated. LC3-I is converted drastically to LC3-II, which is another reason for the increase in LC3-II during autophagy.

Lithium has both inhibitory and activating effects on autophagy. PLCβ plays an important role in the inhibitory effect of autophagic processes initiated by lithium through the generation of IP$_3$. Decreased levels of IP$_3$ or inositol trigger autophagy [46]. Downregulation of PLCε shows drastic autophagy due to enhanced digestion of organelles. The rate at which autophagy proceeds is proportional to the digestion of organelles [47].

When protein gets misfolded, it accumulates in the ER. To ensure quality control of proteins and for protein folding, the ER is important. Phospholipases play a major role in ER stress-induced autophagy. Different reasons for ER stress-induced autophagy can include nutrient deprivation, nutrient excess, inflammation, and alteration in the ER luminal Ca++. The ER stress causes Ca^{++} release through IP$_3$ receptor. PKCθ is a calcium-independent enzyme that plays an important role in ER stress-induced autophagy. The leaked calcium ions tend to produce indirect activation of PKCθ and demonstrate autophagy. PKCθ phosphorylation is a hallmark step that can be analyzed through microscopic techniques, where it resembles a dot structure in the cytoplasm. PLC mediates calcium indirect activation and, upon inhibition of PLC, inhibits the secondary type of programmed cell death (PCD) [48,49].

Autophagy helps some kinds of cell, such as macrophages, to survive. Bacterial organisms like *Listeria* monocytogenes have the tendency to colonize the cytoplasm of macrophages, where they co-localize along with LC3 in vacuoles. The existence of bacteria inside cells automatically triggers autophagy. These localized bacteria acquire the ability to escape into cytoplasm, by way of phospholipases like PI-PLC and PC-PLC. These bacterial phospholipases disrupt the double-membrane autophagosomes during the infection mechanism. Bacterial virulence factors are also contributed by enzyme precursor genes like *plaA* and *plaB*, which are responsible for the escape [50].

As already mentioned, autophagy is evolutionarily conserved. A peculiar study was recently conducted in which, during starvation, female mouse brain cells showed statistically greater survival compared to male cells [51]. Such starving cells use lipid droplet formation as a survival technique. The researchers used PLA$_2$ activity to study the susceptibility to autophagy. Female neurons showed greater PLA$_2$ activity than male ones. PLA$_2$ activity was analyzed because of its major products: FFAs and lysophospholipids. PKC activation reduces lysophospholipids, leading to autophagic cell death [51].

Many researchers are currently looking at induction of autophagy in cancer cells. The main idea is to induce autophagy at a specific target in order to provide therapeutic benefits while reducing collateral damage to normal cells. Mice lacking PLD1/2 show decreased levels of macroautophagy, which is an organ-catabolizing process. This process is vital for tumor-cell spread [52].

19.4 Role of Phospholipases in Necrosis

Necrosis is a major type of cell death that can occur either accidentally or as a result of some pathophysiological processes, such as glutamate excitotoxicity, ischemia/reperfusion, ROS-induced injury, and inflammation. Like, autophagy, necrosis is evolutionarily conserved. Caspases are usually not required, and the cell organelles bulge, membranes undergo damage beyond hope of repair, and there is organelle breakdown. Necrosis is regulated by genes that encode the plasma membrane and ER Ca^{++} channel, calpains, cathspins, other genes, and so on. Most of the time, it occurs as a result of ROS produced in the mitochondria. Mediators of necrosis are lipoxygenase and PLA_2 in general. Phospholipases contribute in pushing the trigger to the necrosis side by helping in organelle degradation. Some new triggers of necrosis have recently been discovered, such as snake and bee venoms and viruses.

Phospholipases participate in lipid hydroperoxidation, leading to cell organelle and membrane disruption. This is one of the important features of necrosis. Phospholipases have been proven to participate in the release of lysophospholipids and AA when a high concentration of bee-venom PLA_2 is administered to cells, leading to necrotic death. The mechanism of action of bee-venom PLA_2 is entirely dependent on interleukin (IL) receptors [53]. One good example to illustrate the effect of phosholipases on cells is given in Figures 19.1 and 19.2. When 25 µg of myotoxin II (MT-II), a Lys49 phospholipase A_2 homolog from the venom of the snake *Bothrops asper*, was administered to lymphoblastoid cells, the cells showed chromatin condensation, as shown in Figure 19.1. But when 50 µg of MT-II was administered, the cell expressed necrotic features such as plasma-membrane and mitochondrial rupture, as shown in Figure 19.2 [54]. Release of ceramide was also seen, in a caspase-independent manner, and the necrotic process was enhanced.

In TNF, chemical- and oxidant-induced necrosis and PLA activity and expression are higher when compared to normal conditions. Ceramide production also regulates the production of ROS in mitochondria, the stimulation of NOS, and lipid peroxidaiton. Ceramide plays an important role in the activation of calpains via the JNK pathway in caspase-independent forms of cell death such as necrosis. As already discussed,

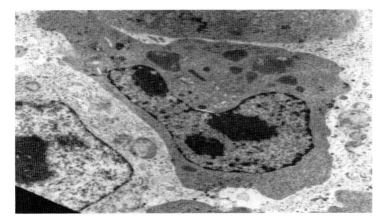

Figure 19.1 Lymphoblastoid cells administered 25 µg myotoxin II (MT-II), showing nuclear condensation. *Source:* Mora et al. 2005 [54]. Reproduced with permission from Elsevier.

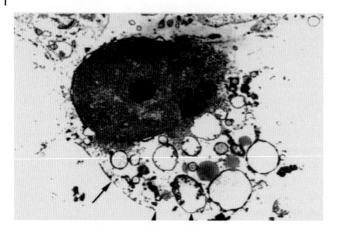

Figure 19.2 Lymphoblastoid cells administered 50 µg myotoxin II (MT-II), showing plasma-membrane and mitochondrial rupture. *Source:* Mora et al. 2005 [54]. Reproduced with permission from Elsevier.

lysosomal rupture by lysosomal phosholipases can change the form of death from apoptosis to necrosis. Release of lysosomal lipases from the ruptured lysosomes strikes every organelle without specificity and hastens the process of necrosis.

19.5 Conclusion

In the three major types of cell death discussed in this chapter, the role of phospholipases is to hydrolyze the phospholipids found in organelles and plasma membranes. PLB is not well studied, and thus we have not discussed its role here. Phospholipases make good therapeutic targets for overcoming diseases like atherosclerosis, developmental defects, cancer, neural-related diseases like schizophrenia, prion-induced encephalitis, Alzheimer's disease (AD), and Parkinson's disease (PD). All variants of phospholipase do not contribute to cell death.

References

1 Aoki J, Inoue A, Makide K, Saiki N, Arai H. Structure and function of extracellular phospholipase A1 belonging to the pancreatic lipase gene family. *Biochimie* 2007;**89**(2):197–204.
2 Baba T, Kashiwagi Y, Arimistsu N, Kogure T, Edo A, Maruyama T, et al. Phosphatidic acid (PA)-preferring phospholipase A1 regulates mitochondrial dynamics. *J Biol Chem* 2014;**289**(16):11 497–511.
3 Najajima K-I, Sonoda H, Mizoguchi T, Aoki J, Arai H, Nagahama M, et al. A novel phospholipase A1 with sequence homology to a mammalian Sec23p-interacting protein, p125. *J Biol Chem* 2002;**277**(13):11 329–35.
4 Balsinde J, Balboa MA. Cellular regulation and proposed biological functions of group VIA calcium-independent phospholipase A2 in activated cells. *Cell Signal* 2005;**17**(9):1052–62.

5 Matsuzawa A, Hattori K, Aoki J, Arai H, Inoue K. Protection against oxidative stress-induced cell death by intracellular platelet-activating factor-acetylhydrolase II. *J Biol Chem* 1997;**272**(51):32 315–20.
6 Schaloske RH, Dennis EA. The phospholipase A2 superfamily and its group numbering system. *Biochem Biophys Acta* 2006;**1761**(11):1246–59.
7 Maury E, Prévost M-C, Nauze M, Redoulès D, Tarroux R, Charvéron M, et al. Human epidermis is a novel site of phospholipase B expression. *Biochem Biophys Res Commun* 2002;**295**(2):362–9.
8 Okada Y, Diogo D, Greenberg JD, Mouassess F, Achkar WA, Fulton RS, et al. Integration of sequence data from a consanguineous family with genetic data from an outbred population identifies PLB1 as a candidate rheumatoid arthritis risk gene. *PLoS One* 2014;**10**(9):e87645.
9 Asano A, Nelson-Harrington JL, Travis AJ. Phospholipase B. Is activated in response to sterol removal and stimulates acrosome exocytosis in murine sperm. *J Biol Chem* 2013;**288**(39):28 104–15.
10 Gresset A, Sondek J, Harden TK. The phospholipase C isozymes and their regulation. *Subcell Biochem* 2012;**58**:61–94.
11 Philip F, Sahu S, Caso G, Scarlata S. Role of phospholipase C-β in RNA interference. *Adv Biol Requl* 2013;**53**(3):319–30.
12 De Filippo MR, Rizzo F, Marchese G, Giurato G, Nassa G, Ravo M, et al. Lack of pathogenic mutations in six patients with MMPSI. *Epilepsy Res* 2014;**108**(2):340–4.
13 Koh HY. Phospholipase C-b1 and schizophrenia-related behaviors. *Adv Biol Requl* 2013;**53**(3):242–8.
14 Bavelloni A, Dmitrienko GI, Goodfellow VJ, Ghavami A, Piazzi M, Blalock W, et al. PLCβ1a and PLCβ1b selective regulation and cyclin D3 modulation reduced by kinamycin F during K562 cell differentiation. *J Cell Physiol* 2015;**230**(3):587–94.
15 Poli A, Faenza I, Chiarini F, Matteucci A, McCubrey JA, Cocco L. K562 cell proliferation is modulated by PLCβ1 through a PKCa-mediated pathway. *Cell Cycle* 2013;**12**(11):1713–21.
16 Rudge SA, Wakelam MJO. Inter-regulatory dynamics of phospholipase D and the actin cytoskeleton. *Biochim Biophys Acta* 2009;**1791**(9):856–61.
17 Oliveira TG, Di Paolo G. Phospholipase D in brain function and Alzheimer's disease. *Biochim Biophys Acta* 2010;**180**(8):799–805.
18 Gao Q, Frohman MA. Roles for the lipid-signaling enzyme MitoPLD in mitochondrial dynamics, piRNA biogenesis, and spermatogenesis. *BMB Rep* 2012;**45**(1):7–13.
19 Schatter B, Jin S, Loffelholz K, Klein J. Cross-talk between phosphatidic acid and ceramide during ethanol-induced apoptosis in astrocytes. *BMC Pharmacol* 2005;**5**:3.
20 Abul-Milh M, Wu Y, Lau B, Lingwood CA, Foster DB. Induction of epithelial cell death including apoptosis by enteropathogenic Escherichia coli expressing bundle-forming pili. *Infect Immun* 2001;**69**(12):7356–64.
21 Sun GY, Xu J, Jensen MD, Simonyi A. Phospholipase A2 in the central nervous system: implications for neurodegenerative diseases. *J Lipid Res* 2004;**45**(2):205–13.
22 Last V, Williams A, Werling D. Inhibition of cytosolic phospholipase A2 prevents prion peptide-induced neuronal damage and co-localisation with beta III tubulin. *BMC Neurosci* 2012;**13**:106.
23 Lei X, Barbour SE, Ramanadham S. Group VIA Ca++-independent phospholipase A 2β(iPLA2β) and its role in β-cell programmed cell death. *Biochimie* 2010;**92**(6):627–37.

24 Chiricozzi E, Ferandez-Fernandez S, Nardicchi V, Almeida A, Bolanos JP, Goracci G. Group IIA secretory phospholipase A2(GIIA) mediates apoptotic death during NMDA receptor activation in rat primary cortical neurons. *J Neurochem* 2010;**112**:1574–83.
25 Ravichandran KS. Beginnings of a good apoptotic meal: the find-me and eat-me signaling pathways. *Immunity* 2011;**35**(4):445–55.
26 Aits S, Jaattela M. Lysosomal cell death at a glance. *J Cell Sci* 2013;**126**:1905–12.
27 Ilić D, Almeida EA, Schlaepfer DD, Dazin P, Aizawa S, Damsky CH. Extracellular matrix survival signals transduction of local adhesion kinase suppress p53-mediated apoptosis. *J Cell Biol* 1998;**143**(2):547–60.
28 Kim HJ, Kim KS, Kim SH, Baek SH, Kim HY, Lee C, Kim JR. Induction of cellular senescence by secretory phospholipase A_2 in human dermal fibroblasts through an ROS-mediated p53 pathway. *J Gerontol A Biol Sci Med Sci* 2009;**64**(3):351–62.
29 Marchi S, Marinello M, Bononi A, Bonora M, Giorgi C, Rimessi A, Pinton P. Selective modulation of subtype III IP_3R by Akt regulates ER Ca++ release and apoptosis. *Cell Death Dis* 2012;**3**:e304.
30 Cifone MG, Roncaioli P, De Maria R, Camarda G, Santoni A, Ruberti G, Testi R. Multiple pathways originate at the Fas/APO-1 (CD95) receptor: sequential involvement of phosphatidylcholine-specific phospholipase C and acidic sphingomyelinase in the propagation of the apoptotic signal. *EMBO J* 1995;**14**(23):5859–68.
31 Plo I, Lautier D, Levade T, Sekouri H, Jaffrézou JP, Laurent G, Bettaïeb A. Phosphatidylcholine-specific phospholipase C and phospholipase D are respectively implicated in mitogen-activated protein kinase and nuclear factor κB activation in tumour-necrosis-factor treated immature acute-myeloid-leukaemia cells. *Biochem J* 2000;**351**(Pt. 2):459–67.
32 Todt JC, Hu B, Curtis JL. The receptor tyrosine kinase MerTK activates phospholipase Cγ2 during recognition of apoptotic thymocytes by murine macrophages. *J Leukoc Biol* 2004;**75**(4):705–13.
33 Chattopadhyay A, Carpenter H. PLC-γ1 is required for IGF-I protection from cell death induced by loss of extracellular matrix adhesion. *J Cell Sci* 2002;**115**:2233–9.
34 Nakamura Y, Hamada Y, Fujiwara T, Enomoto H, Hiroe T, Tanaka S, et al. Phospholipase C-δ1 and -δ3 are essential in the trophoblast for placental development. *Mol Cell Biol* 2005;**25**(24):10 979–88.
35 Nakamura Y, Kanemaru K, Kojima R, Hashimoto Y, Marunouchi T, Oka N, et al. Simultaneous loss of phospholipase C-δ1 and phospholipase C-δ3 causes cardiomyocyte apoptosis and cardiomyopathy. *Cell Death Dis* 2014;**5**:e1215.
36 Huang C, Bruggeman LA, Hydo LM, Miller RT. Shear stress induces cell apoptosis via a c-Src-phospholipase D-mTOR signaling pathway in cultured podocytes. *Exp Cell Res* 2012;**318**(10):1075–85.
37 Yamada M, Banno Y, Takuwa Y, Koda M, Hara A, Nozawa Y. Overexpression of phospholipase D prevents actinomycin D-induced apoptosis through potentiation of phosphoinositide 3-kinase signalling pathways in Chinese-hamster ovary cells. *Biochem J* 2004;**378**(2):649–56.
38 Gao Q, Frohman MA. Roles for the lipid-signaling enzyme MitoPLD in mitochondrial dynamics, piRNA biogenesis, and spermatogenesis. *BMB Rep* 2012;**45**(1):7–13.
39 Yazlovitskaya EM, Linkous AG, Thotala DK, Cuneo KC, Hallahan DE. Cytosolic phospholipase A_2 regulates viability of irradiated vascular endothelium. *Cell Death Differ* 2008;**15**(10):1641–53.

40 Pucer A, Brglex V, Payre C, Pungercar J, Lambeau G, Petan T. Group X secreted phospholipase A_2 induces lipid droplet formation and prolongs breast cancer cell survival. *Mol Cancer* 2013;**12**(1):111.
41 Mirzayans R, Andrais B, Scott A, Murray D. New Insights into p53 signaling and cancer cell response to DNA damage: implications for cancer therapy. *J Biomed Biotechnol* 2012;**2012**:170325.
42 Rubinsztein DC, Codogno P, Levine B. Autophagy modulation as a potential therapeutic target for diverse diseases. *Nat Rev Drug Discov* 2012;**11**(9):709–30.
43 Pivtoraiko VN, Stone SL, Roth KA, Shacka JJ. Oxidative stress and autophagy in the regulation of lysosome-dependent neuron death. *Antioxid Redox Signal* 2009;**11**(3):481–96.
44 Moreau K, Ravikumar B, Puri C, Rubinsztein DC. Arf6 promotes autophagosome formation via effects on phosphatidylinositol 4,5-bisphosphate and phospholipase D. *J Cell Biol* 2012;**196**(4):483–96.
45 Mancuso DJ, Kotzbauer P, Wozniak DF, Sims HF, Jenkins CM, Guan S, et al. Genetic ablation of calcium-independent phospholipase $A_2\gamma$ Leads to alterations in hippocampal cardiolipin content and molecular species distribution, mitochondrial degeneration, autophagy, and cognitive dysfunction. *J Biol Chem* 2009;**284**(51):35 632–44.
46 Fornai F, Longone P, Ferrucci M, et al. Autophagy and amyotrophic lateral sclerosis: the multiple roles of lithium. *Autophagy* 2008;**4**(4):527–30.
47 Cardenas C, Foskett JK. Mitochondrial Ca++ signals in autophagy. *Cell Calcium* 2012;**52**(1):44–51.
48 Sakaki K, Kaufman RJ. Regulation of ER stress-induced macroautophagy by protein kinase C. *Autophagy* 2008;**4**(6):841–3.
49 Sakaki K, We J, Kaufman RJ. Protein kinase CIs required for autophagy in response to stress in the endoplasmic reticulum. *J Biol Chem* 2008;**283**(22):15 370–80.
50 Birmingham CL, Canadien V, Gouin E, Troy EB, Yoshimori T, Cossart P, et al. Listeria monocytogenes evades killing by autophagy during colonization of host cells. *Autophagy* 2007;**3**(5):442–51.
51 Du L, Hickey RW, Bayir H, Watkins SC, Tyurin VA, Guo F, et al. Starving neurons show sex difference in autophagy. *J Biol Chem* 2009;**284**(4):2383–96.
52 Peng X, Frohman MA. Mammalian phospholipase D physiological and pathological roles. *Acta Physiol* 2012;**204**(2):219–26.
53 Van Herreweghe F, Festjens N, Declercq W, Vandenabeele P. Tumor necrosis factor-mediated cell death: to break or to burst, that's the question. *Cell Mol Life Sci* 2010;**67**:1567–79.
54 Mora R, Valverde B, Diaz C, Lomonte B, Gutierrez JM. A Lys49 phospholipase A_2 homologue from Bothrops asper snake venom induces proliferation, apoptosis and necrosis in a lymphoblastoid cell line. *Toxicon* 2005;**45**(5):651–60.

20

TRIAD (Transcriptional Repression-Induced Atypical Death)

Takuya Tamura and Hitoshi Okazawa

Department of Neuropathology, Medical Research Institute, Tokyo Medical and Dental University, Tokyo, Japan

Abbreviations

Aβ	amyloid beta
AD	Alzheimer's disease
ALS	amyotrophic lateral sclerosis
AMA	α-amanitin
APP	amyloid precursor protein
Bax	Bcl-2-associated X protein
BDNF	brain-derived neurotrophic factor
cyt c	cytochrome c
DRPLA	dentate-rubro-pallido-luysian atrophy
ER	endoplasmic reticulum
FL-YAP	full-length YAP
FTLD	frontotemporal dementia
HD	Huntington's disease
HIP	huntingtin-interacting protein
PCD	programmed cell death
PCR	polymerase chain reaction
PD	Parkinson's disease
PHF	paired helical filament
PolII	RNA polymerase II
polyQ	polyglutamine
PUMA	p53-upregulated modulator of apoptosis
SCA1	spinocerebellar ataxia type 1
SCA3	spinocerebellar ataxia type 3
SOD	superoxide dismutase
TAF1	TATA-binding protein-associated factor-1
TEAD	TEA-domain family member
TRIAD	transcriptional repression-induced atypical death
VCP	valosin-containing protein
YAP	Yes-associated protein

Apoptosis and Beyond: The Many Ways Cells Die, First Edition. Edited by James Radosevich.
© 2018 John Wiley & Son Inc. Published 2018 by John Wiley & Son Inc.

20.1 Introduction

Neuronal death is the final phase of neurodegeneration and is irreversible. Therefore, it is a major goal to understand the mechanism of cell death in neurodegenerative disorders in order to discover a treatment for them. The pathological features of neurodegenerative diseases are an aggregation of abnormal protein and cell death. However, recent molecular biology and genetics analyses have elucidated the means of death by possible neuronal dysfunction. These studies prompted us to investigate the molecular mechanism of neuronal death in a lengthy list of neurodegenerative disorders. Neurodegenerative diseases are exacerbated by a gradual progression (2–20 years till bedridden) of brain pathology in human patients. Some of the neurons that are degenerated survive for an extensive period of time. How does neuronal dysfunction induce a slow, progressive cell death? Transcriptional repression-induced atypical death (TRIAD) is a form of type 3 cell death. We consider TRIAD to be a "prototype" of cell death in neurodegenerative disorders. In this chapter, we will review recent concepts of cell death in neurodegeneration and suggest how TRIAD may relate to these.

20.1.1 Cellular Dysfunction in Neurodegenerative Disorders

The characteristics of neurodegenerative disorders are an "aggregation of abnormal proteins" and "neuronal cell death." Studies of neurodegeneration have long been conducted based on the hypothesis that "aggregates induce neuronal death." Typically, Lewy bodies in Parkinson's disease (PD), paired helical filaments (PHFs) in Alzheimer's disease (AD), Bunina bodies in amyotrophic lateral sclerosis (ALS), and polyglutamine (polyQ) aggregates in polyQ diseases are known. There are also nonpathological protein aggregates, such as the marinesco body, which accumulate in neurons. These have not been proven to cause "cell death," but are associated with neurodegeneration. What should happen before cell death? The classical idea suggested there is nothing: cell death is sudden and spontaneous. Recently, however, the idea that "a function inhibition precedes cell death" has begun to become mainstream. Several studies support the idea of "neuronal cellular dysfunction theory." Recent research, including excellent interactome analyses, show that polyQ proteins can physically interact with more than 100 proteins [1–3]. These polyQ-interacting proteins are mainly related to nuclear functions or the protein-degradation pathway. Moreover, transcriptional repression in polyQ disease is becoming a widely accepted theory, based on studies of mutant-polyQ proteins [4–7]. A polyQ-disease gene, huntingtin, interacts with an axonal transport-related protein, huntingtin-interacting protein (HIP), which suggests that abnormal huntingtin can interrupt axonal transportation of brain-derived neurotrophic factor (BDNF) [8,9]. Meanwhile, Prof. Selkoe's group reported that pre-aggregate amyloid beta (Aβ) inhibits a hippocampal long-term potentiation [10]. This report signaled a paradigm shift, and clearly elucidated the neuronal function inhibition that underlies a symptom of AD. They suggested that a soluble degenerative protein can encompass a pathology. Likewise, Prof. Goldstein's group reported a function of amyloid precursor protein (APP) in endoplasmic transportation, which suggests a functional disruption in AD [11,12]. Also, Prof. Sisodia's group reported an abnormal axonal transportation induced by presenilin 1 overexpression [13]. Several polyQ-disease model mice, including spinocerebellar ataxia type 3 (SCA3) and dentate-rubro-pallido-luysian atrophy (DRPLA), did not show a clear cell death in spite of severe phenotypes [14–16]. These reports suggest a pre-cell death neuronal dysfunction that is the cause of symptoms. Altogether, nuclear function that includes transcriptional roles in the pathogenesis of neurodegenerative diseases is still a hot topic in the field [17]. TRIAD can be induced by transcriptional repression. Transcription is the basis of cell-function maintenance, and transcriptional repression affects endoplasmic transportation and/or synaptic transmission.

20.1.2 Forms of Cell Death in Neurodegenerative Disorders

It is an old consensus that cell death in a neurodegeneration disease accompanies poor inflammatory response, and the pathological features suggest a cell disappears quietly. Only ghost tangle-like traces, and not a phagocytosis response, can be observed in AD brains. Such a morphological change without an inflammatory response reminds us of apoptosis. However, it is still controversial whether a cell death in neurodegenerative diseases counts as apoptosis or not, since typical apoptosis is hardly ever observed in a human patient's brain (like anoxia or ischemia at death). A recent study questions the presence of apoptosis even in the fastest-progressing neurodegeneration, prion disease [18]. The percentage of cell death that accompanies typical apoptotic features like apoptotic bodies is extremely low. On the other hand, Shimizu et al. [19] elucidated a non-apoptotic cell death that accompanies autophagy, but autophagy is considered to be a protective mechanism that controls protein quality [20–24]. Thus, autophagic cell death may not be observed in neurodegeneration.

20.1.3 Abnormal Conformation Proteins and Cellular Dysfunction

The significance of aggregates in neurodegeneration has been widely challenged in the last decade. The argument began with polyQ diseases and spread to other research. For example, distributions of aggregate and neuronal death are inconsistent in human patients and in a mouse model of spinocerebellar ataxia type 1 (SCA1) [25]. The authors' group and Prof. Finkbeiner's group independently discovered that a cell with aggregates survives longer after transfection of mutant huntingtin, the causative gene of Huntington's disease (HD) [26,27]. These reports suggest a protective role of aggregates. Similarly, toxicity of soluble amyloid monomer and oligomer in AD is reported [10,11]. Neurodegenerative diseases are also called "conformational diseases" [28]. Misfolded disease proteins make a β-sheet structure and then aggregate via a soluble form. An excellent *in vitro* study demonstrating amyloid fibril suggested that toxicity is mainly derived from the soluble form (monomer or oligomer) [29]. Therefore, a recent hypothesis suggests that a soluble form of misfolded protein is the cause of this toxicity. In response to these results, recent studies are focusing on novel gain of protein–protein interaction among misfolded proteins, which can induce disturbances in a certain molecular pathway, as well as following neurodegeneration and neuron death.

20.2 Induction of TRIAD *In Vitro*

TRIAD is a type of cell death that is induced by transcriptional repression. We demonstrated that two distinct molecules (toxins) can induce TRIAD in rat primary culture neurons [30]. Both α-amanitin (AMA) and actinomycin D are considered to suppress transcription. First, we added 10–250 μg of AMA into a medium of Hela and primary-culture cortical and striatum neurons from rat embryo (E15) and cerebellar neuron from rat pups (P7). We found these cells gradually died over several days. These conditions mimic the original TRIAD induction in the original article. We employed actinomycin D to confirm that this kind of cell death occurs.

20.2.1 α-amanitin

AMA is a bicyclic octapeptide toxin derived from the poisonous mushroom, *Amanitia phalloides*. It specifically inhibits RNA polymerase II (PolII) by tightly binding to the bridge helix structure [31,32]. Thus, it inhibits the elongation step of PolII-dependent transcription [33]. AMA-resistant mutant animals are reported in various species, including hamster [34], rat [35], mouse [36,37], *Drosophila* [38], and *Caenorhabditis elegans* [39]. These mutants carry point mutations in the PolII gene. The resistance is responsible for the lower binding constant of AMA in the presence of mutant PolII. The mutants are resistant to very high concentrations of AMA (100–1000 times higher than wild-type). Therefore, low concentrations of AMA (−10 times wild-type) is a highly specific inhibitor of PolII. We employed AMA to depress mRNA transcription, and found it induced slowly progressive cell death in primary-culture neurons and in Hela cells [30]. The neurons took more than 144 hours to die, which is extremely long – retinoic acid-induced apoptosis occurs within 12 hours. Thus, AMA-induced transcriptional repression leads to slow, progressive neuronal death. This AMA-induced slow, progressive cell death is defined as TRIAD.

20.2.2 Actinomycin D

Actinomycin D is a class of polypeptide antibiotic derived from soil bacteria of the genus *Streptomyces*. It directly binds to double-strand DNA and inhibits transcription and DNA duplication of enzymes [40] by stalling the rapidly moving fraction of PolII [41]. High concentrations of actinomycin D can also inhibit DNA replication, but only low concentrations suppress RNA elongation. Thus, we can confirm that transcriptional repression occurs by a distinct mechanism that can induce slow, progressive cell death. Actinomycin D also induced slow, progressive cell death in primary cultures of cortex, striatum, and cerebellar neurons [30]. Again, it did not induce genomic DNA fragmentation or caspase activation. Most additional molecular analyses were carried out using AMA, not actinomycin D. Thus, the cell death induced by actinomycin D can be called TRIAD, but it is unclear whether AMA-induced and actinomycin D-induced TRIAD are identical in various aspects.

20.3 Morphological Features of TRIAD

Programmed cell death (PCD) is classified into three types, based on morphological differences. Type 1 is apoptosis, type 2 accompanies autophagy, and type 3 accompanies nonlysosomal vacuolar cell death (Table 20.1) [42]. TRIAD does not resemble active cell death (type 1, apoptosis and type 2, autophagic cell death). The authors' group found it is accompanied by unique cytoplasmic (not nuclear) vacuoles. Thus, TRIAD can be classified as a kind of type 3 cell death.

Table 20.1 Types of cell death.

Type of death	Type 1, apoptosis	Type 2, autophagic	Type 3, necrotic	Passive necrosis
Chromatin condensation	+	–	–	–
Nucleus fragmentation	+	–	–	–
DNA fragmentation	+	–	–	–
Caspase activation	+	–	–	–
Autophagosome	–	+	–	–
Nonlysosomal vacuole	–	–	+	+
Plasma-membrane corruption	–	–	+	+

20.3.1 Difference from Apoptosis

The morphological features of apoptosis are fragmentation of chromatin and nuclear condensation. The authors' group carried out electron microscopic analyses to observe the morphology of TRIAD. We found that both AMA and actinomycin D did not induce chromatin fragmentation and nuclear condensation [30]. Therefore, we consider TRIAD is different from typical apoptosis.

20.3.2 Difference from Autophagic Cell Death

Autophagic cell death is another type of PCD that accompanies increased autophagy markers. LC3-positive cytoplasmic vacuoles are employed as a typical marker of autophagosome [43]. Our group used EGFP–LC3 fusion protein to discover whether vacuoles in TRIAD are autophagosomes. We found increased vacuoles in Hela-TRIAD were not labeled with EGFP–LC3, suggesting that TRIAD is different from typical autophagic cell death [30]. LC3-negative vacuoles were increased by AMA treatment in Hela cells.

20.3.3 Difference from Necrosis

The classical definition of necrosis is not active PCD but rather passive cell death induced by irreversible damage. Recently, programmed necrosis (e.g., necroptosis), which has necrosis-like morphological features, was reported. Both passive and active necrosis have the similar morphological features: organelle swelling and corruption of the plasma membrane. An electron-microscopic observation demonstrated that cortical neuron-TRIAD did not accompany a corruption of plasma-membrane or cytoplasimic abnormality or mitochondrial swelling. The authors' group searched for the source of the cytoplasmic vacuole in Hela-TRIAD. Golgi58k, early endosome antigen-1, CD63, and CCO1 are the specific markers of the Golgi apparatus, endosome, lysosome, and mitochondria, respectively. The TRIAD vacuoles were not positively stained by the organelle-specific antibodies, supporting the idea that the cytoplasmic vacuole was not derived from the Golgi apparatus, endosome, lysosome, or mitochondria [30]. Therefore, we suspected that TRIAD is different from typical necrosis.

20.3.4 Original Phenotype: ER Vacuolization

The entity of the unique vacuole remained unclarified. The authors' group finally elucidated that the cytoplasmic vacuole is derived from the endoplasmic reticulum (ER) by using the ER-retention signal peptide, KDEL. KDEL is a C-terminal or N-terminal tetrapeptide signal (lys-asp-glu-leu) that keeps proteins from being secreted from the ER [44]. A secretory protein can leave the ER after the KDEL signal is cleaved out. Fluorescent signal was located in the vacuoles when ECFP–KDEL protein was expressed during Hela-TRIAD, suggesting that vacuoles may be derived from expanded ER. Thus, the characteristic morphological feature of TRIAD is ER-derived cytoplasmic vacuoles. We can clearly discriminate TRIAD as a type 3 cell death (programmed necrosis) based on this unique morphological feature.

20.4 Molecular Features of TRIAD

Established cell death has specific molecular features, but the molecular pathway that underlies TRIAD is not well studied. The authors' group confirmed that TRIAD is not accompanied by the distinct molecular changes of established PCD. Microarray analysis presented three genes specifically upregulated and eight specifically downregulated during TRIAD [30]. The authors' group found that one of the downregulated genes, Yes-associated protein (YAP), can regulate TRIAD.

20.4.1 Difference from Apoptosis

The principle molecular feature of apoptosis is activation of the caspase pathway. An apoptosis-inducing stimulus induces an activation of caspase via cytochrome c (cyt c) release from the mitochondria. Activation of initiator caspases (e.g., caspase 9 or 12) sequentially activates the next caspase, until effector-caspase (e.g., caspase 3 or 7) activation induces apoptosis [45]. Caspase-activated DNase induces genomic DNA fragmentation [46,47]. The authors' group elucidated the difference between apoptosis and TRIAD [30]. The direct electrophoresis method demonstrated no obvious genomic DNA fragmentation induced by TRIAD of P19, Hela, or cerebellar and cortical neurons. Western blot analysis using caspase 3, 7, and 12 antibodies demonstrated no activated (cleaved) form of these caspases that was detected during TRIAD in cerebellar and cortical neurons. Western blot analysis of cell fractions demonstrated that both cytosolic and mitchondrial cyt c were not increased during TRIAD of cerebellar and cortical neurons. Repeatedly, actinomycin D did not induce caspase activation or genomic DNA fragmentation. As expected, the caspase inhibitors z-DEVD-fmk and z-VAD-fmk did not repress AMA-induced cell death in neurons or in HeLa cells. Together with morphological analysis, this made the distinct features of apoptosis and TRIAD clearer.

20.4.2 Difference from Autophagic Cell Death

Autophagy is a basic catabolic mechanism of intracellular proteins. Its main pathway, macroautophagy, is induced by rapamycin, an inhibitor of mTOR. Inactivation of mTOR enhances the subsequent steps of autophagy, and finally forms an autophagosome [48]. As described earlier, an autophagosome is labeled by a specific marker, LC3. Rapamycin treatment of Hela cells induced LC3-positive vacuoles but not LC3-negative

(KDEL-positive) ones [30]. Thus, the typical autophagy mechanism looks not to be involved in vacuole formation during TRIAD.

20.4.3 Difference from Necrosis

Recently, several other type 3 (necrotic) cell deaths have been described, in addition to TRIAD. Paraptosis [49], necroptosis [50], and entosis [51] are caused by intrinsic factors and are considered to be active forms of necrosis [52–54]. Several molecules are reported to be involved in such active necroses (e.g., TNF-R1, Rip1, Rip3) [55]. The changes and effects of the programmed necrosis-related proteins have not yet been addressed in TRIAD. Recent studies have shown that some molecules are commonly employed in multiple types of cell death [56]. The detailed molecular mechanisms of type 3 cell deaths are not clarified, but their importance is increasing daily [57]. Therefore, it is an important question whether molecules of other forms of cell death also participate in TRIAD.

20.4.4 Original Phenotype: Gene-Expression Changes

The authors' group addressed gene-expression change in TRIAD, because the molecular pathway was completely unclear [30]. A microarray analysis was done using RNA 1 hour after AMA treatment to detect initial changes. In cerebellar TRIAD, 114 and 141 genes were up- and downregulated, respectively, by more than twofold. In cortical TRIAD, 14 and 75 genes were up- and downregulated, respectively, by more than twofold. Across both forms of TRIAD, 7 and 12 genes were up- and downregulated, respectively. The authors' group also analyzed gene-expression change in low-potassium-induced apoptosis in cerebellar neuron and found 103 and 98 genes that were up- and downregulated, respectively, by more that twofold. Finally, 3 and 8 genes were specifically up- and downregulated in both forms of TRIAD but were not changed in apoptosis (Table 20.2).

20.4.5 Expression Changes in YAP Isoforms

The authors' group confirmed the downregulation of YAP by AMA treatment using northern blot analysis [30]. Unexpectedly, we identified novel YAP isoforms (splicing variants) during polymerase chain reaction (PCR) cloning with RNA extracted from normal cortical and cerebellar neurons. The novel isoforms of YAP containing 13-, 25-, and 61-nt inserts led to a reading-frame shift, causing truncation of the COOH-terminal transcriptional activation domain (Figure 20.1). Therefore, these isoforms were designated as YAPΔCs. Recently, Sudol et al. [58] classified at least eight YAP splicing variants, but all of them contained the COOH-terminal transcriptional activation domain, making our YAPΔCs the original isoforms. Western blot analysis elucidated different time constants in the decrease between full-length YAP (FL-YAP) and YAPΔCs after AMA treatment [30]. In primary cortical neurons, the expression of YAPΔCs was sustained for 6 days after AMA addition, whereas FL-YAP was repressed within 2 days. Notably, the expression of YAPΔCs was very low in HeLa cells. Therefore, YAPΔCs look to be a neuron-specific variant that has a novel function in TRIAD.

20.4.6 Phosphorylation of p73

YAP is a transcription cofactor. It enhances transcription of target genes by binding to its binding-target transcription factors, one of which is p73 [59]. DNA damage leads

Table 20.2 Genes up- and downregulated in cerebellar and cortical TRIAD but not changed in apoptosis.

Upregulated genes

Spot Index	ID	Name	Definition	Accession Number	Ratio (CTX_Ama −/+)	Ratio (CBLL_Ama −/+)	Ratio (CBLL_apoH/L)
2613	4191	702439960	Incyte EST		0.447	0.303	0.508
2630	4345	701531578	Mouse Naip3 gene, exon 1; neuronal apoptosis-inhibitory protein 1 (Naip1), and general transcription factor IIH polypeptide 2 (Gtf2h2) genes, complete cds	AF242432	0.410	0.332	0.648
5710	4020	701795169	ApolipoproteinL3 (TNF-inducible protein CG12-1)	NW_2354630	0.398	0.393	0.683

Downregulated genes

Spot Index	ID	Name	Definition	Accession Number	Ratio (CTX_Ama −/+)	Ratio (CBLL_Ama −/+)	Ratio (CBLL_apoH/L)
138	18236	702552229	Rat zinc finger, DHHC domain containing 21, mRNA, complete cds	XM_233110.2	2.065	2.058	0.714
8582	6934	700816371	Rat YAP65 mRNA	XM_343347	2.261	2.431	1.876
9202	2431	702073550	Rat apolipoprotein E gene, complete cds	J02582	2.023	2.074	1.467
13695	14518	702141227	Rat mRNA for elongation factor 2 (EF-2)	Y07504	2.282	2.141	1.959
13780	13758	700146774	Rat mRNA for non-neuronal enolase (NNE) (alpha-alpha enolase, 2-phospho-D-glycerate hydrolase EC 4.2.1.11)	X02610	2.068	2.134	1.832
14305	11702	700486110	Mouse mRNA for alpha-tubulin 8 (Tuba8 gene)	AJ245923	2.469	2.161	1.622
14320	10638	700146495	Rattus norvegicus heat shock protein 8 (Hspa8), mRNA	NM_024351	2.059	2.074	1.844
15558	3601	700600666	Rat X-chromosome-linked phosphoglycerate kinase mRNA, complete cds	M31788	2.075	2.026	1.458

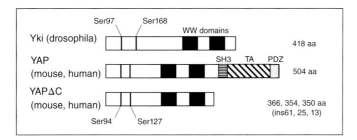

Figure 20.1 Schematic representation of various YAP proteins.

to p73-mediated transactivation of cell-death genes, including Bcl-2-associated X protein (Bax) and, possibly, p53-upregulated modulator of apoptosis (PUMA) [60]. Western blot analysis showed that the amount and phosphorylation of p73 were enhanced in AMA-treated cortical neurons, whereas the total amount of p73 was not changed. It is becoming clear that the various kinases (chk2, ATM, and c-abl) can phosphorylate p73 [61]. Although the role of phosphorylation of p73 in cell death is still a challenging theme, it is a feature of TRIAD. On the other hand, YAP is a transcriptional coactivator of p73 that mediates apoptosis [62]. It promotes nuclear localization and proapoptotic signaling of p73 via its PDZ-binding motif [63]. The authors hypothesized that YAPΔCs can suppress the expression of apoptosis-related genes in a dominant-negative manner. YAP and p73 can form a proapoptotic feedback loop [64]. YAPΔCs may maintain only the proapoptotic function and not the facilitator function of cell death. TRIAD may share certain molecular features with apoptosis because a proapoptotic signal was changed.

20.5 Roles of YAPΔC in TRIAD and Neurodegeneration

YAP/yorkie is a well-conserved protein that functions in cell proliferation, where it plays a proapoptotic role. It was first identified as an associate protein of the SH3 domain of Yes and Src protein-tyrosine kinases [65] and is one of the two main effectors of the Hippo tumor-suppressor pathway [66]. YAP is a transcriptional coactivator of TEAD, which is the target of Hippo pathway, and TEAD activation by YAP upregulates genes that promote cell growth and inhibit apoptosis [67]. YAPΔCs are novel isoforms of YAP that lack C-terminal functional domains. Their overexpression prohibits TRIAD and polyQ-induced cell death. Therefore, YAPΔCs may be a primary controller of TRIAD.

20.5.1 YAPΔCs Suppress TRIAD in Culture Neuron

Adenovirus vector-derived YAPΔCs repressed TRIAD of primary cortical neurons. Transfection of an siRNA targeting a sequence shared by three YAPΔC isoforms but not FL-YAP accelerated TRIAD to 90%, supporting the idea that YAPΔCs suppress the cell-death process in TRIAD at least partially. AMA also accelerated the phosphorylation of p73 without a change in the amount of total p73. This suggests that the suppression of TRIAD by YAPΔCs is due to p73, the target transcription factor of FL-YAP. As FL-YAP decreased prior to YAPΔCs in an early phase of TRIAD, the authors' group hypothesized that YAPΔCs inhibit the function of p73 in cell death, leading to neuronal cell death by antagonizing FL-YAP.

20.5.2 YAPΔCs Suppress Culture Neuron Death Induced by Mutant Huntingtin

YAPΔCs also suppressed a cell death induced by expression of the huntingtin-exon1-111Q-induced cell death of cortical neurons. Consistent with this, YAPΔC-specific siRNA promoted the huntingtin-induced cell death. These data support the concept of TRIAD: "cell death in neurodegeneration." TRIAD can share a molecular pathway with huntingtin-induced cell death. p73 phosphorylation was also observed in the huntingtin-induced cortical neuron death. p73 siRNA repressed the cell death of mutant huntingtin-expressing neurons, as expected. Taken together, p73 can mediate downstream signaling of YAPΔCs to both TRIAD and huntingtin-induced neuronal death.

20.5.3 YAPΔCs Suppress Neuron Death Induced by Mutant Huntingtin in an Animal Model

In order to examine the *in vivo* effect of YAPΔCs on huntingtin-induced neurodegeneration, the authors' group employed a popular *Drosophila* HD model that possesses an eye phenotype [68]. The fly model expressed human Htt120Q under the control of an eye-specific promoter, GMR, and ommatidia structure and photoreceptor neurons were severely disrupted. A transgenic of a YAPΔC with a 61-nt insert (YAPΔC61) markedly preserved the ommatidia structure and maintained the number of photoreceptor neurons in the HD model fly without causing an expression change of Htt120Q. Collectively, these *in vivo* data further suggest the possibility that YAPΔC isoforms might play a protective role against the toxicity of mutant huntingtin in HD pathology.

20.5.4 YAPΔCs as a Final Target of the Hippo Pathway

YAP is the final target molecule of the Hippo pathway [66,69]. Specifically, activation of Hippo signaling phosphorylates YAP and inhibits its translocation into nucleus. The Hippo pathway controls cell proliferation and apoptosis (see Figure 20.1). YAP enhances the cell-proliferation signal via activation of a transcription factor, TEA-domain family member (TEAD) [70]. A cell-proliferation signal thus enhanced can counteract a cell-death signal. YAPΔC lacks an SH3 domain-binding motif at the C terminal, a transactivation domain, and a PDZ domain binding motif (Figure 20.1). It is unclear how these motifs and domains contribute to the function of YAP. In fact, the *Drosophila* YAP homolog, yorkie, which has equivalent function, is more homologous to YAPΔC and lacks these motifs and domains. YAPΔC cannot enhance proliferation of NIH-3T3 cells, although a mutant YAP lacking only the transactivation domain enhanced the proliferation [71]. On the other hand, replacement of Ser94, which is essential for TEAD binding with Ala, disrupted the cell-proliferation activity. Taken together, these results suggest that SH3 and PDZ domain-binding motifs have cell-proliferation activity via TEAD. On the other hand, Ser127, which is the phosphorylation target of Lats and WW domains (which are essential for TEAD binding and Ser94) are conserved in YAP and yorkie, suggesting these domains and motifs are essential for YAP/yorkie functions. Therefore, the function of YAPΔC via TEAD in TRIAD needs to be further addressed. Other YAP target transcription factors, Runx2 and TBX5, are also activated by YAP [72,73]. TRIAD may be related to these factors, but further study is necessary.

20.6 TRIAD *In Vivo*

The therapeutic effect of YAPΔC in a *Drosophila* HD model suggests a possible relationship between TRIAD and cell death in neurodegenerative diseases. There are some similarities between TRIAD and neurodegenerative-disease model animals and human patients. The authors' group elucidated the characteristic changes in TRIAD. From a molecular point of view, an expression change in YAPΔCs and a phosphorylation change are typical in TRIAD. From a morphological one, an ER-derived cytoplasmic vacuole is typical. We summarize these phenomena in neurodegenerative model animals and human patients in this section.

20.6.1 Possible TRIAD in Animal Models

In HD models, several studies have reported atypical cell death with cytoplasmic vacuolization. The authors' group elucidated an increase of phosphorylated p73 in striatal neurons of mutant huntingtin-exon1 transgenic mice (R6/2 mice), although total (nonphosphorylated and phosphorylated) p73 protein levels were not changed. On the other hand, YAPΔCs were expressed in striatal neurons of both normal and R6/2 mice, whereas the signal was relatively stronger in R6/2 mice. Valosin-containing protein (VCP)/p97, a member of the AAA+ family of ATPase proteins interacts with huntingtin, and the mutation itself also cause a frontotemporal lobar dementia. The expression of a mutant VCP leads to cytoplasmic vacuolization, which may be homologous to vacuoles in TRIAD, because they were fused to ER [74]. Thus, we cannot exclude the possibility that TRIAD occurs in HD and related diseases.

TRIAD-related phenomena have also been discovered in another disease. Prof. Abe's group analyzed motor-neuronal death in an ALS model mouse and found a type 3 cell death [75]. Mutation in superoxide dismutase (SOD) is a major cause of familial SOD. The ALS model mouse that the group employed was a mutant SOD (G93A SOD1) transgenic mouse. YAPΔCs decreased in the spinal cords of the SOD mice during disease progression, although FL-YAP was maintained. The total p73 also decreased with age, and the phosphorylated p73 to total p73 ratio increased during the late symptomatic stage. These results suggest that TRIAD, which can be controlled by YAPΔC-p73, is also important in motor-neuronal death in ALS. The expression level of YAPΔC was much lower in ALS than in HD model mice; this may relate to the different progression speeds.

20.6.2 Possible TRIAD in Human Patients

The authors' group elucidated a feature of TRIAD – higher levels of p73 phosphorylation – in HD-patient brains using a western blot analysis. Immunohistostaining also detected a phosphorylation of p73 in striatal neurons of human HD patients. We raised YAPΔC isoform-specific antibodies and showed the existence of YAPΔCs in striatal neurons of human HD patients. The YAPΔC and phosphorylated p73 were co-localized [30]. Levels of phosphorylated p73 and YAPΔCs were very low in normal human brains, suggesting these molecules are specific in pathologies. Thus, TRIAD-like molecular changes also occurred in the brains of human patients. These data suggest that TRIAD may occur in the HD pathology.

20.7 Conclusion

The symptoms of neurodegenerative disease precede the start of cell death, and the cause of symptoms is the functional disturbance of neurons. This functional disturbance shifts to cell death during the long progression in human patients, although the cause of brain atrophy can be explained by volume reduction per single neurons (dendrite, spine loss). No prototype is known of little or no cell death that can directly implicate cell death and functional disturbance. Transcription is the basis of most vital phenomena, such as synaptic transmission, protein degradation, and intracellular transportation. Therefore, TRIAD should be a possible solution to this issue, of use in deciphering the relationship between cell death and functional repression. Recently, TDP-43 was found in a tau-negative inclusion body in frontotemporal dementia (FTLD) [76]. Because TDP-43 is an RNA-binding protein, its accumulation in the cytoplasm can reduce its original distribution in the nucleus. If the reduction of TDP-43 can suppress nuclear functions, FTLD accompanies a TRIAD-like cell death. Moreover, X-linked recessive dystonia, DYT3, is caused by expression repression of TATA-binding protein-associated factor-1 (TAF1) [77]. Repression of TAF1 can induce a repression in general transcription. Thus, DYT3 can accompany a possible TRIAD. To understand the molecular mechanisms of type 3 cell death (including TRIAD), it is important to elucidate the generality of the cell-death mechanism in neurodegenerative diseases.

References

1. Okazawa H. Polyglutamine diseases: a transcription disorder? *CMLS* 2003;**60**(7):1427–39.
2. Stelzl U, Worm U, Lalowski M, Haenig C, Brembeck FH, Goehler H, et al. A human protein-protein interaction network: a resource for annotating the proteome. *Cell* 2005;**122**(6):957–68.
3. Lim J, Hao T, Shaw C, Patel AJ, Szabó G, Rual JF, et al. A protein-protein interaction network for human inherited ataxias and disorders of Purkinje cell degeneration. *Cell* 2006;**125**(4):801–14.
4. Qi ML, Tagawa K, Enokido Y, Yoshimura N, Wada Y, Watase K, et al. Proteome analysis of soluble nuclear proteins reveals that HMGB1/2 suppress genotoxic stress in polyglutamine diseases. *Nat Cell Biol* 2007;**9**(4):402–14.
5. Zoghbi HY, Orr HT. Pathogenic mechanisms of a polyglutamine-mediated neurodegenerative disease, spinocerebellar ataxia type 1. *J Biol Chem* 2009;**284**(12):7425–9.
6. Enokido Y, Tamura T, Ito H, Arumughan A, Komuro A, Shiwaku H, et al. Mutant huntingtin impairs Ku70-mediated DNA repair. *J Cell Biol* 2010;**189**(3):425–43.
7. Barclay SS, Tamura T, Ito H, Fujita K, Tagawa K, Shimamura T, et al. Systems biology analysis of Drosophila in vivo screen data elucidates core networks for DNA damage repair in SCA1. *Hum Mol Genet* 2014;**23**(5):1345–64.
8. Zuccato C, Ciammola A, Rigamonti D, Leavitt BR, Goffredo D, Conti L, et al. Loss of huntingtin-mediated BDNF gene transcription in Huntington's disease. *Science* 2001;**293**(5529):493–8.
9. Gauthier LR, Charrin BC, Borrell-Pàges M, Dompierre JP, Rangone H, Cordelières FP, et al. Huntingtin controls neurotrophic support and survival of neurons by enhancing BDNF vesicular transport along microtubules. *Cell* 2004;**118**(1):127–38.

10 Walsh DM, Klyubin I, Fadeeva JV, Cullen WK, Anwyl R, Wolfe MS, et al. Naturally secreted oligomers of amyloid beta protein potently inhibit hippocampal long-term potentiation in vivo. *Nature* 2002;**416**(6880):535–9.

11 Kamal A, Almenar-Queralt A, LeBlanc JF, Roberts EA, Goldstein LS. Kinesin-mediated axonal transport of a membrane compartment containing beta-secretase and presenilin-1 requires APP. *Nature* 2001;**414**(6864):643–8.

12 Stokin GB, Almenar-Queralt A, Gunawardena S, Rodrigues EM, Falzone T, Kim J, et al. Amyloid precursor protein-induced axonopathies are independent of amyloid-beta peptides. *Hum Mol Genet* 2008;**17**(22):3474–86.

13 Lazarov O, Morfini GA, Pigino G, Gadadhar A, Chen X, Robinson J, et al. Impairments in fast axonal transport and motor neuron deficits in transgenic mice expressing familial Alzheimer's disease-linked mutant presenilin 1. *J Neurosci* 2007;**27**(26):7011–20.

14 Schilling G, Wood JD, Duan K, Slunt HH, Gonzales V, Yamada M, et al. Nuclear accumulation of truncated atrophin-1 fragments in a transgenic mouse model of DRPLA. *Neuron* 1999;**24**(1):275–86.

15 Goti D, Katzen SM, Mez J, Kurtis N, Kiluk J, Ben-Haïem L, et al. A mutant ataxin-3 putative-cleavage fragment in brains of Machado-Joseph disease patients and transgenic mice is cytotoxic above a critical concentration. *J Neurosci* 2004;**24**(45):10 266–79.

16 Bichelmeier U, Schmidt T, Hübener J, Boy J, Rüttiger L, Häbig K, et al. Nuclear localization of ataxin-3 is required for the manifestation of symptoms in SCA3: in vivo evidence. *J Neurosci* 2007;**27**(28):7418–28.

17 Mohan RD, Abmayr SM, Workman JL. The expanding role for chromatin and transcription in polyglutamine disease. *Curr Opin Genet Dev* 2014;**26**:96–104.

18 Liberski PP, Sikorska B, Bratosiewicz-Wasik J, Gajdusek DC, Brown P. Neuronal cell death in transmissible spongiform encephalopathies (prion diseases) revisited: from apoptosis to autophagy. *Int J Biochem Cell Biol* 2004;**36**(12):2473–90.

19 Shimizu S, Kanaseki T, Mizushima N, Mizuta T, Arakawa-Kobayashi S, Thompson CB, Tsujimoto Y. Role of Bcl-2 family proteins in a non-apoptotic programmed cell death dependent on autophagy genes. *Nat Cell Biol* 2004;**6**(12):1221–8.

20 Kegel KB, Kim M, Sapp E, McIntyre C, Castaño JG, Aronin N, DiFiglia M. Huntingtin expression stimulates endosomal-lysosomal activity, endosome tubulation, and autophagy. *J Neurosci* 2000;**20**(19):7268–78.

21 Ravikumar B, Vacher C, Berger Z, Davies JE, Luo S, Oroz LG, et al. Inhibition of mTOR induces autophagy and reduces toxicity of polyglutamine expansions in fly and mouse models of Huntington disease. *Nat Genet* 2004;**36**(6):585–95.

22 Hara T, Nakamura K, Matsui M, Yamamoto A, Nakahara Y, Suzuki-Migishima R, et al. Suppression of basal autophagy in neural cells causes neurodegenerative disease in mice. *Nature* 2006;**441**(7095):885–9.

23 Komatsu M, Waguri S, Koike M, Sou YS, Ueno T, Hara T, et al. Homeostatic levels of p62 control cytoplasmic inclusion body formation in autophagy-deficient mice. *Cell* 2007;**131**(6):1149–63.

24 Boland B, Kumar A, Lee S, Platt FM, Wegiel J, Yu WH, Nixon RA. Autophagy induction and autophagosome clearance in neurons: relationship to autophagic pathology in Alzheimer's disease. *J Neurosci* 2008;**28**(27):6926–37.

25 Watase K, Weeber EJ, Xu B, Antalffy B, Yuva-Paylor L, Hashimoto K, et al. A long CAG repeat in the mouse Sca1 locus replicates SCA1 features and reveals the impact of protein solubility on selective neurodegeneration. *Neuron* 2002;**34**(6):905–19.

26. Arrasate M, Mitra S, Schweitzer ES, Segal MR, Finkbeiner S. Inclusion body formation reduces levels of mutant huntingtin and the risk of neuronal death. *Nature* 2004;**431**(7010):805–10.
27. Tagawa K, Hoshino M, Okuda T, Ueda H, Hayashi H, Engemann S, et al. Distinct aggregation and cell death patterns among different types of primary neurons induced by mutant huntingtin protein. *J Neurochem* 2004;**89**(4):974–87.
28. Carrell RW, Lomas DA. Conformational disease. *Lancet* 1997;**350**(9071):134–8.
29. Ross CA, Poirier MA. Protein aggregation and neurodegenerative disease. *Nat Med* 2004;**10**(Suppl.):S10–17.
30. Hoshino M, Qi ML, Yoshimura N, Miyashita T, Tagawa K, Wada Y, et al. Transcriptional repression induces a slowly progressive atypical neuronal death associated with changes of YAP isoforms and p73. *J Cell Biol* 2006;**172**(4):589–604.
31. Cochet-Meilhac M, Chambon P. Animal DNA-dependent RNA polymerases. 11. Mechanism of the inhibition of RNA polymerases B by amatoxins. *Biochimi Biophys Acta* 1974;**353**(2):160–84.
32. Bushnell DA, Cramer P, Kornberg RD. Structural basis of transcription: alpha-amanitin-RNA polymerase II cocrystal at 2.8 A resolution. *Proc Natl Acad Sci USA* 2002;**99**(3):1218–22.
33. Chafin DR, Guo H, Price DH. Action of alpha-amanitin during pyrophosphorolysis and elongation by RNA polymerase II. *J Biol Chem* 1995;**270**(32):19 114–19.
34. Ingles CJ, Guialis A, Lam J, Siminovitch L. Alpha-amanitin resistance of RNA polymerase II in mutant Chinese hamster ovary cell lines. *J Biol Chem* 1976;**251**(9):2729–34.
35. Somers DG, Pearson ML, Ingles CJ. Isolation and characterization of an alpha-amanitin-resistant rat myoblast mutant cell line possessing alpha-amanitin-resistant RNA polymerase II. *J Biol Chem* 1975;**250**(13):4825–31.
36. Bryant RE, Adelberg EA, Magee PT. Properties of an altered RNA polymerase II activity from an alpha-amanitin-resistant mouse cell line. *Biochemistry* 1977;**16**(19):4237–44.
37. Bartolomei MS, Corden JL. Localization of an alpha-amanitin resistance mutation in the gene encoding the largest subunit of mouse RNA polymerase II. *Mol Cell Biol* 1987;**7**(2):586–94.
38. Greenleaf AL, Weeks JR, Voelker RA, Ohnishi S, Dickson B. Genetic and biochemical characterization of mutants at an RNA polymerase II locus in D. melanogaster. *Cell* 1980;**21**(3):785–92.
39. Sanford T, Golomb M, Riddle DL. RNA polymerase II from wild type and alpha-amanitin-resistant strains of Caenorhabditis elegans. *J Biol Chem* 1983;**258**(21):12 804–9.
40. Sobell HM. Actinomycin and DNA transcription. *Proc Natl Acad Sci USA* 1985;**82**(16):5328–31.
41. Kimura H, Sugaya K, Cook PR. The transcription cycle of RNA polymerase II in living cells. *J Cell Biol* 2002;**159**(5):777–82.
42. Schweichel JU, Merker HJ. The morphology of various types of cell death in prenatal tissues. *Teratology* 1973;**7**(3):253–66.
43. Kabeya Y, Mizushima N, Ueno T, Yamamoto A, Kirisako T, Noda T, et al. LC3, a mammalian homologue of yeast Apg8p, is localized in autophagosome membranes after processing. *EMBO J* 2000;**19**(21):5720–8.
44. Stornaiuolo M, Lotti LV, Borgese N, Torrisi MR, Mottola G, Martire G, Bonatti S. KDEL and KKXX retrieval signals appended to the same reporter protein determine

different trafficking between endoplasmic reticulum, intermediate compartment, and Golgi complex. *Mol Biol Cell* 2003;**14**(3):889–902.
45. Shi Y. Mechanisms of caspase activation and inhibition during apoptosis. *Mol Cell* 2002;**9**(3):459–70.
46. Enari M, Sakahira H, Yokoyama H, Okawa K, Iwamatsu A, Nagata S. A caspase-activated DNase that degrades DNA during apoptosis, and its inhibitor ICAD. *Nature* 1998;**391**(6662):43–50.
47. Sakahira H, Enari M, Nagata S. Cleavage of CAD inhibitor in CAD activation and DNA degradation during apoptosis. *Nature* 1998;**391**(6662):96–9.
48. Pattingre S, Espert L, Biard-Piechaczyk M, Codogno P. Regulation of macroautophagy by mTOR and Beclin 1 complexes. *Biochimie* 2008;**90**(2):313–23.
49. Sperandio S, de Belle I, Bredesen DE. An alternative, nonapoptotic form of programmed cell death. *Proc Natl Acad Sci USA* 2000;**97**(26):14 376–81.
50. Degterev A, Huang Z, Boyce M, Li Y, Jagtap P, Mizushima N, et al. Chemical inhibitor of nonapoptotic cell death with therapeutic potential for ischemic brain injury. *Nat Chem Biol* 2005;**1**(2):112–19.
51. Overholtzer M, Mailleux AA, Mouneimne G, Normand G, Schnitt SJ, King RW, et al. A nonapoptotic cell death process, entosis, that occurs by cell-in-cell invasion. *Cell* 2007;**131**(5):966–79.
52. Degterev A, Hitomi J, Germscheid M, Ch'en IL, Korkina O, Teng X, et al. Identification of RIP1 kinase as a specific cellular target of necrostatins. *Nat Chem Biol* 2008;**4**(5):313–21.
53. Galluzzi L, Kroemer G. Necroptosis: a specialized pathway of programmed necrosis. *Cell* 2008;**135**(7):1161–3.
54. Hitomi J, Christofferson DE, Ng A, Yao J, Degterev A, Xavier RJ, Yuan J. Identification of a molecular signaling network that regulates a cellular necrotic cell death pathway. *Cell* 2008;**135**(7):1311–23.
55. Vandenabeele P, Declercq W, Van Herreweghe F, Vanden Berghe T. The role of the kinases RIP1 and RIP3 in TNF-induced necrosis. *Sci Signal* 2010;**3**(115):re4.
56. Kang R, Zeh HJ, Lotze MT, Tang D. The Beclin 1 network regulates autophagy and apoptosis. *Cell Death Differ* 2011;**18**(4):571–80.
57. Golstein P, Kroemer G. A multiplicity of cell death pathways. Symposium on apoptotic and non-apoptotic cell death pathways. *EMBO Rep* 2007;**8**(9):829–33.
58. Sudol M, Shields DC, Farooq A. Structures of YAP protein domains reveal promising targets for development of new cancer drugs. *Semin Cell Dev Biol* 2012;**23**(7):827–33.
59. Strano S, Munarriz E, Rossi M, Castagnoli L, Shaul Y, Sacchi A, et al. Physical interaction with Yes-associated protein enhances p73 transcriptional activity. *J Biol Chem* 2001;**276**(18):15 164–73.
60. Melino G, Bernassola F, Ranalli M, Yee K, Zong WX, Corazzari M, et al. p73 induces apoptosis via PUMA transactivation and Bax mitochondrial translocation. *J Biol Chem* 2004;**279**(9):8076–83.
61. Conforti F, Sayan AE, Sreekumar R, Sayan BS. Regulation of p73 activity by post-translational modifications. *Cell Death Dis* 2012;**3**:e285.
62. Basu S, Totty NF, Irwin MS, Sudol M, Downward J. Akt phosphorylates the Yes-associated protein, YAP, to induce interaction with 14-3-3 and attenuation of p73-mediated apoptosis. *Mol Cell* 2003;**11**(1):11–23.
63. Oka T, Sudol M. Nuclear localization and pro-apoptotic signaling of YAP2 require intact PDZ-binding motif. *Genes Cells* 2009;**14**(5):607–15.

64. Lapi E, Di Agostino S, Donzelli S, Gal H, Domany E, Rechavi G, et al. PML, YAP, and p73 are components of a proapoptotic autoregulatory feedback loop. *Mol Cell* 2008;**32**(6):803–14.
65. Sudol M. Yes-associated protein (YAP65) is a proline-rich phosphoprotein that binds to the SH3 domain of the Yes proto-oncogene product. *Oncogene* 1994;**9**(8):2145–52.
66. Huang J, Wu S, Barrera J, Matthews K, Pan D. The Hippo signaling pathway coordinately regulates cell proliferation and apoptosis by inactivating Yorkie, the Drosophila homolog of YAP. *Cell* 2005;**122**(3):421–34.
67. Zhao B, Kim J, Ye X, Lai ZC, Guan KL. Both TEAD-binding and WW domains are required for the growth stimulation and oncogenic transformation activity of yes-associated protein. *Cancer Res* 2009;**69**(3):1089–98.
68. Jackson GR, Salecker I, Dong X, Yao X, Arnheim N, Faber PW, et al. Polyglutamine-expanded human huntingtin transgenes induce degeneration of Drosophila photoreceptor neurons. *Neuron* 1998;**21**(3):633–42.
69. Oh H, Irvine KD. Yorkie: the final destination of Hippo signaling. *Trends Cell Biol* 2010;**20**(7):410–17.
70. Bandura JL, Edgar BA. Yorkie and Scalloped: partners in growth activation. *Dev Cell* 2008;**14**(3):315–16.
71. Zhang X, Grusche FA, Harvey KF. Control of tissue growth and cell transformation by the Salvador/Warts/Hippo pathway. *PLoS One* 2012;**7**(2):e31994.
72. Zaidi SK, Sullivan AJ, Medina R, Ito Y, van Wijnen AJ, Stein JL, et al. Tyrosine phosphorylation controls Runx2-mediated subnuclear targeting of YAP to repress transcription. *EMBO J* 2004;**23**(4):790–9.
73. Murakami M, Nakagawa M, Olson EN, Nakagawa O. A WW domain protein TAZ is a critical coactivator for TBX5, a transcription factor implicated in Holt-Oram syndrome. *Proc Natl Acad Sci USA* 2005;**102**(50):18 034–9.
74. Hirabayashi M, Inoue K, Tanaka K, Nakadate K, Ohsawa Y, Kamei Y, et al. VCP/p97 in abnormal protein aggregates, cytoplasmic vacuoles, and cell death, phenotypes relevant to neurodegeneration. *Cell Death Differ* 2001;**8**(10):977–84.
75. Morimoto N, Nagai M, Miyazaki K, Kurata T, Takeshi Y, Ikeda Y, et al. Progressive decrease in the level of YAPdeltaCs, prosurvival isoforms of YAP, in the spinal cord of transgenic mouse carrying a mutant SOD1 gene. *J Neurosci Res* 2009;**87**(4):928–36.
76. Neumann M, Sampathu DM, Kwong LK, Truax AC, Micsenyi MC, Chou TT, et al. Ubiquitinated TDP-43 in frontotemporal lobar degeneration and amyotrophic lateral sclerosis. *Science* 2006;**314**(5796):130–3.
77. Makino S, Kaji R, Ando S, Tomizawa M, Yasuno K, Goto S, et al. Reduced neuron-specific expression of the TAF1 gene is associated with X-linked dystonia-parkinsonism. *Am J Hum Genet* 2007;**80**(3):393–406.

21

Alkylating-Agent Cytotoxicity Associated with O^6-Methylguanine

Latha M. Malaiyandi,[1] Lawrence A. Potempa,[2] Nicholas Marschalk,[1] Paiboon Jungsuwadee,[3] and Kirk E. Dineley[1]

[1]*Chicago College of Osteopathic Medicine, Midwestern University, Downers Grove, IL, USA*
[2]*Roosevelt University College of Pharmacy, Schaumburg, IL, USA*
[3]*School of Pharmacy, Fairleigh Dickinson University, Florham Park, NJ, USA*

Abbreviations

ABVD	adriamycin (doxorubicin), bleomycin, vincristine, dacarbazine
AGT	alkylguanine methyltransferase
AIF	apoptosis-inducing factor
Apaf-1	apoptotic protease activating factor-1
APE-1	apurinic/apyrimidinic endonuclease 1
ATM	ataxia telangiecstasia mutated
Bak	Bcl-2 homologous antagonist killer
BARD1	BRCA1-associated RING-domain protein 1
Bax	Bcl-2-associated X protein
Bcl-2	B-cell lymphoma-2
BEACOPP	bleomycin, etoposide, adriamycin (doxorubicin), cyclophosphomide, oncovin (vincristine), prednisone, procarbazine
BER	base excision repair
BID	BH3-interacting domain death agonist
BRCA1	breast cancer type 1-susceptibility protein
CNS	central nervous system
COPD	chronic obstructive pulmonary disease
CpG	cytosine–phosphate–guanine
CVD	cisplatin, vincristine, dacarbazine
CYP	cytochrome P450
cyt c	cytochrome c
DCLRE1C	DNA crosslink repair 1C
DISC	death-inducing signal complex
DMN	dimethylnitrosamine
DNA-PKcs	DNA protein kinase – catalytic subunit
DTIC	dacarbazine
endo G	endonuclease G
ENU	ethylnitrosourea
FADD	Fas-associated death domain

Apoptosis and Beyond: The Many Ways Cells Die, First Edition. Edited by James Radosevich.
© 2018 John Wiley & Son Inc. Published 2018 by John Wiley & Son Inc.

FasL	Fas ligand
FDA	Food and Drug Administration
GBM	glioblastoma multiforme
G-CIMP	glioma CpG island methylator phenotype
GI	gastrointestinal
GINET	gastrointestinal neuroendocrine tumor
H2AX	histone 2AX
HCC	hepatocellular carcinoma
HNPCC	hereditary nonpolyposis colorectal cancer
HR	homologous recombination
HSV	herpes simplex virus
IHC	immunohistochemistry
LIG3	DNA ligase III
MAM	methylazoxymethanol
MAOI	monoamine oxidase inhibitor
MDC1	mediator of DNA-damage checkpoint 1
MER	methyl excision repair
MGMT	O^6-methylguanine-DNA-methyltransferase
MMR	mismatch repair
MMS	methyl methanesulfonate
MNNG	methyl-N'-nitro-N-nitrosoguanidine
MNU	methylnitrosourea
MOPP	mustargen, oncovin, procarbazine, prednisone
MPG	N-methylpurine-DNA glycosylase
MRE11	meiotic recombination 11
MRN complex	MRE11, RAD50, and NBS1
MSI	microsatellite instability
MSP	methylation-specific polymerase chain reaction
MTIC	5-(3-methyltriazen-1-yl)imidazole-4-carboxamide
N3-meA	N3-methyladenine
N7-meG	N7-methylguanine
NAD^+	oxidized nicotinamide adenine dinucleotide
NDMA	N-nitrosodimethylamine
NER	nucleotide excision repair
NHEJ	non-homologous end joining
NNK	4-(methylnitro-samino)-1-(3-pyridyl)-1-butanone
NOC	N-nitroso compound
Noxa	phorbol-12-myristate-13-acetate-induced protein 1
O^6-BG	O^6-benzylguanine
O^6-BTG	O^6-(4-bromothenyl)guanine (or lomeguatrib)
O^6-meG	O^6-methylguanine
O^4-meT	O^4-methylthymine
OMM	outer mitochondrial membrane
PARG	poly ADP ribose glycohydrolase
PARP-1	poly ADP ribose polymerase 1
PCB	procarbazine
PCD	programmed cell death
PCNA	proliferating cell nuclear antigen
PIDD	p53-induced death domain protein
PKC	protein kinase C
PNET	pancreatic neuroendocrine tumor
Polβ	DNA polymerase β

PUMA	p53-upregulated modulator of apoptosis
ROS	reactive oxygen species
RT-PCR	reverse-transcription polymerase chain reaction
SCID	severe combined immunodeficiency
SMC	structural maintenance of chromosome
STZ	streptozotocin
TMZ	temozolomide
TNF	tumor necrosis factor
TPA	12-*O*-tetradecanoylphorbol-13-acetate
TRAIL	TNF-related apoptosis-inducing ligand
WHO	World Health Organization
XLF	XRCC4-like factor
XRCC1	x-ray repair cross-complementing protein 1

21.1 Introduction

This chapter is organized around four major themes. Section 21.2 reviews the general principles of alkylating agents, including the chemistry of their interactions with DNA. Section 21.3 reviews the cellular repair pathways triggered in response to DNA methylation, including that of the methyltransferase enzyme alkylguanine methyltransferase (AGT), mismatch repair (MMR), base excision repair (BER), homologous recombination (HR), and non-homologous end joining (NHEJ). Section 21.4 reviews the consequences of failed DNA repair (i.e., programmed cell death, PCD). Section 21.5 reviews the pharmacology and medical applications of the four most clinically relevant methylating agents: the three non-classic alkylating agents, procarbazine (PCB), dacarbazine (DTIC), and temozolomide (TMZ), and the methylnitrosourea streptozotocin (STZ). These agents are currently used to treat cancers such as Hodgkin's lymphoma, melanoma, glioma, and colon and pancreatic cancers.

For deeper insights into any of the generalized topics presented here, the reader is referred to many excellent review articles cited throughout.

21.2 General Principles of Alkylating Agents

The primary goal of this chapter is to provide an introduction to the very complicated topic of cell death associated with DNA alkylation. Despite great strides in basic research and clinical experience reaching back a full century, alkylating agent-induced cell death remains mysterious in many ways [1–3]. Assimilation of what is known about alkylating agents is an interdisciplinary challenge requiring considerable expertise in medicinal biochemistry, molecular biology, genetics, apoptosis, and clinical oncology. Many of the most clinically useful alkylating agents are bifunctional; that is, they are capable of crosslinking between two different DNA sites. This greatly complicates the potential number and variety of lesions, which, not surprisingly, sets in motion complicated DNA-repair processes that are poorly understood [4]. For the sake of clarity, this chapter introduces concepts of DNA alkylation using the simplest and perhaps most widely understood reaction: methylation. Although methylating agents react with DNA at many sites, they do not form crosslinks, and most authorities attribute their cytotoxicity to a single type of lesion: addition of a methyl group to the O^6 position of the guanine base. From this uniform beginning, it becomes possible to follow a chain of events involving sequential but overlapping repair systems that attempt to fix that lesion. Should the alkylated lesions prove too numerous and irreparable, controlled cell-death programs override DNA-repair processes, marking the cell for death via pathways that resemble apoptosis. Most fundamentally, the goal of treatment with therapeutic alkylating agents is to enhance activation of apoptotic pathways while minimizing the activity of DNA-repair pathways so that cancerous cells will die.

21.2.1 Classic versus Non-classic Alkylating Agents

The terminology and classification of alkylating agents are used inconsistently even amongst authoritative sources. "Classic" alkylating agents usually refers to agents containing chloroethyl groups that are bifunctional; that is, drugs that have two reactive groups, each of which can covalently bind at two separate locations on their target substrate. Binding sites of classic alkylating agents include a diverse array of positions on nucleic acids and proteins, but the most numerous and likely the most important is the N7 nitrogen atom in the purine ring of guanine (Figure 21.1). Bifunctional agents can form interstrand crosslinks between antiparallel strands of the DNA double helix, which, because of the covalent linkage that prevents DNA from separating, usually yields greater toxicity *in vitro* and *in vivo*. Examples of bifunctional alkylating agents include cyclophosphamide, ifosfamide, and bendamustine [5,6]. As a measure of instructional expedience, many pharmacology texts also group cisplatin and other platinum compounds under the heading "alkylating agents," due to a roughly similar mechanism of cytotoxicity and overlapping clinical uses. Like the classic alkylating agents, platinum compounds form covalent crosslinks between two different positions in DNA strands [7]. However, they form metal adducts, not alkyl ones, so they are not considered true alkylating agents. Methylating agents are usually considered "non-classic" alkylating agents. While they can methylate 13 different positions in DNA, their most important reaction is methylation of the O^6 extracyclic oxygen atom of guanine. Methylating agents are monofunctional; that is, they cannot form inter- or intrastrand crosslinks. While this diminishes their killing power relative to the aforementioned bifunctional reagents, methylating agents nevertheless hold important roles in the treatments of various cancers [8].

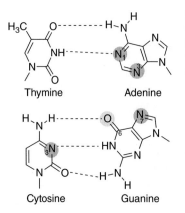

Figure 21.1 Alkylation of various sites on DNA bases. The major alkylation sites on the DNA bases are drawn within boxes. S_N1 reactions methylate nitrogen and oxygen atoms, whereas S_N2 reactions methylate nitrogen atoms almost exclusively.

21.2.2 Lesions Formed by Methylating Agents

The types of DNA lesion formed by an alkylating agent depend importantly on whether the agent reacts by S_N1 or S_N2 nucleophilic substitution. S_N1 reactions follow first-order kinetics, where the reaction rate depends solely on the concentration of the agent. S_N2 reactions proceed by second-order kinetics, where the reaction rate depends on the concentrations of both the alkylating agent and the nucleophile (i.e., the N or O positions of nucleotide bases). The practical upshot is that drugs reacting via the S_N1 mechanism tend to alkylate mostly at nitrogen atoms in DNA, and considerably less at oxygen ones, whereas those reacting by the S_N2 mechanism alkylate almost exclusively at nitrogen atoms. Table 21.1 shows the alkylation patterns achieved by different alkylating agents. Nearly all alkylating and methylating agents of clinical utility react via S_N1. In fact, busulfan is the only clinically prominent alkylating agent that reacts with double-stranded DNA via the S_N2 mechanism [9].

The most cytotoxic lesion for all methylating agents is known to be O^6-methylguanine (O^6-meG), which usually accounts for no more than 5–10% of total methylation adducts. Methylating agents can form lesions at 13 different nitrogen and oxygen positions in DNA, with the precise distribution of preferred methylation positions differing between different agents (see Table 21.1) [10,11]. Regardless of the particular methylating agent used, the most abundant lesions formed are nitrogen methylations, and most of these occur at N7-methylguanine (N7-meG), because its nucleophilic tendency is the highest of all atoms in the DNA bases. The N3 positions in both adenine and guanine are also methylated, but at significantly lower levels. Despite their total abundance, N-methylpurines are together regarded as relatively benign, so long as they are promptly fixed by DNA BER mechanisms. In circumstances where BER is compromised (e.g., by mutation or possibly drug intervention), N-methylpurines do become importantly cytotoxic. This is discussed in greater detail in later sections. The pyrimidine base thymine is also methylated at O^4 in small quantities, making up no more than 1% of

Table 21.1 Relative percentages of total methylated and ethylated adducts isolated from rat and mouse DNA. MMS, methyl methanesulfonate; DMN, dimethylnitrosamine; MNU, methylnitrosourea; MNNG, methyl-N'-nitro-N-nitrosoguanidine; ENU, ethylnitrosourea. Dash indicates not reported; ND indicates reported, but not detected. *Source:* Adapted from Beranek 1990 [11] with permission from Elsevier.

Site of Alkylation		Methylating Agents				Ethylating Agent
		MMS S_N2	DMN S_N1	MNU S_N1	MNNG S_N1	ENU
Adenine	N3	9.4	2.3–5.0	1.1–3.6	8.6	3.2–4.3
Cytosine	O^2	–	ND	ND	–	1.3–2.2
Guanine	N3	0.7	0.6–0.9	0.6–1.6	–	1.4–1.9
	O^6	ND	6.1–7.8	3.6–10.0	9.2	5.5–7.2
	N7	85.5	66.8–74.5	70.0–86.6	82.2	13.5–19.6
Deoxythymidine	O^2	–	0.2	ND	–	5.9–8.9
	O^4	–	ND	1.8	–	0.8–3.0
Phosphotriesters		1.0	9.0–12.0	13.4	–	58.4–61.9

total lesions. It is readily repaired by nucleotide excision repair (NER) processes. O^4-methylthymine (O^4-MeT) is a potent mutagen, but the literature remains uncertain about its role in cell death. Methylating agents also react with the sugar-phosphate backbone of DNA to form a sizeable number of methylphosphotriester adducts. In prokaryotes, methylphosphotriesters may have a role in signaling DNA damage. In mammalian cells, little is known about their role or consequences, or whether they contribute to cytotoxicity.

The toxicity of all alkylating agents is highly biased toward rapidly proliferating cells – an obvious attribute when the target is a neoplastic cell, whose growth is by definition deranged and accelerated. This bias also explains archetypical side effects of alkylating agents arising from toxicity of the bone marrow, reproductive tissues, gastrointestinal epithelium, and hair follicles. Neither classic nor non-classic alkylating agents are considered cell cycle-phase specific. The DNA lesions they produce impair both the DNA replication necessary for cell division and the DNA transcription necessary for the synthesis of RNA. While it is true that a highly synchronous cell population entering S-phase is demonstrably more vulnerable to alkylating agents, the cytotoxicity is not importantly dependent on cell-cycle phase to the degree seen, for example, with taxanes and vinca alkaloids, which are most effective during M-phase, or antimetabolites such as 5-fluorouracil, which are most effective during S-phase [12,13].

21.2.3 DNA Methylation by Endogenous and Environmental Factors

This chapter mainly concerns DNA methylation resulting from therapeutically useful drugs, but it is important to note that methylation is ubiquitous and constant. Certain DNA methylation reactions are in fact a normal process in cellular metabolism, and methylation also occurs in response to environmental and dietary factors. The best-characterized endogenous methylating species is S-adenosylmethionine, which is a methyl donor used by enzymes to methylate DNA promoters as a normal part of epigenetic regulation of transcription. However, S-adenosylmethionine is also known to react with DNA non-enzymatically, generating each day in a typical cell an estimated 4000 N7-meG adducts, 600 N3-methyladenine (N3-meA) adducts, and fewer than 30 O^6-meG adducts [14]. N7-meG is poorly repaired, but is not itself mutagenic or toxic. Methylation does, however, render it more likely to degenerate spontaneously into an apurinic site, which can block DNA replication. In contrast to the relatively benign N7-meG, N3-meA is more cytotoxic. N3-meA lesions can directly block DNA polymerase activity, and, like those of N7-meG, can spontaneously depurinate into an abasic site. In summary, the repair of N3-meA and the secondary lesions of both N3-meA and N7-meG depend on a functional and properly balanced BER system (see Section 21.3.6). Smoking induces DNA methylation that is of relevance to the development of cancer, cardiovascular disease, chronic obstructive pulmonary disease (COPD), and other tobacco-related diseases. The carcinogens 4-(methylnitro-samino)-1-(3-pyridyl)-1-butanone (NNK) and N-nitrosodimethylamine (NDMA) are well-characterized methylating agents in tobacco smoke. Diet is another important source of methylating agents. Colorectal cancer has been associated with diets rich in red meat and nitrate-processed meat, which generate DNA-damaging N-nitroso compounds (NOCs) that harm the gastrointestinal (GI) epithelium. For readers seeking in-depth treatment of DNA methylation arising from endogenous, dietary, and environmental factors, we recommend several excellent reviews [15,17].

21.3 The *MGMT* Gene and AGT Protein Structure

The AGT encoded by the human *MGMT* gene is the most important repair mechanism for O^6-meG, and is in fact the only repair enzyme that can address the O^6-meG primary lesion. In contrast to the other repair systems presented in this chapter, which are all carried out by multiple proteins over multiple steps, AGT repair of O^6-meG is mechanistically far simpler: AGT demethylates O^6-meG in a single step that irreversibly inactivates the enzyme. Due to its station as the solitary mediator of a critical process that lacks redundancy, AGT has been an appealing yet vexing drug target. Furthermore, *MGMT* promoter status is gaining acceptance as a predictive and prognostic indicator of both cancer susceptibility and therapeutic response to methylating agents. For all of these reasons, the *MGMT* gene and its AGT protein product are perhaps the most clinically accessible of the DNA-repair processes.

Alkyltransferase genes and alkyltransferase activity are highly conserved throughout the biota, having been documented in bacteria such as *Escherichia coli* and *Mycobacterium*, yeast, the *Caenorhabditis elegans* roundworm, zebra fish, mice, rats, and humans. The human *MGMT* gene is located on the long arm of chromosome 10, in band q26 [18,19]. It comprises five exons, extending over roughly 300 kb; four of the five are coding, producing a single splice variant that is translated into protein. The promoter contains binding sites for the transcription factors Sp1, activator proteins 1 and 2 (Ap1 and Ap2), and two glucocorticoid response elements. Thus, induction of *MGMT* can be achieved, at least to a modest degree, by cyclic AMP, protein kinase C (PKC), and glucocorticoids. Consistent with its role in DNA repair, *MGMT* is also induced by stressors that damage DNA, such as ionizing radiation and alkylating agents. The *MGMT* promoter region is rich in cytosine–guanine repeats, (so-called cytosine–phosphate–guanine (CpG) islands), which allow gene silencing via methylation. The degree to which certain tumors manifest *MGMT* methylation has become an important topic of consideration in clinical oncology, particularly with respect to gliomas of the central nervous system (CNS), melanoma, and colon cancer (see Section 21.3.5).

The transcribed AGT (EC 2.1.1.63) enzyme is 207 amino acids long, with a molecular weight of ∼20–24 kDa. The protein exhibits a two-domain structure, joined by a zinc-stabilized helical bridge, and possesses a reactive site cysteine [20]. Loss of the zinc atom reduces but does not extinguish enzymatic activity. Synthesized AGT contains a nuclear localization signal, so that most cellular protein is found in the nucleus. AGT can be phosphorylated, which promotes nuclear accumulation and resists degradation, but which also reduces its efficiency for DNA repair [21]. The only known function of AGT is DNA repair. It has a half-life of roughly 1 day.

Hundreds of human *MGMT* gene variants have been described, the clinical significance of which is only starting to be appreciated [22,23]. In some cases, the presence of single amino-acid substitutions is associated with increased or decreased risk for certain cancers. For instance, the I143V variant is more commonly found in individuals of European descent relative to Asians and Africans. Among other predilections, this variant appears to correlate with a lower susceptibility to colorectal and head and neck cancer in women. Another example is the L84F polymorphism. This variant does not differ significantly by race and appears to repair O^6-meG adducts as effectively as wild-type enzyme. However, L84F may be linked to heightened susceptibility to the damaging effects of the nicotine derivative NNK found in cigarettes. A third example is the K178R variant, which is associated with individuals who do not respond as favorably to therapeutic alkylating agents, presumably because this AGT isoform repairs some DNA lesions more effectively than the wild-type enzyme. It is increasingly apparent that certain gene polymorphisms or mutations may substantially change AGT's capacity to mitigate the activity of certain alkylating agents (thus affecting the therapeutic response to chemotherapeutic drugs). This is discussed in more detail in Section 21.3.4.

21.3.1 MGMT Expression Patterns

In evaluating *MGMT* DNA-repair function involving O^6-methyl alkylation reactions, the cellular presence and activity of *MGMT* can be assessed by *MGMT* mRNA levels, AGT protein levels (using either Western Blot or immunohistochemistry (IHC) analysis), and enzyme assays that determine alkytransferase enzyme activity [24,25]. It is important to note that AGT can be hard to detect in cultured human cells. In fact, many immortalized cells do not manifest AGT activity due to transformation protocols that use the SV40 virus, which is known to inactivate *MGMT* [26]. Cells that do not manifest alkyltransferase activity are designated *mer* minus (i.e., negative for methyl excision repair, MER). Newly established lymphoblastic cells usually do show AGT activity. Moreover, the loss of MER function in certain cultured cells may be reversible, through mechanisms that are not yet understood. The synthesis of AGT does not appear to change meaningfully throughout the cell cycle.

Tissue *MGMT* expression is ubiquitous, but varies from organ to organ, being higher in liver, testis, GI tract, and blood, and lower in brain and lung. While *MGMT* expression can vary up to eightfold between individuals, expression in a given tissue does not vary substantially throughout development – with the exception of the liver, where adult AGT activity exceeds that of the fetus by fivefold. AGT activity varies widely in different cancers, with colon cancer showing higher *MGMT* activity and many gliomas lacking AGT. Interestingly, cells within the same tumor may differ substantially in this enzyme activity. In some cancers (e.g., breast and ovarian), AGT activity increases with advancing cancer stage, while in certain brain cancers, it successively declines as they advance to stages 3 and 4 [24].

Upregulation of *MGMT* expression is not easily demonstrated in cultured human cells. In rodent models, however, it can be promoted to a modest degree by exogenous stimuli. Upregulating factors include both endogenous stimuli (e.g., glucocorticoids) and DNA-damaging events (e.g., alkylating agents and other genotoxins, and radiation). This occurs largely through promoter activation. The precise DNA damage stimulus leading to upregulation is unknown, but double-stranded DNA breaks are suspected to be an important common factor.

21.3.2 Mechanism of Action of AGT

The preferred substrate for AGT is double-stranded DNA. However, it can also repair lesions in single-stranded DNA and DNA–RNA hybrids. AGT repairs DNA that is being actively read in transcription or in replication, as well as DNA that is condensed and at rest. It monitors double-stranded DNA via a scanning mechanism whereby bases are flipped one at a time into the active site [10,18]. When O^6-meG is encountered, it is repaired by a single-step demethylation reaction that removes the methyl group from the oxygen in guanine and transfers it directly to the sulfur atom of the active-site cysteine (Figure 21.2). The binding of the methyl group to the cysteine is covalent, and thus the mechanism constitutes a suicide operation that permanently destroys AGT activity. For

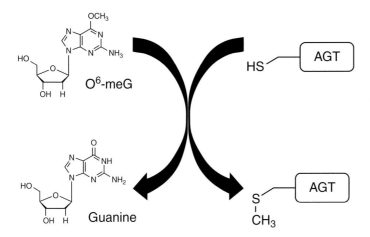

Figure 21.2 Demethylation of a DNA adduct by the enzyme alkylguanine methyltransferase (AGT). Covalent transfer of the methyl group to the cysteine site renders AGT inactive and restores the guanine residue of DNA to the normal, unmethylated form.

this reason, *MGMT* is said to react stoichiometrically, and it is not an enzyme in the most technically precise sense. Once methylated, AGT detaches from the DNA by an asparagine hinge mechanism and is rapidly ubiquitinated and degraded in the cellular proteasome. AGT can repair O^6 alkylations formed by larger groups, including ethyl, propyl, butyl, and chloroethyl; however, these reactions are slower than O^6 demethylation [27]. Also, while only AGT can repair O^6-meG, other NER enzymes can repair adducts formed with the larger alkyl groups.

Failure to repair O^6-meG will cause DNA polymerase to mismatch O^6-meG with thymine bases instead of cytosine bases on the daughter strand during DNA replication. The subsequent round of replication will then introduce an A:T transition mutation at that nucleotide site. At this point, repair of the original O^6-meG can still be effected by AGT, leaving a G:T mismatch and a mutated daughter strand. However O^6-meG:T is recognized and acted upon by additional proteins of the MMR complex, which, over several rounds of DNA replication, repeatedly excise thymine on the nascent strand and reinsert it opposite O^6-meG. This repetitive thymine reinsertion constitutes a so-called "futile MMR cycle," which leads to single-strand gaps that collapse the replication fork, causing double-strand breaks and, eventually, PCD [28].

Methylating agents also form O^4-meT lesions, but much less frequently than O^6-meG lesions. However O^4-meT is considerably more mutagenic on a per mole basis. O^4-MeT resembles cytosine, and so generates T–C base transitions. In humans, NER is probably the most important repair system for O^4-meT [29]. In contrast, human AGT appears limited in its ability to make this repair. In fact, it may be that AGT actually impedes repair by binding to the O^4-meT lesion and blocking access by NER [30]. Interestingly, certain mutated forms of human AGT can exhibit much greater O^4-meT repair activity compared to wild-type.

21.3.3 *MGMT* Transgenic and Knockout Mice

According to Fu et al. [31], mouse knockout (KO) and transgenic models support an important role for *MGMT* in DNA repair [18]. Mouse models engineered for targeted overexpression of AGT in various organs are resistant to tumor induction by alkylating agents. Specifically, mice that overexpress AGT in thymus, liver, or lung are less prone to alkylating agent-induced tumors in those tissues [32–34]. In skin cancer models induced by sequential topical application of the nitrosourea agent methylnitrosourea (MNU) followed by the tumor promoter 12-O-tetradecanoylphorbol-13-acetate (TPA), AGT overexpression protected mice against epidermal papilloma and tumor formation [35]. In C3H mice, transgenic expression of human AGT reduced their tendency toward spontaneous hepatocellular carcinoma (HCC) [36]. Overexpression of AGT can also protect bystander tissue from the toxic effects of certain alkylating agents. The CNS toxicities of the methylating agent methylazoxymethanol (MAM) were markedly reduced in mice that overexpressed AGT [37]. Interestingly – and by way of underscoring the nuanced complexities of these models – AGT overexpression did *not* protect against CNS toxicity caused by nitrogen mustard, a bifunctional alkylating agent.

MGMT undoubtedly serves an important role, but it cannot be regarded as an essential gene, because its targeted disruption causes no obvious phenotype: *MGMT* KO mice are overtly normal, live a normal life span, and reproduce normally. They are, however, more sensitive to the toxic effects of various alkylating agents [38,39]. In various trials using different alkylating agents, *MGMT* null mice exhibited a range of toxicities, including severe myelosuppression and atrophy of the spleen, thymus, and cerebellum. *MGMT* KOs were also more likely to subsequently develop tumors of the lung, liver, and colon in the aftermath of alkylating-agent treatment [18,19]. Human *MGMT* deficiency is associated with a mutator phenotype. Loss of *MGMT* expression correlates with the presence of mutations in the KRAS oncogene and/or the tumor-suppressor p53 in colorectal cancer [40,41]. Furthermore, HCC resulting from aflatoxin, an alkylating agent and a dietary environmental toxin, is more common in individuals with silenced *MGMT* [42].

21.3.4 *MGMT* as a Therapeutic Target

Because one molecule of AGT can repair only a single O^6 methyl lesion, a cell's ability to recover from alkylating stress depends first on the existing pool of AGT and then on the capacity to synthesize additional enzyme. Why an enzyme of such ostensible importance is not reactivated by a demethylation reaction is a matter of curiosity and speculation [43]. This might be explained by a high energetic cost of removing the methyl group. Alternatively, it may result from the simple rapidity with which methylated AGT enters the downstream pathways of degradation. It is further possible that such rapid degradation is necessary in order to avoid the toxic consequences of accumulated methylated AGT. For instance, it has been demonstrated that alkylated AGT can interfere with estrogen-receptor activation [44].

The suicide mechanism of AGT has motivated the use of O^6 guanine derivatives and other similar agents as potential enhancers of clinically used alkylating agents, with the rationale of diminishing levels of functional AGT, in order to enhance the anticancer potency of alkylating drugs. The two most studied compounds are synthetic analogs of guanine: O^6-benzylguanine (O^6-BG) and O^6-(4-bromothenyl)guanine (O^6-BTG; also known as lomeguatrib or PaTrin-2) (Figure 21.3) [45]. These agents have no DNA alkylating activity and are not toxic, but they do strongly inhibit AGT, presumably by an irreversible covalent transfer of a benzyl group to the active site.

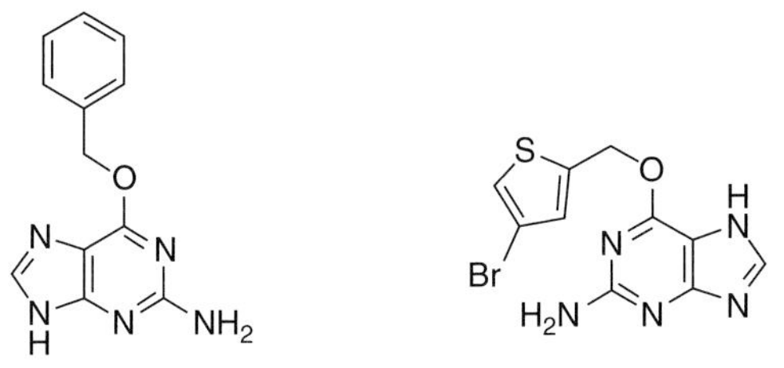

O⁶-benzylguanine (O⁶-BG)   Lomeguatrib (O⁶-BTG)

Figure 21.3 The alkylguanine methyltransferase (AGT) inhibitors O⁶-benzylguanine (O⁶-BG) and lomeguatrib. The clinical use of O⁶-BG and lomeguatrib as potentiators of methylating agent antitumor activity is complicated by severe myelosuppression.

In humans, O⁶-BG inhibits AGT with an IC_{50} of 180 nM. Its plasma half-life is relatively short, due to its hepatic cytochrome P450 (CYP) metabolism, but its major oxidation product, O⁶-benzyl-8-guanine, has a half-life of several hours, also inhibits AGT, and more effectively penetrates the cerebrospinal fluid. Compared to O⁶-BG, O⁶-BTG improves potency by about an order of magnitude; further, while O⁶-BG must be injected, O⁶-BTG can be administered orally [46,47].

Despite a compelling theoretical basis and substantiation in a host of preclinical models showing that AGT inhibition substantially improves the potency of alkylating agents, clinical trials of AGT inhibitors have been disappointing. In phase II clinical studies, O⁶-BG had no impact on clinical outcome in the treatment of various cancers, including multiple myeloma, advanced melanoma, and glioblastoma multiforme (GBM). A critical obstacle to the clinical use of such inhibitors involves reagent toxicity to healthy cells, because AGT inhibitors – like the alkylating agents themselves – do not discriminate between healthy and cancerous cells. Indeed, the combination of AGT inhibition and DNA alkylation has proved to be particularly myelosuppressive, leading to dose-limiting neutropenia, leukopenia, and thrombocytopenia.

The systemic toxicity associated with combined AGT inhibition and DNA alkylation has spurred the development of methods that will more specifically deliver AGT inhibitors to cancer cells. One method of potential application to solid tumors is the direct delivery of an AGT inhibitor into the tumor using a surgically implanted catheter. The basic feasibility of this approach has been demonstrated on a case-study basis, but has not yet been substantiated by formal clinical trials.

Chemical modification of AGT inhibitors has also been explored as a way of targeting tumor cells more selectively. Many cancer types are aberrantly rich in folate (vitamin B9) receptors, which is a selective adaptation that helps aggressively growing cells meet their exaggerated needs for *de novo* synthesis of nucleotide bases. Folate-receptor upregulation is particularly common in ovarian cancer, but also occurs in some brain and colon cancers. The mechanism of folate uptake into cells relies on tight binding of folate to the folate receptor on the external membrane, after which the receptor–ligand complex is absorbed internally. Thus, folate uptake is classic receptor-mediated endocytosis. This mechanism has been exploited as a homing tactic by linking anticancer drugs to folate derivatives [48]. Unfortunately, and like numerous other folate conjugates, the preclinical success of O⁶-BG folate derivatives has not yet been repeated in clinical trial [49].

The Nobel laureate Otto Warburg was the first to appreciate that many cancers rely disproportionately on glucose- and glycolytic-pathway metabolism [50,51]. He hypothesized that metabolic conversion of glucose to primarily fermentative products (i.e., acids and CO_2) instead of oxidative products (i.e., O_2 and water) is in fact a root cause of cancer. Because fermentative metabolism is much less efficient than oxidative metabolism, growing cancer cells require a constant and elevated supply of glucose. Cancer cells are known to upregulate the glucose metabolic pathways and to increase their glucose uptake via increased expression of plasma-membrane glucose transporters. This tendency has been exploited for the preferential delivery of anticancer agents into cancer cells, including glucose-like molecules that can inhibit AGT. This approach was first explored with STZ, which is actually a glucose relative of the methylating agent MNU. STZ has been shown to have anticancer activity and is an important therapy for metastatic cancer of the pancreatic islet cells, which tend to richly express high-capacity glucose transporters. These proof-of-principle experiments motivated the synthesis of glucose conjugates of O^6-BG and O^6-BTG, in the hope of fine-tuning the delivery of AGT inhibitors to tumor cells [52]. *In vitro* experiments show these agents (i) are readily taken up by cancer cells, (ii) are not by themselves toxic, and (iii) potentiate the toxic effects of the methylating agent TMZ and the nitrosourea alkylating agent fotemustine. However, this strategy is so far untested by human clinical trials.

Another potential complexity of AGT-inhibitor therapy relates to polymorphic isoforms of AGT that are resistant to AGT inhibitors. For instance, the G160R polymorphism of AGT is roughly 40 times less sensitive to O^6-BG relative to wild-type [53]. The frequency of G160R isotype is high in Japanese but low in Caucasians. As such, should AGT inhibitor therapy become a highly preferred treatment option, pharmacogenetic prescreening of an individual's *MGMT* gene would almost certainly assume a prominent role in the relevant clinical decision-making.

While AGT inhibition combined with alkylating therapy is limited by myelosuppression, overexpression of AGT can confer nearly complete resistance to clinically achievable doses of certain alkylating agents. Thus, it may be possible to use gene-transfer techniques to selectively reinforce host hematopoietic cells with certain AGT isoforms that are resistant to AGT inhibitors [18]. This strategy has been explored by using viral vectors to deliver genes for inhibitor-resistant *MGMT* DNA into blood stem cells. To date, gene therapy using a few isozymes of AGT has been explored, including *E. coli* alkyltransferase, which is unaffected by O^6-BG, and mutated versions of human AGT, which have a greatly reduced affinity for AGT inhibitors [54,55]. A potential hurdle to all gene-therapy treatment options involves the selection of an ideal viral vector for delivery of the genomic payload. Viral vectors tried so far include retrovirus, foamy virus, and lentivirus. As with the AGT inhibitor-based strategies already discussed, this approach is supported by preclinical data, but has yet to be proven in human clinical trials.

21.3.5 Clinical Oncology and *MGMT*

As previously stated, the *MGMT* promoter region is rich with CpG islands that are susceptible to aberrant methylation, which silences transcriptional gene expression, ultimately reducing the production of AGT. The mechanisms behind *MGMT* promoter methylation are poorly understood but are probably related to the presence of mutated isocitrate dehydrogenase enzymes (IDH1 or IDH2), which are found in approximately 80% of grade 2 and grade 3 gliomas and in secondary GBM [56]. Mutated IDH causes a buildup of 2-hydroxyglutarate, a suspected oncometabolite, which is thought to alter epigenetic regulatory mechanisms. This is not limited to *MGMT*; IDH mutations appear to be a major cause of the glioma CpG island methylator phenotype (G-CIMP), in which numerous tumor-suppressor and repair mediators are silenced by methylation. The potential use of *MGMT* methylation status in glioma diagnosis and therapy is discussed later.

21.3.5.1 *MGMT* Testing in Clinical Practice

Currently, there is great interest in *MGMT* function as it relates to both carcinogenesis and chemotherapy susceptibility. This has been best explored in the context of malignant brain neoplasm. The prevalence of malignant brain tumors is 5/100 000, 80% of which are gliomas [57]. Gliomas are named for the types of glial cells from which they originate (e.g., ependymomas, oligodendromas, and astrocytomas). Prognosis and treatment vary with tumor type and grade. Anaplastic astrocytoma (grade III) and glioblastoma mutliforme (grade IV) have the worst prognosis and fewest treatment options. Current standard of care includes initial surgical resection, if possible, followed by radiation and the methylating agent TMZ. Prior to the establishment of TMZ as the standard of care for GBM, various chemotherapy regimens were offered and implemented, but they had little impact on morbidity and mortality. Unfortunately, even with TMZ treatment, GBM still has the most dire prognosis of any solid tumor. Mean time to recurrence is 7 months, and mean survival is about 15 months. Lower-grade astrocytomas have a better prognosis, but these tumors are very likely to recur after initial treatment, typically with progression from lower to higher grade. Brain neoplasm classification and grading follow the World Health Organization (WHO) criteria, last updated in 2007 [58]. This grading scale uses histopathological tissue-sample characterization for the diagnosis, but it is important to note that many cases do not adhere to a single histological type. In summary, *MGMT* promoter methylation status and other genotypic signatures are likely to join histological evaluation as an integral part of glioma classification and staging [59].

Knowledge of patient *MGMT* promoter status may also have predictive power. Because TMZ's cytotoxicity arises from O6-meG, and because *MGMT* promoter methylation inhibits repair of that lesion, it follows that the extent of *MGMT* promoter methylation in the tumor can be an important predictive factor in determining the efficacy of TMZ and other methylating agents. This rationale is supported by clinical studies. In GBM patients with methylated *MGMT*, treatment with radiation and TMZ yielded a mean survival of 21.7 months, compared to 15.3 months in those with fully expressed *MGMT* [60]. This alludes to the important question of whether TMZ should be administered to patients with hypermethylated *MGMT*, who are unlikely to derive meaningful benefit. Because there are so few validated treatment options for GBM, and because the outcome is so bleak, current clinical practice does not typically interrogate *MGMT* status, and all patients receive TMZ anyway. This has led to a push to develop guidelines for the regular testing of *MGMT* methylation, particularly in elderly GBM patients, in whom *MGMT* hypermethylation is less common than in younger patients [61]. Classifying tumor *MGMT* methylation in patients would provide a means of avoiding futile TMZ therapy in those unlikely to respond, and of establishing who will benefit most from future therapies aimed at depleting AGT or otherwise disabling *MGMT*.

21.3.5.2 Detecting *MGMT* Methylation Status

A variety of laboratory protocols are used to determine *MGMT* methylation status in a glioma tissue sample, including methylation-specific polymerase chain reaction (MSP), pyrosequencing, mRNA sequencing by reverse-transcription polymerase chain reaction (RT-PCR), and Western blot for *MGMT* protein [25]. The best characterized and most commonly used method of determining *MGMT* status is MSP; this is also the simplest and least expensive. Briefly, bisulfite is used to convert cytosine into uracil. In strands with CpG-island methylation, the 5-methylcytosine remains unconverted, and thus unpaired. Primers specific for 5-methylcytosine residues are then used in the PCR process to replicate these sections. The weakness of MSP is that it does not detect heterogeneity throughout the tissue; therefore, non-cancer cells in the sample will distort the results [62]. Pyrosequencing is an alternative method, in which the complementary strand is synthesized one base at a time. As each nucleotide is attached to the growing strand, pyrophosphate is liberated and converted into ATP by sulfurylase. The ATP is then used by a luciferase reaction that converts luciferin to oxyluciferin, yielding a visible-light signal that can be quantified. Pyrosequencing is an option for patients whose MSP result is equivocal. RT-PCR has also been investigated for the characterization of the expression of *MGMT* mRNA, and it shows concordance with pyrosequencing [63]. Other methods, including Western blot and immunofluorescence, are under investigation, but they are currently confined to non-clinical uses. Ultimately, next-generation sequencing and microarray testing will likely become the gold standard. Regardless of the chosen method of *MGMT* typing, a standardized, reproducible protocol with stringent quality control is of utmost importance if it is to be clinically viable.

21.3.5.3 *MGMT* and GI Cancer

An expanding body of evidence suggests that knowledge of *MGMT* methylation can improve treatment of colorectal cancer, particularly for patients who have failed other chemotherapy regimens. *MGMT* methylation is found in about 30–40% of colorectal cancers, and its presence in precancerous aberrant crypt foci suggests that *MGMT* silencing is an early event that may have prognostic value [64]. TMZ has been investigated as a treatment for *MGMT*-methylated colorectal carcinoma that is refractory to other treatments, including fluoropyrimidines, oxaliplatin, and irinotecan. Early results suggest that it may benefit a subset of patients with treatment-resistant disease [65]. While the role of *MGMT* in colorectal cancer is poorly characterized relative to gliomas, *MGMT* methylation status is likely to become an important tool in the classification, prognosis, and guidance of treatment of colorectal cancer.

MGMT methylation in the stomach epithelium is associated, independently, with gastric tumors and with *Helicobacter pylori* infection. A recent report suggests that *H. pylori* eradication therapy can partially reverse *MGMT* methylation in gastric biopsy cells. The study showed that 70% of patients infected with *H. pylori* had *MGMT* methylation, compared to 30% of those who were not infected [66]. After *H. pylori* elimination, 22% of patients regained epithelial AGT protein expression.

21.3.6 BER of N-Methylpurine Lesions

As previously mentioned, BER has an important role in the repair of the major N-methylation adducts (i.e., N7-meG, N3-meA, and N3-meG) [67]. The BER process can be organized into five steps: (i) recognition and removal of the methylated base; (ii) cleavage of the abasic site; (iii) binding of scaffolding proteins to the break; (iv) insertion of the correct base; and (v) sealing of the interruptions in the phosphodiester backbone [68,69]. In the first step, DNA glycosylase is responsible for recognizing and removing the methylated base; in human cells, it is referred to as N-methylpurine-DNA glycosylase (MPG, also known as alkyladenine DNA glycosylase, AAG, EC 3.2.2.20). MPG initiates the process by flipping the nucleotide out of the double helix into its catalytic site and removing it, leaving an abasic site in the DNA strand. Because the abasic site can collapse the transcription machinery, it must be further repaired. In the second step, apurinic/apyrimidinic endonuclease 1 (APE-1) carries out the excision, cleaving the phosphodiester backbone and creating a single-strand break in the process. In the third step, poly ADP ribose polymerase 1 (PARP-1) and x-ray repair cross-complementing protein 1(XRCC1) are recruited to the break to serve as scaffolding proteins capable of stabilizing the structure and to recruit other factors that can repair single-strand breaks. In the fourth step, DNA polymerase β (Polβ) inserts a single correct base, in what is called short patch repair. Finally, in the fifth step, DNA ligase III (LIGIII) anneals the broken phosphodiester backbone. Longer sequences of repair are also possible, in what is called "long-patch" repair, which requires replication machinery – including polymerases δ and ε, proliferating cell nuclear antigen (PCNA), and ligase I – to seal the backbone.

A simplistic view of DNA repair might suggest that maximal expression of repair-machinery components would confer maximal benefit. However, transgenic mice overexpressing *MPG* prove that BER operates in a fine balance, where overactivity can produce intermediates that are equally as toxic – or even more so – compared to the original DNA lesion. In *MPG* overexpressing mice, treatment with the methylating agent methyl methanesulfonate (MMS) causes severe toxicity in numerous organs, including the thymus, spleen, bone marrow, retina, pancreas, and cerebellum [31,70]. Conversely, mice lacking *MPG* display resistance to pancreatic toxicity caused by the methylnitrosourea STZ [71] and to the CNS cerebellar toxicity caused by other methylating agents [37,71]. However, tolerance is not seen in all MPG-deficient tissues; for example, embryonic stem cells and fibroblasts become more sensitive to methylating agents when MPG is lacking [72,73].

The link between BER and cancer is evident in other steps of the BER pathway. For instance, Polβ variations are observed in perhaps one-third of human cancers [74,75]. One of the more common variants results from alternative splicing that produces a deletion variant lacking 28 amino acid residues (Polβ Δ208-236), which appears particularly relevant in colorectal and breast cancers. Other work indicates that injection of mice with Polβ-overexpressing CHO cells results in tumorigenesis. Giving hope for therapeutic potential, a small-molecule inhibitor of Polβ has been shown to potentiate the cytotoxicity of TMZ in cultured cells, and to inhibit tumor growth in mouse xenografts [76]. In addition to Polβ, other components of BER are involved in carcinogenesis, including APE-1, which is elevated in prostate, ovarian, cervical, and colon cancers [77]. Methoxyamine (Figure 21.4), a small-molecule inhibitor of APE-1, is currently under evaluation in several clinical trials as an add-on to TMZ therapy for glioblastoma, advanced solid tumors, relapsed lymphoma, and other solid tumors [78].

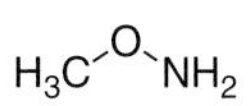

Figure 21.4 Methoxyamine, a small-molecule inhibitor of apurinic/apyrimidinic endonuclease 1 (APE-1). Methoxyamine is currently under evaluation in several clinical trials as an add-on to temozolomide (TMZ) therapy for glioblastoma, advanced solid tumors, relapsed lymphoma, and other solid tumors.

21.3.7 MMR and O^6-meG

Conventional thinking holds that the O^6-meG lesion is not itself directly responsible for the activation of downstream mediators of apoptosis. Rather, the key event is the activation of the MMR complex. MMR is a system of DNA-repair proteins with three major capabilities: **mismatch recognition**, which detects erroneous nucleotide pairs; **excision**, which removes the erroneous nucleotides from the daughter strand; and **resynthesis**, which refills the gap created with the correct nucleotides [78,80]. Compared to AGT, which is a single pseudoenzyme working in a single step, MMR is a substantially more complicated system, comprising numerous factors and reactions. Nevertheless, it is useful to compare their common and salient features. Like *MGMT*, (i) MMR is highly conserved, (ii) MMR deficiency gives rise to a mutator phenotype, and (iii) MMR deficiency is often attributable to promoter hypermethylation. However, it is important to note that in the context of methylating agent-induced cell death, the MMR status within a cell takes precedence over the *MGMT* status; that is, even in the presence of a functional *MGMT* gene and/or protein, cells deficient in MMR become highly tolerant of methylating agents [81]. The reasons for this are explained herein.

MMR begins with the binding of the MutSα complex to a single base mismatch, or a small insertion/deletion. MutSα is itself a heterodimer composed of MSH2 and MSH6 proteins. (MutSβ is also a MutS heterodimer, formed by MSH2 and MSH3 subunits. Compared to MutSα, MutSβ can repair longer sequences and is considered more important in meiosis.) MutS heterodimers have ATP-binding domains, and in fact are members of the ABC-transporter superfamily. The ATP–ADP exchange is used to power a conformational change into a clamp-like structure that complexes with another heterodimer, MutLα. MutLα is composed of MLH1, mutations of which are the most common MMR mutation in Lynch syndrome (see later), and PMS2, an endonuclease that can introduce nicks in the DNA strand. In summary, the MutSα/MutLα structure can slide along DNA in either direction, correcting aberrant sections. In the case of O^6-meG, MMR enters the previously described futile cycle, which eventually produces a double-strand break. Double-strand breaks are thought to be the common signaling point in the initiation of apoptosis resulting from various genotoxic insults, including ionizing radiation, methylating and alkylating agents, and platinum-containing compounds such as cisplatin. A critical distinction between *MGMT* and MMR mechanisms is now apparent: whereas operational *MGMT* confers tolerance to the cytotoxicity of methylating agents, a *functional* MMR system is necessary to convert O^6-meG lesions into cell-killing double-strand DNA breaks [82].

The best-known clinical impact of MMR deficiency is microsatellite instability (MSI): error-prone regions of DNA consisting of short tandem repeats of one to four nucleotides. In the absence of properly functioning MMR, misalignments between the parental and daughter strands cause small insertions or deletions to accumulate in these regions, which in turn can cause frame shifts with potentially huge phenotypic ramifications. Accordingly, MSI is very strongly associated with Lynch syndrome, also known as hereditary nonpolyposis colorectal cancer (HNPCC) [83]. In fact, MSI is evident in more than 90% of Lynch tumors. Lynch syndrome is actually a syndrome of heightened susceptibility to cancer in a plethora of tissues and organs. While endometrial cancer is the most common extracolonic manifestation, Lynch patients are also at increased risk of cancer of the ovary, stomach, small bowel, hepatobiliary system, renal pelvis and ureter, brain, and sebaceous glands.

21.3.8 Double-Strand DNA Break Repair: Homologous Recombination and Non-Homologous End Joining

As the result of a futile MMR cycle, a situation that begins as a simple-seeming methyl lesion can deteriorate into the comparatively serious double-strand DNA break. The cell possesses two major cellular repair pathways for double-strand DNA breaks: HR and NHEJ. HR uses homologous DNA strands (i.e., sister chromatids) as a template to resynthesize DNA in a broken region [84,85]. This can result in sister chromatid exchange, where homologous DNA is exchanged between the two sister chromatid strands. Because it uses a template strand as a reference, HR is a highly accurate process that does not generate errors and is therefore nonmutagenic. However, its timing in the cell cycle is restricted to late the S-phase and the G_2-phase, when an identical sister chromatid is available for template use. HR relies on resection processing of DNA strands in order to create a 3′ overhang, which is used to identify the homologous sequence within the homologous duplex DNA.

In contrast, NHEJ joins the ends of DNA strands without regard to homology and is prone to errors, including chromosomal rearrangements. It can occur throughout the cell cycle, but is preferentially activated during the G_1-phase. The cellular and molecular signals that trigger whether the HR or the NHEJ repair pathway gets activated are not well understood. Cell-cycle status is obviously important, with the precise mechanism involving an antagonistic interplay between breast cancer type 1-susceptibility protein (BRCA1) and tumor suppressor p53-binding protein 1 (53BP1). BRCA1 favors HR, while 53BP1 favors NHEJ [86]. This is clearly an intricately regulated process, with both shared and distinct molecular players between the two pathways. The common stimulus for both, however, appears to be the detection of double-strand breaks in DNA.

21.3.8.1 Homologous Recombination

In HR, the most important sensor of double-strand breaks is the MRN complex, which takes its name from its three major components: **M**RE11, **R**AD50, and **N**BS1 (NBS1 is also known as nibrin). Each component contributes a specific function: MRE11 (meiotic recombination 11) possesses 3′ to 5′ exonuclease and endonuclease activity; RAD50 contributes to the structural maintenance of chromosome (SMC) protein, which hydrolyzes ATP, which in turn contributes to DNA-conformation changes that signal the presence of a double-strand break; and NBS1/nibrin is thought to contribute to the recruitment of the ataxia telangiecstasia mutated (ATM) protein. ATM is a serine/threonine protein kinase that phosphorylates itself and numerous other downstream targets. One of its earliest and most important functions is the phosphorylation of histone 2AX (H2AX), which occurs within minutes of double-strand break induction. Phosphorylated H2AX, in combination with mediator of DNA-damage checkpoint 1 (MDC1) protein, relaxes chromatin so that it is more accessible to repair mechanisms, and serves as a recruitment signal for other mediators such as BRCA1, which is also phosphorylated by ATM.

The ATM phosphorylation cascade eventually leads to the phosphorylation and stabilization of the p53 tumor-suppressor transcription factor. The p53 protein is widely considered the "guardian of the genome," because it occupies such a central and capital position in the decision-making mechanisms the cell uses to choose between DNA repair and apoptosis. When DNA damage is moderate, p53 allows for repair by arresting the cell cycle at G_1 through transcriptional activation of p21. In cases of more severe DNA damage, p53 activates a panel of proapoptotic genes, including Fas, B-cell lymphoma-2 (Bcl-2)-associated X protein (Bax), p53-upregulated modulator of apoptosis (PUMA), phorbol-12-myristate-13-acetate-induced protein 1 (Noxa), apoptotic protease activating factor-1 (Apaf-1), and p53-induced death domain protein (PIDD). The nature of the DNA lesion appears to influence which of these mediators assumes the more important roles. It has been suggested that apoptosis due to methylating agents is particularly dependent on p53-dependent upregulation of Fas, while apoptosis due to ionization radiation or cisplatin relies on Noxa and/or PUMA [87]. The more general point is that the precise apoptotic sequence almost certainly depends on the precise nature of the DNA damage.

21.3.8.2 Non-homologous End Joining

NHEJ doubled-stranded DNA repair is less understood compared to HR [88]. Accumulating evidence suggests it is a prominent repair pathway and may in fact be responsible for repairing the vast majority of double-strand breaks caused by ionizing radiation. Because NHEJ does not require a template in the form of a sister chromatid, it can be active throughout the cell cycle, but it is also prone to insertion, deletion, and even translocation errors when blunt ends are joined incorrectly. The NHEJ machinery must be able to carry out three major functions: (i) identify double-strand breaks; (ii) enzymatically clean DNA termini; and (iii) align and ligate the broken strands. This section will describe just a few of the numerous enzymes and factors known or suspected to be important to NHEJ.

NHEJ senses double-strand breaks via the Ku70–Ku80 heterodimer, which associates with the breaks very quickly after they occur. Ku70–Ku80 is composed of two subunits with similar structure but important functional differences. Ku70 has enzymatic activity that can cleave abasic sites that would otherwise interfere with ligation. Ku80 interacts with DNA protein kinase – catalytic subunit (DNA-PKcs), which is a relative of ATM kinase, but has enzymatic activity specific to Ku-bound DNA ends. More generally, Ku with DNA-PKcs serves as an adapter complex that recruits other factors and enzymes necessary for NHEJ, such as ligase IV and its cofactors, XRCC4 and XRCC4-like factor (XLF), which function to further stabilize the strands with the associated machinery and to confer strand-connecting ligase activity. A variety of nucleases and polymerases may also become involved, particularly in cases where more extensive remodeling of the damaged site is necessary. One of the best studied is the protein DNA crosslink repair 1C (DCLRE1C), also known as Artemis. While DCLRE1C/Artemis appears to have an important role in the repair of certain double-strand breaks caused by ionizing radiation, it is actually best studied in the context of V(D)J recombination, the gene rearrangement process exclusive to developing T and B lymphocytes. V(D)J rearrangement is necessary for the production of a wide variety of immunoglobulin antigen-binding sites and T-cell antigen-receptor binding sites.

In contrast to HR, very little is known about the signaling relationship between NHEJ and p53. However, there is reasonable evidence that p53 is involved in apoptosis subsequent to NHEJ activity. In mice with wild-type p53, targeted disruption of ligase IV causes neuronal apoptosis and embryonic lethality. By contrast, in p53/ligase IV double mutants, the mice are viable and neuronal apoptosis is lessened. A similar outcome has been observed in XRCC4 mutants on varied p53 background [89]. A few additional reports describe associations between p53 and NHEJ mediators, but a direct connection between NHEJ activity and p53 activation remains uncertain.

Numerous diseases and cancer syndromes are associated with defective proteins in the DNA double-strand break-repair systems. Mutations in the ATM and MRN systems are responsible for ataxia-telangiectasia, also referred to as Louis–Bar syndrome, a rare inherited disease marked by neurodegeneration, poor coordination, sensitivity to ionizing radiation, and increased likelihood of cancer [90]. "Telangiectasia" refers to the presence of small, dilated blood vessels, which are typically visible in the sclera of the eye. A mutated DNA ligase IV (*LIG4*) gene causes a related phenotype marked by mental and physical retardation, facial malformation, sensitivity to ionizing radiation, and immunodeficiency [91]. Severe combined immunodeficiency (SCID) results from mutations in the *DCLRE1C* gene [92]. *BRCA1/BRCA2* mutations are infamous risk factors for breast and ovarian cancers, and probably contribute to a host of other cancers, including those of the prostate, pancreas, and colon [93].

The vast array of enzymes and proteins required to repair methylated DNA suggests a surplus of drug targets to be exploited for the treatment of cancer (Figure 21.5) [94]. The reality, however, is that all of the DNA-repair pathways have proved quite difficult to purposefully sabotage, at least in a manner that spares normal cells. Despite decades of

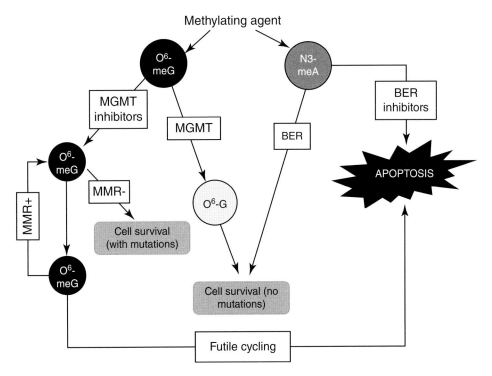

Figure 21.5 Role of repair mechanisms in methylating agent-induced cytotoxicity. Monofunctional alkylating agents generate O^6-methylguanine (O^6-meG) or N3-methyladenine (N3-meA) DNA adducts, which are repaired by different mechanisms. O^6-meG is directly repaired by O^6-methylguanine-DNA methyltransferase (MGMT), which restores normal DNA structure. If the O^6-meG lesion is not repaired by AGT, MMR enters a futile cycle that causes double-strand DNA breaks and, eventually, PCD. The N3-meA DNA adduct is repaired primarily by base excision repair (BER). Clinical trials are currently evaluating poly ADP ribose polymerase 1 (PARP-1) inhibition of BER as a method of enhancing the anticancer activity of methylating agents.

work that has elevated a host of promising compounds into clinical trials, the only DNA repair-inhibiting drug to become US Food and Drug Administration (FDA)-approved for the treatment of any cancer is olaparib, an inhibitor of the PARP-1 enzyme – and it is also worth noting that olaparib was nearly abandoned due to initially disappointing human-trial results.

21.3.9 PARP-1

The enzyme PARP-1 serves important roles in cell-cycle control, DNA-damage repair, and apoptosis [95]. The PARP family has 17 members, but PARP-1 is by far the best studied. After binding to a region of DNA damage, PARP-1 catalyzes the synthesis of long and branched homopolymers of ADP-ribose, using the oxidized form of oxidized nicotinamide adenine dinucleotide (NAD^+). The resulting ADP-ribose chains extend up to 200 units in length and are anchored to particular amino acids in histone proteins and in the PARP protein itself. Because each ADP-ribose unit has two phosphate groups, the polymer structure concentrates negative charges around DNA, which is also negatively charged. The two like-charged structures repel each other, which helps relax DNA and makes it more accessible to repair factors. Poly ADP-ribose also recruits myriad other repair mediators to the damage site. NBS1/nibrin has poly ADP-ribose-binding activity and recruits the MRN complex, paving the way for ATM activation. Poly ADP-ribose also recruits factors active in the HR and NHEJ pathways, such as BRCA1/BRCA1-associated RING-domain protein 1 (BARD1) and LIG4. Thus, PARP-1 has important regulatory effects on both of the major double-strand-break mechanisms.

PARP-1 is rapidly activated when needed: its activity increases by several orders of magnitude within seconds of DNA damage, and the consumption of NAD^+ necessary to fuel it can significantly deplete the intracellular pool of NAD^+. Such a rapid and robust response to DNA damage suggests that PARP-1 is an important component of the early DNA damage-sensing apparatus, similar to the MRN complex and the Ku heterodimers previously discussed. After the damage-repair machinery is activated, poly ADP-ribose is degraded by poly ADP ribose glycohydrolase (PARG), which cleaves ADP-ribose monomers via endo- and exoglycohydrolase activity, thus releasing them for recycling back into NAD^+. This recycling must occur quickly, because the cell cannot sustain long periods in such short supply of NAD^+, lest other important metabolic pathways such as carbohydrate and lipid metabolism become starved.

The rationale for PARP-1-inhibitor therapy is based on the genetic concept of "synthetic lethality," which hypothesizes that mutations in two or more genes will cause cell death, but that mutation of only one gene will have no effect on cell survival [96]. In this scheme, the main point of PARP-1 inhibition is to obstruct DNA single-strand-break repair. Accumulation of single-strand breaks leads to replication failures and, subsequently, to accumulation of double-strand DNA breaks. To repair double-strand DNA breaks, a normal cell favors the highly accurate process of homologous repair. In the tumor cell, however, HR is impaired due to *BRCA1/BRCA2* mutations, leaving only the error-prone NHEJ available for the repair of DNA double-strand breaks. Overreliance on NHEJ then produces widespread genomic instability, eventually culminating in cell death. To summarize, synthetic lethality is achieved in this context by superimposing a pharmacological inhibition of one gene product (PARP-1) on top of a preexisting genetic incapacitation of another (*BRCA1/BRCA2*).

Synthetic lethality also implies a certain degree of anticancer selectivity: because carriers of *BRCA1/BRCA2* mutations are heterozygous, tumor development occurs only after a second genetic hit damages the remaining functional allele. The resulting neoplastic cells should be more sensitive to PARP-1 inhibition, but normal host tissues should remain relatively tolerant, owing to operational homologous repair capabilities conferred by the functional *BRCA1/BRCA2* allele.

Paul Ehrlich, who is regarded by many as the grandfather of pharmacology, famously remarked that the journey from the laboratory to the patient's bedside is extraordinarily arduous. No doubt, this sentiment is familiar to the pioneers of PARP-1 inhibitors, the first of which earned FDA approval more than 50 years after the PARP-1 enzyme was discovered. In December 2014, olaparib was approved as monotherapy for chemotherapy-resistant ovarian cancer. Olaparib inhibits PARP-1 with an IC_{50} ~4.9 nM and is a nicotinamide mimetic with a competitive mode of inhibition, like most other candidate PARP-1 inhibitors (Figure 21.6) [97,98]. It has excellent oral availability, is eliminated by renal and hepatic mechanisms with a plasma half-life of ~6 hours, and is reasonably well tolerated as a single agent. Common adverse effects include nausea, vomiting, and GI distress. In clinical trials, however, olaparib and other PARP-1 inhibitors significantly exacerbated the myelosuppression of other chemotherapeutics (such as the methylating agents DTIC and TMZ), while achieving only minimal therapeutic gain. As is often the case in the drug-development process, real-patient results confounded exceedingly positive preclinical data. Nonetheless, TMZ-olaparib may yet become a particularly effective regimen for MMR-deficient cancers, which are resistant to TMZ and other methylating agents due to their inability to trigger the futile MMR cycle. While some early human-trial reports are disappointing, extensive interest in PARP-1 as a drug target remains. According to ClinicalTrials.gov, over 30 olaparib trials are currently in the recruiting phase, with numerous additional ones evaluating other PARP-1 inhibitors.

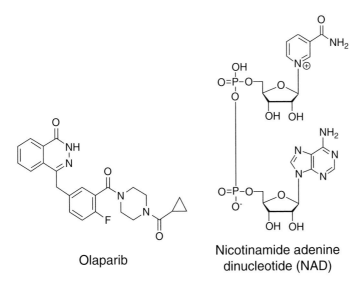

Figure 21.6 Olaparib, a competitive inhibitor of poly ADP ribose polymerase 1 (PARP-1). Most candidate PARP-1 inhibitors mimic the structure of nicotinamide adenine dinucleotide (NAD).

21.4 Methylating Agent-Induced Cell Death

We present here a simplistic summary of some of the major components of the apoptotic program, insofar as it relates to the question of methylating agent-induced cell death. For an in-depth discussion of apoptotic programs and other forms of controlled cell death, the reader should consult other chapters of this book.

21.4.1 Apoptosis and DNA Repair

Conventional thinking holds that apoptosis is a consequence of failed DNA repair. As summarized by Roos and Kaina [99], the causative link between DNA damage and apoptosis is supported by three fundamental observations. First, cells harboring a deficiency in DNA-repair systems are often hypersensitive to the toxic effects of DNA-damaging agents. This is true for deficiencies of *MGMT*, BER, NER, double-strand break repair, and DNA interstrand crosslink repair. Second, the incorporation of false nucleotides into cellular DNA induces apoptosis. Mammalian cells that are modified to use the thymidine kinase enzyme of herpes simplex virus (HSV) incorporate the nucleotide mimetic ganciclovir into their DNA. This causes strand-replication difficulties and then apoptosis. Third, the introduction of restriction enzymes into mammalian cells via electroporation causes single-strand nicks, double-strand breaks, and apoptosis. When attempts to repair DNA damage fail, cells undergo a program of cell death known as apoptosis.

Apoptosis occurs via the activation of caspases (**c**ysteine-dependent **asp**artate-directed prote**ases**), which are enzymes specifically regulated for rapid activation. Proapoptotic caspases are divided into two basic groups, based on their role in a hierarchy of activation. Initiator caspases include caspases 2, 8, 9, and 10. Their main function is to activate (by proteolytic cleavage) the second group: the effector caspases (caspases 3, 6, and 7). Amplification is an important feature of caspase-activation pathways. For example, a single activated caspase 3 can cleave and activate many other effectors.

Activation of a caspase cascade can occur via two canonical pathways: the extrinsic pathway and the intrinsic pathway [100]. Although stimuli for each pathway differ, they are linked at several levels. For instance, activation of the death-receptor pathway activates caspase 8, which can activate the BH3-interacting domain death agonist (BID), which then causes release of cytochrome c (cyt c) from the mitochondria [101]. Likewise, caspase 6 activation in the mitochondrial pathway can cleave and activate caspase 8 in the death-receptor pathway [102].

21.4.2 Extrinsic versus Intrinsic Pathway

The extrinsic pathway begins with the activation of membrane-bound receptors in response to extracellular ligands such as tumor necrosis factor (TNF), Fas ligand (FasL), and TNF-related apoptosis-inducing ligand (TRAIL). Upon receptor–ligand binding, an intracellular adapter protein such as Fas-associated death domain (FADD) is recruited on the cytoplasmic side. FADD has a death-domain motif that recruits pro-caspase 8. The FasR–FADD–pro-caspase 8 conglomerate forms the death-inducing signal complex (DISC), which recruits additional pro-caspase 8 molecules that are activated to caspase 8 by autocleavage. Activated caspase 8 (initiator) then activates caspase 3 (effector) and other downstream mediators. This eventually results in organized proteolytic degradation of the cytoskeleton, organelles, and nuclear membrane, and in the classic hallmarks of apoptosis, such as chromatin condensation.

The intrinsic pathway is the major mechanism of apoptosis and is activated in response to a variety of intracellular stresses, including reactive oxygen species (ROS), excess free calcium, and DNA damage. Such stressors cause the opening of a large, nonspecific pore in the outer mitochondrial membrane (OMM), in an event called mitochondrial permeability transition. Opening of the permeability transition pore causes collapse of the mitochondrial transmembrane gradient and failure of oxidative phosphorylation. Important mitochondrial components are lost through the pore, including apoptosis-inducing factor (AIF), cyt c, and endonuclease G (endo G). Once in the cytosol, cyt c and Apaf-1 form a structure called the apoptosome. The initiator pro-caspase 9 binds to the apoptosome and undergoes autocatalytic processing to generate active caspase 9, which in turn cleaves pro-caspase 3 into active caspase 3, resulting in downstream events that carry out apoptosis. In healthy cells, antiapoptotic proteins of the BCL-2 family prevent mitochondrial permeability transition by blocking proapopotic proteins such as Bax and Bcl-2 homologous antagonist killer (Bak) from inserting into the mitochondrial membrane and causing the release of cyt c. In response to DNA damage, however, p53 upregulates the expression of proapoptotic proteins such as Bax, Bak, Noxa, and PUMA while inhibiting the expression of antiapoptotic proteins, thereby ensuring apoptosis of the cells.

21.4.3 Alkylating Agents and Patterns of Cytotoxicity

Ostensibly, it would seem rather straightforward to question whether apoptosis caused by methylating agents proceeds by the extrinsic or the intrinsic pathway. Unfortunately, this appears to be a very complicated issue, and there is considerable disagreement in the literature (recently reviewed in [87]). For example, Roos and colleagues [103] found that, in human peripheral lymphocytes with wild-type p53, O^6 methylation caused apoptosis via the death receptor (i.e., the extrinsic pathway). In a subsequent study, this time using glioma cells, the same group found that TMZ and methyl-N'-nitro-N-nitrosoguanidine (MNNG) induced the extrinsic pathway in p53 wild-type cells, while in p53-deficient cells death proceeded by the intrinsic pathway [104]. Others used human myeloid precursor cells treated with TMZ and O^6-BG to show that apoptosis proceeded via the mitochondrial pathway, as evidenced by increased levels of p53, p21, and g-H2AX and increased mitochondrial damage. The Fas system was also engaged, although increased Fas protein was not tied to caspase 8 activation [105]. Similarly, Hickman and Samson [106] reported that MNNG induced apoptosis entirely via the intrinsic pathway in TK6 human lymphoblast cells, despite concomitant upregulation of the Fas pathway and caspase activation. Indeed, even the trustworthy generalization that p53 wild-types are more sensitive to DNA damage is challenged by the finding that p53 mutant human melanoma cells are more sensitive to alkylating agents relative to their counterparts with wild-type p53 [107]. Others have gone further, arguing that TMZ induces not apoptosis but senescence in human melanoma cells, and that MNNG in mouse embryo fibroblasts causes a regulated form of necrosis that is independent of p53 and Bcl-2 proteins [108,109]. Some of the differences between these studies may be explained by the time frame of experiments, as suggested by Quiros and colleagues [110], who found that a minimum of two rounds of replication was necessary before apoptosis was observed in MNNG-treated cells.

Given the mechanism of action of alkylating agents, which readily damage not just nuclear but also mitochondrial DNA, combined with the fact that mitochondria have limited DNA-repair capabilities, one might assume that alkylating agents cause cellular apoptosis primarily via the intrinsic pathway. However, not much is known about the effects of alkylating agents on mitochondrial DNA. Interestingly, a recent report showed that MMS caused persistent damage to mitochondrial DNA while only minimally affecting mitochondrial metabolism, over an acute time frame of several hours [111]. Nevertheless, because binding between death receptors and their ligands (i.e., TNF and CD95L, which are inflammatory components of the immune system) is necessary for extrinsic-pathway activation, it is conceivable that alkylating agents could also trigger the extrinsic pathway by activating the immune system and/or inducing the expression of death receptors such as Fas, TRAILR1, and/or TNFR1 in the affected cells. Some studies have shown that in situations where immune components are involved, the extrinsic pathway becomes activated. Whether there is a real connection between immunologic components and activation of the extrinsic pathway remains elusive.

21.5 Pharmacology of Clinically Relevant Methylating Agents

21.5.1 O^6-Methylating Agents in Clinical Use

PCB, DTIC, TMZ, and STZ (Figure 21.7, Table 21.2) are important in therapies for Hodgkin's lymphoma, glioma, melanoma, and colorectal and pancreatic islet cancer. All four agents share a common mechanism of cytotoxicity, which relies on the transfer of a methyl group to various nitrogen and oxygen atoms in DNA. Among the 13 different possible lesions, methylation of O^6 in guanine is considered the most important for achieving cytotoxicity, despite the fact that O^6-meG usually forms no more than 5–10% of

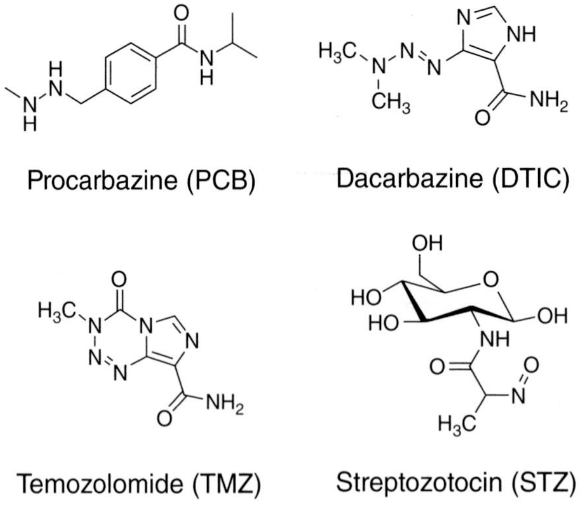

Procarbazine (PCB)

Dacarbazine (DTIC)

Temozolomide (TMZ)

Streptozotocin (STZ)

Figure 21.7 The four methylating agents currently in clinical use: procarbazine (PCB), dacarbazine (DTIC), temozolomide (TMZ), and streptozotocin (STZ).

Table 21.2 Clinical pharmacology of the methylating agents, procarbazine (PCB), dacarbazine (DTIC), temozolomide (TMZ), and streptozotocin (STZ).
BEACOPP, bleomycin, etoposide, adriamycin (doxorubicin), cyclophosphomide, oncovin (vincristine), prednisone, procarbazine; MOPP, mustargen (nitrogen mustard), oncovin (vincristine), procarbazine, prednisone; CVD, cisplatin, vincristine, dacarbazine; MAOI, monoamine oxidase inhibitor; TCA, tricarboxylic acid; GI, gastrointestinal; ABVD, adriamycin (doxorubicin), bleomycin, vincristine, dacarbazine.

Methylating agent	Clinical use and typical regimen	Pharmacokinetics	Drug interactions	Toxicity (in order of severity)
Procarbazine (PCB) Matulane® 1969	Hodgkin's lymphoma (BEACOPP, MOPP), Metastatic melanoma (alone, CVD)	Hepatic activation; $t_{1/2} = 7$ min	Potentiate effects of barbiturates, MAOIs, TCAs Disulfiram reaction with alcohol	Myelosuppression, GI effects, neurotoxicity, cutaneous hypersensitivity
Dacarbazine (DTIC) DTIC-Dome® 1975	Hodgkin's lymphoma (ABVD)	Hepatic activation; $t_{1/2} = 3$ min; poor cerebrospinal fluid permeation	Immune adjuvants may alter activity	Myelosuppression, GI effects, flu-like malaise, alopecia, photosensitivity
Temozolomide (TMZ) Temodar® 1999	Anaplastic astrocytoma, glioblastoma multiforme	Spontaneous activation, pH-dependent; $t_{1/2} = 1.8$ h	Unknown	Myelosuppression, nausea, and vomiting
Streptozotocin (STZ) Zanosar® 1982	Metastatic pancreatic islet-cell carcinoma	$t_{1/2} = 15$ min	Unknown, but used in combination therapy with alkylating agents	Nausea and vomiting, myelosuppression, hepatotoxicity, renal toxicity, hypernociception

the total methylated lesions caused by these drugs. The presence of O^6-meG in double-strand DNA causes DNA-repair mechanisms to falter, introducing DNA double-strand breaks, which trigger apoptosis. Unfortunately, the molecular events transpiring afterward are not easily generalized, due to the great number and variety of apoptotic players known to participate in different model systems, including p53, PARP-1, Fas, caspases, and Bcl proteins. Much current research is aimed at pharmacologically manipulating the relationship between DNA-repair machineries and the apoptotic program, with the goal of tipping the system toward cell death. The long-awaited advent of clinically substantiated PARP-1 inhibition is perhaps the best contemporary example of this.

PCB, DTIC, and TMZ are usually considered "non-classic" alkylating agents. STZ is a nitrosourea compound with unusual structural and functional properties. None of these agents is bifunctional, and hence none can form interstrand crosslinks. All are converted to reactive electrophilic intermediates via pathways that are complicated and, in some cases, poorly understood. Because of their highly reactive nature, the half-life of these compounds is generally short. Figure 21.8 presents simplified activation and reaction pathways for the two triazenes, DTIC and TMZ.

The best-characterized mechanism of resistance to O^6-G methylating agents is the enzyme AGT, encoded by the *MGMT* gene. AGT transfers the methyl group from guanine to itself, repairing DNA but destroying its own activity in the process. Cells possessing robust AGT activity are resistant to methylating-agent cytotoxicity, while cells that do not express the *MGMT* gene (usually due to epigenetic silencing) are more sensitive to such agents. Deficiency of DNA MMR is another clinically relevant

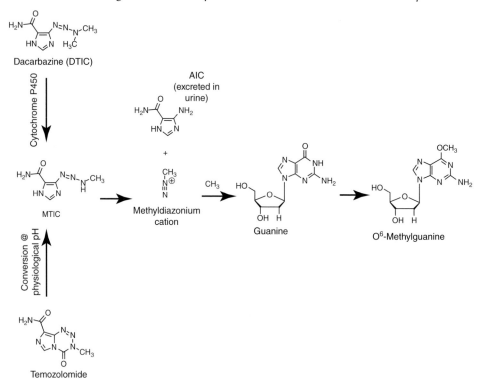

Figure 21.8 Activation and mechanism of action of the triazenes, dacarbazine (DTIC), and temozolomide (TMZ). *Source:* Adapted from Marchesi et al. [118].

resistance mechanism. In contrast to *MGMT*, whose proper function opposes O^6-meG cytotoxicity, functional MMR is *necessary* for O^6-meG cytotoxicity. MMR tries to repair O^6-meG lesions but fails, which causes double-strand DNA breaks and, ultimately, cell death.

21.5.2 Procarbazine

PCB is a methylhydrazine derivative that was initially synthesized as a monoamine oxidase inhibitor (MAOI) but was subsequently found to have anti-tumor activity. It is widely used in Hodgkin's lymphoma, small-cell lung carcinoma, and melanoma. For Hodgkin's lymphoma, PCB is used in a combination chemotherapy known as MOPP (mustargen, oncovin, procarbazine, prednisone) [112].

PCB has excellent oral bioavailability and is eliminated mostly by renal mechanisms. It requires metabolic activation via hepatic CYP enzymes, but this process is poorly understood [113]. The major mode of toxicity is O^6-guanine methylation; other toxic effects include direct chromosomal damage and inhibition of DNA, RNA, and protein synthesis.

Because it potently inhibits monoamine oxidase, PCB can cause severe hypertension when combined with tricyclic antidepressants, sympathomimetics, and fermented foods, which tend to be rich in tyramine [114]. It produces a typical disulfiram-like reaction (i.e., headaches, sweating, and facial flushing). It may also cause CNS depression and enhance the effects of other CNS depressants. The effects of barbiturates, narcotics, sedatives, and other drugs are potentiated due to interactions stemming from CYP metabolism.

An important toxic effect of PCB is mild to moderate myelosuppression, which manifests as leukopenia and thrombocytopenia. Due to immunosuppression, PCB is useful in the treatment of lupus erythematosus and post-bone marrow transplantation. GI effects lead to anorexia, nausea, and vomiting, which are improved when the dosing regimen is built up gradually. Mild neurotoxicities include depression, drowsiness, and some parasthesia, which are more intense with intravenous administration. Some patients experience skin hypersensitivity in the form of rash. PCB is considered gonadotoxic and teratogenic [115,116].

21.5.3 Dacarbazine

The triazene DTIC was proven in German clinical trials to be just as effective as PCB in children with Hodgkin's lymphoma. In the US, DTIC is currently approved for the treatment of Hodgkin's, where it is used in combination with adriamycin (doxorubicin), bleomycin, vincristine, dacarbazine (ABVD). It was shown to be a crucial component of Hodgkin's treatment by studies that omitted it in early-stage disease, which were halted because of disease progression and increased relapses [117]. DTIC is also approved for metastatic malignant melanoma.

DTIC was rationally designed in the 1950s with the goal of inhibiting purine synthesis, but this is not the basis of its anti-tumor activity. The cytotoxicity of DTIC instead results from the methylation of guanine residues at the O^6 position. DTIC is administered either intravenously or intra-arterially. Like PCB, it is metabolized by the hepatic CYP system into the active metabolite 5-(3-methyltriazen-1-yl)imidazole-4-carboxamide (MTIC); however, 50% is recovered in the urine unchanged [118]. It has no known food or drug interactions. The effect of DTIC is potentiated in the presence of other alkylating agents, due to the depletion of *MGMT* levels, which sensitizes cells to the effects of all O^6-guanine methylating agents.

DTIC produces severe nausea and vomiting in 90% of patients. Regimens that include it necessitate the strongest antiemetic therapy, which involves a combination of neurokinin-1 antagonist, 5-HT_3 antagonist, and dexamethasone [119]. Mild-to-moderate myelosuppression is common, whereas a general malaise, alopecia, and renal or hepatic dysfunction are rarely observed. Patients should avoid sunlight, as DTIC induces photosensitivity [120]. Variable immunosuppression has been observed in experimental systems. Even in combination with other methylating agents and/or with radiation treatment, DTIC does not appear to be associated with the formation of secondary malignancy. It is teratogenic in animal models, producing malformations in the urinary, reproductive, skeletal, visual, and cardiovascular systems. No widespread multicenter human trials exist, but in small-scale studies of pregnant women with Hodgkin's lymphoma, the ABVD regimen was well tolerated, with minimal teratogenicity [121]. As of now, DTIC is the standard of care, but further studies are needed.

21.5.4 Temozolomide

TMZ is a second-generation triazene that shows anti-tumor activity in murine tissue and is less toxic than the previous triazenes [118]. It was first approved for the treatment of aggressive glioma in 1999, and is currently the standard of care for essentially all GBM patients in the USA [122]. It is also approved for single-agent treatment of metastatic melanoma, as it has been shown to be just as effective as DTIC, and to possess some advantages over it (oral dosing, penetration to CNS for brain metastasis).

TMZ has several pharmacokinetic features that set it apart from PCB or DTIC: excellent tissue penetration, including of the CNS, and spontaneous conversion to the active form that does not rely on hepatic CYP bioactivation. Like DTIC, TMZ is converted to the active metabolite, MTIC. For TMZ, this conversion heavily relies on a physiological pH, but it is highly stable in acidic conditions, which ensures oral bioavailablity. Spontaneous activation of TMZ reduces CYP-based variability between individuals, and also permits regional or localized application to the site of tumor [123]. TMZ is administered orally or intravenously, and relies on both hepatic and renal elimination.

In addition to *MGMT*- and MMR-related resistance, BER may also confer resistance to TMZ, as suggested by evidence for reversal of resistance with BER inhibitors [124,125].

A major side effect of TMZ is myelosuppression. Roughly 10–15% of patients experience mild-to-moderate nausea and vomiting, which is dose-limiting and is controlled with antiemetic therapy [118]. As with the other agents in this section, TMZ is mutagenic, carcinogenic, and teratogenic, and is contraindicated in pregnant and lactating women.

21.5.5 Streptozotocin (Streptozocin)

STZ is a methyl nitrosourea glucose analog naturally produced by the soil bacterium *Streptomyces achromogenes*, and in fact was originally used as a broad-spectrum antibiotic. Generally speaking, it is a poorly characterized compound. STZ is usually categorized with other nitrosoureas as a classic alkylating agent, but its cytotoxicity arises from mostly O^6-methylation of guanine. Due to its glucose-like structure, STZ preferentially accumulates in pancreatic β cells and is commonly used to induce diabetes in animal models. As an antineoplastic agent, STZ has been incorporated in combination chemotherapy for metastatic pancreatic islet-cell tumors for more than 30 years [126,127]. The typical combination is with doxorubicin and/or fluorouracil. Its use is otherwise limited.

Unlike most of the nitrosourea alkylating agents, which are hydrophobic, STZ is hydrophilic, which limits membrane penetration. It is administered intravenously, undergoes hepatic metabolism, and is excreted primarily in the urine. STZ structurally resembles glucose, making it a good substrate for the low-affinity glucose transporter GLUT2 (encoded by *SLC2A2*), which is highly expressed in pancreatic β cells, as well as in liver and kidney. Cells that do not express GLUT2 are relatively insensitive to STZ [128]. Compared to other methylating agents, STZ's mode of cytotoxicity is less understood, but its primary toxicity is attributed to O^6-meG lesions. STZ also causes oxidative damage to DNA and other subcellular constituents by acting as an NO donor, and by interfering with glucose metabolism. Pancreatic β cells are particularly susceptible because of a shortage of free-radical scavengers believed to be secondary to increased islet-cell nitric oxide, which is involved in the insulin-release pathway [129].

There are no well-known drug–drug interactions for STZ. However, it is used in combination regimens with other antineoplastic agents for pancreatic neuroendocrine tumors (PNETs), which typically originate from pancreatic islet cells. Unfortunately, PNETs are unresectable, so STZ treatment is usually palliative. STZ has also been combined with either cyclophosphamide or 5-fluorouracil for GI neuroendocrine tumors (GINETs, typically known as "carcinoids"). However, due to poor response response rates and renal toxicity, these regimens are now rarely used in this context [126].

STZ causes nausea and vomiting, and some myelosuppression is seen in combination therapies. Its emetogenic effect can be ameliorated with 5-HT$_3$ antagonists. Because liver and kidney cells also express GLUT2, hepatotoxicity or renal toxicity is observed in 30% of patients, who may progress to renal failure that requires dialysis [126]. Although iatrogenic diabetes is possible in patients receiving STZ, it is not clinically relevant, because STZ-including regimens are palliative in nature. The STZ-induced hypernociception demonstrated in animals is a useful model for diabetic neuropathy. STZ is a recognized human carcinogen and is shown to induce tumors in rat kidney, liver, and pancreas.

Acknowledgments

We gratefully acknowledge Ms. Angela Karash, M.S., CPhT, for her assistance with figures and structures.

References

1 Emadi A, Jones RJ, Brodsky RA. Cyclophosphamide and cancer: golden anniversary. *Nat Rev Clin Oncol* 2009;**6**:638–47.
2 DeVita VT, Chu E. A history of cancer chemotherapy. *Cancer Res* 2008;**68**:8643–53.
3 Gilman A. The initial clinical trial of nitrogen mustard. *Am J Surg* 1963;**105**:574–8.
4 Deans AJ, West SC. DNA interstrand crosslink repair and cancer. *Nat Rev Cancer* 2011;**11**:467–80.

5 Fleming RA. An overview of cyclophosphamide and ifosfamide pharmacology. *Pharmacotherapy* 1997;**17**:146S–54S.
6 Kalaycio M. Bendamustine: a new look at an old drug. *Cancer* 2009;**115**:473–9.
7 Basu A, Krishnamurthy S. Cellular responses to cisplatin-induced DNA damage. *J Nucleic Acids* **2010**, 1–16 (2010).
8 Day RS, Ziolkowski CH, Scudiero DA, Meyer SA, Mattern MR. Human tumor cell strains defective in the repair of alkylation damage. *Carcinogenesis* 1980;**1**:21–32.
9 Galaup A, Paci A. Pharmacology of dimethanesulfonate alkylating agents: busulfan and treosulfan. *Expert Opin Drug Metab Toxicol* 2013;**9**:333–47.
10 Shrivastav N, Li D, Essigmann JM. Chemical biology of mutagenesis and DNA repair: cellular responses to DNA alkylation. *Carcinogenesis* 2010;**31**:59–70.
11 Beranek DT. Distribution of methyl and ethyl adducts following alkylation with monofunctional alkylating agents. *Mutat Res* 1990;**231**:11–30.
12 Stanton RA, Gernert KM, Nettles JH, Aneja R. Drugs that target dynamic microtubules: a new molecular perspective. *Med Res Rev* 2011;**31**:443–81.
13 Longley DB, Harkin DP, Johnston PG. 5-fluorouracil: mechanisms of action and clinical strategies. *Nat Rev Cancer* 2003;**3**:330–8.
14 De Bont R, van Larebeke N. Endogenous DNA damage in humans: a review of quantitative data. *Mutagenesis* 2004;**19**:169–85.
15 DeMarini DM. Genotoxicity of tobacco smoke and tobacco smoke condensate: a review. *Mutat Res* 2004;**567**:447–74.
16 Bartsch H, Montesano R. Relevance of nitrosamines to human cancer. *Carcinogenesis* 1984;**5**:1381–93.
17 Jägerstad M, Skog K. Genotoxicity of heat-processed foods. *Mutat Res* 2005;**574**:156–72.
18 Gerson SL. MGMT: its role in cancer aetiology and cancer therapeutics. *Nat Rev Cancer* 2004;**4**:296–307.
19 Kaina B, Christmann M, Naumann S, Roos WP. MGMT: key node in the battle against genotoxicity, carcinogenicity and apoptosis induced by alkylating agents. *DNA Repair* 2007;**6**:1079–99.
20 Daniels DS, Mol CD, Arvai AS, Kanugula S, Pegg AE, Tainer JA. Active and alkylated human AGT structures: a novel zinc site, inhibitor and extrahelical base binding. *EMBO J* 2000;**19**:1719–30.
21 Srivenugopal KS, Mullapudi SR, Shou J, Hazra RK, Ali-Osman F. Protein phosphorylation is a regulatory mechanism for O6-alkylguanine-DNA alkyltransferase in human brain tumor cells. *Cancer Res* 2000;**60**:282–7.
22 Sharma S, Salehi F, Scheithauer BW, Rotondo F, Syro LV, Kovacs K. Role of MGMT in tumor development, progression, diagnosis, treatment and prognosis. *Anticancer Res* 2009;**29**:3759–68.
23 Pegg AE, Fang Q, Loktionova NA. Human variants of O6-alkylguanine-DNA alkyltransferase. *DNA Repair* 2007;**6**:1071–8.
24 Margison GP, Povey AC, Kaina B, Koref M. Variability and regulation of O6-alkylguanine–DNA alkyltransferase. *Carcinogenesis* 2003;**24**(4):625–35.
25 Christmann M, Verbeek B, Roos WP, Kaina B. O(6)-methylguanine-DNA methyltransferase (MGMT) in normal tissues and tumors: enzyme activity, promoter methylation and immunohistochemistry. *Biochim Biophys Acta* 2011;**1816**:179–90.
26 Harris LC, von Wronski MA, Venable CC, Remack JS, Howell SR, Brent TP. Changes in O6-methylguanine-DNA methyltransferase expression during immortalization of cloned human fibroblasts. *Carcinogenesis* 1996;**17**:219–24.

27 Morimoto K, Dolan ME, Scicchitano D, Pegg AE. Repair of O6-propylguanine and O6-butylguanine in DNA by O6-alkylguanine-DNA alkyltransferases from rat liver and E. coli. *Carcinogenesis* 1985;**6**:1027–31.
28 O'Brien V. Signalling cell cycle arrest and cell death through the MMR system. *Carcinogenesis* 2005;**27**:682–92.
29 Klein JC, Bleeker MJ, Roelen HC, Rafferty JA, Margison GP, Brugghe HF, et al. Role of nucleotide excision repair in processing of O4-alkylthymines in human cells. *J Biol Chem* 1994;**269**:25 521–8.
30 Samson L, Han S, Marquis JC, Rasmussen LJ. Mammalian DNA repair methyltransferases shield O4MeT from nucleotide excision repair. *Carcinogenesis* 1997;**18**:919–24.
31 Fu D, Calvo JA, Samson LD. Balancing repair and tolerance of DNA damage caused by alkylating agents. *Nat Rev Cancer* 2012;**12**(2):104–20.
32 Nakatsuru Y, Matsukuma S, Nemoto N, Sugano H, Sekiguchi M, Ishikawa T. O6-methylguanine-DNA methyltransferase protects against nitrosamine-induced hepatocarcinogenesis. *Proc Natl Acad Sci USA* 1993;**90**:6468–72.
33 Dumenco LL, Allay E, Norton K, Gerson SL. The prevention of thymic lymphomas in transgenic mice by human O6-alkylguanine-DNA alkyltransferase. *Science* 1993;**259**:219–22.
34 Liu L, Qin X, Gerson SL. Reduced lung tumorigenesis in human methylguanine DNA–methyltransferase transgenic mice achieved by expression of transgene within the target cell. *Carcinogenesis* 1999;**20**:279–84.
35 Becker K, Dosch J, Gregel CM, Martin BA, Kaina B. Targeted expression of human O(6)-methylguanine-DNA methyltransferase (MGMT) in transgenic mice protects against tumor initiation in two-stage skin carcinogenesis. *Cancer Res* 1996;**56**:3244–9.
36 Zhou ZQ, Manguino D, Kewitt K, Intano GW, McMahan CA, Herbert DC, et al. Spontaneous hepatocellular carcinoma is reduced in transgenic mice overexpressing human O6-methylguanine-DNA methyltransferase. *Proc Natl Acad Sci USA* 2001;**98**:12 566–71.
37 Kisby GE, Olivas A, Park T, Churchwell M, Doerge D, Samson LD, et al. DNA repair modulates the vulnerability of the developing brain to alkylating agents. *DNA Repair (Amst)* 2009;**8**:400–12.
38 Glassner BJ, Weeda G, Allan JM, Broekhof JL, Carls NH, Donker I, et al. DNA repair methyltransferase (Mgmt) knockout mice are sensitive to the lethal effects of chemotherapeutic alkylating agents. *Mutagenesis* 1999;**14**:339–47.
39 Shiraishi A, Sakumi K, Sekiguchi M. Increased susceptibility to chemotherapeutic alkylating agents of mice deficient in DNA repair methyltransferase. *Carcinogenesis* 2000;**21**:1879–83.
40 Esteller M, Toyota M, Sanchez-Cespedes M, Capella G, Peinado MA, Watkins DN, et al. Inactivation of the DNA repair gene O6-methylguanine-DNA methyltransferase by promoter hypermethylation is associated with G to A mutations in K-ras in colorectal tumorigenesis. *Cancer Res* 2000;**60**:2368–71.
41 Esteller M, Risques RA, Toyota M, Capella G, Moreno V, Peinado MA, et al. Promoter hypermethylation of the DNA repair gene O(6)-methylguanine-DNA methyltransferase is associated with the presence of G:C to A:T transition mutations in p53 in human colorectal tumorigenesis. *Cancer Res* 2001;**61**:4689–92.
42 Zhang YJ, Chen Y, Ahsan H, Lunn RM, Lee PH, Chen CJ, Santella RM. Inactivation of the DNA repair gene O6-methylguanine-DNA methyltransferase by promoter

hypermethylation and its relationship to aflatoxin B1-DNA adducts and p53 mutation in hepatocellular carcinoma. *Int J Cancer* 2003;**103**:440–4.

43 Gouws C, Pretorius PJ. O6-methylguanine-DNA methyltransferase (MGMT): can function explain a suicidal mechanism? *Med Hypotheses* 2011;**77**:857–60.

44 Teo AK, Oh HK, Ali RB, Li BF. The modified human DNA repair enzyme O(6)-methylguanine-DNA methyltransferase is a negative regulator of estrogen receptor-mediated transcription upon alkylation DNA damage. *Mol Cell Biol* 2001;**21**:7105–14.

45 Kaina B, Margison GP, Christmann M. Targeting O^6-methylguanine-DNA methyltransferase with specific inhibitors as a strategy in cancer therapy. *Cell Mol Life Sci* 2010;**67**:3663–81.

46 Dolan ME, Mitchell RB, Mummert C, Moschel RC, Pegg AE. Effect of O6-benzylguanine analogues on sensitivity of human tumor cells to the cytotoxic effects of alkylating agents. *Cancer Res* 1991;**51**:3367–72.

47 Shibata T, Glynn N, McMurry TB, McElhinney RS, Margison GP, Williams DM. Novel synthesis of O6-alkylguanine containing oligodeoxyribonucleotides as substrates for the human DNA repair protein, O6-methylguanine DNA methyltransferase (MGMT). *Nucleic Acids Res* 2006;**34**:1884–91.

48 Nelson ME, Loktionova NA, Pegg AE, Moschel RC. 2-amino-O4-benzylpteridine derivatives: potent inactivators of O6-alkylguanine-DNA alkyltransferase. *J Med Chem* 2004;**47**:3887–91.

49 Xia W, Low PS. Folate-targeted therapies for cancer. *J Med Chem* 2010;**53**:6811–24.

50 Kim JW, Dang CV. Cancer's molecular sweet tooth and the Warburg effect. *Cancer Res* 2006;**66**:8927–30.

51 Diaz-Ruiz R, Rigoulet M, Devin A. The Warburg and Crabtree effects: on the origin of cancer cell energy metabolism and of yeast glucose repression. *Biochim Biophys Acta* 2011;**1807**:568–76.

52 Kaina B, Mühlhausen U, Piee-Staffa A, Christmann M, Garcia Boy R, Rösch F, Schirmacher R. Inhibition of O6-methylguanine-DNA methyltransferase by glucose-conjugated inhibitors: comparison with nonconjugated inhibitors and effect on fotemustine and temozolomide-induced cell death. *J Pharmacol Exp Ther* 2004;**311**:585–93.

53 Xu-Welliver M, Leitão J, Kanugula S, Meehan WJ, Pegg AE. Role of codon 160 in the sensitivity of human O6-alkylguanine-DNA alkyltransferase to O6-benzylguanine. *Biochem Pharmacol* 1999;**58**:1279–85.

54 Ragg S, Xu-Welliver M, Bailey J, D'Souza M, Cooper R, Chandra S, et al. Direct reversal of DNA damage by mutant methyltransferase protein protects mice against dose-intensified chemotherapy and leads to in vivo selection of hematopoietic stem cells. *Cancer Res* 2000;**60**:5187–95.

55 Davis BM, Koç ON, Gerson SL. Limiting numbers of G156A O(6)-methylguanine-DNA methyltransferase-transduced marrow progenitors repopulate nonmyeloablated mice after drug selection. *Blood* 2000;**95**:3078–84.

56 Cohen AL, Holmen SL, Colman H. IDH1 and IDH2 mutations in gliomas. *Curr Neurol Neurosci Rep* 2013;**13**:345.

57 Alifieris C, Trafalis DT. Glioblastoma multiforme: pathogenesis and treatment. *Pharmacol Ther* 2015;**152**:63–82.

58 Louis DN, Ohgaki H, Wiestler OD, Cavenee WK, Burger PC, Jouvet A, et al. The 2007 WHO classification of tumours of the central nervous system. *Acta Neuropathol* 2007;**114**:97–109.

59 Vigneswaran K, Neill S, Hadjipanayis CG. Beyond the World Health Organization grading of infiltrating gliomas: advances in the molecular genetics of glioma classification. *Ann Transl Med* 2015;**3**:95.
60 Hegi ME, Diserens AC, Gorila T, Hamou MF, de Tribolet N, Weller M, et al. MGMT gene silencing and benefit from temozolomide in glioblastoma. *N Engl J Med* 2005;**352**:997–1003.
61 Zarnett OJ, Sahgal A, Gosio J, Perry J, Berger MS, Chang S, Das S. Treatment of elderly patients with glioblastoma: a systematic evidence-based analysis. *JAMA Neurol* 2015;**72**:589–96.
62 Wick W, Weller M, van den Bent M, Sanson M, Weiler M, von Deimling A, et al. MGMT testing – the challenges for biomarker-based glioma treatment. *Nat Rev Neurol* 2014;**10**:372–85.
63 Tanaka S, Akimoto J, Narita Y, Oka H, Tashiro T. Is the absolute value of O(6)-methylguanine-DNA methyltransferase gene messenger RNA a prognostic factor, and does it predict the results of treatment of glioblastoma with temozolomide? *J Neurosurg* 2014;**121**:818–26.
64 Inno A, Fanetti G, Di Bartolomeo M, Gori S, Maggi C, Cirillo M, et al. Role of MGMT as biomarker in colorectal cancer. *World J Clin Cases* 2014;**2**:835–9.
65 Pietrantonio F, Ferrone F, de Braud F, Castano A, Maggi C, Bossi I, et al. Activity of temozolomide in patients with advanced chemorefractory colorectal cancer and MGMT promoter methylation. *Ann Oncol* 2014;**25**:404–8.
66 Sepulveda AR, Yao Y, Yan W, Park DI, Kim JJ, Gooding W, et al. CpG methylation and reduced expression of O6-methylguanine DNA methyltransferase is associated with *Helicobacter pylori* infection. *Gastroenterology* 2010;**138**:1836–44.
67 Wyatt MD, Allan JM, Lau AY, Ellenberger TE, Samson LD. 3-methyladenine DNA glycosylases: structure, function, and biological importance. *Bioessays* 1999;**21**:668–76.
68 Krokan HE, Nilsen H, Skorpen F, Otterlei M, Slupphaug G. Base excision repair of DNA in mammalian cells. *FEBS Lett* 2000;**476**:73–7.
69 Dianov GL, Hübscher U. Mammalian base excision repair: the forgotten archangel. *Nucleic Acids Res* 2013;**41**:3483–90.
70 Meira LB, Moroski-Erkul CA, Green SL, Calvo JA, Bronson RT, Shah D, Samson LD. Aag-initiated base excision repair drives alkylation-induced retinal degeneration in mice. *Proc Natl Acad Sci USA* 2009;**106**:888–93.
71 Cardinal JW, Margison GP, Mynett KJ, Yates AP, Cameron DP, Elder RH. Increased susceptibility to streptozotocin-induced beta-cell apoptosis and delayed autoimmune diabetes in alkylpurine-DNA-N-glycosylase-deficient mice. *Mol Cell Biol* 2001;**21**:5605–13.
72 Engelward BP, Weeda G, Wyatt MD, Broekhof JL, de Wit J, Donker I, et al. Base excision repair deficient mice lacking the Aag alkyladenine DNA glycosylase. *Proc Natl Acad Sci USA* 1997;**94**:13 087–92.
73 Engelward BP, Dreslin A, Christensen J, Huszar D, Kurahara C, Samson L. Repair-deficient 3-methyladenine DNA glycosylase homozygous mutant mouse cells have increased sensitivity to alkylation-induced chromosome damage and cell killing. *EMBO J* 1996;**15**:945–52.
74 Starcevic D, Dalal S, Sweasy JB. Is there a link between DNA polymerase beta and cancer? *Cell Cycle* 2004;**3**:998–1001.
75 Lange SS, Takata KI, Wood RD. DNA polymerases and cancer. *Nat Rev Cancer* 2011;**11**:96–110.

76 Jaiswal AS, Banerjee S, Panda H, Bulkin CD, Izumi T, Sarkar FH, et al. A novel inhibitor of DNA polymerase beta enhances the ability of temozolomide to impair the growth of colon cancer cells. *Mol Cancer Res* 2009;7:1973–83.
77 Kelley MR, Fishel ML. DNA repair proteins as molecular targets for cancer therapeutics. *Anticancer Agents Med Chem* 2008;**8**:417–25.
78 Liu L, Gerson SL. Therapeutic impact of methoxyamine: blocking repair of abasic sites in the base excision repair pathway. *Curr Opin Investig Drugs* 2004;**5**:623–7.
79 Modrich P. Mechanisms in eukaryotic mismatch repair. *J Biol Chem* 2006;**281**:30305–9.
80 Hsieh P, Yamane K. DNA mismatch repair: molecular mechanism, cancer, and ageing. *Mech Ageing Dev* 2008;**129**:391–407.
81 Fink D, Aebi S, Howell SB. The role of DNA mismatch repair in drug resistance. *Clin Cancer Res* 1998;**4**:1–6.
82 Koi M, Umar A, Chauhan DP, Cherian SP, Carethers JM, Kunkel TA, Boland CR. Human chromosome 3 corrects mismatch repair deficiency and microsatellite instability and reduces N-methyl-N'-nitro-N-nitrosoguanidine tolerance in colon tumor cells with homozygous hMLH1 mutation. *Cancer Res* 1994;**54**:4308–12.
83 Boland CR, Goel A. Microsatellite instability in colorectal cancer. *Gastroenterology* 2011;**138**(6):2073–87.
84 Helleday T. Pathways for mitotic homologous recombination in mammalian cells. *Mutat Res* 2003;**532**:103–15.
85 Li X, Heyer WD. Homologous recombination in DNA repair and DNA damage tolerance. *Cell Res* 2008;**18**:99–113.
86 Daley JM, Sung P. 53BP1, BRCA1, and the choice between recombination and end joining at DNA double-strand breaks. *Mol Cell Biol* 2014;**34**:1380–8.
87 Roos WP, Kaina B. DNA damage-induced cell death: from specific DNA lesions to the DNA damage response and apoptosis. *Cancer Lett* 2013;**332**:237–48.
88 Lieber MR. The mechanism of double-strand DNA break repair by the nonhomologous DNA end-joining pathway. *Annu Rev Biochem* 2010;**79**:181–211.
89 Gao Y, Ferguson DO, Xie W, Manis JP, Sekiguchi JA, Frank KM, et al. Interplay of p53 and DNA-repair protein XRCC4 in tumorigenesis, genomic stability and development. *Nature* 2000;**404**:897–900.
90 Lavin MF. Ataxia-telangiectasia: from a rare disorder to a paradigm for cell signalling and cancer. *Nat Rev Mol Cell Biol* 2008;**9**:759–69.
91 O'Driscoll M, Cerosaletti KM, Girard PM, Dai Y, Stumm M, Kysela B, et al. DNA ligase IV mutations identified in patients exhibiting developmental delay and immunodeficiency. *Mol Cell* 2001;**8**:1175–85.
92 Moshous D, Callebaut I, de Chasseval R, Corneo B, Cavazzana-Calvo M, Le Deist F, et al. Artemis, a novel DNA double-strand break repair/V(D)J recombination protein, is mutated in human severe combined immune deficiency. *Cell* 2001;**105**:177–86.
93 Roy R, Chun J, Powell SN. BRCA1 and BRCA2: different roles in a common pathway of genome protection. *Nat Rev Cancer* 2012;**12**:68–78.
94 Jekimovs C, Bolderson E, Suraweera A, Adams M, O'Byrne KJ, Richard DJ. Chemotherapeutic compounds targeting the DNA double-strand break repair pathways: the good, the bad, and the promising. *Front Oncol* 2014;**4**:86.
95 Li M, Yu X. The role of poly(ADP-ribosyl)ation in DNA damage response and cancer chemotherapy. *Oncogene* 2015;**34**(26):3349–56.

96 McLornan DP, List A, Mufti GJ. Applying synthetic lethality for the selective targeting of cancer. *N Engl J Med* 2014;**371**:1725–35.
97 Curtin N. PARP inhibitors for anticancer therapy. *Biochem Soc Trans* 2014;**42**:82–8.
98 Lee JM, Ledermann JA, Kohn EC. PARP inhibitors for BRCA1/2 mutation-associated and BRCA-like malignancies. *Ann Oncol* 2014;**25**:32–40.
99 Roos WP, Kaina B. DNA damage-induced cell death by apoptosis. *Trends Mol Med* 2006;**12**:440–50.
100 Fulda S, Debatin KM. Extrinsic versus intrinsic apoptosis pathways in anticancer chemotherapy. *Oncogene* 2006;**25**:4798–811.
101 Gross A, Yin XM, Wang K, Wei MC, Jockel J, Milliman C, et al. Caspase cleaved BID targets mitochondria and is required for cytochrome c release, while BCL-XL prevents this release but not tumor necrosis factor-R1/Fas death. *J Biol Chem* 1999;**274**:1156–63.
102 Cowling V, Downward J. Caspase-6 is the direct activator of caspase-8 in the cytochrome c-induced apoptosis pathway: absolute requirement for removal of caspase-6 prodomain. *Cell Death Differ* 2002;**9**:1046–56.
103 Roos W, Baumgartner M, Kaina B. Apoptosis triggered by DNA damage O6-methylguanine in human lymphocytes requires DNA replication and is mediated by p53 and Fas/CD95/Apo-1. *Oncogene* 2004;**23**:359–67.
104 Roos WP, Batista LF, Naumann SC, Wick W, Weller M, Menck CF, Kaina B. Apoptosis in malignant glioma cells triggered by the temozolomide-induced DNA lesion O6-methylguanine. *Oncogene* 2007;**26**:186–97.
105 Wang H, Cai S, Ernstberger A, Bailey BJ, Wang MZ, Cai W, et al. Temozolomide-mediated DNA methylation in human myeloid precursor cells: differential involvement of intrinsic and extrinsic apoptotic pathways. *Clin Cancer Res* 2013;**19**:2699–709.
106 Hickman MJ, Samson LD. Apoptotic signaling in response to a single type of DNA lesion, O(6)-methylguanine. *Mol Cell* 2004;**14**:105–16.
107 Naumann SC, Roos WP, Jöst E, Belohlavek C, Lennerz V, Schmidt CW, et al. Temozolomide- and fotemustine-induced apoptosis in human malignant melanoma cells: response related to MGMT, MMR, DSBs, and p53. *Br J Cancer* 2009;**100**:322–33.
108 Mhaidat NM, Zhang XD, Allen J, Avery-Kiejda KA, Scott RJ, Hersey P. Temozolomide induces senescence but not apoptosis in human melanoma cells. *Br J Cancer* 2007;**97**:1225–33.
109 Zong WX, Ditsworth D, Bauer DE, Wang ZQ, Thompson CB. Alkylating DNA damage stimulates a regulated form of necrotic cell death. *Genes Dev* 2004;**18**:1272–82.
110 Quiros S, Roos WP, Kaina B. Processing of O 6-methylguanine into DNA double-strand breaks requires two rounds of replication whereas apoptosis is also induced in subsequent cell cycles. *Cell Cycle* 2014;**9**:168–78.
111 Furda AM, Marrangoni AM, Lokshin A, Van Houten B. Oxidants and not alkylating agents induce rapid mtDNA loss and mitochondrial dysfunction. *DNA Repair* 2012;**11**:684–92.
112 Longo DL, Young RC, Wesley M, Hubbard SM, Duffey PL, Jaffe ES, DeVita VT Jr. Twenty years of MOPP therapy for Hodgkin's disease. *J Clin Oncol* 1986;**4**:1295–306.
113 Patterson LH, Murray GI. Tumour cytochrome P450 and drug activation. *Curr Pharm Des* 2002;**8**:1335–47.

114 Kraft SL, Baker NM, Carpenter J, Bostwick JR. Procarbazine and antidepressants: a retrospective review of the risk of serotonin toxicity. *Psychooncology* 2014;**23**:108–13.
115 Howell S, Shalet S. Gonadal damage from chemotherapy and radiotherapy. *Endocrinol Metab Clin North Am* 1998;**27**:927–43.
116 Lee IP, Dixon RL. Mutagenicity, carcinogenicity and teratogenicity of procarbazine. *Mutat Res* 1978;**55**:1–14.
117 Behringer K, Goergen H, Hitz F, Zijlstra JM, Greil R, Markova J, et al. Omission of dacarbazine or bleomycin, or both, from the ABVD regimen in treatment of early-stage favourable Hodgkin's lymphoma (GHSG HD13): an open-label, randomised, non-inferiority trial. *Lancet* 2015;**385**:1418–27.
118 Marchesi F, Turriziani M, Tortorelli G, Avvisati G, Torino F, De Vecchis L. Triazene compounds: mechanism of action and related DNA repair systems. *Pharmacol Res* 2007;**56**:275–87.
119 Basch E, Prestrud AA, Hesketh PJ, Kris MG, Feyer PC, Somerfield MR, et al. Antiemetics: American Society of Clinical Oncology clinical practice guideline update. *J Clin Oncol* 2011;**29**:4189–98.
120 Treudler R, Georgieva J, Geilen CC, Orfanos CE. Dacarbazine but not temozolomide induces phototoxic dermatitis in patients with malignant melanoma. *J Am Acad Dermatol* 2004;**50**:783–5.
121 Cardonick E, Iacobucci A. Use of chemotherapy during human pregnancy. *Lancet Oncol* 2004;**5**:283–91.
122 Stupp R, Dietrich PY, Ostermann Kraljevic S, Pica A, Maillard I, Maeder P, et al. Promising survival for patients with newly diagnosed glioblastoma multiforme treated with concomitant radiation plus temozolomide followed by adjuvant temozolomide. *J Clin Oncol* 2002;**20**:1375–82.
123 Reid JM, Stevens DC, Rubin J, Ames MM. Pharmacokinetics of 3-methyl-(triazen-1-yl) imidazole-4-carboximide following administration of temozolomide to patients with advanced cancer. *Clin Cancer Res* 1997;**3**:2393–8.
124 Boulton S, Pemberton LC, Porteous JK, Curtin NJ, Griffin RJ, Golding BT, Durkacz BW. Potentiation of temozolomide-induced cytotoxicity: a comparative study of the biological effects of poly(ADP-ribose) polymerase inhibitors. *Br J Cancer* 1995;**72**:849–56.
125 Curtin NJ, Wang LZ, Yiakouvaki A, Kyle S, Arris CA, Canan-Koch S, et al. Novel poly (ADP-ribose) polymerase-1 inhibitor, AG14361, restores sensitivity to temozolomide in mismatch repair-deficient cells. *Clin Cancer Res* 2004;**10**:881–9.
126 Weatherstone K, Meyer T. Streptozocin-based chemotherapy is not history in neuroendocrine tumours. *Target Oncol* 2012;**7**:161–8.
127 Eleazu CO, Eleazu KC, Chukwuma S, Essien UN. Review of the mechanism of cell death resulting from streptozotocin challenge in experimental animals, its practical use and potential risk to humans. *J Diabetes Metab Disord* 2013;**12**:60.
128 Elsner M, Guldbakke B, Tiedge M, Munday R, Lenzen S. Relative importance of transport and alkylation for pancreatic beta-cell toxicity of streptozotocin. *Diabetologia* 2000;**43**:1528–33.
129 Spinas GA. The dual role of nitric oxide in islet beta-cells. *News Physiol Sci* 1999;**14**:49–54.

22

Entosis

Jamuna A. Bai and Ravishankar Rai V.

Department of Studies in Microbiology, University of Mysore, Mysore, India

Abbreviations

Atg	autophagy protein
CICS	cell-in-cell structure
cyt c	cytochrome c
E-cadherin	epithelial cadherin
ECM	extracellular matrix
GFP	green fluorescent protein
GTPase	guanosine triphosphatase
HD	Huntington's disease
LAP	LC3-associated phagocytosis
LC3	light chain 3
LPAR2	LPA receptor 2
MCAK	mitotic centromere-associated kinesin
MLC	myosin light chain
MMP	mitochondrial membrane potential
mTOR	mammalian target of rapamycin kinase
mTORC1	mammalian target of rapamycin complex 1
RhoGAP	Rho-GTPase-activating protein
ROCK1	Rho-associated, coiled coil-containing protein kinase 1
VPS34	vacuolar protein sorting 34

22.1 Introduction

In multicellular organisms, homeostasis and cell death are well regulated. However, an imbalance in these processes leads to the increased accumulation of cells due to hyperproliferation and decreased cell death, resulting in the manifestation of diseases such as cancer. Previously, cell death during development and in tumor inhibition was mainly attributed to the process of apoptosis. However, recent studies have shown that other than apoptosis, which has a central role in cell death, certain non-apoptotic cell mechanisms are also involved, either concomitantly with apoptosis or as alternative

Apoptosis and Beyond: The Many Ways Cells Die, First Edition. Edited by James Radosevich.
© 2018 John Wiley & Son Inc. Published 2018 by John Wiley & Son Inc.

processes in its place. An example of one such non-apoptotic cell-death mechanism is the process of entosis. Entosis has been described as a cellular cannibalism and cell-death mechanism wherein a cell engulfs another living cell. This results in the death of the engulfed cell within the phagosome of the host. However, entosis has been observed to occur efficiently in apoptosis- and autophagy-incompetent cells such as MCF-7 breast cancer cells, which lack both caspase 3 and beclin 1. Entosis is not inhibited by the apoptosis inhibitors Bcl-2 or Z-VAD-fmk. The internalized cells initially are alive and healthy, capable of dividing within the engulfing cell, and are sometimes released. However, the majority of the internalized cells die by lysosomal degradation. Hence, the cell-in-cell structural formation of entosis represents a novel cell-death modality. The role of entosis in tumor progression and suppression is currently being explored [1].

22.2 Definition

Entosis is defined as a non-apoptotic cell-death program in which matrix-detached cells are involved in the invasion of one cell into another, leading to a transient state in which a live cell is contained within a host. The internalized cells are either degraded by lysosomal enzymes or released. This cell internalization process is a commonly observed "cell-in-cell" cytological feature in tumor cells [2].

22.3 Discovery of Entosis

The term "entosis" was first used by Overholtzer in 2007 [3] to describe a cell-death mechanism resembling the cell-in-cell structure (CICS) usually observed in the non-phagocytic cells in clinical tumor samples. Such "cell-in-cell" phenotypes were also observed in lymphoblasts from patients with Huntington's disease (HD), but were described there as "cellular cannibalism." Entosis occurred as a result of detachment from the extracellular matrix (ECM). The cells involved did not activate apoptotic markers or signals, and it was observed that entosis proceeded with the aid of E-cadherins and adherens-junction formation. There is no caspase activation during the initialization of entosis, but it requires the activation of the small guanosine triphosphatase (GTPase) Rho and the Rho-associated, coiled coil-containing protein kinase 1 (ROCK1) in the host cell. The internalized cells are stable initially, with some dividing and others getting released. But most disappear eventually, due to cell death associated with the lysosomal hydrolase-mediated degradation of internalized cell (Figure 22.1). Hence, entosis differs from phagocytosis and other cell-death processes. The three major differences are as follows:

- In entosis, the internalized cells are not released from the phagosome and degraded within the lysosome.
- Entosis involves homotypic interactions (the internalized and host cells are of the same type) and the absence of phagocytes.
- Entosis is not affected by chemical and genetic interventions that inhibit caspase-dependent and -independent intrinsic apoptosis (e.g., caspase inhibitors, BCL-2 overexpression) [4].

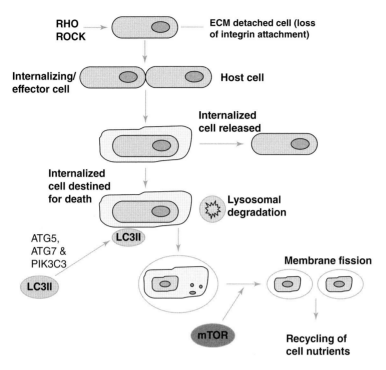

Figure 22.1 Entosis: an overview of a non-apoptotic cell-death pathway. ECM, extracellular matrix; ROCK, Rho-associated, coiled coil-containing protein kinase.

22.4 Entosis: Initiation and Progress

The entosis process starts with the detachment of cells from the ECM and eventually results in the non-apoptotic cell death of the internalized cells (Table 22.1). The internalization of one cell by another requires the formation of adherens junctions between the two. This occurs in the absence of integrin signaling and involves the force-driven invasion of one cell into the other, which requires the Rho GTPase and its downstream effector, ROCK, in the internalizing cell. The absence of integrin–ECM attachment is essential for cell invasion. Further, the generation of an unopposed and unbalanced contractile force results in the invasion and internalization of the cell into its

Table 22.1 Overview of the entosis process [6].

- Cells internalize into their neighbors following detachment from matrix
- Cell internalization occurs independent of apoptotic processes
- Cell internalization requires actin, myosin II, Rho, and ROCK activity
- Cadherins are required for entosis
- Adherens junctions track internalization
- Cells undergo entosis – invasion for cell-in-cell structure formation
- Few internalized cells can survive, and they either divide or get released
- Entosis usually results in cell death mediated by lysosomal degradation
- Internalized cells can also die by alternative mechanisms

neighboring host. Unlike phagocytosis, entosis requires force generation from the actin/myosin cytoskeleton. On invasion, the internalized cell remains stable and viable until it is released by extrication or expulsion from the host cell or it perishes within it. Most internalized cells are marked for cell death by lysosome-mediated degradation and a non-apoptotic process. However, it is not clear that the recognition mode of internalized cells is by lysosomes. The signals in the internalized cells that initiate the process of lysosomal degradation are also not well elucidated. In entosis, the lysosomes do not recruit the autophagy process, which involves sequestering cellular constituents into vesicles and targeting them to the lysosomes for complete degradation [5].

Thus, entosis has been proposed to be another mechanism by which cells are eliminated that have detached from the ECM. By internalizing and degrading cells, entosis eliminates ductal epithelial cells in normal development and inhibits tumor cells.

22.5 Entosis and Other Cell-in-Cell Structures: Differences and Similarities

Live cell engulfment results in the formation of CICSs. CICSs described to date include entosis, emperipolesis, cannibalism, and phagocytosis [7].

22.5.1 Entosis and Phagocytosis

Phagocytosis eliminates dying cells. This process is very much essential for the development and homeostasis of multicellular organisms. If dead cells are not phagocytozed and apoptotic cells are improperly degraded, damage to tissue and inflammation causing developmental defects and autoimmune disease will result. Unlike phagocytosis, entosis is a form of cell engulfment that targets only live cells, not dead ones. Phagocytosis is required for normal development and growth, whereas entosis is predominantly found in tumors [7]. It is involved in tumor-suppressive cell-death mechanisms [8] and in the promotion of tumor progression [9].

In both the phagocytic and the entotic cell, the vacuoles undergo fusion with lysosomes. The lysosomal hydrolytic enzymes degrade the dead cells in phagocytosis and execute the death of live engulfed cells in entosis [8]. After lysosome fusion, phagosome maturation involves sequential lipid phosphorylations and protein recruitment. The autophagy protein light chain 3 (LC3) is targeted to apoptotic cell phagosomes. It promotes the membrane–membrane fusion required for phagosome maturation and facilitates degradation of dead cell components [2,8]. The entotic vacuoles also utilize LC3, wherein the autophagy proteins regulate lysosome fusion and cell death [10].

Entosis, in contrast to phagocytosis, is a homotypic process, involving invasion of a live tumor cell by a similar neighboring cell. The internalized or effector cells are enclosed in the host cell, and the LC3 proteins of the host bind to their entotic vacuole membranes. The entotic vacuoles then fuse with lysosomes present in the host cell, causing degradation of the effector or internalized cells. In the entosis process, the translocation of LC3 involves the autophagy lipidation machinery, including autophagy protein 5 (Atg5), Atg7, and the lipid kinase vacuolar protein sorting 34 (VPS34) – not the autophagosomes. Thus, entosis is a cell-in-cell, non-autophagosome-dependent lysosomal death pathway that differs from phagocytosis [11].

22.5.2 Entosis and Cannibalism

CICSs are frequently observed in highly malignant or metastatic tumors. In self-defense, tumor cells engulf immune cells such as neutrophils and lymphocytes. During starvation, tumor cells phagocytoze their neighbors. This phenomenon, resembling autophagic digestion of cellular organelles, is called cannibalism. Cannibalism is used by tumor cells for survival during starvation and for immune evasion. Entosis is a homogeneous cell-in-cell phenomenon, while cannibalism can be either homogeneous or heterogeneous. Entosis can also occur in epithelial cells apart from tumor cells, but it is homotypic. Entosis involves internalization of living cells, while cannibalism has no restriction regarding dead or live cells. Entosis requires conjugations or adherens junctions for cell-to-cell contact and requires Rho and ROCK activities for internalization. Thus, entosis is an active process, requiring the aid of actin polymerization. Finally, in entosis, the effector cells perish through non-apoptotic lysosome-mediated degradation. Cannibalism is mainly used by tumor cells to kill immune cells such as neutrophils and lymphocytes. Entosis is a homogenous cell-eat-cell phenomenon that happens in both normal and tumor cells, although it has a tumor-suppressive effect in soft agar assay [12].

Metastatic cancer cells undergoing starvation by cannibalism actively "eat" other cells and degrade them in caveosomes, whereas in the internalized cells, they are trapped in the vacuole of host cells (entotic vacuole) and killed by lysosomal degradation. In entosis, internalization favors the accumulation of autophagosomes and autolysosomes within host cells and their fusion with the entotic vacuole, resulting in autophagosome-independent lysosomal death of the internalized cells [3].

22.5.3 Entosis and Emperitosis

Lysosomal degradation-involved entosis and apoptosis-involved emperitosis (killer cell-involved apoptotic cell-in-cell death) are both cell-in-cell death pathways. The difference between these two phenomena is determined by the target cells sensing different in-cell signals triggered by cell-in-cell formation. Entosis undergoes lysosomal degradation, while emperitosis undergoes either caspase- or cytochrome c (cyt c)-triggered apoptosis. In entosis, target cells degrade homogenous cells for nutrition and proliferation, whereas in emperitosisis, a heterotypic CICS is formed in response to an "in-cell danger signal" with the aim of completely eliminating effector cells by apoptosis. Thus, emperitosis and entosis are CICSs with different aims underlying their cellular biological behavior [11].

22.5.4 Entosis and Apoptosis

Apoptosis is a form of cell death that is tightly regulated by a genetic program. Apoptotic cells are engulfed by phagocytesor neighboring cells. Unlike apoptosis, entosis is activated when cells lose their attachment to the ECM. In entosis, the internalized live cells do not have phosphatidylserine residues on their plasma membranes and are devoid of the engulfment signals required for apoptosis of cells. The entotic cell–cell interaction requires cadherin for cell contact. Further, the internalization of the cell requires actin polymerization and myosin II, Rho, and Rho kinase ROCK activity in the effector cells. The host cell uses lysosomes to degrade the intruded cell, unlike the apoptosis process, which involves caspases. Entosis, which usually occurs only in tumors, may also contribute to cell death during development and tissue homeostasis, like apoptosis [13].

22.6 Molecular Mechanism Regulating Entosis

The engulfment mechanism of entosis is controlled by internalization of cells through invasion involving adherens junctions and Rho-mediated contractile force. Once internalized, the cells are enclosed in an entotic vacuole, mature, and become acidified like a phagosome, resulting in their non-apoptotic death. The maturation of entotic vacuoles and the death of internalized cells are dependent on autophagy proteins, in a process resembling LC3-associated phagocytosis (LAP). First, PI3P formation occurs, then – prior to lysosome fusion – green fluorescent protein (GFP)-LC3 is recruited to the entotic vacuoles. GFP-LC3 recruitment depends on the downstream lipidation machinery, including Atg5, Atg7, and VPS34, but does not require the upstream Ulk complex. The GFP-LC3-labeled entotic vacuole has a single-membrane structure. Further, the site of action of autophagy proteins is localized to the host cells. Inhibition of Atg5 in internalized cells doesn't affect LC3 recruitment, whereas Atg5 knockdown decreases the rate of GFP-LC3 recruitment and entotic cell death. Thus, the autophagy proteins play a noncanonical role in a non-cell autonomous death of cells [2].

22.6.1 Actin-Based Cytoskeleton Dynamics

The entosis process depends on the calcium-dependent cell-adhesion protein, epithelial cadherin (E-cadherin). The protein cadherin transduces a contractile force between the actomyosin cytoskeleton and the plasma membrane [14]. Further, myosin is activated by Rho GTPase through the effector kinases ROCK1 and ROCK2 [15]. Entosis begins with the interaction of the two cells – the effector and the host cell. The Rho–ROCK–myosin pathway is required specifically by ingested cells, as these invade their hosts using contractile force [6]. Recent studies show the importance of ezrin in regulating dynamic cytoskeleton–membrane interactions during entosis [16]. Entosis is also regulated by a Par3–Lgl antagonism in the Rho–ROCK–myosin pathway [17]. Thus, actin cytoskeleton dynamics is involved in the regulation of entosis by actin cytoskeleton dynamics [18].

22.6.2 Microtubule-Based Cytoskeleton Dynamics

Entosis has also been implicated in the formation of aneuploidy associated with an aberrant cell-division control. Microtubule plus end-tracking protein TIP150 is involved in the loading of mitotic centromere-associated kinesin (MCAK) on to the microtubule plus ends and arranges microtubule plus-end dynamics during cell division. TIP150, in association with MCAK, controls entosis via a regulatory circuitry that includes Aurora A-mediated phosphorylation of MCAK. MCAK forms an intramolecular association required for TIP150 binding. It cooperates with TIP150 to promote microtubule dynamics and modulate the mechanical rigidity of the cells during entosis. A dynamic interaction of MCAK with TIP150 arranged by Aurora A-mediated phosphorylation governs entosis through regulation of microtubule plus-end dynamics and cell rigidity. The Aurora A-mediated phosphorylation of MCAK affects its intramolecular association, which disturbs the MCAK–TIP150 interaction, resulting in inhibition of entosis.

Thus, Aurora A regulation is involved in the control of microtubule plasticity during cell-in-cell processes [18].

TIP150 is a unique microtubule plus end-tracking protein that interacts with and targets the MCAK at the plus ends, allowing it to regulate kinetochore microtubule plus-end dynamics in cell division [19]. The microtubule plus end-tracking proteins establish a complex structural platform that serves as a molecular machine to regulate microtubule dynamics and thereby orchestrate cellular dynamic events such as cell division and migration [20]. TIP150- and MCAK-dependent microtubule dynamics are essential for entosis and ensure that the interaction of TIP150 with MCAK is modulated by Aurora A phosphorylation. The MCAK–TIP150 interaction is essential for the invasion of effector cells into the host cell. The knockdown or overexpression of TIP150/MCAK affects only the cell-in-cell process and doesn't influence initial cell–cell contacts, implying a critical role for microtubule plus-end dynamics in entosis. Aurora A-mediated phosphorylation of MCAK also has a critical role in entosis, by regulating the MCAK–TIP150 interaction. Thus, a mechanistic link is present with the Aurora A-mediated phosphorylation of MCAK in cell-in-cell processes [18].

A reversible phosphorylation of MCAK is necessary for it to bind to TIP150. This is a potential molecular mechanism for the regulation of its functional activity in cell-in-cell processes. Microtubule depolymerase MCAK binds to TIP150, and this complex tracks microtubule plus ends for the regulation of microtubule plus-end dynamics and cell rigidity during entosis. Aurora A phosphorylates the N-terminal MCAK to release its intramolecular N-C association, which is what inhibits the MCAK–TIP150 interaction. Aurora A-phosphorylated MCAK is no longer associated with its microtubule plus end via TIP150 to exert its depolymerase activity, causing the microtubule plus ends to become static and cell-rigidity regulation to be disturbed. MCAK phosphorylation is dynamically regulated either by the phosphorylation/dephosphorylation cycle or by spatial separation of enzyme–substrate contact. The perturbation of cell rigidity by either hyperstabilization of microtubules or destabilization of microtubules harnesses the progression of entosis. Thus, the dynamic interaction of MCAK–TIP150 orchestrated by Aurora A phosphorylation is essential for entosis. The role of MCAK–TIP150 interaction in cell internalization of the target cell reveals that Aurora A phosphorylation governs entosis by regulating this dynamic interaction. The phosphoregulation of MCAK by Aurora A in cell-in-cell processes demonstrates a critical role for MCAK–TIP150 in regulating microtubule plus-end dynamics related to the cell-in-cell process [18].

22.7 Signaling Mechanism in Entosis

It is known that for a cell to invade into a neighboring cell, Rho-dependent signaling and actin are required [6]. The potential extracellular ligands or cell-surface receptors involved in this migratory process are unknown, but the signal transducer G-protein-coupled LPA receptor 2 (LPAR2) is specifically required by the actively invading cell. G12/13 and PDZ-RhoGEF are required for entotic invasion. This process is driven by blebbing and a uropod-like actin structure present at the end of the invading cell. The RhoA-regulated form in Dia1 is required for entosis downstream of LPAR2. Thus, a G-protein-coupled receptor-signaling process regulates actin dynamics during cell-in-cell invasion [12,13].

22.8 Occurrence of Entosis in Different Types of Cells

Entosis, or the process of forming homotypic CICSs [22], has been observed in many types of tumors, including carcinomas, sarcomas, angiomyolipomas, and melanomas. A number of studies have demonstrated the role of cell-in-cell invasion in tumor progression. Unlike the well-characterized phagocytosis and autophagy, which normally target dead, dying, or pathogenic cells and are driven by phagocytic hosts, entosis occurs when target cells invade nonphagocytic hosts. These cells are usually viable after the internalization [22].

22.9 Entosis and its Implications in Cancer: Pro-tumorigenic Process or Tumor-Suppressive Mechanism?

CICSs – especially "entosis" – are frequently found in human malignancies. A prominent oncogene, activated *Kras*, can stimulate entosis, while a prominent tumor suppressor, E-cadherin, can also induce it. In cancerous cells, the malignant population is genetically and epigenetically heterogeneous. Competition exists among distinct cells, consisting in the engulfment of one by another. Competition by entosis leads to the physical elimination of the effector or internalized cells by non-apoptotic cell death as soon as the phagosome enveloping the engulfed cell is ligated with LC3 and fuses with lysosomes [7].

The host and the effector cells are identified by mechanical deformability, which is controlled by RhoA and actomyosin. The tumor cells, which have high deformability, engulf and outcompete their neighboring cells, which have low deformability in heterogeneous populations. Activated Kras and Rac signaling downregulates contractile myosin in the host cell, allowing for the internalization of neighboring cells, which eventually undergo cell death. Thus, mechanical deformability in the effector and host cells is required for entosis to proceed [23].

Cytoskeletal tension associated with entosis involves RhoA activity and contractile myosin. Thus, entosis results from imbalances of normal cell–cell adhesion-associated forces. The force imbalance activates myosin II contractility and promotes entosis. The activation of RhoA or the introduction of phosphomimetic myosin light chain (MLC) induces the growth of adherens junctions as a result of an increased tugging force by actomyosin. Such tension imbalance, which can promote junctional growth, is capable of enclosing an entire cell. Imbalances in cell deformability may occur normally between adherent cells, but the effect of cell–matrix adhesion prevents engulfment of cells [23].

Entosisis induced by epithelial E- or P-cadherins is associated to the polarized distribution of RhoA and contractile actomyosin. These are dependent on the p190A Rho-GTPase-activating protein (RhoGAP), which is recruited to epithelial junctions. Rho-GTPase and Rho-kinase are not required in engulfing cells but are required in internalizing cells. This makes it clear that the internalized cells are destined for death. The cells undergoing internalization are accompanied by the ROCK-dependent accumulation of actomyosin. The actomyosin contractility within the effector cells constitutes a critical driving force of entosis. The process of entosis confers the host cells with a competitive advantage over aggressive tumor cells, which can salvage nutrients and amino acids for anabolic reaction via their cannibalistic activity or can increase their genomic instability subsequent to mitotic aberrations. However, inducing entosis by re-expression of epithelal E- or P-cadherins can reduce the clonogenic potential of breast cancer cells. Thus, entosis may repress or favor oncogenesis and tumor progression [7].

22.10 Biological Significance

With regard to its biological significance, entosis may have two contradictory biological effects. It "inhibits" tumor metastasis by discarding internalizing cells that detach from the ECM. However, if the effector cell blocks the cytokinesis of the target cells, it can induce the formation of multinucleated or aneuploid target cells. It can also result in chromosome instability, which leads to the further malignancy of target cells through cell fusion. If the target tumor cell with an internalized effector cell is considered as an entity, the killing of internalized cells through entosis can be considered a homeostasis mechanism designed to maintain internal stability [11].

22.11 Overview

The entosis process was first observed in human mammary epithelial MCF10A cells in suspension cultures. It was noted that some cells in small aggregates appeared to be contained inside large vacuoles, implying their internalization within a neighboring cell. The internalization of one cell within another was observed as early as 6 hours in suspension, and within 12 hours, many of the cells were completely internalized. Cell internalization was initiated approximately 3–6 hours after detachment and took several hours to complete. In some cases, complex structures resulting from many internalization events were formed, and some cells appeared to reside in a large vacuole following internalization. Internalizing cells were viable and bore into outer cells, suggesting that this process might involve the active invasion of one cell into another. Examination of engulfed cells from 6-hour suspension cultures did not exhibit any apoptotic or necrotic features. The internalizing cells retained mitochondrial membrane potential (MMP). Thus, it could be concluded that cell internalization in entosis occurs independent of apoptotic processes. Further, it was observed that in the entosis process in MCF10A cells, matrix detachment was not a phagocytic response to an apoptotic program. The overexpression of Bcl2 or treatment of cells with the caspase inhibitor had no effect on cell internalization. Thus, the entosis process was also not associated with caspase activation. The addition of phosphatydilserine liposomes, which inhibit phagocytosis, also had no effect on cell internalization. No engulfment was observed when MCF10A monolayers were incubated with apoptotic cells, suggesting that MCF10A cells cannot readily phagocytize apoptotic cells. Thus, cell engulfment of MCF10A cells in suspension cultures is not regulated by apoptotic processes, meaning it is distinct from the phagocytosis process. The complex mechanisms that control phagosome maturation up to the point of lysosome fusion are clear. Subsequently, degradation of engulfed cell components occurs, involving the export of digested components and the processing of vacuole membranes. The maturation of phagosomes and entotic vacuoles, involving fission, reduces their size and redistributes degraded components into the lysosome network of phagocytes in a mammalian target of rapamycin kinase (mTOR)-regulated manner. Lysosomes continue to fuse with vacuoles even as they undergo shrinkage.

Continual fusion/fission between lysosomes and the shrinking vacuoles disperses the vacuolar contents throughout the lysosome networks during vacuolar shrinkage. The entotic vacuoles have Rab-7, a late endosome marker, at all sizes, representing different stages of shrinkage; this suggests that vacuoles are in fact hybrid late-endosome/lysosome organelles. Large vacuoles such as phagosomes and entotic vacuoles may therefore continually fuse with lysosomes and at the same time undergo fission in order to maintain lysosome-network homeostasis. Macroendocytic vacuoles formed by phagocytosis, or the live-cell engulfment program of entosis, undergo sequential steps of maturation, leading to the fusion of lysosomes that digest internalized cargo. Phagosomes and entotic vacuoles undergo a late maturation step characterized by fission, which redistributes the vacuolar contents into lysosomal networks. Vacuole fission is regulated by the serine/threonine protein kinase mammalian target of rapamycin complex 1 (mTORC1), which localizes to vacuole membranes surrounding engulfed cells. The degrading internalized cells supply the host cells with amino acids to be used in translation, and rescue cell survival and mTORC1 activity in starved macrophages and tumor cells [9–11].

22.12 Conclusion

While apoptosis has been studied extensively for more than 2 decades, non-apoptotic cell-death mechanisms are only beginning to be elucidated. These play a major role in development and tissue homeostasis. Entosis is a mechanism of cell elimination by a non-apoptotic pathway. The invading cells in entosis also provide an advantage to the host cell, through nutrient recycling during metabolic stress and starvation. As entosis is involved in eliminating matrix-detached and cancer cells, perhaps inducing it would be therapeutically advantageous and inhibiting it would be detrimental. However, in-depth studies are required to know whether cancer cells have an advantage in escaping entosis, or whether entosis is deregulated in cancers. Entosis is a new phenomenon that requires an implicit definition, and the genetic factors controlling it have to be understood in order to establish its relevance *in vivo* [5]. The mechanistic elucidation of alternative cell-death mechanisms such as entosis is essential to gaining new insights with major implications for the fields of biology and medicine [13].

References

1 Kroemer G, Galluzzi L, Vandenabeele P, Abrams J, Alnemri ES, Baehrecke EH, et al. Classification of cell death: recommendations of the Nomenclature Committee on Cell Death 2009. *Cell Death Differ* 2009;**16**(1):3–11.

2 Florey O, Overholtzer M. Autophagy proteins in macroendocytic engulfment. *Trends Cell Biol* 2012;**22**(7):374–80.

3 Overholtzer M, Maileux AM, Mouneimne G, Normand G, Schnitt SJ, King RW, et al. A nonapoptotic cell death process, entosis, that occurs by cell-in-cell invasion. *Cell* 2007;**131**(5):966–79.

4 Borghi N, Sorokina M, Shcherbakova OG, Weis WI, Pruitt BL, Nelson WJ, Dunn AR. E-cadherin is under constitutive actomyosin-generated tension that is increased at cell–cell contacts upon externally applied stretch. *Proc Natl Acad Sci USA* 2012;**109**:12 568–73.

5. Florey O, Krajcovic M, Sun Q, Overholtzer M. Entosis. *Curr Biol* 2010;**20**(3):R88.
6. He M-F, Wang S, Wang Y, Wang X-N. Modeling cell-in-cell structure into its biological significance. *Cell Death Dis* 2013;**4**:e630.
7. Galluzzi L, Vitale I, Abrams JM, Alnemri ES, Baehrecke EH, Blagosklonny MV. Molecular definitions of cell death subroutines: recommendations of the Nomenclature Committee on Cell Death 2012. *Cell Death Differ* 2012;**19**:107–20.
8. Krajcovic M, Johnson NB, Sun Q, Normand G, Hoover N, Yao E, et al. A non-genetic route to aneuploidy in human cancers. *Nat Cell Biol* 2011;**13**:324–30.
9. Krajcovic M, Krishna S, Akkari L, Joyced JA, Overholtzer M. mTOR regulates phagosome and entotic vacuole fission. *Mol Biol Cell* 2013;**24**:3736–45.
10. Krajcovic M, Overholtzer M. Mechanisms of ploidy increase in human cancers: a new role for cell cannibalism. *Cancer Res* 2012;**72**:1596–601.
11. Kroemer G, Perfettini J-L. Entosis, a key player in cancer cell competition. *Cell Res* 2014;**24**:1280–1.
12. Overholtzer M, Brugge JS. The cell biology of cell-in-cell structures. *Nat Rev Mol Cell Biol* 2008;**9**:796–809.
13. Purvanov V, Holst M, Khan J, Baarlink C, Grosse C. G-protein-coupled receptor signaling and polarized actin dynamics drive cell-in-cell invasion. *eLife* 2014;**3**:e02786.
14. Qian Y, Shi Y. Natural killer cells go inside: entosis versus cannibalism. *Cell Res* 2009;**19**:1320–1.
15. Rath N, Olson MF. Rho-associated kinases in tumorigenesis: re-considering ROCK inhibition for cancer therapy. *EMBO Rep* 2012;**13**:900–8.
16. Sun Q, Luo T, Ren Y, Florey O, Shirasawas S, Sasazuki T, et al. Competition between human cells by entosis. *Cell Res* 2014;**24**:1299–310.
17. Wan Q, Liu J, Zheng Z, Zhu H, Chu X, Dong Z, et al. Regulation of myosin activation during cell–cell contact formation by Par3-Lgl antagonism: entosis without matrix detachment. *Mol Biol Cell* 2012;**23**:2076–91.
18. Wang S, Guo Z, Xia P, Liu T, Wang J, Li S, et al. Internalization of NK cells into tumor cells requires ezrin and leads to programmed cell-in-cell death. *Cell Res* 2009;**19**:1350–62.
19. Wang Y, Wang X. Cell-in-cell: a virgin land of cell biology. *Oncoimmunology* 2013;**2**(10):e25988.
20. Ward T, Wang M, Liu X, Wang Z, Xia P, Chu Y, et al. Regulation of a dynamic interaction between two microtubule-binding proteins, EB1 and TIP150, by PCAF orchestrates kinetochore microtubule plasticity and chromosome stability during mitosis. *J Biol Chem* 2013;**288**:15 771–85.
21. White E. Entosis: it's a cell-eat-cell world. *Cell* 2007;**131**:840–2.
22. Xia P, Wang Z, Liu X, Wu B, Wang J, Ward T, et al. EB1 acetylation by P300/CBP-associated factor (PCAF) ensures accurate kinetochore-microtubule interactions in mitosis. *Proc Natl Acad Sci USA* 2012;**109**:16 564–9.
23. Xia P, Zhou J, Song X, Wu B, Liu X, Zhang S, et al. Aurora A orchestrates entosis by regulating a dynamic MCAK–TIP150 interaction. *J Mol Cell Biol* 2014;**6**:1–15.
24. Yuan J, Kroemer G. Alternative cell death mechanisms in development and beyond. *Genes Dev* 2010;**24**:2592–602.

23

Mitotic Catastrophe

Raquel De Souza,[1] Lais Costa Ayub,[2] and Kenneth Yip[3]

[1]*University Health Network, Department of Radiation Physics, Pharmaceutical Sciences, University of Toronto, Toronto, ON, Canada*
[2]*Department of Molecular, Genetic and Structural Biology, University of Ponta Grossa, Ponta Grossa, PR, Brazil*
[3]*Department of Biology, University of Toronto, Toronto, ON, Canada*

Abbreviations

AIF	apoptosis-inducing factor
APC	anaphase-promoting complex
ATG5	autophagy-related protein 5
ATM	ataxia telangiectasia mutated
ATR	ataxia telangiectasia mutated-RAD3 related
Bax	Bcl-2-associated X protein
CAK	Cdk-activating kinase
Cdc2	cell-division cycle 2
Cdk1	cyclin-dependent kinase 1
Chk1	checkpoint kinase-1
Chk2	checkpoint kinase-2
cyt c	cytochrome c
DSB	double-strand break
H2AX	histone 2AX
MC	mitotic catastrophe
MCT-1	multiple-copies T-cell malignancy 1
NCCD	Nomenclature Committee on Cell Death
NER	nucleotide excision repair
PAR	poly ADP ribose
PARG	poly ADP ribose glycohydrolase
PARP	poly ADP ribose polymerase
PDT	photodynamic therapy
Plk1	polo-like kinase 1
ROS	reactive oxygen species
TUNEL	terminal deoxynucleotidyl transferase dUTP nick end labeling

Apoptosis and Beyond: The Many Ways Cells Die, First Edition. Edited by James Radosevich.
© 2018 John Wiley & Son Inc. Published 2018 by John Wiley & Son Inc.

23.1 Introduction

The definition of "mitotic catastrophe" (MC) has continuously changed over the years, and a consensus has yet to be reached regarding its precise designation. It has been considered a state that may or may not result in cell death, a distinct case of apoptosis, or a pre-lethal stage [1]. It is currently accepted as a mechanism ensuing from mitotic failure that leads to apoptosis, non-apoptotic cell death, or senescence [2]. Specifically, the 2012 Nomenclature Committee on Cell Death (NCCD) states that MC is triggered during the M phase by damage to the mitotic apparatus, involves mitotic arrest, and triggers either cell death or senescence [2].

MC ensues as a result of premature entry into mitosis when DNA damage or abnormalities in mitotic spindles are present [2]. In fact, the idea of MC first emerged as a way of describing an observation in a mutated *Schizosaccharomyces pombe* strain that lacked *wee1*, a molecule responsible for negatively regulating premature mitotic entry in the presence of DNA damage [3]. Great abnormalities in segregation of chromosomes could be observed [3]. Premature mitotic entry is a result of debilitation of the checkpoints responsible for hindering entry into mitosis without damage being repaired first. Two checkpoints play important roles in MC when deficient. DNA damage or incomplete chromosomal replication is sensed at the G_2/M checkpoint, which prevents entry into mitosis [4]. Once mitosis is underway, the spindle assembly checkpoint ensures that anaphase does not proceed until mitotic spindle formation has occurred properly and chromosomes have secured their bipolar attachment [3]. Upon damage, progression of the cell cycle is halted by these cell-cycle checkpoints. If damage is not repaired, cell-death mechanisms ensue to eliminate the damaged cell [5]. Cells that lack the tumor suppressor p53, for instance – as is the case with more than 50% of human solid tumors – are deprived of a G_2/M checkpoint, and can escape arrest [6]. Aberrant entry into mitosis while the cell is still critically damaged results in MC. This can result in cell death; alternatively, if cell death does not follow, the cells can escape the MC state, and become tetraploid after one aberrant division and endopolyploid after many cycles [3]. Such cells can develop mechanisms for surviving in this state and establish abnormal cell populations [6]. This process can lead to carcinogenesis, as MC-driven micronucleation and polyploidy due to abnormal mitosis are characteristics of tumor progression [7]. Thus, MC leading to cell death can be considered a process that prevents genomic instability in the face of damage to DNA or mitotic spindles [8]. In addition to DNA damage, MC can also ensue as a result of agents that excessively promote or hinder microtubule assembly [9], fusion of mitotic cells with interphase cells, or centrosome overduplication or failure to duplicate [3]. Cell fusion results in MC because active cyclin-dependent kinase 1 (Cdk1)/cyclin B (see later) in the mitotic cell drives interphase cells into premature mitosis [10].

MC has been widely described as a pre-stage of apoptosis [9], but apoptosis is not strictly required for the lethality of MC. For instance, treatment of HeLa cells with etoposide concurrent with Bcl2 suppression did not change the overall clonogenic survival of cells, indicating that although apoptosis was diminished, overall lethality was not [9]. Cell manipulations that would prevent apoptosis upon treatment with DNA damaging or mitotic spindle-poisoning agents, such as caspase inhibitors, do not prevent MC, as giant multinucleated cells can still be observed [3]. However, this does not hold as an argument to support that MC and apoptosis are distinct, since apoptosis can occur in a caspase-independent manner [3]. Morphologically, MC is quite distinct [5]. Chromatin condensation and chromosome segregation are not completed properly, ensuing giant polyploidy cells that contain decondensed chromatin in multiple micronuclei [3,5]. Micronuclei are structures that arise from nuclear envelope formation around mis-segregated chromosomal clusters, and are at most one-third of the size of a normal nucleus [1,9]. Such features are unique to MC and are not observed in cells undergoing apoptosis. The link between MC and other modes of cell death is quite complex, and is discussed in more detail later in the chapter.

Mounting evidence showcases the significant role MC plays in cancer chemotherapy and radiotherapy. Agents that are widely used in the clinical setting, including cisplatin, taxanes, doxorubicin, and etoposide, are continuously shown to induce cell death through MC. Such agents can induce death through MC and/or other pathways of cell death, depending on the cell type or even the dose. As an example, exposure of cells to different doses of doxorubicin yields populations of cells that are very distinct morphologically. Treatment with progressively higher concentrations increasingly induces a morphology characteristic of apoptosis, with cells displaying a small and rounded shape and invaginated membranes. Lower concentrations result in large cells with multiple micronuclei, typical of MC [5]. This phenomenon can be exploited when dealing with cancers resistant to treatment. As cancer cells possess faulty checkpoint controls, they may tolerate genotoxic insults induced by chemotherapeutic agents without undergoing the usual apoptotic pathway. This type of resistance can either be acquired during therapy due to adaptive mutations in checkpoint pathways or it can be inherent to the cells. Certain drugs and other interventions can be employed to overcome such resistance by inducing MC as an alternative route to cell death [11]. The various opportunities presented by MC in the treatment of cancer are discussed in detail in this chapter.

23.2 Mitosis and the Cell Cycle in the Context of MC

The idea of a cell cycle was first proposed in 1952 by Howard and Pelc, who observed fava bean cell division and categorized their observations in terms of phases of the cell cycle: G_1, or gap phase 1; S, when DNA synthesis occurs; G_2, or gap phase 2; and M, or mitosis [12]. Later studies identified the quiescence phase, G_0 [13]. Cells undergo growth during G_1, duplicate their genomes and begin to form mitotic spindles during S, experience additional growth during G_2, and perform mitosis and cytokinesis during M [14]. The cell-cycle progress described in this section is summarized in Figure 23.1.

A simple depiction of mitosis itself can be synopsized into four key stages. (i) During prophase, duplicated chromatin condensation occurs and the nuclear envelope disintegrates. Centrosomes start to move toward the two polar ends of the cell. Kinetochores attach to centromeres, the physical point that joins sister chromatids together. (ii) During metaphase, tubulin units polymerize to form microtubules that connect centromeres, via kinetochores, to centrosomes at either pole of the cell. At this point, chromosomes are aligned at the equator of the cell. (iii) Microtubules then depolymerize during anaphase, causing each pair of sister chromatids to segregate and migrate toward opposite spindle poles. (iv) A new nuclear membrane forms around each set of chromosomes, marking telophase. After mitosis, cell division is completed by cytokinesis. Mammalian cell division, from cytoskeleton reorganization to chromosome segregation and cytokinesis, is a complex process that requires about 1 hour to complete [15].

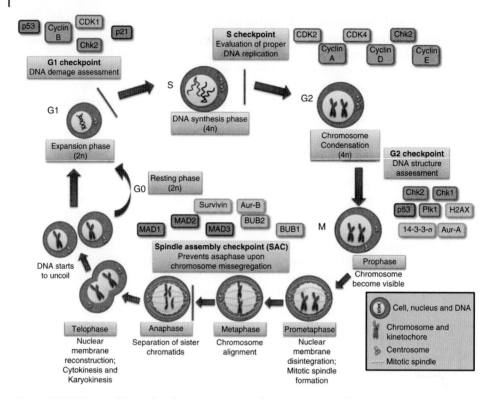

Figure 23.1 Stages of the cell cycle and mitosis, and the main molecular entities involved in the two cell-cycle checkpoints most relevant to mitotic catastrophe (MC).

When a cell transitions from G_1 to S, it commits to completing an entire cycle [16]. To avoid the proliferation of abnormal cells or unnecessary cell death in the case of damage or mistakes during the cycle, cells have quality-control mechanisms that operate throughout, termed cell-cycle checkpoints. These checkpoints operate between or during phases of the cell cycle to ensure proper fidelity of events prior to continued progression through the cycle [14]. If a problem is detected, the cell cycle is halted, or arrested, by molecular players within each checkpoint until repair or other requirements are met [14]. Cell-cycle checkpoint requirements are evolutionarily conserved, as supported by the fact that five of the six genes required for cell-cycle control in the yeast *Saccharomyces pombe* have homologs in vertebrates [17]. Cdks, in complex with their adaptor molecules, the cyclins, are the key participants in cell-cycle progression, and their inhibition is necessary to halt the cell cycle [14,18].

The mitosis-promoting factor is composed of Cdk1 (also termed cell-division cycle 2, Cdc2) bound to cyclin B. When complexed, these two molecules induce cell-cycle progression from G_2 to M phase. The requirement of this mitosis-promoting factor is evolutionarily conserved from yeast to mammals [15]. Cyclin B is synthesized from late S phase and is degraded in metaphase immediately after chromosome alignment [15]. The tumor-suppressor p53, a transcription factor, downregulates *cdk1* expression while increasing the abundance of p21, 14-3-3σ, and Gadd45, which are inhibitors of Cdk1 [19]. If DNA damage is sensed during the G_1 phase, p53 activates p21, which directly inhibits Cdks and causes G_1 arrest [18,20]. During the G_2 phase, Cdk1 is inhibited by Wee1 and Myt1 kinases, while during mitosis, Cdc25 phosphatases and the Cdk-activating kinase (CAK) are responsible for activating it [3,19]. These Cdk1 regulators are controlled by upstream molecules such as Chk1, Chk2, p38, polo-like kinase 1 (Plk1), and Aurora A, which, together, regulate arrest or progression during G_2/M [15]. Cyclin B moves to the nucleus and accumulates so that it can bind to Cdk1 when it becomes available [19]. Transition from prophase to metaphase occurs when Cdc25 dephosphorylates and activates Cdk1, priming it for coupling with cyclin B [21]. Plk1 is responsible for activating Cdc25 [21]. It is important to note, however, that Plk1 has a number of other targets, including the anaphase-promoting complex (APC) and tubulins. Plk1 inhibition leads to G_2/M arrest and eventual apoptosis, demonstrating the importance of its multifactorial role in G_2/M transition [21]. For cells to proceed into anaphase, the APC must disable the activity of Cdk1/cyclin B by triggering cyclin B ubiquitination and subsequent degradation [19,22]. In order for mitosis to proceed, the Cdk1/cyclin B complex initiates a range of signaling cascades, activating processes such as condensation of chromatids, decomposition of the nuclear membrane, and organization of the microtubules [22].

The checkpoints most relevant in the context of MC are the G_2/M checkpoint, which controls entry into mitosis, and the spindle-assembly checkpoint, which ensures appropriate chromosome attachment to kinetochores, centrosome alignment, and spindle formation. Cells that lack a functional G_1 checkpoint, as is the case in some cancer cells, rely heavily on the proper functioning of the G_2/M checkpoint to avoid cell death [14].

23.2.1 The G_2/M Checkpoint

If DNA damage is detected, the G_2/M checkpoint blocks entry into mitosis by hindering the activation of the mitosis-promoting factor Cdk1/cyclin B through activation of Chk1 and Chk2 [6,10]. The sensors HUS7, RAD1, RAD9, and RAD17 detect DNA damage [3], causing the ataxia telangiectasia mutated (ATM) and ataxia telangiectasia mutated-RAD3 related (ATR) protein kinases to activate histone 2AX (H2AX) and the checkpoint kinases Chk1 and Chk2 via phosphorylation [23]. ATM is specifically crucial in the cellular response to DNA damage, while ATR is involved in a broader range of checkpoint functions and is considered a key mediator of the G_2/M checkpoint [17]. The absence of ATR is enough for cells to overcome this checkpoint [17]. The ensuing γH2AX foci recruit proteins associated with DNA repair to the damage site, including RAD51 (which has a crucial role in repair), DNA damage-binding protein 2 (which is needed for nucleotide excision repair, NER), and BARD1 (which interacts with BRCA1) [8]. Meanwhile, active Chk1 inactivates proteins in the Cdc25 family of phosphatases. The Cdc25 phosphatases are responsible for activating Cdk1. When Chk1 inhibits Cdc25 from performing this role, cell-cycle arrest ensues, inhibiting entry into mitosis until DNA damage is repaired. Wee1 also becomes activated, to enhance Cdk1 inhibition [24]. Chk2 functions as an effector that can halt the cell cycle by phosphorylating numerous substrates, including p53, Cdc25 phosphatases, and BRCA1 [3]. It also indirectly activates proapoptotic proteins, such as Apaf-1 [10].

In normal cells, when irreparable DNA damage occurs, cell death takes place during interphase. Alternatively, cells can proceed through interphase, in which case cell-cycle arrest ensues in G_1 or G_2, followed by apoptosis, such that damaged cells never enter mitosis prematurely [25].

23.2.2 The Spindle-Assembly Checkpoint

At the exit of mitosis is yet another checkpoint, termed the spindle-assembly checkpoint, which can block the APC from activation. This molecular complex is required to activate the degradation of cyclin B, which leads to chromatid separation in anaphase and mitotic exit [6]. When APC is inhibited, nuclear division cannot occur [3]. The spindle-assembly checkpoint monitors chromosome segregation to ensure it occurs correctly [8]. Proteins associated with this checkpoint, including BUB1, BUB2, and MAD1, 2, and 3, localize to kinetochores and prevent anaphase from occurring until the kinetochores have correctly attached to microtubules [3,8,26]. Numerous proteins catalyze this attachment, including dynein, which is recruited to kinetochores by dynactin to ensure proper orientation of microtubule attachment [8]. The chromosomal passenger complex of proteins, composed of Aurora B kinase, survivin, INCENP, and Borealin/Dasra-B, is involved in the alignment of chromosomes [26]. Survivin is a substrate of Cdk1 that complexes with Aurora B kinase and the inner centromere-binding protein at the microtubules of the mitotic spindle [3]. Microtubules consist of polymerized alpha and beta tubulin subunits, which form protofilaments [27,28]. The protein RCC1 is responsible for generating a RanGTP gradient that allows tubulin polymerization and depolymerization, necessary for proper microtubule function [26]. As mentioned earlier, Aurora A is one of the upstream regulators of Cdk1. Additionally, it is a key molecular player in centrosome duplication and the formation of the bipolar mitotic spindle [29]. Malfunction of any of these orchestrated molecular interactions is enough for the checkpoint to halt the cell cycle at metaphase, which can result in MC [26]. APC is only activated once cells pass the spindle-assembly checkpoint [24]. Otherwise, cells become arrested between metaphase and anaphase. In cells that have normal p53 activity, a tetraploid state is maintained at G_1, and p21 is activated.

23.3 Molecular Processes in MC Initiation and Execution

In essence, MC occurs when cells prematurely enter or resume mitosis despite the presence of damage or irregularities due to unscheduled activation of the Cdk1/cyclin B complex [10]. MC is quite predominant in cancer cells, in which one of the most common types of change is deregulation of checkpoint control mechanisms [9]. Cells deficient in p53 may undergo MC rather than immediate apoptotic death during interphase [1,30]. A lack of p53 constitutes a hindered G_2/M checkpoint; this is seen in over 50% of human tumors [1]. Cancer cells are also more prone to MC because sensors of DNA damage and regulators of cell-cycle arrest are deregulated [19].

During MC, cells with compromised p53 function can escape mitotic arrest and enter the endocycle, where genome duplication is maintained without cell division [30]. Such cells have more than two sets of chromosomes, a phenomenon termed "polyploidy." A small percentage of these cells are able to transition from the endocycle into endomitosis, starting directly from metaphase, which results in abnormal polyploidy cells that can accumulate a ploidy of up to 64C upon continuous repetition of this process [30,31]. This is a common occurrence in cancer cells. Endocycling depends on prolonged arrest at mitosis, as is seen during MC, and requires the activity of both Cdk1/cyclin B and APC. In addition to mononucleated giant cells, multinucleated giant cells can also form, by overcoming the spindle checkpoint via cyclin B hyperactivity and resuming the endocycle at anaphase [31]. Interestingly, meiosis-specific genes have been implicated in the formation of polyploid cells following MC in p53-deficient cells [32].

23.3.1 Cell-Cycle Arrest Failure upon DNA Damage

As discussed earlier, normal cells do not enter mitosis upon cell-cycle arrest at interphase, G_1, or G_2 [25]. Such cells undergo apoptosis if damage cannot be repaired. Premature entry into mitosis only occurs if a deficiency in cell-cycle checkpoints is present, and such entry is a requirement for MC [25]. This failure to maintain G_2 arrest consists in Cdk1/cyclin B activation when DNA damage issues have not yet been resolved [24]. In general, cells with wild-type p53 undergo apoptosis upon sustaining detrimental DNA damage, whereas p53-deficient cells have a higher tendency to undergo MC [33]. Figure 23.2 shows the various possible scenarios that can follow DNA damage during G_2/M.

If cells enter mitosis with DNA damage (e.g., double-strand breaks, DSBs), the ends of the broken chromosomes will be too distant from each other and too condensed to allow for repair and rejoining, as the mitotic spindles will be in the process of separating chromosome strands [18]. Loss of entire sections of chromosomes is therefore possible, and MC can ensue (followed by apoptosis or necrosis) as a mechanism to avoid genomic instability [18].

The DNA-damage checkpoint activates Chk1 and Chk2, which phosphorylate Cdc25 phosphatases to ensure the Cdk1/cyclin B complex remains inactive, thereby preventing mitotic entry [10,34]. Aberrant overexpression of Cdk1/cyclin B leads to premature condensation of chromatin and MC. For instance, downregulation of 14-3-3σ (see next paragraph) increases Cdk1/cyclin B in the nucleus, resulting in failed G_2/M arrest and MC [10]. Depletion of Chk1 also allows premature mitosis to proceed in the face of DNA damage, due to the failure to inhibit Cdk1/cyclin B, and MC ensues [35]. In this scenario, cell death can be inhibited by the addition of a caspase inhibitor, indicating that cell death that follows MC can be caspase-dependent.

The regulatory molecule 14-3-3σ was discovered to be an integral constituent of the G_2 checkpoint in 1999 [36]. Prior to this, it was known that in normal cells, 14-3-3σ is induced by p53 following DNA damage, and that 14-3-3σ overexpression causes G_2 arrest. Chan and colleagues [36] observed that 14-3-3σ$^{-/-}$ colorectal cancer cells treated with radiation or adriamycin, both of which lead to DNA damage, failed to sustain G_2 arrest and proceeded to MC. This was due to the translocation of Cdk1 and cyclin B from the cytoplasm into the nucleus, driving cells prematurely into mitosis and culminating in MC [10]. Immunoprecipitation confirmed that, in fact, 14-3-3σ is responsible for retaining Cdk1 and cyclin B in the cytoplasm until issues causing G_2 block have been resolved. Since this discovery, cells deficient in 14-3-3σ have been used as a suitable model of MC [37]. When Chk1 phosphorylates Cdc25, 14-3-3σ proteins maintain Cdc25 in the cytoplasm so that it cannot activate Cdk1/cyclin B in the nucleus [17]. A 2011 study employed proteomics and network analysis to show that the 14-3-3σ signaling network was the most involved in MC [38]. The 14-3-3σ molecule was found to ultimately regulate proteins associated with the cell cycle, DNA repair, the cytoskeleton, and chromatin remodeling [38]. The Cdk inhibitor, p21, discussed earlier, is also an integral part of G_2 arrest upon DNA damage, as it inhibits Cdk1 activity [20]. However, colon-cancer cells have been used to show that both 14-3-3σ and p21 functions are not enough to maintain G_2 arrest, and that cells can still enter mitosis

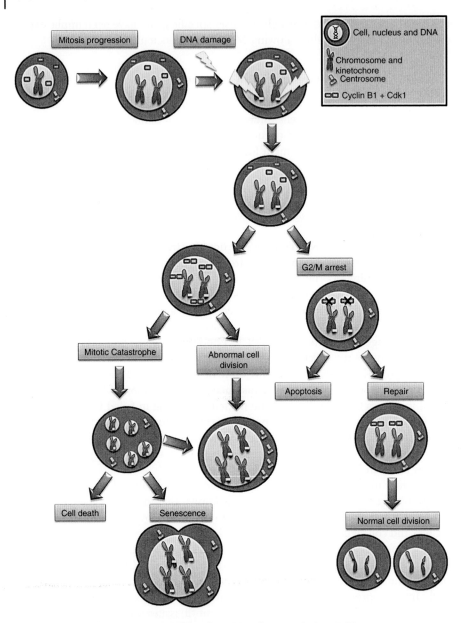

Figure 23.2 Possible scenarios resulting from DNA damage during G_2/M.

with damaged DNA and ultimately yield tetraploid cells [20]. It was concluded that p21 causes a delay in G_2, rather than full arrest, and that 14-3-3σ itself cannot completely inhibit entry into mitosis [20]. Rather than 14-3-3σ, recent evidence suggests that Chk2 (see earlier) may be the main negative regulator of MC [25].

These molecules are some of the key players in MC. The role of other molecules is constantly being elucidated. CyclinA1 is a protein pertinent to meiosis, but it is expressed abnormally in some cancers, such as acute myelogenous leukemia, where it is induced by p53 when DSB repair is needed [39]. CyclinA1 functions to enhance G_1–S progression, but it also promotes G_2 arrest, MC, and apoptosis in ovarian, lung, and renal cancer cells [39]. Recent evidence shows that the absence of a pre-mRNA splicing factor termed Cdc5L, overexpressed in osteosarcoma and cervical cancers, causes multiple aberrations that lead to MC [8]. This study was the first to identify Cdc5L as an important mitosis regulator. When it was knocked down, mitotic arrest and misaligned chromosomes ensued, resulting in MC due to compromised S and G_2/M checkpoints. The authors used microarrays to determine that Cdc25L inhibits mitotic progression by regulating the expression of the proteins responsible for damage repair and progression of mitosis [8].

As mentioned earlier, cells may have the ability to overcome long-term G_2 arrest, reenter mitosis, and proceed with the cell cycle with damaged DNA, resulting in either genomic instability or delayed cell death through MC [16]. To avoid death, a small fraction of cells can undergo a state of quiescence, followed later by G_2 slippage and reentry into the cell cycle, resulting in daughter cells that are endopolyploid – a state in which chromosome duplication occurs without cell division (Figure 23.3) [16]. It is not yet understood how cells enter quiescence from G_2, a transition that normally happens in the G_1 phase. It has recently been observed that cells exposed to high doses of radiation enter a state of long-term G_2 arrest, and that G_2/M transition genes are downregulated [16]. This observation was the first to show that cells arrested at G_2 as a result of DNA damage have the ability to bypass mitosis through G_2 slippage directly into G_1/G_0 quiescence, allowing them to avoid mitotic death [16]. This could be an issue for cancer radiotherapy, suggesting that high doses of radiation could lead to senescent cells that can reenter the cell cycle after some time has passed, potentially leading to recurrence [16].

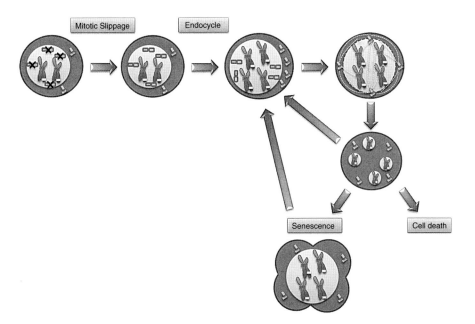

Figure 23.3 Depiction of the endocycle process upon mitotic slippage. Mitotic reentry and progression in the presence of structural damage can cause endoreplication, initiation of cell death, or senescence – situations that can lead to a return of endocycling. The endocycle results in endopolyploid cells, which contain reduplicated chromatin within the nucleus.

An interesting connection between autophagy and MC has recently been demonstrated. Autophagy is a self-digestion process used by cells to degrade unwanted cellular components. It is achieved by sequestering such components in membrane vesicle autophagosomes, which fuse with lysosomes, where the components are degraded [40–42]. Treatment of cells with DNA-damaging agents causes upregulation of autophagy-related protein 5 (ATG5), which translocates to the nucleus and causes MC. Interestingly, the MC thus induced does not lead to cell death, so long as autophagy is not inhibited [40]. This finding contributes to the argument that MC does not necessarily lead to cell death and may, as in this scenario, be a mechanism of cell survival in response to DNA-damaging chemotherapy. Blocking of autophagy causes cell death to follow MC, which may be a promising strategy for increasing cell death following chemotherapy exposure.

23.3.2 Spindle-Assembly Checkpoint Failure

When anaphase is inhibited due to mitotic spindle damage, the cell cycle of such tetraploid cells becomes halted, and the cells are induced into a G_1 state through a process mediated by p53. This may occur because of agents that disrupt mitotic spindle organization [43]. Chemotherapeutic agents that cause spindle damage, such as the taxane paclitaxel, cause chromosomes to form disorganized clusters and decondense, which triggers abnormal nuclear-envelope formation and multiple micronuclei in cells deficient in spindle-checkpoint control, as is the case with many cancer cells [9]. Cell-cycle arrest mediated by the spindle-assembly checkpoint is relatively short-lived. If spindle abnormalities are not corrected, MC and cell death may ensue. However, mitotic slippage may also be seen. Arrest will eventually reoccur, leading to MC and cell death, or to the creation of polyploidy or aneuploidy cells [44]. The latter is more likely to happen if p53 activity is compromised, as p53-dependent apoptotic cell death cannot follow MC [44]. Mitotic slippage has been observed upon prolonged treatment of cells with microtubule-disrupting chemotherapeutic agents [28].

For chromosomes to properly segregate, centrosomes must be correctly placed at opposite poles of the dividing cell to mediate proper nucleation of microtubules [43]. Nucleoporins are the proteins of the nuclear pore complex, which regulate transport of macromolecules to and from the cellular nucleus [45]. In addition to this transport role, they play a part in mitotic spindle function during mitosis. They associate with kinetochores and spindle poles, and prompt appropriate bipolar spindle formation [45]. The nucleoporin RanBP2 is an example of this important role. A lack of RanBP2 expression is enough to cause irregular metaphase and erroneous chromosome segregation, followed by MC [45].

Overduplication of centrosomes (e.g., as a result of radiation) causes multipolarity and induces the formation of multiple spindle poles and consequent micronuclei [9,41]. Among other molecular players, p53 is required to ensure such overduplication does not occur [4]. As already discussed, Aurora A localizes to centrosomes and is a key molecular player in ensuring correct centrosome function [29]. Aberrations in centrosomal proteins (importantly, Aurora A, ninein, TOG, and TACC3) are found in various cancers and can lead to centrosome disruptions and perturbations in the proper organization of microtubules [29,43]. Depletion of any of these centrosomal proteins causes arrest of cells at metaphase, MC, and cell death [43]. The involvement of the spindle assembly checkpoint in this process remains controversial [43]. Such cell death is dependent on phosphorylation of Chk2 (the kinase heavily involved in DNA damage response, as discussed earlier), which interestingly also has an important role in cell death resulting from centrosome disruption [43].

A lack of centrosomal duplication may also occur, hindering chromosomal segregation. Centrosome amplification can ensue from prolonged G_2 delay as a result of DNA damage [4]. A 2007 study showed that after irradiation, 30% of cells showed features of MC, 70% of which showed centrosome overduplication [4]. As p53 is a regulator of centrosome duplication, cells harboring mutations in this transcription factor display abnormal centrosome duplication [46].

A 2012 study identified multiple-copies T-cell malignancy 1 (MCT-1) as a new centrosomal oncogene highly expressed in lymphomas and lung cancers [46]. Its overexpression in p53-deficient cells causes centrosome amplification, as well as missegregation of chromosomes, ultimately leading to either MC or increased malignancy [46].

Inappropriate chromatin condensation also results in MC. The proteins condensin I and II are responsible for ensuring optimal chromosome condensation for mitosis [26]. Entire chromosome shattering has been observed after treatment with doxorubicin [47]. In this scenario, suboptimally condensed chromatin attaches to centromeres, but a metaphase plate cannot form, leading to MC and eventual apoptosis [48].

23.4 Cell Death Resulting from MC and Links to Other Modes of Cell Death

As mentioned earlier, controversy still exists over whether MC should be considered a unique form of cell death, or if it is simply a pre-stage that results in apoptosis, necrosis, senescence, or viable polyploidy [42,49]. In fact, multinucleation as a result of MC has been shown to precede apoptotic and necrotic cell death [33]. An interesting 2011 report provides evidence that mechanical collapse can occur due to the stresses of abnormal mitosis and cause cell death independent of apoptosis and necrosis [41]. The authors used time-lapse microscopy and a cellular model of MC to show that cell death that followed mitotic arrest in p53-deficient cells was a result of mechanical stress imposed by multipolar spindles [41]. Perhaps this unique scenario constitutes cell death that is entirely MC-based.

Apoptosis is arguably the most widely recognized type of cell death. This p53-dependent programmed death involves caspase activation and mitochondrial membrane permeabilization, which causes the release of multiple proapoptotic effectors [41,50]. Apoptosis can also occur independently of caspase activation through the apoptosis-inducing factor (AIF) released from the mitochondria [41]. Significant overlap exists between MC and apoptosis. Cells undergoing MC display characteristics of apoptotic cells, including release of proapoptotic proteins from mitochondria (e.g., cytochrome c (cyt c) and AIF) and caspase activation [3,38]. MC also requires the activation of caspase 2, which precedes the release of mitochondrial proteins [49,50].

According to recent evidence, caspases 2, 3, and 9 are activated during cell death associated with MC, which may indicate that activation of apoptosis occurs subsequent to MC events [25,43]. Mitochondrial membrane permeabilization and cyt c release occur at a point between the activation of caspase 2 and that of caspase 3, suggesting that caspase 2 triggers this release [50,51]. The involvement of proapoptotic Bcl-2 proteins in MC execution and ultimate apoptotic cell death has also been shown [43]. Caspases are important players in the apoptotic pathway, which act by inducing proteolysis of its targets [52]. Initiator caspases, such as caspase 8 –and 9, activate effector caspases, which execute the apoptotic cascade [52]. Caspase 2, in particular, has been identified in multiple studies as a key player in MC-mediated cell death, and is considered its principal executioner [53]. Interestingly, caspase 2 activation is not an event of classic apoptosis [27]. Caspase 8 activation seems to trigger the death-signal cascade; however, apoptosis is not a requirement for MC-induced cell death, as necrosis may also follow MC, as evidenced by a loss of membrane integrity, a hallmark of necrosis [25,53]. For instance, caspase 3 blockage did not inhibit MC [27]. Treatment with the microtubule inhibitor compretastatin-A4 prodrug activated caspase 9 and poly ADP ribose polymerase (PARP, an important caspase target), but only a small percentage of WSU-CLL cells underwent apoptotic cell death [52]. A general caspase inhibitor did not block cell death, showing that caspase 9 activation is not required for MC-induced cell death in this scenario [52] – although another study found that caspase inhibition combined with inhibition of mitochondrial membrane permeabilization did inhibit MC [50]. Caspase-independent apoptosis has also been implicated in MC-related cell death, however – specifically, the lysosomal apoptotic pathway, in which lysosomal proteases trigger an apoptotic response [27]. Activation of the spindle checkpoint also involves caspase-independent apoptosis [43].

A 2012 study investigated the final outcome of p53-deficient and -proficient cells treated with the microtubule-disrupting agent paclitaxel at nanomolar concentrations, which cause MC [33]. In both cases, G_2/M arrest ensued. However, premature entrance into mitosis occurred at a greater frequency in p53-deficient cells, with MC being followed by caspase-independent apoptosis in about half of such cells [33]. The authors observed that p53 deficiency did enhance polyploidy due to aberrant continuation of the cell cycle, but that such cells eventually underwent apoptotic cell death. It was suggested that apoptosis occurs in p53-deficient cells through Bcl-2-associated X protein (Bax)-mediated mitochondrial apoptotic induction, via activation of the spindle-assembly checkpoint rather than the G_2/M checkpoint [33].

Differences between MC and apoptosis are recognizable. Morphologically, MC and apoptosis are quite distinct. While apoptotic cells have condensed chromatin, shrunken cytoplasm, nuclear fragmentation, and plasma-membrane blebbing, MC cells possess uncondensed chromosomes, are very large, are mono- or multinucleated, display multipolar spindles, are polyploidy, and do not display DNA fragmentation detectable by terminal deoxynucleotidyl transferase dUTP nick end labeling (TUNEL) assay [9,41,54]. Aside from these morphological differences, important biochemical dissimilarities also exist. MC does not require p53, which partially explains why this pathway to cell death is commonly observed in cancer cells [50]. The fact that caspase 2 activation precedes mitochondrial membrane permeabilization is quite unique as well [50].

It is still not entirely clear which mechanisms dictate whether MC will result in apoptosis or necrosis. A 2011 study revealed that, although MC itself does not require p53, the tumor suppressor seems to be required for apoptosis to follow MC upon DNA damage [37]. The interplay between γH2AX, ATM, and p53 appears to be essential to this end [37]. When p53 and ATM were depleted, necrosis – rather than apoptosis – was observed to occur post-MC, caused by doxorubicin treatment. Lack of p53 function may therefore determine that cells undergo necrosis rather than apoptosis following MC, as was the case with breast-cancer cells treated with DNA-damaging agents [55]. Cells with a $14\text{-}3\text{-}3\sigma^{-/-}$ genotype, which are commonly used as models of MC, as previously mentioned, showed amplified γH2AX occurrence due to the presence of additional DSBs superfluous to those generated by prior doxorubicin treatment. Cells in the MC state will attempt to repair damaged DNA; however, since chromosomes are densely packed, such repair attempts will actually lead to chromosome breaks and additional DNA damage [37]. This additional damage is what is believed to trigger eventual p53-dependent apoptosis.

Alternative doses of various chemotherapeutic agents known to primarily induce apoptosis can cause cell death via MC, presenting an opportunity to circumvent apoptosis resistance in cancer cells [56]. For instance, lower doses of doxorubicin induce MC in tumor cells, rather than immediate apoptosis [57]. Effective cell killing using this strategy was achieved even in cells overexpressing the antiapoptotic protein Bcl-xL, overcoming resistance to high-dose doxorubicin [57]. High- and low-dose doxorubicin induce different responses in the roles of Cdk1 and Cdk2, which may partially contribute to the different cell-death pathways undertaken [56]. While a high dose is associated with downregulation of the kinase activities of Cdk1 and Cdk2, low doses induce upregulation of such activities in p53-deficient Huh-7 cells [56]. Expression levels of cyclin A and B are down- and upregulated, respectively, in parallel. In the absence of p53 function, the G_2/M checkpoint is not fully induced when DNA damage is present at low levels (e.g., with low doses of doxorubicin). This causes cells to enter mitosis even in the presence of damage, leading to MC [56]. In prostate-cancer cell lines, G_2/M arrest and the degree of immediate apoptosis induced by doxorubicin are both concentration-dependent [51]. The idea of inducing MC presents an opportunity to successfully treat cancers that have become resistant to apoptotic cell death.

Variations in the mode of cell death can also occur if the same agent, alone or in combination, is used to treat cells that are variously sensitive and resistant to that agent. For instance, acute myeloid leukemia cells resistant to cytosine arabinoside, the main agent used to treat the disease, undergo MC when this agent is combined with Aurora A inhibition, whereas cells that are still sensitive to cytosine arabinoside undergo apoptosis when exposed to the same combination [29]. Thus, cancer cells that have become resistant to apoptosis can still be eradicated through the induction of MC. In fact, MC is believed to be an alternative mode of cell death that can evade apoptosis resistance [29]. Multidrug resistance can itself elicit cells to undergo MC rather than apoptosis upon chemotherapeutic treatment. For example, the *mdr1* gene encodes for the well-characterized cell-membrane drug efflux transporter, P-glycoprotein, which is overexpressed in a wide variety of multidrug-resistant cancers. *mdr1* overexpression has been shown to decrease the occurrence of apoptosis in cells, which is compensated for by an increase in MC [9].

23.5 MC Detection Methods

The NCCD advises that two combined techniques are minimally necessary to distinguish modes of cell death, based on morphological and biochemical detection methods [1]. The morphological features of MC are giant cells featuring micro- and multinucleation (resulting from aberrant chromosome fragments), nuclear-envelope formation around such fragments (resulting in large cells containing micronuclei), and abnormal centrosome numbers, as discussed earlier [1,51]. Biochemically, the activity of the DNA-damage checkpoint can be observed through increased γH2AX, reduced Chk1 function (through phosphorylation), and reduced Cdk1 activity [2,23]. Table 23.1 lists a variety of proteins that can be probed, while Table 23.2 identifies methods that can be used in the detection and discernment of MC.

Table 23.1 Molecules quantified by Western blot to determine the occurrence of mitotic catastrophe (MC). Molecules of interest include proteins that act on cell-cycle arrest, cell-cycle progression, the DNA-damage checkpoint, the spindle-damage checkpoint, and mitotic progression.

Molecules of Interest	References
HKL-1	[58]
14-3-3σ	[36]
Akt, phosphor-AKt, α-tubulin, phosphor-FKHR, phosphor-GSK3, cyclin D1, Chk1, p38, Chk2, Cdc25C, Cdc2, P-gp	[59]
β-tubulin, cyclin B1, BubR1, MPM-2, Bcl-2, survivin, PARP, BCL-XL, caspases 2 and 3	[60]
Chk1, cyclin B1 and E1	[23]
Cyclin A and B1, Cdc2	[61]
Caspases 2 and 3, p21WAF1, cyclin D1 and B1, mitotic form of histone H3, GADPH, H3pS10	[33]
Aur-A, β-actin, α-tubulin, PARP, caspases 3 and 9, cyclin B1 and E, Cdc2, MPM2, p38MAPK	[29]
Bcl-xL, capase 3, α-tubulin	[57]
cyclin B1, Aur-A, PLK-1, β-actin	[53]
β-actin, P21waf1, P27kip1, pCdc2	[27]
BubR1, CENP-A, Plk1, Chk1, Chk2, Cdk2, 4 and 6, Cdc2, cyclin A, B, D, E, p27	[56]
p53, p21	[62]
Caspases 3 and 9, PARP, Mcl-1, Bcl-xL, Bcl-2, Bak, Bax, cyclin B1, Cdc2 and p-Cdc2, ubiquitin, β-actin	[63]
p53, p21, G3PDH, MDM2, GADD45, 14-3-3σ, caspases 3 and 9, cyclin B, Cdc2 and 25, Wee1, Chk1	[52]
Caspases 3 and 9, Bcl2, Bax, cyclin B1, Cdc2 and 25, Chk1 and 2, actin, CUCPB2	[64]
Caspases 3 and 7, β-actin, p21 and 27, cyclin B1 and D2, Cdk1, Bcr-Abl	[65]
α-tubulin, actin	[66]
Cyclin A1 B1 and A2, p53, actin, ribosomal protein S9, Cdk1	[39]
Sox2, musashi, nestin, gH2AX, p53 and 21, Mcl-1, Bcl-2, Bcl-xL, DNA-PK, Bax, actin	[67]
securin, Cdc20, tubulin, cyclin B1, Chk1, Chk2, Cdh1, Brca1	[68]
RPA32, β-actin	[69]
PLK1, Cdc25C	[21]
CHK1 and p-CHK1, CHK2 and CHK2, β-actin	[70]
p53, p21, caspases 2, 3, and 8, G3PDH, PARP, γ-H2AX, cyclin B1, Cdc20, BubR1, Mad2	[37]
Caspases 3, 8, and 9, α-tubulin, PARP	[71]
ATM, H2AX, p53, tubulin, DNMT1, DNMT3A, DNMT3B	[72]
Chk1, Cdc2, cyclin B1 and A, histone H3 and H2AX, Chk2, p53, cyt c, VDAC, ATM, ATR	[35]
Cdk1, cyclin A2, b1 AND E2, α-tubulin, Cdc20, Chk1, H3, p31, Cdc27, Mad2, PARP	[24]
Cyclin A, E, and B, p53, p21, ATM, Cdc2 p34, actin, NBS1, MAD2, tubulin	[73]
Phosphorylated H3	[74]
GADPH, Caspase 3, PARP1, Hsp60, Cit C	[75]

Table 23.1 (Continued)

Molecules of Interest	References
Cdc25, Cdc2, α-tubulin, actin, Chk2, PARP, caspase 3	[76]
Survivin, Aur-B, total H3, Ac-tubulin, actin	[77]
ERK MAPK, Aur-2, cyclin B1 and A	[78]
Survivin, Aur-A and -B, γH2AX, Op18, γ-tubulin, α-tubulin, Chk1 and 2, p53, p21	[79]
PARP, caspases 3, 7, and 9, Akt	[80]
β-actin, Cdk1, FLAG, PARP, tubulin, cyclin B1 and A2, histone H3	[81]
Chk2, γH2AX, β-actin	[82]
PARP, caspase 3, γ-actin, cyclin B1 and D1	[83]
β-actin, MAD2, BUBR1, histone H3, PARP, PDS1	[84]
p53, p21, cyclin B1, actin	[85]
Wee1, Cdc2, γH2AX, β-actin, p53	[86]
Cdk1, histone H3, Plk1, PARP, Eg5	[87]
Plk1, actin	[88]
PARP, GEPDH, ERK1/ERK2, Akt, Cdc2 p34, histone H3	[89]
P53, actin, DO-1	[4]
Actin, p53, p21	[90]
Caspases 8 and 9, PARP, actin	[91]
Cyclin B1, actin, p53	[92]
Chk1, Chk2, α-actin	[93]
Caspases 2 and 3, Bax, Bcl-2, p21, actin	[51]
Aur-A, actin, γ-tubulin, PCM-1, TACC3, TOG, ninein	[43]
Caspase 3, p20	[94]
Cdc25C, Cdc 2, cyclin B1, Akt, Chk1, Cdc2 p34, Aur-B, Plk1, Hsp90, Hsp70, β-actin	[95]
Capases 2 and 3a	[50]
β-actin, PARP	[96]
P300, CBP, P/CAF, H2AX, hCAP-D3 and -G, MAD2, BubR1, securin, histone H3, cyclin B1, actin	[97]
α-tubulin, Bax, Cdc25A, Chk1, cyclin E, Mam3, p21, PARP1, pRb	[41]
P21, c-myc, β-tubulin	[55]
Cdc25C, Chk1, Cdc2, p34, Plk1, survivin, β-actin	[98]
14-3-3-σ, Cdk2, Cdc2, α-p21, cyclin B and E	[20]
Cdc6, Mcm2, 3, and 6, GAPDH	[99]
Actin, Cdc25C, Chk1 and 2, Cdk1, cyclin B1	[100]
Cdc2, cyclin B1, actin, p53, Chk1 and 2, Cdc25C, Bac, Bcl	[101]
Cyclin B1, AT8, β-actin	[102]
Actin, Cdc2, Aur-A, Aur-B, Bad, Bak, Bax, Bcl-2, Bcl-xL, caspases 3 and 9, PARP, Puma, Stat3, Bid, Bim, survivin, 14-3-3-σ	[103]
Cdc2, Bcl-2, Bcl-xL, Bax, survivin, actin	[104]

(continued)

Table 23.1 (Continued)

Molecules of Interest	References
P53, 14-3-3-σ, cyclin B1, β-actin	[105]
Aur-A, centrin, actin	[106]
P62, p53, MPM2, β-actin, Keap1, Nqo1, caspase 3, PARP	[44]
Survivin, Aur-B, actin	[26]
Cdk1, MPM2, Cdc20 and 27, BubR1, β-actin	[107]
p53, p21, γH2AX	[34]
H2AX, HSP70, cyclin E, histone H3, γ-tubulin, α-tubulin, Plk1, RCC1, 14-3-3-σ, Chk 1 and 2	[38]
Cdc25C, Cdc2, cyclin B1, β-actin, histone H3, tubulin, caspases 3 and 6	[28]
ATM, actin, PARG	[108]
Survivin, Bcl-2, PARP, XIAP, caspase 3	[11]
Rd2, actin	[109]
Cyclin D1 and B, Rb, Cdk1, ATM, Chk2, γH2AX, wee-1, histone H3	[110]
Caspases 3 and 7, beclin, parkin, LC3-I, -II, β-actin, PINK1, TOM20	[111]
Cyclin D1, actin, Cdc2, Chk2, HuR, CUGBP2, TIA1, VEGF, COX-2	[112]
SLX1, SLX4, MUS81, XPF, Chk1, Chk2, RPA32	[113]
Chk1, p53, caspase 3, histone H3, PARP, GADPH	[114]
Cdc2, PARP, actin	[115]
Actin, Ran, p53, V5, HRP, cyclin B1, Cdc2 p34	[116]
Mad2, α-tubulin, cyclin A and B1, p53, p21, securing, BubR1	[117]
ATM, γH2AX, GADPH, p53, ATF3	[118]
Plk1, β-actin, PARP, cyclin B1	[119]
Cyclin B1, cyclin E, HA-E4orf4, actin, caspases 1, 2, 3, 6, 7, 8, and 9, TOM20, AIF, cyt c	[120]
Chk1, Chk2, GADPH, p53, cyclin B, Cdk1	[10]
MCM3, cyclin B1, Mad2, BubR1, securing, Chk1	[121]
β-actin, γ-tubulin, α-tubulin, cyclin E, V5, Plk1, GST, Cdk2, p53, Cdc2, histone H3, Aur-A, Aur-B, ATR, BubR1	[46]
Bax, caspases 8 and 9, Bcl-xL, PARP, β-actin, cyclin B1	[122]
Bim, α-tubulin, Cox IV, HRP	[123]
Bax, Bcl-2, surivin, β-actin	[124]
P21, p53, actin, β-tubulin	[125]
Cyclin B, Cdk2, Cdk4, β-actin	[126]
P38, caspases 3, 6, 7, and 8, PARP, CENP-A, actin, Plk1, cyt c	[127]
GADPH, β-actin	[32]
Cyclin A, E, and B1, actin, p21, wee1, myt1, Cdc2	[128]

Table 23.2 Techniques used in the identification of mitotic catastrophe (MC).

Method and Rationale	Molecules/Structures of Interest	References
Immunohistochemistry Employed to identify MC upon treatment. For instance, it allows for determination of cell-cycle phase and the presence of MC-related proteins, cytoskeleton analysis, and the study of chromosome compaction during cell division	α- and γ-tubulin G2 or M phase 53BP1 S phase Cyclin B1 and Cdc2 Phosphorylated H3 Cyt c, AIF, tubulin, DAPI MDC1, γH2AX, MRE11, ATM, Cy3, FITC Aur A COX-2, HuR Plk1	[21,39,48,53,58,64,74,76,80, 86,101,106,112,119,126,127,129–131]
Fluorescence videomicroscopy Allows for generation of the cell-cycle profile of cells	Cell morphology	[49,50,132]
Flow cytometry Employed to determine cell-cycle distribution and confirm cell-cycle arrest	Determined by the stage of cell division in question	[5,8,11,18,20,21,24,27–29,32–35,37, 38,40,41,43,44,46,51–53,55,56,58,60–67,69, 70,72,73,75,76,78,79,81–85,87–101, 103–123,125–128,131,133–147]
Immunoelectron microscopy Used to confirm the presence of proteins of interest	Cofilin and actin	[5]
Fluorescence microscopy Used for DNA evaluation and cell-morphology characterization	Nucleus morphology, DNA, and mitotic spindle	[8,10,20,23,26,28,29,32–34,37,38,40, 41,43,44,46,48,50,55,57–62,64–67, 69–71,77–79,81–86,91,93,95–102,105, 107–111,113–115,117–123,126–128, 130,133,136–139,141,143–146, 148–151]

(*continued*)

Table 23.2 (*Continued*)

Method and Rationale	Molecules/Structures of Interest	References
Time-lapse microscopy Used for evaluation of cell division	Dividing cells	[4,8,24,26,34,40,41,43,46,48,71,82,86,87, 94,95,97,106,108,117,121,131,139,140,152]
UV microscopy Used for cell-morphology evaluation	Cellular morphology	[53,92]
Kinase and capase assay Used to measure activity of kinases and caspases pertinent to MC	Cdc2 and Cdk1	[63]
	Cyclin B1 and A1	[35]
	PLK1 and Cdc25C	[21]
	Cdc2	[73,153]
	Histone H1	[81,143]
	Caspase 2, 8 and 9	[91]
	Chk2 and Cdk2	[56,128]
	Caspase 3 and 2	[55,125]
	Caspase 3 and 7	[98]
	Cdk2 Cdc2, cyclin a and B, histone H1	[20]
	Cdk1	[100]
	Plk1	[119]
	Caspase 3	[11,104,115,124]
	Aur-A and -B	[106]
	BubR1 and Plk3	[107]
	Cyclin E and Cdc2	[46]

Costumer-designed low-density array Used for analysis of gene expression relevant to MC	Cdc2 and cyclin B1	[34,116]
	Genes encoding for the following proteins: TNF, KIF20A, STK6, CENP-E, PLK1, BUB1, KPNA2, ACTA2, MYL9, PKC pathway, serine-threonine kinase pathway, TP53 pathway, CCNE1 and 2, CCDN2, CCNB, CDC25 transcripts, MYT1, Cdk inhibitors p21 CDKN1A, CDNK1C, CDKN2C, FGF1 receptor, STAT5A	[132]
Agarose-gel electrophoresis Used to determine whether DNA fragmentation can be observed in MC-related cell death	Fragmented DNA	[56,151]
Fluorogenic assay Used to assess caspase activity	Caspase 3	[71]
Mitotic index Used to identify the phases of the cell cycle during which cells are arrested	Cell cycle	[65,66,71,75,80,95,99,104,105,123, 125,126,135,143,145,147]
Cytofluorimetric assay Used to characterize cell-cycle distribution	Cell cycle	[10,50,71]
Microscopic laser cytometry Used to evaluate nuclear morphology	Nucleus	[35]
Cytology Used in analysis of cell death	Cell morphology	[134]
Transmission electron microscopy Used in ultrastructural cell-morphology analysis	Cell morphology	[28,31,79,85,94,111,118,134,140,151]
Laser-scanning cytometry Used to identify populations of cells at each stage of mitosis	Cell cycle	[102,125,142,145,148]

(continued)

Table 23.2 (*Continued*)

Method and Rationale	Molecules/Structures of Interest	References
Phase-contrast microscopy Used to identify the effects of cytotoxic treatment on cell morphology	Cell morphology	[94]
S-phase checkpoint assays Used in the quantification of cells undergoing S-phase checkpoint activation	Cells with increased DNA synthesis	[34]
Cytokinesis-block proliferation index Used in the quantification of proliferation speed	Cells undergoing mitosis	[108]
Fluorescent in situ hybridization Used in the assessment of metaphase spread and telomere aberrations	Cells undergoing mitosis	[31,108,113,147]
Nuclear morphology analysis Used in the quantification of the nuclear irregularity index	Nucleus	[110]
DNA image cytometry Used in the analysis of the variation of DNA content during metaphase, anaphase, and telophase	Cells undergoing mitosis	[31]
Tubulin GTPase assay Used in cytoskeleton analysis	Functioning cytoskeleton	[146]
Two-photon excitation microscopy Used in 3D cell-organization analysis	Cell	[124]

23.6 MC in the Treatment of Cancer

The induction of MC presents an important opportunity to selectively target cancer cells. Checkpoint function is widely compromised in cancer cells, causing them to be particularly sensitive to MC, as such cells proceed through to mitosis even in the presence of DNA or spindle damage [10]. Further inhibition of checkpoint proteins has the potential to enhance chemo- or radiotherapy, which can cause DNA damage, spindle damage, or both.

23.6.1 Cancer Radiotherapy

Radiation therapy is a highly effective modality for the treatment of solid tumors. MC is a major pathway to cell death following radiotherapy [16]. Ionizing radiation causes the formation of DNA DSBs and, ultimately, cell death if repair cannot be performed, and cell-cycle arrest at G_1/S or G_2/M ensues [70]. As is the case with the DNA-damage response covered earlier, radiation-induced damage triggers a p53-mediated response, in which p21 is activated for G_1/S arrest and Cdk1/cyclin B inhibition occurs to inhibit entry into mitosis [62]. As p53 deregulation is a common feature of cancer cells, cell-cycle checkpoint activation is impaired, causing G_2/M-arrested cells to adapt and resume cycling, with MC ensuing. As discussed earlier, cyclin B accumulation is responsible for this abnormal entry into mitosis [133]. In this scenario, DNA damage continues to accumulate long after irradiation, likely due to additional damage induced by abnormal cell-cycle progression [133]. In addition to DNA damage, radiation causes centrosome overduplication, contributing to the MC phenotype [4]. Delayed cell death occurs, with apoptosis happening 2–4 days after irradiation *in vitro*, and continuing up to 1 week later [62].

Ionizing radiation leads to MC by inducing the activation of ATM or ATR, Chk1 and/or 2 phosphorylation, Cdc25 inactivation, Cdk1 inhibition, and finally cell-cycle arrest at G_2/M [129]. As mentioned earlier, cells can repair DNA damage that ensues from radiation before resuming the cell cycle. In the context of cancer therapy, this is undesirable. Agents that drive cells into mitosis before repair have the potential to drive them toward MC and eventual cell death, thus sensitizing cancer cells to radiation therapy. This can be achieved by inhibiting the cell's DNA-repair machinery or by overcoming cell-cycle arrest before damage can be repaired. One example of the latter strategy involves the inhibition of Chk proteins. In 2011, the Chk inhibitor XL-844 was shown to enhance the sensitivity of HT-29 human colon-cancer cells to radiation if the inhibitor was administered immediately following irradiation [129]. Interestingly, no effect was observed if XL-844 was administered prior to or 24 hours post-irradiation, likely because the inhibitor exerts its effects during the cell's immediate arrest and repair responses to damage. It was demonstrated that, in fact, XL-844 inhibited Chk2 and drove cells prematurely into mitosis. Inhibition of Aurora B, the kinase involved in centrosomal and mitotic spindle function, also leads to radiosensitization through enhanced MC [145]. Interestingly, the effect is only observed if Aurora B inhibition occurs 24 hours pre-radiation.

Inhibition of DNA-repair processes is widely explored as a strategy for radiosensitization. Chromatin remodeling is required for DNA-damage repair after irradiation, and inhibition of the enzymes required for this process hinders repair and enhances radiation-induced damage [70]. For instance, a compound termed "C646" inhibits a histone acetyltransferase (required for chromatin remodeling) and sensitizes non-small-cell lung cancer cells to irradiation by inhibiting Chk1 and Chk2 phosphorylation and inducing MC [70]. Certain cancers that present resistance to radiotherapy show enhanced DNA-repair capabilities. Glioblastoma, for instance, has been shown to highly overexpress wee1, which, as discussed earlier, is a negative regulator of Cdk1 [86]. This overexpression likely enhances G_2/M arrest to allow for sublethal DNA-damage repair, preventing unscheduled entry into mitosis and MC/cell death. It has been shown that wee1 inhibition is a promising target for radiosensitization, as premature entry into mitosis ensues, followed by cell death [86]. Inhibition of Plk1, also overexpressed in glioblastomas, is another promising strategy [88]. Plk1 chemical inhibition in combination with radiation resulted in reduced DNA repair and decreased clonogenicity [88]. The high levels of radioresistance encountered in glioblastoma cases have been partly linked to increased activity of the ATM/ATR pathway upon DNA damage [131]. Kinase inhibitors capable of reducing ATM/ATR activation have been shown to radiosensitize glioblastoma cells to radiation [131]. The PARP enzymes play important roles in detecting DNA repair and inducing signaling to recruit repair molecules to sites of damage. PARP enzymes catalyze the synthesis of poly ADP ribose (PAR), which recruits repair enzymes, while poly ADP ribose glycohydrolase (PARG) is responsible for PAR degradation [108]. PARG depletion increases radiosensitization through slower damage repair and centrosome amplification, leading cells to aberrant mitosis and MC [108].

DNA-damaging or microtubule-disrupting agents, some of which are discussed in the next section, can also be rationally combined with radiotherapy for synergistic cell-killing effects. Nocodazole, for example, is a microtubule-damaging agent that enhances MC induced by high linear-energy transfer radiation in p53-deficient cells [136]. The combination results in a very delayed cell death, indicating that MC is the main pathway leading to eventual apoptotic cell death, as seen by features typical of MC followed by delayed apoptosis [136].

Radiotherapy is less effective in cancers that possess intact p53 function, as they display more robust checkpoints and enhanced DNA-repair responses [85]. While radiation induces G_2/M arrest, cyclin B is elevated in p53-deficient cells, causing them to undergo MC [92]. The glioblastoma cell line U87MG has been used to show that the presence of p53 translates to enhanced DNA-damage repair and resistance to chemo- and radiotherapy [85]. A rational intervention would be the abrogation of the G_2/M checkpoint in such cases, which would enhance MC rates. Employment of UCN-01, a Chk1 inhibitor, shows potential in radiosensitizing p53-competent cancer cells [85]. Induction of cyclin B to trigger mitotic entry and ensuing MC also has potential as a radiosensitizing strategy in such cells [92].

Mild hyperthermia (40.5–42.0 °C) has long been explored as a strategy for radiosensitization, and is arguably one of the most potent used to this end [154]. Hyperthermia itself causes accumulation of cells at S and G_2 phases and premature mitotic entry, although DNA damage has been ruled out as its mechanism of action [135]. Heat shock does, however, affect response pathways to double-stranded DNA breaks induced by radiation [154]. When combined with radiation, hyperthermia leads to a greater frequency of MC due to dramatic increases in cyclin B and enhanced mitotic entry in spite of DNA damage [135].

Radionuclides tagged with antibodies that recognize surface antigens unique to cancer cells are viable options currently being explored in preclinical studies for cancer radiotherapy. Radionuclides that emit beta rays, including ^{131}I and ^{177}Lu, have a

0.7–3.9 mm range in tissues and can be used for solid-tumor targeting [132]. Alpha emitters have much shorter ranges of 50–100 μm but show high linear energy transfer and, thus, have greater relative biological effectiveness on tumors [132]. The alpha emitter ^{213}Bi, targeted to gastric cancer cells, has been shown to induce G_2/M arrest and micronucleation, indicative of MC [132]. Downregulation of Aurora A, Plk1, and other proteins involved in mitotic spindle assembly was observed.

As is the case with other types of DNA damage, MC can be followed by senescence rather than cell death. In response to radiotherapy, cells can resist premature entry into mitosis and remain in a long-term G_2/M arrest. They eventually undergo mitotic slippage and enter a state of senescence [16]. Although this results in tumor growth control, such cells can eventually reenter the cell cycle and produce abnormal polyploidy cells capable of repopulating the tumor, resulting in disease recurrence [16]. It has been shown that high-dose radiation is more likely to induce this senescence phenotype than moderate- or low-dose [16].

23.6.2 Cancer Chemotherapy

Emerging evidence over the past 2 decades and beyond has revealed that some of the most widely used chemotherapeutic agents induce MC as a predominant mode of cell death. This is not surprising, since classic chemotherapy targets highly proliferative cells and interferes with their mitotic machinery. Such agents, some of which are outlined in Table 23.3, cause DNA or spindle damage during the cell cycle. As discussed earlier, many chemotherapeutic agents that are predominantly understood to cause cell death via apoptosis can induce MC if alternative drug concentrations are used. Doxorubicin is one of the most widely used chemotherapeutics clinically. It inhibits topoisomerase II and leads to the formation of free radicals [54]. While high doses cause apoptotic cell death in p53-deficient cells, characterized by cyt c release and caspase 3 activation, low doses lead to MC that is ultimately followed by apoptotic cell death [54].

The important roles of ATM, ATR, Chk1, and Chk2 in hindering mitotic progression when damage is sensed have already been discussed. Targeting these effectors using chemotherapeutic strategies, or in combination with classically used chemotherapeutics that damage DNA or the mitotic spindle, is a rational approach to inducing MC in cancer cells. For instance, the exploratory small molecule UCN-01 inhibits Chk1 and induces mitotic entry by targeting Cdc25C-Cdk1/cyclin B interactions in order to overcome the G_2/M checkpoint [34]. This premature mitotic entry precedes MC in cells that have DNA damage caused by irradiation [24]. In one study, although a large proportion of cells did complete mitosis without undergoing MC, post-mitotic cell death eventually ensued after an abnormally long mitosis, and it was observed that the length of mitosis was a determinant of whether cell death through MC would occur [24]. The identification of targets that affect MC also presents an opportunity for the design of small-molecule inhibitors or enhancers of these targets. For instance, Notch signaling has been shown to increase cyclin B expression through regulation of NF-κB [100]. Stimulation of Notch activity leads to increased cyclin B and Cdk1 activities, resulting in MC in cancer cells [100].

Photodynamic therapy (PDT) is another interesting treatment modality that causes cell death via MC. PDT involves a small-molecule photosensitizer that is activated in the presence of light at specific wavelengths (or of molecular oxygen) and forms reactive oxygen species (ROS) that are damaging to cancer cells [83]. It is known to affect

Table 23.3 Examples of classic chemotherapeutics that induce mitotic catastrophe (MC).

Type of Damage	Mechanism	Drug	Mode of Cell Death	Cancer Cell Lines Studied	References
DNA damage	Topoisomerase II inhibitor, DNA intercalator	Doxorubicin	MC and/or senescence at low concentrations, apoptosis at high concentrations	Liver (Huh-7)	[56,57,127]
				N/A (CHO AA8)	[5,54]
	Topoisomerase II inhibitor	Etoposide	MC in 40–80% at sublethal concentrations	Leukemia (Jukrat T cells)	[40]
	DNA crosslinking agent	Cisplatin	MC in 40–80% at sublethal concentrations	Leukemia (Jukrat T cells)	[40]
		Oxaliplatin	MC and apoptosis; proportion of each is cell line-dependent	Esophageal (TE3 and TE7)	[103]
	DNA methylating agent	Temozolomide	MC	Glioblastoma (U87MG)	[59]
Spindle damage	Microtubule depolymerization inhibitor	Docetaxel	MC predominantly	Breast (CF-10A, MCF-7, MDA-mb-231)	[134]
			MC followed by apoptosis	Prostate (PC3, DU145)	[51]
		Paclitaxel	In acute micromolar concentrations, MC followed by necrosis	Cervical (HeLa)	[138]
			MC in 40–80% at sublethal concentrations	Leukemia (Jukrat T cells)	[40]
			MC followed by apoptosis-like cell death	Colon (HCT116)	[33]
		Nocodazole	MC in 40–80% at sublethal concentrations	Leukemia (Jukrat T cells)	[40]
	Microtubule polymerization inhibitor	Vincristine	MC in apoptosis-resistant cells	Leukemia (HCW-2)	[11]
		Combretastatin CA-4 prodrug	MC predominantly	Non-small-cell lung (H460)	[71,123]
				Leukemia (WSU-CCL)	[52]

microtubule function, leading to arrest by the spindle-assembly checkpoint, MC, and ultimately apoptotic cell death [83]. A glycophthalocyanine photosensitizer, GPh3, has been shown to induce microtubule abnormalities, multipolar spindles, arrest at metaphase, and MC followed by apoptosis in HeLa cervical cancer cells [83].

23.7 Conclusion

Normal cells possess mechanisms that halt their duplication upon taking damage to their DNA or the structures of the mitotic spindle. MC ensues when aberrant mechanisms permit a detrimentally damaged cell to bypass the G2/M or spindle-assembly checkpoint of the cell cycle, allowing abnormal mitosis to begin or to proceed. Such irregular mitotic entry or continuation triggers MC as a mechanism to avoid genomic instability. As cancer cells commonly display deficient checkpoint controls, MC is a common pathway to death in such cells. Failure of MC to ultimately lead to cell death can result in polyploidy and the propagation of genomic aberrations. Controversy remains over whether MC constitutes a unique form of cell death or is simply a pre-stage to apoptosis or necrosis, although mounting evidence suggests the latter. The MC process itself, however, is quite unique and clearly distinguishable, morphologically and biochemically, from other modes of cell death. Therapeutically, strategies have been explored to drive apoptosis-resistant cancer cells toward MC, ultimately leading to death via this alternative pathway. Classical radiotherapy kills cancer cells via MC, and radiosensitizing agents capable of triggering the MC response have been explored in connection with enhancement of the effects of radiation. Although chemotherapeutic agents are widely recognized to result in apoptotic cell death, studies are continuously elucidating the important role of MC-dependent death in the efficacy of chemotherapy. As details pertaining to the mechanisms and applications of MC are revealed, the importance of this remarkable pathway to cell death becomes increasingly evident.

References

1. Caruso R, Fedele F, Lucianò R, Branca G, Parisi C, Paparo D, Parisi A. Mitotic catastrophe in malignant epithelial tumors: the pathologist's viewpoint. *Ultrastruct Pathol* 2011;**35**(2):66–71.
2. Ayscough K, Hayles J, MacNeill SA, Nurse P. Cold-sensitive mutants of p34cdc2 that suppress a mitotic catastrophe phenotype in fission yeast. *Mol Gen Genet* 1992;**232**(3):344–50.
3. Castedo M, Perfettini JL, Roumier T, Andreau K, Medema R, Kroemer G. Cell death by mitotic catastrophe: a molecular definition. *Oncogene* 2004;**23**(16):2825–37.
4. Dodson H, Wheatley SP, Morrison CG. Involvement of centrosome amplification in radiation-induced mitotic catastrophe. *Cell Cycle* 2007;**6**(3):364–70.
5. Grzanka D, Marszałek A, Izdebska M, Gackowska L, Andrzej Szczepanski M, Grzanka A. Actin cytoskeleton reorganization correlates with cofilin nuclear expression and ultrastructural changes in cho aa8 cell line after apoptosis and mitotic catastrophe induction by doxorubicin. *Ultrastruct Pathol* 2011;**35**(3):130–8.
6. Ianzini F, Kosmacek EA, Nelson ES, Napoli E, Erenpreisa J, Kalejs M, Mackey MA. Activation of meiosis-specific genes is associated with depolyploidization of human

tumor cells following radiation-induced mitotic catastrophe. *Cancer Res* 2009;**69**(6):2296–304.
7. Erenpreisa JE, Ivanov A, Dekena G, Vitina A, Krampe R, Feivalds T, et al. Arrest in metaphase and anatomy of mitotic catastrophe: mild heat shock in two human osteosarcoma cell lines. *Cell Biol Int* 2000;**24**(2):61–70.
8. Mu R, Wang YB, Wu M, Yang Y, Song W, Li T, et al. Depletion of pre-mRNA splicing factor Cdc5L inhibits mitotic progression and triggers mitotic catastrophe. *Cell Death Dis* 2014;**5**:e1151.
9. Roninson IB, Broude EV, Chang BD. If not apoptosis, then what? Treatment-induced senescence and mitotic catastrophe in tumor cells. *Drug Resist Updat* 2001;**4**(5):303–13.
10. Castedo M, Perfettini JL, Roumier T, Yakushijin K, Horne D, Medema R, Kroemer G. The cell cycle checkpoint kinase Chk2 is a negative regulator of mitotic catastrophe. *Oncogene* 2004;**23**(25):4353–61.
11. Magalska A, Sliwinska M, Szczepanowska J, Salvioli S, Franceschi C, Sikora E. Resistance to apoptosis of HCW-2 cells can be overcome by curcumin- or vincristine-induced mitotic catastrophe. *Int J Cancer* 2006;**119**(8):1811–18.
12. Howard A, Pelc S. Synthesis of deoxyribonucleic acid in normal and irrradiated cells and its relation to chromosome breakage. *Heredity* 1952;**6**:261–73.
13. Mendelsohn ML, Autoradiographic analysis of cell proliferation in spontaneous breast cancer of C3H mouse. III. The growth fraction. *J Natl Cancer Inst* 1962;**28**:1015–29.
14. Yasutis KM, Kozminski KG. Cell cycle checkpoint regulators reach a zillion. *Cell Cycle* 2013;**12**(10):1501–9.
15. Gheghiani L, Gavet O. Deciphering the spatio-temporal regulation of entry and progression through mitosis. *Biotechnol J* 2014;**9**(2):213–23.
16. Shen Z, Huhn SC, Haffty BG. Escaping death to quiescence: avoiding mitotic catastrophe after DNA damage. *Cell Cycle* 2013;**12**(11):1664.
17. Canman CE. Replication checkpoint: preventing mitotic catastrophe. *Curr Biol* 2001;**11**(4):R121–4.
18. Huang X, Tran T, Zhang L, Hatcher R, Zhang P. DNA damage-induced mitotic catastrophe is mediated by the Chk1-dependent mitotic exit DNA damage checkpoint. *Proc Natl Acad Sci USA* 2005;**102**(4):1065–70.
19. Castedo M, Perfettini JL, Roumier T, Kroemer G. Cyclin-dependent kinase-1: linking apoptosis to cell cycle and mitotic catastrophe. *Cell Death Differ* 2002;**9**(12):1287–93.
20. Andreassen PR, Lacroix FB, Lohez OD, Margolis RL. Neither p21WAF1 nor 14-3-3sigma prevents G2 progression to mitotic catastrophe in human colon carcinoma cells after DNA damage, but p21WAF1 induces stable G1 arrest in resulting tetraploid cells. *Cancer Res* 2001;**61**(20):7660–8.
21. Cogswell JP, Brown CE, Bisi JE, Neil SD. Dominant-negative polo-like kinase 1 induces mitotic catastrophe independent of cdc25C function. *Cell Growth Differ* 2000;**11**(12):615–23.
22. Nigg EA. Mitotic kinases as regulators of cell division and its checkpoints. *Nat Rev Mol Cell Biol* 2001;**2**(1):21–32.
23. Cahuzac N, Studény A, Marshall K, Versteege I, Wetenhall K, Pfeiffer B, et al. An unusual DNA binding compound, S23906, induces mitotic catastrophe in cultured human cells. *Cancer Lett* 2010;**289**(2):178–87.

24 On KF, Chen Y, Ma HT, Chow JP, Poon RY. Determinants of mitotic catastrophe on abrogation of the G2 DNA damage checkpoint by UCN-01. *Mol Cancer Ther* 2011;**10**(5):784–94.

25 Surova O, Zhivotovsky B. Various modes of cell death induced by DNA damage. *Oncogene* 2013;**32**(33):3789–97.

26 Ho CY, Wong CH, Li HY. Perturbation of the chromosomal binding of RCC1, Mad2 and survivin causes spindle assembly defects and mitotic catastrophe. *J Cell Biochem* 2008;**105**(3):835–46.

27 Shen JK, Du HP, Yang M, Wang YG, Jin J. Casticin induces leukemic cell death through apoptosis and mitotic catastrophe. *Ann Hematol* 2009;**88**(8):743–52.

28 Qi M, Yao G, Fan S, Cheng W, Tashiro S, Onodera S, Ikejima T. Pseudolaric acid B induces mitotic catastrophe followed by apoptotic cell death in murine fibrosarcoma L929 cells. *Eur J Pharmacol* 2012;**683**(1–3):16–26.

29 Cheong JW, Jung HI, Eom JI, Kim SJ, Jeung HK, Min YH. Aurora-A kinase inhibition enhances the cytosine arabinoside-induced cell death in leukemia cells through apoptosis and mitotic catastrophe. *Cancer Lett* 2010;**297**(2):171–81.

30 Erenpreisa J, Kalejs M, Cragg MS. Mitotic catastrophe and endomitosis in tumour cells: an evolutionary key to a molecular solution. *Cell Biol Int* 2005;**29**(12):1012–18.

31 Erenpreisa J, Kalejs M, Ianzini F, Kosmacek EA, Mackey MA, Emzinsh D, et al. Segregation of genomes in polyploid tumour cells following mitotic catastrophe. *Cell Biol Int* 2005;**29**(12):1005–11.

32 Kalejs M, Ivanov A, Plakhins G, Cragg MS, Emzinsh D, Illidge TM, Erenpreisa J. Upregulation of meiosis-specific genes in lymphoma cell lines following genotoxic insult and induction of mitotic catastrophe. *BMC Cancer* 2006;**6**:6.

33 Llovera L, Mansilla S, Portugal J. Apoptotic-like death occurs through a caspase-independent route in colon carcinoma cells undergoing mitotic catastrophe. *Cancer Lett* 2012;**326**(1):114–21.

34 Tse AN, Schwartz GK. Potentiation of cytotoxicity of topoisomerase i poison by concurrent and sequential treatment with the checkpoint inhibitor UCN-01 involves disparate mechanisms resulting in either p53-independent clonogenic suppression or p53-dependent mitotic catastrophe. *Cancer Res* 2004;**64**(18):6635–44.

35 Niida H, Tsuge S, Katsuno Y, Konishi A, Takeda N, Nakanishi M. Depletion of Chk1 leads to premature activation of Cdc2-cyclin B and mitotic catastrophe. *J Biol Chem* 2005;**280**(47):39 246–52.

36 Chan TA, Hermeking H, Lengauer C, Kinzler KW, Vogelstein B. 14-3-3sigma is required to prevent mitotic catastrophe after DNA damage. *Nature* 1999;**401**(6753):616–20.

37 Imreh G, Norberg HV, Imreh S, Zhivotovsky B. Chromosomal breaks during mitotic catastrophe trigger gammaH2AX-ATM-p53-mediated apoptosis. *J Cell Sci* 2011;**124** (Pt. 17):2951–63.

38 Zhang B, Huang B, Guan H, Zhang SM, Xu QZ, He XP, et al. Proteomic profiling revealed the functional networks associated with mitotic catastrophe of HepG2 hepatoma cells induced by 6-bromine-5-hydroxy-4-methoxybenzaldehyde. *Toxicol Appl Pharmacol* 2011;**252**(3):307–17.

39 Rivera A, Mavila A, Bayless KJ, Davis GE, Maxwell SA. Cyclin A1 is a p53-induced gene that mediates apoptosis, G2/M arrest, and mitotic catastrophe in renal, ovarian, and lung carcinoma cells. *Cell Mol Life Sci* 2006;**63**(12):1425–39.

40. Maskey D, Yousefi S, Schmid I, Zlobec I, Perren A, Friis R, Simon HU. ATG5 is induced by DNA-damaging agents and promotes mitotic catastrophe independent of autophagy. *Nat Commun* 2013;**4**:2130.
41. Fragkos M, Beard P. Mitotic catastrophe occurs in the absence of apoptosis in p53-null cells with a defective G1 checkpoint. *PLoS One* 2011;**6**(8): e22946.
42. Soares AS, Costa VM, Diniz C, Fresco P. Combination of ClIBMECA with paclitaxel is a highly effective cytotoxic therapy causing mTORdependent autophagy and mitotic catastrophe on human melanoma cells. *J Cancer Res Clin Oncol* 2014;**140**(6):921–35.
43. Kimura M, Yoshioka T, Saio M, Banno Y, Nagaoka H, Okano Y. Mitotic catastrophe and cell death induced by depletion of centrosomal proteins. *Cell Death Dis* 2013;**4**: e603.
44. Bui CB, Shin J. Persistent expression of Nqo1 by p62-mediated Nrf2 activation facilitates p53-dependent mitotic catastrophe. *Biochem Biophys Res Commun* 2011;**412**(2):347–52.
45. Hashizume C, Kobayashi A, Wong RW. Down-modulation of nucleoporin RanBP2/Nup358 impaired chromosomal alignment and induced mitotic catastrophe. *Cell Death Dis* 2013;**4**:e854.
46. Shih HJ, Chu KL, Wu MH, Wu PH, Chang WW, Chu JS, et al. The involvement of MCT-1 oncoprotein in inducing mitotic catastrophe and nuclear abnormalities. *Cell Cycle* 2012;**11**(5):934–52.
47. Stevens JB, Liu G, Bremer SW, Ye KJ, Xu W, Xu J, et al. Mitotic cell death by chromosome fragmentation. *Cancer Res* 2007;**67**(16):7686–94.
48. Hübner B, Strickfaden H, Müller S, Cremer M, Cremer T. Chromosome shattering: a mitotic catastrophe due to chromosome condensation failure. *Eur Biophys J* 2009;**38**(6):729–47.
49. Rello-Varona S, Kepp O, Vitale I, Michaud M, Senovilla L, Jemaà M, et al. An automated fluorescence videomicroscopy assay for the detection of mitotic catastrophe. *Cell Death Dis* 2010;**1**:e25.
50. Castedo M, Perfettini JL, Roumier T, Valent A, Raslova H, Yakushijin K, et al. Mitotic catastrophe constitutes a special case of apoptosis whose suppression entails aneuploidy. *Oncogene* 2004;**23**(25):4362–70.
51. Fabbri F, Amadori D, Carloni S, Brigliadori G, Tesei A, Ulivi P, et al. Mitotic catastrophe and apoptosis induced by docetaxel in hormone-refractory prostate cancer cells. *J Cell Physiol* 2008;**217**(2):494–501.
52. Nabha SM, Mohammad RM, Dandashi MH, Coupaye-Gerard B, Aboukameel A, Pettit GR, Al-Katib AM. Combretastatin-A4 prodrug induces mitotic catastrophe in chronic lymphocytic leukemia cell line independent of caspase activation and poly(ADP-ribose) polymerase cleavage. *Clin Cancer Res* 2002;**8**(8):2735–41.
53. Cives M, Ciavarella S, Rizzo FM, De Matteo M, Dammacco F, Silvestris F. Bendamustine overcomes resistance to melphalan in myeloma cell lines by inducing cell death through mitotic catastrophe. *Cell Signal* 2013;**25**(5):1108–17.
54. Grzanka D, Grzanka A, Izdebska M, Gackowska L, Stepien A, Marszalek A. Actin reorganization in CHO AA8 cells undergoing mitotic catastrophe and apoptosis induced by doxorubicin. *Oncol Rep* 2010;**23**(3):655–63.
55. Mansilla S, Priebe W, Portugal J. Mitotic catastrophe results in cell death by caspase-dependent and caspase-independent mechanisms. *Cell Cycle* 2006;**5**(1):53–60.

56 Park SS, Eom YW, Choi KS. Cdc2 and Cdk2 play critical roles in low dose doxorubicin-induced cell death through mitotic catastrophe but not in high dose doxorubicin-induced apoptosis. *Biochem Biophys Res Commun* 2005;**334**(4):1014–21.

57 Park SS, Kim MA, Eom YW, Choi KS. Bcl-xL blocks high dose doxorubicin-induced apoptosis but not low dose doxorubicin-induced cell death through mitotic catastrophe. *Biochem Biophys Res Commun* 2007;**363**(4):1044–9.

58 Hsu MH, Liu CY, Lin CM, Chen YJ, Chen CJ, Lin YF, et al. 2-(3-methoxyphenyl)-5-methyl-1,8-naphthyridin-4(1H)-one (HKL-1) induces G2/M arrest and mitotic catastrophe in human leukemia HL-60 cells. *Toxicol Appl Pharmacol* 2012;**259**(2):219–26.

59 Hirose Y, Katayama M, Mirzoeva OK, Berger MS, Pieper RO. Akt activation suppresses Chk2-mediated, methylating agent-induced G2 arrest and protects from temozolomide-induced mitotic catastrophe and cellular senescence. *Cancer Res* 2005;**65**(11):4861–9.

60 Wang X, Wu E, Wu J, Wang TL, Hsieh HP, Liu X. An antimitotic and antivascular agent BPR0L075 overcomes multidrug resistance and induces mitotic catastrophe in paclitaxel-resistant ovarian cancer cells. *PLoS One* 2013;**8**(6):e65686.

61 Hyzy M, Bozko P, Konopa J, Skladanowski A. Antitumour imidazoacridone C-1311 induces cell death by mitotic catastrophe in human colon carcinoma cells. *Biochem Pharmacol* 2005;**69**(5):801–9.

62 Eriksson D, Löfroth PO, Johansson L, Riklund KA, Stigbrand T. Cell cycle disturbances and mitotic catastrophes in HeLa Hep2 cells following 2.5 to 10 Gy of ionizing radiation. *Clin Cancer Res* 2007;**13**(18 Pt. 2):5501s–8s.

63 Shen L, Au WY, Wong KY, Shimizu N, Tsuchiyama J, Kwong YL, et al. Cell death by bortezomib-induced mitotic catastrophe in natural killer lymphoma cells. *Mol Cancer Ther* 2008;**7**(12):3807–15.

64 Natarajan G, Ramalingam S, Ramachandran I, May R, Queimado L, Houchen CW, Anant S. CUGBP2 downregulation by prostaglandin E2 protects colon cancer cells from radiation-induced mitotic catastrophe. *Am J Physiol Gastrointest Liver Physiol* 2008;**294**(5):G1235–44.

65 Wolanin K, Magalska A, Mosieniak G, Klinger R, McKenna S, Vejda S, et al. Curcumin affects components of the chromosomal passenger complex and induces mitotic catastrophe in apoptosis-resistant Bcr-Abl-expressing cells. *Mol Cancer Res* 2006;**4**(7):457–69.

66 Jackson SJ, Murphy LL, Venema RC, Singletary KW, Young AJ. Curcumin binds tubulin, induces mitotic catastrophe, and impedes normal endothelial cell proliferation. *Food Chem Toxicol* 2013;**60**:431–8.

67 Firat E, Gaedicke S, Tsurumi C, Esser N, Weyerbrock A, Niedermann G. Delayed cell death associated with mitotic catastrophe in gamma-irradiated stem-like glioma cells. *Radiat Oncol* 2011;**6**:71.

68 Yu X, Chen J. DNA damage-induced cell cycle checkpoint control requires CtIP, a phosphorylation-dependent binding partner of BRCA1 C-terminal domains. *Mol Cell Biol* 2004;**24**(21):9478–86.

69 Ashley AK, Shrivastav M, Nie J, Amerin C, Troksa K, Glanzer JG, et al. DNA-PK phosphorylation of RPA32 Ser4/Ser8 regulates replication stress checkpoint activation, fork restart, homologous recombination and mitotic catastrophe. *DNA Repair (Amst)* 2014;**21**:131–9.

70 Oike T, Komachi M, Ogiwara H, Amornwichet N, Saitoh Y, Torikai K, et al. C646, a selective small molecule inhibitor of histone acetyltransferase p300, radiosensitizes lung cancer cells by enhancing mitotic catastrophe. *Radiother Oncol* 2014;**111**(2):222–7.

71 Vitale I, Antoccia A, Cenciarelli C, Crateri P, Meschini S, Arancia G, et al. Combretastatin CA-4 and combretastatin derivative induce mitotic catastrophe dependent on spindle checkpoint and caspase-3 activation in non-small cell lung cancer cells. *Apoptosis* 2007;**12**(1):155–66.

72 Chen T, Hevi S, Gay F, Tsujimoto N, He T, Zhang B, et al. Complete inactivation of DNMT1 leads to mitotic catastrophe in human cancer cells. *Nat Genet* 2007;**39**(3):391–6.

73 Cherubini G, Petouchoff T, Grossi M, Piersanti S, Cundari E, Saggio I. E1B55K-deleted adenovirus (ONYX-015) overrides G1/S and G2/M checkpoints and causes mitotic catastrophe and endoreduplication in p53-proficient normal cells. *Cell Cycle* 2006;**5**(19):2244–52.

74 Ogawa O, Zhu X, Lee HG, Raina A, Obrenovich ME, Boswer R, et al. Ectopic localization of phosphorylated histone H3 in Alzheimer's disease: a mitotic catastrophe? *Acta Neuropathol* 2003;**105**(5):524–8.

75 Cotugno R, Fortunato R, Santoro A, Gollotta D, Braca A, De Tommasi N, Belisario MA. Effect of sesquiterpene lactone coronopilin on leukaemia cell population growth, cell type-specific induction of apoptosis and mitotic catastrophe. *Cell Prolif* 2012;**45**(1):53–65.

76 Zhang HY, Gu YY, Li ZG, Jia YH, Yuan L, Li SY, et al. Exposure of human lung cancer cells to 8-chloro-adenosine induces G2/M arrest and mitotic catastrophe. *Neoplasia* 2004;**6**(6):802–12.

77 Zhang X, Zhang Z, Chen G, Zhao M, Wang D, Zhang X, et al. FK228 induces mitotic catastrophe in A549 cells by mistargeting chromosomal passenger complex localization through changing centromeric H3K9 hypoacetylation. *Acta Biochim Biophys Sin (Shanghai)* 2010;**42**(10):677–87.

78 Scaife RM. G2 cell cycle arrest, down-regulation of cyclin B, and induction of mitotic catastrophe by the flavoprotein inhibitor diphenyleneiodonium. *Mol Cancer Ther* 2004;**3**(10):1229–37.

79 Shi X, Wang D, Ding K, Lu Z, Jin Y, Zhang J, Pan J. GDP366, a novel small molecule dual inhibitor of survivin and Op18, induces cell growth inhibition, cellular senescence and mitotic catastrophe in human cancer cells. *Cancer Biol Ther* 2010;**9**(8):640–50.

80 Nomura M, Noumra N, Newcomb EW, Lukyanov Y, Tamasdan C, Zagzag D. Geldanamycin induces mitotic catastrophe and subsequent apoptosis in human glioma cells. *J Cell Physiol* 2004;**201**(3):374–84.

81 Chan YW, Chen Y, Poon RY. Generation of an indestructible cyclin B1 by caspase-6-dependent cleavage during mitotic catastrophe. *Oncogene* 2009;**28**(2):170–83.

82 Tominaga Y, Wang A, Wang RH, Wang X, Cao L, Deng CX. Genistein inhibits Brca1 mutant tumor growth through activation of DNA damage checkpoints, cell cycle arrest, and mitotic catastrophe. *Cell Death Differ* 2007;**14**(3):472–9.

83 Soares AR, Neves MG, Tomé AC, Iglesias-de la Cruz MC, Zamarrón A, Carrasco E, et al. Glycophthalocyanines as photosensitizers for triggering mitotic catastrophe and apoptosis in cancer cells. *Chem Res Toxicol* 2012;**25**(4):940–51.

84 Wu YC, Yen WY, Ho HY, Su TL, Yih LH. Glyfoline induces mitotic catastrophe and apoptosis in cancer cells. *Int J Cancer* 2010;**126**(4):1017–28.

85. Ianzini F, Domann FE, Kosmacek EA, Phillips SL, Mackey MA. Human glioblastoma U87MG cells transduced with a dominant negative p53 (TP53) adenovirus construct undergo radiation-induced mitotic catastrophe. *Radiat Res* 2007;**168**(2):183–92.
86. Mir SE, De Witt Hamer PC, Grawczyk PM, Balaj L, Claes A, Niers JM, et al. In silico analysis of kinase expression identifies WEE1 as a gatekeeper against mitotic catastrophe in glioblastoma. *Cancer Cell* 2010;**18**(3):244–57.
87. Chen Y, Chow JP, Poon RY. Inhibition of Eg5 acts synergistically with checkpoint abrogation in promoting mitotic catastrophe. *Mol Cancer Res* 2012;**10**(5):626–35.
88. Tandle AT, Kramp T, Kil WJ, Halthore A, Gehlhaus K, Shankavaram U, et al. Inhibition of polo-like kinase 1 in glioblastoma multiforme induces mitotic catastrophe and enhances radiosensitisation. *Eur J Cancer* 2013;**49**(14):3020–8.
89. Hemstrom TH, Sandstrom M, Zhivotovsky B. Inhibitors of the PI3-kinase/Akt pathway induce mitotic catastrophe in non-small cell lung cancer cells. *Int J Cancer* 2006;**119**(5):1028–38.
90. Gonçalves AP, Máximo V, Lima J, Singh KK, Soares P, Videira A. Involvement of p53 in cell death following cell cycle arrest and mitotic catastrophe induced by rotenone. *Biochim Biophys Acta* 2011;**1813**(3):492–9.
91. Eriksson D, Blomberg J, Lindgren T, Löfroth PO, Johansson L, Riklund K, Stigbrand T. Iodine-131 induces mitotic catastrophes and activates apoptotic pathways in HeLa Hep2 cells. *Cancer Biother Radiopharm* 2008;**23**(5):541–9.
92. Ianzini F, Bertoldo A, Kosmacek EA, Phillips SL, Mackey MA. Lack of p53 function promotes radiation-induced mitotic catastrophe in mouse embryonic fibroblast cells. *Cancer Cell Int* 2006;**6**:11.
93. Evison BJ, Pastuovic M, Bilardi RA, Forrest RA, Pumuye PP, Sleebs BE, et al. M2, a novel anthracenedione, elicits a potent DNA damage response that can be subverted through checkpoint kinase inhibition to generate mitotic catastrophe. *Biochem Pharmacol* 2011;**82**(11):1604–18.
94. Ceelen LM, Haesebrouck F, D'Herde K, Krysko DV, Favoreel H, Vandenabeele P, et al. Mitotic catastrophe as a prestage to necrosis in mouse liver cells treated with Helicobacter pullorum sonicates. *J Morphol* 2009;**270**(8):921–8.
95. Zajac M, Moneo MV, Carnero A, Benitez J, Martínez-Delgado B. Mitotic catastrophe cell death induced by heat shock protein 90 inhibitor in BRCA1-deficient breast cancer cell lines. *Mol Cancer Ther* 2008;**7**(8):2358–66.
96. Rello-Varona S, Stockert JC, Cañete M, Acedo P, Vaillanueva A. Mitotic catastrophe induced in HeLa cells by photodynamic treatment with Zn(II)-phthalocyanine. *Int J Oncol* 2008;**32**(6):1189–96.
97. Ha GH, Kim HS, Lee CG, Park HY, Kim EJ, Shin HJ, et al. Mitotic catastrophe is the predominant response to histone acetyltransferase depletion. *Cell Death Differ* 2009;**16**(3):483–97.
98. Chen CA, Chen CC, Shen CC, Chang HH, Chen YJ. Moscatilin induces apoptosis and mitotic catastrophe in human esophageal cancer cells. *J Med Food* 2013;**16**(10):869–77.
99. Chen S, Wan P, Ding W, Li F, He C, Chen P, et al. Norcantharidin inhibits DNA replication and induces mitotic catastrophe by degrading initiation protein Cdc6. *Int J Mol Med* 2013;**32**(1):43–50.
100. Curry CL, Reed LL, Broude E, Golde TE, Miele L, Foreman KE. Notch inhibition in Kaposi's sarcoma tumor cells leads to mitotic catastrophe through nuclear factor-kappaB signaling. *Mol Cancer Ther* 2007;**6**(7):1983–92.

101 Ramalingam S, Natarajan G, Schafer C, Subramaniam D, May R, Ramachandran I, et al. Novel intestinal splice variants of RNA-binding protein CUGBP2: isoform-specific effects on mitotic catastrophe. *Am J Physiol Gastrointest Liver Physiol* 2008;**294**(4):G971–81.

102 Chen B, Cheng M, Hong DJ, Sun FY, Zhu CQ. Okadaic acid induced cyclin B1 expression and mitotic catastrophe in rat cortex. *Neurosci Lett* 2006;**406**(3):178–82.

103 Ngan CY, Yamamoto H, Takagi A, Fujie Y, Takemasa I, Ikeda M, et al. Oxaliplatin induces mitotic catastrophe and apoptosis in esophageal cancer cells. *Cancer Sci* 2008;**99**(1):129–39.

104 Fujie Y, Yamamoto H, Ngan CY, Takagi A, Hayashi T, Suzuki R, et al. Oxaliplatin, a potent inhibitor of survivin, enhances paclitaxel-induced apoptosis and mitotic catastrophe in colon cancer cells. *Jpn J Clin Oncol* 2005;**35**(8):453–63.

105 Taylor BF, McNeely SC, Miller HL, Lehmann GM, McCabe MJ Jr., States JC. p53 suppression of arsenite-induced mitotic catastrophe is mediated by p21CIP1/WAF1. *J Pharmacol Exp Ther* 2006;**318**(1):142–51.

106 Isham CR, Bossou AR, Negron V, Fisher KE, Kumar R, Marlow L, et al. Pazopanib enhances paclitaxel-induced mitotic catastrophe in anaplastic thyroid cancer. *Sci Transl Med* 2013;**5**(166):166ra3.

107 Xu HZ, Huang Y, Wu YL, Zhao Y, Xiao WL, Lin QS, et al. Pharicin A, a novel natural ent-kaurene diterpenoid, induces mitotic arrest and mitotic catastrophe of cancer cells by interfering with BubR1 function. *Cell Cycle* 2010;**9**(14):2897–907.

108 Ame JC, Fouguerel E, Gauthier LR, Biard D, Boussin FD, Dantzer F, et al. Radiation-induced mitotic catastrophe in PARG-deficient cells. *J Cell Sci* 2009;**122**(Pt. 12):1990–2002.

109 Yu, S-L, Kang M-S, Kim H-Y, Lee SH, Lee S-K. Restoration of proliferation ability with increased genomic instability from Rad2p-induced mitotic catastrophe in *Saccharomyces cerevisiae*. *Mol Cell Toxicol* 2011;**7**:195–206.

110 Filippi-Chiela EC, Thomé MP, Bueno e Silva MM, Pelegrini AL, Ledur PF, Garicochea B, et al. Resveratrol abrogates the temozolomide-induced G2 arrest leading to mitotic catastrophe and reinforces the temozolomide-induced senescence in glioma cells. *BMC Cancer* 2013;**13**:147.

111 Lee SY, Oh JS, Rho JH, Jeong NY, Kwon YH, Jeong WJ, et al. Retinal pigment epithelial cells undergoing mitotic catastrophe are vulnerable to autophagy inhibition. *Cell Death Dis* 2014;**5**:e1303.

112 Subramaniam D, Ramalingam S, Linehan DC, Dieckgraefe BK, Postier RG, Houchen CW, et al. RNA binding protein CUGBP2/CELF2 mediates curcumin-induced mitotic catastrophe of pancreatic cancer cells. *PLoS One* 2011;**6**(2):e16958.

113 Sarbajna S, Davies D, West SC. Roles of SLX1-SLX4, MUS81-EME1, and GEN1 in avoiding genome instability and mitotic catastrophe. *Genes Dev* 2014;**28**(10):1124–36.

114 Du M, Qiu Q, Gruslin A, Gordon J, He M, Chan CC, et al. SB225002 promotes mitotic catastrophe in chemo-sensitive and -resistant ovarian cancer cells independent of p53 status *in vitro*. *PLoS One* 2013;**8**(1):e54572.

115 Skwarska A, Augustin E, Konopa J. Sequential induction of mitotic catastrophe followed by apoptosis in human leukemia MOLT4 cells by imidazoacridinone C-1311. *Apoptosis* 2007;**12**(12):2245–57.

116 Burns TF, Fei P, Scata KA, Dicker DT, El-Deiry WS. Silencing of the novel p53 target gene Snk/Plk2 leads to mitotic catastrophe in paclitaxel (taxol)-exposed cells. *Mol Cell Biol* 2003;**23**(16):5556–71.

117 Nitta M, Kobayashi O, Honda S, Hirota T, Kuninaka S, Marumoto T, et al. Spindle checkpoint function is required for mitotic catastrophe induced by DNA-damaging agents. *Oncogene* 2004;**23**(39):6548–58.
118 Hung JY, Wen CW, Hsu YL, Lin ES, Huang MS, Chen CY, Kuo PL. Subamolide a induces mitotic catastrophe accompanied by apoptosis in human lung cancer cells. *Evid Based Complement Alternat Med* 2013;**2013**:828143.
119 Schmit TL, Zhong W, Setaluri V, Spiegelman VS, Ahmad N. Targeted depletion of Polo-like kinase (Plk) 1 through lentiviral shRNA or a small-molecule inhibitor causes mitotic catastrophe and induction of apoptosis in human melanoma cells. *J Invest Dermatol* 2009;**129**(12):2843–53.
120 Li S, Szymborski A, Miron MJ, Marcellus R, Binda O, Lavoie JN, Branton PE. The adenovirus E4orf4 protein induces growth arrest and mitotic catastrophe in H1299 human lung carcinoma cells. *Oncogene* 2009;**28**(3):390–400.
121 Ingemarsdotter C, Keller D, Beard P. The DNA damage response to non-replicating adeno-associated virus: centriole overduplication and mitotic catastrophe independent of the spindle checkpoint. *Virology* 2010;**400**(2):271–86.
122 Strauss SJ, Higginbottom K, Jüliger S, Maharaj L, Allen P, Schenkein D, et al. The proteasome inhibitor bortezomib acts independently of p53 and induces cell death via apoptosis and mitotic catastrophe in B-cell lymphoma cell lines. *Cancer Res* 2007;**67**(6):2783–90.
123 Cenciarelli C, Tanzarella C, Vitale I, Pisano C, Crateri P, Meschini S, et al. The tubulin-depolymerising agent combretastatin-4 induces ectopic aster assembly and mitotic catastrophe in lung cancer cells H460. *Apoptosis* 2008;**13**(5):659–69.
124 Indovina P, Rainaldi G, Santini MT. Three-dimensional cell organization leads to a different type of ionizing radiation-induced cell death: MG-63 monolayer cells undergo mitotic catastrophe while spheroids die of apoptosis. *Int J Oncol* 2007;**31**(6):1473–83.
125 Mansilla S, Priebe W, Portugal J. Transcriptional changes facilitate mitotic catastrophe in tumour cells that contain functional p53. *Eur J Pharmacol* 2006;**540**(1–3):34–45.
126 Pan J, Hu H, Zhou Z, Sun L, Peng L, Yu L, et al. Tumor-suppressive mir-663 gene induces mitotic catastrophe growth arrest in human gastric cancer cells. *Oncol Rep* 2010;**24**(1):105–12.
127 Eom YW, Kim MA, Park SS, Goo MJ, Kwon HJ, Sohn S, et al. Two distinct modes of cell death induced by doxorubicin: apoptosis and cell death through mitotic catastrophe accompanied by senescence-like phenotype. *Oncogene* 2005;**24**(30):4765–77.
128 Roy RV, Suman S, Das TP, Luevano JE, Damodaran C. Withaferin A, a steroidal lactone from *Withania somnifera*, induces mitotic catastrophe and growth arrest in prostate cancer cells. *J Nat Prod* 2013;**76**(10):1909–15.
129 Riesterer O, Matsumoto F, Wang L, Pickett J, Molkentine D, Giri U, et al. A novel Chk inhibitor, XL-844, increases human cancer cell radiosensitivity through promotion of mitotic catastrophe. *Invest New Drugs* 2011;**29**(3):514–22.
130 Perfettini JL, Nardacci R, Séror C, Raza SQ, Sepe S, Saïdi H, et al. 53BP1 represses mitotic catastrophe in syncytia elicited by the HIV-1 envelope. *Cell Death Differ* 2010;**17**(5):811–20.
131 Minata M, Gu C, Joshi K, Nakano-Okuno M, Hong C, Nguyen CH, et al. Multi-kinase inhibitor C1 triggers mitotic catastrophe of glioma stem cells mainly through MELK kinase inhibition. *PLoS One* 2014;**9**(4):e92546.

132 Seidl C, Port M, Gilbertz KP, Morgenstern A, Bruchertseifer F, Schwaiger M, et al. 213Bi-induced death of HSC45-M2 gastric cancer cells is characterized by G2 arrest and up-regulation of genes known to prevent apoptosis but induce necrosis and mitotic catastrophe. *Mol Cancer Ther* 2007;**6**(8):2346–59.

133 Ianzini F, Mackey MA. Delayed DNA damage associated with mitotic catastrophe following X-irradiation of HeLa S3 cells. *Mutagenesis* 1998;**13**(4):337–44.

134 Morse DL, Gray H, Payne CM, Gillies RJ. Docetaxel induces cell death through mitotic catastrophe in human breast cancer cells. *Mol Cancer Ther* 2005;**4**(10):1495–504.

135 Mackey MA, Ianzini F. Enhancement of radiation-induced mitotic catastrophe by moderate hyperthermia. *Int J Radiat Biol* 2000;**76**(2):273–80.

136 Li P, Zhou L, Dai Z, Jin X, Liu X, Matsumoto Y, et al. High LET radiation enhances nocodazole induced cell death in HeLa cells through mitotic catastrophe and apoptosis. *J Radiat Res* 2011;**52**(4):481–9.

137 Dempe JS, Pfeiffer E, Grimm AS, Metzler M. Metabolism of curcumin and induction of mitotic catastrophe in human cancer cells. *Mol Nutr Food Res* 2008;**52**(9):1074–81.

138 Michalakis J, Georgatos SD, Romanos J, Koutala H, Georgoulias V, Tsiftsis D, Theodoropoulos PA. Micromolar taxol, with or without hyperthermia, induces mitotic catastrophe and cell necrosis in HeLa cells. *Cancer Chemother Pharmacol* 2005;**56**(6):615–22.

139 Lanz HL, Zimmerman RM, Brouwer J, Noteborn MH, Backendorf C. Mitotic catastrophe triggered in human cancer cells by the viral protein apoptin. *Cell Death Dis* 2013;**4**:e487.

140 Wei JH, Seemann J. Nakiterpiosin targets tubulin and triggers mitotic catastrophe in human cancer cells. *Mol Cancer Ther* 2010;**9**(12):3375–85.

141 Sihn CR, Suh EJ, Lee KH, Kim TY, Kim SH. p55CDC/hCDC20 mutant induces mitotic catastrophe by inhibiting the MAD2-dependent spindle checkpoint activity in tumor cells. *Cancer Lett* 2003;**201**(2):203–10.

142 de-Sá-Júnior PL, Pasqualoto KF, Kerreira AK, Tavares MT, Damião MC, de Azevedo RA, et al. RPF101, a new capsaicin-like analogue, disrupts the microtubule network accompanied by arrest in the G2/M phase, inducing apoptosis and mitotic catastrophe in the MCF-7 breast cancer cells. *Toxicol Appl Pharmacol* 2013;**266**(3):385–98.

143 Ianzini F, Mackey MA. Spontaneous premature chromosome condensation and mitotic catastrophe following irradiation of HeLa S3 cells. *Int J Radiat Biol* 1997;**72**(4):409–21.

144 O'Boyle NM, Carr M, Greene LM, Keely NO, Knox AJ, McCabe T, et al. Synthesis, biochemical and molecular modelling studies of antiproliferative azetidinones causing microtubule disruption and mitotic catastrophe. *Eur J Med Chem* 2011;**46**(9):4595–607.

145 Tao Y, Leteur C, Calderaro J, Girdler F, Zhang P, Frascogna V, et al. The aurora B kinase inhibitor AZD1152 sensitizes cancer cells to fractionated irradiation and induces mitotic catastrophe. *Cell Cycle* 2009;**8**(19):3172–81.

146 Holmfeldt P, Larsson N, Segerman B, Howell B, Morbito J, Cassimeris L, Gullberg M. The catastrophe-promoting activity of ectopic Op18/stathmin is required for disruption of mitotic spindles but not interphase microtubules. *Mol Biol Cell* 2001;**12**(1):73–83.

147 Sohn SH, Multani AS, Gugnani PK, Pathak S. Telomere erosion-induced mitotic catastrophe in continuously grown chinese hamster don cells. *Exp Cell Res* 2002;**279**(2):271–6.

148 Nakahata K, Miyakoda M, Suzuki K, Kodama S, Watanabe M. Heat shock induces centrosomal dysfunction, and causes non-apoptotic mitotic catastrophe in human tumour cells. *Int J Hyperthermia* 2002;**18**(4):332–43.

149 Buttner EA, Gil-Krzewska AJ, Rajpurohit AK, Hunter CP. Progression from mitotic catastrophe to germ cell death in *Caenorhabditis elegans* lis-1 mutants requires the spindle checkpoint. *Dev Biol* 2007;**305**(2):397–410.

150 Denison SH, May GS. Mitotic catastrophe is the mechanism of lethality for mutations that confer mutagen sensitivity in *Aspergillus nidulans*. *Mutat Res* 1994;**304**(2):193–202.

151 Yin L, Sit KH. Micrococcal nuclease (endonuclease) digestion causes apoptosis and mitotic catastrophe with interphase chromosome condensation in human Chang liver cells. *Cell Death Differ* 1997;**4**(8):796–805.

152 Chu K, Teele N, Dewey MW, Albright N, Dewey WC. Computerized video time lapse study of cell cycle delay and arrest, mitotic catastrophe, apoptosis and clonogenic survival in irradiated 14-3-3sigma and CDKN1A (p21) knockout cell lines. *Radiat Res* 2004;**162**(3):270–86.

153 Campbell SD, Sprenger F, Edgar BA, O'Farrell PH. Drosophila Wee1 kinase rescues fission yeast from mitotic catastrophe and phosphorylates Drosophila Cdc2 *in vitro*. *Mol Biol Cell* 1995;**6**(10):1333–47.

24

NETosis and ETosis: Incompletely Understood Types of Granulocyte Death and their Proposed Adaptive Benefits and Costs

Marko Radic

Department of Microbiology, Immunology and Biochemistry, University of Tennessee Health Science Center, Memphis, TN, USA

Abbreviations

APC	antigen-presenting cell
CNS	central nervous system
ECM	extracellular matrix
ELISA	enzyme-linked immunosorbent assay
ER	endoplasmic reticulum
ERK	growth factor-regulated extracellular signal-related kinase
GI	gastrointestinal
GM-CSF	granulocyte–macrophage colony-stimulating factor
IFNα	interferon alpha
IFNβ	interferon beta
IKK	inhibitor-kappaB kinase
IL-1β	interleukin 1β
LPS	lipopolysaccharide
LTA	lipoteichoic acid
MEK	mitogen/extracellular signal-regulated kinase
MMP9	matrix metalloprotease 9
MPO	myeloperoxidase
MS	multiple sclerosis
MRSA	methicillin-resistant *S. aureus*
NADP	nicotinamide adenine dinucleotide phosphate
NET	neutrophil extracellular trap
NOX	nicotinamide adenine dinucleotide phosphate oxidase
PAD	peptidylarginine deiminase
PCD	programmed cell death
PKC	protein kinase C
PLCγ	phospholipase gamma
PTM	post-translational modification
RA	rheumatoid arthritis
RIPK1	receptor-interacting protein kinase 1
ROS	reactive oxygen species
SLE	systemic lupus erythematosus

Apoptosis and Beyond: The Many Ways Cells Die, First Edition. Edited by James Radosevich.
© 2018 John Wiley & Son Inc. Published 2018 by John Wiley & Son Inc.

TLR	Toll-like receptor
TNF	tumor necrosis factor
TNFR	tumor necrosis factor receptor
WBC	white blood cell

24.1 Role of Neutrophils in Infection and Inflammation

The contribution of neutrophils to the defense of the body against infection is readily apparent without the need for sophisticated tools. The outward signs of a localized infection and the body's response to it can be easily appreciated. If the body's protective surface layers are breached by an abrasion or a puncture, a microbe from the environment may gain access to the normally sterile tissues beneath. If the microbe establishes a localized infection, the innate immune system mobilizes a rapid response. Cells exit the bloodstream and migrate to the site of the infection in order to contain and combat the intruding pathogens. As a result, the site becomes "inflamed": literally, "on fire." The outward signs of inflammation are elevated temperature, swelling, redness, and pain at the site of the infection. The major cellular actors that initiate the local inflammatory response are the neutrophils. These exit the blood vasculature and migrate toward the site of the infection. The swelling and redness reflect the increased permeability of the blood vessels near the infection site and the influx of the migrating neutrophils along with fluids from the blood plasma. In humans, nearly 60% of the white blood cells (WBCs) are made up of neutrophils, and they represent the most numerous cells in the blood that respond to infection. They respond rapidly, and reach the site of the infection by following a "trail" of foreign molecular traces, which increase in concentration as they get closer to the microbial invaders. The migration toward the elevated levels of microbial signals is called chemotaxis. Neutrophils possess cell-surface receptors that can sense the increased concentration of "chemoattractants" (molecules that stimulate neutrophil migration) and aim their movement toward the infection.

Even the earliest microscopic studies of the cellular response to infectious organisms observed that cells are capable of engulfing ("eating") microbes [1]. Such engulfment is called phagocytosis, and the cells that do it are called phagocytes. The most notable phagocytes are macrophages, but neutrophils are equally capable. In addition, neutrophils generate a true arsenal of chemical compounds, which function to damage and destroy invading pathogens [2]. They produce a range of highly reactive oxygen radicals, which combine with halides or nitrogen and represent the most effective means of destroying microbes, via a chemical attack on their essential macromolecules. The neutrophil arsenal also includes hypochlorite, which is the natural equivalent of household bleach [3]. These chemical weapons are combined with enzymes such as proteases that can further damage the essential molecular pathways that support microbial survival in the host organism.

Clearly, the rapid and abundant production of reactive oxygen species (ROS) and proteolytic enzymes by neutrophils may also have an effect on the host organism [4]. Initial studies of innate immunity observed that activated neutrophils can secrete these compounds from the cell to the exterior and thus contribute to a noxious environment in the host. This process of releasing the contents of neutrophil granules to the outside of the cell was termed "degranulation" and was recognized as a contributing factor in the potential tissue damage that can occur during innate immune responses to an active infection. In the initial research on neutrophil damage to the host, the precise mechanisms that lead to the release of granule contents were not clearly delineated [5]. Thus, a more precise view of these events was needed.

24.2 Apoptosis of Neutrophils

Research on neutrophils revealed that, following their maturation in the bone marrow and exit into the blood circulation, they are short-lived and need to be rapidly replenished in order to maintain high numbers in circulation. Estimates suggest that the life span of circulating neutrophils is from hours to a few days [6]. The rapidly senescing neutrophils are cleared by the spleen and the bone marrow, only to be replaced by newly maturing cells from the marrow [7]. Tens of billions of neutrophils are produced each day, and a similar number are cleared to maintain homeostasis. Various stressful situations affect the levels of neutrophils in the periphery. For example, even vigorous exercise [8] or emotional stress [9] can alter the number of neutrophils in circulation. Thus, neutrophil numbers may fluctuate during the day as a consequence of activity levels and energy demands.

In an infection, additional numbers of neutrophils are rapidly mobilized from the bone marrow to the bloodstream. The main cytokine signal for the departure of neutrophils from the bone marrow is the systemic increase in the levels of granulocyte–macrophage colony-stimulating factor (GM-CSF) [10]. A rapid rise in the numbers of neutrophils, the main leukocytes in the blood, is considered a reliable and early sign of an infection somewhere in the body. The elevated numbers of blood leukocytes (referred to as "leukocytosis") are a strong indication that the patient has an ongoing and acute infection. Alternatively, high circulating levels of neutrophils may indicate a leukemia, which results in massive increases in the levels of leukocytes.

The short life span of neutrophils raises the question whether apoptosis is the mechanism of cell death that accomplishes the removal of senescent neutrophils. Neutrophils, like most other cell types in metazoans, have the machinery for apoptotic cell death and can be triggered to initiate the activation of caspases, following the engagement of cell-surface receptors that belong to the tumor necrosis factor receptor (TNFR) family [11]. Neutrophil apoptosis proceeds through the canonical stages of nuclear condensation and cellular fragmentation, resulting in the appearance of blebs formed by the protrusion of condensed fragments of the nucleus from the cell surface. The nuclear fragments detach from the remnants of the apoptotic cell and become apoptotic bodies, which are cleared by other phagocytic cells nearby. The uptake of neutrophils into tissue macrophage is considered an important step in the return of a tissue to its pre-infection homeostatic balance [12].

In an infection, neutrophils are a cell type that can produce and secrete small chemical messengers that promote inflammation, which are jointly referred to as "proinflammatory cytokines" and "chemokines" [13]. An important consequence of localized proinflammatory cytokine production is the delay of neutrophil apoptosis. In the presence of cytokines such as interleukin 1β (IL-1β), IL-8, and GM-CSF, the life span of neutrophils is extended from the norm [14,15]. In addition, certain bacterial products, such as lipopolysaccharide (LPS) and lipoteichoic acid (LTA), which are cell-wall components of Gram-negative and Gram-positive bacteria, respectively, also delay neutrophil apoptosis [16,17]. A common mechanism for this delay of apoptosis involves the activation of NF-κB.

Because of the important role IL-1β plays in inflammation, researchers examined the mechanisms of its secretion and discovered an intriguing characteristic. Even though IL-1β is a secreted protein, it does not contain a leader peptide, which usually precedes the mature protein in translation-coupled transport into the endoplasmic reticulum (ER) [18]. This mechanism is broadly shared and conserved among the vast majority of secreted proteins, and IL-1β's lack of the leader peptide stimulated a search for an alternative secretion mechanism. Several authors suggested a mechanism that sidesteps the plasma membrane and allows the release of the cytokine from the cytoplasm of the cell. One possible route may involve the opening of plasma-membrane pores [19] or the release of a multimeric protein complex from cells [20]. However, an additional potential mechanism became appreciated once alternative forms of neutrophil cell death were identified.

24.3 Non-apoptotic Death

Researchers have focused attention on the ability of neutrophils to undergo alternative forms of cell death. Two important alternatives are necroptosis [21] and pyroptosis [22], discussed in Chapters 6 and 15, respectively. Both of these types of death are induced as a result of infection or inflammation, and they serve to promote and amplify the inflammatory response. In general, the alternative forms of cell death may occur *in vivo* depending on the specific set of stimuli that the cells are exposed to and the environment, which may provide different extracellular physicochemical conditions. Neutrophils are found in nearly every tissue in the body and patrol mucosal tissues (e.g., the oral cavity, alveolar spaces in the lungs, and the gastrointestinal (GI) tract). Each of these sites provides different oxygen-tension, pH, and ionic conditions, not to mention the presence or absence of other cells of the host or associated forms of the microbiome. Therefore, it is likely that different forms of neutrophil cell death reflect, in part, their anatomic location and the molecular and cellular factors that lead to the neutrophil's demise.

24.4 Discovery of Neutrophil Extracellular Traps

The discovery of neutrophil extracellular traps (NETs) is an example of "seeing is believing." It is hard to imagine that Volker Brinkman, Arturo Zychlinsky, and colleagues [23] were the first to observe the physical demise of a neutrophil that was brought about by the rupture of the plasma membrane and the release of nuclear chromatin to the extracellular space. Instead, it is likely that others had seen NETotic cells in their neutrophil preparations and ignored them or considered them an accident of sample preparation. Thus, only a few publications mentioned this striking type of cell death prior to 2004 [24,25]. The remarkable discovery of NETs was made possible by the logical and persistent pursuit of the initial observation. Brinkman et al. [23] noted that the release of extracellular chromatin occurred more readily in the presence of certain bacterial products, such as the bacterial cell-wall component LPS. Notably, they observed both Gram-positive and Gram-negative bacteria entangled with extracellular fibers composed of DNA and nucleosome-core histones. In addition, they noted that several of the neutrophil-granule proteins did not freely diffuse away from the ruptured neutrophils, but instead associated with the NETs. Therefore, they proposed that NETs serve to capture bacteria and expose them to a high local concentration of bactericidal compounds, limiting the further spread of infection. Decades earlier, Hirsch [26] had identified the bactericidal properties of histone proteins, but the discovery of NETosis provided a context for this characteristic of the nuclear proteins. The release of extracellular chromatin could thereby limit an infection by inflicting structural damage on diverse bacterial species.

The initial observations by Brinkmann et al. [23] also provided a structural basis to account for the effect of bacterial virulence factors. Studies on *Staphylococcus aureus* and many other microbial pathogens had identified bacterial DNases produced during bacterial infections, which enhanced dissemination in the host [27]. Variants lacking such nucleolytic enzymes produced more limited and, thus, less virulent infections [28]. The discovery of NETs could explain the increased virulence of DNase-secreting bacteria, as the nucleolytic activity of the secreted enzymes may free these bacteria by damaging the NETs, allowing them to escape to secondary sites of infection and avoid the bactericidal activity of neutrophil chromatin and granule components [29].

Nevertheless, the initial reports of NETosis brought about skepticism. A series of studies disputed the antimicrobial role of NETs [30,31]. Healthy criticism is necessary and should be welcomed in a newly forged field of scientific inquiry, because caution sharpens the careful scrutiny of new phenomena. Several reports demonstrated that bacteria may be trapped but not killed by NETs [32,33]. Even prolonged incubations failed, in some cases, to reduce the numbers of infectious bacteria. Clearly, complexities found in an infected tissue are not easily reproduced *in vitro*. The field of NETosis investigation responded by looking for detailed evidence that NETosis is regulated and that it provides benefits to the organism in cases of infection.

In addition, NETosis, like most other defense mechanisms devised by evolutionary selection pressures, can be eluded by microbial pathogens. There are several bacterial pathogens that have acquired the ability to neutralize the effects of NETs or to evade their lethal effects [29,32]. Some microbes have even found ways to incorporate the NET chromatin and DNA into their pathogenic mechanisms.

24.5 Testing and Validation of Antimicrobial Functions of NETs

Given the potential life-saving importance of NETosis in innate immune responses to infections, it comes as no surprise that other cells of the myeloid lineage share the capacity for extracellular chromatin release. Thus, it was demonstrated that eosinophils and basophils, close relatives of neutrophil granulocytes, also have the ability to undergo a cell death that results in the release of nuclear chromatin [34–36]. For that reason, the more specific terminology of NETs was broadened to apply to other cell types, and the expression "ETs" was proposed to more generally refer to extracellular traps released from any of these. Intriguingly, an important cell type of the innate immune system that comprises cells of the monocyte/macrophage lineage can also release extracellular traps [37–39]. These extracellular structures are sometimes called "METs." Even though the specific mechanisms may differ and the complement of extracellular chromatin-associated proteins may uniquely reflect the cellular origin of the ETs, each of them consists of DNA fibers that are released from the cells. In a possible variation on the theme of extracellular DNA, studies on basophils and eosinophils have revealed that these cells may expel mitochondrial DNA instead of nuclear chromatin to form ETs [35].

A remarkable diversity of microorganisms can elicit the release of ETs from innate immune cells (Table 24.1). In a growing number of cases, the specific microbial products that elicit NETosis have been identified [40–42]. As already mentioned, various chemical NETosis inducers have been used to characterize the pathways involved in executing the

Table 24.1 Microbes reported to induce NETosis.

Authors	Microorganism
	Bacteria
Lippolis et al. [44]	*Enterococcus faecalis*
Lippolis et al. [44] Grinberg et al. [45] Webster et al. [46]	*Escherichia coli*
Hong et al. [47]	*Haemophilus influenzae*
Hakkim et al. [48]	*Helicobacter pylori*
Papayannopoulos et al. [49]	*Klebsiella pneumoniae*
Ermert et al. [50]	*Listeria monocytogenes*
Ramos-Kichik et al. [51]	*Mycobacterium tuberculosis*
von Kockritz-Blickwede et al. [52]	*Pseudomonas aeruginosa*
Lippolis et al. [44]	*Serratia marcescens*
Brinkmann et al. [23]	*Shigella flexneri*
Brinkmann et al. [23]	*Staphylococcus aureus*
Beiter et al. [53] Crotty Alexander et al. [54]	*Streptococcus pneumoniae*
Buchanan et al. [55]	*Streptococcus pyogenes*
	Fungi
Bruns et al. [56] McCormick et al. [57]	*Aspergillus fumigatus*
Urban et al. [58]	*Candida albicans*
Springer et al. [59]	*Cryptococcus gattii*
Urban et al. [60]	*Cryptococcus neoformans*
	Parasites
Behrendt et al. [61]	*Eimeria bovis*
Guimaraes-Costa et al. [62]	*Leishmania amazonensis*
Bonne-Annees et al. [63]	*Strongyloides stercoralis*
Abi Abdallah et al. [64]	*Toxoplasma gondii*
	Viruses
Saitoh et al. [65]	Human immunodeficiency virus
Raftery et al. [66]	Hentavirus
Narasaraju et al. [67], Tripathi et al. [68]	Influenza A virus
Jenne et al. [69]	Poxvirus

complex morphological transitions that underlie this form of cell death. The more remarkable of these include nitric oxide, hydrogen peroxide, and glucose oxidase, each of which is directly related to the production of reactive oxygen [43]. As neutrophils are capable of generating ROS, it is reasonable to propose that neutrophils themselves can amplify the innate immune response and enhance the production of NETs, as well as the release of ETs from other innate immunity cell types.

The important role of an intracellular calcium spike in the stimulation of NETosis is indicated by the effective induction of NET release from cells that are treated by compounds, which, as artificial pores, facilitate the translocation of calcium ions across the plasma membrane [70,71]. Figures 24.1 shows neutrophils treated with the ionophore A23187 in media containing calcium. A group of cells is shown that includes neutrophils with intact plasma membranes intermingled with cells that have released extracellular chromatin. The released chromatin strongly interacts with antibodies to different histones, suggesting the unhindered access to NETs by proteins that are present in blood, plasma, and lymphatic fluids. By extension, this also suggests that NETs could bind and be taken up by B cells and other antigen-presenting cells (APCs), thus potentially acting as stimuli for the adaptive immune system.

The induction of NETosis/ETosis is stimulated by various cytokines whose secretion is associated with an inflammatory response. These include tumor necrosis factor (TNF) [72], IL-1 [73], IL-8 [74], IL-23 [73], and GM-CSF. It is of interest that some of these cytokines are known to suppress apoptosis of neutrophils. As neutrophils themselves can secrete many such cytokines, it is possible that activation of neutrophils

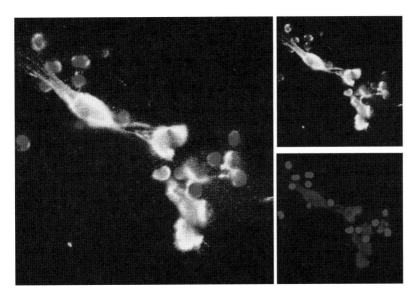

Figure 24.1 **Microscopic image of neutrophils undergoing NETosis.** Purified human blood neutrophils were treated with calcium ionophore A23187 for 1 hour on cover glass slides in Hank's buffered salt solution supplemented with 1 mM calcium chloride. Following the incubation with ionophore, cells were fixed with paraformaldehyde, washed in buffer, and incubated with a monoclonal mouse antibody to histone H1. Bound antibody was detected with a fluorescently tagged secondary antibody cocktail. Cells were examined by confocal laser microscopy to visualize the distribution of histone H1 and DNA. Histone H1 is associated with NETs, whereas neutrophils that retain intact plasma membranes do not allow access of antibody.

generates conditions that favor a shift from apoptosis to NETosis in the presence of an infectious pathogen or in chronic inflammation.

A special role may be attributed to platelets that are well known for forming intimate associations with neutrophils in circulation [75]. The interaction between platelets and neutrophils may be instrumental in guiding the neutrophil response to bacterial products. Thus, the activation of Toll-like receptor 4 (TLR4) on platelets may be an important signal to induce NETosis [76]. Similarly, complement activation and the production of complement fragments such as C5a (also known as anaphylotoxin) may further augment the release of NETs, much like the cytokines interferon alpha and beta (IFNα and IFNβ) [77]. Clearly, the induction of NETosis reflects the intricate nature of the microbial pathogens that may infect the host, and multiple overlapping mechanisms are in place to recognize and respond to the infectious organisms.

24.6 Containment of Microbes versus Escape from NETs

Even though convincing arguments link NETosis to innate immune responses in microbial infections, it would be simplistic to assume that any defense mechanism is failsafe or that evolution did not produce measures and countermeasures on both sides of the host–microbe battle. Thus, a detailed examination of the response of pathogens to ETs reveals mechanisms that allow pathogens to evade extracellular chromatin and even use the nuclear material to enhance colonization. One mechanism for enhanced resistance to damage inflicted by ETs is the expression of extensive capsular polysaccharide coats of several pathogenic bacteria, including many strains of *Staphylococcus* and *Streptococcus* [55,78,79]. *S. aureus*, in addition to degrading DNA from ETs, metabolizes ET DNA to yield deoxyadenosine, a toxic nucleotide that can induce activation of caspase 3 and thereby promote apoptosis in the vicinity of the site of infection [80]. The metabolite forms a zone of immunosuppression surrounding the infection, which isolates the bacteria from the innate immune response [80]. Other NET components, such as the bactericidal peptide LL-37, may also be targeted by bacterial virulence mechanisms, because Streptococcal species produce the M1 protein, which inactivates the LL-37 defensin [81,82]. Countermeasures could act at multiple levels, as human serum proteins inhibit bacterial nucleases and promote clearance of NET-associated bacteria [83].

Further examples of the bacterial adaptations that evade ETs include the potential dissemination of intracellular bacteria by the induction of NETosis in cells that have previously ingested *S. aureus* [84]. In this scenario, the rupture of neutrophils can provide a mechanism for repeated cycles of infection and release from the infected cells. This mechanism may utilize one of several bacterial products known to induce cell lysis [85]. Even more remarkably, the extracellular chromatin released from neutrophils may be of use to bacteria. A potential example is provided by bacteria of the genus *Haemophilus*, which, alongside others, are responsible for difficult-to-treat middle-ear infections. *Haemophilus* may use extracellular DNA as the building block for the tight, extracellular matrix (ECM) that makes up the bacterial biofilms [47]. A similar situation may apply to lung infections mediated by *Pseudomonas*, which also may use extracellular DNA in the construction of biofilms [86]. Biofilms form a fibrous matrix that binds together bacteria of a certain type and allows them to survive conditions that would be inhospitable to single cells of the same bacterium (the so-called "pelagic form"). An essential component of biofilms is extracellular DNA. This is demonstrated by their sensitivity to rapid disassembly following the addition of a DNase [87]. The cells contained in a biofilm can colonize smooth surfaces such as indwelling catheters and, due to the compact nature of biofilms, exhibit an increased resistance to antibiotics and activated complement. A better understanding of the origin of the extracellular DNA may therefore have implications for improved treatment of more resistant forms of biofilm-forming human pathogens.

24.7 Regulation of NETosis

In order to distinguish between different mechanisms of cell death, it is essential to establish specific enzymatic reactions that are regulated differently in different forms of programmed cell death (PCD). To illustrate this point, it is useful to learn about a class of enzymes that carries out the post-translational modification (PTM) of arginine residues in proteins into citrulline residues. This reaction is catalyzed by a small family of enzymes that are conserved in vertebrates, the peptidylarginine deiminases (PADs) [88,89]. The five human PADs are Ca^{++}-dependent enzymes with different tissue distributions that, among other functions, regulate zygotic development, assist in the myelination of neurons, and contribute to the mechanical properties of the epidermis. PAD4 is the PAD that is expressed in the highest levels in granulocytes, monocytes, and mast cells. There is some uncertainty about the cellular location of PAD4, as it has the expression pattern of late granule proteins [90] yet its products are observed in the cytoplasm and nucleus of human neutrophils [91].

PAD4 is the only PAD that contains a nuclear localization sequence and thus can enter the cell nucleus [91]. There, PAD4 deiminates positively charged histone proteins, which organize nuclear genetic material into the compact chromatin of mammalian chromosomes. The basic building block of chromatin is the nucleosome, a structure containing approximately 150 base pairs of DNA that forms two turns around a central core of histones [92]. The core of a nucleosome incorporates two copies each of histones 2A, 2B, 3, and 4. The linker histone H1, which in humans comprises seven protein isoforms, organizes the path of DNA between adjacent nucleosomes by interacting with the intervening linker DNA. H1 thus determines the stacking of nucleosomes into the compact structure of nuclear chromatin [93]. The deimination reaction converts certain positively charged arginine residues in core and linker histones to uncharged citrullines. It is plausible that the loss of positive charges destabilizes histone–DNA interactions and leads to an unraveling of the compact nuclear chromatin. This relaxing of chromatin structure near gene promoters may have an important role during the reprogramming of gene expression, as illustrated by the transient activation of PAD4 during the development of pluripotent precursor cells into differentiated cell lineages [94].

During NETosis, the deimination of histones occurs at a much larger scale, and may promote the release of relaxed DNA from the nucleus to yield the extracellular fibers that make up NETs. The necessity of the deimination reaction in NETosis is indicated by the failure of PAD4-deficient neutrophils to produce NETs [95] and by the suppression of NETosis by PAD4 inhibitors [96]. The activity of PAD4 is induced by stimuli associated with inflammation. In neutrophils, PAD4 catalytic activity is induced by bacterial breakdown products such as LPS [89], LTA, and formylated Met-Leu-Phe (f-MLP), a tripeptide derived by proteolysis from the amino terminus of bacterial proteins. The deimination of histones is also induced following treatment of neutrophils with TNF and hydrogen peroxide, suggesting that innate inflammatory signals and late products of neutrophil activation induce this unique chromatin PTM. Several studies have used the presence of deiminated histones *in vivo* as evidence for the induction of NETosis [97–99], although it may be prudent to confirm this conclusion with other suitable assays or via microscopy. Currently, suitable NET-detection assays include a sandwich enzyme-linked immunosorbent assay (ELISA) that can detect complexes between DNA and myeloperoxidase (MPO) [100]. The visual classification of the different stages of NETosis can be carried out either manually or by computer-assisted image analysis [101].

Because PAD4 activation and NETosis induction are tightly linked, studies of PAD4 activation can help to reveal the biochemical regulatory pathways underlying NETosis. One set of studies from the author's laboratory explored the role of cell-surface receptors and the cytoskeleton in the induction of histone deimination by PAD4 [102]. Adhesion of neutrophils to substrate via integrin receptors and intact actin and microtubulin networks were required for PAD4 activation to occur. This indicated, for the first time, that structural tethering and cytoplasmic force generation are essential to the initiation of NETosis. Other researchers subsequently confirmed that the Mac-1 integrin plays a paramount role in the stimulation of NETosis *in vitro* and *in vivo* [66,103,104].

In general, signals at the cell surface, including adhesion receptors, Fc receptors, and innate pattern-recognition receptors, may contribute to the induction of NETosis. How these signals are appropriately interpreted by the cell remains unknown. However, certain features of NETosis signaling are beginning to be understood. Because the research in different laboratories uses different inducing conditions, certain features of NETosis may not apply in every experimental setting. In addition, there is consensus that more than one mechanism of NETosis may exist. As already mentioned, some indications suggest that mitochondrial DNA, as well as nuclear DNA, may be extruded by different granulocytes and in response to different stimuli. In the classical ("lytic") NETosis, the nuclei swell, resulting first in the release of nuclear chromatin to the cytoplasm, and then in the rupture of the plasma membrane and the discharge of chromatin to the extracellular milieu [105]. In the more rapid "vital" NETosis, on the other hand, neutrophils fragment portions of their nuclei and transport them via vesicles that fuse with the plasma membrane and release portions of nuclear chromatin from the cell. Despite this release of genomic material, the neutrophils remain viable and capable of migration and phagocytosis [106].

The steps involved in signal transduction from the cell-membrane receptors that trigger the rupture of the plasma membrane and the release of NETs are shown in Figure 24.2. Various cell-surface receptors (e.g., TLR4 [107]) may be triggered by bacterial products or intact microbes and organize the signaling complex containing MyD88, TRAF 4 and 6, and IRAK1. This complex may then activate the nicotinamide adenine dinucleotide phosphate (NADP) oxidase (NOX) complex, setting a number of reactive oxygen production steps in motion [3]. Specifically, ROS are produced, which react with halides such as chloride to form perchlorate end products. Separately, signaling from the cell surface may recruit TAK1 to activate inhibitor-kappaB kinases (IKKs) and the mitogen/extracellular signal-regulated kinase (MEK)/Erk pathway [48]. The IKKs catalyze key reaction steps in the activation and nuclear translocation of members of the NF-κB transcription-factor family.

Signaling at the plasma membrane may also activate G-coupled receptors, which can signal to induce phospholipase gamma (PLCγ) and the transient release of calcium ions

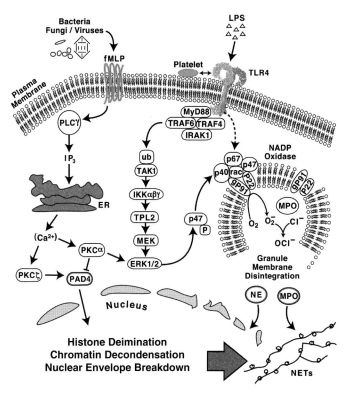

Figure 24.2 Proposed pathways regulating NETosis. Signals acting at the cell surface activate receptors that recognize conserved molecular signals associated with bacterial infections and initiate signaling cascades. These cascades lead to intracellular calcium release, reactive oxygen and perchlorate generation, and activation of transcription factors that induce the inflammatory response. The result is the dissolution of the nuclear envelope and of granule membranes, and the relaxation of chromatin through the action of peptidylarginine deiminase 4 (PAD4). Chromatin escapes the nuclear confines and is released from cells as neutrophil extracellular traps (NETs). For details, see text. *Source:* Adapted from Dwivedi and Radic 2014 [108] with permission from BMJ Publishing Group Ltd.

from intracellular stores [109]. The calcium then acts as a second signal to activate classical protein kinase C (PKC) isoforms. In a series of elegant experiments by Neeli et al. [71], a mutual counterbalance between PKCα and the atypical PKC isoform, PKCξ, was discovered; the authors proposed that this balance participates in the regulation of NETosis. Along a separate pathway, calcium may also contribute to the activation of growth factor-regulated extracellular signal-related kinase (ERK1/2) [48], such that the PKC and ERK pathways may intersect and mutually modulate the outcome of the initiating signals. In a feedforward fashion, the MEK/Erk pathway was shown to regulate PKCξ and matrix metalloprotease 9 (MMP9) expression at podosomes, and it is conceivable that it participates in the activation of NETosis [110].

The outward signs of NETosis activation are the large-scale morphologic transitions that lead to the permeabilization of the nuclear and granule membranes and the release of chromatin and granule components into the cytoplasm [105]. Certain granule proteins, such as elastase and MPO, are thought to strongly attach to chromatin at this stage of NETosis. In fact, NETosis is suppressed by a deficiency of elastase, and resistance to infection is reduced in mice challenged with bacteria [49]. Such associations between nuclear chromatin and granule components presumably enhance the bactericidal properties of NETs. An added feature of the NET chromatin scaffold is that it serves as an anchor for defensins such as LL-37 [82,111], which is known to contribute to the lysis of bacterial cell membranes.

24.8 Similarities and Differences between NETosis and Other Forms of Cell Death

Careful examination of PAD4 activation provided the initial clue that NETosis and apoptosis are mutually exclusive cell fates [70]. Outwardly, the late stages of the two cell forms of death could not be more divergent. In apoptosis, DNA is tightly packaged inside nuclear fragments that are suitable for efficient clearance by phagocytes. In NETosis, DNA is unraveled and dispersed such that it forms a matrix for capture of microbes. At the molecular level, the distinction between apoptosis and NETosis is demonstrated by the fact that PAD4 is strongly inhibited by stimuli that induce neutrophil apoptosis [70]. Camptothecin, a widely used mammalian topoisomerase inhibitor that efficiently induces apoptosis, and staurosporine, another broad apoptosis inducer, prevent histone deimination. Thus, PAD4 activation can differentiate between neutrophil responses to stimuli inducing apoptosis versus innate responses to microbes.

Recently, increased attention has been directed at regulated necrosis, also referred to as "necroptosis" [112,113]. As an alternative to apoptosis, necroptosis has garnered interest because it is associated with inflammation and may serve in microbial defenses. In general, the parallels between NETosis and necroptosis include the facts that both can be induced by physiological stimuli of the TNF pathway [114], that both utilize mechanisms that lead to an increased cytosolic content of calcium ions [115], and that both are associated with ROS production during the execution phase [116]. More importantly, recent experiments with methicillin-resistant *S. aureus* (MRSA) and neutrophils have revealed that a significant proportion of neutrophil-internalized bacteria survive within the phagosome and break out of the granulocytes [117], while the phagocytic neutrophils are less efficiently cleared by activated tissue macrophage. The release of viable *S. aureus* from neutrophils resembles NETosis and shows striking parallels with necroptosis, as the process can be interrupted by inhibition of an important mediator of necroptosis: receptor-interacting protein kinase 1 (RIPK1) [117]. Further shared aspects of signaling pathways that mediate NETosis and necroptosis will need to be explored in order to test the important possibility that NETosis is a particular subtype of programmed necrosis.

24.9 Involvement of NETosis in Autoimmunity

The initial observations by Brinkman, Zychlinsky, and colleagues [23] placed NETosis in the context of innate immunity to microbial infections. Interestingly, the antibodies that the authors used to visualize NETs were previously characterized autoantibodies to histones and chromatin. These antibodies were derived from a strain of mouse that develops a spontaneous autoimmune disease similar to human lupus [118]. The human autoimmune disease, whose full name is systemic lupus erythematosus (SLE), is characterized by the production of a highly diagnostic class of autoantibodies that reacts with DNA or nucleoprotein complexes consisting of DNA and histones [119]. Lupus has puzzled clinicians and researchers for decades, because antibodies to DNA and nucleosomes are highly indicative of lupus and related autoimmune disorders, but it is nearly impossible to raise antibodies to these conserved self-components through immunization of a wild-type mouse [120]. This intrinsic resistance to immunization by self molecules is called immunological tolerance.

Possible clues to mechanisms that can account for the break in self-tolerance and the development of autoimmunity come from studies of PADs, as several PAD substrates are targets of autoimmune reactions [72,121]. In general, autoantibodies often target-specific PTM, including those that are introduced during regulated cell death. It is plausible that PTMs change the sequence, and possibly the conformation, of autoantigens, and thereby convert the previously unreactive self-constituents into targets of the autoimmune response. Clearly, this scenario cannot by itself explain the induction of autoimmunity, because PTMs occur regularly in the course of cellular activation, senescence, and death – all integral aspects of metazoan life itself. The genetic predisposition of particular individuals and the precise circumstances of environmental stimuli that affect the immune system represent other factors that may contribute to the break in tolerance and the development of autoimmunity [108,122].

The precise mechanism of NETosis exhibits an additional feature that is favorable for the stimulation of the adaptive immune system. In viable cells, SLE autoantigens are safely protected in the nucleus, and therefore are not readily accessible to B and T lymphocytes, the main actors in the adaptive immune response. Because adaptive immunity relies on the recognition and uptake of antigens into APCs, the fact that nuclear chromatin is sequestered from cells of the adaptive immune system argues that such reactivity should rarely occur. Apoptotic cells retain most of the nuclear chromatin in tightly packaged apoptotic bodies that are taken up by phagocytic macrophage. Nucleases expressed in the phagosome of these scavenger cells are essential in removing the DNA of apoptotic cells and preventing the development of autoimmunity [123]. However, the release of chromatin from neutrophils in NETosis, coupled with the physical juxtaposition of chromatin and microbial pathogens, which can act as stimuli (adjuvants) of an immune response, represents an intriguing combination of conditions that may lead to the induction of adaptive immunity and the production of auto-antibodies to nuclear SLE autoantigens.

We believe that not all aspects of NETosis in autoimmunity have been adequately explored. In Figure 24.3, we show the result of incubating anti-RNA autoantibodies with NETs. RNA is another macromolecule capable of participating in the formation of an ECM. In parallel with DNA and chromatin, RNA and associated proteins may distribute in an extracellular location and perform important functions in an innate immune response. In addition, autoantibodies to RNA form an important and diagnostic category [124] whose production may be stimulated by the proposed interaction between NETs and the adaptive immune system.

An important piece of evidence linking NETosis to the induction of autoantibody responses is the observation that autoantibodies preferentially react with NETs rather than with unstimulated neutrophils [125]. In addition, as citrullination (deimination) of autoantigens such as histones is greatly increased in NETosis, the preference of autoantibodies for the citrullinated isoforms of histones and other autoantigens strongly

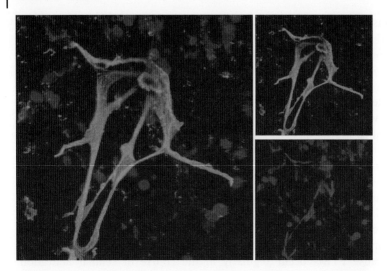

Figure 24.3 Confocal microscopy image of RNA localization in neutrophil extracellular traps (NETs). Purified neutrophils were induced to undergo NETosis, as described in the text. Fixed cells were reacted with a mouse monoclonal that reacts specifically with RNA (BWR4). RNA staining was detected along large fibrous structures that resemble NETs and that partially overlap with DNA.

suggests that B cells are stimulated to react with antigens released during the process of NETosis [108]. It has been observed that rheumatoid arthritis (RA) sera preferentially react with the citrullinated form of a protein called filaggrin, a component of the ECM [126]. However, the relation of these autoantibodies to NETosis remained obscure until the events surrounding NETosis became more clearly understood.

PAD4 itself associates with NETs, and through the process of NETosis gains access to the extracellular space. This observation may be highly relevant, because numerous extracellular proteins have been observed to acquire citrullines in tissues experiencing inflammation. For example, the synovial spaces of arthritis joints contain elevated levels of citrullinated proteins, and deimination is proportional to joint and cartilage destruction [127,128]. These observations are consistent with the massive influx of blood neutrophils into joints during the early inflammatory phases of arthritis. If neutrophils respond to the inflammatory milieu by inducing the expression of PAD4, the release of NETs, and the extrusion of the active enzyme into the synovium, it is likely that proteins of the ECM will serve as PAD4 substrates. Thus, proteins such as filaggrin, fibrinogen, and collagen can acquire elevated levels of citrulline.

Consistent with their contribution to the initiation of adaptive immunity, several components of NETs constitute targets for autoantibodies [108,121,129]. One mechanism that could favor the development of adaptive immune responses involves the uptake of NET components by dendritic cells. When incubated in the presence of NET chromatin or LL-37, dendritic cells secrete IFN [130,131]. In turn, B cells may respond to NETs with the production of most of the typical lupus autoantibodies, such as antibodies to histones and DNA, as well as other NET constituents. It is consistent with experimental evidence to suggest that other autoimmune disorders (e.g., RA [72], Sjogren's syndrome [125,132], multiple sclerosis (MS) [133], and psoriasis [73]) also respond to NET components.

In addition, the exposure of NET antigens in the extracellular milieu provides a suitable substrate that is accessible for binding by circulating autoantibodies to DNA and chromatin. Combined, the antibody–antigen complexes may participate in the formation of extracellular immune deposits. Such immune deposits play a major role in the pathogenesis of lupus, because immune complexes formed by the crosslinking of antibodies and a repetitive, polymeric substrate lodge in various tissues throughout the body [134]. Such immune complexes lead to the systemic inflammation (often involving the skin, kidneys, lungs, pericardium, joints, and central nervous system (CNS)) seen in SLE patients. The participation of NETs in the initial induction of autoreactive immune responses, the progression of autoimmune diseases to a tissue-destructive phase, and the perpetuation of a chronic inflammatory reaction all place NETs at the center of a self-perpetuating cycle that drives ongoing pathogenesis of the major autoimmune diseases.

24.10 Conclusion

At present, NETosis research is in an exponentially growing phase, and the various topics addressed in this chapter are under intense scrutiny, leading to new and surprising discoveries. It is exciting to note that NETosis is an area of basic biological interest as a mechanism of regulated cell death. It also remains a fascinating area of research into innate immune responses to infections by all major classes of pathogens. The mechanism of NETosis is of interest in an evolutionary context because lower eukaryotes [135], and even plants [136], exhibit certain features of extracellular chromatin release. Moreover, there is little doubt that inappropriate induction of NETosis, ineffective clearance of NETs, and defects in pathways regulating NETosis are intimately involved in major human disorders ranging from autoimmunity to cardiovascular diseases [137,138], and even cancer [139]. An exciting development is the observation that NETosis can be seen under intravital microscopy and can be linked with an innate neutrophil response called "swarming" [140]. Neutrophils swarm in response to signals emanating from dying cells – a response that is engaged in infections [141], trauma [142], and wound healing [140]. With an increased understanding of NETosis, inhibitors of steps in NETosis have been developed and tested in animal models of autoimmune diseases. For example, following the demonstration that PAD4 is an important mediator of NETosis *in vivo*, chemical inhibitors of PAD4 have been used in mouse models of SLE [143], collagen-induced arthritis [144], and MS [145]. In each case, the manifestations of the disease improved. Thus, within a short time, NETosis inhibitors should make an important contribution to the treatment of autoimmune diseases in clinics.

References

1 Gordon S. Elie Metchnikoff: father of natural immunity. *Eur J Immunol* 2008;**38**(12):3257–64.
2 Borregaard N. Neutrophils, from marrow to microbes. *Immunity* 2010;**33**(5):657–70.
3 Winterbourn CC, Kettle AJ. Redox reactions and microbial killing in the neutrophil phagosome. *Antioxid Redox Signal* 2013;**18**(6):642–60.
4 Pham CT. Neutrophil serine proteases: specific regulators of inflammation. *Nat Rev Immunol* 2006;**6**(7):541–50.
5 Naucler C, Grinstein S, Sundler R, Tapper H. Signaling to localized degranulation in neutrophils adherent to immune complexes. *J Leukoc Biol* 2002;**71**(4):701–10.

6 Pillay J, den Braber I, Vrisekoop N, Kwast LM, de Boer RJ, Borghans JA, et al. *In vivo* labeling with 2H2O reveals a human neutrophil lifespan of 5.4 days. *Blood* 2010;**116**(4):625–7.
7 Rankin SM. The bone marrow: a site of neutrophil clearance. *J Leukoc Biol* 2010;**88**(2):241–51.
8 Nunes-Silva A, Bernardes PT, Rezende BM, Lopes F, Gomes EC, Marques PE, et al. Treadmill exercise induces neutrophil recruitment into muscle tissue in a reactive oxygen species-dependent manner. An intravital microscopy study. *PLoS One* 2014;**9**(5):e96464.
9 Vitlic A, Khanfer R, Lord JM, Carroll D, Phillips AC. Bereavement reduces neutrophil oxidative burst only in older adults: role of the HPA axis and immunesenescence. *Immun Ageing* 2014;**11**:13.
10 Boettcher S, Gerosa RC, Radpour R, Bauer J, Ampenberger F, Heikenwalder M, et al. Endothelial cells translate pathogen signals into G-CSF-driven emergency granulopoiesis. *Blood* 2014;**124**(9):1393–403.
11 Akgul C, Edwards SW. Regulation of neutrophil apoptosis via death receptors. *Cell Mol Life Sci* 2003;**60**(11):2402–8.
12 Savill J. Apoptosis in resolution of inflammation. *J Leukoc Biol* 1997;**61**(4):375–80.
13 Nathan C. Neutrophils and immunity: challenges and opportunities. *Nat Rev Immunol* 2006;**6**(3):173–82.
14 van Raam BJ, Drewniak A, Groenewold V, van den Berg TK, Kuijpers TW. Granulocyte colony-stimulating factor delays neutrophil apoptosis by inhibition of calpains upstream of caspase-3. *Blood* 2008;**112**(5):2046–54.
15 Hsu LC, Enzler T, Seita J, Timmer AM, Lee CY, Lai TY, et al. IL-1beta-driven neutrophilia preserves antibacterial defense in the absence of the kinase IKKbeta. *Nat Immunol* 2011;**12**(2):144–50.
16 Brach MA, deVos S, Gruss HJ, Herrmann F. Prolongation of survival of human polymorphonuclear neutrophils by granulocyte-macrophage colony-stimulating factor is caused by inhibition of programmed cell death. *Blood* 1992;**80**(11):2920–4.
17 Lotz S, Aga E, Wilde I, van Zandbergen G, Hartung T, Solbach W, Laskay T. Highly purified lipoteichoic acid activates neutrophil granulocytes and delays their spontaneous apoptosis via CD14 and TLR2. *J Leukoc Biol* 2004;**75**(3):467–77.
18 Eder C. Mechanisms of interleukin-1beta release. *Immunobiology* 2009;**214**(7):543–53.
19 Pelegrin P, Surprenant A. Pannexin-1 mediates large pore formation and interleukin-1beta release by the ATP-gated P2X7 receptor. *EMBO J* 2006;**25**(21):5071–82.
20 Baroja-Mazo A, Martin-Sanchez F, Gomez AI, Martínez CM, Amores-Iniesta J, Compan V, et al. The NLRP3 inflammasome is released as a particulate danger signal that amplifies the inflammatory response. *Nat Immunol* 2014;**15**(8):738–48.
21 Newton K, Dugger DL, Wickliffe KE, Kapoor N, de Almagro MC, Vucic D, et al. Activity of protein kinase RIPK3 determines whether cells die by necroptosis or apoptosis. *Science* 2014;**343**(6177):1357–60.
22 Chen KW, Gross CJ, Sotomayor FV, Stacey KJ, Tschopp J, Sweet MJ, Schroder K. The neutrophil NLRC4 inflammasome selectively promotes IL-1beta maturation without pyroptosis during acute *Salmonella* challenge. *Cell Rep* 2014;**8**(2):570–82.
23 Brinkmann V, Reichard U, Goosmann C, Fauler B, Uhlemann Y, Weiss DS, et al. Neutrophil extracellular traps kill bacteria. *Science* 2004;**303**(5663):1532–5.
24 Takei H, Araki A, Watanabe H, Ichinose A, Sendo F. Rapid killing of human neutrophils by the potent activator phorbol 12-myristate 13-acetate (PMA)

accompanied by changes different from typical apoptosis or necrosis. *J Leukoc Biol* 1996;**59**(2):229–40.

25 Suzuki K, Namiki H. Phorbol 12-myristate 13-acetate induced cell death of porcine peripheral blood polymorphonuclear leucocytes. *Cell Struct Funct* 1998;**23**(6):367–72.

26 Hirsch JG. Bactericidal action of histone. *J Exp Med* 1958;**108**(6):925–44.

27 Olson ME, Nygaard TK, Ackermann L, Watkins RL, Zurek OW, Pallister KB, et al. *Staphylococcus aureus* nuclease is an SaeRS-dependent virulence factor. *Infect Immun* 2013;**81**(4):1316–24.

28 Bogni C, Segura M, Giraudo J, Giraudo A, Calzolari A, Nagel R. Avirulence and immunogenicity in mice of a bovine mastitis *Staphylococcus aureus* mutant. *Can J Vet Res* 1998;**62**(4):293–8.

29 Berends ET, Horswill AR, Haste NM, Monestier M, Nizet V, von Kockritz-Blickwede M. Nuclease expression by *Staphylococcus aureus* facilitates escape from neutrophil extracellular traps. *J Innate Immun* 2010;**2**(6):576–86.

30 Menegazzi R, Decleva E, Dri P. Killing by neutrophil extracellular traps: fact or folklore? *Blood* 2012;**119**(5):1214–16.

31 Nauseef WM. Editorial: Nyet to NETs? A pause for healthy skepticism. *J Leukoc Biol* 2012;**91**(3):353–5.

32 Lappann M, Danhof S, Guenther F, Olivares-Florez S, Mordhorst IL, Vogel U. In vitro resistance mechanisms of *Neisseria meningitidis* against neutrophil extracellular traps. *Mol Microbiol* 2013;**89**(3):433–49.

33 Parker H, Albrett AM, Kettle AJ, Winterbourn CC. Myeloperoxidase associated with neutrophil extracellular traps is active and mediates bacterial killing in the presence of hydrogen peroxide. *J Leukoc Biol* 2012;**91**(3):369–76.

34 Schorn C, Janko C, Latzko M, Chaurio R, Schett G, Herrmann M. Monosodium urate crystals induce extracellular DNA traps in neutrophils, eosinophils, and basophils but not in mononuclear cells. *Front Immunol* 2012;**3**:277.

35 Yousefi S, Gold JA, Andina N, Lee JJ, Kelly AM, Kozlowski E, et al. Catapult-like release of mitochondrial DNA by eosinophils contributes to antibacterial defense. *Nat Med* 2008;**14**(9):949–53.

36 Guimaraes-Costa AB, Nascimento MT, Wardini AB, Pinto-da-Silva LH, Saraiva EM. ETosis: a microbicidal mechanism beyond cell death. *J Parasitol Res* 2012;**2012**:929743.

37 Aulik NA, Hellenbrand KM, Czuprynski CJ. *Mannheimia haemolytica* and its leukotoxin cause macrophage extracellular trap formation by bovine macrophages. *Infect Immun* 2012;**80**(5):1923–33.

38 Mohanan S, Horibata S, McElwee JL, Dannenberg AJ, Coonrod SA. Identification of macrophage extracellular trap-like structures in mammary gland adipose tissue: a preliminary study. *Front Immunol* 2013;**4**:67.

39 Liu P, Wu X, Liao C, Liu X, Du J, Shi H, et al. *Escherichia coli* and *Candida albicans* induced macrophage extracellular trap-like structures with limited microbicidal activity. *PLoS One* 2014;**9**(2):e90042.

40 Pilsczek FH, Salina D, Poon KK, Fahey C, Yipp BG, Sibley CD, et al. A novel mechanism of rapid nuclear neutrophil extracellular trap formation in response to *Staphylococcus aureus*. *J Immunol* 2010;**185**(12):7413–25.

41 Cogen AL, Yamasaki K, Muto J, Sanchez KM, Crotty Alexander L, Tanios J, et al. *Staphylococcus epidermidis* antimicrobial delta-toxin (phenol-soluble modulin-

gamma) cooperates with host antimicrobial peptides to kill group A *Streptococcus*. *PLoS One* 2010;**5**(1):e8557.
42. Aulik NA, Hellenbrand KM, Klos H, Czuprynski CJ. *Mannheimia haemolytica* and its leukotoxin cause neutrophil extracellular trap formation by bovine neutrophils. *Infect Immun* 2010;**78**(11):4454–66.
43. Patel S, Kumar S, Jyoti A, Srinag BS, Keshari RS, Saluja R, et al. Nitric oxide donors release extracellular traps from human neutrophils by augmenting free radical generation. *Nitric Oxide* 2010;**22**(3):226–34.
44. Lippolis JD, Reinhardt TA, Goff JP, Horst RL. Neutrophil extracellular trap formation by bovine neutrophils is not inhibited by milk. *Vet Immunol Immunopathol* 2006;**113**(1–2):248–55.
45. Grinberg N, Elazar S, Rosenshine I, Shpigel NY. Beta-hydroxybutyrate abrogates formation of bovine neutrophil extracellular traps and bactericidal activity against mammary pathogenic *Escherichia coli*. *Infect Immun* 2008;**76**(6):2802–7.
46. Webster SJ, Daigneault M, Bewley MA, Preston JA, Marriott HM, Walmsley SR, et al. Distinct cell death programs in monocytes regulate innate responses following challenge with common causes of invasive bacterial disease. *J Immunol* 2010;**185**(5):2968–79.
47. Hong W, Juneau RA, Pang B, Swords WE. Survival of bacterial biofilms within neutrophil extracellular traps promotes nontypeable *Haemophilus influenzae* persistence in the chinchilla model for otitis media. *J Innate Immun* 2009;**1**(3):215–24.
48. Hakkim A, Fuchs TA, Martinez NE, Hess S, Prinz H, Zychlinsky A, Waldmann H. Activation of the Raf-MEK-ERK pathway is required for neutrophil extracellular trap formation. *Nat Chem Biol* 2011;**7**(2):75–7.
49. Papayannopoulos V, Metzler KD, Hakkim A, Zychlinsky A. Neutrophil elastase and myeloperoxidase regulate the formation of neutrophil extracellular traps. *J Cell Biol* 2010;**191**(3):677–91.
50. Ermert D, Urban CF, Laube B, Goosmann C, Zychlinsky A, Brinkmann V. Mouse neutrophil extracellular traps in microbial infections. *J Innate Immun* 2009;**1**(3):181–93.
51. Ramos-Kichik V, Mondragón-Flores R, Mondragón-Castelán M, Gonzales-Pozos S, Muñiz-Hernandez S, et al. Neutrophil extracellular traps are induced by *Mycobacterium tuberculosis*. *Tuberculosis (Edinb)* 2009;**89**(1):29–37.
52. von Kockritz-Blickwede M, Goldmann O, Thulin P, Heinemann K, Norrby-Teglund A, Rohde M, Medina E. Phagocytosis-independent antimicrobial activity of mast cells by means of extracellular trap formation. *Blood* 2008;**111**(6):3070–80.
53. Beiter K, Wartha F, Albiger B, Normark S, Zychlinsky A, Henriques-Normark B. An endonuclease allows *Streptococcus pneumoniae* to escape from neutrophil extracellular traps. *Curr Biol* 2006;**16**(4):401–7.
54. Crotty Alexander LE, Maisey HC, Timmer AM, Rooijakkers SH, Gallo RL, von Köckritz-Blickwede M, Nizet V. M1T1 group A streptococcal pili promote epithelial colonization but diminish systemic virulence through neutrophil extracellular entrapment. *J Mol Med (Berl)* 2010;**88**(4):371–81.
55. Buchanan JT, Simpson AJ, Aziz RK, Liu GY, Kristian SA, Kotb M, et al. DNase expression allows the pathogen group A *Streptococcus* to escape killing in neutrophil extracellular traps. *Curr Biol* 2006;**16**(4):396–400.
56. Bruns S, Kniemeyer O, Hasenberg M, Aimanianda V, Nietzsche S, Thywissen A, et al. Production of extracellular traps against *Aspergillus fumigatus in vitro* and in infected

lung tissue is dependent on invading neutrophils and influenced by hydrophobin RodA. *PLoS Pathog* 2010;**6**(4):e1000873.
57. McCormick A, Heesemann L, Wagener J, Marcos V, Hartl D, Loeffler J, et al. NETs formed by human neutrophils inhibit growth of the pathogenic mold *Aspergillus fumigatus*. *Microbes Infect* 2010;**12**(12–13):928–36.
58. Urban CF, Reichard U, Brinkmann V, Zychlinsky A. Neutrophil extracellular traps capture and kill *Candida albicans* yeast and hyphal forms. *Cell Microbiol* 2006;**8**(4):668–76.
59. Springer DJ, Ren P, Raina R, Dong Y, Behr MJ, McEwen BF, et al. Extracellular fibrils of pathogenic yeast *Cryptococcus gattii* are important for ecological niche, murine virulence and human neutrophil interactions. *PLoS One* 2010;**5**(6):e10978.
60. Urban CF, Ermert D, Schmid M, Abu-Abed U, Goosmann C, Nacken W, et al. Neutrophil extracellular traps contain calprotectin, a cytosolic protein complex involved in host defense against *Candida albicans*. *PLoS Pathog* 2009;**5**(10):e1000639.
61. Behrendt JH, Ruiz A, Zahner H, Taubert A, Hermosilla C. Neutrophil extracellular trap formation as innate immune reactions against the apicomplexan parasite *Eimeria bovis*. *Vet Immunol Immunopathol* 2010;**133**(1):1–8.
62. Guimarães-Costa AB, Nascimento MT, Froment GS, Soares RP, Morgado FN, Conceição-Silva F, Saraiva EM. Leishmania amazonensis promastigotes induce and are killed by neutrophil extracellular traps. *Proc Natl Acad Sci USA* 2009;**106**(16):6748–53.
63. Bonne-Année S, Kerepesi LA, Hess JA, Wesolowski J, Paumet F, Lok JB, et al. Extracellular traps are associated with human and mouse neutrophil and macrophage mediated killing of larval *Strongyloides stercoralis*. *Microbes Infect* 2014;**16**(6):502–11.
64. Abi Abdallah DS, Lin C, Ball CJ, King MR, Duhamel GE, Denkers EY. *Toxoplasma gondii* triggers release of human and mouse neutrophil extracellular traps. *Infect Immun* 2012;**80**(2):768–77.
65. Saitoh T, Komano J, Saitoh Y, Misawa T, Takahama M, Kozaki T, et al. Neutrophil extracellular traps mediate a host defense response to human immunodeficiency virus-1. *Cell Host Microbe* 2012;**12**(1):109–16.
66. Raftery MJ, Lalwani P, Krautkrämer E, Peters T, Scharffetter-Kochanek K, Krüger R, et al. β2 integrin mediates hantavirus-induced release of neutrophil extracellular traps. *J Exp Med* 2014;**211**(7):1485–97.
67. Narasaraju T, Yang E, Samy RP, Ng HH, Poh WP, Liew AA, et al. Excessive neutrophils and neutrophil extracellular traps contribute to acute lung injury of influenza pneumonitis. *Am J Pathol* 2011;**179**(1):199–210.
68. Tripathi S, Verma A, Kim EJ, White MR, Hartshorn KL. LL-37 modulates human neutrophil responses to influenza A virus. *J Leukoc Biol* 2014;**96**(5):931–8.
69. Jenne CN, Wong CH, Zemp FJ, McDonald B, Rahman MM, Forsyth PA, et al. Neutrophils recruited to sites of infection protect from virus challenge by releasing neutrophil extracellular traps. *Cell Host Microbe* 2013;**13**(2):169–80.
70. Neeli I, Khan SN, Radic M. Histone deimination as a response to inflammatory stimuli in neutrophils. *J Immunol* 2008;**180**(3):1895–902.
71. Neeli I, Radic M. Opposition between PKC isoforms regulates histone deimination and neutrophil extracellular chromatin release. *Front Immunol* 2013;**4**:38.
72. Khandpur R, Carmona-Rivera C, Vivekanandan-Giri A, Gizinski A, Yalavarthi S, Knight JS, et al. NETs are a source of citrullinated autoantigens and stimulate inflammatory responses in rheumatoid arthritis. *Sci Transl Med* 2013;**5**(178):178ra140.

73 Lin AM, Rubin CJ, Khandpur R, Wang JY, Riblett M, Yalavarthi S, et al. Mast cells and neutrophils release IL-17 through extracellular trap formation in psoriasis. *J Immunol* 2011;**187**(1):490–500.

74 Gupta AK, Hasler P, Holzgreve W, Gebhardt S, Hahn S. Induction of neutrophil extracellular DNA lattices by placental microparticles and IL-8 and their presence in preeclampsia. *Hum Immunol* 2005;**66**(11):1146–54.

75 Hu H, Varon D, Hjemdahl P, Savion N, Schulman S, Li N. Platelet-leukocyte aggregation under shear stress: differential involvement of selectins and integrins. *Thromb Haemost* 2003;**90**(4):679–87.

76 Clark SR, Ma AC, Tavener SA, McDonald B, Goodarzi Z, Kelly MM, et al. Platelet TLR4 activates neutrophil extracellular traps to ensnare bacteria in septic blood. *Nat Med* 2007;**13**(4):463–9.

77 Martinelli S, Urosevic M, Daryadel A, Oberholzer PA, Baumann C, Fey MF, et al. Induction of genes mediating interferon-dependent extracellular trap formation during neutrophil differentiation. *J Biol Chem* 2004;**279**(42):44 123–32.

78 Wartha F, Beiter K, Albiger B, Fernebro J, Zychlinsky A, Normark S, et al. Capsule and D-alanylated lipoteichoic acids protect *Streptococcus pneumoniae* against neutrophil extracellular traps. *Cell Microbiol* 2007;**9**(5):1162–71.

79 Morita C, Sumioka R, Nakata M, Okahashi N, Wada S, Yamashiro T, et al. Cell wall-anchored nuclease of *Streptococcus sanguinis* contributes to escape from neutrophil extracellular trap-mediated bacteriocidal activity. *PLoS One* 2014;**9**(8):e103125.

80 Thammavongsa V, Missiakas DM, Schneewind O. *Staphylococcus aureus* degrades neutrophil extracellular traps to promote immune cell death. *Science* 2013;**342**(6160):863–6.

81 Lauth X, von Kockritz-Blickwede M, McNamara CW, Myskowski S, Zinkernagel AS, Beall B, et al. M1 protein allows Group A streptococcal survival in phagocyte extracellular traps through cathelicidin inhibition. *J Innate Immun* 2009;**1**(3):202–14.

82 Neumann A, Berends ET, Nerlich A, Molhoek EM, Gallo RL, Meerloo T, et al. The antimicrobial peptide LL-37 facilitates the formation of neutrophil extracellular traps. *Biochem J* 2014;**464**(1):3–11.

83 Schilcher K, Andreoni F, Uchiyama S, Ogawa T, Schuepbach RA, Zinkernagel AS. Increased neutrophil extracellular trap-mediated *Staphylococcus aureus* clearance through inhibition of nuclease activity by clindamycin and immunoglobulin. *J Infect Dis* 2014;**210**(3):473–82.

84 Lu T, Kobayashi SD, Quinn MT, Deleo FR. A NET outcome. *Front Immunol* 2012;**3**:365.

85 Malachowa N, Kobayashi SD, Freedman B, Dorward DW, DeLeo FR. *Staphylococcus aureus* leukotoxin GH promotes formation of neutrophil extracellular traps. *J Immunol* 2013;**191**(12):6022–9.

86 Dwyer M, Shan Q, D'Ortona S, Maurer R, Mitchell R, Olesen H, et al. Cystic fibrosis sputum DNA has NETosis characteristics and neutrophil extracellular trap release is regulated by macrophage migration-inhibitory factor. *J Innate Immun* 2014;**6**(6):765–79.

87 Whitchurch CB, Tolker-Nielsen T, Ragas PC, Mattick JS. Extracellular DNA required for bacterial biofilm formation. *Science* 2002;**295**(5559):1487.

88 Vossenaar ER, Zendman AJ, van Venrooij WJ, Pruijn GJ. PAD, a growing family of citrullinating enzymes: genes, features and involvement in disease. *Bioessays* 2003;**25**(11):1106–18.

89. Rohrbach AS, Slade DJ, Thompson PR, Mowen KA. Activation of PAD4 in NET formation. *Front Immunol* 2012;**3**:360.
90. Rorvig S, Ostergaard O, Heegaard NH, Borregaard N. Proteome profiling of human neutrophil granule subsets, secretory vesicles, and cell membrane: correlation with transcriptome profiling of neutrophil precursors. *J Leukoc Biol* 2013;**94**(4):711–21.
91. Nakashima K, Hagiwara T, Yamada M. Nuclear localization of peptidylarginine deiminase V and histone deimination in granulocytes. *J Biol Chem* 2002;**277**(51):49 562–8.
92. Luger K, Dechassa ML, Tremethick DJ. New insights into nucleosome and chromatin structure: an ordered state or a disordered affair? *Nat Rev Mol Cell Biol* 2012;**13**(7):436–47.
93. Syed SH, Goutte-Gattat D, Becker N, Meyer S, Shukla MS, Hayes JJ, et al. Single-base resolution mapping of H1-nucleosome interactions and 3D organization of the nucleosome. *Proc Natl Acad Sci USA* 2010;**107**(21):9620–5.
94. Christophorou MA, Castelo-Branco G, Halley-Stott RP, Oliveira CS, Loos R, Radzisheuskaya A, et al. Citrullination regulates pluripotency and histone H1 binding to chromatin. *Nature* 2014;**507**(7490):104–8.
95. Li P, Li M, Lindberg MR, Kennett MJ, Xiong N, Wang Y. PAD4 is essential for antibacterial innate immunity mediated by neutrophil extracellular traps. *J Exp Med* 2010;**207**(9):1853–62.
96. Lewis HD, Liddle J, Coote JE, Atkinson SJ, Barker MD, Bax BD, et al. Inhibition of PAD4 activity is sufficient to disrupt mouse and human NET formation. *Nat Chem Biol* 2015;**11**(3):189–91.
97. Munks MW, McKee AS, Macleod MK, Powell RL, Degen JL, Reisdorph NA, et al. Aluminum adjuvants elicit fibrin-dependent extracellular traps *in vivo*. *Blood* 2010;**116**(24):5191–9.
98. Li Y, Liu B, Fukudome EY, Lu J, Chong W, Jin G, et al. Identification of citrullinated histone H3 as a potential serum protein biomarker in a lethal model of lipopolysaccharide-induced shock. *Surgery* 2011;**150**(3):442–51.
99. Martinod K, Demers M, Fuchs TA, Wong SL, Brill A, Gallant M, et al. Neutrophil histone modification by peptidylarginine deiminase 4 is critical for deep vein thrombosis in mice. *Proc Natl Acad Sci USA* 2013;**110**(21):8674–9.
100. Yoo DG, Floyd M, Winn M, Moskowitz SM, Rada B. NET formation induced by *Pseudomonas aeruginosa* cystic fibrosis isolates measured as release of myeloperoxidase-DNA and neutrophil elastase-DNA complexes. *Immunol Lett* 2014;**160**(2):186–94.
101. Brinkmann V, Goosmann C, Kuhn LI, Zychlinsky A. Automatic quantification of *in vitro* NET formation. *Front Immunol* 2012;**3**:413.
102. Neeli I, Dwivedi N, Khan S, Radic M. Regulation of extracellular chromatin release from neutrophils. *J Innate Immun* 2009;**1**(3):194–201.
103. Byrd AS, O'Brien XM, Johnson CM, Lavigne LM, Reichner JS. An extracellular matrix-based mechanism of rapid neutrophil extracellular trap formation in response to *Candida albicans*. *J Immunol* 2013;**190**(8):4136–48.
104. Behnen M, Leschczyk C, Möller S, Batel T, Klinger M, Solbach W, Laskay T. Immobilized immune complexes induce neutrophil extracellular trap release by human neutrophil granulocytes via FcgammaRIIIB and Mac-1. *J Immunol* 2014;**193**(4):1954–65.

105 Fuchs TA, Abed U, Goosmann C, Hurwitz R, Schulze I, Wahn V, et al. Novel cell death program leads to neutrophil extracellular traps. *J Cell Biol* 2007;**176**(2):231–41.

106 Yipp BG, Petri B, Salina D, Jenne CN, Scott BN, Zbytnuik LD, et al. Infection-induced NETosis is a dynamic process involving neutrophil multitasking *in vivo*. *Nat Med* 2012;**18**(9):1386–93.

107 Tadie JM, Bae HB, Jiang S, Park DW, Bell CP, Yang H, et al. HMGB1 promotes neutrophil extracellular trap formation through interactions with Toll-like receptor 4. *Am J Physiol Lung Cell Mol Physiol* 2013;**304**(5):L342–9.

108 Dwivedi N, Radic M. Citrullination of autoantigens implicates NETosis in the induction of autoimmunity. *Ann Rheum Dis* 2014;**73**(3):483–91.

109 Gupta AK, Giaglis S, Hasler P, Hahn S. Efficient neutrophil extracellular trap induction requires mobilization of both intracellular and extracellular calcium pools and is modulated by cyclosporine A. *PLoS One* 2014;**9**(5):e97088.

110 Xiao H, Bai XH, Wang Y, Kim H, Mak AS, Liu M. MEK/ERK pathway mediates PKC activation-induced recruitment of PKCzeta and MMP-9 to podosomes. *J Cell Physiol* 2013;**228**(2):416–27.

111 Lande R, Gregorio J, Facchinetti V, Chatterjee B, Wang YH, Homey B, et al. Plasmacytoid dendritic cells sense self-DNA coupled with antimicrobial peptide. *Nature* 2007;**449**(7162):564–9.

112 Vanden Berghe T, Linkermann A, Jouan-Lanhouet S, Walczak H, Vandenabeele P. Regulated necrosis: the expanding network of non-apoptotic cell death pathways. *Nat Rev Mol Cell Biol* 2014;**15**(2):135–47.

113 Linkermann A, Green DR. Necroptosis. *N Engl J Med* 2014;**370**(5):455–65.

114 Holler N, Zaru R, Micheau O, Thome M, Attinger A, Valitutti S, et al. Fas triggers an alternative, caspase-8-independent cell death pathway using the kinase RIP as effector molecule. *Nat Immunol* 2000;**1**(6):489–95.

115 Cai Z, Jitkaew S, Zhao J, Chiang HC, Choksi S, Liu J, et al. Plasma membrane translocation of trimerized MLKL protein is required for TNF-induced necroptosis. *Nat Cell Biol* 2014;**16**(1):55–65.

116 Shulga N, Pastorino JG. GRIM-19-mediated translocation of STAT3 to mitochondria is necessary for TNF-induced necroptosis. *J Cell Sci* 2012;**125**(Pt. 12):2995–3003.

117 Greenlee-Wacker MC, Rigby KM, Kobayashi SD, Porter AR, DeLeo FR, Nauseef WM. Phagocytosis of *Staphylococcus aureus* by human neutrophils prevents macrophage efferocytosis and induces programmed necrosis. *J Immunol* 2014;**192**(10):4709–17.

118 Losman MJ, Fasy TM, Novick KE, Monestier M. Monoclonal autoantibodies to subnucleosomes from a MRL/Mp(-)+/+ mouse. Oligoclonality of the antibody response and recognition of a determinant composed of histones H2A, H2B, and DNA. *J Immunol* 1992;**148**(5):1561–9.

119 Isenberg DA, Manson JJ, Ehrenstein MR, Rahman A. Fifty years of anti-ds DNA antibodies: are we approaching journey's end? *Rheumatology (Oxford)* 2007;**46**(7):1052–6.

120 Zouali M, Migliorini P, Stollar DB. Murine lupus anti-DNA antibodies cross-react with the hapten (4-hydroxy-5-iodo-3-nitrophenyl)acetyl, but immunization-induced anti-DNA antibodies do not. *Eur J Immunol* 1987;**17**(4):509–13.

121 Darrah E, Andrade F. NETs: the missing link between cell death and systemic autoimmune diseases? *Front Immunol* 2012;**3**:428.

122 Muller S, Radic M. Citrullinated autoantigens: from diagnostic markers to pathogenetic mechanisms. *Clin Rev Allergy Immunol* 2015;**49**(2):232–9.

123 Kawane K, Ohtani M, Miwa K, Kizawa T, Kanbara Y, Yoshioka Y, et al. Chronic polyarthritis caused by mammalian DNA that escapes from degradation in macrophages. *Nature* 2006;**443**(7114):998–1002.

124 Arbuckle MR, McClain MT, Rubertone MV, Schofield RH, Dennis GJ, James JA, Harley JB. Development of autoantibodies before the clinical onset of systemic lupus erythematosus. *N Engl J Med* 2003;**349**(16):1526–33.

125 Dwivedi N, Upadhyay J, Neeli I, Khan S, Pattanaik D, Myers L, et al. Felty's syndrome autoantibodies bind to deiminated histones and neutrophil extracellular chromatin traps. *Arthritis Rheum* 2012;**64**(4):982–92.

126 Schellekens GA, de Jong BA, van den Hoogen FH, van de Putte LB, van Venrooij WJ. Citrulline is an essential constituent of antigenic determinants recognized by rheumatoid arthritis-specific autoantibodies. *J Clin Invest* 1998;**101**(1):273–81.

127 Chang X, Yamada R, Suzuki A, Sawada T, Yoshino S, Tokuhiro S, Yamamoto K. Localization of peptidylarginine deiminase 4 (PADI4) and citrullinated protein in synovial tissue of rheumatoid arthritis. *Rheumatology (Oxford)* 2005;**44**(1):40–50.

128 Romero V, Fert-Bober J, Nigrovic PA, Darrah E, Hague UJ, Lee DM, et al. Immune-mediated pore-forming pathways induce cellular hypercitrullination and generate citrullinated autoantigens in rheumatoid arthritis. *Sci Transl Med* 2013;**5**(209):209ra150.

129 Kaplan MJ, Radic M. Neutrophil extracellular traps: double-edged swords of innate immunity. *J Immunol* 2012;**189**(6):2689–95.

130 Villanueva E, Yalavarthi S, Berthier CC, Hodgin JB, Khandpur R, Lin AM, et al. Netting neutrophils induce endothelial damage, infiltrate tissues, and expose immunostimulatory molecules in systemic lupus erythematosus. *J Immunol* 2011;**187**(1):538–52.

131 Garcia-Romo GS, Caielli S, Vega B, Connolly J, Allantaz F, Xu Z, et al. Netting neutrophils are major inducers of type I IFN production in pediatric systemic lupus erythematosus. *Sci Transl Med* 2011;**3**(73):73ra20.

132 Dwivedi N, Neeli I, Schall N, Wan H, Desiderio DM, Csernok E, et al. Deimination of linker histones links neutrophil extracellular trap release with autoantibodies in systemic autoimmunity. *FASEB J* 2014;**28**(7):2840–51.

133 Mastronardi FG, Wood DD, Mei J, Raijmakers R, Tseveleki V, Dosch HM, et al. Increased citrullination of histone H3 in multiple sclerosis brain and animal models of demyelination: a role for tumor necrosis factor-induced peptidylarginine deiminase 4 translocation. *J Neurosci* 2006;**26**(44):11 387–96.

134 Radic M, Marion TN. Neutrophil extracellular chromatin traps connect innate immune response to autoimmunity. *Semin Immunopathol* 2013;**35**(4):465–80.

135 Robb CT, Dyrynda EA, Gray RD, Rossi AG, Smith VJ. Invertebrate extracellular phagocyte traps show that chromatin is an ancient defence weapon. *Nat Commun* 2014;**5**:4627.

136 Wen F, White GJ, VanEtten HD, Xiong Z, Hawes MC. Extracellular DNA is required for root tip resistance to fungal infection. *Plant Physiol* 2009;**151**(2):820–9.

137 Savchenko AS, Martinod K, Seidman MA, Wong SL, Borissoff JI, Piazza G, et al. Neutrophil extracellular traps form predominantly during the organizing stage of human venous thromboembolism development. *J Thromb Haemost* 2014;**12**(6):860–70.

138 Mangold A, Alias S, Scherz T, et al. Coronary neutrophil extracellular trap burden and deoxyribonuclease activity in ST-elevation acute coronary syndrome are predictors of ST-segment resolution and infarct size. *Circ Res* 2015;**116**(7):1182–92.

139 Demers M, Krause DS, Schatzberg D, Martinod K, Voorhees JR, Fuchs TA, et al. Cancers predispose neutrophils to release extracellular DNA traps that contribute to cancer-associated thrombosis. *Proc Natl Acad Sci USA* 2012;**109**(32):13 076–81.

140 Lämmermann T, Afonso PV, Angermann BR, Wang JM, Kastenmüller W, Parent CA, Germain RN. Neutrophil swarms require LTB4 and integrins at sites of cell death in vivo. *Nature* 2013;**498**(7454):371–5.

141 Chtanova T, Schaeffer M, Han SJ, van Dooren GG, Nollmann M, Herzmark P, et al. Dynamics of neutrophil migration in lymph nodes during infection. *Immunity* 2008;**29**(3):487–96.

142 Roth TL, Nayak D, Atanasijevic T, Koretsky AP, Latour LL, McGavern DB. Transcranial amelioration of inflammation and cell death after brain injury. *Nature* 2014;**505**(7482):223–8.

143 Knight JS, Zhao W, Luo W, Subramanian V, O'Dell AA, Yalavarthi S, et al. Peptidylarginine deiminase inhibition is immunomodulatory and vasculoprotective in murine lupus. *J Clin Invest* 2013;**123**(7):2981–93.

144 Willis VC, Gizinski AM, Banda NK, Causey CP, Knuckley B, Cordova KN, et al. N-alpha-benzoyl-N5-(2-chloro-1-iminoethyl)-L-ornithine amide, a protein arginine deiminase inhibitor, reduces the severity of murine collagen-induced arthritis. *J Immunol* 2011;**186**(7):4396–404.

145 Wei L, Wasilewski E, Chakka SK, Bello AM, Moscarello MA, Kotra LP. Novel inhibitors of protein arginine deiminase with potential activity in multiple sclerosis animal model. *J Med Chem* 2013;**56**(4):1715–22.

25

Parthanatos: Poly ADP Ribose Polymerase (PARP)-Mediated Cell Death

Amos Fatokun

School of Medical Sciences, Faculty of Life Sciences, University of Bradford, Bradford, UK

Abbreviations

AD	Alzheimer's disease
ADP	adenosine diphosphate
AIF	apoptosis-inducing factor
ALS	amyotrophic lateral sclerosis
AP-1	activator protein 1
AP-2	activator protein 2
ARH3	ADP ribose-(arginine) protein hydrolase
ARTD	ADPribosyltransferase
ATP	adenosine triphosphate
Bax	Bcl-2-associated X protein
BER	base excision repair
BRCA1	breast cancer-associated gene 1
BRCA2	breast cancer-associated gene 2
BRCT	breast cancer-associated gene 1 C-terminal
CNS	central nervous system
CypA	cyclophilin A
cyt c	cytochrome c
DSB	double-strand break
endo G	endonuclease G
eNOS	endothelial nitric oxide synthase
FAD	flavine adenine dinucleotide
FDA	Food and Drug Administration
HD	Huntington's disease
Hq	Harlequin
HR	homologous recombination
HSP70	heat-shock protein 70
iNOS	inducible nitric oxide synthase
KO	knockout
MNNG	methyl-N'-nitro-N-nitrosoguanidine
mPT	mitochondrial permeability transition
MPTP	mitochondrial permeability transition pore
NAD^+	oxidized nicotinamide adenine dinucleotide

Apoptosis and Beyond: The Many Ways Cells Die, First Edition. Edited by James Radosevich.
© 2018 John Wiley & Son Inc. Published 2018 by John Wiley & Son Inc.

NADH	reduced nicotinamide adenine dinucleotide
NCCD	Nomenclature Committee on Cell Death
NCX	Na^+/Ca^{++} exchanger
NHEJ	non-homologous end joining
NLS	nuclear location sequence
NMDA	N-methyl-D-aspartate
nNOS	neuronal nitric oxide synthase
NO	nitric oxide
Oct-1	octamer transcription factor 1
OMM	outer mitochondrial membrane
$ONOO^-$	peroxynitrite
PAAN	parthanatos AIF-associated nuclease
PAR	poly ADP ribose
PARG	poly ADP ribose glycohydrolase
PARP	poly ADP ribose polymerase
PD	Parkinson's disease
PLSCR	phospholipid scramblase
RNS	reactive nitrogen species
ROS	reactive oxygen species
SP-1	specificity protein 1
SSA	single-strand annealing
SSB	single-strand break
STAT-1	signal transducer and activator of transcription 1
TCA	tricarboxylic acid
TEF-1	translational elongation factor 1
TNF	tumor necrosis factor
TRAIL	tumor necrosis factor-related apoptosis-inducing ligand
UV	ultraviolet
YY-1	yin-yang-1

25.1 Introduction

Cell-death research has advanced significantly in the last few years, riding fruitfully on the increasing availability of more powerful approaches, especially in molecular and cell biology and imaging, to visualize and/or quantitate even very subtle molecular changes, resulting in the current description of an array of cell-death paradigms – a list that carries the prospect of expanding even further. Beyond the more familiar realms of apoptosis, necrosis, and perhaps autophagic cell death, bona fide cell-death types have now been identified and reasonably characterized, many of which are considered in detail in other chapters. Furthermore, cell-death types such as necrosis, which were previously thought to be completely random or unregulated (as opposed to those such as apoptosis, which were thought to be predominantly programmed), are now considered, at least in certain regards, to encompass some regulated or programmed components [1,2].

In order to avoid the confusion that might arise from an explosion in the choice of terms by which to describe an existing or emerging form of cell death, the Nomenclature Committee on Cell Death (NCCD) has given useful recommendations to unify definitions and descriptions, which advocate preference for the use of measurable biochemical features rather than just morphological attributes [1,3]. This, in a way, captures the relevance of the significant cross-talk – at the biochemical or molecular level – that may exist between one cell death and another, such that the description of a particular cell death should be able to achieve a clear separation of the features it shares with the other form(s) of cell death from those that are unique to it. It is now known that recognition of the reality of shared mechanistic events across different cell-death pathways is critical to the development of novel therapeutics, in view of the fact that most pathological scenarios underlain by cell death display more than one death type, although one may be predominant.

This chapter will consider poly adenosine diphosphate (ADP)-ribose polymerase (PARP)-dependent cell death, otherwise known as "parthanatos." Parthanatos will be described with regard to the evolution of knowledge about its separate identity, its unique biochemical features and mediators, the disease conditions in which it has been shown to be relevant, and its implications for therapeutics development. In presenting this chapter, the author seeks to enable the reader to gain sufficient breadth of knowledge on the profile of parthanatos and its relationships with other forms of cell death, which can be built upon by consulting further material if more depth is required on particular aspects.

25.2 Definition of Parthanatos and its Place in the Current Cell-Death Pantheon

Until around 2008, the coinage "parthanatos" was not in existence and the cell-death phenomenon it represents was simply described as "PARP-mediated cell death." However, progress with the identification of the biochemical mediators of PARP-mediated cell death recognized, among other things, the critical role of its polymer, poly ADP ribose (PAR), in the propagation of the toxic message along the death cascade. PAR polymer was shown to be a death signal, as exposure of neurons directly to it resulted in their death [4]. PARP-dependent cell death therefore obtained a further, mechanistic definition as death resulting from exposure to PAR. As a result, and in order to capture this newborn reality in subsequent descriptions, an expression was coined by the laboratory of Ted and Valina Dawson at the Johns Hopkins University School of Medicine in the USA as a concatenation of "PAR" and "Thanatos," the latter representing the personification of death in Greek mythology. Thus, "PAR-Thanatos," rendered in writing as "parthanatos," is death resulting from the toxic effects of PAR, or PAR-mediated cell death, which would normally result from the overactivation of the enzyme PARP. The term "parthanatos" has been used ever since, and has gained significant traction in the cell-death community [5–9].

As will be discussed later, parthanatos was previously misclassified as necrosis, due to the depletion of cellular energy stores that accompanies it, and as cells deficient in energy are generally unable to die through the energy-dependent route of apoptosis, cells undergoing what is now known as parthanatos were incorrectly thought to succumb to necrotic cell death [10]. Parthanatic cell death manifests several biochemical features that make it distinct from apoptosis, necrosis, autophagic cell death, or any other type of cell death [8,9]. For emphasis, it is known that, unlike apoptosis, parthanatos does not depend on caspases for its execution [11], does not lead to the formation of apoptotic bodies, and causes large- rather than small-scale DNA fragmentation [9]. On the other hand, while it does not share the cell swelling that is characteristic of necrosis, it involves the loss of cell-membrane integrity seen in the latter [11,12]. The identities and sequence of involvement of the mediators uniquely associated with parthanatos are now sufficiently described. Table 25.1 shows the morphological and biochemical features and disease relevance that parthanatos shares with three other relatively well-known types of cell death, namely apoptosis, necrosis, and autophagy (autophagic cell death), as well as its distinct signatures.

25.3 Relevance of PARP and Parthanatos to Disease Conditions

Several disease conditions have been identified in which the enzyme PARP or the parthanatos pathway has been reported to be relevant. As reviewed elsewhere, PARP is involved in the pathogenesis of many human diseases [13]. There is evidence for the involvement of PARP activation in diseases affecting a wide range of organ systems, including neurological [14–16] and neurodegenerative [17,18] conditions, diabetes [19], colitis [20], arthritis [21], liver toxicity [22], and uveitis [23]. Parthanatos was first demonstrated in the central nervous system (CNS) [24] and a role has been found or suggested for PARP-1 in experimental models of trauma [16], stroke [25], Parkinson's disease (PD) [18,26,27], Alzheimer's disease (AD) [17], Huntington's disease (HD) [28], amyotrophic lateral sclerosis (ALS) [29], and spinal cord injury [30]. Many more conditions have been described, including multiorgan failure, in which PARP plays a role with or without accompanying cell death (parthanatos) [31–35].

25.4 Known Mediators of Parthanatos and their Cross-Talk

Biochemical events unique to parthanatos include rapid activation of PARP, synthesis and accumulation of PAR, and translocation of AIF from the mitochondria to the nucleus, all of which occur in a highly choreographed fashion. We will consider these mediators in the order of their recruitment into – or participation in – parthanatos. Where relevant, their normal physiological roles are also discussed, and related to the pathological role switch that implicates them in parthanatic cell death.

25.4.1 Poly ADP-Ribose Polymerases

The PARPs (EC 2.4.2.30) are a family of 17 proteins that are also, although less commonly, referred to as poly ADP ribose synthetases (synthases) or poly ADP ribose transferases. Some have argued that the nomenclature "PARP," suggesting that the enzymes only catalyze poly ADP ribosylation, should be reviewed, as there is now evidence that some recently identified members of the family do catalyze mono ADP ribosylation. In that light, it has been suggested the PARPs be called Diphtheria toxin-like ADPribosyltransferase (ARTD) enzymes [36]. However, as many current expert sources still use the "PARP" designation, we will retain it in this book in order to avoid any confusion. The most important and well characterized of the PARPs is PARP-1, which is so recognized because it alone produces 95% of the PAR polymer. This is why PARP-mediated cell death may once have been considered synonymous with PARP-1-mediated cell death, although the importance of the contributions of other isoforms, especially PARP-2, is now being intensely studied. PARP-1 is a nuclear enzyme that is

Table 25.1 Similarities and differences between apoptosis, necrosis, parthanatos, and autophagy. The morphological and biochemical features shared by the four types of cell death, as well as those that are unique to each, are listed with a view to comparing and contrasting them. To illustrate disease relevance, the context of neurological and neurodegenerative conditions is used to demonstrate the involvement of each cell death type in human pathologies. Parthanatos is distinguished by PARP activation, PAR polymer formation, the mitochondrial release of apoptosis-inducing factor (AIF) and its translocation into the nucleus, and large-scale ($\approx$50 kb) DNA fragmentation. PS, phophatidylserine; cyt c, cytochrome c; AIF, apoptosis-inducing factor; ATP, adenosine triphosphate; PARP, poly ADP ribose polymerase; PAR, poly ADP ribose; AD, Alzheimer's disease; PD, Parkinson's disease; ALS, amyotrophic lateral sclerosis; HD, Huntington's disease.
Source: Wang et al. 2009 [9]. Reproduced with permission from Elsevier.

	Apoptosis	Necrosis	Autophagy	Parthanatos
Membrane	Apoptotic bodies;	Disrupted,	Formation of double-membrane	Loss of integrity:
	Blebbing;	Loss of integrity;	bound autophagosomes	PS externalization
	PS externalization	Blebbing,		
Cytoplasm	Morphologically intact;	Vacuolation of the cytoplasm	Autophagic vacuoles,	Condensation
	Condensation	Disrupted; Cell swelling	Lysosomal degradation	
Mitochondria	Depolarization;	Loss of mitochondria ultrastructure	Degradation	Depolarization;
	Cyt c release to cytosol;			AIF release to nucleus;
Nuclei	Chromatin condensation;	Morphologie changes;	–	Chromatin condensation;
	Shrink	Chromatin digestion		Shrink
DNA	DNA fragmentation	DNA hydrolysis	–	Large DNA fragmentation
	(DNA ladder)	(smear)		($\approx$50 kb)
ATP	Energy-dependent	No energy requirement	Energy-dependent	Energy-independent
Key mediators	Subset of caspases	–	Atg6/Beclin-1	PARP activation;
				PAR polymer formation;
				Caspase-independent
Propidium iodide	–	+	+	+
Annexin V	+	–	–	+
Inflammatory	–	+	–	–
Neurologic disorders	Excitotoxicity, ischemia, stroke, trauma, AD, PD, ALS, HD, ataxias	Excitotoxicity, stroke, ischemia, AD, PD. ALS, HD, epilepsy	AD, PD, ALS, HD, prion disease	Excitotoxicity, stroke, ischemia, PD

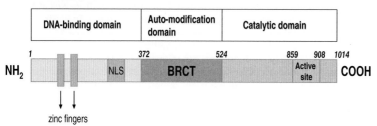

Figure 25.1 Primary structure of poly ADP ribose polymerase 1 (PARP-1), showing its three domains. (i) An N-terminal DNA-binding domain with two zinc-finger motifs and a nuclear location sequence (NLS). (ii) A central automodification domain, with a breast cancer-associated gene 1 (*BRCA1*) C-terminal (BRCT) motif containing phosphorylation sites for the regulation of PARP-1 activity. (iii) A C-terminal catalytic domain, containing the nicotinamide adenine dinucleotide (NAD)-binding site and the poly ADP ribose (PAR)-synthesizing domain. *Source:* Fatokun et al. 2014 [8]. Reproduced with permission from John Wiley & Sons.

physiologically important to the cell, in which it regulates homeostasis and genomic stability [37–39], among other possible or related functions [38,40–43]. PARP-1 is able to sense mild DNA damage in the form of single-strand nicks and breaks, and through the synthesis of PAR polymer, can facilitate the recruitment of the base excision repair (BER) machinery. PARP-1 is a 116 kDa protein and, as shown in Figure 25.1, is endowed with three major domains [34,38,44,45]: (i) an N-terminal domain (42 kDa), which has two zinc-finger motifs and a nuclear location sequence (NLS) for DNA binding; (ii) a central automodification domain (16 kDa); and (iii) a C-terminal catalytic domain (55 kDa), which features the nicotinamide adenine dinucleotide (NAD)-binding site and the PAR-synthesizing domain.

There is evidence of significant conservation of PARP-1 in eukaryotes, with about 92% homology (highest for the catalytic domain) between human and mouse PARP-1 [34]. With regard to its physiological roles, PARP-1 has the capacity to increase its activity up to 500-fold in response to mild DNA damage, and it can produce PAR polymer by using oxidized nicotinamide adenine dinucleotide (NAD$^+$) generated from adenosine triphosphate (ATP). Residues of the PAR polymer (usually about 50–200 in number), in a process called parylation, may attach either to PARP itself (autoribosylation) or to several other nuclear proteins (heteroribosylation) (e.g., histones, topoisomerases I and II, DNA polymerases, DNA ligase-2, high-mobility-group proteins, and transcription factors) [38,46,47]. It should be noted that parylation is a post-translational modification that occurs in both mitotic and post-mitotic mammalian cells [46,48].

A pathological context is found for the process of parylation when DNA damage is no longer mild but is instead profound. Excessive activation of PARP-1 follows, leading to the production of a vast number of PAR polymers, which are (unlike in the case of mild DNA damage) rather long-chained and branched. This represents the initiation of the toxic parthanatic cascade, and the sequence of the untoward molecular events that leads to parthanatic cell death begins to unravel. It is very important to distinguish between the degree of DNA damage that induces PARP-1 to produce levels and types of PAR that facilitate, rather than impair, DNA repair and the degree of damage that is so characteristically detrimental as to produce PAR of a sufficient number and complexity to direct the impacted cells toward death. As will be discussed later, the repair of DNA by activated PARP-1 can be cleverly impaired in cancer cells, leading to their eventual demise. This is particularly relevant in cancers in which there is loss-of-function mutation in certain genes, especially breast and ovarian cancers that harbor mutations in the *BRCA1* and *BRCA2* genes – a liability that renders them highly sensitive to the effects of PARP inhibitors through synthetic lethality (see later). In contrast, overactivation of PARP-1 in functional cells unfortunately leads to their death, and the therapeutic objective in this context will be the prevention of lethality by the blocking of excessive activation of PARP-1. It is this aspect of the DNA-damage response underpinning parthanatos that will now be discussed.

Many toxic stimuli and conditions can induce excessive activation of PARP-1, leading to parthanatos; these include oxidative stress caused by reactive oxygen species (ROS) (e.g., hydrogen peroxide, hydroxy radical), nitrosative stress induced by reactive nitrogen species (RNS) (e.g., nitric oxide (NO), peroxynitrite (ONOO$^-$, formed from the reaction of nitrogen with the superoxide anion)), hypoxia, hypoglycemia, ischemia, ischemic reperfusion injury, and alkylating agents such as methyl-N′-nitro-N-nitrosoguanidine (MNNG). This is the basis for the use of these sources of toxicity in experimental models of parthanatos. Exposure of neurons to N-methyl-D-aspartate (NMDA) can also lead to parthanatic cell death through excessive stimulation of the glutamate NMDA receptor, resulting in intracellular calcium overload, generation of ROS, and generation of ONOO$^-$, a potent inducer of DNA damage that causes PARP-1 overactivation [12,13,24,25,49–54].

25.4.2 Poly ADP-Ribose Polymer

As already emphasized, parthanatos involves, around its initiation stage, the synthesis and accumulation of a vast amount of complex and highly toxic PAR polymer, occasioned by PARP overactivation. While PARP itself is unable to leave the nucleus, the PAR polymer has the ability to exit it to travel to the cytoplasm and subsequently to the mitochondria. In a study of the role of the PAR polymer in parthanatos, protection was achieved against cell death caused by exposure of mouse primary neurons to NMDA when cytosolic PAR was neutralized with PAR-specific antibodies. Neurons died when exposed directly to the PAR polymer, with the extent of the resulting toxicity being directly proportional to the increase in the dose and/or the complexity of the polymer [4]. PAR has also been shown to interact with myriad proteins in order to regulate their physiological functions [39,43]. These and other reports provide strong evidence that the polymer is a credible death signal and serves as a death messenger of PARP.

The mitochondria have long been known to be a source of death mediators, and in parthanatos the PAR polymer serves as the agent of communication between PARP-1 and mitochondrial mediators of parthanatos. The advent of the polymer in the mitochondria can therefore bring about molecular events therein that are shared by other cell-death types, such as phosphatidylserine externalization (which also occurs in apoptosis), dissipation of mitochondrial membrane potential, and mitochondrial permeability transition (mPT) (resulting from the opening of the mitochondrial permeability transition pore, MPTP), as well as other mitochondrial events that are exclusive to parthanatos, most strikingly the release and translocation of AIF from the mitochondria to the nucleus, leading to large-scale ($\approx$50 kb) DNA fragmentation – again, a unique characteristic of parthanatos. The possibility of late caspase activation in parthanatos has been argued, but this is not in any way necessary for its occurrence [11].

The unraveling of details of the molecular mechanisms by which PAR induces mitochondrial changes in parthanatos provided a strong rationale for scrutinizing the validity of the suicide hypothesis that suggested energy depletion as the primary cause of death. PARP-1 overactivation causes massive depletion of cellular NAD^+, a cofactor in glycolysis and the tricarboxylic acid (TCA) cycle, and consequently of ATP, as the synthesis of each molecule of NAD^+ requires four molecules of ATP. Cells were then thought to die from the energy deficiency that resulted over time, in a manner that appeared consistent with the execution of necrosis [10,55–58]. More recently, it was shown that the energy depletion occurs via inhibition of glycolysis [59]. However, direct evidence to support the suicide hypothesis has been lacking, and some studies have shown, for example, that cellular energy stores are not different between PARP-1 knockout (KO) mice and wild-type following focal ischemic injury – although infarct volume was smaller in the former [60] – and that in the absence of NAD^+ depletion, PARP-1 activation in cells lacking poly ADP ribose glycohydrolase (PARG) causes parthanatos [61]. While there is a resurgence of significant interest in delineating the firm roles of cellular energy depletion in parthanatos, it is now unequivocally resolved that the main culprit event at the mitochondrial level is the PAR polymer inducing the release of AIF and its translocation from the mitochondria to the nucleus [4] (see later).

25.4.3 Poly ADP Ribose Glycohydrolase

PARG is an endogenous enzyme that helps to regulate levels of the PAR polymer by catalyzing its degradation, following its synthesis by PARP [44,62]. PAR polymers are hydrolyzed to free ADP ribose units through its endoglycosidic and exoglycosidic activity [63]. Found in humans and other species [64–66], PARG exists in alternatively spliced isoforms [67–69], with its full length localizing to the nucleus and the splice variants showing diverse localization patterns [65,67–70]. Many studies of the effects of PARG in parthanatos have strongly demonstrated that regulation of PAR polymer levels is critical to cell survival in this type of cell death [61]. For example, overexpression of PARG has been shown to protect against PAR-mediated cell death [4], hydrogen peroxide-induced cell death [71], excitotoxicity, and stroke [4,72]. Conversely, its deletion led to increased toxicity [4], except when a PARP inhibitor was present [73], and cells from PARG KOs were significantly more sensitive to MNNG and menadione – agents that activate PARP-1 significantly – causing accumulation of PAR [73]. In fact, in mice, whole-body knockout of PARG is known to be embryonically lethal at about day E3.5. In all these studies, the basis for the toxicity (or non-survival) resulting from the absence of PARG was suggested to be the excessive accumulation of the synthesized PAR polymer. Another enzyme, ADP ribose-(arginine) protein hydrolase (ARH3), also displays PARG-like activity, but it does not seem to have an appreciable role in cell death, not being protective against death or PAR accumulation [73,74].

25.4.4 Apoptosis-Inducing Factor

When excessive levels of the toxic PAR polymer are not curbed, the course of parthanatos progresses and the free polymer reaches the mitochondria (although protein-conjugated PAR may also be involved), where it causes the translocation to the nucleus of the AIF [75]. To avoid any confusion, it should be promptly recognized that, in retrospect, the name "AIF" is rather a misnomer, as it was given when this protein was thought to participate predominantly in apoptosis, whereas it is now known that its role is more related to parthanatos, which is caspase-independent. Its unlikely function in caspase-dependent cell death (classical apoptosis) is generally regarded as merely epiphenomenal, providing an alternative route by which cells in which there is caspase activation are able to die [9,76].

AIF is a flavoprotein that is localized to the intermembrane space of the mitochondria [77]. However, there is evidence that some 30% of the protein may be loosely associated with the cytosolic side of the outer mitochondrial membrane (OMM), which may be responsible for the rapid release of a small initial pool of AIF in certain conditions inducing parthanatos, such as NMDA toxicity [78]. The *AIF* gene is found on chromosome X and has 16 exons [77]. Its synthesis as a 67 kDa precursor occurs in the cytoplasm, followed by its importation into the mitochondria, where processing to the mature 62 kDa form occurs [79,80]. AIF has five isoforms in humans [81–84], and the solved structures of human and mouse AIFs [85,86] reveal strong, positive electrostatic potential at their surfaces, which may bind to DNA [86]. The use of the Harlequin (*Hq*) mouse, in which there is approximately 80% downregulation of AIF expression [87], has furnished insights into AIF's physiological roles. AIF is now known to support cell survival when localized to the mitochondria, as oxidative stress and progressive degeneration of terminally differentiated cerebellar and retinal neurons are seen in this mouse model [87]. Other functions have also been reported [88–90].

There are three suggested functional domains in AIF. The first is an N-terminal flavine adenine dinucleotide (FAD)-binding domain and the second is a central reduced nicotinamide adenine dinucleotide (NADH)-binding domain, both of which underlie AIF's oxidoreductase activity. The third is the C-terminal domain, which is the locus of its cell death-mediating ability [77]. Interestingly, AIF is capable of inducing cell death in the absence of its oxidoreductase activity [86], and mutation to the PAR-binding site on AIF (which is separate from the DNA-binding site) [91] renders the mutant AIF – which retains its NADH oxidase activity and binds FAD or DNA – incapable of release from the mitochondria or of translocation to the nucleus to mediate cell death (parthanatos) in response to the activation of PARP-1.

Countless studies have provided compelling evidence that PARP-1 overactivation leads to the translocation of AIF to the nucleus. These studies induced cellular injury by exposing cells to insults known to cause DNA damage, including excitotoxicity, oxidative stress, nitrosative stress, and DNA-alkylating agents. The pathological scenarios led to nuclear translocation of AIF, followed by large-scale DNA fragmentation (≈ 50 kb) – characteristic of parthanatos – and also to chromatin condensation, culminating in cell death [11,92–94]. Furthermore, NMDA and purified PAR polymer were less toxic to *Hq* mouse neurons [75] (and the mice showed reduced lesions) compared with wild-type when both groups were exposed to excitotoxicity and stroke [4,88,91,95–97]. However, when AIF expression was forced in the *Hq* neurons, NMDA-induced cell death was raised to the same level as in wild-type [75]. Exposure of cells to exogenous PAR induced AIF translocation [75]. With AIF having been identified as a PAR polymer-binding protein [43], a physical interaction between PAR and AIF is considered a requirement before the latter can be released from the mitochondria [91].

Intriguingly, this release precedes that of cytochrome c (cyt c), as well as caspase activation (in parthanatos, caspase activation happens at the late stage, if at all). AIF translocation is now regarded as the commitment point (point of no return) in parthanatic cell death.

Upon its entry into the nucleus, AIF is believed to bind to DNA in order to effect cell death [86], but other mechanisms may also exist, since mouse AIF, unlike human AIF, does not have recognizable structural motifs for DNA binding [85,86]. As it is not certain that AIF has any intrinsic endonuclease activity, how it causes DNA fragmentation is still a subject of intense research, but the process might involve recruitment of endogenous proteases or nucleases, which cause the fragmentation; on the other hand, AIF may interact with the DNA to make it more susceptible to the activity of proteases or nucleases. An example in this regard is cyclophilin A (CypA), which has been shown to interact with AIF to form a DNA-degradation complex, albeit in the context of apoptosis-associated chromatinolysis [98]. However, neither of the two interacting partners has a nuclease domain, so the manner of their interaction is unknown. Mammalian endonuclease G (endo G) has also been suggested to interact with AIF, based purely on observed DNA-degrading interactions between CPS-6, the ortholog of endo G in *Caenorhabditis elegans*, and WAH-1, AIF's ortholog in *C. elegans* [99]. However, no evidence exists for a role for endo G in mammalian parthanatos [100,101]. A frantic search is therefore underway for a parthanatos AIF-associated nuclease (PAAN) in vertebrates.

Figure 25.2 summarizes the sequence of events leading to parthanatos in neurons and in other cell types. The decrease in cellular energy levels (NAD^+, ATP) caused by the induction of parthanatos is shown as a secondary contributor to the cell-death process, while the primary mechanistic cause is PAR-induced AIF translocation.

25.5 Putative Mediators of Parthanatos

While there is convincing evidence for the involvement of PAR and AIF in PARP-1-mediated cell death (parthanatos), there is some controversy over the likely involvement of some other mediators. For example, the Bcl-2 proteins have been suggested. However, no protection against parthanatos was found when *Bcl-2* was overexpressed, although cell-death onset was delayed [11]. A separate study showed AIF as essential for the neuronal death induced by Bcl-2-associated X protein (Bax)-dependent and Bax-independent mechanisms [88]. With regard to calpains, which are calcium-activated cysteine proteases, they have been shown to perhaps favor excitotoxicity by cleaving and hence inactivating the Na^+/Ca^{++} exchanger (NCX), thus promoting a feedforward loop of Ca^{++} overload [102], and this has been suggested to have some potential implications for the release of AIF in parthanatos. Other studies have revealed that calpain 1 is critical to the mitochondrial release of AIF from isolated mitochondria [103] and that AIF-mediated cell death involves a sequential activation of PARP-1, calpains, and Bax, but does not involve cathepsins [104]. In a similar vein, calpain 1 activity has been found to be required for AIF translocation after ischemic neuronal injury, with overexpression of the endogenous calpain inhibitor, calpastatin, inhibiting cell death following PARP-1 activation [79]. In contrast, however, in a separate study, no requirement for calpain 1 was found for AIF translocation in parthanatos, as the protease cleaved recombinant AIF in a cell-free system but not in intact cells, where AIF was released and translocated to

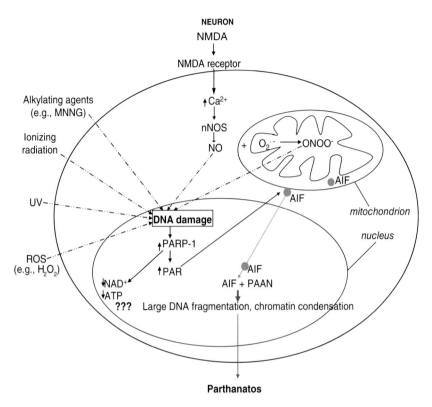

Figure 25.2 The parthanatos cascade. Identities and sequences of choreographed molecular events in parthanatos, as it occurs in a typical cell (neuronal, non-neuronal). Activation of the N-methyl-D-aspartate (NMDA) receptors in neurons causes increased calcium influx, resulting in the activation of calcium-dependent neuronal nitric oxide synthase (nNOS), which produces nitric oxide (NO). While NO at high levels may induce DNA damage directly, it more commonly reacts with the superoxide anion ($O_2^{\bullet-}$) in the mitochondria to produce peroxynitrite ($ONOO^-$), a very potent inducer of DNA damage. Production of NO through the activation of endothelial nitric oxide synthase (eNOS) or inducible nitric oxide synthase (iNOS) may be more important in non-neuronal cells, again leading to $ONOO^-$ generation (not shown). Some stimuli can induce DNA damage directly, including reactive oxygen species (ROS), such as hydrogen peroxide (H_2O_2), alkylating agents (e.g., methyl-N'-nitro-N-nitrosoguanidine, MNNG), ultraviolet (UV) radiation, and ionizing radiation. Stimuli inducing DNA damage are shown by broken arrows. DNA damage causes poly ADP ribose polymerase 1 (PARP-1) overactivation, leading to poly ADP ribose (PAR) polymer synthesis and accumulation. PARP-1 overactivation depletes the cellular pool of oxidized nicotinamide adenine dinucleotide (NAD^+) and adenosine triphosphate (ATP), but this is not the primary cause of cell death, as indicated by "???." PAR polymer signals to the mitochondria and directly binds the PAR polymer-binding site on the apoptosis-inducing factor (AIF), inducing its mitochondrial release and nuclear translocation. Upon entry into the nucleus, AIF causes large-scale ($\approx$50 kb) DNA fragmentation and chromatin condensation through an as yet unidentified parthanatos AIF-associated nuclease (PAAN). This is the molecular route that underlies parthanatos. Events after AIF release from the mitochondria are depicted by solid arrows. *Source:* Fatokun et al. 2014 [8]. Reproduced with permission from John Wiley & Sons.

the nucleus in its uncleaved (62 kDa) form and NMDA induced AIF translocation under conditions that did not activate calpain [105]. Taken together, it remains to be unequivocally resolved whether or not these mediators have obligatory roles in parthanatos.

25.6 Protein–Protein Interactions of Potential Relevance in Parthanatos

Studies are now beginning to establish the likely roles of protein–protein interactions and protein kinases in parthanatos. PARP has been shown to have an interaction network [106]. At the transcriptional level, it regulates the expression of several proteins, including inducible nitric oxide synthase (iNOS), cytokines, chemokines, and adhesion molecules. NF-κB is a major transcription factor that regulates these proteins, and PARP functions as its co-activator [107]. PARP is also involved in the activation of transcription factors associated with inflammation, including activator protein 1 (AP-1), AP-2, translational elongation factor 1 (TEF-1), specificity protein 1 (SP-1), octamer transcription factor 1 (Oct-1), yin-yang-1 (YY-1), and signal transducer and activator of transcription 1 (STAT-1) [32,42,108]. Similarly, multiple apoptogenic proteins may be involved in the nuclear translocation of AIF during focal cerebral ischemia [79,95,109–112]. In fact, AIF is thought to interact with a myriad of proteins other than calpains, cathepsins, and Bax along its nuclear translocation path. Heat-shock protein 70 (HSP70) blocks chromatin condensation of purified nuclei induced by AIF in a cell-free system [113], possibly through physical interactions between the two proteins [113,114]. This beneficial (protective) effect of HSP70 may be counteracted by granzyme-b [112], a serine protease that can mediate cell death independent of caspase activation [115]. Phospholipid scramblases (PLSCRs), which have been reported to promote bidirectional lipid scrambling [116–118], have also been shown to interact with AIF. There are four human scramblases (PLSCR1–4), while eight orthologs (SCRM1-8) have been identified in *C. elegans*. SCRM1 shares the highest homology with the human PLSCRs. It has been shown that the *C. elegans* AIF ortholog, WAH-1, following release from the mitochondria, is able to promote plasma-membrane phosphatidylserine externalization through the plasma-membrane protein SCRM-1, acting as its downstream effector [119]. Whether and how the PLSCRs interact with AIF in the context of parthanatos in mammals is still unclear, although these mediators have been reported to be involved in mammalian apoptosis induced by the tumor necrosis factor (TNF)-related apoptosis-inducing ligand (TRAIL) [120].

25.7 Parthanatos and the Development of Novel Therapeutics

The course of parthanatos (see earlier) reveals a pathway that should lend itself to therapeutic exploitation for the development of drugs for conditions in which cell death plays an important role. Figure 25.3 shows the therapeutic opportunities available at the different levels of the cascade, based on our current understanding of the molecular mechanisms involved. Significant success in this endeavor has already been realized for the topmost level – PARP activation – especially for the treatment of cancers [107,121–125]. Other potential interventions targeting events downstream of PARP activation are still being tackled at the experimental stage.

25.7.1 PARP Inhibitors

Inhibition of PARP presents an enormous advantage in the development of chemotherapeutics, whether as adjuvant drugs for use with mainline anticancer therapies

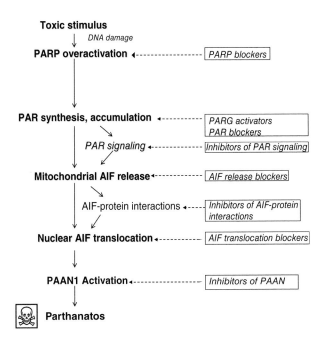

Figure 25.3 Existing or potential molecular therapeutic targets in parthanatos. Discrete levels within the parthanatos cascade are available for therapeutic exploitation. The direction of the solid arrows depicts pathway progression. Existing or potential therapeutic interventions are shown in solid rectangular boxes, each linked with broken arrows to its relevant step in parthanatos. Various poly ADP ribose polymerase (PARP) inhibitors are currently undergoing clinical trial or already in clinical use, but potential interventions downstream of PARP activation are still largely only experimentally explored. *Source:* Fatokun et al. 2014 [8]. Reproduced with permission from John Wiley & Sons.

(chemotherapy or radiotherapy) or as mainline drugs in cancer patients with certain inherited mutations. The rationale for the use of PARP inhibitors as adjuvant anticancer agents is that, as PARP is involved in the repair of single-strand DNA breaks, its inhibition will block – or at least impair – the repair of DNA damage induced by chemotherapy or radiotherapy in cancer cells (the goal of these therapies is to damage the DNA of the target cell) [126]. This can help sensitize cancer cells to mainline anticancer agents (improved efficacy) or help overcome their resistance to anticancer agents.

PARP inhibitors can also be used as mainline anticancer agents. It is known that certain cancers harbor loss-of-function mutations in a number of genes, which can be used to induce their demise through the application of relevant anticancer drugs. For example, some breast, ovarian, and prostate cancers carry mutations in the tumor-suppressor genes *BRCA1* and *BRCA2*, which are involved in the homologous recombination (HR) repair of DNA double-strand breaks (DSBs). Cancer cells with *BRCA1* or *BRCA1* loss of function harbor a deficiency in HR, which means they should be incompetent in repairing their DSBs. This vulnerability presents a golden opportunity for therapeutic intervention through PARP inhibition, since a PARP inhibitor will prevent the repair of single-strand breaks (SSBs), the accumulation of which will result in DSBs, which the HR-incompetent cancer cells will be incapable of repairing.

Accumulation of DSBs will lead to their deaths (other mechanisms of DSB repair that the cells may attempt to use, such as non-homologous end joining (NHEJ) and single-strand annealing (SSA), are highly error-prone). This is the mechanism of synthetic lethality, as illustrated in Figure 25.4.

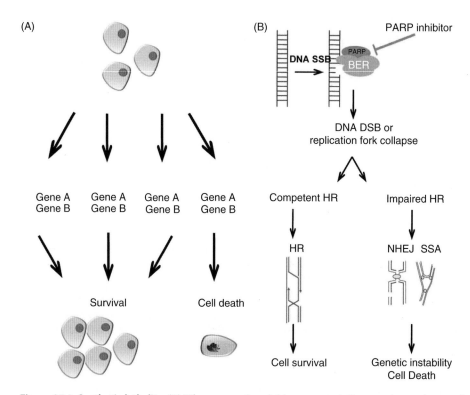

Figure 25.4 Synthetic lethality. (A) When genes *A* and *B* have no mutation, or when only one of them has a mutation (left three columns), the phenotype of the cell is viable (cell survival). However, when there is a mutation (or concurrent inhibition) in both (right column), cell death will occur (cell non-survival). This is the principle of synthetic lethality. (B) PARP inhibitors cause selective cell death in cells with impaired homologous recombination (HR) DNA repair, such as ovarian- and breast-cancer cells carrying *BRCA1* or *BRCA2* loss-of-function mutations. The base excision repair (BER) machinery, of which PARP is a key component, will normally repair single-strand breaks (SSBs) in the DNA, with PARP having the ability to localize to the site of breakage and act as a molecular beacon for the other components of the machinery. Therefore, if PARP is inhibited, BER is not recruited and so SSBs are not repaired. Accumulation of SSBs then causes replication fork stalling and collapse, and allows for the formation of double-strand breaks (DSBs). In a normal cell, DSBs are repaired either by the error-free HR mechanism or by an error-prone mechanism such as non-homologous end joining (NHEJ) or single-strand annealing (SSA). Cells with mutations such as the *BRCA1* or *BRCA2* mutation are HR-incompetent and so cannot repair their DSBs (or can only repair them using error-prone pathways). This engenders genomic instability and cell death. A PARP inhibitor used in these cancers should therefore block BER recruitment, causing the death of the cancer cells through synthetic lethality. This is the basis for the use of PARP inhibitors as monotherapy. Although not illustrated here, a PARP inhibitor used in combination with a mainline anticancer drug will also potentiate the effect of the mainline drug (increased efficacy) and block or impair the process by which cancer cells exposed to chemo- or radiotherapy gain resistance to such therapy by repairing the DNA damage they have suffered. *Source:* Dedes et al. 2011 [127]. Reproduced with permission from Taylor & Francis.

First-generation PARP inhibitors include nicotinamide, benzamide, and 3-aminobenzamide. Second-generation inhibitors are analogs of benzamides. Third-generation inhibitors have diverse chemical structures, including derivatives of imidazopyridine, imidazoquinolinone, and isoquinolindione [128]. Many other classes of inhibitor are currently in development [107]. It is of considerable interest that recent research has identified several natural compounds (mainly the flavone family of flavonoids) as PARP inhibitors [129,130], some of which have been shown to be neuroprotective [130]. Current efforts are focused on developing isoform-specific PARP inhibitors [124,125,131–133]. Over the last decade and a half, there has been considerable interest in the use of PARP inhibitors in clinical medicine, and several clinical trials of a number of PARP inhibitors as single agents (e.g., KU59436, BSI-201, ABT-888, INO-1001, rucaparib, olaparib, BGB-290) or in combination with other agents (e.g., INO-1001, GPI 21016, ABT-888, BMN-673, E7449, AZD2281, PF-01367338, olaparib) have been conducted, or are currently being recruited into [84,107,108,126,134–137]. In late 2014, olaparib was approved by the US Food and Drug Administration (FDA) as oral monotherapy for patients with germline *BRCA*-mutated advanced ovarian cancer that has already been treated with three or more lines of chemotherapy [138].

While there is a possibility of developing PARP inhibitors as neuroprotective agents, this may be significantly more challenging [139], and it might not be ultimately desirable to inhibit PARP for a prolonged period in neurons. This could limit the usefulness of such future drugs, say for the treatment of chronic neurodegenerative conditions. A more desirable approach to the exploitation of the parthanatos pathway for novel neurotherapeutics development will therefore rationally seek potential druggable targets downstream of PARP activation.

25.7.2 Inhibitors of Parthanatos Downstream of PARP Activation

The desirability and applicability of direct PARP inhibitors in anticancer therapy have been sufficiently demonstrated. Downstream of PARP activation are a number of other molecular events that may constitute therapeutic targets, from reduced PARG activity to PAAN activation (see Figure 25.4). To date, these are all still at the experimental stage. In relation to exploiting the next potential intervention step following PARP-1 activation – by means of clearing excessive levels of the PAR polymer – an endogenous inhibitor of parthanatos (like PARG) has been discovered. Named "Iduna" (RNF146), this is a PAR-dependent E3 ubiquitin ligase [140] that binds PAR. It is induced by NMDA-receptor activation and protects against glutamate NMDA receptor-mediated excitotoxicity *in vitro* and *in vivo* and against stroke, by interfering with parthanatos [141]. Another step in the parthanatos cascade beyond PARP activation that has received significant attention as a potential therapeutic target is the translocation of AIF from the mitochondria to the nucleus, perhaps due to its identification as the commitment point for cell death. Efforts are underway to develop high-throughput platforms capable of screening small molecules for compounds that block the AIF translocation event. These compounds will find exceptional relevance in CNS drugs development, as they can be developed as disease-modifying neuroprotectants for the management of neurological and neurodegenerative diseases in which parthanatos plays a significant role. Identification of AIF translocation blockers or of compounds targeting one or more of the other downstream steps in parthanatos that are devoid of any direct inhibition of the enzyme PARP will constitute a landmark development.

25.8 Conclusion

PARP-1-mediated cell death, or parthanatos, is a distinct cell-death pathway that is relevant to the pathophysiology of a range of conditions (as well as to their therapeutics development), including cancer and neurodegenerative diseases. The pathway uniquely features rapid activation of PARP-1, followed in sequence by synthesis and accumulation of the PAR polymer and translocation of AIF from the mitochondria to the nucleus, with the eventuality of death. Knowledge of the temporal and spatial regulation of parthanatos is now substantial, although there are aspects of its mechanisms that have yet to be unequivocally clarified. Future cell-death research will showcase enormous interest in the cross-talk that exists between parthanatos and other forms of cell death, and will use the knowledge gained to inform the development of better therapeutics for the treatment of conditions in which cell death is implicated.

References

1. Galluzzi L, Vitale I, Abrams JM, Alnemri ES, Baehrecke EH, Blagosklonny MV, et al. Molecular definitions of cell death subroutines: recommendations of the Nomenclature Committee on Cell Death 2012. *Cell Death Differ* 2012;**19**(1):107–20.
2. Nagley P, Higgins GC, Atkin JD, Beart PM. Multifaceted deaths orchestrated by mitochondria in neurones. *Biochim Biophys Acta* 2010;**1802**(1):167–85.
3. Galluzzi L, Bravo-San Pedro JM, Vitale I, Aaronson SA, Abrams JM, Adam D, et al. Essential versus accessory aspects of cell death: recommendations of the NCCD 2015. *Cell Death Differ* 2015;**22**(1):58–73.
4. Andrabi SA, Kim NS, Yu SW, Wang H, Koh DW, Sasaki M, et al. Poly(ADP-ribose) (PAR) polymer is a death signal. *Proc Natl Acad Sci USA* 2006;**103**(48):18 308–13.
5. Harraz MM, Dawson TM, Dawson VL. Advances in neuronal cell death 2007. *Stroke* 2008;**39**(2):286–8.
6. David KK, Andrabi SA, Dawson TM, Dawson VL. Parthanatos, a messenger of death. *Front Biosci* 2009;**14**:1116–28.
7. Andrabi SA, Dawson TM, Dawson VL. Mitochondrial and nuclear cross talk in cell death: parthanatos. *Ann NY Acad Sci* 2008;**1147**:233–41.
8. Fatokun AA, Dawson VL, Dawson TM. Parthanatos: mitochondrial-linked mechanisms and therapeutic opportunities. *Br J Pharmacol* 2014;**171**(8):2000–16.
9. Wang Y, Dawson VL, Dawson TM. Poly(ADP-ribose) signals to mitochondrial AIF: a key event in parthanatos. *Exp Neurol* 2009;**218**(2):193–202.
10. Ha HC, Snyder SH. Poly(ADP-ribose) polymerase is a mediator of necrotic cell death by ATP depletion. *Proc Natl Acad Sci USA* 1999;**96**(24):13 978–82.
11. Yu SW, Wang H, Poitras MF, Coombs C, Bowers WJ, Federoff HJ, et al. Mediation of poly(ADP-ribose) polymerase-1-dependent cell death by apoptosis-inducing factor. *Science* 2002;**297**(5579):259–63.
12. Wang H, Yu SW, Koh DW, Lew J, Coombs C, Bowers W, et al. Apoptosis-inducing factor substitutes for caspase executioners in NMDA-triggered excitotoxic neuronal death. *J Neurosci* 2004;**24**(48):10 963–73.
13. Pacher P, Szabo C. Role of the peroxynitrite-poly(ADP-ribose) polymerase pathway in human disease. *Am J Pathol* 2008;**173**(1):2–13.

14. Endres M, Wang ZQ, Namura S, Waeber C, Moskowitz MA. Ischemic brain injury is mediated by the activation of poly(ADP-ribose)polymerase. *J Cereb Blood Flow Metab* 1997;**17**(11):1143–51.
15. Joashi UC, Greenwood K, Taylor DL, Kozma M, Mazarakis ND, Edwards AD, Mehmet H. Poly(ADP ribose) polymerase cleavage precedes neuronal death in the hippocampus and cerebellum following injury to the developing rat forebrain. *Eur J Neurosci* 1999;**11**(1):91–100.
16. LaPlaca MC, Zhang J, Raghupathi R, Li JH, Smith F, Bareyre FM, et al. Pharmacologic inhibition of poly(ADP-ribose) polymerase is neuroprotective following traumatic brain injury in rats. *J Neurotrauma* 2001;**18**(4):369–76.
17. Love S, Barber R, Wilcock GK. Increased poly(ADP-ribosyl)ation of nuclear proteins in Alzheimer's disease. *Brain* 1999;**122**(Pt. 2):247–53.
18. Mandir AS, Przedborski S, Jackson-Lewis V, Wang ZQ, Simbulan-Rosenthal CM, Smulson ME, et al. Poly(ADP-ribose) polymerase activation mediates 1-methyl-4-phenyl-1 2,3,6-tetrahydropyridine (MPTP)-induced parkinsonism. *Proc Natl Acad Sci USA* 1999;**96**(10):5774–9.
19. Yamamoto H, Okamoto H. Protection by picolinamide, a novel inhibitor of poly (ADP-ribose) synthetase, against both streptozotocin-induced depression of proinsulin synthesis and reduction of NAD content in pancreatic islets. *Biochem Biophys Res Commun* 1980;**95**(1):474–81.
20. Zingarelli B, Szabo C, Salzman AL. Blockade of poly(ADP-ribose) synthetase inhibits neutrophil recruitment, oxidant generation, and mucosal injury in murine colitis. *Gastroenterology* 1999;**116**(2):335–45.
21. Miesel R, Kurpisz M, Kroger H. Modulation of inflammatory arthritis by inhibition of poly(ADP ribose) polymerase. *Inflammation* 1995;**19**(3):379–87.
22. Stubberfield CR, Cohen GM. NAD+ depletion and cytotoxicity in isolated hepatocytes. *Biochem Pharmacol* 1988;**37**(20):3967–74.
23. Mabley JG, Jagtap P, Perretti M, Getting SJ, Salzman AL, Virág L, et al. Anti-inflammatory effects of a novel, potent inhibitor of poly (ADP-ribose) polymerase. *Inflamm Res* 2001;**50**(11):561–9.
24. Zhang J, Dawson VL, Dawson TM, Snyder SH. Nitric oxide activation of poly(ADP-ribose) synthetase in neurotoxicity. *Science* 1994;**263**(5147):687–9.
25. Eliasson MJ, Sampei K, Mandir AS, Hurn PD, Traystman RJ, Bao J, et al. Poly(ADP-ribose) polymerase gene disruption renders mice resistant to cerebral ischemia. *Nat Med* 1997;**3**(10):1089–95.
26. Outeiro TF, Grammatopoulos TN, Altmann S, Amore A, Standaert DG, Hyman BT, Kazantsev AG. Pharmacological inhibition of PARP-1 reduces alpha-synuclein- and MPP+-induced cytotoxicity in Parkinson's disease *in vitro* models. *Biochem Biophys Res Commun* 2007;**357**(3):596–602.
27. Lee Y, Kang HC, Lee BD, Lee YI, Kim YP, Shin JH. Poly (ADP-ribose) in the pathogenesis of Parkinson's disease. *BMB Rep* 2014;**47**(8):424–32.
28. Vis JC, Schipper E, de Boer-van Huizen RT, Verbeek MM, de Waal RM, Wesseling P, et al. Expression pattern of apoptosis-related markers in Huntington's disease. *Acta Neuropathol* 2005;**109**(3):321–8.
29. Hivert B, Cerruti C, Camu W. Hydrogen peroxide-induced motoneuron apoptosis is prevented by poly ADP ribosyl synthetase inhibitors. *Neuroreport* 1998;**9**(8):1835–8.
30. Maier C, Scheuerle A, Hauser B, Schelzig H, Szabó C, Radermacher P, Kick J. The selective poly(ADP)ribose-polymerase 1 inhibitor INO1001 reduces spinal cord injury

during porcine aortic cross-clamping-induced ischemia/reperfusion injury. *Intensive Care Med* 2007;**33**(5):845–50.
31. de la Lastra CA, Villegas I, Sanchez-Fidalgo S. Poly(ADP-ribose) polymerase inhibitors: new pharmacological functions and potential clinical implications. *Curr Pharm Des* 2007;**13**(9):933–62.
32. Gero D, Szabo C. Poly(ADP-ribose) polymerase: a new therapeutic target? *Curr Opin Anaesthesiol* 2008;**21**(2):111–21.
33. Szabo C. Roles of poly(ADP-ribose) polymerase activation in the pathogenesis of diabetes mellitus and its complications. *Pharmacol Res* 2005;**52**(1):60–71.
34. Virag L, Szabo C. The therapeutic potential of poly(ADP-ribose) polymerase inhibitors. *Pharmacol Rev* 2002;**54**(3):375–429.
35. Zhao H, Ning J, Lemaire A, Koumpa FS, Sun JJ, Fung A, et al. Necroptosis and parthanatos are involved in remote lung injury after receiving ischemic renal allografts in rats. *Kidney Int* 2015;**87**(4):738–48.
36. Hottiger MO, Hassa PO, Luscher B, Schuler H, Koch-Nolte F. Toward a unified nomenclature for mammalian ADP-ribosyltransferases. *Trends Biochem Sci* 2010;**35**(4):208–19.
37. Smith S. The world according to PARP. *Trends Biochem Sci* 2001;**26**(3):174–9.
38. Hong SJ, Dawson TM, Dawson VL. Nuclear and mitochondrial conversations in cell death: PARP-1 and AIF signaling. *Trends Pharmacol Sci* 2004;**25**(5):259–64.
39. Krietsch J, Caron MC, Gagné JP, Ethier C, Vignard J, Vincent M, et al. PARP activation regulates the RNA-binding protein NONO in the DNA damage response to DNA double-strand breaks. *Nucleic Acids Res* 2012;**40**(20):10 287–301.
40. D'Amours D, Desnoyers S, D'Silva I, Poirier GG. Poly(ADP-ribosyl)ation reactions in the regulation of nuclear functions. *Biochem J* 1999;**342**(Pt. 2):249–68.
41. Chiarugi A. Poly(ADP-ribose) polymerase: killer or conspirator? The "suicide hypothesis" revisited. *Trends Pharmacol Sci* 2002;**23**(3):122–9.
42. Kraus WL, Lis JT. PARP goes transcription. *Cell* 2003;**113**(6):677–83.
43. Gagné JP, Isabelle M, Lo KS, Bourassa S, Hendzel MJ, Dawson VL, et al. Proteome-wide identification of poly(ADP-ribose) binding proteins and poly(ADP-ribose)-associated protein complexes. *Nucleic Acids Res* 2008;**36**(22):6959–76.
44. Kameshita I, Matsuda Z, Taniguchi T, Shizuta Y. Poly (ADP-ribose) synthetase. Separation and identification of three proteolytic fragments as the substrate-binding domain, the DNA-binding domain, and the automodification domain. *J Biol Chem* 1984;**259**(8):4770–6.
45. Sousa FG, Matuo R, Soares DG, Escargueil AE, Henriques JA, Larsen AK, Saffi J. PARPs and the DNA damage response. *Carcinogenesis* 2012;**33**(8):1433–40.
46. Shall S, de Murcia G. Poly(ADP-ribose) polymerase-1: what have we learned from the deficient mouse model? *Mutat Res* 2000;**460**(1):1–15.
47. Smulson ME, Simbulan-Rosenthal CM, Boulares AH, Yakovlev A, Stoica B, Iyer S, et al. Roles of poly(ADP-ribosyl)ation and PARP in apoptosis, DNA repair, genomic stability and functions of p53 and E2F-1. *Adv Enzyme Regul* 2000;**40**:183–215.
48. Kanai M, Uchida M, Hanai S, Uematsu N, Uchida K, Miwa M. Poly(ADP-ribose) polymerase localizes to the centrosomes and chromosomes. *Biochem Biophys Res Commun* 2000;**278**(2):385–9.
49. Dawson VL, Dawson TM, Bartley DA, Uhl GR, Snyder SH. Mechanisms of nitric oxide-mediated neurotoxicity in primary brain cultures. *J Neurosci* 1993;**13**(6):2651–61.

50 Dawson VL, Dawson TM, London ED, Bredt DS, Snyder SH. Nitric oxide mediates glutamate neurotoxicity in primary cortical cultures. *Proc Natl Acad Sci USA* 1991;**88**(14):6368–71.
51 Dawson VL, Kizushi VM, Huang PL, Snyder SH, Dawson TM. Resistance to neurotoxicity in cortical cultures from neuronal nitric oxide synthase-deficient mice. *J Neurosci* 1996;**16**(8):2479–87.
52 Gonzalez-Zulueta M, Ensz LM, Mukhina G, Mukhina G, Lebovitz RM, Zwacka RM, et al. Manganese superoxide dismutase protects nNOS neurons from NMDA and nitric oxide-mediated neurotoxicity. *J Neurosci* 1998;**18**(6):2040–55.
53 Xia Y, Dawson VL, Dawson TM, Snyder SH, Zweier JL. Nitric oxide synthase generates superoxide and nitric oxide in arginine-depleted cells leading to peroxynitrite-mediated cellular injury. *Proc Natl Acad Sci USA* 1996;**93**(13):6770–4.
54 Stone TW, Addae JI. The pharmacological manipulation of glutamate receptors and neuroprotection. *Eur J Pharmacol* 2002;**447**(2–3):285–96.
55 Berger NA. Poly(ADP-ribose) in the cellular response to DNA damage. *Radiat Res* 1985;**101**(1):4–15.
56 Berger NA, Berger SJ. Metabolic consequences of DNA damage: the role of poly(ADP-ribose) polymerase as mediator of the suicide response. *Basic Life Sci* 1986;**38**:357–63.
57 Berger NA, Berger SJ, Catino DM, Petzold SJ, Robins RK. Modulation of nicotinamide adenine dinucleotide and poly(adenosine diphosphoribose) metabolism by the synthetic "C" nucleoside analogs, tiazofurin and selenazofurin. A new strategy for cancer chemotherapy. *J Clin Invest* 1985;**75**(2):702–9.
58 Berger NA, Sims JL, Catino DM, Berger SJ. Poly(ADP-ribose) polymerase mediates the suicide response to massive DNA damage: studies in normal and DNA-repair defective cells. *Princess Takamatsu Symp* 1983;**13**:219–26.
59 Andrabi SA, Umanah GK, Chang C, Stevens DA, Karuppagounder SS, Gagné JP, et al. Poly(ADP-ribose) polymerase-dependent energy depletion occurs through inhibition of glycolysis. *Proc Natl Acad Sci USA* 2014;**111**(28):10 209–14.
60 Goto S, Xue R, Sugo N, Sawada M, Blizzard KK, Poitras MF, et al. Poly(ADP-ribose) polymerase impairs early and long-term experimental stroke recovery. *Stroke* 2002;**33**(4):1101–6.
61 Zhou Y, Feng X, Koh DW. Activation of cell death mediated by apoptosis-inducing factor due to the absence of poly(ADP-ribose) glycohydrolase. *Biochemistry* 2011;**50**(14):2850–9.
62 Whitacre CM, Hashimoto H, Tsai ML, Chatterjee S, Berger SJ, Berger NA. Involvement of NAD-poly(ADP-ribose) metabolism in p53 regulation and its consequences. *Cancer Res* 1995;**55**(17):3697–701.
63 Davidovic L, Vodenicharov M, Affar EB, Poirier GG. Importance of poly(ADP-ribose) glycohydrolase in the control of poly(ADP-ribose) metabolism. *Exp Cell Res* 2001;**268**(1):7–13.
64 Ame JC, Apiou F, Jacobson EL, Jacobson MK. Assignment of the poly(ADP-ribose) glycohydrolase gene (PARG) to human chromosome 10q11.23 and mouse chromosome 14B by in situ hybridization. *Cytogenet Cell Genet* 1999;**85**(3–4):269–70.
65 Meyer-Ficca ML, Meyer RG, Coyle DL, Jacobson EL, Jacobson MK. Human poly(ADP-ribose) glycohydrolase is expressed in alternative splice variants yielding isoforms that localize to different cell compartments. *Exp Cell Res* 2004;**297**(2):521–32.

66 Winstall E, Affar EB, Shah R, Bourassa S, Scovassi AI, Poirier GG. Poly(ADP-ribose) glycohydrolase is present and active in mammalian cells as a 110-kDa protein. *Exp Cell Res* 1999;**246**(2):395–8.

67 Bonicalzi ME, Vodenicharov M, Coulombe M, Gagne JP, Poirier GG. Alteration of poly(ADP-ribose) glycohydrolase nucleocytoplasmic shuttling characteristics upon cleavage by apoptotic proteases. *Biol Cell* 2003;**95**(9):635–44.

68 Haince JF, Ouellet ME, McDonald D, Hendzel MJ, Poirier GG. Dynamic relocation of poly(ADP-ribose) glycohydrolase isoforms during radiation-induced DNA damage. *Biochim Biophys Acta* 2006;**1763**(2):226–37.

69 Meyer RG, Meyer-Ficca ML, Whatcott CJ, Jacobson EL, Jacobson MK. Two small enzyme isoforms mediate mammalian mitochondrial poly(ADP-ribose) glycohydrolase (PARG) activity. *Exp Cell Res* 2007;**313**(13):2920–36.

70 Burns DM, Ying W, Kauppinen TM, Zhu K, Swanson RA. Selective down-regulation of nuclear poly(ADP-ribose) glycohydrolase. *PLoS One* 2009;**4**(3):e4896.

71 Blenn C, Althaus FR, Malanga M. Poly(ADP-ribose) glycohydrolase silencing protects against H2O2-induced cell death. *Biochem J* 2006;**396**(3):419–29.

72 Cozzi A, Cipriani G, Fossati S, Faraco G, Formentini L, Min W, et al. Poly(ADP-ribose) accumulation and enhancement of postischemic brain damage in 110-kDa poly(ADP-ribose) glycohydrolase null mice. *J Cereb Blood Flow Metab* 2006;**26**(5):684–95.

73 Koh DW, Lawler AM, Poitras MF, Sasaki M, Wattler S, Nehls MC, et al. Failure to degrade poly(ADP-ribose) causes increased sensitivity to cytotoxicity and early embryonic lethality. *Proc Natl Acad Sci USA* 2004;**101**(51):17 699–704.

74 Hanai S, Kanai M, Ohashi S, Okamoto K, Yamada M, Takahashi H, Miwa M. Loss of poly(ADP-ribose) glycohydrolase causes progressive neurodegeneration in Drosophila melanogaster. *Proc Natl Acad Sci USA* 2004;**101**(1):82–6.

75 Yu SW, Andrabi SA, Wang H, Kim NS, Poirier GG, Dawson TM, Dawson VL. Apoptosis-inducing factor mediates poly(ADP-ribose) (PAR) polymer-induced cell death. *Proc Natl Acad Sci USA* 2006;**103**(48):18 314–19.

76 Cregan SP, Fortin A, MacLaurin JG, Callaghan SM, Cecconi F, Yu SW, et al. Apoptosis-inducing factor is involved in the regulation of caspase-independent neuronal cell death. *J Cell Biol* 2002;**158**(3):507–17.

77 Susin SA, Lorenzo HK, Zamzami N, Marzo I, Snow BE, Brothers GM, et al. Molecular characterization of mitochondrial apoptosis-inducing factor. *Nature* 1999;**397**(6718):441–6.

78 Yu SW, Wang Y, Frydenlund DS, Ottersen OP, Dawson VL, Dawson TM. Outer mitochondrial membrane localization of apoptosis-inducing factor: mechanistic implications for release. *ASN Neuro* 2009;**1**(5): pii:e00021.

79 Cao G, Xing J, Xiao X, Liou AK, Gao Y, Yin XM, et al. Critical role of calpain I in mitochondrial release of apoptosis-inducing factor in ischemic neuronal injury. *J Neurosci* 2007;**27**(35):9278–93.

80 Otera H, Ohsakaya S, Nagaura Z, Ishihara N, Mihara K. Export of mitochondrial AIF in response to proapoptotic stimuli depends on processing at the intermembrane space. *EMBO J* 2005;**24**(7):1375–86.

81 Delettre C, Yuste VJ, Moubarak RS, Bras M, Lesbordes-Brion JC, Petres S, et al. AIFsh, a novel apoptosis-inducing factor (AIF) pro-apoptotic isoform with potential pathological relevance in human cancer. *J Biol Chem* 2006;**281**(10):6413–27.

82. Delettre C, Yuste VJ, Moubarak RS, Bras M, Robert N, Susin SA. Identification and characterization of AIFsh2, a mitochondrial apoptosis-inducing factor (AIF) isoform with NADH oxidase activity. *J Biol Chem* 2006;**281**(27):18 507–18.
83. Loeffler M, Daugas E, Susin SA, Zamzami N, Metivier D, Nieminen AL, et al. Dominant cell death induction by extramitochondrially targeted apoptosis-inducing factor. *FASEB J* 2001;**15**(3):758–67.
84. Lorenzo HK, Susin SA. Therapeutic potential of AIF-mediated caspase-independent programmed cell death. *Drug Resist Updat* 2007;**10**(6):235–55.
85. Maté MJ, Ortiz-Lombardía M, Boitel B, Haouz A, Tello D, Susin SA, et al. The crystal structure of the mouse apoptosis-inducing factor AIF. *Nat Struct Biol* 2002;**9**(6):442–6.
86. Ye H, Cande C, Stephanou NC, Jiang S, Gurbuxani S, Larochette N, et al. DNA binding is required for the apoptogenic action of apoptosis inducing factor. *Nat Struct Biol* 2002;**9**(9):680–4.
87. Klein JA, Longo-Guess CM, Rossmann MP, Seburn KL, Hurd RE, Frankel WN, et al. The harlequin mouse mutation downregulates apoptosis-inducing factor. *Nature* 2002;**419**(6905):367–74.
88. Cheung EC, Melanson-Drapeau L, Cregan SP, Vanderluit JL, Ferguson KL, McIntosh WC, et al. Apoptosis-inducing factor is a key factor in neuronal cell death propagated by BAX-dependent and BAX-independent mechanisms. *J Neurosci* 2005;**25**(6):1324–34.
89. Joza N, Oudit GY, Brown D, Bénit P, Kassiri Z, Vahsen N, et al. Muscle-specific loss of apoptosis-inducing factor leads to mitochondrial dysfunction, skeletal muscle atrophy, and dilated cardiomyopathy. *Mol Cell Biol* 2005;**25**(23): 10 261–72.
90. Pospisilik JA, Knauf C, Joza N, Benit P, Orthofer M, Cani PD, et al. Targeted deletion of AIF decreases mitochondrial oxidative phosphorylation and protects from obesity and diabetes. *Cell* 2007;**131**(3):476–91.
91. Wang Y, Kim NS, Haince JF, Kang HC, David KK, Andrabi SA, et al. Poly(ADP-ribose) (PAR) binding to apoptosis-inducing factor is critical for PAR polymerase-1-dependent cell death (parthanatos). *Sci Signal* 2011;**4**(167): ra20.
92. Cregan SP, Dawson VL, Slack RS. Role of AIF in caspase-dependent and caspase-independent cell death. *Oncogene* 2004;**23**(16):2785–96.
93. Virag L. Structure and function of poly(ADP-ribose) polymerase-1: role in oxidative stress-related pathologies. *Curr Vasc Pharmacol* 2005;**3**(3):209–14.
94. Yu SW, Wang H, Dawson TM, Dawson VL. Poly(ADP-ribose) polymerase-1 and apoptosis inducing factor in neurotoxicity. *Neurobiol Dis* 2003;**14**(3): 303–17.
95. Culmsee C, Zhu C, Landshamer S, Becattini B, Wagner E, Pellecchia M, et al. Apoptosis-inducing factor triggered by poly(ADP-ribose) polymerase and Bid mediates neuronal cell death after oxygen-glucose deprivation and focal cerebral ischemia. *J Neurosci* 2005;**25**(44):10 262–72.
96. Yuan M, Siegel C, Zeng Z, Li J, Liu F, McCullough LD. Sex differences in the response to activation of the poly (ADP-ribose) polymerase pathway after experimental stroke. *Exp Neurol* 2009;**217**(1):210–18.
97. Zhu C, Wang X, Deinum J, Huang Z, Gao J, Modjtahedi N, et al. Cyclophilin A participates in the nuclear translocation of apoptosis-inducing factor in neurons after cerebral hypoxia-ischemia. *J Exp Med* 2007;**204**(8):1741–8.

98. Cande C, Vahsen N, Kouranti I, Schmitt E, Daugas E, Spahr C, et al. AIF and cyclophilin A cooperate in apoptosis-associated chromatinolysis. *Oncogene* 2004;**23**(8):1514–21.
99. Wang X, Yang C, Chai J, Shi Y, Xue D. Mechanisms of AIF-mediated apoptotic DNA degradation in *Caenorhabditis elegans*. *Science* 2002;**298**(5598): 1587–92.
100. David KK, Sasaki M, Yu SW, Dawson TM, Dawson VL. EndoG is dispensable in embryogenesis and apoptosis. *Cell Death Differ* 2006;**13**(7):1147–55.
101. Xu Z, Zhang J, David KK, Yang ZJ, Li X, Dawson TM, et al. Endonuclease G does not play an obligatory role in poly(ADP-ribose) polymerase-dependent cell death after transient focal cerebral ischemia. *Am J Physiol* 2010;**299**(1):R215–21.
102. Bano D, Young KW, Guerin CJ, Lefeuvre R, Rothwell NJ, Naldini L, et al. Cleavage of the plasma membrane Na+/Ca2+ exchanger in excitotoxicity. *Cell* 2005;**120**(2):275–85.
103. Polster BM, Basanez G, Etxebarria A, Hardwick JM, Nicholls DG. Calpain I induces cleavage and release of apoptosis-inducing factor from isolated mitochondria. *J Biol Chem* 2005;**280**(8):6447–54.
104. Moubarak RS, Yuste VJ, Artus C, Bouharrour A, Greer PA, Menissier-de Murcia J, Susin SA. Sequential activation of poly(ADP-ribose) polymerase 1, calpains, and Bax is essential in apoptosis-inducing factor-mediated programmed necrosis. *Mol Cell Biol* 2007;**27**(13):4844–62.
105. Wang Y, Kim NS, Li X, Greer PA, Koehler RC, Dawson VL, Dawson TM. Calpain activation is not required for AIF translocation in PARP-1-dependent cell death (parthanatos). *J Neurochem* 2009;**110**(2):687–96.
106. Droit A, Hunter JM, Rouleau M, Ethier C, Picard-Cloutier A, Bourgais D, Poirier GG. PARPs database: a LIMS systems for protein-protein interaction data mining or laboratory information management system. *BMC Bioinformatics* 2007; **8**:483.
107. Peralta-Leal A, Rodríguez-Vargas JM, Aguilar-Quesada R, Rodríguez MI, Linares JL, de Almodóvar MR, Oliver FJ. PARP inhibitors: new partners in the therapy of cancer and inflammatory diseases. *Free Radic Biol Med* 2009;**47**(1):13–26.
108. Jagtap P, Szabo C. Poly(ADP-ribose) polymerase and the therapeutic effects of its inhibitors. *Nat Rev Drug Discov* 2005;**4**(5):421–40.
109. Zhu C, Qiu L, Wang X, Hallin U, Candé C, Kroemer G, et al. Involvement of apoptosis-inducing factor in neuronal death after hypoxia-ischemia in the neonatal rat brain. *J Neurochem* 2003;**86**(2):306–17.
110. Zhu C, Wang X, Huang Z, Qiu L, Xu F, Vahsen N, et al. Apoptosis-inducing factor is a major contributor to neuronal loss induced by neonatal cerebral hypoxia-ischemia. *Cell Death Differ* 2007;**14**(4):775–84.
111. Li X, Nemoto M, Xu Z, Yu SW, Shimoji M, Andrabi SA, et al. Influence of duration of focal cerebral ischemia and neuronal nitric oxide synthase on translocation of apoptosis-inducing factor to the nucleus. *Neuroscience* 2007;**144**(1):56–65.
112. Chaitanya GV, Babu P.P. Multiple apoptogenic proteins are involved in the nuclear translocation of Apoptosis Inducing Factor during transient focal cerebral ischemia in rat. *Brain Res* 2008;**1246**:178–90.
113. Ravagnan L, Gurbuxani S, Susin SA, Maisse C, Daugas E, Zamzami N, et al. Heat-shock protein 70 antagonizes apoptosis-inducing factor. *Nat Cell Biol* 2001;**3**(9):839–43.

114 Gurbuxani S, Schmitt E, Cande C, Parcellier A, Hammann A, Daugas E, et al. Heat shock protein 70 binding inhibits the nuclear import of apoptosis-inducing factor. *Oncogene* 2003;**22**(43):6669–78.
115 Sutton VR, Davis JE, Cancilla M, Johnstone RW, Ruefli AA, Sedelies K, et al. Initiation of apoptosis by granzyme B requires direct cleavage of bid, but not direct granzyme B-mediated caspase activation. *J Exp Med* 2000;**192**(10):1403–14.
116 Zhou Q, Zhao J, Stout JG, Luhm RA, Wiedmer T, Sims PJ. Molecular cloning of human plasma membrane phospholipid scramblase. A protein mediating transbilayer movement of plasma membrane phospholipids. *J Biol Chem* 1997;**272**(29):18 240–4.
117 Wiedmer T, Zhou Q, Kwoh DY, Sims PJ. Identification of three new members of the phospholipid scramblase gene family. *Biochim Biophys Acta* 2000;**1467**(1):244–53.
118 Frasch SC, Henson PM, Kailey JM, Richter DA, Janes MS, Fadok VA, Bratton DL. Regulation of phospholipid scramblase activity during apoptosis and cell activation by protein kinase Cdelta. *J Biol Chem* 2000;**275**(30):23 065–73.
119 Wang X, Wang J, Gengyo-Ando K, Gu L, Sun CL, Yang C, et al. *C. elegans* mitochondrial factor WAH-1 promotes phosphatidylserine externalization in apoptotic cells through phospholipid scramblase SCRM-1. *Nat Cell Biol* 2007;**9**(5):541–9.
120 Ndebele K, Gona P, Jin TG, Benhaga N, Chalah A, Degli-Esposito M, Khosravi-Far R. Tumor necrosis factor (TNF)-related apoptosis-inducing ligand (TRAIL) induced mitochondrial pathway to apoptosis and caspase activation is potentiated by phospholipid scramblase-3. *Apoptosis* 2008;**13**(7):845–56.
121 Bryant HE, Schultz N, Thomas HD, Parker KM, Flower D, Lopez E, et al. Specific killing of *BRCA2*-deficient tumours with inhibitors of poly(ADP-ribose) polymerase. *Nature* 2005;**434**(7035):913–17.
122 Farmer H, McCabe N, Lord CJ, Tutt AN, Johnson DA, Richardson TB, et al. Targeting the DNA repair defect in *BRCA* mutant cells as a therapeutic strategy. *Nature* 2005;**434**(7035):917–21.
123 McCabe N, Lord CJ, Tutt AN, Martin NM, Smith GC, Ashworth A. BRCA2-deficient CAPAN-1 cells are extremely sensitive to the inhibition of poly (ADP-ribose) polymerase: an issue of potency. *Cancer Biol Ther* 2005;**4**(9):934–6.
124 Peralta-Leal A, Rodriguez MI, Oliver FJ. Poly(ADP-ribose)polymerase-1 (PARP-1) in carcinogenesis: potential role of PARP inhibitors in cancer treatment. *Clin Transl Oncol* 2008;**10**(6):318–23.
125 Turner NC, Lord CJ, Iorns E, Brough R, Swift S, Elliott R, et al. A synthetic lethal siRNA screen identifying genes mediating sensitivity to a PARP inhibitor. *EMBO J* 2008;**27**(9):1368–77.
126 Shall S, Gaymes T, Farzaneh F, Curtin N, Mufti GJ. The use of PARP inhibitors in cancer therapy: use as adjuvant with chemotherapy or radiotherapy; use as a single agent in susceptible patients; techniques used to identify susceptible patients. In Tulin AV (ed.). *Poly(ADP-Ribose) Polymerase*. Berlin: Springer.
127 Dedes KJ, Wilkerson PM, Wetterskog D, Weigelt B, Ashworth A, Reis-Filho JS. Synthetic lethality of PARP inhibition in cancers lacking *BRCA1* and *BRCA2* mutations. *Cell Cycle* 2011;**10**(8):1192–9.
128 Eltze T, Boer R, Wagner T, Weinbrenner S, McDonald MC, Thiemermann C, et al. Imidazoquinolinone, imidazopyridine, and isoquinolindione derivatives as novel and potent inhibitors of the poly(ADP-ribose) polymerase (PARP): a comparison with standard PARP inhibitors. *Mol Pharmacol* 2008;**74**(6):1587–98.

129. Geraets L, Moonen HJ, Brauers K, Gottschalk RW, Wouters EF, Bast A, Hageman GJ. Flavone as PARP-1 inhibitor: its effect on lipopolysaccharide induced gene-expression. *Eur J Pharmacol* 2007;**573**(1–3):241–8.
130. Fatokun AA, Liu JO, Dawson VL, Dawson TM. Identification through high-throughput screening of 4′-methoxyflavone and 3′,4′-dimethoxyflavone as novel neuroprotective inhibitors of parthanatos. *Br J Pharmacol* 2013;**169**(6):1263–78.
131. Ishida J, Yamamoto H, Kido Y, Kamijo K, Murano K, Miyake H, et al. Discovery of potent and selective PARP-1 and PARP-2 inhibitors: SBDD analysis via a combination of X-ray structural study and homology modeling. *Bioorg Med Chem* 2006;**14**(5):1378–90.
132. Iwashita A, Hattori K, Yamamoto H, Ishida J, Kido Y, Kamijo K, et al. Discovery of quinazolinone and quinoxaline derivatives as potent and selective poly(ADP-ribose) polymerase-1/2 inhibitors. *FEBS Lett* 2005;**579**(6):1389–93.
133. Pellicciari R, Camaioni E, Costantino G, Formentini L, Sabbatini P, Venturoni F, et al. On the way to selective PARP-2 inhibitors. Design, synthesis, and preliminary evaluation of a series of isoquinolinone derivatives. *ChemMedChem* 2008;**3**(6):914–23.
134. Green DR, Kroemer G. Pharmacological manipulation of cell death: clinical applications in sight? *J Clin Invest* 2005;**115**(10):2610–17.
135. Jagtap PG, Baloglu E, Southan GJ, Mabley JG, Li H, Zhou J, et al. Discovery of potent poly(ADP-ribose) polymerase-1 inhibitors from the modification of indeno[1,2-c]isoquinolinone. *J Med Chem* 2005;**48**(16):5100–3.
136. Pacher P, Szabo C. Role of poly(ADP-ribose) polymerase 1 (PARP-1) in cardiovascular diseases: the therapeutic potential of PARP inhibitors. *Cardiovasc Drug Rev* 2007;**25**(3):235–60.
137. Rouleau M, Patel A, Hendzel MJ, Kaufmann SH, Poirier GG. PARP inhibition: PARP1 and beyond. *Nat Rev Cancer* 2010;**10**(4):293–301.
138. Anon. Olaparib approved for advanced ovarian cancer. *Cancer Discovery* 2015;**5**(3):218–20.
139. Graziani G, Szabo C. Clinical perspectives of PARP inhibitors. *Pharmacol Res* 2005;**52**(1):109–18.
140. Kang HC, Lee YI, Shin JH, Andrabi SA, Chi Z, Gagné JP, et al. Iduna is a poly(ADP-ribose) (PAR)-dependent E3 ubiquitin ligase that regulates DNA damage. *Proc Natl Acad Sci USA* 2011;**108**(34):14 103–8.
141. Andrabi SA, Kang HC, Haince JF, Lee YI, Zhang J, Chi Z, et al. Iduna protects the brain from glutamate excitotoxicity and stroke by interfering with poly(ADP-ribose) polymer-induced cell death. *Nat Med* 2011;**17**(6):692–9.

26

Methuosis: Drinking to Death

Madeeha Aqil

Department of Oral Medicine and Diagnostic Sciences, University of Illinois at Chicago, Chicago, IL, USA

Abbreviations

AIF	apoptosis-inducing factor
cyt c	cytochrome c
ER	endoplasmic reticulum
ERK	growth factor-regulated extracellular signal-related kinase
GBM	glioblastoma multiforme
GTPase	guanosine triphosphatase
LMP	lysosomal membrane permeabilization
MOMP	mitochondrial outer-membrane permeabilization
PARP	poly ADP ribose polymerase
PCD	programmed cell death
PI3-K	phosphatidylinositol 3-kinase

26.1 Introduction

Apoptosis is the most well characterized form of programmed cell death (PCD) [1,2]. However, a number of non-apoptotic cell-death pathways are now recognized as playing important roles in embryonic development, neurological disorders, and cancer treatment [3]. Most of these pathways function independently of caspase activation [4,5]. The most widely studied form of non-apoptotic cell death is autophagic cell death [6]. Its characteristic morphologic feature is accumulation of autophagosomes and lysosomes, which engulf organelles and cytoplasm, ultimately eating up the whole cell [7]. Recently, the Maltese group described an unusual form of cell death termed "methuosis" in glioblastoma cells, which is caused by macropinocytosis dysregulation [8]. Macropinocytosis is a physiologic process by which cells internalize extracellular fluid into vesicles called macropinosomes [9]. However, overstimulation of macropinocytosis, as well as defects in endocytic-vesicle trafficking and recycling, can lead to the accumulation of progressively larger vacuoles, eventually disrupting membrane integrity and leading to cell death [10–12]. This chapter describes the discovery of methuosis, details

Apoptosis and Beyond: The Many Ways Cells Die, First Edition. Edited by James Radosevich.
© 2018 John Wiley & Son Inc. Published 2018 by John Wiley & Son Inc.

its mechanisms, compares it with other types of cell death, and discusses its importance for the development of new cancer therapeutics.

26.2 Discovery of Methuosis

The first report of non-apoptotic cell death induced by the ectopic expression of the constitutively active oncoprotein, H-Ras (G12V), was made in 1999, by Chi et al. [4]. Instead of cell proliferation, these authors observed that glioblastoma and gastric carcinoma cells accumulated multiple large, phase-lucent cytoplasmic vacuoles, and eventually underwent caspase-independent cell death. They described this as "autophagic cell death" [4]. However, there were no confirmatory studies to show that the vacuoles induced by activated Ras are autophagosomes. Recently, the Maltese group characterized these vacuoles using electron microscopy and established their origin from plasma-membrane projections [8]. They do not exhibit double-membrane morphology or degradative contents, as would be expected for autophagosomes. In fact, the Ras-induced vacuoles are derived from macropinosomes and continually expand, leading to cell rupture. Immunoflourescence studies show that epitope-tagged G12V localizes to their membranes. Studies with fluid phase-fluorescent markers tracking vacuole movement inside the cell revealed that they do not recycle back to the cell surface or fuse with lysosomes [8,13]. Based on these findings, the Maltese group proposed a model wherein constitutive activation of Ras oncoprotein leads to extensive endosomal vacuolization, not just because of the increase in macropinocytosis, but also because of the disruption in normal endocytic trafficking and lysosomal fusion of macropinsosmes [14].

Glioblastoma cells expressing G12V lose viability over a period of 4–6 days, with morphological features distinct from those of apoptosis. In particular, the cytoplasmic space is almost completely occupied by massive fluid-filled vacuoles, and there is no chromatin condensation or fragmentation in the nuclei [15]. Interestingly, inhibiting the expression of autophagy proteins and treating the cells with caspase or necroptosis inhibitor did not prevent Ras-induced cell death. Thus, based on these findings, Maltese and Overmeyer [15] proposed this to be a novel form of cell death associated with hyperstimulation of fluid uptake and accumulation of swollen macropinosomes. Since it is linked to macropinocytosis, often referred to as "cell drinking," they coined the term "methuosis," derived from the Greek *methuo* ("drink to intoxification"), to describe this form of cell death.

26.3 Mechanism of Methuosis

The key downstream mediator in activated Ras-induced cell death is a member of the Rho guanosine triphosphatase (GTPase) family, namely Rac1 GTPase, as demonstrated by Bhanot et al. [13]. Rac 1 is known to play an important role in the initial steps of macropinosome formation and trafficking. As the ectopic Ras expression exceeds the level required for activation of the canonical growth-stimulatory phosphatidylinositol 3-kinase (PI3-K) and growth factor-regulated extracellular signal-related kinase (ERK) pathways, there is an increase in levels of endogenous activated Rac1, and extensive cytoplasmic vacuolization is observed. On the other hand, inhibiting Rac 1 activation delays the accumulation of Ras-induced vacuoles. Moreover, the morphological effects of activated H-Ras expression are observed with ectopic expression of activated Rac-1 [8,15]. Bhanot et al. further demonstrated that Arf6, a GTPase involved in macropinosome recycling, is also affected by constitutive Ras expression and is involved in the vacuolar cytopathology observed in methuosis [13]. They found that in glioma cells expressing H-Ras, there is an inverse relationship between the expression of activated Rac1 and that of Arf6: as the relative amount of Rac1 GTP increased the amount of Arf-6 GTP decreased [13]. This was regulated by Rac-1-mediated activation of GIT-1, which controls Arf6 levels. Thus, it is proposed that Ras-induced extreme vacuolization is a result of the combined effect of an increase in macropinososme formation and a decrease in macropinosome recycling. However, the inability of LAMP-1-positive vacuoles to fuse with lysosome and eventually degrade also suggests additional defects at the late endosome–lysososome fusion step [14]. A working model of methuosis is depicted in Figure 26.1.

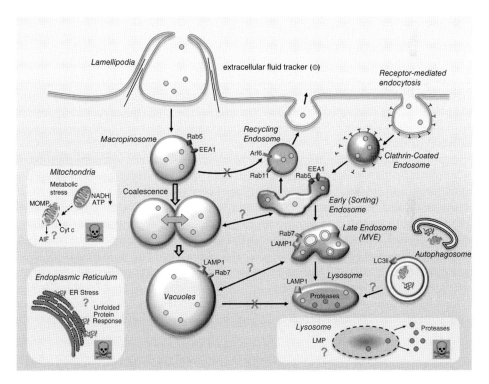

Figure 26.1 A working model of methuosis. Large fluid-filled vacuoles are formed from nascent macropinosomes, as they coalesce after entering the cell. These vacuoles are unable to recycle back like normal macropinosomes and acquire some characteristics of late endosomes (Rab7 and LAMP1). However, they do not label with acidotropic markers (lysotracker and acridine orange) or fuse with pre-existing lysosomes. It is unknown if the vacuoles can disrupt clathrin-dependent protein trafficking, merge with early or late endosomes, or affect macroautophagy inside the cell. A key question is how the accumulation of vacuoles eventually leads to cell death. Some contributing factors might be perturbed mitochondrial energy metabolism and mitochondrial outer-membrane permeabilization (MOMP), endoplasmic reticulum (ER) stress, lysosomal membrane permeabilization (LMP), apoptosis-inducing factor (AIF), and cytochrome c (cyt c). *Source:* Maltese and Overmeyer 2014 [14]. Reproduced with permission from Elsevier.

Apart from glioma cells, cell death induced by Ras overexpression is also observed in gastric carcinoma [4] and osteosarcoma [13] cells *in vitro*, suggesting that this mechanism is not specific to brain cancers [16]. Finally, studies of *Aspergillus fumigatus* show that expression of activated Ras causes extreme vacuolar expansion and lysis of fungal hyphae [17]. Hence, it is believed that an analogous Ras-mediated methuosis pathway might be conserved in lower eukaryotes [18].

26.4 Methuosis Induced by Synthetic Small Molecules: Indole-Based Chalcones

After initial reports describing methuosis in glioblastoma multiforme (GBM) cells, studies were done to identify small molecules that could induce methuosis. In 2011, Overmeyer et al. [19] described a synthetic indole-based chalcone termed MIPP: an acronym for 3-(2-methyl-1H indol-3-yl)-1-(4-pyridinyl)-2-propen-1-one. MIPP triggered rapid accumulation of phase-lucent cytoplasmic vacuoles at low micromolar concentrations in GBM cells and other cell lines [19]. Subsequently, the researchers developed a more potent analog of MIPP termed MOMIPP, with a 5-methoxy group added to the indole ring [20]. As revealed by time-lapse microscopy experiments, macropinosmes start entering cells between 13 and 80 minutes after MIPP addition and rapidly coalesce to form larger vacuoles [20]. Further experiments with fluorescent fluid phase tracers confirmed their endocytic origin. The morphological characteristics of dying cells were similar to Ras-induced methuosis and supported a non-apoptotic cell-death mechanism [21].

Structural studies with these synthetic molecules revealed that altering the position of the nitrogen in the pyridinyl moiety from para to meta configuration eliminated the vacuolization and cytotoxicity effects [21]. This implies that methuosis induced by these molecules results from their interaction with one or more specific protein targets: plausibly, the regulatory and structural components of the early or late endocytic pathways. As with Ras, the vacuoles induced by MIPP are distinct from early endosomes, autophagosomes, and recycling endosomes, and acquire late-endosomal markers Rab7 and LAMP1 [19]. However, MIPP-induced vacuolization ceases after a few hours as larger vacuoles accumulate, suggesting a disruption in new macropinosome generation, in addition to recycling and lysosomal fusion steps. Further, the activation states of Rac1 and Arf6 do not change, suggesting that MIPP probably targets the downstream molecules [21].

26.5 Comparison with Other Cell-Death Mechanisms

Several distinct non-apoptotic forms of cell death have been characterized. These include the type II or autophagic cell deaths oncosis, paraptosis, and necroptosis. In this section, we will briefly discuss how methuosis compares with some of these other mechanisms.

26.5.1 Comparison with Autophagy

As already discussed, the morphological features of activated Ras-induced vesicles are distinct from those of autophagosomes. Specifically, these vacuoles are phase- and electron-lucent and are bounded by a single membrane, rather than the characteristic double membrane of autophagosomes [8]. Interestingly, autohpagy is upregulated in activated Ras-expressing GBM cells, as revealed by autophagy-marker LC3II localization and expression studies conducted using immunoflourescence microscopy [10]. Nevertheless, the vesicles labeled with LC3II are smaller and are spatially separate from the larger phase-lucent vacuoles. Further inhibiting autophagy-protein Beclin-1 expression does not prevent cell death in these GBM cells [8]. As a result, autophagy is believed to be a compensatory stress response, rather than a cell-death mechanism, in this scenario.

26.5.2 Comparison with Apoptosis

Some features of activated Ras-induced cell death are common with apoptosis, including activation of caspase 3 and cleavage of caspase substrates, poly ADP ribose polymerase (PARP), and lamin A/C. However, other typical features, including chromatin condensation, cell shrinkage, and plasma-membrane blebbing, are missing [22]. Moreover, the loss of cell viability is not prevented by caspase inhibitor, and caspase activation is not an obligatory step in the Ras-induced cell-death pathway [8].

26.5.3 Comparison with Other Cell-Death Pathways

Electron-microscopy and fluorescence-microscopy studies were carried out in order to study the membrane origin of Ras-induced vacuoles, using markers for the endoplasmic reticulum (ER), mitochondria, endosomes, and lysosomes. These studies demonstrated that the Ras-induced vacuoles did not originate from swollen ER membranes, as in paraptosis, or from the mitochondria, as in oncosis [14]. Membrane integrity loss occurs similarly to that in necroptosis, but no protection is observed by necrostatin, distinguishing methuosis from necroptosis [15].

26.6 Conclusion

Current cancer chemotherapeutic regimes rely mostly on drugs that activate apoptotic cell-death pathways [23]. However, many tumors that initially respond to chemotherapeutic drugs eventually develop multidrug resistance by upregulating drug-efflux mechanisms or DNA-repair pathways [24,25]. Moreover, tumor cells generally possess mutations in genes that control apoptosis, making them relatively insensitive to apoptotic cell death [26]. Together, these factors contribute to the chemoresistance [27] frequently encountered in relapsing tumors [28]. As a result, there is growing interest in identifying alternative cell-death pathways and small molecules capable of inducing non-apoptotic cell death in drug-resistant and recurrent cancers [5,29]. Lately, the Maltese group has discovered a novel caspase-independent cell-death pathway termed "methuosis" *in vitro* in GBM cells, triggered by overexpression of activated Ras protein [10]. GBM is the most common primary brain tumor in adults, and is resistant to most forms of chemotherapy [30]. The therapeutic regime marginally increases a patient's life span, but there is no curative treatment [31]. As reported by Maltese and Overmeyer [14], expression of activated Ras leads to hyperstimulation of fluid uptake and progressive accumulation of fluid-filled cytoplasmic vacuoles derived from macropinosomes and non-clathrin-coated endosomes. Eventually, there metabolic breakdown occurs, and the cell loses its membrane integrity. Further, the Maltese group identified small molecules that can induce methuosis in a therapeutic context and carried out structure–activity relationship studies to identify key residues for methuosis induction [20]. Interestingly, these studies led to the identification of compounds capable of maintaining methuosis or uncoupling cell death from vacuolization, which can potentially serve as a prototype of a new class of drugs against resistant tumors [32]. Triggering methuosis in cell culture requires ectopic expression of artificially high levels of activated Ras protein. Gaining a better insight into the mechanism of this excessive vacuolar cytopathology (followed by cell death) may lead to the development of better pharmacological strategies capable of inducing methuosis against notoriously resistant cancers in a therapeutic context.

References

1. Schultz DR, Harrington WJ Jr. Apoptosis: programmed cell death at a molecular level. *Semin Arthritis Rheum* 2003;**32**(6):345–69.
2. Flusberg DA, Sorger PK. Surviving apoptosis: life-death signaling in single cells. *Trends Cell Biol* 2015;**25**(8):446–58.
3. Galluzzi L, Vitale I, Abrams JM, Alnemri ES, Baehrecke EH, Blagosklonny MV, et al. Molecular definitions of cell death subroutines: recommendations of the Nomenclature Committee on Cell Death 2012. *Cell Death Differ* 2012;**19**(1):107–20.
4. Chi S, Kitanaka C, Noguchi K, Mochizuki T, Nagashima Y, Shirouzu M, et al. Oncogenic Ras triggers cell suicide through the activation of a caspase-independent cell death program in human cancer cells. *Oncogene* 1999;**18**(13):2281–90.
5. Kornienko A, Mathieu V, Rastogi SK, Lefranc F, Kiss R. Therapeutic agents triggering nonapoptotic cancer cell death. *J Med Chem* 2013;**56**(12):4823–39.
6. El-Khattouti A, Selimovic D, Haikel Y, Hassan M. Crosstalk between apoptosis and autophagy: molecular mechanisms and therapeutic strategies in cancer. *J Cell Death* 2013;**6**:37–55.
7. Fulda S, Kögel D. Cell death by autophagy: emerging molecular mechanisms and implications for cancer therapy. *Oncogene.* 2015;**34**(40):5105–13.
8. Overmeyer JH, Kaul A, Johnson EE, Maltese WA. Active Ras triggers death in glioblastoma cells through hyperstimulation of macropinocytosis. *Mol Cancer Res* 2008;**6**(6):965–77.
9. Levin R, Grinstein S, Schlam D. Phosphoinositides in phagocytosis and macropinocytosis. *Biochim Biophys Acta* 2015;**1851**(6):805–23.
10. Kaul A, Overmeyer JH, Maltese WA. Activated Ras induces cytoplasmic vacuolation and non-apoptotic death in glioblastoma cells via novel effector pathways. *Cell Signal* 2007;**19**(5):1034–43.
11. Naumann U, Wischhusen J, Weit S, Rieger J, Wolburg H, Massing U, et al. Alkylphosphocholine-induced glioma cell death is BCL-X(L)-sensitive, caspase-independent and characterized by massive cytoplasmic vacuole formation. *Cell Death Differ* 2004;**11**(12):1326–41.
12. Li C, Macdonald JI, Hryciw T, Meakin SO. Nerve growth factor activation of the TrkA receptor induces cell death, by macropinocytosis, in medulloblastoma Daoy cells. *J Neurochem* 2010;**112**(4):882–99.
13. Bhanot H, Young AM, Overmeyer JH, Maltese WA. Induction of nonapoptotic cell death by activated Ras requires inverse regulation of Rac1 and Arf6. *Mol Cancer Res* 2010;**8**(10):1358–74.

14 Maltese WA, Overmeyer JH. Methuosis: nonapoptotic cell death associated with vacuolization of macropinosome and endosome compartments. *Am J Pathol* 2014;**184**(6):1630–42.
15 Maltese WA, Overmeyer JH. Non-apoptotic cell death associated with perturbations of macropinocytosis. *Front Physiol* 2015;**6**:38.
16 Nara A, Aki T, Funakoshi T, Uemura K. Methamphetamine induces macropinocytosis in differentiated SH-SY5Y human neuroblastoma cells. *Brain Res* 2010;**1352**:1–10.
17 Fortwendel JR. Ras-mediated signal transduction and virulence in human pathogenic fungi. *Fungal Genom Biol* 2012;**2**(1):105.
18 Fortwendel JR, Juvvadi PR, Rogg LE, Steinbach WJ. Regulatable Ras activity is critical for proper establishment and maintenance of polarity in *Aspergillus fumigatus*. *Eukaryot Cell* 2011;**10**(4):611–15.
19 Overmeyer JH, Young AM, Bhanot H, Maltese WA. A chalcone-related small molecule that induces methuosis, a novel form of non-apoptotic cell death, in glioblastoma cells. *Mol Cancer* 2011;**10**:69.
20 Robinson MW, Overmeyer JH, Young AM, Erhardt PW, Maltese WA. Synthesis and evaluation of indole-based chalcones as inducers of methuosis, a novel type of nonapoptotic cell death. *J Med Chem* 2012;**55**(5):1940–56.
21 Trabbic CJ, Dietsch HM, Alexander EM, Nagy PI, Robinson MW, Overmeyer JH, et al. Differential induction of cytoplasmic vacuolization and methuosis by novel 2-indolyl-substituted pyridinylpropenones. *ACS Med Chem Lett* 2014;**5**(1):73–7.
22 Chen M, Wang J. Initiator caspases in apoptosis signaling pathways. *Apoptosis* 2002;**7**(4):313–19.
23 Li-Weber M. Targeting apoptosis pathways in cancer by Chinese medicine. *Cancer Lett* 2013;**332**(2):304–12.
24 Fokas E, Prevo R, Hammond EM, Brunner TB, McKenna WG, Muschel RJ. Targeting ATR in DNA damage response and cancer therapeutics. *Cancer Treat Rev* 2014;**40**(1):109–17.
25 Groth-Pedersen L, Jaattela M. Combating apoptosis and multidrug resistant cancers by targeting lysosomes. *Cancer Lett* 2013;**332**(2):265–74.
26 Ghavami S, Hashemi M, Ande SR, Yeganeh B, Xiao W, Eshraghi M, et al. Apoptosis and cancer: mutations within caspase genes. *J Med Genet* 2009;**46**(8):497–510.
27 Abdullah LN, Chow EK. Mechanisms of chemoresistance in cancer stem cells. *Clin Transl Med* 2013;**2**(1):3.
28 Smolewski P, Robak T. Inhibitors of apoptosis proteins (IAPs) as potential molecular targets for therapy of hematological malignancies. *Curr Mol Med* 2011;**11**(8):633–49.
29 Kreuzaler P, Watson CJ. Killing a cancer: what are the alternatives? *Nat Rev Cancer* 2012;**12**(6):411–24.
30 Pan H, Alksne J, Mundt AJ, Murphy KT, Cornell M, Kesari S, et al. Patterns of imaging failures in glioblastoma patients treated with chemoradiation: a retrospective study. *Med Oncol* 2012;**29**(3):2040–5.
31 Grossman SA, Ye X, Piantadosi S, Desideri S, Nabors LB, Rosenfeld M, et al. Survival of patients with newly diagnosed glioblastoma treated with radiation and temozolomide in research studies in the United States. *Clin Cancer Res* 2010;**16**(8):2443–9.
32 Trabbic CJ, Overmeyer JH, Alexander EM, Crissman EJ, Kvale HM, Smith MA, et al. Synthesis and biological evaluation of indolyl-pyridinyl-propenones having either methuosis or microtubule disruption activity. *J Med Chem* 2015;**58**(5):2489–512.

27

Oncosis

Priya Weerasinghe,[1] Sarathi Hallock,[2] Robert Brown,[1] and L. Maximilian Buja[1]

[1]*Department of Pathology and Laboratory Medicine, University of Texas Health Science Center at Houston, Houston, TX, USA*
[2]*Memorial University of Newfoundland, AMC Cancer Research Center, University of Texas Health Science Center at Houston, Houston, TX, USA*

Abbreviations

ATO	arsenic trioxide
ATP	adenosine triphosphate
Bax	Bcl-2-associated X protein
BCD	blister cell death
cyt c	cytochrome c
DCS	diffuse cell swelling
ER	endoplasmic reticulum
ERK	growth factor-regulated extracellular signal-related kinase
MAPK	mitogen-activated protein kinase
MOMP	mitochondrial outer-membrane permeabilization
MPTP	mitochondrial permeability transition pore
mTOR	mammalian target of rapamycin
NK	natural killer
PARP	poly ADP ribose polymerase
PI3-K	phosphatidylinositol 3-kinase
PORIMIN	pre-oncosis receptor-induced membrane injury
PS	phosphatidylserine
RC	ruptured cell
RISK	reperfusion injury salvage kinase
STP	Society of Toxicologic Pathology
UCP-2	uncoupling protein 2

27.1 Introduction

Studies on how cells die have shown that there are two primary mechanisms: apoptosis and non-apoptotic cell death [1,2]. One report suggested that the best term for cell death other than by apoptosis would be "accidental cell death" [1]. According to the authors, the term "necrosis" is inappropriate for non-apoptotic cell death, as it does not indicate a form of death, but merely refers to the changes that take place following it. One well-known form of accidental cell death is ischemic cell death. The term "oncosis" has been proposed as a major contributor for this, as it is accompanied by cell swelling [1,3]. Oncosis and apoptosis are different forms of cell death. "Oncosis" is a much older term. "Apoptosis" has been known by several designations, including "single-cell necrosis" and "shrinkage necrosis." Oncosis is the pre-lethal phase that follows a lethal cell injury such as complete ischemia or exposure to chemical toxins [1,2]. While oncosis has been studied, research on apoptosis, from a molecular genetics standpoint, has been much more extensive. Both types of cell death are "programmed," in that the genetic information on many of the factors and enzymes pre-exists in the cell [4,5]. Reports outlining the relationship between cell injury and cell death indicate that apoptotic and oncotic mechanisms can proceed together with oncotic morphology, dominating the end steps of irreversible injury in solid organs [6].

Increasing attention is being devoted to elucidating the differences in cell death under pathological and physiological conditions. As new terms such as "apoptosis" have appeared and the study of new and additional forms of cell death has been pursued, some conceptual and semantic strains have developed. In fact, the Society of Toxicologic Pathology (STP) has long attempted to forge a consensus on the nomenclature of the various forms of cell death [5]. How cells die is very important to the practice of diagnosis and research in toxicological pathology [4].

Following injury, cells undergo a series of responses that collectively form what we recognize as a disease process [7]. Many injuries to cells are sublethal and result in altered or new steady states in which the cells are able to survive (e.g., vacuolization from dilatation of lysosomes) [7]. Injuries that are lethal to cells (e.g., complete ischemia), in comparison, lead to cell death following a period of reversible reactions. After the cell dies, a series of degradative reactions take place that restore the environment to equilibrium. These postmortem cellular alterations are referred to as "necrosis" [1,8]. Therefore, cell death and necrosis define clearly separate entities. The terms "apoptosis" and "oncosis" refer to cellular processes as the result of injury prior to cell death. They are distinguished primarily on the basis of nuclear morphology and cell volume [4]. During apoptosis, cells shrink, have multiple cytoplasmic protrusions or blebs, preserve their membrane permeability, exhibit nuclear fragmentation with marked nuclear chromatin condensation, and ultimately fragment into what are known as apoptotic bodies [1]. On the other hand, oncosis is related to energy depletion, which causes impairment of ionic pumps of the cell membrane, clearing of the cytosol, cell swelling, dilatation of the Golgi and endoplasmic reticulum (ER), mitochondrial condensation following swelling, nuclear chromatin clumping, and formation of cytoplasmic blebs or blisters that are free of any organelles [1,6].

It is known that necrotic cells lose plasma-membrane integrity early in the process, allowing dyes such as trypan blue and propidium iodide to enter into them [9]. The exclusion of propidium iodide and trypan blue by cells undergoing oncosis indicates that oncosis is not identical to early necrosis, but rather precedes it. The morphology of blister formation induced by sanguinarine resembles that of "oncosis," as previously discussed [1,10]. Oncosis has been described as a kind of accidental cell death that has hallmarks of blebbing, cellular swelling, organelle swelling, and increased membrane permeability caused by the failure of the ionic pumps of the plasma membrane [1].

Reports have shown that oncosis (blister formation) increases the concentration of cytosolic calcium with rearrangement of cytoskeletal proteins [4,8].

In this chapter, we explore the pathobiology of oncosis in comparison to apoptosis based on the analysis of several specific settings where cells manifest oncosis or a combination of oncosis and apoptosis.

27.2 Oncosis Induced by Anticancer Agents

K562 erythroleukemia cells, when exposed to sanguinarine (a benzophenanthridine alkaloid) for 2 hours using concentrations of 1.5 µg/mL or 12.5 µg/mL, displayed two different morphologies, corresponding to two different modalities of cell death: at 1.5 µg/mL, >90% of cells showed formation of apoptotic bodies over the entire cell surface (Figure 27.1), consistent with the current criteria for apoptosis; at 12.5 µg/mL, >90% of cells had the morphology of large blister formation (Figures 27.1C, 27.2A, and 27.3C) (commonly, a single blister, and usually not more than two) or oncosis. Treatment of cells with concentrations of sanguinarine above 12.5 µg/mL resulted in ruptured blisters (Figure 27.2D). Several reports have found K562 cells to be rather resistant to the induction of apoptosis [11]; sanguinarine, on the other hand, is able to effectively overcome this resistance via the induction of apoptosis and oncosis. A detailed dose–response of K562 cells has been established (Figure 27.4).

The molecular mechanisms of the sanguinarine-induced direct tumor cell killing of K562 human erythroleukemia cells and CEM-T4 human lymphoblastic cells via the induction of oncosis – in comparison to apoptosis – have been studied. Sanguinarine was found to induce homogenous populations of apoptosis and oncosis; these could be induced in the same cell line using similar experimental conditions but different drug concentrations. This, in our view, makes sanguinarine-induced apoptosis and oncosis an ideal model for the study of both these cell-death phenomena. Other chemotherapeutic agents, such as arsenic trioxide (ATO), cisplatin, doxorubicin, and etoposide, have also been shown to induce apoptosis and oncosis [12,13].

27.3 Morphological Variants of Oncosis

It has been reported that K562 human erythroleukemia cells, when treated with sanguinarine at concentrations of 1.5 or 12.5 µg/mL for a short period of time (2 hours),

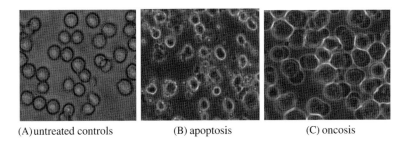

(A) untreated controls　　　(B) apoptosis　　　(C) oncosis

Figure 27.1 Light micrographs of K562 cells (at original magnification of ×282) show (A) untreated and (B,C) treated with sanguinarine. Treatment of cells with sanguinarine concentrations of 1.5 and 12.5 µg/mL for 2 hrs resulted in the morphologies of (B) apoptosis and (C) oncosis, respectively.

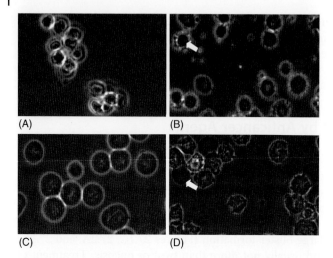

Figure 27.2 Light micrographs of untreated K562 (original magnification ×282). (A) parent cells and (B,C,D) cells treated with sanguinarine. Exposure of cells with sanguinarine at concentrations of 1.5, 6.25, and 12.5 µg/mL for 2 hours resulted in the morphology of (B) apoptosis, (C) oncosis-diffuse cell swelling (DCS), and oncosis-blister cell death (BCD) (D), respectively. The arrow in (B) indicates an apoptotic cell, while the arrow in (D) shows a cell undergoing oncosis-BCD.

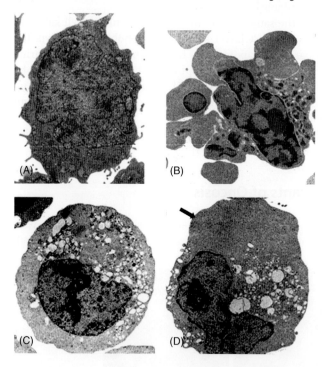

Figure 27.3 Electron micrographs (original magnification ×14 100) showing (A) an untreated control K562 cell, (B) apoptosis, (C) oncosis-diffuse cell swelling (DCS), and (D) oncosis-blister cell death (BCD). The arrow in (D) indicates a blister in cells that were exposed to sanguinarine at concentrations of 1.5, 6.25, and 12.5 µg/mL. Apoptosis in these cells shows classic apoptotic bodies, nuclear fragmentation, and chromatin condensation. The oncosis-BCD micrograph reveals a classic-looking blister formation, with patchy chromatin condensation and increased vacuolization. Cells treated with an intermediate concentration of sanguinarine (6.25 µg/mL) reveal a morphology of oncosis-DCS, along with patchy chromatin condensation and vacuolization.

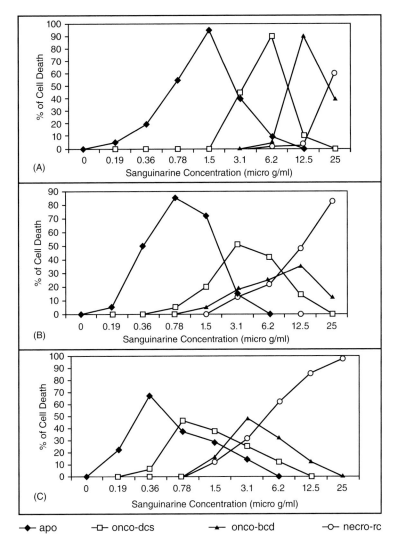

- ◆ apo
- ─□─ onco-dcs
- ─▲─ onco-bcd
- ─○─ necro-rc

Figure 27.4 Several dose–response curves for sanguinarine, using K562 cells at various time points: (A) 2 hours, (B) 12 hours, and (C) 24 hours. A significant result in this study is the shift to the left of the curves as the duration of drug exposure is increased. (A) Results of a quantitative morphological analysis of K562 cells exposed to increasing concentrations (0.19, 0.36, 0.78, 1.5, 3.1, 6.2, 12.5, and 25.0 µg/mL) of sanguinarine for 2 hours. Maximum apoptosis (apo) is noted at 1.5 µg/mL, whereas maximum oncosis-diffuse cell swelling (DCS) is seen at 6.25 µg/mL. The greatest oncosis-blister cell death (BCD) is noted at 12.5 µg/mL Ruptured cells not releasing trypan blue dye is considered to be cell death by necrosis; this is greatest when the concentration of sanguinarine is at 25 µg/mL. (B) Quantitative morphology of cells exposed to sanguinarine had the most apoptosis, at 0.78 µg/mL (12 hours). On the other hand, the greatest oncosis-DCS was found at 3.1 µg/mL. Maximal oncosis-BCD was found at 12.5 µg/mL, while ruptured cells were seen with concentrations greater than 3.1 µg/mL. For each cell population exposed to sanguinarine for 12 hours, the morphologies of cell death were more heterogenous compared with those exposed for only 2 hours. Therefore, with 12 hours of exposure, mixed populations of apoptosis, oncosis-DCS, oncosis-BCD, and ruptured cells were seen for all drug concentrations. (C) Quantitative morphology after 24 hours of drug treatment revealed maximum apoptosis at 0.36 µg/mL, maximum oncosis-DCS at 0.78 µg/mL, and maximum oncosis-BCD at 3.1 µg/mL. Ruptured cells were noted at concentrations greater than 1.5 µg/mL. Each data point is the mean from three independent experiments.

express to morphologies of two different kinds of cell death. At the lower concentration (1.5 µg/mL), the morphology is that of apoptosis, while at the higher concentration (12.5 µg/mL), it is that of single blister formation or oncosis-blister cell death (BCD) [14]. This dual cell-death modality induced by sanguinarine is termed "bimodal cell death" [15,16]. At an intermediate sanguinarine concentration of 6.25 µg/mL cells displayed diffuse cell swelling (DCS), which might represent a variant of oncosis resembling the morphology of autophagy [17]. The terms "BCD" and "DCS" are used to describe the primary morphological features found in each of the two forms of oncosis: namely, the single blister formation used to identify oncosis-BCD and the diffuse swelling used to identify oncosis-DCS. Exposure of cells with concentrations of sanguinarine greater than 12.5 µg/mL resulted in blisters that ruptured.

27.4 Biochemical Mechanisms of Oncosis

Several cell lines, including K562 (leukemia), CEM-VLB and CEM-T4 (lymphoma), prostate DU145 and LNCap (prostate), MDA-MB-435 and MCF-7 (breast), and some cervical and neuroblastoma lines, exhibit the classical pattern of bimodal cell death when treated with sanguinarine. Sanguinarine can induce oncosis and has been found to be biochemically, morphologically, and genetically different than apoptosis. Briefly, during apoptosis, cells shrink and show multiple cytoplasmic protrusions or blebs, marked nuclear chromatin condensation, karryohexis, and ultimately fragmentation. However, oncosis induced by sanguinarine is characterized by karryolysis, cell swelling, dilatation of the Golgi and ER, clearing of the cytosol, mitochondrial floccular condensation followed by swelling, cytoplasmic blebs that are free of organelles, and nuclear chromatin clumping (Table 27.1).

A critical event in apoptosis is mitochondrial outer-membrane permeabilization (MOMP), which promotes release of cytochrome c (cyt c) and other molecules, resulting in caspase activation. Conversely, a critical mitochondrial event in oncosis is the opening of the mitochondrial permeability transition pore (MPTP), which is located in the inner membrane: this occurs without cyt c release. With the opening of the MPTP, there is an immediate loss of the electrical potential difference across the inner membrane ($\Delta\Psi$m), causing cessation of adenosine triphosphate (ATP) synthesis and an influx of solute, which leads to mitochondrial swelling. These mitochondrial changes result from the activation of "mitochondrial death channels" [18]. The important role of the mitochondria in this process has also been shown in ischemic and toxic liver injury [19–22]. The interrelationships between oncosis (also sometimes called "oncotic necrosis") and apoptosis are increasingly being recognized. They include models characterized by the initial activation of apoptosis followed by oncosis, and vice versa [23]. This has given rise to the use of new terms, including "programmed death," "type III necrosis," and "necroptosis programmed necrosis" [20,24–28]. To conceptualize this, certain oncotic events – such as the loss of the mitochondrial membrane potential – occur in close temporal relationship to the release of cyt c from the mitochondria. Just as with oncosis, secondary necrotic changes follow if the apoptotic bodies are not rapidly removed, as in acute myocardial infarction [24]. Elucidation of the key role of the mitochondria and of the mechanisms of mitochondrial dysfunction has enabled new avenues of research into the creation of novel pharmacological agents for the treatment of disease processes involving oncosis, such as salvaging ischemic myocardium due to reperfusion

Table 27.1 Characteristics of apoptosis versus oncosis. BCD, blister cell death; PS, phosphatidylserine; Bax, Bcl-2-associated X protein.

Criterion	Apoptosis	Oncosis-BCD
Morphology as studied by light and electron microscopy	Formation of apoptotic bodies containing organelles, nuclear fragmentation, and chromatin condensation	Formation of blisters (often one, but rarely two) devoid of organelles, patchy chromatin condensation, increased vacuolization
Cell injury as measured by ^{51}Cr release assay	Relatively lower degree of injury, approximately 20% of 51Cr released	Relatively higher degree of injury, approximately 40% of ^{51}Cr released
Cell viability and plasma-membrane integrity as measured by trypan blue and propidium iodide dye exclusion assays	Excludes both dyes after 2-hour exposure, demonstrating viability (ceases to exclude after 18 hours)	Excludes both dyes after 2-hour exposure, demonstrating viability (ceases to exclude after 6 hours)
Membrane PS flip as detected by the annexin-V-assay	PS flipping	No PS flipping
DNA nicking as measured by the TdT end-labeling method	DNA breaks	No DNA breaks
Bax protein expression as measured by Western blotting and flow cytometry	Increase in Bax expression	No change in Bax expression
Bcl-2 protein expression as measured by Western blotting and flow cytometry	No significant change in Bcl-2 expression	No significant change in Bcl-2 expression
NF-κB protein expression as measured by flow cytometry	Decrease in NF-κB expression	Increase in NF-κB expression
Caspase 3 involvement/activati-on as measured by Western blotting and fluorimetric assay	Caspase 3 activation involved	Caspase 3 activation not involved

injury [29–31]. These interventions may interact with a newly discovered reperfusion injury salvage kinase (RISK) pathway, part of which involves phosphatidylinositol 3-kinase (PI3-K) acting on Akt and the mammalian target of rapamycin (mTOR), and another part mitogen-activated protein kinase (MAPK) and p42/p44 growth factor-regulated extracellular signal-related kinase (ERK). The two arms of the system converge on p70s6 kinase, which activates glycogen synthase kinase β, which in turn acts to prevent opening of the MPTP. The activation of the RISK pathway thus acts to inhibit the MPTP from opening, preserve mitochondrial membrane potential, enhance uptake of calcium from the sarcoplasm into the sarcoplasmic reticulum, and activate anti-apoptotic and antioncotic processes. Thus, the RISK pathway appears to be a mechanism of programmed cell survival [29,31].

Our understanding of the mechanisms of cell death is increasing rapidly. Efforts are directed at elucidating the biochemical mechanisms of cell death, with a view to many different end goals. An important biochemical event leading to oncosis – and not to apoptosis – is a rapid loss of intracellular ATP ($[ATP]_i$) [4]. Stopping ATP synthesis rapidly causes a deactivation of Na^+, K^+-ATPase at the cell membrane, leading to an increase in $[Na^+]_i$ and $[Cl^-]_i$ with a water influx, cellular swelling, and a rapid increase in $[Ca^{++}]_i$ [32,33]. Such findings are corroborated by recent reports showing that a modest increase in the expression level of uncoupling protein 2 (UCP-2) results in a rapid and significant fall in mitochondrial membrane potential and a reduction in intracellular ATP, and thus in the morphological changes seen in oncosis [34].

Certain biochemical changes during apoptosis, such as phosphatidylserine (PS) flipping and DNA fragmentation, are not noted in oncosis. O_2 consumption studies show that during apoptosis induced by sanguinarine, there is an initial increase in O_2 consumption, followed by a gradual decline. In contrast, oncosis consistently shows a rapid decline in O_2 consumption. These findings are corroborated by mitochondrial potential studies using the Rh 123 assay. Mitochondrial potential during apoptosis induced by sanguinarine remained largely unchanged during the first 2 hours, whereas that during oncosis showed a clear reduction compared to untreated controls (unpublished data). These biochemical parameters found in cell-death patterns induced by sanguinarine are in general agreement with those associated with oncosis and apoptosis reported in the literature.

27.5 Genetic and "Programmed" Nature of Oncosis

Recent literature has revealed some of the molecular pathways of alternate non-apoptotic forms of cell death. Each of these is thought to share some – but not all – of the characteristics of classical apoptotic pathways [35]. For example, it has been reported that there are less well defined molecular cell-death pathways that do not require caspase activation [36,37]. However, recent reports also support the idea that the cell-death process of oncosis is caspase 1-dependent [38,39] and that the transcription factor NF-κB protects cells against oncosis [40]. Several other reports indicate that apoptosis may share common molecular pathways with other types of cell death at an early stage [41]. Our findings show that tumor cell death may be induced by oncosis – a hitherto less understood form of cell death that is caspase 3-independent. It has also been shown that oncosis, like apoptosis, can be activated by triggering cell-surface receptors such as the pre-oncosis receptor-induced membrane injury (PORIMIN) receptor [42].

In order to elucidate the possible role of Bcl-2 in oncosis, the response to the anticancer agent sanguinarine was measured in the low Bcl-2-expressing K562 and CEM-T4 cells and in the high Bcl-2 expressing JM1 cells. These cells were treated with sanguinarine concentrations of 1.5 and 12.5 µg/mL (i.e., concentrations that induce apoptosis and oncosis, respectively, in K562 and CEM-T4 cells), and multiple parameters of their effects were studied. In general, it was found that JM1 cells treated with sanguinarine failed to undergo either apoptosis or oncosis. Thus, the overexpression of antiapoptotic Bcl-2 may have prevented sanguinarine from inducing apoptosis and oncosis in JM1 cells. These results indicate that the resistance of JM1 cells to the alkaloid sanguinarine may have been due to an antioncosis role played by Bcl-2, in addition to its widely reported antiapoptotic role. We believe that the chemoresistance caused by the increase in Bcl-2 might be associated with its dual role of antiapoptosis and antioncosis [15].

Evidence of the involvement of other Bcl-2 family proteins was compared in apoptosis and oncosis. The expression of proapoptotic Bax protein was found to be increased in sanguinarine-induced apoptosis as compared to untreated control K562 and CEM-T4 cells. In contrast, sanguinarine-induced oncosis was not accompanied by any significant increase in Bax protein expression. Also the Bax/Bcl-2 ratio was found to correlate with the susceptibility of cells to apoptosis but not to oncosis. The expression of proapoptotic Bak, Bad, Bik/Nbk, Bcl-XS, Bid, and p53 and antiapoptotic Bcl-XL proteins was not significantly affected by sanguinarine in either cell line at either low or high level [16]. Studies also showed that sanguinarine-induced apoptosis is associated with a decrease in the nuclear transcription factor NF-κB, which corroborates similar reports in the literature [43,44]. However, sanguinarine-induced oncosis is associated with an increase in NF-κB [16]. Studies comparing apoptosis and oncosis with regard to the protease cascade showed that unlike in apoptosis, caspase 3 activation may not be a necessary element in the execution of oncosis [14,36,37,45,46]. A related study found that in multidrug-resistant cervical cancer cells, sanguinarine-induced apoptosis led to caspase 3 activation and the cleavage of poly ADP ribose polymerase (PARP) and other caspase 3 substrates, whereas the blockage of this pathway led to oncosis [47]. Several reports in the literature suggest that apoptosis shares common molecular pathways with other types of cell death at an early stage [41].

Sanguinarine, a known anticancer agent, was found to cause tumor cell death by oncosis and apoptosis, as well as specific and dose-dependent changes at the transcriptome level. Microarray analysis using Illumina Beadchips of RNAs prepared from cultured mouse cardiomyocytes support earlier observations of the involvement of genetic and transcriptional events in oncosis [48]. Sanguinarine, at higher oncosis-inducing concentrations, altered the expression of 2514 probes, and at lower apoptosis-inducing concentrations, altered the expression of 1643 probes at a level of significance of $p < 0.001$. The increase in the number of probes altered at the higher dose suggests that there are multiple biochemical or signaling pathways involved. The activation of probes associated with post-translational modification, cell-cycle regulation, protein synthesis, and other functional groupings revealed by pathway analysis suggests widespread cellular activation at the high versus the low dose. Activation of transcription-related groupings, with RNA polymerase II occupying a nodal position, was observed at the higher sanguinarine dose using Ingenuity's network function. These findings corroborate active cellular functioning at the higher dose, including the activation of major canonical pathways. Perforin, a cytolytic protein expressed in the granules of CD8 T cells and natural killer (NK) cells, is induced more than 11-fold during oncosis [48].

27.6 Incidence of Oncosis and Relevance to Disease

Reports on oncosis are steadily increasing in the literature. Examples include: killing of human monocyte-derived macrophages by virulent *Shigella flexneri* [49]; killing of neutrophils and macrophages by *Pseudomonas aeruginosa* [50]; macrophage killing by strains of *Escherichia coli* [51]; cell killing by *Bacillus anthracis* toxin [52]; radiation exposure-related cell death [53]; cell death due to atherosclerotic plaques [54]; cell death due to stroke and ischemic heart disease [4,55,56]; and cell death induced by environmental toxins like alkylating agents [57]. Thus, the mounting evidence of oncosis in the literature ranges from bacterial survival strategies allowing the pathogen to escape and colonize host tissue, resulting in escape from host immune mechanisms, to toxicity related to a variety of environmental situations. Furthermore, it has been reported that laboratory-altered and natural isolates of HIV-1 result in T-cell death through oncosis rather than apoptosis [58,59], which is contrary to what was previously believed. Moreover, it is also now known that other chemotherapeutic agents, such as ATO, cisplatin, doxorubicin, and etoposide, kill cells by oncosis in addition to apoptosis [12,13]. Thus, the study of oncosis, like the study of apoptosis, will significantly improve our understanding of disease processes. It has other potential implications in the fight against cancer, cardiovascular disease, stroke, AIDS, and infectious diseases, too. A molecular understanding of oncosis could help us better utilize the molecular basis of the action of anticancer agents, and thus could be used as an approach to identifying novel targets for improving therapeutic outcome.

27.7 Oncosis in Myocardial Ischemia

A large body of evidence indicates that profound myocardial ischemia following coronary occlusion leads to a pattern of cardiomyocyte damage that is characterized by mild cell swelling and progressive membrane damage – in other words, oncosis. This can progress to rapid cell swelling when reperfusion starts [60,61].

Recently, reports of apoptosis have been documented in models of myocardial ischemia and reperfusion [60,62]. Studies using caspase inhibitors and/or mutant mouse models indicate that treated or mutant animals subjected to myocardial ischemia have a reduced amount of infarction, but not complete protection from it, suggesting that the mechanism of cell injury and death in early myocardial ischemia involves both oncosis and apoptosis in approximately equal degree [63,64]. Other studies have shown that cardiomyocytes contain the molecular machinery required to progress to cell death via oncosis and apoptosis (Figure 27.5) [48]. The rate and amount of ATP depletion appears to be an important influence on the progression of cell damage via oncosis, in that rapid ATP depletion favors this pathway, whereas ATP preservation favors aptotosis. Also, apoptotic and oncotic mechanisms can operate in the same ischemic myocytes. Autophagy, resembling the morphology of the oncosis variant DCS, has been shown to be another type of cell modulation that can cause various forms of ischemic injury [65,66]. Thus, cell injury and death induced by severe blood flow reduction represents a hybrid pattern best termed "ischemic cell injury and cell death." This discussion highlights that apoptosis and oncosis are mediated by distinct but overlapping pathways involving cell-surface death receptors, mitochondria, and ER. A major role for

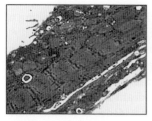 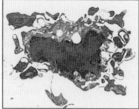 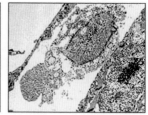

Figure 27.5 Electron micrographs (original magnification ×10 000) showing mouse cardiomyocytes. Left panel: untreated. Middle panel: treated with a concentration of sanguinarine that induces apoptosis. Right panel: treated with a concentration of sanguinarine that induces oncosis.

the mitochondria has been identified in both myocardial ischemia and other forms of injury [24,28].

It is hypothesized that oncotic cell death involves progressive membrane injury with three separate phases [6,67]. In phase 1, the cell becomes committed to oncosis due to selective membrane injury that leaks ions and water as a result of ATP depletion. This leads to cell swelling without a generalized increase in cell membrane permeability. In phase 2, the cell passes beyond the point of reversibility, with the cell membrane becoming leaky to trypan blue and propidium iodide, suggesting the development of a nonselective increase in membrane permeability. During phase 3, the eventual physical disruption of the cell membrane places it in the necrotic phase. It is known that early necrotic cells lose plasma-membrane integrity, permitting the entry of trypan blue and propidium iodide [9]. The exclusion of these dyes by oncotic cells shows that oncosis is not representative of early necrosis. It has been reported that oncosis is a form of accidental cell death that involves cellular swelling, organelle swelling, blebbing, and increased membrane permeability caused by the failure of the ion pumps [1].

27.8 Oncosis in Chemotherapeutic Cardiotoxicity

Chemotherapeutic cardiotoxicity is emerging as an important issue among cancer survivors and is an area of much significance to the rapidly growing field of cardio-oncology. Chemotherapeutic agents of concern include the anthracyclines. One such anthracycline, doxorubicin, is an effective antineoplastic drug that is widely used in the treatment of solid and hematologic tumors. Its efficacy, however, is reduced in the face of cumulative cardiotoxic adverse effects that are dose-related, progressive, and irreversible [68]. Among the many cardiotoxic complications of doxorubicin therapy are heart failure, myocardial ischemia, hypertension, thromboembolism, and arrhythmia [69–72]. In the past, cardiotoxicity was less evident, but it has increased in incidence due to the use of combination and adjuvant therapies. The aging of the populations of developed countries and the higher probability that a patient may have both cancer and cardiovascular disease make this population group even more vulnerable to chemotherapy-induced cardiomyocyte injury. Cardiomyocyte injury results in a characteristic pattern of metabolic and ultrastructural changes that can lead to irreversible injury to the heart and other organs. Our preliminary findings, as well as reports in the literature [73], show that doxorubicin can induce at least two forms of cell injury leading to death: oncosis and apoptosis. Dose–response experiments have shown that cardiomyocytes, when treated with doxorubicin at low concentrations, mainly exhibit the classic morphology of apoptosis. At higher concentrations, they exhibit that of oncosis. Doxorubicin cardiotoxicity is most likely the result of cardiomyocytes that have been injured by both oncosis and apoptosis. Many other chemotherapeutic agents, such as as cisplatin, ATO, and etoposide, have been shown to induce both oncosis and apoptosis [12,13,74]. Recent advances in studies pertaining to the relationship between cell injury and cell death suggest that apoptotic and oncotic mechanisms can proceed together and thereby lead to a hybrid type of oncotic mechanism that dominates the final phase of irreversible cell injury [6]. Therefore, the pathways and modes of cell injury are not only of key importance to the practice of diagnostic cardiac pathology, but also form the basis for mechanism-based treatment and prevention options for chemotherapeutic cardiotoxicity.

References

1. Majno G, Joris I. Apoptosis, oncosis and necrosis. An overview of cell death. *Am J Pathol* 1995;**146**(1):3–15.
2. Majno G, Joris I. Cell injury and cell death. In *Cells Tissues and Disease: Principles of General Pathology*, 2nd edn. New York: Oxford University Press. 2004.
3. Darzynkiewicz Z, Juan G, Li X, Gorczyca W, Murakami T, Traganos F. Cytometry in cell necrobiology: analysis of apoptosis and accidental cell death (necrosis). *Cytometry* 1997;**27**:1–20.
4. Trump BF, Berezesky IK, Chang SH, Phelps PC. The pathways of cell death: oncosis, apoptosis and necrosis. *Toxicol Pathol* 1997;**25**(1):82–8.
5. Levin S, Bucci TJ, Cohen SM, Fix AS, Hardisty JF, LeGrand EK, et al. The nomenclature of cell death: recommendations of an ad hoc committee of the society of toxicologic pathologists. *Toxicol Pathol* 1999;**27**:484–90.
6. Buja LM. Myocardial ischemia and reperfusion injury. *Cardiovasc Pathol* 2005;**4**:170–5.
7. Trump BF, Ginn FL. The pathogenesis of subcellular reaction to lethal injury. In: Bajusz E, Jasmin G (eds.). *Methods and Achievements in Experimental Pathology*, Vol. IV. Karger: Basel. 1969.
8. Trump BF, Berezesky IK. The role of cytosolic Ca++ in cell injury, necrosis and apoptosis. *Curr Opin Cell Biol* 1992;**4**:227–32.
9. O'Brien MC, Healy SF Jr, Raney SR, Hurst JM, Avner B, Hanley A, et al. Discrimination of late apoptotic/necrotic cells (Type III) by flow cytometry in solid tumors. *Cytometry* 1997;**28**:81–9.
10. Phelps PC, Smith MW, Trump BF. Cytosolic ionized calcium and bleb formation after acute cell injury of cultured rabbit renal tubule cells. *Lab Invest* 1989;**60**(5):630–41.
11. Kobayashi T, Ruan S, Clodi K, Kliche KO, Shiku H, Andreeff M, Zhang W. Overexpression of Bax gene sensitizes K562 erythroleukemia cells to apoptosis induced by selective chemotherapeutic agents. *Oncogene* 1998;**16**:1587–91.
12. Zhu J, Okumura H, Ohtake S, Nakamura S, Nakao S. The molecular mechanism of arsenic trioxide-induced apoptosis and oncosis in leukemia/lymphoma cell lines. *Acta Haematol* 2003;**110**(1):1–10.
13. Gonzalez VM, Fuertes MA, Alonso C, Perez JM. Is cisplatin-induced cell death always produced by apoptosis? *Mol Pharmacol* 2001;**59**(4):657–63.
14. Weerasinghe P, Hallock S, Tang S, Liepins A. Role of Bcl-2 family proteins and caspase-3 in sanguinarine-induced bimodal cell death. *Cell Biol Toxicol* 2001;**17**:371–81.
15. Weerasinghe P, Hallock S, Tang S, Liepins A. Sanguinarine induces bimodal cell death in K562 but not in high Bcl-2 expressing JM1 cells. *Pathol Res Pract* 2001;**197**:717–26.
16. Weerasinghe P, Hallock S, Liepins A. Bax, Bcl-2 and NF-κB expression in sanguinarine induced bimodal cell death. *Exp Mol Pathol* 2001;**71**:89–98.
17. Hallock S, Tang SC, Buja LM, Trump BF, Liepins A, Weerasinghe P. Aurintricarboxylic acid inhibits protein synthesis independent, sanguinarine-induced apoptosis and oncosis. *Toxicol Pathol* 2007;**35**(2):300–9.
18. Webster KA. Mitochondrial death channels. *Am Sci* 2009;**97**(5):384–91.
19. Hinson JA, Roberts DW, James LP. Mechanisms of acetaminophen-induced liver necrosis. *Handb Exp Pharmacol* 2010; (196):369–405.
20. Jaeschke H, Lemasters JJ. Apoptosis versus oncotic necrosis in hepatic ischemia/reperfusion injury. *Gastroenterology* 2003;**125**(4):1246–57.

21 Kass GE. Mitochondrial involvement in drug-induced hepatic injury. *Chem Biol Interact* 2006;**163**(1-2):145–59.
22 McGill MR, Sharpe MR, Williams CD, Taha M, Curry SC, Jaeschke H. The mechanism underlying acetaminophen-induced hepatotoxicity in humans and mice involves mitochondrial damage and nuclear DNA fragmentation. *J Clin Invest* 2012;**122**:1574–83.
23 Whelan RS, Kaplinskiy V, Kitsis RN. Cell death in the pathogenesis of heart disease: mechanisms and significance. *Ann Rev Physiol* 2010;**72**:19–44.
24 Whelan RS, Konstantinidis K, Wei AC, Chen Y, Reyna DE, Jha S, et al. Bax regulates primary necrosis through mitochondrial dynamics. *Proc Nat Acad Sci USA* 2012;**109**:6566–71.
25 Golstein P, Kroemer G. A multiplicity of cell death pathways. A symposium on apoptotic and non-apoptotic cell death pathways. *EMBO Rep* 2007;**8**:829–33.
26 Golstein P, Kroemer G. Cell death by necrosis: toward a molecular definition. *Trends Biochem Sci* 2007;**32**:37–43.
27 Konstantinidis K, Whelan RS, Kitsis RN. Mechanisms of cell death in heart disease. *Arterioscler Thromb Vasc Biol* 2012;**32**:1552–62.
28 Kung G, Konstantinidis K, Kitsis RN. Programmed necrosis, not apoptosis, in the heart. *Circ Res* 2011;**108**:1017–36.
29 Bell RM, Yellon DM. Conditioning the whole heart – not just the cardiomyocyte. *J Mol Cell Cardiol* 2012;**53**:24–32.
30 Oerlemans MI, Koudstall S, Chamuleau SA, de Kleijn DP, Doevendans PA, Sluijter JP. Targeting cell death in the reperfused heart: pharmacological approaches for cardioprotection. *Int J Cardiol* 2013;**165**(3):410–22.
31 Yellon DM, Hausenloy DJ. Myocardial reperfusion injury. *N Engl J Med* 2007;**357**:1121–35.
32 Trump BF, Berezesky IK. Cellular and molecular pathobiology of reversible and irreversible injury. In: Tyson CA, Frazier JM (eds.). *Methods in Toxicology, Vol. 1B: In Vitro Toxicity Indicators*. Academic Press: New York. 1994.
33 Trump BF, Berezesky IK. Calcium mediated cell injury and cell death. *FASEB J* 1995;**9**:219–28.
34 Mills EM, Xu D, Fergusson MM, Combs CA, Xu Y, Finkel T. Regulation of cellular oncosis by uncoupling protein 2. *J Biochem* 2002;**277**(30):27 385–92.
35 Igney FH, Krammer PH. Death and anti-death: tumour resistance to apoptosis. *Nat Rev* 2002;**2**:277–90.
36 Borner C, Monney L. Apoptosis without caspases: an inefficient molecular guillotine? *Cell Death Diff* 1999;**6**:497–507.
37 Sperandio S, de Belle I, Bredesen DE. An alternative, nonapoptotic form of programmed cell death. *Proc Nat Acad Sci USA*. 2000;**97**:14 376–81.
38 Sun GW, Lu J, Pervaiz S, Cao WP, Gan YH. Caspase-1 dependent macrophage death induced by *Burkholderia pseudomallei*. *Cell Microbiol* 2005;**10**:1447–58.
39 Thumbikat P, Dileepan T, Kannan MS, Maheswaran SK. Mechanisms underlying *Mannheimia haemolytica* leukotoxin-induced oncosis and apoptosis of bovine alveolar macrophages. *Microb Pathog* 2005;**38**:161–72.
40 Franek WR, Morrow DM, Zhu H, Vancurova I, Miskolci V, Darley-Usmar K, et al. NF-kappaB protects lung epithelium against hyperoxia-induced nonapoptotic cell death-oncosis. *Free Radic Biol Med* 2004;**37**:1670–9.
41 Shirai T. Commentary: Oncosis and apoptosis: two faces of necrosis in a new proposal to clear up confusion regarding cell death. *Toxicol Pathol* 1999;**27**:1495–6.

42 Ma F, Zhang C, Prasad KV, Freeman GJ, Schlossman SF. Molecular cloning of porimin, a novel cell surface receptor mediating oncotic cell death. *Proc Nat Acad Sci USA* 2001;**98**:9778–83.

43 Kajino S, Suganuma M, Teranishi F, Takahashi N, Tetsuka T, Ohara H, et al. Evidence that de novo protein synthesis is dispensable for anti-apoptotic effects of NF-kB. *Oncogene* 2000;**19**:2233–9.

44 Kumar AP, Garcia GE, Orsborn J, Levin VA, Slaga TJ. 2-methoxyestradiol interferes with NF kappa B transcriptional activity in primitive neuroectodermal brain tumors: implications for management. *Carcinogenesis* 2003;**24**:209–16.

45 Kojima H, Talanian RV, Alnemmi ES, Wong WW, Keefe DW. Activation of the CPP 32 protease in apoptosis induced by 1-∃-D arabinofuranosylcytosine and other DNA damaging agents. *Blood* 1996;**88**:1936–43.

46 Irabo AM, Huang Y, Fang G, Liu L, Bhalla K. Overexpression of Bcl2 or BclXL inhibits ara-c-induced CPP32/Yama protease activity and apoptosis of human acute myelogenous leukemia HL-60 cells. *Cancer Res* 1996;**56**:4743–8.

47 Ding Z, Tang SC, Weerasinghe P, Yang X, Pater A, Liepins A. The alkaloid sanguinarine is effective against multidrug resistance in human cervical cells via bimodal cell death. *Biochem Pharmacol* 2002;**63**(8):1415–21.

48 Weerasinghe, P, Hallock S, Brown RE, Loose DS, Buja LM. A model for cardiomyocyte cell death: insights into mechanisms of oncosis. *Exp Mol Pathol* 2013;**94**(1):289–300.

49 Fernandez-Prada CM, Hoover DL, Venkatesan MM. Human monocyte derived macrophages infected with virulent shigella flexneri *in vitro* undergo a rapid cytolytic event similar to oncosis but not apoptosis. *Infect Immun* 1997;**65**(4):1486–96.

50 Dacheux D, Goure J, Chabert J, Usson Y, Attree, I. Pore-forming activity of type III system-secreted proteins leads to oncosis of *Pseudomonas aeruginosa*-infected macrophages. *Mol Microbiol* 2001;**40**(1):76–85.

51 Fernandez-Prada CM, Tall BD, Elliott SE, Hoover DL, Nataro JP, Venkatesan MM. Hemolysin-positive enteroaggregative and cell detaching *Echerichia coli* strains causes oncosis of human monocyte derived macrophages and apoptosis of murine J774 cells. *Infect Immun* 1988;**66**(8):3918–24.

52 Lin GE, Kao YT, Liu WT, Huang HH, Chen KC, Wang TM, Lin HC. Cytotoxic effects of anthrax lethal toxin on macrophage-like cell line J774A.1. *Curr Microbiol* 1996;**33**(4):224–7.

53 Cornelissen M, Thierens H, De Ridder L. Interphase death in human peripheral blood lymphocytes after moderate and high doses of low and high LET radiation: an electron microscopic approach. *Anticancer Res* 2002;**22**(1A):241–5.

54 Crisby M, Kallin B, Thyberg J, Zhivotovsky B, Orrenius S, Kostulas V, Nilsson J. Cell death in human atherosclerotic plaques involves both oncosis and apoptosis. *Atherosclerosis* 1997;**130**(1-2):17–27.

55 Ohno M, Takemura G, Ohno A, Misao J, Hayakawa Y, Minatogushi S, et al. "Apoptotic" myocytes in infarct area in rabbit hearts may be oncotic myocytes with DNA fragmentation: analysis by immunogold electron microscopy combined with *in situ* nick end-labeling. *Circulation* 1998;**98**:1422–30.

56 Muller M, Ballanyi K. Dynamic recording of cell death in the *in vitro* dorsal vagal nucleus of rats in response to metabolic arrest. *J Neurophysiol* 2003;**89**(1):551–61.

57 el Affar B, Shah RG, Dallaire A, Castonguay V, Shah G. Role of poly (ADP-ribose) polymerase in rapid intracellular acidification induced by alkylating DNA damage. *PNAS* 2002;**99**(1):245–50.

58 Lenardo MJ, Angleman SB, Bounkeua V, Dimas J, Duvall MG, Graubard MB, et al. Cytopathic killing of peripheral blood CD4(+) T lymphocytes by human immunodificiency type1 appears necrotic rather than apoptotic and does not require env. *J Virol* 2002;**76**(10):5082–93.

59 Bolton DL, Hahn B, Park EA, Lehnhoff LL, Hornung F, Lenardo MJ. Death of CD4 T cell lines caused by human immunodeficiency virus type 1 does not depend on caspases or apoptosis. *J Virol* 2002;**76**:5094–107.

60 Buja LM. Modulation of the myocardial response to ischemia. *Lab Invest* 1998;**78**:1345–73.

61 Reimer KA, Ideker RE. Myocardial ischemia and infarction: anatomic and biochemical substrates for ischemic cell death and ventricular arrhythmias. *Hum Pathol* 1987;**18**:462–75.

62 Buja LM, Vela D. Cardiomyocyte death and renewal in the normal and diseased heart. *Cardiovasc Pathol* 2008;**17**:349–74.

63 Buja LM, Weerasinghe P. Unresolved issues in myocardial reperfusion injury. *Cardiovasc Pathol* 2010;**19**(1):29–35.

64 Foo RS, Mani K, Kitsis RN. Death begets failure in the heart. *J Clin Invest* 2005;**115**:565–71.

65 Gottlieb RA. Cell death pathways in acute I/R injury. *J Cardiovasc Pharmacol Ther* 2011;**16**:233–8.

66 Huang C, Yitzhaki S, Peng CN, Liu W, Giricz Z, Mentzer R.N. Jr. Gottlieb RA. Autophagy induced by ischemic preconditioning is essential for cardioprotection. *J Cardiovasc Transl Res* 2010;**3**:365–73.

67 Buja LM, Eigenbrodt ML, Eigenbrodt EH. Apoptosis and necrosis. Basic types and mechanisms of cell death. *Arch Pathol Lab Med* 1993;**117**:1208–14.

68 Barry E, Alvarez JA, Scully RE, Miller TL, Lipshultz SE. Anthracycline-induced cardiotoxicity: course, pathophysiology, prevention and management. *Expert Opin Pharmacother* 2007;**8**:1039–58.

69 Albini A, Pennesi G, Donatelli F, Cammarota R, De Flora S, Noonan DM. Cardiotoxicity of anticancer drugs: the need for cardio-oncology and cardio-oncological prevention. *J Natl Cancer Inst* 2010;**102**(1):14–25.

70 Minotti G, Salvatorelli E, Menna P. Pharmacological foundations of cardio-oncology. *J Pharmacol Exp Ther* 2010;**334**:2–8.

71 Xue D, Zhang W, Liang T, Zhao S, Sun B, Sun D. Effects of arsenic trioxide on the cerulein-induced AR42J cells and its gene regulation. *Pancreas* 2009;**38**:183–9.

72 Yeh ET, Bickford CL. Cardiovascular complications of cancer therapy: incidence, pathogenesis, diagnosis, and management. *J Am Coll Cardiol* 2009;**53**:2231–47.

73 Selleri S, Arnaboldi F, Vizzotto L, Balsari A, Rumio C. Epithelium–mesenchyme compartment interaction and oncosis on chemotherapy-induced hair damage. *Lab Invest* 2004;**84**:1404–17.

74 Chanan-Khan A, Srinivasan S, Czuczman MS. Prevention and management of cardiotoxicity from antineoplastic therapy. *J Support Oncol* 2004;**2**:251–6.

28

Autoschizis: A Mode of Cell Death of Cancer Cells Induced by a Prooxidant Treatment *In Vitro* and *In Vivo*

J. Gilloteaux,[1] J.M. Jamison,[2] D. Arnold,[3] and J.L. Summers[2]

[1] Department of Anatomical Sciences, St. Georges' University International School of Medicine, KB Taylor Global Scholar's Program at Northumbria University, Newcastle upon Tyne, UK
[2] Department of Urology and Apatone Development Laboratory, Summa Health System, Akron, OH, USA
[3] Anesthesiology Unit, Belfast, ME, USA

> *[Sumus] quasi nanos gigantium humeris insidentes.*
> From Bernard de Chartres, in Jean de Salisbury, *Metalogicon*, III

This chapter is dedicated to the memory of our colleague and friend, Henryk S. Taper, M.D., Ph.D., who initiated and progressed this research with us, and who died in 2009.

Abbreviations

5-FU	5-fluouracil
ATG5	autophagy-related gene 5
ATP	adenosine triphosphate
CD_{50}	50% cytotoxic dose
CNS	central nervous system
cyt c	cytochrome c
DES	diethylstilbestrol
DFR	dense fibrillar region
DMNQ	2,3-dimethoxy-naphthoquinone
ECM	extracellular matrix
ER	endoplasmic reticulum
FFR	fine fibrillar region
G6PD	glucose-6-phosphate dehydrogenase
GAPDH	glyceraldehyde-3-phosphate dehydrogenase
GSH	glutathione
GSK-3	glycogen synthase kinase-3
HIF-1	hypoxia-inducible factor 1
HK	hexokinase
IG	interchromatin granule
LEI	leukocyte-elastase inhibitor

Apoptosis and Beyond: The Many Ways Cells Die, First Edition. Edited by James Radosevich.
© 2018 John Wiley & Son Inc. Published 2018 by John Wiley & Son Inc.

LM	light microscopy
NAD⁺	oxidized nicotinamide adenine dinucleotide
NADH	reduced nicotinamide adenine dinucleotide
NADPH	reduced nicotinamide adenine dinucleotide phosphate
NCCD	Nomenclature Committee on Cell Death
NOR	nucleolar organizer region
OMM	outer mitochondrial membrane
PBS	phosphate-buffered saline
PCD	programmed cell death
PS	phosphatidylserine
RER	rough endoplasmic reticulum
ROS	reactive oxygen species
SADA	serum alkaline DNase activity
SCID	severe combined immunodeficiency
SEM	scanning electron microscopy
SER	smooth endoplasmic reticulum
SOD	superoxide dismutase
TEM	transmission electron microscopy
TNF	tumor necrosis factor
VC	vitamin C, ascorbate
VC:VK3	Apatone (combination ascorbate and menadione)
VK3	vitamin K3, menadione bisulfite

28.1 Introduction

As early as 1954, several studies originating from the Institut du Cancer de Montréal, Université de Montreal, Montréal, Canada, showed that spontaneous and experimental human tumor tissues lack or are deficient in DNase and RNase activities [1–5]. This could be the result of some unknown blocking molecules [1,3,5,6]. After spending a sabbatical doing research in Montréal, Henryk Taper pursued a series of experiments based on the premise that any nontoxic compound capable of unblocking or reactivating nuclease activities would either favor injurious defects to cancer cells, enhancing their attack by anticancer treatment, or else eventually make malignant cells more manageable by removing their more lethal characters, making them less aggressive than benign ones or, eventually – and better – killing them entirely.

This hypothesis led Taper and collaborators to publish a series of studies between 1967 and his unexpected death in 2009. Much of the revelavant data can be found in his PhD thesis, published by the Université de Montreal's affiliated press [7]. In all his work, using *in vivo* methodology, Henryk demonstrated that human tumor implants or induced tumors with carcinogens can be treated with a combination of prooxidant vitamins and some oligofructose in the diet, in order to starve the tumor cells. More significantly, a combined treatment of vitamin C (VC, sodium ascorbate) and vitamin K3 (VK3, menadione bisulfite) – abbreviated VC:VK3, and named "Apatone" – following irradiation or in combination with chemotherapy improved life expectancy in a rodent model. To verify these data, tumor-bearing organs were prepared for histochemistry, and it was shown in a series of antioxidant organic compounds that CK3 reactivated both nucleases, alkaline and acid deoxyribonucleases, and the ribonucleases [8,9]. The data were also verified using biochemical techniques [7,10,11]. Most tumor tissues were shown to be deficient in nucleases, but not those treated by VC alone, VK3 alone, or combined VC:VK3, which indeed were reactivated [7,9]. This mixture is nontoxic and inexpensive, so it could be used to assist classic oncologic treatments against several tumors, especially the malignant types.

The incidence of carcinomas in the digestive system [2,11] and central nervous system (CNS) [2,12,13], as well as in carcinogenesis, was found to be in a direct relationship with the repressed activity of nucleases (DNases and RNases) in tumors [14–18]. Hence, a possible reactivation of both alkaline and acid nucleases could be expected to decrease the resistance of cancer cells to radiation and chemotherapy [17,18]. It was also found that the coadministration of VC and VK3 in a variety of carcinoma cell lines resulted in tumor-specific antitumor activity of the aforementioned therapies at doses that were 10–50 times lower than when either vitamin was administered alone. Additionally, studies using a murine ascites transplantable liver tumor model have shown that the VC:VK3 combination is an effective chemosentisizer that induces little systemic or major organ pathology [19,20]. Similarly, studies carried out *in vivo* provide similar data to those *in vitro*, reinforcing this contention when a combined vitamin treatment allowed us to show the degradation of the proliferating tumors in nude mice [9,21].

28.2 About Cell Thanatology

Over the last 45 years, innumerable studies in experimental cell biology have been carried out in the field of cell death. These studies have led to detailed descriptions and characterizations of several types of cell death, named by diverse research groups. A recent self-elected committee is now apparently deciding whether a "cell death name" in published research is acceptable, based on a perplexing comment recently published by the Nomenclature Committee on Cell Death (NCCD). They propose to only recognize a cell death name through its biochemical characterization, and not its morphology [22]. The most current publication by the NCCD group, as of 2015, proposes to abandon the "morphologic catalogue of cell death" in favor of a new classification based on quantifiable biochemical parameters [23]. According to the NCCD – without morphologic evidence – a cell death name should be created or given based on a histogram, curves of enzymatic defect, or other nonrepresentational evidence. That is like having a corpse but avoiding dissection because you don't have the time to use an electron microscope or complementary morphological tools, and so ignoring molecular findings [22–26]. This imposition on discovery by a non-official, self-elected committee seems to go against scientific freedom. As of now, our group has published more than 21 papers and two book chapters, and so far has not even been recognized by the NCCD. Several members have contacted us in the past, both online and by telephone, and have received our reprints showing not only morphological but also biochemical data proving the validity of this new mode of cell death. Why, then, is it being ignored? All our published studies have been accessible through public information sites on the Internet. It is interesting to find that even though it does not see cell-structure changes as a clue for progression toward a descriptive illustration of a mode of cell death, the same French group directing the NCCD has undermined itself by publishing a report on organelles and cell death [27].

Necrosis and apoptosis are both well-known cell-death mechanisms. They cause processes that are classically taught in most biomedical lectures, colleges, and universities. They are extensively copied and illustrated, with less and less imagery, but certainly with some biochemical markers, according to the NCCD. Necrosis was eventually also named "oncosis" [28]. Apoptosis, originally defined morphologically by Kerr and others, is currently based on morphologic descriptions [23–32]. Several other modes of cell death have been described in many published studies, based on subtle biochemical metabolome changes or triggers. This chapter aims to clarify and describe more fully these less known modes. It is based on the two classic cell deaths in which autoschizis is mentioned [33].

We are grateful and honoured for the invitation by Dr. James A Radosevich, chief editor, to submit this chapter. Combining morphology with other techniques to ensure progress, it is hoped that this may stimulate further investigations to characterize autoschizis cell death. It is important to note that autoschizis cell death was discovered serendipitously in 1995. It was then rediscovered in 1998, and abundantly documented with morphologic and biochemical data, both *in vitro* and *in vivo* in the form of xenotransplants obtained in nude mice [34,35]. It holds great promise as it involves the killing of cancer cells and may have application in some forms of cancer treatment.

28.3 Background

Many tumors arise from the epithelium. These are known as carcinomas, adenocarcinomas, and adenomas. In order to corroborate the cytotoxic effects of prooxidant-induced damages (which eventually kill tumor cells), let us first make a succinct survey of the general ultrastructural aspects of these carcinoma cells. We will do this by describing their most common characteristics, using studies made in our laboratories and verified by other laboratories around the world.

28.3.1 Tumor Cell *In Vitro*

Viewed by scanning electron microscopy (SEM) or light microscopy (LM), all human and animal carcinoma cells – exemplified by human bladder (T24 and RT4), ovarian (MDAH 2774), and prostate (DU145) cell lines and by the murine prostate (TRAMP) cell line – used in our studies commonly appear pleiomorphic with a polyhedral-like profile in most preparations. In our studies, preparations for transmission electron microscopy (TEM) culminate in the centrifugation of samples, unless they can be fixed and cut *in situ* after being embedded in the same site. Otherwise, the collection of cells involves detaching them from their cultivated sites, whether on cover slips or in Petri dishes and flasks. Centrifugation of treated cells is carried out in a table-top centrifuge simultaneously with sham-treated ones, in order to allow meaningful comparisons to be made.

28.3.1.1 Carcinoma Cell Morphology

Figure 28.1 shows T24 cells (A,B) and DU145 cells (C) on a Petri dish support following SEM preparation. The DU145 cells are shown after a careful 1–2-minute slow centrifugation (500–1000 rev/min) in a table-top microfuge; the cell shape imparted by the centrifugal force changes the cells' aspect and renders them obviously more spherical. In addition, the cell diameter can vary according to the cell line studied. After each

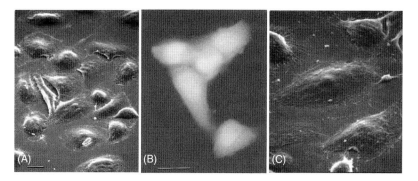

Figure 28.1 SEM view of (A,B) sham human bladder (T24) cells and (C) a prostate (DU145) cell growing on a small Petri dish support coated with polylysine display. Their flattened aspect where the centrally-located nuclear swelling allows nucleolar bulges to be noted. The smooth upper surface contrasts with the cell edges showing microvilli, filopodia and short lamellipodia. In B: LM field of view showing few clustered sham T24 cells after stained with FITC-phalloidin immunofluorescence revealing the entire sham cells are filled with a homogeneous content of actin, including the nucleus. Scales equal 10 μm in A or B for both micrographs; C is enlarged trice of A.

treatment, cells were centrifuged with the same apparatus and at the same time in the same fluid milieu, so that a comparison can be made between cells after collection and treatments.

Fixed *in situ* (on their cultivation surfaces), carcinoma cells display polymorphisms. They show cells frequently overlapping their margins or crawling over one another (Figure 28.1A–C), especially when they are confluent [36,39]. They can loosely aggregate as polygonal, squamous-like cells that overlap and show no contact inhibition, revealing some spherical shapes (due to cell division) with average diameters ranging from 28 to 30 μm: much larger than ovary (MDAH 2774) and prostate carcinoma (DU145 or TRAMP) cells. Tumor cells display a smooth to rugged or corrugated peau d'orange-like surface, decorated by abundant but diversely shaped microvilli. They show branching aspects that interdigitate their surfaces and are often decorated by a delicate coat of innumerable, irregular, oddly shaped and sized microvilli [40–42] with some multifurcate aspects [34,43,44] and some long, delicate filopodia or microspikes (0.5–2.5 μm in length); these especially appear when cells are fixed on cultivation surfaces. Cell contact inhibition can be demonstrated in cultivation by observing overlapping cells, caused by their surface charge and fibrolytic activities [45,46], even if they are fixed and processed in nonconfluent cultures [34,35,47,48]. When few contacts are made, they can appear similar morphologically to the zonulae adherentes between some cytoplasmic extensions <0.5 μm in length [43]. However, if the morphologic aspect appears typical, then, in most tumor cells, the expressed macromolecular components will be flawed by mutations (cadherins, catenin, etc.) [49,50].

Based on observations with SEM and TEM, the tumor-cell membranes of control or sham-treated cells do not show any damage due to centrifugation or cell processing. It is when they are treated by menadione that small foamy whorls (<0.3 μm) are sometime seen along some cell profiles [51]. An LM view of cultivated T24 cells following immunolabeling with phalloidin (for an actin pattern) demonstrates a full nucleic and cytoplasmic content of green-yellowish immunofluorescence (Figure 28.1B).

28.3.1.2 Organelles

28.3.1.2.1 Nucleus

Based on ultrastructural views and random sectioning, most carcinoma cells (e.g., DU145 cells, sham cells) possess a main ovoid-shaped, large euchromatic nucleus and scant cytoplasm. The heterochromatin appears as a dispersed "marbled" contrast among the nucleoplasm, which also includes peripheral patches along the inner membrane of its envelope [52,53]. The overall nucleoplasm is usually described as a "fibrogranular network," where fibrillar proteins of diverse sorts and some ribonucleoproteins occupy a space in which many macromolecules are still unknown in the normal nucleus; these are probably even more diverse in the neoplastic cell nucleus.

The tumor-cell literature describes the nuclear membrane as exhibiting "unpredictable" and unexplainable irregularities [53–55]. These invaginations are often diagnostic of poorly differentiated cells [56]. The carcinoma nuclear profile can typically possess a deep or shallow indentation known as a "pocket," "bleb," "canal," "tubular invagination," or "infolding" [57,58]. This often shows as one or more variable-sized but typical shallow indented grooves in a 1 µm-thick section, with a wide, distorted coffee-bean profile *in vitro* (Figure 28.2) [59] and *in vivo*. This is likely caused by the arrangement of the nuclear cytoskeletal lamins [60] and other interactions with the cytoplasmic cytoskeleton [61,62], which have also been associated with alterations in specific transduction activities [63,64] and with tumor-suppressor gene activity [65]. In some cancer cells, these irregularities in lamins have been studied for their mechanistic alterations [66], as well as for their impact on chromosomal abnormalities, resulting from lamin defects; these can include aneuploidy and chromosome instability [67,68]. Such nuclear coves or cytoplasmic invaginations are often rich in F-actin [69,70] and cytokeratins [71]; they can act as pathways along which transcripts move into the cytoplasm [72]. The channel-like structure can increase the surface area-to-volume ratio of the nucleus and facilitates the sustenance of the nucleoplasmic reticulum [58], as well as enhancing cytoplasmic–nuclear and, especially, cytoplasmic–nucleolar exchanges. This is because the nucleoli are often localized near the termini of these invaginations of the nuclear envelope. There are linkages between the nuclear envelope and cellular defects that are common in cancer cells; some nuclear-envelope proteins often show aberrant expression in tumors. As a consequence, the nuclear-envelope components are a potential tool for use as diagnostic and prognostic markers in cancers [73–76].

Tumor cells usually contain one large, reticulated, and branching nucleolus, sometime adjacent to or associated with the nuclear indented zone, forming thick strands and complex anastomoses in a network-like pattern containing the typical nucleolar components: dense and fine fibrillar components fashioning several nucleolar organizer regions (NORs), the granular component, interstices, and the associated heterochromatin. Due to its content in charged histones, the latter component is usually more electron densely contrasted than the others. The nucleoli often contain multiple NORs [77–86], which are characteristic of extremely active (and suggest poorly differentiated) cells [87–90]. They are usually associated with aneuploidy [83,91–94]. In an enlargement of Figure 28.2 (shown as Figure 28.3), one can see fine fibrillar centers and spherical grainy patches of granular regions dispersed in the large, complex, branching masses of heterochromatin, extended with a complex architecture of dense fibrillar component into the nucleolar mass. In the nucleus and its nucleolus, the

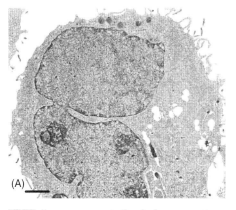
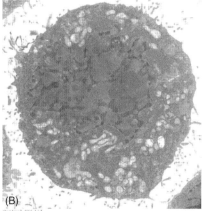
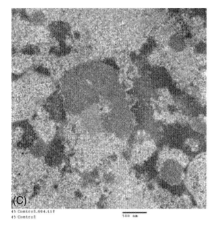

Figure 28.2 TEM aspects of human carcinoma cells. (A) A large, coffee bean-shaped nucleus of an ovarian carcinoma (MDAH 2774) cell, with its small cytoplasm crowded by electron densely contrasted ribonucleoproteins and containing a few fatty deposits and contrasted vesicular lysosomes (top area). A branching nucleolus appears in this section as three contrasted patches in the lower nucleoplasm. Scale = 2 µm. (B) A human prostate carcinoma (DU145) cell, with its large nucleus and small cytoplasm containing its typical pale contrasted mitochondria with disorganized cristae. Some smooth–rough endoplasmic reticulum (SER–RER) can be seen adjacent to or among these mitochondria, as well as a few superficial glycogen and fatty droplets. The enormous nucleolus contains many nucleolar organizer region (NORs) among the marbled aspect taken by the heterochromatin. Note the irregularly size and oddly shaped microvilli coating the cell surface. Scale = 2 µm. (C) Enlarged view of (B), where a nuclear wedge contains part of the enormous nucleolus attached to the nuclear envelope. Its zonation of dense fibrillar regions (DFRs) as extensions of the heterochromatin accounts for this marbled, high-contrast, but complex aspect. Several fine fibrillar regions (FFRs) are located in the center of the micrograph, surrounded by an arch of granular components (RNA with ribonucleoproteins), which can be seen as small spherical patches surrounded by loosely shaped DFRs. Scale = 0.5 µm.

heterochromatin is always the most highly contrasted structure following conventional fixation and staining for electron microscopy.

More than one nucleolus can be found in cancer cells, but they are exceptional in normal cells, even if they have been described in some LM observations made in rodent tissues [95]. Knowing that TEM only shows a narrow slice of cells, one has to be careful to claim what is only a partial slice of no more than one large, branching complex

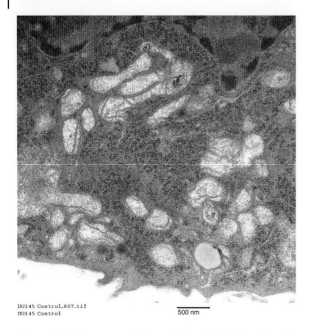

Figure 28.3 Human prostate carcinoma (DU145) cell perikaryon, demonstrating many pale-contrasted mitochondria, the dystrophic cristae of which have diverse orientations and flocculent matrices. The cytoplasm is evidently congested by polyribosomes and/or a few rough–smooth endoplasmic reticulum (RER–SER) cisternae. A few small, dense lysosomal bodies can be seen near the bottom part of the cytoplasm. Scale = 0.5 µm.

nucleolus per cell in the cell lines studied (Figures 28.2 and 28.3). It is possible that the thickness of sectioning may influence such an account. In fact, Figure 28.2A shows a section of the same nucleus seen later in Figure 28.28, showing three nucleoli that are actually part of one large branching one. A note about there being more than one nucleolus: some of our SEM micrographs suggest that possibly one may see two nucleoli in a tumor cell (Figure 28.1C; see also Figure 28.15B, later). In all carcinoma cells, the nuclear-to-cytoplasmic ratio (N:C) is characteristically different to that in normal cells, which maintain one N:C ratio per amount of DNA [93,96–101]. It has been shown that the higher the ratio, the better the prognosis, especially in glioblastomas [102].

28.3.1.2.2 Mitochondria

In tumor cells, the mitchondria appear dispersed, and often occupy a large cytoplasmic region facing the nuclear indent with all the other cell organelles. When fixed on their cultivation site, the cytoplasmic region spreads them outward in a fan-like fashion. This distribution can be suggested by microtubule components originating from a centrosome core, located adjacent to the nuclear concave zone that organizes their position throughout the cells. The population of these organelles varies among tumor cells, but can understandably be sparse, since it is likely associated to their poor aerobic metabolism [103]. The mitochondria range between 1.0 and 3.5 µm in length and are 0.5 µm in width, with a pale contrast as they stand out in the cytoplasm (Figure 28.4). Some have typical, elongated profiles, narrow and oblong in shape; these are often located in the perikaryon. Many others, more distally located in the cytoplasm, can be pleiomorphically large,

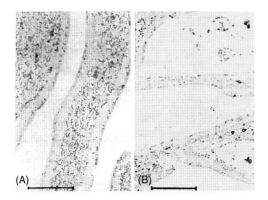

Figure 28.4 Histochemical localization of catalase activity in (A) sham epithelium and (B) pre- and postnatally diethylstilbestrol (DES)-treated Syrian hamster endometrium. Scale = 100 μm. *Source:* Gilloteaux and Steggles 1982 [129]. Reproduced with permission from John Wiley & Sons.

reaching 4–5 μm in length. Some rare tumor cells can display numerous mitochondria as oncocyte-like cells; these are the topic of another publication [21].

In tumor cells, the mitochondria have evolved and adapted with an intrinsic molecular makeup through the cell genome. They have also undergone mtDNA mutations [104,105]. Changes from normal cells accumulated during the formation of the tumor mitochondria affected (i) the composition of inner-membrane lipids and (ii) the enzymes located in those membranes and in the matrix. These include the antioxidant protection mechanisms that may influence the typical functional homeostasis maintained by ionized calcium [106–109]. No less than 77 mtDNA mutations have been reported for tumor cells [104,110].

Modifications of their inner-membrane lipids and enzyme components – along with a lack of cytochrome c (cyt c) oxidase – have been found in some tumor cells [111,112]. Morphologically, some of these alterations may be reflected by the positioning and altered topology of the cristae, which are oriented oblique, wavy, or parallel to the mitochondrial long axis instead of perpendicular to it, as is recognized in their classic topology (Figures 28.2 and 28.4) [113]. A fine, fuzzy-like dispersed fine granular material can be noted in the matrices, even without the use of phosphate buffers. This seems to be related to dysfunction in their ionic regulation, suggesting an excessive cationic uptake that would make them poorly preserved for ultrastructural viewing. Often, this mitochondrial fine view is referred to as "cristolytic" or "in cristolysis" [114]. These atypical features fit their altered tumor-cell metabolic status and defects [115,116].

Further supporting this, defects in ion channels [104], a poor expression or lack of mitochondrial cyt c oxidase [111,117], and a deficiency of antiperoxidative protective enzymes such as superoxide dismutase (SOD), glutathione (GSH) S-transferase, and thiols can be accounted for by many other changes. Other depleted enzymes include those controlling the fatty acid oxidation metabolism, located in the matrix of the mitochondrion, and those dealing with part of the urea cycle, gluconeogenesis, which regulates ion uptake (e.g., phosphate and calcium). Lower oxidative phosphorylation may well induce a low output of adenosine triphosphate (ATP), but this can be compensated for by a glycolytic adaptation, where fumarate acts as a significant fuel for these cancer cells [113,118–120]. By surviving with an adapted, glycolytic modified metabolism, many tumor cells, including DU145 cells, produce high levels of reactive oxygen species (ROS); DU145 cells show more than a fourfold increased ROS production as compared to normal prostate cells, due to their high level of glycerophosphate dehydrogenase and activity [121]. These excesses can cause lipid and mtDNA breaks.

Because cancer cells have been shown to have little or no mitochondrial SOD activity [122], this relative absence of ROS detoxifying enzymes should suggest that they are even more susceptible to the cytotoxic effects of peroxides. One of the histochemical landmarks of cancer cells and oxygen insensitivity is a significantly increased glucose-6-phosphate dehydrogenase (G6PD) activity, which is enhanced by hypoxia-inducible factor 1 (HIF-1) [123]. The poor antioxidant defence of such cells is supportive of the hypothesis that prooxidant compounds can assist in protecting active cells and help in killing cancer cells [2,7,20,124–126].

28.3.1.2.3 Peroxisomes

Peroxisomes and their activities in tumor cells is a subject neglected in cancer biology. Catalase activity (catalatic one) has been found elevated in some forms of renal carcinogenesis [10,127]. Most tumor cells have some population of peroxisomes wherein their associated catalatic and peroxidatic activities are usually significantly lower than what is found in normal cells. In our experience in carcinogenesis – induced through hormonal treatment in hamster endometrium and kidneys (see Figures 28.5 and 28.6A, B) – this is also true for other cancer cells in other organs [128,129]. Catalatic activity is low to negligible compared with normal, untreated tissues. This fact alone, probably forgotten by most clinicians, suggests that an appropriate prooxidant treatment would injure dysplastic cells toward cell death more easily than it would normal cells. Peroxisomes can be potent antioxidant armamentaria as organelles of normal cells. However, cancer cells contain 10–100-fold lower catalase activity than normal cells [130–138].

28.3.1.2.4 Rough Endoplasmic Reticulum

The RER loosely meanders as a few long, entwined cisterns among the other organelles. It sometime closely associates with mitochondria with small, narrow, elongated or even sinuous, curved cistern shapes in continuity with the SER.

28.3.1.2.5 Golgi Apparatus

The Golgi apparatus reveals minute cisterns or components in the perikaryal zone, and in the carcinoma cells used for experiments, is difficult to detect and, when found, appears small and without associated secretory vesicles in the heavily particulate

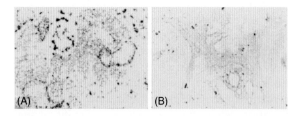

Figure 28.5 **(A) Control kidney showing staining for catalatic activity in the cortex, with strong distal convoluted tubule content.** Glomeruli present no catalase activity besides an obvious contrast caused by erythrocyte trapped in the rete mirabile. **(B) Diethylstilbestrol (DES)-treated hamster kidney showing poor or no staining for catalase activity.** Some disorganization of cortex tissue. Scale = 100 μm. *Source:* Gilloteaux and Steggles 1983 [130]. Reproduced with permission from Elsevier.

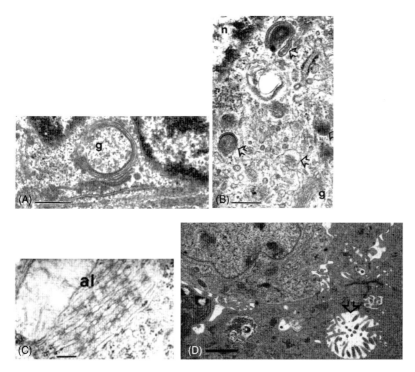

Figure 28.6 Examples of fine structures of human carcinoma organelles in bladder carcinoma (RT4) cells. (A) Example of small Golgi apparatus located in the perikaryon. (B) Lysosome-processing endocytosis in sham tumor cell (open arrows). Scale = 1 μm. (C) Annulate lamellae from a human prostate (DU145) carcinoma cell adjacent to a single mitochondrion. Scale = 100 nm. (D) Intracellular lumina (arrowed), lined by typical microvilli in a prostate (DU145) tumor cell. Scale = 1 μm. g, Golgi apparatus; n, nucleus. *Source:* Part A from Gilloteaux et al. 2012 [43]. Reproduced with permission from Taylor & Francis Group. Part B from Gilloteaux et al. 2014 [148]. Reproduced with permission from Taylor & Francis Group. Part C from Gilloteaux et al. 2013 [44]. Reproduced with permission from John Wiley & Sons. Part D from Gilloteaux et al. 2014 [114]. Reproduced with permission from Taylor & Francis Group.

cytoplasm (Figure 28.6A) [43,137]. However, its poor or absence of apparent secretory products is supportive of the dedifferentiating status of tumor cells.

28.3.1.2.6 Lysosomes

Lysosomes and their activities are suggested by viewing small, electron-dense onion complexes, heterogeneous vacuoles, or membranous components encircling organelles, confirming small but nondeleterious cell maintenance as autophagocytotic events occur in the cytoplasm of active sham-treated carcinoma cells [21,51,100], probably linked to peroxidative events linked in turn to accumulations of ROS (see earlier). The lysosomal bodies are more numerous than those of normal, active epithelial cells. This perinuclear zone clusters lysosomes in some carcinoma cells; this may be caused by the recruitment or alteration of normal functions in motor proteins. Lysosomes can be scarce – although their activities in tumor cells are not [139,140]. However, in some aggressive tumors, they can be abundant; this may assist in tumor penetration and growth, as they are released or captured by the surrounding matrix [141]. Primary lysosomes are poorly represented, as suggested, but they have not been characterized through cytochemistry in this cell line and they seem scattered among the smooth and short endoplasmic reticulum (ER) cisterns.

28.3.1.2.7 Annulate Lamella
The annulate lamella (Figure 28.6C) is a typical organelle landmark for tumor cells, found only on gametes, embryonal and fetal tissues, and cancer cells [142–146]. They can be seen as short to long lamellae among the perikaryal cytoplasm. In most tumor cells, as in gametes and embryonic cells, these peculiar structures support active synthesis of membranes of the nuclear envelope and the RER [35,43].

Another typical structure of tumor cells, but one that is not restricted to carcinoma, is the **confronting cisternae** noted in DU145 cells [43,147].

Some carcinoma cells position in their perikaryal areas one small to large **intracytoplasmic lumina** (Figure 28.6D), whose internal surface is usually coated with microvilli [113,148]. These surface specializations can be abraded as smooth surfaces when mucus or other secretory materials fill them. They probably arise in malignant cells, as they evolve with a loss of cell polarity.

28.3.1.2.8 Inclusions and Storage Components
The cytosol of carcinoma cells is often filled by innumerable ribonucleoproteins and/or ribosomes, a feature that is also found with liver carcinogenesis [114,149]. This may reflect the loss or inactivation of RNases during transformation and in tumor cells, as shown in Figure 28.3 (see also Figure 28.38C and previous data [9,12,13,113]), and it includes xenotransplants [43,44,150,151]. An abundance of ribosomes or polysomes, admixed with glycogen-like particles or often aggregates, forms most of the fine inclusions distributed quasi-evenly throughout the cytosol. These are distributed among all the other organelles of these active tumor cells, and are easily found even after staining with toluidine blue and viewed by LM. Glycogen accumulates and appears with a poor contrast in the peripheral locations of the cytoplasm, where it is detected as patches or dispersed aggregates. It can also be formed from adjacent small lipid droplets (Figures 28.2 and 28.4). Glycogen accumulation in most carcinomas, such as DU145 cells grown *in vitro* or *in vivo* [43,113,152], is the energetic reserve that is associated with glycolysis and the high glycogen synthase kinase-3 (GSK-3) activity found in androgen-independent tumors, such as DU145 [153–155]. This is because, like most cancer cells, these have a modified metabolism in which fumarate can become a significant fuel source [103,115,156–158]. The fine-structure observations made in DU145 have also been made in other human prostate tumors [159]. Our data describing the paucity of organelles are mirrored in solid DU145 tumors xenotransplants [43,44,150,151].

28.3.1.3 Other Considerations
Finally, contributions by mutations and aneuploidy result in a tumor-cell genome that harbours an altered structure, modified growth-factor functionss [160], proto-oncogenes [161], and tumor-suppressor activities [162]. Cell-signal transduction changes cell-cycle regulation; for example, p53 is overexpressed and/or mutated in more than 50% of prostate tumors [163,164]. A similar thing is seen in breast-cancer cells [165]. Mutation of p53 in conjunction with splicing variants of autophagy-related gene 5 (ATG5) prevents autophagic cell death in these cells [166,167]. Unlike in normal cells, where p53 tumor-suppressor proteins can induce apoptosis or some other cell-death pathway, the use of cyt c as an intracellular critical key signal is not possible in abnormal

cells, because p53 protein is mutated in many carcinomas [168,169]. Being deficient in cyt c expression makes a signal for "death" deficient, which either allows these tumor cells to survive after injury, thriving and carrying on growth by replicating their genomic diversity instead of being controlled by their genetic makeup, or eventually makes (DU145) cells less likely to die from apoptosis, because their Fas-ligand receptor of programmed cell death (PCD) is dysfunctional [170]. These characteristics make DU145 likely to move toward another cell-death pathway as a result of a cytotoxic prooxidant treatment.

28.3.2 Tumor Cells *In Vivo*: Implanted Carcinomas

Human carcinoma cells can be implanted and grown in nude mice. For example, implanted in the abdominal cavity, DU145 tumor cells were found to attach to many areas of the peritoneum after some pioneer carcinoma cells had removed the original mesothelial cells of the mesothelium covering through a sort of bulldozing action; they replaced them with an aggressive growth pattern. These tumors were described using SEM and TEM [21,43]. They grew as solid masses and invaded the subjacent tissues of the organs they colonized. For example, they invaded into the diaphragm muscle through the stroma and constituted at least two invasive phenotypical growths, including both solid carcinomas and adenoid, cribriform, or adenoid carcinomas (Figures 28.7–28.9).

28.3.2.1 Xenotransplanted Carcinoma Cell Morphology

Implanted carcinomas are composed of two cell types: (i) a poorly differentiated, pale eosinophilic cell type with a diameter ranging from 10 to 15 μm, which grows tumors

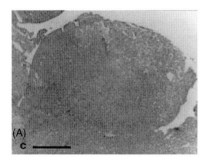

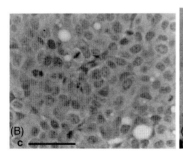
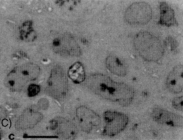

Figure 28.7 Xenotransplant of prostate (DU145) cells in severe combined immunodeficiency (SCID) nude mice (nu/nu) growing on diaphragm as a solid tumor that has not yet vascularized. The compact tissue shows pleiomorphic cells, with poor intercellular spaces, some mitotic stages, and cytophagic activities. Scale: (A) = 100 μm; (B) = 50 μm; (C) = 10 μm.

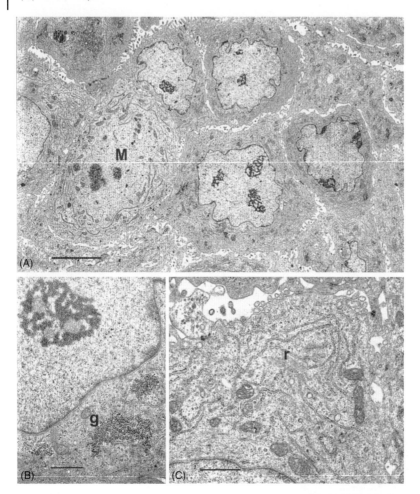

Figure 28.8 TEM views from the mid region of a dome-shaped, DU145 carcinomatous tumor area growing as a dome-shaped solid carcinoma on the nu/nu mouse diaphragm. (A) General aspects of a field of tumor cells, with their large nuclei and active nucleoli. One mitotic cell is shown (M). The DU145 cells are closely associated, without any fibrillar matrix. However, the narrow intercellular spaces are filled with microvilli and filopodia-like cell extensions and are connected with narrow zonulae adherentes. Some examples are enlarged in Figure 28.5. Scale = 5 µm. (B) High magnification of the nucleolus and its nucleolar organizer regions (NORs), glycogen (g) patches near small rough endoplasmic reticulum (RER) segments. Scale = 200 nm. (C) Scattered mitochondria among a bunch of RER wavy tubular components. Note the intercellular spaces. Scale = 200 nm. *Source:* Gilloteaux et al. 2012 [43]. Reproduced with permission from Taylor & Francis.

intraperitoneally; and (ii) a large, basophilic cell type, 8–25 µm in diameter, which is able to invade the peritoneal stroma of organs (i.e., the diaphragm).

28.3.2.2 Organelles and Organization of Xenotransplanted Cells

The basophilic cells are pleiomorphic, but have a larger nuclear:cytoplasmic ratio than the pale cells. They can be found predominately near the base of the tumor. In LM, the 1 µm-thick sections show that the small tumors are quite solid and do not have much extracellular space (Figures 28.7–28.9).

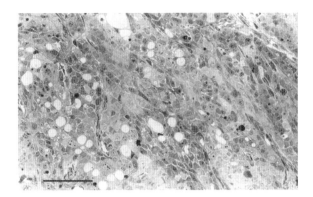

Figure 28.9 LM micrograph of a 1 μm-thick epoxy section (stained by toluidine blue) of a 9 mm-diameter DU145 xenotransplanted tumor growing on the diaphragm of nude mice for at least 10 weeks after implantation. Note the many signet cells formed out of intracellular lumina, producing small to large cystic-like spaces throughout the heterogeneous but solid carcinomatous tissue organization. Scale = 100 μm.

DU145 implants growing on the peritoneal surfaces of nude mice appear with a similar histology to that shown in Figure 28.7A, and are also similar to those described in other laboratories, such as TLT hepatoma, where the same prooxidant activities of VC:VK3 were studied [7,9,17–19,171–173]. Examined using TEM during its early growth, as shown in Figure 28.8 [43], this type of solid tumor cell shows intercellular spaces and tiny or narrow intercellular junctions that resemble the zonulae adherentes. With SEM, most cells show odd shapes and pleiomorphic microvilli-like features on their surfaces, with short, long, or even branching types.

When DU145 cells are attached to a collagenous substratum or to the extracellular matrix (ECM) of the submesothelium of implanted nude mice, they are pleiomorphic, mainly polygonal in shape, and often closely adjacent to reduced attachments. All the tumor cells examined appear poorly differentiated and contain large, spherical nuclei with at least one nuclear indentation and a small cytoplasm that is characteristic of them. They evidently show a high nuclear:cytoplasmic ratio (Figures 28.7B,C and 28.8A). Each nucleus contains a large, extended, and branching nucleolus with multiple NORs (Figures 28.2A,B and 28.8A,B).

The DU145 cytoplasm is crowded with microfilaments, ribosomes, and glycogen-like particles, in which a few other organelles are dispersed. For example, small round to oblong mitochondria among whorls of wavy or twisted concentric or winding cisterns of RER are revealed (Figure 28.8C) where any part of a Golgi apparatus is difficult to detect in ultrathin sections. Small, poorly contrasted vesicles, ranging between 50 and 150 nm, can be seen near the margins of the cells, representing primary lysosomes. Intracellular lumina, as well as aggregates of glycogen, are visible among the ER and some lysosomes [21,43,51]. Other lysosomal bodies enclose heterogeneous content, as they remodel parts of cells as autophagosomes (Figure 28.6B) Using LM, large digestive vacuoles can be seen, without detail; they include damaged cells, making them cytophagosomes (Figure 28.7B,C) [43]. The same can be found in larger tumors where signet cell-like cells seem to originate from intracytoplasmic lumina (Figure 28.9).

Intracytoplasmic lumina found in xenotransplants are similar in morphology to those observed *in vitro*. A malignant tumor cell may contain one or more intracellular lumen. This type of apparent cytoplasmic void forms out of coalescing cytoplasmic vesicles, which become coated with a rich glycocalyx associated with microvilli [43] and are probably formed from a loss of cell polarity among the tumor cells [142,148,174]. After toluidine blue staining, a few of these intracellular cavities reveal deposits with a fine granular contrast – especially after they have lost their microvillar surfaces – and may contain mucus. In this case, the tumor is considered quite aggressive, because these carcinoma cells are known as signet cells. The development of signet cells occurs during the enlargement of the solid tumors. It could be that, through their further enlargement, these cells are able to evolve into crypts and form "cribriform" carcinomas and, from them, adenocarcinoma aspects. These can be noted especially in prostatic carcinomas, but also in other types, as illustrated in Figure 28.9 and hypothesized in xenotransplant evolution and development [43,44].

Carcinomas as tumor cells derived from epithelia can be characterized by:

1) **pleiomorphism with loss of contact inhibition and polarity;**
2) **epigenetic anomalies in cells and mitochondria:**
   a) **high nuclear:cytoplasmic ratio and aneuploidy;**
   b) **large nucleolus with numerous NORs and annulate lamellae;**
   c) **mtDNA with mutations adapting mitochondria to aerobic and anaerobic metabolisms according to their tumor location;**
   d) **abundant lysosomes but defective peroxisomes, with Golgi;**
   e) **often poorly developed apparatus;**
   f) **lack or deficiency of DNases and RNases.**

28.4 Early Observations of a New Mode of Cell Death: Autoschizis

Based on the findings by Henryk Taper concerning the nuclease reactivation in cancer cells [7,18,175], the antitumor activity of VC and VK3 combinations against several human urologic carcinoma cell lines was evaluated using cytotoxicity tests and morphologic and biochemical investigations. It was observed that these tumor cells, like the xenotransplant tumors, did not appear to die through either apoptosis or necrosis [34,35,176].

Initially, one human prostate carcinoma (DU145) cell line was studied, first using *in vitro* techniques, and then SEM and TEM. These early observations noted cytotoxic changes that suggested cell enucleation, with peculiar cuts suggestive of self-mutilation [34,177,178]. However, due to the poor technical quality of this study, we were unconvinced by the data, and decided to repeat it using two other tumor cell lines, namely human bladder carcinoma (T24) and ovarian (MDAH 2774) cells, and later again with TRAMP cells. The first cell line provided us with many surprising morphological details via LM, SEM, and TEM techniques [35,47,179,180], as well as flow cytometry [176,181] and some biochemical observations [176–178]. The data collected from the T24 cell line later gave us almost similar cytoplasmic changes concerning cell size and injuries leading to cell death. There were cell-cycle disruptions, characterized by blocks in the G1/S and G2/M stages [176,181], and specific nuclear damages that included membrane, metabolic, and DNA changes with degradation [47,176–178,182,183].

Following these promising preliminary studies, we examined other cancer cell lines (J82, SCaBER, TCCSUP, and RT4) again replicated most of our original data in terms of cytotoxicity [176–179]. At this point, a careful review of all of the data made us realize that we had discovered a new mode of cell death, which we first called "autoschizis necrosis," then "autoschizis" – especially because of the difference in injurious nuclear changes observed when compared with apoptosis [35,179,180]. This terminology will be justified later in the chapter. Further data from DU145 and other cell lines [15,34,35,43,51,71,139,179,180,183–188], as well as from xenotransplants [9,21], helped clarify the details concerning this type of cell death.

28.5 Cytotoxicity of VC:VK3 on Tumor Cells: The *Causa Mortis*

Details of our observations were obtained in both human prostate carcinoma (DU145) cells [176,181] and bladder carcinoma (T24) cells [188,189]. Recent data obtained in T24 cells are outlined in the following section and summarized in Table 28.1. In this table, the fractional inhibitory concentration (FIC) index is used to evaluate the synergism of the vitamins. A FIC < 1.0 indicates a combination is synergistic, while a FIC > 1.0 indicates it is antagonistic. A FIC = 1.0 indicates the combination is indifferent.

28.5.1 Bladder Carcinoma (T24) Cells

Vitamin treatment of T24 cells resulted in 50% cytotoxic dose (CD_{50}) values of 1492 ± 141 mM for VC alone and 13.1 ± 0.01 mM for VK3 alone. When the cells were exposed to Apatone, the CD_{50} values of decreased to 212 ± 7.6 and 2.13 ± 0.06 mM, respectively (Table 28.1). These results represent a fivefold decrease in the CD_{50} of VC and a sixfold decrease in that of VK3. The FIC for the T24 cells was 0.158, indicating a synergistic interaction between the vitamins. The results of another study demonstrated that human foreskin fibroblasts were four- to sixfold less sensitive to the cytotoxic action of the vitamins than were the tumor cells [190].

Because the endpoints in an MTT assay can be a function of a number of factors, including antimetabolic activity, cell death, and cell-cycle blockage, flow cytometry was employed to determine whether vitamin treatment affected the cell cycle of T24 cells. Detached cells in the supernatant were pooled with adherent cells and then analyzed by flow cytometry. A histogram summarized the cytotoxicity against T24 cells.

Data were complemented by other data collected from other cell lines using similar techniques; we have also attached another example with human prostate carcinoma (DU145) cells using the same vitamin concentrations (Table 28.1).

28.5.2 Prostate Carcinoma (DU145) Cells

Prostate carcinoma (DU145) cells treated by the combined vitamins resulted in CD_{50} values of $2,455 \pm 28$ µM for VC alone and 12.9 ± 0.8 µM for VK3 alone. When the vitamins were combined, the CD_{50} values for VC and VK3 decreased to 122 ± 15.6 and 1.22 ± 0.16 µM, respectively (Table 28.1). These results represent a tenfold decrease in the CD_{50} of VC and a fivefold decrease in that of VK3. The FIC for continuously treated DU145 cells was 0.145, indicating a synergistic interaction between the vitamins. Human foreskin fibroblasts are four- to sixfold less sensitive to the cytotoxic action of the vitamins than are the tumor cells [188]. At the end of the experiment, and following data collection, the enhanced antitumor activity was 25-fold increased 1 hour post-treatment and 20-fold increased 5 days post-treatment using VC:VK3 compared to each vitamin alone [34,177,178].

Table 28.1 Antitumor activity of vitamins after 1-hour and 5-day co-incubation. Measurements of antitumor activity obtained by microtetrazolium assay following 1-hour and 5-day exposure to vitamin C (VC, ascorbate), vitamin K3 (VK3, menadione bisulfite), or a vitamin combination with VC:VK3 ratio 100:1. Values are the means ± standard error of three experiments, with six readings per experiment. *Source:* Venugopal et al. 1996 [178]. Reproduced with permission from Elsevier.

		Antitumor Activity of Vitamins				
		Vitamins Alone		Vitamin Combinations		
Cell Line	Exposure Time	Vit C 50% Cytotoxic Dose (μM)	Vit K$_3$ 50% Cytotoxic Dose (μM)	Vit C 50% Cytotoxic Dose (μM)	Vit K$_3$ 50% Cytotoxic Dose (μM)	Fractional Inhibitory Concentration (F.I.C.)
Caki-1	5 day	376 ± 1	19.3 ± 0.33	89 ± 2.9	0.90 ± 0.01	0.283
	1 hour	2,777 ± 41	56.1 ± 4.2	379 ± 1.6	3.79 ± 0.02	0.206
DU 145	5 day	2,455 ± 28	12.9 ± 0.8	122 ± 15.6	1.22 ± 0.16	0.145
	1 hour	6,009 ± 8	77.2 ± 2.9	312 ± 4.0	3.13 ± 0.06	0.093
J82	5 day	376 ± 1	8.2 ± 0.15	79 ± 1.0	0.79 ± 0.01	0.301
	1 hour	914 ± 8	19.5 ± 0.4	188 ± 1.9	1.87 ± 0.02	0.296
RT4	5 day	2,427 ± 28	12.8 ± 0.03	110 ± 9.6	1.10 ± 0.10	0.136
	1 hour	4,737 ± 27	60.7 ± 4.0	267 ± 4.9	2.68 ± 0.05	0.100
SCaBER	5 day	327 ± 6	9.7 ± 0.15	71 ± 1.0	0.71 ± 0.01	0.291
	1 hour	1,782 ± 80	23.5 ± 0.1	303 ± 0.6	3.04 ± 0.02	0.206
T24	5 day	1,492 ± 141	13.1 ± 0.01	212 ± 7.6	2.13 ± 0.06	0.158
	1 hour	4,974 ± 246	73.0 ± 5.9	120 ± 7.0	1.20 ± 0.07	0.040
TCCSUP	5 day	297 ± 7	6.3 ± 0.06	48 ± 0.1	0.489 ± 0.01	0.238
	1 hour	675 ± 8	40.5 ± 0.7	154 ± 2.7	1.54 ± 0.03	0.266
Tera-2	5 day	1,312 ± 10	21.8 ± 0.73	195 ± 0.7	1.96 ± 0.01	0.239
	1 hour	5,515 ± 114	90.3 ± 0.6	447 ± 5.0	4.47 ± 0.05	0.130

FIC = $CD_{50}^{A\,comb}/CD_{50}^{A\,alone} + CD_{50}^{B\,comb}/CD_{50}^{B\,alone}$, where $CD_{50}^{A\,alone}$ and $CD_{50}^{B\,alone}$ are 50% cytopathic doses of each vitamin alone; $CD_{50}^{A\,comb}$ and $CD_{50}^{B\,comb}$ are the 50% cytopathic doses of the vitamins administered together.
Antitumor activity was measured by a MTT assay following a 5 day or 1 hour exposure to vitamin C, vitamin K$_3$ or a VC/VK$_3$ combination with VC:VK$_3$ ratio of 100:1. Values are the mean ± the Standard error of the mean of three experiments with six readings per experiment.

28.6 Flow Cytometry

28.6.1 Bladder Carcinoma (T24) Cells

Human foreskin fibroblasts were mixed with either DU145 [176] or T24 cells in an effort to determine the channel number of the true diploid G0/G1 peak (Figure 28.10) [189]. Looking at the most recent data from T24 mixed with fibroblasts, a true diploid peak (G0/G1 for the fibroblasts) is located in channel 59, and its corresponding G2/M peak in channel 118. Conversely, the T24 cells exhibit a G0/G1 peak in channel 108 and a G2/M peak in channel 214. These observations are in agreement with karyology studies that describe T24 cells as hypo- to hypertetraploid.

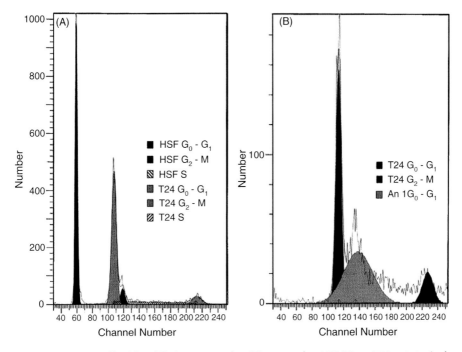

Figure 28.10 T24 cells either (A) sham-treated or (B) exposed to VC:VK3 at 90% cytotoxic doses for 1 hour. The vitamins were removed, and the cells washed twice with phosphate-buffered saline (PBS) and overlain with culture medium. After 24 hours, the cells were harvested, and DNA ploidy and cell-cycle analysis were performed on an Ortho Cytoron flow cytometer. Data were collected from 2×10^4 cells (when possible), stored, and analyzed using ModFit Cell Cycle Analysis. Sham-treated T24 cells served as the negative control, while human foreskin fibroblasts served as the diploid control. (A) Flow cytometry trace of a mixture of diploid human foreskin fibroblasts (HSF) and bladder cancer cells (T24), which allows for the determination of the ploidy of the T24 cells with respect to normal diploid cells, as well as the cell-cycle distribution of both cell types following sham treatment. (B) Flow cytometry trace of T24 following Apatone treatment, showing the enrichment in the G0/G1 and the G2/M populations, as well as the appearance of an aneuploid peak indicative of autoschizis. *Source:* Jamison et al. 2010 [189]. Reproduced with permission from SAGE.

In the case of **sham treatment**, 72% of the cells were in G0/G1, 18% were in S phase, and 10% were in G2/M. Forward scatter of Apatone-treated cells (data not shown) revealed the presence of two populations. The first (primarily adherent cells) had the same dimensions as the sham-treated T24 cells, whereas the second (primarily detached cells) was smaller in size. Both populations also differed with respect to their DNA content, with the detached population having less DNA than the adherent. This corresponds with the observations made with the Feulgen DNA stain [183].

In the case of **VC:VK3 treatment** (Figure 28.2) [189], 47% of the cells counted were adherent, aneuploid, and of the same size as controls (Table 28.1) [189]. However, following VC:VK3 or Apatone treatment, 0% of the cells in the aneuploidy population was blocked in late G1 to early S phase. The remaining 53% of the cells counted were detached and of a smaller size compared to control cells (10–15 μm); 79% of these were in G0/G1 and 21% in G2/M (Figure 28.9, Table 28.1) [189]. The lack of cells in S phase suggests that either tumor cells were arrested in late G1 phase or they were arrested in both G1 and G2/M (with the cells in G1 arrested in G1 and those that had passed the G1

Table 28.2 Cell-cycle distribution of the percentage of T24 cells measured. T24 cells were exposed to both vitamins at their 90% cytotoxic doses for 1 hour, incubated for 24 hours, and then harvested. DNA ploidy and cell-cycle analysis were performed on an Ortho Cytoron flow cytometer, and the data were analyzed using ModFit cell-cycle analysis. Sham-treated T24 cells served as the negative control. Human foreskin fibroblasts served as the diploid control. VC, vitamin C; VK3, vitamin K3. *Source:* Jamison et al. 2010 [189]. Reproduced with permission from SAGE.

Vitamin	G0/G1 phase	S phase	G2/M phase
VC and VK3			
Small cells	79	0	20
Large cells	100	0	0
Total cells	89	0	11
Sham/control	70	19	11

checkpoint arrested in G2/M). When the cell-cycle distributions of both populations were calculated using weighted averages, 89% of the cells were in G0/G1 ((79 × 0.53) + (47 × 1.00)) and 11% were in G2/M (Table 28.2).

28.6.2 Prostate Carcinoma (DU145) Cells

Human foreskin fibroblasts were mixed with DU145 in an effort to determine the channel number of the true diploid G0/G1 peak (Figure 28.11), which was located in

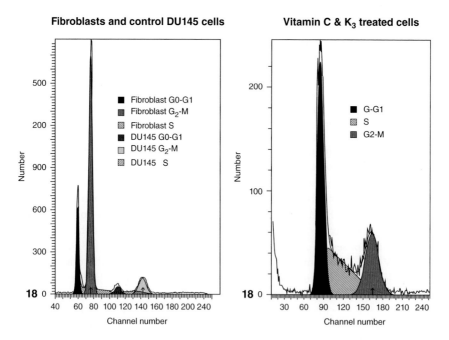

Figure 28.11 DU145 cells exposed to vitamins VC and VK3 at 90% cytotoxic doses for 24 hours and then harvested. DNA ploidy and cell-cycle analysis were performed on a FACScan flow cytometer and analyzed using MofFit Cell Cycle Analysis. Sham-treated Du145 cells served as the negative control. Human foreskin fibroblasts served as the diploid control.

Table 28.3 Cell-cycle distribution in percentage (%) of DU145 cells after treatments.
Source: Jamison et al. 1996 [177]. Reproduced with permission from Elsevier.

Treatment	G0/G1 phase	S phase	G2/M phase
Ascorbate alone (VC)	46	40	14
Menadione alone (VK3)	51	43	6
Combination of VC:VK3	44	33	23
Sham-control	84	5	11

channel 59, with its corresponding G2/M peak located in channel 118. Conversely, T24 cells exhibited a G0/G1 peak in channel 119 and a G2/M peak in channel 156. These observations are in agreement with karyology describing the DU145 aneuploid cell line, which is hypotriploid. The distribution among the phases is given in Table 28.3.

VC:VK3 treated cells (Figure 28.12B) show a G0-G1 peak in channel 85 and a G2-M peak in channel 164. Sub G0-G1 multi-cut debris is also shown in Figure 28.12B. As a consequence of this treatment, the proportion of DU145 cells in the S phase and G2-M phase are 33 and 23% respectively, compared with 5% and 11% for sham – control DU145 cell (Figure 28.11A,B)

As a result of a prooxidant treatment of carcinoma cells, flow cytometry suggests significant:

1) **cytotoxicity induces disturbance of the cell cycle, with blocks in G1/S and G2/M phases through cytoskeletal damage (i.e., are functional check points still active in the treated tumor cells?);**
2) **aneuploid nuclei and data suggestive of the appearance of tumor-cell corpses.**

28.7 Autoschizis or Self-Excisions of Tumor Cells Caused By Prooxidants

Alterations induced in tumor cells by the combination of VC and VK3 versus sham treatment were surveyed and found to include LM, SEM, and TEM morphologic changes affecting the plasma-membrane and superficial cytoplasm, nucleus and nucleolus, and organelles, as well as injuries or damage progressing toward cell death, especially autoschizis and autoschizic cell death. Those changes induced by the prooxidant treatment will be further explained in view of the biochemical data in order to justify the new name given to this form of cell death.

Using human carcinoma cells [bladder (T24 and RT4), ovarian (MDAH 2774), prostate (DU145), and TRAMP cell lines] and a murine prostate carcinoma (Table 28.4), measurements of treated tumor cells demonstrate that they decrease in average size. This is reflected and confirmed by pieces of cell or cell debris detected in the intercellular spaces. Additionally, the cells also change shape. Other illustrations complement these data (see also Table 28.3), as they reflect measurements from the same populations of carcinoma cells analyzed: one of two human bladder (i.e., T24 cells in Figure 28.12),

Table 28.4 Carcinoma cell-size alterations after different treatments (measurements do not include cell corpses and debris size). Dsem, average diameter in µm ± s.e.m; n, number of measurements; Vcalc, average volume calculated in $µm^3$; ±, percentage increase or decrease of calculated volume; –%, percentage volume decrease compared to sham-treated cells. In the RT4 line (*), measurements were limited to low TEM micrographs. 1-hour sham was used as baseline for comparisons and calculations.

Treatment		1-hour	2-hour	4-hour
DU145				
SHAM	Dsem	15.6 ± 3.2 (n = 267)	15.5 ± 3.4 (n = 452)	15.9 ± 2.1 (n = 336)
	Vcalc	3479	3535	3053
VC:VK3	Dsem	14.2 ± 4.1 (n = 216)	11.9 ± 4.9 (n = 245)	10.9 ± 2.9 (n = 278)
	Vcalc	3208	2487	1379
	±	–9%	–70%	–45%
MDAH 2774				
SHAM	DsemVcalc	17.5 ± 3.1 (n = 345)	16.9 ± 3.3 (n = 347)	17.2 ± 3.6 (n = 298)
		4577	4316	5425
VC:VK3	Dsem	14.7 ± 3.8 (n = 321)	12.8 ± 2.4 (n = 267)	10.5 ± 1.8 (n = 300)
	Vcalc	3315	1839	974
	±	–28%	–57%	–80%
T24				
SHAM	Dsem	29.4 ± 3.7 (n = 250)	29.5 ± 3.0 (n = 219)	30.0 ± 3.3 (n = 235)
	Vcalc	18988	17975	19335
VC:VK3	Dsem	24.1 ± 4.6 (n = 160)	16.7 ± 6.8 (n = 213)	15.4 ± 2.1 (n = 183)
	Vcalc	13240	6795	2807
	±	–30%	–62%	–85%
RT4 *				
SHAM	Dsem	20.5 ± 3.2 (n = 12)	n/a	n/a
	Vcalc	6970		
VC:VK3	Dsem	19.0 ± 4.0 (n = 11)	12.5 ± 2.0 (n = 11)	6.0 ± 1.2 (n = 19)
	Vcalc	6371	1596	195
	±	–9%	–73%	–97%
TRAMP				
SHAM	D_{sem}	8.6 ± 2.5 (n = 944)	8.8 ± 2.5 (n = 909)	8.8 ± 2.8 (n = 838)
	V_{calc}	717	756	792
VC:VK3	Dsem	6.7 ± 2.4 (n = 574)	5.1 ± 2.6 (n = 667)	5.0 ± 3.3 (n = 649)
	Vcalc	395	239	299
	±	–45%	–68%	–62%

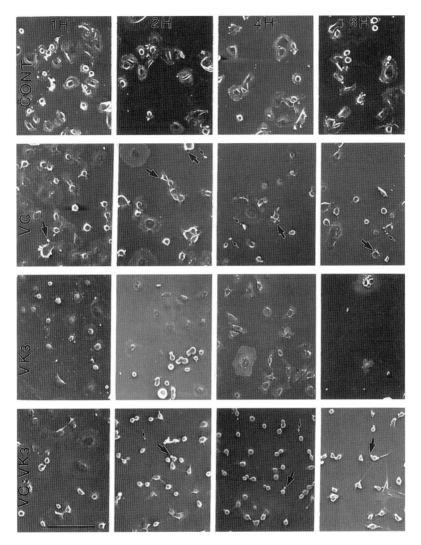

Figure 28.12 SEM of T24 cells cultivated without attaining confluence. Gallery of representative fields of observation, in which the rows are the treatments (sham or control, VC alone, VK3 alone, and combined VC:VK3) and the columns are the treatment durations (1, 2, 4, and 6 hours). Note the general alterations of tumor cell morphology, especially the diminished cell size. A treatment lasting more than 4 hours, especially VK3 and VC:VK3, resulted in a remaining scant population density in a sequence of apparent surviving cells as VK3 < VC:VK3 < VC < sham treatments. Scale = 100 μm for all SEM micrographs. Source: Gilloteaux et al. 2006 [184]. Reproduced with permission from John Wiley & Sons.

ovarian (MDAH 2774; Figure 28.13), and murine prostate (TRAMP) carcinoma cells (Figure 28.14).

Table 28.4 shows that, after a 1-hour VC:VK3 treatment, the average cell diameter is reduced in all tumor cell lines when compared with 1-hour sham-treatment: by 9, 28, 30, and 45% in the case of DU145, MDAH 2774, T24, and TRAMP, respectively. A 2-hour treatment reduces the cell volume more – by 57 (MDAH 2774), 62 (T24), 68 (TRAMP) and 70% (DU145) – showing some plateaued consistency. Finally, a 4-hour treatment reduces

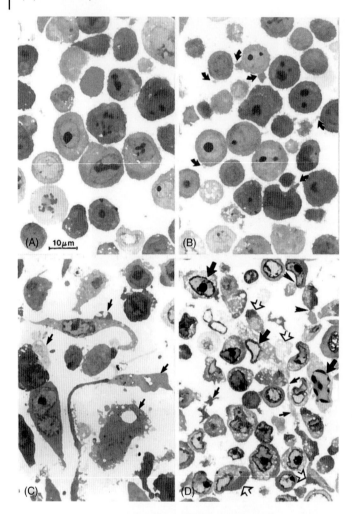

Figure 28.13 LM of MDAH 2774 cells following treatment obtained from 1 μm-thick epoxy sections stained by toluidine blue. (A) Control-sham displaying examples of typical tumor cells, with a large nucleus and large nucleolus and stages of mitotic activity. (B) VC-treated cells demonstrating blebs (curved arrows) and hairy aspects. (C) VK3-treated cells revealing induced pleiomorphism, vacuolization, and self-excisions of cytoplasmic pieces (small arrows) associated with other superficial damages (cytoplasmic blebs). Note the large nucleoli and the indented nuclear shapes with binucleate aspects. (D) VC:VK3-treated cells demonstrating their diminished size while maintaining a large nucleus and vacuolated cytoplasm. These appear with elongated shapes and vacuoles, and are poorly contrasted by toluidine blue. Most can be viewed in the process of self-excision (open arrows). Innumerable cell debris (small arrows) results from the cytoplasmic pieces (same open arrows) that are scattered throughout the intercellular field of observation. Nuclei are changed, and display a milky pink to rosy hue, revealing condensation of heterochromatin along the inner membrane, with some oblong patches (thick arrows). Scale = 10 μm. *Source:* Gilloteaux et al. 2003 [392]. Reproduced with permission from John Wiley & Sons.

it even more: by 45 (DU145), 62 (TRAMP), 80 (MDAH 2774), and 85% (T24). These diminished volumetric values (as calculated), even if they are exaggerated due to processing, reflect that prolonged treatment makes tumor cells lose not only size, but also number, as shown in the panels of micrographs in all published data. A 6-hour treatment of T24 cells

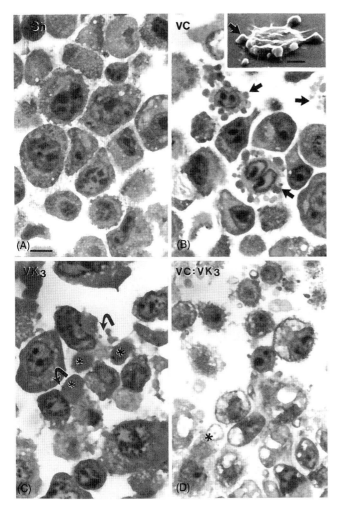

Figure 28.14 Comparative photomontage of LM views of 1 μm-thick epoxy sections (stained by toluidine blue) of TRAMP carcinoma cells after 2-hour (A) sham, (B) VC alone, (C) VK3 alone, and (D) VC:VK3 combination treatments. SEM view added as insert in (B). In (A), sham cells show polygonal-like aspects with obvious pleiomorphism. In all of the treatments, nuclei depict poor contrast of the nucleoplasm that suggests some chromatin degradation and a rounding of the nucleoli into ribonucleoproteins masses. Following VC treatment, many cells demonstrate superficial but notable blister-like excisions (arrows), as verified with the SEM insert view. VK3-treated cells reveal large metachromatic cell-piece self-excisions (curved arrows). A poor contrast of the nucleoplasm, appearing as "milky pink" instead of red-violet, indicates some ongoing degradation of the chromatin, with only a rim of chromatin on the inner nuclear membrane and hypercondensation of the nucleolus. After VC:VK3 combined treatment, TRAMP cells exaggerate cytotoxic damage, with more self-excisions and dramatic vacuolizations, which significantly decrease cell and nucleus sizes with lytic defects. Some resisting cytoskeleton (keratin) forms circling patterns that enclose damaged cytoplasm contents and nuclei; even some self-cut pieces appear vacuolated (*). Scale = 5 μm.

with SEM verified that a tumor-cell population was drastically decreased (as well as cell size), but the number of cells was so low after VK3 and VC:VK3 treatments that it was not possible to accomplish meaningful measurements. Many cell corpses and debris were viewed by SEM, and there were certainly no cells left to be measured.

28.7.1 Bladder Carcinoma (T24) Cells

The composite pane of Figure 28.12 illustrates and summarizes SEM views of those T24 cells that have been treated with phosphate-buffered saline (PBS) as sham/control, VC alone, VK3 alone, or VC:VK3 for 1, 2, 4, and 6 hours. Similar views of the tumor cells as shown in Figure 28.12, along with LM micrographs of 1 µm-thick epoxy sections taken from slow centrifugation (when cells are spherical), allowed us to measure the cell diameters of the treated cells in other cell lines and to calculate their respective average cell volumes [183].

Other measurements taken *in vitro* were verified; these were significant across all treatments, as reported by K. McGuire (thesis, pers. comm.).

All cells were processed similarly and at the same time by the same solutions, so shrinkage caused by SEM processing was equal and only related to cell damage. A similarly large percentage of volume decrease occurred after these different cells were treated with the combined prooxidative vitamin mixture.

28.7.2 Ovarian Carcinoma (MDAH 2774) Cells

The cytotoxic actions of VC, VK3, and VC:VK3 on ovarian endometroid carcinoma MDAH 2774 cells and sham-control untreated cells can be seen in Figure 28.13, where four representative illustrations show the dramatic changes caused by the combination of the prooxidant compounds used as an anti-cancer drug (namely, Apatone).

Sham pictures were obtained from random 1 µm-thick epoxy sections cut through several embedments of collected samples obtained using the same technique of slow centrifugation as the treated ones, and were centrifuged at the same time and for the same 1-minute duration.

LM sham ovarian carcinoma (MDAH 2774) cells showed their morphology was spherical to ovoid (Figure 28.13A) and their content consisted in a round nucleus with one large but narrow indentation. The nucleus usually has a branching or round eccentric nucleolus, but these sham/control-treated cells had numerous small vacuoles of fatty deposits and intracellular lumina with accumulations of glycogen and other carbohydrates. Most of the cytoplasm was filled by ribonucleoproteins, giving the sham cells a more or less deep purple-blue hue after toluidine blue stain. Some sham cells seen with a pale, euchromatic nucleoplasm are indicative of imminent cell division. A summary of measurements of the sham cells is given in Table 28.4. According to Table 28.2, sham treatment showed that $7.2 \pm 2.6\%$ of total cells ($n = 541$) were undergoing mitosis, a few were undergoing necrosis/oncosis ($1.3 \pm 0.5\%$), and some were apoptotic ($3.5 \pm 0.6\%$). Autoschizis was found in $2.7 \pm 0.55\%$ of the cell population counted. Incidentally, a few cells (0.9%) were noted to have two nuclei, and these binucleated cells were predominantly in telophase.

As a result of **VC treatment**, a representative view of the MDAH 2774 cells is illustrated in Figure 28.13B, following review of a number of them ($n = 562$). This treatment induced many cells to undergo autoschizis cell death (19%). Other parameters (i.e., mitosis, necrosis, and apoptosis) were within the same range as in other carcinoma cells (i.e., DU145 [148], T24 [35,47,48], or TRAMP cells [192]) in this chapter (Figure 28.14); they show the typical formation of blisters from small to large, with eventually many of them decorating the cell outlines, the aspect of which usually appears as tiny oblong spheroids or subspherical blebs that range between 0.3 and 2.5 µm in diameter.

Cell debris can be seen in the intercellular spaces, while most of the treated cells show large compacted nucleoli.

After **menadione sulfite (VK3) treatment** (Figure 28.13C), of 546 cells counted, about 3% underwent mitosis, 0% apoptosis, 0.9% necrosis, and 32% autoschizic cell death. The carcinoma cells became obviously pleiomorphic as a result of the treatment, which seemed to deeply alter the cells' cytoskeleton, providing it with a new shape (i.e., cytokeratin). Fewer cells per field of view could be seen compared with both aforementioned treatments (sham and VC), and they had deeply altered spherical cell shapes; even elongated cells, as if they are exaggerated stretched cells, can be seen throughout the observed microscopic fields. In addition, the intercellular spaces appeared with scattered pieces of cells or debris. Following treatment, large basophilic, cytoplasmic appendages appeared to hang from each tumor cell. These could be as little as 0.5 or as large as 5 µm in diameter. They could be seen with vacuolated aspects and were in the process of self-excising, producing the cell debris observed in large numbers in the intercellular spaces. In addition, tumor cells revealed a smaller size than the sham or VC-treated cells, due to losing pieces of their cytoplasm (Table 28.5).

After a **combined VC:VK3 treatment**, the population of MDAH 2774 cells counted following SEM surveys is reduced tremendously, leaving corpses or cell debris in any field of view, as in Figure 28.13 for T24 cells. In order to verify the damage to those cells surviving from the initial population, treated cells left in culture were centrifuged. The sample wrongly gave the impression of a crowded space, even though one could see a significant reduction of cells. These 1 µm-thick sections revealed the main spherical shape was favored, due to the mode of collection, but the cell changes revealed the elongated distortions already noted after menadione treatment. Even though the tumor cells were all altered as a result of this combined treatment, they retained a mostly oblong shape, with small to large appendages of proceeding self-excisions. Following this combined treatment, tumor cells were significantly smaller compared to other cell types (n = 717 in Table 28.1). About 7% of mitotic cells were found in this population, with about 1.9% undergoing necrotic oncosis, 3% apoptosis, and a huge 43% autoschizic cell injury and death.

The reduction of size is caused by the multiple self-excisions, some of which are still attached by narrow bridges that appear as "appendages." Increased condensation of the nucleoli in the nucleoplasm of quasi-cells appears as a pale pink to rosy, milky color after toluidine blue staining. The heterochromatin enhances the contrast of the inner membrane of the nuclear envelope, as if highlighted in a cartoon fashion, with a small number of oblong, compacted patches (thick arrows in Figure 28.13D). The cytoplasm has more contrast in the cell peripheral zones, which contain glycogenic reserves, as compared with the perikaryal zones, which show damaged organelles and vesicle

Table 28.5 Percentage ($\pm$ s.e.m.) of ovarian carcinoma (MDAH 2774) cells undergoing mitosis and cell death *in vitro* after 2 hours of each treatment. n, number of cell counted.

Treatment	Mitosis	Oncosis	Apoptosis	Autochizis
Sham/control (n = 541)	7.2 ± 2.6	1.3 ± 0.6	3.5 ± 0.6	2.7 ± 0.5
VC (n = 562)	2.9 ± 1.5	2.5 ± 0.7	2.6 ± 0.5	19.0 ± 2.5
VK3 (n = 546)	3.1 ± 1.3	0	0.9 ± 0.1	32.1 ± 3.5
VC:VK3 (n = 717)	7.0 ± 2.1	1.9 ± 0.5	3.0 ± 0.5	43.0 ± 6.5

appearing as vacuoles. With TEM, one can show that mitochondria have been damaged by osmotic shock and remain along with parts of the SER as small vacuolated spaces.

Examining cells after cultivation shows that other cell deaths occurred during cultivation and, according to the treatment stimuli, reveals that the major cytotoxic activity is effected by prooxidation and autoschizis cell death.

28.7.3 Murine Prostate Carcinoma (TRAMP) Cells

A photomontage of LM observations taken from 1 μm-thick epoxy sections of a treated TRAMP murine model of prostate carcinoma stained with toluidine blue is shown in Figure 28.14. These representative illustrations show quite elegantly the main events that occur following each vitamin treatment. In LM, micrographs can illustrate many of the self-excising events (i.e., autoschizic events) leading tumor cells to their demise.

Figure 28.14A shows **sham-treated cells** after a mild centrifugation (500 rpm/min), revealing their somewhat polygonal aspects, which almost cover the entire section of the pellet in epoxy. Note the large central nucleus with huge nucleolus or nucleoli occupying the largest part of the cell, confirming the high nuclear:cytoplasm volume ratio.

Figure 28.14B shows **VC-treated** cells, demonstrating many bulging blebs or blisters of variable size (some as large as 1.5–2.0 μm). An example of a SEM view inserted in the same illustration is quite clear, revealing melted edges frozen in place by the fixative as if to suggest superficial ongoing small self-cuts (2 μm scale). This micrograph also demonstrates a cell "doublet" undergoing blebbing, and at least one mitotic block. A cell block is noted for the cell cycle in T24, as well as for DU145 [176,181,188]; this may provide an example of cell-cycle arrest.

Figure 28.14C shows **VK3-treated cells**, which also demonstrate dramatic self-cuts, of the largest size noted. These are densely stained by toluidine blue, showing a more purplish hue, which is suggestive of a dramatic excision of cytoplasmic pieces or debris of between <1 and almost 2 μm diameter; these can be found in the intercellular spaces.

Figure 28.14D shows TRAMP cells following **combined VC:VK3 treatment**, demonstrating great size reduction and a significantly reduced diameter. This can be seen as small blisters and pieces of cell that are connected with the mother cell. Tumor cells have vacuoles and a rim of cytoskeleton around the nucleus and within the cell. Most cells show a vacuolated cytoplasm, suggesting severe damage. The nuclei of these treated cells have an enormous single nucleolus. There is a ring of superficial fibers, which seem to support the entire cell, and separate the nucleus from its subservient organelles and cytoplasm. Among this cohort of treated cells, many vacuolated bodies can be viewed, which originate from self-excisions and are often vacuolated themselves.

28.7.4 Other Human Carcinoma Cell Lines

Data from published sources concerning the cytotoxicity of prooxidant treatments of other cancer cell lines, including other carcinoma and adenocarcinoma cell lines, are given in Tables 28.1 and 28.4 [177,178].

The prooxidant treatment consisting in a mixture of VC:VK3 at a ratio of 100:1 induces carcinoma cells *in vitro* to self-cut pieces of their cytoplasm and significantly change their morphology, as seen by a:

1) **decrease of their size;**

2) **alteration in their shape via cytoskeletal damage (hence, "pleiomorphism");**
3) **high death rate among the cells (i.e., autoschizis).**

28.8 Injuries Induced by VC:VK3 Combination Treatment Lead to Tumor Cell Death

Several tumor cell lines were studied as sham-controls using LM, SEM, and TEM and described alongside treated cells in order to verify the changes resulting from treatment. We shall examine as examples human bladder carcinoma (T24), ovarian (MDAH 2774), and prostate carcinoma (DU145) cells. In addition, we present xenotransplants from several of our published manuscripts [35,179,180,183,185,186], along with some unpublished illustrations.

28.8.1 Cell Shape and Size: Hockey Puck-Shaped Cells as Corpses

After treatment by VK3 of more than 2 hours or by VC:VK3 of 1 hour, we observed that *in vitro*-treated carcinoma cells underwent peculiar morsellation. Repeating the same experiments using SEM on DU145, we saw that the treated cells enucleated [34]. However, because the cells were grown on gold grids, there was poor resolution and the data were confused. We repeated the experiment again using T24 cells grown on Petri dishes. We cut the bottoms out of the dishes and prepared these makeshift slides for SEM, which enabled us to investigate large surfaces and facilitate the acquisition of a large number of micrographs. Together with the results of SEM conducted at Penn State Behrend in Erie, PA, these assisted us in understanding most of the morphologic changes associated with the various treatments. This allowed us to confirm that tumor-cell injuries induced by prooxidant treatment undergo dramatic changes, as revealed by early data in the *Scanning* journal in 1998, hinting at a new mode of cell death (i.e., autoschizis) [35].

With 1-hour or longer treatment with VC:VK3, or 2 hours or longer with VK3 alone, T24 cells demonstrated really astonishing views, which in some ways repeated those found in DU145 cells. Again, enucleating-like or self- or auto-mutilation of cells could be seen, in which oblong to spherical pieces exited a quasi-squamous cell to become first spherical, then flattened and "puck-shaped," or else somewhat biconcave, almost likes erythrocytes. Figure 28.15 provides a general view of sham T24 cells (A) and 2-hour VC:VK3-treated T24 cells (C). Figure 28.15B reveals flattened cells and extrusions, with an elongated bulging cell (bottom right) with an almost complete extrusion (middle right), a spherical-shaped cell (top left), and a flat, puck-like cell (middle edge). A copy of our published micrographs from the *Scanning* journal (Figure 28.15C) displays a cluster of these hockey puck-like cells in a field of view [35]. Following this, a 20–30-minute lag period shows a cell morphology reduced to spheroids, with short "foot-like" appendages (Figure 28.15D).

28.8.2 Membrane Defects: Annexin IV-FITC

During the process of injury caused by prooxidant treatment, self-excision can occur in many tumor cells, leading to changes in their nuclei. As treatment creates oxidative species, plasma membranes can be investigated for changes in their phospholipid composition, while it is at the same time verified whether the observed cell injuries

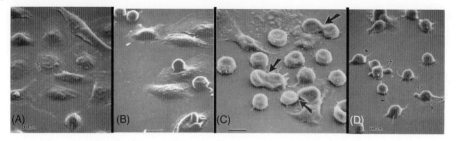

Figure 28.15 T24 human bladder carcinoma cells viewed with SEM. (A) 2-hour sham-treated cells, showing a nonconfluent aspect and a lack of contact inhibition. (B) Field of T24 cells treated for 1 hour with a VC:VK3 combination or for 2 hours by VK3 alone. This view shows sham-like cells with extruding and extruded perikarya. (C,D) Clusters of cells obtained after 2-hour treatment by VC:VK3, showing (C) either a hockey-puck or a biconcave shape, resolving later into (D) a spherical shape, as diminutive cells are reduced to nucleus and narrow perikaryal. Scale = 10 μm. *Source:* Gilloteaux et al. 2006 [184]. Reproduced with permission from John Wiley & Sons.

and death correspond either to necrosis/oncosis or to apoptosis. Annexin IV-FITC labeling can detect specific chemical changes invovled in the phosphatidylserine (PS) redistribution into the outer leaflet of the cell membrane in case of injurious changes toward apoptosis, but not autoschizis. This peculiar distribution of annexin IV-FITC labeling – different from that found in apoptosis, but not specifically the same as that in autoschizis – was included in a presentation we gave on cell death in Orlando, FL. *Science News* interviewed our group, and put an image on the cover of its June 2001 issue illustrating the tumor-cell perikaryon separating from the cytoplasm.

LM observations of VC:VK3-treated T24 cells reveal a complementary marker in support of the discovery of this new mode of cell death. T24 cells exposed to the vitamin combination at their 90% cytotoxic doses for 1, 2, or 4 hours, plus annexin IV-FITC (added to a final concentration of 1 μg/mL plus 1 μg/mL propidium iodide), were incubated for 15 minutes in the dark at room temperature and induced to die as three populations [193–195] – confirming the data in Table 28.2 for an ovarian carcinoma cell line:

1) the first population was dual-labeled (green membrane and red-orange nucleus) as either necrotic or oncotic cells;
2) the second population was annexin IV-FITC-labeled (green membranes), without staining of the nucleus structure, as apoptotic cells; and
3) the third population was PI-labeled (red-orange nucleus either alone or with tiny, adherent pieces of yellow or green membrane) as autoschizic cells [35,47,188].

It is possible that cells in the second population that had "lost" their nucleus and were now undergoing autoschizis were counted as apoptotic cells, as PI was shown to stain apoptotic cells along with membrane. Figure 28.16 provides micrographs illustrating these observations [193].

28.8.3 Ultrastructural Damage Caused by Prooxidant Treatment

28.8.3.1 Self- or Auto-excision-Induced Damage: Autoschizis

Further details can be gained regarding SEM views by analyzing the similar views produced by TEM and LM fluorescence of phalloidin-stained cells, used to detect the network of actin or stress filaments during prooxidant "stressed" treatment.

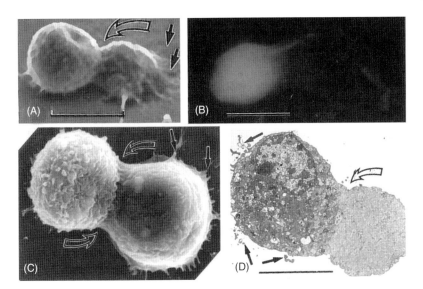

Figure 28.16 T24 cells observed after 2-hour VC:VK3-treatment. Tumor cells significantly decrease the size and stage of self-splitting cells due to their "doublet" aspects. (A) LM phalloidin-FITC, demonstrating an actin network well represented in the nucleus and perikaryon, as compared to the extended cytoplasm, where no or only traces of actin can be seen at the edge. (B) Profile side-view of a self-excising cell. (C) SEM polar view of a self-cutting cell, where the largest spherical-shaped part of the T24 cell contains the nucleus. (D) TEM view, showing the seceding cytoplasmic piece still attached to the perikaryon. A curved arrow suggests the split region. Scale = 10 μm. *Source:* Gilloteaux et al. 2006 [184]. Reproduced with permission from John Wiley & Sons.

Because T24 cells can become spherical and of almost identical shape (after self-cuts), we found cells that illustrate this type of cytoplasmic degradation, leaving them as smaller pieces, as suggested by previous micrographs. This is shown in Figure 28.16A. The associated Figure 28.16B–D illustrates the same process, with corresponding micrographs taken from LM (Figure 28.16B), SEM (Figure 28.17C), and TEM (Figure 28.16D). The LM fluorescence view clearly shows a dense cell-actin network distributed asymmetrically by a fluorescent yellowish-green hue, which is mainly perikaryal and intranuclear. The long cytoplasmic extension (later to be self-excised) lacks fluorescence, save a narrow edge. This suggests a profound rearrangement of the stress (actin) filaments as they organize the perikaryal zone. Compare this pattern with that of sham-treated T24 cells (Figure 28.1B).

Further views of the ongoing self-cutting activities of the treated T24 cells are provided in Figures 28.16 and 28.17. Comparison between these TEM and SEM complementary aspects shows that the process of auto-cut or autoschizis initially consists of a single large piece of cytoplasm, which appears to separate from these treated cells. At first, the "doublet" aspect consists of an enucleate portion and a nucleate portion (Figure 28.16A,B). One can see that this peculiar event involves a major part of the treated cytoplasm, which is attached to the substratum by filopodia and lamellipodia, while the nucleus and its perikaryon secede from it.

The remodelled aspect after self-excision in this tumor cell type provides a perikaryal area with a collapsed shape that contributes to provide a biconcave morphology resembling an erythrocyte-like morphology. At this time, one can only hypothesize that it is caused by the collapse of or defects in the cytoskeleton – especially the

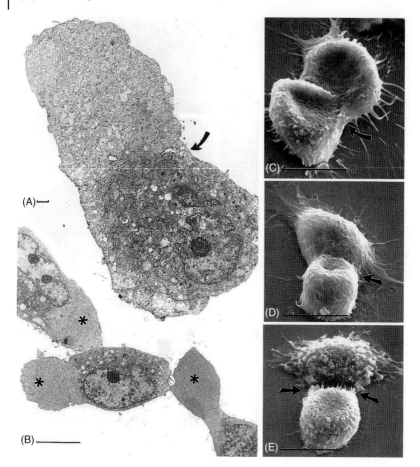

Figure 28.17 Combined VC:VK3 treatment of T24 cells, lasting for 1 and 2 hours. TEM views (A,B) complement the SEM data (C–E). All illustrations show cells undergoing alterations as SEM "doublets," while TEM views demonstrate that these "doublets" consist of two different cell portions: an enucleate (often attached to the substrate) and a nucleate (biconcave) part. The "schizic" or "schizous" perikarya parts keep contact with the cultivation support through loose extensions while progressively seceding. Scale: (A) = 1 µm; (B) = 5 µm; (C–E) = 10 µm. Source: Gilloteaux et al. 2006 [184]. Reproduced with permission from John Wiley & Sons.

microtubule and cytokeratin-lamins architectural support, which usually provides the cell shape [196], giving a hockey-puck shape after such large self-mutilation [179,180,183].

Using Petri dishes, as noted in Section 28.8.1 for SEM surveys of the cultivation surface, we were able to view large populations of cells. Following VC:VK3 treatment, a frequency of T24 cell self-excisions was obtained that was systematically a function of treatment duration: after 1-hour incubation, only 10–15% of T24 cells revealed self-cuts; after 2 hours, more than 35% showed cuts; and after 4 hours, more than 50% showed ultrastructural aspects consistent with autoschizis cell death [35,179–181,183,188].

This process of self-mutilation or self-cuts is therefore not an isolated event. We have collected more than 100 micrographs showing diverse stages of self-excision, and we have mounted a series of TEM views (along with two SEM aspects) that show that this unique discovery is supportive of a new mode of cell death in tumor cells (see Figure 28.18). This is useful in showing not only that the self-excising processes is characteristic of the prooxidant treatment of T24, but also that alterations occur simultaneously within the first hour post-treatment at the level of the nuclei and their content, especially chromatin alterations and the nucleolar defects or nucleolysis that accompanies nucleoplasm damage.

Figure 28.18A,C provides two SEM views that link to the previous pane, where the extrusion of a nucleus occurs with a small rim of cytoplasm – a bit like what is found in erythropoiesis [197,198]. These views show the nucleus with a rim erupting from a left-out cytoplasm, which can remain attached to the cultivation substratum. Figure 28.18B, D–L provides TEM views showing diverse aspects of the segregation of the perikaryal zone and cytoplasm by self-cuts.

During these changes, the nuclear content loses its typical nucleolus and NORs early on. It becomes less and less chromatic, and the nucleolus appears as a compacted mass that reduces in size with the severity and duration of self-excisions, leaving a nucleus with a pale-contrasted heterochromatin that has lost its marbled heavy contrast. Most of the nucleoplasm is reduced to finer and finer granulation. As part of this injurious process, the nucleus surrounds itself with a rim of cytoplasm, crowded by organelles that are "excluded" from the other side of the treated cells; this is constituted by a cytoplasm emptied of organelles but still containing particulate ribonucleoproteins and dispersed glycogen.

The main excision process, as viewed throughout the submitted micrographs, can be summarized by Figure 28.19. This micrograph seems to shed some interesting light on how the excision process occurs, as it may involve a local rearrangement of organelles "distributed along a trough or furrow," which forms the edge of a major cytoplasm piece with the early phase of self-excision. This TEM micrograph appears to suggest the stressed RER alignment with a defective cytoskeleton may favor such excision activity. However, further histochemical and immunochemistry data need to be obtained in order to verify the cytoskeletal damages claimed.

Finally, in order to further convince our readers of the self-excision or autoschizis mechanism (which has already been shown in many views of carcinoma cells treated by VC:VK3), we have added a few examples of micrographs of T24 cells treated for a duration of 4 hours, as shown in Figure 28.18A–E. In these views, one can see many puck-like or erythrocyte-shaped tumor cells undergoing further and further self-excisions of their rimmed cytoplasms while and after organelles are defectuous.

The remaining nucleus, and its loss of perikaryon, can be matched by SEM and TEM views in Figure 28.20. The nucleus seen with TEM is noted to have lost most of its chromatin, which is left as patches of granules and a quasi-empty nucleoplasm clearly displaying chromatolysis. Figure 28.20 shows other events related to this extreme self-cutting, reducing the T24 into minute cells.

It is interesting to share some observations from 1 µm sections and other autoschizic forms of cell death, as noted in TRAMP cells (Figure 28.14D), where, after VC:VK3 treatment, the perikaryal zone is preserved against the cytoplasm, with or without many organelles. This suggests an encircling bundle of cytokeratin maintains the perikaryal zone away from a "lost" or "sacrificed" cytoplasm to maintain the nucleus and its "court crowd" of organelles, which remain preserved until cell death, as shown in other cell lines (e.g., DU145).

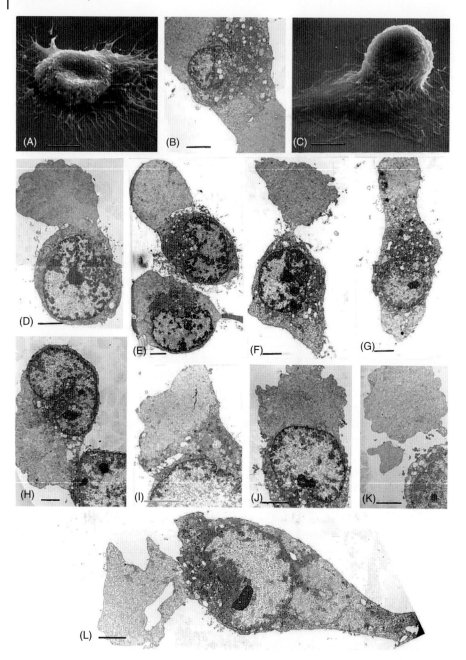

Figure 28.18 Human bladder carcinoma (T24) cells observed after 2 hours' treatment with VC: VK3. This series of micrographs demonstrates aspects of the self-excision progression and activity encountered in these treated cells. Two SEM aspects are shown in (A) and (C), while the remainder of the pane (B,D–L) consists in TEM aspects. Note the nuclear chromatin becoming less and less contrasted. Scale = 1 μm. *Source:* Parts from Gilloteaux et al. 2001 [181,184]. Reproduced with permission from Taylor & Francis Group.

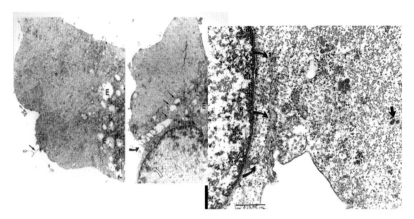

Figure 28.19 TEM views of cytoplasmic excisions in treated T24 cells, shown at high magnification. The cytoplasmic pieces appear to contain a finely granular cytoplasm that includes innumerable ribonucleoproteins and some swollen, hydropic mitochondria (m), with matrices often emptied and containing a fuzzy inner layer instead of the inner limiting membrane; these sometimes appear as interrupted lines within intramatrical deposits of variable size. Alignments of damaged mitochondria (straight arrows), with or without other cytoplasmic vesicles, can delineate an excision groove (curved arrow in Figure 28.8) or rudimentary microvillus (small arrow in Figure 28.7). Right panel shows an enlarged view of another self-excision of the same cell line undergoing cell death. Scale = 200 nm. *Source:* Parts from Gilloteaux et al. 2001 [181]. Reproduced with permission from Taylor & Francis Group.

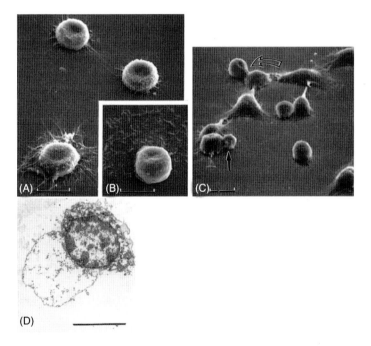

Figure 28.20 SEM and TEM views of T24 cells after 3-hour treatment by VC:VK3, showing the "puck-like" cells under further self degradation. Note that the puck-cell cytoplasm left around the nucleus becomes vacuolated, while the nucleus undergoes chromatolysis at this stage of diminution, only a few moments before cell death through autoschizis. In (C), some of the same carcinoma cells can be seen clinging to their short survival after self-excision, before undergoing further degradation as in (B). Scale = 1 μm.

28.8.3.2 Organelle Damage during Autoschizis

Specific alterations or injuries of cell organelles caused by prooxidant treatment accompany cell death by autoschizis.

28.8.3.2.1 Nucleus

Reviewing all the panes with TEM views displayed with the aforementioned descriptions, as well as those published on T24 cell demise, it is interesting to note the patterns of electron density in the chromatin and nucleolar structures of the nuclei of the cells that were treated [182,183]. It is clear that the treated cells show noticeable and progressive changes associated with chromatolysis and the segregation of their nucleolar components, followed by fragmentation and compaction of their granular components. Figure 28.21 provides another collection of bladder (T24)-treated cells undergoing irreversible cell injuries as a result of VC:VK3 treatment, showing again both the nucleus and cytoplasmic changes. The cytoplasm self-excisions are obvious, and other examples are provided in Figures 28.11–28.20. However, one would like to further illustrate in this pane the not-so-obvious nuclear changes that encompass chromatolysis and nucleolar changes, as can be seen more clearly in this panel.

Figure 28.21A provides an example of a sham T24-treated cell, where the nucleus content characterizes tumor cells by the high N:C ratio, the number of NORs, and the typical multiple-like nucleoli or "branching" nucleolus (see the general description of carcinoma cells in Figure 28.1). Figure 28.21B–H demonstrates autoschizis changes. In Figure 28.21B, the self-cut process has started, with a nucleus that shows a granulation and a peripheral heterochromatin layer that is more thickened than in sham, where the nucleolus appears as a densely contrasted mass without NORs. In Figure 28.21C, condensation of heterochromatin and, perhaps, patches of nucleolar structures remain. The nucleoplasm is finely granular, with a few lumps of chromatin decorating it. In Figure 28.21D, the nucleus further depicts poorer contrast, with nucleolar bodies highly contrasted with marbled or heavy reticulated aspects of the granular component, while the heterochromatin remains in patches of granules that are more abundant in the subenvelope regions than the main nucleoplasm. The cytoplasm is further self-excising, and organelles are poorly represented due to plane sectioning. In Figure 28.21F,G, while further degradation by self-cuts can be seen, diminishing the cell size, the nuclei are even less contrasted, leaving heterochromatin granular patches with more and more dilution; in (G), the process of cell degradation consists also in further shrinkage of the nucleoli, made of lesser and lesser clumped proteins, with intrinsic spaces. The damaged organelles located in the perikaryon can be viewed and appreciated in both (F) and (G). Finally, and at a further stage of cell demise, Figure 28.21H shows a damaged cell and a cell that appears to follow the sequence toward cell death: the T24 cell has decreased in size tremendously, and the small nucleus is filled with more or less uniformly fine granulations. A leftover sieve-like nucleolus can be seen, with whatever remains of the intranucleoplasm structure. Note that the injured organelles surround the nucleus where further cytoplasmic vacuolizations have occurred.

28.8.3.2.2 Karyolysis in Bladder Carcinoma (T24) Cells

In Figure 28.21, a series of photomicrographs made after a 4-hour VC:VK3 treatment (Figures 28.21A–D) suggest a progressive segregation of the nucleolar components, where the electron densely contrasted chromatin (DNA-labeled) associates and interacts with the other nucleolar components, as does chromatin associated with the inner

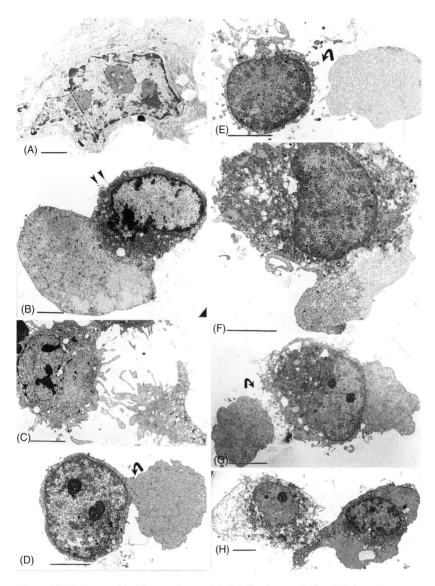

Figure 28.21 Human bladder carcinoma (T24) cells observed after (A) sham treatment and (B) 2-hour VC:VK3 treatment, prepared from flat-bottom cultivation in Petri dishes. These cells show diverse aspects of the cell self-excision that characterizes autoschizis, along with nuclear changes during a progressive karyolysis that includes nucleolus fragmentation, with the granular components massing into morsels. Scale = 5 μm. *Source:* Gilloteaux et al. 2001 [181]. Reproduced with permission from Taylor & Francis.

nuclear membrane (see Figures 28.22B,C). The nucleolonema somewhat remains as a main mass of granular component, unlabelled by DNA antibody, first with small interstices and then forming a compact aggregate. The dense component is discrete and is detected only in the area shown in Figure 28.22B. From a series of chromatin fingers interacting with and penetrating the nucleolus (Figures 28.22A,B), the granular

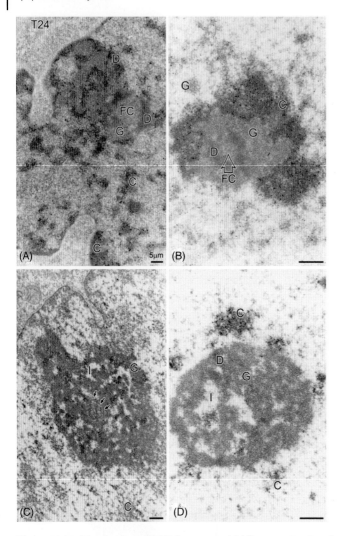

Figure 28.22 Distribution of DNA immunogold (diameter 5–10 nm) within the nucleoli of T24 bladder carcinoma cells treated for 4 hours with PBS (sham), VK3, VC, or VC:VK3. (A) Example of a sham-treated T24 cell, showing a large reticulated nucleolus with a dispersed dense fibrillar component (D) and several fibrillar centers (FC) scattered throughout the granular region (G). The chromatin (C) is characteristically labeled, and DNA can be detected throughout the nucleolus, forming a large network from one side to the other of the nuclear envelope. (B) Following menadione treatment, the nucleolus acquires a more compact morphology, and the dense fibrillar component forms a large, condensed arched structure, surrounded by the granular component. The outermost electron-dense component is the chromatin, which is properly labeled. Both fibrillar centers show a few labels adjacent to the dense component. (C) An ascorbate (VC)-treated T24 nucleolus, showing DNA labels in packets in narrow interstices (I) of the reticulated and compact mass made by the granular region and throughout the nucleoplasm. Small arrows denote a thread-like dense fibrillar component. (D) T24 cells treated by apatone demonstrate the most compact nucleolus, with large interstices throughout the granular component. Only a few labeled strands of chromatin appear centrifugal to the compact nucleolar mass. No labels are found in the granular component. Rare DNA labels can be found in the nucleoplasm. Scale = 0.5 μm. *Source:* Jamison et al. 2010 [189]. Reproduced with permission from SAGE.

component becomes more and more compacted, whereas the chromatin component appears partially to totally extruded and exits the granular component (Figure 28.22D).

The TdT method, in conjunction with immunogold labeling, has been used to visualize the DNA of ultrathin sections of equivalent fields of view (Figure 28.23) of T24 cells (1- and 4-hour treatments). After random sectioning, sham-treated T24 cells exhibit many of the characteristic features of tumor cells. For example, they possess complex indented nuclei with large nucleoli. These nucleoli contain large nucleolonemal masses with reticulated, dense fibrillar components that usually occupy a large area of the nucleoplasm. In addition, the nucleolar mass is often closely associated with the inner nuclear envelope near one or more of its nuclear indentations. The dense fibrillar component of the nucleolus is contrasted, and appears as thick strands or coarse anastomosing loops.

A few large fibrillar centers are visible (reaching diameters up to 1.5 mm; see Figure 28.23A), and are surrounded by the dense fibrillar component. Some of the fibrillar centers near the periphery of the nucleolus are labeled by the DNA antibody. Small interstices are also apparent, while the granular component forms narrow islands intertwined between the thick, thread-like networks of the dense fibrillar component. The perinucleolar chromatin is loosely associated with the nucleolonema, and appears as electron-dense patches dispersed throughout the nucleoplasm and as patches of heterochromatin located along the inner nuclear membrane. This component is also specifically labeled by the immunogold-labeled DNA antibody. The other components of the nucleolus, including the dense fibrillar component and the granular component, are visible as aggregates of 12–15 nm particles and are not labeled. No DNA label was found in the cytoplasm.

As a result of VC treatment (Figure 28.23C), the nucleolus appears as an elongated compact mass whose components are segregated. The majority of the perinucleolar chromatin lies outside the nucleonema. One dense fibrillar region (DFR) can be detected within the large granular component, which is not labeled by the DNA immunogold antibody. A dense fibrillar component surrounds the fine fibrillar areas and can be detected inside the compact aggregates of the granular component. Typically, a small amount of label is associated with the fibrillar center region. The majority of the DNA immunogold label covers the electron-dense intranucleolar chromatin as well as the perinucleolar chromatin.

Following VK3 treatment, the nucleolus (Figure 28.23B) appears as a compact mass whose components are segregated. The perinucleolar chromatin lies outside the nucleolonema and does not exhibit interstices. The majority of the DNA immunogold label covers the electron-dense intranucleolar chromatin as well as the perinucleolar chromatin. After this treatment, the dense fibrillar component that surrounds the fine fibrillar areas can be detected inside the compact aggregates of the granular component. Typically, a small amount of label is associated with the fibrillar center region. One DFR can be detected within the large granular component and is not labeled by the DNA immunogold antibody.

Following a 1-hour VC:VK3 treatment (Figure 28.23D), the nucleoli appear as round, centrally located nuclear masses that have lost contact with the nuclear envelope. These nucleoli are almost completely composed of the granular component and are not labeled by the immunogold. The granular component is surrounded by perinucleolar chromatin aggregates. The nucleoli also contain a few variably sized interstices. A small amount of immunogold label can be detected in the most peripheral of these interstices. Some of

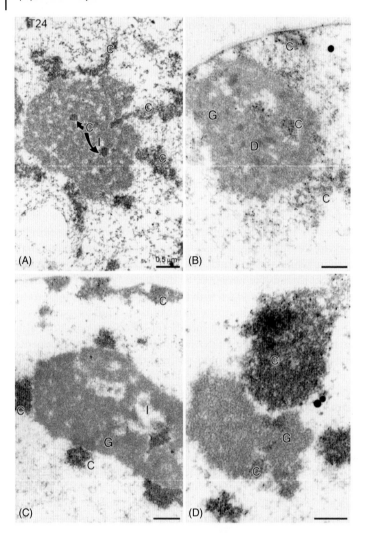

Figure 28.23 Series of photomicrographs made after 1-, 2-, 3-, and 4-hour Apatone treatment, showing the progressive segregation of the nucleolar components. The electron densely contrasted chromatin (C) (DNA-labeled) associates and interacts with the other nucleolar components, along with chromatin associated with the inner nuclear membrane (B,C). The nucleolonema remains somewhat as a main mass of granular component (G), unlabelled by DNA antibody, first with small interstices (I), and later forming a compact aggregate (D). The dense component is discrete and is detected in (B) only. From a series of chromatin fingers interacting with and penetrating the nucleolus (A,B), the granular component becomes more and more compacted, whereas the component named "chromatin associated with nucleolus" appears to be partially to totally extruded and to exit the granular component (D) as a result of treatment duration. This process leaves the ribonucleoproteins as a nucleolus remnant. Scale = 0.5 μm. *Source:* Jamison et al. 2010 [189]. Reproduced with permission from SAGE.

these interstices possess a small amount of delicate, electron-dense chromatin that is intertwined with the nucleolonema and extends beyond the outermost limit of the perinucleolar chromatin. A minute amount of unlabelled dense fibrillar material can be detected at the edge of the large nucleolonema. No fine fibrillar component is evident.

These nuclear changes, especially at the level of the nucleolus structure, involve the machinery for RNA transcription and production of its rRNAs, which is seriously impaired and disappears as a result of the combined VC:VK3 treatment. These changes make it even more important that the DNA is degraded by DNases and the nucleus with its nucleoplasm shows a process of karyorrhexis and karyolysis. As an osmotic end point, the tumor-cell demise has a nucleus that swells and bursts as soon as most of the cytoplasm simultaneously loses all of its organelles and the plasma membrane becomes filled with enormous gaps. During karyolysis, nucleoli became condensed and fragmented early in the process of autoschizic cell death, while chromatin became progressively less electron-dense.

Thus, VC and VK3 treatment induce only minor alterations in nucleolar structure, while treatment with the VC:VK3 combination produces marked alterations. These changes are characterized by the redistribution of nucleolar components, the formation of ring-shaped nucleoli, condensation (an increase in the proportion of perinucleolar chromatin), and the enlargement of nucleolar fibrillar centers (as shown in Figures 28.22 and 28.23) [189]. Immunogold DNA labeling of the compact nucleoli of VC:VK3-treated DU145 cells encompasses label localized predominantly on the peripheral chromatin, with a small amount detected in the peripheral area of the fibrillar region (Figure 28.9). Immunogold labeling of the RNA of the nucleoli of both sham- and vitamin-treated DU145 cells shows large areas of label at the boundary between the granular and fibrillar components, with only a few gold particles on the fibrillar centers. Vitamin treatment also alters the fibrillarin staining pattern, from one that matches with the fibrillar centers and adjacent regions to a more homogeneous one covering the entire nucleolus, with the VC:VK3 combination exhibiting the most pronounced effect. These fibrillarin changes are consistent with those observed following DNase I treatment.

28.8.3.2.3 Karyolysis in Bladder Carcinoma (RT4) Cells

Sham RT4 cells possess variable-sized, ovoid, euchromatic nuclei with shallow and narrow indented grooves that give them coffee-bean profiles. Viewed in random sections, the nuclei contain at least one large, reticulated, and branching nucleolus, which is usually adjacent to or associated with the indented zone of the nucleus. Within the nucleolus, the dense fibrillar and granular components form a heavily contrasted network of anastomosing strands. The nucleoplasm is punctuated by a fine grainy network that contains patches of chromatic materials of between 40 and 350 nm in diameter, while the heterochromatin decorating the inner nuclear membrane is a quasi-uniform layer of 50–150 nm thickness interrupted by the nuclear pores (Figure 28.24A).

VC-treated cells contain ovoid, indented nuclei that appear more spherical than the ones described in sham-treated cells. After 2 hours of treatment, the inner membrane of the nuclear envelopes is outlined by a series of 100–250 nm-thick, heterochromatic, densely contrasted layers interrupted by numerous nuclear pores. The remaining nucleoplasm consists of fine, electron-dense specks characteristic of a euchromatic nucleus, while the nucleoli are strongly contrasted condensed bodies. A close examination of the nucleoli reveals aggregation of their granular components, which mainly consist of ribonucleoproteins and maturing rRNA precursors; a number of large nucleolar interstitial spaces form a nucleolonema (Figure 28.24B). After this VC treatment, the remaining heterochromatin component of these nucleoli appears as grossly branching extensions contiguous with the heterochromatin that lines the inner

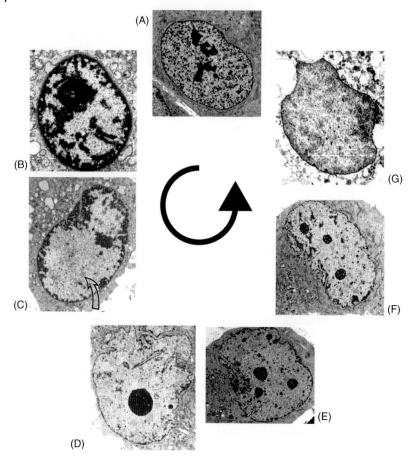

Figure 28.24 Bladder carcinoma (RT4) cells undergoing autoschizis cell death. This figure shows the progressive, ongoing nucleus and nucleolar alterations associated with this mode of cell death, from (A) sham-treated to (B–G) condensation of peripheral chromatin and accumulation of the rRNA granular component as the nucleolus becomes dysfunctional. The heterochromatin leaves the nucleolonema, and the nucleolus that remains is the granular component made of ribonucleoproteins (see Figure 28.25). The karyolytic progression pursues its course, and as a result the nucleoplasm becomes less and less electron-contrasted as its fibrogranular component dissapears and is replaced by a fine granulation. The reduction and fragmentation of the nucleolus granular component may be affected by RNase- and proteolytic-induced degradations. The cell demise can be viewed while the entire chromatin content is lyzed into sort of tiny, fuzzily contrasted, loosely diffuse distributions in the nucleoplasm. The nucleus maintains its envelope until osmotic shock breaks it to kill the tumor cell [51]. Scale = 1 μm.

nuclear envelope (Figure 28.24B), and it appears that the nucleolar bodies are tethered to this heterochromatin.

After 4 hours, VC treatment exacerbates the nuclear and nucleolar damage. Specifically, the heterochromatin is found in a thinner, 50–80 nm electron-dense heterochromatic layer along the inner nuclear envelope surface. In addition, within the finely contrasted nucleoplasm, clusters of aggregated chromatic dots create 0.3–0.6 μm oblong to circular patches of interchromatin granules (IGs) that circumscribe the nucleolar masses. These can be called satellite nucleoli, and their mere presence suggests that the nucleoli lack polymerases or even transcripts and that the cell is destined to proceed toward cell death [199–201]. Some chromatin can be detected via its higher contrast compared to the ribonucleoproteins in the central interstitium (Figure 28.24B).

Following **menadione treatment of 1-hour duration**, as well as in other cytotoxic treatments, most organelles become segregated into large cytoplasmic sectors that face the concave side of the indented nuclei (Figure 28.24C,D). After 4 hours' treatment, the nuclear/cytoplasmic ratio appears to have significantly increased. This overall appearance may be the result of a series of self-excisions of the cytoplasm. This type of event has been seen in T24 bladder cancer cell lines and dramatically reduces the cytoplasmic volume, leaving the nucleus surrounded by only a narrow rim of cytoplasm (Figure 28.24C–E).

After 4-hour VK3 treatment, the carcinoma cells display a thicker-than-100 nm band of heterochromatin along the inner envelope, but with longer treatment, the thickness of the electron-dense heterochromatin layer becomes interrupted; it decreases further throughout the nucleus section and the nucleoplasm becomes paler (Figure 28.24E,F). As a result of this treatment, nucleoli are fragmented into multiple variable-sized compact masses that can reach up to 2 μm in diameter. These nucleoli parts are composed of granular masses with punctate nucleolar interstitial spaces surrounded by innumerable 80–150 nm ribonucleoproteins, which are the remnants of the granular component. As the treatment time increases from 1 to 4 hours, several loose aggregates of chromatic dots appear as interchromatic granules in the nucleoplasm around the compacted nucleolar masses. It is possible to count as many as 20 of these electron-dense aggregates in one nucleus (Figure 28.24C,D) [51,153]. At higher magnification, the spherical nucleolar masses can be resolved into aggregations of ribosomal particles of 12–20 nm diameter. Rarely, a small amount of peripheral heterochromatin contiguous with the dense layer along the inner nuclear envelope may surround the nucleoli. In addition, a few small aggregates of heterochromatin can be seen on the periphery of these nucleolar masses.

After **short VC:VK3 treatment**, the nucleus displays a narrow peripheral heterochromatic layer along the internal nuclear envelope and a euchromatic nucleoplasm. This heterochromatic layer is marbled in appearance and variable in thickness. In the smallest cells, the nuclei remain centrally located and the perikaryal cytoplasm is reduced to a thin rim of less than 2 μm diameter [51]. The nucleoplasm encompasses one or more oblong to spherical, highly contrasted, compact nucleolar masses. The integrity of the nuclear envelope is apparently retained until cell death. Throughout the cell-demise process, the condensation and fragmentation of the chromatin can be viewed through finer and finer granulations appearing in the nucleoplasm and, along with the condensation, segregation and fragmentation of the nucleoli as described for other cells undergoing irreversible injuries leading to chromatolysis [83]. Figure 28.25 demonstrates a condensate of ribonucleoproteins as what remains of a nucleolus. During the degenerative process of nucleolysis, it is common to find IGs clustered around the nucleolar masses; these are caused by chromatin degradation after leaving the nucleolus, and likely result from reactivations of DNases and RNases, as shown in Figures 28.24 and 28.25. A reconstruction of chromatolysis is attempted in Figure 28.24A–F.

28.8.3.2.4 LM:Histochemical Detection of DNA Appearance with Feulgen Stain

Further confirmation of such chromatolysis, as suggested by so many TEM views, can be found in a Feulgen histochemical test verifying the DNA content of treated cells. This

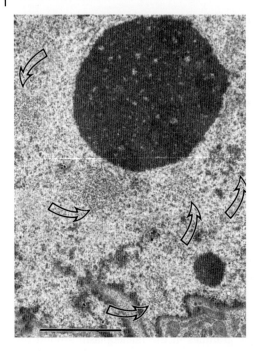

Figure 28.25 The formation of a compacted mass of particulate ribonucleoproteins as a remnant of the nucleolus after chromatin has left (see Figures 28.22 and 28.23). Interchromatin granules (IGs) can be seen as chromatin deteriorates (curved arrows). Scale = 1 µm.

technique is based on a stoichiometric attachment of the Schiff base with DNA purines and pyrimidines, and it allows a densitometry check on the stained nuclei.

Measurements confirmed the progressive chromatolysis occurring during anticancer treatment, with significant changes in densitometry in the sequence from highest-degraded DNA from VC:VK3 > VK3 > VC > sham-control T24 cells [183].

Figure 28.26 shows a gallery of Feulgen-stained T24 cells that display changes in DNA content as a result of 1-, 2-, or 4-hour treatment (columns) with culture medium alone (CONT as sham), VC alone, VK3 alone, or combined VC:VK3 (rows). At all three time points, the nuclei stained by the **sham-treated cells** (CONT) are uniformly strongly contrasted in dark magenta. All T24 cells display large nuclei, some of which are coffee bean-shaped and surrounded by a nonspecific pale lilac-hued cytoplasm. Some pale oval nucleoli can be viewed amid the well-contrasted nuclei, due to their absence of contrast. During mitotic stages of mitosis (e.g., anaphase; see Figure 28.24 CONT 2 hours), the chromosomes are darkly stained, as expected, while the cytoplasm retains a background pale lilac stain. Nuclear DNA staining reflectance intensities are not significantly different across 1-, 2-, and 4-hour sham treatments, being 29.9 ± 2.7, 30.0 ± 2.7, and 29.6 ± 2.2, respectively (Figure 28.27).

The nuclei of **VK3-treated cells** are less contrasted by Feulgen stain compared to VC and sham treatments (Figure 28.27 VK3 1, 2, and 4 hours). Likewise, cell size is reduced by the loss of cytoplasm caused by autoschizic excisions, and the cells become smaller than sham- and VC-treated cells. Many twin-like cells can be seen as pairs of small cells of equal or unequal size scattered throughout the field of view. During this altered cytokinesis, as noted with flow cytometry, several asymmetric cell divisions can be seen, resembling those seen by TEM and SEM (Figure 28.26 VK3 1 and 2 hour inserted views). Following cytokinesis, the nucleate portion exhibits a very narrow perikaryal cytoplasm around a smaller nucleus size with poor contrast and noted nucleolar masses, which are

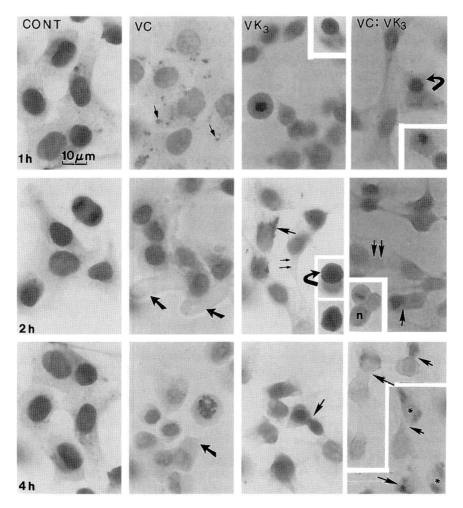

Figure 28.26 Feulgen-stained treated T24 human bladder carcinoma cells. Columns represent duration of treatment (1, 2, and 4 hours), and rows treatment type (sham-control (CONT), VC alone, VK3 alone, combined VC:VK3). In all micrographs, "n" and "*" represent the nuclei, in which DNA runs from dark purplish-red- to pink-stained. Small arrows in VC 1 hour indicate the cytoplasmic localization of accumulated dyes. In all other panes, small arrows mark selected examples of cytoplasmic self-excisions and curved arrows indicate perikaryal excisions. Note that asterisks also mark the locations of quasi-unstained nuclei. Scale = 10 μm. *Source:* Gilloteaux et al. 2006 [184]. Reproduced with permission from John Wiley & Sons.

difficult to see in the illustration, but can be viewed easily by fine-focusing the microscope slides. The anucleate portion appears to be elongated, stretched, and delineated by irregular edges. Self-excising perikarya areas can be seen in Figure 28.26 VK3 1 and 2 hours. Finally, minute cells appear with paler and paler contrast for their nuclei. Well-contrasted DNA-rich material is visible in the pale-stained cytoplasm and can be interpreted as DNA leaking from the nucleus into lysosomes. Large cytoplasmic pale pieces can be seen self-excising from these cells (Figure 28.26 VK3 4 hours).

VC-treated cells clearly show a lesser degree of nuclear staining than sham-control cells, due to a diminished DNA content. Large and poorly stained round to oblong

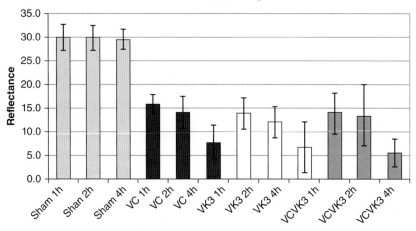

Figure 28.27 Histogram of averaged (s.e.m.) densitometry reflectance values obtained from T24 cells stained by Feulgen after different treatments. Treatments are either sham (1, 2, and 4 hours), VC only (1, 2, and 4 hours), VK3 only (1, 2, and 4 hours), or combined VC:VK3 (1, 2, and 4 hours), as shown in a Table 28.3. *Source:* Gilloteaux et al. 2006 [184]. Reproduced with permission from John Wiley & Sons.

nuclear masses can be detected in the background of the magenta-stained chromatin. Some contrasted cytoplasmic deposits are noted, which are probably dilated RER and mitochondria, or else lysosomes displaying autophagy. After 2- and 4-hour VC treatment, many cells are smaller than sham cells, due to some reduction of their cytoplasm. In the 2-hour group, treated cells poorly stained with cytoplasmic lamellipodia-like extensions can be seen without any inclusions. In the same fields of view, one can see a large cell in advanced prophase, in which chromonema can be detected (Figure 28.26 VC 2 hours). Pale oval-shaped nucleolar bodies can be seen among the chromatic material. After 4 hours of treatment, cells contain fine chromatic material and appear to be blocked in the early stage of prophase (Figure 28.24 VC 4 hours); all nuclei are smaller than in sham-treated nuclei, and the population attached to the culture support is less numerous than for 1 and 2 hour-treated cells.

After **1 hour's treatment with VC:VK3** (Figure 28.27 VC:VK3 1 hour), cells exhibit changes similar to those seen following 4-hour VC or 2-hour VK3 treatment. The ongoing self-excisions and the resulting progressive reduction of cell size are characteristic of this treatment, as can be seen in Figure 28.26 VC:VK3 1–4 hours. A few isolated cells with poorly contrasted nuclei and a narrow rim of cytoplasm can be discerned. The cell cycle is still evident as blocked in some cases (e.g., Figure 28.26 VC:VK3 2 hour insert depicting metaphase). Many cells show their shearing parts of the cytoplasm. After a 4-hour treatment, it can be seen that the nuclei have only a minute amount of DNA left, which makes them appear washed out, and staining is condensed away from the nuclei (*) into lysosomal bodies.

28.8.3.2.5 DNA Gel Electrophoretic Pattern from Bladder (T24) Karyolytic Damages

The total DNA extracted from sham-, VC-, VK3-, and VC:VK3-treated T24 cells was resolved electrophoretically with the aim of verifying whether DNA was cleaved by the vitamin prooxidant treatments, as was suggested by the Feulgen staining technique. In all the vitamin-treated cells, DNA gel patterns showed an increased spread according to treatment time, which was visible after treatment as short as 1 hour. During DNA processing, the viscosity of the treated cell extracts was greatly reduced compared to sham-treated cells, extracts of which were always more viscous and fibrous. This apparent lower viscosity found in treated cells suggests that their DNA is cleaved and progressively degraded. It also confirms the observations of the Feulgen staining technique, as well as those of densitometry (i.e., a low contrast and the quasi-disappearance of strong magenta contrast in nuclei compared to the same pale hue as seen in the cytoplasm). DNA-degradation analyses support the finding that VC alone, menadione alone, or their combination induced a form of cell death that resembled necrosis/oncosis, where DNA degrades into smears instead of the laddering pattern found with apoptotic dying cells.

28.8.3.2.6 Karyolysis in Ovarian Carcinoma (MDAH 2774) Cells

The human ovarian carcinoma (MDAH 2774) cells shown in Figures 28.28 and 28.29 provide representative TEM micrographs of the damage in the nucleus and some parts of the cytoplasm as a result of prooxidant treatment, as reported in at least three previous publications [71,139,185,202].

Figure 28.28 illustrates typical sham-control treated MDAH 2774 cells, which are round and poorly differentiated, with diameters ranging between 15 and 23 µm. They have a large nucleus, containing an enormous nucleolus. The random cut sections make it look as if the cell has three nucleoli, but in fact the nucleolus is a huge, branching structure with many NORs, and is therefore shown in three parts.

Many of these tumor cells undergo changes in normal cultivation that allow them to die of either apoptosis, some rare form of necrosis/oncosis, or, following prooxidant treatment in our experiments, mainly autoschizis (see Table 28.6). This inducement is extreme after treatment that encompasses VC:VK3; in this case, autoschizis cell deaths are more numerous. They are most obvious after 2 hours of treatment, as large numbers of cells undergo not only self-excision but also chromatolysis, as revealed in Figure 28.28.

Most of the cytoplasm appears relatively electron-dense (after 1 hour's treatment), due to the abundant distribution of the ribonucleoprotein granules; the peripheral cytoplasm also contains numerous glycogen-like particles. A few fat-containing vacuoles can be detected by the presence of a residual osmiophilic content. A large number of rough and smooth endoplasmic cisternae are distributed throughout the remaining cytoplasm. The large nuclei appear ovoid, or else are deeply indented by a deep, narrow groove. Because of its large size, only the edge of the intranuclear branching structure of the nucleolus is usually detected (Figure 28.28A). The overall nucleoplasm is euchromatic, with a delicate thickening of the inner nuclear membrane clearly interrupted by nuclear pores.

Following VK3 treatment lasting longer than 2 hours, or after 1 hour of VC:VK3 treatment, 1.0–1.5% of cells are multinucleate, probably due to cell-cycle block [176,187]. The majority of the cells are uninucleate and appear either subspherical, with fusiform to elongate regions, or spindle-shaped (Figure 28.28). Pieces of cytoplasm of 1.5–5.0 µm diameter and blisters of varying sizes are in the process of being excised. Unlike in apoptosis, all of these cytoplasmic fragments do not contain detectable membranous organelles or any nuclear fragments (such as apoptotic bodies). As a consequence of these progressive self-excisions, cells diminish in size by more than one-third, leaving organelles clustered around the nuclei in a thin rim of cytoplasm. Figure 28.28 shows the progressive nucleoplasmic alterations that result from changes

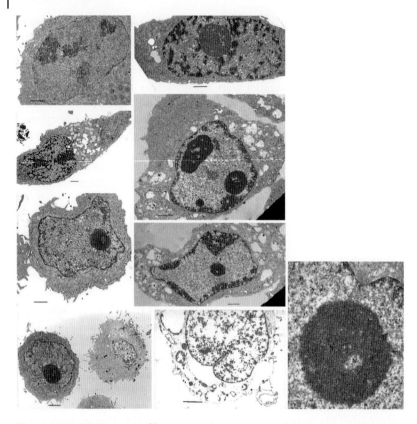

Figure 28.28 TEM aspects of human ovarian carcinoma (MDAH 2774) cell alterations during the process of cell death by autoschizis following 2-hour VC:VK3 treatment. Top-left image shows a sham tumor cell, with its large nucleus and reticulated nucleolus. In the rest of the images, the nucleus has chromatin redistributed as a peripheral coating of the inner nuclear membrane and extracted from the nucleolus, making a "marbled" patchy thicker outer nucleoplasm that compacts to progressively disappear through chromatolysis. At the same time, the nucleolus compacts as a granular mass with interstices left by retiring chromatin, which become entirely compacted and fragment into smaller size. The treated carcinoma cells decrease in size before chromatolysis with deleterious swelling of the nucleus and bursting of the envelope accompanies autoschizis cell death. Scale = 1 µm. *Source:* Gilloteaux et al. 2003 [71]. Reproduced with permission from Cambridge University Press.

in the chromatin. The majority of the cells exhibit well-contrasted chromatin, which decorates the inner nuclear envelope in a layer 0.2–0.5 µm thick, at first quite narrow but becoming thicker, although not compacted; in fact, it is more like a marbled dissociated network. At first, the chromatin remains associated with and penetrates a huge round nucleolus. Subsequently, the heterochromatin leaves the nucleolus, which becomes a small to large, highly contrasted mass of ribonucleoprotein (Figure 28.29). As shown in Figure 28.28, the chromatin associated with the nucleolus leaves it and becomes a series of eccentrically located chromatic bodies. These become distant from the nucleolar mass and appear only as electron-dense patches along the nuclear envelope. The final stage in the process before karyorrhexis and karyolysis is the swelling of the remaining organelles and nuclear envelope, which disintegrates due to osmotic defects.

Following VC:VK3 treatment, the described alterations in cell ultrastructure occur in all treated cells – especially those treated for more than 2 hours, due to the exacerbated

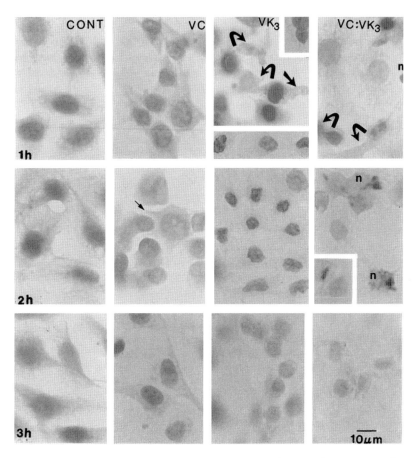

Figure 28.29 Feulgen-stained ovarian (MDAH 2774) endometroid carcinoma cells. Rows represent period of treatment (1, 2, and 3-hour), and columns treatment type (sham (CONT), VC, VK3, and combined VC:VK3). In all the micrographs, "n" and "*" represent the nucleus. Small arrows in VC 1 hour indicate the cytoplasmic localization of accumulated dyes. In all others, they mark selected examples cytoplasmic self-excisions. Curved arrows indicate perikaryal excisions. Stars indicate unstained nuclei. Scale = 10 μm. *Source:* Gilloteaux et al. 2004 [185]. Reproduced with permission from Elsevier.

Table 28.6 Antitumor activity of vitamins.

Cell line	Vitamins alone		Vitamins together		Fractional inhibitory concentration (FIC)
	Vitamin C 50% cytotoxic dose (μM)	Vitamin K_3 50% cytotoxic dose (μM)	Vitamin C 50% cytotoxic dose (μM)	Vitamin K_3 50% cytotoxic dose (μM)	
MDAH	1528 ± 29	41.8 ± 0.62	165 ± 1.5	1.65 ± 1.53	0.147
MRHF	8000 ± 246	500 + 5.0	1000 ± 28	10 ± 0.2	0.145

damage. During these progressive changes, the majority of the mitochondria exhibit condensed bodies in their matrices, while their inner and outer membranes apparently remain intact. The intermediate filaments of the cytoskeleton appear in bundles surrounding the nucleus and the organelles. Finally, the RER first undergoes degranulation in the case of cell injury, and then is maintained intact until near the completion of autoschizis cell death, as explained through many of the aforementioned micrographs.

28.8.3.2.7 LM:Histochemical Detection of DNA Appearance with Feulgen Stain

Figure 28.29 provides representative micrographs of sham-, VC-, VK3-, and VC:VK3-treated MDAH 2774 cells after Feulgen staining, in order to survey their relative DNA content. The rows of mounted micrographs display changes in Feulgen staining patterns following 1, 2, and 3-hour treatment (columns) with culture medium (CONT), VC alone, VK3 alone, or the VC:VK3 combination. The reason for limiting our investigation to 3 hours' duration is because there were so few cells left intact that the number that could be evaluated became questionable, as tumor cells were dying very efficiently due to prolonged treatment. At all three time points chosen, Feulgen-stained sham (CONT) cells can be viewed as uniformly contrasted. The cells display ovoid magenta nuclei surrounded by pale-lilac cytoplasm nonspecifically stained by the Schiff base during dehydration of the cells. The abundant cytoplasm of the control cells enables one to see the multipolar shape of most of the MDAH cells. Cells are superimposed on top of one another, which is indicative of their absence of contact inhibition. In some favorable cases, when the control cells are viewed through immersion oil, many elongated particulate structures (probably mitochondria) are visible, as their DNA is also stained.

Following VC treatment, MDAH cells reveal a lesser degree of nuclear staining than do sham-treated cells. This indicates a decrease of DNA content. After 1 hour's VC treatment, well-stained marginated chromatin and large, oblong, poorly stained nucleolar masses are visible. Following 2 hours' VC treatment, many cells seem smaller than in control, due to a reduction in the cytoplasm (see Table 28.4). However, a small number of cells with large diameters and very large nuclei can also be seen. After 3 hours' VC treatment, all nuclei appear smaller than control nuclei, and the population of cells attached to the coverslips is less numerous than for the 1- and 2-hour VC treatments. Exposure of the MDAH cells to VC results in two populations of cells: those undergoing autoschizis and those that have likely been blocked at G1/S in the cell cycle [176]. By 3 hours' VC treatment, many of the cells undergoing autoschizis have detached from the coverslip, leaving only those that are blocked. These blocked cells exhibit far fewer defects in DNA content than the autoschizic cells [71,185,202].

After 1 hour of VK3 treatment, the nuclei are more intensely stained than was the case for 1-hour VC treatment. While the nuclei of the 1- and 2-hour VK3 treatment groups are still well stained, cell size is obviously reduced by the loss of cytoplasm via autoschizic excisions, as can be seen through some ongoing events captured by the fixation procedure in action (i.e., Figure 28.29 VK3 1 hour and inserts). Some of these cells display asymmetrical cytokinesis and clear-cut dotted margination of chromatin (Figure 28.29 VK3 1 hour, top insert). After 2 hours' VK3 treatment, most cells definitively appear smaller than both sham- and VC-treated cells, due to the excision of cytoplasmic pieces. Many of these cells exhibit marginated chromatin with a reticulated pattern (Figure 28.29 VK3 2 hour). The large, pale, ovoid masses in these nuclei may be the segregated masses of nucleolar ribonucleoproteins shown on TEM, which are surrounded by blocks of DNA-containing chromatin. Finally, after 3 hours of VK3 treatment, the smallest MDAH cells are seen, consisting of poorly stained nuclei surrounded by well-delineated but narrow rims of cytoplasm.

Following treatment with the VC:VK3 combination, MDAH cells exhibit a variety of cell shapes. A time-dependent reduction in the number of adherent cells and of staining contrast for chromatin is also apparent. After 1 hour of treatment (Figure 28.29 VC:VK3 1 hour), the ongoing self-excisions and the progressive reduction of cell size that characterize autoschizis are evident. A few cells are reduced to their nuclei, surrounded by narrow rims of cytoplasm. After 2 hours of VC:VK3 treatment, most of the remaining cells that have been reduced in size appear clustered in small groups. The nuclei of these cells are stained in a pale lilac hue, demonstrating a visibly significant loss of DNA. As with other tumor cells tested with Feulgen stain after similar treatment, some stained material is located in the cytoplasm that surrounds these poorly stained nuclei; these are either artifacts or DNA captured by autophagocytosis. The inset of Figure 28.29 VC:VK3 2 hours demonstrates intensely stained clumps of DNA among the unstained cytoplasm of an MDAH cell that is undergoing autoschizis. This DNA degradation and autoschizis is increased in the 3-hour treatment group, where minute cells can be seen to be undergoing further cytoplasmic self-excisions, as well as nuclear diminution and degradation. In this treatment group, the nuclear mass is always detected only with a poorly stained, pinkish chromatic material. Only 5% of cells are left after 4-hour VC:VK3 treatment – too few to count – so the pane is completed with 3 hours' treatment.

28.8.3.2.8 Densitometry Measurements

In order to quantify the staining patterns that obviously demonstrate DNA degradation, we tried to measure the differences between treatments. The measurements of the DNA in the nuclei of treated MDAH cells shown in Figures 28.30 and 28.31 were only made

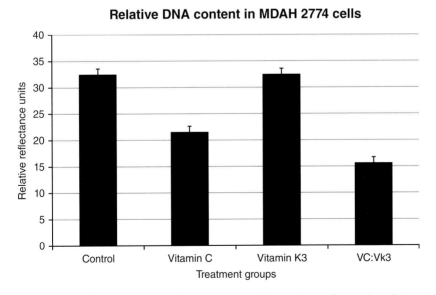

Figure 28.30 Histogram summarizing the relative DNA contents of control- and vitamin-treated MDAH 2774 cells. Bars represent densitometry measurements of the intensity of reflectance, based on the illustrations in Figure 28.1 minus the background reflectance. *Source:* Gilloteaux et al. 2004 [185]. Reproduced with permission from Elsevier.

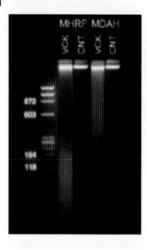

Figure 28.31 Gel electrophoresis of the total DNA extracted from sham- and VC:VK3-treated human foreskin fibroblasts (MHRF) and MDAH 2774 cells. Lane 1 contains a molecular-weight marker, lane 2 contains the DNA from VC:VK3-treated MHRF cells (VCK), lane 3 contains the DNA from sham-treated MHRF cells (CNT), lane 4 contains the DNA from vitamin-treated MDAH cells (VCK), and lane 5 contains the DNA from sham-treated MDAH cells (CNT).
Source: Gilloteaux et al. 2004 [185]. Reproduced with permission from Elsevier.

after a 1-hour vitamin exposure, due to the abundance of cell death, detachment, and disintegration following longer exposures. As shown in the histogram of Figure 28.30, the DNA-staining intensity of sham-treated (CONT cells) is 32.5 ± 1.127. The intensity values for VC and VK3 are 21.6 ± 1.127 and 32.6 ± 1.127, respectively, while that for VC:VK3 is 15.6 ± 1.127. Note that the average DNA levels following VK3 are the same as in the control group. By contrast, the DNA level following VC or VC:VK3 treatment is significantly lower than in controls ($p < 0.0001$). In addition, the DNA intensity level following VC:VK3 treatment is significantly different from that following VC treatment ($p < 0.03$). The standard-error values are all identical, because they are based on a pooled variance estimate.

28.8.3.2.9 DNA Gel Electrophoresis Pattern from Ovarian (MDAH 2774) Karyolytic Damage

In an effort to further confirm biochemically that DNA was cleaved and eventually degraded during VC:VK3 treatment through the efficient activities of DNases, total DNA was extracted from MDAH-treated cells and human foreskin fibroblasts (MHRF) following sham treatment or 4–hour VC:VK3 treatment. The DNA was then resolved electrophoretically, as shown in Figure 28.31. Lanes 2 and 3 contain the DNA of VC:VK3-treated (VCK) and sham-treated (CNT) MHRF cells. The DNA from the sham-treated MHRF cells appears as a single high-molecular-weight band slightly below the well in which it was loaded. Conversely, the DNA from the VC:VK3-treated MHRF cells appears as a low-molecular-weight smear at or below 200 bases. The DNA from the sham-treated MDAH cells also appears as a single high-molecular-weight band slightly below the well in which it was loaded, while the DNA from the vitamin-treated MDAH cells exhibits a spread pattern. The absence of laddering in the vitamin-treated lanes strongly suggests that VC:VK3-induced cell death is not a form of apoptosis.

The diffence in treatment with different cytotoxicities is summarized here for the sake of completion and comparison [185]. Table 28.6 shows a close toxicity, but the dose required to achieve cytotoxicity in MHRF cells is five times more elevated than for cancer cells, indicating that cancer cells are the first to be damaged, in a sequence indicated as VC:VK3 > VK3 > VC > sham treatment, as expected. This sequence can be verified as shown. In the aforementioned experiments, a new mode of cell death by self-excision – not just of cytoplasm, but also of DNA – has occured.

28.8.3.2.10 Karyolysis and Other Damages in Prostate Carcinoma (DU145) Cells

Human prostate carcinoma (DU145), viewed by LM from *in vitro* cultivation, demonstrates that the anticancer treatment used with a combined VC:VK3 is more efficient than VK3 alone, or even than VC alone, in terms of cytotoxicity (Table 28.4) and biochemical data [177,178]. Table 28.1 confirms, along with LM, SEM, and TEM aspects [34], that, generally, VK3 and VC:VK3 treatments result in a diminished cell size by making suggestive self-cuts into a more spherical cell morphology with a diameter less than 15 µm (i.e., smaller than sham- or control-untreated cells, as shown in Figures 28.1 and 28.31A) [34,113,148]. After a treatment with VC:VK3 of longer than 2 hours, the carcinoma cells become subspherical to spherical in shape. Some of them still display appendice-like extensions, exhibiting widely diverse shapes (Figure 28.31B) [113,148]. Differently sized pieces can be auto- or self-excised by the cell body, which remains as a narrow perikaryal cytoplasmic area. When fixed on cell-culture surfaces, self-cut extensions appear like petals falling off a flower (Figures 28.31B and 28.32), resulting in the final small, subspherical, puck-like cells ablated from the cytoplasm. These can be shown with both SEM and TEM in this cell line, as well as in other tumor cell lines.

Following this treatment, the cell population of *in vitro* cultivated cells (Petri dishes) is reduced compared to sham, VK3, or VC treatment. In fact, the 1 µm-thick sections stained by toluidine blue also have significantly fewer remaining DU145 cells compared to the other treatments. In addition, DU145 displays the smallest cell size recorded (5.0–8.5 µm, as noted in Table 28.4). Many pieces of cells or corpses are found in intercellular spaces to confirm these observations. After a slow sedimentation, and before fixing these samples, some supernatant of this treatment – not studied ultrastructurally – always appears cloudy, suggesting a content of innumerable suspended cell fragments. These cells reveal a more spherical aspect, with the narrowest cytoplasm of all the treatment types, and more of the similar but complex injuries described in both VC- and VK3-treated cells.

As was found in the initial SEM micrographs, SEM and TEM micrographic collections of this carcinoma cell demonstrate a greater reduction in cell size due to self-excisions following exposure to VC:VK3 as compared to VC or VK3 alone [34,148]. Likewise, there is an increased prevalence of cytoplasmic vacuoles with other organelle damages.

28.8.3.2.11 Oxidative Stress of DU145 Cells Favors Formation of Phagophores

It is interesting to see that, as a result of this treatment, the outer nuclear membrane and the adjacent ER of the perikaryal area display a reticulated-like, tubular aspect that can

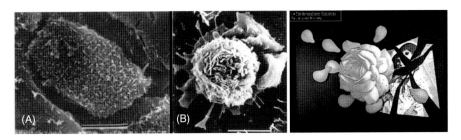

Figure 28.32 (A) SEM view of a DU145 cell fixed on a gold grid, as cultivation support at the confluence. (B) DU145 cells undergoing autoschizis, which resembles the painting "La Fin de Madame Gardenia" by Jacques Monory (at right). *Source:* Gilloteaux et al. 1995 [34]. Gilloteaux et al. 2014 [114]. Reproduced with permission from Taylor & Francis.

associate with the lysosomal bodies constructing certain autophagocytic structures. Out of this network appears a series of flat, well-contrasted vesicles displaying stiffed-like membranes, revealing apposed, curved, or end-curved rods similar to collapsed Birbeck bodies, each extremity of which displays a small dilated internum. These peculiar structures also look like the isolation membranes surveyed as precursors of autophagosomes, known as phagophores, and accumulate as preying small segments of cytoplasm with organelles (Figures 28.33A and 28.34).

Most of the time, few or no intact recognizable organelles remain against a heterogeneous, sometimes finely granular background, where one finds contrasted lysosomes as large autophagocytic whorls or bodies and abundant vacuoles of diverse sizes, similar to

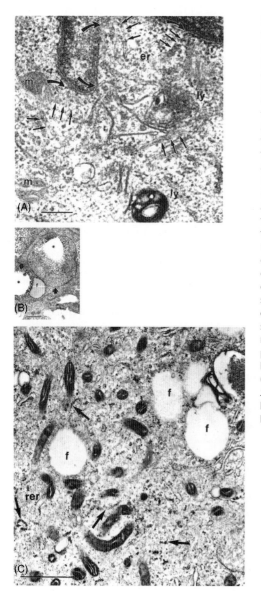

Figure 28.33 (A) Perikaryal area after 1-hour treatment by VC:VK3-treated DU145 cell. The area shows loosely amassed, damaged endoplasmic reticulum (ER), contiguous and extending from the outer nuclear membrane (thick curved arrows). It is flat, appears like stiff cisterns with small dilated luminal extremities that mingle among damaged mitochondria and lysosomal bodies (one markedly so). Note the loose, poorly contrasted cisterns that seem to be issued from the outer nuclear membrane and eventually extend as rough endoplasmic reticulum (RER) (sets of straight arrows).
(B) Large glycogen patch among immummerable ribonucleoproteins. This patch displays void areas (*) adjacent to an apparently intact fatty droplet (f) and is almost entirely surrounded by a phagophore-phagosome, suggesting that glycogen and its cytoplasm are captured. **(C) Superficially, most peripheral cytoplasm still appears to be decorated with small fatty droplets and small, elongated, but contrasted mitochondria.** Numerous polysomes and segments of the sinuous, loose RER network can be seen [43]. The abundant polyribosomes may reflect the unusual, poor RNase activity of this tumor cell type. Scale: (A) = 500 nm; (B) = 500 nm; (C) = 1 μm.
Source: Gilloteaux et al. 2014 [114]. Reproduced with permission from Taylor & Francis.

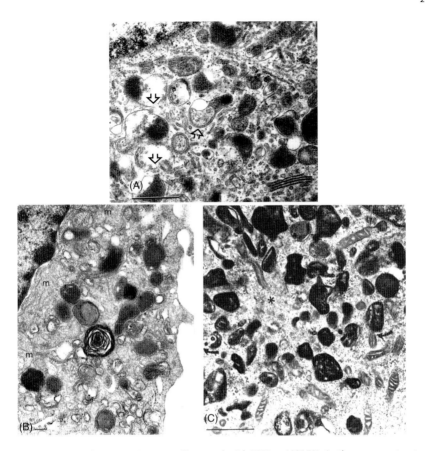

Figure 28.34 (A) RT4 carcinoma cell treated with VK3 or VC:VK3. In these cases, treatment induces lysosome leakiness. This can be seen in the enlarged area with accumulated lysosomal bodies, which shows their leaky envelopes (open arrowheads). **(B,C) VK3-treated DU145 cells.** Overall aspect of a nuclear cove containing a large congregation of electron-dense bodies – mitochondria and lysosomes irradiating from a central field (*), where the centrosome locates and organizes a supportive microtubule network. Some mitochondria appear elongated, distorted with thinning aspects, and serpent-shaped. Some make contact with the lysosomal bodies and/or self-excise through a tenuous part (straight arrows) while contacting tubular ER structures. Contiguous or continuous aspects can be found within the lysosomal bodies (curved arrow). Scale = 1 μm. *Source:* Gilloteaux et al. 2010 [51]. Reproduced with permission from Taylor & Francis.

those noted in VC-treated cells (Figures 28.34, 28.35, 28.36B, and 28.37A,B). When recognized, the defectuous mitochondria can reveal high contrast or condensation of some disorganized membranes in the free cytoplasm, often adjacent to enormous, curved phagophores or inside autophagosomes as remains (Figures 28.33B,C and 28.36D).

In other remaining cytoplasmic areas, VC:VK3 treatment showed isolation membrane-capturing patches of glycogen, in the vicinity of stored fatty deposits (Figure 28.33B).

28.8.3.2.12 Karyolytic Aspects of VC:VK3 Treatment

Most cells that survive treatment and are injured contain a large nucleus with altered chromatin and an altered nucleolus (Figure 28.38). After combined vitamin treatment,

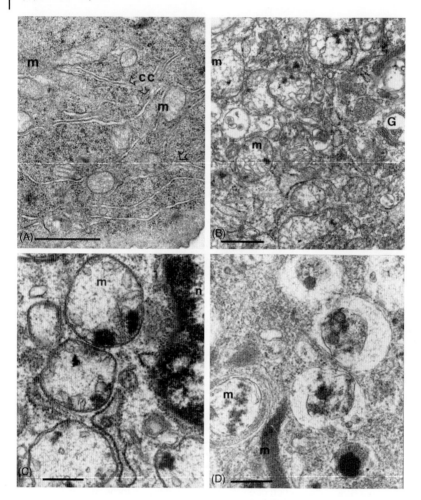

Figure 28.35 DU145-cell mitochondria after (A) sham, (B) VC, (C) VK3, and (D) VC:VK3 treatment. The comparative display demonstrates that from a cristolytic morphology (A), the matrix is changed due to swelling and deposition of highly contrasted material, as well as damages. It is after VC:VK3 treatment (D) that the mitochondria themselves – as well as the adjacent space – are swollen and become vacuole-like spaces among organelles amassed in the perikaryon. Scale = 1 μm.

nuclei are usually eccentrically located, have a small diameter, and contain a poorly contrasted nucleoplasm encompassing one or more oblong to spheroid, highly contrasted nucleolar masses, caused by their extreme compaction (Figure 28.38). At first, a thin nucleolus and heterochromatin go from a somewhat reduced contrast to clusters of fine chromatic punctate dots, before melting down into a fine to indistinguishable, poorly contrasted, homogeneous aspect, without showing any compaction on nucleoids as in apoptosis or apoptotic bodies. In the smallest cells, the nucleus remains in a central location and the cytoplasm is reduced to a rim of <2 μm, as seen on LM, TEM, and SEM views [51,148]. In these small and dying cells, the narrow perikaryal cytoplasmic sector is filled with innumerable strings of ribosomal particles, which remain attached as beads on strings to mRNAs shaped like short, winding, straight to sinuous hair loops (Figure 28.33C).

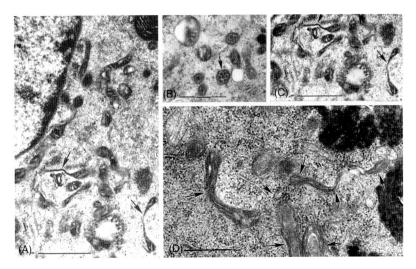

Figure 28.36 VK3-treated Du145 cells, with enlarged views of the aggregated, distorted mitochondria, somewhat contrasted in (D). The altered mitochondria demonstrate their electron contrast, and a unique stretched, twisted-like morphology (arrows) near onion (phagosomes) bodies; some dilated intermembranous spaces can be vacuolated. (C) shows a field of altered mitochondria, albeit with tubular cristae in their stretched cross-sections (arrow). In (D), adjacent to or in contact with large, heavily contrasted lysosomal bodies, a few twisted, distorted mitochondria (arrows) are linked with smooth endoplasmic reticulum (SER) and these lysosomes via some extended endomembranes. The cytoplasm contains innumerable particles (free ribosomes mixed with b-glycogen) and scattered pieces of endoplasmic cistern. Scale: (A–C) = 1 µm; (D) = 200 nm.
Source: Gilloteaux et al. 2014 [114]. Reproduced with permission from Taylor & Francis.

Figure 28.38B,C shows nuclei undergoing the initial steps toward chromatin degradation into chromatolysis. The nucleoli progressively change from a compacted aspect to a fragmented one, in which a thick layer of peripheral heterochromatin coats the inner nuclear membrane, before being thickened into interrupted contrasted patches (Figure 28.32C). With VC:VK3 treatment (Figure 28.38), the nucleus – with its nucleolar alterations – follows several similar sequences of degeneration to those seen in the T24 and MDAH 2774 cells described anteriorly.

The nucleoplasm shows fine granulation, resulting from heterochromatin dilution and disaggregation. Some cells found in the 1 µm-thick sections are nucleated, confirming the blockages in the cell cycle described by Jamison and colaborators [176,188]. Figure 28.38B,C already shows a nucleolus with segregated components, because a progressive condensation of the granular component has occurred, producing a large, amassed, accumulated structure, electron-densely contrasted and with a few sieve-like perforations caused by the nucleolus interstices left where the associated chromatin was either extracted or not (Figures 28.38C and following), in a similar way to that observed in T23-treated cells.

Further nucleolus self-accumulation of granular components can be seen in Figure 28.38, where several cells are seen in a progressive sequence of events. The accumulated nucleolar masses can be dismantled and fragmented, leaving one mass of condensate ribonucleoproteins further shrinking the nucleoplasm and losing its chromaticity. A chromatic peripheral rim alongside the inner nuclear membrane also appears underlined by this deposit (Figure 28.38).

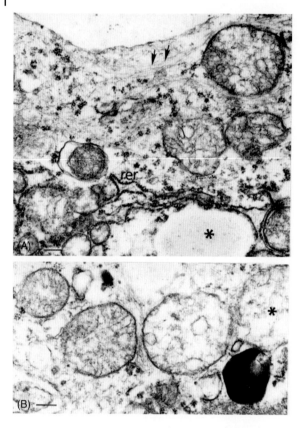

Figure 28.37 3-hour VC:VK3-treated DU145-cell cytoplasms. (A) Remaining cell cytoplasm with swollen mitochondria (bulbous cristae) among small, well-contrasted clumps of ribosomes and polyribosomes, associated or not with swollen rough endoplasmic reticulum (RER) membranes and dispersed remnants of cytokeratin filaments (small arrows). Endoplasmic reticulum (ER) cisterns may be swollen (*), and mitochondria can be completely surrounded by an isolation membrane or autophagosome. (B) Example of a cytoplasm of a dying cell adjacent to a flocculent, finely granular deposit, left over from an open nucleus envelope as a result of osmotic damage. Shows damaged organelles, including an electron-dense bursting lysosome body. Scale = 500 nm.

Further damage of the nucleolar remainder at its nadir produces a smaller and smaller structure that eventually disintegrates upon cell demise (Figure 28.38), or else is found among the intercellular spaces if released as large fragments.

In the meantime, chromatolysis pursues its course (Figure 28.38I). In Figure 28.38, an advanced chromatin degradation (chromatolysis) can be seen, with only small heterochromatin patches persisting. first attached to the inner nuclear membrane. From the nucleoplasm and this rim of heterochromatin, a fine granulation seeds the nucleoplasm, of the same size as the IGs found earlier [148]. Possibly, these IGs are actually the products of further nucleolar fragmentation (Smetana et al., pers. comm.) [51,183]. The significant and progressive decrease in the contrast of these granules into "soft" contrast granules appears to further support some ongoing DNase-mediated degradation of this substrate, as verified in other tumor cell lines by Feulgen staining (Figure 28.25) and DNA gel electrophoresis (Figures 28.31). In some cases, one can clearly verify an end of karyorrhexis after VC or VK3 treatment, which can be caught by the fixation procedure

[148]. However, after VC:VK3 treatment, only the end of karyorrhexis can be captured (Figure 28.38B). Ultimately, small remnants of nucleolus and IG, as well as a diminished contrast of chromatin, can be viewed, due to progressive DNA lysis or chromatolysis, indicating cell demise by autoschizis.

DU145 cell death occurs while the nuclear envelope remains apparently intact, until the perikaryal cytoplasm is self-excised or bursts due to osmotic swelling of the nucleocytoplasmic remnants (Figure 28.38D,E). Rupture of the nuclear envelope occurs following its oncotic burst. In this event, the outer nuclear-membrane pores and the membrane itself show a series of blisters (Figure 28.38E).

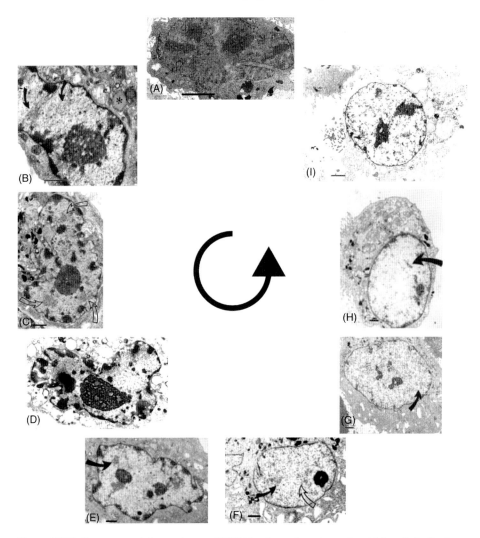

Figure 28.38 Human prostate carcinoma (DU145) cells undergoing autoschizis cell death. Arrow shows progression from sham treatment (A) to 1- (B) to 4-hour (C–I) VC:VK3 treatment, demonstrating the ongoing nuclear injuries that occur throughout the process. Note the chromatin leaving the nucleolus (C), and the karyorrhexis and progressive karyolysis producing a clearing of the nucleoplasm, which become flocculent with almost total washout (I). Nucleolar damage encompasses segregation of the cellular components and fragmentation of their ribonucleoprotein masses. Scale = 1 μm.

Figure 28.34 compares the morphology of the sham DU145 cell (Figure 28.34A) with that of a VC:VK3-treated DU145 cell undergoing cell death by autoschizis (Figure 28.34B). A comparison can be made between this death morphology and the artistic "death" of a gardenia flower illustrated in Figure 28.32C.

28.8.3.2.13 Other Organelle Injuries: Mitochondria and Lysosomes with SER and Autophagy

After menadione or VK3 treatment, DU145 cells show many defects, with worse damage than after VC treatment. Among these defects, autophagosomes arise from interactions between injured endomembranes (SER, RER, and Golgi) and lysosomes, resulting in small to large contrasted onion bodies (1.5–2.5 µm diameter). In the same cell areas, the altered mitochondria are captured or linked with lysosomes through membranous liners and become constituents of the amassed autophagosomes [113,148]. In the vicinity of these pile-ups, ER cisterns appear to cluster into "isolation membranes": heavily contrasted, scythe- or curved-shaped membrane leaflets, which are the precursors of phagophores and autophagosomes. Some ER-phagosomes with elongated vacuoles bear flattened edges that can sometimes reach the plasmalemma.

It is of interest to try to work out (based on ultrastructural observations that show phagophores and autophagocytotic events) which structures produce these endomembranes. The ER itself seems to be cut or carved into small pieces throughout the treated cells. However, in some few micrographs collected from observations of the perikaryon of certain cells, the nuclear external membrane appears to produce membrane extensions that burst into phagophores (Figure 28.33A). These display a typical curly aspect, either by themselves or in continuity with the outer-membrane extensions of the lysosomal bodies; they range between 1.0 and 2.0 µm in diameter. These structures can evolve into phagosomes with increased segregation of altered organelles and segments of cytoplasm, and the vacuoles can eventually link and fuse altogether, contributing to the self-excision of pieces of cell or to autoschizic cell degradation, which involves the shedding of cell pieces or corpses.

28.8.3.2.14 Lysosomes, SER, and Prooxidant Cytotoxicity

Most of the organelle defects that encompass changes in mitochondria are also accompanied by endomembrane. In particular, the ER (more often SER than RER) is accompanied by lysosomal activity favoring cell damage, due to leakiness resulting from VK3 treatment (Figure 28.34). It is worth noting that, even with TEM, injured mitochondria cannot always be distinguished from lysosomal bodies, as they attain an enormous size and can acquire a swollen matrix with huge osmiophilic deposits.

Numerous and sometime prominent autophagosomes can be seen adjacent to or as an extension of the nuclear envelope. Diverse stages of autophagocytosis with defects in lysosomal membranes can be observed alongside or associated with other injuries that ultimately kill – in this case – these carcinoma cells. While the tumor cell cycle is blocked after this combined treatment (Figure 28.11A,B and Table 28.3), the heterochromatin changes strongly suggest random degradation as the cause of the same injured morphology in other tumor cells (Figures 28.26–28.31). This induced process of tumor-cell demise is the most frequently observed *in vitro*, and is what is termed "autoschizis cell death" [35]. This process has recently been reviewed and summarized [113].

The leakiness of lysosomes and their hydrolases caused by VK3 treatment can be seen as the treated cells lose their reducing agents and exhibit an increased internal acidity.

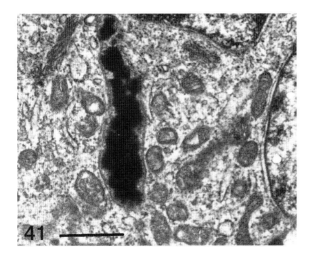

Figure 28.39 RT4 cells exposed to VC:VK3 for 4 hours. Among the perikaryal organelles, small mitochondria with minute matrical condensation and variable-sized, electron-dense autophagosomes can be seen. Inclusions suggestive of large mitochondria, including a large one adjacent to the nuclear envelope, are also visible (black arrowhead). A large portion of the cytoplasm containing minute smooth endoplasmic reticulum (SER) components entwined with SER-like tubular aggregates and loose membranes can be seen in the lower portion of the figure. Scale = 1 μm.

Most organelles with proton pumps are now degraded by heavy contrast (mitochondria, lysosomes, and phagosomes), as in Figure 28.35A–C [51].

Mitochondria, in most fields of view, are altered and reveal a small size and dimorphic shape compared with sham-treated cells. Following VK3 treatment, they display some swelling and a higher electron contrast compared to sham (Figure 28.35A) and VC-treated (Figures 28.35B) cells. In the case of DU145, mitochondria treated by VK3 and VC:VK3 reveal peculiar distortions in pulled, straight, and twisted regions. Long, narrow regions taper, hook, and apparently blend with SER membranes through twisted and stretched extremities, in which the cristae develop a tubular appearance. The mitochondria in these regions seem "melted," as if they belong in a Salvador Dalí painting. On many occasions, they merge with – and even contain – a lipid droplet or vacuole-like space, as shown in VC:VK3 (Figure 28.35A–D). In some similarly treated carcinoma cells, such as RT4 [51], the organelle's matrix can be entirely filled by menadione and other defectuous (probably lipochrome and cation) components (Figure 28.39) [51].

The cove-like areas around the nucleus contain congregations of a few intact-like organelles, along with many osmiophilic bodies. They appear as if they have been corralled or amassed in the remainder cytoplasmic area, as if blocked along suggestive lines of damaged microtubules issued from a remaining centrosome [113]. Most of the injured organelles seem to be congregated, with an irradiating, fan-like spread (Figures 28.34A–C and 28.40A).

Some lysosomal bodies can be formed from damaged SER–RER and/or damaged mitochondria following long-duration treatments. These findings suggest that ER, mitochondria, and lysosomal bodies can interact with one another to produce autophagosome bodies out of phagophore complexes. After this treatment, most of the surviving tumor cells will have a high nuclear:cytoplasmic ratio.

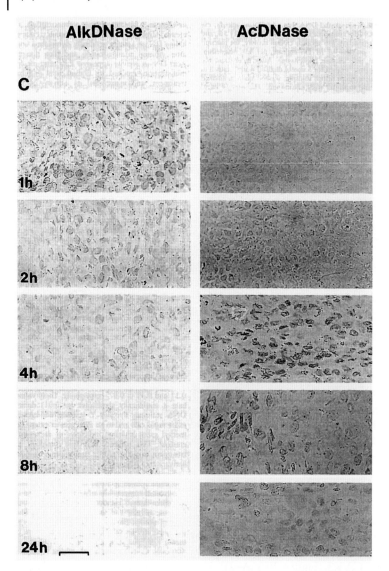

Figure 28.40 Reactivations of DNases in an implanted human prostate tumor after single-dose administration of intraperitoneal and oral VC and VK3. Left column: Little if any alkaline (DNase I) activity is found in the sham-treated tumor (C). However, there is distinct alkaline DNase activity, primarily in the nuclei, 1 hour after injection. This activity decreases during the second hour (2 hours) and becomes difficult to appreciate at later time points (4, 8, and 24 hours). Right column: AcDNase (DNase II) activity observed in the tumor after the same vitamin treatment. Little or no activity is observed in the sham-treated tumor (C). Progressive activation of DNase occurs more than 2 hours after vitamin treatment, reaching a maximum after 4 hours. This activity decreases after 8 hours but is still evident after 24. Scale = 20 μm. *Source:* Taper, 2001. Reproduced with permission from SAGE Publishing.

Following **VC treatment alone**, DU145 cells display organelles aggregated in the perikaryal area, while small, round, fatty deposits are outwardly located in the cytoplasm along with often enormous glycogen patches [113]. Cytoplasmic damage includes continuous small auto- or self-excisions (autoschizis), recognizable by the cellular debris scattered among the remaining cells. These excised pieces confirm the smaller diameter of the examined cells (Table 28.1) by significantly reducing the cytoplasm to a narrow, perikaryal area that becomes congested with mostly injured organelles as a result of lysosomal activities. Most of the mitochondria bear wavy, washout to quasi-inexistent cristae and some swelling, with poorly contrasted matrix containing one or more eccentric, irregularly shaped membrane whorls and flocculent osmiophilic deposit-like patches of irregular blotches of deposit (Figure 28.35B).

Following **VC:VK3 treatment**, most tumor cells reveal the same sorts of damage imparted by both VK3 and VC alone. With time, the mitochondrial injuries become vacuolated, first showing intramatrical heterogeneous deposits that resemble those noted after VK3 treatment alone. After >2 hours' treatment, most mitochondria are replaced by vacuolated bodies, which seem to have loose onion-like aspects concentric to and enclosing a heterogeneous content (Figure 28.35D). Alternatively, some cells may be more resistant to the prooxidant treatment. Near the time of cell demise by autoschizis, some mitochondria can still be recognized; these are swollen, and are eventually surrounded by phagophores, with remnants of cristae still showing. They appear as if in osmotic shock during oncotic/necrosis and are replaced by vacuoles (Figure 28.37A,B).

In vitro, most carcinoma cells undergoing a prooxidant treatment endure significant injuries that show on LM, SEM, and TEM, leading to autoschizis cell death:

1) **surviving, diminished, and damaged cells with a peculiar shape undergo progressive disintegration of their nuclear content through karyolysis and cytoplasmic damage;**
2) **membrane and cytoskeletal defects are visualized with no or little debris by annexin-IV FITC, making the nucleus totally visible, unlike in apoptosis;**
3) **nuclear chromatolysis verified by histochemistry and gel electrophoresis demonstrates DNA being degraded by random cuts resembling those of necrosis; degradation forms IGs;**
4) **nucleolar segregation, fragmentation, and/or compaction of rRNA pieces can be seen;**
5) **inducement of phagophores occurs, with lysosomal activation of autoph-agocytoses;**
6) **oxidative mitochondrial damage is seen, with energetic reserves of phagocytosis.**

28.9 Xenotransplanted Prostate Carcinoma (DU145) in Nude Mice Treated with VC:VK3

28.9.1 Sequential Reactivations of Nucleases, Inducers of *Causa Mortis*

Based on Taper and collaborators' studies, it was demonstrated that the activity of alkaline DNase (DNase I; EC 3.1.21.1) and acid DNases (DNase II; EC 3.1.22.1) can be inhibited in non-necrotic cells of malignant tumors in humans and experimental animals, as well as during the early stages of experimental carcinogenesis [7,10,12,13,124,149]. Because these DNases' reactivations can happen in spontaneous induced tumor necrosis and regression [7,8,10], compounds capable of reactivating DNase in tumor cells should offer great potential for therapeutic intervention in cancer.

It has been found that VK3 (2-methyl-1-4-naphthoquinone) selectively reactivates alkaline DNase in malignant tumor cells, whereas VC exclusively reactivates acid DNase or DNase I [10]. A combined administration of VC:VK3 (at a single IP dose of VC = 1 g/

kg body weight and VK3 = 0.01 g/kg) in ascites tumor-bearing mice synergistically increased the life span of the mice by 45%. Separate administration of VC increased life span by 14.7% and that of VK3 by 1.07% [16]. Combined VC:VK3 treatment also potentiated the effects of chemotherapy induced by six different cytotoxic drugs [16], sensitized TLT carcinomas from implants in nude mice resistant to vincristine [17], and potentiated the therapeutic effects of radiotherapy [203].

The tumor growth-inhibiting and chemosensitizing effects of VC:VK3 were confirmed by *in vitro* experiments with different lines of human tumor cells [18,19,204]. These *in vitro* studies were extended to a battery of human urological tumor cell lines, including DU145, an androgen-independent prostate carcinoma cell line [8,34,177,178,205]. In this study, we have investigated the therapeutic effect of VC: VK3 on human prostate carcinoma cells (DU145) transplanted into 5-week-old male NCr athymic nude mice. The histochemical data published in support of the xeno-transplanted prostate carcinoma treatment are summarized in Figure 28.41, extracted from an article published by Taper and our laboratories in 2001 [9].

The photomontage of Figure 28.40 rightfully summarizes the observations made in detecting histochemical DNase I and DNase II activities in DU145 implants after VC:VK3 treatment. The sham-treated tumors exhibited little if any DNase I or DNase II activity (Figure 28.40, column C). However, after combined VC:VK3 treatment, DNase I activity appeared significantly sooner than DNAse II activity, because a positive detection showed as early as 1 hour after vitamin administration, and decreased slightly thereafter until 2 hours after treatment. DNase I activity gradually diminished until 8 hours after vitamin administration and then disappeared. DNase II (acid DNase) activity appeared 2 hours after vitamin administration, reached its highest level 4–8 hours after treatment, and maintained itself for 24 hours after treatment. It is obvious that if both DNases have a combined activity and reactivate enzymes, it will peak at 2 hours' duration and afterwards. The data presented in this chapter seem to be supported by these critical observations.

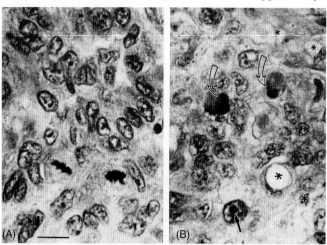

Figure 28.41 Hematoxylin- and eosin-stained implanted tumor. (A) Micrographic field of sham-treated tumor cells showing pleomorphic cell and nuclear morphology. Note the two mitotic figures in the field. (B) Micrographic field of a vitamin-treated tumor, showing cells with pyknotic nuclei (curved arrows). Other nuclei with marginated chromatin exhibit condensed and enlarged nucleoli (small arrow). Many vacuolated cells (stars) are evident throughout the tumor mass. Scale = 25 μm. *Source:* Taper, 2001. Reproduced with permission from SAGE Publishing.

Following treatment with the combined vitamins as prooxidants, a reactivation of RNases was noted, but this was not further studied. The only observation we made was of the dysmorphology and segregation of the nucleoli of tumor cells, with ultimate compaction of ribonucleoproteins after chromatin left the nucleolus, leaving empty interstices. This was followed by compaction, and later fragmentation, of the amassed granular structures among the nucleoplasm. In some cases (MDAH 2774), the ultimate cell demise leaves corpses and either large or small, fragmented masses of the leftover nucleoli in the intercellular spaces.

28.9.2 Autoschizic Injuries and Deaths Induced by Prooxidant (VC:VK3) Treatment

VC or VK3 treatment alone causes a significant reduction in the cell diameter of tumor cells as compared to sham-treated cells (Table 28.4). Both isolated treatments affect the cytoskeleton (VK3 > VC), as they favor self-excisions or superficial bursts of the cytoplasm free of organelles. These excisions are confirmed by cell-distorted lamellipodia, edges, and debris of all sizes throughout the intercellular space, whether viewed by SEM or TEM (Figures 28.2A and 28.3B) [35,43,51,113,148,173,182,183,185,186,192]. After each treatment, nuclei display a narrow 50–150 nm heterochromatic electron-dense layer covering their inner envelope. Prolonged treatment of either VK3 or VC leads to later condensation of the nucleolus into one or more compact osmiophilic masses, with segregation of their components. Up to 20 nucleoplasm aggregates of separated chromatic dots create ovoid to circular patches of 0.3–0.6 μm diameter circumscribing the remaining small and large nucleolar masses; these are referred to as "satellite nucleoli," and their presence alongside the compacted altered nucleoli indicates that the latter are remodeled, as some of their protein components (e.g., fibrillarin) can migrate without polymerases or transcripts decreasing or totally shutting down all cell-synthesis or cell-repair mechanisms [189]. All these nuclear events are indicators of irreversible cell injuries preceding the demise of the tumor cells by autoschizis. With immunolabeling and other biochemical investigations, specific steps and differences can be further elucidated.

Again, in both treatment regimens, most organelles are segregated into a large cytoplasmic sector facing the concave side of the indented nucleus. Among them, mitochondria, SER, and RER show swelling and other damage associated with intracellular lysosomal leakages, as seen with prostate DU145 (Figure 28.35A) [113,148] and RT4 bladder carcinoma cells (Figure 28.34) [51]. Based on the knowledge that VK3 is a faster reactivator of DNase I than VC is of DNase II, let's review the qualities of each of these components of VC:VK3 treatment.

The consequences of the reactivation of both DNase I and DNase II can be seen in tumor cells 24 hours after VC and VK3 treatment. When the nucleic acids of sham-treated tumor cells were stained with methyl green-pyronin Y, intense methyl green staining was visible in the nucleus and intense pyronin Y staining in the cytoplasm (Figure 28.42). In vitamin-treated tumor cells, the methyl green staining was greatly diminished, indicating a decrease in DNA content, in line with studies carried out *in vitro* (Figures 28.13 and 28.26). Densitometry analysis of the staining intensity of control and vitamin-treated tumor cells produced optical density values of 0.63 ± 0.11 and 0.37 ± 0.07, respectively. These values indicate a statistically significant ($p < 0.01$) decrease in DNA staining. Pyronin Y staining diminished at a more rapid rate and to an even greater extent than methyl green staining, indicating a decreased RNA

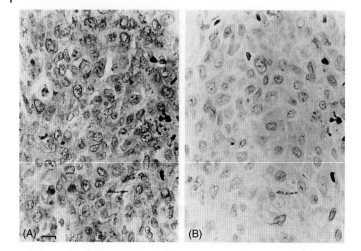

Figure 28.42 Methyl green-pyronin Y staining of sham- and vitamin-treated implanted tumors.
(A) Sham-treated tumor cells demonstrate intense intranuclear staining of DNA, as well as intense staining of cytoplasmic and nucleolar RNA. (B) Vitamin-treated tumor cells stained 24 hours after vitamin treatment exhibit weak staining of both nucleic acids as a consequence of DNase (and possibly RNase) reactivation. Scale = 10 μm. *Source:* Taper, 2001. Reproduced with permission from SAGE Publishing.

content. This suggests that this combined vitamin treatment may also reduce the transcription or induce the activation of RNases.

New data concerning DU145 implants, obtained while writing this chapter, provide further support for the claim that reactivation of nucleases can facilitate the degradation of tumor cells. Accordingly, we shall provide here some preliminary, still unpublished observations (Figure 28.43).

In addition to the histochemical detection of nucleases noted in the following paragraphs, a few unpublished observations can be found in Figure 28.41. The presence of mitotic figures indicates that cells in this tumor were still actively dividing. Sections of the control tumor exhibited an average of 5 ± 1 mitotic figures/500 cells. Conversely, the cells in the vitamin-treated section (Figure 28.41B) averaged 1.23 ± 0.84 mitotic figures/500 cells. Although the incidence of mitosis was small, the difference between control and vitamin-treated cells was statistically significant ($p < 0.01$). The nuclei of some cells appeared pyknotic, while others showed marginated chromatin and predominant condensed nucleoli. This section was also characterized by the presence of many vacuolated cells and of cell necroses.

Figure 28.44B shows a solid tumor disrupted by cryptic erosion following VC:VK3 treatment, which displays numerous freed cells undergoing cell death mainly through autoschizis. The injured and dying tumor cells have detached from their adjacent cells, having obliterated edge structures to provide some intercellular contacts [43], as well as undergoing some minute or large self-excisions (Figure 28.43A,B). Figure 28.43D provides a detailed view of such a self-excision. Through these erosions, the cells freed themselves from the solid tumor that was making cribriform areas [43] (see Figure 28.9) become vacuolated while the nuclei altered their contents in a similar manner to that seen *in vitro* by LM (see Figures 28.13D and 28.14D).

These results show that VC:VK3 treatment is capable of stimulating the degradation of nuclear content through nucleases.

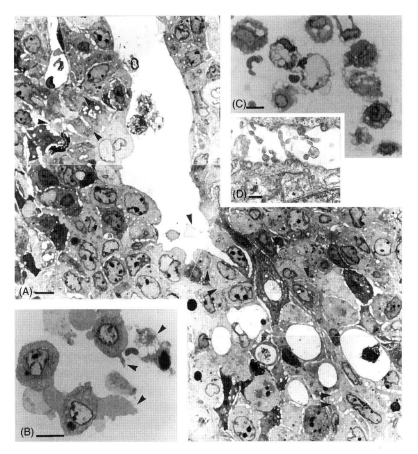

Figure 28.43 LM and TEM of a region of xenografted DU145 prostate tumor growing on a diaphragm. (A–C) Pleomorphism is shown by the heterogeneity of the cell components. Note the high nuclear:cytoplasm ratio and the fact that there are some cryptic structures among the tissue, created by the signet cells and cell deaths, which contribute to shrinking the tumor after treatment. One can clearly see that some injured cells are detached from the bulk tissue growth and are showing autoschizic self-excisions caused by protruding appendages extended away from the cell bodies, which resemble those found *in vitro* after treatment with VC:VK3. (D) A small area revealed by TEM shows a self-excising cell liberating cytoplasm pieces by creating an intracellular gap during the cutting process.

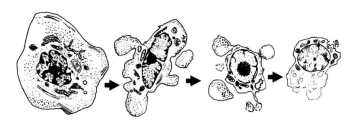

Figure 28.44 Diagrammatic representation of tumor cells undergoing autoschizic cell death.
Source: Gilloteaux et al. 2010 [51]. Reproduced with permission from Taylor & Francis.

In xenotransplants, most carcinoma cells undergoing prooxidant treatment endure significant injuries leading to autoschizis cell death, as seen on LM, SEM, and TEM:

1) reduced cell sizes and peculiar shapes undergoing progressive degradation of the nuclear content and cytoplasm self-excision;
2) **karyolysis of VC:VK3 > VK3 > VC**, favored by DNase and RNase activation, inducing injurious events, as seen by LM, TEM, and gel electrophoresis:
    1) membrane and cytoskeletal defects contributing to cell-morphology changes;
    2) peroxidations triggering phagophores and lysosomal leaks, causing cytoplasmic and concommittant organelle damage (karyolysis and nucleolysis, mitochondria), and autophagosomes capturing damaged parts;
3) **cytotoxicity of treated tumors**, including shrinkage through dismantling of the carcinoma organization by an alteration of cellular organization, disruption of cell size by self-cuts, emigration, and cell death by accrued autoschizis.

28.9.3 DNases I and II and RNases

Most tissues contain both DNase I and DNase II, or else acid DNase (E.C. 3.1.22.1) nucleases [7,207,208], with a nomenclature often based on the organ from which the enzymes were first extracted or on their type of action on nucleic acids. Details concerning nucleases can be found in several published works [209–211], reviewed in [7].

DNase I and II and RNases are diversely affected in their activities according to the presence of sulfhydryl groups, especially at lysosomal sites [212,213].

The nature and biological mechanisms of these nucleases are often associated with the notion that alkaline types (e.g., DNase I) can be found in secretions (with extracellular function), while acid types (DNAse II) and RNases are located in lysosomes, suggesting their intracellular functions [207,208,214].

DNase I and RNase I in liver and other mammalian organs are found in the intermembranous space of the mitochondria [215–220], as well as in the nuclear fraction [221,222]. All sorts of pathologic conditions can alter these enzymes (viral, bacterial, degeneration, autolysis, regeneration, "fasting" in liver, etc.) [7].

RNase is ubiquitous [223], and it is generally accepted that the DNAse II and RNase II contents of normal cells mainly belong in lysosomes [216,224], although some DNase II has also been seen in the nuclear fraction [139,208,225–229], such as in HeLa cells [228,229]. Like DNase II, RNase II is found in the soluble fraction of rat liver [230,231], chiefly in the nuclear fraction [217,232,233] associated with the nucleolus [225].

Natural inhibitors of DNases and RNases were investigated long ago, but little study has been made of how they relate to oncology. Lindberg and Skoog [234] and Roth [235] have detected an alkaline DNase (DNase I) that can protect polyribosome integrity [236]. In human urine, it is possible to detect proteins that inhibit DNAse II [237], which are damaged by sulfhydryl reagents such as VK3 [235,238,239]. Some proteinaceous components of the lysosomal membranes and latent enzymes of the lysosomes or their membrane attachments depend on the status of sulfhydryl groups [213]. The significance of nuclease inhibitors in the homeostasis of cells was reviewed by Taper [7], who found, like Hakim [240], that nuclease inhibitors acted as "enzyme pace makers," capable of influencing the functional activities of substrates (DNA and RNA), because nucleases were segregated in packets by membranous compartments [241]. One can image (based on the research accomplished with VC:VK3) these membranes being damaged, the "packages" liberated, and, hence, reactivity occuring, favoring the death of tumor cells!

Wroblewski and Bodansky [242] have described how the average alkaline DNase activity in the serum of 50 patients with different types of malignant tumors was significantly lower than that of a group of normal individuals or of patients with noncancerous diseases. Others have not only confirmed this observation, but found that the majority of cancer-bearing patients have a complete absence of alkaline DNase activity in serum [243]. Even though it did not isolate any inhibitor of this nuclease in the serum, an additional experiment found a decrease of activity of alkaline DNase introduced in tissue-culture medium during the incubation of cancerous versus normal tissues. This suggests that DNase was absorbed by the inhibitor present in the incubated neoplastic tissue. Recent advances in the field suggest mechanisms like these involve nuclease blocks.

According to the earliest work using histochemistry [1], a deficiency of nuclease activity was seen in more than 60 different animal and human tumors. The decrease and even disappearance of such activity during experimental liver carcinogenesis has also been reported [244]. It has been confirmed by both histochemical [7,245–247] and biochemical [243,248–255] methods, as well as by immunochemical technique approaches [256].

Taper demonstrated that the activity of both DNase I (EC 3.1.21.1) and DNase II (EC 3.1.22.1) was inhibited early in experimental carcinogenesis, before the phenotypic signs of malignancy appeared (Table 28.7) [7,9–11,124,149,257].

Furthermore, DNase reactivation occurred during the early stages of spontaneous and/or induced tumor-cell necrosis [257]. In fact, the variations in serum alkaline DNase activity (SADA) may serve as a possible test for the prognosis of cancer therapy. In

Table 28.7 Nuclease properties.

Nucleases	Optimal pH	Intracellular Locations	Extracellular Locations	Other
DNase I or alkaline DNase (EC 3.1.4.5)	7.0–8.0	Nucleus and intermembranous space of mitochondria	Secreted digest, exogenous DNA	Activated by menadione or Mg^{++} Inhibitedby Zn^{++} Serum CSF Pancreatic juice Urine Thermostable
RNase I or alkaline RNase (EC 2.7.7.17)	7.5–8.5	Nucleus, nucleolus, and intermembrane space of mitochondria	-	Most cells
DNAse II oracid DNAse (EC 3.1.4.6)	4.2–5.5	Nucleus, lysosomes	-	dsDNa cuts Most cells
RNase II or acid RNase (EC 2.7.7.17)	5.5	Nucleus, nucleolus	Against infectious RNAs (including viral)	Liver and many other tissues

positive responders, SADA levels decreased in the days immediately following therapy, before increasing again a few weeks after treatment to levels equal to or higher than at the start. The maintenance of high SADA levels for several months accompanied remission. Conversely, a sudden decrease in SADA levels preceded the recurrence of cancer by several weeks. Negative responders to cancer treatment did not demonstrate these specific SADA variations [258].

Because DNase reactivation appeared to be linked to tumor necrosis and regression, elucidating compounds capable of reactivating DNases in tumors may offer great therapeutic potential. However, VC and VK3 are of particular interest as therapeutic agents because of their low systemic toxicity and because VK3 has been shown to selectively reactivate alkaline DNase in malignant tumor cells, while VC exclusively reactivates DNase II [7,9,11,20,257]. The reactivation of VC:VK3 appears to combine the reactivations of DNases and RNases of both types in tumor cells. This affects cancer cell integrity, through membrane and protein damage to these structures. The –SH groups cannot be repaired, evidently facilitating irreversible injurious events to cells on their way toward cell death.

Because endonuclease activation is one of the earliest changes denoting irreversible commitment to cell death, it is generally believed to be involved in the triggering of cell death, rather than the result of it. Several candidate deoxyribonuclease molecules have been identified in various cell lines and tissues, including the caspase-activated DNA fragmentation factor caspase 3-activated DNase (DFF40/CAD) nuclease, DNase I, and DNase II. DFF endonuclease is primarily responsible for mediating DNA laddering during apoptosis after its activation by caspases (mainly caspase 3). DFF is composed of two subunits: a 40 kD caspase 3-activated nuclease and its 45 kD inhibitor. This protein complex resides in the nucleus and is activated by caspase-3 cleavage of the inhibitor and its subsequent dissociation from the endonuclease. The activated endonuclease exhibits a pH optimum of 7.5, requires Mg^{2+} (not Ca ions), and is inhibited by Zn^{2+} [260,261]. Although the pH optimum, nuclear localization, and other factors suggest that the DFF40/CAD nuclease may be the reactivated DNase I, a number of other factors argue that another DNase is primarily responsible for the cleavage observed during vitamin-induced tumor-cell death. First, autoschizis rather than apoptosis is the predominant form of vitamin-induced tumor-cell death, and the DFF40/CAD nuclease is associated with apoptotic cell death. Second, preliminary studies conducted in our laboratories indicate that caspase-3 activation by the vitamin combination is marginal and only occurs 3–4 hours after vitamin treatment [173,204,262].

Since nucleases, DNases, and RNases are reactivated after prooxidant treatment, this raises the question of whether DNA is degraded along with other verifiable cancer-cell structures.

28.10 Summarizing the Data

The data surveyed in this chapter, initiated by the pioneering research studies from the late Henryk Taper [7,20], and further obtained from our collaborations with him in United States and from the Louvain laboratory [173,204,262], demonstrate that sham-treated carcinoma cells and solid tumors (Figure 28.42A,B) are essentially devoid of DNase activity. After administration of a VC:VK3 combination, both alkaline DNase (DNase I) and acid DNase (DNase II) activities are reactivated and can be detected histochemically in cryo-sections of tumor specimens by a modified Gomori's lead nitrate method (Figure 28.40) [7,9]. This corresponds to a significant degradation of the tumors and to survival of the implanted mice [7,9,20,204].

While ascorbate, in the form of ascorbic acid or sodium salt (VC), is traditionally perceived as an antioxidant, it may also act as a prooxidant, and increase DNA damage to induce cell death [263]. Menadione (VK3) is an oxidant that exhibits antitumor activity against a variety of tumor cell lines, as well as human explants that are resistant to other types of chemotherapy [264]. When VK3 is combined with VC, the interaction fosters redox cycling [265], which increases oxidative stress through a rapid production of H_2O_2 [266] and ROS [267]. Within the first hour following combined VC:VK3 treatment, this oxidative stress decreases cellular thiol levels to less than half those of sham-treated cells. The loss of protection against ROS is accompanied by the oxidation and subsequent disruption of cellular caspases (including caspase 3), as well as microtubules and other cytoskeletal proteins [35,181]. This cytoskeletal disorganization is reflected by blister and bleb formation, as well as by acute distortions in tumor-cell shape [35,51,173,182,183,186]. After 1 hour of vitamin exposure, significant levels of lipid peroxidation and damage to the cell membrane occur, suggesting that wholesale indiscriminate lipid peroxidation is a late event in the cell death process. However, TEM micrographs demonstrate that the vitamin combination rapidly alters mitochondrial architecture, and induces ultrastructural changes in both the SER and the RER as well. As a consequence of these changes, Ca^{++} transport systems of the mitochondria, SER, and RER are perturbed, and there is an increase in intracellular Ca^{++} levels, leading to the reactivation of DNases [7,9]. While lipid peroxidation and the subsequent loss of membrane integrity may be responsible for the release of Ca^{++} into the cytoplasm, the fact that ATP production by the mitochondria increases 1 hour after VC treatment and 3 hours after VC:VK3 combined treatment suggests that the Ca^{++} release occurs via modulation of the voltage-dependent anion channel [19]. A number of other cellular processes are also affected by the presence of ascorbic acid, and especially dehydroascorbate, including: modulation of signal transduction, cell-cycle arrest, inhibition of glycolytic respiration, and inhibition of metastasis.

Taken together, these results indicate that autoschizis (the type of cell death induced by the vitamin combination) entails the coordinated modulation of cell signalling and metabolism by VC and VK3 in their various redox states, coupled with an attack of H_2O_2 and ROS on cellular thiols, membranes, cytoskeleton, and DNA, which continues until cell death by self-morsellation.

In this process of cell death, termed "autoschizis," DNase I activity appears as early as 1 hour after vitamin administration, decreases after 2 hours, and disappears again by 8 hours post-administration [9]. DNase II activity appears 2 hours after vitamin administration, reaches its highest level at 4–8 hours, and is maintained until 24 hours post-treatment. Methyl green staining indicates that DNase activity is accompanied by a decrease in the DNA content of the tumor cells; this has been confirmed by Feulgen stain in T24 and MDAH 2774 carcinoma [183,185]. Examination of 1 μm sections of the vitamin-treated tumors indicates that the prooxidant activities of both vitamins induce DNase reactivation through cell injuries [9].

Peroxidative activities initiate membrane and cytoskeletal catastrophic breakdown, leading to cell self-excisions – hence the name, "cell death by autoschizis." Concomitant damages include defects of energetic organelles left unprotected by the depletion of antioxidative metabolic mechanisms, along with degradations of nuclear DNA and its nucleolus (observed through chromatolysis and nucleololysis), with fragmentation by cathepsins. The auto- or self-excisions typically continue while karyolysis progresses until the perikaryon (consisting of an apparently intact nuclear envelope surrounded by a thin rim of cytoplasm containing damaged organelles) has been damaged [35,47,51,113,139,148,179,180,183–185]. This cell death is the main form observed in the tumor tissues. Although both necrotic and apoptotic cell death are observed in the same treated tissue, only a few cells show these processes.

In vivo studies designed to determine the effect of vitamin administration on the life span of nude mice demonstrated that mice receiving both oral and intraperitoneal vitamins lived significantly longer ($p < 0.01$) than controls. The results of additional *in vivo* studies demonstrated that administration of clinically attainable doses of oral vitamins *ad libitum* in drinking water could significantly reduce the growth rate of solid tumors in nude mice ($p < 0.05$). Finally, nude mice receiving the vitamin combination did not exhibit any significant bone-marrow toxicity, changes in organ weight, or pathological changes in these organs. Additional experiments using transgenic mice that more closely mimic the development and metastases of prostate cancer in humans are underway.

Among the different mechanisms that may be involved in this therapeutic effect of combined vitamins against cancer, the most plausible is a stimulation of a redox-cycling system that produces H_2O_2 and other active oxygen species involved in cell-membrane lipid peroxidation, DNase activation, and DNA breaks leading to cell death [34,176–178,205]. The most important fact is that this action of VC:VK3 appears to be selective for cancer cells [17]. Unlike normal organs and tissues, cancer cells are usually deficient in catalase, SOD, and/or GSH peroxidase, which constitute the cellular armamentarium system/defense system against free radicals [268]. This hypothesis is supported by the fact that simultaneous administration of VC:VK3 with catalase suppressed the potentiating and sensitizing effect of these vitamins [19].

Because the vitamin combination is a chemosensitizer [16] and a radiosensitizer [203], combined VC:VK3 administration can be considered a new, nontoxic adjuvant cancer therapy, which can be easily introduced into the classical protocols of clinical cancer therapy without any supplementary risk for patients.

When VC is combined with VK3, the interaction fosters a one-electron reduction to produce the long-lived semi-quinone and ascorbyl radical. This increases the rate of redox cycling of the quinone, forming H_2O_2 and other ROS capable of damaging lipids, along with the protective mechanisms against the induced ROS. This results in damage of membranes, associated superficial cytoskeleton, mitochondria, and lysosomes, activation of autophagocytosis, and injurious defects of nuclear content through the reactivation of DNases and RNases. Additionally, ROS appear to favor starvation through inhibition of the regulation of tumor-adapted glycolytic and aerobic metabolic activities.

28.10.1 Terminology: Justification of Cell Death by Autoschizis

Obire, non novus mori, sed nove

("To die, not a new thing, but a new mode")

At first, the data were quite puzzling. They initially seemd to describe a kind of necrotic mechanism, but with a study on bladder carcinoma (T24) cells, collating observations of morphology, flow cytometry, biochemistry, and cytotoxicity in at least 12 human cell lines (bladder (T24), prostate (DU145), ovarian (MDAH 2774) and TRAMP (murine prostate carcinoma)), both *in vitro* and *in vivo*, it was found that cells were self-cutting as a main first appearance *in vitro*. This was thus a new form of cell death.

"Autoschizis" is a neologism that was created from the Greek root *autos* ("self") and *schizis* or *schisis* ("division"or "cut"). Thus, the term "autoschizis" signifies "self-cutting" or "self-dividing.," The major steps and events historically observed in this mode of cell death were initiated by superficial membrane defects triggered by auto- or self-excisions or by self-morsellation (from Latin *morsus*) due to cytoskeletal-induced defects.

Autoschizic cell death has been seen as the cytotoxicity resulting from the prooxidant actions of VC:VK3 on several carcinoma cell lines, including renal carcinoma (Caki-1), ovarian MDAH 2774 carcinoma [35,137,179,180,184], and a murine prostatic carcinoma (TRAMP) [192]. A number of independent laboratories have also reported observations of autoschizic cell death [173,204,262]. Recently, we have detailed tumor-cell damage by describing in some detail organelle changes in human prostate carcinoma DU145 cell lines and nuclear changes in the nucleolus (as shown in Figures 28.19, 28.22, 28.25, 28.30, and 28.36) [21], as well as damge to other organelles, especially the mitochondria and lysosomes, culminating in self-cannibalism [148]. These events were also accompanied by entosis, with further cannibalism contributing to removal of corpses and to cell survival via larger cells eating smaller ones [113].

This chapter has aimed at providing comparisons of some of the data collected in several of these cancer cell lines *in vitro* and in xenotransplants *in vivo*. Autoschizis cell death, as we have revealed, has been studied with the goal of supporting a new anticancer strategy using the VC:VK3 combination as an adjuvant or treatment, because, according to animal studies, the propensity of the mixture is to preferentially kill neoplastic cells while leaving normal tissue cells unharmed.

28.10.2 Autoschizis: A Mode of Self-Excision that Can also be Observed in Differentiated Normal Cells and Tissues

Future investigations could consider autoschizis as a mode of cell self-cutting that, via analogous cytotoxic actions, can eventually result in cell death or in alteration of cells without death, as seen in the following mammalian examples: cardiac development [269], erythrocyte maturation [197,198], muscle-tissue metamorphosis [270], functional prostate epithelium [271], and Clara cell differentiation [272]. In some cells, the autoschizic self-excisions do not necessarily lead to cell death, but do result in self-cutting, in a manner similar to apocrine secretion events. Thus, one can document normal cell cytoskeleton rearrangement favoring apocrine self-cuts or other similar types of cut. These cell-regulating self-excisions may bear similar intracellular macro-molecular interactions to those found in autoschizis cell-death processes, but in such cases the preservation of cell-organelle integrity is paramount. Cytoskeletal components may not entail associated damage because certain regulatory enzymes (caspases) are not activated through mitochondrial damage signals.

28.10.3 Autoschizis and Other Cell Deaths

In view of the morphologic injuries observed with VC and VK3 alone and with combined VC:VK3, one can see that the injurious events triggered by prooxidative cytotoxicity are initiated by membranous and cytoskeletal damage, triggering self-excising cytoplasm and dramatically decreasing cell size. A short duration (1–3 hours) of DNase I and RNase treatment damages lysosomes and mitochondrial DNA and abrogates most of the already poor metabolic armamentarium of the tumor cells (GSH, H_2O_2, and ROS protectors, especially peroxisomes). With this energetic poisoning, it also prevents any repairs of damage incurred by the initial oxidative stress. However, the reactivation of a second wave of DNase II and RNases further degrades the nucleus, with the accompaniment of cytoplasmic cathepsins from leaky lysosomes. Concomitantly, these components assist in disintegrating the DNA and nucleolar components while at the same time removing (by autophagocytosis) the organelles rendered defectuous by the persisting oxidative reactions. These autophagocytic activities merely capture pieces of the injured cells for energy salvage through recycling, however; ultimately, the tumor cells self-destruct. The ultimate osmotic overload caused by karyolysis also favors tumor-cell demise.

Under normal cellular metabolism, controlled by an intact genome, cells function in the least oxidative environment possible, through maintenance of a neutral to poorly alkaline homeostatic state of functioning. All protective systems are meanwhile ready in case of invaders or internal toxicants (e.g., DNases, RNases, peroxisomes, lysosomes, SODs, cytochromes P450, cell-cycle checkpoints, etc.).

Transient changes occurring in normal cells during interphasic to replicative activities are supposedly regulated by arrays of proto-oncogenic tumor-suppressor genes and their products. This process maintains the optimal balance of transduction inputs. Many growth factors are implicated in cell replication and progression toward a normal phenotype, which results in organization of the cells into tissues, and their integration into organ systems. These events are cyclically possible in a typical "controlled cell division." Normal cells with epigenetic anomalies will be committed to apoptosis (e.g., p53, APC, ATG, etc.).

Normal cells without any altered genome suffer necrosis if invaded by toxins or starved of oxygen and nutrients. This is usually an oncotic cell demise; that is, it can be called "oncosis" [28].

Any disruptive epigenetic changes caused in control systems that do not have built-in redundancy or secondary checkpoints can become detrimental to their homeostastic functionality. This disruption can favor uncontrolled cell division and evolve into an anomaly under carcinogenetic conditions, characterized by the initiation, promotion, and progression steps, and exacerbated by a prooxidant environment. Cancer cells can evade the inducement of suicide through mutations of their mitochondrial Bcl proteins and some cell cycle-controlling checkpoints (i.e., p53 mutations, BRCA1, *myc*, *jun*, etc.). The only way to control or eventually kill these cells is through an exaggerated oxidative environment. This is because even though they adapt their metabolism, they lack or have only poor antiperoxidative defenses compared to nomal cells (e.g., catalase). Through reactivation of lowered DNAses and RNases, cells can favor their self-injuries through lysosomal leaks and nuclear damage; that is, they favor a non-PCD that is different from apoptosis, such as autoschizis.

Autoschizis should therefore be considered a regulated necrosis or PCD that encompasses several subtypes, namely necroptosis, pathanos, pyroptosis, and oxytosis. These have all been reviewed and illustrated in a recent monograph on the topic, in which autoschizis was also briefly explained [273]. Morphologically, as shown throughout this chapter, and by a survey of our data, tumor cells undergo autoschizis cell death while maintaining an apparently intact nuclear envelope containing degraded chromatin, in a similar manner to that described in necrosis/oncosis. The cells, however, maintain a narrow perikaryal cytoplasm, containing damaged organelles. There is a continuing reduction of size by self-excision of the cytoplasm and removal of its damaged organelles via autophagy. In autoschizis, the auto- or self-excised cell fragments do not contain nuclear fragments or organelles like those seen in apoptotic bodies or in extrusion of nuclear RNA packets. Finally, in autoschizis, cathepsins and not caspases act as the cell executioners [204].

Similar morphologic alterations with self-excisions can be found in the literature; these have mistakenly been described as a form of apoptotic cell death or as some type of PCD due to the quasi-universal call for this mode of cell death. Of course, with the new "biochemical terminology" reinvented by the CNND group [23], one can expect more names to be coined according to the biochemical molecules or transduction mechanisms altered, with unfortunately no specific, comprehensive, illustrative imagery to help identify each new form of cell death.

28.11 Conclusion

Autoschizis is a unique, non-apoptotic form of cell death induced by prooxidants. It has been attributed to redox cycling of vitamins and to the generation of ROS, followed by alterations in membrane lipids and organelles, and by nuclease (DNase and RNases) reactivations. Following concomitant nuclear and cytoplasmic damage, including lysosomal membrane defects involving autophagocytosis and energetic blocks, cell death occurs via a unique process of self-excision of cytoplasmic pieces with no organelle content (autoschizic bodies). Degradation of the protective metabolome against H_2O_2, including GSH and other antioxidants, induces the cell to die.

This form of cell death was termed "autoschizis" as a result of the phenotypic and genotypic self-cuts it involves. Due to the intracellular biochemical damages measured, as well as the recent suggestions made by the NCCD, autoschizis cell death may encompass necroptosis, paranosis, pyroptosis, and oxytosis.

28.12 Addendum

28.12.1 Biochemical Cytotoxic Alterations Consequent on Prooxidant Treatment

28.12.1.1 Antitumor Activity of Menadione or VK3

Menadione, as VK3, is a synthetic derivative of VK1 whose antitumor activities have been studied for over 65 years [274–277]. The VK3 used in this treatment is the promoter form [15], because it is a well-characterized prooxidant or oxidant that can stimulate growth, trigger apoptosis, or cause necrosis depending on the dose and time of exposure [278]. For example, 5 mM VK3 stimulates DNA synthesis and promotes mitogenesis, 10 mM VK3 exerts antiproliferative effects, 20 mM VK3 can induce apoptosis, and 100 mM VK3 induces necrosis [279]. Menadione-induced defects are dose- and time-dependent, and can include many intracellular defects reviewed in previous studies [43,51,113,148,176–179,181–183,185,186,188,189,192].

VK3 exhibits a broad spectrum of selective antitumor and cytotoxic activity *in vitro* against rodent cell lines [276,277,280–283], as well as against human breast, cervix, colon, leukemia, liver, lung, lymphoma, and nasopharynx cell lines [19,259,264,282,283]. VK3 has also been shown to be effective against multiple drug-resistant leukemia cell lines and against adriamycin-resistant leukemia cells in rats [281,284]. Weekly intraperitoneal administration of VK3 (10 mg/mL) to hepatoma-bearing rats for a 4-week duration increased survival to 60 days for test rats as compared to 17 days for controls, with 5 out of 16 long-term survivors [285]. VK3 at doses between 150 and 200 mg/day intravenously has been shown to be a radiosentizing agent in patients with inoperable bronchial carcinoma and a chemosensitizer when combined with chemotherapeutic agents [286]. When VK3 was added to human oral epidermoid carcinoma (KB) cell cultures with chemotherapeutic agents, synergism was observed with bleomycin, cisplatin, dacarbazine, and 5-fluouracil (5-FU). An additive effect was observed when actinomycin D, cytarabine, doxorubicin, hydroxyurea, mercaptopurine, mitomycin C, mitoxantrone, thiotepa, vincristine, or VP-16 was used [282]. Synergistic activity was also observed between VK3 pretreatment and doxorubicin, 5-FU, and vinblastine in nasopharyngeal carcinoma cells, and between it and doxorubicin or mitomycin in MCF-7 breast cancer cells [287–289]. A study with rats showed that the combination of methotrexate (0.75 mg/kg/day) and VK3 (250 mg/kg/day) resulted in a 99% inhibition of tumor growth; decreasing the dosage of VK3 to 225 mg/kg/day led to an 84% inhibition. In addition, circulating levels of VK3 as low as 1 µM induced synergism with the methotrexate [290,291].

28.12.1.2 Nuclear Damage Caused by Menadione or VK3

VK3's cytotoxicity can be imparted to its antitumor activity through its ability to induce cell-cycle arrest or delay at both the G1/S and the G2/M phase [176,282]. The majority of its antitumor properties are attributed to its prooxidant qualities and its poisonous tumor nucleoside metabolism [259,282].

The histochemical detection of DNases and RNases [7,9] demonstrated a rapid reactivation of DNase I and RNases following VK3 treatment in solid tumors: as soon as within 1 hour post treatment, lasting no longer than 2 hours. Histochemical Feulgen staining patterns verified by densitometry [183,185,186] confirmed that VK3 induces some DNA degradation or karyolysis in several cell lines, as exemplified earlier in this chapter with bladder and ovarian carcinoma cells (see, e.g., Figures 28.26 and 28.29). In addition to karyolitic defects, nucleolysis damage induced by VC:VK3 oxidative treatment destroy the transcriptive machinery, along with some of the mitochondria. This is probably because VK3 alone – and in cooperation with VC later – damages these organelles due to DNase I being bound to G-actin in a 1:1 complex, which represses it. When it separates from the complex to make F-actin [292,293], it can be speculated that de-repressed (i.e., free) DNase I also remodels actin's stress filaments and functions by altering its bonds with other macromolecules, as well as going along with degrading mtDNA and nuclear DNA [294,295]. Actin is part of the nucleoplasm and is involved in transcriptional activity [296]; it also interacts with tubulin associated with the nucleolus [65]. Outside of cytoskeletal damage, actin alone (or in conjunction with the altered lamins that tumor cells contain) causes alterations of the "integrity" of nucleolar NORs [83,298] and of fibrillarin [189]. The vacuous interstices in the nucleolar masses following VK3 treatment demonstrate that the chromatin (DNA) is quasi-absent from this nucleolus or else is degraded following such treatment [148,188,189]. Nuclear chromatin changes reflect DNA degradation through a decreased electron density of the nucleoplasm, which accompanies an initial layer on the inner nuclear membrane, and segregation of the nucleolus components, as viewed diagrammatically [148]. Breakdown of the chromatin can lead to the production of IGs [51,148,183]; this has been noted by Kerr and collaborators [30] during apoptosis, as confirmed by Smetana (pers. comm.) [51].

Chromatin and nucleolar alterations observed after treatment with VK3 alone are consistent with a composite of morphological changes seen following cell-cycle arrest in G1/S or G2/M through disruption of the tumor nuclear cytoskeleton [176,297,299,300], and with activation of DNases, actin [301,302], and myosin I, as noted with RT4 cells [51].

28.12.1.3 Peroxidation of Proteins Caused by Menadione or VK3

The cycling of VK3, like that of VC (reviewed in [51,113,177,178,185,188]) results in the production of ROS, which can damage antioxidant defenses by drawing down cell thiols that interfere with protein folding and protein functions by removing their critical steric shape, alter signal transduction [303], and promote degradation of histones and keratin [304]. As a promoter, the prooxidant activity of VK3 can be reduced intracellularly via one- or two-electron transfer [305]. DT-diaphorase is the principal enzyme that catalyzes the two-electron reduction of the promoter to hydroquinone, which may form nontoxic conjugates or slowly autoxidize to reform the promoter. The autoxidation of the hydroquinone generates the superoxide radical and other ROS (including the hydroxyl radical), and is the rate-limiting step in the redox cycling of the promoter. While two-electron reduction has been considered a mechanism for detoxification, recent evidence suggests that it can also cause redox cycling. The single-electron reduction of the promoter to the semi-quinone can be catalyzed by a number of promoter enzymes, including reduced nicotinamide adenine dinucleotide phosphate (NADPH) cytochrome P-450 reductase, NADPH cytochrome b5 reductase, and NADPH ubiquinone oxidoreductase [306]. Subsequently, the semi-quinone reduces oxygen (O_2) to superoxide, and in the process regenerates the promoter. Redox cycling ensues, and large amounts of superoxide are produced and can dismutate to form H_2O_2 or more toxic ROS via intracellular metal-catalyzed reactions [307].

The redox cycling generates oxidative stress and leads to a complex stress response that results in structural damage to the tumor cells [282]. Tumor cells are very sensitive to any strong oxidative stress, because they typically have lower levels of the enzymes necessary to detoxify ROS than do normal, untransformed cells [308]. The coordinated attack of ROS may induce membrane damage in the exoplasmic regions in the form of leakiness and blebs, and initiates progressive cell injuries leading to cell cytolysis [309]. VK3 predominately affects the cytoskeleton of the tumor cells, because the treated cells show dramatic changes in their shape and undergo self-excision, including of their perikaryal areas and nucleus, with "puck-shaped" cells or corpses as shown in Figure 28.15. Damage of actin has been noted – especially nuclear actin [299,300]. These cytoskeletal alterations contribute to internal and external phenotypical morphological changes, favoring odd cell shapes. This includes "dwarfing": surviving tumor cells following self-excision form populations of small cells. Shedding of these cells leads to the creation of autoschizic bodies (pieces of cytoplasm without organelles – in contrast to apoptotic bodies, which may contain both organelles and pieces of nucleus with chromatin compacted without the envelope).

The cell cytoskeleton includes not only some regulatory proteins, but also intrinsic, superficial ones (i.e., spectrin [276], filamins), as well as the cytokeratin that caused the changes in cell morphology (i.e., drastic reduction in cell size into a biconcave shape and odd or pleiomorphic shapes with lamellar extensions in the process of self-excisions) and led the cell to accumulate organelles in one particular area [21,34,51,71,113,148,179,180,183,185,186].

28.12.1.4 Peroxidations of Organelles Caused by Menadione or VK3

Mitochondrial dysmorphology is often present following VK3 treatment. The mitochondria appear to have been stretched or squeezed and to have extended in an effort to make contact with or become part of the autophagosomes of the damaged cells. The continuity between the outer mitochondrial membrane (OMM) and the ER [113,148,301] could certainly facilitate ionic and enzymatic cooperative events as a result of damage to the mitochondrial envelope by ROS, which can upset Ca^{++} homeostasis [302]. This could happen through disruption in the transition pore control, eventually triggering the release of cyt c to the cytoplasm and the induction of apoptosis [311] or other cell-death pathways, including autoschizis [173,204]. It appears that the majority of VK3-induced cell death takes the form of autoschizis [34,35,43,71,176–178,182,183,186,187,189]. In addition, defects in caspase 3 [204] may create different internal signaling events that ultimately induce a different cell death than apoptosis [148].

In an effort to assess the role of the covalent binding of VK3 to protein thiols and primary amines in the cytotoxicity of vitamins, Buc Calderon and coworkers [15,148,188,204] performed experiments with 2,3-dimethoxy-naphthoquinone (DMNQ), a structural analog of VK3 that contains no arylation sites. When DMNQ was combined with VC instead of VK3, the profile of cytotoxicity observed was the same as when VK3 was used. These results indicate that the diverse effects of VK3 are caused by reactions derived from redox cycling rather than covalent binding to essential macromolecules [15,148,188,204].

At first, no major defects in Ca^{++} uptake by tumor cells after treatment by VK3 deposits were found, and the longer the treatment, the more dense a form of accumulation occurred, especially in some carcinoma cell lines, as in this descending sequence: RT4 > T24 > DU145 > MDAH 2774 > TRAMP. Damaged lipids and cysteine groups allow the inner thiol content to acquire spotty and peculiar osmiophilic deposits, caused by calcium ion exchange with osmium salts, forming osmate salts, which are similar to those observed in ZIO staining [312]. In addition, menadione can diffuse in the mitochondrial matrix to also become part of these electron-dense deposits [113,315].

As a recycling or feeding resource activity, the damaged organelles are part of the autophagosome bodies crowding the perikaryal zones. Mitochondria can accumulate, and are seen to interact with the defectuous endoplasmic membranes [113] in a series of degradative events, as suggested by starvation [310]. Increased lysosomal activity through membrane peroxidation can be induced by VK3, as lysosomes possess sulfhydryl groups – key to maintaining their membrane integrity [213]; it can also be caused by other complex secondary transduction signals. These cell injuries result from repetitive autophagic events related to lipid peroxidation-induced membrane damage [313] and from other events related to disrupted permeability of the mitochondria and ER, as described in the previous paragraph. The huge osmiophilic bodies seen after tumor cells are treated by VK3 confirm the ongoing autophagocytotic activities. The autophagosomes accumulate with all sorts of membrane peroxidative damages and with activation of SER–RER stresses upon the formation of phagophores, facilitating defectuous mitochondria and other organelles to undergo capture as a result of their continuity with the SER [301]. They can be stockpiled along with other autophagotic events in perikaryal zones similar to the lipofuscin bodies found in cardiac cells or neurons as a result of the low-lipase activities of lysosomes [314,315]. In these freely moving tumor cells, the accrual of these "recycled" structures is poorly dispersed or secreted due to cytoskeletal microtubule damages caused by the treatment. This results in their hoarding, accumulating at a location of the centrosome. The nuclear cove is unable to organize and allow lysosomal bodies to reach distal regions of the cytoplasm to be expelled from the injured cells [43,51,113,148,186,316].

Abrogation of all of the peroxidative actions caused by menadione can be blocked and verified by addition of catalase in experimental settings [9,176,177,188,297]; among them, the specific oxidative actions against thiols can be inhibited by the addition of L-cysteine in the culture medium [304].

28.12.1.5 VK3 and Signal Transduction

In addition to oxidative stress, VK3 activity causes the modulation of transcription factors by ROS, tyrosine kinases, and tyrosine phosphatases and the subsequent transcription of growth factors, inflammatory cytokines, and apoptotic factors, which can result in cell-cycle arrest, cell death, and the regulation and/or differentiation of cell growth (Figure 28.5) [188]. For example, addition of 50 µM VK3 to *in vitro* malignant cells has been found to produce a transient increase in *c-fos* expression within 1 hour and elevated expression of *c-myc* for up to 9 hours [264]. The *c-myc* gene product is the nuclear protein transcription factor, c-Myc, which forms a complex with Max that activates other genes involved in transformation, immortalization, cell differentiation, and induction of apoptosis [320,321]. The *c-fos* gene product, c-Fos, is a nuclear protein that is involved in growth-related transcriptional control. c-Fos and the *c-jun* gene product, c-Jun, dimerize in order to bind to the AP-1 recognition site and regulate growth and tumor-promoter stimuli [264,320]. The proto-oncogene bcl-2 has been shown to protect cells against apoptotic cell death in a large number tumor of models. It is localized to intracellular sites of oxygen free-radical generation, including the mitochondria, ER, and nuclear membranes. While the addition of bcl-2 to lymphoblastoid cell cultures (2B4 and FL5.12) blocks cell death caused by the oxidative burst of 50–200 µM VK3, it cannot decrease VK3-induced ROS production. These results suggest that ROS acts as second messenger that signals downstream transcription factors like NF-κB and Fos/Jun [322]. NF-κB is involved in stress-induced FasL expression – Fas is one of the important death receptors in the tumor necrosis factor (TNF) superfamily, and the gene encoding its ligand, designated FasL, activates it by trimerization of the receptor. Activation of apoptotic cell death is commonly mediated by the Fas/FasL system, and VK3 induces both Fas and FasL expression [290,323]. However, VK3 is known to induce oxidative stress via the production of ROS, which are also known to induce FasL mRNA expression and to rapidly decrease GSH during apoptosis induced by crosslinking of Fas receptor [324,325]. Therefore, in some cells, Fas activation could include tumor cells as a consequence of VK3-induced oxidative stress.

DU145 cells die through non-apoptotic cell death, as they do not express Fas/FasL. Similarly, VK3 has been shown to inhibit the growth and induce the apoptosis of HR stomach cancer cells in a dose-dependent fashion. Growth inhibition is attributed to sulfhydryl arylation of the cysteines that mediate protein tyrosine phosphorylation [326]. The inhibition of cell growth by the VK3-induced tyrosine phosphorylation of hepatocyte growth-factor receptors (*c-met*) and epidermal growth-factor receptors (EGFRs) activates the Ras signaling pathway. Ras can also be activated by a number of other factors, including G-protein coupled receptors (GPCRs), Janus kinase I (Jak), and increased intracellular calcium levels (such as those observed during VC:VK3 treatment). Once activated, Ras is able to stimulate many effector proteins, including pl20 GTPase-activating protein, guanine nucleotide exchange factors such as RaiGDS, and a number of protein kinases, such as phosphatidylinositol 3-kinase (PI3-K), protein kinase C (PKC), mitogen-activated protein kinase kinase (MAP2K), mitogen-activated protein kinase kinase (MAP3K), c-Jun NH2 terminal kinase (JNK), and extracellular signal-regulated kinase (ERK). Therefore, Ras activation can trigger a diverse range of responses, including cell death, cell survival, and remodeling of the cell's actin cytoskeleton, depending on the effector protein with which it interacts (L 182). VK3 also produces a sustained phosphorylation of ERK, which is associated with cellular signal transduction, proliferation, and apoptosis. It further induces both protein tyrosine kinase activation from the receptor pathway and inhibition of the ERK protein tyrosine phosphatases [290,326].

VK3 cytotoxicity can be summarized as inducing:

1) **a significant depletion of reduced GSH, ATP, and SH-reducing species through oxidation of sulfhydryl groups. Damages then affect the cytoskeletal proteins and associated macromolecules, as well as lysosomal integrity;**
2) **reactivation of DNase I and the subsequent degradation of tumor-cell and mitochondrial DNA;**
3) **deregulation of cellular Ca2+ sequestration, impeding the phosphorylation of ERK and/or other regulating factors controlling the cell cycle.**

28.12.2 Ascorbate (VC)

For almost 70 years, the use of VC alone has been proposed for the prevention and treatment of many forms of cancer, due to its antioxidant properties [199,327,328]. Humans cannot synthesize VC and must obtain it through their diets. While various mechanisms have been proposed to account for its antitumor properties, no evidence has been presented to adequately explain why it preferentially attacks primary tumor cells or their metastases, leaving normal cells alone [15,20,325]. Some explanations are reviewed in several of our recent publications [9,34,35,51,71,173,176–178,183,185,186,192,204]. One key point is its preferential accumulation in tumor cells [15,20,188] versus normal cells, which incorporate only a small amount.

28.12.2.1 How Does VC Accumulate in Tumor Cells?

Tumor cells can take up VC *in vitro* and bioconcentrate it to levels that are 50- to 100-fold greater than the concentrations in the adjacent normal tissue [330] or the extracellular milieu [331–334]. This excess accumulation of VC occurs intracellularly in its **oxidized form, dehydroascorbate (DHA)**, taken up and transported through Na+-dependent ascorbic acid transporters [335] or through facilitative glucose transporters (GLUT-1, 3, and 4 channels). This was shown experimentally in carcinoma cells, especially *in vitro* bladder and prostatic carcinoma cells [297,331–334,336]. Solid tumors can accumulate VC, since they usually have a higher VC concentration than the cells in the adjacent normal tissue [330]. Once DHA is inside the cancer cells, it is reduced back into VC, which **cannot be transported back across the cell membrane** by the bidirectional GLUTs [333,334].

A few types of specialized cell transport ascorbic acid directly through the VC cotransporter and DHA, which enters most cell monocytes, T-lymphocyte cell lines, prostate and breast cancer cells, and hematopoietic human xenografted tumors through facilitative glucose transporters [333,337,338]. In some tumors, such as myeloid leukemia and HL-60 engrafts, as well as neutrophils, ascorbic acid can be oxidized to dehydroascorbic acid directly so that it may enter the cell [333,339–341]. Epithelial tumors appear to rely on superoxide, which is produced constitutively via NADPH oxidase of non-neoplastic stromal cells to oxidize the ascorbic acid [342,343]. Because VC is taken up intracellularly in excess by cancer cells, it becomes an internal cytotoxic agent [263]. It can act on melanoma [344], leukemia [345], neuroblastoma [275], ascites [346], acute lymphoblastic leukemia, epidermoid carcinoma, and fibrosarcoma cell lines [258].

VC is also selectively toxic to tumor cells *in vivo* [14,17,177] and can kill cells in solid primary tumors and metastases [14,16,177,181]. The enrichment of the intracellular VC content of tumor cells by the addition of exogenous VC was announced by Cameron and Pauling [347] and verified in some other studies [348–350]. Intracellular DHA–VC cycling induces metabolic peroxidative activities, wherein irreversible injuries cause diverse organelle stress and likely succeed in killing the most susceptible cells. VC in excess can inhibit metastases and their potential invasion by inducing a marked decrease in metalloproteinases (MMP-2 and -9) via a post-transcriptional inhibition of MMP proenzyme production [351]. Finally, VC is a chemosensitizing and radiosensitizing agent [7,14,349].

28.12.2.2 VC is an Antioxidant that can Become an Oxidant

VC treatment in solution becomes DHA in blood [204,333,337]. In normal cells, VC functions as an antioxidant and reduces cytotoxicity and mutagenicity through its ability to scavenge free radicals and to induce G_2/M cell-cycle arrest, which allows DNA repair to occur [352].

Conversely, if VC is added during oxidative stress, it acts as a prooxidant and increases cytotoxic and mutagenic effects via a mechanism that implies and demands the involvement of two signals: (i) increased intracellular DHA levels and (ii) increased DNA damage [352]. In this case, DHA accumulates (DHA signal) but is not readily reduced to ascorbic acid. Likewise, intracellular levels of reduced GSH are greatly reduced. This results in excess levels of intracellular H_2O_2, which damages DNA and generates the DNA damage signal. Additional cytotoxicity and mutagenicity occur because of an increase in intracellular levels of oxidized products of ascorbic acid, an increase in intracellular oxidative stress due to reduced intracellular GSH levels and partial G_2/M cell-cycle arrest, and incomplete DNA repair [352].

The antitumor effects of VC are therefore related to its ability to induce cell-cycle arrest and the redox cycle. Intracellularly, ascorbic acid as VC in its ascorbate bioconcentrated form can be oxidized by either single- or two-electron transfer and converted intracellularly back to VC by reduced nicotinamide adenine dinucleotide (NADH)-dependent semidehydroxyascorbate reductase or GSH-dependent DHA reductase [201,334,349]. On discovering VC, Szent-Györgyi [353] observed that it reacted with GSH and thiols; this led others to find that VC metabolites can act as an ROS generator that depletes protective thiol residues [177,178,200,350]. Such depletion favors membrane peroxidations by generating more ROS with intracellular H_2O_2, which contributes to further depleting cellular thiol levels and inducing plasma- and organelle-membrane injuries (including of mitochondria and lysosomes). Protein integrity [354] is then affected by ROS and initiates and provide continuous oxidative stress, eventually resulting in tumor cell death, especially when tumor cells are depleted of catalytic activities [129,200]. VC redox cycling can account for lipid peroxidation. These injuries, and the increased fluidity of the cell membranes, are partly responsible for the elongation of the viewed filopodia and for the deformations of the superficial cytoplasm, as shown via LM, SEM, and TEM for treatment by VC alone at the earliest observed time points [35,71,182,183,355]. In view of the Feulgen data, VC can also participate as a prooxidant in the degradation of cellular DNA [355,356]. The presence of small pieces of cellular debris in the intercellular spaces in many micrographs supports cytoplasmic as well as DNA damage, and these contentions agree with the flow-cytometry data [176,297].

Taper and his associates showed that DNase II (EC 3.1.22.1) may be responsible for the ultrastructural damage caused by VC exposure [7,15–18]. In the cytoplasm, DNase II is bound in a 1:1 complex with a repressor protein, which can be activated after a delay of at least 2 hours following the activation of DNAse I by VK3 when it is used as an anticancer agent. Acidification of the tumor cell reactivates DNase II by releasing it from its repressor. Recent studies suggest that DNase II is combined with an anti-apoptotic serine protease inhibitor, leukocyte-elastase inhibitor (LEI), and that acidification allows for the separation of LEI from DNase II, which creates an active lysosomal DNase II [357]. The acidification of the cytoplasm required to reactivate DNase II suggests permeabilization of the lysosomal membrane with the release of lytic enzymes, including cathepsins [191,297]. This was also seen with TEM, following similar treatment of an RT4 bladder carcinoma cell [51].

In most VC-treated cells, the nucleolus acquires a compact aspect, visualized as either one spherical electron-dense body in contact with the nuclear envelope or as a thick electron-dense patch in contact with a rim of chromatin in a diminutive nucleus. Such nucleolar changes are consistent with a late G2 block of the cell cycle in which the dense and fine fibrillar components are excluded from a mass of synthesized ribonucleoproteins following damage of the intrinsic nucleoskeletal components caused by peroxidation and DNase action [81,148,189]. It has also been noted that IGs can be seen in the nucleoplasm of most VC:VK3-treated cells, with concentrations consistently proportional with the amount of nuclear damage caused.

After at least a 2-hour treatment, VC can induce a reduction in nuclear/nucleolar volume. The remaining nucleolar structure results in a dense aggregate of its granular component (ribonuclear proteins). A prolonged (at least 4-hour) treatment of VC for any carcinoma cells increases the number of cells with the changed morphology and exacerbates death, as this is a typical delay following addition of VC in cultivated cells in order to validate the enhanced reactivation of DNase II, which is visible after 2-hour histochemistry (Figure 28.43) [7,9].

28.12.2.3 In Tumor Cells, VC can Control the Metabolism and Expression of Specific Growth Receptors

The HIF transcriptional cascade is activated during cancer and leads to increased angiogenesis, enhanced glycolysis, and other processes that promote tumor growth [358]. The activity of HIF hydroxylases is dependent on dioxygen, and their diminished activity in hypoxia allows HIF-α subunits to escape proteolysis and become transcriptionally active. In addition, hydroxylases – including the HIF hydroxylases – are dependent on VC for their full catalytic activity *in vitro* [359].

In one study, 25 µM VC was shown to strikingly suppress the levels of HIF-1α and HIF transcriptional targets in a human androgen-independent prostate adenocarcinoma cell line (PC-3), an ovarian carcinoma cell line (OVCAR3), and a breast-cancer cell line (HS578T) under nornoxic conditions. Similar results were obtained with iron supplementation. VC exposure also substantially reduced the level of vascular endothelial growth factor (VEGF) mRNA and inhibited VEGF secretion into the medium following insulin-like growth factor and insulin treatment. These results indicate that responses – including cell survival and remodeling of the cells' actin cytoskeleton, as well as cell death – depend on the effector protein with which Ras interacts [360].

Although the antitumor effects of VC alone were disappointing in the aforementioned clinical applications [361], it was found to potentiate the cytotoxic effects of a variety of chemotherapeutic drugs [349,362,363]. Furthermore, VC was found to exert a radiosensitizing activity [364]. In a randomized trial with 50 human subjects, which looked at the effect of concurrent VC (five daily doses of 1 g each) and radiotherapy on different tumor types, better complete responses to radiation were noted in the VC group at 1 month (87 vs. 55%) and 4 months (63 vs. 45%) post treatment. Side effects were lesser in the VC-treated subjects as well [274].

With the cycling of VC between ascorbate and DHA, and its concomitant oxidation of GSH and other thiol-rich compounds, the nuclear cytoskeleton is likely to become damaged [296,365], along with numerous membrane peroxidations [354]. Through unrepaired membrane defects, cascades of certain intracellular signals, as well as ionic defects, will eventually lead to tumor cell death [200,204].

28.12.2.4 VC, Mitochondria-SER, and Intracellular Ca^{++}

Mitochondrial membrane alterations after fixation are detected as small whorl-like changes following a short treatment duration. Longer treatments favor additional damages, such as intramatrical deposits and swelling, making the mitochondria appear as vacuoles due to their significant internal damage [34,51,183,192,205]. Likewise, in other aggressive carcinoma cells, the mitochondria can be adapted to a salvage energetic pathway using lactate [118–120]. Treated by VC, the injured mitochondria is rendered dysfunctional, with increased malonaldehyde levels, and ATP production is reduced by deregulation of the respiratory-chain complexes involved in the tumor-peculiar metabolic homeostasis.

These cytotoxic defects contribute to devastate cancer cells' precarious adaptive glycolytic homeostasis and accentuate further deterioration of DU145 cells and other carcinomas. More of the same mitochondrial structural defects have also been noticed in this and other cancer cell lines, along with increased autophagocytotic activity [51,113,199].

Ca^{++} dysregulation led to exposure of PS on the outer surface of the plasma membrane and activated a number of enzymes (phopholipases, proteases, and endonucleases). As shown in Figure 28.16, co-incubation of tumor cells with an annexin V-FITC/PI cocktail following vitamin exposure revealed three populations of dying cells: dual-labeled cells were necrotic or oncotic, annexin-labeled cells were apoptotic, and PI-labeled cells were autoschizic. These latter, minute cells were in the majority. In addition, flow cytometry of annexin-V/PI-labeled cells following VC:VK3 treatment verified the presence of autoschizis instead of apoptosis [173,195].

Because Ca^{++} dysregulation is also known to activate DNases – especially DNase I – that have previously been damaged by VK3 and then VC, detrimental cuts in mtDNA and the nuclear DNA staining pattern of Feulgen contribute to demonstrate the time-dependent decrease of tumor-cell DNA caused by both decreased synthesis and degradation via peroxidation and DNases. Electrophoretic analysis of DNA after VC: VK3 treatment of tumor cells (MDAH 2774, T24, and DU145) revealed, in every case, a degradation of chromatin DNA in the sequence sham < VC < VK3 < VC:VK3. As a result, the total DNA extracted for gel electrophoresis displayed spread patterns instead of the ladder patterns characteristic of apoptosis. *In vivo* data collected both *in vitro* and from xenotransplants in nude mice as solid tumors treated by the vitamins also proved a spread pattern resulting from the caspase 3-independent sequential reactivation of DNase I and II [7,9,173,204,297].

The intracellular vitamin-induced decrease in thiols and the increase in hydrogen peroxides and other ROS produced by this are accompanied by the oxidation and subsequent disruption of microtubule and other cytoskeletal proteins. Taken together, these results indicate that autoschizis entails the coordinated injury of thiols, membranes, the cytoskeleton, and the DNA by ROS and other species, which continues until cell death by self-excision and the degradation of organelles through osmotic shock, along with degradation and salvage caused by autophagocytosis events [113].

28.12.2.5 Summary
VC becomes a prooxidant because it is taken up in excess by cancer cells. Intracellular DHA–VC cycling induces metabolic peroxidative activities involving irreversible injuries caused by diverse organelle stress. This was first announced by Cameron and Pauling [65], and verified in many other studies.

28.12.3 Cytotoxic Injuries Found after VC:VK3 Treatment

By accumulating phagosomes in the perikaryon [368], along with tumor glycolytic metabolism [200,348,350], the VC:VK3 combined treatment accomplishes what one vitamin alone cannot do: it kills tumor cells quickly. After 3 hours of combined treatment, 80% inhibition of glycolysis of DU145 cells followed a 30% inhibition of glyceraldehyde-3-phosphate dehydrogenase (GAPDH). Additionally, there was a nearly 100% depletion of cellular oxidized nicotinamide adenine dinucleotide (NAD^+) levels within the first 2 hours of treatment [51,177,178,188].

28.12.3.1 VC:VK3 Induces the Formation of Hydrogen Peroxide and ROS, and can Alter Signal Transduction Pathways in Tumor Cells
Taper and collaborators showed that hydrogen peroxide was necessary to provide the antitumor activity of the VC:VK3 combination. Likewise, the data collected in our laboratories confirmed that the combined vitamin treatment led to the generation of hydrogen peroxide and ROS, which together affect the lipid components of tumor cells more easily than they do normal cells, as well as decreasing cell defenses by lowering the thiol content. Thiol buffers intracellular hydrogen peroxide and ROS levels, so reducing it leaves prooxidant-treated cells susceptible to peroxide damage. Coadministration of exogenous catalase destroyed the anti-tumor activity of the vitamins and implicated hydrogen peroxide in its mechanism [177,178]. The addition of SOD or mannitol to the tumor cells did not modify these anti-tumor effects. Superoxide and hydroxyl radicals did not appear to be the primary mediators of vitamin-induced cytotoxicity.

H_2O_2, linked to the synergistic cytotoxicity of the VC:VK3 combination, may also be involved in signal transduction. When Buc Calderon evaluated the VC:VK3 combination in conjunction with a series of protein kinase and protein phosphatase inhibitors, only sodium orthovanadate (a tyrosine phosphatase inhibitor) was able to suppress the cytotoxicity of the vitamin combination [173,204]. The rather specific inhibitory effect of orthovanadate, and the fact that H_2O_2 can derepress NF-κB by inducing the phosphorylation of I-κB, led some authors to evaluate the role of NF-κB in autoschizis [262]. Under physiological conditions, NF-κB is present in its inactive state in the cytoplasm as a complex with I-κB. This latter is degraded by phosphorylation and ubiquitination, allowing the activation of NF-κB and its translocation to the nucleus, where its binds to the promoter region of DNA and activates genes that mediate carcinogenesis and metastasis [262].

Future studies should progress toward exploring the potential relationships between activation of NF-κB, kinase cascades, and autoschizis.

28.12.3.2 VC:VK3 Treatment as a Strategy to Starve Tumor Cells

When VC and VK3 are combined in 100:1 ratio, the combination forms a redox pair and redox cycle. The VC:VK3 interaction not only fosters a single-electron reduction, producing long-lived semi-quinone and ascorbyl radicals and increasing the rate of redox cycling of the quinone to form H_2O_2 and other ROS, but also undergoes a two-electron transfer, which ensures ascorbic acid and DHA are present at pharmacologic levels for a protracted period of time. Three key glycolytic enzymes – hexokinase (HK), GAPDH, and G6PD – are known to exhibit redox sensitivity and can be inhibited by DHA when purified enzymes or erythrocyte lysates are employed [369]. While the inhibitory effect of DHA on all three enzymes can be abrogated by adding sufficiently high concentrations of the enzymes' specific substrates, the DHA-mediated inhibition of HK persists in intact erythrocytes, because the critical concentration of its substrate (glucose) cannot be achieved as glucose diffuses out of the erythrocytes. These results suggest that as DHA levels in the tumor cell increase as a consequence of redox cycling and cellular thiol levels decrease, DHA may inhibit HK either directly by binding in the active site or indirectly by binding near the binding site and preventing the approach of the ATP.

TEM micrographs have, however, demonstrated that mitochondrial architecture is rapidly altered by the vitamins' treatment, through lipid and protein damage, and is involved in the triggering mechanisms of autoschizis. Immunofluorescence and confocal microscopy studies with a Alexa fluor-488-labeled antibody to cyt c illustrated a change from punctate staining of the mitochondria in sham cells to a diffuse staining pattern in VC:VK3-treated cells. This altered pattern of staining is indicative of some cyt c release from the mitochondria, at least in the cells studied [188].

With respect to the mitochondrial defects induced, intracellular ATP levels were measured to determine whether VC:VK3-induced damage could be related to significant ATP depletion of the cell charge and might favor this newly discovered mode of cell death. Surprisingly, the measurements revealed that ATP levels first dropped and then rebounded to some higher level than control [188]. However, a prolonged exposure to VC:VK3 significantly depleted ATP [204].

DHA has been shown to inhibit the phosphorylation of other important signal-transduction and transcription factors, including NF-κB and p38 MAPK by preventing the binding of ATP [369–371]. Specifically, NF-κB is of great interest because it is constitutively expressed in tumor cells and promotes tumor progression while inhibiting tumor-cell death. Under physiological conditions, NF-κB is present in its inactive state in the cytoplasm as a complex with I-κB. The latter is degraded by the proteasome following phosphorylation and ubiquitination reactions. This allows the release of

NF-κB and its translocation to the nucleus, where it binds to the promoter region of DNA and activates genes that mediate carcinogenesis and metastasis. NF-κB can be released from I-κB repression by ROS-mediated oxidation of the inhibitor of kappa β kinase (IKK), IKKβ, or by phosphorylation of I-κB by IKKβ that has been phosphorylated by ATP [266,372]. VC quenches ROS signaling induced by the cytokine–receptor interaction, preventing the activation of ROS-mediated responses [370], while DHA functions as a kinase inhibitor of IKKα, IKKβ, and p38 MAPK by preventing phosphorylation [372]. The precise mechanism of kinase inhibition by DHA is not certain, but it appears to entail the binding of DHA to a pocket near the active site of the kinase, preventing the binding of ATP. The results of Verrax and coworkers [173,204,262,373] are consistent with those of Cárcamo and coworkers [372]. A constitutive activation of NF-κB was observed in untreated cells, as well as in cells exposed to VC or VK3 alone. While H_2O_2 enhanced the activation of NF-κB, VC:VK3 produced a strong inhibition of NF-κB. The lack of inhibitory activity with VC alone compared to VC:VK3 suggests the latter produces increased levels of DHA and inhibits the activation of IKKβ. DHA has also been shown to inactivate some subcomponents of the intracellular protein degradation machinery, such as cathepsin B [374], altering cell-membrane integrity (and inducing chemical modifications, cross-linking, and browning of proteins) by reacting with nucleophilic groups in the proteins [375].

During hepatic carcinogenesis, Bannash [114] noted that glycogen deposits appear moth-eaten due to the increased glycogen catabolism associated with aerobic glycolysis [103,115]. However, the cytotoxicity effects are suggested by the rapid phenotypic alteration of the tumor cells' morphology, as revealed in several previous caspase 3-independent reactivations of DNase I within the first hour of treatment [9,204]. DNA degradation appears as heterochromatic changes, involving clearing of nucleoplasm, while the nucleolus sees its components segregated, one of them by fibrillarin [189].

28.12.3.3 Lysosomes, Injurious Defects of Organelles, and Autophagosomes in Treated Carcinoma Cells

Lysosomal activities can reflect the ultimate and desperate strategy of injured tumor cells: isolation and digestion, in an attempt to survive despite their heavily mutilated nucleus. At this stage of damage, however, DU145 cells are doomed to die by autoschizis death [113,148,176,188] like TRAMP cells [192] – as seen in treated DU145 xenotransplants [9].

In combined VC:VK3 treatment, menadione has potent injurious cytotoxicity. It acts within 1 hour, while adding to and catalyzing injuries caused by DHA. VC cycling is delayed 2–3 hours both *in vitro* and *in vivo* [9,15,34,176]; that is, both VC and VK3 are necessary to accomplish better and more complete anticancer action, after a lag of at least 3 hours following treatment [7,9,20].

Most lysosomal activities result from the oxidant-linked injuries caused by VC:VK3, due to leakage. Defectuous mitochondria and other endomembranes have also been shown in another carcinoma cell line and in xenotransplants [9,113,348], as well as cytochemically [2,7,191,297]. Here, leaky lysosomes liberate their hydrolytic enzymes, some reaching the nucleus (i.e., ribonuclease, K and L cathepsins [297], and DNase I and II [1,7,9,15]), where they degrade nuclear structure and DNA [148,188].

The induced injuries spread throughout the cell and cannot be repaired, as they associate with lysosomal fragilization [51,191] (as supported by the earliest descriptions [366]), culminating with cytolysis [367] and, in some cases, cell death.

Autophagosomes accumulate in all treatments, with increasing number in the sequence VC:VK3 > VK3 > VC > sham, indicating ongoing injury of organelles, especially with regard to the mitochondria and endomembranes [51,122,177,178].

28.12.3.4 Biochemical Data Supportive of Tumor-Cell Damage

28.12.3.4.1 Lipid Peroxidations
Changes caused by lipid damage can account for:

1) the increased fluidity of the cell membrane, causing elongation of the filopodia without microspikes and other superficial alterations [35,51,179,180,183];
2) the destruction of Ca^{++} homeostasis [376–378]. These changes in turn lead to potentiation of nucleases [15,378], increased thiol oxidations, and poisoning of the mitochondria, which further lead to ATP depletion and lower the safe level of cell energy charge [379] that ultimately favors cell damage through irreversible repair and fatal cytotoxicity, even though tumor cells are adapted to a peculiar glycolytic-aerobic metabolism [103,113,118–121]. However, VC treatment triggers defects less rapidly than does VK3. The latter worsens ongoing damage via additional nuclease reactivations, with DNAse I acting faster than DNAse II and favoring lysosomal leakage, acting with hydrolases against nucleolar structures and all sorts of RNA defects [7,9].

VC:VK3 administration-induced hydrogen peroxide can induce lipid peroxidation through the Fenton reaction, which amplifies itself throughout lipids, phospholipids, and some single-DNA breaks [380]. The amount of peroxidized lipid was evaluated by the thiobarbituric acid method, which revealed that these vitamins must remain in contact with the cell for a long period of time (2–3 hours) before significant levels of lipid peroxidation and damage to the cell membranes can occur.

28.12.3.4.2 Nucleic Acids and Proteins Synthesis
Tritiated thymidine, radiolabeled uridine, and (32) S-cysteine incorporation assays were employed to evaluate vitamin-induced changes in DNA, RNA, and protein synthesis. A continuous exposure to vitamins depressed the rate of synthesis of nucleic acids. With histology surveys, the optical density of Feulgen-stained tumor cells also showed a significant treatment time-dependent decrease in DNA content, with the VC:VK3 combination producing the highest decrease in staining intensity due to the caspase 3-independent reactivation of DNAse I and DNAse II in bladder, ovarian, and other carcinoma cells [297].

Figure 28.45 shows the results of treating bladder carcinoma (RT4) cells [297] with sham, VC, VK3, and VC:VK3 for 1–5 hours.

Tritiated thymidine, radiolabeled uridine, and (32) cysteine incorporation assays were employed to evaluate vitamin-induced changes in DNA, RNA, and protein synthesis. Continuous exposure was seen to depress the rate of synthesis of all macromolecules [15,173,178,204,381]. In addition, gel electrophoresis of total DNA from T24 and MDAH 2774 cells treated by prooxidants demonstrated a random degradation. It has long been known that comparing nuclear content analyses of tumor and normal cells can show very different results, with the amount of DNA being greater in tumor cells but most other cellular components being similar [382]. Following treatment with VC:VK3, gel electrophoreses showed a random degradation of DNA – not a laddering type as seen in apoptosis, but instead resembling necrosis. It is as if the tumor cells

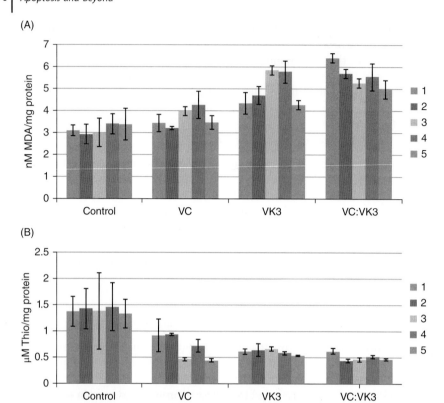

Figure 28.45 (A) Vitamin-induced lipid peroxidation. RT-4 cells were treated with the vitamins at their CD$_{90}$ doses (VC, 8750 µM; VK3, 90 µM; VC:VK3, 520:5.2 µM), harvested at 1-hour intervals over 5 hours, and assayed for lipid peroxidation using the thiobarbituric acid method. Malondialdehyde (MDA) production was monitored fluorimetrically, and data were expressed as nM MDA/mg protein and calculated based on an MDA standard curve. Values are the mean ± standard error of three experiments, with three readings per experiment, compared to control. VC:VK3 treatments resulted in significant amounts of lipid peroxidation versus control ($p \ll 0.0022$). **(B) Vitamin-induced alterations in the thiol content of RT4 cells.** RT4 cells were treated for 1 hour with the vitamins at their CD$_{90}$ doses (VC, 8750 µM; VK3, 90 µM; VC:VK3, 520:5.2 µM), harvested at 1-hour intervals over 5 hours, and assayed for cellular thiol content by monitoring absorbance following reaction with Ellman's reagent. Data are expressed as µM thiol/mg protein, calculated on the basis of a GSH standard curve. Values are the mean ± standard error of the mean of three experiments, with three readings per experiment. VC:VK3 causes significant depletion of cellular thiols compared to control ($p \ll 0.0022$).

started with an "apoptotic"-like initiation (DNase stimulated activation, self-excisions) but continued as "oncosis" or "necrosis" [28].

Using **gel electrophoresis**, total DNA isolation from several cell lines following vitamin treatments always reveals a smearing pattern, with a subdued laddering embedded in it, occuring as early as 1 hour after vitamin treatment [47,183,185]. These results show a DNA degradation pattern that is inconsistent with DFF40/CAD nuclease activation and which precedes caspase 3 activation. Although these observations do not eliminate DFF40/CAD participation in later stages of the process, they do suggest that other DNases are involved. DNase I and DNase II are of particular interest due to their broad tissue distribution and because they have been implicated as possible effectors of apoptosis [383,384].

DNase I has a pH optimum around 7.5, and its enzymatic activity occurs in a pH with an interval range of 5.5–9.0. Although DNase I requires micromolar concentrations of both Ca^{++} and Mg^{++} for optimal enzymatic activity, the enzyme may be activated by 1 mM Mg^{++} alone. Subcellular fractionation studies have shown that DNase I is localized primarily in the mitochondria, with activity also being detected in the nucleus [216]. A more recent immunolocalization study indicates that substantial amounts of DNase I can be located in the ER/nuclear envelope [384].

In contrast to DNase I, DNase II (acid DNase) does not require divalent cations for its activity [207,383]. It has an acidic pH optimum at about 5.0, with enzymatic activity occurring in a pH range of 3.0–7.0. Subcellular fractionation studies demonstrate that although the specific activity of DNase II is highest in the lysosomes [139], over 50% of the total cellular DNase II is located in the nuclear fraction. The presence of nuclear DNase II has also been confirmed in fixed tissue sections by histochemical detection using the lead nitrate method [245]. Nuclear DNase II is very stable, being only slightly depleted after 24 hours, even in the presence of cycloheximide [386].

The activity of DNase I and DNase II is inhibited in non-necrotic cells of malignant tumors in humans and experimental animals [7,13,14,124,149]. Reversal of this inhibition in malignant tumors has been linked to necrotic events and to efficient tumor therapy [7,8,387]. In a study employing transplantable murine hepatoma cells, DNases were shown to be reactivated in necrotic tumor cells following *in vivo* irradiation or *in vitro* treatment with a variety of different compounds.

Compounds that reactivate DNase I and II activities demonstrate a distinct enzymatic specificity [7,8,388]. Sulfhydryl oxidants, including VK3, exclusively reactivate DNase I. Reducing agents, including VC, reactivate DNase II without having any effect on DNase I. The reactivation of DNase I occurs rapidly (within 1 hour), with a shorter duration than that of DNase II (Figure 28.40) [8]. The reversal of DNase II deficiency is delayed compared to that of DNAse I. However, once DNase II activity is restored, its half-life is much longer than that of DNase I (Figure 28.43) [7–9].

Although the relatively rapid reversibility of the deficiency of alkaline DNase argues against a real decrease in enzyme content and a subsequent *de novo* synthesis of the DNase, the later time course of acid DNase activity could involve activation of both preexisting and *de novo* acid DNase synthesis [8]. At least in the case of the alkaline DNase, these observations point to a masking of enzymatic activity, either by structural latency (i.e., the release of sequestered DNases from other intracellular compartments) or by the presence of enzyme inhibitors [389]. Furthermore, the lack of reactivation of DNases in tumor cells after treatment with phospholipases, as well as the inhibition of usually active DNases in normal rat liver by tumor homogenate, suggests that DNase deficiency in malignant cells results from the presence of inhibitors and is not linked to structural latency [7,8].

Protein nuclease inhibitors that bind to DNase molecules in a 1:1 ratio have been found in normal tissues [234,390], and may even be increased in tumor cells [243]. In the case of DNase I, monomeric actin has been implicated as the inhibitory molecule [383,391]. A monomeric natural protein inhibitor of DNase II has been purified to homogeneity from bovine liver [390]. Small shifts in pH have been shown to affect the interaction between the inhibitor and DNase II, and to lead to recovery of nuclease activity [227], whereas DNase I is reactivated by sulfhydryl reagents [388]. However, if the nuclease reactivation by both VC and VK3 of the DNase inhibitor complexes facilitates cancer cells in being killed by autoschizis, then its biochemical mechanisms or controls (markers) have not yet been elucidated.

> *Born of undesirable cells,*
>
> *Tumors construct harmful and rebellious islands,*
>
> *Where points of flaws would favor their demises*
>
> *And out of these exits new life paths can be . . .*
>
> J. Gilloteaux

Acknowledgments

The authors want to acknowledge support from the American Institute for Cancer Research, the Hess-Roth-Kaminsky Urological Foundation, Erie, PA, St. George's University School of Medicine, Grenada, WI and Bay Shore, NY, the Summa Research Foundation, Akron, OH, and IC Med Tech, San Diego, CA, USA. Large parts of this chapter were submitted at the 1st International Symposium on Innovative Anticancer Drugs and Strategies, held in Newcastle upon Tyne, UK on June 3–4, 2010; this event was organized as a memorial for Henryk S. Taper, M.D., Ph.D. In addition, the authors wish to recognize Steve Getch, Communication Specialist, Summa, Akron, OH, USA and Ms. Kate Adamson, Graphic Department of Northumbria University, Newcastle upon Tyne, UK for their assistance with the electronic transfers of the original micrographies and graphics illustrating this chapter. Ms. Alanna Hallett, SGU-NU is also thanked for her skillful typographic assistance in helping make an accurate list of citations.

References

1. Daoust R, Cantero A. The distribution of deoxyribonuclease in normal, cirrhotic and neoplastic rat livers. *J Histochem Cytochem* 1959;**7**:139–43.
2. Fort L, Taper HS, Brucher JM. Nucleases activity in different segments of the human digestive tube compared to the incidence of carcinomas (histochemical study). *Histochemie* 1969;**20**:150–8.
3. Daoust R, Calamai R. Hyperbasophilic foci as sites of neoplastic transformation in hepatic parenchyma. *Cancer Res* 1971;**31**:1290–6.
4. Murthy SM, Daoust R. The distribution of acid and alkaline ribonuclerase activities in preneoplastic and neoplastic rat livers. *J Histochem Cytochem* 1977;**25**(2):115–21.
5. Dabrowska W, Copper EJ, Laskowski M. A specific inhibitor for esoxyribonuclease. *J Biol Chem* 1949;**177**:991–2.
6. Daoust R. Loss of ribonuclease activity in preneoplastic liver. *Proc Amer Assoc Cancer Res* 1971;**12**:18.
7. Taper HS. *The Relation between the Histochemical Activity of Nucleases and Neoplasms in Rat and Man.* Louvain: Vander. 1975.

8   Taper HS, Deckers CO, Dekers-Passau LO. Increase in nuclease activity as a possible mean for detecting tumor cell sensitivity to anticancer agents. *Cancer* 1981;**47**:523–9.
9   Taper HS, Jamison JM, Gilloteaux J, Gwin CA, Gordon T, Summers JL. In vivo reactivation of DNAses in implanted human prostate tumors after administration of vitamin C/K3 combination. *J Histochem Cytochem* 2001;**49**:109–19.
10 Taper HS. Reversibility of acid and alkaline deoxyribonuclease deficiency in malignant tumor cells. *J Histochem Cytochem* 1980;**29**:1053–60.
11 Taper HS. La détection histochimique de la désoxyribonucléase alcaline. *Ann Histochim* 1968;**13**:301–17.
12 Taper HS, Brucher JM, Fort L. Activity of alkaline and acid nucleases in tumors of the human central nervous system. Histochemical study. *Cancer* 1971;**28**(2):482–90.
13 Taper HS, Fort L, Brucher JM. Histochemical activity of alkaline and acid nucleases in the rat liver parenchyma during N-nitrosomorpholine carcinogenesis. *Cancer Res* 1971;**31**(6):913–16.
14 Taper HS, Keyeux A, Roberfroid M. Potentiation of radiotherapy by non toxic pretreatment with combined vitamin C and K3 in mice bearing solid, transplantable tumor. *Anticancer Res* 1996;**16**:499–504.
15 Taper HS, Jamison JM, Gilloteaux J, Summers JL, Buc Calderon P. Inhibition of the development of metastases by dietary vitamin C: K3 combination. *Life Sci* 2004;**75**:955–67.
16 Taper HS, de Gerlache J, Lans M, Roberfroid M. Non-toxic potentiation of cancer chemotherapy by combined C and K3 vitamin pretreatment. *Int J Cancer* 1987;**40**:575–9.
17 Taper HS, Roberfroid M. Non-toxic sensitization of cancer chemotherapy by combined C and K3 vitamin pretreatment in a mouse tumor resistant to oncovin. *Anticancer Res* 1992;**12**:1651–4.
18 De Loecker W, Janssens J, Bonte J, Taper HS. Effects of sodium ascorbate (vitamin C) and 2-methyl-1,4-naphtoquinone (vitamin K3) treatment on human tumor cell growth in vitro. II. Synergism with combined chemotherapy action. *Anticancer Res* 1993;**13**:103–6.
19 Noto V, Taper HS, Jiang YH, Janssens J, Bonte J, De Loecker W. Effects of sodium ascorbate (vitamin C) and 2-methyl-1,4-naphtoquinone (vitamin K3) treatment on human tumor cell growth in vitro. I. Synergism of combined vitamin C and K3 action. *Cancer* 1989;**63**:901–6.
20 Taper, HS. Altered deoxyribonuclease activities in cancer cells and its role in non-toxic adjuvant cancer therapy with mixed vitamins C and K3. *Anticancer Res* 2008;**28**:2727–32.
21 Gilloteaux J, Jamison JM, Arnold D, McGuire K, Summers JL. Autoschizis: a new cell death found in vitro and in xenotransplanted tumors induced by an oxidative stress mechanism lead by Apatone©. Workshop on Cancer Cell Death, European Cancer Research Conference, Amsterdam, NL. 2014.
22 Galluzzi L, Vitale I, Abrams JM, Alnemri ES, Baehrecke EH, Blagosklonny MV, et al. Molecular definitions of cell death subroutines: recommendations of the Nomenclature Committee on Cell Death. *Cell Death Differ* 2012;**19**:107–20.
23 Galluzzi L, Bravo-San Pedro JM, Vitale I, Aaronson SA, Abrams JM, Adam D, et al. Essential versus accessory aspects of cell death: recommendations of the NCCD 2015. *Cell Death Differ* 2015;**22**:58–63.

24 Galluzzi L, Maiuri MC, Vitale I, Zischka H, Castedo M, Zitvogel L, et al. Cell death modalities: classification and pathophysiological implications. *Cell Death Differ* 2007;**14**:1237–43.

25 Green DR, Galluzzi L, Kroemer G. Cell biology. Metabolic control of cell death. *Science* 2014;**345**(6203):1250256.

26 Kroemer G, Galluzzi L, Vandenabeele P, Abrams J, Alnemri ES, Baehrecke EH, et al. 2009. Classification of cell death: recommendations of the Nomenclature Committee on Cell Death 2009. *Cell Death Differ* 2009;**16**(1):3–11.

27 Galluzzi L, Bravo-San Pedro JM, Kroemer G. Organelle-specific initiation of cell death. *Nat Cell Biol* 2014;**16**:728–36.

28 Majno G, Joris I. Apoptosis, oncosis, and necrosis. An overview of cell death. Fine structural organization of the interphase nucleus in some mammalian cells. *J Ultrastruct Res* 1995;**27**:266–88.

29 Wyllie AH, Kerr JF, Currie AR. Cell death: the significance of apoptosis. *Int Rev Cytol* 1980;**68**:251–306.

30 Wyllie AH, Beattie GJ, Hargreaves AD. Chromatin changes in apoptosis. *Histochem J* 1981;**13**:681–92.

31 Falcieri E, Stuppia L, Di Baldassarre A, Mariani AR, Cinti C, Columbaro M, et al. Different approaches to the study of apoptosis. *Scanning Microsc* 1996;**10**:227–36, disc. 235-7.

32 Häcker G. The morphology of apoptosis. *Cell Tissue Res* 2000;**301**:5–17.

33 Saikumar P, Venkatachalam MA. Apoptosis and cell deaths. In: Cagle PT, Allen TC (eds.). *Basic Concepts of Molecular Pathology*. Dordrecht: Springer Verlag. 2009.

34 Gilloteaux J, Jamison JM, Venugopal M, Giammar D, Summers JL. Scanning electron microscopy and transmission electron microscopy aspects of synergistic antitumor activity of vitamin C-vitamin K3 combinations against human prostatic carcinoma cells. *Scan Microsc Int* 1995;**9**:159–73.

35 Gilloteaux J, Jamison JM, Arnold D, Ervin E, Eckroat L, Docherty JJ, et al. Cancer cell necrosis by autoschizis: synergism of antitumor activity of vitamin C:vitamin K3 on human bladder carcinoma T24 cells. *Scanning* 1998;**20**:564–75.

36 Brehmer B, Riemann JF, Bloodworth J.M. Jr., Madsen PO. Electron microscopic appearance of cells from carcinoma of the prostate in monolayer tissue culture. *Urol Res* 1973;**1**:27–31.

37 Celesk RA, Pollard M. Ultrastructural cytology of prostate carcinoma cells from Wistar rats. *Invest Urol* 1976;**14**:95–9.

38 Chakraborty J, Von Stein GA. Pleomorphism of human prostatic cancer cells (DU 145) in culture – the role of cytoskeleton. *Exp Mol Pathol* 1986;**44**(2):235–45.

39 Carruba G, Pavone C, Pavone-Macaluso M, Mesiti M, d'Aquino A, Vita G, et al. Morphometry of *in vitro* systems. An image analysis of two human prostate cancer cell lines (PC3 and DU-145). *Pathol Res Pract* 1989;**185**(5):704–8.

40 Gonda MA, Aaronson SA, Ellmore N, Zeve VH, Nagashima K. Ultrastructural studies of surface features of human normal and tumor cells in tissue culture by scanning and transmission electron microscopy. *J Natl Cancer Inst* 1976;**56**(2):245–63.

41 Domagala W, Koss LG. Configuration of surfaces of human cancer cells in effusions. A scanning electron microscopic study of microvilli. *Virchows Arch B Cell Pathol* 1977;**26**(1):27–42.

42 Renshaw AA, Nappi D, Cibas ES. Cytology of metastatic adenocarcinoma of the prostate in pleural effusions. *Diagn Cytopathol* 1996;**15**:103–7.

43 Gilloteaux J, Jamison JM, Neal DR, Summers JL. Xenotransplanted human prostate carcinoma (DU145) cells develop into carcinomas and cribriform carcinomas: ultrastructural aspects. *Ultrastruct Pathol* 2012;**36**:294–311.

44 Gilloteaux J, Jamison JM, Arnold D, Summers JL. Human prostate DU145 carcinoma cells implanted in nude mice remove the peritoneal mesothelium to invade and grow as carcinomas. *Anat Rec A Disc Mol Cell Evol Biol* 2013;**296**:40–55.

45 Abercrombie M, Ambrose EJ. The surface properties of cancer cells: a review. *Cancer Res* 1962;**22**:525–48.

46 Ossowski L, Quigley JP, Reich E. Fibrinolysis associated with oncogenic transformation. Morphological correlates. *J Biol Chem* 1974;**249**:4312–20.

47 Ervin E, Jamison JM, Gilloteaux J, Docherty JJ, Summers JL. Characterization of the early events in vitamin C and K3-induced death of human bladder tumor cells. *Scanning* 1998;**20**:210–11.

48 Gilloteaux J, Jamison JM, Von Greuningen V, Arnold D, Summers JL. Autoschizis in ovarian carcinoma cells MDAH as a result of treatment by combined vitamins C and K3 treatment. *Scanning* 2000;**22**:119–20.

49 Hu X, Ruan Y, Cheng F, Yu W, Zhang X, Larré S. p130Cas, E-cadherin and β-catenin in human transitional cell carcinoma of the bladder: expression and clinicopathological significance. *Int J Urol* 2011;**18**(9):630–7.

50 Ning XH, Guo R, Han L, Zhang AL, Liu X, Li ZX, et al. [DZNep raises miR-200c expression to delay the invasion and migration of MGC-803 gastric carcinoma cells]. *Sheng Li Xue Bao* 2015;**67**(1):83–9.

51 Gilloteaux J, Jamison JM, Neal DR, Loukas M, Doberstyn T, Summers JL. Cell damage and death by autoschizis in human bladder (RT4) carcinoma cells resulting from treatment with ascorbate and menadione. *Ultrastruct Path* 2010;**34**:140–60.

52 Cremer M, Küpper K, Wagler B, Wizelman L, von Hase J, Weiland Y, et al. Inheritance of gene density – related higher order chromatin arrangements in normal and tumor cell nuclei. *J Cell Biol* 2003;**162**:809–20.

53 Zink D, Fisher AH, Nickerson JA. Nuclear structure in cancer cells. *Nat Rev Cancer* 2004;**4**:677–87.

54 Frost JK. The cancer cell: criteria of malignancy. Lilienfeld AM, Gifford AJ (eds.). *Chronic Diseases and Public Health*. Baltimore, MD: Johns Hopkins Press, 1966.

55 Frost JK. The cell in health and disease. An evaluation of cellular morphologic expression of biologic behaviour. 2nd, revised edn. *Monogr Clin Cytol* 1986;**2**:1–304.

56 Johnson N, Krebs M, Boudreau R, Giorgi G, LeGros M, Larabell C. Actin-filled nuclear invaginations indicate degree of cell de-differentiation. *Differentiation* 2003;**71**:414–24.

57 Burns ER, Soloff BL, Hanna C, Buxton DF. Nuclear pocket associated with the nucleolus in normal and neoplastic cells. *Cancer Res* 1971;**31**:159–65.

58 Lagace TA, Ridgway ND. The rate-limiting enzyme in phosphatidylcholine synthesis regulates proliferation of the nucleoplasmic reticulum. *Mol Biol Cell* 2005;**16**:1120–30.

59 Coene E, Van Oostveldt P, Willems K, van Emmelo J, De Potter CR. BRCA1 is localized in cytoplasmic tube-like invaginations in the nucleus. *Nat Genet* 1997;**16**:122–4.

60 Starr DA. Communication berween the cytoskeleton and the nuclear envelope to position the nucleus. *Mol Biosyst* 2007;**3**(9):583–9.

61. Starr DA, Fridolfsson HN. Interactions between nuclei and the cytoskeleton are mediated by SUN-KASH nuclear envelope bridges. *Ann Rev Cell Dev Biol* 2010;**26**:421–4.
62. Tapley EC, Starr DA. Connecting the nucleus to the cytoskeleton by SIN-KASH bridges across the nuclear envelope. *Curr Opin Cell Biol* 2013;**25**:57–62.
63. Dahl KN, Ribeiro AJS, Lammerding J. Nuclear shape, mechanics, and mechanotransduction. *Circ Res* 2008;**102**:1307–18.
64. Gorjánácz M. Nuclear assembly as a target for anti-cancer therapies. *Nucleus* 2014;**5**:1–9.
65. Hořesjši B, Vinopal S, Stádková V, Dráberová E, Sulimenko V, Sulimenko T, et al. Nuclear γ-tubulin associates with nucleoli and interacts with tumor suppressor protein p53. *J Cell Physiol* 2012;**227**:367–82.
66. Frankhouser CM, Skepnek R, Shimi T, Goldman AE, Goldman RD, Olivera de la Cruz M. Mechanical model of blebbing in nuclear lamin meshworks. *Proc Natl Acad Sci USA* 2013;**110**:3248–58.
67. Capo-chichi CD, Cai KQ, Simpkins F, Ganjei-Azar P, Godwin AK, Xu XX. Nuclear envelope structural defects cause chromosomal Numerical instability and aneuploidy in ovarian cancer. *BMC Med* 2011;**9**:28.
68. Helfand BT, Wang Y, Pfeghaar K, Shimi T, Taimen P, Shumaker DK. Chromosomal regions associated with prostate cancer risk localize to lamin B-deficient microdomains and exhibit reduced gene transcription. *J Pathol* 2012;**226**:735–45.
69. Clubb BH, Locke M. 3T3 cells have nuclear invaginations containing F-actin. *Tissue Cell* 1998;**30**:684–91.
70. Nickerson JA. Nuclear dreams: the malignant alteration of nuclear architecture. *J Cell Biochem* 1998;**70**(2):172–80.
71. Gilloteaux J, Jamison JM, Arnold D, Taper HS, Von Gruenigen VE, Summers JL. Microscopic aspects of autoschizic cell death in human ovarian carcinoma (2774) cells following vitamin C, vitamin K3 or vitamin C:K3 treatment. *Microsc Microanal* 2003;**9**:311–29.
72. Bourgeois CA, Hubert J. Spatial relationship between the nucleolus and the nuclear envelope: structural aspects and functional significance. *Int Rev Cytol* 1988;**111**:1–52.
73. Bussolati G. Proper detection of the nuclear shape: ways and significance. *Rom J Morphol Embryol* 2008;**49**(4):435–9.
74. Bussolati G, Maletta F, Asioli S, Annaratone L, Sapino A, Marchio C. "To be or not to be a good shape": diagnostic and clinical value of nuclear shape irregularities in thyroid and breast cancer. *Adv Exp Med Biol* 2014;**773**:101–21.
75. De Las Heras JI, Batrakou DG, Schirmer EC. Cancer biology and the nuclear envelope: a convoluted relationship. *Semin Cancer Biol* 2013;**23**:125–37.
76. Fisher AH. The diagnostic pathology of the nuclear envelope in human cancers. *Adv Exp Med Biol* 2014;**773**:49–75.
77. Green EU. On a special condition of the interphase nucleus in normal and cancerous cells. *Cancer Res* 1949;**9**(5):267–76.
78. DeRobertis E. Advances in the ultrastructure of the nucleus and chromosomes. *Natl Cancer Inst Monogr* 1964;**14**:33–55.
79. Smetana K, Busch H. Studies on the ultrastructure of the nucleoli of the Walker tumor and rat liver. *Cancer Res* 1964;**24**:537–57.
80. Leak LV, Caulfield JB, Burke JF, McKhann CF. Electron microscopic studies on a human fibromyxosarcoma. *Cancer Res* 1967;**27**(2):261–85.

81 Hernandez-Verdun D. Assembly and disassembly of the nucleolus during the cell cycle. *Nucleus* 2011;**2**:189–94.
82 Wachtler F, Schwarzacher HG, Ellinger A. The influence of the cell cycle on structure and number of nucleoli in cultured human lymphocytes. *Cell Tissue Res* 1982;**225**:155–63.
83 Goessens G. Nucleolar structure. *Int Rev Cytol* 1984;**87**:107–58.
84 Ploton D, Beorchia A, Menager M, Jeannesson P, Adnet JJ. The three-dimensional ultrastructure of interphasic and metaphasic nucleolar argyrophilic components studied with high-voltage electron microscopy in thick sections. *Biol Cell* 1987;**59**:113–20.
85 Trerè D, Ceccarelli C, Danova M, Derenzini M. *In vivo* bromodeoxyuridine labeling index, AgNOR protein expression and DNA content in human tumors. *Eur J Histochem* 1996;**40**(1):17–26.
86 Raska I, Shaw PJ, Cmarko D. New insights into nucleolar architecture and activity. *Int Rev Cytol* 2006;**255**:177–234.
87 Kram N, Nessim S, Geller SA. A study of colonic adenocarcinoma, with comparison of histopathology, DNA flow cytometric data, and number of nucleolar organizer regions (NORs). *Mod Pathol* 1989;**2**(5):468–72.
88 Derenzini M, Trerè D, Chieco P, Melchiorri C. Interphase AgNOR quantity is not related o DNA content in 11 established human cancer cell lines. *Exp Cell Res* 1994;**211**(2):282–5.
89 Smetana K, Gyorkey F, Gyorkey P, Busch H. Comparative studies on the ultrastructure of nucleoli in human lymphosarcoma cells and leukemic lymphocytes. *Cancer Res* 1970;**30**(4):1149–55.
90 Derenzini M, Betts CM, Ceccarelli C, Eusebi V. Ultrastructural organization of nucleoli in benign and malignant melanomas. *Virchows Arch B Cell Pathol Incl Mol Pathol* 1986;**5**:342–52.
91 Mirré C, Knibiehler B. A re-evaluation of the relationships between the fibrillar centres and the nucleolus-organizing regions in reticulated nucleoli: ultrastructural organization, number and distribution of the fibrillar centres in the nucleolus of the mouse Sertoli cell. *J Cell Sci* 1982;**55**:247–59.
92 Seed J. The relations between DNA, RNA, and protein in normal embryonic cell nuclei and spontaneous tumor cell nuclei. *J Cell Biol* 1964;**80**:17–23.
93 Tavares AS, Costa J, de Carvalho A, Reis M. Tumor ploidy and prognosis in carcinomas of the bladder and prostate. *Br J Cancer* 1966;**20**:438–41.
94 Simard R, Langelier Y, Mandeville R, Maestracci N, Royal A. Inhibitors as tools in elucidating the structure and function of the nucleus. In: Busch H (ed.). *The Cell Nucleus*, Vol. 3 New York: Academic Press. 1974.
95 Schwarzacher HG, Wachtler F. The nucleolus. *Anat Embryol (Berl)* 1993;**188**(6):515–36.
96 Shea JR, Leblond CP. Number of nucleoli in various cell types of the mouse. *J Morphol* 1966;**119**:425–34.
97 Wilson EB. *The Cell in Development and Heredity*. New York: McMillan. 1928.
98 Molnar F, Daoust R. Nucleocytoplasmic ratios in different populations of rat liver parenchymal cells during azo dye carcinogenesis. *Cancer Res* 1965;**25**(8):1213–18.
99 Frost JK. An evaluation of cellular morphologic expression of biologic behavior. *Monogr Clin Cytol* 1961;**2**:1–142.

100 Burchardt P. [Electron microscopy of prostatic carcinoma cells (author's transl)]. *Urologe A* 1980;**19**(6):379–84.
101 Srigley JR, Hartwick WJ. Selected ultrastructural aspects of urothelial and prostatic tumors. *Ultrastruct Pathol* 1988;**12**:49–65.
102 White FH, Jin Y, Yang L. An evaluation of the role of nuclear cytoplasmic ratios and nuclear volume densities as diagnostic indicators in metaplastic, dysplastic and neoplastic lesions of the human cheek. *Histol Histopathol* 1997;**12**:69–77.
103 Nafe R, Franz K, Schlote W, Schneider B. Morphology of tumor cell nuclei is significantly related with survival time of patients with glioblastomas. *Clin Cancer Res* 2005;**11**:2141–8.
104 Warburg O. On respiratory impairment in cancer cells. *Science* 1956;**124**:269–70.
105 Dakubo GD, Parr RL, Costello LC, Franklin RB, Thayer RE. Altered metabolism and mitochondrial genome in prostate cancer. *J Clin Pathol* 2006;**59**:10–16.
106 Lin CS, Lee HT, Lee SY, Shen YA, Wang LS, Chen YJ, Wei YH. High mitochondrial DNA copy number and bioenergetic function are associated with tumor invasion of esophageal squamous cell carcinoma cell lines. *Int J Mol Sci* 2012;**13**(9):11 228–46.
107 Chatterjee A, Mambo E, Sidransky D. Mitochondrial DNA mutations in human cancer. *Oncogene* 2006;**25**:4663–73.
108 Chatterjee A, Dasgupta S, Sidransky D. Mitochondrial subversion in cancer. *Cancer Prev Res (Phila)* 2011;**4**:638–54.
109 Cook CC, Higuchi M. The awakening of an advanced malignant cancer: an insult to the mitochondrial genome. *Biochim Biophys Acta* 2012;**1820**(5):652–62.
110 Singh AK, Pandey P, Tewari M, Pandey HP, Shukla HS. Human mitochondrial genome flaws and risk of cancer. *Mitochondrial DNA* 2014;**25**(5):329–34.
111 Jerónimo C, Nomoto S, Caballero OL, Usadel H, Henrique R, Varzim G, et al. Mitochondrial mutations in early stage prostate cancer and bodily fluids. *Oncogene* 2001;**20**(37):5195–8.
112 Herrmann PC, Gillespie JW, Charboneau L, Bichsel VE, Paweletz CP, Calvert VS, et al. Mitochondrial proteome: altered cytochrome c oxidase subunit levels in prostate cancer. *Proteomics* 2003;**3**:1801–10.
113 Petros JA, Baumann AK, Ruiz-Pesini E, Amin MA, Sun CQ, Hall J, et al. mtDNA mutations increase tumorigenicity in prostate cancer. *Proc Natl Acad Sci USA* 2005;**102**:719–24.
114 Gilloteaux J, Jamison JM, Summers JL. Pro-oxidant treatment of human prostate carcinoma (DU145) cells induces autoschizis cell death: autophagosomes build up out of injured endomembranes and mitochondria. *Ultrastruct Pathol* 2014;**38**:315–28.
115 Arismendi-Morrilo G. Electron microscopy morphology of the mitochondrial network in human cancer. *Int J Biochem Cell Biol* 2008;**41**:2062–8.
116 Warburg O, Wind F, Negelein E. The metabolism of tumors in the body. *J Gen Physiol* 1927;**8**(6):519–30.
117 Modica-Napolitano JS, Kulawiec M, Singh KK. Mitochondria and human cancer. *Curr Mol Med* 2007;**7**:121–31.
118 Kimbro KS, Simons JW. Hypoxia-inducible factor-1 in human breast and prostate cancer. *Endocr Relat Cancer* 2006;**13**:739–49.
119 Féron O. Pyruvate into lactate and back: from the Warburg effect to symbiotic energy fuel exchange in cancer cells. *Radiother Oncol* 2009;**92**:329–33.
120 Porporato PE, Dhup S, Dadhich RK, Copetti T, Sonveaux P. Anticancer targets in the glycolytic metabolism of tumors: a comprehensive review. *Front Pharmacol* 2011;**2**:49.

121 Dhup S, Dadhich RK, Porporato PE, Sonveaux P. Multiple biological activities of lactic acid in cancer: influences of tumor growth, angiogenesis and metastasis. *Curr Pharmacol Design* 2012;**18**:1319–30.

122 Chowdhury SKR, Gemin A, Singh G. High activity of mitochondrial glycerophosphate dehydrogenase and glycerophosphate-dependent ROS production in prostate cancer cell lines. *Biochem Biophys Res Comm* 2005;**333**:1139–45.

123 Dionisi O, Galeotti T, Terranova T, Azzi A. Superoxide radicals and hydrogen peroxide formation in mitochondria from normal and neoplastic tissues. *Biochem Biophys Acta* 1975;**403**:292–300.

124 Bos R, Zhong H, Hanrahan CF, Mommers ECM, Semenza GL, Pinedo HM, et al. Levels of hypoxia-inducible factor1 during breast carcinogenesis. *J Natl Cancer Inst* 2001;**93**(4):309–14.

125 Fort L, Taper HS, Brucher JM. Gastric carcinogenesis induced in rats by methylnitrosurea (MNU). Morphology and histochemistry of nucleases. *Z Krebsforsch* 1974;**81**:51–62.

126 Fort L, Taper HS, Brucher JM. Morphological alterations and focal deficiency of the histochemical activity of acid and alkaline nucleases in rat liver after chronic administration of phenobarbital. *Beitr Pathol* 1977;**161**(4):363–75.

127 Vo TK, Druez C, Delzenne N, Taper HS, Roberfroid M. Analysis of antioxidant defense systems during rat hepatocarcinogenesis. *Carcinogenesis* 1988;**9**:2009–13.

128 Inoue K, Kawahito Y, Tsubouchi Y, Kohno M, Yoshimura R, Yoshikawa T, Sano H. Expression of peroxisome proliferator-activated receptor gamma in renal cell carcinoma and growth inhibition by its agonists. *Biochem Biophys Res Commun* 2001;**287**(3):727–32.

129 Gilloteaux J, Steggles AW. Alterations in hamster uterine catalase levels as a result of estrogen treatment. A histoenzymatic analysis. *Ann NY Acad Sci* 1982;**386**:546–8.

130 Gilloteaux J, Steggles AW. Histoenzymatic alterations in kidney catalase activity following hormonal treatment of Syrian hamster. *Cell Biol Int Rep* 1983;**7**:31–3.

131 Benade L, Howard T, Burk D. Synergistic killing of Ehrlich ascites carcinoma cells by ascorbate and 3-amino-2, 3, 4-triazole. *Oncology* 1969;**23**:33–43.

132 Sima AAF. Peroxisomes (microbodies) in human glial tumors. *Acta Neuropathol (Berl)* 1980;**51**:113–17.

133 Cablé S, Keller JM, Colin S, Haffen K, Kédinger M, Parache RM, Dauca M. Peroxisomes in human colon carcinomas. A cytochemical and biochemical study. *Virchows Arch B Cell Pathol Mol Pathol* 1992;**62**:221–6.

134 Keller JM, Cablé S, el Bouhtoury F, Heusser S, Scotto C, Armbruster-Ciolek E, et al. Peroxisome through cell differentiation and neoplasia. *Biol Cell* 1993;**77**:77–88.

135 Lauer C, Völkl A, Riedl S, Fahimi HD, Beier K. Impairment of peroxisomal biogenesis in human colon carcinoma. *Carcinogenesis* 1999;**20**(6):985–9. Erratum in: *Carcinogenesis* 1999; **20**(10):2037.

136 Van Noorden CJ. A simple histochemical assay to detect cancer cells. *Folia Histochem Cytobiol* 2000;**38**(3):99–102.

137 Frederiks WM, Bosch KS, Hoeben KA, van Marle J, Langbein S. Renal cell carcinoma and oxidative stress: the lack of peroxisomes. *Acta Histochem* 2010;**112**(4):364–71.

138 Frederiks WM, Vreeling-Sindelárová H, Van Noorden CJ. Loss of peroxisomes causes oxygen insensitivity of the histochemical assay of glucose-6-phosphate dehydrogenase activity to detect cancer cells. *J Histochem Cytochem* 2007;**55**(2):175–81.

139 Gilloteaux J, Jamison JM, Arnold D, Taper HS, Summers JL. Autoschizis: another cell death for cancer cells facilitated by vitamins C and K3 treatment. *Mol Biol Cell* 1999;**10**:44a.

140 Wattiaux-De Coninck S, Van Dijck JM, Morris HP, Wattiaux R. Lysosomes in hepatomas. *Biochem J* 1969;**115**(5):52P.

141 Allison AC. Lysosomes in cancer cells. *J Clin Pathol. Suppl (Roy Coll Path)* 1974;**7**:43–50.

142 Poole AR. Tumor lysosomal enzymes and invasive growth. In: Dingle TJ (ed.). *Lysosomes in Biology and Pathology*, Vol. 3 Amsterdam: North Holland. 1973.

143 Ghadially F. (ed.). *Ultrastructural Pathology of the Cell and Matrix*, Vols. 1 and 2. London: Butterworth Scientific. 1988.

144 Kessel RG. Annulate lamellae. *J Ultrastruct Res* 1968;**10**:1–82.

145 Kessel RG. The structure and function of annulate lamellae: porous cytoplasmic and intranuclear membranes. *Int Rev Cytol* 1983;**82**:181–303.

146 Kessel RG. The annulate lamellae – from obscurity to spotlight. *Electron Microsc Rev* 1989;**2**(2):257–348.

147 Kessel RG. Annulate lamellae: a last frontier in cellular organelles. *Int Rev Cytol* 1992;**133**:43–120.

148 Gilloteaux J, Jamison JM, Neal DR, Summers JL. Synergistic antitumor cytotoxic actions of ascorbate and menadione on human prostate (DU145) cancer cells *in vitro*: nucleus and other injuries preceding cell death by autoschizis. *Ultrastruct Pathol* 2014;**38**:116–40.

149 Bannash P. The cytoplasm of hepatocytes during carcinogenesis. In: Rentchnick P (ed.). *Recent Results in Cancer Research*. New York: Springer Verlag. 1968.

150 Taper HS, Bannasch P. Histochemical correlation between glycogen, nucleic acids and nucleases in preneoplastic lesions of rat liver after short-term administration of N-nitrosomorpholine. *Z Krebsforsh* 1976;**87**:53–65.

151 Mickey DD, Stone KR, Wunderli H, Mickey GH, Vollmer RT, Paulson DF. Heterotransplantation of a human prostatic adenocarcinoma cell line in nude mice. *Cancer Res* 1977;**37**:4049–58.

152 Mickey DD, Stone KR, Wunderli H, Mickey GH, Paulson DF. Characterization of a human prostate adenocarcinoma cell line (DU 145) as a monolayer culture and as a solid tumor in athymic mice. *Prog Clin Biol Res* 1980;**37**:67–84.

153 Schnier JB, Nishi K, Gumerlock PH, Gorin FA, Bradbury EM. Glycogen synthesis correlates with androgen-dependent growth arrest in prostate cancer. *BMC Urol* 2005;**5**:6.

154 Mazor M, Kawano Y, Zhu H, Waxman J, Kypta RM. Inhibition of glycogen synthase kinase-3 represses androgen receptor activity and prostate cancer growth. *Oncogene* 2004;**23**:7882–92.

155 Rinnab L, Schütz SV, Diesch J, Schmid E, Küfer R, Hautmann RE, et al. Inhibition of glycogen synthase kinase-3 in androgen-responsive prostate cancer cell lines: are GSK inhibitors therapeutically useful? *Neoplasia* 2008;**10**:624–34.

156 Darrington RS, Campa VM, Walker MM, Bengoa-Vergniory N, Gorrono-Etxebarria I, Uysal-Onganer P, et al. Distinct expression and activity of GSK-3α and CSK-3β in prostate cancer. *Int J Cancer* 2012;**131**:E872–83.

157 Wu R, Racker E. Regulatory mechanisms in carbohydrate metabolism. IV. Pasteur effect and Crabtree effect in ascites tumor cells. *J Biol Chem* 1959;**234**:1036–41.

158 Dang CV, Semenza GL. Oncogenic alterations of metabolism. *Trends Biochem Sci* 1999;**24**(2):68–72.
159 Sonveaux P, Vegran P, Schroeder T, Wergin MC, Verrax J, Rabbani ZN, et al. Targeting lactate-fueled respiration selectively kills hypoxic tumor cells in mice. *J Clin Invest* 2008;**118**:3930–40.
160 Mao P, Nakao K, Angrist A. Human prostatic carcinoma: an electron microscope study. *Cancer Res* 1966;**26**:955–73.
161 Aaronson SA. Growth factors and cancer. *Science* 1991;**254**:1146–53.
162 Weinstein IB, Joe AK. Mechanisms of disease: oncogene addiction-a rationale for molecular targeting in cancer therapy. *Nat Clin Pract Oncol* 2006;**3**:448–57.
163 Sherr CJ. Principles of tumor suppression. *Cell* 2004;**116**:235–46.
164 Thomas DJ, Robinson M, King P, Hasan T, Charlton R, Martin J, et al. p53 expression and clinical outcome in prostate cancer. *Br J Urol* 1993;**72**(5 Pt. 2):778–81.
165 Cheng L, Sebo TJ, Cheville JC, Pisansky TM, Slezak J, Bergstrahl EJ, et al. p53 protein overexpression is associated with increased cell proliferation in patients with locally recurrent prostate carcinoma after radiation therapy. *Cancer* 1999;**85**:1293–9.
166 Ziyaie D, Hugo TR, Thompson AM. p53 and breast cancer. *Breast* 2000;**9**:239–46.
167 Liu EY, Ryan KM. Autophagy and cancer – issues we need to digest. *J Cell Sci* 2012;**125**(Pt. 10):2349–58.
168 Chen X, Chen J, Gan S, Guan H, Zhou Y, Ouyang Q, Shi J. DNA damage strength modulates a bimodal switch of p53 dynamics for cell-fate control. *BMC Biol* 2013;**11**:73.
169 Grogno DJ, Caplan R, Sarkar FH, Lawton CA, Hammond EH, Pilepich MV, et al. p53 status and prognosis of locally advanved prostate adenocarcinoma: a study based on RTOG 8610. *J Natl Cancer Inst* 1997;**89**:158–65.
170 Zhou S, Kachhap S, Singh KK. Mitochondrial impairment in p53-deficient human cancer cells. *Mutagenesis* 2003;**18**:287–92.
171 Srikanth S, Franklin CC, Duke RC, Kraft RS. Human DU145 prostate cancer cells overexpressing mitogen-activated protein kinase phosphatase-1 are resistant to Fas ligand-induced mitochondrial perturbations and cellular apoptosis. *Mol Cell Biochem* 1999;**199**:169–78.
172 Taper HS, Wooley GW, Teller MN, Lardis MP. A new transplantable mouse liver tumor of spontaneous origin. *Cancer Res* 1966;**26**:143–8.
173 Verrax J, Pedrosa RC, Beck R, Dejeans N, Taper H, Buc Calderon P. In situ modulation of oxidative stress: a novel and efficient strategy to kill cancer cells. *Curr Med Chem* 2009;**16**(15):1821–30.
174 Verrax J, Cadrobbi J, Delvaux M, Jamison JM, Gilloteaux J, Summers JL, et al. The association of vitamin C and K3 kills cancer cells mainly by autoschizis, a novel form of cell death. Basis for their potential use as coadjuvants in anticancer therapy. *Eur J Med Chem* 2003;**38**:451–7.
175 Ghadially FN. *Ultrastructural Pathology of the Cell and Matrix*, 4th edn. Boston, MA: Butterworth-Heinemann. 1997.
176 Jamison JM, Gilloteaux J, Taper HS, Summers JL. Evaluation of the *in vitro* and *in vivo* antitumor activities of vitamin C and K3 combinations against human prostate cancer. *J Nutr* 2001;**131**:158S–60S.
177 Jamison JM, Gilloteaux J, Venugopal M, Koch JA, Sowick C, Shah R, Summers JL. Flow cytometric and ultrastructural aspects of the synergistic antitumor activity of

vitamin C-vitamin K3 combinations against prostatic carcinoma cells. *Tissue Cell* 1996;**28**:687–701.
178 Venugopal M, Jamison JM, Gilloteaux J, Koch JA, Summers M, Hoke J, et al. Synergistic antitumor activity of vitamins C and K3 against human prostate carcinoma cell lines. *Cell Biol Intl* 1996;**20**:787–97.
179 Venugopal M, Jamison JM, Gilloteaux J, Koch JA, Summers JL. Synergistic antitumor activity of vitamin C and K3 on human urologic tumor cell lines. *Life Sci* 1996;**59**:1389–400.
180 Gilloteaux J, Jamison JM, Arnold D, Summers JL. Autoschzis: another cell death for cancer cells induced by oxidative stress. *Ital J Anat Embryol* 2001;**106**:79–92.
181 Gilloteaux J, Jamison JM, Arnold D, Taper HS, Summers JL. Ultrastructural aspects of autoschizis: a new cancer cell death induced by the synergistic action of ascorbate: menadione on human bladder carcinoma cells. *Ultrastruct Pathol* 2001;**25**:183–92.
182 Jamison JM, Gilloteaux J, Nassiri MR, Venugopal M, Neal DR, Summers JL. Cell cycle arrest and autoschizis in human bladder carcinoma cell line following vitamin C and vitamin K3 treatment. *Biochem Pharmacol* 2004;**67**:337–51.
183 Gilloteaux J, Jamison JM, Taper HS, Summers JL. The antitumor activity of Apatone against human prostate cancer correlates with its ability to reactivate DNases and induce autoschizis. *J Men's Health* 2009;**6**:248.
184 Gilloteaux J, Jamison JM, Arnold D, Neal DR, Summers JL. Morphology and DNA degeneration during autoschizic cell death in bladder carcinoma T24 cells induced by ascorbate and menadione treatment. *Anat Rec A Discov Mol Cell Evol Biol* 2006;**288**:58–83.
185 Gilloteaux J, Jamison JM, Lorimer HE, Jarjoura D, Taper HS, Buc Calderon P, et al. Autoschizis: a new form of cell death for human ovarian carcinoma cells following ascorbate:menadione treatment. Nuclear and DNA degradation. *Tissue Cell* 2004;**36**:197–209.
186 Gilloteaux J, Jamison JM, Arnold D, McGuire K, Loukas M, Sczepaniak JP, et al. Autoschizis: a new cell death induced found in tumor cells induced by oxidative stress mechanism. In: Mendez-Villas A, Alvarez JD (eds.). *Microscopy, Science, Technology, Application and Education.* Bajadoz, Spain: Formatex Research Center, Microscopy. 2011.
187 Jamison JM, Gilloteaux J, Taper HS, Buc Calderon P, Summers JL. Autoschizis: a novel cell death. *Bochem Pharmacol* 2002;**63**:1773–83.
188 Jamison JM, Gilloteaux J, Taper HS, Buc Calderon P, Perlaky L, Thiry M, et al. The in vitro and in vivo antitumor activity of vitamin C:K3 combinations against prostate cancer. Lucas JN (ed.). *Trends in Prostate Cancer Research.* Hauppauge, NY: Nova Sci Publishers. 2005.
189 Jamison JM, Gilloteaux J, Perlaky L, Thiry M, Smetana K, Neal D, et al. Nucleolar changes and fibrillarin redistribution following Apatone treatment of human bladder carcinoma cells. *J Histochem Cytochem* 2010;**58**:635–51.
190 Zhang W, Negoro T, Satoh K, Jiang Y, Hashimoto K, Kikuchi H, et al. Synergistic cytotoxic action of vitamin C and vitamin K3. *Anticancer Res* 2001;**21**:3439–44.
191 McGuire K, Jamison JM, Neal D, Gilloteaux J, Summers JL. Elucidating the pathway of Apatone induced DNAse II reactivation during autoschizic cell death. *Proc Microsc Microanal* 2009;**15**(Suppl. 2):888.
192 Gilloteaux J, Jamison JM, Neal DR, Summers JL. Cell death by autoschizis in TRAMP prostate carcinoma cells as a result of treatment by ascorbate:menadione combination. *Ultrastruct Pathol* 2005;**23**:221–36.

193 Gerke V, Moss SE. Annexins: from structure to function. *Physiol Rev* 2002;**82**:331–71.
194 Segawa K, Suzuki J, Nagata S. Constitutive exposure of phosphatidylserine on viable cells. *Proc Natl Acad Sci USA* 2011;**108**(48):19 246–51.
195 Takatsu H, Tanaka G, Segawa K, Suzuki J, Nagata S, Nakayama K, Shin HW. Phospholipid flippase activities and substrate specificities of human type IV P-type ATPases localized to the plasma membrane. *J Biol Chem* 2014;**89**:33 543–56.
196 Ingber DE. Tensegrity I. Cell structure and hierarchical systems biology. *J Cell Sci* 2003;**116**:1157–73.
197 Tavassoli M, Crosby WH. Fate of the nucleus of the marrow erythroblast. *Science* 1973;**179**:912–13.
198 Sonoda Y, Sasaki K, Suda M, Itano C, Iwatsuki H. Effects of colchicine on the enucleation of erythroid cells and macrophages in the liver of mouse embryos: ultrastructural and three-dimensional studies. *Anat Rec* 1998;**251**:290–6.
199 Leung PY, Miyashita K, Young M, Tsao CS. Cytotoxic effect of ascorbate and its derivatives on cultured malignant and nonmalignant cell lines. *Anticancer Res* 1993;**13**:475–80.
200 Maramag C, Menon M, Blaji KC, Reddy PG, Laxmanan S. Effect of vitamin C on prostate cancer cells *in vitro*: effect on cell number, viability, and DNA synthesis. *Prostate* 1997;**32**:188–95.
201 Menon MM, Maramag C, Malhotra RK, Seethalakshmi L. Effect of vitamin C on androgen independent prostate cancer cells (PC3 and Mat-Ly-Lu) *in vitro*: involvement of reactive oxygen species-effect on cell number, viability and DNA synthesis. *Cancer Biochem Biophys* 1998;**16**:17–30.
202 Von Gruenigen VE, Jamison JM, Gilloteaux J, Lorimer HE, Summers M, Pollard RR, et al. The *in vitro* antitumor activity of vitamins C and K3 against ovarian carcinoma. *Anticancer Res* 2003;**23**:3279–87.
203 Verrax J, Stockis J, Tison A, Taper HS, Buc Calderon P. Oxidative stress by ascorbate/menadione association kills K562 human chronic myelogenous leukaemia cells and inhibits its tumor growth in nude mice. *Biochem Pharmacol* 2006;**72**:671–80.
204 Buc Calderon P, Cadrobbi J, Marques C, Hong-Ngoc N, Jamison JM, Gilloteaux J, et al. Potential therapeutic application of the association of vitamins C and K3 in cancer treatment. *Curr Med Chem* 2002;**9**(24):2271–85.
205 Jamison JM, Gilloteaux J, Koch JA, Nicastro E, Docherty JJ, Hoke J, et al. Vitamin C and K3-induced oxidative stress in human prostate tumor cells: mitochondrial ultrastructural alterations. *Microsc Microanal* 1997;**3**(Suppl. 2):23–4.
206 Connolly JM, Rose DP. Angiogenesis in two human prostate cancer cell lines with differing metastatic potential when growing as solid tumors in nude mice. *J Urol* 1998;**160**:932–6.
207 Alffrey V, Mirsky AE. Some aspects of deoxyribonuclease activities of animal tissues. *J Gen Physiol* 1952;**36**:227–41.
208 Siebert G, Lang K, Lucius-Lang S, Herkert L, Starr G, Rossmuller G, Jockel H. Eigenschaften und histologische Bedeutung der Deoxyribonuclease aus Säugetiereorganen. *Hoppe-Seyler's Z Physiol Chem* 1953;**295**:229–43.
209 Laskowski M. DNAses and their use in the studies of primary structure of nucleic acids. *Adv Enzymol* 1967;**29**:165–220.
210 Privat de Garilhe M. *Enzymes in Nucleic Acid Research*. Paris: Hermann. 1967.

211 Shugar D, Sierakowska H. Mammalian nucleolytic enzymes and their localization. In: Davidson JN, Cohn WE (eds.). *Progress in Nucleic Acid Research and Molecular Biology*, Vol. 7 New York: Academic Press. 1967.

212 Verity MA, Reith A. Effects of mercurial compounds on structure-linked latency of lysosomal hydrolases. *Biochem J* 1967;**105**:685–90.

213 Verity MA, Caper R, Brown WJ. Effect of cations on structure-linked sedimentability of lysosomal hydrolases. *Biochem J* 1968;**109**:149–54.

214 Lieberman MW, Sullivan RJ, Shull KH, Liang H, Farber E. Partial characterization and cellular localization of two deoxyribonuclases in the small intestine of the rat. *Canad J Biochem* 1971;**49**:38–43.

215 Roth JS. Ribonuclease. VI. Partial purification and characterization of the ribonucleases of rat liver mitochondria. *J Biol Chem* 1957;**227**:591–604.

216 Beaufay H, Bendall DS, Baudhuin P, de Duve C. Tissue fractionation studies. 12. Intracellular distribution of some dehydrogenases, alkaline deoxyribonuclease and iron in rat liver tissue. *Biochem J* 1959;**73**:623–8.

217 De Lamirande G, Allard C, Cantero A. Intracellular distribution of deoxyrinonucleo-depolymerase in normal rat liver, liver tumor and liver of animals fed p-dimethylaminoazobenzene. *Can J Biochem* 1954;**32**:35–40.

218 Van Lancker JL, Holtzer RL. Tissue fractionation studies of mouse pancreas. *J Biol Chem* 1959;**234**:2359–63.

219 Girija NS, Sreenivasan A. Characterisation of ribonucleases and ribonuclease inhibitor in subcellular fractions of rat adrenals. *Biochem J* 1966;**98**:562–6.

220 Morais R, Blackstein M, de Lamirande G. Partial purification and some properties of a 5'-RNase from rat liver mitochondrial fraction. *Arch Biochem* 1967;**121**:711–16.

221 Keir HM, Smellie RMS, Siebert G. Intracellular location of DNA nucleotidyltransferase. *Nature* 1962;**196**:752–4.

222 Smith J, Keir HM. DNA nucleotidyltrasnferase in nuclei and cytoplasm prepared from thymus tissue in non aqueous media. *Biochim Biophys Acata* 1963;**68**:578–88.

223 Roth JS. Some observations on the assay and properties of ribonucleases in normal and tumor tissues. In: Bush H (ed.). *Methods in Cancer Research*. New York: Academic Press. 1967.

224 De Duve C, Wattiaux H. Functions of lysosomes. *Ann Rev Physiol* 1966;**28**:435–92.

225 Siebert G, Villalobos J, Ro TS, Steele WJ, Lindenmayer G, Adams H, Busch H. Enzymatic studies on isolated nucleoli of rat liver. *J Boil Chem* 1966;**241**:71–8.

226 Swingle KF, Cole LJ. Acid deoxyribonuclease in rat liver cell nuclei isolated in the presence of calcium ions. *J Histochem Cytochem* 1964;**12**:442–7.

227 Lesca P. Age variations of acid deoxyribonuclease activity in mouse nuclei. *Nature* 1968;**220**:76–7.

228 Slor H, Lev T. Acid deoxyribonuclease activity in purified calf thymus nuclei. *Biochem J* 1971;**123**:993–5.

229 Ling G, Waxman DJ. DNase I digestion of isolated nulcei for genome-wide mapping of DNase hypersensitivity sites in chromatin. *Methods Mol Biol* 2013;**977**:21–33.

230 Wattiaux R, Baudhuin P, Berleur AM, De Duve C. Tissue fractionation studies. 8. Cellular localization of bound enzymes. *Biochem J* 1956;**63**:608–12.

231 Futai M, Miyata S, Mizumo D. Acid ribonucleases of lysosomal and soluble fractions from rat liver. *J Biol Chem* 1969;**244**:4951–60.

232 Brown KD, Jacobs G, Laskowski M. The distribution of nucleodepolymerases in calf thymus fractions. *J Biol Chem* 1952;**194**:445–54.

233 Reid E, El-Aaser AB, Turner MK, Siebert G. Enzymes of ribonucleic acid and ribonucleotide metabolism in rat liver nuclei. *Hoppe-Seyler's Z Physiol Chem* 1964: **339**:139–49.

234 Lindberg MU, Skoog L. Purification from calf thymus of an inhibitor of deoxyribonuclease I. *Eur J Biochem* 1970;**13**:326–35.

235 Roth JS. Ribonuclease VII. Partial purification and characterisation of a ribonuclease inhibitor in rat liver supernatant fraction. *J Biol Chem* 1958;**321**:1085–95.

236 Sirakov LM, Kochakian CD. The stability of ribonuclease inhibitor from guinea pig liver. *Biochim Biophys Acat (Amst)* 1969;**195**:572–5.

237 Kowlessar OD, Okada S, Potter JL, Altman KJ. Occurrence of an inhibitor for deoxyribonuclease II in human urine. *Arch Biochem* 1957;**68**:231–3.

238 Kurnick NB. Deoxyribonuclease activity of sera of man and some other species. *Arch Biochem* 1953;**43**:97–107.

239 Roth JS. Ribonuclease V. Studies on the properties and distribution of ribonuclease inhibitor in the rat. *Biochim Biophys Acta (Amst)* 1956;**21**:34–43.

240 Hakim AA. Enzyme pacemakers IV and V. *Enzym Biol Clin* 1959;**20**:249–60.

241 Köenig H, Guines D, Donald T, Gray R, Scott J. Studies of brain lysosomes. I. Subcellular distribution of five acid hydrolases, succinate dehydrogenase and gamgliosides in rat brain. *J Neurochem* 1964;**11**:729–43.

242 Wrolewski F, Bodansky O. Presence of deoxyribonuclease activity in human serum. *Proc Sox Exp Biol Med (NY)* 1950;**74**:443–5.

243 Loiselle JM, Carrier R. Désoxyribonucléase sérique chez les cancéreux. *Rev Cancer Biol* 1963;**22**:341–6.

244 Amano H, Daoust R. The distribution of ribonuclease activity in rat liver during azo-dye carcinogenesis. *J Histochem Cytochem* 1961;**9**:161–4.

245 Vorbrodt A. Histochemical studies on the intracellular localization of acid deoxyribonuclease. *J Histochem Cytochem* 1961;**9**:647–55.

246 Steigleder GK, Fisher J. Uber die Lokalisierung von Ribonuclease (RNase) und Deoxyribonuclease (DNase). Aktivität in normaler, in entzündlich veränderter Haut und bei Hauttumoren. *Arch Klin Exp Derm* 1963;**217**:553–62.

247 Amano H, Amano M. Histochemical studies on deoxyribonuclease activity in normal and azo-dye -fed rat liver by the modified phosphatase method. *Gann* 1968;**59**:19–24.

248 Ledoux L, Pileri A, Vanderhaeghe F. Activité de la ribonucléase et teneur en acide ribonucléique des tissus normaux et néoplasmes. *Arch Physiol* 1957;**66**(1):124–5.

249 Ledoux L, Brandly S. Ribonuclease activity and ribonucleic acid content of homologous human normal and cancer cells. *Nature* 1958;**181**:913–14.

250 Ellem KA, Colter JS, Kuhn J. Ribonucleases of mouse tissues and of the Ehrlich ascites tumor. *Nature* 1959;**184**(Suppl. 13):984–5.

251 Colter JS, Kuhn J, Ellem KA. The ribonucleases of mouse ascite tumors. *Cancer Res* 1961;**21**:48–51.

252 Ambellan E, Hollander VP. Ribonuclease inhibitor in lymphosarcoma P 1798: effects of steroid and 5-FU. *Proc Sox Exp Biol Med* 1968;**127**:482–4.

253 Chakravorty AK, Busch H. Alkaline ribonuclease and ribonuclease inhibitor in nuclear and nucleolar preparations from normal and neoplastic tissues. *Cancer Res* 1967;**27**:789–92.

254 de Lamirande G. Intracellular distribution of 5'-ribonuclease and 5'-phosphodiesterase in the Novikoff hepatoma ascite cells. *Cancer Res* 1967;**27**:1722–3.

255 Arora DJS, de Lamirande G. Ribonuclease activity of ribonucleoprotein particles isolated from livers of rats fed 4-dimethylamino-azobenzene, primary liver tumors and Novikoff hepatoma. *Cancer Res* 1968;**28**:225–7.

256 Gordon J. Ribonucleases of the rat. Immunochemical properties. *Arch Biochem* 1965;**112**:429–35.

257 Taper HS, Lans M, Economidou-Karaoglou A, de Gerlache J, Roberfroid M. Variations in serum alkaline DNase activity: apossible clinical test for therapeutic prognosis of human tumors. *Anticancer Res* 1986;**6**:949–56.

258 Comporti M. Three modes of free radical-induced cell injury. *Chem Biol Interact* 1989;**72**:1–56.

259 Economidou-Karaoglou A, Opsomer M, Lans M, Taper HS, Deckers C, Roberfroid MB. Predictive value of serum alkaline DNase activity variations in treatment of head and neck cancer. *Acta Oncol* 1990;**29**(2):163–6.

260 Inohara N, Koseki T, Chen S, Benedict MA, Nunez G. Identification of regulatory and catalytic domains in the apoptosis nuclease DFF/CAD. *J Biol Chem* 1999;**274**:270–4.

261 Widlak P, Li P, Wang X, Garrad WT. Cleavage preferences of the apoptotic endonucleaase DFF40 (acspase acticated Dnase or nuclease) on naked DNA and chromatin substrates. *J Biol Chem* 2000;**275**:8226–32.

262 Verrax J, Cadrobbi J, Delvaux M, Jamison JM, Gilloteaux J, Taper HS, Buc Calderon P. The association of vitamins C and K3 kills cancer cells mainly by autoschizis, a novel form of cell death. Basis for their potential use as coadjuvants in anticancer therapy. *Eur J Med Chem* 2003;**38**:415–57.

263 Podmore ID, Griffiths HR, Herbert KE, Mistry N, Mistry P, Lunec J. Vitamin C exhibits pro-oxidant properties. *Nature* 1998;**392**:559.

264 Parekh HK, Mansuri-Torshizi H, Srivastava TS, Chitnis MP. Circumvention of Adriamycin resistance: effect of 2-methyl-1,4-naphtoquinone (vitamin K3) on drug cytotoxicity in sensitive and MDR P388 leukemia cells. *Cancer Lett* 1992;**61**:147–56.

265 Jarabak R, Jarabak J. Effect of ascorbate on the DT-diaphorase mediated redox cycling of 2-methyl-1,4-naphtoquinone. *Arch Biochem Biophys* 1995;**318**:418–23.

266 Cárcamo JM, Pedraza A, Borquez-Ojeda O, Zhang B, Sanchez R, Golde DW. Vitamin C is a kinase inhibitor: dehydroxyascorbate acid inhibits IkappaBalpha kinase beta. *Mol Cell Biol* 2004;**24**:6645–52.

267 Gant TW, Rao DN, Mason R, Cohen GM. Redox cycling and sulphydryl arylation; their relative importance in the mechanism of quinone cytotoxicity to isolated hepatocytes. *Chem Biol Interact* 1988;**65**(2):157–73.

268 Sinha Bk, Mimnaugh EG. Free radicals and anticancer drug resistance. *Free Rad Biol Med* 1968;**8**:301–17.

269 Velkey JM, Bernanke DH. Apoptosis during coronary artery orifice development in the chick embryo. *Anat Rec* 2001;**262**:310–17.

270 Lockshin RA, Zakeri Z. Programmed cell death: early changes in metamorphosing cells. *Biochem Cell Biol* 1994;**72**:589–96.

271 Aumüller G, Seitz J, Riva A. Functional morphology of prostate gland. In: Riva A, Testa-Riva F, Motta PM (eds.). *Ultrastructure of Male Urogenital Glands*. Boston, MA: Kluwer Academic. 1994.

272 Verity MA, Reith A. Effect of mercurial compounds on structure-linked latency of lysosomal hydrolases. *Biochem J* 1967;**105**:685–90.

273 Vanden Berghe T, Linkermann A, Jouan-Lanhouet S, Walczak H, Vandenabeele P. Regulated necrosis: the expanding network of non-apoptotic cell death pathways. *Nat Rev Mol Cell Biol* 2014;**15**(2):135–47.

274 Mitchell JS, Simon-Reuss I. Combination of some effects of x-radiation and a synthetic vitamin K substitute. *Nature* 1947;**159**(4055):98.

275 Prasad KN, Edwards-Prasad J, Sakamoto A. Vitamin K3 (menadione) inhibits the growth of mammalian tumor cells in culture. *Life Sci* 1979;**29**:1387–92.

276 Akman SA, Dietrich M, Chlebowski RT, Limberg P, Block JB. Modulation of cytotoxicity of menadione sodium bisulfite versus leukemia L1210 by the acid-soluble thiol pool. *Cancer Res* 1985;**45**:5257–62.

277 Chlebowski RT, Dietrich M, Akman S, Block JB. Vitamin K3 inhibition of malignant cell growth and human tumor colony formation. *Cancer Res* 1985;**45**:5257–62.

278 Dypbukt JM, Ankerona M, Burkitt, M, Sjöholm, Å, Ström K, Orrenius S, Nicotera P. Different prooxidant levels stimulate growth, trigger apoptosis, or produce necrosis of insulin-secreting RINm5F cells. *J Biol Chem* 1994;**269**:30 553–60.

279 Markovits J, Sun TP, Juan CC, Ju CG, Wu FY. Menadione (vitamin K3) enhances the mitogenic signal of epidermal growth factor via extracellular signal-regulated kinases. *Int J Oncol* 1998;**13**:1163–70.

280 Waxman S, Bruckner H. The enhancement of 5-fluoouracil anti-metabolic activity by leucovorin, menadione and alpha-tocopherol. *Eur J Cancer Clin Oncol* 1982;**18**:685–92.

281 Nutter LM, Cheng AL, Hung HL, Hsieh RK, Ngo EO, Liu TW. Menadione: spectrum of anticancer therapy and effects on nucleoside metabolism in human neoplastic cell lines. *Biochem Pharmacol* 1991;**1241**:1283–92.

282 Ngo EO, Sun TP, Chang JY, Wang CC, Chi KH, Cheng AL, Nutter LM. Menadione-induced DNA damage in a human tumor cell line. *Biochem Pharmacol* 1991;**42**:1961–8.

283 Wu FY, Liao WC, Chang HM. Comparison of antitumor activity of vitamins K1, K2 and K3 on human tumor cells by two (MTT and SRB) cell viability assays. *Life Sci* 1993;**52**:1797–804.

284 Wu FY, Chang NT, Chen WJ, Juan CC. Vitamin K3-induced cell cycle arrest and apoptotic cell death are accompanied by altered expression of c-fos and c-myc in nasopharyngeal carcinoma cells. *Oncogene* 1993;**8**:2237–44.

285 Mitchell JS, Brinkley D, Haybittle JL. Clinical trial of radiosensitizers, including synkavit and oxygen inhaled at atmospheric pressure. *Acta Radiol Ther Phys Biol* 1965;**3**:329–41.

286 Liao WC, Wu FY, Wu CW. Binary/ternary combined effects of vitamin K3 with other antitumor agents in nasopharyngeal carcinoma CG1 cells. *Int J Oncol* 2000;**17**:323–8.

287 Gold J. In vivo synergy of vitaminK3 and methotrexate in tumor-bearing animals. *Cancer Treat Rep* 1986;**70**:1433–5.

288 Tetef M, Matgolin K, Ahn C, Akman S, Chow W, Leong L, et al. Mitomycin C and menadione for the treatment of lung cancer: a phase II trial. *Invest New Drugs* 1995;**13**:157–62.

289 Tetef M, Margolin K, Ahn C, Akman S, Chow W, Coluzzi P, et al. Mitomycin C and menadione for the treatment of advanced gastrointestinal cancers: aphse II trial. *J Cancer Res Clin Oncol* 1995;**121**:103–6.

290 Akman S, Carr B, Leong LA. Phase I trial of menadiol sodium diphosphate (SYNKAYVITE) (M) in advanced cancer. *Proc Am Soc Clin Oncol* 1988;**7**:76.

291 Lamson DW, Plaza SM. The anticancer effects of vitamin K. *Altern Med Rev* 2003;**8**:303–18.
292 Kabsch W, Mannherz HG, Suck D. Three-dimensional structure of the complex of actin and DNAse I at 4.5 Å resolution. *EmBO J* 1985;**4**:2113–18.
293 Khaitlina SY, Strzelecka-Golaszewska H. Role of DNase-I-binding loop in dynamic properties of actin filament. *Biophys J* 2002;**82**(1 Pt. 1):321–34.
294 Burtnick LD, Chan KW. Protection of actin against proteolysis by complex formation with deoxyribonuclease I. *Can J Biochem* 1980;**58**:1348–54.
295 Malorni W, Iosi F, Mirabelli F, Bellomo G. Cytoskeleton as a target in menadione-induced oxidative stress in cultured mammalian cells: alterations underlying surface bleb formation. *Chem Biol Interact* 1991;**80**(2):217–36.
296 Philimonenko W, Janácek J, Harata M, Hozák P. Transcription-dependent rearrangements of actin and nuclear myosin I in the nucleolus. *Histochem Cell Biol* 2010;**134**:243–9.
297 McGuire K, Jamison JM, Gilloteaux J, Summers JL. Vitamin C and K3 combination causes enhanced anticancer activity against RT-4 bladder cancer cells. *J Cancer Sci Ther* 2013;**4**:7–19.
298 Thiry M, Lafontaine DL. Birth of a nucleolus: the evolution of nucleolar compartments. *Trends Cell Biol* 2005;**15**(4):194–9.
299 Rando OJ, Zhao K, Crabtree G. Searching for a function for nuclear actin. *Trends Cell Biol* 2000;**10**(3):92–7.
300 Spencer VA, Costes S, Inman JL, Xu R, Chen J, Hendzel MJ, Bissell MJ. Depletion of nuclear actin is a key mediator of quiescence in epithelial cells. *J Cell Sci* 2011;**24**:123–32.
301 Franke WW, Kartenbeck J. Outer mitochondrial membrane continuous with endoplasmic reticulum. *Protoplasma* 1971;**73**:35–41.
302 Baumgartner HK, Gerasimenko JV, Thorne C, Ferdek P, Pozzan T, Tepikin AV, et al. Calcium elevation in mitochondria is the main Ca^{2+} requirement for mitochondria permeability transition pore (mPTP) opening. *J Biol Chem* 2009;**284**:20 796–803.
303 Di Monte D, Bellomo G, Thor H, Nicotera P, Orrenius S. Menadione-induced cytotoxicity is associated with protein thiol oxidation and alteration in intracellular Ca2+ homeostasis. *Arch Biochem Biophys* 1984;**235**:343–50.
304 Scott GK, Atsriku C, Karminker P, Held J, Gibson B, Baldwin MA, Benz CC. Vitamin K3 (menadione)-induced oncosis associated with keratin 8 phosphorylation and histone H3 arylation. *Mol Pharmacol* 2005;**68**:606–15.
305 O'Brien PJ. Molecular mechanisms of quinone cytotoxicity. *Chem Biol Interact* 1991;**80**:1–41.
306 Rizzuto R, Pitton G, Azzone GF. Effect of Ca2+, peroxides, SH reagents, phosphate and aging on the permeability of mitochondrial membranes. *Eur J Biochem* 1987;**162**:239–49.
307 Fenton HJH. LXXIII. – Oxidation of tartaric acid in presence of iron. *J Chem Soc Trans* 1894;**65**:899–910.
308 Townsend DM, Tew KD, Tapiero H. The importance of glutathione in human disease. *Biomed Pharmacother* 2003;**57**:145–55.
309 Arrick BA, Griffith OW, Cohn ZA. Glutathione depletion sensitizes tumor cells to oxidative cytolysis. *J Biol Chem* 1982;**257**:1231–7.
310 Criddle DN, Gillies S, Baumgartner-Wilson HK, Jaffar M, Chinje EC, Passmore S, et al. Menadione-induced reactive oxygen species generation via redox cycling

promotes apoptosis of murine pancreatic acinar cells. *J Biol Chem* 2006;**281**:40 485–92.

311 Gilloteaux J, Naud J. The zinc iodide-osmium tetroxide staining-fixative of Maillet. Nature of the precipitate studied by x-ray microanalysis and detection of Ca^{2+}-affinity subcellular sites in a tonic smooth muscle. *Histochemistry* 1979;**63**:227–43.

312 Ide T, Tsutsui H, Hayashidani S, Kang D, Suematsu N, Nakamua K, et al. Mitochondrial DNA damage and dysfunction associated with oxidative stress in failing hearts after myocardial infarction. *Circ Res* 2001;**88**:529–35.

313 Hamasaki M, Furuta N, Matsuda A, Nezu A, Yamamoto A, Fujita N, et al. Autophagosomes form at ER-mitochondria contact sites. *Nature* 2013;**495**:389–93.

314 Kim HK, Park WS, Kang SH, Warda M, Kim N, Ko JH, et al. Mitochondrial alterations in human gastric carcinoma cell line. *Am J Physiol Cell Physiol* 2007;**293**:C761–71.

315 Wattiaux R. Biochemistry and function of lysosomes. In: Lima-de-Faria A (ed.). *Handbook of Molecular Cytology*. Amsterdam: North Holland. 1969.

316 Hallman A, Milczarek R, Lippinski M, Kossowska E, Spodnik JH, Wozniak M, et al. 2004. Fast perinuclear clustering of mitochondria in oxidatively stressed human choriocarcinoma cells. *Folia Morphol* **63**:407–12.

317 Bellomo G, Mirabelli F, Vairetti M, Iosi F, Malorni W. Cytoskeleton as a target in menadione- induced oxidative stress in cultured mammalian cells. *J Cell Physiol* 1990;**143**:118–28.

318 Gerasimenko JV, Gerasimenko OV, Petersen OH. Membrane repair: Ca(2+)-elicited lysosomal exocytosis. *Curr Biol* 2001;**11**(23):R971–4.

319 Osada S, Yoshida K. A novel strategy for advanced pancreatic cancer – progression of molecular targeting therapy. *Anticancer Agents Med Chem* 2009;**9**(8):877–81.

320 Cole MD, McMahon SB. The Myc oncoprotein: a critical evaluation of transactivation and target gene regulation. *Oncogene* 1999;**18**:2916–24.

321 Hockenberry DM, Oltvai ZN, Yin XM, Milliman CL, Korsmeyer SJ. Bcl-2 functions in an antioxidant pathway to prevent apoptosis. *Cell* 1993;**75**:241–51.

322 Liu GY, Frank N, Bartsch H, Lin JK. Induction of apoptosis by thiuramdisulfides, the reactive metabolites of dithiocarbamates, through coordinative modulation of NFkappaB, c-fos/c-jun, and p53 proteins. *Mol Carcinog* 1998;**22**(4):235–46.

323 Hug H, Strand S, Grambihler A, Galle J, Hack V, Stremmel W, et al. Reactive oxygen intermediates are involved in the induction of CD95 ligand mRNA expression by cytostatic drugs in hepatoma cells. *J Biol Chem* 1997;**272**(45):28 191–3.

324 van den Dobbelsteen DJ, Nobel CSI, Schlegel J, Cot-greave IA, Orrenius S, Slater AF. Rapid and specific efflux of reduced glutathione during apoptosis induced by anti-Fas/APO-1 antibody. *J Biol Chem* 1996;**271**:15 420–37.

325 Osada S, Saji S, Osada K. Critical role of the extracellular signal-regulated kinase phosphorylation on menadione (vitamin K3) induced growth inhibition. *Cancer* 2001;**91**:1156–65.

326 Vojtek AB, Ders CJ. Increasing complexity of the Ras signalling pathway. *J Biol Chem* 1998;**273**:19 925–8.

327 Frei B, England L, Ames BN. Ascorbate is an outstanding antioxidant in human blood plasma. *Proc Natl Acad USA* 1989;**86**:6377–81.

328 Pavelić K, Kos Z, Spaventi S. Antimetabolic activity of L-ascorbic acid in human and animal tumors. *Int J Biochem* 1989;**21**(8):931–5.

329 Sharma P, Mongan PD. Ascorbate reduces superoxide production and improves mitochondrial respiratory chain function in human fibroblasts with electron transport chain deficiencies. *Mitochondrion* 2001;1(2):191–8.

330 Langemann H, Torhorst J, Kabiersch A, Krenger W, Honegger CG. Quantitative determination of water- and lipid-soluble antioxidants in neoplastic and non-neoplastic human breast tissue. *Int J Cancer* 1989;43(6):1169–73.

331 Baader SL, Bruchelt G, Trautner MC, Boschert H, Niethammer D. Uptake and cytotoxicity of ascorbic acid and dehydroascorbic acid in neuroblastoma (SK-N-SH) and neuroectodermal (SK-N-LO) cells. *Anticancer Res* 1994;14(1A):221–7.

332 Spielholz C, Golde DW, Houghton AN, Nualart F, Vera JC. Increased facilitated transport of dehydroascorbic acid without changes in sodium-dependent ascorbate transport in human melanoma cells. *Cancer Res* 1997;57(12):2529–37.

333 Agus DB, Gambhir SS, Pardridge WM, Spielholz C, Baselga J, Vera JC, Golde DW. Vitamin C crosses the blood-brain barrier in the oxidized form through the glucose transporters. *J Clin Invest* 1997;100(11):2842–8.

334 Liang W-J, Johnson D, Jarvis SM. Vitamin C transport systems of mammalian cells. *Mol Membr Biol* 2001;18:87–95.

335 Tsukaguchi H, Tokui T, Mackenzie B, Berger UV, Chen XZ, Wang Y, et al. A family of mammalian Na+-dependent L-ascorbic acid transporters. *Nature* 1999;399:70–5.

336 Shah S, Nath N. Metabolism of L-ascorbic acid in the prostate of normal and scorbutic guinea pigs. *Metabolism* 1985;34(10):912–16.

337 Vera JC, Rivas CI, Fishbarg J, Golde DW. Mammalian facilitative hexose transporters mediate the transport of dehydroascorbatic acid. *Nature* 1993;364:79–82.

338 Vera JC, Rivas CI, Velasquez FV, Zhang RH, Concha H, Golde DW. Resolution of the facilitated transport of dehydroascorbic acid from its intracellular accumulation as ascorbic acid. *J Biol Chem* 1995;270:23 706–12.

339 Thomas EL, Learn DB, Jefferson MM, Weatherred W. Superoxide-dependent oxidation of extracellular reducing agents by isolated neutrophils. *J Biol Chem* 1988;263:2178–86.

340 Washko PW, Wang Y, Levine M. Ascorbic acid recycling in human neutrophils. *J Biol Chem* 1993;268:15 531–5.

341 Vera JC, Rivas CI, Zhang RH, Farber CM, Golde DW. Human HL-60 myeloid leukemia cells transport dehydroascorbic acid via the glucose transporters and accumulate reduced ascorbic acid. *Blood* 1994;84:1628–34.

342 Matsubara T, Ziff M. Increased superoxide anion release from human endothelial cells in response to cytokines. *J Immunol* 1986;137:3295–8.

343 Meier B, Cross AR, Hancock JT, Kaup FJ, Jones OT. Identification of a superoxide-generating NAPDH oxidase system in human fibroblasts. *Biochem J* 1991;275:241–5.

344 Bram S, Froussard P. Guichard M, Jasmin C. Augery Y, Sinoussi-Barre F, Wray W. Vitamin C preferential toxicity for malignant melanoma cells. *Nature* 1980;284:629–31.

345 Park CH, Amare M, Savin MA. Hoogstrntcn B. Growth suppression of human leukemic cells *in vitro* by L-ascorbic acid. *Cancer Res* 1980;40:1062–5.

346 Liotti FS, Menghini AR, Guerrieri P, Talesa V, Bodo M. Effects of ascorbic acid and dehydroascorbic acid on the multiplication of tumor ascites cells *in vitro*. *J Cancer Res Clin Oncol* 1984;108:230–2.

347 Cameron E, Pauling L, Liebovitz B. Ascorbic acid and cancer. A review. *Cancer Res* 1979;39:663–81.

348 Yamafuji K, Nakamura Y, Omura H, Soeda T, Gyotoku K. Antitumor potency of ascorbic acid, dehydroascorbic or 2,3-diketogulonic acid and their action on deoxyribonucleic acid. *Z Krebsforsch Onkol Cancer Res Clin Oncol* 1971;**76**:1–7.

349 Koch CJ, Biaglow JE. Toxicity, radiation sensitivity modification and metabolic effects of dehydroascorbate and ascorbate in mammalian cells. *J Cell Physiol* 1978;**94**:299–306.

350 De Laurenzi V, Melino G, Savini I, Annicchiarico-Petruzzelli M, Finazzi-Agrò A, Avigliano L. Cell death by oxidative stress and ascorbic acid regeneration in human neuroectodermal cell lines. *Eur J Cancer* 1995;**31A**(4):463–6.

351 Nagao N, Nakayama T, Etoh T, Saiki I, Miwa N. Tumor invasion is inhibited by phosphorylated ascorbate via enrichment of intracellular vitamin C and decreasing of oxidative stress. *J Cancer Res Clin Oncol* 2000;**126**(9):511–18.

352 Bijur GN, Briggs, Hichcock CL, Williams MV. Ascorbic acid dehydroascorbate induces cell cycle arrest at G2/M DNA damage checkpoint during oxidative stress. *Environ Mol Mutagen* 1999;**33**:144–52.

353 Szent-Györgyi A. Observations on the function of peroxidase systems and the chemistry of the adrenal cortex: description of a new carbohydrate derivative. *Biochem J* 1928;**22**(6):1387–409.

354 Desai ID, Tappel AL. Damage to proteins by peroxidized lipids. *J Lipid Res* 1963;**4**:204–7.

355 Lupulescu A. Vitamin C inhibits DNA, RNA and protein synthesis in epithelial neoplastic cells. *Int J Vitam Nutr Res* 1991;**61**:125–9.

356 Rosin MP, San RH, Stich HF. Mutagenic activity of ascorbate in mammalian cell cultures. *Cancer Lett* 1980;**8**(4):299–305.

357 Tardy C, Codogno P, Autefage H, Levade T, Andrieu-Abadie N. Lysosomes and lysosomal proteins in cancer cell death (new players of an old struggle). *Biochim Biophys Acta* 2006;**1765**:101–25.

358 Epstein ACR, Gleade GM, McNeill LA, Hewitson KS, O'Rourke J, Mole DR, et al. C. elegans EGL-9 and mammalian homologues define a family of dioxygenases that regulate HIF by prolyl hydroxylation. *Cell* 2001;**107**:43–54.

359 Knowles HJ, Raval RR, Harris AL, Ratcliffe PJ. Effect of ascorbate on the activity of hypoxia-inducible factor in cancer cells. *Cancer Res* 2003;**63**(8):1764–8.

360 Lin TH, Chen Q, Howe A, Juliano RL. Cell anchorage permits efficient signal transduction between ras and its downstream kinases. *J Biol Chem* 1997;**272**(14):8849–52.

361 Dunham WB, Zuckerkandl E, Reynolds R, Willoughby R, Marcuson R, Barth R, Pauling L. Effects of intake of L-ascorbic acid on the incidence of dermal neoplasms induced in mice by ultraviolet light. *Proc Natl Acad Sci USA* 1982;**79**(23):7532–6.

362 Zaizen Y, Nakagawara A, Ikeda K. Patterns of destruction of mouse neuroblastoma cells by extracellular hydrogen peroxide formed by 6-hydroxydopamine and ascorbate. *J Cancer Res Clin Oncol* 1986;**111**(2):93–7.

363 Prasad SB, Giri A, Arjun J. Use of subtherapeutical dose of cisplatin and vitamin C against murine Dalton's lymphoma. *Pol J Pharmacol Pharm* 1992;**44**(4):383–9.

364 Hanck AB. Vitamin C and cancer. *Prog Clin Biol Res* 1988;**259**:307–20.

365 Pedersen T, Aebi U. Nuclear actin extends, with no contraction in sight. *Mol Biol Cell* 2005;**16**:5055–60.

366 Novikoff AB, Shin WY. Endoplasmic reticulum and autophagy in rat hepatocytes. *Proc Natl Acad Sci USA* 1978;**75**(10):5039–42.

367 Van Caneghem P. Influence of substances with thiol functions and of their reagents on the fragility lysosomes. *Biochem Pharmacol* 1972;**21**(18):2417–24.
368 Ericsson JLE. Mechanism of cellular autophagy. In: Dingle JT, Fell HB (eds.). *Lysosomes in Biology and Pathology*. Amsterdam: North-Holland. 1969.
369 Fiorani M, De Sanctis R, Scarlatti F, Stocchi V. Substrates of hexokinase, glucose-6-phosphate dehydrogenase, and glyceraldehyde-3-phosphate dehydrogenase prevent the inhibitory response induced by ascorbic acid/iron and dehydroascorbic acid in rabbit erythrocytes. *Arch Biochem Biophys* 1998;**356**:159–66.
370 Pethig R, Gascoyne PR, McLaughlin JA, Szent-Györgyi A. Ascorbate-quinone interactions: electrochemical, free radical, and cytotoxic properties. *Proc Natl Acad Sci USA* 1983;**80**(1):129–32.
371 Fiorani M, De Sanctus R, Scarlatti F, Vallorani L, De Bellis R, Serafini G, et al. Dehydroascorbic acid irreversibly inhibits hexokinase activity. *Mol Cell Biochem* 2000;**209**:145–53.
372 Cárcamo JM, Bórquez-Ojeda O, Golde DW. Vitamin C inhibits granulocyte macrophage-colony-stimulating factor-induced signaling pathways. *Blood* 2002;**99**(9):3205–12.
373 Verrax J, Delvaux M, Beghein N, Taper H, Gallez B, Buc Calderon P. Enhancement of quinone redox cycling by ascorbate induces a caspase-3 independent cell death in human leukaemia cells. An in vitro comparative study. *Free Radic Res* 2005;**39**(6):649–57.
374 Lockwood TD. Inactivation of intracellular proteolysis and cathepsin B enzyme activity by dehydroascorbic acid and reactivation by dithiotreitol in perfused rat heart. *Biochem Pharmacol* 1997;**54**:669–75.
375 Dunn JA, Ahmed MU, Murtiashaw MH, Richardson JM, Walla MD, Thorpe SR, Baynes JW. Reaction of ascorbate with lysine and protein under autoxidizing conditions: formation of N epsilon-(carboxymethyl)lysine by reaction between lysine and products of autoxidation of ascorbate. *Biochemistry* 1990;**29**(49):10 964–70.
376 Orrenius S, McConkey DJ, Nicotera P. Role of calcium in toxic and programmed cell death. *Adv Exp Med Biol* 1991;**283**:419–25.
377 Farber JL. The role of calcium in lethal cell injury. *Chem Res Toxicol* 1990;**3**(6):503–8.
378 Sakagami H, Satoh K. Modulating factors of radical intensity and cytotoxic activity of ascorbate. *Anticancer Res* 1997;**17**:3513–20.
379 Atkinson DE. The energy charge of the adenylate pool as regulatory parameter. Interaction with feedback modifiers. *Biochemistry* 1968;**7**:4030–4.
380 Filho ACM, Meneghini R. *In vivo* formation of single-strand breaks in DNA by hydrogen peroxide is mediated by the Haber-Weiss reaction. *Biochim Biophys Acta* 1984;**781**:56–63.
381 Clopton DA, Saltman P. Low-level oxidative stress causes cell-cycle specific arrest in cultured cells. *Biochem Biophys Res Commun* 1995;**210**(1):189–96.
382 Zbarsky IB, Dmitrieva NP, Yermolayeva LP. On the structure of tumor cell nuclei. *Exp Cell Res* 1962;**27**:573–6.
383 Peitsch MC, Polzar B, Tschopp J, Mannherz HG. About the involvement of deoxyribonuclease I in apoptosis. *Cell Death Differ* 1994;**1**(1):1–6.
384 Krieser RJ, Eastman A. The cloning and expression of human deoxyribonuclease II. A possible role in apoptosis. *J Biol Chem* 1998;**273**(47):30 909–14.
385 Peitsch MC, Polzar B, Stephan H, Crompton T, MacDonald HR, Mannherz HG, Tschopp J. Characterization of the endogenous deoxyribonuclease involved in nuclear

DNA degradation during apoptosis (programmed cell death). *EMBO J* 1993;**12**(1):371–7.

386 Barry MA, Eastman A. Identification of deoxyribonuclease II as an endonuclease involved in apoptosis. *Arch Biochem Biophys* 1993;**300**(1):440–50.

387 Patutina OA, Mironova NL, Ryabchikova EI, Popova NA, Nikolin VP, Kaledin VI, et al. Tumoricidal activity of RNase A and DNase I. *Acta Naturae* 2010;**2**(1):88–94.

388 Festy B, Paoletti C. [Demonstration and properties of 1 (or several) natural inhibitor(s) of pancreatic neutral deoxyribonuclease]. *C R Hebd Seances Acad Sci* 1963;**257**:3682–5.

389 Sierakowska H, Shugar D. Mammalian nucleolytic enzymes. *Prog Nucleic Acid Res Mol Biol* 1977;**20**:59–130.

390 Lesca P. Protein inhibitor of acid deoxyribonucleases. Improved purification procedure and properties. *J Biol Chem* 1976;**251**(1):116–23.

391 Lacks SA. Deoxyribonuclease I in mammalian tissues. Specificity of inhibition by actin. *J Biol Chem* 1981;**256**(6):2644–8.

392 Gilloteaux J, Jamison JM, Arnold D, Summers JL. Autoschizis of human ovarian carcinoma cells: scanning electron and light microscopy of a new cell death induced by sodium ascorbate: menadione treatment. *Scanning* 2003;**25**:137–49.

29

Programmed Death 1 (PD1)-Mediated T-Cell Apoptosis and Cancer Immunotherapy

Chandi Charan Mandal,[1] Jatin Mehta,[2] and Vijay K. Prajapati[1]

[1]*Department of Biochemistry, Central University of Rajasthan, Rajasthan, India*
[2]*National Institute of Pathology, ICMR, Safdarjang Hospital, New Delhi, India*

Abbreviations

ACT	adoptive cell transfer
ADCC	antibody-dependent cell-mediated cytotoxicity
AKT	protein kinase B
ALK	anaplastic lymphoma kinase
APC	antigen-presenting cell
BCR	B-cell receptor
CLL	chronic lymphocytic leukemia
CTL	cytotoxic T lymphocyte
CTLA4	cytotoxic T lymphocyte-associated protein
EAE	experimental autoimmune encephalomyelitis
EGFR	endothelial growth-factor receptor
ERK	growth factor-regulated extracellular signal-related kinase
FDA	Food and Drug Administration
FDC	follicular dendritic cell
Foxop3	forkhead box P3
GM-CSF	granulocyte–macrophage colony-stimulating factor
HLA	human leukocyte antigen
HPV	human papillomavirus
ICOS	inducible T-cell COStimulator
IFNα	interferon alpha
IFNγ	interferon gamma
IgM	immunoglobulin M
IgSF	immunoglobin superfamily
IL-1	interleukin 1
IRF-1	interferon regulatory factor 1
ITIM	immunoreceptor tyrosine-based inhibitory motif
ITSM	immunoreceptor tyrosine-based switch motif
JAK	Janus kinase
LCK	lymphocyte-specific protein tyrosine kinase
MEK	mitogen/extracellular signal-regulated kinase
MHC	major histocompatibility complex

Apoptosis and Beyond: The Many Ways Cells Die, First Edition. Edited by James Radosevich.
© 2018 John Wiley & Son Inc. Published 2018 by John Wiley & Son Inc.

MS	multiple sclerosis
NK	natural killer
NOD	nucleotide-binding oligomerization domain
NPM	nucleophosmin
OSCC	oral squamous cancer cell
PBMC	peripheral blood mononuclear cell
PCD	programmed cell death
PD1	programmed death 1
PD-L1	programmed death ligand 1
PD-L2	programmed death ligand 2
PI3-K	phosphatidylinositol 3-kinase
PTEN	phosphatase and tensin homolog
RA	rheumatoid arthritis
SLE	systemic lupus erythematosus
SNP	single-nucleotide polymorphism
STAT	signal transducer and activator of transcription
TCR	T-cell receptor
TKI	tyrosine kinase inhibitor
TLR	Toll-like receptor
UTR	untranslated region

29.1 Introduction

Programmed cell death (PCD) or apoptosis is essential not only for the development of the human body, but also for the maintenance of cellular homeostasis. Unhealthy conditions are known to arise due to disturbances in the balance between cell formation and cell death, leading to the development of several pathophysiologic diseases, such as neurological disorders and cancers. The immune system primarily acts as a sentinel, protecting us from infectious and foreign substances by inducing clonal expansion and activation of T and B cells to fight them. The immune system selectively eliminates autoreactive T cells during their maturation to ensure tolerance to self-antigens. Failure of this tolerance mechanism may lead to tissue damages and the development of autoimmune diseases, whereas excessive apoptosis of reactive T cells or inactivation of T cells will fail to defend against infection [1]. Central peripheral tolerance mechanisms can destroy most of the autoreactive T cells in the thymus. However, if some escape this selective elimination, these same mechanisms promote either T-cell anergy, exhaustion, inactivation, or apoptosis. Cytotoxic T cells recognize tumor antigens displayed along with class I major histocompatibility complex (MHC) on their surface, and initiate killing of them by means of T-cell granule content or interferon gamma (IFNγ) [2,3]. However, tumor cells often escape being killed by the immune system by altering the normal immune response, by inducing apoptosis, anergy, or exhaustion of active T cells, or by other ingenious strategies of immune evasion [4]. Chen et al. [3] have proposed that oncogenic signals of cancer cells release antigen peptide, which is captured by antigen-presenting cells (APCs), and that these APCs travel to lymphoid organs and present antigen–MHC complex to T cells. These interactions help in the priming and activation of effector T cells, which attack cancer cells by recognizing the antigen peptide–MHC complex they display. Effective T-cell activation requires a minimum of two signals provided by the APCs [5]. The most essential signal is the interaction between T cells and APCs brought about by the T-cell receptor (TCR)–antigen peptide–MHC complex. However, clonal activation and expansion of T cells requires the APCs to provide a second signal to the T cells by engaging their co-stimulatory ligands (B7-1, B7-2) with the CD28 receptor on the cell surfaces. These two positive signals jointly activate the T cells and ensure their clonal expansion. Instead of co-stimulatory signals, APCs can also provide co-inhibitory molecules (programmed death ligand 1 and 2, PD-L1 and PD-L2), which bind to the programmed death 1 (PD1) receptor of activated T cells, and subsequently inhibit T-cell function.

Several reports have outlined the roles of PD-L1, PD-L2, and PD1 in T-cell apoptosis [6,7]. In this chapter, we discuss how this PD-L–PD1 pathway is involved in T-cell apoptosis, cell-cycle inhibition, and proliferation. Emerging evidence from several clinical trials shows a very promising role for monoclonal antibodies (e.g., Nivolumab, Prenbrolizumab) that target the PD-L–PD1 pathway as a treatment for cancers. We describe how these antibodies manifest anti-tumor activity. We also examine whether the engineering of T cells targeting the PD-L–PD1 pathway is a good strategy for treating cancer patients and whether microRNA can be used to modulate the activity of T cells for use in immune cancer therapy.

29.2 T-Cell Development, Differentiation, Function, and Degradation

T cells are produced in the bone marrow. The immature T cells, called prothymocytes, travel to the thymus gland [8], where they are processed and become mature (i.e., functionally competent). They then travel to peripheral lymphoid compartments, including the spleen, kidney, and blood [8].

T-cell development and degradation are briefly sketched in Figure 29.1. In the early phase, immune-response cells, APCs, and dendritic cells, which carry antigens, leave the site of tissue infection or cancer tissue and travel to secondary lymphoid organs to activate naïve T cells. There, APCs present antigen through the MHC to naïve T cells. The T cells recognize the MHC–antigen complex through TCR. They can differentiate into different subsets depending on inflammatory signals and other environmental factors. For example, interleukin 12 (IL-12), T-bet, and Stat4 convert to Th1-type differentiation, whereas IL-4, GATA3, and Stat6 convert to Th2-type, and TGFβ, forkhead box P3 (Foxop3), and IL-2 convert to Treg cells (Figure 29.2). APCs also express stimulatory ligands such as B7-1 and B7-2, which bind to CD28 on the T cells; this ligand–receptor interaction may increase TCR signaling to activate and differentiate T cells. Activated T cells travel from the secondary lymphoid organs and penetrate to the sites of tissue inflammation, where they interact with them and eventually kill the target cells.

29.3 Structure of PD1

The cell-surface receptor PD1, also known as CD279 protein, belongs to the immunoglobin superfamily (IgSF). It was first isolated from the cDNA subtraction library, when stimulated murine T cells (immature T lymphocytes, a T-cell hybridoma 2B4.11)

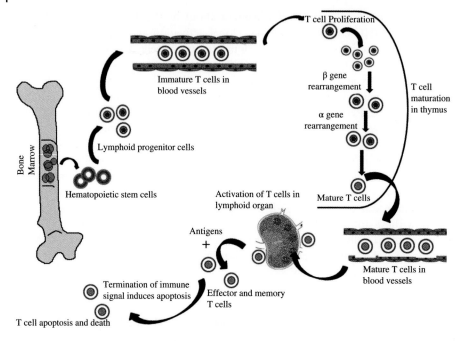

Figure 29.1 Development, activation, and degradation of T cells.

were undergoing apoptosis [9]. It was reported that the longest open reading frame encodes 288 amino acids, which have two hydrophobic regions. PD1 is a type I transmembrane glycoprotein. The hydrophobic region present in the middle part is the transmembrane domain, and the N-terminus hydrophobic region is a signal

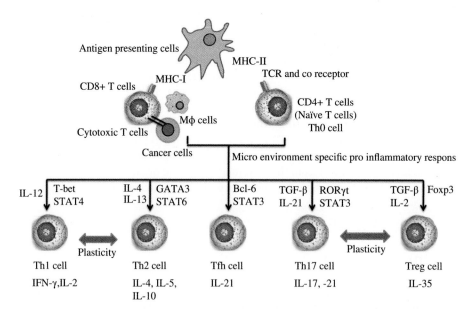

Figure 29.2 CD4+ T-cell (Th0) differentiation into different T subpopulations.

peptide [9]. After cleavage of the signal peptide, the predicted mature protein is 268 amino acids, with an extracellular domain (147 amino acids), a transmembrane region (27 amino acids), and a cytosolic domain (94 amino acids) [9] Multiple sequence alignments have documented a large degree of similarity between the extracellular domain of this PD1 and other IgSF proteins, such as TCR Vα, Vβ, CD4 (I), and CD8(β); thus, it is referred as an IgSF protein [9]. In humans, PD1 is encoded by the *pdcd1* gene, which is located on chromosome 2. Mouse PD1 is located on chromosome 1. The *pdcd1* gene has five exons, where exons 1 and 2 encode a signal peptide and an extracellular immunoglobin domain, respectively, exon 3 encodes stalk- and membrane-spanning regions, and exons 4 and 5 form the cytoplasmic tail [10]. Exon 5 also encodes 3′ untranslated regions (UTRs). There is 64% amino acid identity between mouse and human PD1 in the protein sequence [7,11,12].

PD1 shares almost 30% amino acid sequence homology with CD28, cytotoxic T lymphocyte-associated protein (CTLA4), and inducible T-cell COStimulator (ICOS) (other IgSF proteins), and it is thus referred to as an extended family member of these proteins [13]. Its predicted molecular weight is 29.3 kD, but experimentally this has been shown to be 50–55 kD when protein samples of stimulated murine T cells are used, suggesting that PD1 is a highly glycosylated protein [14]. Although x-ray crystallographic data show that the extracellular domain of PD1 has a similar structure to that of CTLA4, both have two layers of β-sandwich secondary structure (Figure 29.3). The cytoplasmic domain contains both an immunoreceptor tyrosine-based inhibitory motif (ITIM) and an immunoreceptor tyrosine-based switch motif (ITSM). It is evident that the tyrosine residues (phosphorylation site) present in both ITIM and ITSM modulate PD1 signaling in T cells. The cytoplasmic tail of PD1 binds to the tyrosine phosphatases SHP1/SHP2

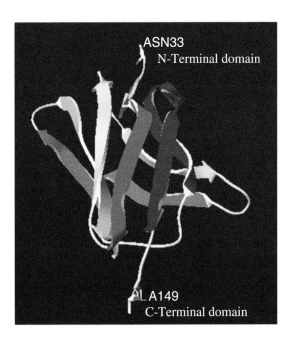

Figure 29.3 Three-dimensional (3D) structures of programmed death 1 (PD1) (PDB: 3RRQ). Images made using SPDBV 4.1.0. This extracellular human PD1 (A) domain contains ten β sheets.

thorough this ITSM sequence, and these phosphatases may remove the phosphate group of key proteins of TCR signaling to shut down PD1 signaling [5,15–17].

29.4 Expression and Functions of PD1

It has been documented that genetic ablation of the PD1 gene generates many immunological disorders, such as the spontaneous autoimmune diseases arthritis, gastritis, and cardiomyopathy [18–20]. Moreover, the presence of polymorphisms within the *pdcd1* gene is associated with multiple sclerosis (MS), rheumatoid arthritis (RA), type 1 diabetes, and systemic lupus erythematosus (SLE) [21–23]. Thus, expression of PD1 prevents autoimmune diseases by suppressing the activity of the immune system. PD1 is expressed following stimulation of many types of hematopoietic cell, including T and B lymphocytes, monocytes, and dendritic cells. It is exclusively found during thymus development [10,14]. The expressions of the family members CD28, CTLA4, and ICOS are limited mainly to T cells [24].

A considerable amount of basal PD1 expression has been reported in a murine T-cell hybridoma 2B4.11, which was markedly increased when cells were stimulated with phorbol myristate acetate (PMA)/ionomycin or anti-CD3 [14]. Double-negative $CD4^-CD8^-$ T cells show PD1 expression. This varies with the early T-lymphocyte markers CD44 and C25, and is augmented by signaling by the TCR complex and by TCRβ rearrangement in T cells [25].

PD1 expression has also been detected primarily in the T-cell zone of the spleen, and is co-localized with CD3 expression. In the thymus, it is mainly found on thymocytes in the medulla, and less so in the cortex. It is detected on small round cells, suggestive of germinal-center B cells. At 9 weeks of age, expression of PD1 and its ligand PD-L1 is detected on the infiltrating mononuclear cells in the islet of pancreas. However, no staining of PD1 or its ligands is observed, regardless of the initial infiltration of mononuclear cells into the pancreas, in 3-week-old nucleotide-binding oligomerization domain (NOD) mice [26]. Expression of PD1 is also found in the germinal center of human tonsils [27]. Many intrinsic and extrinsic factors or agents may stimulate PD1 expression on T cells. For example, estrogen, TCR- and B-cell receptor (BCR)-signaling, chronic viral infection, and antigens may upregulate the expression of PD1 on T cells or APC cells to control the immune response [14,28,29]. The common gamma-chain cytokines IL-2, IL-7, IL-15, and IL-21, which have key roles in T-cell expansion and survival, can also induce PD1 expression on T cells [30]. Similarly, ligation of Toll-like receptor 2 (TLR2), TLR3, TLR4, or NOD upregulates PD1 expression, but this expression is inhibited by IL-4 and TLR9 ligation [31]. However, proinflammatory cytokines (IL-1β, IL-6, IL-8, TNF-α), immunosuppressive cytokines (TGF-β, IL-10), and certain immunoregulatory cytokines (IL-4, IL-18) do not show any effect on PD1 expression [30]. Bioinformatic analysis shows the presence of NFATc1 binding sites on the upstream region of the *pdcd1* gene. Moreover, treatment of $CD8^+$ T cells with cyclosporine A, calcineurin inhibitor, or VIVIT peptide, a NFATc1-specific inhibitor, shows a significant reduction in basal and induced PD1 expression [32]. On T-cell activation, calcineurin phosphatase dephosphorylates NFATc1, which translocates to the nucleus, but cyclosporine A inhibits NFATc1 nuclear translocation by blocking calcineurin activity, resulting in inhibition of PD1 expression in T cells.

29.5 Structure and Expression of the PD1 Ligands PD-L1 and PD-L2

Researchers had hypothesized that the ligand of PD1 could be a member of the B7 family, since PD1s have a large degree of similarity with CTLA4 and CD28 [9]. Later, it was discovered by two independent research groups that PD-L1 belongs to the B7 family, and its cDNA was isolated based on a homology search of that family [7,33].

PD-L1 is also known as B7-H1 and CD274. The *pd-l1* gene is located on chromosomes 9 and 19 in humans and mice, respectively. PD-L1 is a type 1 transmembrane glycoprotein, with 290 amino acids [34]. It contains a signal sequence, an extracellular IgV and IgC domain, and a transmembrane domain with a short cytoplasmic tail, like other B7 family proteins, including B7-1 and B7-2. Human PD-L1 has 70% amino acid identity with murine PD-L1 [7,11,12]. The crystal structures of PD-L1 alone and in combination with PD1 have been determined (Figure 29.4).

Similarly, through a gene bank search, Freeman's research group [35] identified the second ligand of PD1, known as PD-L2. The same group successfully cloned PD-L2 cDNA from human placental tissues [35]. PD-L2 is also known as B7DC and CD273. Like *pd-l1*, the *pd-l2* gene is located on chromosomes 9 and 19 of humans and mice, respectively. Human PD-L2 is also a type 1 transmembrane glycoprotein with 274 amino acids; this is 26 amino acids longer than mouse PD-L2, which has a premature stop codon [34,35]. There is 38% amino acid identity between the protein sequence of PD-L1 and that of PD-L2 [11,12]. Like PD-L1, PD-L2 has a signal sequence, an extracellular IgV and IgC domain, and a transmembrane domain with a short cytoplasmic tail. The cytoplasmic tail of PD-L2 is not conserved much between the mouse and the human, whereas that of PD-L1 is greatly conserved [35]. The crystal structures of PD-L2 alone and in combination with PD1 have alsp been solved (Figure 29.5) [34].

Expression of PD-L1 mRNA has been detected in a variety of tissues, including heart, lung, and placenta [26], whereas the expression of B7-1 (CD80) and B7-2 (CD86) is

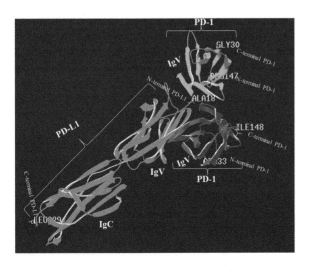

Figure 29.4 Three-dimensional (3D) structure of PD1–PD-L1 complex (PDB: 3BP5). Image made using SPDBV 4.1.0. Both PD1 domains shown contain 11 β sheets. The N-terminal domain of PD-L1 contains eight β sheets and one helix. The C-terminal domain of PD-L1 also contains eight β sheets.

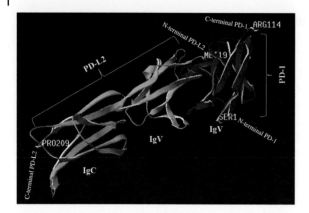

Figure 29.5 Three-dimensional (3D) structure of PD1–PD-L2 complex (PDB: 3BP5). Image made using SPDBV 4.1.0. The PD1 domain contains 11 β sheets. The N-terminal domain of PD-L2 contains nine β sheets and one helix. The C-terminal domain of PD-L2 contains ten β sheets.

limited primarily to lymphoid cells. The human tissue-specific expression pattern of PD-L1 is almost the same as that of mouse. Several reports have detected PD-L1 protein expression on cardiac endothelium, pancreatic islet cells, mononuclear cells within the lamina propria, alveolar macrophages, and syncyciotrophoblasts within the placenta [26]. Moreover, PD-L1 is expressed diffusely throughout the white pulp, with strong expression in the marginal zone, while PD-L2 expression is not detected in the spleen. The expressions of both of these ligands have been found in thymus dendritic cells [26].

Like PD-L1, PD-L2 mRNA expression is detected in tissues such as lung, brain, kidney, and placenta [26,35]. However, PD-L2 protein exists at below detectable levels in these tissues in humans, although lymphoid tissues show positive expression of PD-L2 protein [26]. Brown et al. [36] reported that mouse non-lymphoid tissues such as cardiac endothelium, placental endothelium, and myocardium show expression of PD-L2 protein [36]. However, no expression of PD-L1 or PD-L2 protein has been found in mouse germinal-center B cells or follicular dendritic cells (FDCs), although PD-L1 and PD-L2 protein expression has been found on human FDCs [26,36]. Research studies conducted in an experimental autoimmune encephalomyelitis (EAE) model show that both PD-L1 and PD-L2, including PD1, have been found to be expressed on infiltrating cells of the endothelium and brain tissues [26]. In brief, mouse hematopoietic cells (e.g., T cells, B cells, macrophages, dendritic cells, and bone marrow-derived mast cells) and non-hematopoietic cells (e.g., endothelial, epithelial, and muscle cells) show constitutive expression of PD-L1 but not of PD-L2. However, PD-L2 expression is induced on macrophages, dendritic cells, and bone marrow-derived mast cells [37]. This constitutive expression of PD-L1 but not PD-L2 on non-hematopoietic tissues indicates that PD-L1 may function more dominantly than PD-L2 in regulating tissue-directed inflammatory responses [26].

It has been experimentally validated that both PD-L1 and PD-L2 bind to PD1 receptor. However, PD-L1 does so with greater affinity. It has also been demonstrated, by Butte et al. [37], that PD-L1 binds with B7-1, another B7 family member, but that PD-L2 does not. These authors proposed that B7-1 may compete with PD1 for binding to PD-L1, since they have a partial overlap of binding sites [37], and that the binding of B7-1 to

Table 29.1 Co-stimulatory and co-inhibitory ligands that interact with their receptors to modulate T-cell function.

Characteristic	Ligand Name	Receptor Name	Ligand–Receptor/ Ligand–Ligand Interaction	Function
Co-stimulators	B7-1 (CD80)	CD28	B7-1-CD28	Proliferation, survival, memory formation, differentiation
	B7-2 (CD86)	CD28	B7-2-CD28	
	B7-H2 (CD275)	CD28	B7-H2-CD28	
	B7-H2 (CD275 or ICOSL)	ICOS (CD278)	B7-H2-ICOS	
Co-inhibitors	B7-1 (CD80)	CTLA-4 (CD152)	B7-1-CTLA-4	Cell-cycle inhibition, apoptosis, tolerance, exhaustion, inhibition of effector formation
	B7-2 (CD86)	CTLA-4 (CD152)	B7-1-CTLA-4	
	B7-H1 (CD274 or PD-L1)	PD-1 (CD279)	B7-H1-PD-1	
	B7DC (CD273 or PD-L2)	PD-1 (CD279)	B7DC-PD-1	
	B7-1 (CD80)		B7-1-B7-H1	
	B7-H1 (CD274 or PD-L1)	Unknown	B7-H1-Unknown	-
	B7DC (CD273 or PD-l1)	Unknown	B7DC-Unknown	-

PD-L1 may prevent T-cell activation instead of inducing it. The ligand–receptor interactions and functions of T cells are summarized in Table 29.1.

It has been documented that *pdcd1*-deficient T cells show less proliferation and less production of the cytokines IL-2 and IFNγ when incubated with PD-L1 and CD3 beads as compared to immunoglobin-plus-CD3 beads [37]. PD-L1-plus-CD3 beads also showed inhibition of wild-type T cells, similar to PD1-deficient T cells. In contrast, *pdcd1* knockout (KO) T cells did not show inhibition of cell proliferation when they were incubated with PD-L2-plus-CD3 beads. However, PD-L2-plus-CD3 beads showed inhibition of the expansion of wild-type T cells [37]. These results indicate that there may be an inhibitory counter-receptor for PD-L1 other than PD1 [37]. Moreover, this study demonstrated that PD-L1 can exert an inhibitory effect on T cells through either B7-1 or PD1.

29.6 Regulation of PD-L1 and PD-L2 Expression

A research study by Liang et al. [26] showed that signal transducer and activator of transcription 4 and 6 ($STAT4^{-/-}$ and $STAT6^{-/-}$) dendritic cells did not alter the expression of PD-L1, but that this expression was slightly reduced in $p50^{-/-}p65^{+/-}$ (NF-κB) dendritic cells when they were treated with IL-4 or IFNγ. However, gel mobility-shift and chromatin-immunoprecipitation experiments revealed that the transcription factor

STAT3 binds to the PD-L1 promoter to transcriptionally upregulate its expression [38]. Moreover, chimeric nucleophosmin (NPM)/anaplastic lymphoma kinase (ALK) can upregulate PD-L1 expression in T lymphoma cells. Similarly, reporter assays using PD-L1 promoter have revealed that interferon regulatory factor 1 (IRF-1) is primarily responsible for the constitutive expression of PD-L1, as well as for the IFNγ-mediated PD-L1 overexpression in human lung-cancer A549 cells [39]. Parsa et al. [40] reported that the phosphatidylinositol 3-kinase (PI3-K)/protein kinase B (AKT) pathway induces PD-L1 expression by activating mTOR/S6K1 signaling in glioma cells. However, other investigators failed to show the inducing effect of the PI3-K, mTOR, or mitogen/extracellular signal-regulated kinase (MEK)/growth factor-regulated extracellular signal-related kinase (ERK) pathways in the regulation of PD-L1 expression when studying lung and hepatocellular cancer cells [39]. Loss of phosphatase and tensin homolog (PTEN) expression upregulates the PI3-K/AKT pathway to increase PD-L1 expression [5], while the Janus kinase (JAK)-STAT signaling pathway also upregulates PD-L1 expression. All these results indicate that endogenous expression of PD-L1 can be regulated by multiple signaling pathways, and that its regulation may vary in different cellular contexts.

Expression of PD-L2 is not regulated by NF-κB, but it might positively regulate PD-L1 expression [26]. Several reports have shown that many cytokines, including IL-2, IL-7, IL-15, and IL-21, can induce PD-L1 expression on T cells and APCs [30]. Interferons, IL-4, and granulocyte–macrophage colony-stimulating factor (GM-CSF) may stimulate expression of PD-L2 on dendritic cells, while the common gamma-chain cytokines can induce greater PD-L1 expression as compared to PD-L2 on human monocytes/macrophages [30]. Similarly, inflammatory macrophages show basal expression of PD-L1, while INFγ stimulation increases induction of PD-L1 expression. On the other hand, inflammatory macrophages do not show expression of PD-L2 until cells are stimulated with IL-4 [41]. Moreover, PD-L1 expression depends on STAT1 activation, whereas PD-L2 depends on that of STAT6. However, constitutive expression of PD-L1 is independent of STAT1 activation. In addition, Th1 cells enhance PD-L1 expression on macrophages, but PD-L2 expression in inflammatory macrophages is induced by Th2 cells [41]. These differential expression patterns indicate that PD-L1 and PD-L2 may have largely non-overlapping functions in regulating immune responses and autoimmunity.

29.7 Expression Patterns of PD1, PD-L1, and PD-L2 in Cancer Tissues and Tumor-Infiltrating Lymphocytes

Recent studies have documented that a large number of cancer tissue samples show expression of PD-L1 in many cancer types. For example, more than 50% of cancer patients studied by Salih et al. [42] showed higher levels of PD-L1 expression in leukemia cancer cells, and almost all leukemia cancer cell lines showed significant expression levels of PD-L1. CD8+ and CD4+ T cells isolated from chronic lymphocytic leukemia (CLL) patients contained an increased expression of PD1 receptor [43]. An elevated level of PD-L1 expression was also found on Treg cells isolated from the peripheral blood cells of melanoma cancer patients [44]. A greater number of CD8+PD1+ T cells were found in peripheral blood mononuclear cells (PBMC) of oral squamous cancer cells (OSCCs) than in healthy controls. In addition, OSCCs showed an increased level of PD-L1 compared to controls [45]. Similarly, human papillomavirus (HPV) types 6 and 11-induced respiratory papillomas exhibited an increased level of PD-L1 expression as compared to healthy laryngeal tissues [46]. Shen et al. [47] reported that a subset of osteosarcoma cancer tissues showed increased levels of PD-L1 and, moreover, that this expression was positively associated with the number of tumor-infiltrating cells present in the tumor tissue environment. A subset of T cells of peripheral blood showed higher expression of PD1 in the case of non-small-cell lung-cancer patients; this PD1 expression can be used as a prognostic marker [48]. Recent evidence shows that the mutation pattern of oncogenes may positively correlate with expression of PD1 or PD-L1 [49]. For instance, the presence of oncogene KRAS mutations shows positive association with PD1-positive non-small-cell patients, whereas the presence of an endothelial growth-factor receptor (EGFR) mutation shows a positive correlation with PD-L1-positive adenocarcinoma [49]. It has also been reported that smoking status positively correlates with PD1-positive non-small-cell patients. These data suggest that smoking creates mutations in lung-cancer tissues, and that these mutations provide oncogenic signals that upregulate the PD-L–PD1 pathway. Another study provided evidence that a large percentage of non-small-cell lung-cancer specimens show increased levels of both PD1 and PD-L1 [50]. This same study showed that EGFR–tyrosine kinase inhibitor (TKI) therapy provides a significantly greater disease control rate, increased progression-free survival, and greater overall survival in PD-L1-positive cancer patients as compared to other forms of cancer.

Several lines of evidence show a differential expression of PD-L1 in breast cancer. For example, Ghebeh et al. [51] reported that 50% of breast-cancer tissues contain high level of PD-L1. In addition, 44% of tumor-infiltrating lymphocytes (TILs) showed PD-L1 expression, which was positively correlated with tumor size, while intra-tumor expression of PD-L1 was positively associated with tumor grade III, ER-negative, and PR-negative status [51]. Similarly, another research group showed that almost 60% of breast-cancer tissues had PD-L1 mRNA expression, which was associated with TILs, and with longer recurrence-free survival [52]. Soliman et al. [53] reported that the basal type of breast-cancer cell line showed high expression of PD-L1 as compared to other cell types. However, they noted that increased levels of PD-L1 expression were obtained in 20% of breast cancer samples, while basal and mesenchymal breast-cancer tissues showed higher expression of PD-L1 as compared to other cancer types. It was demonstrated that the level of PD-L1 expression positively correlates with better survival and response to chemotherapy in case of aggressive breast-cancer types [54]. All these data together suggest that PD-L1 is highly expressed in a large number of breast-cancer patients. Moreover, PD-L1 expression correlates with aggressiveness of breast cancer. However, chemotherapy treatment works better if cancer tissues have a high level of PD-L1 expression.

Similarly, the presence of T cells expressing PD1 within follicular lymphoma is referenced as a marker for the transformation of this follicular lymphoma [55]. Kim et al. [56] reported that levels of PD-L1 could be a prognostic marker for soft-tissue carcinoma. There is substantial evidence that an elevated level of PD-L1 expression in cancer tissue or in TILs is a novel potential biomarker, and correlates with longer survival of cancer patients.

29.8 Role of PD1, PD-L1, and PD-L2 in Apoptosis

In 1992, Ishida et al. [9] isolated for first time the PD1 gene by using a cDNA subtraction library method on a stimulated murine T-cell hybridoma. They observed that *de novo* RNA synthesis is required for classical apoptotic cell death. DNA fragmentation, a classical apoptotic feature, appeared when murine hybridoma 2B4.11 cells were stimulated with ionomycin and PMA. Similarly, other hematopoietic LyD9 cells showed DNA fragmentation when cells were deprived of IL-3. Ishida et al. found that addition of actinomyocin D at the time of cell stimulation inhibits DNA fragmentation, but that DNA fragmentation was not inhibited (or even partially inhibited) when actinomycin D was added some time after cell stimulation. From these experiments, they proposed that *de novo* RNA synthesis is required for a classical apoptotic death. Thus, in order to discover a novel gene with responsibility for apoptotic death, they performed cDNA subtractive hybridization experiments using stimulated hybridoma 2B4.11 cells and IL-3-deprived LyD9 cells. In these experiments, they prepared a cDNA probe capable of hitting cell death-specific cDNAs that are common between stimulated 2B4.11 and IL-3-deprived LyD9 cells. Ultimately, they identified PD1 as a subtractive clone, and sequenced the *pdcd1* gene. Amino acid sequence similarity defines PD1 it as a member of IgSF. Ishida et al. also reported that extensive DNA fragmentation and augmentation of PD1 mRNA were found when anti-CD3 monoclonal antibody was injected into the thymus of mice. They concluded that PD1 might play a significant role in the process of induction of apoptotic cell death, but that additional factors might modulate its function. They hypothesized that PD1 expression is a cell-death inducer [9].

PD1-deficient C57BL/6 mice exhibited severe autoimmune features such as a lupus-like arthritis and glomerulonephritis. Moreover, PD1 augments the B-cell proliferative response upon immunoglobulin M (IgM) stimulation. Similarly, PD1 KO BALB/C mice showed several autoimmune diseases, such as dilated cardiomyopathy with excessive thrombosis [14,20,35]. These lines of evidence suggest that PD1 might have an inhibitory response to the immune system. Later, two independent research groups identified and isolated the *pd-l1* gene. Dong et al. [33] showed that PD-L1 does not bind to the receptors CD28 and CTLA4, but Freeman et al. [7] revealed that it does bind to PD1 receptor. Freeman et al. noted that neither B7-1 nor PD-L1 was detected in unstimulated blood monocyte cells, but found that their expressions were increased when these cells were treated with IFNγ. They also showed that PD-L1 inhibits anti-CD3 monoclonal antibody-induced T-cell proliferation in $PD1^{+/+}$ murine T cells, but that it has no such effect in the case of $PD1^{-/-}$ murine T cells. They observed that PD-L1 inhibits the proliferation of anti-CD3 monoclonal antibody-stimulated CD4+ T cells isolated from human PBMCs, with simultaneous inhibition of IFNγ and IL-10 production. However, they also found that PD-L1 ligand can inhibit cell proliferation when CD4+ T cells are stimulated with high concentrations of anti-CD3 antibody in the absence of CD28 stimulation [7]. These observations suggest that engagement of PD1 by PD-L1 can suppress the cell proliferation of weak TCR-activated T cells. Similarly, Dong et al. [6] showed that tumor-associated PD-L1 ligand induces apoptosis of human CD8+ T cell when PD-L1-transfected 624 melanoma cells are co-cultured with CD8+ T cells. They also revealed that PD-L1 reduces T-cell proliferation in a mice model. Moreover, soluble PD1Ig inhibits the apoptosis of M15 T cells, suggesting that PD1 could be a receptor of PD-L1, which delivers inhibitory signals to T cells. The PD-L1-transfected 624 melanoma cells did not show any expression of FasL or TRAIL. These observations suggest that the PD-L1-mediated apoptosis of T cells could be independent of the FasL/TRAIL pathway, and that nonredundant regulation of T-cell apoptosis may occur via the PD-L1–PD1 and Fas–FasL pathways [6]. A study by Shi et al. [57] demonstrated that PD-L1 expression is induced in hepatocellular HepG2.2.15 cells when these cells are co-cultured with CD8+ T cells or treated with IFNγ. Here, CD8+ T cell-induced PD-L1 expression of cancer cells is mediated through IFNγ. Induction of apoptosis of CD8+ T cells is seen when these cells are co-cultured with IFNγ-treated HepG2.2.15 cells, but blocking PD-L1 markedly inhibits this. These data support the idea that PD-L1 enhances apoptosis of T cells [57]. Similarly, Latchman et al. [35] found that incubation of cells on PD-L2 immunoglobin-coated beads in the presence of anti-CD3 shows inhibition of T-cell proliferation and simultaneous inhibition of IL-10 and IL-4 production. They concluded that PD-L2–PD1 interaction inhibits TCR-mediated T-cell proliferation in the absence of CD28 stimulation [35]. Butte et al. [37] were the first to report that PD-L1 interacts with the co-stimulatory B7-1 molecule to inhibit T-cell response. They had noticed that co-stimulatory B7-1 increases the proliferation of wild-type T cells when these cells are incubated with B7-1 immunoglobin and CD3 beads, but that cell proliferation is inhibited when $CD28^{-/-}CTLA4^{-/-}$ T cells are used in this experiment [35]. It was also observed that PD-L1 inhibits the proliferation of PD1 KO T cells, as well as wild-type T cells. These results suggest that the interaction of PD-L1 and B7-1 may have a T cell-inhibitory response. Contrasting findings have also been reported by several investigators. For instance, incubation of T cells (isolated from human PBMCs) on PD-L1 immunoglobin-coated culture plate in the presence of a suboptimal dose of anti-CD3 shows a marked increase of cell proliferation as compared with negative control [33]. Moreover, cytokines such as IL-10 and INFγ are also increased when cells are cultured PD-L1 immunoglobin-coated culture plate [33]. Similarly, PD-L1 derived from leukemia cells does not show any influence of T-cell proliferation, T-cell activation, or cytokine production [42]. However, production of melanoma antigen-specific cytotoxic T lymphocytes (CTLs) is increased, due to blockade of PD1 [44]. Thus, most research studies – but not all of them – suggest that the induction and interaction of PD1, PD-L1, and PD-L2 reduces T-cell proliferation, either by inducing apoptosis or by inhibiting cell-cycle progression. However, the effect of the PD-L–PD1 pathway in T-cell survival or proliferation may vary depending on the cytokine milieu, antigen specificity, allogenic or syngeneic nature, and tissue-specific qualities. In brief, at low TCR activation, this pathway may show inhibitory response to T-cell proliferation, but at high TCR activation, it may not be able to counterattack the proliferative signals produced by B7-1/B7-2–CD28 stimulation. On the other hand, high CD28 stimulation may weaken the apoptosis of the PD-L–PD1 pathway.

29.9 Co-inhibitory Signal of PD-L1/PD-L2–PD1 Inhibits TCR Signaling

The activated T cell recognizes the antigen peptide–MHC complex displayed by APC/dendritic cells through TCR and commences TCR signaling. This first signal is a prerequisite for TCR signaling-mediated T-cell activation, but it is not sufficient to boost T cells for its activation. For that, a second stimulatory signal must be supplied upon interaction of the TCR–antigen peptide–MHC complex. Several co-stimulatory ligands and their TCRs have been reported (see Table 29.1), among which B7-1 and B7-2

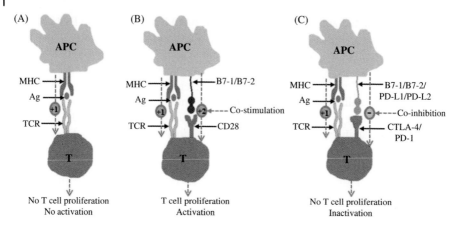

Figure 29.6 Activation of T cells depends on second co-stimulatory or co-inhibitory signals. (A) Positive signal 1, caused by the interaction between the major histocompatibility complex (MHC), an antigen peptide, and a T-cell receptor (TCR). (B) Two positive signals. The second is caused by the interaction between CD28 and a co-stimulatory ligand (such as B7-1 and B7-2). (C) One positive and one negative signal. The former is caused by the interaction of the MHC, an antigen peptide, and a TCR, the latter by the interaction between a co-inhibitory ligand and its receptor (e.g., B7-1 and CTLA4, B7-2 and CTLA4, PD-L1 and PD1, or PD-L2 and PD1). The +1 and +2 arrows represent signals 1 and 2, the − arrow represents the inhibitory signal.

are the most potent, binding to CD28 to potentiate TCR signaling (Figure 29.6). B7-1 and B7-2 may also provide an inhibitory signal if they bind to the co-inhibitory receptor CTLA4. The newly discovered PD-L1 and PD-L2, expressed on the surfaces of APCs (dendritic or cancer cells), bind to the PD1 receptor expressed on the surfaces of activated cells. This engagement of the extracellular domain of PD1 receptor through PD-L1 or PD-L2 provides an inhibitory signal to T cells, which may eventually inhibit the active TCR signaling caused by positive signal 1 and positive signal 2. Cancer cells often escape from immune attack. Substantial data suggest that most cancer types express PD-L1 protein on their surface (see earlier). Tumor or epithelial cells may display antigen peptide to MHC on their surface, and this antigen peptide–MHC complex can be recognized by T cells [2]. Since cancer cells often express inhibitory ligand PD-L1, which can bind to PD1 to abrogate TCR signaling, resulting in T-cell apoptosis, cell-cycle inhibition, and T-cell inactivation (Figure 29.7), it has been suggested that inflammatory signals from the microenvironment might modulate cancer cell-mediated inhibitory signaling of T cells. For example, microenvironment-released or -activated T cell-derived IFNγ may upregulate PD-L1 expression on APCs and cancer cells [2,7,39,41,58] and may increase PD1 on T cells (Figure 29.8). An increase in PD1 and PD-L1 may prevent the T-cell activation by blocking TCR signaling.

We have described the basics of TCR signaling, based on several studies (Figure 29.9) [5,24,59]. Co-stimulatory signals caused by the interaction of CD28 with B7-1/B7-2 activate lymphocyte-specific protein tyrosine kinase (LCK). Activated LCK, in the presence of antigen-induced TCR signaling, activates the central ZAP70 protein, which directly or indirectly activates many key signaling molecules, including PI3-K, GRB2, and PLCγ. These signaling molecules eventually activate many different transcription factors, including NFAT, NF-κB, and ELK. Moreover, activated PI3-K activates AKT signaling. Thus, TCR signaling regulates many biological functions, including enhancement of cell proliferation, cell division, cytokine production, cell survival, and protein

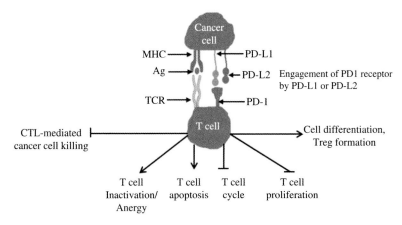

Figure 29.7 Engagement of the PD1 receptor by PD-L1/PD-L2 of cancer cells, inducing T-cell apoptosis and inactivation. Cancer cell-expressed PD-L1 binds with the PD1 receptor of the T cell. This engagement inhibits TCR signaling, resulting in inactivation of T cells, induction of T cells, inhibition of the cell cycle, and induction of T-cell proliferation. Moreover, PD1–PD-L1 interaction induces conversion of T cells to T-repressor cells. Finally, T cells are unable to show cytotoxic T lymphocyte (CTL)-mediated cancer-cell killing.

synthesis (Figure 29.9). However, the engagement of PD1 by PD-L1 or PD-L2 increases the phosphorylation of tyrosine residues present in the ITIM and ITSM of PD1 [24] and activates phosphatase SHP2. Activated SHP2 removes the phosphorylation of key signaling molecules that were activated by TCR signaling [5,15–17]. Thus, increased

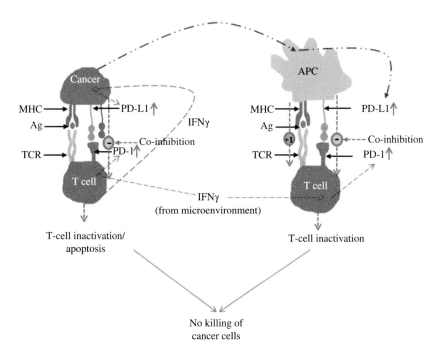

Figure 29.8 Promotion of the production of inhibitory factors (PD-L1 and PD1) by cancer cells and microenvironment signals, in order to prevent the cancer-immunity cycle. Up arrows indicate increased production.

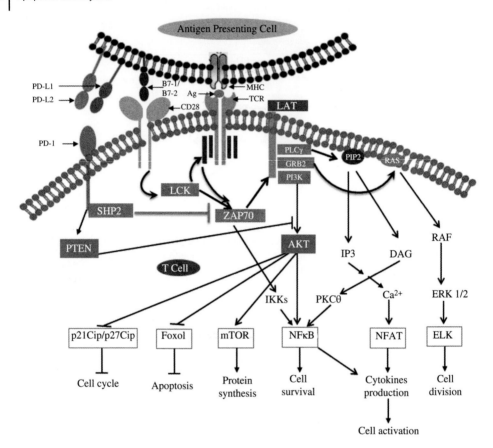

Figure 29.9 Schematic diagram of the molecular mechanism by which the PD-L–PD1 pathway inhibits T-cell receptor (TCR) signaling.

expression of PD-L1 in cancer cells may abrogate activated TCR signaling to inhibit T-cell immune function.

29.10 MicroRNA and PD1 Signaling

MicroRNAs are small, ~22-nucleotide-long, single-stranded, non-coding RNA molecules that bind with the 3′ UTRs of target mRNAs to inhibit protein production, either by suppressing translation or by augmenting mRNA cleavage [60,61]. The role of microRNAs in cancer progression and metastasis has already been established. Dysregulation of microRNAs promotes many pathological functions. Some microRNAs (such as miR-34 and miR-122) are now being used in clinical trials. In the near future, microRNAs could be used as a potential therapy for the treatment of a number of different diseases. In fact, the effects of several microRNAs have been implicated in the regulation of the PD-L1–PD1 signaling pathway. For example, miR-513 binds to the 3′ UTR of PD-L1 mRNA and inhibits its expression in epithelial cholangiocyte cells. In fact, basal mRNA expression – but not the protein of PD-L1 – is found in resting cholangiocyte cells, but INFγ treatment downregulates miR-513 to increase the PD-L1 protein level [62].

Similarly, in retinoblastoma cancer, overexpression of miR-513a downregulates PD-L1 mRNA and protein by directly interacting with the 3′ UTR [63]. However, the expression of miR-513 does not show any correlation with the expression of PD-L1 protein in vestibular schwannoma tumors [64]. One study reported that miR-20b, miR-21, and miR-130b inhibit PTEN expression in order to upregulate PD-L1 expression in advanced colorectal cancer [65]. Moreover, single-nucleotide polymorphisms (SNPs) have been reported on the miR-570 binding site of the 3′ UTR of PD-L1 mRNA in gastric cancers. It was seen stated that this polymorphosim disrupted the binding between miR-570 and the 3′ UTR of PD-L1 to upregulate PD-L1 expression [66]. It has also been documented that miR-4717 expression downregulates PD1 expression in lymphocytes from chronic hepatitis B virus patients [67]. The same study further documented that miR-4717 directly interacts with the 3′ UTR of PD1 mRNA to inhibit its expression. It was previously reported that PD1 deficiency increases autoimmune diseases, whereas miR-155 KO mice show a high resistance to autoimmune encephalomyelitis. However, double-knockout miR-155 and PD1 again show susceptibility to autoimmune diseases [68]. These data indicate that this relationship between miR-155 and PD1 may play a significant role in modulating T-cell function. Several reports provide more evidence that $PD1^{-/-}$ antigen-specific CD4+ T cells show upregulation of miR-21 expression. Moreover, siRNA-mediated downregulation of PD1 increases miR-21 expression in CD4+ T cells [69]. It is established from a number of different studies that miR-21 binds to the 3′ UTR of PDCD4 to inhibit apoptosis. Thus, it could be the case that the blocking of PD1 in T cells might upregulate miR-21 to prevent apoptosis of T cells by downregulating PDCD4. In the near future, researchers may discover some special microRNAs that inhibit the PD-L1–PD1 pathway, as well as some key oncogenic signaling molecules that enable these microRNAs to block the oncogenic signaling of tumor cells and inhibit the PD1–PD-L1 pathway of T cells. Thus, in Table 29.2, we have listed all predicted microRNAs present in the 3′ UTRs of human PD1, PD-L1, and PD-L2.

Table 29.2 Putative microRNA binding sites in the 3′ UTRs of human PD1, PD-L1, and PD-L2.

PD-1 3′ UTR (Human)		PD-L1 3′ UTR (Human)		PD-L2 3′ UTR (Human)	
Micro RNA	Binding Site	Micro RNA	Binding Site	Micro RNA	Binding Site
hsa-miR-374b	1121,1132	hsa-miR-15a	554	hsa-miR-519d	83 110, 223
hsa-miR-23b*	898	hsa-miR-15b	554	hsa-miR-93	84,110, 223
hsa-miR-23a*	901	hsa-miR-16	553	hsa-miR-106b	85, 113, 225
hsa-miR-374a	1121	hsa-miR-424	553	hsa-miR-20a	84, 110, 223
hsa-miR-4290	786, 843	hsa-miR-195	555	hsa-miR-20b	84, 110, 223
hsa-miR-205*	913	hsa-miR-497	555	hsa-miR-106a	84, 111, 223
hsa-miR-670	547	hsa-miR-145	583	hsa-miR-17 84	111, 223
hsa-miR-3184	734	hsa-miR-383	1	hsa-miR-340	1034, 1283
hsa-miR-423-5p	735	hsa-miR-339-5p	241	hsa-miR-130a	80
hsa-miR-4253	614	hsa-miR-429	420	hsa-miR-130b	83

(continued)

Table 29.2 (Continued)

PD-1 3′ UTR (Human)		PD-L1 3′ UTR (Human)		PD-L2 3′ UTR (Human)	
Micro RNA	Binding Site	Micro RNA	Binding Site	Micro RNA	Binding Site
hsa-miR-544b	49	hsa-miR-200b	420	hsa-miR-758	622
hsa-miR-661	185,612	hsa-miR-200c	416	hsa-miR-19a	78, 888
hsa-miR-185*	146	hsa-miR-422a	398	hsa-miR-19b	78, 888
hsa-miR-220c	1110	hsa-miR-378	400	hsa-miR-301a	79
hsa-miR-608	723	hsa-miR-384	303	hsa-miR-301b	79
hsa-miR-4256	556	hsa-miR-141	404	hsa-miR-455-5p	1089
hsa-miR-2861	693	hsa-miR-200a	404	hsa-miR-329	350
hsa-miR-4302	1027	hsa-miR-125a-3p	23	hsa-miR-362-3p	350
hsa-miR-4265	177	hsa-miR-324-5p	83	hsa-miR-454	81
hsa-miR-4296	179	hsa-miR-873	341	hsa-miR-182	384
hsa-miR-4322	176	hsa-miR-544	342	hsa-miR-203	790
hsa-miR-140-3p	1113	hsa-miR-194	523	hsa-miR-28-5p	1136
hsa-miR-671-5p	1	hsa-miR-421	500	hsa-miR-708	1137
hsa-miR-147	1022	hsa-miR-503	552	hsa-miR-150	1
hsa-miR-612	609	hsa-miR-513a-5p	238	hsa-miR-186	1030, 1283
hsa-miR-1285	612	hsa-miR-513b	487, 504	hsa-miR-520a-3p	111
hsa-miR-942	536			hsa-miR-520d-3p	109
hsa-miR-4313	784			hsa-miR-520b	110
hsa-miR-4270	279			hsa-miR-520c-3p	109
hsa-miR-1908	988			hsa-miR-520e	112
hsa-miR-663	991			hsa-miR-373	108
hsa-miR-18a*	839			hsa-miR-302a	110
hsa-miR-4267	681			hsa-miR-302b	110
hsa-miR-3151	147			hsa-miR-302c	110
hsa-miR-548s	1			hsa-miR-302e	116
hsa-miR-944	1129			hsa-miR-372	110
hsa-miR-1256	172			hsa-miR-302d	110
hsa-miR-342-5p	1029			hsa-miR-339-5p	462
hsa-miR-588	822			hsa-miR-410	1288
hsa-miR-3189	775			hsa-miR-539	1213
hsa-miR-492	244			hsa-miR-145	507
hsa-miR-1915	881			hsa-miR-143	1209
				hsa-miR-590-3p	700
				hsa-miR-149	1
				hsa-miR-411	1296
				hsa-miR-876-5p	946

29.11 Targeting of the PD-L–PD1 Pathway as a Promising Cancer Immunotherapy

Immunotherapy has been used as a treatment for cancer patients for a few decades. However, the results are not very promising, and this therapy is restricted to specific cancer types. For example, interferon alpha (IFNα) and IL-2 have been used to treat metastatic renal and melanoma cancer patients, but toxicity and the low response of these agents limit their application [70,71]. The immune system of the cancer microenvironment differs greatly from the normal physiologic immune system. Cancer microenvironment-released signal-driven oncogenic signaling in cancer cells largely disturbs the normal physiologic function of immune cells present in the tumor microenvironment. Normally, cancer tissues produce some factors that reduce the immunogenic load to cancer tissues. Cancer cells may evade the immune attack by inducing T-cell anergy, T-cell nonresponsiveness, T-cell exhaustion, suppressor T cells, and T-cell apoptosis, or by inhibiting T-cell proliferation. Binding of PD-L1 to the PD1 receptor provides an inhibitory signal to T cells for its inactivation. This provides a molecular shield to cancer cells against the cytotoxicity of CD8+ T cells. Moreover, all types of cancer tissue may express PD-L1 ligand on the cell surface. Thus, targeting of the PD-L1–PD1 pathway seems a promising immunotherapy for cancer treatment. In fact, many drugs (monoclonal antibodies) that block the interaction of PD-L1 and PD1 are currently being used in clinical trials (Table 29.3). Early results of these trials show a very promising response in different types of cancer patients [2,72,73]. In 2010, the Brahmer research group was the first to use an anti-PD-L1 antibody (fully human IgG4-blocking monoclonal antibody against PD1, BMS-936559, Nivolumab) in a human clinical (phase I) trial for a few cancer patients. Safety and pharmacodynamics studies showed that patients tolerated this antibody well, and this treatment had quite a good response [74].

Based on these results, Brahmer et al. [77] conducted the same study using a greater number of types of advanced cancer, and their results were quite promising (trial number: NCT00729664). In 2014, they reported, based on this continuation study, that responses lasted for longer, persisting even after the discontinuation of the drug, and that long-term safety was assured [81]. They observed that 28 and 27% of melanoma and renal cancer patients, respectively, responded to nivolumab, but that response rates were lower in the case of non-small-cell lung cancer. Robert et al. [82] reported, based on a phase III clinical trial in melanoma patients without BRAF mutation, that nivolumab showed better response and lower toxic effects as compared to dacarbazine. They also found that the observed response rate was greater in the case of PD-L1-positive cancer patients as compared to PD-L1-negative patients.

Similarly, anti-PD1 antibody (humanized IgG1-blocking monoclonal antibodies against PD1, Pidilizumab, CT-011) was first used in a phase I clinical trial with hematologic malignancy patients; this study reported that patients tolerated different doses of this antibody well [83]. A phase II clinical trial showed that its response rate was relatively lower than that of other antibodies, such as nivolumab [76].

Pembrolizumab (MK-3475), a humanized PD1 IgG4 isotype antibody, is currently in clinical trial. A follow-up study reported a large (37–38%) response rate in advanced melanoma patients [75,84]. Engineered human IgG1 anti-PD-L1 monoclonal antibody is also in clinical trial. It shows less antibody-dependent cell-mediated cytotoxicity (ADCC) compared to nivolumab and pebbrolizumab [72,78].

Table 29.3 Monoclonal antibodies of PD1 and PD-L1 currently in clinical trial.

Drugs	Target Protein	Nature/Type of Antibody	Company Name	Approval/Clinical Trial	References
Nivolumab, BMS-936558, MDX-1106, ONO-4538	PD-1	Fully human IgG4, mutation S228P	Bristol-Myers Squibb	Melanoma Squamous non-small-cell lung cancer Hodgkin's lymphoma	[74]
Pembrolizumab (MK-3475)	PD-1	Humanized IgG4 kappa, mutation S228P	Merck	Melanoma Non-small cell lung cancer	[75]
Pidilizumab, CT-011	PD-1	Humanized IgG1-kappa	CureTech	Hematological malignancy patients (clinical trial)	[76]
AMP-224	PD-1	Fusion protein, blocks PD-L–PD1 interaction	GSK	Metastatic colorectal cancer	[72,73]
BMS-936559	PD-L1	Human IgG4, S228P	Bristol-Myers Squibb	Patients with advanced cancer (melanoma, renal-cell cancer, non-small-cell lung cancer)	[77]
MSB-0010718C	PD-L1	Fully human IgG1	Merck	Merkel cell carcinoma	[72]
MPDL3280A	PD-L1	Engineered human IgG1 (fragment crystallizable (Fc) domain)	Roche	Metastatic melanoma, non-small-cell lung cancer	[78,79]
MEDI-4736	PD-L1	Human IgG1	MedImmune, AstraZeneca	Non-small-cell lung cancer, melanoma	[72,80]

29.12 Adoptive Cell Transfer for Suppression of Tumor Growth

In an adoptive cell-transfer (ACT) method, tumor-specific T cells are often isolated from tumor-infiltrating lymphocytes cells, especially from melanoma cancer patients. These isolated T cells are activated and expanded *in vitro*, and subsequently transplanted to the patients [4].

ACT of T lymphocyte-engineered cells has been implicated in the treatment of cancers [85–87]. T-lymphocyte cells are the main player in the human immune system, and orchestrate cell-mediated immunity by the expression of cytokines in response to a diseased microenvironment and cancer conditions. B-lymphocyte cells are associated with the production of antibodies and are characterized by the presence of CD19 surface marker over the B cells. In certain leukemia and lymphoma conditions, B-lymphocyte cell-surface markers show defects in the organization of CD19 protein and cause immunodeficiency characterized by hypogammaglobulinemia and the inability to mount an antibody response against an antigen or against cancerous cells. CD19 cell-surface markers over B immune-cell deficiency lead to the development of cancers such as acute lymphoblastic leukemia (ALL), B-cell non-Hodgkin's lymphoma, and CLL [88]. In diseased B-cell conditions, T cells become the workhorses of the immune system and develop cell-mediated immunity. In cancer, B cells navigate detection by T cells; therefore, cell-mediated immunity is not sufficient to kill cancerous B cells. To fight against this condition, T cell-based immunotherapy is the most efficient therapeutic option. In T-cell immunotherapy, patient T cells are genetically engineered in such a way that they can recognize and bind to CD19 protein on the surface of cancerous B cells. The genetically engineered T cells are called "chimeric antigen-receptor T cells." When such chimeric antigen-receptor T cells are kept back in the patient peripheral blood, they can recognize rapidly dividing cancerous B cells by forming receptor–ligand bonds; as a result, the cancerous B cells are killed and removed from the patient blood system. Increasing scientific evidence in support of T cell-based immunotherapy has the potential to control blood cancer in human beings.

Recognition of malignant cells by CTL and natural killer (NK) cells depends upon the human leukocyte antigen (HLA) class. It has been shown that downregulation of HLA class I in malignant cells can help cancer cells to escape from the T-lymphocyte cells and from cell-mediated immunity. Such downregulation can be the result of several factors, including mutation in HLA class I or a defect in the antigen-presentation or the proteosomal protein-degradation process. These factors negatively affect the interaction between immune cells (e.g., NK cells, dendritic cells, CTL cells) and malignant cells. With the help of genetic engineering, a patient T cell can be modified in such a way that its binding with immune cells is enhanced; this process can allow control over tumor formation [89]. Malignant tissues are capable of expressing a number of weak signals in the form of antigens and self-antigens. These antigens are controlled by the activity of self-tolerance of CD4+ CD25+Foxp3+ regulatory T (Treg) cells. Treg cells reduce the immunity of T cells against self- and tumor-associated antigen signals. The population of Treg cells increases at the time of carcinoma formation in the peripheral blood. Therefore, inhibiting the activity of Treg cells using a chemical inhibitor such as cyclophosphamide can help in the development of T cell-based immunotherapy and so provide protection from malignant tissues [90]. Fully activated T cells may sometimes recognize the self-diversity of antigens. The activation of T cells depends on the interaction between TCR stimulation and the activation of co-stimulatory signals such as CD28 and CTLA4 molecules [91]. Here, CD28 is a positive and CTLA4 is a negative co-stimulatory signal for the interaction of T cells with APCs. Therefore, blockage of CTLA4 enhances T-cell responses in various ways, providing protection against the tumor microenvironment.

Several lines of evidence show that T-cell engineering based on PD-L–PD1 signaling holds great promise for T cell-based adoptive immunotherapy [87]. For instance, Ankri et al. [92] used an engineered T cell in which the extracellular part of PD1 was fused with the signaling part of the co-stimulatory CD28 molecule. Such human T cell-expressing PD1/CD28 chimeric molecules were co-cultured with tumor cells, and this co-culture showed increased cytokine secretion. Moreover, these engineered T cells showed superior anti-tumor function against human melanoma tumors [92]. On the other hand, it is also documented that CTLA4- or PD1-based antigen-specific inhibitory chimeric antigen receptors restrict cytokine secretion and proliferation induced by endogenous T cell-receptor signaling, so that these engineered T cells may be used in the clinical application of autologous and allogeneic T-cell therapies [93]. In brief, changing the specificity and function of T cells through engineering seems to be a promising avenue of immunotherapy against a number of diseases, including cancer.

29.13 Conclusion

Immunotherapy-based treatment strategies have been implicated in the treatment of several diseases, including cancer, viral and bacterial infection, and autoimmune diseases. Many clinical trials using anti-PD1 and anti-PD-L1 monoclonal antibodies are now ongoing. Early results of phase I, phase II, and phase III studies hold great promise for the treatment of many cancers, including melanoma, renal, and lung. Because of the large positive response rate, some of these monoclonal antibodies, such as nivolumab and pembrolizumab, have already been approved by the US Food and Drug Administration (FDA) for the treatment of several types of cancer, including advanced melanoma, non-small-cell lung cancer, and Hodgkin's lymphoma (Table 29.3).

All these successes are coming after the identification of the PD-L–PD1 pathway as a co-inhibitor of TCR signaling. Most of the studies, but not all, showed that engagement of the PD1 receptor of the T cell by PD-L or treatment of PD1 ligands on activated T cells led to induction of apoptosis, inhibition of proliferation, and cell-cycle progression of T cells. Some studies show contrasting results, where the PD-L–PD1 pathway increases T-cell proliferation. Many have shown that PD1 expression on the T-cell surface is increased only upon T-cell activation. Many indicate that the PD-L–PD1 pathway can inhibit cell proliferation or T-cell exhaustion only when T cells are suboptimally active (e.g., in the presence of anti-CD3 and the absence of anti-CD28). This pathway may fail to induce apoptosis if a T cell is optimally active (in the presence of both anti-CD3 and anti-CD28). The activity of this pathway depends on the cytokine milieu present in the tumor microenvironment, and varies from tissue type to tissue type.

Many studies have been carried out in order to discover the correlation between expression of PD-L1 and prognosis of cancer patients. A large percentage of cancer tissues show an elevation of PD-L1 expression, which positively correlates with cancer progression if patients are treated with an anti-PD1- or anti-PD-L1-based antibody. Several clinical trials have been restricted to PD-L1-positive patients, since they may give a better objective response rate as compared to PD-L1-negative patients. However, the standard level (range) of PD-L1 expression is not yet established and the best PD-L1 antibody to use in immune histochemistry has not been determined. Such a standard level is needed so that every pathology laboratory can effectively report the positive or negative effects of PD-L1 on cancer tissues. One of the leading questions concerns the development of a biomarker that can predict the response of these therapies.

The role of a few microRNAs, such as miR-513 and miR-570, in inhibiting the PD-L–PD1 pathway has also been discussed. Future study will confirm whether microRNAs that target this pathway can be used in cancer therapeutics. It would be best to select those microRNAs (see Table 29.2) that inhibit the PD-L–PD1 pathway by directly targeting PD1 or PD-L 3′ UTR, and which inhibit key oncogenic signaling, as they will better prevent tumorigenesis by activating T cells and inhibiting the oncogenic signal of cancer cells. Engineered T cells expressing these microRNAs can be used to provide ACT for the suppression of tumor growth.

It will be important to determine whether treatment involving blockade of the PD-L–PD1 pathway induces autoimmune diseases. Future studies will confirm whether this treatment strategy can be successfully used for those cancer patients who have developed autoimmune diseases such as diabetes type I and arthritis.

In brief, blocking of the PD-L–PD1 pathway of activated T cells inhibits apoptosis and induces T-cell activation, which may promote CTL-mediated cancer cell killing.

References

1. Feig C, Peter ME. How apoptosis got the immune system in shape. *Eur J Immunol* 2007;**37**(S1):S61–70.
2. Sharma P, Allison JP. The future of immune checkpoint therapy. *Science* 2015;**348**(6230):56–61.
3. Chen DS, Mellman I. Oncology meets immunology: the cancer-immunity cycle. *Immunity* 2013;**39**(1):1–10.
4. Kershaw MH, Westwood JA, Darcy PK. Gene-engineered T cells for cancer therapy. *Nat Rev Cancer* 2013;**13**(8):525–41.
5. Francisco LM, Sage PT, Sharpe AH. The PD-1 pathway in tolerance and autoimmunity. *Immunol Rev* 2010;**236**(1):219–42.
6. Dong H, Strome SE, Salomao DR, Tamura H, Hirano F, Flies DB, et al. Tumor-associated B7-H1 promotes T-cell apoptosis: a potential mechanism of immune evasion. *Nat Med* 2002;**8**(8):793–800.
7. Freeman GJ, Long AJ, Iwai Y, Bourgue K, Chernova T, Nishimura H, et al. Engagement of the PD-1 immunoinhibitory receptor by a novel B7 family member leads to negative regulation of lymphocyte activation. *J Exp Med* 2000;**192**(7):1027–34.
8. Reinherz EL, Schlossman SF. The differentiation and function of human T lymphocytes. *Cell* 1980;**19**(4):821–7.
9. Ishida Y, Agata Y, Shibahara K, Honjo T. Induced expression of PD-1, a novel member of the immunoglobulin gene superfamily, upon programmed cell death. *EMBO J* 1992;**11**(11):3887.
10. Keir ME, Butte MJ, Freeman GJ, Sharpe AH. PD-1 and its ligands in tolerance and immunity. *Annu Rev Immunol* 2008;**26**:677–704.
11. Lin DY, Tanaka Y, Iwasaki M, Gittis AG, Su HP, Mikami B, et al. The PD-1/PD-L1 complex resembles the antigen-binding Fv domains of antibodies and T cell receptors. *Proc Natl Acad Sci USA* 2008;**105**(8):3011–16.
12. Lázár-Molnár E, Yan Q, Cao E, Ramagopal U, Nathenson SG, Almo SC. Crystal structure of the complex between programmed death-1 (PD-1) and its ligand PD-L2. *Proc Natl Acad Sci USA* 2008;**105**(30):10 483–8.
13. Okazaki T, Iwai Y, Honjo T. New regulatory co-receptors: inducible co-stimulator and PD-1. *Curr Opin Immunol* 2002;**14**(6):779–82.
14. Agata Y, Kawasaki A, Nishimura H, Ishida Y, Tsubata T, Yagita H, Honjo T. Expression of the PD-1 antigen on the surface of stimulated mouse T and B lymphocytes. *Int Immunol* 1996;**8**(5):765–72.
15. Okazaki T, Maeda A, Nishimura H, Kurosaki T, Honjo T. PD-1 immunoreceptor inhibits B cell receptor-mediated signaling by recruiting src homology 2-domain-containing tyrosine phosphatase 2 to phosphotyrosine. *Proc Natl Acad Sci USA* 2001;**98**(24):13 866–71.

16 Chemnitz JM, Parry RV, Nichols KE, June CH, Riley JL. SHP-1 and SHP-2 associate with immunoreceptor tyrosine-based switch motif of programmed death 1 upon primary human T cell stimulation, but only receptor ligation prevents T cell activation. *J Immunol* 2004;**173**(2):945–54.

17 Sheppard K-A, Fitz LJ, Lee JM, Benander C, George JA, Wooters J, et al. PD-1 inhibits T-cell receptor induced phosphorylation of the ZAP70/CD3ζ signalosome and downstream signaling to PKCθ. *FEBS Lett* 2004;**574**(1):37–41.

18 Nishimura H, Minato N, Nakano T, Honjo T. Immunological studies on PD-1 deficient mice: implication of PD-1 as a negative regulator for B cell responses. *Int Immunol* 1998;**10**(10):1563–72.

19 Nishimura H, Nose M, Hiai H, Minato N, Honjo T. Development of lupus-like autoimmune diseases by disruption of the PD-1 gene encoding an ITIM motif-carrying immunoreceptor. *Immunity* 1999;**11**(2):141–51.

20 Nishimura H, Okazaki T, Tanaka Y, Nakatani K, Hara M, Matsumori A, et al. Autoimmune dilated cardiomyopathy in PD-1 receptor-deficient mice. *Science* 2001;**291**(5502):319–22.

21 Kroner A, Mehling M, Hemmer B, Rieckmann P, Toyka KV, Mäurer M, Wiendl H. A PD-1 polymorphism is associated with disease progression in multiple sclerosis. *Ann Neurol* 2005;**58**(1):50–7.

22 Hatachi S, Iwai Y, Kawano S, Morinobu S, Kobayashi M, Koshiba M, et al. CD4+ PD-1+ T cells accumulate as unique anergic cells in rheumatoid arthritis synovial fluid. *J Rheumatol* 2003;**30**(7):1410–19.

23 Nielsen C, Hansen D, Husby S, Jacobsen B, Lillevang ST. Association of a putative regulatory polymorphism in the PD-1 gene with susceptibility to type 1 diabetes. *Tissue Antigens* 2003;**62**(6):492–7.

24 Sharpe AH, Freeman GJ. The B7–CD28 superfamily. *Nat Rev Immunol* 2002;**2**(2):116–26.

25 Nishimura H, Agata Y, Kawasaki A, Sato M, Inamura S, Minato N, Yagita H, et al. Developmentally regulated expression of the PD-1 protein on the surface of double-negative (CD4–CD8–) thymocytes. *Int Immunol* 1996;**8**(5):773–80.

26 Liang SC, Latchman YE, Buhlmann JE, Tomczak MF, Horwitz BH, Freeman GJ, Sharpe AH. Regulation of PD-1, PD-L1, and PD-L2 expression during normal and autoimmune responses. *Eur J Immunol* 2003;**33**(10):2706–16.

27 Iwai Y, Okazaki T, Nishimura H, Kawasaki A, Yagita H, Honjo T. Microanatomical localization of PD-1 in human tonsils. *Immunol Lett* 2002;**83**(3):215–20.

28 Polanczyk MJ, Hopke C, Vandenbark AA, Offner H. Estrogen-mediated immunomodulation involves reduced activation of effector T cells, potentiation of treg cells, and enhanced expression of the PD-1 costimulatory pathway. *J Neurosci Res* 2006;**84**(2):370–8.

29 Freeman GJ, Wherry EJ, Ahmed R, Sharpe AH. Reinvigorating exhausted HIV-specific T cells via PD-1–PD-1 ligand blockade. *J Exp Med* 2006;**203**(10):2223–7.

30 Kinter AL, Godbout EJ, McNally JP, Sereti I, Roby GA, O'Shea MA, Fauci AS. The common γ-chain cytokines IL-2, IL-7, IL-15, and IL-21 induce the expression of programmed death-1 and its ligands. *J Immunol* 2008;**181**(10):6738–46.

31 Yao S, Wang S, Zhu Y, Luo L, Zhu G, Flies S, et al. PD-1 on dendritic cells impedes innate immunity against bacterial infection. *Blood* 2009;**113**(23):5811–18.

32 Oestreich KJ, Yoon H, Ahmed R, Boss JM. NFATc1 regulates PD-1 expression upon T cell activation. *J Immunol* 2008;**181**(7):4832–9.

33. Dong H, Zhu G, Tamada K, Chen L. B7-H1, a third member of the B7 family, co-stimulates T-cell proliferation and interleukin-10 secretion. *Nat Med* 1999;**5**(12):1365–9.
34. Pentcheva-Hoang T, Corse E, Allison JP. Negative regulators of T-cell activation: potential targets for therapeutic intervention in cancer, autoimmune disease, and persistent infections. *Immunol Rev* 2009;**229**(1):67–87.
35. Latchman Y, Wood CR, Chernova T, Chaudhary D, Borde M, Chernova I, et al. PD-L2 is a second ligand for PD-1 and inhibits T cell activation. *Nat Immunol* 2001;**2**(3):261–8.
36. Brown JA, Dorfman DM, Ma FR, Sullivan EL, Munoz O, Wood CR, et al. Blockade of programmed death-1 ligands on dendritic cells enhances T cell activation and cytokine production. *J Immunol* 2003;**170**(3):1257–66.
37. Butte MJ, Keir ME, Phamduy TB, Sharpe AH, Freeman GJ. Programmed death-1 ligand 1 interacts specifically with the B7-1 costimulatory molecule to inhibit T cell responses. *Immunity* 2007;**27**(1):111–22.
38. Marzec M, Zhang Q, Goradia A, Raghunath PN, Liu X, Paessler M, et al. Oncogenic kinase NPM/ALK induces through STAT3 expression of immunosuppressive protein CD274 (PD-L1, B7-H1). *Proc Natl Acad Sci USA* 2008;**105**(52):20 852–7.
39. Lee S-J, Jang B-C, Lee S-W, et al. Interferon regulatory factor-1 is prerequisite to the constitutive expression and IFN-γ-induced upregulation of B7-H1 (CD274). *FEBS Lett* 2006;**580**(3):755–62.
40. Parsa AT, Waldron JS, Panner A, Crane CA, Parney IF, Barry JJ, et al. Loss of tumor suppressor PTEN function increases B7-H1 expression and immunoresistance in glioma. *Nat Med* 2007;**13**(1):84–8.
41. Loke PN, Allison JP. PD-L1 and PD-L2 are differentially regulated by Th1 and Th2 cells. *Proc Natl Acad Sci USA* 2003;**100**(9):5336–41.
42. Salih HR, Wintterle S, Krusch M, Kroner A, Huang YH, Chen L, Wiendl H. The role of leukemia-derived B7-H1 (PD-L1) in tumor–T-cell interactions in humans. *Exp Hematol* 2006;**34**(7):888–94.
43. Riches JC, Davies JK, McClanahan F, Fatah R, Iqbal S, Agrawal S, et al. T cells from CLL patients exhibit features of T-cell exhaustion but retain capacity for cytokine production. *Blood* 2013;**121**(9):1612–21.
44. Wang W, Lau R, Yu D, Zhu W, Korman A, Weber J. PD1 blockade reverses the suppression of melanoma antigen-specific CTL by CD4+ CD25Hi regulatory T cells. *Int Immunol* 2009: **21**(9):1065–77.
45. Malaspina TS, Gasparoto TH, Costa MR, de Melo EF. Jr. Ikoma MR, Damante JH, et al. Enhanced programmed death 1 (PD-1) and PD-1 ligand (PD-L1) expression in patients with actinic cheilitis and oral squamous cell carcinoma. *Cancer Immunol Immunother* 2011;**60**(7):965–74.
46. Hatam LJ, DeVoti JA, Rosenthal DW, Lam F, Abramson AL, Steinberg BM, Bonagura VR. Immune suppression in premalignant respiratory papillomas: enriched functional CD4+ Foxp3+ regulatory T cells and PD-1/PD-L1/L2 expression. *Clin Cancer Res* 2012;**18**(7):1925–35.
47. Shen JK, Cote GM, Choy E, Yang P, Harmon D, Schwab J, et al. Programmed cell death ligand 1 expression in osteosarcoma. *Cancer Immunol Res* 2014;**2**(7):690–8.
48. Waki K, Yamada T, Yoshiyama K, Terazaki Y, Sakamoto S, Matsueda S, et al. PD-1 expression on peripheral blood T-cell subsets correlates with prognosis in non-small cell lung cancer. *Cancer Sci* 2014;**105**(10):1229–35.

49. D'Incecco A, Andreozzi M, Ludovini V, Rossi E, Capodanno A, Landi L, et al. PD-1 and PD-L1 expression in molecularly selected non-small-cell lung cancer patients. *Br J Cancer* 2015;**112**(1):95–102.
50. Lin C, Chen X, Li M, Liu J, Qi X, Yang W, et al. Programmed death-ligand 1 expression predicts tyrosine kinase inhibitor response and better prognosis in a cohort of patients with epidermal growth factor receptor mutation-positive lung adenocarcinoma. *Clin Lung Cancer* 2015;**16**(5):e25–35.
51. Ghebeh H, Mohammed S, Al-Omair A, Qattan A, Lehe C, Al-Qudaihi G, et al. The B7-H1 (PD-L1) T lymphocyte-inhibitory molecule is expressed in breast cancer patients with infiltrating ductal carcinoma: correlation with important high-risk prognostic factors. *Neoplasia* 2006;**8**(3):190–8.
52. Schalper KA, Velcheti V, Carvajal D, Wimberly H, Brown J, Pusztai L, Rimm DL. In situ tumor PD-L1 mRNA expression is associated with increased TILs and better outcome in breast carcinomas. *Clin Cancer Res* 2014;**20**(10):2773–82.
53. Soliman H, Khalil F, Antonia S. PD-L1 expression is increased in a subset of basal type breast cancer cells. *PloS One* 2014;**9**(2):e88557.
54. Sabatier R, Finetti P, Mamessier E, Adelaide J, Chaffanet M, Ali HR, et al. Prognostic and predictive value of PDL1 expression in breast cancer. *Oncotarget* 2015;**6**(7):5449–64.
55. Smeltzer JP, Jones JM, Ziesmer SC, Grote DM, Xiu B, Ristow KM, et al. Pattern of CD14+ follicular dendritic cells and PD1+ T cells independently predicts time to transformation in follicular lymphoma. *Clin Cancer Res* 2014;**20**(11):2862–72.
56. Kim JR, Moon YJ, Kwon KS, Bae JS, Wagle S, Kim KM, et al. Tumor infiltrating PD1-positive lymphocytes and the expression of PD-L1 predict poor prognosis of soft tissue sarcomas. 2013;**8**(12):e82870.
57. Shi F, Shi M, Zeng Z, Qi RZ, Liu ZW, Zhang JY, et al. PD-1 and PD-L1 upregulation promotes CD8+ T-cell apoptosis and postoperative recurrence in hepatocellular carcinoma patients. *Int J Cancer* 2011;**128**(4):887–96.
58. Houman Nourkeyhani SG. Immune checkpoint inhibitors for renal cell carcinoma. *J Targeted Ther Cancer* 2014;**3**(5):46–50.
59. Chen L, Flies DB. Molecular mechanisms of T cell co-stimulation and co-inhibition. *Nat Rev Immunol* 2013;**13**(4):227–42.
60. Bartel DP. MicroRNAs: genomics, biogenesis, mechanism, and function. *Cell* 2004;**116**(2):281–97.
61. Lai EC. Micro RNAs are complementary to 3′ UTR sequence motifs that mediate negative post-transcriptional regulation. *Nat Genet* 2002;**30**(4):363–4.
62. Gong A-Y, Zhou R, Hu G, Li X, Splinter PL, O'Hara SP, et al. MicroRNA-513 regulates B7-H1 translation and is involved in IFN-γ-induced B7-H1 expression in cholangiocytes. *J Immunol* 2009;**182**(3):1325–33.
63. Wu L, Chen Z, Zhang J, Xing Y. Effect of miR-513a-5p on etoposide-stimulating B7-H1 expression in retinoblastoma cells. *J Huazhong Univ Sci Technolog Med Sci* 2012;**32**:601–6.
64. Archibald DJ, Neff BA, Voss SG, Splinter PL, Driscoll CL, Link MJ, et al. B7-H1 expression in vestibular schwannomas. *Otol Neurotol* 2010;**31**(6):991–7.
65. Zhu J, Chen L, Zou L, Yang P, Wu R, Mao Y, et al. MiR-20b, -21, and -130b inhibit PTEN expression resulting in B7-H1 over-expression in advanced colorectal cancer. *Hum Immunol* 2014;**75**(4):348–53.

66 Wang W, Li F, Mao Y, Zhou H, Sun J, Li R, et al. A miR-570 binding site polymorphism in the B7-H1 gene is associated with the risk of gastric adenocarcinoma. *Hum Genet* 2013;**132**(6):641–8.

67 Zhang G, Li N, Li Z, Zhu Q, Li F, Yang C, et al. microRNA-4717 differentially interacts with its polymorphic target in the PD1 3′ untranslated region: a mechanism for regulating PD-1 expression and function in HBV-associated liver diseases. *Oncotarget* 2015;**6**(22):18 933–44.

68 Zhang J, Braun MY. PD-1 deletion restores susceptibility to experimental autoimmune encephalomyelitis in miR-155-deficient mice. *Int Immunol* 2014: **26**(7):407–15.

69 Iliopoulos D, Kavousanaki M, Ioannou M, Boumpas D, Verginis P. The negative costimulatory molecule PD-1 modulates the balance between immunity and tolerance via miR-21. *Eur J Immunol* 2011;**41**(6):1754–63.

70 Medical Research Council Renal Cancer Collaborators. Interferon-α and survival in metastatic renal carcinoma: early results of a randomised controlled trial. *Lancet* 1999;**353**(9146):14–17.

71 Rosenberg SA, Yang JC, Topalian SL, Schwartzentruber DJ, Weber JS, Parkinson DR, et al. Treatment of 283 consecutive patients with metastatic melanoma or renal cell cancer using high-dose bolus interleukin 2. *JAMA* 1994;**271**(12):907–13.

72 Mahoney KM, Freeman GJ, McDermott DF. The next immune-checkpoint inhibitors: PD-1/PD-L1 blockade in melanoma. *Clin Ther* 2015;**37**(4):764–82.

73 Kim JW, Eder JP. Prospects for targeting PD-1 and PD-L1 in various tumor types. *Oncology (Williston Park)* 2014;**28**(Suppl. 3):15–28.

74 Brahmer JR, Drake CG, Wollner I, Powderly JD, Picus J, Sharfman WH, et al. Phase I study of single-agent anti–programmed death-1 (MDX-1106) in refractory solid tumors: safety, clinical activity, pharmacodynamics, and immunologic correlates. *J Clin Oncol* 2010;**28**(19):3167–75.

75 Hamid O, Robert C, Daud A, Hodi FS, Hwu WJ, Kefford R, Wolchok JD, et al. Safety and tumor responses with lambrolizumab (anti-PD-1) in melanoma. *N Engl J Med* 2013;**369**(2):134–44.

76 Atkins MB, Kudchadkar RR, Sznol M, McDermott DF, Lotem M, Schachter J, et al. Phase 2, multicenter, safety and efficacy study of pidilizumab in patients with metastatic melanoma. J Clin Oncol 32(15):9001.

77 Brahmer JR, Tykodi SS, Chow LQ, Hwu W-J, Topalian SL, Hwu P, et al. Safety and activity of anti-PD-L1 antibody in patients with advanced cancer. *N Engl J Med* 2012;**366**(26):2455–65.

78 Hamid O, Sosman JA, Lawrence DP, Sullivan RJ, Ibrahim N, Kluger HM, et al. Clinical activity, safety, and biomarkers of MPDL3280A, an engineered PD-L1 antibody in patients with locally advanced or metastatic melanoma (mM). *J Clin Oncol* **31**(15):9010.

79 Soria J, Cruz C, Bahleda R, Scoria JC, Cruz C, Delord JP, et al. Clinical activity, safety and biomarkers of PD-L1 blockade in non-small cell lung cancer (NSCLC): additional analyses from a clinical study of the engineered antibody MPDL3280A (anti-PDL1). *Eur J Cancer* 2013.

80 Lutzky J, Antonia SJ, Blake-Haskins A, Li X, Robbins PB, Shalabi AM, et al. A phase 1 study of MEDI4736, an anti-PD-L1 antibody, in patients with advanced solid tumors. *J Clin Oncol* **32**(15):3001.

81 Topalian SL, Sznol M, McDermott DF, Kluger HM, Carvajal RD, Sharfman WH, et al. Survival, durable tumor remission, and long-term safety in patients with advanced melanoma receiving nivolumab. *J Clin Oncol* 2014;**32**(10):1020–30.

82 Robert C, Long GV, Brady B, Dutriaux C, Maio M, Mortier L, et al. Nivolumab in previously untreated melanoma without BRAF mutation. *N Engl J Med* 2015;**372**(4):320–30.

83 Berger R, Rotem-Yehudar R, Slama G, Landes S, Kneller A, Leiba M, et al. Phase I safety and pharmacokinetic study of CT-011, a humanized antibody interacting with PD-1, in patients with advanced hematologic malignancies. *Clin Cancer Res* 2008;**14**(10):3044–51.

84 Kefford R, Ribas A, Hamid O, Robert C, Daud A, Wolchok JD, et al. Clinical efficacy and correlation with tumor PD-L1 expression in patients (pts) with melanoma (MEL) treated with the anti-PD-1 monoclonal antibody MK-3475. *J Clin Oncol* **32**(15):3005.

85 Hawkins RE, Gilham DE, Debets R, Eshhar Z, Taylor N, Abken H, et al. Development of adoptive cell therapy for cancer: a clinical perspective. *Hum Gene Ther* 2010;**21**(6):665–72.

86 Restifo NP, Dudley ME, Rosenberg SA. Adoptive immunotherapy for cancer: harnessing the T cell response. *Nat Rev Immunol* 2012;**12**(4):269–81.

87 Daniel-Meshulam I, Ya'akobi S, Ankri C, Cohen CJ. How (specific) would you like your T-cells today? Generating T-cell therapeutic function through TCR-gene transfer. *Front Immunol* 2012;**3**:186.

88 Wang K, Wei G, Liu D. CD19: a biomarker for B cell development, lymphoma diagnosis and therapy. *Exp Hematol Oncol* 2012;**1**(1):36.

89 Hicklin DJ, Marincola FM, Ferrone S. HLA class I antigen downregulation in human cancers: T-cell immunotherapy revives an old story. *Mol Med Today* 1999;**5**(4):178–86.

90 Ghiringhelli F, Larmonier N, Schmitt E, Parcellier A, Cathelin D, Garrido C, et al. CD4+ CD25+ regulatory T cells suppress tumor immunity but are sensitive to cyclophosphamide which allows immunotherapy of established tumors to be curative. *Eur J Immunol* 2004;**34**(2):336–44.

91 Chambers CA, Kuhns MS, Egen JG, Allison JP. CTLA-4-mediated inhibition in regulation of T cell responses: mechanisms and manipulation in tumor immunotherapy. *Ann Rev Immunol* 2001;**19**(1):565–94.

92 Ankri C, Shamalov K, Horovitz-Fried M, Mauer S, Cohen CJ. Human T cells engineered to express a programmed death 1/28 costimulatory retargeting molecule display enhanced antitumor activity. *J Immunol* 2013;**191**(8):4121–9.

93 Fedorov VD, Themeli M, Sadelain M. PD-1- and CTLA-4-based inhibitory chimeric antigen receptors (iCARs) divert off-target immunotherapy responses. *Sci Transl Med* 2013;**5**(215):215ra172.

Index

A0 *see* thioxotriazole copper(II) complex
A20, necroptosis 118
14-4-4σ, mitotic catastrophe 481–482

a

AA *see* arachidonic acid
accumulation
  ascorbate 662–663
  glutamate 199
ACD *see* autophagic cell death
acetyl-CoA synthetases (ACS) 19
acetyl coenzyme A (acetyl-CoA) 19
ACL *see* ATP-citrate lyase
ACS *see* acetyl-CoA synthetases
actin, entosis 468
actinomycin D 414
activating transcription factors (ATF) 303
activation
  cytokines 325
  DNases 188–189, 645–652, 669–672
  eryptosis 372–373
  T cells 703, 707–710
acute phenoptosis 241–248
  cancer 246–247
  diabetes 247–248
  invertebrates 242–244
  plants 241–242
  septic shock 245–246
  vertebrates 244–248
adaptive immunity 370–371
adenine nucleotide translocase (ANT) 21, 33–34, 382
adenosine triphosphate/adenosine diphosphate (ATP/ADP) ratios 18–19
adherens junctions (AJs) 154, 160–161
adhesomes 150–151, 153
adipose necrosis 94
adoptive cell-transfer (ACT) 714–716

aging
  as chronic phenoptosis 248–250
  dietary restriction 263–267
  evolvability 256–257
  heavy hyaluronan 256
  living conditions 270
  neoteny 253–256
  pharmacological targets 274–275
  physical activity 267–269
  prevention 261–272
  psychological factors 270–272
  small antioxidant doses 274
  stochastic injury 260–261
  vascular, autophagy 79–81
agranulocytes, structure 370
AGT *see* alkylguanine methyltransferase
AIF *see* apoptosis-inducing factor
AIM-2 like receptors (ALRs) 322–324, 328
AIPs *see* inhibitors of apoptosis proteins
AJs *see* adherens junctions
Akt
  anoikis 152–153, 160–161
  T cell signaling 708–710
ALD *see* apoptotic-like death pathway
Alix-CT *see* C-terminal half of AIP1/Alix
alkylating agents
  apoptosis 448–450
  base excision repair 440–441
  classic vs. non-classic 430
  endogenous and environmental 431
  general principles 429–433
  homologous recombination 443–444
  methylative lesion formation 431–432
  O^6-methylguanine cytotoxicity 427–462
  mismatch recognition 442
  non-homologous end joining 444–446
  PARP-1 446–447

alkylating agents (*Continued*)
  pharmacology 450–455
  sites of action 430
alkylguanine methyltransferase
    (AGT) 433–447
  inhibition 436–438
  mechanism of action 434–435
  structure 433
α-amantin (AMA) 414
α-amino-3-hydroxy-5-methyl-isoxazole-4-
    proprionate (AMPA)
    receptors 198
alpha-enolase (ENOA) 356
α-syn 31
α-tubulins 354–355
ALRs *see* AIM-2 like receptors
ALS *see* amyotrophic lateral sclerosis
Alzheimer's disease (AD)
  amyloid plaques 413
  cyclophilin D-dependent necrosis 386
  excitotoxicity 201–202
  mitochondria 31
AMA *see* α-amantin
α-amantin (AMA) 414
American ginseng 358–359
α-amino-3-hydroxy-5-methyl-isoxazole-4-
    proprionate (AMPA)
    receptors 198
amyloidosis 89
amyloid plaques 413
amyloid precursor proteins (APP) 31, 202
amyotrophic lateral sclerosis (ALS) 202
anemia 373
animal models of TRIAD 421
annexin IV-FITC 611–612
annulate lamella 594
anoikis 145–168
  cancer 165–167
  core features 146–154
  coroner's conclusions 161–162
  diseases 165–167
  epidermolysis bullosae 165
  extracellular matrix 149–155
  FAK/Src signaling 150–152
  induction 155–159, 161
  intestinal epithelial cells 162–165
  muscular dystrophies 165
ANT *see* adenine nucleotide translocase
Antechinus stuartii 244
anthracosis 89–90
anticancer drugs
  autophagic responses 76–77

paraptotic 351, 353–354, 356–362
  see also Apatone; cancer; pharmacological
    targets
antigen presentation 370
antiholins 61–63
anti-inflammatory cytokines 212–214
antioxidants
  small doses and aging 274
  see also Apatone; ascorbate
Apaf-1 *see* apoptotic protease activating
    factor-1
Apatone (combined ascorbate/menadione
    treatment/VC:VK3)
  cell cycle effects 600–603
  cytotoxicity 666–672
  DNA effects 669–672
  flow cytometric evaluation 600–603
  karyolysis 637–642
  lipid peroxidation 669
  morphological alterations 598–599,
    603–645
    cell shape and size 611
    membrane defects 611–612
  protein synthesis effects 669–672
  starvation effects 667–668
  ultrastructural damage 612–645
  see also ascorbate; autoschizis; menadione
ApoE *see* apolipoprotein E
apolipoprotein E (ApoE), Wallerian
    degeneration 214
apoptosis
  activation 2
  avoidance in cancer 30–31
  caspases 320–321
  cell-surface signaling 320–321
  ceramides 400
  chemotactic signaling 29
  core concepts 2–3
  crosstalk 118, 330–332
  cytoplasmic pathways 3–7
  definitions 2
  deregulation 30–33
  detachment-induced *see* anoikis
  detection 8
  lysosomal permeabilization 401
  methylating agents 448–450
  mitochondria 1–6, 18–29, 30–36
    deregulation 30–33
    incoming signals 20–25
    metabolism 18–19
    outgoing signals 25–29
  mitotic catastrophe 485–487

neutrophils   513–514
nuclear features   7–8
phospholipase A   399–401
phospholipases   400–403
programmed death 1-mediated   695–697, 706–716
shear stress   402
vs. cornification   192
vs. entosis   467–468
vs. methuosis   563
vs. oncosis   573
vs. parthanatos   539
vs. pyronecrosis   231
vs. TRIAD   415, 416
apoptosis-associated speck-like protein containing a CARD (ASC/PYCARD)   228, 231–232
apoptosis-inducing factor (AIF)   26, 29
  oxytosis   300–302
  parthanatos   542–544
  pharmacological interventions   549
apoptosis signal-regulating kinase-1 (ASK-1)   157, 294–295
apoptosomes   148–149
apoptotic-like death (ALD) pathway   59–60
apoptotic protease activating factor-1 (Apaf-1)   3–4, 26–27
Arabidopsis thaliana (*A. thaliana*)   241–242
arachidonic acid (AA)   199, 293, 400
Argiope aurantia, phenoptosis   243
ASC *see* apoptosis-associated speck-like protein containing a CARD
ascorbate (VC)
  accumulation   662–663
  calcium overload   665–666
  cytotoxicity   662–666
  growth receptor regulation   664–665
  menadione combination therapy cytotoxicity   599–600, 666–672
    see also Apatone
  organelle effects   665–666
  pro-oxidant property development   663–664
  tumor cytotoxicity   599–600
ASK-1 *see* apoptosis signal-regulating kinase-1
Aspergillus fumigates, pyroptosis   328
ataxia telangiectasia mutated (ATM)   479
ataxia telangiectasia mutated-RAD3 related (ATR)   479
ATF *see* activating transcription factors
ATG4 see autophagy-specific gene 4
ATG5/ATG7   78

ATM *see* ataxia telangiectasia mutated
ATP/ADP ratios   18–19
ATP-citrate lyase (ACL)   19
ATP-synthase
  apoptosis avoidance   30
  cyclophilin D-dependent necrosis   382–383
  paraptosis   355
ATR *see* ataxia telangiectasia mutated-RAD3 related
atrophy, cell injury   84–85
autoimmunity
  NETosis   522–525
  programmed death 1   700
autophagic cell death (ACD)   74–77
  vs. cornification   192
  vs. TRIAD   415, 416–417
autophagosomes   668–669
autophagy   71–82
  autoschizis   635–637, 642
  cancer treatment   76–77
  cardiovascular disease   77–81
  cell death   74–77
  chaperone-mediated   72, 74
  classifications   72–74
  infections   77
  macro   73
  micro   73–74
  phospholipases   403–405
  stages of   75
  starvation   73–74, 77
  vascular aging   79–81
  vs. methuosis   562–563
  vs. parthanatos   539
autophagy-specific gene 4 (ATG4)   23
autoschizis   583–693
  annexin IV-FITC   611–612
  autophagy   635–637, 642
  biochemical alterations   657–672
    Apatone   666–672
    ascorbate effects   662–666
    menadione effects   657–662
  cell shape/size effects   611
  definitions   654–655
  DU145 cells   595–604, 635–652
  early observations   598–599
  endoplasmic reticulum   642–645
  flow cytometry measurements   600–603
  karyolysis   618–642
  lysosomes   642–643, 668–669
  MDAH 2774 cells   606, 608–610, 629–634

autoschizis (*Continued*)
  membrane defects   611–612
  mitochondria   642–643
  morphology   598–599, 603–645
  RT4 cells   623–625
  T24 cells   600–602, 608, 613–623, 625–629
  TRAMP cells   604, 607, 610
  ultrastructural damage   612–645
  VC:VK3 cytotoxicities   599–600
  vs. other PCD forms   655–657
  xenotransplanted prostate carcinoma   645–652
axonal injuries
  neuropathic pain   215–216
  traumatic   208–209
  Wallerian degeneration   205–223

b

Bacillus anthracis (*B. anthracis*)   63, 234, 326–327
Bacillus subtilis (*B. subtilis*)   51–54, 58–59
bacteria   49–70
  biofilms   60–63
  extrinsic pathway   3, 5–6
  necroptosis   117
  NET evasion   518–519
  neutrophil extracellular traps   514–519
  oncosis   575–576
  phenoptosis   239–240
  pyroptosis   324–330
  sporulation   51–54
  toxin–antitoxin systems   50, 54–60
baculoviral IAP repeat (BIR) domain   27
baicalein   294
Bak *see* Bcl-2 homologous antagonist killer
B. anthracis   63
BARD1   479
base excision repair (BER)   431, 440–441
basophils   370–371
Bax *see* Bcl-2 associated X protein
Bazan effect   400
Bc *see* bromocriptine
B-cell lymphoma-2 (Bcl-2) family   147
  mitochondrial signaling   22–23, 30, 33–34
  as therapeutic target   33–34
  tumor resistance   30
  unfolded protein response   24
B cells, NETosis   524
Bcl-2 associated X protein (Bax)   320, 380
  oncosis   573–574
  oxytosis   298

Bcl-2 homologous antagonist killer (Bak)   320, 380
Bcl-2 proteins
  apoptotic caspases   320
  MOMP   296–299
  oncosis   573–574
Bcl-xL   35–36
BDNF *see* brain-derived neurotrophic factor
BECN1   78–81
O^6-benzylguanine (O^6-BG)   436–437
benzyloxycarbonyl-valyl-alanyl-aspartyl fluoromethyl ketone (zVAD-fmk)   348–350
BER *see* base excision repair
β-subunit of ATP-synthase   355
β-tubulins   354–355
O^6-BG *see* O^6-benzylguanine
BH3-interacting domain (BID)   3–5, 6, 148, 149, 297–298
BID *see* BH3-interacting domain
bifunctional alkylating agents, definition   430
bifunctional proteins in phenoptosis   258–259
big potassium (BK) channels   348–350
big potassium calcium (BKCa) channels   348–350
bimolecular nucleophilic substitution (S_N2 reaction)   431
bioenergetics of mitochondria   20
biofilms
  NET evasion   518–519
  programmed cell death   50, 60–63
  sporulation signaling   53–54
BIR domain *see* baculoviral IAP repeat domain
BK *see* bongkrekic acid
BKCa channels *see* big potassium calcium channels
BK channels *see* big potassium channels
Bladder carcinoma (RT4) cells, autoschizis   623–625
bladder carcinoma (T24) cells
  Apatone cytotoxicity   599–600
  autoschizis   600–602, 608, 613–623, 625–629
  cell cycle stages   600–602
"blebs"   3
B lymphocytes
  functions   371
  maturation   371
  structure   370

bongkrekic acid (BK)   21
bowhead whales   251
brain-derived neurotrophic factor (BDNF),
        Wallerian degeneration   215
brain injuries
    excitotoxicity   200
    necroptosis   114
    see also nerve injuries; neurodegeneration
brefeldin A   303
bristlecone pine (*Pinus longaeva*)   250–251
bromocriptine (Bc)   357
B. subtilis see *Bacillus subtilis*
Buffalo, co-evolution   257
Burkholderia Spp.   329

c

CAD see caspase-activated DNase
Caenorhabditis elegans (*C. elegans*)   264
C. albicans see *Candida albicans*
calcification, dystrophic   95–96
calcium overload
    ascorbate cytotoxicity   665–666
    excitotoxicity   198–199
    parthanatos   544–545
calcium release-activated calcium channel
        protein 1 (ORAI1)   305
calpain I   200
camptothecin   522
CAMs see cell-adhesion molecules
cancer
    adoptive cell-transfer   714–716
    AGT inhibition   437–438
    anoikis   165–167
    Apatone therapy
        cytotoxicity   599–600
        DNA/protein synthesis
            effects   669–672
        lipid peroxidation   669
        organelle effects   668–669
        starvation   667–668
        starvation effects   667–668
    apoptotic resistance   30–31, 708–710,
            713–714
    ascorbate cytotoxicity   599–600, 662–666
        accumulation   662–663
        growth receptor regulation   664–665
        organelle effects   665–666
        pro-oxidant property
            development   663–664
    autophagy-inducing drugs   76–77
    autoschizis   583–693
    cell cycle stages   600–603

entosis   470–472
ferroptosis   139
granzyme A   5
growth receptors, ascorbate
        effects   664–665
immune avoidance   708–710, 713–714
immunotherapy   362, 713–716
membrane macrophage colony-stimulating
        factor   361–362
menadione cytotoxicity   599–600,
        657–662
    necrosis   657–658
    nuclear damage   658–659
    organelle peroxidation   660–661
    protein peroxidation   659–660
    signal transduction effects   661–662
methylating agents   432
MGMT gene   437–440
mismatch recognition   442
mitotic catastrophe induction   495–499
morphology   586–598
    in vitro   586–595
    in vivo   595–598
paraptotic medicines   351, 353–354,
        356–362
phenoptosis   246–247
programmed death ligand 1/2   704–705
synthetic lethality   446–447
Candida albicans (*C. albicans*),
        pyroptosis   325–326
cannibalism
    B. subtilis sporulation   51–53
    vs. entosis   467
CAPS see CIAS1-associated periodic
        syndromes
carbidopa (Sinemet)   202
carbon monoxide (CO)   348–350
carbonyl cyanide *m*-chlorophenylhydrazone
        (CCCP)   22, 35
carcinomas
    autoschizis   583–693
    characterization   598
    in vitro morphology   586–595
    in vivo morphology   595–598
    see also cancer; individual cell lines...;
        tumors
CARD see caspase-associated recruitment
        domain
cardiotoxicity, chemotherapeutic
        oncosis   577
cardiovascular development,
        autophagy   77–78

cardiovascular disease
   autophagy 77–81
   necroptosis 112
caseous necrosis 92–93
caspase 1 228, 321–328
caspase 2 156, 159
caspase 3 4, 6, 7–8, 28, 274
caspase 5 228
caspase 7 28, 159
caspase 8
   activation 3–4, 7
   anoikis 156, 158–159
   necroptosis 102, 106
   oxytosis 297–298
caspase 9 7, 26–27, 160
caspase 10 4, 7
caspase 11 328–330
caspase 14 188, 192
caspase-activated DNase (CAD) 2, 7–8, 26, 188–189
caspase-associated recruitment domain (CARD) 26, 321–322, 324–325
caspases (cysteine-dependent aspartate-specific proteases)
   anoikis 145–168
   apoptotic 320–321
   inflammasomes 322–330
     canonical 322–328
     non-canonical 328–330
   inflammatory 321–322
   mitochondrial pathways 2–3, 25–29
   phenoptosis 258–259
   pyroptosis 320–330
cathespin B inhibitor 231
CCAT/enhancer binding protein (C/EBP) homologous protein (CHOP) 303
CCCP *see* carbonyl cyanide *m*-chlorophenylhydrazone
CCL2 *see* monocyte chemoattractant protein-1
C_c^-Cys/Glu antiporter 291, 304
Cdc25 479, 481–483
Cdk-activating kinase (CAK) 479
Cdks 478–480, 481–482
C. elegans see Caenorhabditis elegans
cell-adhesion molecules (CAMs) 154
cell–cell contacts, cornification 191
cell cycle
   mitotic catastrophe 477–485
   tumor stages 600–603
cell death receptors 102
cell–extracellular matrix interactions 149–153

   see also integrins
cell-in-cell (CIC) structures
   cancer 467, 470–472
   cannibalism 467
   emperitosis 467
   entosis 463–473
   phagocytosis 466
cell injury 83–98
   atrophy 84–85
   caseous necrosis 92–93
   coagulative necrosis 92
   dystrophic calcification 95–96
   fat necrosis 94
   fibrinoid necrosis 94–95
   gummatous necrosis 93
   hydropic swelling 84
   hyperplasia and hypertrophy 85–86
   intracellular accumulations 87–90, 95–96
   irreversible 90–96
   liquefactive necrosis 92
   metaplasia 86–87
   necrotic nuclear changes 90–91
   necrotic tissue patterns 91–95
   nuclear changes 90–91
   reversible 84–90
cell survival
   anoikis 147–148
   phospholipase A 403
cell thanatology 585–586
cellular FLICE-inhibitory protein (c-FLIP) 5, 102, 148
cellular inhibitor of apoptosis protein (cIAP) 102
cellular metabolism, mitochondria 18–19
cellular morphology
   Apatone effects 611–612
   autoschizis 598–599, 603–645
   carcinomas 586–598
     in vitro 586–595
     in vivo 595–598
   erythrocytes 368–369
   mitochondria 17–18, 590–592
   mitotic catastrophe 486
   oncosis 569–572
   paraptosis 346–347
   pyronecrosis 227
   TRIAD 414–416
centrosomes in mitotic catastrophe 477, 484–485
ceramides 400
cerulein-induced pancreatitis 111, 112
c-FLIP *see* cellular FLICE-inhibitory protein

c-fos 661
CFU-GM *see* colony-forming-unit–granulocyte–macrophages
chalcones 562
chaperone-mediated autophagy (CMA) 72, 74
chaperones and oxidative stress 304
characterization of carcinomas 598
checkpoint kinase-1 (Chk1) 481
checkpoint kinase-2 (Chk2) 479, 484–485
checkpoints of the cell cycle 477–480
chemokines in Wallerian degeneration 214
chemotactic signaling in apoptosis 29
chemotherapy
  mitotic catastrophe 487, 497–499
  oncosis 569, 577
  paraptosis 351, 353–354, 356–362
  PARP inhibitors 547–549
  pyroptosis 326–327
Chk1 *see* checkpoint kinase-1
Chk2 *see* checkpoint kinase-2
cholesterol, intracellular accumulation 87–88
CHOP *see* CCAT/enhancer binding protein (C/EBP) homologous protein
chromatin condensation 7–8
chromosome segregation 480, 484–485
Chx *see* cyclohexamine
cIAP *see* cellular inhibitor of apoptosis protein
CIAS1 see cold-induced autoinflammatory syndrome 1 gene
CIAS1 *see* cryopyrin
CIAS1-associated periodic syndromes (CAPS) 229, 325–326
CIC *see* cell-in-cell structures
cicadas, phenoptosis 243–244
ciclopirox olamine (CPX) 140
CidAB 61–63
ciliary neurotrophic factor (CNTF) 215
Cisplatin (CisPt) 357
CisPt *see* Cisplatin
citrate as a predictor of death 272–273
C-Jun N-terminal kinase (JNK)
  anoikis 156–158
  -dependent-phosphorylated BID 158
  non-canonical inflammasomes 328–329
  oxytosis 294–295
classic alkylating agents 430
classifications
  alkylating agents 430
  autophagy 72–74

toxin–antitoxin systems 55–57
clodronate 34
ClpAP 57
ClpXP 58
CLRs *see* C-type lectin receptors
CNTF *see* ciliary neurotrophic factor
coagulative necrosis 92
co-evolution 257
co-inhibitory ligands 703, 707–710
cold-induced autoinflammatory syndrome 1 (CIAS1) gene 228–229
colon cancer 166, 357–358
colony-forming-unit–granulocyte–macrophages (CFU-GM) 371
colorectal cancer 357–358
combined ascorbate/menadione treatment *see* Apatone
condensation OF chromatin 7–8
confronting cisternae 594
copper complexes 357–358
corneodesmosomes 185, 191
cornification 183–192
  cell–cell contacts 191
  cellular remodelling 186–188
  in comparison 191–192
  differentiation 183–186
  lysosomes 187–188
  nuclear lysis 188–189
  protein crosslinking 189–190
coroner's perspectives, anoikis 161–162
corticosteroids 352
co-stimulatory ligands 703, 707–710
COX *see* cyclooxygenase
CP *see* [Cu (thp) 4][PF6]
CpG islands *see* cytosine–phosphate–guanine islands
CPX *see* ciclopirox olamine
crosslinking in cornification 189–190
crosstalk
  apoptosis 330–332
  inflammasomes 330–332
  integrins/receptor tyrosine kinases 152
  necroptosis 118, 330–332
  parthanatos 538–544
  pyroptosis 118, 330–332
cryopyrin (CIAS1/NLRP3) 228–229, 322–323, 325–326
C-terminal half of AIP1/Alix (Alix-CT) 351
CTLs *see* cytotoxic T lymphocytes
C-type lectin receptors (CLRs) 228–229
curcumin 359–360

[Cu (thp) 4][PF6] (CP)   357–358
CwlC/H   53–54
CX3CL1 *see* fractalkine
cyclin A1   482–483
cyclin B   481–482
cyclohexamine (Chx) induction   8
cyclooxygenase (COX)   293
cyclophilin A (CypA)   300–302, 377
cyclophilin D (CypD)   21
  -dependent necrosis   375–394
    cyclophilins   376–379
    disease roles   383–386
    future prospects   388–389
    inhibition   386–388
    mechanism   379–383
    myocardial infarction   383–385
    neurological disorders   385–386
  physiological roles   378–379
  structure   378
  as a therapeutic target   33–34
cyclophilins   376–379
cyclosporine A (CsA)   21, 386–388, 700
CYLD *see* cylindromatosis
cylindromatosis (CYLD)   102
CypA *see* cyclophilin A
CypD *see* cyclophilin D
cysteine-dependent aspartate-specific
    proteases (caspases)
  anoikis   145–168
  apoptotic   320–321
  inflammasomes   322–330
    canonical   322–328
    non-canonical   328–330
  inflammatory   321–322
  mitochondrial pathways   2–3, 25–29
  phenoptosis   258–259
  pyroptosis   320–330
cysteine/glutamate antiporter   135
cysteine uptake inhibition   300
cytochrome c (cyt c) release   3–5, 25–26
cytokines
  NETosis   517–518
  programmed death ligand effects   703
  pyroptosis   325
  Wallerian degeneration-mediated recovery   211–214
cytoplasm
  apoptotic pathways   3–7
    mitochondrial signaling   20–25
cytosine–phosphate–guanine (CpG)
    islands   433
cytoskeletal dynamics, entosis
    regulation   468–469
cytotoxicity
  alkylating agents
    apoptosis   448–450
    DNA repair mechanisms   440–446
    general principles   429–433
  Apatone   599–600, 666–672
  ascorbate   662–666
  menadione   657–662
  methylating agents
    alkylguanine
      methyltransferase   433–447
    apoptosis   448–450
    DNA repair mechanisms   440–446
    general principles   431–432
    pharmacology   450–455
  O^6-methylguanine   427–462
    general principles   429–433
    lesion formation   431–432
    pharmacology   450–455
cytotoxic T lymphocytes (CTLs)   3, 5–6

d

DAPK *see* death-associated protein kinases
DCS *see* diffuse cell swelling
DDR *see* DNA damage response
death-associated protein kinases
    (DAPK)   158
death effector domains (DED)   3–4, 5
death-inducing signal complex (DISC)   3–5, 6, 148–149
death receptor pathway *see* extrinsic pathway
death receptors
  apoptotic caspases   320
  core concepts   2–3
  KILLER   25
death signals in mitochondria   19
decarbazine (DTIC)   450–453
DED *see* death effector domains
degradation, T cells   697–698
dehydroascorbic acid (DHA)   662–665
demethylation   434–435
demyelination   211–215
deregulation of apoptosis   30–33
desquamation   191
detachment-induced apoptosis *see* anoikis
detection of methylation   439–450
development
  microbial   50–51
  T cells   697–698
DHA *see* dehydroascorbic acid

Dia1   469
diabetes   247–248
DIABLO   26–27
diagnostic techniques for mitotic catastrophe   488–494
dietary restriction and aging   263–267
diferuloylmethane (curcumin)   359–360
differentiation
   epidermal keratinocytes   183–186
   hair follicles   185–186
   intestinal epithelial anoikis   162–165
   T cells   697–698
diffuse cell swelling (DCS)   576
   see also oncosis
dimerization of death effector domains   3–4, 5
DISC see death-inducing signal complex
discovery
   entosis   464–465
   methuosis   560
   NETs   514–515
diseases, anoikis   165–167
disulfide bonds in cornification   190
DNA
   alkylating agents
      general principles   429–433
      methylation   431–432
      O^6-methylguanine   427–462
      mismatch recognition   442
      sites of action   430
   Apatone treatment effects   669–672
   base excision repair   440–441
   demethylation   434–435
   extracellular   50
   G_2/M Checkpoint   479
   homologous recombination   443–444
   hydrolysis   2, 3, 7–8, 188–189
   lesion formation   431–432
   methylation
      alkylguanine methyltransferase   433–447
      apoptosis   448–450
      base excision repair   440–441
      endogenous/environmental causes   431
      lesion formation   431–432
      mismatch recognition   442
      pharmacology   450–455
      repair mechanisms   440–446
   microsatellite instability   442
   mismatch recognition   442
   mitotic catastrophe   479, 481–484, 495–497
   non-homologous end joining   444–446
   PARP-1   446–447
   parthanatos   540–541
DNA damage response (DDR)   25
DNases
   ascorbate effects   665
   autoschizis   645–652, 669–672
   cornification   188–189
   see also karyolysis
double strand breaks (DSB)   443–444, 482–483
downregulation, TRIAD   417
downstream mechanisms of oxytosis   300–302
doxorubicin cardiotoxicity   577
Drosophila melanogaster   264
DRP1 see dynamin-related protein-1
DSB see double strand breaks
DTIC see decarbazine
DU145 cells
   autoschizis   595–603, 635–652
   cell cycle stages   602–603
   morphology   595–598
   VC:VK3 treatment   599–603, 645–652
   xenotransplanted   595–598, 645–652
dynamin-related protein-1 (DRP1)   298–299
dystrophic calcification   95–96

e

ears, necroptosis   113
E-cadherins see epithelial cadherins
ECM see extracellular matrix
E. coli see *Escherichia coli*
eDNA see extracellular DNA
effector caspases in apoptosis   320–321
EGFRs see epidermal growth factor receptors
eicosanoid biosynthesis   293
elastase, NETosis   521–522
ELK   708, 710
embryonic development   108–109, 111
emperitosis vs. entosis   467
EMT see epithelial–mesenchymal transition
endo G see endonuclease G
endogenous methylating agents   432
endonuclease G (endo G)   4, 8, 26, 29
endonucleases, basic concepts   2
endoplasmic reticulum (ER)
   autoschizis   642–645
   stress vs. oxytosis   302–305
   vacuolization in TRIAD   416
endopolyploidy   483–484
endotoxins, phenoptosis   245–246

ENOA *see* alpha-enolase
entosis   463–473
　biological significance   471–472
　cancer   470–471
　cell-in-cell structures   466–468
　definition   464
　discovery   464–465
　initiation   466–467
　progression   466–467
　regulation   468–469
　signaling   469
　vs. apoptosis   467–468
　vs. phagocytosis   466
　see also necrosis
environmental sensing by mechanotransduction   154–155
eosinophils   370, 371
epidermal growth factor (EGF)   352
epidermal growth factor receptors (EGFRs)   661–662
epidermal keratinocytes   183–192
　see also cornification
epidermis
　differentiation   183–186
　see also skin
epidermolysis bullosae   165
epilepsy   198
epithelial cadherins (E-cadherins)   154, 160–161
epithelial cells, anoikis   162–165
epithelial–mesenchymal transition (EMT)   166–167
ER *see* endoplasmic reticulum
Erastin   134–136, 138
ERKs *see* extracellular signal-related kinases
eryptosis   372–373
erythrocytes   367–374
　eryptosis   372–373
　hematopoiesis   368–369
　life cycle   369
　proerythroblasts   369
　structure and function   368–369
erythropoiesis   368–369
Escherichia coli (*E. coli*)   57–60, 400
ETosis *see* NETosis
eukaryotic initiation factor 2 (eIF2α)   303–304
evolution
　aging   256–257
　phenoptosis   248–250
excercise and aging   267–269
excitotoxicity   197–204
　Alzheimer's disease   201–202
　brain injuries   200
　calcium overload   198–199
　glutamate   198–199
　Huntington's disease   199, 201
　intracellular toxicity   199
　oxidative stress   200
　radiologic observations   202–203
exonucleases   2
expression
　MGMT   434
　programmed death ligands   700, 701–705
　TRIAD   417
extracellular death factors (EDF)   54–60, 514–519
extracellular DNA (eDNA)   50, 61
extracellular matrix (ECM)
　anoikis   149–155
　biofilms   50
　cell interactions   149–150
　integrins   150–153
extracellular signal-related kinases (ERKs)
　integrin signaling   153
　menadione cytotoxicity   662
　methuosis   560–561
　NETosis   521
　oncosis   573–574
　oxytosis   294
　paraptosis   350–351
extrinsic pathway   2–3, 4–6, 148–149
　inhibition   5
　link to intrinsic pathway   4, 8
　methylating agents   448–450
eyes, necroptosis   113
ezrin   468

f

F680/F684   387–388
FADD *see* Fas-associated death domain protein
FAK *see* focal adhesion kinase
familial Mediterranean fever   323
Fas-associated death domain protein (FADD)   3–4, 5, 102, 105, 158–159
FasL *see* fatty acid synthase ligand
FasR *see* fatty acid synthase receptor
fasting, aging   263–267
fat necrosis   94
fatty acid synthase ligand (FasL)   5, 6, 400–401
fatty acid synthase receptor (FasR)   5, 6
FcyR-mediated phagocytosis   72

Fenton reaction 131
ferroptosis 129–140
    clinical significance 139–140
    Erastin 134–138
    induction/inhibition 138–139
Feulgen staining 625–628, 631–633
FFAs *see* free fatty acids
F_0–F_1 ATPase *see* ATP-synthase
fibrinoid necrosis 94–95
filaggrin 185, 188
FK506-binding proteins (FKBPs) 377
FKBPs *see* FK506-binding proteins
flow cytometry, autoschizis 600–603
fMLP *see* formylated Met–Leu–Phe
focal adhesion kinase (FAK) 150–152, 155–156, 401
food restriction 263–267
formation of phagolysosomes 73
formylated Met–Leu–Phe (fMLP) 519–521
four-punch hit 155, 160
fractalkine (CX3CL1) 29
Francisella tularensis 328
free fatty acids (FFAs) 94
free radicals 130–133
functional recovery from nerve injuries 209–211
fusion, lysosomes 72

g

Gal-3 *see* galectin-3
galectin-3 (Gal-3) 214
γ-interferon inducible protein 16 (IFI16) 322–323, 328
gastrointestinal (GI) cancer, *MGMT* gene 440
gel electrophoresis 628–629, 633–634
geroprotection
    dietary restriction 263–267
    living conditions 270
    physical activity 267–269
    psychological factors 270–272
GhoS 56
GhoT 56
GI cancer *see* gastrointestinal cancer
GluN2D subunit 253–255
glutamate 198–199
glutamine starvation 24
glutathione (GSH) 19, 35, 135, 292–293
glutathione peroxidases (GPX) 35, 292–293
4-(*N*-(*S*-glutathionylacetyl)amino) (GSAO) 33–34
glycogen storage diseases 89

G_2/M Checkpoint 479, 480–484
GM-CSF *see* granulocyte–macrophage colony-stimulating factor
Golgi apparatus, carcinoma morphology 592–593
GPX *see* glutathione peroxidases
Gracilinanus microtarsus 244
Gram negative bacteria
    definition 50
    MazEF 57–60
    programmed cell death 57–60
Gram positive bacteria
    biofilm development 60–63
    definition 50
    sporulation 51–54
granular layer cornification 186–188
granulocyte–macrophage colony-stimulating factor (GM-CSF) 513, 517–518, 704
granulocytes
    maturation 371
    NETosis 511–534
    structure 369–370
granulomas, necrosis 92–94
granzyme A (GzmzA) 3–4, 5–6
granzyme B (GzmzB) 3–4, 6
granzyme K (GzmzK) 5
granzyme M (GzmzM) 5
granzymes
    formation 3
    see also Perforin/granzyme pathway
growth-factor withdrawal 24
growth receptors, ascorbate effects 664–665
GSAO *see* 4-(*N*-(*S*-glutathionylacetyl)amino)
GSH *see* glutathione
GTPases *see* Rho guanine triphosphatases
gummatous necrosis 93
Gzmz N... *see* granzyme N...

h

Haber–Weiss reaction 131
hair follicle differentiation 185–186, 191
Harlequin (*Hq*) mutation 302
HCC *see* hepatocellular carcinoma
HCNP *see* hippocampal cholinergic neurostimulating peptide
heart failure (HF) and autophagy 78–79
heavy hyaluronan 256
hedonic factors to life expectancy 270–272
Helicobacter pylori (*H. pylori*) 440
hematopoiesis 109, 368
hematopoietic stem cells (HSCs) 368

hemidesmosomes 155, 161
hemosiderin 90
hemoxygenase-2 348
hepatitis, necroptosis 113
hepatocellular carcinoma (HCC) 139, 362
hereditary nonpolyposis colorectal cancer (HNPCC) 442
Heterocephalus glaber (naked mole rats) 251–256
hexokinase (HK) 30
HF *see* heart failure
hIECs *see* human intestinal epithelial cells
HIF cascade, ascorbate 664–665
high-mobility group protein B1 (HMGB1) 232
HIP *see* huntingtin-interacting protein
hippocampal cholinergic neurostimulating peptide (HCNP) 355–356
Hippo pathway 420
histones 5–6, 519–522
HMGB1 *see* high-mobility group protein B1
HNPCC *see* hereditary nonpolyposis colorectal cancer
holins 61–63
homeostasis and autophagy 71–82
homologous recombination 443–444, 482–483
homology, programmed death 1 699
honokiol (HNK) 353–354
H. pylori see Helicobacter pylori
Hq see Harlequin mutation
HSCs *see* hematopoietic stem cells
HtrA2 26
human intestinal epithelial cells (hIECs) 162–165
huntingtin-interacting protein (HIP) 412
huntingtin protein 412–413
Huntington's disease (HD)
  excitotoxicity 199, 201
  polyQ aggregates 413
  TRIAD 413, 419–421
  YAPΔCs 419–421
hydrogen peroxide 130, 666–667
hydrolysis of DNA 2, 3, 7–8, 188–189
hydropic swelling 84
hydroxyl radicals 130
hyperplasia 85–86
hypertrophy 85–86

i

ICAD *see* inhibitor of caspase-activated DNase
Iduna 549
IFI16 *see* γ-interferon inducible protein 16
IFNs *see* interferons
IKKs *see* inhibitor-kappaB kinases
IL-1α *see* interleukin 1α
IL-1β *see* interleukin 1β
IL-6 *see* interleukin 6
IL-10 *see* interleukin 10
IL-18 *see* interleukin 18
ILK *see* integrin-linked kinase
immunity
  adaptive 370–371
  cancerous evasion 708, 710, 713–714
  extrinsic pathway 3, 5–6
  innate 370
  pyronecrosis 225–236
immunity-related guanosine triphosphatases (IRG) 233
immunophilins 377
  see also cyclophilin...
immunotherapy 713–716
inclusions, carcinoma *in vitro* 594
indole-based chalcones 562
induction
  anoikis 155–159
  eryptosis 372–373
  ferroptosis 138–139
  necroptosis 100–106
  necrosis by menadione 657–658
  neuronal protection 207–208
  parthanatos 538–545
  pyronecrosis 232–234
  TRIAD
    in vitro 413–414
    in vivo 421
infections
  autophagy 77
  bacterial 3, 5–6, 117, 322–328, 512–517, 573–574
  extrinsic pathway 3, 5–6
  granulomas 92–94
  necroptosis 117
  neutrophil extracellular traps 514–519
  neutrophils 512–519
  oncosis 575–576
  oxidative stress 24–25
  phenoptosis 245–246
  pyronecrosis 225–236
  pyroptosis activation 324–330
  viral 3, 5–6, 24–25, 117, 576
inflammasomes 228
  AIM2 322–324, 328

canonical   322–328
crosstalk   330–332
IFI16   322–323, 328
NAIP/NLRC4   322–323, 327–328
NLRP1   322–323, 326–327
NLRP3   322–323, 325–326
non-canonical   328–330
pyroptosis   322–330
structural features   324
inflammation
caspases   321–322
cell death mechanisms   228–229
neuropathic pain   215–216
neutrophils   512–513
parthanatos   546
Wallerian degeneration   212–214
inflammatory cell death
caspases   320–322
pyroptosis   317–341
inflammatory diseases
necroptosis   110–117
NETosis   522–525
Influenza   325–326
inhibition
alkylguanine methyltransferase   436–438
cyclophilin D-dependent
   necrosis   386–388
cysteine uptake   300
extrinsic pathway   5
ferroptosis   138–139
glutathione peroxidase-4   292–293
necroptosis   117–119
PARP-1   446–447
poly ADP ribose polymerases   546–549
T cell receptor signaling   707–710
inhibitor of caspase-activated DNase
   (ICAD)   7–8
inhibitor-kappaB kinases (IKKs)   520–521
inhibitors of apoptosis proteins (AIPs)   27
initiation
entosis   466–467
eukaryotic factor 2   303–304
NETosis   519–521
innate immunity   370
insulin-like growth factor 1 receptor
   (IGF1R)   350–351
integrin-linked kinase (ILK)   152–153
integrins   149–153
adhesomes   150, 153
anoikis induction   155–160
β1   151
receptor tyrosine kinase crosstalk   152

γ-interferon inducible protein 16
   (IFI16)   322–323, 328
interferons (IFNs)   104–106
interleukin 1α (IL-1α)   212
interleukin 1β (IL-1β)   212, 216, 322, 325, 513–514
interleukin 6 (IL-6)   212–214, 215
interleukin 10 (IL-10)   212–214
interleukin 18 (IL-18)   325
intestinal diseases and necroptosis   114
intestinal epithelium, anoikis   162–165
intracellular accumulations of cell
   injury   87–90, 95–96
intracellular receptors, necroptosis   106
intracellular toxicity, excitotoxicity   199
intracytoplasmic lumina   594, 597–598
intrinsic pathway   149
link to extrinsic pathway   4, 8
methylating agents   448–450
invertebrates, phenoptosis   242–244
in vivo
xenotranslated carcinoma
   autoschizis   645–652
   morphology   595–598
I/R *see* ischemia/reperfusion
IRG *see* immunity-related guanosine
   triphosphatases
iron
anemia and eryptosis   373
ferroptosis   135–137
human distribution   135, 136
reactive oxygen species   134
irreversible cell injury   90–96
intracellular accumulations   95–96
nuclear changes   90–91
tissue patterns   91–95
see also necrosis
ischemia/reperfusion (I/R)
autophagy   78
Bazan effect   400
oncosis   572, 576–577

j

JAK-STAT signaling, necroptosis   104–106
jBID *see* JNK-dependent-phosphorylated BID
JNK *see* C-Jun N-terminal kinase
JNK-dependent-phosphorylated BID
   (jBID)   158

k

K5 *see* keratin 5
kallikreins (KLK)   191

karyolysis 91
  Apatone treatment 637–642
  autoschizis 618–642
  Feulgen staining 625–628, 631–633
  gel electrophoresis 628–629, 633–634
karyorrhexis 91
keratin 5 (K5) 185
keratin-associated proteins (KRTAPs) 185
keratinocytes
  cornification 183–192
  differentiation 183–186
kidneys, necroptosis 113–114
killifish 250
kinetochores 477, 480
KLK *see* kallikreins
knockout (KO) mice, *MGMT* 435–436
KRTAPs *see* keratin-associated proteins

l

LC3, lipidation 72
LCK *see* lymphocyte-specific protein tyrosine kinase
leaderless mRNA 59
LEKTI *see* lympho-epithelial Kazal-type related inhibitor
lesions, methylating agents 431–432
leukocytes 369–372
  agranulocyte structure 370
  functions 370–371
  granulocyte structure 369–370
  hematopoiesis 368
  maturation 371
Levodopa 202
LgrAB 61–63
LIF 215
life cycle
  erythrocytes 369
  macrophages 370
  see also phenoptosis
life expectancy
  dietary restriction 263–267
  living conditions 270
  physical activity 267–269
  psychological factors 270–272
linear ubiquitin-chain assembly complex (LUBAC) 102
Lions, co-evolution 257
lipidation, LC3 72
lipid peroxidation 293–294, 669
lipids, intracellular accumulation 87
lipofucsin 90

lipopolysaccharide (LPS) 328–330
lipoxygenase (LOX) 293–294
liquefactive necrosis 92
Listeria monocytogenes 328
liver, necroptosis 112–113
living conditions, geroprotection 270
lomeguatrib 436–437
lonidamine 33–34
LOX *see* lipoxygenase
LP *see* lysosomal permeabilization
LPAR2 *see* LPA receptor 2
LPA receptor 2 (LPAR2) 469
LPC *see* lysophosphatidylcholine
LPS *see* lipopolysaccharide
LUBACC *see* linear ubiquitin-chain assembly complex
lungs, anthracosis 89–90
lupus *see* systemic lupus erythematosus
lymphocytes
  degradation 697–698
  functions 370–371
  maturation 697–698
  NETosis 511–534
  programmed death 1-mediated apoptosis 695–722
  structure 370
  T cell receptor signaling 703, 707–710
lymphocyte-specific protein tyrosine kinase (LCK) 708, 710
lympho-epithelial Kazal-type related inhibitor (LEKTI) 191
Lynch syndrome 442
lysophosphatidylcholine (LPC) 29
lysosomal permeabilization (LP) 401
lysosomes
  Apatone effects 668–669
  autoschizis 642–643, 668–669
  carcinoma morphology 593–594
  cornification 187–188
  fusion 72
LytC 53–54
LytRS 61–62

m

mAbs *see* monoclonal antibodies
macroautophagy 73
macrophage inflammatory protein 1α (MIP-1α) 214
macrophages
  function 370
  life cycle 370
  METs 515–516

traumatic nerve injury 208–209, 211–212, 214
macropinosomes 560–562
maitotoxin 234
major histocompatibility complex (MHC) 370, 707–710
MAM *see* methylazoxymethanol
mammalian target of rapamycin (mTOR) 152, 187, 708, 710
MAP2K *see* mitogen-activated protein kinase kinase
MAP3K *see* mitogen-activated protein kinase kinase kinase
MAPKs *see* mitogen-activated protein kinases
marinesco bodies 412
matrix metalloproteases (MMP) 19, 521
maturation
  leukocytes 371
  T cells 697–698
MazEF 55, 57–60
MCAK *see* mitotic centromere-associated kinesin
MCP-1 *see* monocyte chemoattractant protein-1
MD *see* muscular dystrophy
MDAH 2774 (ovarian carcinoma) cells 606, 608–610, 629–634
Mdivi-1 36
mechanotransduction 154–155
MEK *see* mitogen/extracellular signal-related kinase
melanin, cell injury 90
Melanoplus spretus (Rocky Mountain locust) 257
membrane defects of Apatone treatment 611–612
membrane macrophage colony-stimulating factor (mM-CSF) 348–349, 361–362
menadione (VK3)
  ascorbate combination therapy cytotoxicity 599–600, 666–672
    see also Apatone
  cytotoxicity 657–662
  necrosis induction 657–658
  nuclear damage 658–659
  organelle peroxidation 660–661
  protein peroxidation 659–660
  signal transduction effects 661–662
  tumor cytotoxicity 599–600
O^6-meT *see* O^6-methylguanine
metaplasia 86–87

metastasis, anoikis 165–167
methionine-free diets 264–265
methuosis 559–565
  discovery 560
  indole-based chalcones 562
  pathophysiology 560–562
  vs. apoptosis 563
  vs. autophagy 562–563
methylating agents
  endogenous and environmental 431
  general concepts 430
  lesion formation 431–432
  pharmacology 450–455
  see also non-classic alkylating agents
methylation of DNA
  alkylguanine methyltransferase 433–447
  apoptosis 448–450
  base excision repair 440–441
  CpG islands 433
  detection, *MGMT* gene 439–450
  homologous recombination 443–444
  mismatch recognition 442
  non-homologous end joining 444–446
  PARP-1 446–447
methylazoxymethanol (MAM) 436
O^6-methylguanine-DNA-methyltransferase (*MGMT*) gene
  clinical onclogy 438–439
  expression patterns 434
  GI cancer 440
  knockout/transgenic mice studies 435–436
  methylation detection 439–450
  O^6-methylguanine toxicity 433–447
  as a therapeutic target 436–438
O^6-methylguanine (O^6-meG) 427–462
  general principles 429–433
  lesion formation 431–432
  MGMT gene 433–447
  pharmacology 450–453
3-(2-methyl-1H indolyl-3-yl)-(4-pyridinyl)-2-propen-1-one (MIPP), methuosis 562
methylnitrosourea (MNU) 438
methyl-N'-nitro-N-nitrosoguanidine, parthanatos 542, 545
1-methyl 4-phenyl-1,2,3,6-tetrahydropyridine 35
O^4-methylthymine (O^4-meT) 432
MGMT see O^6-methylguanine-DNA-methyltransferase *gene*
MHC *see* major histocompatibility complex

microautophagy 73–74
microbes
  apoptotic-like death pathway 59–60
  biofilms 50, 60–63
  MazEF 57–60
  NET evasion 518–519
  phenoptosis 239–240
  programmed cell death 49–70
  sporulation 51–54
  toxin–antitoxin systems 50, 54–60
microRNAs, programmed death 1
    regulation 710–712
microsatellite instability (MSI) 442
microtubule plus end-tracking
    proteins 468–469
microtubules
  entosis regulation 468–469
  paraptosis 354–355
  spindle-assembly checkpoint 480, 484
MIP-1α see macrophage inflammatory
    protein 1α
MIPP see 3-(2-methyl-1H indolyl-3-yl)-
    (4-pyridinyl)-2-propen-1-one
mismatch recognition (MMR) 442
mitochondria 13–47
  aging 261–263
  apoptotic pathways 1–6, 20–29
    deregulation 30–33
    incoming signals 20–25
    outgoing signals 25–29
  autoschizis 642–643
  bioenergetics 20
  cellular metabolism 18–19
  cyclophilin D-dependent
      necrosis 379–383
  death signals 19
  dysfunction
    ascorbate 665–666
    oxytosis 295–300
      downstream mechanisms 300–302
      upstream mechanisms 293–295
  morphology 17–18, 590–592
  neurodegeneration 31–33
  oncosis 572–575
  paraptosis 355
  parthanatos 541–542
  as a therapeutic target 33–36
  tumor resistance 30–31
mitochondrial fission 1 protein (FIS1) 298
mitochondrial outer-membrane
    permeabilization (MOMP) 3–4, 6
  Bcl-2 proteins 296–299
  cancerous inhibition 30–31
  cyclophilin D-dependent
      necrosis 379–383
  oncosis 572–573
  outgoing signals 25–29
  oxyxtosis 295–299
mitochondrial permeability transition pore
    (MPTP) 376, 379–383, 572
  components 21
  states 21–22
  as a therapeutic target 33–34
mitochondrial reactive oxygen species
    (mROS)
  dietary restriction 264–265
  functions 299–300
  regulation and phenoptosis 273
mitofusins 298
mitogen-activated protein kinase kinase 1
    (MEKK1) 159–160
mitogen-activated protein kinase kinase
    kinase (MAP3K) 294
mitogen-activated protein kinase kinase
    (MAP2K) 294
mitogen-activated protein kinases (MAPKs)
  menadione cytotoxicity 662
  mitochondria 23–24
  oncosis 573–574
  oxytosis 294
mitogen/extracellular signal-related kinase
    (MEK) 153, 521
Mito Q 35
mitotic catastrophe (MC) 475–509
  cancer treatment 495–499
  cell cycle 477–480
  cell death mechanisms 485–487
  chemotherapy 487, 497–499
  detection methods 487–494
  general concepts 476–477
  G_2/M Checkpoint 479, 480–484
  morphology 486
  quiescence 483–484
  radiotherapy 495–497
  spindle-assembly checkpoint 480,
      484–485
mitotic centromere-associated kinesin
    (MCAK) 468–469
mixed lineage kinase domain-like (MLKL)
    protein 107, 110–117
MLC see myosin light chain
mM-CSF see membrane macrophage colony-
    stimulating factor
MMP see matrix metalloproteases

MMR *see* mismatch recognition
MOMP *see* mitochondrial outer-membrane permeabilization
monoclonal antibodies (mAbs)   362
monocyte chemoattractant protein-1 (MCP-1/CCL2)   214
monocytes
  functions   371
  maturation   371
  METs   515–516
  structure   370
morphology
  autoschizis   598–599, 603–645
  carcinomas   586–598
    in vitro   586–595
    in vivo   595–598
  erythrocytes   368–369
  mitochondria   17–18, 590–592
  mitotic catastrophe   486
  oncosis   569–572
  paraptosis   346–347
  pyronecrosis   227
  TRIAD   414–416
MPTP *see* mitochondrial permeability transition pore
mRNA, leaderless   59
MrpC   60
MSI *see* microsatellite instability
mTOR *see* mammalian target of rapamycin
multistep differentiation of epidermis   184–185
murine prostate carcinoma cells   604, 607, 610
muscular dystrophy (MD)   165
MutLα   442
MutSα   442
Mycobacterium tuberculosis (*M. tuberculosis*)   60
Mycobacterium xanthus (*M. xanthus*)   60
myocardial infarction
  Bazan effect   400
  cyclophilin D-dependent necrosis   383–385
myocardial ischemia, oncosis   576–577
myosin, entosis   468
myosin light chain (MLC), anoikis   157
Myt1   479

n

N:C *see* nuclear-to-cytoplasmic ratios
NA^+/Ca^{++} exchanger (NCX)   544–545
NACHT *see* nucleotide binding and oligomerization domains
$NAD^+/NADH$ *see* nicotinamide adenine dinucleotide
NADPH450 reductase *see* nicotinamide dinucleotide phosphate 450 reductase
$NADP^+/NADPH$ *see* nicotinamide adenine dinucleotide phosphate
NAIP *see* neuronal apoptosis inhibitory protein
naked mole rats (*Heterocephalus glaber*)   251–256
natural killer (NK) cells
  extrinsic pathway   3, 5–6
  functions   370–371
  maturation   371
  structure   370
natural selection, phenoptosis   248–250, 258–259
NCX *see* NA^+/Ca^{++} exchanger
necroptosis   99–119
  cell death receptors   102
  crosstalk   118, 330–332
  definitions   100
  embryonic development   108–109, 111
  hematopoiesis   109
  history of   101
  induction   100–106
  infections   111–117
  inflammatory diseases   110
  inhibition   118–119
  interferons   104–106
  intracellular receptors   106
  mechanisms   106–107
  mixed lineage kinase domain-like protein   107, 110–117
  neutrophils   514
  outcomes   107–117
  RIPK1   106, 110–117
  RIPK3   102–104, 106–107, 110–117
  toll-like receptors   102–104
  vs. NETosis   522
necrosis
  cyclophilin D-dependent   375–394
    cyclophilins   376–379
    disease roles   383–386
    future prospects   388–389
    inhibition   386–388
    mechanism   379–383
  dystrophic calcification   95–96
  intracellular accumulations   95–96
  menadione   657–658
  mitochondria   23–24

necrosis (*Continued*)
  mitotic catastrophe 486–487
  nuclear changes 90–91
  phospholipases 405–406
  tissue patterns 91–95
  vs. parthanatos 539
  vs. pyronecrosis 229–230
  see also entosis
necrosomes 23–24, 100, 102–104, 106–107, 109–111
Neisseria gonorrhoeae (*N. gonorrhoeae*) 233, 325–326
neoplasia 86
neoteny 253–256
NER *see* nucleotide excision repair
nerve growth factor (NGF) 215–216
nerve injuries
  demyelination 211–215
  recovery pathway activation 209–211
  Wallerian degeneration 205–223
    immune responses 208–209
    neuroprotection induction 207–208
    neuroprotective mechanism 206–207
    recovery pathway activation 209–211
NETosis 511–534
  autoimmunity 522–525
  NETs 514–519
  regulation 519–522
  vs. other forms of PCD 522
NETs *see* neutrophil extracellular traps
neurodegeneration
  amyloidosis 89
  cellular dysfunction 412–413
  conformational protein abnormalities 413
  cyclophilin D-dependent necrosis 385–386
  excitotoxicity 197–204
  ferroptosis 139
  forms of cell death 413
  mitochondria 31–33
  necroptosis 112
  nitric oxide synthase 199
  oxytosis 289–316
  protection by aggregates 413
  TRIAD 411–426
  YAPΔCs 419–421
neurokinin-1 receptor (NK1R) 352
neuronal apoptosis inhibitory protein (NAIP) 322–323, 327–328
neuropathic pain 215–216
neuroradiology 202–203

neurotrophic factors in Wallerian degeneration 215
neutrophil extracellular traps (NETs) 514–519
  discovery 514–515
  microbe adaptations 518–519
  testing and validation 515–518
neutrophils
  apoptosis 513–514
  functions 370
  infection/inflammation 512–513
  maturation 371
  NETosis 511–534
  non-apoptotic death 514
  structure 369
NFAT 708, 710
NF$_\kappa$B, T cell signaling 708, 710
NGF *see* nerve growth factor
N. gonorrhoeae see *Neisseria gonorrhoeae*
nicotinamide adenine dinucleotide (NAD$^+$/NADH), NADPH ratios 19
nicotinamide adenine dinucleotide phosphate (NADP$^+$/NADPH), NAD$^+$ ratios 19
nicotinamide dinucleotide phosphate 450 (NADPH450) reductase 348
Niemann–Pick disease 87–88
nigericin 234
nitric oxide synthase (NOS) 199–200
nitro-PAHs *see* nitro-polycyclic aromatic hydrocarbons
nitro-polycyclic aromatic hydrocarbons (nitro-PAHs) 352–353
1-nitropyrene (1-NP) 352–353
Nivolumab 713–714
NK1R *see* neurokinin-1 receptor
NK *see* natural killer cells
NLRC4 322–323, 327–328
NLRP1 322–323, 326–327
NLRP3 *see* cryopyrin
NLRs *see* NOD-like receptors
NMDA *see* N-methyl-D-aspartate
N-methyl-D-aspartate (NMDA) receptors 198–199, 253–255, 543–545, 549
N-methylpurine lesions 440–441
Nmnat1 206–207
Nmnat2 207–208
NOD *see* nucleotide-binding oligomerization domain
NOD-like receptors (NLRs)
  inflammasomes 322–334
  pyronecrosis 228–229

non-aging living beings   250–256
non-classic alkylating agents   430
   see also methylating agents
non-homologous end joining   444–446
NOS see nitric oxide synthase
Nothobranchius Spp.   250
1-NP see 1-nitropyrene
NT-3 see neurotrophin 3
NT-4 see neurotrophin 4
Nuc-1 see nuclease protein
nuclear-to-cytoplasmic ratios (N:C)   590
nuclease protein (Nuc-1)   7
nucleolar organizer regions
         (NORs)   588–589
nucleolus, carcinomas   589–590
nucleophilic substitutions   431–432
nucleoporins in mitotic catastrophe   484
nucleotide binding and oligomerization
         (NACHT) domains   324
nucleotide-binding oligomerization (NOD)
         domain   700
nucleotide excision repair (NER)   431
nucleus
   apoptotic pathways   7–8
      mitochondrial signaling   25
   autoschizis   618–642
   cornification   188–189
   endonucleases   2
   exonucleases   2
   in vitro carcinoma morphology   588–590
   karyolysis   91
   karyorrhexis   91
   menadione effects   658–659
   necrosis   90–96
   pyknosis   91

o

Oct-1 see octamer transcription factor 1
octamer transcription factor 1 (Oct-1)   546
Octopus hummelincki, phenoptosis   243
OGD see oxygen–glucose deprivation
oligomycin sensitivity-conferring protein
         (OSCP)   383
Omi   26
OMM see outer mitochondrial membrane
oncology, *MGMT* gene   438–439
oncosis   567–581
   cardiotoxicity   577
   chemotherapy   569, 577
   genetics   574–575
   morphologies   569–572
   myocardial ischemia   576–577

pathophysiology   572–575
vs. apoptosis   573
OP-A see Ophiobolin A
Ophiobolin A (OP-A)   360–361
ORAI1 see calcium release-activated calcium
         channel protein 1
organelles
   Apatone treatment effects   668–669
   in vitro carcinoma morphology   588–595
   in vivo carcinoma morphology   596–598
   peroxidation, menadione   660–661
OSCP see oligomycin sensitivity-conferring
         protein
outer mitochondrial membrane (OMM)
         permeabilization   295–299
outside–in signaling   150–153
ovarian carcinoma (MDAH 2774) cells   606,
         608–610, 629–634
oxidative stress
   chaperones   304
   excitotoxicity   200
   glutathione   292–293
   oxytosis   289–316
   unfolded protein response   304–305
   viral infection   24–25
OXPHOS machinery   18–19, 30
oxygen–glucose deprivation (OGD)   295
oxytosis   289–316
   downstream mechanisms   300–302
   ERKs   294
   glutathione   292–293
   JNKs   294–295
   lipid peroxidation   293–294
   lipoxygenase   293–294
   MAPKs   294
   mitochondrial dysfunction   295–300
   MOMP   295–299
   p38 kinases   295
   unfolded protein response   302–304
   upstream mechanisms   293–295
   vs. ER stress   302–305

p

p38, anoikis   156–158
p38 kinases, oxytosis   295
p53
   mitochondrial death signaling   25
   mitotic catastrophe   478, 481–485, 486–487
p53-induced death domain protein 1
         (PIDD)   149
p53-upregulated modulator of apoptosis
         (PUMA)   25

p73, TRIAD 417–421
Pacific Salmon 244
paclitaxel *see* taxol
PADs *see* peptidylarginine deaminases
P. aeruginosa see Pseudomonas aeruginosa
PAHs *see* polycyclic aromatic hydrocarbons
pain, Wallerian degeneration 215–216
Panax quinquefolius (American ginseng) 358–359
pancreatic necroptosis 111, 112
Panton–Valentine leukocidin (PVL) 233
PAR *see* poly ADP-ribose
parakeratosis 192
paraptosis 343–366
  1-nitropyrene-mediated 352–353
  cancer immunotherapy 362
  corticosteroid-mediated 352
  definition 345–346
  EGF-mediated 352
  honokiol-induced 353–354
  IGF1R-induced 350–351
  mechanisms 348–350
  medical implications 356–362
  mitochondria 355
  morphology 346–347
  NK1R-mediated 352
  pathways 350–354
  proteomics 354–356
  TAJ/TROY-mediated 351–352
  taxol 353, 356
  VR1-mediated 351
  yessotoxin-induced 353
PARG *see* poly ADP ribose glycohydrolase
PARK2 gene 31
Parkinson's disease (PD)
  cyclophilin D-dependent necrosis 386
  excitotoxicity 202
  mitochondria 31–33
PARP-1 *see* poly ADP ribose polymerase 1
PARPs *see* poly ADP ribose polymerases
parthanatos 535–558
  apoptosis-inducing factor 542–544
  crosstalk 538–544
  definition 537–538
  induction 538–545
  inflammation 546
  pharmaceutical targets 546–549
  poly ADP ribose glycohydrolase 542
  poly ADP ribose polymerases 538–541
  poly ADP-ribose polymers 541–542
  protein–protein interactions 546
  putative mediators 544–545

vs. other PCD forms 539
pathophysiology
  excitotoxicity 198–200
  methuosis 560–562
  oncosis 572–575
  parthanatos 538
  TRIAD 416–419
PaTrin-2 436–437
PCB *see* procarbazine
PCD *see* programmed cell death
PD1 *see* programmed death 1
PD *see* Parkinson's disease
PDCD5 351–352
PD-L1/PD-L2 *see* programmed death ligand 1/2
PEBP *see* phosphatidylethanolamine-binding protein
Pembrolizumab 713–714
peptidoglycan (PG) 50
peptidylarginine deaminases (PADs) 519–525
peptidyl-prolyl cis-trans-isomerases (PPIase) 376–379
Perforin/granzyme (PFN/Gzm) pathway 2–3, 5–6
perforin (PFN) 5
peripheral blood mononuclear cells (PBMC) 704–705
peripheral nervous system (PNS)
  functional recovery 209–211
  Wallerian degeneration 208–215
permeabilization, outer mitochondrial membrane 295–299
peroxidation
  lipids
    Apatone effects 669
    oxytosis 293–294
  organelles, menadione 660–661
  proteins, menadione 659–660
peroxisomes 592
PFN *see* perforin
PFN/Gzm *see* Perforin/granzyme pathway
PG *see* peptidoglycan
PGMB *see* peripheral blood mononuclear cells
phage, holins 61
phagocytes, functions 370
phagocytosis
  neutrophils 512
  vs. entosis 466
phagolysosomes, formation 73
phagophores, autoschizis 635–637

pharmacological targets
  aging   274–275
  alkylguanine methyltransferase   436–438
  autophagy   76–77
  ferroptosis   139–140
  MGMT gene   436–439
  mitochondria   33–36
  necroptosis   118–119
  paraptosis   356–362
  PARP-1   446–447
  parthanatos   546–549
  PD1/PD-L1 immunotherapies   713–716
phenoptosis   237–288
  aging as   248–250
  bifunctional proteins   258–259
  cancer   246–247
  citrate as a predictor of death   272–273
  definition   238
  diabetes   247–248
  evolvability   256–257
  food restriction   263–267
  invertebrates   242–244
  mROS regulation   273
  neoteny   253–256
  non-aging living beings   250–256
  pharmacological perspectives   274–275
  plants   240–242
  plastoquinonyl decyltriphenylphosphonium   261–263, 265–266, 274
  rationale for   259
  reproduction   241–244
  septic shock   245–246
  stochastic injury   260–261
  unicellular organisms   239–240
  vertebrates   244–248
phorbol myristate acetate (PMA)   700
phosphatidylethanolamine   400
phosphatidylethanolamine-binding protein (PEBP)   355–356
phosphatidylinositol 3-kinase (PI3-K)
  anoikis   152, 160–161
  methuosis   560–561
  T cell activation   708–710
phosphatidylserine (PS)   320–321
phospholipase A (PLA)
  apoptosis   399–401
  autophagy   404
  cell survival   403
  excitotoxicity   199
  necrosis   405–406
  structure and functions   396–398

phospholipase B (PLB)   399
phospholipase C (PLC)
  apoptosis   401–402
  autophagy   404
  general properties   399
  NETosis   520–521
phospholipase D (PLD)
  apoptosis   402–403
  autophagy   403–404
  general properties   399
phospholipases   395–409
  apoptosis   400–403
  autophagy   403–405
  general properties   396–399
  necrosis   405–406
phosphorylation, p73 in TRIAD   417–421
photodynamic therapy   497–499
physical activity, aging   267–269
phytotoxins, paraptosis   360–361
PI3-K *see* phosphatidylinositol 3-kinase
PIDD *see* p53-induced death domain protein 1
PIDDosomes   149
Pidilizumab   713–714
pigments, intracellular accumulation   89–90
pinching off   2, 3–4
PINK1   31–33
Pinus longaeva (bristlecone pine)   250–251
PKR *see* protein kinase R
PLA *see* phospholipase A
plants, phenoptosis   240–242
Plasmodium Spp.   325–326
plastoquinonyl decyltriphenylphosphonium (SkQ1)   261–263, 265–266, 274
  diabetes   247
platelet-activating factor   199
PLB *see* phospholipase B
PLC *see* phospholipase C
PLD *see* phospholipase D
Plk1 *see* polo-like kinase 1
PNS *see* peripheral nervous system
polo-like kinase 1 (Plk1)   479
poly ADP ribose glycohydrolase (PARG)   542
poly ADP-ribose (PAR) polymers   541–542
poly ADP ribose polymerase 1 (PARP-1)   446–447
poly ADP ribose polymerases (PARPs)
  disease profiles   538
  inhibition   546–549
  parthanatos   535–558
    definition   537–538
    disease profiles   538

poly ADP ribose polymerases (PARPs) (*Continued*)
  pathophysiology 538–541
  pharmaceutical targets 546–549
  protein–protein interactions 546
 pyroptosis 320–322
 synthetic lethality 547–548
polycyclic aromatic hydrocarbons (PAHs) 352–353
polyQ diseases 412–413
 see also Huntington's disease
PORIMIN receptor *see* pre-oncosis receptor-induced membrane injury receptor
post-translational modifications (PTM) 519
potassium channels in paraptosis 348–350
ppGpp, MazEF signaling 57
PPIF gene 378
praying mantis 243
predator–victim co-evolution 257
pre-oncosis receptor-induced membrane injury (PORIMIN) receptor 574
presenelins 202
prevention, aging 261–272
procarbazine (PCB) 450–453
pro-caspase 8 3–4
proerythroblasts 369
progeria, phenoptosis 244–245
programmed cell death (PCD)
 autophagy 74–77
 biofilms 50, 60–63
 definition 49
 detection 8
 inflammation 228–229
 MazEF signaling 57
 microbial 49–70
 mitotic catastrophe 485–487
 reactive oxygen species 56, 60
 sporulation 51–54
 toxin–antitoxin systems 50, 54–60
 see also apoptosis
programmed death 1 (PD1) 695–722
 apoptotic roles 706–707
 autoimmunity 700
 cancer-related expression 704–705
 expression 700
 functions 700
 homology 699
 microRNAs 710–712
 monoclonal antibodies 713–714
 regulation 700, 710–712
 structure 697–700

programmed death ligand 1/2 (PD-L1/-L2) 701–710
 apoptotic roles 706–707
 cancer-related expression 704–705
 expression 701–705
 microRNAs 710–712
 monoclonal antibodies 713–714
 regulation 703–704
 structure 701–703
progression of entosis 466–467
prooxidants *see* ascorbic acid; autoschizis; menadione; reactive oxygen species
prostate carcinoma (DU145) cells
 autoschizis 595–603, 635–652
 cell cycle stages 602–603
 flow cytometry measuring autoschizis 602–603
 morphology 595–598
 VC:VK3 treatment 599–603, 645–652
 xenotransplanted 595–598, 645–652
prostate carcinoma (TRAMP) cells 604, 607, 610
protection, protein aggregates in neurodegeneration 413
protein crosslinking in cornification 189–190
protein kinase C (PKC) 199, 521
protein kinase R (PKR) 105
protein phosphatase 1b (PPM1B) 102
protein–protein interactions in parthanatos 546–547
proteins
 intracellular accumulation 89
 peroxidation 659–660
 synthesis, Apatone effects 669–672
proteomics of paraptosis 354–356
PS *see* phosphatidylserine
Pseudomonas aeruginosa (*P. aeruginosa*) 58–59
psychological factors, aging 270–272
PTM *see* post-translational modifications
PUMA *see* p53-upregulated modulator of apoptosis
PVL *see* Panton–Valentine leukocidin
PYCARD *see* apoptosis-associated speck-like protein containing a CARD
pyknosis 91
pyrin 323–324
pyronecrosis 225–236
 cryopyrin 228–229
 definition 227
 induction 232–234

morphology   227
NLRs   228–229
pathways   231–232
vs. apoptosis   231
vs. necrosis   229–230
vs. pyroptosis   230–231
pyroptosis   317–341
  canonical inflammasomes   322–328
  caspases   320–322
  crosstalk   118, 330–332
  cytokine activation   325
  neutrophils   514
  non-canonical inflammasomes   328–330
  vs. pyronecrosis   230–231
pyruvate oxidase   63

q
quiescence and mitotic catastrophe   483–484

r
RA *see* rheumatoid arthritis
Rac1 GTPase   560–562
radiology, excitotoxicity   202–203
radiotherapy and mitotic
      catastrophe   495–497
Raf, integrin signaling   153
Raf kinase-inhibitor protein (RKIP) *see*
      phosphatidylethanolamine-binding
      protein
RAGE *see* receptor for advanced glycation
      end products
RAIDD *see* RIPK-associated protein with a
      death domain
rapamycin-insensitive companion of
      mammalian target of rapamycin
      (RICTOR)   152
Ras
  integrin signaling   153
  menadione cytotoxicity   661–662
  methuosis   560–562
rate-limiting steps, glutathione   292–293
RBCs (red blood cells) *see* erythrocytes
RCD *see* regulated cell death
reactive nitrogen species (RNS)   19
reactive oxygen species (ROS)   130–133
  Apatone therapy   666–667
  ascorbate   665–666
  cellular sources   131–132
  cyclophilin D-dependent necrosis   384
  excitotoxicity   200
  food restriction   264–265
  impacts   132–133

iron   131, 135–137
JNK activation   295
mitochondria   23
NETosis   520–521
programmed cell death, microbial   56, 60
regulation and phenoptosis   273
receptor for advanced glycation end products
      (RAGE)   232
receptor-interacting protein kinase-1
      (RIPK1)   23–24
  necroptosis   100, 102, 106–107, 110–117
  NETosis   522
receptor-interacting protein kinase-3
      (RIPK3)   23–24
  inhibition   116–117
  necroptosis   102–104, 106–107, 110–117
receptor-interacting protein (RIP)   3–4, 5
receptor tyrosine kinases (RTK)   152
red blood cells (RBCs) *see* erythrocytes
redox cycling, ascorbate treatment   664–665
regulated cell death (RCD)
  anoikis   146–168
    core features   146–154
    definition   146–147
    tissue integrity   155
regulation
  entosis   468–469
  erythropoiesis   369
  NETosis   519–522
  programmed death 1   700, 710–712
  programmed death ligand 1/2
      expression   703–704
renal diseases, necroptosis   113–114
repair
  DNA
    base excision repair   440–441
    homologous recombination   443–444
    mismatch recognition   442
    non-homologous end joining   444–446
    PARP-1   446–447
reperfusion injury salvage kinase
      (RISK)   573–574
reperfusion shock   78
reproduction, phenoptosis   241–244
RER *see* rough endoplasmic reticulum
retinoic acid-inducible gene 1 (RIG-1)   106
reversible cell injury   84–90
  atrophy   84–85
  hydropic swelling   84
  hyperplasia and hypertrophy   85–86
  intracellular accumulations   87–90
  metaplasia   86–87

rheumatoid arthritis (RA)   524
RhoA   157
Rho-associated coiled coil-containing protein kinase 1 (ROCK1)   8, 157, 468
Rho guanine triphosphatases (GTPases)   560–562
ribosomal RNA (rRNA), MazF   59–60
RICTOR *see* rapamycin-insensitive companion of mammalian target of rapamycin
RIG-1 *see* retinoic acid-inducible gene 1
RIP *see* receptor-interacting protein
RIPK1 *see* receptor-interacting protein kinase-1
RIPK-associated protein with a death domain (RAIDD)   149
RISK *see* reperfusion injury salvage kinase
RKIP *see* phosphatidylethanolamine-binding protein
RNA
  MazEF pathway   55, 59–60
  microbial PCD   55, 59
RNases, autoschizis   650–652
RNA viruses, oxidative stress   24–25
RNS *see* reactive nitrogen species
ROCK1 *see* Rho-associated coiled coil-containing protein kinase 1
Rocky Mountain locust *see* Melanoplus spretus
ROS *see* reactive oxygen species
rough endoplasmic reticulum (RER), morphology   592
rRNA *see* ribosomal RNA
RT4 cells, autoschizis   623–625
RTK *see* receptor tyrosine kinases

S
S1P *see* sphingosine-1-phosphate
S_N1 reaction *see* unimolecular nucleophilic substitution
S_N2 reaction *see* bimolecular nucleophilic substitution
Salmonella Spp., pyroptosis   327–330
sanglifehrins   387–388
sanguinarine   569–572, 575
saponification   94
S. aureus see Staphylococcus aureus
Schwann cells   209, 211–212
self-excision *see* autoschizis
self-tolerance *see* autoimmunity
semelparous vertebrates   244–245
Sendai virus   106

septic shock   245–246
sequestosome-1 (SQSTM1/p62)   79–80
S. flexneri see Shigella flexneri
SHARPIN   111, 118
shear stress   402
Shigella flexneri (*S. flexneri*)   232
S. hygroscopicus see Streptomyces hygroscopicus
signaling
  ascorbate treatment   664–665
  entosis   469
  menadione:ascorbate combination therapy   666–667
  NETosis   520–521
  T cell receptors   703, 707–710
signal transducer and activation of transcription 1 (STAT-1)   546
Sinemet *see* carbidopa
SKF *see* sporulation killing factor
skin
  anoikis   165
  cornification   183–192
  differentiation   183–186
  necroptosis   112
  see also epidermis; epithelial cells
SkQ1 *see* plastoquinonyl decyltriphenylphosphonium
SLE *see* systemic lupus erythematosus
Smac *see* small mitochondria-derived activator of caspases
small mitochondria-derived activator of caspases (SMAC)   26–27, 102
smooth endoplasmic reticulum (SER), autoschizis   642–645
SOCE *see* store-operated calcium entry
SOD *see* superoxide dismutase
Sorafenib   139
sources of reactive oxygen species   131–132
SPD *see* sporulation-delaying protein
SpdI   53
sphingosine-1-phosphate (S1P)   29
spindle-assembly checkpoint   480, 484–485
splicing variants of Yes-associated protein   417, 419–421
S. pneumoniae see Streptococcus pneumoniae
Spo0A   51–53
sporulation   51–54
sporulation-delaying protein (SPD)   51–53
sporulation killing factor (SKF)   51–53
SpxB   63
SQSTM1/p62 *see* sequestosome-1
Src

E-cadherin signaling 154
integrin signaling 150–152, 155–156, 158–159
SS tetrapeptide (SS-31) 35
Staphylococcus aureus (*S. aureus*) 60–63, 232
starvation
glutamine 24
menadione:ascorbate combination therapy 667–668
microautophagy 73–74, 77
STAT-1 *see* signal transducer and activation of transcription 1
sterile tissue damage in necroptosis 110–111
stochastic injury in aging 260–261
store-operated calcium entry (SOCE) 304–305
Streptococcus pneumoniae (*S. pneumoniae*) 325–326
Streptomyces hygroscopicus (*S. hygroscopicus*) 234
streptozotocin (STZ) 438, 450–454
structural features
inflammasomes 324
programmed death 1 697–700
programmed death ligand 1/2 701–703
S. typhimurium *see* *Salmonella typhimurium*
β-subunit of ATP-synthase 355
sulfur-containing proteins and reactive oxygen species 133
superoxide dismutase (SOD) 35, 591
α-syn 31
synthetic lethality 446–447, 547–548
syphilis 93
systemic lupus erythematosus (SLE) 523, 525

t

T24 cells
autoschizis 608, 613–623, 625–629
cell cycle stages 600–602
VC:VK3 cytotoxicity 599–600
TA *see* toxin–antitoxin systems
TAJ/TROY 351–352
taxol (paclitaxel) 35, 353, 356
tazarotene-induced gene 3 (TIG3) 189
tBID *see* truncated BID protein
T-cell receptors (TCR) 703, 707–710
T cells
activation 703, 707–710
adoptive cell-transfer 714–716

apoptosis 695–697, 706–716
degradation 697–698
maturation 697–698
programmed death 1-mediated apoptosis 695–722
programmed death ligand 1/2 701–710
TEF-1 *see* translational elongation factor 1
temozolomide (TMZ) 450–455
terminal deoxynucleotidyl transferase dUTP nick end labeling (TUNEL) 8
termites, phenoptosis 242–243
testosterone 368
TGases *see* transglutaminases
T. gondii *see* *Toxoplasma gondii parasitophorous*
thanatology 585–586
thapsigargin 303–304
therapeutic targets *see* pharmacological targets
thioredoxin reductase (TRXR2) 35
thioxotriazole copper(II) complex (A0) 358
TIG3 *see* tazarotene-induced gene 3
TIP150, entosis regulation 468–469
tissue integrity, anoikis 155
tissue patterns, necrosis 91–95
TLR2 *see* toll-like receptor 2
TLR3 *see* toll-like receptor 3
TLR4 *see* toll-like receptor 4
T lymphocytes 370–371
TMZ *see* temozolomide
TNF α *see* tumor necrosis factor alpha
TNFR1 *see* tumor necrosis factor receptor 1
TNFR *see* tumor necrosis factor receptors
TNF receptor-associated death domain (TRADD) 3–4, 5, 102
TNF receptor-associated factor 2 (TRAF2) 102
tocotrienols, paraptosis 359
toll-like receptor 2 (TLR2) 232, 700
toll-like receptor 3 (TLR3) 102–104, 700
toll-like receptor 4 (TLR4) 102–104, 232, 520–521, 700
toxin–antitoxin (TA) systems 50, 54–60
toxI/ToxN 55
Toxoplasma gondii parasitophorous (*T. gondii*) 233, 326–327
TRADD *see* TNF receptor-associated death domain
TRAF2 *see* TNF receptor-associated factor 2
TRAIL *see* tumor necrosis factor-related apoptosis-inducing ligand

TRAMP (murine prostate carcinoma) cells   604, 607, 610
transcriptional repression-induced atypical death (TRIAD)   411–426
 actinomycin D   414
 α-amantin   414
 cellular dysfunction   412–413
 conformational protein abnormalities   413
 ER vacuolization   416
 expression patterns   417
 Hippo pathway   420
 induction
  in vitro   413–414
  in vivo   421
 molecular features   416–419
 morphology   414–416
 p73   417–419
 vs. apoptosis   415, 416
 vs. autophagic cell death   415, 416–417
 vs. necrosis   415, 417
 YAP   416–421
 YAPΔCs   419–421
transgenic mice, MGMT   435–436
transglutaminases (TGases)   188, 189–190
transient receptor potential cation channel subfamily V member 1 (TRPV1)   247
translational elongation factor 1 (TEF-1)   546
traumatic nerve injury
 demyelination   211–215
 recovery pathway activation   209–211
 Wallerian degeneration   208–215
TRIAD see transcriptional repression-induced atypical death
TRIM/SUP7 complex   102
tropomyosin   354–355
TRPV1 see transient receptor potential cation channel subfamily V member 1
truncated BID (tBID) protein   5, 148, 149, 297–298
TRXR2 see thioredoxin reductase
α-tubulins, paraptosis   354–355
β-tubulins, paraptosis   354–355
TUCAN see tumor upregulated CARD-containing antagonist of caspasenine
tumor-infiltrating lymphocytes   704–705
tumor necrosis factor alpha (TNFα)   3–4, 212, 216, 400–401

tumor necrosis factor receptor 1 (TNFR1)   3–4, 102–103
tumor necrosis factor receptors (TNFR)   351–352, 401–402, 513–514
tumor necrosis factor-related apoptosis-inducing ligand (TRAIL)   25, 320
tumor necrosis factor (TNF)   102–103, 517–518
tumors
 adoptive cell-transfer   714–716
 AGT inhibition   437–438
 anoikis   165–166
 Apatone therapy
  cytotoxicity   599–600
  DNA/protein synthesis effects   669–672
  lipid peroxidation   669
  organelle effects   668–669
  starvation effects   667–668
 apoptotic resistance   30–31, 708–710, 713–714
 ascorbate cytotoxicity   599–600, 662–666
  accumulation   662–663
  growth receptor regulation   664–665
  organelle effects   665–666
  pro-oxidant property development   663–664
 autophagy-inducing drugs   76–77
 autoschizis   583–693
 cannibalism   466
 cell cycle stages   600–603
 entosis   470–472
 ferroptosis   139
 growth receptors, ascorbate effects   664–665
 immune avoidance   708–710, 713–714
 immunotherapy   362, 713–716
 membrane macrophage colony-stimulating factor   361–362
 menadione cytotoxicity   599–600, 657–662
 methylating agents   432
 MGMT gene   437–440
 mitotic catastrophe induction   495–499
 morphology   586–598
  in vitro   586–595
  in vivo   595–598
 paraptotic medicines   351, 353–354, 356–362
 programmed death 1, expression   704–705

    programmed death ligand 1/2,
        expression   704–705
    starvation, VC:VK3 therapy   667–668
    synthetic lethality   446–447
tumor upregulated CARD-containing
        antagonist of caspasenine
        (TUCAN)   228
TUNEL *see* terminal deoxynucleotidyl
        transferase dUTP nick end labeling
tunicamycin   303–304
turquoise killifish   250

u

U251 glioma cells   348–349
Ube4b   207
uncoupling protein 2 (UCP2)   574
unfolded protein response (UPR)   24,
        302–304
unicellular organisms, phenoptosis   239–240
unimolecular nucleophilic substitution (S_N1
        reaction)   431
untranslated regions (UTRs)   699
UPR *see* unfolded protein response
upstream mechanisms, oxytosis   293–295

v

vacuolization, in TRIAD   416
vanilloid receptor subtype 1 (VR1)   351
vascular aging   79–81, 115
vascular endothelial growth factor
        (VEGF)   665
VC:VK3 (combined ascorbate/menadione
        treatment) *see* Apatone
VDACs *see* voltage-dependent anion
        channnels
vegetative growth   241–242
VEGF *see* vascular endothelial growth factor
vertebrates, phenoptosis   244–248
v-FLIPs *see* viral homolog of c-FLIPs
viral homolog of c-FLIPs (v-FLIPs)   111
viral infection
    extrinsic pathway   3, 5–6
    necroptosis   117
    oncosis   576
    oxidative stress   24–25
Visomitin   274
vitamin C (VC) *see* ascorbate
VK3 *see* menadione
voltage-dependent anion channnels
        (VDACs)   21
    apoptotic avoidance   30

    ferroptosis   134–136
    as therapeutic target   33–34
VR1 *see* vanilloid receptor subtype 1

w

Wallerian degeneration   205–223
    cellular changes   210
    chemokines   214
    cytokines   211–214
    definition   206
    demyelination   211–215
    immune responses   208–209
    neuropathic pain   215–216
    neuroprotection induction   207–208
    neuroprotective mechanism   206–207
    neurotrophic factors   215
    recovery pathway activation   209–211
WBCs (white blood cells) *see* leukocytes
Wee1   479
western blotting, mitotic catastrophe   488–490
white blood cells (WBCs) *see* leukocytes
Wld   206–208

x

X-box binding protein 1 (XBP1)   303
XBP1 *see* X-box binding protein 1
xenotransplanted carcinoma
    autoschizis   645–652
    morphology   595–598
X-linked inhibitor of apoptosis protein
        (XIAP)   27, 30, 102

y

YAP *see* Yes-associated protein
YAPΔCs   417, 419–421
ydcDE   58–59
yeast, phenoptosis   239–240
Yes-associated protein (YAP)   416–421
yessotoxin (YTX)   353
Yfhl/YknWXYZ   53
yin-yang-1 (YY-1)   546
YopJ protein   118
yorkie   419
    see also Yes-associated protein
YY-1 *see* yin-yang-1

z

Z-VAD-fmk   231
zVAD-fmk *see* benzyloxycarbonyl-valyl-
        alanyl-aspartyl fluoromethyl ketone
Zwf   58